Flora Australiensis

A Description of the Plants of the Australian Territory

Volume 3: Myrtaceae to Compositae

George Bentham
Ferdinand von Mueller

CAMBRIDGE UNIVERSITY PRESS

Cambridge, New York, Melbourne, Madrid, Cape Town,
Singapore, São Paolo, Delhi, Tokyo, Mexico City

Published in the United States of America by Cambridge University Press, New York

www.cambridge.org
Information on this title: www.cambridge.org/9781108037402

This edition first published 1866
This digitally printed version 2011

ISBN 978-1-108-03740-2 Paperback

CAMBRIDGE LIBRARY COLLECTION

Books of enduring scholarly value

Life Sciences

Until the nineteenth century, the various subjects now known as the life sciences were regarded either as arcane studies which had little impact on ordinary daily life, or as a genteel hobby for the leisured classes. The increasing academic rigour and systematisation brought to the study of botany, zoology and other disciplines, and their adoption in university curricula, are reflected in the books reissued in this series.

Flora Australiensis

George Bentham (1800–84) was one of Britain's most influential botanists, whose own collection of plant specimens numbered more than 100,000. Although he donated his herbarium to the Royal Botanic Gardens, Kew in 1854, he continued to make significant contributions to the field, including this exhaustive, seven-volume work detailing the plant life of Australia, which was published from 1863 to 1878. It was part of a series of works commissioned by the British government to document the flora in its colonies. Using the extensive numbers of specimens at Kew – and with the help of Ferdinand Mueller (1825–96), a German botanist in Australia – Bentham was able to compile descriptions of more than 8,000 species of Australian plants, making these volumes the first completed compendium of the flora of any large continental area. Volume 3, published in 1866, describes 14 orders of dicotyledon flora in the subclasses polypetalae and monopetalae.

Cambridge University Press has long been a pioneer in the reissuing of out-of-print titles from its own backlist, producing digital reprints of books that are still sought after by scholars and students but could not be reprinted economically using traditional technology. The Cambridge Library Collection extends this activity to a wider range of books which are still of importance to researchers and professionals, either for the source material they contain, or as landmarks in the history of their academic discipline.

Drawing from the world-renowned collections in the Cambridge University Library, and guided by the advice of experts in each subject area, Cambridge University Press is using state-of-the-art scanning machines in its own Printing House to capture the content of each book selected for inclusion. The files are processed to give a consistently clear, crisp image, and the books finished to the high quality standard for which the Press is recognised around the world. The latest print-on-demand technology ensures that the books will remain available indefinitely, and that orders for single or multiple copies can quickly be supplied.

The Cambridge Library Collection will bring back to life books of enduring scholarly value (including out-of-copyright works originally issued by other publishers) across a wide range of disciplines in the humanities and social sciences and in science and technology.

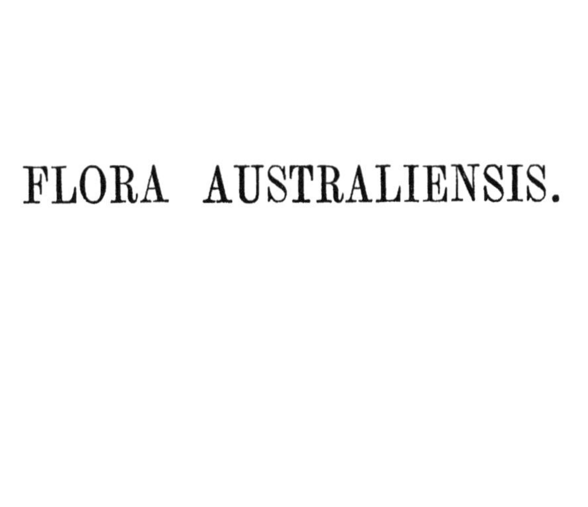

FLORA AUSTRALIENSIS.

FLORA AUSTRALIENSIS:

A DESCRIPTION

OF THE

PLANTS OF THE AUSTRALIAN TERRITORY.

BY

GEORGE BENTHAM, F.R.S., P.L.S.,

ASSISTED BY

FERDINAND MUELLER, M.D., F.R.S. & L.S.,

GOVERNMENT BOTANIST, MELBOURNE, VICTORIA.

VOL. III.

MYRTACEÆ TO COMPOSITÆ.

PUBLISHED UNDER THE AUTHORITY OF THE SEVERAL GOVERNMENTS
OF THE AUSTRALIAN COLONIES.

LONDON:
LOVELL REEVE & CO., 5, HENRIETTA STREET, COVENT GARDEN.
1866.

J. E. TAYLOR AND CO., PRINTERS,
LITTLE QUEEN STREET, LINCOLN'S INN FIELDS.

CONTENTS.

CONSPECTUS OF THE ORDERS CONTAINED IN THE THIRD VOLUME.

CLASS I. DICOTYLEDONS.

SUBCLASS I. POLYPETALÆ.

SERIES III. CALYCIFLORÆ.

(Continued from Vol. II.)

XLVIII. MYRTACEÆ. Trees or shrubs, very rarely undershrubs. Leaves opposite or alternate, without stipules, usually dotted. Flowers regular or nearly so. Calyx-lobes usually imbricate. Petals imbricate. Stamens indefinite or rarely definite; anthers opening in longitudinal slits or rarely in terminal pores. Ovary inferior, 2- or more-celled with 2 or more ovules in each cell, or rarely 1-celled with 1 placenta. Style undivided. Seeds without albumen. Cotyledons flat or folded, not convolute.

XLIX. MELASTOMACEÆ. Shrubs or rarely trees or herbs. Leaves opposite, not dotted, without stipules. Flowers regular or nearly so. Petals contorted. Stamens definite; anthers opening in terminal pores, very rarely in longitudinal slits. Ovary inferior or enclosed in the calyx, 2- or more-celled, with 2 or more ovules in each cell, or rarely 1-celled with a central placenta. Style undivided. Seeds without albumen. Cotyledons flat or folded, not convolute.

L. LYTHRARIEÆ. Herbs shrubs or trees. Leaves opposite or alternate, without stipules. Flowers regular or nearly so. Calyx-lobes valvate. Petals usually crumpled in the bud. Stamens definite or rarely indefinite. Ovary usually enclosed in the calyx-tube, 2- or more-celled, with few or many ovules in each cell. Style undivided. Seeds without albumen. Cotyledons not convolute.

LI. ONAGRARIEÆ. Herbs (in the Australian genera). Leaves opposite or alternate, without stipules. Flowers regular or nearly so, usually 4-merous. Calyx-lobes valvate. Petals imbricate. Stamens definite. Ovary inferior, 2- or more-celled, very rarely 1-celled. Style undivided. Seeds without albumen.

LII. SAMYDACEÆ. Trees or shrubs. Leaves alternate. Stipules small or none. Flowers regular or nearly so. Petals and sepals nearly similar. Stamens indefinite or alternating with small scales or glands. Ovary 1-celled with parietal placentas. Style entire or branched. Seeds albuminous.

LIII. PASSIFLOREÆ. Climbers (in the Australian genera). Leaves alternate, with stipules. Flowers regular. Petals persistent with the calyx-lobes and often resembling them. Stamens definite. Ovary stalked, 1-celled, with parietal placentas. Style branched. Seeds albuminous.

LIV. CUCURBITACEÆ. Herbs either prostrate or climbing with tendrils. Leaves alternate, without stipules. Flowers unisexual, regular. Stamens 3 or 5. Ovary inferior, at first 1-celled, the (3) parietal placentas soon meeting in the axis and dividing the cavity into 3 or 6 cells or remaining 1-celled with 1 placenta. Style entire or branched. Seeds without albumen.

LV. FICOIDEÆ. Herbs or rarely undershrubs, sometimes succulent. Leaves alternate or rarely opposite, without any or with minute scarious stipules. Petals none or indefinite and narrow. Stamens indefinite or rarely definite. Ovary inferior or superior, several-celled (rarely reduced to 1 cell). Placentas basal or nearly so. Styles free or united at the base. Embryo curved in a mealy albumen.

LVI. UMBELLIFERÆ. Herbs or rarely shrubs. Leaves alternate, often dissected, without any or rarely with scarious stipules. Calyx-teeth small or obsolete. Stamens as many as petals, and inserted with them round an epigynous 2-lobed disk. Ovary inferior, usually 2-celled with 1 pendulous ovule in each cell. Styles 2. Fruit dividing into 2 small dry 1-seeded nuts. Seeds albuminous with a minute embryo.

LVII. ARALIACEÆ. Trees shrubs or very rarely tall herbs. Leaves alternate, without stipules. Flowers of *Umbelliferæ*, except that the ovary-cells are often more than two. Fruit drupaceous, the endocarp hardened into 2 or more distinct, 1-seeded pyrenes, the epicarp fleshy, succulent, or rarely dry and thin. Seeds albuminous with a minute embryo.

LVIII. CORNACEÆ. Trees shrubs or very rarely herbs. Leaves opposite or very rarely (as in the Australian genus) alternate, without stipules. Petals valvate. Stamens as many or twice as many as petals. Ovary inferior, 1- or 2-celled with 1 pendulous ovule in each cell. Style simple. Seeds albuminous, the embryo nearly as long as the albumen.

SUBCLASS II. MONOPETALÆ.

Petals united into a single lobed corolla (exceptionally free in a few *Loranthaceæ*).

LIX. LORANTHACEÆ. Parasitical shrubs or trees. Leaves opposite or alternate, without stipules. Stamens opposite the corolla-lobes or petals. Ovary inferior, 1-celled, with 1 erect ovule, not perceptible till the flowering is over. Seeds albuminous.

LX. CAPRIFOLIACEÆ. Trees, shrubs, or climbers, rarely herbs. Leaves opposite (pinnate in the Australian genus) without real stipules. Stamens as many as corolla-lobes and alternate with them. Ovary inferior, 2- or more-celled. Seeds albuminous.

LXI. RUBIACEÆ. Trees, shrubs, or herbs. Leaves opposite, with interpetiolar or sheathing stipules. Stamens as many as corolla-lobes and alternate with them. Ovary inferior, 2- or more-celled, very rarely reduced to 1 cell. Seeds albuminous.

LXII. COMPOSITÆ. Herbs shrubs or rarely trees. Leaves opposite or alternate, without stipules. Flowers or florets collected in heads, each head surrounded by a calyx-like involucre, the true calyx of each floret wanting or reduced to a pappus. Stamens as many as corolla-lobes and alternate with them. Ovary inferior, 1-celled, with 1 erect ovule. Seed without albumen.

FLORA AUSTRALIENSIS.

ORDER XLVIII. MYRTACEÆ.

Calyx-tube adnate to the ovary at the base or up to the insertion of the stamens; limb more or less divided (usually to the base) into 4 or 5, very rarely 3 or more than 5, lobes or teeth, or reduced to a narrow border, or entirely wanting; lobes usually imbricate or open in the bud. Petals usually as many as calyx-lobes, very much imbricate in the bud, the external one sometimes larger than the others, but usually all nearly equal when expanded, sometimes all concrete and falling off in a single operculum, or rarely entirely wanting. Stamens indefinite, usually numerous or rarely few and definite, inserted in one or several rows on a disk, either thin and lining the calyx-tube above the ovary and forming a thickened ring at its orifice, or thicker and forming a ring close round the summit of the ovary; filaments free or rarely united into a ring or tube at the base, or into as many bundles as there are calyx-lobes; anthers 2-celled, versatile or attached by the base, the cells opening in longitudinal slits, or rarely in terminal pores. Ovary inferior or rarely almost superior, but enclosed in the calyx-tube, sometimes 1-celled, with a placenta attached to the base or adnate to one side, more frequently 2- or more celled, with the placentas in the inner angle of each cell, very rarely 1-celled, with 2 parietal placentas. Style simple, with a small or a capitate or peltate, very rarely lobed stigma. Ovules 2 or more to each placenta, in 2 or more rows, or very rarely solitary, erect pendulous or laterally attached, anatropous or amphitropous. Fruit inferior, adnate to the calyx-tube, and crowned by the persistent limb, or marked by its scar when deciduous, or very rarely half or almost wholly superior, and surrounded at the base by the persistent calyx-tube, either capsular and opening loculicidally at the summit, in as many valves as cells, or indehiscent, dry, and 1-seeded, or succulent and indehiscent. Perfect seeds usually very few or solitary in each cell, even when the ovules are numerous, or rarely numerous and perfect; teeth either thin and membranous, or crustaceous, fleshy or bony. Albumen none, or very scanty near the hilum. Embryo straight or variously curved, fleshy, with minute cotyledons at one end, or with large, flat, or variously folded cotyledons, or with thick fleshy distinct or consolidated cotyledons, and an exceedingly short radicle, or rarely apparently homogeneous, the cotyledons inconspicuous before germination. Abortive ovules in many capsular genera, enlarged without being fertilized, and simulating the seeds, but of a hard, nearly homogeneous, woody, or granular consistence.—Trees or shrubs, very rarely undershrubs, Leaves simple, entire or rarely obscurely crenate-toothed, opposite or less frequently alternate, more or less dotted in all but the *Lecythideæ*, with small resinous glands, either pellucid or black and superficial, often scarcely visible when the leaf is thick. Stipules none, or rarely very minute and fugacious. Flowers solitary or in racemes panicles or cymes, axillary or apparently terminal from the terminal bud not growing out till after the flowering is over. Bracts solitary at the base of the peduncles, or forming an imbricate involucre from the abortion of the lower flowers. Bracteoles 2 at the base of or on the pedicel, sometimes very small or abortive, and often exceedingly deciduous.

The fleshy-fruited genera of the Order are widely spread over the tropical regions both of the New and the Old World, including many of the largest forest trees, and are, in Australia, almost limited to the tropics, a very few species extending into N. S. Wales, and only one into Victoria. The capsular genera are either entirely or chiefly Australian; four of the larger ones, represented by a few species in New Caledonia and the Indian Archipelago, one *Xanthostemon*, represented by more species in New Caledonia than in Australia, two small ones are in New Caledonia, and not yet found in Australia, one *Eucalyptus*, is represented in Timor, if not in the Moluccas, but is not in New Caledonia, another, *Metrosideros*, is more abundant in the Pacific islands than in Australia, and extends also to the Malayan peninsula, and in anomalous forms (perhaps not strictly congeners) to S. Africa and S. America. Two of the widest-spread genera, *Leptospermum* and *Metrosideros*, are also in New Zealand.

TRIBE I. **Chamælaucieæ.**—*Ovary* 1-*celled, with a single placenta. Fruit indehiscent, dry, with* 1 *or rarely* 2 *seeds. Shrubs often heath-like. Leaves small. Flowers solitary, or very rarely* 2 *together in the axils of the leaves or bracts, scattered along the branches, or forming a terminal head.*

Stamens definite, in a single series, more or less united in a ring at the base, and often alternating with staminodia.
- Stamens 8, without staminodia. Flowers 4-merous, the outer ones of the head enlarged and sterile . . . . 1. ACTINODIUM.
- Stamens 10, alternating with as many staminodia (very minute or wanting in one species of *Darwinia* and one of *Verticordia*).
  - Calyx-lobes 5, subulate, entire . . . . 3. HOMORANTHUS.
  - Calyx-lobes 5, broad, entire or shortly ciliolate.
    - Anthers globose or didymous, opening in terminal pores or short slits. Style usually long . . . . 2. DARWINIA.
    - Anther-cells parallel, opening in longitudinal slits. Style short . . . . 6. CHAMÆLAUCIUM.
  - Calyx-lobes 5 or 10, deeply divided into subulate, plumose, or hair-like lobes . . . . 4. VERTICORDIA.
- Stamens 20, without staminodia. Calyx-lobes 10, entire . . . 5. PILEANTHUS.

Stamens indefinite, without staminodia, numerous, or, if few, not regularly alternate or opposite to the calyx-lobes.
- Calyx-lobes persistent, or rarely falling off with the upper portion of the tube. Ovules 2, on a filiform placenta attached both to the base and summit of the ovary.
  - Calyx-lobes terminating in a long bristle or rarely in a short point . . . . 7. CALYTHRIX.
  - Calyx-lobes truncate or retuse, not pointed . . . . 8. LHOTZKYA.
- Calyx-lobes short, deciduous. Ovules 2, the placenta basal or adnate to one side of the ovary . . . . 9. HOMALOCALYX.

Stamens 5 or 10, regularly alternate with or opposite to the calyx-lobes, quite distinct and without staminodia.
  Ovules 2 or more, ascending or attached to a lateral placenta. Stamens, when 5, alternate with the petals . . . . 10. THRYPTOMENE.
  Ovules 2 or 4, pendulous from the summit of a filiform placenta. Stamens, when 5, opposite to the petals . . . . 11. MICROMYRTUS.

TRIBE II. **Leptospermeæ.**—*Ovary divided into 2 to 5, or rarely more cells. Capsule opening at the summit in as many valves as there are cells, or very rarely indehiscent, with 1 or 2 seeds.*

Stamens in a single row, definite or indefinite, shorter than or rarely shortly exceeding the petals, free or united in bundles, alternating with the petals. Leaves small or narrow.
  Leaves opposite.
    Ovules 2 in each cell, superposed or solitary. Flowers small, in axillary cymes, or rarely solitary . . . . 12. SCHOLTZIA.
    Ovules several in each cell, in 2 rows or in a ring round a peltate placenta, or if 2, collateral. Flowers axillary, solitary or rarely few, on a common peduncle.
      Stamens free, rarely exceeding 20, and usually much fewer. Flowers small . . . . 13. BÆCKEA.
      Stamens united in bundles, alternating with the petals. Flowers small . . . . 14. ASTARTEA.
      Stamens numerous, often united in a ring at the base . . . 15. HYPOCALYMNA.
      Stamens numerous, free. Calyx large, red, urceolate . . . 16. BALAUSTION.
  Leaves alternate.
    Stamens free, definite, or if indefinite none opposite the centre of the petals. Flowers in globular sessile heads . . . . 17. AGONIS.
    Stamens numerous, in a continuous series. Flowers solitary or crowded, but not in heads . . . . 18. LEPTOSPERMUM.
Stamens exceeding the petals, indefinite, either free or united in bundles, opposite the petals. Leaves small or narrow, or rarely large and many-nerved. Flowers closely sessile (except in some species of *Kunzea*).
  Anthers versatile, with parallel cells, opening longitudinally.
    Stamens free (almost in 5 bundles in 1 species of *Callistemon*).
      Calyx-lobes usually persistent. Ovary 2- to 5-celled. Seeds pendulous. Flowers in heads or solitary, or rarely in short spikes . . . . 19. KUNZEA.
      Calyx-lobes usually deciduous. Ovary 3- or 4-celled. Seeds ascending. Flowers in spikes, terminal or crowned by the year's shoot . . . . 20. CALLISTEMON.
    Stamens united in 5 bundles opposite the petals (almost free in 1 species of *Melaleuca*).
      Staminal bundles united high up in a tube . . . . . . 21. LAMARCHEA.
      Staminal bundles distinct or scarcely united at the base.
        Ovules several in each cell. . . . . 22. MELALEUCA.
        Ovules solitary in each cell . . . . 23. CONOTHAMNUS.
  Anthers erect, attached by the base. Stamens united in bundles opposite the petals, or nearly free in some species of *Eremæa* and *Phymatocarpus*.
    Ovules 1 to 4 in each cell, peltate and laterally attached.
      Anther-cells opening at the top in transverse valves. Ovules 1 in each cell . . . . 24. BEAUFORTIA.
      Anther-cells placed back to back, and opening in outward longitudinal slits. Ovules 4 in each cell . . . . 25. REGELIA.
    Ovules 2 or more in each cell, erect or ascending, linear or cuneate.

Anthers obovoid, the cells back to back, opening in outward transverse valves. Ovules 2 to 4 in each cell. Leaves small, opposite . . . . 26. PHYMATOCARPUS.
Anthers oblong or linear, the cells parallel, turned inwards, opening in longitudinal slits. Ovules several. Leaves long, alternate. Flowers lateral. . . . . 27. CALOTHAMNUS.
Anthers obovoid, the cells back to back, opening in outward longitudinal slits. Ovules several. Leaves small, scattered. Flowers 1 to 3, nearly terminal . . . . 28. EREMÆA.
Stamens numerous, free or obscurely united at the base. Petals attached by a broad base, free or consolidated into an operculum. Leaves usually large. Flowers in umbels heads or cymes, rarely sessile on the stem.
Calyx-teeth distinct, distant. Petals free . . . . 29. ANGOPHORA.
Calyx truncate, entire or with 4 minute teeth. Petals united in an operculum . . . . 30. EUCALYPTUS.
Stamens exceeding the petals, indefinite, free, or rarely united in bundles opposite the petals. Leaves large or myrtle-like, penniveined. Flowers in pedunculate heads cymes or corymbs, or rarely solitary and pedicellate.
(Stamens scarcely exceeding the petals in some species of *Tristania.*)
Stamens united in 5 bundles. Leaves alternate or in one species opposite . . . . 31. TRISTANIA.
Outer stamens with reniform sterile anthers. Leaves opposite, narrow . . . . 33. LYSICARPUS.
Flowers in globular pedunculate heads. Leaves opposite. Stamens of *Metrosideros*. . . . . 32. SYNCARPIA.
Flowers in cymes. Stamens free, all perfect.
Ovules numerous, horizontal or ascending, covering the placenta. Leaves opposite . . . . 34. METROSIDEROS.
Ovules in a ring round a club-shaped or peltate placenta. Leaves alternate . . . . 35. XANTHOSTEMON.
Ovules pendulous or recurved. Calyx-lobes almost petal-like. Leaves opposite . . . . 36. BACKHOUSIA.
Stamens indefinite, free. Fruit dry, indehiscent. Ovary perfectly or imperfectly 2-celled or 1-celled by abortion.
Calyx-lobes almost petal-like. Petals 4, shorter than or scarcely exceeding the calyx-lobes. Flowers in cymes heads or umbels 36. BACKHOUSIA.
Calyx-lobes 8. Petals none. Flowers solitary, sessile . . . . 37. OSBORNIA.
Calyx-lobes 5, narrow. Petals 5. Flowers solitary, pedicellate . 41. FENZLIA.

TRIBE III. **Myrteæ.**—*Ovary divided into* 2 *or more cells, or very rarely* 1-*celled, with* 2 *parietal placentas. Fruit an indehiscent berry or drupe.*

Ovary 1-celled, with 2 parietal placentas. Leaves 3-nerved . . . 40. RHODAMNIA.
Ovary 2-celled (or 1-celled by abortion), with 2 or 3 superposed ovules in each cell. Leaves white underneath . . . . 41. FENZLIA.
Ovary with 2, 4, or 6 rows of superposed ovules, separated by vertical septa, the ovules themselves separated by transverse septa (1-, 2-, or 3-celled, with double rows of ovules in each cell, all separated by spurious septa.) Leaves sometimes 3-nerved . . . . 38. RHODOMYRTUS.
Ovary 2- or 3-celled, with several ovules in each cell, without spurious dissepiments.
Embryo long and narrow, curved, circular, or spiral, with small cotyledons. Flowers 5-merous or rarely 4-merous, solitary or racemose . . . . 39. MYRTUS.
Embryo thick and fleshy, either indivisible or with 2 thick fleshy cotyledons and a short radicle. Flowers 4-merous or rarely 5-merous, solitary or in trichotomous cymes or panicles . . . . 43. EUGENIA.

Ovary 5- or rarely 4-celled, with 2 to 6 ovules in each cell. Embryo of *Myrtus* . . . . 42. NELITRIS.
(One species of *Kunzea*, has the fruit succulent and pulpy.)

TRIBE IV. **Lecythideæ** (SUBTRIBE **Barringtonieæ**).—*Ovary divided more or less completely into 2 or more cells. Fruit indehiscent, hard and fibrous or fleshy. Leaves alternate or crowded at the ends of the branches, large, not dotted. Calyx often nearly valvate.*

Stamens all perfect. Fruit angular, fibrous, with a single seed . . 44. BARRINGTONIA.
Outer or inner stamens, or both without anthers. Fruit ovoid or globular, not angular, fleshy, with several seeds enveloped in pulp 45. CAREYA.

(*Bartlingia*, Ad. Brongn., referred by Schauer to *Chamælaucieæ*, proves to be *Pultenæa obovata*, described above, Vol. II. p. 123, having been originally examined in a state of very young bud, before the irregularity of the petals was developed.)

TRIBE I. CHAMÆLAUCIEÆ.—Ovary 1-celled. Fruit 1- or rarely 2-seeded, indehiscent. Shrubs often heath-like, with small leaves. Flowers usually small, solitary or very rarely 2 or 3 together in the axils of the leaves or bracts, either along the branches or in terminal heads, the floral leaves either like the stem-leaves, or dilated and bract-like, or forming an involucre.

The first two subtribes of *Chamælaucieæ* have a peculiar habit, which had induced their being proposed as a distinct Order, but some of the third subtribe (*Thryptomeneæ*) pass so gradually into the *Leptospermeæ*, as only to be distinguishable from *Bæckea* by the examination of the ovary.

SUBTRIBE I. EUCHAMÆLAUCIEÆ.—Stamens twice as many as petals, with intervening staminodia rarely wanting, or 4 times as many as petals without staminodia, the filaments more or less distinctly united in a ring at the base. Ovules 2 to 10, attached to an excentrical basal placenta, or in 2 rows, on a short lateral placenta. Embryo, where known, consisting of a thick radicle, the shape of the seed, with a slender neck lying on the summit, apparently entire or with 2 minute cotyledons at the end.

1. **ACTINODIUM,** Schauer.

(Triphelia, *R. Br.*)

Calyx-tube acutely 4-angled; lobes 4, petal-like, entire. Petals 4, as long as the calyx. Stamens 8, in a single row, those opposite the sepals more inflected in the bud; anthers nearly globular, opening in 2 minute pores; staminodia none. Ovary 1-celled, with a single ovule, erect from a short basal placenta; style exserted, with a terminal oblong stigma. Fruit . . .—Shrub, with the habit of *Darwinia*. Leaves heath-like, scattered. Flowers small, in terminal heads, the outer barren flowers with elongated calyx-lobes, petals, and petal-like bracts and bracteoles forming a ray, within an involucre of coloured floral leaves or bracts.

The genus is limited to the single Australian species.

1. **A. Cunninghamii,** *Schau. in Lindl. Introd. Nat. Syst. ed.* 2. 440, *Myrt. Xeroc.* 24. *t.* 1 B, *and in Pl. Preiss.* i. 96. An erect glabrous heath-like shrub of 1 to 2 feet, with slender virgate branches. Leaves scat-

tered, sessile, erect or slightly spreading, linear-terete and channelled above or triquetrous, obtuse or mucronulate, either slender and distant, or short, thick, and almost imbricate. Perfect flowers apparently pink or white, very small and numerous, in a dense hemispherical terminal head, each flower in the axil of a lanceolate or linear, almost petal-like bract, with 2 similar bracteoles under the calyx, the outer flowers of the head usually barren, pedicellate, the bracts, bracteoles, calyx-lobes, and petals, all linear and petal-like, and growing out to 3 or even 4 lines, forming an apparently white ray to the head, and the whole surrounded by a short involucre of more or less coloured, oblong or obovate, acuminate, imbricate bracts or floral leaves passing into the stem-leaves. Calyx 1 to 1½ lines long; the lobes about as long as the tube. Petals narrow, entire or toothed at the end. After the flowering is over, either the central shoot grows out, leaving the old receptacle as a thickening of the branch, or 2 or 3 new shoots grow out from under the head.—*Triphelia brunioides*, R. Br.; Endl. in Hueg. Enum. 48; *Actinodium proliferum*, Turcz. in Bull. Mosc. 1849, ii. 17.

W. Australia. King George's Sound and adjoining districts, and eastward to Cape Riche, *R. Brown*, and others; *Drummond*, 3*rd Coll. n.* 211, 4*th Coll. n.* 43 *and* 44, 5*th Coll. n.* 102; *Preiss, n.* 223; Moir's Inlet, *Maxwell.*

2. DARWINIA, Rudge.

(Genetyllis, *DC.*; Hedaroma, *Lindl.*; Polyzone, *Endl.*; Schuermannia, *F. Muell.*; Cryptostemon, *F. Muell.*; Francisia, *Endl.*)

Calyx-tube nearly cylindrical, turbinate or hemisphærical, the lower adnate part more or less distinctly 5- or rarely 10-ribbed, the upper disk-bearing free portion scarcely ribbed; lobes 5, scarious or petal-like, often very minute. Petals 5, entire. Stamens 10, alternating with as many staminodia, very shortly united at the base in a single ring, or rarely the staminodia when broad forming an outer row; anthers globular, opening in 2 minute pores near the scarcely prominent connective. Ovary 1-celled, with 2, 3, and very rarely (except in *D. micropetala*) 4 ovules, inserted on a very short basal, usually excentric placenta. Style exserted, usually long, and more or less bearded towards the end; stigma terminal, minute or capitate. Fruit formed of the slightly-enlarged and somewhat hardened calyx. Seed usually solitary, filling the fruit, testa very thin. Embryo consisting of a homogeneous mass or thick radicle of the shape of the seed, with a rather slender neck lying along the flattened apex, entire, or perhaps divided at the point into two minute cotyledons.—Shrubs with usually a heath-like or *Diosma*-like habit. Leaves small, opposite or scattered, entire. Flowers small, nearly sessile, or shortly pedicellate in the upper axils, or in terminal heads, the floral leaves or bracts either large and coloured, or small like the stem-leaves. Bracteoles thin and scarious, concave, and keeled, enclosing the young bud, and very deciduous, or small, narrow, and more persistent.

The genus is limited to Australia. Perfect seeds have been examined only in very few species.

SECTION I. **Genetyllis.**—*Calyx-lobes not exceeding half the length of the petals, and often very minute. Flowers in single terminal heads, rarely becoming lateral by the elongation of the central axis.*

A. *Flower-heads usually nodding, surrounded by a campanulate or ovoid involucre of coloured imbricated bracts longer than the flowers, and enclosing them.*

Leaves scattered, oblong, ½ in. long or more, with recurved margins. Involucral bracts obovate, oblong.
  Leaves and bracts entire.
    Leaves elliptical-oblong. Inner bracts obovate, streaked. Calyx 10-ribbed at the base . . . . 1. *D. macrostegia.*
    Leaves linear-oblong. Inner bracts broadly oblong, one-coloured. Calyx 5- or rarely 7-ribbed at the base . . . 2. *D. Hookeriana.*
  Leaves and bracts ciliate. Calyx without prominent ribs . . 3. *D. fimbriata.*
Leaves opposite, small, erect, concave. Involucral bracts broadly oblong, entire. Calyx prominently 5-ribbed at the base . . . 4. *D. speciosa.*
Leaves linear, semiterete or triquetrous, scattered, crowded.
  Involucral bracts ovate-oblong, entire. Flowers numerous in the head. Bracteoles narrow. Calyx-lobes very small . . 5. *D. Meissneri.*
  Involucral bracts ovate-lanceolate, numerous, entire. Flowers 4 in the head. Bracteoles orbicular. Calyx-lobes broad, half as long as the petals . . . . 6. *D. helichrysoides.*
  Involucral bracts linear, numerous, ciliate. Flowers numerous in the head. Bracteoles linear. Calyx-lobes very small . 7. *D. œderoides.*

B. *Flower-heads erect or nodding, surrounded by an involucre of coloured bracts scarcely exceeding the flowers, or shorter than them and usually spreading.*

Leaves scattered and crowded, not opposite. Flowers numerous in the head.
  Leaves semiterete or triquetrous, 3 to 4 lines long. Calyx-tube irregularly glandulose-rugose . . . . 8. *D. virescens.*
  Leaves oblong, rarely above 2 lines long. Calyx-tube marked with parallel rings of glandular papillæ.
    Leaves with recurved ciliate margins. Bracteoles short. Calyx fully 3 lines, with 2 or 3 rings. Staminodia lanceolate 9. *D. Oldfieldii.*
    Leaves entire. Bracteoles narrow. Calyx about 2 lines, with 5 or 6 rings. Staminodia filiform . . . . 10. *D. purpurea.*
Leaves mostly opposite, oblong, ¼ to ½ in. long. Flowers 4 in the head. Calyx smooth, with an obscure glandular ring . . . . 11. *D. citriodora.*

C. *Flowers in terminal heads or in the upper axils, the floral leaves or bracts not very different from the stem leaves.*

Leaves mostly opposite, linear lanceolate or falcate. Stems diffuse or prostrate.
  Leaves with revolute margins. Flowers 4 to 8 in the head. Petals with narrow concave coloured tips. Staminodia prominently glandular . . . . 12. *D. thymoides.*
  Leaves triquetrous or laterally flattened. Flowers 2 to 4 in the head. Petals ovate. Staminodia small and subulate . . . 13. *D. taxifolia.*
Leaves crowded, not opposite. Erect bushy shrubs.
  Leaves obovate or oblong, often imbricate. Flowers distinctly pedicellate, often becoming lateral by the elongation of the shoot.
    Flowers numerous, scarcely 2 lines long. Calyx narrow . . 14. *D. vestita.*
    Flowers few, nearly 4 lines long. Calyx broadly turbinate . 15. *D. pauciflora.*
Leaves semiterete or triquetrous. Flowers sessile or nearly so, the heads always terminal.
  Calyx marked with numerous rings of glandular tubercles or papillæ . . . . 16. *D. diosmoides.*
  Calyx 5-ribbed, otherwise smooth . . . . 17. *D. fascicularis.*

SECTION II. **Schuermannia,** *F. Muell.—Calyx-lobes as long as the petals, or longer.*

Flowers in the axils of the upper leaves, few or forming compound heads or corymbs, or rarely simple heads.

Flowers in dense terminal simple heads. Inner bracts broad, thin, and coloured, but short. Leaves linear, slender, crowded. Staminodia broad . . . . 18. *D. pinifolia.*
Flowers in compound heads (contracted corymbs). Leaves mostly opposite.
Compound heads hemispherical. Bracts ovate, coloured, but shorter than the flowers. Leaves linear or lanceolate, often ½ in. long, with ciliate edges. Staminodia lanceolate . . . 19. *D. sanguinea.*
Compound heads small, globular, without coloured bracts. Leaves triquetrous, about 1 line long. Staminodia minute . 20. *D. micropetala.*
Flowers few in the upper axils. Calyx glabrous. Leaves opposite.
Leaves linear-triquetrous. Flowers nearly sessile. Petals as long as the calyx-lobes. Ovules 2 . . . . 21. *D. Schuermanni.*
Leaves obovate. Flowers pedicellate. Petals half as long as the calyx-lobes. Ovules 6 . . . . 23. *D. Thomasii.*
Flowers in broad leafy corymbs. Calyx hemispherical, softly villous. Leaves opposite, linear-triquetrous . . . . 22. *D. verticordina.*

SECTION I. GENETYLLIS.—Calyx-lobes not exceeding half the length of the petals, and often very minute. Flowers in simple terminal heads, rarely becoming lateral by the elongation of the central axis.

In the whole of this section the inflorescence is quite simple,—a reduced spike or raceme, each flower being sessile or very shortly pedicellate in the axil of a floral leaf or bract, with a pair of concave bracteoles close under the calyx. In the first group, comprising the first 7 species, the terminal shoot is wholly arrested, the flowers forming a strictly terminal head on a club-shaped obovoid, globular, or broad and disk-shaped receptacle, the floral leaves within the head reduced to small scarious bracts, those subtending the external flowers, with more or less of the stem-leaves next to the head much enlarged, coloured, and petal-like, forming a campanulate or ovoid involucre completely enclosing the flowers. In the second group, comprising the species 8 to 11, the flower-heads are as compact or nearly so, but the involucres are short, more or less spreading, and do not conceal the flowers. In the third, comprising the species 12 to 17, the heads are smaller and looser, the terminal shoot occasionally grows out from the centre, the receptacle is but slightly thickened, the floral leaves differ but little from those of the stem, and the flowers are sometimes pedicellate, thus showing the connection with the axillary inflorescence of those species of the following section, where it is simple.

1. **D. macrostegia,** *Benth. in Journ. Linn. Soc.* ix. 179.—Erect, attaining 2 or 3 ft. Leaves scattered, elliptical-oblong or slightly cuneate, very obtuse, ½ to ¾ in. long, with recurved entire margins. Involucres campanulate, nearly 1½ in. long, the petal-like inner bracts broadly obovate, pale yellow streaked with red, quite entire, a few outer ones shorter and redder, and 2 or 3 of the lowest passing into the stem-leaves. Flowers rather numerous. Bracteoles acuminate, as long as the flowers, deciduous. Calyx-tube marked in the adnate part with 10 prominent ribs and transversely wrinkled between them, the free part smooth; lobes very small, obovate. Petals white, about 1½ lines long. Stamens short; staminodia short, linear-clavate. Style nearly as long as the involucre, bearded towards the end. Ovules 2.—*Genetyllis macrostegia*, Turcz. in Bull. Mosc. 1849, ii. 18; Kipp. in Journ. Linn. Soc. i. 51; *Hedaroma tulipifera*, Lindl. in Gardn. Chron. 1854, 323; *Genetyllis tulipifera*, Hook. Bot. Mag. t. 4858.

W. Australia, *Drummond, 4th Coll. n.* 40, *5th Coll. n.* 97. Stirling range and E. Mount Barren, *Maxwell.*

2. **D. Hookeriana,** *Benth. in Journ. Linn. Soc.* ix. 179.—Very nearly resembles *D. macrostegia*, but is usually smaller, more slender, and less twiggy. Leaves scattered, linear-oblong, ½ to ¾ in. long, with recurved entire margins. Involucres ovoid, about 1¼ in. long, the inner bracts broadly oblong, of a uniform pink colour, or slightly white at the edge, and not streaked, the outer bracts short and recurved, but otherwise like the stem leaves. Flowers like those of *D. macrostegia*, but rather smaller, and the base of the calyx-tube has only 5 or very rarely 6 or 7 prominent ribs, and is only slightly tuberculate between them. Stamens, staminodia, and style as in *D. macrostegia*, or the style rather stouter.—*Genetyllis macrostegia*, Hook. Bot. Mag. t. 4860, not of Turcz.; *G. Hookeriana*, Meissn. in Journ. Linn. Soc. i. 37.

W. Australia, *Drummond, 5th Coll. n.* 98; *Maxwell.*

3. **D. fimbriata,** *Benth. in Journ. Linn. Soc.* ix. 179.—A bushy shrub of 1 to 2 feet. Leaves scattered, often crowded, oblong-elliptical, very obtuse, 2 to 3 lines, or on the main branches 4 lines long, the margins recurved and strongly ciliate-denticulate. Involucres ovoid, about ¾ in. long or rather more, the inner bracts petaloid, pink, broadly oblong or almost cuneate and very obtuse, the outer ones short, broad, and squarrose but coloured, and all ciliate. Flowers rather numerous. Bracteoles rather shorter than the flowers. Calyx about 3 lines long, without prominent ribs; lobes minute or quite inconspicuous. Petals triangular, about 1 line long. Staminodia filiform, nearly as long as the filaments. Styles thick, often as long as the involucre, shortly bearded towards the end.—*Genetyllis fimbriata*, Kipp. in Journ. Linn. Soc. i. 49; Hook. Bot. Mag. t. 5468.

W. Australia. Stirling range, E. extremity, *Drummond, 5th Coll. n.* 99.

4. **D. speciosa,** *Benth. in Journ. Linn. Soc.* ix. 179.—A small shrub with numerous short ascending or erect branches, not above 6 in. in our specimens. Leaves all opposite, erect, narrow-oblong, obtuse, concave, 2 to 3 lines long, or rather more on the main stems. Involucres ovoid, above 1 in. long, apparently red; inner bracts ovate-oblong, entire, a few outer ones much shorter, but not squarrose. Bracteoles lanceolate, shorter than the calyx. Calyx 2 to 2½ lines long, the adnate part prominently 5-ribbed; lobes lanceolate or acuminate, often nearly half as long as the petals. Petals 1 to 1½ lines long. Staminodia small.—*Genetyllis speciosa*, Meissn. in Journ. Linn. Soc. i. 36.

W. Australia. Between Moore and Murchison rivers, *Drummond, 6th Coll. n.* 34.

5. **D. Meissneri,** *Benth. in Journ. Linn. Soc.* ix. 179.—An erect heath-like shrub. Leaves scattered, crowded, linear, mostly 3 to 4 lines long, convex underneath, but furrowed next to the margin. Involucres broadly campanulate, ¾ to above 1 in. long; inner bracts ovate or ovate-oblong, shortly acuminate or mucronulate, apparently red, entire; outer ones short, ovate, with green leaf-like points. Flowers about 8 to 10 in the head. Bracteoles narrow, often exceeding the calyx. Calyx about 3 lines long, the

adnate part without prominent ribs, but with a granular surface; lobes ovate, not ¼ line long. Petals triangular, rather above 1 line. Staminodia small. Style variable in length.—*Genetyllis Meissneri*, Kipp. in Journ. Linn. Soc. i. 49.

W. Australia. Middle Mount Barren, *Drummond, 5th Coll. n.* 100, and with rather paler smaller involucres, n. 101.

6. **D. helichrysoides,** *Benth. in Journ. Linn. Soc.* ix. 179.—Slender and erect, often under 1 ft. high. Leaves scattered, rather crowded, linear-triquetrous or semiterete, spreading, 2 to 3 lines long. Involucres narrow, nearly 1 in. long; bracts numerous, ovate-lanceolate, acute, mostly with a prominent midrib, the inner ones coloured, passing gradually into the short broad outer ones. Flowers about 4 in the head. Bracteoles very broadly orbicular. Calyx above 3 lines long, the adnate part without prominent ribs, but the surface granular; lobes broad, very obtuse, thicker than in any other species, streaky, and half as long as the petals. Petals about 1 line long. Staminodia rather thick, capitellate.—*Genetyllis helichrysoides*, Meissn. in Journ. Linn. Soc. i. 37.

W. Australia. Between Moore and Murchison rivers, *Drummond, 6th Coll. n.* 35.

7. **D. œderoides,** *Benth. in Journ. Linn. Soc.* ix. 179.—Low and much branched. Leaves scattered, crowded, linear-triquetrous or semiterete, spreading, 2 to 3 lines long. Involucres ovoid, nearly 1 in. long, with numerous linear or linear-lanceolate, imbricate bracts, the outer ones short and entire like the stem-leaves, passing gradually into the inner long coloured ones, which are elegantly ciliate with rather long hairs. Flowers numerous, on a flat receptacle of 4 or 5 lines diameter. Bracteoles linear, ciliate. Calyx nearly 3 lines long, the adnate part obtusely 5-angled; lobes very small. Petals at least 1½ lines long. Staminodia slender.—*Genetyllis œderoides*, Turcz. in Bull. Mosc. 1849, ii. 18.

W. Australia. King George's Sound, *M'Lean;* Southern districts? *Drummond, 4th Coll. n.* 41.

8. **D. virescens,** *Benth. in Journ. Linn. Soc.* ix. 179.—A decumbent shrub, the bark of the young branches rather thick and white. Leaves scattered, crowded, linear, semiterete or triquetrous, obtuse, mostly 3 to 4 lines long. Flower-heads dense, hemispherical, often above 1 in. diameter. Involucral bracts numerous, but not exceeding the flowers, lanceolate or ovate-lanceolate, scarcely coloured, the inner ones narrower and shorter. Flowers numerous, each on an exceedingly short thick turbinate pedicel, but the broad flat receptacle not otherwise divided. Bracteoles ovate, shorter than the flowers. Calyx about 4 lines long, the tube glandular and obscurely 5-ribbed; lobes ovate, scarious, about one-fourth the length of the petals. Petals nearly 2 lines long, obtuse. Staminodia slightly clavate.—*Genetyllis virescens*, Meissn. in Journ. Linn. Soc. i. 38.

W. Australia. Between Moore and Murchison rivers, *Drummond, 6th Coll. n.* 37; Port Gregory, *Oldfield.*

9. **D. Oldfieldii,** *Benth. in Journ. Linn. Soc.* ix. 180.—Erect and bushy, attaining 3 to 4 ft. Leaves scattered, crowded, oblong, obtuse,

scarcely above 2 lines long, the margins recurved, and shortly ciliate-denticulate. Flower-heads dense, hemispherical. Involucral bracts numerous, not exceeding the flowers, imbricate but squarrose, ovate, ciliate, more or less coloured. Flowers 10 to 12 or more. Bracteoles narrow. Calyx fully 3 lines long, the adnate part not ribbed, granular at the base, and separated from the smooth free part by 2 or 3 rings of prominent glandular papillæ; lobes very small and scale-like. Petals ovate, nearly $1\frac{1}{2}$ lines long. Staminodia lanceolate.

W. Australia. Murchison river, *Oldfield.* Nearly allied to *D. purpurea,* but differs in its ciliate leaves, larger flowers, shorter bracts, fewer rings to the calyx, etc.

10. **D. purpurea,** *Benth. in Journ. Linn. Soc.* ix. 180.—Erect and much branched. Leaves scattered, crowded, and almost imbricate, linear, obtuse, 1 to 2 lines long, convex underneath, flat or concave above, the edges entire, or very minutely denticulate-ciliate. Flowers numerous, in dense hemispherical heads. Involucral bracts numerous, more or less coloured, imbricate, but somewhat spreading, rather longer than the flowers, the outer ones ovate, passing into the inner obovate or spathulate ones. Bracteoles rather narrow. Calyx about 2 lines long, the adnate part 5-ribbed at the base, the upper half encircled by 5 or 6 rings of glandular papillæ, the free part smooth; lobes very small and scale-like. Petals about 1 line long. Staminodia filiform or slightly clavate.—*Polyzone purpurea,* Endl. in Ann. Wien. Mus. ii. 191; *Genetyllis purpurea,* Schau. Myrt. Xeroc. 27. t. 2 B.

W. Australia. In the interior, *J. S. Roe (Herb. Wien. Mus.)*

11. **D. citriodora,** *Benth. in Journ. Linn. Soc.* ix. 180.—A diffuse shrub of 1 to 2 ft., the young branches with 2 prominent angles under the leaves. Leaves nearly opposite, from narrow-oblong to almost ovate-lanceolate, obtuse, $\frac{1}{4}$ to $\frac{1}{2}$ in. long, or longer on the main branches, the margins recurved or revolute. Flowers usually 4, in small terminal heads; involucre scarcely exceeding the flowers, consisting usually of 4 outer leaf-like bracts, and 4 inner ovate ones more or less coloured. Bracteoles broad and short. Calyx about 3 lines long, the adnate part obtusely 5-angled, with occasionally an obscure ring of glandular papillæ at the base of the smooth free part; lobes ovate, about half as long as the petals. Staminodia spathulate.—*Genetyllis citriodora,* Endl. in Hueg. Enum. 47; Schau. Myrt. Xeroc. 31. t. 2 C, and in Pl. Preiss. i. 97; *Hedaroma latifolium,* Lindl. Swan Riv. App. 7. t. 2 B; *Genetyllis pimeleoides,* F. Muell. Fragm. ii. 169.

W. Australia. Swan River to King George's Sound, and eastward to Cape Riche, *Baudin's Expedition; Huegel; Drummond,* 1*st Coll. n.* 148; *Preiss, n.* 2014, and others.

12. **D. thymoides,** *Benth. in Journ. Linn. Soc.* ix. 180. Low, diffuse, slender, and much branched. Leaves mostly opposite, linear or lanceolate, obtuse, 3 to 4 lines long, the margins revolute, the upper and floral ones sometimes longer. Flowers sessile, 4 to 8 together in terminal heads, the outer bracts or floral leaves sometimes slightly exceeding them but not coloured; inner bracts (within the head) very small and narrow. Bracteoles very broad, much shorter than the flowers, and falling away very early. Calyx rather slender, 2 to 3 lines long, strongly 5-ribbed, otherwise smooth,

lobes narrow-ovate, scarcely ⅓ the length of the petals. Petals about 1 line long, rather narrow, concave, with a deep-coloured spot at the tip. Staminodia linear-lanceolate, bordered by 3 to 5 prominent tubercular glands. Style bearded towards the end as in the other species, but the hairs very deciduous.—*Hedaroma thymoides*, Lindl. Swan Riv. App. 7; *Genetyllis thymoides*, Schau. Myrt. Xeroc. 33; *Darwinia brevistyla*, Turcz. in Bull. Mosc. 1847, i. 155.

W. Australia. Swan River, *Drummond*, 1*st Coll.*, *also n.* 53 *and* 149, 3*rd Coll. n.* 23, 4*th Coll. n.* 42.

13. **D. taxifolia,** *A. Cunn. in Field N. S. Wales*, 352. A straggling or decumbent shrub, or when luxuriant almost arborescent. Leaves mostly opposite, linear-falcate, triquetrous or laterally compressed, acute, ¼ to ½ in., or in very luxuriant specimens all above ½ in. long, almost petiolate, the floral ones not enlarged. Flowers 2 to 4 together at the ends of the branchlets, not exceeding the leaves. Bracteoles broad, acute, as long as the flowers. Calyx 2½ lines long, prominently 5-ribbed, the adnate part slightly rugose between the ribs; lobes very small and scale-like. Petals ovate, ½ line long or rather more. Staminodia very small and subulate.—*D. laxifolia*, Schau. Myrt. Xeroc. 38.

N. S. Wales. Rocky declivities of the Blue Mountains, *A. Cunningham.* Moist sandy heaths between Sydney and South Head, *R. Brown.*

Var. *grandiflora.* Calyx fully 3 lines long, the lobes nearly half as long as the petals.—Illawarra, *Herb. F. Mueller.*

Schauer was mistaken in supposing that A. Cunningham's specific name of *taxifolia* was a misprint; it was intended to allude to the peculiar bifarious arrangement of the leaves in luxuriant branches.

14. **D. vestita,** *Benth. in Journ. Linn. Soc.* ix. 180. Erect, bushy, with short and rigid or long and virgate branches. Leaves scattered, mostly crowded, from obovate or oblong to almost linear, 1 to 2 lines long, almost imbricate on the smaller branches, concave above, strongly keeled underneath. Flowers on pedicels of about 1 line, in globular terminal umbels or heads, becoming sometimes lateral verticils by the elongation of the central shoot, the floral leaves like the stem ones or rather broader. Bracteoles nearly as long as the calyx, but very deciduous. Calyx not 2 lines long, the adnate part 5-ribbed, otherwise smooth, the free part obscurely 10-ribbed; lobes very small and scale-like. Petals white, above 1 line long. Staminodia subulate, rather longer than the filaments. Style not twice as long as the petals.—*Genetyllis vestita*, Endl. in Hueg. Enum. 47; Schau. Myrt. Xeroc. 30; and in Pl. Preiss. i. 96.

W. Australia. King George's Sound, *Baxter, Huegel*, and others; eastward to Cape le Grand, *Maxwell*; and thence to Swan River, *Preiss, n.* 433; *Drummond*, 4*th Coll. n.* 161, 5*th Coll. n.* 103; Cape Naturaliste, *Oldfield.*

15. **D. pauciflora,** *Benth. in Journ. Linn. Soc.* ix. 180. Apparently tall and bushy, with numerous short branches. Leaves scattered, erect or spreading, obovate or oblong, very obtuse, 1 to 2 lines long, imbricate on the smaller branches, concave above, convex underneath, but scarcely keeled, usually glaucous, entire or slightly serrulate-ciliate; the floral ones similar. Flowers shortly pedicellate in the upper axils, either forming a terminal head of 3 to

6, or more frequently lateral by the elongation of the central shoot. Bracteoles very broad, shorter than the calyx, and very deciduous. Calyx-tube broadly turbinate, nearly 2 lines long, the adnate part 5-ribbed, otherwise smooth; lobes very minute or scarcely conspicuous. Petals white, ovate, nearly as long as the calyx-tube, slightly serrulate. Staminodia slender. Style shortly exceeding the petals, bearded under the broadly-capitate stigma.

W. Australia. Between Moore and Murchison rivers, *Drummond, 6th Coll. n.* 38; S. Hutt River, *Oldfield.* Nearly allied to *D. vestita*, but the few flowers, broad calyx, and large petals give it a very different aspect.

16. **D. diosmoides,** *Benth. in Journ. Linn. Soc.* ix. 180. An erect bushy shrub of 2 or 3 feet, with the aspect of a heath or a Diosma. Leaves scattered, crowded, linear, semiterete or triquetrous, thick or slender, obtuse, 1 to 2 or rarely 3 lines long. Flowers numerous, in compact terminal globular heads of 3 or 4 lines diameter, the floral leaves on the outside not different from the stem ones. Bracteoles oblong-lanceolate, shorter than the calyx. Calyx about 1½ lines long, the adnate part obscurely 5-ribbed, and covered nearly from the base by glandular papillæ more or less distinctly arranged in 6 to 8 prominent parallel rings, the free part short and smooth; lobes very small and scale-like. Petals white, about ¾ line long. Staminodia slender, about as long as the filaments. Style exserted.—*Genetyllis diosmoides*, DC. Prod. iii. 209, and Mem. Myrt. t. 2 (incorrect as to the stamens); Schau. Myrt. Xeroc. 28. t. 2 A (the staminodia too broad), and in Pl. Preiss. i. 96; *G. Drummondii*, Turcz. in Bull. Mosc. 1847, i. 155 (a short-leaved form).

W. Australia. Common in rocky places and on the seacoast at King George's Sound and adjoining districts, *R. Brown* and others, *Drummond, 4th Coll. n.* 22, *5th Coll. suppl. n.* 21; *Preiss, n.*. 223.

Genetyllis affinis, Turcz. in Bull. Mosc. 1847. i. 155 is said to differ in the calyx quite smooth, but in Drummond's specimens, 4th Coll. n. 21, quoted by him, it has certainly the parallel rings of the species.

17. **D. fascicularis,** *Rudge in Trans. Linn. Soc.* xi. 299. t. 22. An erect much-branched heath-like shrub. Leaves scattered, often crowded, linear, slender, semiterete or obtusely triquetrous, subulate-pointed, mostly 4 to 5 lines long, shortly petiolate, the floral ones not different or slightly longer. Flowers about 6 to 12 together in terminal heads within the last leaves. Bracteoles narrow and short. Calyx slender, not 3 lines long, the adnate part prominently 5-ribbed, otherwise smooth; lobes very small and scale-like. Petals broad, about ½ line long. Staminodia short and filiform. Style long and slender.—Schau. Myrt. Xeroc. 36. t. 2 D.

N. S. Wales. Port Jackson, *R. Brown* and others.

Cryptostemon ericæus, F. Muell., published by Miquel in Nederl. Kruidk. Arch. iv. 115, from F. Mueller's description taken from a garden specimen of a N. S. Wales plant of which no specimen is preserved, is most probably *Darwinia fascicularis,* with which the description agrees in every respect except that the staminodia are not mentioned. These, however, may well have been overlooked. *Francisia,* Endl. Gen. Pl. 1226, proves from the investigation of Dr. Fenzl to have been established on a drawing of Ferd. Bauer's, n. 1226, representing *D. fascicularis,* the stamens by some error, possibly of the press, being described as 20 instead of 10.

SECTION II. SCHUERMANNIA, *F. Muell. Fragm.* iv. 57. (*Schnermannia*, probably from a typographical error.)—Calyx-lobes as long as the petals or longer. Flowers in the axils of the upper leaves, few, or forming compound heads or corymbs, or rarely simple heads.

18. **D. pinifolia,** *Benth. in Journ. Linn. Soc.* ix. 181. Erect and much-branched, closely resembling *D. fascicularis* in habit, foliage, and inflorescence, but with very different calyx and staminodia. Leaves scattered, crowded, linear, slender, semiterete or triquetrous, acute or mucronate, often ½ in. long, attenuate at the base but not petiolate. Flowers in dense terminal heads, the outer floral leaves like those of the stem, the bracts within the heads shorter, broader, and thin. Bracteoles ovate-oblong or spathulate, shorter than the flowers, Calyx slender, nearly 3 lines long, the adnate part 5-ribbed, the free part broader and 10-ribbed; lobes broadly ovate, about the same length and consistence as the broad obtuse petals. Staminodia broadly obtuse, more or less outside the stamens.—*Hedaroma pinifolium*, Lindl. Swan Riv. App. 7; *Genetyllis pinifolia*, Schau. Myrt. Xeroc. 34.

W. Australia. Swan River, *Mylne.*

19. **D. sanguinea,** *Benth. in Journ. Linn. Soc.* ix. 181. Apparently diffuse. Leaves opposite, often crowded, linear-oblong or lanceolate, 2 to 3 lines long, with rather thin recurved ciliolate margins. Flowers crowded in a dense terminal hemispherical compound head of ½ to 1 in. diameter, consisting of several partial heads of about 4 flowers each. Bracts or floral leaves ovate, usually coloured, but shorter than the flowers. Bracteoles very broad, mucronulate, shorter than the calyx. Calyx about 3 lines long, the adnate part prominently 5-ribbed and granular-tuberculate between the ribs; lobes cordate-ovate, nearly 1 line long. Petals ovate, about as long as the calyx-lobes. Staminodia lanceolate.—*Genetyllis sanguinea*, Meissn. in Journ. Linn. Soc. i. 38.

W. Australia. Between Moore and Murchison rivers, *Drummond, 6th Coll. n.* 36.

20 **D. micropetala,** *Benth. in Journ. Linn. Soc.* ix. 181. Erect and bushy, with slender branches. Leaves opposite or alternate, not crowded, linear, triquetrous, obtuse, 1 or rarely 2 lines long, the floral ones rather larger but scarcely otherwise different. Flowers in small terminal compound almost globular heads, 2 to 4 together in each partial head. Bracteoles nearly as long as the flowers. Calyx about 1½ lines long, the adnate part of the tube 5-ribbed, otherwise smooth; lobes petal-like, and as long as or rather longer than the petals. Staminodia very minute. Style not twice as long as the petals. Ovules 4.—*Genetyllis micropetala*, F. Muell. Fragm. i. 12.

S. Australia. Kangaroo Island, *Bannier.*

21. **D. Schuermanni,** *Benth. in Journ. Linn. Soc.* ix. 181. Procumbent and much branched. Leaves opposite, linear, triquetrous, shortly acute or mucronulate, ¼ to ½ in. long, the floral ones not different. Flowers solitary in the upper axils of short branchlets, on very short pedicels. Bracteoles broad, about as long as the calyx. Calyx nearly 4 lines long, the adnate part prominently and obtusely 5-ribbed, otherwise smooth; lobes lanceolate, petal-like, about the length of the ovate petals. Staminodia subulate.

Style long. Ovules 2.—*Schuermannia homoranthoides*, F. Muell. in Linnæa, xxv. 387; *Genetyllis Schuermanni*, F. Muell. Fragm. i. 12.

S. Australia. Near Boston Point, Port Lincoln, *Wilhelmi.*

22. **D. verticordina,** *Benth. in Journ. Linn. Soc.* ix. 181. Erect and densely bushy. Leaves opposite, linear, semiterete or triquetrous, mostly about 3 lines long. Flowers in the upper axils of the short branchlets forming a dense flat-topped leafy corymb, the pedicels 1 to 2 lines long. Calyx-tube hemispherical, 1½ lines diameter, softly pubescent, more villous at the base with a dense ring of white hairs as in many species of *Verticordia*; lobes ovate, scarious, nearly 2 lines long, very shortly and irregularly denticulate-ciliate. Petals ovate-lanceolate, rather shorter than the calyx-lobes entire, scarious with a broad dark-coloured central line. Stamens united for nearly 1 line above the calyx-tube; staminodia lanceolate-subulate, forming a distinct outer series. Style very long. Ovules 2.—*Chamælaucium verticordinum*, F. Muell. Fragm. iv. 57; *Verticordia integrisepala*, F. Muell. Herb.

W. Australia. Rocks near Cape le Grand, *Maxwell,* to the eastward of King George's Sound, *Baxter.*

Although this plant has, as observed by F. Mueller, the calyx-tube, and some other characters of *Verticordia,* yet, on the whole, he appears to have referred it more correctly to his section *Schuermannia,* at first proposed as a distinct genus, then reduced to *Genetyllis,* and afterwards transferred to *Chamælaucium.* The anthers and ovary are those of *Darwinia* (*Genetyllis*), and not of *Chamælaucium,* and the affinity with the former genus is still further indicated by the remarkably long style which is common in *Darwinia,* rare in *Verticordia,* and unknown in *Chamælaucium.*

23. **D. Thomasii,** *Benth. in Journ. Linn. Soc.* ix. 181. Slender and somewhat glaucous. Leaves opposite, obovate-falcate, very oblique, the midrib near the shorter edge, and terminating in a short recurved point or acute angle, the floral ones not different. Flowers large, pink, on pedicels of 3 lines or more in the upper axils. Bracteoles persistent, almost petal-like, obtuse, with a sharp point about 3 lines long. Calyx-tube rather narrow, about as long as the bracteoles; lobes petal-like, obovate-oblong, about 2 lines long, minutely denticulate. Petals orbicular, entire, about half as long as the calyx-lobes. Stamens shorter than the petals; anthers globular, the cells opening in oblong pores; staminodia rather shorter, adnate at the base to the filaments of the petaline stamens. Style twice as long as the calyx, shortly bearded below the stigma. Ovules 6.—*Chamælaucium Thomasii*, F. Muell. Fragm. iv. 137. t. 30.

Queensland. Sandstone country, head of Cape River, *Bowman.*

3. HOMORANTHUS, A. Cunn.

Calyx-tube narrow, the adnate part 5-ribbed; lobes 5, subulate, longer than the petals. Petals 5, entire. Stamens 10, alternating with as many staminodia, and united with them very shortly at the base in a single ring. Anthers globular, opening in 2 minute pores near the scarcely prominent connective. Ovary 1-celled, with about 4 ovules inserted on a short basal placenta. Style exserted, bearded towards the end; stigma terminal, minute.

Fruit . . .—Shrub. Leaves opposite. Flowers 2 to 4 together at the ends of the branches.

The genus is limited to a single species, only differing from *Darwinia* in the subulate calyx-lobes.

1. **H. virgatus,** *A. Cunn. in Schau. Myrt. Xeroc.* 41. t. 3 A. Spreading or diffuse, closely resembling *Darwinia taxifolia* in aspect. Leaves linear, slender, triquetrous, often falcate, obtuse or shortly acute, $\frac{1}{4}$ to $\frac{1}{2}$ in. long, the floral ones not different. Flowers 2 to 4 together at the ends of the branches, nearly sessile. Bracteoles broad, concave, keeled, scarious, enclosing the young bud but falling off long before the flower expands. Calyx-tube $2\frac{1}{2}$ lines long, prominently 5-ribbed, and the adnate part somewhat rugose between the ribs. Petals broad, about $\frac{3}{4}$ line long. Staminodia filiform. Ovules in all the flowers examined 4, according to Schauer 4 to 8. Style not very long.—*H. flavescens,* A. Cunn. in Schau. l. c. 40. t. 3 B.

Queensland. Islands of Moreton Bay, *A. Cunningham, Fraser, F. Mueller.*
N. S. Wales. Forest lands skirting Liverpool Plains and Mohe's Rivulet below Wellington Valley, *A. Cunningham;* Cape Brown, *C. Moore.*
I can discover no difference whatever between the two supposed species.

4. VERTICORDIA, DC.

(Chrysorrhoe, *Lindl.*)

Calyx-tube hemispherical turbinate or rarely cylindrical, the adnate part 5 or 10-ribbed, or smooth; lobes 5, spreading, deeply divided into digitate pectinate or ciliate lobes, or into numerous long, simple, hair-like lobes or cilia, with, in some species, accessory lobes, alternating with and outside the principal ones, scarious, reflexed on the tube, with long cilia turned up again from the base of the calyx; occasionally also 5 herbaceous appendages reflexed on the tube under the primary lobes. Petals 5, entire fringed or digitate. Stamens 10, alternating with as many staminodia, more or less united at the base in a ring or broad tube in a single row, or the staminodia when broad forming an outer series; anthers either globular and 2-porose, as in *Darwinia*, or with 2 parallel cells opening in longitudinal slits, as in *Chamælaucium.* Ovary 1-celled, either with 2 or 4 ovules on a small excentric placenta, or about 8 or 10 on a more or less peltate placenta. Style included or shortly exserted, rarely elongated; stigma terminal, small, or capitate or peltate. Fruit formed by the hardened base of the slightly-enlarged persistent calyx. Seed usually solitary, testa very thin; embryo consisting of a homogeneous mass of the shape of the seed, with a slender neck lying along the flattened apex, entire or perhaps divided at the point into two minute cotyledons.—Shrubs with usually a heath-like or Diosma-like aspect, glabrous except the cilia on the edges of the leaves. Leaves small, opposite or rarely (in *V. serrata*) alternate, entire. Flowers usually pedicellate in the upper axils, forming often broad terminal leafy corymbs, or simple leafy spikes or racemes below the ends of the branches; the elegantly plumose radiating calyx-lobes often coloured, the floral leaves resembling the upper stem-leaves, but in some species all the upper leaves short, broad, and concave, whilst the lower ones are slender and triquetrous. Bracteoles thin and scarious, folded over each other or enclosing the flower-bud, but very deciduous, or rarely connate at the base and persistent, the keel often terminating in a point at or below the apex, very variable in length even in the same species.

The genus is limited to Australia. It is characterized by the calyx. In other respects the first section has the characters of *Darwinia*, the second those of *Chamælaucium*. In the few seeds which I have seen ripe I could find no notch in the slender end of the embryo, and it is therefore still uncertain whether that or the thick homogeneous mass is the radicular end.

SECTION 1. **Euverticordia.**—*Anthers nearly globular, opening in 2 almost dorsal pores; connective either small and inconspicuous or more or less thickened or produced into a concave or hooded appendage, concealing the pores. Ovules 2 or rarely 4 or 1, on a small or stalk-like placenta.*

A. *Calyx-tube narrow, 5-ribbed, glabrous; lobes 5, erect; with 3 to 5 long simple hair-like divisions.*

Flowers small, in umbel-like corymbs, the floral leaves reduced to small bracts . . . . 1. *V. Wilhelmii.*

B. *Calyx-tube hemispherical, smooth or ribbed, pubescent or with a tuft of spreading hairs round the base; primary lobes 5, spreading, deeply divided into 5, 7, or rarely 9 linear or subulate pectinate-ciliate digitate lobes. Petals entire denticulate or ciliate. Connective small.*

Calyx-lobes contracted into a short broad claw, the tube hirsute with long hairs at the base, glabrous or pubescent above.
  Flowers white or pink. Leaves rather slender . . . . . . 2. *V. densiflora.*
  Flowers yellow. Leaves very short and thick . . . . . . 3. *V. stelluligera.*
Calyx-lobes digitate from the base.
  No staminodia. Flowers very small . . . . 4. *V. minutiflora.*
  Staminodia linear.
    Divisions of the calyx-lobes flat and scarious, pectinate-pinnatifid; tube pubescent all over.
      Flowers white or pink. Style scarcely exserted . . . . 5. *V. Fontanesii.*
      Flowers yellow. Style very long . . . . 6. *V. helichrysantha.*
    Divisions of the calyx-lobes subulate, the pectinate cilia long.
      Calyx-tube hirsute only with a ring of hairs at the base, otherwise glabrous. Flowers very numerous, in broad terminal leafy corymbs, often pedunculate.
        Leaves small, obovate or oblong . . . . 7. *V. Brownii.*
        Leaves linear, triquetrous, slender . . . . 17. *V. polytricha.*
      Calyx-tube hirsute all over. Flowers very small in the upper axils of the densely-tufted branches. Leaves linear, triquetrous . . . . 8. *V. conferta.*
  Staminodia lanceolate, petal-like, fringed . . . . 9. *V. Harveyi.*

C. *Calyx-tube turbinate or hemispherical, glabrous, primary lobes 5, spreading, deeply and digitately divided into 5, 7, or rarely 9 linear or subulate pectinate-ciliate lobes. Connective thickened and usually produced into a concave appendage. Flowers yellow.*

Petals deeply fringed. Staminodia fringed. Corymbs small, few-flowered. Bracteoles persistent . . . . 10. *V. fimbrilepis.*
Petals rigid, denticulate. Staminodia entire. Corymbs broad, many-flowered. Bracteoles very deciduous.
  Leaves serrulate-ciliate. Staminodia broad. Connective-appendage short . . . . 11. *V. serrata.*
  Leaves entire. Staminodia subulate. Connective-appendage large and hood-shaped . . . . 12. *V. nitens.*
Petals digitately divided into 5, 7, or 9 rigid subulate lobes.
  Staminodia oblong or lanceolate. Bracteoles persistent.

Connective-appendage large, with 2 horns turned down over the cells. Staminodia variously toothed or fringed . . . 13. *V. grandiflora.*
Connective-appendage short, not horned. Staminodia entire.
Stem-leaves linear-triquetrous, rigid, often above ½ in. long. Flowers few, large . . . . 14. *V. chrysantha.*
Stem-leaves slender, those of the corymb often small and ovate. Flowers numerous, small . . . . 15. *V. Preissii.*
Staminodia ciliate or fringed . . . . 16. *V. acerosa.*

D. *Calyx-tube with a ring of long hairs at the base, otherwise glabrous, pubescent, or shortly villous; primary lobes* 5, *spreading or partially reflexed; divided into numerous subulate or hair-like lobes or long cilia. Connective small.*

Divisions or cilia of the calyx-lobes all horizontally spreading. Flowers very small.
Stems erect. Flowers in broad dense leafy corymbs, often pedunculate. Petals slightly ciliate . . . . 17. *V. polytricha.*
Stems procumbent. Petals entire . . . . 18. *V. demissa.*
Lateral divisions or cilia of the calyx-lobes outermost in imbrication, more or less reflexed on the calyx-tube (but without real accessory reflexed lobes).
Stems diffuse or prostrate. Style very long.
Petals entire. Style glabrous . . . . 19. *V. humilis.*
Petals ciliate. Style bearded at the top with long purple hairs 20. *V. penicillaris.*
Stems erect. Flowers corymbose. Petals fringed with fine cilia.
Staminodia entire. Stigma capitate . . . . 21. *V. multiflora.*
Staminodia fringed. Stigma broadly peltate . . . . 22. *V. Huegelii.*

E. *Calyx with* 5 *spreading primary lobes, either digitate with pectinate divisions or divided into very numerous hair-like lobes, and* 5 *accessory lobes, alternating with the primary ones on the outside, very thin and scarious, closely reflexed on the tube, divided into numerous fine cilia, and turned up again from the base of the tube. Connective small. Lower leaves laterally compressed or triquetrous.*

Staminodia fringed or ciliate.
Flowers rather large. Petals fringed . . . . 23. *V. insignis.*
Flowers rather small. Petals entire . . . . 24. *V. habrantha.*
Staminodia subulate, entire.
Flowers rather large, numerous, in a broad corymb. Hair-like divisions of the calyx exceedingly numerous. Petals fringed with numerous long cilia. Style straight, glabrous . . . 25. *V. monadelpha.*
Flowers few, in small corymbs, or in the upper axils. Divisions of the spreading calyx-lobes few, ciliate. Petals fringed at the end with a few irregular lobes. Style bent and bearded towards the end . . . . 26. *V. Lehmanni.*

SECTION 2. **Catocalypta.**—*Anthers ovoid or oblong, with* 2 *parallel cells, adnate to a more or less thickened connective, and opening in longitudinal slits. Ovules several, usually* 8 *or* 10, *in* 2 *rows, on an obliquely peltate, oblong or rarely stalk-like placenta.*

A. *Calyx-lobes* 5, *spreading, without reflexed accessory lobes or herbaceous appendages. Racemes short, mostly terminal, almost corymbose. Leaves linear-triquetrous or semiterete.*

Leaves mostly ½ in. long, or more. Lateral cilia of the calyx-lobes reflexed on the tube. Petals fringed or denticulate, shorter than the calyx-segments . . . . 27. *V. Cunninghamii.*
Leaves mostly under ¼ in. Calyx-lobes without reflexed divisions or cilia. Petals entire, longer than the calyx-lobes . . . . 28. *V. picta.*
Leaves mostly 1 to 2 lines long. Lateral cilia of the calyx-lobes reflexed on the tube . . . . 29. *V. pennigera.*

B. *Calyx-lobes* 5, *spreading, without reflexed accessory lobes, but with* 5 *herbaceous reflexed appendages on the tube under the segments. Flowers usually forming oblong racemes or spikes below the ends of the branches. Leaves small.*

Leaves linear or lanceolate, serrulate. Calyx-appendages shorter than broad, sometimes scarcely conspicuous . . . . 29. *V. pennigera.*
Leaves obovate or oblong, not 2 lines long. Calyx-appendages ovate, usually half as long as the tube.
Leaves spreading or loosely imbricate . . . . 30. *V. Drummondii.*
Leaves closely appressed and imbricate . . . . 31. *V. pholidophylla.*

C. *Calyx-lobes* 5, *spreading, with subulate plumose divisions,* 5 *accessory lobes outside and alternating with the primary ones, thin and transparent, reflexed on the tube, fringed or densely ciliate and turned up again from the base of the tube, and* 5 *herbaceous reflexed appendages on the tube under the spreading lobes. Flowers forming racemes or spikes below the ends of the branches or rarely short terminal racemes. Leaves obovate or orbicular, usually glaucous.*

Leaves mostly about 1 line long, closely imbricate.
Reflexed accessory calyx-lobes ovate-lanceolate, fringed with long cilia. Petals with long cilia . . . . 32. *V. spicata.*
Reflexed accessory calyx-lobes orbicular, shortly fringed. Petals entire or scarcely denticulate . . . . 33. *V. lepidophylla.*
Leaves obovate, mostly keeled, 2 to 3 lines long. Herbaceous calyx-appendages very short and broad. Flowers white or pink, in short racemes . . . . 34. *V. ovalifolia.*
Leaves obovate or orbicular, 2 to 3 lines long. Herbaceous calyx-appendages half as long as the tube. Flowers yellow, in long racemes below the ends of the branches . . . . 35. *V. chrysostachya.*
Leaves orbicular, 3 to 4 lines long. Flowers large, red or blue.
Plumose divisions of the spreading calyx-lobes white and scarious. Petals fringed with long cilia . . . . 36. *V. oculata.*
Plumose divisions of the spreading calyx-lobes hair-like and purple or red. Petals bordered by short teeth . . . . 37. *V. grandis.*

SECTION 1. EUVERTICORDIA. Anthers nearly globular, opening in two almost dorsal pores; connective either small and inconspicuous, or more or less thickened, or produced into a concave or hooded appendage concealing the pores. Ovules 2 or rarely 4 or 1, on a small or stalk-like placenta.

This section, with the anthers and ovary of *Darwinia* and *Homoranthus,* is only distinguished from them by the calyx.

A. Calyx-tube narrow, 5-ribbed, glabrous; primary lobes 5, erect, each divided into 3 to 5 long, simple, hair-like lobes.

This single species differs from all others of the genus in inflorescence and the shape of the calyx, and in its lobes forms an approach to those of *Homoranthus.*

1. **V. Wilhelmii,** *F. Muell. in Trans. Vict. Inst.* 122. Erect and bushy. Leaves linear, semiterete or triquetrous, slender, obtuse or mucronulate, 2 to 3 lines long, crowded on the smaller shoots. Flowers small, in small dense terminal corymbs on very short pedicels, the floral leaves in the corymb all reduced to small bracts. Bracteoles very thin and deciduous. Calyx-tube narrow, above 1 line long, 5-ribbed, glandular-rugose between the ribs, lobes 5, erect, thin, short, and broad, terminating in 3 or 5 long hair-like divisions, the middle ones exceeding the petals. Petals ovate-oblong, entire, about 1 line long. Stamens very short with small globular anthers, the connective not prominent. Staminodia minute, subulate. Style exserted,

slender, bearded towards the end. Ovules 2 or 4, one of them speedily enlarging.

S. Australia. Boston Point, Port Lincoln, *Wilhelmi.*

B. Calyx-tube hemispherical, smooth or ribbed, pubescent or with a tuft of spreading hairs round the base; primary lobes 5, spreading, deeply divided into 5, 7, or rarely 9 linear or subulate pectinate-ciliate digitate lobes. Petals entire denticulate or ciliate. Connective small.

This group has the calyx-tube of D. with the segments of C.

2. **V. densiflora,** *Lindl. Swan Riv. App.* 6. Erect and bushy, attaining 2 or 3 ft. Leaves linear, semiterete or triquetrous, slender, $\frac{1}{4}$ to $\frac{1}{2}$ in. long, crowded on the short side-branches so as to form axillary tufts as in *V. Fontanesii,* but usually more remote on the main stems, the floral ones in the corymb often lanceolate. Flowers white or pink, on pedicels rarely exceeding 2 lines, in dense terminal leafy corymbs, usually pedunculate. Calyx-tube hemispherical, hirsute with long hairs at the base only, primary lobes 5, spreading to 4 or 5 lines in diameter, each one contracted at the base into a broad ciliate claw, and divided to below the middle into 3 or 5 pectinate-ciliate digitate lobes. Petals short, nearly orbicular, fringed with numerous cilia. Stamens very shortly united; anthers globular, 2-porose; connective small. Staminodia lanceolate-subulate, entire, slightly glandular. Ovules 2 (or 1?). Style exserted, bearded towards the end.—Schau. Myrt. Xeroc. 50, and in Pl. Preiss. i. 98.

W. Australia. Swan River, *Drummond, 1st Coll. n.* 174; *Preiss, n.* 180; Blackwood River, *Oldfield;* Salt River and south coast, *Maxwell.*

V. cæspitosa, Turcz. in Bull. Mosc. 1847, i. 157, described from Gilbert's specimens, n. 330, which I have not seen, appears, from the character given, to be the same as *V. densiflora.*

3. **V. stelluligera,** *Meissn. in Journ. Linn. Soc.* i. 38. Very near *V. densiflora,* but the broader leaves and small corymbs of yellow flowers give it a very different aspect. Branches divaricate, slender. Leaves from linear-triquetrous to oblong and concave, very obtuse, rarely 3 lines long. Flowers small, yellow, in numerous small but dense leafy corymbs, often pedunculate. Calyx-tube hemispherical, hirsute with long hairs at the base only, contracted at the top; primary lobes 5, scarcely spreading to 3 lines diameter, each contracted into a short broad claw and divided into 3 or 5 linear flat but pectinate digitate lobes. Petals ovate, shorter than the calyx-lobes, fringed with fine cilia. Stamens and lanceolate-subulate staminodia of *V. densiflora.* Style exserted, bearded towards the end. Ovule 1.

W. Australia. Between Moore and Murchison rivers, *Drummond, 6th Coll. n.* 50.

4. **V. minutiflora,** *F. Muell. Fragm.* iv. 58. Erect and bushy, with the habit of *V. Fontanesii,* but more slender, with smaller finer leaves, much smaller flowers, and no staminodia. Leaves linear, semiterete or triquetrous, slender, obtuse, 2 to 3 lines long, crowded on the short lateral branches. Flowers very small on very short pedicels, in small terminal leafy corymbs, or in the upper axils. Bracteoles ovate, more persistent than in the allied species. Calyx-tube ovoid-globular, hirsute, scarcely above $\frac{1}{2}$ line long; primary lobes 5; spreading scarcely to a diameter of 2 lines, digitately

divided from near the base into 3 or 5 white scarious lobes bordered by a few long scarious cilia. Petals ovate, entire, as long as the calyx-lobes. Staminal disk truncate, with 10 short equidistant stamens, and no staminodia; anthers globular, 2-porose, the connective slightly prominent. Ovules 2 (or 1?). Style filiform, slightly bearded towards the end.

W. Australia. Towards the Great Bight, *Maxwell.*

5. **V. Fontanesii,** *DC. Prod.* iii. 209. Erect and bushy, attaining 3 or 4 ft. Leaves linear, semiterete or triquetrous, usually slender, obtuse or mucronate, 3 to 4 lines long, and densely crowded on the short lateral shoots, rarely looser and ½ in. long, or short, thick, and closely decussate. Flowers white or pink, on slender pedicels, rarely above ¼ in. long, in terminal leafy corymbs or rounded dense panicles, close above the stem-leaves or shortly pedunculate. Calyx-tube hemispherical, without prominent ribs, densely and softly hirsute all over, the adnate part and ovary exceedingly short; primary lobes 5, spreading to a diameter of 3 to 4 lines in the common form, divided nearly to the base into about 5 or 7 digitate, linear, but flat and scarious lobes, pectinate-ciliate or pinnatifid, or sometimes toothed only towards the end. Petals oblong or ovate, as long as the calyx-lobes, slightly pubescent, entire or slightly ciliate towards the base. Stamens shortly united; anthers globular, 2-porose, the connective inconspicuous; staminodia oblong-linear, obtuse, glandular. Ovules 2 or 4. Style filiform, scarcely exserted, more or less bearded towards the end.—Schau. Myrt. Xeroc. 47, and in Pl. Preiss. i. 98; *Chamælaucium plumosum*, Desf. in Mem. Mus. Par. v. 42. t. 4; *Verticordia Sieberi*, Diesing in Schau. Myrt. Xeroc. 49.

W. Australia. King George's Sound and adjoining districts, *A. Cunningham* and others; *Drummond, 3rd Coll. n.* 30, *Preiss, n.* 172, 174.

Var. *grandiflora.* Calyx-tube with shorter hairs; lobes spreading to about ½ in. diameter —*Drummond, 5th Coll. n.* 110.

Var. (?) *parviflora.* Flowers small, the petals ovate-lanceolate.—Lucky Bay, *R. Brown.*

6. **V. helichrysantha,** *F. Muell. Herb.* An erect shrub, with the habit, foliage and inflorescence of the large-flowered variety of *V. Fontanesii*, but the flowers appear to be yellow and the style much longer. Leaves linear, semiterete or triquetrous, very obtuse, 2 to 4 lines long, crowded on the smaller shoots. Flowers on pedicels as long as the leaves, in small, terminal, leafy corymbs, or in the upper axils. Calyx-tube hemispherical, scarcely ribbed, softly hirsute all over; primary lobes 5, spreading to nearly ½ in. diameter, each one digitately divided into 5 or 7 linear, flat, scarious, pectinate-ciliate lobes. Petals ovate-oblong, entire, pubescent, as long as the calyx. Stamens and staminodia of *V. Fontanesii.* Style subulate, ¾ in. long, shortly bearded towards the end. Ovules 2.

W. Australia. Phillips Range and Cape Riche, *Maxwell.*

7. **V. Brownii,** *DC. Prod.* iii. 209. Erect, bushy, and much branched. Leaves obovate or oblong, very obtuse, keeled or triquetrous, mostly 1 to 1½ lines long, almost imbricate and decussate on the short barren branches. Flowers small, on pedicels of 1 to 2 lines, and very numerous, in dense, broad, terminal, leafy corymbs, more or less pedunculate. Bracteoles distinct. Calyx-tube hemispherical or almost disk-shaped, contracted at the top, not

ribbed, glabrous, except a ring of spreading hairs round the base; primary lobes 5, spreading to about 3 lines diameter, deeply divided into about 7 long hair-like lobes, all pectinately fringed by long cilia chiefly below the middle. Petals shorter than the calyx-lobes, orbicular, entire or obscurely denticulate, glabrous, connivent. Stamens very shortly united in a ring; anthers globular, 2-porose, with a small connective; staminodia lanceolate-linear, glandular inside. Ovule usually 1. Style shortly exserted.—Schau. Myrt. Xeroc. 52; *Chamælaucium Brownii*, Desf. in Mem. Mus. Par. v. 271. t. 19.

W. Australia. Lucky Bay (Cape Le Grand), *R. Brown;* ranges to the eastward of King George's Sound, *Baxter, Maxwell, Drummond n.* 176, 5*th Coll. n.* 50.

8. **V. conferta,** *Benth.* Rigid and erect, with numerous short branches, forming apparently a low, dense, flat-topped bush in Drummond's specimens, with more of the habit of *V. Harveyi* in Maxwell's. Leaves linear, semiterete or triquetrous, obtuse, mostly 2 to 3 lines long, all densely crowded. Flowers very small, on very short pedicels in the upper axils. Calyx-tube scarcely above ½ line long, 10-ribbed, shortly hirsute; primary lobes 5, spreading to little more than 2 lines diameter; each one deeply divided into 3 or 5 subulate, more or less pectinate-ciliate lobes, with a few simple cilia. Petals ovate, obtuse, entire, rather firm and glabrous, or scarcely pubescent. Staminal disk very broad; filaments short; anthers globose, 2-porose, with a scarcely prominent connective; staminodia subulate, slender. Ovule 1 (or 2?). Style exserted, subulate, bearded or glabrous.

W. Australia, *Drummond,* 5*th Coll. n.* 114; swampy places, near E. Mount Barren, *Maxwell.*

9. **V. Harveyi,** *Benth.* Erect, with slender branches, often nearly leafless below the corymbs, as in *V. densiflora* and *V. polytricha.* Leaves linear-subulate, semiterete or triquetrous, ¼ to ½ in. long, crowded in the corymbs, and some of them exceeding the flowers. Flowers small, numerous, on pedicels rarely exceeding 1 line, in dense leafy corymbs. Calyx-tube hemispherical, obscurely ribbed, softly pubescent; primary lobes 5, spreading to about 3 lines diameter, digitately divided into about 7 subulate, pectinate-ciliate lobes. Petals ovate or broadly lanceolate, fringed with rather long cilia or rarely entire. Stamens very short; anthers globular, 2-porose, the connective slightly thickened but small. Staminodia lanceolate, petal-like and fringed, or linear-lanceolate and entire. Style shortly exserted, bent and bearded towards the end. Ovules 2.

W. Australia. Near Cape Riche, *Harvey, Maxwell.*
Var. *nudipetala.* Petals scarcely fringed. Staminodia entire.—W. Australia, *Maxwell.*

C. Calyx-tube turbinate or hemispherical, ribbed, glabrous; primary lobes 5, spreading, deeply divided into 5, 7, or rarely 9 linear or subulate pectinate-ciliate lobes. Connective thickened and usually produced into a concave appendage. Flowers yellow. Bracteoles often persistent.

This group corresponds nearly to Schauer's section *Chrysoma.* The calyx-segments are as in B, but the tube is different, the petals more rigid, and the connective often curiously developed.

10. **V. fimbrilepis,** *Turcz. in Bull. Mosc.* 1847, i. 158. Erect, with slender branches. Leaves linear, semiterete or triquetrous, obtuse or mucro-

nulate, mostly 2 to 3 lines long, clustered on the smaller branches. Flowers rather small, on pedicels scarcely above 2 lines long, in small, terminal, leafy corymbs, or rarely ovoid, leafy panicles. Bracteoles usually persistent and connate at the base. Calyx-tube almost hemispherical, 10-ribbed, glabrous, not above 1 line diameter; primary lobes 5, spreading to about 4 lines diameter, each one digitately divided into 5 or 7 linear, pectinate-ciliate (white ?) lobes. Petals ovate, striate, deeply fringed, nearly as long as the calyx-lobes. Stamens short; anthers globular, 2-porose, the connective thickened and produced into a short obtuse appendage, sometimes exceeding the cells; staminodia petal-like, fringed. Ovules 2. Style very short, glabrous.

W. Australia, *Drummond, 3rd Coll. n.* 24.

11. **V. serrata,** *Schau. Myrt. Xeroc.* 70. Stout, erect, and rigid. Leaves mostly alternate, from linear-lanceolate acute and ½ to ¾ in. long, to oblong or obovate, and 2 to 4 lines long, all rigid, concave or keeled, the margins ciliate with short stiff hairs. Flowers yellow, on slender pedicels, in dense, terminal, leafy corymbs. Bracteoles very deciduous. Calyx-tube glabrous; primary lobes 5, spreading to 4 or 5 lines diameter, each with a broadly cordate or auriculate base, deeply divided into digitate, plumose-ciliate, subulate lobes. Petals ovate, rather thick, concave, more or less toothed at the end. Stamens shortly united; anthers globular, 2-porose, the connective produced into an ovate or lanceolate, obtuse, concave appendage, exceeding the cells; staminodia petal-like, oblong, entire, connivent over the stamens. Style subulate, glabrous. Ovules 2.—Hook. Journ. Bot. ii. t. 13; *Chrysorrhoe serrata*, Lindl. in Swan Riv. App. 6.

W. Australia, *Drummond.* His specimens comprise four forms, which might almost be taken for distinct species:—*a.* Leaves short and broad. Petals broad, rather deeply toothed. Appendage of the connective broad, very obtuse. *1st Coll. n.* 145, *3rd Coll. n.* 169.—*b.* Leaves long and narrow. Petals ovate, less deeply toothed. Appendage of the connective rather narrow and sometimes acuminate. *3rd Coll. n.* 168.—*c.* Leaves of *b.* Flowers of *a.* *5th Coll. n.* 107.—*d.* Leaves of *a.* Flowers of *b.* (*4th Coll. ?*) *n.* 47.

12. **V. nitens,** *Schau. Myrt. Xeroc.* 71, *t.* 4 B, *and in Pl. Preiss.* i. 102. Small, erect, and corymbosely branched. Leaves linear, semiterete, rather slender, acute or mucronulate, mostly ½ to ¾ in., but the lower ones sometimes above 1 in. long. Flowers golden-yellow, on slender pedicels, in a broad terminal corymb. Bracteoles very deciduous. Calyx-tube shortly turbinate, glabrous; primary lobes 5, spreading to about 4 lines diameter, divided nearly to the base into 7 to 11 digitate, subulate, pectinate-plumose lobes. Petals ovate or obovate, glabrous, about as long as the calyx, thick and striate in the centre, thin at the edge, and fringed with short irregular teeth. Stamens scarcely united at the base; anthers globular, 2-porose, but often almost entirely enclosed in the large concave or hood-shaped appendage of the connective, which is usually obtusely 2-lobed at the top; staminodia short, subulate. Style filiform, glabrous. Ovules 2.—Bot. Mag. t. 5286; *Chrysorrhoe nitens*, Lindl. in Comp. Bot. Mag. ii. 357, and Swan Riv. App. t. 1.

W. Australia. Swan River, *Drummond, 1st Coll., 3rd Coll. n.* 166; *Oldfield; Preiss, n.* 173.

13. **V. grandiflora,** *Endl. in Ann. Wien. Mus.* ii. 195. Erect, rigid, and rather stout, 1 to 2 ft. high or rather more. Leaves from linear, semi-

terete or triquetrous, and ½ to 1 in. long, to oblong or obovate, concave or keeled, and 2 to 3 lines long, usually thick, obtuse or mucronulate, crowded on the short lateral shoots. Flowers yellow, on slender pedicels, in a rather loose, terminal leafy corymb. Bracteoles persistent, connate at the base. Calyx-tube glabrous, broadly turbinate; primary lobes 5, often spreading to a diameter of ¾ in., but sometimes smaller, each deeply divided into 5 to 9 digitate, subulate, pectinate-ciliate, or plumose lobes. Petals deeply divided into 7 to 11 or even more digitate, subulate, entire lobes. Stamens slightly united at the base; anthers globose, 2-porose, the connective thickened into a concave or hooded appendage, much longer than the cells, with 2 long horn-like points turned down over them; staminodia lanceolate or oblong, petal-like, more or less toothed or fringed, connivent outside the stamens. Style subulate, glabrous. Ovules 2.—Schau. Myrt. Xeroc. 75; Hook. Journ. Bot. ii. t. 14; *V. heliantha*, Lindl. Swan Riv. App. 6; *V. nobilis*, Meissn. in Journ. Linn. Soc. i. 39.

W. Australia. Swan River, *Drummond*, 1*st Coll.*; Murchison river and Irwin river, Champion Bay, *Oldfield*, *Drummond*, 6*th Coll. n.* 47.

14. **V. chrysantha,** *Endl. in Ann. Wien. Mus.* ii. 195. Very nearly allied to *V. grandiflora*, with the same foliage, persistent bracteoles, calyx and petals. Inflorescence usually looser, the flowers fewer and rather smaller, yet larger than in *V. Preissii*. Anthers with the connective erect and projecting beyond the cells, larger than in *V. Preissii* and in *V. acerosa*, but not 2-horned, as in *V. grandiflora*; staminodia oblong-lanceolate, petal-like and connivent outside the stamens, entire or slightly toothed.—Schau. Myrt. Xeroc. 73, and in Pl. Preiss. i. 102; *V. Gilbertii*, Turcz. in Bull. Mosc. 1847, i. 160.

W. Australia. In the interior, *Roe*, *Preiss*, *n.* 178, *Gilbert*; Oldfield Range, *Maxwell*. The specimens are none of them very satisfactory.

15. **V. Preissii,** *Schau. in Pl. Preiss.* i. 101. An erect shrub of 1 to 1½ ft. Stem-leaves linear, semiterete or triquetrous, acute or mucronate, rarely above ½ in. long, usually densely crowded on the short lateral branches, the upper ones below the corymb often more distant, those in the corymb shorter and lanceolate, or in some specimens small, ovate or even orbicular, thick and concave. Flowers yellow, on slender pedicels, in a compact terminal corymb. Bracteoles persistent, connate at the base. Calyx-tube broadly turbinate, 10-ribbed, glabrous; primary lobes 5, spreading to a diameter of about 4 lines, digitately divided into 7 or 9 subulate, pectinate-ciliate or plumose lobes. Petals deeply divided into 7 to 11 digitate, subulate, entire lobes, about as long as the calyx. Stamens very shortly united; anthers globular, 2-porose, the connective slightly thickened and produced into a concave appendage, very shortly exceeding the cells; staminodia oblong, petal-like, entire, connivent outside the stamens. Style filiform, glabrous. Ovules 2.—*V. Endlicheriana*, Schau. in Pl. Preiss. i. 101.

W. Australia. Between King George's Sound and Swan River, *Oldfield*, *Drummond*, (4*th Coll.?*) *n.* 65, 5*th Coll. n.* 112, *A. C. Gregory*; rocky heights, near Halfway House, Gordon river, and foot of Konkoberup hills towards Cape Riche, *Preiss*, *n.* 175, 179, 181; Young river, *Maxwell*. The species is very near on the one hand to *V. chrysantha*, but with much smaller densely corymbose flowers, and on the other to *V. acerosa*, but with perfectly entire staminodia.

16. **V. acerosa,** *Lindl. Swan Riv. App.* 6. Erect, attaining 1 to 3 ft., the branches usually virgate and rather slender. Leaves varying from linear-triquetrous, slender, mucronate and nearly ¾ in. long, to ovate or ovate-lanceolate, concave and 2 to 3 lines long, those crowded on short lateral shoots usually the longest and most slender, but sometimes all long and narrow, or all short and broad. Flowers yellow, rather small, in trichotomous terminal corymbs, on pedicels of 3 lines or more. Calyx-tube turbinate, strongly 10-ribbed, glabrous; primary lobes 5, spreading to 4 or 5 lines diameter, each deeply divided into 5 or 7 digitate, subulate, pectinate-ciliate or plumose lobes. Petals divided almost to the base, into 5, 7 or 9 subulate, digitate lobes, rigid and entire. Stamens very shortly united; anthers globose, 2-porose, the connective with a short obtuse appendage, scarcely exceeding the cells; staminodia lanceolate or oblong, petal-like, fringed or pinnatifid, connivent over the stamens. Style rather short, glabrous. Ovules 2.—Schau. Myrt. Xeroc. 68, and in Pl. Preiss. i. 101.

W. Australia. Swan River, *Drummond, 1st Coll. n.* 164; Darling Range, *Preiss, n.* 176.

D. Calyx-tube hemispherical or turbinate, with a ring of long hairs at the base, otherwise glabrous, pubescent or shortly villous; primary lobes 5, spreading or partially reflexed, divided into numerous subulate or hair-like lobes or long cilia, either all simple or some of them slightly branched or ciliate near the base. Connective small.

This group has the calyx-tube of B, with the lobes of some species of E, and the reflexed marginal cilia may be occasionally mistaken for the accessory lobes of the latter group, but in D these reflexed cilia never really proceed from distinct lobes, alternating with the spreading ones.

17. **V. polytricha,** *Benth.* Erect and bushy, with the habit and foliage nearly of *V. Harveyi*, but with a different calyx. Leaves linear, semiterete or triquetrous, slender, obtuse, mostly 3 to 4 lines long, very densely crowded on the short shoots, more distant below the corymb. Flowers small, on short pedicels, in broad, compact, terminal, leafy corymbs, often pedunculate. Calyx-tube hemispherical, with long dense hairs at the base, glabrous and contracted at the top; primary lobes 5, spreading to 3 or nearly 4 lines diameter, deeply divided into very numerous, long, simple cilia, the margins not reflexed. Petals short, ovate, pubescent, slightly ciliate. Stamens very shortly united; anthers very small, globular, 2-porose; connective small; staminodia lanceolate, acute, longer than the stamens. Style shortly exserted, slender, incurved and bearded at the end. Ovule 1 (or 2?).

W. Australia. Murchison river, *Oldfield, Drummond, 6th Coll. n.* 49.

18. **V. demissa,** *F. Muell. Herb.* Procumbent and rigid, with numerous short ascending branches. Leaves linear, semiterete or triquetrous, obtuse, rarely above 2 lines long, crowded on the short branches. Flowers small, on short thick pedicels in the upper axils, numerous, but scarcely corymbose. Calyx-tube nearly 1 line diameter, the short adnate part faintly 10-ribbed and densely hirsute with long spreading hairs, the free part broad, smooth, and glabrous; primary lobes 5, spreading to a diameter of about 3 lines, each one divided to the base into long simple cilia or subulate lobes, of

which 2 or 3 thicker and longer than the others. Petals ovate, very thin, densely pubescent, entire or nearly so. Stamens very shortly united; anthers globose, 2-porose; connective thickened, but not exceeding the cells; staminodia lanceolate-subulate, entire. Style rigid, subulate, exceeding the corolla by 3 or 4 lines, bearded with a few stiff hairs near the point. Ovules 2.

W. Australia, *Drummond, 5th Coll. n.* 113; Fitzgerald ranges, *Maxwell.* The rigid spreading bristles of the calyx, the almost globular pubescent corolla in the centre, with the long protruding style give the flowers a peculiar aspect.

19. **V. humilis,** *Benth.* Apparently a small slender procumbent shrub. Leaves linear, triquetrous, or laterally compressed, distinctly petiolate, 2 to 3 lines long, not crowded in our specimen. Flowers rather large, often on very short pedicels in the upper axils. Calyx-tube very broad and short, 10-ribbed, hirsute with spreading hairs near the base, otherwise pubescent; primary lobes 5, spreading to about 5 lines diameter, divided at the base into numerous purple subulate or hair-like lobes, the inner ones more rigid and entire, the outer ones more slender and plumose with a few long cilia. Petals ovate, pubescent, entire, connivent. Stamens very shortly united; anthers globular, 2-porose, with a very small connective. Staminodia lanceolate-subulate, entire. Style very long and subulate, not bearded. Ovules 2 (or sometimes 1?).

W. Australia. *J. S. Roe.*

20. **V. penicillaris,** *F. Muell. Fragm.* i. 226. Rigid, diffuse or prostrate and much branched. Leaves linear, concave, keeled or triquetrous, obtuse or mucronate, 1 to 2 lines long, crowded. Flowers large for the size of the plant, on short pedicels in the upper axils, forming a broad irregular leafy corymb. Calyx-tube hemispherical, densely hirsute with long rigid hairs at the base, otherwise pubescent; primary lobes 5, spreading to a diameter of above ½ in., divided to the base into numerous long hair-like simple or branched cilia, a few more rigid than the others, the marginal ones reflexed on the tube. Petals ovate, fringed with numerous cilia, very fine, but rarely longer than the breadth of the lamina. Stamens very shortly united; anthers nearly globular, 2-porose, the connective not prominent. Style very long, purple, bearded near the end with remarkably long spreading hairs. Ovules 2.

W. Australia. Table Hill, Champion Bay, *Oldfield.*

21. **V. multiflora,** *Turcz. in Bull. Mosc.* 1847, i. 159. Branches apparently divaricate. Leaves linear, thick, semiterete or triquetrous, obtuse, rarely exceeding 2 lines, crowded on the short lateral shoots. Flowers (yellow) rather small, on pedicels attaining 2 to 3 lines, in small dense terminal corymbs. Calyx-tube densely hirsute at the base, the free part broad and glabrous; primary lobes 5, spreading to about 4 lines diameter, deeply divided into very numerous long cilia, of which 3 to 5 thicker and subulate, and a few of the marginal ones sometimes forming auricles reflexed on the tube. Petals ovate, shorter than the calyx-lobes, fringed with numerous cilia. Stamens very shortly united; anthers globular, 2-porose, the connective inconspicuous. Staminodia subulate, longer than the stamens. Style shortly exserted, bearded from below the middle with a few long hairs;

stigma broadly capitate. Ovules 2.—*V. brachypoda*, Turcz. in Bull. Mosc. 1847, i. 158.

W. Australia, *Drummond, 3rd Coll. n.* 26 *and* 28; *5th Coll. n.* 111. Plantagenet, Stirling, and Fitzgerald ranges, *Maxwell.* The species is allied to *V. Huegelii*, but the leaves are shorter and thicker, the cilia of the calyx much less numerous, the stigma much smaller, and the staminodia different.

22. **V. Huegelii,** *Endl. in Hueg. Enum.* 16. Erect with slender branches. Leaves linear, rather slender, semiterete or triquetrous, obtuse, 2 to 4 lines long, crowded on the short lateral shoots. Flowers (white or pink?), on pedicels of 3 to 4 lines, in small loose terminal corymbs or in the upper axils. Calyx-tube strongly 10-ribbed, densely hirsute at the base, otherwise pubescent; primary lobes 5, spreading to a diameter of 4 to 5 lines or rather more, and divided into exceedingly numerous fine cilia, forming a dense globular tuft, a few of the inner ones more rigid and subulate, some of the outer ones occasionally branched, and several on the outer margins closely reflexed on the tube so as often almost to cover it, but without any distinct accessory lobes. Petals much shorter than the calyx-lobes, fringed with numerous fine cilia. Stamens very shortly united; anthers globular, 2-porose, the connective inconspicuous. Staminodia lanceolate, connivent over the stamens, more or less fringed with long cilia, but very variable as to breadth. Style shortly exserted, bearded; stigma peltate, larger than in any other species. Ovules 2.—Schau. Myrt. Xeroc. 61, and in Pl. Preiss. i. 99; *V. fimbripetala*, Turcz. in Bull. Mosc. 1849, ii. 19.

W. Australia. Swan River, *Drummond, 1st Coll. n.* 175; Darling Range, *Preiss, n.* 177; Harvey river, *Oldfield;* Kalgan river, *Maxwell.*

V. stylosa, Turcz. in Bull. Mosc. 1847, i. 160, is founded on specimens of Gilbert's n. 327, which I have not seen. The essential characters given are precisely those of *V. Huegelii*, except the colour of the flowers and the appendage to the anthers, which belong to the group C, in which are never found the other characters given. It is probable, therefore, that some fragments of *V. serrata* or its allies may have got mixed with the author's specimens of *V. Huegelii*, and the *V. stylosa* made up of both.

E. Calyx-tube various; primary lobes 5, spreading, either digitate with pectinate lobes, or divided into very numerous hair-like lobes or cilia, and 5 accessory ones alternating with them on the outside, very thin and scarious, closely reflexed on the tube, divided into numerous fine cilia, and turned up again from the base of the tube. Connective small. Lower leaves laterally compressed or triquetrous.

This group has the accessory calyx-lobes, but not the herbaceous appendages to the tube of the group C of *Catocalypta*, and the anthers and ovary are quite those of *Euverticordia.*

23. **V. insignis,** *Endl. in Hueg. Enum.* 47.—An erect shrub of 1 to 2 ft., branching from the base. Leaves from broadly ovate to oblong, very obtuse or almost mucronate, 2 to 4 lines long, the lower ones and those of the barren branches often laterally compressed or vertical, others with the upper edge dilated, and the upper ones often concave and keeled only. Flowers on pedicels often exceeding 1 in., in loose irregular terminal leafy corymbs. Bracteoles deciduous. Calyx-tube 10-ribbed, about 2 lines long, quite concealed by the accessory lobes, which are closely reflexed to the base, and there turned up again and divided into numerous long cilia, which appear

to form a fringe to the base of the tube; primary lobes spreading to about ½ in. diameter, deeply divided into 5 to 9 digitate linear lobes fringed with long cilia. Petals orbicular, fringed with cilia, inserted on the staminal tube near its base. Stamens united into a broad short tube above the calyx, filaments longer than the petals; anthers small, globular, 2-porose; staminodia fringed with long cilia. Style glabrous with a capitate stigma. Ovules 2. —Lindl. Swan Riv. App. t. 2 A; Schau. Myrt. Xeroc. 65, and in Pl. Preiss. i. 100.

W. Australia. Swan River, *Huegel; Drummond*, 1*st Coll.*; *Preiss, n.* 167 *and* 168, and others.

V. compta, Endl. in Ann. Wien. Mus. ii. 194, and *V. Roei*, Endl. l. c., appear to be only a small-leaved variety of *V. insignis*; the specimens are in a bad state, but the staminodia are certainly fringed in both. Preiss's specimens, referred by Schauer to *V. compta*, appear to me to be a very common form of *V. insignis*.

24. **V. habrantha,** *Schau. in Pl. Preiss.* i. 100. A shrub of 2 or 3 ft., with slender often virgate branches. Lower leaves and those of the short side branches often laterally compressed, falcate-oblong, dilated on the upper edge or triquetrous, and attaining 3 to 4 lines, the upper ones in the corymb are sometimes nearly all obovate or oblong, concave with a prominent keel and not 2 lines long. Flowers rather small, on pedicels of ¼ to ½ in., in irregular terminal leafy corymbs. Bracteoles deciduous. Calyx-tube turbinate, 10-ribbed, pubescent at the base; primary lobes 5, spreading to a diameter of about 4 lines, deeply divided into subulate simple or forked lobes fringed below the middle with long cilia; and 5 accessory outer lobes reflexed on the tube, turned up again from the base, and deeply divided into numerous long fine cilia. Petals ovate, entire or obscurely denticulate, contracted at the base. Stamens shortly united above the calyx; anthers globular, 2-porose; connective small; staminodia rather broad, fringed with a few long cilia. Style exceedingly short, glabrous, with a capitate stigma. Ovules 2.—*V. umbellata*, Turcz. in Bull. Mosc. 1847, i. 159; *V. brachystylis*, F. Muell. Fragm. i. 164.

W. Australia, *Drummond*, 3*rd Coll. n.* 25; 5*th Coll. n.* 108 *and* 109 (the latter with large flowers); Gordon river, *Preiss, n.* 169; Kalgan, Gordon, and Tone rivers, *Oldfield;* Gardiner river and Mount Manypeak, *Maxwell.* Turczaninow must either have mistaken the anther-cells for a cucullate connective, and the small persistent base of the bracteoles for the bracteoles themselves, or to have mixed up his description of this species with that of *V. nitens*.

25. **V. monadelpha,** *Turcz. in Bull. Mosc.* 1847, i. 158. Erect and much branched. Leaves linear, triquetrous or laterally compressed, mostly mucronate, rather thick, often above ½ in. long. Flowers rather large, pink or white, in broad or loose terminal leafy corymbs, each flower having the appearance of a dense globular tuft of hairs of at least ½ in. diameter. Calyx-tube about 2 lines long, broadly turbinate, 10-ribbed and hairy at the base, the free part very broad and glabrous; primary lobes 5, spreading, and 5 accessory outer ones reflexed on the calyx-tube and turned up from its base, all deeply divided into exceedingly numerous long cilia. Petals short, ovate, fringed with long cilia, adnate to the staminal tube to about half its length. Stamens united in a broad tube for about a line above the calyx; filaments

often exceeding the petals; anthers globular, 2-porose, with a minute scale-like appendage to the small connective; staminodia lanceolate-subulate, entire. Style rather short, glabrous; stigma capitate. Ovules 2.—*V. callitricha*, Meissn. in Journ. Linn. Soc. i. 39.

W. Australia. *Drummond, 3rd Coll. n.* 27; Murchison river, *Drummond, 6th Coll. n.* 48; *Oldfield.*

26. **V. Lehmanni,** *Schau. in Pl. Preiss.* i. 99. Slender, erect, and slightly branched, usually from 1 to 1½ ft. high. Leaves mostly in distant pairs, linear-oblong or falcate, laterally compressed or triquetrous, obtuse or mucronate, 3 to 4 lines long, the upper ones near the flowers not half so long, oblong or almost ovate and concave. Flowers rather small, on pedicels of 1 to 2 lines, few in small compact terminal corymbs, or in more luxuriant specimens axillary below the ends of the loosely corymbose upper branches. Calyx-tube 1½ lines long, the adnate part shortly villous at the base, the free part prominently 10-ribbed and glabrous; primary lobes 5, spreading, deeply divided into about 5 subulate lobes, with several long cilia between them, 5 accessory outer ones closely reflexed on the tube and turned up from its base, thin and transparent, deeply divided into numerous cilia. Petals ovate, very thin, irregularly lobed or ciliate at the end, inserted near the top of the staminal tube. Stamens united in a broad short tube; anthers globular, 2-porose, with a slightly-thickened connective; staminodia lanceolate-subulate, slightly glandular. Style shortly exserted, incurved towards the end and bearded at the bend. Ovules 2.

W. Australia, *Drummond, n.* 15; Molloy's Plains, Sussex district, *Preiss, n.* 166.

SECTION 2. CATOCALYPTA.—Anthers ovoid or oblong, with parallel cells adnate to a more or less thickened connective, and opening in longitudinal slits. Ovules several, usually 8 or 10, in 2 rows on an obliquely peltate or rarely stalk-like placenta.

This section, with the anthers and ovary of *Chamælaucium,* is only distinguished from it by the calyx. I have adopted Schauer's name for it, although somewhat differently limited.

A. Calyx-lobes 5, spreading, without reflexed accessory lobes or herbaceous appendages. Racemes short, mostly terminal, almost corymbose. Leaves linear-triquetrous or semiterete.

The two species here inserted have not the herbaceous appendages to the calyx which characterize the rest of the section, and in inflorescence they show an approach to *Euverticordia,* but the anthers and some other points indicate a closer affinity with *Catocalypta.*

27. **V. Cunninghamii,** *Schau. Myrt. Xeroc.* 55. A tall erect shrub. Leaves linear, triquetrous or concave, obtuse or mucronate, mostly ½ in. but sometimes ¾ in. long. Flowers on pedicels of about ¼ to ½ in. in the upper axils, forming short terminal almost corymbose racemes arranged in a long leafy panicle. Calyx-tube hemispherical, 10-ribbed; primary lobes 5, spreading to ½ in. diameter, each one deeply divided into long digitate pectinate-ciliate lobes, the lateral ones reflexed on the tube, but no accessory lobes. Petals much shorter than the calyx-lobes, ovate, fringed with irregular teeth. Stamens shortly united above the calyx; anther-cells parallel, opening longitudinally, adnate to a connectivum, thickened at the end into a small fleshy

appendage; staminodia linear, entire. Style shortly exserted, with a ring of hairs round the capitate stigma. Ovules 8 or 10.

N. Australia. York Sound, *A. Cunningham;* Victoria river, *Bynoe*, islands of the Gulf of Carpentaria, *A. Brown;* Macadam range, *F. Mueller;* Port Essington, *Armstrong*.

28. **V. picta,** *Endl. in Ann. Wien. Mus.* ii. 194. Branches spreading, rather slender. Leaves linear, semiterete or triquetrous, obtuse or mucronate, mostly 2 to 4 lines long. Flowers white or pink, rather large, on pedicels of 3 or 4 lines, in loose terminal corymbs or short leafy panicles. Calyx-tube hemispherical, glabrous, obscurely 10-ribbed in the free part; primary lobes 5, spreading to about 5 lines diameter, deeply divided into 7, 9, or 11 digitate linear pectinate-ciliate scarious lobes. Petals inserted on the staminal tube shortly above the calyx, broadly ovate, entire, longer than the calyx-lobes. Stamens united in a broad tube; filaments short; anthers oblong, with parallel cells opening longitudinally; staminodia lanceolate-subulate, entire. Style shortly bearded below the stigma. Ovules about 10, appended to as many marginal lobes of a somewhat peltate, excentric placenta.—Schau. Myrt. Xeroc. 53.

W. Australia, *Roe*; Swan River, Drummond, 1*st Coll. n.* 170; S. Hutt and Murchison rivers, *Oldfield*.

V. pentandra, Turcz. in Bull. Mosc. 1847, i. 157, described from Gilbert's specimens, n. 329, which I have not seen, appears from the character given not to differ from *V. picta*.

B. Calyx-lobes 5, spreading, without accessory reflexed segments, but with 5 herbaceous reflexed appendages on the tube under the lobes. Flowers usually forming oblong racemes or spikes below the ends of the branches. Leaves small.

The reflexed herbaceous appendages which distinguish this group from A are rather variable, in *V. pennigera* occasionally reduced to a slight gibbosity under the lobes, sometimes in that species extending ¼ down the tube, in others halfway down or nearly to the base, always closely appressed to the tube between the ribs, and sometimes shortly adnate to it.

29. **V. pennigera,** *Endl. in Hueg. Enum.* 46. Stems in some specimens short and erect from a thick stock, in others slender, spreading, or virgate. Leaves linear and semiterete or triquetrous, or oblong and concave, obtuse or mucronate, 1 to 2 lines long, crowded on the small lateral shoots, the margins more or less ciliate. Flowers on short pedicels in the upper axils, forming leafy racemes, sometimes collected into thyrsoid panicles. Calyx-tube turbinate, 5-ribbed; primary lobes 5, spreading to a diameter of 4 or 5 lines, deeply divided into subulate plumose lobes, with a few long lateral cilia closely reflexed on the tube, without accessory lobes, but with herbaceous adnate appendages reflexed on the tube under the lobes, very short and broad and sometimes scarcely more than broad gibbosities. Petals obovate-oblong, striate, toothed or fringed at the end, connivent over the stamens. Stamens shortly united above the calyx; anther-cells parallel, opening longitudinally, the connective not much thickened. Style slightly bearded. Ovules about 6.—Schau. Myrt. Xeroc. 59, and in Pl. Preiss. i. 99; *V. setigera*, Lindl. Swan Riv. App. 7.

W. Australia. Swan River, *Drummond,* 1*st Coll. Preiss, n.* 182; Murchison, Gordon, and Kalgan rivers, *Oldfield;* Dirk Hartog's Island, *Martin;* Gardner ranges and Colt river, *Maxwell.*—The species differs slightly from *V. Drummondii* in the ciliate leaves, the shortness of the calyx-appendages, and the longer more striate petals.

30. **V. Drummondii,** *Schau. Myrt. Xeroc.* 56, *and in Pl. Preiss.* i. 98. A shrub with virgate or divaricate branches, and much the aspect of *Erica vulgaris*. Leaves obovate or oblong, very obtuse, rather thick and concave, 1 to 2 lines long, imbricate on the short lateral shoots, entire or minutely or obscurely denticulate-ciliate. Flowers on short pedicels in the upper axils, forming oblong leafy racemes or dense thyrsoid panicles. Calyx-tube turbinate, 5-ribbed; primary lobes 5, spreading to 4 or 5 lines diameter, deeply divided into subulate plumose lobes, the marginal ones sometimes reflexed on the tube; no accessory lobes, but 5 herbaceous appendages reflexed under the lobes, and often half as long as the tube. Petals ovate, connivent, striate and fringed at the end. Stamens shortly united above the calyx; anther-cells parallel, opening longitudinally; connective not much thickened; staminodia lanceolate-subulate, bordered by prominent glands. Style slightly bearded. Ovules about 6.—*V. carinata,* Turcz. in Bull. Mosc. 1849, ii. 19.

W. Australia. Swan River, *Drummond,* 1*st Coll.*; *Preiss, n.* 171; between Tone and Gordon rivers, *Oldfield.*

Var. *Lindleyi.* Leaves broader and less imbricate, often distant and spreading, and quite entire. *V. Lindleyi,* Schau. Myrt. Xeroc. 58, and in Pl. Preiss. i. 98; *Drummond,* 4*th Coll. n.* 46; Irwin river, *Preiss, n.* 170.

31. **V. pholidophylla,** *F. Muell. Fragm.* i. 227. A shrub of 1 to 2 ft., with spreading branches, very closely allied to *V. Drummondii.* Leaves ovate or obovate, thick, concave, obtuse, rarely above 1 line long, closely imbricate on the smaller branches. Flowers on very short axillary pedicels below the ends of the branches, often assuming a slight yellowish tinge. Calyx-tube turbinate, 5-ribbed; primary lobes 5, spreading to about 4 or 5 lines diameter, deeply divided into 7 or 9 subulate plumose or ciliate lobes, a few of the lateral cilia reflexed on the tube; no accessory lobes, but 5 herbaceous reflexed appendages under the lobes, about half as long as the tube. Petals ovate, ciliate-fringed, about as long as the calyx-segments, inserted on the staminal tube. Stamens united at the base into a short broad tube; anther-cells parallel, opening longitudinally, connective somewhat thickened; staminodia linear-subulate, short. Style incurved and bearded towards the end. Ovules 6 to 8.

W. Australia. Coalcurda, north of Murchison river and sandy plains south of Oolingara, *Oldfield*; Roebuck Bay, *Marten.*

C. Calyx with 5 primary lobes spreading, each one divided into subulate plumose lobes, 5 accessory lobes outside and alternate with the primary one, thin and transparent, reflexed on tube, fringed or densely ciliate and turned up again from the base of the tube, and 5 herbaceous reflexed appendages on the tube between the ribs and under the lobes. Flowers forming racemes or spikes below the ends of the branches, or rarely short terminal racemes. Leaves obovate or orbicular, usually glaucous.

This group has the appendages to the calyx-tube of the preceding one, and in addition the accessory lobes of the group E of *Euverticordia.*

32. **V. spicata,** *F. Muell. Fragm.* i. 226. Much resembling some forms of *V. Drummondii*, but with a different calyx. Leaves obovate or orbicular, concave, obtuse, not 1 line long, minutely denticulate-ciliate, and very closely imbricate except the floral ones, which are twice as large and looser. Flowers nearly sessile, forming dense spikes below the summits of the branches. Calyx-tube 5-ribbed; primary lobes 5, spreading to a diameter of nearly ½ in., deeply divided into 5 to 9 linear-subulate plumose-ciliate lobes, 5 accessory reflexed external lobes ovate-lanceolate, transparent, fringed with a few long cilia, and 5 herbaceous appendages reflexed between the ribs under the primary lobes, and nearly as long as the tube. Petals ovate, thin, fringed with long cilia. Stamens united at the base in a very short broad tube; anthers oblong, the cells parallel, opening longitudinally, and adnate to a broad thick connective; staminodia linear, rather thick. Style shortly exserted, bearded towards the end. Ovules about 8, in two rows, on a rather long stalk-like placenta.

W. Australia. Murchison river, *Oldfield.*

33. **V. lepidophylla,** *F. Muell. Fragm.* i. 228. Erect, attaining 3 or 4 ft., with spreading branches, resembling *V. pholidophylla*, but with a different calyx. Leaves obovate-orbicular, concave, obtuse, rarely above 1 line long, thick, entire or minutely denticulate-ciliate, imbricate on the smaller branches. Flowers on very short pedicels, axillary below the ends of the branches. Calyx-tube nearly hemispherical, 5-ribbed; primary lobes 5, spreading to a diameter of 4 or 5 lines, digitately divided into 7 to 9 linear plumose-ciliate lobes, 5 accessory external lobes closely reflexed on and covering the tube, orbicular, transparent, fringed at the edges, and 5 herbaceous appendages reflexed between the ribs, but exceedingly short and broad. Petals as long as the calyx-lobes, entire or minutely denticulate, attached near the summit of the staminal tube. Stamens united for nearly a line above the calyx; anthers ovoid, with parallel cells opening longitudinally; staminodia spathulate, fringed at the end. Style exserted, bearded towards the end. Ovules about 6.

W. Australia. Murchison river, *Oldfield.*

34. **V. ovalifolia,** *Meissn. in Journ. Linn. Soc.* i. 40. Branches slender, virgate. Leaves obovate, concave, erect, squarrose, or spreading, mostly 2 to 3 lines long. Flowers on pedicels shortly exceeding the leaves, not numerous, in a short terminal corymbose raceme. Calyx-tube about 2½ lines long, the 5 ribs not very prominent; primary lobes 5, spreading to nearly ¾ in. diameter, deeply divided into 8 to 10 long plumose lobes; 5 accessory external lobes, thin and transparent, closely reflexed, and almost covering the tube with their long marginal cilia; 5 herbaceous appendages under the primary lobes reflexed on the tube, but exceedingly short and broad. Petals broad, inserted on the staminal tube shortly above the calyx, irregularly divided into 5 or 6 more or less fringed lobes. Stamens united nearly a line above the calyx; anther-cells parallel, opening longitudinally, connective thick; staminodia slightly clavate at the end. Style bearded below the stigma. Ovules about 8.

W. Australia. Between Moore and Murchison rivers, *Drummond, 6th Coll. n.* 45.

Meissner appears to have overlooked the appendages to the calyx-tube, which, although much shorter than in either of the following species, certainly exist in our specimens.

35. **V. chrysostachya,** *Meissn. in Journ. Linn. Soc.* i. 41. Erect, with virgate branches, glaucous like the allied species or assuming a yellow hue, at least in the dried specimens. Leaves obovate or orbicular, erect, and concave or nearly flat, squarrose or spreading, rather thick and almost nerveless. Flowers yellow, on pedicels rarely exceeding the leaves below the ends of the branches. Calyx-tube about 2 lines long, 5-ribbed; primary lobes 5, spreading to a diameter of about ½ in., deeply divided into plumose lobes; 5 accessory external lobes closely reflexed and turned up from the base of the tube, completely covering it with their numerous cilia, and 5 herbaceous appendages shorter than the tube reflexed upon it between the ribs from under the primary lobes. Petals inserted on the staminal tube, broadly cordate, fringed with long cilia. Stamens united above the calyx in a short broad tube; anther-cells parallel, opening longitudinally, on a thickened connective; staminodia subulate, thickened at the base. Style bearded below the stigma with short hairs. Ovules 6 to 8.

W. Australia. Between Moore and Murchison rivers, *Drummond, 6th Coll. n.* 46.

36. **V. oculata,** *Meissn. in Journ. Linn. Soc.* i. 41. A glaucous shrub attaining 5 or 6 ft., but often flowering when under 2 ft., with slender or spreading branches. Leaves orbicular, stem-clasping, 3 to 5 lines diameter, faintly 3- or 5-nerved or quite nerveless, with thin edges. Flowers lilac or pale, with a dark centre, on pedicels shorter or longer than the leaves below the ends of the branches. Calyx-tube about 3 lines long, 5-ribbed; primary lobes 5, spreading to a diameter of nearly 1 in., deeply divided into long plumose lobes of a shining white, 5 thin transparent accessory lobes reflexed on the tube and turned up from the base, deeply divided into numerous cilia, and 5 herbaceous appendages alternate with the ribs, reflexed on the tube from under the primary lobes, thinner than in the allied species. Petals short and broad, fringed with 10 to 12 long subulate lobes or cilia, inserted on the staminal tube. Stamens united in a broad tube above the calyx; anther-cells parallel, opening longitudinally, the connective not much thickened; staminodia subulate-pointed but very irregular. Style exserted, the stigma surrounded by a tuft of long hairs. Ovules about 8.

W. Australia. Sandy plains between Hutt and Murchison rivers, *Drummond, 6th Coll. n.* 43.

37. **V. grandis,** *Drumm. in Hook. Kew Journ.* v. 119. A stout glaucous shrub of 3 to 6 ft., with erect or spreading branches. Leaves orbicular and half-stem-clasping, 3 to 6 lines diameter, faintly 5- or 7-nerved, with thin edges. Flowers axillary along the virgate branches, each forming when fully out a densely plumose crimson tuft of at least 1 in. diameter. Calyx-tube turbinate, 5-ribbed, about 4 lines long; primary lobes 5, spreading, divided into numerous long plumose lobes; 5 accessory lobes reflexed on the tube and turned up from the base, fringed with fine cilia, and 5 herbaceous appendages between the ribs reflexed from under the primary lobes and nearly as long as the tube. Petals orbicular, fringed with short teeth, inserted on the staminal tube considerably above the calyx. Stamens united

at the base into a broad tube; anther-cells parallel, opening longitudinally, adnate to a thick connective; staminodia subulate. Style exserted, slightly bearded above the middle. Ovules 8 to 10.—Meissn. in Journ. Linn. Soc. i. 42.

N. Australia. Lagrange Bay, N.W. Coast, *Marten.*

W. Australia. Sandy plains, Hill river, *Drummond, 6th Coll. n.* 44; Irwin river, *Oldfield.*

5. PILEANTHUS, Labill.

Calyx-tube turbinate or campanulate, 10-ribbed, lobes 10, spreading, all equal, broad, petal-like, entire. Petals 5, exceeding the calyx, spreading, shortly ciliate. Stamens 20, in a single row, the filaments dilated at the base, and shortly united; anther-cells parallel, opening longitudinally, either contiguous and adnate to the thickened end of the filament or separately attached to the branches of the forked filament. Ovary 1-celled, with 6 to 10 ovules in 2 rows, on an erect free excentric basal placenta. Style filiform, glabrous, with a small terminal stigma. Fruit usually 1-seeded, formed by the hardened base of the persistent calyx, but not seen ripe.—Heath-like shrubs, glabrous except the flowers. Leaves mostly opposite, linear-terete or triquetrous. Flowers in the upper axils forming terminal leafy corymbs. Bracteoles scarious, united, and enclosing the bud, circumsciss at or below the middle, and falling off together.

The genus is limited to West Australia.

Anther-cells contiguous on the clavate end of the filament . . . . 1. *P. peduncularis.*
Filaments forked, each branch bearing one anther-cell.
  Leaves linear-clavate, thick, 2 to 3 lines long. Pedicels short . . 2. *P. Limacis.*
  Leaves mostly linear-terete or triquetrous, 3 to 6 lines long . . 3. *P. filifolius.*

1. **P. peduncularis,** *Endl. in Ann. Wien. Mus.* ii. 196. An erect shrub, more or less corymbosely branched. Leaves linear-terete or triquetrous, obtuse, mostly 2 to 3 lines long and rather thick. Flowers in the upper axils on pedicels of ½ to 1 line long. Bracteoles circumsciss about the middle, leaving a turbinate truncate persistent base, 2 to 3 lines long, almost covering the calyx-tube. Calyx silky-pubescent, tube 2 to 3 lines long; lobes broadly ovate, very obtuse, 1 to 1¼ lines long. Petals obovate, exceeding the calyx. Stamens shorter than the petals; filaments slightly clavate at the end; anther-cells contiguous and adnate.—Schau. Myrt. Xeroc. 29. t. 5 B; *P. vernicosus,* F. Muell. Fragm. i. 225.

W. Australia, *J. S. Roe, Drummond, 4th Coll. n.* 48. Owing to the badness of the original specimen, the petals were by mistake described by Endlicher as shorter than the calyx-lobes.

2. **P. Limacis,** *Labill. Pl. Nov. Holl.* ii. 11. t. 149. Leaves linear-clavate, semiterete, very obtuse, 2 to 3 lines long, smooth or glandular-tuberculate and slightly ciliate. Flowers in the upper axils, on pedicels shorter than the leaves or slightly exceeding them. Bracteoles circumsciss rather below the middle or near the base, leaving a turbinate truncate cup much shorter than the calyx-tube. Calyx-tube above 2 lines long, broadly turbinate, silky-pubescent; lobes petal-like, nearly glabrous, minutely denticulate, shorter than the tube. Petals longer than the calyx-lobes. Filaments forked at the end, each branch bearing one of the anther-cells.—Desf. in Ann. Mus. Par. v. t. 3; DC. Prod. iii. 209; Schau. Myrt. Xeroc. 77. t. 5 A.

W. Australia. Sea-coast, *Labillardière*; Géographe Bay, *Baudin's Expedition* (*Herb. R. Brown*).

3. **P. filifolius,** *Meissn. in Journ. Linn. Soc.* i. 45. Erect and branching, but much less corymbose than *P. peduncularis*. Leaves linear-terete or triquetrous, obtuse, in some specimens rather thick minutely ciliate and 2 to 3 lines long, in others slender, smooth, and ½ in. long or more. Flowers in the upper axils on pedicels often attaining ½ in. Bracteoles circumsciss near the base, leaving the whole calyx-tube exposed. Calyx silky-pubescent, tube about 2 lines long; lobes yellow, nearly as long, obovate, slightly denticulate, Petals pink, more than twice as long as the calyx; lobes obovate, shortly fringed. Stamens shorter than the petals; filaments forked at the end, each branch bearing one of the anther-cells, the alternate stamens rather larger.

W. Australia. Murchison river, *Drummond, 6th Coll. n.* 42, *Oldfield.* Very near *P. Limacis*, and perhaps a slender-leaved variety, with longer pedicels, and the persistent base of the bracteoles usually much shorter.

6. CHAMÆLAUCIUM, Desf.

(Decalophium, *Turcz.*)

Calyx-tube tubular-campanulate or turbinate, 10-ribbed, or only 5-ribbed in the adnate part; lobes 5, spreading, petal-like or ciliate. Petals 5, orbicular, longer than the sepals. Stamens 10, alternating with as many staminodia, very shortly united in a ring in a single row; anthers ovoid or globular, the cells parallel, opening longitudinally, and adnate to a more or less thickened connective. Ovary 1-celled, with 6 to 10 ovules in 2 rows on an erect free excentric basal placenta. Style shorter than the petals or rather longer, thickened at the base, glabrous or fringed with spreading hairs under the capitate stigma. Fruit formed by the hardened base of the persistent calyx. Seeds 1 or 2 (not seen ripe).—Heath-like shrubs. Leaves opposite or rarely (in *C. Drummondii*) scattered, small, narrow, and sessile. Flowers sessile or shortly pedicellate in the axils of the upper stem-leaves, or few in a terminal cluster with the floral leaves reduced to small bracts. Bracteoles broad, thin, scarious, enclosing the young bud, but falling off in most species long before flowering, or rarely persistent.

The genus is limited to Australia. It differs from *Darwinia* in the anthers and in the more numerous ovules, and generally in its rather larger fewer flowers and shorter style.

Style not bearded. Calyx narrow.
  Flowers axillary. Calyx-lobes deeply fringed. Staminodia oblong . . . 1. *C. ciliatum.*
  Flowers in short terminal racemes. Calyx-lobes entire or nearly so. Staminodia slender.
    Calyx-tube about 2 lines long. Filaments all slender; connective scarcely thickened . . . . 2. *C. gracile.*
    Calyx-tube scarcely 1½ lines. Petaline filaments winged at the base; connective large and thick, with globular cells on the top . . . . 3. *C. heterandrum.*
Style with a ring of hairs (sometimes deciduous) under the stigma. Calyx broadly campanulate or turbinate.
  Leaves scattered or crowded, not opposite, ciliate . . . . . . 4. *C. Drummondii.*
  Leaves opposite, not ciliate.
    Bracteoles persistent, covering the calyx-tube. Flowers terminal.

Leaves mostly ¾ in. long . . . . 5. *C. virgatum.*
Leaves mostly ¼ in. long or less . . . . 6. *C. brevifolium.*
Bracteoles falling off long before flowering.
Calyx-lobes entire, very short and broad . . . . . . 7. *C. uncinatum.*
Calyx-lobes ovate, minutely ciliate. Flowers terminal . . 8. *C. megalopetalum.*
Calyx-lobes cordate, ciliate. Flowers terminal, at least when first opening . . . . 9. *C. pauciflorum.*
Calyx-lobes orbicular, minutely ciliate. Flowers axillary . 10. *C. axillare.*

1. **C. ciliatum,** *Desf. in Mem. Mus. Par.* v. 40. *t.* 3. Erect and bushy, about 2 ft. high. Leaves opposite, crowded on the smaller branches, linear-terete or slightly triquetrous, obtuse, mostly 3 to 4 lines long on some specimens, much smaller on others. Flowers axillary below the ends of the branches, or almost terminal on the short side-branches, on pedicels much shorter than the calyx. Bracteoles scarious, cohering, but falling off in a calyptra from the very young bud. Calyx-tube under 2 lines long, narrow-turbinate, prominently ribbed; lobes orbicular, petal-like, fringed, not half so long as the petals. Petals obovate, above 1 line long, quite entire or minutely fringed under a strong lens. Staminodia ovate-oblong, more connate with the petaline than with the sepaline stamens. Connective of the anthers much thickened. Style shorter than the petals, quite glabrous.—DC. Prod. iii. 209; Schau. Myrt. Xeroc. 43, and in Pl. Preiss. i. 97; *Genetyllis pauciflora*, Turcz. in Bull. Mosc. 1849, ii. 17.

W. Australia. King George's Sound and adjoining districts, *Labillardière, R. Brown, Preiss, n.* 360, *Drummond, 2nd Coll. n.* 54, *3rd Coll. Suppl. n.* 13, *4th Coll. n.* 45, *5th Coll. n.* 106; and eastward to Cape Arid, *Maxwell.* The eastern specimens mostly with smaller leaves.

2. **C. gracile,** *F. Muell. Fragm.* iv. 62. Branches slender, divaricate. Leaves opposite, not crowded, linear-terete or slightly triquetrous, obtuse or with a short point, mostly ½ to ¾ in. long, but smaller on the lateral branchlets. Flowers 2 to 6, on pedicels of scarcely 1 line, in short, loose, terminal, corymbose racemes, the floral leaves reduced to small bracts. Bracteoles very deciduous. Calyx-tube slender, narrow-turbinate, about 2 lines long, prominently 10-ribbed, but only 5 ribs reaching to the base; lobes very short, broadly semiorbicular, entire or scarcely fringed. Petals obovate-orbicular, ¾ line long. Stamens nearly as long, the connective scarcely or not at all thickened; staminodia slender. Style often shortly exserted, quite glabrous, with a broad stigma.

W. Australia. Murchison river, *Oldfield, Drummond, 6th Coll. n.* 39, *also* (*5th Coll.?*) *n.* 22.

3. **C. heterandrum,** *Benth.* Bushy and much branched. Leaves opposite, slender, linear-terete or slightly channelled above, obtuse or nearly so, about 2 to 3 lines long. Flowers small, in short, loose, axillary, almost corymbose racemes, with the floral leaves small linear and bract-like. Pedicels solitary in each axil, slender, 1 to 2 lines long. Bracteoles already fallen from the youngest buds seen. Calyx-tube narrow-turbinate or almost cylindrical, prominently ribbed, nearly 1½ lines long; lobes exceedingly short and broad, entire. Petals orbicular, entire, rather more than ½ line diameter. Stamens shorter, quite free; filaments of the sepaline ones filiform; those of the petaline stamens rather longer and more or less dilated at the base or to the

middle into a wing-like appendage on each side; connective of the anthers thick, obovoid or almost turbinate, with 2 globular cells at the top, quite distinct, as in *Thryptomene,* but opening longitudinally and nearly parallel; staminodia minute, inflected, often almost concealed by the appendages of the filaments. Ovules about 6. Style glabrous, with a capitate stigma.

W. Australia, *Drummond (5th Coll.?), n.* 135.

4. **C. Drummondii,** *Meissn. in Journ. Linn. Soc.* i. 44. Branches virgate. Leaves scattered or crowded, not opposite, linear, obtusely keeled, obtuse or scarcely mucronate, 3 to 4 lines long, ciliate with long hairs. Flowers nearly sessile, in terminal heads or clusters, usually of about 6 to 10. Calyx-tube broadly turbinate, about 2 lines long, prominently ribbed; lobes broadly ovate, 1 line long or rather more, shortly ciliate. Petals 1½ line long, minutely fringed. Connective of the anthers thickened into a glandular appendage; staminodia linear, obtuse. Style with a ring of rather long hairs under the broadly capitate stigma.

W. Australia. Sandy plains, near Colbourn springs, N. of Swan River, *Drummond, 6th Coll. n.* 41; and a smaller variety with shorter leaves and fewer flowers, *Drummond (2nd Coll.?), n.* 58.

5. **C. virgatum,** *Endl. in Ann. Wien. Mus.* ii. 193. Apparently larger than any other species, with rigid virgate branches. Leaves opposite, linear, terete or nearly so, obtuse, ¾ to 1 in. long. Flowers on short pedicels, 2 to 4 together at the ends of the branches in the axils of the last leaves, the uppermost pair reduced to small bracts. Bracteoles very broad, brown and scarious, persistent and enveloping the calyx after the flower is expanded. Calyx-tube broadly turbinate, nearly 3 lines long, obtusely ribbed; lobes orbicular, ciliate, about half as long as the petals. Petals orbicular, 1½ lines long, minutely fringed. Anthers with a thickened globose connective and small parallel cells; staminodia linear.—Schau. Myrt. Xeroc. 44. t. 4 A.

W. Australia. E. from New York, *J. S. Roe.* Of this I have only seen the single specimen described by Endlicher and Schauer, which is a very imperfect one, but sufficient to show the remarkable persistent bracteoles, different from those of all other species except *C. brevifolium.*

6. **C. brevifolium,** *Benth.* Branches long and virgate. Leaves opposite, linear, concave or semiterete, mostly erect, appressed and 2 to 3 lines long on the flowering branches, more slender and rather long, with smaller ones clustered in their axils on the main branches. Flowers few, on very short thick pedicels in the upper axils, forming a terminal head or short corymbose raceme. Bracteoles very broad, truncate, brown and scarious, persistent and enveloping the calyx after the flower is expanded. Calyx-tube ovoid, about 2 lines long, the adnate part obtusely 5-ribbed, the free part broader and obscurely 10-ribbed; lobes short, broad, scarious, fringed-ciliate. Petals orbicular, entire, about 1½ lines broad. Connective of the anthers much thickened. Staminodia linear. Style bearded under the stigma with deciduous hairs. Ovules about 6.

W. Australia, *Drummond* (*2nd Coll.?*), *n.* 52. It is possible that this may prove to be a variety of *C. virgatum,* but independently of the foliage, the flowers are smaller and the shape of the calyx appears to be different.

7. **C. uncinatum,** *Schau. in Pl. Preiss.* i. 97. Erect and bushy.

Leaves opposite, linear-triquetrous, usually with a hooked point, from under ½ in. to above ¾ in. long, much attenuate below the middle. Flowers 2 or 4, on pedicels of 2 to 3 lines, in small terminal corymbs, the floral leaves reduced to small bracts. Bracteoles exceedingly deciduous. Calyx-tube thick, full of oily receptacles, broadly turbinate, nearly 3 lines long; lobes very short and broad, quite entire. Petals orbicular, 1½ lines diameter or rather more when fully out. Connective of the anthers thick and globular. Staminodia small, linear or clavate. Style short, with a ring of rigid hairs under the stigma. Ovules 6 to 8.—*C. affine*, Meissn. in Journ. Linn. Soc. i. 45.

W. Australia, *Drummond* (*4th Coll.?*), *n.* 52; Flinders Bay, *Collie;* seacoast, near Fremantle, *Preiss, n.* 359; Swan River, *Gilbert;* between Moore and Murchison rivers, *Drummond, 6th Coll. n.* 40.

Var. *leptophyllum.* Leaves slender, linear-terete, mucronate, but not hooked. Foliage almost of *C. gracile,* with the flowers of *C. uncinatum.* Murchison river, *Oldfield.*

8. **C. megalopetalum,** *F. Muell. Herb.* Bushy and rather rigid. Leaves opposite, oblong-linear, thick, obtuse, mostly 2 to 3 lines long, convex underneath, flat or concave above. Flowers 2 or 4 at the ends of the branches, on pedicels of 1½ to 3 lines. Bracteoles very deciduous. Calyx-tube, in the original form, broadly campanulate, 3 to 4 lines long, the ribs not prominent; lobes broadly ovate, with a rather broad sinus between them, very obtuse, 1 to 1½ lines long, minutely ciliate. Petals from rather less to more than twice as long as the calyx-lobes, quite entire. Connective of the anthers thickened into a semicircular appendage. Staminodia linear. Style with a few stiff hairs in a ring under the stigma. Ovules 8 to 10.

W. Australia. Eastward of King George's Sound, *J. S. Roe, Drummond, Maxwell.* We have four different forms of this plant, which, however, may not be constant enough to establish distinct varieties, viz. 1, with large flowers and crowded leaves, from the interior, *J. S. Roe;* Kojonerup and E. Mount Barren, *Maxwell;* 2, with large flowers and short leaves, erect in distant pairs, from E. Mount Barren, *Maxwell;* 3, with smaller flowers, and leaves short and rather distant, from *Drummond, 5th Coll. n.* 105; and 4, with small flowers, crowded decussate leaves, and the calyx-lobes more distinctly ciliate and separated by a narrower sinus, from *Drummond, 5th Coll. n.* 104.

9. **C. pauciflorum,** *Benth.* An erect shrub, of 1 to 2 ft., with virgate branches. Leaves opposite, erect or slightly spreading, linear or linear-oblong, very obtuse, mostly 2 or rarely 3 lines long, thick, concave, narrowed at the base. Flowers few, rather large, nearly sessile in the upper axils or about 4 in a loose terminal head, the floral leaves broader and shorter than the others, with thin or scarious margins. Bracteoles fallen off from all our specimens. Calyx-tube nearly 3 lines long, turbinate-campanulate, 10-ribbed, the secondary ribs much widened upwards; lobes orbicular-cordate, about ¾ line long, fringed with short cilia. Petals orbicular, about 2 lines diameter, entire. Connective of the anthers thick and ovoid. Staminodia linear-subulate, thickened at the base. Style with a few spreading hairs under the broad stigma. Ovules about 8.—*Decalophium pauciflorum,* Turcz. in Bull. Mosc. 1847, i. 154.

W. Australia, *Drummond, 3rd Coll. n.* 31.

10. **C. axillare,** *F. Muell. Herb.* Rigid virgate and somewhat glaucous. Leaves opposite, linear-triquetrous, mostly mucronate, ½ to ¾ in. long, atte-

nuate at the base. Flowers few, rather large, on pedicels of 1½ to 2 lines in the upper axils, the floral leaves like the others. Bracteoles already fallen from our specimens. Calyx-tube broadly campanulate, about 3 lines long, not very prominently 10-ribbed; lobes broadly ovate or orbicular, minutely ciliate, 1 to 1½ lines long. Petals not twice as long as the calyx-lobes, orbicular, entire. Connective of the anthers thickened into a small appendage. Staminodia linear-lanceolate, often with a rudimentary anther. Style with a ring of stiff hairs under the stigma.

W. Australia. Gales Brook and Russell Range, *Maxwell.*

SUBTRIBE II. CALYTHRICEÆ.—Stamens indefinite, few or numerous, free, in several rows, the inner ones shorter, without staminodia. Ovules 2, collaterally attached to a filiform placenta, extending from the base to the summit of the cavity. Embryo straight, very shortly divided into 2 small cotyledons at the summit.

7. CALYTHRIX, Labill.

(Calycothrix, *Endl.*)

Calyx-tube elongated, usually slender, 10-ribbed, adnate to the ovary at the base or its whole length; lobes 5, spreading, short, with scarious margins, the midrib produced into a long rigid or hair-like awn, or rarely tapering into a shorter point. Petals 5, entire, spreading, deciduous. Stamens indefinite, numerous or rarely 7 to 12, in several rows, the inner ones shorter, deciduous; filaments filiform, quite free; anthers small, versatile; cells parallel, opening in longitudinal slits, connective with a small globular gland-like appendage, rarely thickened or conical and larger than the cells. Ovary 1-celled; ovules 2, collaterally erect, on a filiform placenta attached to the base and to the summit of the cavity, and sometimes continuous with the style. Style filiform, glabrous, with a small capitate stigma. Fruit formed by the lower, usually fusiform, part of the calyx-tube, and usually crowned by the persistent remainder of the calyx. Seed solitary, cylindrical; testa very thin; embryo of the shape of the seed, quite straight, very shortly 2-lobed at the upper end.—Heath-like shrubs. Leaves scattered (not opposite), small, semiterete or 3- or 4-angled or rarely flat and rigid, entire, with occasionally minute hair-like deciduous stipules. Flowers usually shortly pedicellate, solitary in the upper axils, either in terminal leafy heads or more frequently below the ends of the branches. Bracteoles persistent, rigid, continuous with the thickened pedicels, and often united at the base into a turbinate cup, and in the free part overlapping each other and enclosing the base or nearly the whole of the calyx-tube.

The genus is limited to Australia. It has been divided by some according to the presence or absence of stipules, but this character is wholly unavailable in practice. The stipules, when present, are rudimentary only, and so minute and fugacious, that it is often impossible to discover them in some specimens of species where they are occasionally the most conspicuous. Other botanists again have, from the number of stamens, distributed the species into decandrous and icosandrous, or even given in the diagnosis stamens 8, 10, 20 or about 40, but I have found them to vary in this respect in all the species. The majority have above 30 stamens, whilst in the few supposed to be decandrous, the number varies from 7 to about 15, and are not arranged in any regular relation to the sepals and petals, as in the genera with definite stamens. The colour of the flowers appears to be constant in individual species, yellow in some, pink or lilac in others, white in *C. tetragona*, but not of sufficient importance to be available for sectional grouping. The most tangible character I have found, lies in the shape of the calyx-tube and its relation to the ovary, although it is often difficult to verify it without a careful analysis, and, in habit, the majority of the species are very much alike.

A. *Calyx-tube slender, slightly fusiform and adnate to the ovary below the middle, the upper part slender, terete, solid inside, with a convex disk closing the orifice, on which is inserted the style, usually deciduous as well as the stamens.*

Bracteoles free or scarcely united at the base. Flowers yellow.
Glabrous or minutely pubescent. Leaves oblong, erect. Flowers in dense, terminal, leafy heads. Bracteoles usually as long as the calyx-tube . . . . 1. *C. aurea.*
Softly pubescent. Leaves linear, flat or concave. Flowers in terminal heads. Bracteoles acuminate . . . . 2. *C. puberula.*
Glabrous. Leaves semiterete or triquetrous. Bracteoles much shorter than the calyx-tube.
Bracteoles acuminate or subulate-pointed . . . . 3. *C. flavescens.*
Bracteoles broader upwards, herbaceous and obtuse . . . . 4. *C. asperula.*
Bracteoles free or scarcely united at the base. Flowers pink or lilac.
Bracteoles, at the time of flowering, nearly or quite as long as the calyx-tube.
Pubescent. Flowers in terminal heads, the floral leaves lanceolate-villous as well as the bracteoles . . . . 5. *C. sapphirina.*
Glabrous. Flowers in leafy spikes, terminal or below the ends of the branches . . . . 6. *C. breviseta.*
Bracteoles from the first much shorter than the calyx-tube. Stem and leaves slender.
Slender part of the calyx-tube solid . . . . 7. *C. simplex.*
Slender part of the calyx-tube closely enclosing the style, but free from it . . . . 22. *C. tenuiramea.*
Bracteoles connate ¼ to ½ their length. Flowers pink or lilac.
Hirsute. Floral leaves ovate. Bracteoles loose . . . . . . 8. *C. empetroides.*
Glabrous.
Bracteoles nearly or quite as long as the calyx-tube . . . . 9. *C. variabilis.*
Bracteoles much shorter than the calyx-tube.
Leaves oblong-linear, thick, mostly 3 to 4 lines long. Bracteoles forming a loose obovoid-oblong involucre . . . . 10. *C. muricata.*
Leaves slender, 2 to 3 lines long. Bracteoles narrow, appressed . . . . 11. *C. gracilis.*
Leaves very obtuse and thick, under 2 lines long.
Stamens numerous.
Leaves mostly above 1 line long. Bracteoles broad, forming a loose involucre. Calyx-tube 6 to 8 lines . 12. *C. brevifolia.*
Leaves mostly under 1 line. Bracteoles appressed. Calyx-tube 3 to 4 lines . . . . 13. *C. brachyphylla.*
Stamens 7 to 15 (usually about 10) . . . . 14. *C. Leschenaultii.*

B. *Calyx-tube slender, slightly fusiform and adnate to the ovary below the middle, the upper slender part terete, free, enclosing the base of the style, which is usually persistent, the staminal disk forming a ring round it, but free from it.*

Western species.
Bracteoles connate to about the middle.
Bracteoles under 2 lines, much shorter than the calyx-tube . . 15. *C. Oldfieldii.*
Bracteoles 4 to 6 lines, nearly as long as the calyx-tube. Flowers lilac . . . . 16. *C. glutinosa.*
Bracteoles about 3 lines, rather shorter than the calyx-tube. Flowers yellow . . . . 17. *C. angulata.*

Bracteoles free or scarcely united at the base.
Calyx-tube scarcely exceeding the bracteoles and floral leaves.
Low densely-branched shrub. (Flowers yellow ?) . . . . 18. *C. depressa.*
Calyx-tube exceeding the bracteoles, but shorter than the floral leaves. Low densely-branched shrub. Flowers pink or lilac 19. *C. tenuifolia.*
Calyx strigose-pubescent or hirsute, the tube more or less exceeding the bracteoles and floral leaves . . . . 20. *C. strigosa.*
Calyx glabrous, the tube much longer than the bracteoles and floral leaves.
Leaves thick, 2 to 4 lines long. Connective conical . . . 21. *C. decandra.*
Leaves 2 to 3 lines long. Bracteoles acuminate. Connective small . . . . 22. *C. tenuiramea.*
Leaves rarely above 2 lines. Bracteoles obtuse . . . . . 23. *C. Fraseri.*
Leaves very spreading, thick and very obtuse, all under 1 line 24. *C. granulosa.*
Tropical or eastern subtropical species. Bracteoles much shorter than the calyx-tube, connate, from $\frac{1}{4}$ to $\frac{1}{2}$ their length.
Leaves from under $\frac{1}{2}$ line to about 1 line long, minutely ciliate and usually acute and prominently keeled. Petals narrow, acute . 25. *C. microphylla.*
Leaves mostly about 2 lines long, acutely keeled and often minutely ciliate . . . . 26. *C. longiflora.*
Leaves slender, semiterete, 2 to 4 lines long, crowded, not ciliate . 27. *C. leptophylla.*
Leaves oblong-lanceolate, acute, 4 to 6 lines long, not ciliate . . 28. *C. megaphylla.*

C. *Calyx-tube slender, slightly fusiform and adnate to the ovary below the middle, the upper slender part terete, solid inside, terminating in a short broadly campanulate or turbinate free portion.*

Flowers white, usually in terminal leafy heads or short spikes . . 29. *C. tetragona.*

D. *Calyx-tube cylindrical, attenuate at the base, but not contracted above the ovary, the free part scarcely longer than broad.*

Bracteoles more than half as long as the calyx-tube. Calyx-lobes short and broad, with a long hair-like awn . . . . 30. *C. conferta.*
Bracteoles not half as long as the calyx-tube. Calyx-lobes ovate-lanceolate, acuminate, tapering into a short awn . . . . 31. *C. arborescens.*

F. *Calyx-tube pubescent, oblong, more or less contracted above the ovary, the free part short; lobes with very short awns or points.*

Calyx-tube 2 to $2\frac{1}{2}$ lines long, slightly contracted above the ovary . 32. *C. brachychæta.*
Calyx-tube 1 line long, much contracted above the ovary . . . . 33. *C. achæta.*
Calyx-tube nearly glabrous, $1\frac{1}{2}$ lines long, slightly contracted above the ovary. Leaves very fine, 2 to 3 lines long . . . . 34. *C. laricina.*

1. **C. aurea,** *Lindl. Swan Riv. App.* 5. *t.* 3 B. Erect, rather stout and rigid, not much branched, glabrous or minutely pubescent. Leaves erect or rarely spreading, elliptical-oblong or the floral ones ovate-lanceolate, thick, concave, obtuse, mostly 3 to 4 lines long, more or less ciliate on the edges and midrib, or rarely quite glabrous. Flowers yellow, nearly sessile in dense terminal leafy heads. Bracteoles 3 to 5 lines long, free, narrow, acutely keeled, broader upwards and tapering to a fine point. Calyx-tube rarely exceeding the bracteoles and often shorter, the lower fusiform portion 3-angled, the upper slender part solid; lobes short, broad, with spreading awns much longer than the petals. Petals obtuse, 3 to $3\frac{1}{2}$ lines long. Stamens numerous; connective gland globular. Style inserted on the staminal disk, deciduous.—Schau. Myrt. Xeroc. 106, and Pl. Preiss. i. 107.

W. Australia. Swan River, *Drummond*, 1*st Coll.*; sandy plains, Canning river, *Preiss, n.* 184.

2. **C. puberula,** *Meissn. in Journ. Linn. Soc.* i. 48. Much smaller than *C. aurea*, erect, much branched, softly pubescent or villous. Leaves linear, rather flat, but the midrib very prominent underneath and often above also, obtuse or mucronulate, 2 to 3 lines long, the floral ones scarcely larger. Flowers yellow, nearly sessile in the upper axils, but not in such compact heads as in *C. aurea*. Bracteoles free, narrow, hirsute, acuminate or mucronulate, about 3 lines long. Calyx-tube rather longer, fusiform and obscurely triangular, the slender upper portion short, solid; lobes obovate, with an awn not twice as long as the petals. Petals rather obtuse, about 3 lines long. Stamens numerous; connective-gland prominent. Style inserted on the staminal disk, deciduous.

W. Australia. Between Moore and Murchison rivers, *Drummond, 6th Coll. n.* 51.

3. **C. flavescens,** *A. Cunn. in Bot. Mag. under n.* 3323. Rather slender, often under 1 ft. high and simple or nearly so, and from that to above 2 ft. and more or less branched, usually glabrous. Leaves linear-triquetrous and slender, or the floral ones lanceolate and flat, obtuse or scarcely mucronulate, mostly 3 to 4 lines long in the normal form. Flowers yellow, nearly sessile in the upper axils, forming ovoid or oblong terminal leafy spikes, rarely lateral by the elongation of the shoots, or a few short branches forming a compact corymb. Bracteoles free, narrow, 3 to 4 lines long, keeled, and tapering into a fine awn-like point. Calyx-tube 6 to 8 lines long, slightly fusiform and 3- or 5-angled below the middle, the slender upper portion solid inside; lobes 1 to 1½ lines long, truncate or shortly acuminate with an awn 2 or 3 times the length of the petals. Petals obtuse, 3 to 3½ lines long. Stamens numerous; connective-gland globular. Style inserted on the staminal disk, deciduous.—Schau. Myrt. Xeroc. 105, and in Pl. Preiss. i. 106; Field. and Gardn. Sert. Pl. t. 38 (the analysis not correct); *C. luteola*, Schau. in Pl. Preiss. i. 106.

W. Australia. Swan River, *Drummond, 1st Coll.; 2nd Coll. n.* 52; *Preiss, n.* 186, 187, 193; Moore river, *Oldfield;* Kojonerup and Tone river, *Maxwell.*

Var. *Drummondii.* Stouter and more rigid in all its parts. Leaves crowded, 4 to 6 lines long. Flowers rather larger and more numerous below the ends of the branches.—*C. Drummondii*, Meissn. in Journ. Linn. Soc. i. 47. Between Moore and Murchison rivers, *Drummond, 6th Coll. n.* 52.

Var. *tenella.* Apparently diffuse and slender. Flowers smaller and more distant. Bracteoles less pointed.—*C. tenella*, Meissn. in Journ. Linn. Soc. i. 47. Between Moore and Murchison rivers, *Drummond, 6th Coll. n.* 55. This appears to me rather an etiolated form than a distinct race.

Var. *curtophylla.* Leaves short and not very slender, mostly spreading. Flowers rather small. Bracteoles short and narrow. *C. curtophylla*, A. Cunn. in Bot. Mag. under n. 3323, not of Schauer. Swan River, *Fraser*, and southern districts, *Baxter.*—*C. tetragonophylla*, Meissn. in Journ. Linn. Soc. i. 47, from Moore and Murchison rivers, *Drummond, 6th Coll. n.* 54, only differs from this form in a very slight pubescence on the branches and upper leaves.

Amongst all the above forms this species is readily recognized by the bracteoles always finely acuminate, as in *C. simplex* and in *C. tenuiramea*, both of which have pink or lilac flowers.

4. **C. asperula,** *Schau. in Pl. Preiss.* i. 106. Loosely branched and quite glabrous, 1 to 3 ft. high. Leaves more or less spreading, linear or linear-oblong, obtusely triquetrous, rather thick, very obtuse, 1 to 2 lines or

rarely longer. Flowers pale yellow, nearly sessile below the ends of the branches. Bracteoles free, about 2 lines long, narrowed at the base, broader and herbaceous upwards, obtuse or with a short spreading point. Calyx-tube 4 to 5 lines long, slightly fusiform and angular below the middle, the slender upper portion solid inside; lobes broadly obovate, the long awn dilated towards the base. Petals 3 to 3½ lines long, rather acute. Stamens numerous; anthers small, didymous, the cells opening deeply in 2 valves; connective-gland small. Style inserted on the staminal disk, deciduous.

W. Australia. Sandy and stony places, King George's Sound, and to the eastward towards Cape Riche, W. Mount Barren, etc., *Baxter, Preiss, n.* 194, *Oldfield, Maxwell;* Swan River (?), *Drummond,* 1*st Coll.* The bracteoles distinguish this species from all others independently of the colour of the flowers, yellow.

Var. *gracilis.* Leaves slender, flowers small, *Baxter* (*Herb. R. Br.*).

5. **C. sapphirina,** *Lindl. Swan Riv. App.* 5. Erect and nearly simple when first flowering, growing into a straggling shrub of 2 to 3 ft. more or less pubescent. Leaves usually spreading, linear-triquetrous, obtuse or mucronate, 2 to 3 or rarely 4 lines long, the floral ones often lanceolate acuminate and softly villous. Flowers pink or purple, nearly sessile in dense terminal globular or ovoid leafy heads. Bracteoles free, 3 lines long or rather more, keeled, scarious but softly pubescent. Calyx-tube at first scarcely exceeding the bracteoles, but lengthening to about 5 lines, fusiform and pubescent below the middle, the slender upper portion glabrous and solid; lobes truncate with fine awns longer than the petals, and minutely ciliate. Petals about 2 lines long. Stamens numerous; connective-gland small. Style inserted on the staminal disk, glabrous.—Schau. Myrt. Xeroc. 103, and in Pl. Preiss. i. 105; *C. lasiostachya,* F. Muell. Fragm. i. 224.

W. Australia. Swan River, *Drummond,* 1*st Coll. and* 2*nd Coll. n.* 154; *Preiss, n.* 189; Murchison river, *Oldfield.*

6. **C. breviseta,** *Lindl. Swan Riv. App.* 5. Glabrous with erect and virgate or spreading and branched stems. Leaves erect or spreading, linear, semiterete or triquetrous, obtuse or mucronate, 2 to 3 lines long, usually rather slender, the floral ones scarcely broader. Flowers pink or lilac, nearly sessile in the upper axils, but mostly below the ends of the branches. Bracteoles free, rather firm and brown, smooth or more or less glandular-muricate, about 3 lines long. Calyx tube rarely exceeding the bracteoles, fusiform below the middle, the slender upper portion solid; lobes broad, about a line long, the hair-like awn rather longer than the petals, often minutely ciliate. Petals acute, about 4 lines long. Stamens numerous; connective-gland small. Style inserted on the staminal disk, deciduous.—Schau. Myrt. Xeroc. 99; *C. cuspidata,* Turcz. in Bull. Mosc. 1847, i. 162 (from the descr.).

W. Australia. Swan River, *Drummond,* 1*st Coll.*, 5*th Coll. n.* 115.—This has the habit and large bracteoles of *C. variabilis,* but they are free almost or quite to the base. The name is not very appropriate.

7. **C. simplex,** *Lindl. Swan Riv. App.* 5. Glabrous or slightly pubescent, simple, erect, and under 6 in., or taller, with slender spreading branches, as in *C. tenuiramea.* Leaves rather slender, semiterete or triquetrous, 3 to 4

lines long. Flowers pink or lilac, on very short pedicels; axillary below the ends of the branches. Bracteoles free or scarcely united at the base, about 2 lines long, acuminate or almost awned like those of *C. flavescens.* Calyx-tube about 4 lines long, fusiform below the middle, the slender upper portion solid; lobes short, acute, with fine awns slightly dilated at the base and exceeding the petals. Petals about 3 lines long, obtuse. Stamens numerous; the connective small and globular, or in the inner stamens larger and more prominent. Style inserted on the staminal disk, glabrous.—Schau. Myrt. Xeroc. 101, and in Pl. Preiss. i. 105.

W. Australia. Swan River, *Drummond,* 1*st Coll.;* stony hills, Tweed river, *Oldfield;* near Albany, *Preiss, n.* 199, whose specimens, however, I have not seen.—The simple tufted stems which suggested the specific name are by no means constant. The species often assumes the aspect of *C. tenuiramea,* of which it has also the bracteoles, but I never find the style penetrating into the calyx-tube as in that species. *C. simplex* is also very near *C. flavescens,* differing chiefly in the colour of the flowers.

8. **C. empetroides,** *Schau. Myrt. Xeroc.* 102, *and in Pl. Preiss.* i. 105. Low, diffuse or prostrate, much branched, pubescent or hirsute. Leaves oblong-linear, spreading, obtuse or scarcely mucronate, rarely above 2 lines long, the midrib prominent underneath, the floral ones usually ovate and more hirsute. Flowers nearly sessile in the upper axils, or below the ends of the branches. Bracteoles membranous, obtuse, hirsute, connate above the middle into a loose involucre, about 1½ lines long. Calyx-tube about 2½ lines long, slightly hirsute, fusiform below the middle, the slender upper portion solid; lobes broad, almost acute, the hair-like awn minutely ciliate and about as long as the petals. Petals (pink or lilac) about 3 lines long. Stamens numerous; anthers small, with a globular gland-like connective. Style inserted on the staminal disk, deciduous.

W. Australia, *J. S. Roe;* gravelly sides of Mount Bakewell, *Preiss, n.* 195.

C. ciliata, Turcz. in Bull. Mosc. 1847, i. 161, which I have not seen, is probably, from the description given, a variety of *C. empetroides* with narrower leaves.

9. **C. variabilis,** *Lindl. Swan Riv. App.* 5. Quite glabrous, branches usually erect and virgate, sometimes loosely spreading, Leaves linear-triquetrous and slender, 3 to 6 lines long, or thickly linear-oblong and 2 to 3 lines long, those of the flowering branches often much shorter and broader than the others. Flowers (pink or lilac) on very short pedicels in the upper axils below the ends of the branches. Bracteoles 3 to 4 lines long, connate from ¼ to nearly ½ their length, broader upwards, acuminate, the midrib scarcely prominent. Calyx-tube rarely exceeding the bracteoles, slightly fusiform below the middle, the slender upper portion solid; lobes tapering into a slender awn, exceeding the petals and often minutely ciliate. Petals about 3 lines or sometimes nearly 4 lines long, rather acute. Stamens numerous; connective-gland rather large. Style inserted on the slightly concave staminal disk.—Schau. Myrt. Xeroc. 100, and in Pl. Preiss. i. 105.

W. Australia. Swan River, Darling range and neighbourhood, *Collie, Drummond,* 1*st Coll. n.* 157; *Preiss, n.* 197, *Oldfield.*

10. **C. muricata,** *F. Muell. Fragm.* i. 224. Quite glabrous, the branches erect, rigid, and rather stout. Leaves linear or linear-oblong, thick, triquetrous or keeled, obtuse, mostly 3 to 4 lines long, the floral ones and a

few at the base of the shoots often shorter, ovate or ovate-lanceolate. Flowers pink, on stout pedicels of 1 or even 2 lines in the upper axils below the ends of the branches. Bracteoles firm, attaining 3 or 4 lines, connate to about the middle, usually glandular-muricate. Calyx-tube fully ½ in. long, and often lengthening to nearly ¾ in., slightly fusiform below the middle, the upper slender portion solid; lobes broad, nearly 1½ lines long, shortly tapering into very long fine awns. Petals rather broad, above 3 lines long. Stamens numerous. Style inserted on the convex staminal disk, deciduous.

W. Australia. Sandy places, Murchison river, *Oldfield.*—In the ripe fruit the upper portion of the calyx usually falls off, leaving only the fusiform portion enclosing the seed and included in the bracteoles.

Var. *parvifolia.* Leaves mostly under 2 lines. Murchison river, *Oldfield.*

11. **C. gracilis,** *Benth.* Glabrous and very heath-like, with rather slender, short, but virgate branches. Leaves linear, slender, semiterete, erect, obtuse, 2 to 3 lines long or rather more. Flowers (pink or purple) shortly pedicellate in the upper axils. Bracteoles narrow, connate to about the middle, appressed on the calyx-tube, about 2 lines long, obtuse, with the midrib produced into a short point, smooth. Calyx-tube very slender, about 4 lines long, slightly fusiform below the middle, the slender upper portion solid; lobes broad, about 1 line long, shortly acuminate, with a fine awn scarcely exceeding the petals. Petals about 3 lines long, acute. Stamens numerous; anthers small. Style inserted on the convex staminal disk, deciduous.

W. Australia. Murchison river, *Oldfield.*—Allied to *C. muricata,* but the slender leaves, smaller flowers, narrow smooth bracteoles, and slender calyx-tube, give it a very different aspect.

12. **C. brevifolia,** *Meissn. in Journ. Linn. Soc.* i. 46. Possibly a larger, stouter variety of *C. brachyphylla.* Leaves spreading or almost reflexed, oblong-triquetrous or almost ovate, thick, very obtuse, 1 to 1½ lines long or rarely more. Flowers (pink ?) on turbinate or clavate pedicels of 1 to 2 lines below the end of the branches. Bracteoles about 3 lines long, connate to near the middle, the free part broadly cordate-ovate, shortly acuminate, forming a loose involucre. Calyx-tube 6 to 8 lines long, shortly fusiform and very obtusely ribbed below the middle, the very slender upper portion solid; lobes broad, truncate, with a fine awn much longer than the petals. Petals nearly 4 lines long. Stamens numerous; connective-gland small. Style inserted on the staminal disk, deciduous.

W. Australia. Between Moore and Murchison rivers, *Drummond, 6th Coll. n.* 58.

13. **C. brachyphylla,** *Turcz. in Bull. Mosc.* 1847, i. 161. Quite glabrous and much branched. Leaves imbricate or spreading, ovate to oblong-linear, very thick and obtuse, rarely exceeding 1 line and often not above ½ line long. Flowers (pink or lilac) rather small, nearly sessile in the upper axils, either forming short terminal leafy corymbs or all below the ends of the branches. Bracteoles about 1½ to nearly 2 lines long, connate to near the middle, much closer round the calyx than in *C. brevifolia,* rather firm, smooth or glandular-muricate. Calyx-tube slender, 3 to 4 lines long, fusiform below the middle, the slender portion solid; lobes short, truncate,

almost emarginate, the awn hair-like, minutely ciliate, shortly exceeding the petals. Petals $2\frac{1}{2}$ to 3 lines long, acute. Stamens numerous; anthers small, the connective-gland small in the outer stamens, larger and almost conical in the inner ones. Style inserted on the staminal disk, deciduous.

W. Australia. Swan River, *Drummond,* 1*st Coll. n.* 156, *Gilbert;* Murchison river, *Oldfield;* in the interior, *J. S. Roe;* common about King George's Sound, *Milne;* eastward to Cape Arid, *Maxwell;* Moannoka, *Walcott.*

14. **C. Leschenaultii,** *Schau. in Pl. Preiss.* i. 104. Apparently more erect and less branched than in the last two species, which this one otherwise resembles. Leaves oblong-triquetrous, rather thick, very obtuse, 1 to 2 lines long. Flowers (pink or purple) nearly sessile in the upper axils at or below the ends of the branches. Bracteoles 2 to 3 lines long, connate to near the middle, the free part not very broad, usually shortly mucronate. Calyx-tube 4 to nearly 5 lines long, fusiform below the middle, the very slender upper portion solid; lobes usually truncate, deeply coloured with scarious margins, but sometimes almost tapering into the fine awn, which is much longer than the petals. Petals acutely acuminate, about 3 lines long. Stamens 7 to 10 or rarely more; connective-gland small. Style inserted on the staminal disk, deciduous.—*C. curtophylla,* Schau. Myrt. Xeroc. 90, and in Pl. Preiss. i. 104 (from the character given), but not of A. Cunn.

W. Australia. Darling range, Gordon river, and near Albany, *Preiss, n.* 191, 192; Kalgan river and Mount Elphinstone, *Oldfield.*—The specimens which I have seen of *C. curtophylla,* Schau., are past flower; they look more like *C. brachyphylla,* but Schauer says there are only 8 or 9 stamens. *C. curtophylla,* A. Cunn., is a variety of *C. flavescens.*

15. **C. Oldfieldii,** *Benth.* Very nearly allied to *C. brachyphylla,* with the same habit and foliage, but besides the broader bracteoles, the style is persistent and free within the slender portion of the calyx. Quite glabrous. Leaves oblong-triquetrous, thick and very obtuse, mostly about 1 line long. Flowers (pink?) nearly sessile in the upper axils, more corymbose than is usual in *C. brachyphylla.* Bracteoles $1\frac{1}{2}$ to nearly 2 lines long, connate to about the middle, the free part broad and obtuse. Calyx-tube 4 lines long or rather more, fusiform below the middle, the slender upper portion not quite solid, but leaving a deep narrow cavity round the style; lobes short and broad, with fine awns scarcely exceeding the petals. Petals about 3 lines long, obtuse. Stamens numerous.

W. Australia. S. Hutt river, *Oldfield.*

16. **C. glutinosa,** *Lindl. Swan Riv. App.* 5. Erect, with rather stout not much branched stems of 1 to 2 ft. Leaves erect, linear-terete, rather thick, mucronate, $\frac{1}{4}$ to $\frac{1}{2}$ in. long, the floral ones often much shorter, acute, flat and dilated at the base into a short stipule-like lobe on each side. Flowers rather large (purple?), on short pedicels in ovoid terminal heads, or sometimes lateral by the elongation of the shoot. Bracteoles about $\frac{1}{2}$ in. long, connate to the middle, acuminate and almost aristate, keeled, more or less glutinous. Calyx-tube not exceeding the bracteoles at first, but rather longer when in fruit, fusiform below the middle, the upper slender portion free, enclosing the style; lobes usually truncate, the rigid prominent midrib produced into an awn much exceeding the petals. Petals about 4 lines

long. Stamens in some specimens about 10, in others nearly twice as many, inserted round an annular disk; connective-gland small.—Schau. Myrt. Xeroc. 91, and Pl. Preiss. i. 104.

W. Australia. Swan River, *Drummond, 1st Coll., Gilbert;* Darling range, *Preiss, n.* 196.

17. **C. angulata,** *Lindl. Swan Riv. App.* 6. Glabrous, the young branches angular. Leaves spreading, mostly linear-triquetrous, rather thick, obtuse, 2 to 3 lines long, but often shorter and broader on the lateral shoots, and a few ovate, concave, keeled, about 1½ lines long. Flowers apparently yellow, shortly pedicellate in the upper axils below the ends of the branches. Bracteoles about 3 lines long, or rather more, united to the middle, enlarged upwards, keeled, somewhat acute or mucronate, smooth or glandular-muricate. Calyx-tube very slender, scarcely exceeding the bracteoles at first, half as long again when in fruit, slightly fusiform below the middle, the slender upper portion free, enclosing the style; lobes truncate, with a long hair-like awn. Petals about 3 lines long. Stamens numerous; connective-gland globular. —Schau. Myrt. Xeroc. 104, and in Pl. Preiss. i. 106.

W. Australia. Swan River, *Collie, Drummond, 1st Coll. n.* 161, *Turner;* Canning river, *Preiss, n.* 185.

18. **C. depressa,** *Turcz. in Bull. Mosc.* 1847, i. 162. Glabrous, very densely branched and under 1 ft. high. Leaves crowded, linear-triquetrous, obtuse or very shortly mucronate, rarely above 2 lines long, rather thick or the floral ones almost lanceolate and concave. Flowers (yellow ?) nearly sessile in the upper axils. Bracteoles free, about 3 lines long, keeled, narrowed at the base and shortly tapering into a short point. Calyx-tube rarely exceeding the bracteoles, slightly fusiform below the middle, the upper slender free portion enclosing the style; lobes small, broad, acute, with a fine awn scarcely exceeding the petals. Petals about 3 lines long, acute. Stamens numerous; connective thickened into a conical appendage nearly as long as the cells.

W. Australià, *Drummond, 3rd Coll. n.* 24.

19. **C. tenuifolia,** *Meissn. in Journ. Linn. Soc.* i. 46. Glabrous, with numerous erect branches. Leaves crowded on the smaller branches, erect or slightly spreading, linear, slender, semiterete or triquetrous, obtuse or mucronulate, 3 to 4 lines long in the original form. Flowers (pink ?) nearly sessile in the upper axils. Bracteoles free, 2 to 2½ lines long, acuminate. Calyx-tube about 3 lines long, or rarely lengthening to 4 lines in fruit, slightly fusiform below the middle, the upper free portion very slender and enclosing the style; lobes orbicular, about 1 line long, with a fine awn slightly exceeding the petals. Petals acuminate, 2½ to 3 lines long. Stamens numerous; connective-gland small.

W. Australia. Murchison river, *Oldfield, Drummond, 6th Coll. n.* 57.

Var. *rigidior.* Stouter, leaves more rigid and rather longer; flowers rather larger.—*C. rosea,* Meissn. in Journ. Linn. Soc. i. 46.—*Drummond, 6th Coll. n.* 56.

20. **C. strigosa,** *A. Cunn. in Bot. Mag. under n.* 3323. A low bushy shrub, more or less pubescent, or quite glabrous except the flowers. Leaves erect or spreading, linear-oblong, rather thick, subterete or triquetrous, obtuse,

1 to 2 lines long, the floral ones sometimes broadly oblong or ovate and concave. Flowers (pink or lilac?) nearly sessile at or below the ends of the branches, forming sometimes a dense leafy corymbose panicle. Bracteoles free, 2 to nearly 4 lines long, cuneate, obtuse or scarcely acuminate. Calyx-tube slender, pubescent or hirsute, 4 to 6 lines long, slightly fusiform below the middle, the upper slender free portion enclosing the style; lobes ovate-lanceolate at the base, gradually tapering into shortly plumose awns not much exceeding the petals. Petals 3 to 5 lines long. Stamens numerous round a distinct annular disk; connective-glands small.—Schau. Myrt. Xeroc. 108; *C. lasiantha*, Meissn. in Journ. Linn. Soc. i. 46.

W. Australia. Sharks Bay and Dirk Hartog's Island, *A. Cunningham;* Murchison river, Champion Bay, *Drummond, n.* 158, *3rd Coll. n.* 178, and *6th Coll. n.* 53, *Oldfield, Walcott, and others.*

21. **C. decandra,** *R. Br. Herb.; DC. Prod.* iii. 208. Small, erect, and quite glabrous. Leaves crowded, linear, triquetrous or concave, acute or obtuse, rather thick, 2 to 3 or rarely 4 lines long. Flowers large, pink, on short thick pedicels in the upper axils. Bracteoles free, rather narrow, acuminate, about 4 lines long. Calyx-tube 6 to 7 lines long when in flower, still longer afterwards, slightly fusiform below the middle, the long slender upper portion free and enclosing the style; lobes scarious, shortly tapering into the long awn, not at all ciliate. Petals acuminate, fully 5 lines long. Stamens about 10, very unequal; connective thick and obliquely conical, larger than the cells, with a small globular gland in a dorsal cavity.—*C. Candolleana*, Schau. Myrt. Xeroc. 92; *C. conanthera*, F. Muell. Fragm. i. 146.

W. Australia. Lucky Bay, *R. Brown, Baxter;* Eyre's Range, M'Callum's Inlet, Stokes Inlet, E. Mount Barren, *Maxwell.*

22. **C. tenuiramea,** *Turcz. in Bull. Mosc.* 1849, ii. 20. Glabrous, with slender divaricate branches, from under 1 ft. to 2 ft. high. Leaves not crowded, semiterete or triquetrous, obtuse or scarcely mucronate, from under 2 to above 3 lines long. Flowers (pink or lilac) on short axillary pedicels below the ends of the branches. Bracteoles free or shortly united at the base, rather narrow, acuminate or almost awned, 2 to 2½ lines long. Calyx-tube 3 to 3½ lines long, slightly fusiform below the middle, the upper very slender portion apparently solid but not quite so, leaving a very narrow cavity in which the style is free; lobes small, acute, the long rigid awn dilated at the base, fine at the end. Petals rather acute, 3 to 3½ lines long, Stamens numerous; connective gland small.

W. Australia, *Drummond, 4th Coll. n.* 50; towards Cape Riche, *Harvey;* sandy plains, Cape Riche, Gordon river, and near Mount Barker, *Maxwell.* Except in the style free within the calyx-tube, this species much resembles the slender branching forms of *C. simplex.* The so-called stipules are sometimes very conspicuous.

23. **C. Fraseri,** *A. Cunn. in Bot. Mag. under n.* 3323. Quite glabrous. 1 to 2 ft. high, with spreading branches. Leaves spreading or recurved, oblong or linear, keeled or triquetrous, obtuse, rather thick, rarely above 2 lines long. Flowers few in the upper axils below the ends of the branches, rather large, lilac or purple. Bracteoles free, not 2 lines long, narrow-cuneate, very obtuse. Calyx-tube about ½ in. long, slightly fusiform below

the middle, the long slender upper portion free, enclosing the style; lobes truncate, with very fine hair-like awns. Petals very acute, about 4 lines long. Stamens numerous; connective-gland globular.—Schau. Myrt. Xeroc. 98, and in Pl. Preiss. i. 105.

W. Australia. Swan River, *Fraser, Drummond,* 1*st Coll.,* 2*nd Coll. n.* 159, *Preiss, n.* 190, 198; Murchison river, *Oldfield.*

24. **C. granulosa,** *Benth.* Scrubby, glabrous, with tortuous divaricate branches. Leaves very spreading, oblong or ovate, thick and very obtuse, all under 1 line long. Flowers (pink?) on slender pedicels of nearly 1 line, below the ends of the branches. Bracteoles quite free, linear-cuneate, very obtuse, scarcely 1½ line long. Calyx-tube 3 or at length nearly 4 lines long, slightly fusiform below the middle, the upper free portion as long and scarcely more slender, enclosing the style and slightly dilated at the mouth; lobes small, truncate, with fine awns much longer than the petals. Petals about 2½ to 3 lines long, obtuse. Stamens numerous; connective-gland small, globular.

W. Australia. Murchison river, *Oldfield.* This has some resemblance with *C. brachyphylla,* but is more rigid and scrubby, and the bracteoles and calyx are quite different.

25. **C. microphylla,** *A. Cunn. in Bot. Mag. under n.* 3323. A tall shrub, or, on banks of streams, a small tree, with numerous small branchlets covered with imbricated leaves, as in *C. arborescens.* Leaves thick and triquetrous, from under ½ line long and almost obtuse, to above 1 line long and acute, more or less ciliate with very short rigid hairs, or when luxuriant quite glabrous. Flowers (of a rich red?) on thick pedicels of about a line in the upper axils of the short branchlets, forming showy corymbose or oblong leafy panicles. Bracteoles about 2 lines long, setaceous-acuminate, connate at the base. Calyx-tube scarcely 3 lines long when first flowering, but lengthening to 5 lines, slightly fusiform below the middle, the slender upper portion free, enclosing the style; lobes ovate, acuminate, with hair-like awns from half the length of, to longer than the petals. Petals narrow, acute, 4 to 5 lines long. Stamens numerous; connective-gland small.—Schau. Myrt. Xeroc. 89; *C. exstipulata,* DC. Prod. iii. 208, according to Schauer; *C. cupressifolia,* A. Rich. Sert. Astrol. 41. t. 16 (*C. cupressoides,* A. Rich. l. c. 43).

N. Australia. Glenelg river and Roebuck Bay, N. W. Coast, *Marten*, Victoria river and Arnhem's Land, *F. Mueller;* Melville Island (not Port Macquarrie), *Fraser;* islands of the Gulf of Carpentaria, *R. Brown;* exposed cliffs of Port Essington, *A. Cunningham, Armstrong.*

Var.? *longifolia.* Leaves less imbricate, 1 to 1½ line long, mucronate, ciliolate; bracteoles short, calycine awns longer than the petals.—*M'Douall Stuart's Expedition,* lat. 17° 43′.—These small specimens seem almost to connect this species with the following.

26. **C. longiflora,** *F. Muell. Fragm.* i. 12. A tall handsome shrub, quite glabrous. Leaves oblong-linear or cuneate, obtuse or shortly mucronate, 1½ to 2 lines long, or nearly 3 lines on luxuriant shoots, rigid with acute denticulate-ciliate margins, and a very prominent acute keel. Flowers large (pink?) on short thick pedicels, in the axils of small floral leaves, forming terminal heads on the short branchlets. Bracteoles about 1½ lines long, connate at the base, truncate and finely mucronate. Calyx-tube attaining 6

lines, cylindrical, the adnate portion scarcely fusiform, the upper free portion about as long, not more slender, enclosing the style; lobes short, broad, with long hair-like awns. Petals acute, 4 to 5 lines long. Stamens very numerous; connective-gland small.

Queensland. In the interior, *Mitchell;* Suttor river, *F. Mueller.*

27. **C. leptophylla,** *Benth.* Quite glabrous. Leaves crowded on the short branchlets, slender, linear, semiterete or triquetrous, obtuse or scarcely mucronate, mostly about 2 lines. Flowers (pink?) nearly sessile in the upper axils, much smaller than in *C. longiflora*, but otherwise similar. Bracteoles connate below the middle, acutely acuminate, about 2 lines long. Calyx-tube slender, about 4 lines long, the lower portion scarcely fusiform, the upper cylindrical portion free, enclosing the style. Petals and stamens not seen.

Queensland. Newcastle Range, *F. Mueller;* a single specimen snatched in breaking through the scrub and communicated under the name of *C. tenuifolia,* which is now, however, preoccupied by a species of Meissner's. It is evidently very near *C. longiflora* and *C. microphylla,* but can scarcely be considered as conspecific with either.

28. **C. megaphylla,** *F. Muell. Fragm.* i. 146. Quite glabrous. Leaves linear-oblong or lanceolate, acute, narrowed at the base, the larger ones fully ½ in. long, coriaceous, somewhat concave, slightly keeled, the margins not ciliate. Flowers large (deep pink?) nearly sessile in the upper axils. Bracteoles about 3 lines long, shortly connate at the base, acuminate, with long fine points. Calyx-tube fully 8 lines long, slightly fusiform below the middle, the upper portion scarcely more slender, free, enclosing the style; lobes broadly obovate, with awns exceeding the petals. Petals 4 to 5 lines long. Stamens numerous.

N. Australia. M'Adam Range, *A. C. Gregory.* Differs from *C. longiflora* chiefly in the foliage, but the specimens are few and small.

29. **C. tetragona,** *Labill. Pl. Nov. Holl.* ii. 8. *t.* 146. An elegant shrub, usually of 2 or 3 ft. but sometimes drawn up to a much greater height, glabrous pubescent or hirsute with short rigid hairs, the branches virgate or spreading. Leaves erect or spreading, linear, triquetrous or convex underneath, obtuse or mucronulate, mostly 2 to 3 lines long, or when luxuriant nearly twice as long, the stipules which have been chiefly observed in this species so minute and deciduous as to be rarely seen. Flowers white or pink, nearly sessile in the upper axils, forming dense terminal short or oblong leafy heads, becoming lateral by the elongation of the shoots, especially in poor cultivated specimens. Bracteoles free, scarious, keeled, about 2 lines long. Calyx-tube about 2 lines at the time of flowering, lengthening out to 4 lines or even more, the lower portion fusiform, produced into a long slender solid neck or stipes to the short campanulate or turbinate free part; lobes ovate, with fine awns longer than the petals. Petals obtuse, about 2 lines long. Stamens usually above 20; connective gland small. Style inserted on the summit of the solid neck of the calyx.—F. Muell. Fragm. iv. 36; *C. glabra*, R. Br. in Bot. Reg. t. 409; Lodd. Bot. Cab. t. 586; Hook. f. Fl. Tasm. i. 127; *C. glabra*, *C. tetraptera*, and *C. scabra*, DC. Prod. iii. 208; Mem. Myrt. t. 1; *C. ericoides*, A. Cunn. in Field, N. S. Wales, 350;

C. virgata, A. Cunn. in Bot. Mag. t. 3323; *C. brunioides*, A. Cunn. in Bot. Mag. under the same n.; *C. Billardierii*, *C. virgata*, *C. scabra*, and *C. brunioides*, Schau. Myrt. Xeroc. 93 to 97; *C. Brownii*, Schau. l. c. 108, and probably *C. Baueri*, Schau. l. c. 109; *C. pubescens*, Sweet in G. Don, Gen. Syst. ii. 811; *C. Behriana*, Schlecht. Linnæa, xx. 650; *C. Schlechtendahlii*, *C. rosea*, *C. leucantha*, *C. squarrosa*, *C. monticola*, and *C. Muelleri*, Miq. in Nederl. Kruidk. Arch. iv. 116 to 119.

N. S. Wales. Port Jackson to the Blue Mountains, *R. Brown*, *Sieber*, *n.* 285, and others, and in the interior on the Macquarrie, Lachlan, Darling, Murrumbidgee, etc., *A. Cunningham*, *F. Mueller*, and others.

Victoria. Common in the desert tracts on the Murray and Wimmera, ascending in the Australian Alps to 4000 ft., *F. Mueller*.

Tasmania. Sandy places and rocky coasts, frequent, sometimes growing in water and then very tall, *J. D. Hooker*.

S. Australia. From Spencer's and St. Vincent's Gulf, and Kangaroo Island to the Murray, *F. Mueller*, *Whittaker*, and others.

W. Australia. King George's Sound and neighbouring districts, and eastward to the Great Australian Bight, *J. C. Roe*, *Oldfield*, *Maxwell*, *Drummond*, *2nd Coll.*, *n.* 46; *3rd Coll. n.* 53; *5th Coll. n.* 116.

This is undoubtedly a variable species, and individual specimens often exhibit very striking differences, but the numerous species founded upon it have been chiefly distinguished by the degree of pubescence, by the size and direction of the leaves, the length of the calyx-tube, and other characters, often dependent on age, luxuriance, or local circumstances, and which, in the large mass of specimens I have examined, show such insensible gradations that I have in vain sought to class them in distinct varieties by any tangible characters. Amidst all these variations, this species is readily distinguished by the short free part of the calyx always much broader than the neck of the adnate part, although it varies from narrow-campanulate to very broadly turbinate.

30. **C. conferta,** *A. Cunn. in Bot. Mag. under n.* 3323. A tall, erect, glabrous shrub, with very numerous small branches. Leaves imbricate, acutely triangular, acute or mucronate, $\frac{1}{2}$ to $\frac{3}{4}$ line long. Flowers (pink?) on short pedicels in the upper axils. Bracteoles connate below the middle, broader upwards, rather firm, obtuse or shortly acuminate, about 2 lines long. Calyx-tube cylindrical, 2 to nearly 3 lines long, attenuate at the base, but not at all above the ovary, the free part scarcely longer than broad; lobes broad, scarcely acuminate, with fine awns about as long as the petals. Petals 3 to 4 lines long, acute. Stamens numerous; connective-gland globular.—Schau. Myrt. Xeroc. 88. t. 6 B.

N. Australia. Port Keath, Cambridge Gulf, *A. Cunningham;* N. W. Coast, *Bynoe*.

Near *C. arborescens*, with nearly the same shaped calyx-tube, but the bracteoles and calycine lobes quite different.

31. **C. arborescens,** *F. Muell. in Trans. Phil. Inst. Vict.* iii. 42. A tall shrub or small tree, with exceedingly numerous short slender branchlets covered with the imbricate scale-like leaves like those of a tamarisk. Leaves ovate-lanceolate, concave, prominently keeled, scarcely $\frac{1}{2}$ line long, glabrous. Flowers (white?) on thickened pedicels of about 1 line in the upper axils. Bracteoles connate at the base, acuminate, almost aristate, about 1 line long. Calyx-tube cylindrical but thicker than in most species, about 3 lines long, the free part very short and slightly campanulate; lobes ovate-lanceolate, gradually tapering into a short awn rarely exceeding the petals. Petals about 4 lines long. Stamens numerous; connective-gland globular.

N. Australia. Arnhem's Land, *F. Mueller;* Port Essington, *Armstrong.*

32. **C. brachychæta,** *F. Muell. in Trans. Phil. Inst. Vict.* iii. 43. A much-branched shrub of 6 to 12 ft., more or less pubescent. Leaves crowded or imbricate, linear-triquetrous, obtuse or mucronate, 1 to 1½ or rarely 2 lines long. Flowers (white?) nearly sessile below the ends of the branches. Bracteoles free, deciduous, shortly acuminate or mucronate, about as long as the adnate part of the calyx. Calyx-tube oblong, pubescent or hirsute, 2 to 2½ lines long, much attenuate at the base, the free part nearly as long as the adnate portion, cylindrical or contracted upwards; lobes ovate-lanceolate, acuminate or shortly awned, ciliate, about 2 lines long. Petals scarcely as long. Stamens about 20; connective-gland globular.

N. Australia. Sandstone table land, Arnhem's Land, *F. Mueller.* F. Mueller's herbarium comprises also some glabrous specimens from dry stony ridges near the Fitzmaurice River, and others, apparently in an abnormal state, from the Victoria river, with smaller mostly imperfect flowers. The seed in this species is thicker upwards than in most others, but the embryo appears to be the same.

Var. ? *tenuifolia.* Habit nearly of *Lhotzkya ericoides.* Leaves slender, triquetrous, densely crowded, 3 to 4 lines long. Islands of the Gulf of Carpentaria, *R. Brown.*

33. **C. achæta,** *F. Muell. in Trans. Phil. Inst. Vict.* iii. 43. Diffuse or prostrate, with numerous short crowded branches, more or less sprinkled with spreading hairs. Leaves imbricate, oblong-linear, triquetrous, obtuse, mostly under 1 line long. Flowers nearly sessile in the upper axils of the numerous flowering branchlets. Bracteoles broad, scarious, truncate, nearly as long as the calyx-tube. Calyx-tube ovoid-oblong, about 1 line long, very hirsute, the free part contracted and almost as long as the adnate portion; lobes ovate or ovate-lanceolate, tapering into a short awn or point almost concealed by the long hairs fringing the lobe. Petals about 1 line long. Stamens about 20; connective almost didymous besides the small globular gland. Seeds solitary and obovoid, or sometimes 2, nearly hemispherical; embryo of the same shape, but apparently straight, homogeneous and obscurely 2-lobed at the top.—*Lhotzkya cuspidata,* F. Muell. in Hook. Kew Journ. viii. 324.

N. Australia. Upper Glenelg river, N. W. Coast, *Marten;* Victoria river and gullies, and low stony ridges on Fitzmaurice river, *F. Mueller.* This and the preceding species are evidently closely allied to each other, and notwithstanding the shortness of the calyx-awns and thickness of the embryo (which I have scarcely seen perfect) appear to be better referred to *Calythrix* than to any other genus.

34. **C. laricina,** *R. Br. Herb.* A much-branched shrub, spreading, and scarcely 1½ ft. high in barren open places, attaining 6 or 7 ft. in moist situations. Leaves linear-subulate, slender, triquetrous, mucronate, 2 to 3 lines long, crowded on the smaller branchlets. Flowers small, nearly sessile, crowded below the ends of the branches. Bracts truncate or shortly acuminate, ciliate, much shorter than the calyx-tube. Calyx-tube about 1½ lines long, pubescent or nearly glabrous, the free part scarcely contracted; lobes at first broadly lanceolate, ciliate, not so long as the tube, the short awn scarcely exceeding the cilia, after flowering the lobes are longer and taper into a more prominent awn.

N. Australia. Arnhem's Land and islands of the Gulf of Carpentaria, *R. Brown.*

8. LHOTZKYA, Schau.

Calyx-tube elongated, cylindrical or narrow-turbinate, 10- or rarely 5-ribbed; lobes 5, scarious, spreading, short, broad, very obtuse. Petals 5, entire, spreading, deciduous. Stamens indefinite, usually numerous, in several rows, the inner ones shorter, deciduous; filaments filiform, quite free; anthers small, versatile; cells parallel, opening in longitudinal slits, connective terminating in a globular gland-like appendage. Ovary 1-celled; ovules 2, collaterally erect on a filiform placenta attached to the base and to the summit of the cavity. Style filiform, glabrous, with a small capitate stigma. Fruit formed by the lower usually fusiform part of the calyx-tube, and crowned by the persistent remainder of the calyx. Seed solitary, cylindrical, or slightly thickened upwards; testa very thin; embryo of the shape of the seed, quite straight, very shortly 2-lobed at the upper end.—Heath-like shrubs. Leaves scattered or rarely opposite, small, semiterete or 3- or 4-angled, rigid, entire, glabrous or pubescent. Flowers sessile or shortly pedicellate, solitary in each axil along the branches or forming terminal leafy heads. Bracteoles scarious, at least on the margin, the keel often green, persistent, continuous with the rigid pedicel, often united at the base into a turbinate cup, and in the free part overlapping each other, enclosing the base or nearly the whole of the calyx-tube.

The genus is limited to Australia. It is closely allied to *Calythrix*, with which F. Mueller proposes to unite it, but the constant want of any awn or point to the calyx-segments gives it so distinct an aspect that we may be justified in maintaining it as distinct.

Calyx-tube shortly produced above the ovary into a concave disk.
Flowers (pink) small, in terminal heads becoming lateral by the elongation of the shoot.
Quite glabrous. Calyx-tube very narrow-turbinate . . . . 1. *L. glaberrima.*
Pubescent. Calyx-tube cylindrical, hirsute . . . . 2. *L. genetylloides.*
Calyx-tube adnate to the top.
Calyx-tube narrow-turbinate, pubescent or hirsute.
Leaves scattered. Flowers in terminal heads. Bracteoles broad, villous, very conspicuous . . . . 3. *L. violacea.*
Leaves mostly opposite. Flowers lateral. Bracteoles narrow 4. *L. ciliata.*
Calyx-tube after flowering narrowed into a short slender neck.
Flowers small, lateral . . . . 5. *L. brevifolia.*
Calyx-tube cylindrical, glabrous, or slightly scabrous-pubescent, not narrowed at the top.
Flowers violet or purple. Bracteoles broad, obtuse . . . 6. *L. purpurea.*
Flowers white or yellowish. Bracteoles acuminate.
Calyx-tube 7- to 10-ribbed, 1 to $1\frac{1}{2}$ lines long. Leaves mostly 3 to 4 lines long . . . . 7. *L. ericoides.*
Calyx-tube 5-ribbed, about 2 lines long. Leaves mostly 4 to 6 lines long . . . . 8. *L. acutifolia.*

1. **L. glaberrima,** *F. Muell. Fragm.* i. 13. Small, with slender erect branches and quite glabrous. Leaves scattered, linear or oblong, triquetrous or concave and keeled, obtuse, mostly 1 to $1\frac{1}{2}$ lines long, more or less spreading. Flowers pink, nearly sessile in the upper axils but below the ends of the branches. Bracteoles concave, with green keels, shorter than the calyx-tube. Calyx-tube narrow-turbinate, 10-ribbed at the top, about 1 line long or rather more; lobes orbicular, about $\frac{1}{2}$ line long. Petals oblong,

obtuse, nearly 2 lines long. Stamens numerous, anthers without any conspicuous gland.

S. Australia. Kangaroo Island, *Bannier.*

2. **L. genetylloides,** *F. Muell. in Trans. Phil. Soc. Vict.* i. 16. Erect and bushy, glabrous pubescent or hirsute with short hairs. Leaves scattered, crowded, spreading, linear or oblong, usually flat above and convex or keeled underneath, obtuse or mucronate, 1 to 2 or rarely nearly 3 lines long. Flowers (pink or white?) on very short pedicels, in small terminal leafy heads rarely becoming lateral by the elongation of the shoot. Bracteoles obovate, slightly ciliolate, about as long as the calyx-tube. Calyx-tube oblong, slightly narrowed at the top, 10-ribbed, glabrous or hirsute at the top, about $1\frac{1}{2}$ lines long, slightly produced above the ovary into a concave disk; lobes obovate, $\frac{1}{2}$ to $\frac{3}{4}$ line long. Petals oval-oblong, about 2 lines. Stamens often not above 20. Fruiting-calyx about 2 lines long, 5-angled, the secondary ribs disappearing.—*Genetyllis alpestris,* Lindl. in Mitch. Three Exped. ii. 178.

Victoria. Grampian mountains, *Mitchell, F Mueller.*

S. Australia. Scrub of the S.E. portion of the colony, *J. E. Woods.*

Var. *bracteosa.* Floral leaves or bracts ovate or orbicular, very prominent as well as the broad bracteoles.—On the Glenelg, *Robertson.*

3. **L. violacea,** *Lindl. Swan Riv. App.* 7. Erect, bushy, more or less pubescent or hirsute with short hairs. Leaves alternate or scattered, oblong, very obtuse, 2 to 3 lines long, concave above, convex underneath. Flowers (purple?) in the upper axils, forming dense terminal heads, the floral leaves short and broad, the uppermost ones scarious on the edge. Bracteoles obovate, very obtuse, scarious with a herbaceous villous keel and base. Calyx-tube villous, 10-ribbed, narrow-turbinate, tapering to a stalk-like base, adnate to the top; lobes very short and broad. Petals obovate-oblong, nearly 3 lines long. Stamens numerous.—Schau. Myrt. Xeroc. 85.

W. Australia. Swan River, *Drummond,* 1*st Coll. n.* 162; *Gilbert.*

4. **L. ciliata,** *F. Muell. Herb.* Apparently a small species, with slender erect pubescent branches. Leaves mostly opposite, appressed, oblong-lanceolate or linear, almost acute, concave and obtusely keeled, 1 to $1\frac{1}{2}$ lines long, glabrous except the margin, which is ciliate with short soft hairs. Flowers (purple?) nearly sessile below the ends of the branches. Bracteoles narrow, as long as the calyx-tube, connate to the middle. Calyx-tube narrow turbinate, pubescent, obscurely ribbed, rather above 1 line long; lobes broad, truncate, scarious, about $\frac{1}{2}$ line long. Petals about 2 lines. Stamens numerous.

W. Australia. Oldfield river, Plantagenet and Stirling ranges, *Maxwell.*

5. **L. brevifolia,** *Schau. in Pl. Preiss.* i. 103. Branches rather slender, erect, virgate, more or less pubescent or rarely glabrous. Leaves scattered, linear, triquetrous or concave and keeled, obtuse, rarely above 2 lines long, glabrous, minutely ciliate or pubescent. Flowers small, nearly sessile along the branches as in *L. ericoides.* Bracteoles free or scarcely cohering at the base. Calyx-tube at first almost concealed by the bracteoles, but after

flowering attaining 2 lines and exceeding the bracts by about half its length, 10-ribbed, more or less contracted at the top into a short slender web; the lobes not longer than the breadth of the tube, broad, truncate, and slightly emarginate. Petals about 2 lines long. Stamens about 20.

W. Australia. Swan River, *Drummond, 1st Coll. n.* 163; *Preiss, n.* 2638; *Turner*. The small flowers and slender neck of the calyx-tube distinguish this from all others. I have not seen Preiss's specimens, but Schauer's description evidently refers to this species.

6. **L. purpurea,** *F. Muell. Fragm.* i. 224. Erect, bushy, 1 to 1½ ft. high, quite glabrous. Leaves scattered, crowded, linear-triquetrous, obtuse, 3 to 4 lines long. Flowers purple, nearly sessile along the branches. Bracteoles shortly united at the base, broad, concave, obtuse, shorter than the calyx-tube. Calyx-tube slender, 2 to 2½ lines long, adnate to the top, 10-nerved and scarcely contracted above the middle, narrower and scarcely above 5-nerved at the base; lobes scarcely ½ line long, broad, truncate, emarginate. Petals above 2 lines long. Stamens numerous.

W. Australia. Sandy hills, Champion Bay, *Oldfield*. Scarcely differs from *L. ericoides*, except in the broader bracteoles and the larger flowers of a different colour.

7. **L. ericoides,** *Schau. in Lindl. Introd. Nat. Syst. ed.* 2. 439. An erect heath-like shrub of 2 to 4 ft., glabrous or slightly scabrous-pubescent. Leaves scattered, rather crowded, linear-triquetrous, slender, obtuse or mucronulate, 3 to 4 lines long. Flowers (white?) nearly sessile along the branches. Bracteoles lanceolate-acuminate, scarcely exceeding the calyx-tube. Calyx-tube glabrous or minutely pubescent, cylindrical, prominently 7- to 10-ribbed, rather more than 1 line long, wholly adnate and not contracted at the top; lobes ovate, obtuse, scarious, about ½ line long. Petals narrow, above 2 lines long. Stamens numerous.—Schau. Myrt. Xeroc. 83, and in Pl. Preiss. i. 103; *L. scabra*, Turcz. in Bull. Mosc. 1862, ii. 324; *L. hirta*, Regel, Gartenfl. 1863, 337. t. 415 (from the description and figure).

W. Australia. King George's Sound and adjoining districts, *R. Brown, A. Cunningham, Drummond, Preiss, n.* 222, and others.

8. **L. acutifolia,** *Lindl. Swan Riv. App.* 7. Erect with virgate branches, often pubescent, and with the general habit and characters of *L. ericoides*, but stouter. Leaves scattered, crowded, linear, mucronate, mostly 4 to 6 lines long, rigid and prominently keeled underneath. Flowers (white or yellowish?) nearly sessile along the branches, larger than in *L. ericoides*. Bracteoles as long as the calyx-tube, keeled and acute, but scarcely acuminate. Calyx-tube cylindrical, 5-ribbed, about 2 lines long, wholly adnate and not contracted at the top; lobes not ½ line long. Petals (white or yellowish?) above 3 lines long. Stamens numerous.—Arn. in Hook. Journ. Bot. ii. 380. t. 15; Schau. Myrt. Xeroc. 84, and in Pl. Preiss. i. 103.

W. Australia. Swan River, *Drummond, 1st Coll.;* Mount Melville, near Albany, *Preiss, n.* 224. Possible a variety only of *L. ericoides*.

SUBTRIBE III. THRYPTOMENEÆ.—Stamens 5, 10, or indefinite, free, in one or several rows, without staminodia. Ovules 2, rarely 4 to 10, attached in 2 rows to a placenta either basal or adnate to the side of the cavity or extending to the summit of the cavity. Embryo where known very thick, with a slender neck inflected and divided at the end into 2 small cotyledons.

The three genera here included have the habit and embryo of *Bæckea* with the 1-celled ovary of *Euchamælaucieæ*.

9. HOMALOCALYX, F. Muell.

Calyx-tube cylindrical or turbinate, the upper free part short and broad; lobes 5, petal-like, entire, deciduous. Petals 5, entire, deciduous. Stamens indefinite, few or many, free, deciduous, the inner ones shorter, filaments filiform; anthers with two parallel cells opening longitudinally, the connective thickened into a terminal gland. Ovary 1-celled, with 2 ovules on a short basal excentrical placenta. Style filiform, glabrous with a small capitate stigma. Fruit . . .—Heath-like glabrous shrubs. Leaves scattered (not opposite), usually crowded, small, entire. Flowers nearly sessile along the branches, solitary in the axils of the leaves. Bracteoles broad, usually persistent.

A small genus, limited to Australia, allied to *Lhotzkya* in its petals and stamens, and in some measure to *Thryptomene* in the ovary, differing from both in the deciduous calyx-lobes. The ripe fruit is unknown, but in the farthest advanced state that I have seen there is no tendency to the hardening of the endocarp as in *Thryptomene*.

Leaves linear, mucronate. Calyx-tube cylindrical, lobes and petals acute. Stamens 9 to 15 . . . . 1. *H. ericæus.*
Leaves oblong-triquetrous, obtuse. Calyx-tube broadly turbinate, lobes and petals broad, obtuse. Stamens 20 to 30 . . . . . . 2. *H. polyandrus.*

1. **H. ericæus,** *F. Muell. in Hook. Kew Journ.* ix. 309. A small shrub, erect, with slender virgate branches, or spreading and almost procumbent. Leaves crowded, linear, rigid, acutely triquetrous or concave, mucronate, ¼ to ½ in. long. Flowers nearly sessile, or shortly pedicellate in the upper axils. Bracteoles broad, much shorter than the calyx-tube, veined, scarious only at the edges. Calyx-tube oblong-cylindrical, 1½ lines long in flower, longer afterwards, the free part short; lobes lanceolate, acute, about 1 line long, very deciduous. Petals similar to the calyx-lobes, but rather longer. Stamens 9 to 15; anthers small. Ovules 2, collateral, erect on a short basal excentrical placenta, which does not appear to be continued beyond the ovules. Young fruit 1-seeded, enclosed in the enlarged truncate calyx-tube. *Thryptomene homalocalyx*, F. Muell. Fragm. iv. 63.

N. Australia. Islands of the Gulf of Carpentaria, *R. Brown;* elevated table-land between the Roper and Limmen Bight rivers, *F. Mueller.*

2. **H. polyandrus,** *F. Muell. Herb.* Leaves erect, oblong, triquetrous, very obtuse, 1½ to 2 lines long. Flowers on very short pedicels in the upper axils. Bracteoles persistent, very broad, keeled, scarious, forming a truncate cup enclosing the calyx. Calyx-tube very short, broadly turbinate; lobes orbicular, nearly 1½ lines diameter. Petals about the same, and apparently falling off with them. Stamens 20 to 30, crowded almost into a single row; the filaments all short, but the inner ones still shorter and more inflexed; anthers small with a rather large gland to the connective. Ovary very short and broad in the base of the calyx-tube, with 2 ovules erect on a short basal placenta.—*Thryptomene polyandra*, F. Muell. Fragm. iv. 77.

N. Australia (or **Queensland?**). *Leichhardt*, no station given. The specimens are not good, most of the flowers injured or deformed by insects, but the best appear to have the calyx-lobes, petals, and stamens very deciduous, as in *H. ericæus*, leaving a truncate fruiting calyx concealed within the persistent bracteoles.

10. THRYPTOMENE, Endl.

(Paryphanthe, *Schau.*; Astræa, *Schau.*; Eremopyxis, *Baill.*)

Calyx-tube hemispherical turbinate ovoid or shortly cylindrical, adnate to the top or the free part broader; lobes 5, persistent (unless the free part of the calyx falls off), petal-like or scarious, spreading, entire. Petals 5, persistent, usually connivent over the stamens. Stamens 5, alternate with the petals, or 10, often inserted within the margin of the disk; filaments short; anther-cells globular or obovoid, separately inserted on the connective and usually pendulous, either smooth and opening by pores or furrowed and opening by pores or short slits in the furrow. Ovary inferior, 1-celled, the cavity usually small near the top of the calyx-tube, with 2 or rarely 4 ovules on a short basal placenta either excentrical or adhering to one side of the cavity, or rarely the cavity occupies the greater part of the tube, with several ovules in 2 rows on a lateral placenta. Style short, glabrous, with a small capitate stigma. Fruit, where known, formed by the hardened but scarcely enlarged base of the calyx crowned by the persistent calyx-lobes and petals; endocarp cartilaginous or hardened, usually globular, indehiscent or separating into 2 cocci open on the inner face, containing either 1 globular or 2 hemispherical or slightly reniform seeds; testa very thin; embryo folded, the radicular end very thick, the other fold much shorter, narrow with ovate cotyledons.—Heath-like glabrous shrubs. Leaves opposite, small, entire. Flowers axillary, solitary, or rarely 2 or 3 in the same axil, small, nearly sessile or pedicellate. Bracteoles 2 under the calyx, scarious or green in the centre, usually small and in many species so deciduous as to be rarely found on the specimens.

The genus is limited to Australia. With the habit of *Bæckea*, it has most of the characteristics of the *Chamælaucieæ*, with peculiar anthers. The hardened endocarp appears also to be characteristic, but perfect fruits have only been seen in a very few species, and very frequently the seeds are abortive, although enlarged and converted into a hard granular apparently homogeneous mass. In some species, where the cavity of the ovary is very small and quite at the summit of the calyx-tube, the ovules, although really arising from the base of the cavity, appear as they enlarge into the lower part of the tube to be pendulous, but when examined at the time of flowering I have never found them to be really pendulous as in *Micromyrtus*.

A. *Calyx-tube turbinate,* 10-*ribbed or rarely* 5-*ribbed.* *Stamens* 10.

Ovules 6 to 10, in 2 rows on a lateral placenta. Leaves broadly obovate, 1 to 2 lines long . . . . 1. *T. mucronulata.*
Ovules 4 to 6, on a short basal placenta sometimes adhering to one side. Leaves narrow, convex underneath, 1 to 2 lines.
Calyx, free part very short. Filaments twice as long as the anthers.
Calyx-tube 10-ribbed, 1 line diameter . . . . 2. *T. australis.*
Calyx-tube 5-ribbed, ½ line diameter . . . . 3. *T. tenella.*
Calyx, free part as long as the ovary. Filaments short . . . 4. *T. prolifera.*
Ovules 2, on a short basal placenta. Leaves flat, obovate-oblong, 1½ to 3 lines long. Ribs of the calyx-tube rugose . . . . 5. *T. saxicola.*

B. *Calyx-tube broad, slightly turbinate, 15-ribbed. Stamens 10.*

Ovules 2, on a short basal placenta. Leaves broad, thick, scarcely 1 line long . . . . 6. *T. Johnsonii.*

C. *Calyx-tube hemispherical or shortly campanulate, rugose and pitted, without prominent ribs.*

Stamens 10. Ovules 2, on a short basal placenta.
Leaves obovate-oblong, flat, 1½ to 3 lines long . . . . . . 5. *T. saxicola.*
Leaves ovate obovate or oblong, concave, under 1½ lines long.
Flowers nearly sessile, or pedicels under 1 line (except *T. strongylophylla*).
Calyx-lobes minutely denticulate . . . . 8. *T. denticulata.*
Calyx-lobes entire.
Petals about as long as the calyx-lobes . . . . 7. *T. racemulosa.*
Petals twice as long as the calyx-lobes.
Leaves keeled, ½ to 1 line long. Flowers nearly sessile below the ends of the branches . . . . 9. *T. bæckeacea.*
Leaves concave, not keeled, mostly under ½ line, closely imbricate. Flowers at the ends of the branches, on pedicels usually as long as the leaves . . . . 10. *T. strongylophylla.*
Leaves narrow. Flowers on slender pedicels of 2 lines or more 11. *T. hyporhytis.*
Stamens 5. Ovules about 6, in 2 rows, on a lateral placenta.
Leaves obovate, thick, less than 1 line . . . . 12. *T. Maisonneuvii.*

D. *Calyx-tube ovoid turbinate or cylindrical, not rugose, often ribbed. Stamens 5.*

Calyx-tube ovoid-campanulate, not ribbed, shortly produced above the ovary. Leaves flat . . . . 13. *T. Mitchelliana.*
Calyx-tube cylindrical, ribbed, not produced above the ovary.
Leaves flat. Flowers 1½ to 2 lines long . . . . 14. *T. Miqueliana.*
Leaves flat. Flowers very slender, about 1¼ lines long . . . 15. *T. micrantha.*
Leaves triquetrous. Flowers slender, 1½ to nearly 2 lines long 16. *T. ericæa.*
Calyx-tube broadly turbinate, prominently ribbed, not produced above the ovary.
Leaves obovate or broadly oblong, flat. Flowers spreading to 2 lines diameter . . . . 17. *T. oligandra.*
Leaves linear-oblong or cuneate, convex underneath. Flowers not 1 line diameter . . . . 17. *T. oligandra*, var. *parviflora.*

1. **T. mucronulata,** *Turcz. in Bull. Mosc.* 1847, i. 156. Branches numerous, slender, virgate. Leaves erect or somewhat spreading, broadly obovate, flat, with the midrib inconspicuous or slightly prominent towards the end, with a small recurved point, 1 to 2 lines long. Flowers solitary in the axils below the ends of the branches, nearly sessile. Bracteoles obovate, concave, very thin. Calyx-tube turbinate, obtusely 10-ribbed, the short free part forming a broad concave disk; lobes spreading to rather less than 2 lines diameter. Petals about as long as the calyx-lobes, connivent. Stamens 10, inserted just within the margin of the disk; filaments short; anther-cells globular, divaricate, the connective-gland half as large as the cells. Ovules 6 to 10, in 2 rows, along a lateral placenta extending from the base to the summit of the cavity of the ovary, which is larger than in most species.

W. Australia, *Drummond, 3rd Coll. n.* 33.

2. **T. australis,** *Endl. in Ann. Wien. Mus.* ii. 192. Branches slender, erect, virgate. Leaves erect, linear, semiterete or concave, somewhat thickened upwards, with a short fine recurved point rarely wanting, about 2

or rarely 3 lines long. Flowers axillary, on very short pedicels. Bracteoles concave, keeled, very deciduous. Calyx-tube turbinate, 10-ribbed, almost entirely adnate, 1 to 1¼ line long; lobes spreading to nearly 3 lines diameter. Petals about as long as the calyx-lobes, very broad, connivent. Stamens 10, inserted within the margin of the disk; filaments filiform, much longer than the anthers; anther-cells nearly globular, furrowed, opening in pores or short slits in the furrows; connective-gland at least half as large as the cells. Ovules 4 or 6, on a short excentrical placenta at the base of a small cavity near the top of the calyx-tube. Style short with a broad stigma.—Schau. Myrt. Xeroc. 81. t. 6 A.

W. Australia. E. of New York, *J. S. Roe;* Phillips range and Salt river, *Maxwell.*

3. **T. tenella,** *Benth.* Branches virgate, very slender. Leaves erect or spreading, linear, semiterete or concave, slightly thickened upwards, with fine recurved points, as in *T. australis*, but more slender. Flowers axillary, on short pedicels, not half the size of those of *T. australis*, but only seen in fruit. Calyx-tube thin, nearly globular, 5-ribbed, ½ line diameter; lobes spreading to little more than 1 line diameter. Petals not longer, connivent. Stamens all fallen away from the specimens. Endocarp crustaceous, containing 2 hemispherical seeds, or 1 nearly globular and the other small and abortive.

W. Australia, *Drummond, 5th Coll. Suppl. n.* 24.

4. **T. prolifera,** *Turcz. in Bull. Mosc.* 1862, ii. 324. Branches numerous, erect and slender. Leaves erect or slightly spreading, linear or linear-oblong, obtuse or with a short recurved point, 1 to 2 lines long. Flowers on very short pedicels in the upper axils, usually forming a small tuft at the base of the young shoots. Bracteoles small, obovate, concave, narrowed at the base. Calyx-tube turbinate, obtusely 5-ribbed at the base, the free part 10-ribbed and very broad; lobes orbicular, about 1 line broad. Petals about the size of the calyx-lobes. Stamens 10; filaments short; anther-cells globular, divaricate; connective-gland small. Ovary readily separating from the calyx-tube, with 4 ovules on a small erect basal placenta.

W. Australia. Between Moore and Murchison rivers, *Drummond, 6th Coll. n.* 62.

5. **T. saxicola,** *Schau. in Pl. Preiss.* i. 102. Erect with virgate branches, attaining 3 or 4 ft. (rarely diffuse or prostrate?). Leaves obovate-oblong, flat, with the midrib scarcely conspicuous, obtuse or slightly acute, 1½ to 3 lines long. Flowers small, on slender pedicels of 1 to 1½ lines in the upper axils. Bracteoles lanceolate, very deciduous. Calyx-tube not ½ line long, turbinate, irregularly 10-ribbed, with the ribs more or less wrinkled, or entirely wrinkled without distinguishable ribs, the free part very short and broad; lobes broad and very obtuse, about ½ line long. Petals orbicular, nearly 1 line long, connivent. Stamens 10; filaments short; anther-cells pendulous, not furrowed, opening in pores or short slits. Ovary small, near the top of the calyx-tube, with 2 ovules on a short basal placenta. Seeds usually 2.—*Bæckea saxicola*, A. Cunn. in Bot. Mag. t. 3160; *Astræa saxicola*, Schau. in Linnæa, xvii. 239; *Eremopyxis camphorata*, Baill. Adans. ii. 329 (but not *Bæckea camphorata*, R. Br.); *Scholtzia decandra*, F. Muell. Fragm. iv. 75.

W. Australia. King George's Sound, and eastward towards Cape Riche, *A. Cunningham; Drummond, 5th Coll. n.* 126; *Oldfield,* and others.

6. **T. Johnsonii,** *F. Muell. Fragm.* iv. 77. Erect and densely branched. Leaves obovate-orbicular, thick, with a prominent keel, obtuse or with a small recurved point, rarely 1 line long. Flowers almost sessile in the upper axils. Bracteoles obovoid-orbicular, spreading, nearly flat, scarious, above 1 line long. Calyx-tube broadly turbinate, 15-ribbed; lobes very thin and scarious, broadly orbicular, about $\frac{3}{4}$ line long. Petals orbicular, about as long as the calyx-lobes. Anther-cells obovoid-globular, pendulous and divergent, opening in terminal pores. Ovules 2, on a very short basal placenta in a small cavity at the top of the calyx-tube. Fruit with a hard globular endocarp enclosing 1 globular or 2 hemispherical seeds.

W. Australia. Probably Murchison river, *Herb. F. Mueller.*

7. **T. racemulosa,** *Turcz. in Bull. Mosc.* 1847, i. 156. Erect and bushy, with very numerous rather slender branches. Leaves obovate, erect, or slightly spreading, thick and concave or keeled, obtuse, 1 to $1\frac{1}{2}$ lines long. Flowers on very short pedicels in the upper axils. Bracteoles small, broad, concave, keeled, spreading, connate at the base. Calyx-tube hemispherical, not keeled, very rugose and pitted; lobes spreading to 2 lines diameter, somewhat enlarged in fruit. Petals rather shorter than the calyx-lobes Anther-cells obovoid, divergent, deflexed, not furrowed, opening in small terminal pores; connective small. Ovules 2, on a short basal placenta. Seeds 1 or 2 in a hard globular endocarp.

W. Australia, *Drummond, 2nd Coll. n.* 58, *3rd Coll. n.* 32.

8. **T. denticulata,** *Benth.* Much branched. Leaves ovate, obovate or almost oblong, thick, concave, keeled, loosely imbricate and decussate on the smaller branches, obtuse, about 1 line long. Flowers shortly pedunculate in the upper axils. Bracteoles ovate or oblong, concave, spreading, usually persistent even after the flowers have fallen. Calyx-tube turbinate or at length hemispherical, rugose; lobes ovate or orbicular, minutely denticulate or almost entire, rather rigid, at least half as long as the petals. Petals nearly $1\frac{1}{2}$ lines long, entire. Stamens 10; filaments very short; anther-cells obovoid, quite distinct, opening in terminal pores; connective-gland small. Ovules 2, on a short basal placenta, adnate to one side of the cavity.—*Scholtzia denticulata,* F. Muell. Fragm. iv. 75.

W. Australia. Murchison river, *Oldfield, Burgess.* Very nearly allied to *T. bæckeacea,* and perhaps a variety only, with larger less imbricate leaves and larger flowers, the calyx-lobes usually larger in proportion to the petals. Some specimens of Drummond's in the 3rd Coll. n. 33, appear almost to connect the two.

9. **T. bæckeacea,** *F. Muell. Fragm.* iv. 65. A very densely branched shrub of 1 to 2 ft. Leaves obovate or oblong, triquetrous, imbricate and decussate on the branchlets, obtuse, rarely 1 line long. Flowers small in the upper axils, nearly sessile or the pedicels shorter than the leaves, usually below the ends of the branches. Bracteoles concave, keeled, green, with only the margins scarious. Calyx-tube broadly turbinate, rugose, not ribbed; lobes petal-like, orbicular, about $\frac{1}{2}$ line long. Petals orbicular, twice as long as the calyx-lobes. Stamens 10, very short; anther-cells obovoid-globular, pen-

dulous, divergent, opening in small terminal pores; connective-gland small. Ovary very short in the bottom of the calyx-tube, with 2 ovules attached to a lateral placenta.—*Bæckea micrantha*, DC. Prod. iii. 230; Mem. Myrt. t. 14.

W. Australia. Rocky places near the Murchison river, *Oldfield;* Sharks Bay, *Herb. Mus. Par.* (*in Herb. R. Brown and Sonder*).

10. **T. strongylophylla,** *F. Muell. Herb.* Nearly allied to *T. bæckeacea*, and perhaps a small variety. Leaves closely imbricate and decussate on the branchlets, orbicular, concave, not keeled, very obtuse, rarely exceeding ½ line diameter, the floral ones smaller. Flowers in the upper axils, on pedicels of ½ to ¾ line, forming little terminal leafy corymbs. Bracteoles ovate, small, very deciduous. Calyx-tube broadly turbinate, rugose, not ribbed; lobes petal-like, orbicular, not ½ line diameter. Petals orbicular, about twice as long as the calyx-lobes. Stamens 10, short; anther-cells obovoid, pendulous, divergent, opening in small terminal pores. Ovary short, with 2 ovules on a short basal somewhat lateral placenta.

W. Australia. Murchison river, *Oldfield.*

11. **T. hyporhytis,** *Turcz. in Bull. Mosc.* 1862, ii. 324. Apparently a small shrub, with numerous erect branches, not exceeding 6 in. in our specimens. Leaves linear or almost oblong, erect, thick and concave, very obtuse, 2 to 3 lines long, or those immediately about the flowers often much shorter. Flowers small, on slender pedicels of 2 lines or more. Bracteoles short, broad, very concave and keeled. Calyx-tube hemispherical or very broadly turbinate, rugose and pitted; lobes petal-like, orbicular, ¾ line broad. Petals not twice as long as the calyx-lobes. Stamens 10; anther-cells obovoid, divergent, pendulous, opening in terminal pores; connective-gland small. Ovules 2, on a short basal placenta, adnate to one side of the cavity.

W. Australia. Between Moore and Murchison rivers, *Drummond, 6th Coll. n.* 63.

12. **T. Maisonneuvii,** *F. Muell. Fragm.* iv. 64. Very much branched, with the aspect and foliage of *T. Johnsonii.* Leaves obovate, thick, prominently keeled, very obtuse, rarely 1 line long. Flowers nearly sessile in the upper axils. Calyx-tube shortly campanulate, rugose and pitted, not ribbed; lobes very short, thick and triangular, with lateral, divaricate, scarious auricles. Petals orbicular, fully 1 line diameter, and apparently spreading. Stamens 5, inserted outside the broad thick disk; anther-cells distinct, on a thick connective. Overy 1-celled, with 4 to 6 ovules on a lateral placenta in the upper portion of the cavity,

N. Australia. Fincke river, *M'Douall Stuart's Expedition.* The flowers are very far advanced, and I do not feel confident of having exactly ascertained some of the details of their structure.

13. **T. Mitchelliana,** *F. Muell. Fragm.* i. 11. Tall and bushy, with slender virgate branches. Leaves oblong or slightly cuneate, flat, with the midrib scarcely prominent, obtuse or mucronate, ¼ to ½ in. long, or the floral ones or rarely nearly all shorter and broader. Flowers solitary or 2 or 3 together in the upper axils, on pedicels rarely as long as the calyx. Bracteoles falling off so early as to be rarely seen. Calyx-tube ovate-campanulate, about 1 line long, inconspicuously ribbed, produced above the ovary, the free part

sometimes circumsciss and deciduous: lobes orbicular, petal-like, nearly as long as the tube. Petals orbicular, nearly as long as the calyx-lobes. Stamens 5; filaments short; anther-cells distinct, globular, pendulous, furrowed, opening in short slits; connective-gland small. Ovules 2, erect, on a short basal placenta, in a very small cavity at the top of the adnate part of the calyx-tube. Fruiting-calyx slightly enlarged. Seeds usually 2.—*Bæckea calycina*, Lindl. in Mitch. Three Exped. ii. 190; *Paryphanthe Mitchelliana*, Schau. in Linnæa, xvii. 235.

Victoria. Mount Arapiles, *Mitchell;* in the Grampians generally, *F. Mueller*, *Wilhelmi.* Some specimens have the leaves almost of *T. Miqueliana*, from which it is readily distinguished by the calyx. In one specimen, from the Grampians, the pedicels are rather longer than the calyx.

14. **T. Miqueliana,** *F. Muell. Fragm.* i. 11. Leaves flat or slightly concave, from obovate and about 2 lines long to oblong or somewhat cuneate and 3 lines, obtuse or almost acute. Flowers mostly solitary, on short pedicels in the upper axils. Bracteoles very small. Calyx-tube cylindrical or slightly turbinate, 10-ribbed, 1 to 1¼ line long, adnate to the top; lobes petal-like, about ¾ line long. Petals rather smaller. Stamens 5; anther-cells globular, distinct, furrowed, opening in oblong pores or short slits; connective-gland globular. Ovules 2, erect in a very small cavity at the top of the calyx-tube, but, as they enlarge, occupying the greater part of the tube.—*T. saxicola*, Miq. in Nederl. Kruidk. Arch. iv. 116, not of Schauer.

N. S. Wales, *Herb. F. Mueller.*
S. Australia. Spencer's Gulf, *Warburton.*
Very near *T. Mitchelliana* in foliage, but with a different calyx.

15. **T. micrantha,** *Hook. f. in Hook. Kew Journ.* v. 299. *t.* 8, *and Fl. Tasm.* i. 128. A small shrub, with slender, virgate, or spreading branches. Leaves linear-oblong, flat or the margins slightly recurved, obtuse, 2 to 3 lines long. Flowers on exceedingly short pedicels, solitary or 2 or 3 together in the axils along the branchlets of the year. Bracteoles very small. Calyx-tube nearly cylindrical, ¾ line long, 10-ribbed, adnate to the top; lobes concave, not half as long as the tube. Petals still shorter. Stamens 5; anther-cells nearly globular, distinct, slightly furrowed, opening in very short slits; connective-gland nearly as large as the cells. Ovules 2, attached to a nearly basal lateral placenta, in a small cavity near the top of the calyx-tube.

Tasmania. Banks of sand and oyster-shells, Schouten Island, Bass's Straits, *Gunn.*

16. **T. ericæa,** *F. Muell. Fragm.* i. 12. Branches slender, virgate. Leaves linear, semiterete, obtuse or mucronulate, 2 to 3 lines long. Flowers on short pedicels, solitary in each axil below the ends of the branchlets. Bracteoles very small. Calyx-tube cylindrical or slightly turbinate, 10-ribbed, above 1 line long, adnate to the top; lobes obovate, about ¾ line long. Petals rather broader and shorter. Stamens 5; anther-cells globular, distinct, divaricate, furrowed, opening in short slits; connective-gland prominent. Ovules 2, on a basal almost lateral placenta, in a small cavity near the top of the calyx-tube.

S. Australia. Kangaroo Island, *Bannier*, *Waterhouse.* Differs from *T. micrantha* chiefly in the foliage.

17. **T. oligandra,** *F. Muell. Fragm.* i. 11. Arborescent, with numerous slender rigid branchlets. Leaves spreading, broadly ovate or obovate, flat with the midrib and often the primary veins conspicuous underneath, very obtuse, 2 to 3 lines long. Flowers almost sessile, solitary or 2 or 3 together in each axil along the branchlets. Bracteoles orbicular, small. Calyx-tube turbinate, prominently 10-ribbed; lobes petal-like, spreading to about 2 lines diameter. Petals rather shorter than the calyx-lobes, connivent. Stamens 5; anther-cells globular, distinct, furrowed, opening in short slits; connective-gland prominent. Ovules 2, on a lateral almost basal placenta in a small cavity near the top of the calyx-tube.

N. Australia. Islands of the Gulf of Carpentaria, *R. Brown.*

Queensland. Endeavour river, *Banks and Solander, A. Cunningham;* sandy table-land on the Suttor, *F. Mueller;* Lizard Island, *M'Gillivray.*

Var. (?) *parviflora*, F. Muell. Leaves linear-oblong or cuneate, erect or spreading at the top, obtuse or mucronulate, 1 to 2 lines long, concave above, convex underneath, without any prominent midrib. Flowers very small, nearly sessile and solitary in the upper axils. Bracteoles ovate, very deciduous. Calyx-tube scarcely ½ line long, the flowers otherwise as in *T. oligandra.*

N. Australia. Barren places, Gilbert river, Gulf of Carpentaria, *F. Mueller.* The foliage, like that of some *Epacrideæ*, and the very small flowers seem almost sufficient to characterize a distinct species.

11. MICROMYRTUS, Benth.

Calyx-tube cylindrical or turbinate, 5- or 10-ribbed; lobes small, petal-like or scarious, persistent, sometimes reduced to a narrow or scarcely distinguishable border. Petals 5, obovate or orbicular, deciduous or rarely persistent and spreading. Stamens 5 opposite the petals, or 10, those opposite the sepals inserted usually within the margin of the disk; anther-cells distinct, almost globular, opening in parallel divergent or divaricate slits. Ovary adnate, 1-celled; style short, filiform, glabrous, with a capitate stigma; ovules 2, or rarely 4 to 8, collaterally attached at or near the summit of a filiform placenta extending from the base to the top of the cavity. Fruit enclosed in the hardened scarcely enlarged calyx-tube and crowned by the limb, indehiscent. Seed solitary, filling the fruit; testa thin; embryo of the shape of the seed, consisting chiefly of the thick fleshy clavate radicular portion with a short slender neck turned up against one side, and rather deeply divided into 2 linear cotyledons.—Glabrous shrubs, with the habit of the smaller-leaved or more slender species of *Bæckea.* Leaves opposite, small, entire. Flowers small, white or pink, solitary and shortly pedicellate or almost sessile in the axils of the leaves. Bracteoles 2, scarious, close under the calyx, often enclosing the bud, but very deciduous.

The genus is limited to Australia. It is nearly allied to *Thryptomene*, but differs essentially in the ovules and in the placentation, and in most cases in the very deciduous petals. The stamens also, when 5, are opposite the petals, not alternate with them, and the fruit never appears to have the hardened endocarp observable in many species of *Thryptomene.*

Stamens 10. Petals very deciduous. Ovules 2 (Western species).
Calyx nearly cylindrical. Leaves mostly narrow.
Calyx-limb reduced to a very narrow ring . . . . 1. *M. elobata.*
Calyx-lobes distinct, orbicular, not ¼ as long as the petals . 2. *M. racemosa.*
Calyx turbinate. Leaves obovate, keeled. Calyx-lobes at least ⅓ as long as the petals . . . . 3. *M. imbricata.*

Stamens 5.
Petals very deciduous. Ovules 2 (Western species) . . . . 4. *M. Drummondii.*
Petals often persistent. Ovules 2, 4, or more (Southern and Eastern species).
Calyx-tube ovate-turbinate, not exceeding 1 line. Ovules 4. . 5. *M. microphylla.*
Calyx-tube narrow, scarcely exceeding ½ line. Ovules 2 . . 6. *M. minutiflora.*
Calyx-tube narrow, exceeding 1 line. Ovules 6 to 8 . . . 7. *M. leptocalyx.*

1. **M. elobata,** *Benth.* Branches slender, erect and virgate, complete specimens often under 1 ft. high. Leaves erect or slightly spreading, oblong or linear, thick, triquetrous or concave, obtuse or mucronate, 1 to 2 lines long. Flowers nearly sessile in the upper axils, forming small terminal leafy heads becoming lateral by the elongation of the shoot. Bracteoles scarious, with a thick midrib. Calyx-tube above 1 line long, narrow-turbinate or cylindrical, 10-ribbed; limb reduced to an exceedingly narrow ring or border, sometimes scarcely prominent. Petals broadly ovate, about 1 line long, deciduous. Stamens 10, in 2 rows; anther-cells globular, opening in divaricate or transverse slits; connective broad, tipped by a globular gland. Ovules 2.—*Thryptomene elobata*, F. Muell. Fragm. iv. 63.

W. Australia. Sandy plains inland of Israelite Bay, *Maxwell.*

2. **M. racemosa,** *Benth.* Allied to *M. elobata*, but the branches more slender, almost filiform, and the calyx different. Leaves oblong or linear, erect, thick, concave or semiterete, very obtuse, rarely above 1 line long. Flowers on very short axillary pedicels below the ends of the branches. Bracteoles not seen. Calyx-tube slender, cylindrical or slightly turbinate, about 1 line long, with 10 scarcely prominent ribs; lobes orbicular, short and broad but distinct, scarious and minutely denticulate. Petals about ½ line long, deciduous. Stamens 10, in 2 rows; filaments very short; anther-cells globular, distinct, opening in parallel or divergent slits; connective-gland globular. Ovules 2.—*Thryptomene racemosa*, F. Muell. Herb.

W. Australia. *Drummond, 2nd Coll. n.* 235; Murchison river, *Oldfield.*

3. **M. imbricata,** *Benth.* Erect, 1 to 2 ft. high, with numerous slender virgate branches. Leaves obovate or nearly oblong, concave and keeled or triquetrous, often imbricate and decussate on the smaller branches, obtuse, 1 to 1½ or rarely 2 lines long. Flowers small in the upper axils, on pedicels often exceeding the leaves. Bracteoles narrow, very deciduous. Calyx-tube turbinate, 5- or 10-ribbed, about ¾ line long; lobes less than half as long as the petals, broad, obtuse, scarious. Petals broadly obovate, ¾ line long. Stamens 10, those opposite the sepals inserted much within the margin of the disk; anther-cells globular, distinct, opening in divergent or transverse slits. Ovules 2.

W. Australia. Sandy places, Termination Granite, *Maxwell.*

4. **M. Drummondii,** *Benth.* Branches slender, virgate. Leaves obovate or oblong, rather thick, concave and keeled, very obtuse, ½ to 1 line long or rarely more. Flowers very small, on very short axillary pedicels along the branches. Bracteoles short, concave, very deciduous. Calyx-tube turbinate, 5-ribbed, about ½ line long; lobes very small, scarious, entire. Petals obovate-orbicular, at least as long as the calyx. Stamens 5, opposite

the petals; filaments very short; anther-cells globular, distinct, opening in short parallel or diverging slits. Ovules 2. Embryo with the slender 2-lobed portion as long as the thick radicular end.

W. Australia, *Drummond*, 5*th Coll. Suppl. n.* 23.

5. **M. microphylla,** *Benth.* Erect or diffuse and much-branched. Leaves usually obovate-triquetrous, rather thick, very obtuse, and under 1 line long, but sometimes passing from that to nearly linear, semiterete and nearly 2 lines long, decussate on the smaller branches, the upper ones sometimes minutely dentate-ciliate. Flowers nearly sessile in the upper axils, usually forming numerous little almost corymbose leafy racemes on the smaller branches. Bracteoles short, concave, keeled. Calyx-tube ovoid-turbinate, prominently 5-ribbed, about 1 line long; lobes orbicular, scarious, ½ to ¾ line long. Petals orbicular, spreading, deciduous or sometimes persistent, about 1 line diameter. Stamens 5, opposite the petals; filaments filiform, rather thick; anther-cells parallel, opening longitudinally; connective tipped with 1 or 2 globular glands, rarely both wanting. Ovules 4, suspended in pairs from the top of the cavity. Embryo with the slender portion half as long as the thick radicular end and deeply 2-lobed.—*Imbricaria ciliata*, Sm. in Trans. Linn. Soc. iii. 259; *Stereoxylon ciliatum*, Poir. Dict. Suppl. v. 247; *Escallonia ciliata*, Rœm. and Schult. Syst. v. 329; *Bæckea microphylla*, Sieb. in Spreng. Syst. Cur. Post. 149, DC. Prod. iii. 230, F. Muell. Fragm. i. 30; *B. plicata*, F. Muell. Fragm. i. 30; *Thryptomene plicata*, F. Muell. Fragm. iv. 63.

N. S. Wales. Port Jackson to the Blue Mountains, *R. Brown*, *Sieber*, *n.* 282, *and others;* northward to Hunter's River, *Herb. Mueller;* and southward to Argyle county, *Mossman* and others, and probably to the Victorian frontier.

Victoria. Rocky declivities of the Grampians and in the deserts of the Murray and Wimmera, *F. Mueller*, *Dallachy*, and others.

S. Australia. Tattiara country, *J. E. Woods.*

6. **M. minutiflora,** *Benth.* A shrub with slender virgate branches. Leaves erect, linear-triquetrous, decussate and imbricate on the smaller branches as in *M. microphylla*, but more slender. Flowers very small and nearly sessile in the axils below the ends of the branches. Bracteoles very small. Calyx-tube narrow, scarcely above ½ line long, prominently 5-ribbed; lobes short, orbicular, petal-like, minutely ciliate. Petals orbicular, rather more than ½ line diameter. Stamens 5, opposite the petals, with the anthers of *M. microphylla*. Ovules 2, collaterally suspended from the top of the cavity. Fruit not seen.—*Thryptomene plicata*, var. *minutiflora*, F. Muell. Herb.

N. S. Wales. New England, *C. Stuart;* near Richmond, *Wilhelmi.* F. Mueller thinks this is a variety only of the preceding species, but in all the specimens I have examined it appears, like the following *M. leptocalyx* to differ constantly from *M. microphylla* in the form of the flower and the number of ovules.

7. **M. leptocalyx,** *Benth.* A bushy shrub, attaining about 6 ft. Leaves linear-triquetrous, decussate and imbricate on the smaller branches as in *M. microphylla*, but rather longer. Flowers larger than in that species, on pedicels either exceedingly short or sometimes attaining 1 line. Calyx-tube narrow-turbinate, attaining 1¼ line; lobes orbicular, scarious, about half as

long as the petals. Stamens 5, opposite the petals. Ovary 1-celled with a cluster of 6 to 8 ovules suspended from the top of the cavity on a filiform placenta arising from the base as in *M. microphylla.*—*Bæckea leptocalyx*, F. Muell. Fragm. i. 30.

Queensland. Near Mount Pluto, *Mitchell.*

TRIBE II. LEPTOSPERMEÆ.—Ovary divided into 2 to 5 or more cells. Fruit dry, capsular, opening at the top loculicidally in as many valves as cells, or very rarely 1- or 2-seeded and indehiscent.

SUBTRIBE 1. BÆCKEÆ.—Leaves opposite, usually small. Flowers usually small, pedicellate or subsessile, solitary or few in a small cyme umbel or head in the axils of the leaves, sometimes forming a terminal head with the floral leaves reduced to bracts. Stamens definite, or if indefinite usually in a single row, free or united at the base in a ring, or into clusters alternating with the petals (not opposite the petals as in other subtribes), and usually shorter than the petals. Ovules usually in 2 rows. Embryo with a thick radicle, produced at the opposite end into a slender incurved neck or into a short point with very small, often minute cotyledons.

The most constant character of this subtribe is probably that derived from the embryo, in which, so far as known, the cotyledons are always minute, whilst in the following subtribes they are as long as or longer than the radicle. There are still many species, however, where the embryo has not yet been observed. Generally speaking also the subtribe is distinguished by opposite leaves, and the stamens shorter than the petals, but to these there are a few exceptions.

12. SCHOLTZIA, Schau.

(Piptandra, *Turcz.*)

Calyx-tube turbinate, ovoid or hemispherical, adnate to the ovary, the free part short and broad or reduced to a narrow ring; lobes 5, spreading, petal-like or scarious, shorter than the petals, persistent or deciduous. Petals 5, obovate or orbicular, spreading, usually deciduous. Stamens 5 to 10 or rarely as many as 20 or even more, inserted in a single row on the margin of the disk, those opposite the centre of the petals usually wanting, all free, deciduous; filaments filiform or slightly dilated; anther-cells either united and opening in terminal pores, or distinct furrowed and opening into slits; connective usually thickened and tipped by a globular gland. Ovary inferior, flat-topped or slightly convex, with a tubular depression in the centre round the style, 2- or rarely 3-celled, with 2 superposed ovules or rarely 1 or 3 ovules in each cell, attached to a small axile placenta; style filiform, short, glabrous, with a truncate or capitate stigma. Capsule adnate to the hardened persistent but scarcely enlarged calyx-tube and crowned by its lobes, 2-celled, opening on the flat or convex summit, or almost indehiscent or separating into 2 cocci. Seeds 1 or 2 in each cell, filling the cell and shaped accordingly; testa thin; embryo of the shape of the seed, consisting chiefly of a thick fleshy clavate or truncate radicular portion, with a short slender neck turned up against one side and more or less divided into 2 cotyledons. —Glabrous shrubs, with the habit of *Bæckea*. Leaves, opposite, small, entire or minutely denticulate-ciliate. Flowers small, white or pale pink, in little cymes or rarely umbels of 3 or more, or sometimes solitary on short axillary peduncles. Bracts at the base of the peduncles and pedicels or of the sessile calyces small and deciduous.

The genus is confined to Western Australia. It forms the passage, as it were, from *Thryptomene* and *Micromyrtus*, on the one hand, to *Bæckea* on the other. Some species only differ from the section *Babingtonia* of the latter by the number and position of the ovules, and in others the apparently incomplete dissepiment is an approach to the filiform placenta of *Micromyrtus*. The prevalent cymose inflorescence gives most of the species an aspect different from either.

Anthers broadly obovoid or obcordate, the cells more or less united, and opening in terminal pores.
  Flowers sessile or nearly so in the cyme, or within the bracts on the peduncle.
    Leaves closely sessile, orbicular-cordate or reniform. Cymes rather loose . . . . 1. *S. uberiflora.*
    Leaves obovate or almost rhomboidal narrowed at the base. Cymes dense or few-flowered.
      Calyx-tube smooth or slightly rugose.
        Peduncles not exceeding the leaves, 3- to 5-flowered. Calyx-segments entire. Stamens about 20. Ovary 2-celled . . . . 2. *S. obovata.*
        Peduncles exceeding the leaves, with 5 to 9 large flowers. Calyx-segments entire. Stamens 10–12. Ovary 3-celled . . . . 3. *S. spathulata.*
        Peduncles short, with 1 to 3 small flowers. Calyx-segments ciliate. Stamens under 10. Ovary 3-celled . 4. *S. ciliata.*
        Peduncles exceeding the leaves, with 1 to 3 small flowers. Calyx-segments entire. Ovary 3-celled . . . . . 11. *S. Drummondii.*
      Calyx-tube densely pitted and rugose. Flowers small, numerous in the cyme. Ovary 2-celled . . . . . 5. *S. capitata.*
    Leaves linear-terete . . . . 12. *S. teretifolia.*
  Flowers pedicellate, in pedunculate umbels.
    Leaves narrow-cuneate, concave or keeled. Ovules 2 in each cell . . . . 6. *S. umbellifera.*
    Leaves obovate-cuneate, flat or nearly so. Ovules solitary in each cell . . . . 7. *S. laxiflora.*
Anther-cells distinct, either deeply furrowed and opening in slits, or opening in oblong pores.
  Leaves small, obovate or orbicular.
    Flowers in dense cymes. Calyx-tube narrow-turbinate. Ovules 2 in each cell. Stamens 8 to 10 . . . . 8. *S. leptantha.*
    Peduncles 1- to 3-flowered. Calyx-tube broad. Stamens 5.
      Peduncles not exceeding the leaves. Ovary 2-celled.
        Petals about ½ line diameter. Ovules usually solitary . 9. *S. parviflora.*
        Petals about 1 line diameter. Ovules 2 in each cell . . 10. *S. oligandra.*
      Peduncles exceeding the leaves. Petals 1½ lines diameter. Ovary usually 3-celled with 2 ovules in each cell . . . 11. *S. Drummondii.*
  Leaves slender, linear-terete, about 2 lines. Peduncles short, 1- to 3-flowered. Stamens about 20 . . . . 12. *S. teretifolia.*

1. **S. uberiflora,** *F. Muell. Fragm.* iv. 74. A straggling shrub, of about 6 ft. Leaves closely sessile, spreading or reflexed, orbicular-cordate or almost reniform, very obtuse, mostly 2 to 3 lines diameter, rigid and prominently veined. Flowers white or pale pink, smaller than in *S. obovata*, in a small cyme or head on a common peduncle considerably longer than the leaves. Bracts small, ovate, falling off from the very young buds. Calyx-tube tur-

binate and under 1 line long in flower, at length ovoid-campanulate and nearly 1½ lines long; segments not ⅓ as long as the petals. Petals deciduous, above 1 line diameter. Stamens usually 6 to 8; anthers broadly obcordate, the cells opening in terminal pores, connective-gland prominent. Ovary 2-celled, with 2 or rarely 3 ovules in each cell. Seed ovoid; testa crustaceous; embryo with the slender deeply 2-lobed neck not half as long as the thick radicular portion, and folded against it.

W. Australia. Murchison river, *Oldfield.*

2. **S. obovata,** *Schau. in Linnæa*, xvii. 241, *and in Pl. Preiss.* i. 109. A rigid spreading or almost decumbent shrub of 1 to 2 ft., or when luxuriant twice that size. Leaves from obovate to oblong-cuneate or almost rhomboidal, much narrowed at the base, rigid, somewhat concave, obtuse or almost acute, 2 to 3 lines long. Peduncles very short or rarely as long as the leaves, bearing each a cyme of 3 to 5 nearly sessile flowers. Bracts lanceolate, very fugacious. Calyx-tube hemispherical, smooth, about 1½ lines diameter; lobes petal-like, broad, half as long as the petals. Petals about 1½ lines diameter. Stamens about 20; anthers obcordate, the cells opening in small terminal pores, connective without any prominent gland. Ovary flat-topped, the central depression not deep, 2-celled with 2 superposed ovules in each cell.—F. Muell. Fragm. iv. 74; *Bæckea involucrata,* Endl. in Hueg. Enum. 51.

W. Australia. Swan River, *Fraser, Preiss, n.* 343, *Drummond, n.* 147, *2nd Coll. n.* 76, and others; Murchison river, *Oldfield.*

Bæckea obovata, DC. Prod. iii. 230, is referred by Schauer to this species. The diagnoses given will refer equally well to several other species of *Scholtzia,* but from French specimens in Herb. R. Br. it is more probably the *S. leptantha.*

3. **S. spathulata,** *Benth.* Very near *S. obovata*, but independently of the stamens and ovary it is known at once by its longer peduncles and rather larger flowers. Leaves broadly obovate-spathulate, much narrowed at the base, erect or spreading, 2 to 3 lines long or rather more. Peduncles longer than the leaves, bearing each a cyme of 5 to 9 nearly sessile flowers similar to those of *S. obovata*, but with apparently only about 10 stamens; in all the specimens, however, the flowers are far advanced and most of the stamens are fallen off. Ovary convex after flowering, 3-celled in all the flowers examined, with 2 superposed ovules in each cell.—*Piptandra spathulata,* Turcz. in Bull. Mosc. 1862, ii. 324.

W. Australia. Murchison river, *Oldfield, Drummond, 6th Coll. n.* 59.

4. **S. ciliata,** *F. Muell. Fragm.* iv. 76. A spreading much-branched shrub of about 4 ft., the branchlets sometimes almost spinescent. Leaves obovate to nearly orbicular, spreading or recurved at the top, thick, concave, obtuse, mostly about 1 line long, the upper ones often denticulate-ciliate. Peduncles shorter than the leaves, bearing each 1 or 3 almost sessile flowers. Bracts denticulate-ciliate. Calyx-tube rather broad, rugose; lobes ovate-orbicular, ciliate, not half as long as the petals, and often deciduous. Petals about 1 line diameter, deciduous. Stamens 6 to 9; anthers broadly obcordate, the cells opening in terminal pores or short almost confluent slits. Ovary 3-celled, with 2 superposed ovules in each cell.

W. Australia. Murchison river, *Oldfield.* This species much resembles *Thryptomene denticulata,* but the deciduous petals and calyx-lobes and immersed style readily distinguish it, independently of the structure of the ovary.

5. **S. capitata,** *F. Muell. Herb.* A twiggy shrub of 8 to 10 ft. Leaves broadly obovate or almost rhomboidal, obtuse or almost acute, narrowed into a short petiole, concave, thick and rigid, rarely attaining 2 lines. Flowers rather small, white, in a dense almost capitate cyme on a peduncle considerably exceeding the leaves. Calyx-tube ovoid-campanulate, densely pitted and rugose as in some *Thryptomenes;* lobes petal-like, nearly half as long as the petals. Petals scarcely 1 line diameter. Stamens apparently few, but more or less fallen from the flowers examined, all very far advanced. Ovary 2-celled, with 2 superposed ovules in each cell.

W. Australia. Murchison river, *Oldfield, Drummond, n.* 134. F. Muell., Fragm. iv. 75, observes that this may be a variety of *S. uberiflora,* but it has not the peculiar foliage of that species, and in the calyx it is different both from that and from *S. obovata,* which it resembles in some respects.

6. **S. umbellifera,** *F. Muell. Fragm.* iv. 75. A small shrub, with slender, erect, virgate branches. Leaves narrow-cuneate, erect and recurved, thick, concave or keeled, obtuse, about 1 line long, often minutely denticulate-ciliate. Peduncles longer than the leaves, bearing each an umbel of 3 to 6 small white flowers on short slender pedicels. Bracts at the base of the pedicels small and deciduous. Calyx-tube turbinate, not $\frac{1}{2}$ line long; lobes petal-like, not half as long as the petals. Petals orbicular, about $\frac{3}{4}$ line diameter. Stamens 6 to 9; filaments filiform; anthers broadly obcordate or nearly globular, the cells united nearly to the top and opening in terminal pores. Ovary more or less perfectly 2-celled, with 2 superposed ovules in each cell or on each side of the placenta; style very shortly immersed.

W. Australia. Flinders Bay, *Collie;* Champion Bay, *Walcott.*

7. **S. laxiflora,** *Benth.* Apparently a tall shrub with slender virgate branches. Leaves obovate-cuneate, much narrowed at the base, flat or slightly concave, obtuse or mucronulate, 1 to 2 lines long. Flowers small, in little umbels of 3 to 5 or rarely solitary, on filiform peduncles exceeding the leaves. Calyx-tube broadly turbinate, scarcely rugose, about $\frac{1}{2}$ line long or rather more; lobes broad, short, petal-like. Petals orbicular, about 1 line diameter. Stamens 10; anther-cells united, opening in large oblong pores. Ovary 2- or rarely 3-celled, with one ascending ovule in each cell, style very shortly immersed.

W. Australia. Between Moore and Murchison rivers, *Drummond, 6th Coll. n.* 64.

8. **S. leptantha,** *Benth.* A low bushy shrub with virgate branches, allied to *S. capitata* and *S. parviflora,* but readily distinguished by the calyx. Leaves from narrow-obovate to almost orbicular, erect and recurved, rigid, nearly flat or concave, obtuse or with a slightly prominent midrib, 1 to $1\frac{1}{2}$ lines long. Peduncles longer than the leaves, bearing each a dense capitate cyme of small white flowers. Bracts small, deciduous. Calyx-tube narrow-turbinate, obscurely ribbed, about 1 line long; lobes usually short and broad and rarely exceeding $\frac{1}{3}$ of the petals. Petals about $\frac{3}{4}$ line diameter. Stamens about 8 to 10; anther-cells globular, very small, quite distinct, opening in

short longitudinal slits. Ovary more or less perfectly 2-celled, with 2 superposed ovules in each cell or on each side of the placenta; style shortly immersed.

W. Australia. Seashore, Sharks' Bay, *Milne,* also in Herb. R. Brown from *Herb. Mus. Par.* This is, therefore, probably the true *Bæckea obovata,* DC. Prod. iii. 230.

9. **S. parviflora,** *F. Muell. Fragm.* iv. 76. A spreading shrub of 6 to 8 ft. Leaves from broadly-obovate or almost spathulate and much narrowed at the base to nearly orbicular, mostly about or under 1 line long, rarely nearly 2 lines on the main branches. Peduncles very short, bearing usually 3 small white flowers. Calyx-tube broad, slightly rugose, not ½ line long; lobes broad, entire, not half as long as the petals. Petals spreading, scarcely above ½ line diameter. Stamens about 5; anther-cells ovoid-globular, distinct, opening almost to the base in longitudinal nearly parallel slits; connective-gland small. Ovary flat-topped, 2-celled, with 1 ascending ovule in each cell; style immersed in a deep tubular central depression. Young fruit apparently separating into 2 cocci.

W. Australia, *Drummond, 2nd Coll. n.* 75 (*4th Coll.?*), *n.* 56; Murchison river, *Oldfield.* In Drummond's specimens the leaves are smaller, more spreading, more orbicular, and less narrowed at the base than in Oldfield's.

10. **S. oligandra,** *F. Muell. Herb.* A spreading densely-branched shrub of about 4 or 5 ft. Leaves spreading, decussate on the smaller branches, obovate-orbicular, thick, flat or concave, nerveless, very obtuse, 1 line or rather more in diameter. Flowers small, solitary or 3 together, sessile on a short peduncle articulate at the top, with minute orbicular exceedingly deciduous bracts. Calyx-tube turbinate, nearly ¾ line long, lobes short and broad, petal-like, entire. Petals nearly 1 line long. Stamens about 5; filaments short; anther-cells deeply furrowed and opening in the furrows, having the appearance of 4 globular collateral cells. Ovary nearly flat, with 2 superposed ovules in each cell; the style not very deeply immersed. Fruit separating into 2 hard usually 1-seeded cocci.

W. Australia, *Drummond* (*5th Coll?*), *n.* 147; Murchison river, *Oldfield.* Drummond's specimens have numerous flowers, but far advanced, and have lost their stamens. In Oldfield's, the flowers are very few, but more perfect; both appear, however, to belong to one species.

11. **S. Drummondii,** *Benth.* Much branched and rigid. Leaves obovate or orbicular, spreading, thick, flat or concave, very obtuse, mostly about 1 line long. Peduncles rather slender but rigid, longer than the leaves, bearing 1 or rarely 3 flowers, sessile at the top, and much larger than in *L. oligandra.* Calyx-tube hemispherical, smooth or scarcely rugose, lobes petal-like, about half as long as the petals. Petals persistent, spreading, 1½ lines diameter. Stamens not seen. Ovary after flowering convex, 2- or more frequently 3-celled, with 2 superposed ovules in each cell. Style very shortly immersed.

W. Australia, *Drummond* (*3rd Coll.?*), *n.* 38.

12. **S. teretifolia,** *Benth.* Stems in our specimens numerous, erect, 6 to 8 in. high. Leaves linear, terete or channelled above, obtuse, not exceeding 2 lines and mostly clustered as in *Astartea.* Flowers solitary or 2 or 3 together on very short peduncles in the upper axils. Calyx-tube broadly

turbinate, somewhat rugose, about 1 line diameter; lobes broad, scarious, denticulate-ciliate. Petals twice as long as the calyx-lobes, nearly 1½ lines diameter. Stamens above 20; filaments rather long; anthers broadly obcordate or the cells almost distinct, opening in large oblong pores. Ovary very convex or almost free, 2-celled with 2 superposed ovules in each cell; style immersed in a central tubular depression.

W. Australia, *Drummond, n.* 136. The foliage gives this plant a very different aspect from that of the other species, yet the floral characters are entirely those of *Scholtzia.*

13. BÆCKEA, Linn.

(Jungia, *Gærtn.;* Imbricaria, *Sm.;* Schidiomyrtus, Rinzia, Euryomyrtus, Camphoromyrtus, Tetrapora, Harmogia, and Oxymyrrhine, *Schau.;* Babingtonia, *Lindl.;* Ericomyrtus, *Turcz.*)

Calyx-tube turbinate or hemispherical, adnate to the ovary at the base, the free part broad and open; lobes 5, imbricate, continuous with the tube or more or less scarious, usually persistent. Petals 5, broadly obovate or orbicular, longer than the calyx-lobes, spreading. Stamens rarely exceeding 20 and often under 10, free, in a single row round the margin of the disk, and usually horizontally inflected in the bud. Filaments filiform or flat; anther-cells united or distinct, opening in longitudinal slits or in small pores. Ovary adnate to the lower part of the calyx-tube or enclosed in it, and either more or less convex at the top or semiadnate or free except the broad base, 2- or 3-celled, with 2 collateral or several ovules in each cell, in 2 rows or in a ring round a more or less peltate placenta; style filiform, glabrous, inserted in a deep tubular or rarely shallow depression in the centre of the ovary; stigma capitate or peltate. Capsule partially or wholly superior, enclosed in the scarcely enlarged calyx-tube, opening at the top loculicidally in 2 or 3 valves. Seeds either 1 or 2 in each cell and reniform, or several and more or less angular; testa thin or slightly crustaceous; embryo filling the seed, the radicular portion thick and clavate, with a slender short neck folded against the side and shortly divided into 2 ovate or oblong cotyledons.—Heath-like glabrous shrubs. Leaves small, opposite, entire. Flowers small, white or pink, either solitary in the axils on a peduncle articulate at, above, or rarely below the middle, with two small bracteoles at the articulation, or several together on a short common peduncle with a small bract at the base of each pedicel.

The genus is chiefly Australian, but one of the common East Australian species extends into New Caledonia, and 2 or 3 others not Australian are found in New Caledonia or in the Indian Archipelago and S. China.

Much as several of the species differ from each other in the stamens as well as in the ovary, it is exceedingly difficult to distribute the whole into good sections, for the different forms appear either to pass into each other by almost insensible gradations, or to be strictly monotypic, and none have appeared to me to be sufficiently accompanied by differences in habit or by any combination of characters to justify the adoption of any of the long list of separate genera proposed by Schauer and others. The presence or absence of the five stamens opposed to the petals is perhaps the most marked, but even that appears to be uncertain in the few cases where the stamens exceed 20. The anthers of the first sections are very different from those of the last, but those of *Harmogia* and *Oxymyrrhine* show a gradual passage from the one to the other.

A. *Stamens* 10 *or more, of which* 5 *(often larger than the others) opposite the centre of the petals. Anther-cells parallel, opening longitudinally.*

SECTION 1. **Rinzia.**—*Filaments all, or at least those opposite the petals, distinctly flattened.*

Stamens 10.
  Ovary superior except the broad base, with 2 or rarely 3 ovules in each cell. Pedicels exceeding the leaves.
    Leaves ovate or oblong, thick, 1 to 1½ lines long . . . . . 1. *B. platystemona.*
    Leaves linear, 1 to 3 lines long . . . . 2. *B. Fumana.*
  Ovary convex, but almost entirely inferior, with usually 3 ovules in each cell. Pedicels shorter than the small erect leaves . 3. *B. dimorphandra.*
  Ovary very convex or half-superior, with 4 to 6 or more ovules in each cell. Pedicels much longer than the leaves.
    Leaves linear or lanceolate, flat. . . . . 4. *B. schollerifolia.*
    Leaves ovate or orbicular, thick and very convex . . . . . 5. *B. oxycoccoides.*
Stamens 15 to 20. Leaves linear, semiterete. . . . . 6. *B. Drummondii.*

SECTION 2. **Euryomyrtus.**—*Filaments filiform or very slightly dilated.*

Stamens 10. Ovules 2, 3, or rarely 4 in each cell.
  Leaves linear, loose or spreading. Pedicels longer than the leaves.
    Ovary convex . . . . 7. *B. diffusa.*
  Leaves thick, obovoid to linear-terete. Pedicels not exceeding the leaves. Ovary convex . . . . 8. *B. crassifolia.*
  Leaves under 1 line long, imbricate and decussate. Ovary flat-topped . . . . 9. *B. tetragona.*
Stamens 15. Ovules 2 in each cell. Leaves rarely above 1 line long.
  Leaves very small, imbricate and decussate. Flowers nearly sessile 10. *B. ericæa.*
  Leaves above ½ line long, rather loose. Flowers shortly pedicellate 8. *B. crassifolia*, var.
Stamens 20 to 30. Ovary convex, with about 8 ovules in each cell.
  Leaves linear-terete, 1 to 2 lines long. Pedicels short. Calyx-lobes petal-like . . . . 11. *B. polystemona.*

(*B. polyandra* with numerous stamens has the petaline ones present, but the anthers are very different from those of *Euryomyrtus*.)

B. *Stamens few or numerous, but none opposite the centre of the petals, excepting very rarely, when there are more than* 20. *Ovules several in each cell.*

SECTION 3. **Schidiomyrtus.**—*Anther-cells distinct, parallel, opening longitudinally to the base. Flowers solitary. Ovary 2- or rarely (in* B. astarteoides*) 3-celled.*

Leaves broadly ovate or orbicular, flat or concave, 1½ to 3 lines long 12. *B. crenulata.*
Leaves thick, triquetrous, decussate, ½ to 1 line long . . . . . 13. *B. brevifolia.*
Leaves concave, from narrow-obovate and 1 line to linear-cuneate and 3 lines long.
  Calyx-lobes entire . . . . 14. *B. Gunniana.*
  Calyx-lobes denticulate-ciliate . . . . 15. *B. diosmifolia.*
Leaves linear, semiterete or triquetrous, or subulate, 2 to 9 lines long.
  Calyx-tube broadly turbinate.
    Leaves linear, concave, rigid. Stamens about 5.
      Flowers about 1 line long . . . . 16. *B. leptocaulis.*
      Flowers about ½ line long . . . . 17. *B. arbuscula.*
    Leaves semiterete or triquetrous, usually short and clustered in the axils. Stamens 6 to 8 . . . . 18. *B. astarteoides.*
    Leaves linear-subulate, usually long. Stamens 10 to 15. . . . 19. *B. linifolia.*
  Calyx-tube narrow-turbinate. Leaves linear-subulate, usually long.
    Flowers very small. Stamens about 5 . . . . 20. *B. stenophylla.*

SECTION 4. **Harmogia.**—*Anther-cells distinct, nearly globular, deeply furrowed, parallel or divergent, and opening more or less in longitudinal slits in the furrows. Ovary 3-celled, with several ovules in each cell. Eastern species.*

Leaves flat. Flowers often clustered or umbellate.
  Leaves oblong-cuneate or nearly linear, under 3 lines long.
    Pedicels short thick, 1-flowered but often 2 or 3 in the axils . . 21. *B. camphorata.*

Leaves linear-lanceolate or narrow-oblong, 4 lines to 1 in. long.
Flowers mostly in pedunculate umbels . . . . 22. *B. virgata.*
Leaves ovate or oblong, crenulate, 2 to 3 lines long. Flowers 1 to 3 on a slender peduncle . . . . 23. *B. crenatifolia.*
Leaves orbicular, thick, about 1 line long. Flowers 1 to 3 on a short peduncle . . . . 24. *B. Cunninghamii.*
Leaves semiterete or triquetrous, obtuse or with recurved points. Slits of the anther-cells almost shortened to pores.
Leaves slender, mostly imbricate-decussate or short, the recurved points minute or none. Filaments not clavate . . . . . 25. *B. densifolia.*
Leaves rather thick, the recurved points prominent. Filaments clavate under the anther . . . . 26. *B. Behrii.*

SECTION 5. **Oxymyrrhine.**—*Anther-cells more or less united at the base, didymous, deeply furrowed and opening in the furrows, giving the whole anther the appearance of 4 collateral globular cells, either all equal or the 2 central ones smaller. Ovary 3-celled with numerous ovules in each cell. Western species.*

Leaves semiterete, triquetrous, or scarcely flattened, mostly 1½ to 3 lines long.
Leaves recurved at the end. Flowers mostly in threes. Calyx-tube not ribbed, lobes obtuse. Stamens about 8 . . . . 27. *B. uncinella.*
Leaves mostly obtuse. Flowers solitary. Calyx-tube prominently 5-ribbed, lobes acute. Stamens 20 or more . . . . . 28. *B. polyandra.*
Leaves mostly thick and clavate or curved. Flowers 1 to 3. Calyx-lobes acute. Stamens 6 to 8 . . . . 29. *B. corynophylla.*
(See also 25. *B. densifolia,* and 26. *B. Behrii.*)
Leaves mostly obovate or orbicular, 1 to 1½ line long. Flowers solitary. Calyx-lobes short, obtuse.
Leaves thick, spreading. Stamens about 5 . . . . 30. *B. pachyphylla.*
Leaves very concave, erect or imbricate. Stamens 15 to 20 . 31. *B. crispiflora.*

SECTION 6. **Babingtonia.**—*Anther-cells united into an obcordate or almost globular anther, and opening in terminal pores or short slits. Western species.*

Leaves slender, short, semiterete, crowded on the lateral branchlets (except in *B. pygmæa*).
Flowers mostly in clusters, forming a long, unilateral, leafy raceme. Stamens about 10 . . . . 32. *B. camphorosmæ.*
Flowers small, mostly solitary, on slender pedicels. Stamens 20 to 30 . . . . 33. *B. pulchella.*
Flowers very small, solitary or 2 together, on slender pedicels. Stamens about 10 . . . . 34. *B. pygmæa.*
Leaves small, obovate. Flowers small, usually 2 to 5 on each peduncle, in a short corymbose raceme. Calyx-tube not angled.
Stamens 10 to 15. Petals about ½ line diameter . . . . . 35. *B. corymbulosa.*
Stamens usually 10. Petals above 1 line diameter . . . . 36. *B. floribunda.*
Stamens about 5. Petals above 1 line diameter . . . . . 37. *B. pentandra.*
Leaves small, obovate-triquetrous. Flowers solitary. Calyx-tube acutely 5-angled. Stamens 15 to 20 . . . . 38. *B. pentagonantha.*
Leaves rigid, often 2 lines long, or more. Flowers rather large, solitary. Calyx-tube very open, not angled, at least 2 lines diameter. Stamens 15 to 30.
Leaves broadly obovate, cuneate or truncate, scarcely 2 lines long . . . . 41. *B. subcuneata.*
Leaves oval to oblong-linear, concave, 2 to 4 lines long . . . 40. *B. ovalifolia.*
Leaves thick, almost fleshy, oblong or oblong-linear, 2 to 4 lines long . . . . 39. *B. robusta.*
Leaves slender, linear-terete . . . . 42. *B. grandiflora.*

Section I. Rinzia.—Stamens 10 or more, of which 5 opposite the centres of the petals; filaments all, or at least those opposite the petals, much flattened and broad, often notched at the top; anther-cells distinct and parallel, opening longitudinally to the base. Ovary 3-celled. Flowers pedicellate, solitary or rarely 2 or 3 together on an exceedingly short common peduncle.

1. **B. platystemona,** *Benth.* Apparently low and diffuse, with the habit and foliage of *B. crassifolia*, but with very different stamens. Leaves ovate-oblong, thick, concave, very obtuse, 1 to 1½ lines long. Flowers solitary, on pedicels of 2 to 3 lines, with a pair of small coloured bracts at or near the base. Calyx-tube very broadly campanulate, almost introrse at the base, nearly 2 lines diameter; lobes semi-orbicular, continuous with the tube. Petals twice as long as the calyx-lobes, about 1½ lines diameter, very deciduous. Stamens 10; filaments flat, those opposite the petals very broad, emarginate at the end with the connective-gland in the notch, the others smaller and more tapering, not notched; anthers in front of the filament, the cells distinct and parallel, opening longitudinally. Ovary very convex, almost superior, 3-celled, with 2 collateral ovules in each cell; style shortly immersed. Seeds rather large, with a crustaceous granulate testa, as in *B. diffusa.*

W. Australia, *Drummond, 4th Coll. n.* 148; *5th Coll. n.* 122.

2. **B. Fumana,** *F. Muell. Fragm.* iv. 68. A small shrub, with numerous erect or diffuse branches, often not above 6 in. high, attaining 1 ft. in our larger specimens. Leaves loosely imbricate on the smaller branches, or if distant not appressed, linear or linear-oblong, concave or semiterete, obtuse, 1 to 3 lines long or rarely more. Flowers apparently white, on pedicels from rather longer than, to twice as long as the floral leaves, with 2 minute bracts at the base. Calyx broadly hemispherical, almost introrse at the base, about 1½ lines diameter, with broad obtuse lobes nearly as long as the tube, and slightly scarious at the edges. Petals about 1½ lines diameter. Stamens 10; filaments flat, those opposite the petals very broad, emarginate at the end, with the connective-gland in the notch, the others smaller, more tapering, and not notched; anthers in front of the filament, the cells distinct and parallel, opening longitudinally. Ovary nearly globular, adnate by the broad base, but otherwise superior, 3-celled, with 2 collateral ovules in each cell.—*Rinzia Fumana,* Schau. in Linnæa, xvii. 239, and in Pl. Preiss. i. 108.

W. Australia. Swan River, *Drummond, 1st Coll.;* also *n.* 77 *and* 140; King George's Sound, *Preiss, n.* 164.

3. **B. dimorphandra,** *F. Muell. Herb.* A small shrub of ½ to 1 ft., with numerous slender erect virgate branches. Leaves appressed, linear, semiterete or concave, obtuse, 1 to 2 lines long. Flowers solitary, or 2 or 3 together on very short pedicels, with 2 or 3 bracts at the base. Calyx-tube broad; lobes ovate, about as long as the tube, not scarious. Petals nearly 1½ lines diameter. Stamens 10; filaments flat, those opposite the petals broad, 2-lobed at the top, with the connective-gland in the notch, the others smaller, tapering and not notched; anthers in front of the filament; the cells distinct

and parallel, opening longitudinally. Ovary convex, but almost entirely inferior, 3-celled, with usually 3 ovules in each cell. Style shortly immersed.

W. Australia. Sandy places near Cape le Grand, *Maxwell*. Near *B. Fumana*, but besides the differences in the foliage and ovary, the flowers in the dried specimen are of a rich pink, whilst in *B. Fumana* they appear to be white or nearly so.

4. **B. schollerifolia,** *Lehm. in Pl. Preiss.* ii. 369. Small, slender, and diffuse or procumbent. Leaves oblong or lanceolate, flat with slightly recurved margins, obtuse or rather acute, 1½ to 3 lines long. Flowers solitary, on pedicels of ½ to 1 in., not articulate, with a coloured deciduous bract at the base. Calyx-tube broadly turbinate, soon becoming hemispherical, rather above 1 line diameter; lobes broad, rather thick, with thin scarious minutely ciliate margins. Petals above 2 lines diameter. Stamens 10; filaments flat, those opposite the petals broader and emarginate, with the connective-gland in the notch, the others tapering and entire; anthers in front of the filament; the cells distinct, parallel, and opening longitudinally. Ovary very convex, 3-celled, with 4 to 6 ovules in each cell on a small placenta; style shortly immersed.

W. Australia, *Drummond, n.* 63, *and* 5*th Coll. n.* 125, *or in some sets n.* 121; near Seven-mile Bridge, Plantagenet district, *Preiss, n.* 2015.—The habit and foliage are nearly those of *B. diffusa*, but the stamens and ovary are different.

5. **B. oxycoccoides,** *Benth.* Branches prostrate or trailing. Leaves ovate or orbicular, thick, very convex with recurved margins, obtuse, under 2 lines diameter. Flowers pink, solitary, on pedicels of ½ to 1 in., not articulate, with a coloured deciduous bract at the base. Calyx-tube very broadly turbinate, fully 2 lines diameter; lobes very short and broad, with scarious minutely ciliate margins. Petals fully 2½ lines diameter. Stamens 10; filaments flat, thick, erect, all tapering at the end, with small globular connective-glands, or those opposite the petals rather broader and slightly emarginate; anthers in front of the filament; the cells distinct, parallel, and opening longitudinally. Ovary convex, 3-celled, with 10 to 12 or even more ovules in each cell; style shortly immersed.

W. Australia, *Drummond,* 5*th Coll. n.* 120.

6. **B. Drummondii,** *Benth.* Branches apparently divaricate, elongated and rather rigid. Leaves linear, semiterete or triquetrous, rather thick, obtuse or mucronate, mostly 2 to 4 lines long. Flowers rather large, solitary, on short pedicels with 2 small bracts at the base, or on short axillary branchlets with 1 or 2 pairs of small leaves towards the base. Calyx-tube hemispherical, about 2 lines diameter; lobes short and broad, thick at the base, with broad petal-like margins. Petals about 2 lines diameter. Stamens 15 to 20, closely packed in a single row; filaments all flat and broad, but usually quite distinct, those opposite the centre of the petals the largest, all entire, with a small connective-gland; anthers in front of the filaments; the cells almost connate, parallel, opening longitudinally. Ovary convex, 3-celled, with 3 or 4 ovules in each cell on a peltate placenta; style shortly immersed.

W. Australia, *Drummond,* 5*th Coll. n.* 123.

SECTION II. EURYOMYRTUS.—Stamens 10 or more, of which 5 opposite the centre of the petals; filaments filiform or very slightly flattened; anther-cells distinct, parallel, and opening longitudinally to the base. Ovary 3-celled. Flowers pedicellate or nearly sessile, solitary in each axil.

7. **B. diffusa,** *Sieb. in DC. Prod.* iii. 230. Prostrate or diffuse, with slender branches, often attaining a considerable length. Leaves linear, flat, or thick and concave, smooth or striate, acute or almost obtuse, from 2 to 4 or even 5 lines long. Flowers solitary, on slender axillary pedicels of 2 or 3 lines, with a small bract at the base and a pair of bracteoles usually about the middle. Calyx-tube broadly turbinate or hemispherical, at least 1½ lines diameter; lobes short and broad, minutely ciliate. Petals nearly 1½ lines diameter. Stamens 10; filaments filiform or slightly dilated, especially those opposite the petals, but much less so than in the *Rinzias;* anther-cells distinct and parallel, with a rather large obovoid-globular connective-gland. Ovary convex, 3-celled, with 3 or 4 ovules in each cell; style very shortly immersed. Capsule half-superior. Seeds usually 2 in each cell, collateral, rather large, with a lateral hilum; testa crustaceous. Embryo with the slender cotyledonar end transversely flexuose or twisted.—Hook. f. Fl. Tasm. i. 142; F. Muell. Fragm. iv. 67; *B. alpina,* Lindl. in Mitch. Three Exped. ii. 178; *B. thymifolia,* Hook. f. in Hook. Ic. Pl. t. 284, and in Fl. Tasm. i. 141; *B. affinis* and *B. prostrata,* Hook. f. in Hook. Ic. Pl. t. 284; *Euryomyrtus diffusa, E. alpina,* and *E. thymifolia,* Schau. in Linnæa, xvii. 239; *Euryomyrtus parviflora* and *E. Stuartiana,* F. Muell.; Miq. in Ned. Kruidk. Arch. iv. 149.

N. S. Wales. Port Jackson to the Blue Mountains, *R. Brown; Sieber, n.* 276, and others.

Victoria. Alpine and subalpine heights, Mounts William, Buller, Barkly, Liger, etc., *F. Mueller.*

Tasmania. Derwent river and Port Dalrymple, *R. Brown.*—Abundant on heaths, especially on river banks, *J. D. Hooker.*

The forms originally distinguished by J. D. Hooker as species, are now shown by a number of intermediate specimens to run so much into each other as not to be easily separable as varieties.

8. **B. crassifolia,** *Lindl. in Mitch. Three Exped.* ii. 115. Low and much-branched, often diffuse. Leaves spreading, thickly obovoid or oblong, very obtuse, and ½ to 1 line long, or rarely linear-terete, almost acute and 2 lines or rather longer. Flowers on pedicels of ½ to 1 line, solitary in each axil, and usually 2 or 3 only on short lateral branchlets. Bracteoles so deciduous as to be rarely seen. Calyx-tube broadly turbinate or hemispherical, nearly 1 line diameter; lobes broad, obtuse, nearly as long as the tube, with petal-like margins. Petals twice as long as the calyx-lobes. Stamens 10; filaments all filiform; anther-cells parallel, opening longitudinally, with conspicuous connective-glands. Ovary convex, but not much so, 3-celled with 2 collateral ovules in each cell; style very shortly immersed. Capsule nearly half-superior. Seeds and embryo nearly as in *B. diffusa,* but not so large.—F. Muell. Fragm. iv. 66.

N. S. Wales. Deserts of the Darling and Murrumbidgee, *F. Mueller* (I have not seen the specimens).

Victoria. On the Murray and in the Wimmera district, *F. Mueller* and others.

S. Australia. Sandy deserts from Spencer's and St. Vincent's Gulfs to the Murray, *F. Mueller* and others; Kangaroo Island, *Waterhouse.*

Var. (?) *icosandra,* F. Muell. Stamens usually 1 opposite each petal, and 2 or sometimes 3 in the intervals.

W. Australia. Limestone cliffs towards the Great Australian Bight, *Maxwell.*—Notwithstanding the difference in foliage and habit, it is possible that these specimens may be a form rather of *B. ericœa* than of *B. crassifolia.*

The linear-leaved specimens of *B. crassifolia* are from the Botanical Garden of Melbourne.

9. **B. tetragona,** *F. Muell. Herb.* Branchlets small, numerous, erect. Leaves imbricate and decussate, ovate or oblong, thick, concave or keeled, obtuse, $\frac{3}{4}$ to nearly 1 line long. Flowers solitary, almost sessile, with 2 concave very deciduous bracts under the calyx. Calyx-tube turbinate, 5-angled, above 1 line long; lobes broadly ovate and petal-like, or 2 outer ones narrower and greener, half as long as the tube. Petals larger than the calyx-lobes. Stamens 10; filaments filiform or slightly dilated at the base; anther-cells parallel, opening longitudinally; connective-gland globular. Ovary flat-topped, 3-celled, with 2 or rarely 3 ovules in each cell. Style shortly immersed. Seeds nearly as in *B. diffusa.*

W. Australia. E. of King George's Sound, *Baxter;* Middle Mount Barren, *Maxwell;* Lucky Bay, *R. Brown.*

10. **B. ericæa,** *F. Muell. Fragm.* i. 34. Small and very much branched, closely resembling the smaller specimens of *Micromyrtus microphylla,* but quite different in the structure of the flowers. Leaves oblong or linear, thick, concave or keeled, very obtuse, $\frac{1}{2}$ to 1 line long, appressed and distant on the larger branches, imbricate and decussate on the smaller ones. Flowers small, solitary, sessile, with broad scarious bracts under the calyx. Calyx-tube very broad, about $\frac{1}{2}$ line long; lobes short, broad, entire, with coloured scarious margins. Petals twice as long as the calyx-lobes. Stamens 15, of which 5 opposite the petals; filaments short, filiform; anther-cells short, parallel, opening longitudinally; connective rather thick. Ovary very convex, 3-celled, with 2 collateral ovules in each cell. Seeds apparently as in *B. diffusa,* but not seen ripe.

Victoria. In the Murray scrub, *F. Mueller;* Wimmera, *Dallachy.*

11. **B. polystemona,** *F. Muell. Fragm.* ii. 124. Leaves crowded, linear or slightly clavate, terete or concave, obtuse, 1 to 2 lines long. Flowers solitary, on short pedicels in the upper axils, with a pair of very deciduous bracts at the base of the pedicels. Calyx-tube very broad, about 2 lines diameter; lobes short, broad, petal-like. Stamens 20 to 30, of which 5 opposite the centre of the petals; filaments filiform; anther-cells parallel, opening longitudinally; connective-gland rather large. Ovary convex, 3-celled, with about 8 ovules in each cell; style shortly immersed.

N. Australia. Brindley's Bluff, *M'Douall Stuart's Expedition.* Described from a single small specimen in Herb. F. Muell. The more numerous ovules and indefinite stamens might refer it to the following section, but that there appears always to be a stamen opposite the centre of each petal as in *Euryomyrtus.*

SECTION III. SCHIDIOMYRTUS.—Stamens few or numerous, but none opposite the centres of the petals; filaments filiform; anther-cells distinct, parallel, and opening longitudinally to the base. Ovary 2-celled, or in *B. astarteoides* 3-celled, with several ovules in each cell. Flowers solitary in each axil.

12. **B. crenulata,** *DC. Prod.* iii. 230. Branches virgate. Leaves broadly obovate or orbicular, flat or concave, obtuse or almost acute, usually minutely denticulate-ciliate, often imbricate, 1½ to nearly 3 lines long, the floral ones mostly longer than the others. Flowers nearly sessile along the branches, solitary in each axil, shorter than or scarcely exceeding the leaves. Bracteoles ovate-lanceolate, concave, deciduous. Calyx-tube about 1 line long, the adnate part narrow-turbinate, the free part broad; lobes ovate, ½ line long, slightly scarious on the edges. Petals shortly exceeding the calyx-lobes. Stamens 10 or fewer, not opposite the centre of the petals; anthers small, didymous, the cells opening longitudinally; connective-gland inconspicuous. Ovary 2-celled, with 6 to 10 ovules in each cell; style shortly immersed. Seeds obovoid, more or less angular; testa thinly crustaceous; thin end of the embryo closely folded against the radicle, otherwise straight.—R. Br. in Flind. Voy. App. 548; F. Muell. Fragm. iv. 65; *Jungia imbricata*, Gærtn. Fruct. i. 175. t. 35 (incorrect as to the details); *Mollia imbricata*, Gmel. Syst. Veg. 420; *Imbricaria crenulata*, Sm. in Trans. Linn. Soc. iii. 259; *Stereoxylon crenulatum*, Poir. Dict. Suppl. v. 246; *Escallonia crenulata*, Rœm. and Schult. Syst. v. 329; *Bæckea diosmoides*, Sieb. in DC. Prod. iii. 230; *Schidiomyrtus crenulata* and *S. Sieberi*, Schau. in Linnæa, xvii. 237.

N. S. Wales. Port Jackson, *R. Brown; Sieber n.* 277, *and Fl. Mixt. n.* 611, and others, and southward to Illawarra, *Shepherd.*

Var. *tenella.* Leaves smaller; flowers very much smaller, but not otherwise different. *Jungia tenella*, Gærtn. Fruct. i. 175. With the larger variety from most collectors.

13. **B. brevifolia,** *DC. Prod.* iii. 230. Branches numerous, slender, erect. Leaves decussate, triquetrous, thick and very obtuse, ½ to 1 line long. Flowers solitary in the upper axils, on pedicels of from ½ to nearly 2 lines. Bracteoles very deciduous. Calyx-tube turbinate, 1 to 1½ lines long, very broad at the top; lobes short, broad, not scarious. Petals about 1 line long. Stamens about 15, none opposite the centre of the petals; anther-cells parallel, opening longitudinally; connective-gland small. Ovary nearly flat-topped, 2-celled, with 8 to 10 ovules in each cell; style not very deeply immersed.—*Leptospermum brevifolium*, Rudge in Trans. Linn. Soc. viii. 299. t. 14; *Bæckea carnosula*, Sieb. in Spreng. Syst. Cur. Post. 149.

N. S. Wales. Port Jackson, *R. Brown; Sieber, n.* 278, and others.

14. **B. Gunniana,** *Schau. in Walp. Rep.* ii. 921. A densely-branched shrub, either low and prostrate or erect and bushy, attaining 5 or 6 ft. Leaves spreading, flat or concave, from obovate-oblong and scarcely 1 line long, to linear or linear-cuneate, and 3 or even 4 lines long, obtuse or scarcely mucronate-acute. Flowers solitary in the upper axils, on pedicels of 1 to nearly 2 lines. Bracteoles under the calyx so deciduous as to be rarely seen. Calyx-tube turbinate, about 1 line long, very broad at the top; lobes not half as long as the tube, petal-like, obtuse, separated by rather broad sinuses. Petals above 1 line diameter. Stamens 10 or fewer, none opposite the centres of the petals; filaments filiform; anther-cells parallel, opening longitudinally; connective-gland rather prominent. Ovary flat-topped, 2-celled, with 10 to 12 ovules in each cell. Seeds obovoid, more or less angular; testa thinly coriaceous; slender end of the embryo folded against the radicle,

but otherwise straight.—Hook. f. Fl. Tasm. i. 142; F. Muell. Fragm. iv. 66; *B. micrantha*, Hook. f. in Hook. Ic. Pl. t. 309, not of DC.; *B. utilis*, F. Muell.; Miq. in Ned. Kruidk. Arch. iv. 150; *Tetrapora Gunniana*, Miq. l. c.

N. S. Wales. Mount Mitchell, *Beckler*.

Victoria. Common in boggy places in the Australian Alps, *F. Mueller*.

Tasmania. Summit of Table Mountain, *R. Brown;* abundant in alpine places, *J. D. Hooker*.

Var. *latifolia*. Leaves ovate-oblong, 3 to 4 lines long. Baw-Baw Mountains, *F. Mueller*.

15. **B. diosmifolia,** *Rudge in Trans. Linn. Soc.* viii. 298. *t.* 13. Erect or diffuse and much branched, from a thick woody stock. Leaves linear, narrow, oblong, or somewhat cuneate, concave or semiterete, obtuse or mucronulate-acute, more or less denticulate-ciliate, 1 to 2 lines long. Flowers nearly sessile and solitary in the upper axils. Bracteoles obovate-cuneate, concave, as long as the calyx-tube, very deciduous. Calyx-tube turbinate, about 1½ lines long; lobes ovate, denticulate-ciliate. Petals about 1 line diameter. Stamens 7 to 10, none opposite the centre of the petals; filaments filiform; anther-cells parallel, opening longitudinally; connective-gland globular. Ovary small, 2-celled, with about 4 ovules in each cell.—DC. Prod. iii. 230; F. Muell. Fragm. i. 29.

N. S. Wales. Port Jackson, *R. Brown, F. Mueller*, and others; Paramatta, *A. Cunningham, Woolls*.—In all the flowers I examined of Woolls's specimens, the ovary was in a monstrous state, with the ovules all abortive, but with several more or less perfect stamens on the walls of the cavity.

16. **B. leptocaulis,** *Hook. f. in Hook. Ic. Pl. t.* 298, and *Fl. Tasm.* i. 141. Branches erect, from a thick woody base, slender, 1 to 2 ft. high. Leaves linear, concave or semiterete, obtuse or with a short erect point, mostly 3 to 4 lines long. Flowers solitary in the upper axils, on pedicels at least as long as the calyx-tube, with 2 small very fugacious bracteoles at the base. Calyx-tube turbinate, under 1 line long, tapering into the pedicel; lobes small, ovate or oblong, usually separated by marked intervals. Petals nearly 1 line diameter. Stamens about 5, none opposite the centre of the petals; filaments filiform; anther-cells parallel, opening longitudinally; connective-gland very small or none. Ovary 2-celled, with 8 to 10 ovules in each cell.

Tasmania. Abundant on Loddon Plains, on the road to Macquarrie Harbour; top of Rocky Cape, *Gunn*.—This plant is very nearly allied to the narrow-leaved forms of *B. Gunniana*.

17. **B. arbuscula,** *R. Br. Herb.* A slender, erect, bushy, heath-like shrub, scarcely exceeding 6 in., with very numerous filiform branches, quite glabrous. Leaves slender, linear-terete or slightly flattened, 1 to 2 lines long. Pedicels axillary, solitary, 1-flowered, about ½ line long. Flowers the most minute in the Order. Calyx about ¼ line long, turbinate, with 5 lobes, not one-third as long as the tube. Petals not twice as long as the calyx-lobes, spreading. Stamens 5 or fewer; anthers with distinct nearly globular cells opening longitudinally. Ovary 2-celled?

W. Australia. King George's Sound, *R. Brown*. I do not feel certain of having correctly ascertained the structure of the ovary, but the species is evidently allied to *B. leptocaulis*, although the excessive minuteness of the flowers gives it a very different aspect.

18. **B. astarteoides,** *Benth.* A shrub of 2 or 3 ft., with elongated branches. Leaves linear, semiterete or triquetrous, often slightly clavate, obtuse, 2 to 3 lines long, densely clustered on the short axillary branchlets. Flowers small, pink, solitary, on short pedicels, articulate about or above the middle. Bracteoles all fallen from the specimens seen. Calyx-tube turbinate when young, at length hemispherical, scarcely above 1 line diameter; lobes short, broad, with scarious margins. Petals less than 1 line long, much narrowed at the base. Stamens about 6 to 8, none opposite the centre of the petals; filaments filiform; anther-cells parallel, opening longitudinally; connective-gland small. Ovary 2- or 3-celled, with about 8 ovules in each cell; stigma rather broad. Capsule slightly convex. Seeds not angled.

W. Australia. King George's Sound, *R. Brown, A. Cunningham*: along the coast from Bremer Bay, to Experience Bay and inland from Cape Le Grand, *Maxwell*; Lucky Bay, *R. Brown* (with a much smaller style and stigma).—This much resembles *Astartea fascicularis*, but the filaments are quite distinct and distant from each other.

19. **B. linifolia,** *Rudge in Trans. Linn. Soc.* viii. 297. *t.* 12. Tall and erect, with slender virgate branches. Leaves very narrow-linear, semiterete or concave, acute, in some specimens all above ½ in., attaining ¾ or even 1 in., in others mostly ¼ to ½ in. long. Flowers small, solitary in the upper axils, almost sessile or on pedicels rarely attaining 1 line. Calyx-tube turbinate or at length almost urceolate; lobes very broad and short, scarcely pointed. Petals about 1 line diameter. Stamens 10 to 15, none opposite the centre of the petals; filaments filiform; anther-cells parallel, opening longitudinally; connective-gland small. Ovary flat-topped, 2-celled, with 15 to 20 ovules in each cell round an orbicular almost peltate placenta. Capsule separating readily from the calyx-tube. Seeds small, angular. Embryo with the slender cotyledonar end closely folded against the radicle, but otherwise straight.—DC. Prod. iii. 229; F. Muell. Fragm. iv. 71; *B. trichophylla*, Sieb. in Spreng. Syst. Cur. Post. 149.

N. S. Wales. Port Jackson to the Blue Mountains, *R. Brown, Sieber, n.* 280, and others.

Var. (?) *brevifolia*, F. Muell. Fragm. iv. 72. Leaves shorter and more rigid.

Victoria. Boggy places, near Mount Imlay, *F. Mueller*. I refer this here, on the authority of F. Mueller. The specimens appear to me to have rather the aspect, foliage, and calyx of *B. leptocaulis*, but being only in a far advanced fruiting stage, they cannot be accurately determined.

20. **B. stenophylla,** *F. Muell. Fragm.* i. 13. Branches slender, virgate. Leaves slender, linear, semiterete, obtuse, 2, 3 or rarely 4 lines long, mostly crowded or clustered on the short axillary shoots. Flowers very small, shortly pedicellate, solitary in each axil, but often forming little leafy corymbs, on short axillary shoots. Bracteoles narrow, at the base of the pedicel. Calyx-tube narrow-turbinate, ¾ line long; lobes small, broad, very obtuse. Petals about ½ line diameter. Stamens 5 or 6, none opposite the centre of the petals; filaments filiform; anther-cells parallel, opening longitudinally; connective-gland inconspicuous. Ovary 2-celled, with 16 to 20 ovules closely packed round an oblong somewhat peltate placenta.

Queensland. Moreton Island, *F. Mueller*. This differs chiefly from *B. linifolia* in its slender habit and foliage and small narrow flowers. Both are nearly allied to *B. frutes-*

cens, Linn., a common species in the Eastern Archipelago and S. China, distinguished from them chiefly by the more open calyx, and the ovary almost always 3-celled.

SECTION IV. HARMOGIA.—Stamens few or numerous, but none opposite the centres of the petals; filaments filiform or rarely clavate; anther-cells distinct, nearly globular, parallel or divergent, deeply furrowed and opening more or less in longitudinal slits in the furrows. Ovary 3-celled, with several ovules in each cell. Flowers solitary or 2, 3 or more together on a short common peduncle.

The anthers in this and the following section are intermediate, as it were, between those of *Euryomyrtus* and of *Babingtonia*. In the first four species of *Harmogia*, they are very nearly those of *Euryomyrtus*, except that the cells are more globular and do not open quite so deeply in 2 valves; in *B. densifolia* and *B. Behrii*, the slits are almost shortened into pores. In *Oxymyrrhine*, the dehiscence is nearly the same, but the furrows of the cells are as deep as those which separate the cells, so that the anthers appear to have 4 cells similar to the two of *Babingtonia*, and the slits are shortened into pores. In some specimens, however, it is often very difficult to draw any marked line between the several modifications.

21. **B. camphorata,** *R. Br. in Bot. Mag. t.* 2694. Erect, with somewhat virgate branches. Leaves from linear-oblong or slightly cuneate to broadly oblong or almost obovate, flat, obtuse or nearly so, $1\frac{1}{2}$ to 3 lines long. Flowers rather small, solitary or in clusters of 2 or 3, on short pedicels with concave very deciduous bracteoles at the base, without any conspicuous common peduncle. Calyx-tube campanulate, not 1 line long; lobes small, broadly ovate, petal-like, half as long as the petals. Petals about $1\frac{1}{4}$ line long, almost clawed. Stamens about 15, none opposite the centre of the petals; filaments filiform; anther-cells nearly globular, but parallel and opening more or less deeply in longitudinal slits. Ovary flat-topped, 3-celled, with 10 to 20 ovules in each cell round a small slightly peltate placenta; style inserted in a deep tubular central depression. Capsule slightly convex.—DC. Prod. iii. 230; F. Muell. Fragm. iv. 70; *Leptospermum imbricatum*, Sm. in Trans. Linn. Soc. vi. 300; *Camphoromyrtus Brownii*, Schauer in Linnæa, xvii. 240.

N. S. Wales. Paramatta, *R. Brown, Woolls.*

22. **B. virgata,** *Andr. Bot. Rep. t.* 598. Usually tall erect and loosely branched, attaining 10 to 12 ft., rarely low and diffuse. Leaves from linear-lanceolate to narrow-oblong, flat and often 1- or 3-nerved, usually acute and $\frac{1}{2}$ to 1 in. long, but in some specimens all under $\frac{1}{2}$ in. long, and occasionally some or nearly all obtuse both in the short- and long-leaved forms. Flowers small in the upper axils, usually several together in a loose umbel, on a common peduncle of 2 to 4 lines, the pedicels varying from 1 to 3 lines. Calyx-tube turbinate, at length hemispherical, about $1\frac{1}{2}$ lines diameter; lobes short and broad, the midrib more or less produced into a conical point or protuberance. Petals about $1\frac{1}{2}$ lines diameter. Stamens 5 to 15, none opposite the centre of the petals; filaments filiform; anthers didymous, the cells globular, furrowed, opening in short slits; connective thickened into a gland almost as long as the cells. Ovary 3-celled, with 15 to 20 ovules in each cell round a peltate placenta. Capsule nearly flat-topped. Seeds usually angular. Embryo with the slender inflected end very short, with 2 small ovate cotyledons.—DC. Prod. iii. 229; Bot. Mag. t. 2127; Lodd. Bot. Cab. t. 341; Colla, Hort. Ripul. t. 6; F. Muell. Fragm. iv. 69; *Leptospermum*

virgatum, Forst. Char. Gen. 48; *Melaleuca virgata*, Linn. fil. Suppl. 343; *Harmogia virgata*, Schau. in Linnæa, xvii. 238; *Camphoromyrtus pluriflora*, F. Muell. in Trans. Vict. Inst. i. 123; *Harmogia umbellata*, F. Muell. Fragm. ii. 31; *Bæckea umbellata*, F. Muell. Fragm. iv. 69; *Babingtonia virgata*, F. Muell. Fragm. iv. 74.

N. Australia. Sandstone precipices, Victoria river, rare, *F. Mueller.*

Queensland, *Bidwill;* Upper Brisbane river, *F. Mueller;* Moreton Bay, *C. Stuart;* Pine river, *Fitzalan;* Rockhampton, *Dallachy.*

N. S. Wales. Grose and Hawkesbury rivers, *R. Brown;* Blue Mountains, *A. Cunningham;* northward to Macleay river, *Beckler.*

Victoria. On the Snowy and Tambo rivers, *F. Mueller.*

The species is also in New Caledonia.—*B. parvula*, DC. Prod. iii. 229 (*Leptospermum parvulum*, Labill. Sert. Austr. Caled. 62. t. 61. *Harmogia parvula*, Schauer in Linnæa, xvii. 238), also from New Caledonia, is a slight variety, only differing in the shorter more obtuse leaves. The same variety, with even still shorter oblong leaves, is amongst the Queensland specimens communicated by Bidwill.

23. **B. crenatifolia,** *F. Muell. Fragm.* iv. 70. A tall shrub, attaining 10 ft., with erect or pendulous branches. Leaves ovate obovate oblong or almost orbicular, flat, obtuse, minutely crenulate, 2 to 3 lines long. Flowers nearly of *B. virgata*, usually 2 or 3 together, on a common slender peduncle of 2 or 3 lines; pedicels also slender, with minute, very fugacious bracteoles at their base. Calyx-tube hemispherical, fully 1½ lines broad; lobes very short and broad, without any or only a very small dorsal protuberance. Petals fully 1½ lines diameter. Stamens 10 to 15, none opposite the centre of the petals; filaments thickened into a gland at or a little below the top; anthers globose, didymous, the cells opening in slits in the deep furrows. Ovary 3-celled, with 15 to 20 ovules in each cell round a peltate placenta; style rather deeply immersed.—*Camphoromyrtus crenulata*, F. Muell. in Trans. Vict. Inst. i. 123; *Harmogia crenulata*, F. Muell.; Miq. in Nederl. Kruidk. Arch. iv. 148.

Victoria. Along springs and rivulets, Buffalo Range, *F. Mueller.*

24. **B. Cunninghamii,** *Benth.* Branches slender and apparently diffuse. Leaves obovate or orbicular, thick, but flat or nearly so, and very obtuse, mostly under 1 line diameter. Flowers small, either solitary, on slender pedicels of about 1 line, with 2 small fugacious bracteoles at the base, or 2 or 3 together, on a short common peduncle, with a similar bracteole at the base of each pedicel. Calyx-tube at first turbinate, but soon hemispherical; lobes short and broad, with a thick conical point or protuberance either dorsal or nearly terminal. Petals about 1 line diameter. Stamens about 5, none opposite the centre of the petals; filaments slightly thickened near the end; anthers globose, didymous, the cells opening in short slits in the deep furrows. Ovary 3-celled, with above 10 ovules in a ring round the placenta; style deeply immersed. Seeds angular, but not seen ripe.—*Harmogia Cunninghamii*, Schau. in Walp. Rep. ii. 921.

N. S. Wales. Bushy forest country, W. of Wellington Valley, *A. Cunningham.*

25. **B. densifolia,** *Sm. in Trans. Linn. Soc.* iii. 260. Branches rather slender, but rigid and virgate. Leaves crowded and decussate on the smaller branches, linear, slender, semiterete or concave, obtuse or with a minute re-

curved point, mostly 2 to 3 lines long. Flowers solitary in the upper axils, often forming short terminal leafy racemes or corymbs. Pedicels 1 to 2 lines long, with a pair of small deciduous bracteoles below the middle. Calyx-tube broadly turbinate; lobes short, broadly triangular. Petals about 1½ lines diameter. Stamens usually 8 or 9, but sometimes as many as 12, none opposite the centre of the petals; filaments filiform; anthers nearly globular, the cells unequally furrowed and opening in the furrows in short slits; connective-gland conspicuous or small, or wholly disappearing. Ovary flat-topped, usually 3-celled, with about 8 ovules in each cell in the ordinary form; style shortly immersed. Seeds angular; embryo with the slender cotyledonar end short and appressed against the radicle, otherwise straight.—DC. Prod. iii. 230; F. Muell. Fragm. iv. 71; *B. fasciculata*, Sieb. in Spreng. Syst. Cur. Post. 149; *Harmogia densifolia*, Schau. in Linnæa, xvii. 238; *Babingtonia densifolia*, F. Muell. Fragm. iv. 74; *Harmogia Baueriana*, Schau. in Walp. Rep. ii. 921, from the character given.

N. S. Wales. Port Jackson to the Blue Mountains, *R. Brown*, *Sieber*, *n.* 279, and others; New England, *C. Stuart.*

The structure of the anthers in this species is so nearly that of the following section, that I feel doubts as to having correctly placed it in the present one. It varies much in the size of the flower, the length of the pedicel, and attenuate base of the calyx, and the number of ovules. *Harmogia propinqua*, Schau. in Walp. Rep. ii. 921, has smaller flowers, the calyx-tube almost close above the bracteoles, and the connective-gland very small or none. *Bæckea Novo-anglica*, F. Muell. Fragm. iv. 71, or *Babingtonia Novo-anglica*, F. Muell. l.c. 74, has rather larger flowers, the calyx attenuate into a pedicel more distinct than in *H. propinqua*, shorter than in the common form, the connective-gland small or none, and ovules more numerous than usual; the stamens also vary in number and in the degree of dehiscence of the anther-cells; but I find, after examining a considerable number of specimens, that these differences pass so gradually one into the other, that I am unable to characterize the several forms even as distinct varieties.

26. **B. Behrii,** *F. Muell. Fragm.* iv. 68. A tall handsome shrub, with erect virgate branches. Leaves erect or spreading, rather distant, linear, semiterete or triquetrous, with a rather thick recurved point, 2 to 4 lines long. Flowers solitary and pedicellate, or very rarely 2 on a common peduncle, one pedicellate, the other sessile. Bracteoles so fugacious as to be rarely seen. Calyx-tube turbinate; lobes exceedingly short and broad. Petals fully 1½ lines diameter. Stamens 8 to 15, none opposite the centre of the petals; filaments clavate; anthers didymoûs, the cells unequally furrowed and opening in the furrows in short slits. Ovary flat-topped, 3-celled, with 10 to 15 ovules in each cell round a somewhat peltate placenta; style immersed in a rather deep tubular depression and scarcely projecting above it. —*Camphoromyrtus Behrii*, Schlecht. Linnæa, xx. 651.

N. S. Wales. On the Lachlan, Murrumbidgee, and Darling, according to *F. Mueller.* (I have not seen the specimens.)

Victoria. Murray Desert, *F. Mueller;* Wimmera, *Dallachy.*

S. Australia. Port Lincoln, *R. Brown;* St. Vincent's and Spencer's Gulf to the Murray, *Behr, F. Mueller*, and others; and inland to Lake Gillies, *Burkett.*

W. Australia. Specimens from Lucky Bay, *R. Brown*, exactly like the S. Australian ones, and apparently distinct from *B. uncinella.*

The species is nearly allied to *B. densifolia*, differing in foliage, in its rather larger flowers, the remarkably short calyx-lobes, and the clavate filaments, and the slits of the anther-cells, apparently shorter, bringing it still nearer to the following section.

SECTION V. OXYMYRRHINE.—Stamens few or numerous, but none opposite the centre of the petals (except in *B. polyandra*); filaments filiform. Anther-cells more or less united at the base, didymous, deeply furrowed and opening in pores in the furrows, giving the whole anther the appearance of 4 collateral globular cells, either all equal or the 2 central ones smaller, the connective-gland sometimes appearing like a fifth. Ovary 3-celled, with numerous ovules in each cell.

This section might almost be united with the previous one, but the anthers appear to me to form a nearer approach to those of *Babingtonia.* The species are all western, whilst the *Harmogias* are eastern.

27. **B. uncinella,** *Benth.* Branches rather slender. Leaves erect or spreading, rather distant, linear or linear-cuneate, semiterete or triquetrous, with short recurved points, 2 to 3 lines long. Flowers usually 3 together, on short, slender peduncles, the pedicels longer than the calyx-tube. Bracteoles none or exceedingly fugacious. Calyx-tube turbinate, about 1 line long; lobes short and obtuse. Petals little more than 1 line diameter. Stamens about 8, none opposite the centre of the petals; filaments short, not clavate; anther-cells shortly united, deeply furrowed, giving the appearance of 4 collateral lobes of the anther, the 2 central ones smaller than the others, and opening in the furrows in pores or very short slits; connective-gland small. Ovary 3-celled, with 15 to 20 ovules in each cell, round a small peltate placenta; style rather deeply immersed.

W. Australia. Plains E. of Stokes Inlet, *Maxwell.* F. Mueller, Fragm. iv. 69, thinks that this may be a western variety of *B. Behrii,* but it appears to me to differ in inflorescence, in stamens, and in the number of ovules, as well as in some points in the calyx and general aspect. I have, however, only seen two specimens.

28. **B. polyandra,** *F. Muell. Fragm.* iv. 72. Branches slender but rigid. Leaves linear, semiterete or triquetrous, decussate on the smaller branches, obtuse or with a minute recurved point, rarely exceeding 2 lines. Flowers solitary, on pedicels of 1 to 2 lines, articulate with lanceolate bracteoles close under the calyx or at very little distance from it. Calyx-tube turbinate-campanulate, 5-ribbed; lobes short, erect, acute, herbaceous or slightly scarious on the margin and denticulate at the base. Petals above 1½ lines diameter. Stamens 20 to 25, in a single row, those opposite the centre of the petals present and rather larger than the others; filaments thick; anther-cells deeply furrowed, opening in pores in the furrows, the whole anther showing 4 globular, collateral lobes round the more or less prominent connective-gland. Ovary 3-celled, with numerous ovules in each cell in a ring round the peltate placenta; style immersed in a rather deep central depression.—*Oxymyrrhine gracilis,* Schau. in Linnæa, xvii. 240; *Babingtonia gracilis,* F. Muell. Fragm. iv. 74.

W. Australia. King George's Sound, or to the eastward, *R. Brown, Baxter;* sea-coast, E. of Stokes Inlet, and inland from Orleans Bay, *Maxwell.*

29. **B. corynophylla,** *F. Muell. Fragm.* iv. 72. Branches apparently loose and elongated. Leaves not crowded, linear-clavate or cuneate, thick, channelled above, more or less recurved at the end and often laterally compressed, very obtuse, 1½ to nearly 3 lines long. Peduncles short, crowded

at the ends of the branches, bearing each 2, 3 or rarely only 1 flower, on pedicels of 1 line or rather more, the bracteoles very small and narrow. Calyx-tube slightly turbinate, about 1 line long; lobes longer and less obtuse than in most allied species. Petals nearly 1½ lines diameter. Stamens 6 to 8, none opposite the centres of the petals; anther-cells deeply furrowed, opening in pores or short slits in the furrows, the whole anther showing 4 globular, collateral lobes round the globular connective. Ovary flat-topped, 3-celled, with about 10 to 12 ovules in each cell round a slightly peltate placenta; style rather deeply immersed.—*Harmogia corynophylla*, F. Muell. Fragm. ii. 30; *Babingtonia corynophylla*, F. Muell. Fragm. iv. 74.

W. Australia, *Drummond, 5th Coll. n.* 127; Fitzgerald ranges, *Maxwell.*

30. **B. pachyphylla,** *Benth.* Branches apparently loose and elongated. Leaves not crowded, obovate-oblong, very thick and obtuse, 1 to 1½ lines long. Pedicels 1½ to 2 lines long or rather more, solitary or 2 or 3 together on a short common peduncle, with small very fugacious bracteoles at their base. Calyx-tube turbinate or at length nearly globular, about 1 line long; lobes short, broad, obtuse, scarious only at the margin. Petals about 1½ lines diameter. Stamens 5 or fewer, none opposite the centre of the petals; filaments filiform; anther-cells united at the base, deeply furrowed, opening shortly in the furrows, giving the appearance of 4 collateral lobes, the 2 central ones smaller than the lateral ones as in *D. uncinella.* Ovary 3-celled with 8 to 10 ovules in each.

W. Australia. In the interior from the south coast, *Maxwell.* A single specimen in Herb. F. Mueller, which I am unable to refer to any other species. The structure of the flowers is nearly that of *B. corynophylla*, with the foliage and habit more of *B. floribunda.*

31. **B. crispiflora,** *F. Muell. Fragm.* iv. 72. Much branched and often somewhat glaucous. Leaves broadly ovate or obovate, erect or spreading, very concave and obtuse or the midrib slightly produced into a point, 1 to 1½ lines long. Flowers solitary, on pedicels much longer than the leaves, articulate above the middle with a pair of linear leaf-like bracteoles. Calyx-tube urceolate, about 1 line long; lobes short, broad, and rather thick. Petals rather above 1 line diameter, usually much undulate on the margin. Stamens 15 to 20, none opposite the centre of the petals; filaments thickened at the base; anther-cells deeply furrowed, opening shortly in the furrows, giving the appearance of 4 collateral equal lobes round the globular connective as in *D. polyandra.* Ovary 3-celled, with numerous ovules in each cell round a peltate placenta; style rather deeply immersed. Seeds angular; embryo with the short slender cotyledonar end appressed against the radicle, but otherwise straight.—*Harmogia crispiflora*, F. Muell. Fragm. ii. 31; *Babingtonia crispiflora*, F. Muell. Fragm. iv. 74.

W. Australia, *Drummond, 1st Coll. and 3rd Coll. n.* 38.

Section VI. Babingtonia.—Stamens few or more frequently numerous, none opposite the centre of the petals, or rarely forming a complete ring when above 20, filaments filiform or clavate; anther-cells united into an obcordate or almost globular anther, and opening in terminal pores or short slits. Ovary 3-celled with several, often numerous, ovules round a more or less peltate placenta. Flowers solitary or umbellate.—All Western species.

32. **B. camphorosmæ,** *Endl. in Hueg. Enum.* 51. Either low and spreading, or erect and attaining 2 ft. or more; branches usually long and virgate, with numerous short branchlets. Leaves crowded on the branchlets, in some specimens occasionally alternate, linear, semiterete or triquetrous, obtuse or with a minute straight point, mostly 1½ to 2 lines long, or those on the main branches longer and distant. Flowers white or pink, on very short pedicels, solitary or more frequently clustered on a very short common peduncle, with a small deciduous bracteole at the base of each pedicel, the clusters usually forming a long terminal usually one-sided leafy raceme. Calyx-tube broadly turbinate or at length urceolate, about 1 line long; lobes broad, short, scarious, and minutely denticulate, the thick centre sometimes produced into a short conical point. Petals above 1½ lines diameter. Stamens 10, none opposite the centre of the petals; filaments thick, continuous with the connective; anthers thick, obcordate or almost didymous, the cells not furrowed, opening in small terminal pores. Ovary 3-celled, with about 10 ovules in each cell on a placenta ascending from the base; style immersed in a deep tubular central depression. Embryo with very minute ovate cotyledons.—*Babingtonia camphorosmæ*, Lindl. Bot. Reg. 1842, t. 10; Schau. in Pl. Preiss. i. 109.

W. Australia. King George's Sound to Swan River, *Fraser, Drummond*, 1*st Coll.*, *Preiss, n.* 347, 349, and others. Vasse River, *Preiss, n.* 348. One of the grounds on which the genus *Babingtonia* was formed was on the supposed perforation of the ovary in the centre through which the style passed in direct continuation of the placenta, but this appears to be a mistake. The style in this and many other capsular *Myrtaceæ* is ventrally or almost basally attached to the carpels, as in *Labiatæ, Chrysobalaneæ*, many *Rutaceæ*, etc., but the carpels are united so as to form a ring or slender tube close round the style but free from it.

33. **B. pulchella,** *DC. Prod.* iii. 230, *and Mem. Myrt.* t. 13? Erect with numerous slender branches. Leaves slender, semiterete or triquetrous, mostly about 1 line long, crowded or decussate on the smaller branches. Flowers small, mostly solitary, on slender pedicels longer than the leaves and articulate below the middle with a pair of leaf-like bracteoles. Calyx-tube rather broad, about ¾ line long; lobes short, obtusely triangular. Petals about 1 line diameter. Stamens 25 to 30 in a single dense ring; filaments rather thick; anthers nearly globular, the cells united, furrowed, opening in short slits in the furrows. Ovary flat-topped, 3-celled, with many ovules in each cell round a peltate placenta; style rather deeply immersed.—*Ericomyrtus Drummondii*, Turcz. in Bull. Mosc. 1847. i. 155.

W. Australia, *Drummond*, 3*rd Coll. n.* 36.
I have not seen authentic specimens of De Candolle's plant, but this is the only species I have found to agree with his short diagnosis and figure in everything except the bracteoles, which, however, De Candolle may have considered as stem leaves.

34. **B. pygmæa,** *R. Br. Herb.* Slender and erect or spreading, from a few inches to nearly 1 ft. high. Leaves linear-terete, obtuse or almost acute, 1 to 1½ or rarely 2 lines long. Peduncles in the upper axils filiform, bearing 1 or 2 very small flowers on filiform pedicels usually exceeding the leaves. Calyx about ½ line long, with 5 short herbaceous teeth. Petals about ¼ line diameter. Stamens about 10, none opposite the centre of the petals; anthers

nearly globular, the cells connate and opening at the top in short pores. Ovary 3-celled, with several ovules in each cell; style deeply immersed.

W. Australia. King George's Sound, *R. Brown.* Near *R. pulchella,* but with very much smaller flowers and fewer stamens.

35. **B. corymbulosa,** *Benth.* Small, with numerous slender branchlets. Leaves oblong or almost linear-cuneate, rather thick, concave, obtuse, mostly under 1 line long. Flowers very small, in little terminal leafy corymbs, solitary or 2 or 3 together in each axil on a slender peduncle shortly exceeding the leaves, the pedicels often nearly as long as the common peduncle Calyx-tube ovoid, about ½ line long; lobes very short and obtuse. Petals about ½ line diameter. Stamens 10 to 15, none opposite the centre of the petals; anthers nearly globular, the cells united nearly to the top and opening in oblong pores or short slits. Ovary nearly flat-topped, 3-celled, with numerous ovules in each cell round a peltate placenta; style deeply immersed.

W. Australia, *Drummond,* 5*th Coll. Suppl. n.* 25.

36. **B. floribunda,** *Benth.* Nearly allied to *B. pentandra*, and perhaps a variety, but the leaves are much shorter and thicker, oblong or almost obovoid, and under 1 line long except on the main luxuriant branches, where they are linear-distant and appressed. Peduncles longer than the leaves, bearing 1 to 5 flowers, larger than in *B. pentandra*, with similar stamens, but there appear to be always 10, none, however, opposite the centre of the petals. Ovary flat-topped, 3-celled, with 8 to 10 ovules in each cell round a peltate placenta; style deeply immersed.

W. Australia, *Drummond, n.* 9, 138, *and* 3*rd Coll. n.* 37.

37. **B. pentandra,** *F. Muell. Fragm.* iv. 72. Erect, with rather slender virgate branches of 1 to 2 ft., and numerous small branchlets. Leaves linear, semiterete or triquetrous, very obtuse, ¾ to 1½ lines long, decussate on the smaller branches. Peduncles short, bearing 1 to 3 flowers on pedicels of about 2 lines, with small linear deciduous bracteoles at their base. Calyx-tube short, about ¾ line diameter; lobes broad, very obtuse, scarious with thickened centres. Petals at least 1 line diameter. Stamens 5 or sometimes 6, not opposite the petals; filaments short; anthers nearly globular, the cells united nearly to the top and slightly furrowed, opening in terminal pores; connective thick. Ovary flat-topped, 3-celled, with 4 to 6 ovules in each cell round a small placenta; style deeply immersed in a tubular depression of the ovary.—*Harmogia pentandra*, F. Muell. Fragm. ii. 31; *Tetrapora Preissiana*, Schau. in Linnæa, xvii. 283, and in Pl. Preiss. i. 107; *Babingtonia pentandra* and *B. Preissiana*, F. Muell. Fragm. iv. 74.

W. Australia, *Drummond,* 5*th Coll. n.* 117, *Preiss. n.* 345; Gardiner's River, Plantagenet and Stirling ranges, *Maxwell.*

38. **B. pentagonantha,** *F. Muell. Fragm.* iv. 73. A large bushy shrub of 6 to 8 ft., with numerous small erect branches. Leaves decussate on the smaller branchlets, broadly ovate or orbicular, rather thick, concave and keeled, very obtuse, rarely exceeding 1 line. Flowers solitary in the upper axils on very short pedicels, articulate with linear bracteoles about the middle. Calyx-tube 1½ lines long, very prominently 5-angled or almost

winged; lobes short, broad, with scarious margins. Petals not above 1 line diameter. Stamens 15 to 20, none opposite the centres of the petals; anther-cells united, globular, opening in terminal pores or short slits, connective-gland globular and prominent. Ovary flat or concave at the top, 2-celled, with 8 to 10 ovules in each cell round a peltate placenta; style not very deeply immersed.—*Babingtonia pentagonantha*, F. Muell. Fragm. iv. 74.

W. Australia. Murchison river, *Oldfield*, and apparently the same species but the specimens not in flower, Sharks' Bay, *Denham;* Dirk Hartog's island, *Milne.* The prominent angles of the calyx are much more conspicuous in this than in *B. polyandra*, and readily distinguish the species from all others.

39. **B. robusta,** *F. Muell. Fragm.* iv. 72. A straggling shrub of 3 to 6 ft. Leaves linear or oblong, semiterete or concave, thick, very obtuse, mostly 2 to 4 lines long. Flowers solitary on a pedicel of 2 to 4 lines, articulate with 2 deciduous bracteoles about the middle, or rarely 2 or 3 together on a short common peduncle. Calyx-tube nearly 2 lines long, turbinate, smooth or obscurely angled, with short broad rather thick lobes. Petals not large. Stamens 10 to 20, none opposite the centre of the petals; filaments tapering at the end below the thick connective; anthers obcordate or almost didymous, the cells opening in rather large oblong terminal pores. Ovary 3-celled (or rarely 2-celled?) with 6 to 8 ovules in each cell round a small peltate placenta; style deeply immersed.—*Babingtonia robusta*, F. Muell. Fragm. iv. 74.

W. Australia. Sandy plains, Murchison river, *Oldfield, Drummond, 6th Coll. n.* 61.

40. **B. ovalifolia,** *F. Muell. Fragm.* iv. 72. Erect, attaining about 3 ft., with rather short virgate branches. Leaves erect or spreading, ovate, oblong or broadly linear, concave, thick, obtuse, 1½ to 3 lines long. Flowers large for the genus, solitary on pedicels of 2 to 3 lines, articulate about or above the middle, with 2 linear or oblong concave deciduous bracteoles. Calyx-tube very broadly turbinate or hemispherical, about 2 lines diameter, more or less rugose; lobes short, broad, very obtuse, thick in the centre with broad scarious margins. Petals 2½ to 3 lines diameter. Stamens 15 to 20 or even more, those opposite the centre of the petals often wanting; filaments filiform or slightly flattened, the inflated summit continuous with the thickened connective; anthers broadly clavate, the cells scarcely distinct, opening in terminal pores. Ovary convex, 3-celled, with 8 to 10 or sometimes more ovules in each cell round a small peltate placenta; style immersed to half the depth of the ovary. Capsule very convex.—*Harmogia ovalifolia*, F. Muell. Fragm. ii. 32; *Babingtonia ovalifolia*, F. Muell. Fragm. iv. 74.

W. Australia, *Drummond, 5th Coll. n.* 124; E. Mount Barren, *Maxwell.* Drummond's specimens have longer leaves and larger flowers than the single one of Maxwell's.

41. **B. subcuneata,** *F. Muell. Fragm.* iv. 73. Erect, attaining 4 to 6 ft., with virgate branches. Leaves erect or slightly spreading, broadly obovate-cuneate, concave or folded, obtuse or the midrib slightly produced, rather thick, mostly 1½ to 2 lines long. Flowers solitary on short thick pedicels, with a pair of very deciduous bracteoles below the middle. Calyx-tube turbinate-campanulate or hemispherical, rather thick; lobes short and broad, thick, with more or less scarious margins. Petals about 1½ lines

diameter. Stamens about 20, none opposite the centre of the petals; filaments inflated at the summit and continuous with the thickened connective; anthers broadly obcordate, the cells opening in terminal pores. Ovary very convex, 3-celled, with several ovules in each cell; style shortly immersed.—*Babingtonia subcuneata*, F. Muell. Fragm. iv. 74.

W. Australia. Sandy plains, Murchison river, *Oldfield.*

42. **B. grandiflora,** *Benth.* Branches elongated, with numerous small branchlets. Leaves linear, semiterete or triquetrous, obtuse or scarcely mucronate, clustered or decussate on the smaller branches, 2 to 4 lines long, the floral ones distant. Flowers large, solitary, on pedicels of 2 to 4 lines, articulate above the middle with a pair of linear bracteoles. Calyx-tube very open, above 2 lines diameter, truncate, with 5 prominent angles or short teeth. Petals nearly 3 lines diameter. Stamens 15 to 20; filaments thick and dilated, forming an uninterrupted ring, but quite free from each other; anthers large and thick, ovoid or oblong, truncate at the top, the connective forming a short protuberance at the base; the cells quite united, opening in small terminal pores. Ovary flat, 3-celled, with numerous ovules in each cell; style shortly immersed.

W. Australia. Between Moore and Murchison rivers, *Drummond, 6th Coll. n.* 60.

(*B. spinosa,* Sieb. in Spreng. Syst. Cur. Post. 149, is unknown to me, and probably no *Bæckea.*)

14. ASTARTEA, DC.

Calyx-tube turbinate or hemispherical, adnate to the ovary at the base, the free disk-bearing part broad and open; lobes 5, imbricate, continuous with the tube, scarious on the edges, persistent. Petals 5, broadly obovate or orbicular, spreading. Stamens usually above 20 in a single row, more or less united at the base into 5 clusters opposite the calyx-lobes or into a ring scarcely interrupted opposite the petals; anther-cells distinct, opening in longitudinal or transverse slits. Ovary 3-celled, with several ovules in each cell in 2 rows or in a ring round a more or less peltate placenta; style filiform, inserted in a slight or shortly tubular depression in the centre of the ovary; stigma capitate. Capsule almost entirely inferior, opening at the top loculicidally in 3 valves. Seeds more or less angular, with a thin testa; embryo probably as in *Bæckea*, but not seen perfect.—Heath-like glabrous shrubs. Leaves small, opposite, narrow, entire. Flowers small, white or pink, solitary in the axils, nearly sessile, or on a peduncle or pedicel articulate near the base, with 2 small bracteoles at the articulation.

The genus is entirely Australian, only differing from the section *Schidiomyrtus* of *Bæckea* in the stamens more or less united at the base opposite the calyx-lobes, not opposite the petals as in *Melaleuca* and its allies.

Flowers distinctly pedicellate. Anthers opening longitudinally.
  Flowers rather large. Filaments dilated, forming a nearly complete ring at the base . . . . 1. *A. ambigua.*
  Flowers small. Filaments scarcely dilated, united at the base into 5 distinct clusters . . . . 2. *A. fascicularis.*
Flowers nearly sessile. Anthers opening in transverse slits . . 3. *A. intratropica.*

1. **A. ambigua,** *F. Muell. Fragm.* ii. 32. An erect or spreading shrub of 3 or 4 ft. Leaves linear, linear-cuneate or here and there almost lanceo-

late, rigid, concave, obtuse, or with a small often recurved point, 2 to 3 or rarely 4 lines long. Flowers large for the genus, on pedicels 2 to 3 lines long, articulate with 2 minute bracteoles near the base. Calyx-tube broad, almost hemispherical, about 2 lines diameter; lobes semiorbicular. Petals 2 lines diameter. Stamens about 20; filaments of unequal length but all short, dilated, more or less united in a ring either complete or broken opposite the centre of the petals; anther-cells parallel, opening longitudinally in front of the summit of the filament; connective-gland globular. Ovary nearly flat, with 6 to 8 ovules in each cell; style in a very slight central depression.

W. Australia. E. Mount Barren, Mount Bland, and Phillips Ranges, *Maxwell.*

2. **A. fascicularis,** *DC. Prod.* iii. 210. An erect heath-like shrub attaining 8 to 10 ft., rarely low and diffuse. Leaves linear, semiterete or triquetrous, obtuse or mucronulate, usually 2 or 3 lines long and rather slender, but varying from under 2 lines to above 4 lines, thick or slender and almost filiform, often densely clustered in the axils but sometimes distant. Flowers small, on pedicels of 1 to 2 or rarely 3 lines, articulate with a pair of small bracteoles above the middle. Calyx-tube broadly turbinate-campanulate, about 1 line diameter, or rather more when in fruit; lobes scarious on the margin, the centre thickened, and sometimes produced into a conical protuberance or point. Petals usually about 1½ lines diameter, but variable in size. Stamens in 5 distinct clusters, usually of 5 or 6 each, but sometimes only 3 or 4, or 7 or 8 in each cluster; anthers small, didymous, the cells parallel, opening in broad longitudinal slits. Ovary flat or slightly convex, with 6 to 10 ovules in each cell; style in a short central tubular depression.—*Melaleuca fascicularis*, Labill. Pl. Nov. Holl. ii. 29. t. 170; *Leptospermum dubium*, Spreng. Syst. ii. 492; *Bæckea affinis*, Endl. in Hueg. Enum. 51, acccording to Schau. *Astartea leptophylla*, *A. fascicularis*, *A. laricifolia*, *A. scoparia*, *A. aspera*, *A. glomerulosa*, *A. corniculata*, and *A. Endlicheriana*, Schau. in Pl. Preiss. i. 113 to 115.

W. Australia. King George's Sound, Lucky Bay, *R. Brown*, *Labillardière.* Common from the S. coast to Swan and Murchison rivers, *Fraser* and others; *Preiss, n.* 150, 156, 158, 159, 162, 163, 165, 361; *Drummond, 1st. Coll.; 2nd Coll. n.* 60, 70; *3rd Coll. n.* 35; *4th Coll. n.* 52; *5th Coll. n.* 125, 128.

The species is certainly variable as to the size of the flowers, the thick or fine leaves, the greater or less prominence of the appendage or thickening of the calyx-lobes, the number of stamens, etc., but I have been quite unable to sort the very numerous specimens before me into distinct varieties. Some from King George's Sound, *Harvey*, have remarkably small flowers, the petals under 1 line diameter; other small-flowered specimens have the dorsal point of the calyx-lobes much elongated, but do not otherwise differ. In some of Drummond's and Maxwell's specimens from the districts east of King George's Sound the flowers are altogether much larger with the petals nearly 2 lines diameter. Labillardière's figure represents a coarse form, with the leaves less clustered than usual. Drummond's 2nd Coll. n. 60, are like it, but more etiolated.

3. **A. intratropica,** *F. Muell. Fragm.* i. 83. A shrub of several ft. with erect virgate branches. Leaves linear, triquetrous or semiterete, obtuse, rather thick, mostly 3 to 4 lines long, narrowed at the base, not clustered. Flowers almost sessile, with 2 narrow very deciduous bracteoles. Calyx-tube turbinate-campanulate, about 1½ lines diameter, glandular-rugose; lobes broad, very obtuse, thickened in the centre, but without any appendage.

Petals above 1 line diameter. Stamens in 5 distinct clusters of 6 to 8 each; anther-cells distinct, globular, opening in transverse slits, the connective gland nearly as large as each cell. Ovary with numerous ovules in each cell.

N. Australia. Ravines of the sandstone table-land at the head of the Roper and Limmen Bight rivers, *F. Mueller.*

15. HYPOCALYMMA, Endl.

Calyx-tube broadly turbinate or almost flat, adnate to the ovary at the base; lobes 5, broad and obtuse, more or less scarious, shorter than the petals. Petals 5, broadly obovate or orbicular, spreading, often persistent. Stamens numerous, not exceeding the petals, very shortly united in a single ring; filaments filiform, in 1 or more rows, persistent; anthers ovate or oblong, the cells parallel, opening longitudinally. Ovary in the bottom of the calyx, inferior half-inferior or wholly superior except the broad base, with or without a central depression round the style, 2- or 3-celled, with 1, 2 or rarely more ovules in each cell, laterally attached or pendulous; style filiform, with a small or capitate stigma. Capsule more or less inferior or enclosed in the calyx-tube, opening loculicidally at the top or in the whole free portion. Seeds solitary or few in each cell, ovoid-oblong, with an oblong lateral hilum; testa usually crustaceous, with a thin inner membrane round the embryo (or wholly membranous ?). Embryo, where known, straight, filling the seed, quite entire, with a small sometimes slightly incurved papilla at the smallest upper end.—Shrubs, either glabrous or with pubescent branches. Leaves opposite, usually larger than in *Bæckea*, entire or with crisped edges. Flowers axillary, in pairs or rarely 3 or 4 together in each axil, sessile or shortly pedunculate, with 3 scarious bracts or bracteoles under each flower, 1 at the top of the common peduncle and 2 under the calyx.

The genus is limited to Western Australia. It connects, in some measure, *Bæckea* and its allies with *Leptospermum*, but differs from both in the staminal arrangement, and, as far as known, in the embryo. The *H. strictum* has sometimes almost the aspect of *Kunzea pauciflora*, which moreover has very frequently many of the leaves opposite, but is readily distinguished by the 5-celled ovary and capsule.

SECTION I. **Eucalymma.**—*Ovary 2- or 3-celled, with 2 or 3 ovules in each cell, the style continuous with the prominent ridges, without any central depression. Flowers in pairs, sessile or on a very short common peduncle.*

Branches pubescent. Leaves oblong-cuneate, very obtuse. Flowers yellow. Ovary 3-celled, slightly prominent . . . . . . 1. *H. xanthopetalum.*
Glabrous. Ovary 2-celled. Flowers not yellow.
Leaves linear-lanceolate, rigid, acute. Ovary scarcely prominent 2. *H. robustum.*
Leaves triquetrous, 1½ to 2½ in. long. Capsule very convex . 3. *H. longifolium.*
Leaves terete or sulcate, under ¾ in. long. Ovary very prominent, free, except the broad base . . . . 4. *H. strictum.*

SECTION II. **Astrocalymma.**—*Ovary 3-celled, with 1 ovule in each cell, prominently 3-angled, the style inserted in a central depression. Flowers closely sessile, in pairs.*

Leaves broadly oblong, very obtuse. (Flowers white or pink ?) 5. *H. tetrapterum.*
Leaves linear-oblong, obtuse or rather acute. Flowers yellowish 6. *H. linifolium.*
Leaves semiterete or triquetrous, 3 or 4 times as long as the small flowers . . . . 7. *H. angustifolium.*
Leaves semiterete or triquetrous, not exceeding the large flowers. 8. *H. ericifolium.*

SECTION III. **Cardiomyrtus.**—*Ovary 3-celled, with 2 or more ovules in each cell, without prominent ridges, the style inserted in a central depression. Flowers pedunculate.*

Ovules 2 in each cell. Leaves orbicular-cordate.
Leaves with crisped or denticulate recurved margins. Ovary slightly convex . . . . 9. *H. cordifolium.*
Leaves flat and entire. Ovary very prominent, free, except the broad base . . . . 10. *H. boroniaceum.*
Ovules 6 to 12 in each cell.
Branches pubescent. Leaves cordate, ovate . . . . 11. *H. Phillipsii.*
Glabrous. Leaves elliptical, rounded or narrowed at the base . 12. *H. hypericifolium.*

SECTION I. EUCALYMMA, Schau.—Ovary 2- or 3-celled, with 2 or 3 ovules in each cell, the style continuous with the prominent raised angles or ridges of the ovary, without any central depression. Flowers in pairs, sessile or on a very short common peduncle.

The want of the central depression of the ovary round the style is exceptional in the first three subtribes of *Leptospermeæ.*

1. **H. xanthopetalum,** *F. Muell. Fragm.* ii. 29. Erect or diffuse, not much branched, attaining 1 or 2 ft., the branches pubescent. Leaves from narrow-oblong to broadly oblong-cuneate or almost obovate, obtuse, minutely denticulate-ciliate, ½ to ¾ in. long, narrowed at the base, but sessile or half stem-clasping. Flowers yellowish, in closely sessile pairs. Bracts orbicular, scarious, covering the calyx-tube. Calyx-tube nearly 2 lines diameter, the lobes half as long as the petals, entire or denticulate-ciliate. Petals persistent, about 1½ lines diameter. Stamens numerous, the filaments almost 2-seriate. Ovary only slightly prominent at the top, with 3 raised angles continuous with the style without any central depression, 3-celled, with 2 ovules in each cell, but 2 of the cells often very small, with semiabortive ovules.—*H. cuneatum,* Turcz. in Bull. Mosc. 1862, ii. 325.

W. Australia. Murchison river and adjoining districts, *Drummond, 6th Coll. n.* 67, *Oldfield.*

H. ciliatum, Turcz. in Bull. Mosc. 1862, ii. 325, is a slight variety with narrower leaves.

2. **H. robustum,** *Endl. in Hueg. Enum.* 50 (under *Leptospermum*). An elegant shrub, of 1 to 2 or 3 ft., with erect, rigid, virgate branches, quite glabrous. Leaves linear or linear-lanceolate, spreading, rigid, acute, ½ to 1 in. long, with a thick broad midrib, but otherwise nearly flat. Flowers peach-coloured, sessile in pairs or very rarely 3 or 4 together, on a very short, thick, common peduncle. Bracts small, lanceolate, concave. Calyx-tube rugose, 1½ to 2 lines diameter; lobes orbicular, scarious, about 1 line diameter. Petals twice as long as the calyx-lobes. Stamens 30 to 40, nearly as long as the petals. Ovary flat-topped, with 2 prominent ridges continuous with the style, without any central depression, 2-celled, with 3 ovules in each cell.—Lindl. Bot. Reg. 1843, t. 8; Schau. in Pl. Preiss. i. 110.

W. Australia. Swan River, *Huegel, Drummond,* 1*st Coll. n.* 141, *Harvey, Oldfield, Preiss, n.* 342.

3. **H. longifolium,** *F. Muell. Fragm.* ii. 28. Very near *H. strictum,* and perhaps a variety. Branches rigid, virgate, glabrous. Leaves linear-

triquetrous, rigid, tapering into a slightly recurved point, 1½ to 2½ in. long. Flowers sessile, in pairs, on an exceedingly short, thick, common peduncle. Fruiting-calyx very flat and broad, nearly 3 lines diameter, the lobes very short and broad. Petals not seen. Capsule very convex, 2-celled. Seeds not seen.

W. Australia. Murchison river, *Oldfield.*

4. **H. strictum,** *Schau. in Pl. Preiss.* i. 111. A bushy glabrous shrub, of 1 to 2 ft., with numerous, erect, virgate branches. Leaves erect or spreading, linear-terete or sulcate, either obtuse and all under ½ in., or rather longer and more acute. Flowers 2 to 4 together, sessile on an exceedingly short, thick, common peduncle, much smaller than in *H. robustum.* Calyx-tube but little above 1 line diameter, the semiorbicular lobes about half as long. Petals rather above 1 line diameter. Stamens usually rather longer than the petals. Ovary very convex, almost free, except the broad base, with 2 prominent ridges continuous with the style, without any central depression, 2-celled, with 2 or 3 ovules in each cell.—*H. Cunninghamii* and *H. asperum*, Schau. l. c.

W. Australia. King George's Sound and adjoining districts, *R. Brown, A. Cunningham, Fraser*, and others, *Drummond, 4th Coll. n.* 53, *Preiss, n.* 331, 332, 334, 335.

Var. *pedunculatum.* Branches more slender and elongated. Leaves slender. Common peduncles 1 to 1½ lines long.—*Drummond, 3rd Coll. n.* 34.

SECTION II. ASTROCALYMMA, Schau.—Ovary 3-celled, with 1 ovule in each cell, prominently 3-angled, but with a central depression in which the style is inserted. Flowers closely sessile, in pairs.

5. **H. tetrapterum,** *Turcz. in Bull. Mosc.* 1862, ii. 325. Apparently a tall shrub, with virgate, more or less 4-angled branches, and quite glabrous. Leaves closely sessile or half stem-clasping, broadly oblong-cuneate, obtuse, mostly about ½ in. long or rather more. Flowers in closely sessile pairs, not so yellow when dry as in the allied species. Bracts broad, shorter than the calyx. Calyx-tube very open, about 2 lines diameter; lobes semiorbicular, half as long as the petals. Petals persistent, about 1½ lines diameter. Stamens almost 2-seriate. Ovary free, except the broad base, prominently 3-angled, with a short depression round the style, 3-celled, with 1 ovule in each cell or rarely a second abortive one. Capsule exceeding the calyx-tube, but enclosed in the persistent petals. Seeds oblong-reniform, with a large lateral hilum; testa crustaceous; embryo apparently entire.

W. Australia. Between Moore and Murchison rivers, *Drummond, 6th Coll. n.* 68.

6. **H. linifolium,** *Turcz. in Bull. Mosc.* 1862, ii. 325. Stems slightly branched, virgate, 1 to 2 ft. high, quite glabrous. Leaves closely sessile, oblong-linear, thick and rigid, obtuse or mucronate-acute, 4 to 8 lines long. Flowers in closely sessile pairs, apparently yellowish. Bracts orbicular, shorter than the calyx. Calyx-tube very open, about 1½ lines diameter; lobes broad, petal-like, fully half as long as the petals. Petals about 1½ lines diameter. Stamens almost 1-seriate. Ovary prominent, broadly and shortly pyramidal, prominently 3-angled, with a central depression round the style, 3-celled, with 1 ovule in each cell.

W. Australia. Between Moore and Murchison rivers, *Drummond, 6th Coll. n.* 65.

7. **H. angustifoliun,** *Endl. in Hueg. Enum.* 50 (under *Leptospermum*). An erect, bushy, glabrous shrub, from about 1 to 3 ft. high. Leaves narrow-linear, rigid, channelled above or semiterete, rarely rather broader and concave, obtuse or acute, ½ to 1 in. long. Flowers white or pale pink, in sessile pairs, but often in the axil of one only of each pair of leaves. Bracts ovate-cordate, scarious, about 1 line long. Calyx-tube broad and flat, nearly 2 lines diameter, with a slightly contracted rim; lobes broad, from ⅓ to ½ as long as the petals. Petals about 1½ lines diameter. Stamens about as long as the petals, in a single row. Ovary pyramidal at the top, with 3 prominent angles and a short tubular depression round the style, 3-celled, with 1 ovule or very rarely a second abortive one in each cell; stigma small. Seeds like those of *H. tetrapterum*, but the embryo not seen perfect.—Schau. in Pl. Preiss. i. 112; *H. suave*, Lindl. Bot. Reg. 1844, Misc. 27.

W. Australia. Swan River to the S. coast, *Huegel; Drummond, 1st Coll. n.* 137, 142; *Preiss, n.* 333, 336, 338, 339, 340, 341, and others.

Var. *densiflorum*. Leaves shorter, inflorescence more dense, almost spicate; flowers smaller; stamens shorter.—*H. scariosum*, Schau in Pl. Preiss. i. 111.—King George's Sound, *Preiss, n.* 330, *Oldfield*. Some of Drummond's specimens, described as *H. suave*, closely connect this variety with the form originally described as *H. angustifolium*.

8. **H. ericifolium,** *Benth.* Glabrous, with erect virgate branches. Leaves linear or linear-clavate, thick, obtusely triquetrous or channelled above, 2 to 4 lines long. Flowers in sessile pairs, much larger than in the allied species, concealing the floral leaves when several together. Bracts broad, about as long as the calyx-tube. Calyx-tube very broad and flat, about 2 lines diameter, the lobes not above ⅓ as long as the petals. Petals above 2 lines diameter. Ovary broadly pyramidal on the top, with 3 prominently raised angles, and a rather deep central depression round the style, 3-celled, with 1 ovule in each cell.

W. Australia. Champion Bay and Vasse River, *Oldfield*.

SECTION III. CARDIOMYRTUS, Schau.—Ovary 3-celled, with 2 or more ovules in each cell, without prominent angles or ridges, and with a central depression in which the style is inserted. Flowers pedicellate, solitary, clustered or 2 or more together on a common peduncle.

9. **H. cordifolium,** *Lehm.; Schau. in Pl. Preiss.* i. 112. A glabrous shrub, of 2 or 3 ft., with long, loose branches, more or less 4-angled, the angles sometimes dilated under the leaves into denticulate wings. Leaves closely sessile, very broadly orbicular-cordate or almost triangular, the margins recurved and more or less crisped or denticulate, all under ¼ in. diameter in some specimens, about ½ in. in others. Pedicels slender, solitary or more frequently 2 or 3 together, on a short, slender, common peduncle, but the proportions of the peduncle to the pedicels very variable, the whole inflorescence rarely as long as the leaf. Bracts very small and narrow. Calyx-tube very flat and open, about 1 line diameter; lobes herbaceous, orbicular, as long as the tube. Petals about twice the calyx-lobes. Ovary slightly convex, with a central depression round the style, but without prominent ridges, 3-celled, with 2 ovules in each cell.

W. Australia. King George's Sound and to the eastward, *R. Brown, Preiss, n.* 154,

Milne, Harvey, Oldfield, Drummond, n. 41, *2nd Coll. n.* 59, *3rd Coll. n.* 55, *4th Coll. n.* 54.

10. **H. boroniaceum,** *F. Muell. Herb.* Stems several, from a woody stock, simple or slightly branched, mostly about 1 ft. high, glabrous. Leaves closely sessile, orbicular-cordate, flat and quite entire, very obtuse, ¼ to ½ in. diameter. Pedicels slender, usually several together in an axillary cluster or on a very short common peduncle. Bracts very small, concave and coloured, at the base of the pedicels; bracteoles under the calyx sometimes rather larger. Calyx-tube broad and flat, about 1 line diameter; lobes richly coloured, 1½ lines long or more. Petals of a rich red when dry, 3 to 4 lines long. Stamens exceedingly numerous, in more than one row. Ovary obovoid, very much raised, free except the broad base, without raised angles, but with a central depression round the style, 3-celled with 3 collateral ovules in each cell.

W. Australia, *Drummond, 5th Coll. n.* 119, also in *Herb. F. Muell.* from *Dutton.* With the exception of the colour of the flower, the specimens remind one of the European *Hypericum nummularifolium.*

11. **H. Phillipsii,** *Harv. in Nat. Hist. Rev.* v. 296. *t.* 22. Branches scarcely angular, softly pubescent. Leaves closely sessile, cordate-ovate, very obtuse, ¾ to above 1 in. long, glabrous. Flowers large, white, solitary or clustered in the axils, the pedicels rather thick, 1 to 2 lines long, with a small bract at the base, and larger ovate deciduous bracteoles under the calyx. Calyx-tube broad and flat, nearly 2 lines diameter; lobes ovate-orbicular, 1½ lines long. Petals 4 to 5 lines. Stamens very numerous, in more than 1 row. Ovary much raised, obtusely 3-lobed, with a small central depression round the style, 3-celled, with 10 to 12 ovules in each cell.

W. Australia. Raised in the Botanic Garden of Dublin from seeds received from the neighbourhood of King George's Sound.

12. **H. hypericifolium,** *Benth.* Branches erect, elongated, slightly 4-angled, glabrous. Leaves elliptical or almost ovate, obtuse or nearly so, narrowed or rounded at the base, mostly ¾ to 1 in. long. Flowers white, not so large as in *H. Phillipsii,* usually clustered in the axils, the pedicels very short but slender, with a small bract at the base, and 2 rather larger ovate concave bracteoles under the calyx. Calyx-tube very flat, about 1½ lines diameter; lobes ovate-orbicular, about 1½ lines long. Petals twice as large. Stamens numerous, in more than 1 row. Ovary half-superior, broad, obtusely 3-lobed, with a central depression round the style, 3-celled, with 6 to 8 ovules in each cell.

W. Australia, *Drummond, 5th Coll. n.* 118.

16. BALAUSTION, Hook.

(Cheynia, *J. Drumm.*)

Calyx-tube urceolate, adnate to the ovary at the base; lobes 5, broad and obtuse, continuous with the tube. Petals 5, orbicular, spreading. Stamens numerous, free, not exceeding the petals, inserted in a single row round the prominent annular disk; anther-cells versatile; the cells parallel, opening longitudinally. Ovary in the bottom of the calyx, wholly inferior, flat-topped with a central depression round the style, 3-celled, with several ovules in each cell, imbricate in 2 rows on a peltate placenta; style filiform, with a slightly dilated capitate stigma. Capsule opening loculicidally, but not near ripe in our specimens.—Shrub. Leaves opposite, entire. Flowers large, axillary, pedicellate with 2 bracteoles under the calyx.

The genus is limited to the single Australian species.

1. **B. pulcherrimum,** *Hook. Ic. Pl. t.* 852. A low glabrous shrub, with a short thick trunk and numerous decumbent or prostrate stems, extending to about 1 ft. Leaves petiolate, linear-concave and keeled or triquetrous, rigid, acute or mucronate, mostly under ½ in. long. Flowers of a rich red, solitary in the axils below the ends of the branches, on pedicels of 1 to 3 lines. Bracteoles small, ovate. Calyx-tube nearly ½ in. long; lobes about 1 to 1½ lines, coloured like the tube. Petals about 5 lines diameter, with a very short broad claw. Stamens about 30, the filaments somewhat dilated, with a callous protuberance at the base inside. Style long.—*Cheynia pulchella*, J. Drumm. in Hook. Kew Journ. vii. 56.

W. Australia. Northern districts, *Drummond*, 5*th Coll. Suppl. n.* 26.

Subtribe II. Euleptospermeæ.—Leaves scattered or rarely opposite, small or narrow and coriaceous, 1- or more nerved, rarely penniveined. Flowers solitary in the axils of the leaves or bracts, closely sessile except in a very few species. Stamens indefinite, in one or more rows, free or united in bundles opposite the petals, or very rarely definite. Anthers versatile, with distinct parallel cells. Ovules in 2 or more rows in each cell of the ovary. Embryo straight or slightly incurved, the cotyledons usually longer than the radicle.

17. AGONIS, DC.

(Billiottia, *DC.*)

Calyx-tube turbinate or campanulate, adnate to the ovary at the base, the free part broad; lobes 5, ovate, usually scarious, imbricate or open. Petals 5, orbicular, spreading, exceeding the calyx-lobes. Stamens free, not exceeding the petals, either 10 regularly opposite the petals and calyx-lobes, or 20 or more without any opposite the centre of the petals; filaments filiform; anthers versatile, the cells parallel, opening longitudinally; connective with a small globular gland. Ovary inferior, 3-celled, with 2 or 4 ovules in each cell erect from a small nearly basal placenta; style filiform, inserted in a deeply tubular depression in the centre of the ovary, being attached almost to the base of the carpels; stigma capitate or peltate. Capsule opening at the top loculicidally in 3 valves, shorter than the calyx-tube. Seeds oblong or cuneate; testa thin; embryo straight; cotyledons plano-convex, much longer than the radicle.—Shrubs or small trees. Leaves alternate, often crowded on the smaller branchlets, either small or long and narrow, entire. Flowers rather small, closely sessile, in globular axillary or terminal heads, usually surrounded by imbricate scale-like bracts, with 2 smaller bracteoles under each flower, the white persistent petals usually very conspicuous.

The genus is limited to West Australia. Formerly considered as a section of *Lepto-*

spermum on account of its alternate leaves and stamens not exceeding the petals; it is much nearer allied to *Melaleuca* in inflorescence and in the ovary and seeds, whilst the arrangement of the stamens shows a connection with *Bæckea* and its allies. The seeds have been examined in three species only.

SECTION I. **Taxandria.**—*Stamens* 10, *regularly opposite the calyx-lobes and petals. Ovules* 2 *in each cell.*

Leaves spathulate, obovate or oblong-cuneate.
  Leaves mostly obovate, thick, nerveless, rarely above ¼ in. long.
    Bracts not exceeding the calyx-tube . . . . 1. *A. spathulata.*
  Leaves oblong-cuneate, mucronate-acute, 1- or 3-nerved, ½ to 1 in.
    long. Bracts covering the calyx-tube . . . . 2. *A. floribunda.*
  Leaves obovate-oblong, obtuse, rigid, 3-nerved, bordered with silky
    hairs, ½ to 1 in. long. Bracts not exceeding the calyx-tube . . 3. *A. marginata.*
Leaves linear or linear-lanceolate.
  Leaves ½ to 1 in. long, obtuse or acute, not pungent. Bracts and
    calyx-lobes obtuse . . . . 4. *A. linearifolia.*
  Leaves ¼ to ½ in., mucronate-acute and mostly pungent. Bracts
    and calyx-lobes more or less acute . . . . 5. *A. juniperina.*
  Leaves densely clustered, ¼ in. or under, obtuse or rarely acute.
    Bracts and calyx-lobes obtuse . . . . 6. *A. parviceps.*

SECTION II. **Ataxandria.**—*Stamens* 20 *to* 30 (*except in* A. grandiflora), *but none opposite the centres of the petals. Ovules* 4 *to* 6 *in each cell.*

Leaves linear-lanceolate to oblong-cuneate, 1 to 6 in. long. Bracts
  and calyx-lobes obtuse . . . . 7. *A. flexuosa.*
Leaves obovate or oblong-cuneate, undulate, mostly ½ in. long.
  Bracts acuminate. Calyx-lobes acute . . . . 8. *A. undulata.*
Leaves ovate, almost cordate, about ½ in. long. Bracts and calyx-
  lobes obtuse . . . . 9. *A. theæformis.*
Leaves linear. Flowers large, solitary, or 2 to 4 in the head. Bracts
  and calyx-lobes large and scarious . . . . 10. *A. grandiflora.*

SECTION I. TAXANDRIA.—Stamens 10, regularly opposite the calyx-lobes and petals, as in the first two sections of *Bæckea*. Ovules 2 in each cell.

1. **A. spathulata,** *Schau. in Pl. Preiss.* i. 117. A densely-tufted, bushy, or diffuse shrub of 1 to 2 ft., glabrous, or with a few long soft hairs about the upper leaves and inflorescence. Leaves obovate, spathulate, or almost orbicular, narrowed into a distinct petiole, very obtuse, thick, concave, and almost nerveless, mostly 1½ to 3 lines long. Flowers snow-white, in closely sessile terminal or axillary heads of 12 to 20. Outer bracts broadly orbicular, granular-tuberculate, covering the calyx-tube, inner ones obovate; bracteoles narrow, concave. Calyx-tube turbinate, about 1 line long; lobes about half as long as the tube, scarious and ciliate. Petal-claws as long as the calyx-lobes; lamina orbicular, 1 line diameter. Stamens 10, regularly opposite the calyx-lobes and petals; filaments somewhat dilated, especially those opposite the petals. Ovules 2 in each cell of the ovary.

W. Australia. Lucky Bay, *R. Brown;* Kalgan river, *Oldfield;* barren rocky wastes at the foot of the Konkoberup hills, *Preiss, n.* 324; also *Drummond,* 5*th Coll. n.* 131.

Var. *angustifolia.* Leaves longer, narrower, and less obtuse, sometimes almost linear-cuneate and ½ in. long. Flowers rather larger, with shorter and broader claws to the petals.—E. Mount Barren, *Maxwell.*

2. **A. floribunda,** *Turcz. in Bull. Mosc.* 1849, ii. 20. Branches rigid,

flexuose, apparently spreading, the young shoots loosely silky-hairy. Leaves crowded on the smaller branchlets, almost whorled under the flower-heads, oblong-cuneate, acute or mucronate, undulate, much narrowed towards the base, 1- or 3-nerved, from about $\frac{1}{4}$ to above $\frac{1}{2}$ in. long. Flower-heads terminal, or below the ends of the branches after the growth of the axis, very dense, but few-flowered. Imbricate bracts broad, rigid, completely enveloping the calyx. surrounded by a whorl of floral leaves. Calyx-tube pubescent; lobes ciliate. Petal-claws coloured, as long as the calyx-lobes; lamina orbicular, very white. Stamens 10, regularly opposite the calyx-lobes and petals. Ovules 2 in each cell of the ovary.

W. Australia, *Drummond, 4th Coll. n.* 56.—The species is allied to *A. spathulata*, differing chiefly in the narrower leaves and in the bracts. No. 55, 4th Coll. of Drummond, may be the same species in very young bud.

3. **A. marginata,** *Schau. in Pl. Preiss.* i. 117. A tall shrub, the branches and young shoots clothed with soft silky hairs. Leaves obovate-oblong, narrowed into a short petiole, obtuse, or minutely mucronate, $\frac{1}{2}$ to 1 in. long, 3- or rarely 5-nerved, bordered by a rim of dense appressed hairs, which at length wear off. Flower-heads terminal or axillary, of about 12 to 20 flowers. Imbricate bracts broadly orbicular, the inner ones obovate, concave. Calyx-tube rather above 1 line long; lobes about half as long, softly ciliate, and silky-hairy. Petals snow-white, orbicular, above 1 line diameter, the claw shorter than the calyx-lobes, or scarcely any. Stamens 10, opposite the petals and calyx-lobes. Ovules 2 in each cell of the ovary.—*Leptospermum marginatum*, Labill. Pl. Nov. Holl. ii. 10. t. 148; DC. Prod. iii. 226; *Billiottia marginata*, G. Don, Gen. Syst. ii. 827; *Fabricia stricta*, Lodd. Bot. Cab. t. 1219.

W. Australia. King George's Sound, *R. Brown*, *Labillardière*, and adjoining districts, *A. Cunningham; Preiss, n.* 141; *Baxter*, and others.

4. **A. linearifolia,** *Schau. in Pl. Preiss.* i. 118. A tall shrub, attaining 12 ft. or more in some situations, the young shoots loosely and softly hairy, otherwise glabrous. Leaves linear, linear-lanceolate, or somewhat cuneate, mostly acute and narrowed at the base, nerveless, or 1- or 3-nerved, $\frac{1}{2}$ to 1 in. long. Flower-heads small, all axillary. Calyx-lobes short, ovate, obtuse, pubescent, ciliate. Petals with very short broad claws. Stamens 10, regularly opposite the calyx-lobes and petals, the filaments broad at the base. Ovules 2 in each cell of the ovary.—*Leptospermum linearifolium*, DC. Prod. iii. 227; Mem. Myrt. t. 12; *Billiottia linearifolia*, G. Don, Gen. Syst. ii. 827; *Agonis conspicua* and *A. angustifolia*, Schau. in Pl. Preiss. i. 118.

W. Australia. King George's Sound and Lucky Bay, *R. Brown*, and thence to Swan River, apparently common, *A. Cunningham*, and others; *Drummond, 1st Coll. n.* 156; *3rd Coll. n.* 42; *4th Coll. n.* 57; *Preiss, n.* 142, 145, and in some sets, n. 151, which in others is *Leptospermum firmum*. Some specimens from Cape Le Grand, *Maxwell*, and from *Drummond*, *5th Coll. n.* 143, and *5th Coll. n.* 132, have remarkably narrow leaves. Others from Capelrice, *Oldfield*, have very short leaves, almost connecting the species with the long-leaved forms of *A. spathulata*. Preiss's n. 149 and 150 (*A. angustifolia* and *A. conspicua*) have large flowers.

5. **A. juniperina,** *Schau. in Pl. Preiss.* i. 118. A tall shrub, or sometimes a tree of 40 ft. or even more, with rigid branches more or less

pubescent or hirsute when young. Leaves linear-lanceolate, clustered in the axils or on short branchlets, concave, rigid, mucronate-acute or almost pungent, $\frac{1}{4}$ to nearly $\frac{1}{2}$ in. long. Flower-heads globular, terminating short lateral branchlets. Bracts rather small, very concave, mucronate or acute. Calyx-tube softly pubescent, about 1 line long; lobes much shorter, ovate-triangular, acute. Petals about 1 line diameter, on claws nearly as long as the calyx-lobes. Stamens 10, regularly opposite the calyx-lobes and petals. Ovules 2 in each cell of the ovary.

W. Australia, *Drummond, 2nd Coll. n.* 79, *4th Coll. n.* 58; barren gravelly places near Cape Riche, *Preiss, n.* 314, Blackwood river, and by lagoons, Princess Royal Harbour, *Oldfield;* shores of Lake Leven, *Maxwell.*—The species is very closely allied to *A. parviceps,* and some specimens from Hay river, *Maxwell,* with the foliage and larger flowers of *A. juniperina,* have the calyx-lobes and bracts scarcely acute.

6. **A. parviceps,** *Schau. in Pl. Preiss.* i. 119. A much-branched bushy rigid shrub of from 2 to 3 ft. to twice that height. Leaves from linear-spathulate, and under 2 lines, to linear-lanceolate, and 3 lines long or rather more, densely clustered in the axils and on the smaller branchlets, rigid, concave, spreading or recurved, obtuse or slightly mucronate. Flowers small, in small heads in the cluster of leaves, or terminating short branchlets. Bracts obtuse, not exceeding the calyx-tube. Calyx-tube pubescent, rarely above $\frac{3}{4}$ line long; lobes shorter than the tube, ovate, obtuse. Stamens usually 10, regularly opposite the calyx-lobes and petals, those opposite the calyx-lobes smaller and perhaps sometimes wanting; filaments short. Ovules 2 in each cell of the ovary.

W. Australia. Moist boggy ground, King George's Sound, *R. Brown,* and adjoining districts, *A. Cunningham* and others; *Drummond, 2nd Coll. n.* 78; *Preiss, n.* 160, 161; Vasse river, *Oldfield.*

SECTION II. ATAXANDRIA.—Stamens 20 to 30, in a single series, but usually (except in *A. grandiflora*) interrupted opposite the centre of each petal, as in the last four sections of *Bæckea.* Ovules 4 to 6 in each cell of the ovary.

7. **A. flexuosa,** *Schau. in Pl. Preiss.* i. 116. A tall shrub or tree attaining 40 ft., the young shoots often silky-pubescent, at length glabrous. Leaves lanceolate or linear-lanceolate, acute, narrowed at both ends, 3-nerved, 2 to 6 in. long. Flower-heads all axillary. Bracts not numerous, broad, very obtuse, shorter than the calyx. Calyx softly pubescent, the tube 1 to $1\frac{1}{2}$ lines long, the lobes much shorter, scarious, and fringed or ciliate at the edges. Petals obovate, fully 2 lines long. Stamens usually 20, 4 opposite each calyx-lobe, none opposite the petals. Ovules 6 in each cell of the ovary.—*Metrosideros flexuosa,* Willd. Enum. Hort. Berol. 514; *Leptospermum flexuosum,* Spreng. Nov. Prov. according to DC. Prod. iii. 226; Colla, Hort. Ripul. App. t. 2; *Billiottia flexuosa,* G. Don, Gen. Syst. ii. 827; *Leptospermum resiniferum,* Bertol. Amœn. Ital. 29; *L. glomeratum,* Wendl. fil. in Flora, 1819, 678, as corrected in Wendl. Beitr. ii. 22.

W. Australia. King George's Sound, *R. Brown,* and thence to Swan River, *A. Cunningham, Fraser,* and others; *Drummond, n.* 18, 54, *and 2nd Coll. n.* 77; *Preiss, n.* 136, 137, 138, 139, 140, 147.

Var. *latifolia,* Schau. Branches stouter and more rigid; leaves shorter, broader, obtuse, and rigid; flowers larger and more numerous in the head; calyx-tube nearly 2 lines; petals 4 lines long; stamens about 6 opposite each calyx-lobe.—King George's Sound, and to the eastward, *Preiss*, *n*. 144; *Drummond*, *5th Coll. n*. 133; *Maxwell*.

8. **A. undulata,** *Benth*. Branches rigid, almost spinescent, our specimens entirely glabrous. Leaves from obovate to oblong-cuneate, obtuse or mucronulate and often emarginate, much narrowed at the base, undulate, 1- or 3-nerved, rarely exceeding ½ in. Flower-heads terminal or axillary, or sometimes below the ends of the branches, the axis growing out as in *Melaleuca;* flowers not numerous in the head. Bracts acuminate, pubescent, nearly as long as the calyx. Calyx silky, the tube about 1 line long; lobes rather shorter, acute. Petals obovate, not twice as long as the calyx-lobes. Stamens about 4 opposite each calyx-lobe, none opposite the petals. Ovules 4 in each cell of the ovary.

W. Australia, *Drummond*, *n*. 6.—Allied to *A. theæformis*, but differing in foliage and in the calyx-lobes.

9. **A. theæformis,** *Schau. in Pl. Preiss*. ii. 223. A tall shrub with rather slender branches, the young shoots loosely and softly hairy, becoming glabrous when full grown. Leaves ovate or broadly elliptical, acute or obtuse, cordate or truncate at the base, under ½ in. long, and sometimes not above ¼ in., 1-nerved and penniveined, often undulate, with a recurved point. Flower-heads all axillary, of 6 to 12 flowers. Bracts fringed-ciliate. Calyx-tube broad; lobes broad, obtuse, with scarious edges. Petals white, but drying of a yellowish hue, 1 to 1½ lines diameter. Stamens about 20, none opposite the centres of the petals. Ovules about 4 in each cell of the ovary, but only 1 appears to enlarge, the perfect seed has, however, not been seen. —*A. hypericifolia*, Schau. in Pl. Preiss. i. 117.

W. Australia. Moist sandy places and rocks, King George's Sound, *R. Brown*, and adjoining districts, *A. Cunningham; Baxter; Drummond*, *3rd Coll. n*. 41; *Preiss*, *n*. 152, 153; *Oldfield*.

10. **A. grandiflora,** *Benth*. Glabrous, or the young shoots hairy. Leaves densely clustered, linear, concave, obtuse or mucronate, about ½ in. long. Flowers large, solitary, or in heads of 2 to 4, sessile in the clusters of leaves. Bracts scarious, imbricate, covering the calyx. Calyx-tube 1½ lines long, pubescent, with appressed hairs; lobes at least as long as the tube, ovate, scarious. Petals about 4 lines long, obovate, narrowed into a claw. Stamens 20 to 30, rather closer together opposite the calyx-lobes than opposite the petals, but forming a complete ring without any distinct vacancy opposite the centre of the petals, anthers large, with oblong parallel cells and a conspicuous connective-gland. Ovary entirely as in the rest of the section, with 4 to 6 erect ovules in each cell.

W. Australia. Near Hampden, *W. Clarke*.

18. LEPTOSPERMUM, Forst.

(Fabricia, *Gærtn.;* Macklottia, *Korth.;* Homalospermum, *Schau.;* Pericalymma, *Endl.*)

Calyx-tube broadly campanulate or rarely turbinate, adnate to the ovary at the base, free part broad; lobes 5, ovate, herbaceous or membranous, imbricate or open. Petals 5, orbicular, spreading, exceeding the calyx-lobes. Stamens numerous, free, not exceeding the petals, inserted on the margin of the disk in a single row; filaments filiform; anthers versatile, the cells parallel, opening longitudinally; connective with a small globular gland. Ovary inferior or half-superior, enclosed in the calyx-tube, usually 5- or more celled, rarely 3- or 4-celled, with either numerous ovules in each cell densely covering a peltate placenta and horizontal or recurved, or few and recurved in two rows; style filiform, inserted in a slight or deep depression in the centre of the ovary, often short, with a capitate or peltate stigma. Capsule opening at the top loculicidally, either protruding from the calyx-tube or rarely shorter. Seeds either linear-cuneate and wingless or more or less angular with transparent wings or cilia along the angles, but usually only few in each cell or a single one perfect, the others sterile often hard and always wingless.—Shrubs or rarely small trees, glabrous silky-pubescent or hoary. Leaves alternate, small, rigid, entire, nerveless or 1- or 3-nerved. Flowers usually white, sessile or rarely shortly pedicellate, solitary or 2 or 3 together at the ends of short branchlets or in the axils of the leaves. Bracts broad, scarious, 2 or 3 outer ones usually imbricate, but falling off from the very young bud, 2 inner ones or bracteoles opposite and close under the calyx often more persistent.

The genus is common to Australia and New Zealand and the Indian Archipelago. Of the Australian species one is found in New Zealand also, and another in the Indian Archipelago, the remainder are endemic. The species are very difficult to distinguish. The whole of those with 5-celled ovaries, from *L. lanigerum* to *L. erubescens*, different as some of them appear at first sight, pass so gradually one into the other that they might be readily admitted as varieties of one species, whilst on the other hand many of the varieties here enumerated have been distinguished as species by R. Brown, whose herbarium contains a beautiful series of well-selected specimens, as well as by other eminent botanists whose opinions are entitled to great weight. The genus requires, therefore, much further study on the part of those who have the opportunity of observing it in its native stations. From the dried specimens, whether of the species here admitted or of the varieties or races, I have been unable to discover any positive discriminating characters.

Most authors describe the calyx-lobes of *Leptospermum* as valvate; I have always found them decidedly imbricate in the young bud, even in the Javanese specimens communicated by Blume.

SECTION I. **Fabricia.**—*Ovary usually* 6- *to* 10- *or* 4-*celled. Ovules numerous. Seeds when perfect rather broad, fringed or winged at the angles (as far as known). Flowers closely sessile.*

Ovary usually 6- to 10-celled. Eastern and South-eastern species.
  Calyx villous. Capsule half-exserted . . . . 1. *L. Fabricia.*
  Calyx glabrous. Capsule slightly protruding above the calyx-tube . . . . 2. *L. lævigatum.*
Ovary usually 4-celled. Calyx glabrous. Western species . . . 3. *L. firmum.*

SECTION II. **Euleptospermum.**—*Ovary usually* 5-*celled, or here and there* 4-*celled, or* 3-*celled in the last two species. Ovules numerous. Seeds, both perfect and sterile, narrow-linear.*

Calyx-tube glabrous. Ovary 5- or rarely 4-celled.
  Leaves flat or with recurved margins, obtuse or scarcely pointed (except in the large variety) . . . . 4. *L. flavescens.*
  Leaves flat or concave, pungent-pointed, narrow or small . . . 5. *L. scoparium.*
Calyx-tube pubescent or villous. Ovary 5-, rarely 4-celled.

Branches spinescent. Leaves mostly oblong. Calyx-tube loosely villous.
Capsule not prominent. Flowers large. Western species . 7. *L. spinescens.*
Branches not spinescent. Flowers sessile or nearly so. Eastern species.
Calyx broad and obtuse at the base, woolly, loosely villous, or closely tomentose.
Leaves linear, concave, pungent pointed . . . . . . 6. *L. arachnoideum.*
Leaves obovate, oblong or elliptical, flat or with recurved margins, obtuse or shortly mucronate . . . . . . 8. *L. lanigerum.*
Leaves very small (mostly under 2 lines) obovate or oblong, flat, obtuse. Flowers small . . . . 9. *L. parvifolium.*
Calyx usually attenuate at the base, at least when young, silky with appressed hairs.
Calyx-lobes appressed-silky, usually persistent . . . . 10. *L. stellatum.*
Calyx-lobes silky, but thin and deciduous. Stems prostrate 13. *L. rupestre.*
Calyx-lobes membranous, deciduous. Erect or spreading shrubs . . . . 12. *L. myrtifolium.*
Branches not spinescent, often flexuose. Flowers and leaves small. Calyx silky, the *lobes very small.* Capsule not prominent. Southern species . . . . 14. *L. myrsinoides.*
Branches not spinescent. Flowers pedicellate. Calyx silky. Capsule not prominent.
Eastern species. Calyx usually attenuate at the base . . . 11. *L. attenuatum.*
Western species. Branches often flexuose. Flowers and leaves small. Calyx obtuse at the base . . . . 15. *L. erubescens.*
Ovary 3-celled.
Flowers small, glabrous. Capsule shorter than the calyx-tube.
Eastern and tropical species . . . . 16. *L. abnorme.*
Flowers rather large. Calyx densely villous. Western species 17. *L. Roei.*

SECTION III. **Pericalymma.**—*Ovary usually 3-celled. Ovules few* (4 *to* 8 *in each cell*). *Branchlets flexuose and dichotomous. Western species.*

Tall erect shrubs, the trunk not turgid.
Flowers rather large. Calyx clothed with long hairs, the lobes as long as the tube . . . . 18. *L. floridum.*
Flowers rather small. Calyx shortly silky, the lobes much shorter than the tube . . . . 19. *L. ellipticum.*
Dwarf shrub, the base of the stem thickened, almost fusiform . . 20. *L. crassipes.*

L. obliquum, Colla, Hort. Ripul. App. 2. 351, described in leaf only, is not now to be determined. It is probably *L. lanigerum* or *L. flavescens*. *L. tortuosum* and *L. buxifolium*, Dehnh. Rivist. Napol. and *L. ciliolatum*, *L. hypericifolium*, *L. cupressinum*, and *L. cuneiforme*, Otto and Dietr. Allgem. Gart. Zeit., described from garden specimens and quoted with short diagnoses in Walp. Rep. ii. 169, are all unknown to me. They are probably, as well as numerous names of *Leptosperma*, taken from garden catalogues or herbaria by Steudel or by Schauer, and which, being otherwise unpublished, are here omitted, nearly all of them forms of *L. flavescens*, *L. lanigerum*, or *L. scoparium*.

SECTION I. FABRICIA.—Ovary usually 6- to 10-celled or 4-celled. Ovules numerous. Seeds usually 1 or 2, perfect in each cell, rather broad, fringed or winged at the angles, the remainder sterile, slender or flat. Flowers closely sessile.

1. **L. Fabricia,** *Benth.* A shrub or tree resembling the larger specimens of *L. lævigatum*, but the branches often loosely hairy. Leaves from oblong-lanceolate to almost obovate, $\frac{3}{4}$ to $1\frac{1}{2}$ in. long, obtuse or slightly mucronate, 3- or 5-nerved. Flowers larger than in *L. lævigatum*, mostly termi-

nating short leafy branchlets, surrounded by orbicular imbricate deciduous bracts. Calyx more or less tomentose-villous, the tube hemispherical, the lobes nearly as long as the tube, orbicular, very obtuse, silky or villous outside. Stamens numerous. Ovary usually 10-celled. Capsule very prominent above the calyx-rim, the free part usually as long as the enclosed portion. Seeds not seen quite perfect, but in the apparently ripe capsules already burst open the enlarged ovules of each cell are readily detached in a mass with the placenta, the whole assuming the shape represented by Gærtner as that of the seed; enlarged ovules or young seeds very flat, obliquely obovate-oblong, the upper ones falcate, very differently shaped from those of *L. lævigatum*, not winged or very slightly so at the base.—*Fabricia myrtifolia*, Gærtn. Fruct. i. 175. t. 35.

Queensland. Endeavour river, *Banks and Solander;* Haggerstone and Lizard Islands, *A. Cunningham;* Cape York, *W. Hill.* The Banksian specimens described by Gærtner are in the same state, with unripe seeds only, as A. Cunningham's.

N. S. Wales? Some flowering specimens of *Vicary's*, without the precise locality, appear to belong to this species.

2. **L. lævigatum,** *F. Muell. Ann. Rep.* 1858, *and Fragm.* iv. 60. A tall shrub, often arborescent and attaining 20 to 30 ft., glabrous and somewhat glaucous, the young shoots often slightly silky. Leaves from obovate-oblong to oblong-cuneate or narrow-oblong, obtuse, mostly ½ to ¾ in. long, but sometimes 1 in. or even more, more or less conspicuously 3-nerved. Flowers axillary, solitary and sessile or nearly so, or very rarely 2 together on a very short common peduncle. Bracts imbricate, bracteoles cohering, but all very deciduous. Calyx glabrous; tube at first broadly turbinate, at length nearly hemispherical; lobes triangular, much shorter than the tube, usually persistent for a long time but falling off from the ripe fruit. Stamens numerous round a broad very flat disk. Ovary flat-topped, usually 10-celled; style short in a central depresssion; stigma broadly peltate. Capsule nearly flat and scarcely prominent above the calyx-border. Perfect seeds usually 1 or very few in each cell, linear-oblong, more or less compressed, incurved, fringed all round with a transparent wing which readily splits up into cilia. Embryo filling the seed, the cotyledons ovate-oblong, broader and longer than the radicle.—*Fabricia lævigata*, Gærtn. Fruct. i. 175; Bot. Mag. t. 1304; Hook. f. Fl. Tasm. i. 141; *F. myrtifolia*, Sieb. Pl. Exs., not of Gærtn.

N. S. Wales. Port Jackson, *R. Brown, Sieber n.* 309; near the sea, *Woolls;* northward to Hastings river, *Beckler;* and southward to Gabo Island, *Maplestone.*

Victoria. Port Phillip, *R. Brown;* on the seacoast, *Robertson, F. Mueller.*

Tasmania. King's Island, *R. Brown;* maritime sands, common in some parts of the N. and N.W. coast and islands of Bass's Straits, *J. D. Hooker.*

Var. ? *minus*, F. Muell. Branches slender. Leaves oblong-cuneate, mucronate-acute. Flowers much smaller than in the common form, the calyx-lobes more petal-like. Ovary usually 6- to 8-celled, with fewer ovules than in the common form and the capsules more convex. Seeds, according to F. Mueller, with or without wings.—*Fabricia coriacea*, F. Muell., Miq. in Ned. Kruidk. Arch. iv. 147. Perhaps a distinct species.

N. S. Wales. Darling river, *Victorian Expedition.*

Victoria. N.W. desert, *Lockhart Morton, Dallachy;* scrub near the mouth of the Murray, *F. Mueller.*

S. Australia. St. Vincent's and Spencer's Gulfs to the Murray, *F. Mueller* and others.

3. **L. firmum,** *Benth.* A tall erect glabrous shrub, with virgate branches. Leaves linear or linear-lanceolate, acute or rather obtuse, narrowed to the base, rigid, ½ to 1 in. long. Flowers rather large, closely sessile. Bracts small, broad, truncate, persistent. Calyx glabrous, tube very broad; lobes short, broad, membranous, at length deciduous. Ovary 4-celled or rarely 3- or 5-celled, with numerous closely-packed but short ovules in each cell. Fruit hard, usually almost cubical or triquetrous, the capsule protruding from the calyx-tube. Seeds usually 1 or 2 perfect in each cell, obovate-oblong, somewhat flattened, more or less surrounded by a thin wing breaking up into cilia as in *L. lævigatum*, and embryo also as in that species; barren seeds very numerous, small, often irregularly winged.—*Homalospermum firmum*, Schau. in Linnæa, xvii. 242, and in Pl. Preiss. i. 119.

W. Australia. King George's Sound, *R. Brown;* chiefly in marshy places, from the south coast to Swan River, *A. Cunningham, Drummond, 1st Coll. n.* 139, *Preiss, n.* 143 *and* 148, and others.

SECTION II. EULEPTOSPERMUM.—Ovary usually 5-celled or, especially in the last two species, 4- or 3-celled. Ovules numerous. Seeds, both perfect and sterile, narrow-linear, without wings.

4. **L. flavescens,** *Sm. in Trans. Linn. Soc.* iii. 262. Usually a tall shrub, quite glabrous or the young parts minutely silky-hoary. Leaves from narrow-oblong or linear-lanceolate to broadly oblong or even obovate, obtuse or scarcely acute, rigid, flat, nerveless or 1- or 3-nerved, attaining ¾ in. in the largest forms but usually under ½ in. and sometimes all very small. Flowers solitary, terminating the branchlets or axillary and nearly sessile, as variable in size as in *L. lanigerum*, and of the same shape. Calyx quite glabrous, the tube broadly campanulate or hemispherical; lobes ovate, as long as the tube, membranous or thickened in the centre. Ovary 5-celled, more or less convex on the top, with a short central depression round the style. Capsule prominent above the calyx-tube. Seeds all narrow-linear, without wings.—DC. Prod. iii. 227; Hook. f. Fl. Tasm. i. 139; *Melaleuca trinervia*, White, Trav. 229. t. 24?; *Leptospermum polygalifolium*, Salisb. Prod. 350; *L. Thea*, Willd. Spec. Pl. ii. 949, and (on his authority) *Melaleuca Thea*, Wendl. Sert. Hannov. 24. t. 13; *L. tuberculatum*, Poir. Dict. Suppl. iii. 338 (from the character given).

Queensland. Abundant about Brisbane river and Moreton Bay, *A. Cunningham, F. Mueller*, and others; Percy Island, *A. Cunningham;* Port Denison, *Fitzalan.*

N. S. Wales. Port Jackson to the Blue Mountains, *R. Brown, Sieber, n.* 315, *and Fl. Mixt. n.* 549, and others; in the interior, *Fraser;* New England, *C. Stuart;* Illawarra, *A. Cunningham.*

Victoria. Buffalo Range, Yarra, Goulbourn, and Ovens rivers, *F. Mueller.*

Tasmania. Abundant on banks of rivers, etc. *J. D. Hooker.*

This species, which extends also into the Indian Archipelago and Malacca, is scarcely to be distinguished from *L. lanigerum* except by the absence of all hairs or down from the calyx, and is equally variable, the extreme forms being at first sight so dissimilar that it requires the examination of a large number of specimens to believe in their specific identity, and at the same time it is almost impossible to draw a precise line of demarcation between this and several others. The following are the varieties which appear to be the most prominent and distinct.

a. *commune.* Leaves narrow, from under ½ in. to ¾ in. long. Flowers middle-sized.—Bot. Mag. t. 2695; *L. porophyllum*, Cav. Ic. iv. 17. t. 330. f. 2 (from the fig. and descr.); *L. amboinense*, DC. Prod. iii. 229, at least the specimens so named by Miquel and Blume;

Macklottia amboinensis, Korth. in Ned. Kruidk. Arch. i. 196.—From Tasmania to Queensland, and in the Indian Archipelago.

b. *obovatum*, F. Muell. Leaves from broadly obovate to obovate-oblong, under ½ in. long.—*L. obovatum*, Sweet, Fl. Austral. t. 36; *L. micromyrtus*, Miq. in Ned. Kruidk. Arch. iv. 145 (from the character given); N. S. Wales and Victoria, the Port Jackson specimens with rather thin and 3-nerved leaves, the southern ones with much thicker rigid nerveless leaves. *L. emarginatum*, Wendl. in Spreng. Syst. ii. 491, has the leaves narrow as in *a*, but very obtuse or emarginate as in *b*.

c. *grandiflorum*. Leaves rather large. Flowers larger than in any other variety.—*L. grandiflorum*, Lodd. Bot. Cab. t. 514; *L. virgatum*, Schau. in Linnæa, xv. 410; *L. nobile*, F. Muell.; Miq. in Ned. Kruidk. Arch. iv. 145.—Paramatta, *Woolls;* Blue Mountains, *A. Cunningham;* Tasmania, *C. Stuart*.

d. *microphyllum*. Leaves flat, oblong or lanceolate, ¼ to ½ in. long.—Chiefly in Queensland.

e. *minutifolium*, F. Muell. Leaves all under ½ in. and mostly under 2 lines long, obovate or oblong, concave and recurved. Flowers very small.—New England, *C. Stuart*. This may prove sufficiently distinct to be considered as a species.

5. **L. scoparium,** *Forst. Char. Gen.* 48. A rigid very much branched shrub, in alpine situations low and almost prostrate, more usually erect, and attaining sometimes 10 to 12 ft., the young shoots often silky, the adult foliage mostly glabrous. Leaves from ovate to linear-lanceolate or linear, rigid, concave, acute and pungent-pointed, mostly under ½ in. long. Flowers axillary, sessile and solitary, or rarely terminating short lateral branchlets. Calyx quite glabrous, as variable in size as in *L. flavescens*, and the flowers and fruit otherwise precisely as in that species.—Sm. in Trans. Linn. Soc. iii. 262; Andr. Bot. Rep. t. 622; DC. Prod. iii. 227; Bot. Mag. t. 3419; Hook. f. Fl. Tasm. i. 138; Schau. in Linnæa, xv. 424; *L. floribundum*, Salisb. Prod. 349, and *L. recurvifolium*, Salisb. l. c. 350 (from the characters given); *L. juniperinum* (with narrow leaves), Sm. in Trans. Linn. Soc. iii. 263; Vent. Jard. Malm. t. 89; Schau. in Linnæa, xv. 431; *L. multiflorum*, Cav. Ic. Pl. iv. 17. t. 331. f. 1; *L. juniperifolium*, Cav. l. c. 18. t. 331. f. 2; *L. squarrosum*, Sieb. Pl. Exs.; *L. rubricaule*, Link, Enum. Hort. Berol. ii. 25; *L. styphelioides*, Schau. in Linnæa, xv. 423; *L. aciculare*, Schau. l. c. 429; *L. oxycedrus*, Schau. l. c. 432; *L. baccatum*, Schau. l. c. 433, not of Sm. including according to Schau. *L. persiciflorum*, Reichb. Hort. Bot. iii. 8. t. 220; *L. divaricatum*, Schau. in Walp. Rep. ii. 923 (a starved small-leaved form).

Queensland. Moreton Bay, *Murray*, according to Schauer.

N. S. Wales. Port Jackson to the Blue Mountains, *R. Brown*, *Sieb. n.* 310, 311, *and Fl. Mixt. n.* 547, 548, *A. Cunningham*, and others; northward to Clarence river, *Beckler;* and southward to Illawarra, *A. Cunningham;* and Twofold Bay, *F. Mueller*.

Victoria. Common in heaths and moist situations, *Robertson*, *F. Mueller*.

Tasmania. Very abundant throughout the colony, *R. Brown*, *J. D. Hooker*, etc.

S. Australia. Moist localities, St. Vincent's and Spencer's Gulfs, *F. Mueller;* Kangaroo island, *Waterhouse*.

The species is also in New Zealand.

6. **L. arachnoideum,** *Sm. in Trans. Linn. Soc.* iii. 263. A rigid much branched shrub, with the habit of the narrow-leaved forms of *L. scoparium*, and the same pungent crowded rigid concave linear leaves, but with the flowers of *L. lanigerum*, mostly on short lateral leafy branches, closely surrounded by floral leaves. Calyx broad, rather large, loosely woolly-hairy. Capsule shortly protruding from the calyx-tube, 5-celled or very rarely 3- or 4-celled.—DC. Prod. iii. 228; *L. arachnoides*, Gærtn. Fruct. i. 175. t. 35; *L. triloculare*, Vent. Jard. Malm. t. 88; Lodd. Bot. Cab. t. 791.

N. S. Wales. Port Jackson, *R. Brown, A. and R. Cunningham*, and others.

L. baccatum, Sm. in Trans. Linn. Soc. iii. 264, is a form with much less woolly calyxes, almost connecting this species with *L. scoparium*. Some specimens from *C. Moore* are quite like the one in Smith's herbarium.

7. **L. spinescens,** *Endl. in Hueg. Enum.* 51. A very rigid shrub with stout divaricate branches, the smaller ones spinescent. Leaves from obovate to cuneate-oblong or oblanceolate, mostly obtuse, thick and rigid, $\frac{1}{4}$ to $\frac{1}{2}$ in. long, 1-nerved or obscurely 3-nerved. Flowers rather large, solitary and closely sessile. Calyx-tube broadly hemispherical, densely woolly-tomentose, 3 to 4 lines diameter; lobes ovate, tomentose, much shorter than the tube, persistent. Petals scarcely above 2 lines diameter. Stamens about 20; connective gland of the anthers particularly large. Ovary flat-topped, with a very slight central depression, in many flowers rudimentary or completely abortive, 5-celled; ovules very numerous. Capsule hard, nearly flat and not produced above the calyx-tube.

W. Australia. Lucky Bay, *R. Brown;* King George's Sound or to the eastward, *Huegel, Drummond, 1st Coll. n.* 146 *or* 148, *Baxter, Roe.*

8. **L. lanigerum,** *Sm. in Trans. Linn. Soc.* iii. 263. A tall erect shrub, sometimes growing into a small tree, rarely low and bushy, the branchlets usually softly pubescent. Leaves from obovate-oblong to elliptical or narrow-oblong, exceedingly variable in size and indumentum, in some luxuriant specimens attaining $\frac{3}{4}$ in. or even more, but naturally not above $\frac{1}{2}$ in. and in some varieties all very much smaller, obtuse or mucronate-acute, more or less hoary silky or hairy underneath or on both sides, rarely glabrous except a few silky hairs on the margin, when broad and thin showing 1, 3 or 5 nerves, more frequently rigidly coriaceous, the nerves scarcely prominent or concealed by the indumentum. Flowers solitary, terminating very short leafy branchlets, or rarely sessile on the branches without intervening leaves. Calyx broad, more or less densely clothed with silky or woolly hairs; lobes triangular, often as long as the tube. Petals twice as long, broad, distinctly clawed. Stamens about 20 to 30, in a single series. Ovary 5-celled, convex, with a central depression, with numerous ovules in each cell. Capsule nearly globular but depressed at the top, more or less protruding from the calyx-tube, the lobes wearing off, varying from under 3 to above 4 lines diameter. Seeds linear without wings; cotyledons as long as or rather longer than the radicle.—DC. Prod. iii. 227; Hook. f. Fl. Tasm. i. 139; *L. australe*, Salisb. Prod. 350; *Melaleuca trinervia*, White, Journ. 229. t. 24 (quoted by Smith and DC. as *L. trinerve*), is either this or *L. flavescens*.

N. S. Wales. Port Jackson to the Blue Mountains, *R. Brown*, and others, northward to Mount Mitchell, *Beckler*, southward to Illawarra, *A. Cunningham*, and Twofold Bay, *F. Mueller*, and in the interior to Macquarrie and Cox's rivers, *Fraser, A. Cunningham.*

Victoria. Gipps' Land and mountainous districts generally, neighbourhood of Melbourne, Port Phillip, etc., *F. Mueller* and others.

Tasmania. Abundant throughout the island in many soils and situations, *J. D. Hooker.*

S. Australia. Rivoli Bay, mouth of the Glenelg, Port Adelaide, Onkaparinga range. *F. Mueller.*

This exceedingly variable species has the calyx sometimes nearly glabrous, and then passes almost into *L. flavescens*, whilst the smaller-flowered forms are closely connected with several of the following species; the most marked varieties are:—

a. Flowers large. Leaves coriaceous with a very short point, shining above, silky-hairy underneath, with recurved margins. Bracts large and often persistent.—*L. grandifolium*, Sm. in Trans. Linn. Soc. vi. 299.

b. Flowers large. Leaves broad, about ½ in. long, silky or nearly glabrous, the latter including *L. nitidum*, Hook. f. Fl. Tasm. i. 139, and only differing from *L. flavescens* in the hairy calyx.

c. Flowers large. Leaves narrower, ½ to 1 in. long.—*L. grandifolium*, Bot. Mag. t. 1810; Lodd. Bot. Cab. t. 701; *L. tonsum*, Schau. in Linnæa, xv. 422 (from the description).

d. Flowers large. Leaves rather narrow, ¾ to 1 in., long, very rigid, with almost pungent points, connecting the species with *L. scoparium*. Interior of N. S. Wales and Victoria.

e. Flowers smaller. Leaves mostly under ½ in., often silky on both sides, the commonest form in Tasmania, Victoria, and S. Australia.—*L. pubescens*, Lam. Dict. iii. 466; *L. villosum*, Otto and Dietr.; Walp. Rep. ii. 169; *L. Cunninghamii*, Schau. in Linnæa, xv. 420; *L. glaucescens*, Schau. l. c. 421; *L. Candollei*, Schau. l. c. 441; *L. microphyllum*, F. Muell.; Miq. in Ned. Kruidk. Arch. iv. 142.

f. Flowers small. Leaves small, obovate or orbicular. Mountains of Victoria and probably also the specimens from Mount Mitchell, which are however imperfect.

L. pilosum, Schau. in Walp. Rep. ii. 923, is described from Tasmanian specimens of Cunningham's n. 84. I have not found this n. in his herbarium, but the only Tasmanian species to which Schauer's diagnosis is applicable is *L. lanigerum* in some of its numerous forms. *L. splendens*, Schau. l. c. seems to refer to one of the larger varieties of *L. lanigerum*.

9. **L. parvifolium,** *Sm. in Trans. Linn. Soc.* iii. 263. A shrub with slender branches, pubescent or woolly when young. Leaves obovate or oblong, very obtuse, thick, nerveless or faintly 3-nerved, flat, 1, 2 or rarely 3 lines long. Flowers small, solitary or rarely 2 together at or near the end of lateral leafy branchlets. Calyx-tube rather broadly campanulate, but rarely above 1½ lines diameter, loosely and softly villous, lobes ovate, membranous, glabrous or slightly pubescent, nearly as long as the tube. Ovary 5-celled, short, slightly convex.—DC. Prod. iii. 228; *L. eriocalyx*, Sieb. Pl. Exs.

N. S. Wales. Port Jackson to the Blue Mountains, *R. Brown*, *Sieber*, *n.* 313, and others; and westward to Liverpool Plains, *A. Cunningham;* near Richmond, *C. Moore.* Very near the small-leaved forms of *L. lanigerum*, of which F. Mueller considers it a variety.

10. **L. stellatum,** *Cav. Ic.* iv. 16. *t.* 330. *f.* 1 *(from the figure and description).* Much-branched and erect, from 2 or 3 to several feet high, the young shoots silky-pubescent, the adult foliage glabrous or nearly so. Leaves from rather broadly elliptical-oblong to oblong-linear or linear-lanceolate, mostly from ¼ to ½ in. long, obtuse or mucronate-acute, rather rigid, more or less conspicuously 1- or 3-nerved. Flowers rather small, sessile or very shortly pedicellate in the upper axils or terminating short leafy shoots and then often two together. Calyx-tube broadly turbinate, densely silky-pubescent; lobes silky, more acute and more persistent than in *L. myrtifolium.* Ovary flat-topped or concave. Capsule level with the margin of the calyx or scarcely protruding.—*L. sericatum*, Lindl. in Mitch. Trop. Austr. 298.

Queensland. Moreton Island, *M'Gillivray;* Logan river, *Fraser;* near Lake Salvator, *Mitchell;* Rockingham Bay, *Dallachy* (specimens in fruit only, and doubtful. Capsules very small).

N. S. Wales. Port Jackson to the Blue Mountains, *A. and R. Cunningham* and others.

Var. *grandiflorum.* Flowers larger, the calyx-tube fully 2 lines long.—*L. gnidiæfolium* of German gardens, but scarcely of DC. Queensland, *Bowman;* Port Jackson, *Herb. F. Mueller.*

11. **L. attenuatum,** *Sm. in Trans. Linn. Soc.* iii. 262. Very near *L. stellatum,* differing only in the pedicellate flowers. Branches usually slender and loose. Leaves mostly narrow-oblong and about ½ in. long, but varying from broadly oblong and ¼ in. to linear and above 1 in. long. Flowers usually small, solitary in the axils or 2 together on short leafy branchlets, on pedicels of 1 to 2 lines. Calyx-tube densely silky-pubescent, contracted at the base, lobes usually persistent. Capsule scarcely prominent above the calyx-rim.—*L. pendulum,* Sieb. Pl. Exs.; *L. gnidiæfolium,* DC. Prod. iii. 228?; *L. brevipes,* F. Muell. in Trans. Vict. Inst. 1855, 125.

Queenand. Northumberland islands, *R. Brown* (with small leaves and flowers); Ranges near Peak Downs, *F. Mueller* (with long narrow leaves).

N. S. Wales. Port Jackson to the Blue Mountains, *Sieber, n.* 312; Hastings river, *Beckler;* New England, *C. Stuart:* (Leaves small or middle sized, often somewhat cuneate, almost passing into *L. stellatum.*) Bent's Basin, *Woolls* (with very narrow leaves).

Victoria. Avon, Mitta-Mitta, Ovens, and other rivers in Gipps' Land, *F. Mueller.*

12. **L. myrtifolium,** *Sieb. in DC. Prod.* iii. 238. A tall shrub attaining 8 to 10 ft. but flowering when only 1 to 2 ft. high, the branches usually more slender than in *L. lanigerum,* glabrous or silky. Leaves usually small and rarely ½ in. long, obovate or oblong, flat or concave, nerveless or 1- or 3-nerved, glabrous or silky-white. Flowers rather small, all or nearly all solitary, sessile and axillary. Bracts none or already fallen from the very young bud. Calyx-tube turbinate, silky with appressed hairs, rarely above 2 lines diameter, lobes shorter than the tube, glabrous or slightly silky, membranous and much more deciduous than in *L. lanigerum.* Ovary flat-topped or concave, with a central depression round the style, 5-celled. Capsule flat-topped, on a level with or scarcely protruding from the calyx-rim.—Hook. f. Fl. Tasm. i. 140; *Eriostemon? trinerve,* Hook. Journ. Bot. i. 254; *L. multicaule,* A. Cunn. in Field, N. S. Wales. 349; Schau. in Walp. Rep. ii. 923.

Queensland. Moreton Island, *M'Gillivray* (like a var. of *L. lanigerum* from the same place, but with the calyx of *L. myrtifolium*); Rockhampton, *Thozet* (leaves narrow and glabrous).

N. S. Wales. Port Jackson to the Blue Mountains, *Sieber, n.* 314; in the S.W. interior, *Fraser* (all with narrow canescent leaves); near Bathurst, *A. Cunningham* (leaves small and silky).

Tasmania. Sandhills near the sea in the northern parts of the island, *J. D. Hooker* (glabrous with small leaves). The species appears to pass on the one hand into *L. stellatum,* and on the other into *L. lanigerum.*

13. **L. rupestre,** *Hook. f. in Hook. Ic. Pl. t.* 308, *and Fl. Tasm.* i. 140. *t.* 30. A procumbent or prostrate shrub, closely allied to the var. *obovatum* of *L. flavescens,* but connecting as it were that species with some forms of *L. myrtifolium* and *L. lanigerum,* and may be almost considered as an alpine variety of either of them. Leaves obovate to oblong, narrowed into a petiole, thick and usually nerveless, obtuse or nearly so, rarely exceeding 3 lines. Flowers small, sessile in the axils or terminating short leafy branchlets. Calyx-tube broad, loosely silky, lobes membranous and deciduous, but more or less silky. Capsule prominent above the calyx-rim as in *L. flavescens.*

Tasmania. Common on the tops of mountains at an elevation of 3 to 5000 ft., *J. D. Hooker.*

14. **L. myrsinoides,** *Schlecht. Linnæa,* xx. 653. A dense bushy shrub, glabrous or the young shoots silky-pubescent, approaching sometimes in habit *L. scoparium*, but the leaves not pungent, or *L. erubescens*, but with more sessile flowers, and sometimes with flexuose branches almost as in the section *Pericalymma*, and distinguished from all by the shortness of the calyx-lobes. Leaves from obovate to oblong-linear or cuneate, obtuse or obscurely mucronate-acuminate, rigid, concave, 3-nerved, often recurved, mostly 2 to 3 lines but sometimes ½ in. long. Flowers often polygamous, small, almost all terminating very short leafy branchlets. Calyx-tube silky-white with appressed hairs, campanulate, somewhat turbinate, under 2 lines diameter, the free margin often glabrous; lobes exceedingly short, membranous, glabrous. Ovary 4- or 5-celled, nearly flat-topped, with a small central depression. Capsule 2 to 3 lines diameter, scarcely projecting above the calyx.

Victoria. Common in heathy tracts in the western districts, *Robertson, F. Mueller,* and others; Snowy River, *F. Mueller;* N.W. portion, *L. Morton;* Wimmera, *Dallachy.*

S. Australia. Sandy districts between Gawler and Light rivers, *Behr;* St. Vincent's Gulf, *Whittaker, Blandowski;* Kangaroo Island, *Waterhouse.*

15. **L. erubescens,** *Schau. in Pl. Preiss.* i. 121. A spreading much-branched shrub of several feet, the branchlets rather slender but rigid and often flexuose, the young shoots silky, or at length glabrous or nearly so. Leaves from obovate and scarcely 2 lines, to oblong and nearly ½ in. long, rather thick, obscurely 1- or 3-nerved, often concave and recurved, especially when short. Flowers usually shortly pedicellate, axillary and solitary or two together on short lateral branchlets. Calyx-tube broadly turbinate, 1 to 1½ lines long, silky or rarely nearly glabrous; lobes ovate-triangular, persistent, more or less fringed-ciliate. Stamens usually 15 to 20, but sometimes fewer; filaments slightly dilated. Ovary 5- or rarely 4-celled at first, flat-topped; ovules numerous in each cell. Capsule usually more convex, but rarely protruding from the calyx-rim.

W. Australia. Gordon river, *Preiss, n.* 133; Gardner ranges, *Maxwell,* also *Drummond,* 1*st Coll. n.* 145, *Roe.*

Var. *stricta.* Branches straighter, leaves longer, *Drummond,* 5*th Coll. n.* 130, and *Suppl. n.* 28 (the latter intermediate); Phillips and Oldfield rivers, *Maxwell.*

Var. *psilocalyx.* Calyx glabrous or nearly so, and distinctly ribbed.—*Kunzea podantha,* F. Muell. Fragm. ii. 28; W. Australia, *Drummond,* 5*th Coll. n.* 129; Mount Barker, *Oldfield;* Phillip's Flat, Fitzgerald ranges, *Maxwell.*

L. sericeum, Schau. in Pl. Preiss. i. 121; from the Quangen plains, *Preiss, n.* 135, appears, from the poor specimens seen, to be this species, and does not at all agree with Labillardière's figure, which represents *Kunzea sericea.*

16. **L. abnorme,** *F. Muell. Herb.* A tall shrub with rather slender virgate branches, glabrous or the young shoots minutely silky. Leaves linear-lanceolate, acute or mucronate, mostly 1 to 2 in. long, prominently 1-nerved, with 1 or 2 faint lateral nerves on each side. Flowers nearly sessile, rather small, axillary or several together in a compact sessile terminal corymbose raceme. Bracts very deciduous. Calyx-tube turbinate, glabrous, nearly 1½ lines long; lobes ovate-triangular, persistent, with petal-like margins. Petals about twice as long as the calyx-lobes, less contracted at the base than in most *Leptospermums.* Stamens about 25, crowded opposite

the sepals, solitary opposite the petals. Ovary 3-celled, slightly convex, with a deep central depression; ovules numerous. Capsule convex, but shorter than the calyx-tube.—*Kunzea brachyandra*, F. Muell. Fragm. ii. 27.

N. Australia. Port Essington, *Armstrong*. (Flowers small and mostly imperfect.)
Queensland. Northumberland Island, *R. Brown*; Duck Creek, *Dallachy*. (Leaves in both only about ½ in. and the specimens in fruit only and therefore doubtful.)
N. S. Wales. Hastings river, *A. Cunningham*, *Dallachy*; Severn river, *C. Stuart*.

17. **L. Roei,** *Benth.* Branches slender, virgate, silky-pubescent. Leaves obovate-oblong, obtuse, narrowed at the base, flat, 3 to nearly 6 lines long, thick, silky-white, or at length glabrous. Flowers rather large, nearly sessile and axillary. Bracts small. Calyx-tube turbinate, rather broad, densely villous, with white silky hairs, about 1½ lines long; lobes very silky, persistent, about 1 line long. Ovary 3-celled, with numerous ovules in each cell.

W. Australia. In the interior, *Roë*.

SECTION III. PERICALYMMA.—Ovary usually 3-celled. Ovules few (4 to 8 in each cell). Seeds often solitary in each cell, not winged. Branches usually dichotomous and flexuose.

The following three species may possibly prove to be varieties of a single one.

18. **L. floridum,** *Benth.* An erect shrub, attaining 10 ft., but flowering when still small, with the habit, flexuose dichotomous branchlets, foliage and inflorescence of *L. ellipticum*, only differing in the larger flowers, the calyx and bracts clothed with long silky hairs, the calyx-lobes larger, usually about as long as the tube. Stamens numerous.—*Pericalymma floridum*, Schau. in Pl. Preiss. i. 121.

W. Australia. Swan River, *Preiss*, *n.* 131, *Drummond*, 1*st Coll. n.* 144.

19. **L. ellipticum,** *Endl. in Hueg. Enum*, 51. A tall erect glabrous shrub, the smaller branchlets flexuose and dichotomous. Leaves from obovate-elliptical to narrow-cuneate, obtuse acute or mucronate, usually narrowed at the base, concave and recurved at the end, 2 to 3 lines long or rarely more. Flowers rather small, solitary, sessile in the upper axils and often appearing almost terminal, surrounded by 3 or 4 imbricate scarious bracts. Calyx-tube turbinate, about 1 line long; lobes small, ovate, persistent. Petals obovate, often 2 lines long. Stamens about 15. Ovary 3-celled, with 5 or 6 ovules in each cell, the style in a deep central depression; stigma peltate. Fruiting calyx ovoid, 1½ to 2 lines long, crowned by the erect persistent lobes. Capsule much shorter than the calyx-tube, 3-celled with a hardened endocarp. Seeds solitary in each cell; testa thin; embryo straight, the cotyledons much longer than the radicle.—*Pericalymma ellipticum*, Schau. in Pl. Preiss. i. 120.

W. Australia. King George's Sound to Vasse and Swan rivers, *Huegel*, *Drummond*, 2*nd Coll. n.* 80, *Baxter*, *Preiss*, *n.* 132 and 157.

20. **L. crassipes,** *Lehm. Ind. Sem. Hort. Hamb.* 1842, *according to Schauer.* A small shrub, from a few inches to nearly a foot high, the base of the stem much thickened and almost fusiform, otherwise the tortuous dichotomous branches, foliage and inflorescence are those of *L. ellipticum.* Leaves usually small. Flowers much smaller than in *L. ellipticum.* Calyx-tube glabrous, turbinate, ¾ line long; lobes nearly as long. Petals about 1 line

long. Stamens about 10. Ovary 3-celled with 4 or 5 ovules in each cell, shorter than in *L. ellipticum.*—*Pericalymma crassipes*, Schau. in Pl. Preiss. i. 120.

W. Australia. King George's Sound, *R. Brown;* boggy ground near Albany, *Preiss, n.* 155, also *Drummond, n.* 220. Possibly an abnormal state rather than a variety of *L. ellipticum.*

19. KUNZEA, Reichb.

(Salisia, *Lindl.;* Pentagonaster, *Klotzsch.*)

Calyx-tube ovoid or globular, adnate to the ovary at the base, the free part rarely dilated; lobes 5, small, imbricate or open, usually erect, green or scarious at the edges only. Petals 5, small, orbicular, spreading. Stamens longer than the petals, indefinite, free, in 1 or several series; filaments filiform; anthers small, versatile; cells parallel, opening in longitudinal slits, the connective with a small globular gland. Ovary 2- to 5-celled, usually glabrous on the top, with 2 or more frequently numerous ovules in each cell, horizontal or pendulous from a more or less peltate placenta; style filiform, inserted in a slight central depression of the ovary; stigma small or capitate. Capsule wholly inferior, not woody, and in one species fleshy, crowned by the persistent scarcely hardened free portion of the calyx, opening at the top loculicidally. Seeds pendulous, oblong or obovoid; testa thin or firm; embryo straight; cotyledons plano-convex, longer than the superior radicle.—Shrubs, often heath-like. Leaves alternate or very rarely here and there opposite, small, entire. Flowers sessile or rarely pedicellate in the upper axils, or more frequently in terminal heads, rarely an oblong spike below the end of the branch, with a broad scale-like bract, and 2 smaller bracteoles under each flower, and sometimes several empty bracts imbricate round the head.

The genus is limited to Australia. Formerly included in *Metrosideros*, it differs in habit, inflorescence, and structure of the ovary, much nearer allied to *Leptospermum*, but readily distinguished by the exserted stamens; it also passes into *Callistemon*, through *K. Baxteri.* In *K. pauciflora* some of the leaves are often exceptionally opposite, so as almost to connect the genus with *Hypocalymma.*

Section I. **Eukunzea.**—*Ovules not numerous* (2 *to* 12) *in each cell, pendulous, in* 2 *rows.*

Ovary 2- or 3 celled, with 2 to 4 ovules in each cell.
  Leaves linear-cuneate, flat, rigid, 2 to 4 lines long. Flowers small, numerous, in dense globular heads, glabrous or nearly so . . . 1. *K. micrantha.*
  Leaves semiterete, crowded, about 2 lines long. Flowers few, rather large. Calyx densely woolly-white . . . . 2. *K. eriocalyx.*
Ovary 2- or 3-celled, with about 8 ovules in each cell. Leaves semiterete, crowded. Flowers small, in dense heads . . . . . 3. *K. Muelleri.*
Ovary 5-celled, with 8 to 12 ovules in each cell. Western species.
  Flowers in globular terminal heads.
    Leaves linear, mostly acute, 3 to 4 lines long. Flowers greenish-yellow (pink in the following species) . . . . 4. *K. ericifolia.*
    Leaves oblanceolate or linear, obtuse, flat, rigid, mostly 2 to 4 lines. Calyx softly villous . . . . 5. *K. Preissiana.*
    Leaves obovate to linear-cuneate, obtuse, rigid, 2 to 3 lines. Calyx usually glabrous . . . . 6. *K. recurva.*
    Leaves obovate or oblong, 1 to 1½ lines. Flowers small. Calyx softly villous or glabrous . . . . 7. *K. micromera.*

Flowers 2 to 4 together, scarcely capitate. Leaves semiterete, crowded. Calyx glabrous . . . . 8. *K. pauciflora.*

SECTION II. **Salisia.**—*Ovules very numerous in each cell, covering a peltate placenta.*

Ovary 3-celled or rarely 2- or 4-celled. Eastern species.
Flowers axillary or in loose or ovoid heads. Bracts lanceolate or none.
Leaves oblong or linear, 1 line or less. Flowers small, sessile . 9. *K. parvifolia.*
Leaves linear or linear-lanceolate, ¼ to ½ in. or more.
Flowers pedicellate . . . . 10. *K. pedunculatis.*
Flowers sessile . . . . 11. *K. corifolia.*
Flowers sessile, in globular terminal heads. Bracts broad.
Leaves from obovate to linear-cuneate. Bracts leafy or small and scarious. Fruiting-calyx ovoid, dry . . . . 12. *K. capitata.*
Leaves ovate or orbicular. Bracts very broad, as long as the calyx. Fruiting-calyx globular, fleshy . . . . 13. *K. pomifera.*
Ovary 5-celled. Flowers rather long. Western species.
Leaves obovate, silky. Flowers polygamous; males in a loose terminal cluster or short raceme, the perfect ones often solitary . . 14. *K. sericea.*
Leaves linear-oblong or lanceolate. Flowers in a dense spike below the end of the branch, with long crimson stamens . . . . . 15. *K. Baxteri.*

K. trinervia, Turcz. in Bull. Mosc. 1862, ii. 326, said to be from Norfolk Island, 'Reliquiæ Cunninghamianæ,' n. 110, must be founded on some mistake. Cunningham's Norfolk Island collection contains no such plant. The collection sold some years since with "Norfolk Island" printed labels, consisted chiefly of common N.S. Wales species.

SECTION I. EUKUNZEA.—Ovules not numerous (2 to 12) in each cell, pendulous, in 2 rows on an oblong or peltate placenta.

1. **K. micrantha,** *Schau. in Pl. Preiss.* i. 125. Apparently small and erect, quite glabrous or with a very slight pubescence about the young flower-heads. Leaves linear or linear-cuneate, erect or slightly recurved, flat, rigid, obtuse, 1-nerved, 2 to 4 lines long. Flowers numerous, in dense, terminal, globular heads. Bracts broadly ovate-acuminate or rhomboidal, scarious or almost coriaceous, nearly as long as the calyx-tube; bracteoles narrow. Calyx-tube about 2 lines long, but much narrower than in most species, especially at the base, and oblique or incurved, usually glabrous; lobes very small, ovate, obtuse. Petals nearly 1 line long. Stamens not numerous, from a little longer than the petals to twice as long. Ovary 2- or 3-celled, with 2 to 4 pendulous ovules in each cell; stigma small.

W. Australia, *Drummond,* 1*st Coll., also* 5*th Coll. n.* 139 *or* 159; *Preiss;* Salt River, *Maxwell* (in Herb. Oldfield).

The species was originally described by Schauer, from specimens with the flowers so young that he had afterwards misgivings about it and suppressed it (Pl. Preiss. ii. 223), but Drummond's specimens show it to be a very distinct species, quite different from *K. micromera*, Schau., with which Preiss's specimens had been mixed, and which it resembles in foliage. In this respect it resembles also the var. *præstans* of *K. recurva*, but the flowers are different.

2. **K. eriocalyx,** *F. Muell. Fragm.* ii. 28. Apparently a small heath-like plant, glabrous except the inflorescence. Leaves crowded, linear, semiterete, channelled above, obtuse, about 2 lines long. Flowers few, in terminal heads or sometimes solitary. Bracts ovate or rhomboidal, mucronate or acuminate; bracteoles narrow. Calyx-tube nearly 2 lines long, ovoid, densely clothed with white woolly hairs; lobes short, ovoid, obtuse. Petals deep

pink, about 1½ lines diameter. Stamens 12 to 16, from a little longer than the petals to twice as long. Ovary 2-celled, with 2 collateral pendulous ovules in each cell; stigma small, not capitate.

W. Australia. Middle Mount Barren, *Maxwell*, a single small specimen in Herb. F. Muell.

3. **K. Muelleri,** *Benth.* A low heath-like bushy shrub, more or less pubescent. Leaves scattered, occasionally opposite, clustered and almost decussate on the smaller branchlets, linear, concave or semiterete, mostly 2 to 3 lines long. Flowers (yellow ?) sessile, in small, dense, softly villous heads at or just below the ends of the branches. Inner bracts narrow, scarious; bracteoles ovate-lanceolate, acute, keeled, nearly as long as the calyx-tube. Calyx-tube about 1½ lines long; lobes from broadly ovate to lanceolate-triangular, shorter than the petals. Petals about 1 line diameter. Stamens very numerous, in more than one series, not exceeding twice the length of the petals. Ovary 2- or 3-celled, with about 8 ovules in each cell, horizontal or reflexed, on a peltate placenta; stigma small, but capitate. Fruiting-calyx scarcely enlarged. Seeds pendulous, but not seen perfect.—*K. ericifolia*, F. Muell. in Trans. Vict. Inst. 1855, 123, not of Reichb.

Victoria. Haidinger Range, Mount Wellington, and Munyong mountains, at an elevation of 4 to 6000 ft., *F. Mueller.*

4. **K. ericifolia,** *Reichb. Consp.* 175. A tall shrub with virgate branches, more or less pubescent or densely hirsute, as well as the leaves, or rarely nearly glabrous. Leaves linear, spreading or recurved, flat, concave or almost triquetrous, mostly mucronate-acute, rather rigid, 2 to 4 lines long. Flowers (greenish-yellow) in dense globular heads, the rhachis woolly-villous. Bracts obovoid or rhomboidal, acute or almost obtuse, from half as long to nearly as long as the calyx, deciduous; bracteoles smaller. Calyx-tube obovoid or turbinate, glabrous or pubescent, nearly 2 lines long; lobes short, erect. Petals rather above 1 line diameter. Stamens numerous, in several series, 2 or 3 times as long as the petals. Ovary 5-celled, with about 10 ovules in each cell, in 2 rows. Fruiting-calyx considerably enlarged. Seeds black, obovoid-oblong, pendulous; cotyledons twice as long as the radicle.—*Metrosideros ericifolia*, Sm. in Rees, Cyclop. xxiii.; DC. Prod. iii. 225; *Kunzea vestita*, Schau. in Pl. Preiss. i. 126. ii. 224.

W. Australia. King George's Sound, *R. Brown*, and adjoining districts, chiefly in low wet places near the sea, *Menzies, Drummond, 1st Coll. n.* 131; *Preiss, n.* 272, and many others.

Var. *glabrior.* Generally less villous and sometimes nearly glabrous.—Swan River, *Preiss, n.* 271; Gordon river, *Maxwell.*—*Metrosideros propinqua*, Endl. in Hueg. Enum. 50; *Kunzea propinqua*, Schau. in Pl. Preiss. i. 126.

5. **K. Preissiana,** *Schau. in Pl. Preiss.* i. 125. A rather rigid much-branched shrub, from 1 to 3 or 4 ft. high, the young shoots and inflorescence more or less villous, the older leaves nearly glabrous. Leaves oblanceolate or linear-oblong, obtuse, rigid, flat, or slightly concave, 2 to 4 lines long. Flower-heads globular, not very large. Bracts broad, obtuse or shortly acuminate, not exceeding the calyx-tube; bracteoles smaller and narrow. Calyx-tube softly villous or silky, about 1½ lines long; lobes short, ovate, obtuse or scarcely acute. Petals pink, rather above 1 line diameter. Stamens not very

numerous, from a little longer than the petals to nearly twice as long. Ovary short, 5-celled, with 8 to 12 ovules in two rows in each cell, or sometimes 1 or 2 of the cells abortive; stigma small.

W. Australia, *Drummond*, 1*st Coll.*, *Preiss*, *n.* 276, *Maxwell*, *Sandford*; Kalgan river, *Oldfield*. The species is, as it were, intermediate between *K. ericifolia* and *K. recurva*.

Var. *villiceps*. Whole plant villous, with soft spreading hairs. Flowers rather larger, the heads densely villous.—Sandy places, Gordon river, *Preiss*, *n.* 275; Stirling Range, *Oldfield*, *also Drummond*, 3*rd Coll. n.* 39 *or* 49.—*K. villiceps*, Schau. in Pl. Preiss. i. 125.

6. **K. recurva,** *Schau. in Pl. Preiss.* i. 125. A tall shrub with rigid branches, the young shoots slightly pubescent, otherwise glabrous. Leaves obovate or almost spathulate, spreading or recurved, narrowed at the base, obtuse or with a minute recurved point, mostly 2 to 3 lines long. Flower-heads dense, globular, the rhachis usually woolly. Bracts very broad, rigid and dry, with scarious margins, as long as the calyx, but very deciduous; bracteoles smaller. Calyx-tube obovoid, about 2 lines long in the larger forms, glabrous or nearly so; lobes ovate, obtuse. Petals above 1 line diameter. Stamens numerous, in several series, 2 or 3 times as long as the petals. Ovary 5-celled, with 10 to 15 ovules in 2 rows in each cell; stigma scarcely dilated. Fruiting-calyx enlarged, often urceolate.

W. Australia. Swan River, *Drummond*, 1*st Coll.*; Darling Range, *Preiss*, *n.* 290; also *Drummond*, *n.* 24, *and* 5*th Coll. n.* 136 *and* 137.

Var. *melaleucoides*, F. Muell. Leaves smaller, nearly orbicular, sessile. Flowers smaller, deeper coloured. Bracts smaller.—Tone and Vasse rivers, *Oldfield*; Bald Island and Cape Riche, *Maxwell*.

Var. *præstans*. More glabrous. Leaves narrower, but varying from almost obovate, and 2 lines long to linear-cuneate and 5 lines long.—*K. præstans*, Schau. in Pl. Preiss. i. 124.—*Drummond*, 1*st Coll. and* 5*th Coll. n.* 138.

Metrosideros sororia, Endl. in Hueg. Enum. 49, referred by Schauer to his *Melaleuca Endlicheriana* (*M. seriata*, Lindl., var.), seems, from the character given, to be the same as *Kunzea recurva*.

7. **K. micromera,** *Schau. in Pl. Preiss.* ii. 223. A rigid shrub, attaining 2 to 4 ft., with spreading branches, but often quite low and diffuse, glabrous, except the inflorescence. Leaves narrow-obovate or oblong, thick, very obtuse, 1 to 1½ or nearly 2 lines long. Flower-heads numerous, but small and often few-flowered. Bracts obovate, shorter than the calyx-tube, scarious, ciliate villous or nearly glabrous. Calyx-tube scarcely above 1 line long, broadly ovoid or almost globular, softly villous or nearly glabrous; lobes ovate, about as long as the tube. Petals about 1 line diameter. Stamens not numerous, from rather longer than the petals to nearly twice as long. Ovary 5-celled, with 8 to 10 ovules, in 2 rows, in each cell; stigma small but capitate.

W. Australia, *Drummond*, 5*th Coll. n.* 135, *Preiss* (a fragment in Herb. Sonder); Kalgan river, *Oldfield*; Gardiner ranges, *Maxwell*.

8. **K. pauciflora,** *Schau. in Pl. Preiss.* i. 124. A bushy shrub, of 2 or 3 ft., with numerous erect branchlets, glabrous or nearly so. Leaves rather crowded, and in some specimens many of them opposite, erect, narrow-linear, semiterete, obtuse or scarcely mucronate, 2 to 4 lines long. Flowers large for the plant, sessile in the upper axils, sometimes solitary, but usually 2 or 3, rarely up to 6 together, in a terminal head. Bracts broad, scarious, shorter than

the calyx-tube, either with a short or a long leaf-like point. Calyx-tube ovoid, about 2 lines long, glabrous; lobes lanceolate, acute or acuminate, often half as long as the tube. Petals deep pink, fully 2 lines long. Stamens numerous, some scarcely exceeding the petals, others twice as long. Ovary 5-celled, with 8 to 10 ovules in 2 rows in each cell; stigma capitate. Fruiting-calyx slightly enlarged, urceolate.

W. Australia. Gravelly base of the Konkoberup hills towards Cape Riche, *Preiss, n.* 259; *Drummond* (*4th Coll.?*) *n.* 56, *5th Coll. n.* 134; *Maxwell;* base of Mount Bland, *Maxwell.*

SECTION II. SALISIA.—Ovules very numerous in each cell, covering the surface of a peltate placenta.

9. **K. parvifolia,** *Schau. in Pl. Preiss.* i. 124. A shrub, of several ft., with slender divaricate branches and numerous branchlets, softly pubescent when young. Leaves oblong or almost linear, erect or recurved at the end, concave, obtuse or mucronate, rarely above 1 line long. Flowers small, few in terminal heads, becoming lateral by the elongation of the shoot. Bracts lanceolate, acute. Petals and stamens not seen. Fruiting-calyx nearly globular, about 1½ lines diameter, crowned by the short acute teeth. Capsule adnate to about half the calyx-tube, but very convex, so as nearly to fill it, 3-celled, the thick peltate placentas covered with the scars of very numerous ovules. Seeds not seen.

N. S. Wales. Argyle County, *Huegel* (specimen not seen), near Berrima, Illawarra, *M'Arthur.*
Victoria. Buffalo Range, *F. Mueller.*

10. **K. peduncularis,** *F. Muell. in Trans. Vict. Inst.* 1855, 124, *and in Hook. Kew Journ.* viii. 67. A tall shrub or sometimes a small tree, the branchlets virgate, glabrous or very slightly silky when young. Leaves linear or linear-lanceolate, concave, acute, mostly about ½ in., but varying from ¼ to nearly 1 in. long. Flowers small, shortly pedicellate, in the upper axils, forming either short terminal leafy corymbs, or long interrupted leafy racemes. Bracteoles scarious, but falling off from the very young bud. Calyx glabrous, about 1½ lines long; lobes ovate, with scarious margins. Petals obovate, not exceeding 1 line. Stamens above 30, in a single series, from half as long again to twice as long as the petals. Ovary about half as long as the calyx-tube, 3-celled or very rarely 4-celled, with numerous ovules in each cell on a peltate placenta. Fruiting-calyx slightly enlarged. Seeds usually only one perfect in each cell.—*Bæckea phylicoides,* A. Cunn.; Schau. in Walp. Rep. ii. 921; *Kunzea leptospermoides,* F. Muell.; Miq. in Ned. Kruidk. Arch. iv. 146.

N. S. Wales. Banks of rivers, Argyle County, *A. Cunningham.*
Victoria. Snowy River and Macalister river, mountains near Brighton, *F. Mueller.*
Var. *brachyandra,* F. Muell. Stamens shorter, but still exceeding the petals. Leaves oblong-linear. Summits of the White Rock Mountains, Mount Aberdeen, and sources of the Genoa river, *F. Mueller.*

11. **K. corifolia,** *Reichb. Consp.* 175. A tall shrub, glabrous or the young shoots pubescent. Leaves usually crowded on the branchlets or clustered in the axils, linear or linear-lanceolate, rigid, concave, obtuse or mucronate-acute, ¼ to ½ in. long. Flowers white, nearly sessile, solitary in the upper axils of very short leafy branchlets, which are often very numerous

along the main branches. Bracts none besides the floral leaves. Calyx usually glabrous; lobes small, ovate or ovate-lanceolate. Petals rarely above 1 line diameter. Stamens numerous, in 2 or 3 irregular series, at least twice as long as the petals. Ovary 3-celled, rarely 4-celled, with very numerous ovules in each cell covering a broad peltate placenta; stigma capitate or peltate. Fruiting-calyx more or less enlarged, with erect persistent lobes.—Schau. in Pl. Preiss. i. 124; Hook. f. Fl. Tasm. i. 130; *Metrosideros corifolia*, Vent. Jard. Malm. t. 46; DC. Prod. iii. 225; *Leptospermum ambiguum*, Sm. in Trans. Linn. Soc. iii. 264, and Exot. Bot. t. 59; Lodd. Bot. Cab. t. 1998.

N. S. Wales. Port Jackson, *R. Brown*, *Sieber*, *n.* 324.

Victoria. Maritime rocks, Wilson's Promontory, Genoa river, *F. Mueller;* Glenny islands, *Wilhelmi.*

Tasmania. Islands of Bass's Straits, *R. Brown;* granite rocks, Gun-carriage and Flinders islands, *Backhouse*, *Gunn;* Schouten Island, *Herb F. Mueller.*

In the Port Jackson specimens the leaves are more slender than in the Tasmanian ones, which have also slightly tomentose calyxes and constitute the *K. pelagia*, F. Muell.; Miq. in Ned. Kruidk. Arch. iv. 145.

12. **K. capitata,** *Reichb. Consp.* 175. Branches and young shoots more or less villous with long soft hairs. Leaves obovate oblong or linear-cuneate, erect and recurved at the end, rigid, concave, obtuse or with short recurved points, mostly 2 to 4 lines long, 1- or 3-nerved. Flowers in small terminal heads, often becoming lateral by the development of the axis, the floral leaves herbaceous, but usually smaller than the others, or the inner ones reduced to scarious bracts; bracteoles cuneate, scarious, shorter than the calyx-tube. Calyx-tube rather narrow, softly villous, about 2 lines long; lobes short, lanceolate, acute. Petals scarcely exceeding the calyx-lobes. Stamens 2 or 3 times as long as the petals. Ovary 3- or rarely 4-celled, with very numerous ovules in each cell, covering a broad peltate placenta. Fruiting-calyx lengthened sometimes to 3 lines. Seeds ovoid, incurved; testa thin; cotyledons broad and rather thick, tapering into a very short radicle.—*Metrosideros capitata*, Sm. in Trans. Linn. Soc. iii. 273; DC. Prod. iii. 225; *Callistemon (Callistemma) capitatus*, Reichb. Icon. Exot. i. 59. t. 84; *Melaleuca eriocephala*, Sieb. in Spreng. Syst. iii. 336; *Kunzea Schaueri*, Lehm.; Schau. in Pl. Preiss. i. 124; *K. hirsuta*, Turcz. in Bull. Mosc. 1862, ii. 326 (from the character given).

N. S. Wales. Port Jackson to the Blue Mountains, *R. Brown*, *Sieber*, *n.* 322, *Fl. Mixt. n.* 609, and others; northward to Hastings river, *Fraser*, *Beckler.*

Var. (?) *glabrescens.* Branches slender, divaricate. Leaves 2 to 3 lines long. Flowers few in the head. Calyx glabrous or nearly so.—Between Port Jackson and Sydney, *R. Brown.* This variety almost connects the species with *K. parvifolia.*

13. **K. pomifera,** *F. Muell. in Trans. Vict. Inst.* 1855, 124, *and in Hook. Kew Journ.* viii. 66. A rigid prostrate shrub, glabrous or the young shoots slightly pubescent. Leaves ovate, from nearly orbicular and almost cordate, to narrow and acute at the base, rigid, spreading, obtuse or recurved-pointed, mostly 2 to 3 lines long, or 4 lines on luxuriant shoots. Flowers white or yellowish, sessile, not numerous, but forming dense terminal heads, and becoming lateral by the elongation of the shoot. Bracts very broadly orbicular, pubescent, coloured, as long as the calyx-tube, deciduous. Calyx-tube ovoid, silky-pubescent, 1½ to 2 lines long at the time of flowering; lobes small.

Petals scarcely 1 line diameter. Stamens numerous, 3 or 4 times as long as the petals. Ovary very short, 3-celled, with very numerous ovules in each cell, covering a broad peltate placenta. Fruiting-calyx enlarged and succulent, forming a blue berry of 3 to 4 lines diameter, crowned by the lobes. Capsule small, in the base of the calyx. Seeds ovate; testa almost crustaceous; cotyledons thick, ovate, with a very short radicle.

Victoria. Sandhills, on Lakes Nepo and Hindmarsh, Wimmera, *Dallachy*; seabeach, Portland Bay, *Allitt.*

S. Australia. Sandy shore and rocks of St. Vincent's Gulf and Rivoli Bay, *F. Mueller.*

14. **K. sericea,** *Turcz. in Bull. Mosc.* 1847, i. 162. A tall shrub with very rigid tortuose or divaricate branches, tomentose when young. Leaves obovate, crowded on the short branchlets, very obtuse or minutely mucronate, $\frac{1}{4}$ to $\frac{1}{2}$ in. long, or on luxuriant branches narrower almost spathulate and $\frac{3}{4}$ in. long, very rigid and silvery-white on both sides even when old. Flowers large, polygamous, on very short pedicels, the perfect ones often (perhaps always) solitary, the males several together in a terminal cluster or very short raceme. Bracts membranous, broad, concave, very deciduous. Calyx broadly campanulate, 3 to 4 lines diameter; lobes lanceolate, thick, tomentose, shorter than the tube. Petals 2 to $2\frac{1}{2}$ lines diameter. Ovary adnate to the base of the calyx, rudimentary or abortive in the male flowers, 5- or 6-celled in the perfect ones, with very numerous ovules in each cell on a peltate placenta; stigma small, slightly capitate. Capsule filling the slightly enlarged calyx-tube, but not projecting beyond it. Seeds angular, cuneate; testa thin; cotyledons obovate, plano-convex, much longer than the radicle. —*Leptospermum sericeum*, Labill. Pl. Nov. Holl. ii. 9. t. 147; *Salisia pulchella*, Lindl. Swan Riv. App. 10.

W. Australia. Lucky Bay, *R. Brown*; eastward of King George's Sound, *Baxter, Drummond, 5th Coll. Suppl. n.* 27; Swan River, *Fraser, Drummond, 1st Coll. also 3rd Coll. n.* 40 (*4th Coll.?*) *n.* 54.

Schauer appears to have ascertained that *Leptospermum sericeum* of Labillardière was from the "Terre van Leeuwin," not from Tasmania; but he refers to it specimens of Preiss's which evidently belong to *Leptospermum erubescens.* Labillardière's own specimen in Herb. R. Brown, in fruit only, and exactly corresponding with the fruiting specimens figured, is certainly *K. sericea*, but Labillardière's description of the flower refers to a true *Leptospermum*, taken, perhaps, from some Tasmanian specimen of *L. lanigerum.*

15. **K. Baxteri,** *Schau. in Pl. Preiss.* i. 123. A rigid shrub of several feet, minutely silky-pubescent, or the foliage at length glabrous. Leaves crowded, linear-oblong or lanceolate, flat, obtuse or somewhat acute, $\frac{1}{4}$ to $\frac{1}{2}$ in. long, usually bordered by short dense silky hairs. Flowers large, like those of a *Callistemon*, in dense terminal oblong cylindrical spikes of 1 to 2 in., the rhachis and calyxes pubescent or rarely glabrous. Calyx-tube broadly campanulate, $2\frac{1}{2}$ to 3 lines long; lobes leafy, lanceolate or linear, erect, nearly as long as the tube. Petals of a rich red, not exceeding the calyx-lobes. Stamens crimson, $\frac{1}{2}$ to $\frac{3}{4}$ or sometimes nearly 1 in. long; anthers yellow. Ovary concave at the top, glabrous or slightly silky, 5-celled, with very numerous ovules in each cell on a small peltate placenta; stigma slightly clavate. Fruiting-calyx thick and somewhat enlarged, the lobes persistent and erect, capsule about half as long as the tube, wholly adnate.—*Pentagonaster Baxteri,*

Klotzsch in Otto and Dietr. Allg. Gartenzeit. iv. 113 (according to Schauer); *Callistemon macrostachyum*, Lindl. Bot. Reg. 1838, t. 7; *Callistemon Hainesii*, F. Muell. Fragm. iii. 153.

W. Australia. To the eastward of King George's Sound, *Baxter;* between Cape Arid and Cape Le Grand, *Maxwell.* This species has the somewhat ascending ovules, inflorescence, and long richly-coloured stamens of *Callistemon*, but the calyx and ovary are much more those of *Kunzea*, thus closely connecting the two genera.

20. CALLISTEMON, R. Br.

Calyx-tube ovoid, campanulate or urceolate, adnate to the ovary at the base, the free part erect or contracted; lobes 5, imbricate, more or less scarious, deciduous. Petals 5, orbicular, spreading, longer than the calyx-lobes. Stamens much longer than the petals, indefinite, usually in several series, free or very rarely collected in clusters or very shortly united opposite the petals, or all very shortly united in a continuous ring; anthers versatile, the cells parallel, opening longitudinally. Ovary villous on the top, usually convex, with a slight depression round the style, 3- or 4-celled, with very numerous ovules in each cell, horizontal or ascending and covering a peltate placenta; style filiform with a small terminal often scarcely conspicuous stigma. Fruiting-calyx more or less hardened and enlarged, with a truncate orifice; capsule enclosed in and more or less adnate to the calyx, opening loculicidally. Seeds linear or linear cuneate, testa thin; cotyledons plano-convex, longer than the radicle.—Tall shrubs or small trees. Leaves scattered, terete, linear or lanceolate, entire, coriaceous, nerveless or with a prominent midrib and nerve-like margins and pinnate veins. Flowers showy, pale yellow or crimson, in dense oblong or cylindrical spikes, at first terminal, but the axis very soon growing out into a leafy shoot, the lower leaves of the new shoot usually reduced to dry very deciduous scales, each flower closely sessile or slightly immersed in the woody rhachis. Bracts none or dry and deciduous, rarely here and there more persistent and leaf-like. Stamens in most species ½ to 1 in. long or even more.

The genus is confined to Australia. As originally observed by R. Brown, it passes gradually into *Melaleuca*, with which F. Mueller proposes to reunite it, the *C. speciosum* being, as it were, intermediate between the two. On the other hand, it is as closely connected with *Kunzea* through *K. Baxteri*, and that genus again passes into *Leptospermum*. Yet the great majority of species of each of the four groups are separated by characters so marked and prominent that it appears more convenient to retain the four genera as generally admitted.

The species of *Callistemon*, as thus limited, have a remarkable similarity in their floral characters, scarcely differing but in the breadth and consistence of their leaves and in the length and colour of the stamens. They might, indeed, almost be considered as varieties of one species.

Leaves lanceolate.
  Stamens red.
    Western species.
      Leaves thick, penniveined. Flower-spikes dense, large, usually villous. Stamens obscurely or very shortly 5-adelphous . . . . 1. *C. speciosus.*
      Leaves thick but obscurely veined. Flower-spikes not very dense, usually glabrous. Stamens not clustered . . . . 2. *C. phœniceus.*

Eastern species. Leaves usually penniveined. Spikes glabrous or pubescent.
Spikes rather loose. Anthers dark coloured . . . . . . 3. *C. lanceolatus.*
Spikes short dense. Anthers usually yellow . . . . . . 4. *C. coccineus.*
Stamens greenish-yellow. Eastern species. Spikes usually glabrous 5. *C. salignus.*
Leaves linear. Eastern species.
Stamens red.
Leaves flat, penniveined . . . . 6. *C. rigidus.*
Leaves concave, nerveless or 1-nerved . . . . 7. *C. linearis.*
Stamens yellow or greenish . . . . 5. *C. salignus.*
Leaves linear-subulate, terete. Eastern species.
Leaves mostly above 2 in. long. Flowers large. Stamens above ½ in.
Leaves channelled above. Stamens yellowish-green, glabrous . 8. *C. pinifolius.*
Leaves quite terete. Stamens red, filaments hairy . . . . . 9. *C. teretifolius.*
Leaves under 1½ in. long, spreading, pungent. Stamens red, scarcely exceeding 4 lines . . . . 10. *C. brachyandrus.*

C. Sieberi, DC. Prod. iii. 223, was described from Sieber's specimens, n. 637, which I have not seen. In foliage the short character agrees with *C. linearis*, but the species is placed amongst those with yellowish stamens, and these are said to be only a little longer than the petals, which would remove the plant from all the species known to me.

1. **C. speciosus,** *DC. Prod.* iii. 224. A tall bushy shrub or small tree, glabrous except the inflorescence, or the young shoots silky-hairy. Leaves narrow-lanceolate, obtuse or with a callous point, narrowed at the base, mostly 3 to 4 in. long, penniveined, with a prominent midrib and nerve-like margins as in *C. lanceolatus*, but much thicker and more rigid. Flowers large, of a rich red, in dense cylindrical spikes of 3 to 5 or even 6 in., the rhachis and calyx usually pubescent or hirsute. Calyx-tube often 3 lines long; lobes 1 to 1½ lines diameter. Petals 2 to 3 lines. Stamens usually about 1 in. long, of a rich red, more or less distinctly collected in clusters or very shortly united in bundles opposite the petals. Fruiting-calyx globular, about 3 lines diameter, with a broad open truncate orifice. Capsule usually considerably shorter.—Schau. in Pl. Preiss. i. 122; *Metrosideros speciosa*, Sims, Bot. Mag. t. 1761; Lodd. Bot. Cab. t. 285; *Metrosideros glauca*, Bonpl. Jard. Malm. 86. t. 34; *Callistemon glaucus*, F. Muell. Fragm. i. 14; *Melaleuca paludosa*, R. Br. in Ait. Hort. Kew. ed. 2. iv. 410; DC. Prod. iii. 212, not of Schlecht.

W. Australia. King George's Sound and adjoining districts, *Baudin's Expedition*, *R. Brown*, *Preiss*, *n.* 351, *Drummond*, *3rd Coll. n.* 62, and others. I have followed De Candolle in preferring Sims's specific name to Bonpland's, for, although the *Pl. Rar. Malm.* bears the date of 1813 on the title page, the later parts were not published till 1816.

2. **C. phœniceus,** *Lindl. Swan Riv. App.* 10. Very closely allied to *C. lanceolatus*. Leaves narrower, 2 to 4 in. long and rarely 3 lines wide, very thick, rigid, with prominent midrib and nerve-like margins, but the pinnate veins usually quite inconspicuous. Flowers large, the spikes not dense and usually glabrous. Stamens of a rich red, about 1 in. long, not at all clustered; anthers dark or rarely light-coloured. Fruiting-calyx more contracted at the orifice than in *C. lanceolatus*.—Schau. in Pl. Preiss. i. 123; F. Muell. Fragm. iv. 53.

W. Australia. Swan River, *Drummond*, 1*st Coll.*, *Preiss*, *n.* 352, 353; Murchison river, *Oldfield.*

3. **C. lanceolatus,** *DC. Prod.* iii. 223. Usually a tall shrub, but sometimes said to be low and bushy and at others to attain 30 ft., the young shoots silky or loosely hairy and the inflorescence usually pubescent, otherwise glabrous. Leaves lanceolate, variable in breadth, usually acute and 1½ to 2 in. long but varying from 1 to 3 in., rather rigid, more or less distinctly penniveined, the margins often nerve-like. Flower-spikes 2 to 4 in. long, not very dense, the rhachis and calyxes pubescent hirsute or rarely glabrous; occasionally, especially in cultivation, the flowers are more distant and a few of them in the axils of leaf-like bracts. Calyx-tube usually about 2 lines long; lobes broad and very obtuse. Petals greenish or reddish, from 1½ to nearly 3 lines diameter. Stamens red, in some specimens deeply coloured and 1 in. long, in others much paler, more slender and scarcely above ½ in., quite free or very shortly united in a ring at the base. Fruiting-calyx not much enlarged, the truncate orifice usually open.—F. Muell. Fragm. iv. 53; *Metrosideros lanceolata*, Sm. in Trans. Linn. Soc. iii. 272; *M. citrina*, Curt. Bot. Mag. t. 260; *M. lophantha*, Vent. Jard. Cels. t. 69; *M. marginata*, Cav. Ic. iv. 18. t. 332; *Callistemon marginatus*, DC. Prod. iii. 224; *C. scaber*, Lodd. Bot. Cab. t. 1288; *M. rugulosa*, Sieb. Pl. Exs. n. 321, but perhaps not of Willd.; *M. semperflorens*, Lodd. Bot. Cab. t. 523.

Queensland. Shoalwater Bay, *R. Brown;* Brisbane river, Moreton Bay, *Fraser*, *W. Hill*, and others; Burdekin river, *F. Mueller;* Bowen river, *Bowman;* Edgecombe Bay, *Dallachy;* Condamine river and other stations in the interior, *Leichhardt;* Pine river, *Fitzalan* (with the stamens united at the base).

N. S. Wales. Port Jackson to the Blue Mountains, *R. Brown*, *Sieber*, *n.* 321, and others; northward to Hastings river, *Fraser*, *Beckler;* New England, *C. Stuart;* southward to Bango, *M'Arthur.*

Victoria. Eastern Gipps' Land, *F. Mueller.*

4. **C. coccineus,** *F. Muell. Fragm.* i. 13. Very closely allied to *C. lanceolatus* and *C. salignus.* Leaves nearly of the var. *hebestachyus* of the latter, but more rigid, almost pungent, 1 to 1½ in. long, the midrib and nerve-like margins prominent, the pinnate veins inconspicuous, the under surface often and sometimes both surfaces glandular-scabrous. Flowers rather large, the spikes not very dense, the rhachis and calyxes pubescent or glabrous. Calyx-tube 2 to 2½ lines long; lobes short and broad. Petals 2 to 3 lines diameter. Stamens ¾ to 1 in. long, red with yellow anthers, numerous, quite free. Fruiting spikes dense, the calyx more contracted at the orifice than in *C. lanceolata.*—*C. rugulosus*, Miq. in Ned. Kruidk. Arch. iv. 141, but scarcely of DC.

S. Australia. From Spencer's and St. Vincent's Gulf to the Murray and Encounter Bay and in Kangaroo Island, *Behr*, *F. Mueller*, and others.

5. **C. salignus,** *DC. Prod.* iii. 223. A tall shrub or small tree attaining sometimes 30 to 40 ft., and often undistinguishable in foliage and inflorescence from *C. lanceolatus*, the leaves are however usually more acute, more distinctly penniveined, and the nerve-like margins often more prominent; in some forms, however, the venation is, on the contrary, more obscure. Spikes in the common form glabrous, more rarely the rhachis and calyxes pubescent or villous. Flowers generally rather smaller than in *C. lanceolatus*, the

calyx-lobes more ovate. Stamens pale yellow or rarely light pink, usually rather under ½ in. long. Fruiting-calyx and capsule as in *C. lanceolatus.*—Hook. f. Fl. Tasm. i. 131; F. Muell. Fragm. iv. 54; *Metrosideros saligna*, Sm. in Trans. Linn. Soc. iii. 272; Vent. Jard. Cels. t. 70; Bonpl. Pl. Malm. t. 4; Bot. Mag. t. 1821; *Metrosideros pallida*, Bonpl. Pl. Malm. 101. t. 41; *Callistemon pallidus*, DC. Prod. iii. 223; *C. lophanthus*, Lodd. Bot. Cab. t. 1302.

Queensland. Brisbane river, Moreton Bay, *F. Mueller.*

N. S. Wales. Port Jackson to the Blue Mountains, *R. Brown, Sieber, n.* 320, and others; Hastings river, *Beckler.*

Victoria. Common on the Yarra, Ovens, Goulburn, and other rivers, *F. Mueller,* and others.

Tasmania. Derwent river, etc., *R. Brown;* abundant on river banks in all parts of the island. *J. D. Hooker.*

S. Australia. River banks and dry beds of streams towards St. Vincent's Gulf, *Behr, F. Mueller,* and others.

Var. *australis.* Leaves usually smaller (1 to 2 in.), calyx and rhachis glabrous.—*Melaleuca paludosa*, Schlecht. Linnæa, xx. 653, not of *R. Br.; C. paludosus*, F. Muell. Fragm. i. 14. To this belong the majority of the Victorian, Tasmanian, and S. Australian specimens.

Var. *hebestachyus.* Leaves rather small. Calyx and rhachis pubescent or villous.—*C. lophanthus*, Sweet, Fl. Austral. t. 29, but not the syn. of Ventenat quoted.—Victoria and Tasmania. *C. leptostachyus*, Sweet, Fl. Austral. under n. 29, is probably a weak form of the same variety.

Var. *angustifolia.* Leaves linear-lanceolate, very rigid, almost pungent, 1 to 2 in. long. Flowers glabrous.—N. W. interior of N. S. Wales, *A. Cunningham;* New England, *C. Stuart.*

Var. *viridiflora*, F. Muell. Fragm. iv. 53. Leaves rarely exceeding 1 in., narrow-lanceolate, crowded, very rigid, the veins obscure. Flowers rather large, glabrous, the stamens rather above ½ in. long, greenish-yellow.—*Metrosideros viridiflora*, Sims, Bot. Mag. t. 2602; *Callistemon viridiflorus*, DC. Prod. iii. 223; Hook. f. Fl. Tasm. i. 131.—Tasmania, often ascending to 4000 ft., *J. D. Hooker;* Gipps' Land, *F. Mueller.*

Var. *Sieberi*, F. Muell. l. c. Leaves almost linear, crowded, linear, ½ to ¾ in. long. Flowers small, in short spikes.—*C. Sieberi*, DC. Prod. iii. 223, according to F. Mueller, but scarcely agreeing with the character given.—Shoalhaven in N. S. Wales, *Woolls;* Australian Alps, *F. Mueller.*

Melaleuca pithyoides, F. Muell. Herb., from Buffalo Range, enumerated doubtfully under *Callistemon* by Miq. in Ned. Kruidk. Arch. iv. 142, must remain uncertain until the flowers are known. F. Mueller, Fragm. iv. 54, refers it to *C. saligna*, but the leaves are semiterete and pungent as in *Melaleuca nodosa* and *M. pungens;* the fruits, which may be those of a *Melaleuca* or of a *Callistemon*, form a dense cylindrical spike of about 1 in.

6. **C. rigidus,** *R. Br. in Bot. Reg. t.* 393. Very near *C. lanceolatus*, with the same habit, inflorescence and flowers. Leaves linear or very narrowly linear-lanceolate, flat, rigid, acute and almost pungent-pointed, penniveined with the midrib and nerve-like margins prominent, 2 to 5 in. long, and rarely above 2 lines wide. Flowers at least as large as in *C. lanceolatus*, in a dense spike, the rhachis and calyxes pubescent or villous. Stamens often above 1 in long, red with dark coloured anthers. Fruiting-calyx truncate with a thick open orifice exceeding the capsule.—DC. Prod. iii. 223.

N. S. Wales. Lane Cove, *R. Brown.* The specimens in other herbaria are all cultivated. Intermediate between *C. lanceolatus* and *C. linearis.* F. Mueller is disposed (Fragm. iv. 54), and perhaps correctly, to unite it with the latter. To me, however, it appears to be more nearly allied to the former, the leaves being constantly flat. The flower is the same in all three.

C. linearifolius, DC. Prod. iii. 223 (*Metrosideros linearifolia*, Link, Enum. Hort. Berol.

ii. 26) and *C. rugulosus*, DC. l. c. (*Metrosideros rugulosa*, Willd. in Link, l. c. 27; *M. scabra*, Coll. Hort. Ripul. 91; *M. glandulosa*, Desf. Cat. Hort. Par. 407; *M. macropunctata*, Dum. Cours., according to DC. l. c.) are apparently garden varieties further connecting *C. rigidus* with *C. lanceolatus*.

7. **C. linearis,** *DC. Prod.* iii. 223. Considered by F. Mueller as a variety of *C. rigidus*, it differs in the leaves all much narrower; they are quite linear, 2 to 5 in. long, concave or rarely almost flat, obtuse or acute, nerveless or with the midrib scarcely prominent and the lateral veins quite inconspicuous. Flowers large, the rhachis of the spike and calyxes usually pubescent or villous. Stamens about 1 in. long, dark or pale red, or, according to Fraser, sometimes greenish. Fruiting-calyx about 4 lines diameter, more globular and more contracted at the orifice than in *C. lanceolatus* and *C. rigidus*.

N. S. Wales. Port Jackson to the Blue Mountains, *R. Brown*, *Fraser*, *A. and R. Cunningham*, and others. When the leaves are very narrow, the specimens can scarcely be distinguished except in the colour of the stamens from *C. pinifolius*.

8. **C. pinifolius,** *DC. Prod.* iii. 223. A tall shrub, usually quite glabrous, even the inflorescence. Leaves linear-subulate, terete, more or less distinctly channelled above, rigid, obtuse, acute or pungent-pointed, 2 to 4 in. long. Flowers rather large, like those of *C. lanceolatus* except in colour. Stamens ½ to ¾ in. long, of a dull yellowish-green including the anthers.—*Metrosideros pinifolia*, Wendl. Collect. i. 53. t. 16; *C. acerosus*, Tausch in Flora, 1836, 411.

N. S. Wales. Port Jackson, *R. Brown;* Paramatta, *Woolls;* Hunter's River, *A. Cunningham*.

9. **C. teretifolius,** *F. Muell. in Linnæa*, xxv. 387. A spreading shrub of several feet, the young shoots silky, the adult foliage glabrous. Leaves linear-subulate, quite terete and not channelled, rigid, mostly acute, 3 to 4 in. long. Flowers large, in dense spikes, the rhachis and calyxes glabrous or slightly pubescent. Petals fully 2 lines broad. Stamens red or yellow, quite free, ¾ in. long; filaments bearded towards the base with long soft hairs. Fruiting-calyx about 4 lines diameter, nearly globular, much contracted at the orifice.

S. Australia. Rocky mountains of Elders Range, *F. Mueller*. Appears to be constantly distinct from *C. pinifolius* in the leaves not sulcate and the hairy filaments.

10. **C. brachyandrus,** *Lindl. in Journ. Hort. Soc.* iv. 112. A tall stiff bushy shrub or small tree, the young shoots softly hairy, and sometimes soft loose spreading hairs persistent on the older branches and foliage. Leaves linear-subulate, terete and channelled above, rigid and pungent-pointed, mostly ¾ to 1½ in. long. Spikes loose and interrupted or sometimes dense, rarely 2 in. long, the rhachis and calyxes loosely hairy. Calyx-tube broadly campanulate, 1 to 1½ lines long; lobes broad, ciliate, more or less scarious. Petals about 1½ lines diameter, glabrous or pubescent. Stamens quite free and scarcely above 4 lines long; filaments deep red; anthers yellow or pale.—F. Muell. Fragm. iv. 52; *C. arborescens*, F. Muell. in Linnæa, xxv. 388; *C. acerosus*, Miq. in Nederl. Kruidk. Arch. iv. 141, not of Tausch.

N. S. Wales. Darling river and towards the Barrier Range, *Victorian Expedition*.

Victoria. Murray desert, *F. Mueller.*
S. Australia. On the Murray, *F. Mueller.*
C. pithyoides, Miq. in Nederl. Kruidk. Arch. iv. 142, mentioned above as only known in fruit, if a *Callistemon* at all, appears to be nearer this species than to the *C. salignus.*

21. LAMARCHEA, Gaud.

Calyx-tube ovoid-globular, adnate to the ovary at the base, the free part contracted; lobes 5, ovate, leaf-like, deciduous. Petals 5, oblong, spreading. Stamens indefinite, much longer than the petals, united in 5 bundles, distinct above the middle and opposite the petals, but all united, at least to the middle, into a single tube; anthers narrow, versatile, the cells parallel, opening longitudinally. Ovary inferior, slightly convex and densely villous on the top, 3-celled, with numerous ovules in each cell descending from a peltate placenta; style filiform, with a slightly clavate stigma. Fruiting-calyx hardened and enlarged, nearly globular with a truncate orifice. Seeds . . . .—A shrub or small tree, with the habit and foliage of *Melaleuca.*

The genus is limited to a single species, differing from *Melaleuca* only in the monadelphous stamens.

1. **L. hakeæfolia,** *Gaud. in Freyc. Voy.* 484. *t.* 110. A tall shrub or small tree, glabrous or the young shoots glaucous or hoary. Leaves alternate, oblong-linear or lanceolate, rigid, almost pungent-pointed, narrowed at the base, 3-nerved, 1 to 2 in. long. Flowers large, almost sessile, singly scattered along the old wood. Bracts not seen. Calyx-tube glabrous or minutely pubescent, about 2 lines diameter, leaf-like lobes as long as the tube. Petals about twice as long as the calyx-lobes. Stamens (red?) about 1 in. long, from 9 to 15 in each bundle, the common tube more or less incurved and hairy. Fruiting-calyx closely sessile, hard and very smooth, 4 to 5 lines diameter.

W. Australia. Seashore and sandhills, Sharks' Bay, *Milne, M. Brown.*

22. MELALEUCA, Linn.

(Gymnagathis, *Schau.;* Asteromyrtus, *Schau.*)

Calyx-tube campanulate or urceolate, adnate to the ovary at the base, the free part erect contracted or scarcely dilated; lobes 5, imbricate or open, herbaceous or more or less scarious, and then occasionally irregularly confluent. Petals 5, orbicular, spreading. Stamens indefinite, much longer than the petals, united in 5 distinct bundles opposite the petals; the united part or *claw* usually flattened, from very short and broad to long and linear, the filaments (or free parts) filiform, either pinnately arranged along the margin of the claw, or clustered or digitate at the end, or covering also the inner face; anthers versatile, the cells parallel, opening longitudinally. Ovary enclosed in the calyx-tube, inferior or semi-inferior, the convex summit villous (except in *M. calycina*) with a central depression round the style; 3-celled, with indefinite ovules in each cell, either numerous and closely packed on the outer surface of a peltate placenta or few and ascending on a short peltate or 2-fid placenta; style filiform with a peltate capitate or frequently very small stigma. Capsule enclosed in the enlarged and hardened calyx, crowned by the cup-shaped or annular free part of the tube, the lobes rarely persistent, opening loculicidally at the top in 3 valves, and occasionally separable from the calyx into 3 cocci. Seeds more or less cuneate, the perfect ones usually few, testa thin; embryo straight or scarcely curved; cotyledons flat, plano-convex or folded and embracing each other, longer than the radicle.—Shrubs or trees. Leaves alternate or in a few species opposite, entire, usually coriaceous, flat concave or semiterete, 1- 3- or several nerved, very rarely thinner with recurved margins. Flowers red white or yellow, closely sessile and solitary within each bract or floral leaf, in heads or spikes, or rarely solitary and scattered, the axis of the spike usually growing out during or after the flowering, the fruiting spike forming the base of the new branch. Bracts usually scale-like and often imbricate in the young spike, but usually deciduous long before flowering. Bracteoles usually small and deciduous, or sometimes none.

The genus is probably entirely Australian, for the few supposed species common in the Indian Archipelago appear to be varieties of a single one which is also widely dispersed over tropical and Eastern Australia. It is also, generally speaking, a well-defined group, readily distinguished from *Callistemon* by the 5-adelphous stamens, from *Conothamnus* by the ovules and seeds, and from *Beaufortia* and its allies by the anthers. The only exceptions are one or two species in which the claws of the staminal bundles are so short as to connect the genus with *Callistemon*, of which one species (*C. speciosus*) has the stamens almost or quite 5-adelphous, but single transitionary species appear scarcely to justify the union of very large groups otherwise well characterized.

The great similarity of structure throughout the genus prevents the establishing any definite subdivisions, the specific distinctions resting chiefly on habit, foliage, and inflorescence, neither the opposite leaves of some species, nor even the deciduous calyx-rim of the few *Asteromyrti*, having any other character in common to justify their separation as sections. The following series, therefore, although the best I have been able to devise, will be found in many instances to pass gradually one into the other.

SERIES I. **Callistemoneæ.**—*Flowers large, red or rarely greenish-yellow, in oblong or cylindrical dense spikes, glabrous or slightly pubescent, lateral on the old wood or forming the base of leafy branches. Calyx broad at the base. Stamens above ½ in. long (not exceeding ½ in. in any other series).*

Leaves alternate.
  Leaves lanceolate, 1 to 1½ in. long. Claws of the staminal bundles long . . . . 1. *M. longicoma.*
  Leaves lanceolate, 3 to 4 in. long. Flowers large in a long spike. Claws of the staminal bundles very short . . . . *Callistemon speciosus.*
  Leaves linear or semiterete, erect or scarcely spreading, mostly about ½ in. long. Claws of the staminal bundles short.
    Leaves flat or concave, acute . . . . 2. *M. lateritia.*
    Leaves semiterete or slightly flattened, obtuse . . . . . 3. *M. calothamnoides.*
  Leaves very spreading, all under ½ in. long.
    Leaves linear or linear-lanceolate, acute, under ¼ in. long. Staminal claws long . . . . 4. *M. blæriæfolia.*
    Leaves ovate-lanceolate or oblong, obtuse, 3 to 5 lines long. Staminal claws short . . . . 5. *M. diosmifolia.*
    Leaves cordate, pungent. Staminal claws long. . . . 17. *M. cardiophylla var.*
Leaves opposite. Staminal claws long.
  Leaves elliptical-ovate, thick, flat, very obtuse, under ½ in. long 6. *M. elliptica.*
  Leaves lanceolate or oblong, with recurved margins and prominent midrib, ¾ to 1½ in. long . . . . 7. *M. hypericifolia.*
  Leaves linear-concave, almost nerveless, ¾ to 1 in. long . . . 8. *M. fulgens.*
  Leaves linear or linear-lanceolate, ¼ to ½ in. long. Flowers scarcely spicate . . . . 14. *M. Wilsonii.*

SERIES II. **Decussatæ.**—*Glabrous bushy shrubs. Leaves opposite, small, flat or concave, nerveless or 1- or 3-nerved. Flowers pink or rarely white, in small heads or clusters along the previous year's stems, or forming short loose spikes at the base of the new shoot already grown out before the flower expands. Rhachis and calyx glabrous.*

In *M. violacea,* the male flowers occasionally, although rarely, form terminal heads.

(The opposite-leaved species of the series *Spicifloræ* differ in the dense, many-flowered spikes and those of the *Capitatæ* in the flowers, whether in heads or solitary, being always at the ends of the branches at the time of expanding.)

Calyx-lobes more or less scarious and deciduous or wearing off when in fruit.
  Calyx rounded at the base, never immersed in the rhachis.
    Flowers in axillary or lateral clusters, the axis not growing out. Leaves acute, often pungent . . . . 9. *M. acuminata.*
    Flowers in small heads or clusters, forming the base of lateral leafy shoots.
      Flowers few, very small. Stamens 2 or 3 in each bundle 10. *M. leptoclada.*
      Flowers rather numerous. Stamens 20 to 30 in each bundle . . . . 11. *M. basicephala.*
  Calyx attached by the broad base, more or less immersed when in fruit in the thickened rhachis.
    Leaves ovate or obovate, rarely 3 lines long . . . . . . 12. *M. gibbosa.*
    Leaves oblong-lanceolate or linear, 3 to 6 lines long . . . 13. *M. decussata.*
Calyx-lobes herbaceous, persistent, and thickened when in fruit.
  Leaves linear. Filaments clustered at the end of the staminal claws . . . . 14. *M. Wilsonii.*
  Leaves oblong-lanceolate or almost linear, nearly nerveless.
    Filaments pinnate along the upper half of the staminal claws 15. *M. thymifolia.*
  Leaves cordate-ovate or ovate-lanceolate, 3-nerved . . . . 16. *M. violacea.*

SERIES III. **Laterales.**—*Leaves alternate. Flowers usually small, in axillary or lateral clusters, the axis very rarely growing out, the rhachis woolly-pubescent or rarely glabrous.*—Gymnagathis, *Schau.*

Leaves many-nerved, pungent-pointed, under ½ in. long.
  Leaves cordate-ovate or cordate-lanceolate . . . . 17. *M. cardiophylla.*
  Leaves lanceolate, narrowed at the base . . . . 18. *M. undulata.*
Leaves under ½ in. long, obscurely veined, not pungent.
  Leaves flat, ovate, very spreading, under 2 lines long. . . . 19. *M. elachophylla.*
  Leaves flat, obovate to lanceolate, 2 to 4 lines long . . . . 20. *M. lateriflora.*
  Leaves linear, semiterete.
    Flowers pink, immersed in the fissures of the corky branch. 21. *M. exarata.*
    Flowers white. Branches not corky . . . . 22. *M. fasciculiflora.*
Leaves above 1 in. long, obscurely veined, not pungent.
  Leaves linear-subulate, terete . . . . 23. *M. teretifolia.*
  Leaves lanceolate, flat. Calyx glabrous . . . . 24. *M. alsophila.*
  Leaves broadly-oblong to lanceolate, flat, obtuse or scarcely mucronate. Calyx pubescent . . . . 25. *M. acacioides.*

SERIES IV. **Circumscissæ.**—*Leaves alternate (usually above 1 in. long). Flowers in axillary, lateral, or rarely terminal globular heads. Calyx-tube circumsciss at the top of the ovary after flowering, and falling off with the lobes (persistent in the other series). Fruits more or less cohering in a globular head.*—Asteromyrtus, *Schau.*

Leaves obovate-oblong, 3-nerved . . . . 26. *M. Baxteri.*
Leaves oblong or lanceolate, 5- or more nerved.
  Bracts shorter than the calyx-tube. Heads mostly lateral . . 27. *M. symphyocarpa.*
  Outer bracts exceeding the calyx-tube. Heads mostly terminating short leafy branchlets. . . . . 28. *M. angustifolia.*

SERIES V. **Spicifloræ.**—*Leaves alternate or opposite. Flowers either solitary or few* *and distinct, or in more or less interrupted oblong-cylindrical or elongated spikes, sometimes at first terminal but the axis usually growing out before the flowering is over, rarely in dense lateral cylindrical spikes. Rhachis glabrous pubescent or villous.*

A few species, such as *M. eleutherostachya*, with rather dense spikes, almost pass into the yellow-flowered *Capitatæ*, in which the perfect spikes are sometimes elongated.

Leaves mostly opposite.
Leaves lanceolate, about ⅓ in. long, flat with recurved margins, 1-nerved. Flowers few, small . . . . 29. *M. pauciflora.*
Leaves ovate-cordate or ovate-lanceolate, rarely ½ in. long, 5- or 7-nerved. Spikes rather dense. . . . . 30. *M. squarrosa.*
Leaves narrow, rarely exceeding ½ in., nerveless or faintly 1-nerved. Spikes rather dense, lateral, the axis rarely growing out . . . . 31. *M. eleutherostachya.*
Leaves narrow, ¾ to 1½ in. long, nerveless or faintly 1-nerved.
Leaves flat or concave. Spikes loose. Calyx small . . . 32. *M. linariifolia.*
Leaves with involute margins. Flowers distant. Calyx large 33. *M. radula.*
Leaves small, scattered, crowded. Flowers solitary or very few. Calyx large.
Leaves oblong or ovate, squarrose, under 2 lines long . . . 34. *M. pulchella.*
Leaves linear or semiterete, erect, about 2 lines long . . . . 35. *M. conferta.*
Leaves mostly alternate. Flowers usually numerous.
Leaves flat, often vertical, several-nerved, mostly above 1 in. long. Spikes interrupted.
Leaves 2 to 8 in. long, broad or narrow. Stamens glabrous, 5 to 9 in each bundle . . . . 36. *M. leucadendron.*
Leaves 1 to 2 in. long, narrow. Stamens pubescent, 12 to 20 in each bundle . . . . 37. *M. lasiandra.*
Leaves flat, concave or undulate, several-nerved, acute or pungent-pointed, under ½ in. or rarely ¾ in. long.
Leaves flat or undulate, finely striate, mostly about ½ in. long.
Leaves linear-lanceolate or lanceolate. Spikes usually interrupted. Calyx-lobes triangular . . . . 38. *M. genistifolia.*
Leaves ovate-acuminate or ovate-lanceolate. Spikes rather dense. Calyx-lobes very acute . . . . 39. *M. styphelioides.*
Leaves concave, mostly under ¼ in. long, stem-clasping, 3- or 5-nerved. Spikes long, slender, rather dense . . . 40. *M. Huegelii.*
Leaves flat or semiterete, narrow, obscurely 1- or 3-nerved.
Leaves lanceolate or linear-lanceolate, flat, very acute, ¾ to 1 in. long.
Calyx about 1 line long . . . . 41. *M. dissitiflora.*
Calyx about ½ line long . . . . 42. *M. linophylla.*
Leaves lanceolate or oblong-linear, thick, flat, rather crowded, erect or recurved, mostly under ½ in. long. Spikes interrupted near the ends of the branches . . . . . . 43. *M. Preissiana.*
Leaves oblong or linear, narrowed at the base. Spikes rather short, glabrous, mostly at the base of the branches . . 44. *M. crassifolia.*
Leaves narrow-linear or semiterete.
Leaves crowded with small fine recurved points.
Leaves mostly above ½ in. long. Eastern species . . 45. *M. armillaris.*
Leaves mostly under ½ in. long. Western species . . 46. *M. hamulosa.*
Leaves obtuse or with straight points.
Leaves rather crowded, often flat. Flowers pink.
Leaves mostly ½ in. long or more . . . . 47. *M. brachystachya.*
Leaves mostly ¼ in. long or less . . . . 48. *M. glaberrima.*
Leaves rather distant, terete, ½ to 1 in. long or more. Flowers white . . . . 49. *M. rhaphiophylla.*

SERIES VI. **Capitatæ.**—*Leaves alternate or opposite. Flowers, at least the males, in* *terminal globular heads, the perfect ones occasionally in oblong or cylindrical dense spikes, the axis not growing out until after the flowering is over, the rhachis usually woolly hirsute. Fruiting spikes very dense, globular or oblong, rarely reduced to* 2 *or* 3 *fruits.*

SUBSERIES I. **Oppositifoliæ.**—*Leaves opposite. Flowers often few or almost solitary.*

Leaves 1 line long, thick. Spikes glabrous, ovoid-oblong. Bracts small or none . . . . 12. *M. gibbosa.*
Leaves thick, convex underneath, oblong or linear. Bracts imbricate.
Flowers several in a head, glabrous. Leaves under 3 lines long, very obtuse. Staminal claws very short . . . . 50. *M. cymbifolia.*
Flowers 1 to 3 together. Staminal claws as long as the petals.
Leaves mostly above ¼ in. long. Flowers glabrous . . . 51. *M. cuticularis.*
Leaves mostly under ¼ in. long, very obtuse. Flowers and young shoots hoary-pubescent . . . . 52. *M. sparsiflora.*
Leaves ovate-cordate or ovate-lanceolate, flat or concave, about ½ in. long.
Flowers rather large, 1 to 3 together. Bracts imbricate . . 53. *M. calycina.*
Flowers in terminal globular heads, and in lateral clusters . . 16. *M. violacea.*

SUBSERIES II. **Nervosæ.**—*Leaves alternate or scattered, flat, thick,* 3- *to* 7-*nerved, and rarely under* ½ *in. long.*

(75. *M. squamea,* has 3-nerved leaves, but very small; 77. *M. thymoides;* and 78. *M. striata,* have linear-lanceolate, acute, 3- or 5-nerved leaves.)

Laaves broadly ovate-cordate or orbicular, 3- or 5-nerved, ¼ to ½ in. long . . . . 54. *M. cordata.*
Leaves obovate-oblong or broadly oblanceolate, 5- or 7-nerved, 1 to 2 in. long. . . . . 55. *M. globifera.*
Leaves obovate-orbicular or obovate-oblong, prominently 3- or 5-nerved, 4 to 8 lines long . . . . 56. *M. megacephala.*
Leaves obovate-oblong to oblanceolate, ½ to 1 in. long, obscurely 3- or 5-nerved.
Leaves obtuse.
Flower-heads large. Stamens 4 lines long, numerous in each bundle . . . . 57. *M. nesophila.*
Flower-heads small. Stamens 2 lines long, few in each bundle . . . . 64. *M. pentagona var.*
Leaves pungent-pointed . . . . 58. *M. Oldfieldii.*

SUBSERIES III. **Longifoliæ.**—*Leaves linear, terete or flat, mostly above* 1 *in. long. Flowers usually white or yellow.*

Leaves linear, terete or rarely flat, with hooked points or rarely obtuse. Fruit spikes mostly ovoid . . . . 59. *M. uncinata.*
Leaves linear, flat, with straight points. Fruit-spikes mostly ovoid . . . . 60. *M. concreta.*
Leaves linear-terete, with straight points. Fruit-spikes rather large, ovoid . . . . 61. *M. filifolia.*
Leaves terete, with straight points. Fruit-heads small, globular . 62. *M. hakeoides.*
Leaves flat, with straight points. Fruit-heads very small. globular 63. *M. glomerata.*

SUBSERIES III. **Erythrocephalæ.**—*Leaves either linear and under* 1 *in., or if ovate or obovate under* ½ *in. long,* 1- *or obscurely* 3-*nerved. Flowers red pink or rarely orange or yellow, all in globular heads. All Western species except* M. squamea.

Leaves linear, terete or flat, rigid, pungent-pointed or rarely obtuse, 1-nerved, ½ to ¾ in. long. Flower-heads axillary and terminal . . . . 64. *M. pentagona.*
Leaves small, ovate, obovate-oblong or lanceolate, prominently 1-nerved.

Leaves obovate-orbicular, 2 to 3 lines long. Flower-heads large, orange . . . . 65. *M. ciliosa.*
Leaves ovate, thick, rarely above 2 lines. Flower-heads small, pale coloured . . . . 76. *M. densa.*
Leaves ovate-lanceolate or oblong, rigid, 3 to 4 lines long. Stamens few and short. Flower-heads pink . . . . . . 66. *M. polycephala.*
Leaves nerveless or obscurely 1-nerved. Flower-heads terminal, pink (except *M. urceolaris ?*).
Leaves obovate, 2 to 3 lines long. Stamens few and short . . 67. *M. spathulata.*
Leaves linear-oblong or cuneate, thick, hoary, 2 to 4 lines long. Calyx densely hoary-tomentose . . . . 68. *M. eriantha.*
Leaves semiterete or slightly flattened. Calyx glabrous or loosely villous.
Leaves very thick and obtuse, 2 to 3 lines long . . . . . 69. *M. subtrigona.*
Leaves thick, very obtuse, but flat, mostly 3 to 4 lines . . 70. *M. seriata.*
Leaves mostly semiterete or terete, obtuse, 3 to 6 lines.
Flowers pink ?
Calyx-lobe scarious. Western species . . . . 71. *M. scabra.*
Calyx-lobe herbaceous. Eastern species . . . . . . 83. *M. ericifolia var.*
Flowers yellow ? . . . . 72. *M. urceolaris.*
Leaves slender, mostly above ½ in. long, hirsute with fine spreading hairs, or glabrous . . . . 73. *M. trichophylla.*
Leaves rather slender, often ½ in. or more, hoary-tomentose or silky-villous . . . . 74. *M. holosericea.*
Leaves small, ovate-acuminate, rigid, 3-nerved . . . . 75. *M. squamea.*

SUBSERIES IV. **Pallidifloræ.**—*Leaves either linear-subulate and under* 1 *in. or broader and under* ½ *in. long, nerveless or rarely prominently nerved. Flowers white or pale yellow, rarely pale pink, in dense terminal heads or spikes, the males often globular, the perfect ones ovoid or oblong, rarely globular, the rhachis tomentose. Fruits in dense heads or spikes.*

Leaves ovate, acuminate, 3-nerved, about 3 lines long . . . . 75. *M. squamea.*
Leaves ovate, flat or concave, thick, imbricate or squarrose, 1- or 3-nerved, 2 to 3 lines long . . . . 76. *M. densa.*
Leaves flat, linear-lanceolate or oblong, rigid, acute, 3- or 5-nerved.
Branches often spinescent. Leaves under ⅓ in. long. Flowers white or yellow. Stamens 3 to 4 lines long . . . . . . 77. *M. thymoides.*
Branches not spinescent. Leaves about ½ in. long. Flowers white or pale pink. Stamens 5 to 6 lines long . . . . . 78. *M. striata.*
Leaves flat or concave, linear or lanceolate, acute, not thick, ¼ to ½ in. long.
Glabrous or nearly so . . . . 79. *M. polygaloides.*
Hoary-tomentose or villous . . . . 80. *M. incana.*
Leaves linear-subulate, rigid, pungent-pointed, ½ to 1 in. long.
Flower-heads globular or shortly ovoid. Eastern species . . 81. *M. nodosa.*
Flower-heads ovoid or oblong. Western species . . . . . . 82. *M. pungens.*
Leaves narrow-linear, concave or semiterete, not pungent.
Leaves ½ in. long or more. Flowers yellow . . . . 82. *M. pungens var.*
Leaves 4 to 6 lines long. Flowers white or pale.
Staminal claws as long as the petals.
Leaves obtuse or with a short straight point. Eastern species . . . . 83. *M. ericifolia.*
Leaves with recurved points or obtuse. Western species . 84. *M. viminea.*
Staminal claws very short, the filaments irregularly united. Western species . . . . 89. *M. acerosa.*
Leaves mostly 2 to 3 lines long.

Leaves very spreading, obtuse. Flowers rather large, in spikes of above 1 in. . . . . 85. *M. microphylla.*
Leaves slender, acute. Flowers small, in spikes of about ½ in. 86. *M. tenella.*

SUBSERIES V. **Pauciflorae.**—*Leaves under ½ in. long. Flowers white or pale coloured, very few in the head, or the males rarely more numerous. Fruits in clusters of* 2, 3, *or* 4.

Leaves oblong, flat, thick. Fruits 3 lines diameter, thick, and very smooth . . . . 87. *M. leiopyxis.*
Leaves oblong to linear, obtuse, concave. Calyx-lobes as long as the tube. Stamens numerous. Fruits 2 lines diameter . . 88. *M. pustulata.*
Leaves linear, slender, acute.
Flowers yellowish-white, in globular heads, the males rather numerous, rhachis tomentose . . . . 89. *M. acerosa.*
Flowers (pale pink?) 4 to 6 together, rhachis and calyx glabrous . . . . 90. *M. pauperiflora.*
Flowers and leaves white-tomentose . . . . 91. *M. aspalathoides.*

SERIES VII. **Peltatae.**—*Leaves very small, often scale-like, more or less peltately attached. Flowers small, in dense heads or spikes.*

Branchlets not excavated. Leaves mostly opposite, the points spreading, or not closely appressed to the branch.
Leaves very thick, obtuse, spreading, 1 to 2 lines long . . . 92. *M. deltoidea.*
Leaves finely pointed, erect, under 1 line long . . . . 93. *M. minutifolia.*
Branchlets excavated for the scale-like, peltate, closely appressed leaves.
Leaves mostly opposite. Flowers 3 or 4 in the heads. Calyx-lobes and petals striate. Stamens numerous in each bundle . 94. *M. foliolosa.*
Leaves mostly in whorls of 3. Flowers strictly diœcious, in globular heads. Stamens few in each bundle . . . . . . 95. *M. micromera.*
Leaves mostly alternate.
Flowering and fruiting spikes ovoid-globular or shortly oblong 96. *M. thyoides.*
Flowering and fruiting spikes oblong-cylindrical . . . . . 97. *M. tamariscina.*

(The leaves of 40. *M. Huegelii*, when small, have some resemblance to those of this section, but are attached at the base, not above it.)

M. imbricata, Link, Enum. ii. 272, *M. taxifolia*, Schlecht. in Spreng. Syst. Veg. iii. 336, and *M. ternifolia*, F. Muell.; Miq. in Ned. Kruidk. Arch. iv. 123, which I have not seen, are not sufficiently described to be recognizable, but probably belong to some of the above species. There are also numerous names in Steudel's 'Nomenclator' taken up from garden lists, etc., and not otherwise published, which are therefore here omitted.

SERIES I. CALLISTEMONEÆ.—Flowers large, red, or rarely greenish-yellow, in oblong or cylindrical dense spikes, glabrous or slightly pubescent, lateral on the old wood, or forming the base of leafy branches. Calyx broad at the base. Stamens above ½ in. long.

The inflorescence and the length of the stamens give many of the species of this series the aspect of *Callistemon*, but the stamens are always very distinctly 5-adelphous.

1. **M. longicoma,** *Benth.* Apparently a tall shrub, glabrous, except sometimes the inflorescence. Leaves alternate, oblong-lanceolate, mostly acute, much narrowed at the base, 1 to 1½ or even 2 in. long, flat or concave, 1- or 3-nerved, the lateral nerves, when present, close to the margin. Flowers large, of a rich red, in oblong-cylindrical spikes, of 1 to 2 in., forming the base of the young leafy branches, the rhachis and calyxes glabrous or pubescent. Calyx-tube about 1 line long or rather more; lobes ovate, nearly as

long as the tube, the margins slightly scarious. Petals 2 to nearly 3 lines long. Staminal bundles above ¾ in. long, the claws narrow, much longer than the petals, each with a cluster of 20 to 30 filaments at the end; anthers small. Ovules exceedingly numerous in each cell, covering a broad peltate placenta; stigma small. Fruit not seen.

W. Australia, *Drummond, Suppl. to 5th Coll. n.* 32.

2. **M. lateritia,** *Otto in Allgem. Gart. Zeit.* ii. 257, *according to Schau. in Pl. Preiss.* i. 141. A glabrous shrub, of several feet, with virgate branches. Leaves alternate, linear, acute, narrowed at the base, flat or concave, nerveless or obscurely 1-nerved, mostly about ½ in., rarely ¾ in. long, often drying of a bluish colour. Flowers large, of a rich scarlet, in oblong or cylindrical spikes, often 2 to 3 in. long, the axis usually growing out into a leafy shoot before the flowering is over Calyx-tube ovoid; about 1 line long; lobes half as long, very obtuse. Petals scarious, above 1 line diameter. Stamens fully ¾ in. long, very shortly but distinctly united in clusters of 7 to 11; anthers oblong. Ovules very numerous in each cell, covering a broad peltate placenta; stigma slightly dilated.—*M. callistemonea*, Lindl. Swan Riv. App. 8.

W. Australia. Lucky Bay, *R. Brown*; Swan River, *Fraser, Drummond, 1st Coll., Preiss, n.* 354, and others; Harvey and Gordon rivers, *Oldfield;* Bald Island, *Maxwell.*

3. **M. calothamnoides,** *F. Muell. Fragm.* iii. 114. A bushy shrub, of several feet, glabrous, except the inflorescence, and often glaucous, with virgate branches. Leaves scattered, usually crowded, linear-terete or slightly flattened, very obtuse, 4 to 6 lines long or very rarely more. Flowers large, red, in dense cylindrical spikes of 1 to 2 in., lateral on the old wood, and often reflexed, the axis growing out either before the flowering is over or shortly after, the rhachis and calyxes tomentose or nearly glabrous. Calyx-tube ovoid, rather above 1 line long; lobes ovate, about half the length of the tube. Petals 1 to 1½ lines long. Stamens 7 to 8 lines long, shortly united in bundles of about 7; anthers small, ovate. Ovules very numerous in each cell, covering a peltate placenta; stigma scarcely dilated. Fruiting-calyx urceolate, often above 2 lines diameter. Cotyledons not folded.

W. Australia. Rocks near Oolingara, Murchison river, *Oldfield.*

4. **M. blæriæfolia,** *Turcz. in Bull. Mosc.* 1847, i. 165. Glabrous and very densely branched. Leaves scattered, very numerous and all about the same size, spreading, linear or lanceolate, acute or somewhat obtuse, flat, obscurely 1-nerved, 2 to nearly 3 lines long. Flowers rather large (red?), in dense oblong-cylindrical spikes, sessile on the old wood, the axis apparently not growing out, the rhachis and calyxes glabrous. Calyx-tube ovoid, nearly 2 lines long; lobes obtusely triangular, erect, nearly 1 line long. Petals 2 lines diameter. Staminal bundles ½ in. long or rather more, the claw narrow, exceeding the petals, with 10 to 15 or even more filaments, pinnately arranged along the upper half; anthers ovoid. Ovules numerous, covering a peltate placenta; stigma broad.

W. Australia, *Drummond, 3rd Coll. n.* 45. This has the foliage of *M. brachyphylla*, but the calyx and stamens are quite different.

5. **M. diosmifolia,** *Andr. Bot. Rep. t.* 476. A tall, glabrous, rigid

shrub. Leaves scattered, spreading, crowded, ovate-lanceolate or oblong, obtuse, rigid, 1-nerved, numerous, and all about the same size, 3 to 4 lines long in some specimens, nearly ½ in. in others. Flowers rather large, greenish-yellow, in dense oblong or cylindrical spikes below the ends of the branches. Calyx-tube broad, about 2 lines diameter; lobes rounded, very obtuse, about 1 line. Petals about 2 lines. Staminal bundles ½ in. long or rather more, the claws shorter than the petals, divided each into 3 to 5 or rarely 7 filaments. Ovules numerous in each cell, covering a peltate placenta. Fruiting-calyx depressed-globular, very thick and hard, often nearly ½ in. diameter.—DC. Prod. iii. 212; *M. chlorantha*, Bonpl. Pl. Malm. 22. t. 8; *M. foliosa*, Dum. Cours, according to DC.

W. Australia. King George's Sound, *Menzies*, and to the eastward, *Baudin's Expedition*, *A. Cunningham*, *Drummond*. The leaves in Andrews's figure are unusually narrow; Boupland's represents the more ordinary form.

6. **M. elliptica,** *Labill. Pl. Nov. Holl.* ii. 31. *t.* 173. Glabrous, except sometimes the inflorescence; branches divaricate. Leaves opposite, oval, very obtuse, flat, rather thick, ¼ to nearly ½ in. long, faintly 1-nerved, with thickened margins, more or less glaucous. Flowers large and showy, red, in oblong-cylindrical lateral spikes of 2 to 3 in., the axis rarely growing out, the rhachis and calyxes glabrous or slightly pubescent. Calyx-tube thick, 1½ to nearly 2 lines long; lobes ovate, thick, about 1 line. Petals about 2 lines. Staminal bundles nearly 1 in. long, the claws linear, very much exceeding the petals, each with a cluster of 20 to 30 filaments at the ends; anthers ovate. Ovules exceedingly numerous in each cell, covering the back of a hirsute placenta. Fruiting-calyxes densely packed, at least 4 lines diameter, the lobes more persistent than in the allied species, and connivent.—DC. Prod. iii. 215.

W. Australia, *Drummond, 5th Coll. n.* 140, 187; *J. S. Roe;* Young river, *Maxwell.*

7. **M. hypericifolia,** *Sm. in Trans. Linn. Soc.* iii. 279. A tall glabrous shrub. Leaves mostly opposite, lanceolate or elliptical-oblong, obtuse or mucronate, ¾ to 1½ in. long, flat or with recurved margins, the midrib prominent underneath. Flowers large, of a rich red, in dense spikes of about 2 in., forming the base of leafy branches. Calyx-tube sessile by its broad base, about 1 line long; lobes broad, obtuse, herbaceous, about as long as the tube. Petals broad, concave, contracted at the base, about 2 lines long. Staminal bundles at least ¾ in. long, the slender claws much longer than the petals, each with 15 to 20 filaments at the end. Ovules exceedingly numerous in each cell, covering the broad peltate placenta.—DC. Prod. iii. 214; Andr. Bot. Rep. t. 200; Vent. Jard. Cels. t. 10; Lodd. Bot. Cab. t. 199; *Metrosideros hypericifolia*, Salisb. Prod. 351.

N. S. Wales. In swampy places, Port Jackson, *Burton;* the other specimens I have seen, in Smith's and several other herbaria, are all cultivated, unless it be one in Herb. F. Mueller, of doubtful origin, but found by him amongst some *Callistemons*, from Moreton Bay. The leaves of this species, rather thin, with a tendency to a recurved margin, differ in this respect from all others, except *M. pauciflora.*

8. **M. fulgens,** *R. Br. in Ait. Hort. Kew. ed.* 2. iv. 415. A tall glabrous shrub. Leaves mostly opposite, linear or linear-lanceolate, acute or

obtuse, narrowed at the base, very concave, nerveless or 1-nerved, mostly ¾ to 1 in. long, the glandular dots usually very conspicuous and black. Flowers large and showy, of a rich red, in rather loose oblong lateral spikes, the axis only growing out after flowering. Calyx-tube urceolate-globular, glabrous, 2 lines diameter or rather more; lobes short, broad, scarious, with thick centres. Petals 2 to 3 lines diameter. Staminal bundles often above 1 in. long, the claws usually exceeding the petals, with numerous filaments at the end. Ovules very numerous in each cell, covering the peltate placenta; stigma slightly dilated. Fruiting-calyx, when full grown, thick, hard, and nearly ½ in. diameter, but in many specimens remaining small, although apparently ripe. Cotyledons broad and folded.—DC. Prod. iii. 214; Bot. Reg. t. 103; Lodd. Bot. Cab. t. 378.

W. Australia. Dry gravelly ridges, King George's Sound, and to the eastward, *Fraser, Baxter, Maxwell.* The foliage is that of *M. radula,* but the flowers are very different.

SERIES II. DECUSSATÆ.—Glabrous bushy shrubs. Leaves opposite, small, flat or concave, nerveless or 1- or 3-nerved. Flowers pink or rarely white, in small heads or clusters along the previous year's stems, or forming short loose spikes at the base of the new shoot already grown out before the flowers expand, rarely (in *M. violacea*) a few males also in terminal heads. Rhachis and calyxes glabrous.

9. **M. acuminata,** *F. Muell. Fragm.* i. 15. Glabrous, with rather slender virgate branches. Leaves mostly opposite, lanceolate or elliptical, acute or acuminate, and sometimes pungent-pointed, narrowed at the base, mostly 3 to 4 lines long. Flowers whitish, few together, in lateral clusters on the previous year's branches. Calyx-tube ovoid, rounded at the base, above 1 line long; lobes very short and obtuse. Petals about 1 line diameter. Staminal bundles 3 to 4 lines long, the claws exceeding the petals, each divided at the end into about 9 to 15 filaments. Ovules numerous, on a short usually bifid placenta. Fruiting-calyx nearly globular, truncate, about 2 lines diameter.

Victoria. Wimmera and Murray Desert, *Dallachy.*

S. Australia. Port Lincoln, *R. Brown;* Mount Baker Creek, *L. Fischer;* Kangaroo Island, *R. Brown, Waterhouse.*

W. Australia. Murchison river, *Oldfield* (with whitish flowers, as in the S. Australian specimens); in the interior, *J. S. Roe* (with flowers apparently reddish, and leaves rather narrower).

10. **M. leptoclada,** *Benth.* Glabrous, with spreading elongated almost filiform branches. Leaves opposite, often distant, elliptical oblong-linear or lanceolate, acute or rather obtuse, flat or concave, in some specimens under 2 lines, in others 3 to 6 lines long, nerveless or obscurely 1-nerved. Flowers very small and few at the base of the leafy branches, growing out long before flowering, the rhachis and calyxes quite glabrous. Calyx-tube ovoid, about ¾ line long; lobes herbaceous or with scarious margins, obtuse, half as long as the tube. Petals nearly 1 line diameter. Stamens rather longer than the petals, shortly united in bundles of 2 or 3. Ovules not very numerous, erect, on a small placenta; style rather thick, with a small stigma. Fruiting-calyx attaining 1½ lines diameter.

W. Australia. King George's Sound, *R. Brown*; also *Drummond, 3rd Coll. n.* 65; *4th Coll. n.* 66.

11. **M. basicephala,** *Benth.* Glabrous, with rather slender virgate branches. Leaves opposite or rarely alternate on luxuriant shoots, elliptical-oblong or oblanceolate, obtuse or rather acute, narrowed at the base, flat or concave, 3 to 6 lines long, nerveless or 1-nerved. Flowers deep pink, in globular heads, closely sessile on the old wood, but forming the base of leafy lateral branches grown out long before the flowers expand, the rhachis and calyxes glabrous. Calyx-tube ovoid-globular, about 1 line diameter. Stamens attaining 2 lines, shortly united in bundles of 20 to 30; anthers ovate. Ovules rather numerous, erect, on a small placenta; style rather thick, with a small stigma.

W. Australia, *Drummond, 3rd or 4th Coll. n.* 48.

12. **M. gibbosa,** *Labill. Pl. Nov. Holl.* ii. 30. *t.* 172. An erect glabrous shrub, of 6 to 12 ft., either dense and bushy, or with loose slender branches. Leaves mostly opposite, ovate or obovate, spreading or recurved at the top, obtuse or mucronulate, in some specimens 2 to 3 or even 4 lines long, in others much smaller, concave, with the midrib slightly prominent underneath and sometimes 2 smaller lateral nerves. Flowers red, rather small, not numerous, in short ovoid or almost globular lateral heads or spikes, often forming the base of leafy branches. Calyx-tube shorter than broad, scarcely 1 line diameter, closely sessile by its broad base; lobes short and broad, with thick centres and petal-like margins. Petals scarcely above 1 line diameter. Stamens about 3 lines long or rather more, shortly united in bundles of 10 to 15. Ovules rather numerous, but less so than in *M. thymifolia*, erect, on a short thick placenta. Fruiting spikes ½ to 1 in. long, the calyxes somewhat enlarged, truncate, more or less immersed in the thickened woody rhachis.—DC. Prod. iii. 215; Hook. f. Fl. Tasm. i. 129.

Victoria. Marshy places, between the Grampians and Victoria ranges, *F. Mueller*, and others; on the Glenelg, *Robertson*; Portland, *Allitt*.

Tasmania. Derwent river, *R. Brown*; common in the northern parts of the island, near the sea, and in the interior, *J. D. Hooker*.

S. Australia. Moist places, Kangaroo Island, *Waterhouse*.

13. **M. decussata,** *R. Br. in Ait. Hort. Kew. ed.* 2. iv. 415. A tall glabrous shrub, attaining sometimes 20 ft. Leaves mostly opposite, from oblong lanceolate to almost linear or very rarely broad, obtuse or acute, mostly ¼ to ½ in. long, rigid, concave, nerveless or obscurely 1- or 3- nerved, erect or recurved, often decussate on the smaller branches. Flowers rather small, either in oblong or almost globular lateral heads or spikes and then usually barren, or, when fertile, in oblong or cylindrical interrupted spikes forming the base of leafy branches, the rhachis and calyxes glabrous. Calyx-tube closely sessile by the broad base, scarcely 1 line diameter; lobes short and broad. Petals above 1 line diameter. Stamens not above 3 lines long, very shortly united in bundles of 10 to 15. Ovules rather numerous, erect on a short thick placenta. Fruiting-calyx not much enlarged, truncate, more or less embedded in the thickened woody rhachis.—DC. Prod. iii. 214; Bot. Mag. t. 2268; Colla, Hort. R pul. t. 15; Lodd. Bot. Cab. t. 1208;

M. parviflora, Reichb. Iconogr. Exot. t. 31; *M. oligantha*, F. Muell.; Miq. in Ned. Kruidk. Arch. iv. 123; *M. tetragona*, Otto; DC. Prod. iii. 215.

Victoria. Summit of Mount William and others of the Grampians, *F. Mueller* (a form with low depressed stems and narrow leaves).

S. Australia. S. coast, *R. Brown;* Port Lincoln and Marble Range, *Wilhelmi;* Onkaparinga range, *F. Mueller;* St. Vincent's Gulf, *Blandowski.*

M. elegans, Hornsch., from the diagnoses in Walp. Rep. ii. 162, is most probably a garden variety of *M. decussata.*

14. **M. Wilsonii,** *F. Muell. Fragm.* ii. 124. *t.* 15. A tall elegant shrub, glabrous or the young shoots slightly pubescent. Leaves opposite, almost imbricate on the smaller branches, linear or linear-lanceolate, erect or scarcely spreading, mostly acute, ¼ to ½ in. long, thick, convex underneath and obscurely 1- or 3-nerved. Flowers red, solitary or 2 or 3 together in the axils of stem-leaves, often numerous along the principal branches without forming regular spikes, or rarely the upper ones in an irregular terminal head or spike. Calyx-tube glabrous, ovoid, rounded at the base, about 1 line long; lobes lanceolate, acute, at least as long as the tube. Petals very concave, narrowed at the base, above 1 line long. Staminal bundles about ½ in. long, the claw slender and much longer than the petals, each with 15 to 20 filaments at or near the end. Ovules rather numerous in each cell, covering a laterally attached peltate placenta.

Victoria. Around Lake Hindmarsh, Wimmera, *Dallachy.*

S. Australia. Desert of the Tattiara country, *J. E. Woods;* Port Lincoln, *R. Brown* (with more obtuse and somewhat spreading leaves).

15. **M. thymifolia,** *Sm. in Trans. Linn. Soc.* iii. 278 *and Exot. Bot. t.* 36. A low glabrous shrub, rarely above 2 ft. high, but very spreading and gregarious, often covering acres of ground. Leaves mostly opposite, lanceolate elliptical-oblong or almost linear, nearly acute, ¼ to ½ in. long or rarely more, rigid, concave, the midrib scarcely conspicuous. Flowers red, not numerous, in short ovoid or oblong lateral spikes, the axis often growing out into a leafy shoot at the time of flowering, the rhachis and calyxes glabrous. Calyx-tube ovoid, rounded at the base, about 1½ lines long; lobes much shorter, thick and obtuse. Petals nearly 2 lines long. Staminal bundles ½ in. long, the claws exceeding the petals, each with numerous filaments pinnately arranged along the upper half with a few on the inner face; anthers very small. Ovules exceedingly numerous in each cell, densely covering the peltate placenta; style rather long, the stigma slightly dilated. Fruiting-calyx not immersed in the rhachis, crowned by the persistent lobes.—DC. Prod. iii. 214; Bot. Mag. t. 1868; Lodd. Bot. Cab. t. 439; *Metrosideros calycina*, Cav. Ic. iv. 20. t. 336 (from the fig. and descr.); *Melaleuca coronata*, Andr. Bot. Rep. t. 278; *M. gnidiæfolia*, Vent. Jard. Malm. t. 4; *M. discolor*, Reich. in Spreng. Syst. iii. 337; Iconogr. Exot. t. 113; *Metrosideros gracilis*, Salisb. Prod. 352?

N. S. Wales. Port Jackson to the Blue Mountains, *R. Brown, Burton, Sieber, n.* 323, and others; Hastings river, *Beckler.*

Victoria? Some specimens from Churchill Island, Port Phillip, *Gunn*, have the calyx of this species, but rather smaller, in other respects they are more like and perhaps a variety of *M. decussata.*

In some of R. Brown's specimens the leaves are all narrow-linear.

16. **M. violacea,** *Lindl. Swan Riv. App.* 8. A low spreading glabrous shrub, the flowering branches often corky. Leaves opposite, sessile, spreading, cordate-ovate or ovate-lanceolate, acute or almost obtuse, rigid, 3-nerved, $\frac{1}{4}$ to $\frac{1}{2}$ in. long. Flowers purple-red, either in terminal globular heads with a few large bracts or in small axillary or lateral clusters with very few or small bracts, the rhachis and calyxes glabrous. Calyx-tube $\frac{3}{4}$ to 1 line long; lobes ovate, nearly as long. Petals $1\frac{1}{2}$ to above 2 lines diameter. Staminal bundles 3 to 4 lines long, but always so much incurved as to appear short; claws narrow, exceeding the petals, with short not very numerous filaments pinnately arranged from near the base. Ovules rather numerous on a shortly bifid placenta; style rather thick, with a dilated stigma. Fruiting-calyx often corky, with prominent persistent lobes.—Schau. in Pl. Preiss. i. 146.

W. Australia. King George's Sound and to the eastward towards Cape Riche, *Preiss, n.* 273, 274, *Oldfield, Maxwell, Drummond, Suppl. to 4th Coll. n.* 6, 7; (Swan River, *Drummond?*).

Var. *petiolata.* Leaves smaller, shortly petiolate, rounded at the base.—*Drummond, 5th Coll. Suppl. n.* 29.

Series III. Laterales.—Leaves alternate. Flowers usually small, in axillary or lateral clusters, the axis rarely growing out, the rhachis woolly pubescent or rarely glabrous.

17. **M. cardiophylla,** *F. Muell. Fragm.* i. 225. A tall bushy shrub, with rigid although often slender divaricate branches, the young shoots and inflorescence often pubescent, otherwise glabrous. Leaves alternate, from very broadly cordate-ovate and 2 or 3 lines long to ovate-lanceolate and nearly $\frac{1}{2}$ in. long, often stem-clasping, ovate, acuminate or pungent-pointed, rigid, striate with many nerves. Flowers usually rather small, white, in small lateral clusters. Calyx-tube thick, above 1 line long, striate; lobes short, broad, continuous with the tube, occasionally with narrow scarious margins. Petals above 1 line long, rigid in the centre with scarious margins. Staminal bundles about 3 lines long; the claws spathulate, exceeding the petals, each with numerous filaments pinnately arranged along the upper half. Ovules rather numerous, erect on a short placenta; style short, stigma not dilated. Fruiting-calyxes distinct, nearly globular, very smooth, about 4 lines diameter.

W. Australia. Murchison river, *Oldfield*, including a variety with lanceolate leaves.

Var. *parviflora.* Flowers small, the calyx woolly.—Swan River, *Drummond, 1st Coll.*; Murchison river, *Oldfield*; Sharks' Bay, *Denham*; Dirk Hartog's Island, *Milne*; Bay of Rest, N.W. Coast, *A. Cunningham.*

Var.? *longistaminea,* F. Muell. Staminal bundles $\frac{3}{4}$ in. long, the claws much longer than the petals.—Murchison river, *Oldfield.*

18. **M. undulata,** *Benth.* A very rigid shrub, with thick often tortuous branches, usually glabrous except the inflorescence. Leaves alternate, lanceolate or ovate-lanceolate, acuminate, acute and pungent-pointed, narrowed at the base, rigid, undulate but otherwise flat, more or less distinctly several-nerved, $\frac{1}{4}$ to nearly $\frac{1}{2}$ in. long. Flowers white, much larger than in *M. lateriflora,* not numerous, in lateral clusters of which the upper ones rarely form an irregular terminal spike, the rhachis glabrous or pubescent. Calyx-

tube ovoid, above 1 line long; lobes short, broad, thick, striate, with narrow scarious edges. Staminal bundles 3 to 4 lines long, the claws rather narrow, longer than the petals, each with numerous filaments clustered at the end with a few on the inner face. Ovules not very numerous, erect on a short placenta; stigma small. Fruiting-calyx thick, hard and smooth, about 3 lines diameter.

W. Australia, *Drummond, 3rd or 4th Coll. n.* 45; between Cape Arid and Lucky Bay, *Baxter;* towards the Great Bight, *Maxwell.* This species has the leaves sometimes of *M. styphelioides*, but a very different inflorescence and calyx. A specimen from the Melbourne Botanic Garden has the young shoots hairy.

Var. *minor.* Leaves narrower nearly those of *M. lateriflora*, var. *acutifolia*, but the flowers of *M. undulata.*—*Drummond, 5th Coll. n.* 172.

19. **M. elachophylla,** *F. Muell. Fragm.* iii. 120. A bushy glabrous shrub of several feet. Leaves scattered, spreading, ovate, obtuse, 1 to nearly 2 lines long, flat or slightly concave, thick, rigid, nerveless. Flowers pink or purple, few in small loose globular heads, at first lateral, but the axis soon growing out into a leafy branch, the rhachis and calyxes glabrous. Calyx-tube ¾ line long; lobes short, obtuse, herbaceous or slightly scarious. Petals above 1 line diameter. Staminal bundles 3 to 4 lines long, the claws about as long as the petals, each with 7 to 11 filaments. Ovules not very numerous, on a peltate placenta. Fruiting-calyx about 2 lines diameter.

W. Australia, *Drummond, 5th Coll. n.* 153; Fitzgerald river, *Maxwell.*

20. **M. lateriflora,** *Benth.* Glabrous except the slightly pubescent young shoots and the inflorescence. Leaves alternate, broadly obovate-spathulate in the primary form, obtuse or mucronulate, 2 to 4 lines long, flat, rigid, obscurely several-nerved, narrowed into a distinct petiole. Flowers small (white?), in globular clusters, axillary or lateral on the old wood, the rhachis pubescent. Calyx-tube glabrous, ovoid, about 1 line long; lobes not half so long, obscurely striate in the centre with broad scarious margins. Petals stiff, about 1 line diameter. Staminal bundles 3 to 4 lines long, the claws longer than the petals, each with 15 to 20 or even more filaments crowded at the end. Ovules rather numerous in each cell, erect on a small bifid placenta; style, when perfect, long, with a small stigma. Fruiting-calyx very smooth, about 1½ lines diameter, crowned by the short persistent lobes.

W. Australia, *Drummond* (*4th Coll.?*), *n.* 75.

Var. *elliptica.* Leaves mostly broadly elliptical-oblong, 4 to 6 lines long, with a few of the lower ones only of each branch more or less obovate-spathulate.—*Drummond* (*4th Coll.?*), *n.* 62.

Var. *acutifolia.* Leaves oblong-lanceolate, oblanceolate or slightly cuneate, acute, nerveless.—*Drummond, 5th Coll. n.* 140. The form of the leaves in this variety is so different that it seems difficult to unite it with that first described, but the inflorescence and flowers are precisely the same, and the var. *elliptica* is intermediate as to foliage.

21. **M. exarata,** *F. Muell. Fragm.* iii. 114. A low spreading shrub, glabrous except sometimes the inflorescence, the bark of the flowering branches very corky and deeply furrowed. Leaves scattered, crowded, linear, thick, concave or semiterete, obtuse, 2 to 3 lines long, often much tuberculate. Flowers small, red, irregularly scattered along the previous year's branches or forming long cylindrical but not dense spikes, and usually inserted in the

furrows of the cork, the rhachis and calyxes glabrous or rarely pubescent. Calyx-tube campanulate, about 1 line long; lobes ovate, rather shorter than the tube. Petals above 1 line long, very spreading or deflexed. Stamens scarcely above 3 lines long, shortly united in bundles of 7 to 11. Ovules erect, on a short bifid placenta; stigma small. Fruiting-calyx about 2 lines diameter, more or less corky, and often half-immersed in the cork of the branch.—*Calothamnus (?) suberosa*, Schau. in Pl. Preiss. i. 156.

W. Australia, *R. Brown;* towards Cape Riche, *Preiss, n.* 206 *b, Drummond, n.* 17, 43, *and* 5*th Coll. n.* 161, *Maxwell.*

Drummond's specimens, 5th Coll. n. 162, 168 (partly) have the bright red flowers and other characters of *M. exarata*, except that the branches are not at all corky.

22. **M. fasciculiflora,** *Benth.* Glabrous, except sometimes the inflorescence, the branches not at all or only slightly corky. Leaves scattered, often crowded, linear, thick, concave or semiterete, obtuse, in some specimens all under ¼ in. in others ¼ to ½ in. long. Flowers apparently white, in lateral or axillary clusters along the previous year's branches, or rarely the males in small terminal heads, the axis not growing out, the very short rhachis glabrous or pubescent. Calyx-tube glabrous, ovoid-campanulate, about 1 line long; lobes half as long, obtusely triangular. Petals scarcely 1 line long, usually reflexed. Stamens nearly 3 lines long, very shortly united in bundles of 7 to 11. Ovules not numerous, erect on a small bifid placenta; stigma small. Fruiting-calyx thick, about 2 lines diameter, scarcely corky, densely clustered.

W. Australia, *Drummond,* 5*th Coll. n.* 159, 164, 168 (partly); Gordon river, *Maxwell.*

23. **M. teretifolia,** *Endl. in Hueg. Enum.* 49.. A tall erect shrub with long rigid branches, quite glabrous. Leaves alternate, linear-subulate, terete, smooth or slightly sulcate above, rigid, acute, mostly 1½ to 2 in. long. Flowers white, rather small, in sessile axillary or lateral heads, the rhachis glabrous. Calyx-tube ovoid or campanulate, about 1 line long; lobes short, obtuse, herbaceous or with very narrow scarious margins. Petals about 1 line diameter. Staminal bundles nearly 3 lines long, the claws about as long as the petals, each with 7 to 11 filaments at the end. Ovules rather numerous on a peltate placenta; style rather thick, with a broad stigma. Fruiting-calyx about 1½ lines diameter, urceolate or nearly globose.—*M. hakeacea*, F. Muell. Fragm. iii. 117; *Gymnagathis teretifolia*, Schau. in Linnæa, xvii. 243, and in Pl. Preiss. i. 133.

W. Australia. Marshes or moist sandy places. Swan River, *Huegel, Drummond,* 1*st Coll. and* (2*nd Coll. ?*) *n.* 49; Woodman's Point and Hester Point, *Preiss, n.* 268, 269; Harvey river, *Oldfield;* Hampden, *Clark.*

24. **M. alsophila,** *A. Cunn. Herb.* Quite glabrous. Leaves alternate, mostly vertical, oblanceolate, acute or rarely obtuse, much narrowed at the base, 1½ to 2½ in. long, flat, thick, rigid, obscurely 3- or 5-nerved. Flowers small, in small sessile axillary or lateral clusters or heads. Calyx almost urceolate, quite glabrous, the tube about ¾ line long; lobes about half as long, ovate, obtuse. Petals small, exceedingly deciduous. Staminal bundles about 3 lines long, the claws exceeding the petals, each with 7 to 11 fila-

ments at the end. Ovules few in each cell, erect, on a small placenta; stigma small.

N. Australia. N.W. Coast, Usborne's Harbour, *Voyage of the Beagle;* Liverpool river and Cambridge Gulf, *A. Cunningham.*

25. **M. acacioides,** *F. Muell. Fragm.* iii. 116. A small tree, of a pale green, nearly glabrous or the young shoots and inflorescence pubescent. Leaves alternate, from broadly oblong and under 1 in. to lanceolate or almost linear and 2 in. long, obtuse or scarcely mucronate, narrowed at the base, often vertical, flat, thick, faintly 3- or 5-nerved. Flowers small, in small dense sessile globular heads, mostly axillary or lateral, the rhachis and calyxes pubescent. Calyx-tube nearly globular, about $\frac{3}{4}$ line diameter; lobes short and broad. Petals about $\frac{1}{2}$ line diameter. Staminal bundles about 3 lines long, the claws much longer than the petals, unequally divided at the end each into 5 to 7 filaments. Ovules few in each cell, rather large, erect on a short thick placenta. Fruiting-calyx often scarcely above 1 line diameter.

N. Australia. Pandanus swamps, Else's Creek, Arnhem's Land, and dry plains at the sources of the Roper river, *F. Mueller.*

SERIES IV. CIRCUMSCISSÆ.—Leaves alternate. Flowers in axillary lateral or rarely terminal globular heads. Calyx-tube circumsciss at the top of the ovary after flowering and falling off with the lobes, the adnate portion alone persistent. Fruits more or less cohering in a globular head.

This series might be considered as a distinct section under Schauer's name, *Asteromyrtus*, characterized by the circumsciss calyx-tube, which has not been observed in any other species. The inflorescence, however, the cohering fruiting-calyxes, and other accessory characters occur in other groups of the genus.

26. **M. Baxteri,** *Benth.* Very rigid, the young shoots softly tomentose. Leaves alternate, obovate-oblong, obtuse, rigidly coriaceous, 3-nerved, 1 to nearly 2 in. long. Flowers not seen. Fruiting-heads lateral, sessile, densely globular, the fruits almost truncate, about 3 lines diameter. Calyx truncate to the level of the capsule as in *M. symphyocarpa* and *M. angustifolia.*

W. Australia. King George's Sound or to the eastward, *Baxter.*

27. **M. symphyocarpa,** *F. Muell. in Trans. Phil. Inst. Vict.* iii. 44. Glabrous and glaucous. Leaves alternate, oblong, obtuse, narrowed at the base, $1\frac{1}{2}$ to $2\frac{1}{2}$ in. long, mostly vertical, flat, rigid, many-nerved. Flowers in dense globular lateral heads, sessile on the former year's branches. Bracts shorter than the calyx-tube. Calyx-tube campanulate, often angular by pressure, about 2 lines long, glabrous or pubescent; lobes short, broad, orbicular. Petals rather above 1 line diameter. Staminal bundles 4 to 5 lines long, the claws narrow, much longer than the petals, each with a tuft of slender filaments at the end. Ovules rather numerous, erect on a short thick placenta. Fruiting-heads $\frac{1}{2}$ in. diameter, the fruits closely appressed or connate, the calyx-tube circumsciss and deciduous, leaving the adnate part truncate on a level with the capsule.

N. Australia. Islands of the Gulf of Carpentaria, *R. Brown;* on the mainland,

F. Mueller. Very near *M. angustifolia*, the veins of the leaves more numerous and slender, the inflorescence mostly lateral, and the bracts smaller.

28. **M. angustifolia,** *Gærtn. Fruct.* i. 172. *t.* 35. Glabrous or the young shoots slightly silky. Leaves alternate, narrow-oblong, often narrowed at the base, mostly 1½ to 2 in. long, flat, often vertical, distinctly 5-nerved. Flowers in dense terminal globular sessile heads. Bracts broad, imbricate, scale-like, usually exceeding the calyx-tube and persistent. Calyx-tube broad, nearly 2 lines diameter, silky-pubescent or villous; lobes short and broad. Petals 1½ lines diameter. Staminal bundles 4 to 5 lines long, the claws united in a ring at the base, narrow, exceeding the petals, each with a tuft of numerous short slender filaments at the end; anthers very small. Ovules several in each cell, erect on a short placenta. Fruiting-heads about ½ in. diameter, the fruits very closely appressed but scarcely connate, the calyx-tube circumsciss and deciduous, leaving the adnate part truncate on a level with the tube.—*Asteromyrtus Gærtneri*, Schau. in Linnæa, xvii. 243.

Queensland. Endeavour river, *Banks and Solander*, *A. Cunningham*.

In some heads the ovary remains small and abortive; the calyx, enlarging much after flowering, becomes broadly campanulate, bordered by the persistent hardened claws of the staminal bundles.

Gærtner figures the seeds as winged, but it is doubtful whether he had them perfect, otherwise he would have seen the embryo.

SERIES V. SPICIFLORÆ.—Leaves alternate or opposite. Flowers either solitary, or few and distant, or in more or less interrupted, oblong cylindrical or elongated spikes, sometimes at first terminal, but the axis usually growing out before the flowering is over, rarely in dense lateral cylindrical spikes. Rhachis glabrous pubescent or villous.

29. **M. pauciflora,** *Turcz. in Bull. Mosc.* 1847, i. 166. A tree of 60 to 80 ft., the young shoots silky-pubescent, the older foliage glabrous. Leaves opposite, spreading, from oblong-elliptical and obtuse or mucronate to lanceolate and acute, ¼ to ½ in. long, rather thin with the margins often recurved and the midrib prominent underneath as in *M. hypericifolia*. Flowers (white?) few, small, in short terminal spikes, the axis growing out before the flowering is over, the rhachis pubescent. Calyx glabrous, campanulate; the tube scarcely above ½ line long; lobes nearly as long, ovate, obtuse. Petals about 1 line diameter. Staminal bundles 3 to 4 lines long, the claws usually shorter than the petals but variable, each with 7 to 15 filaments at the end; anthers small. Ovules rather numerous in each cell, erect on a small peltate placenta; style long with a small stigma. Fruiting-calyx rather above 1 line diameter, crowned by the persistent lobes, but not seen very perfect.

N. S. Wales. "East Australia, *Gilbert*, *n.* 40;" *n.* 221 of Sydney woods sent to Paris in 1854, *M'Arthur*. I have not seen Gilbert's specimens, but M'Arthur's quite agree with Turczaninow's description, and the peculiar foliage occurs in no other species except *M. hypericifolia*.

Victoria? A specimen without flower from the muddy banks of the Yarra in *Herb. F. Mueller*, may belong to this species.

30. **M. squarrosa,** *Sm. in Trans. Linn. Soc.* vi. 300. A handsome erect shrub, usually from 6 to 10 ft., but sometimes attaining twice that height, glabrous or the young shoots and inflorescence pubescent or villous.

Leaves mostly opposite or nearly so, from broadly ovate-cordate to ovate-lanceolate, 5 or 7-nerved, rigid, acute, almost pungent, mostly 3 to 4 lines and rarely ½ in. long. Flowers yellowish-white, sessile in oblong or cylindrical spikes of about 1 to 2 in., at first terminal, but the axis often growing out before the flowering is over, or the flowers from the first much below the ends of the branches, the floral-leaves or bracts sometimes almost like the stem-leaves, but usually shorter, broader and sometimes reduced to small coloured bracts. Calyx-tube ovoid, above 1 line long; lobes very short, herbaceous, obtuse. Petals scarcely 1 line long. Stamens rarely above 3 lines long, very shortly united in bundles of 8 to 12 or almost free; anthers oblong. Ovules not very numerous in each cell, erect on a short 2-lobed placenta. Fruiting-spikes rather dense, but not closely compact as in the *Capitatæ.*—DC. Prod. iii. 215; Labill. Pl. Nov. Holl. ii. 28. t. 169; Bot. Mag. t. 1935; Hook. f. Fl. Tasm. i. 129; Lodd. Bot. Cab. t. 1130; *M. myrtifolia*, Vent. Jard. Malm. t. 47.

N. S. Wales. Port Jackson to the Blue Mountains, *Fraser*, *R. Cunningham*, and others; Illawarra, *Shepherd.*

Victoria. Marshes of the Yarra, *F. Mueller*; moist heaths on the Glenelg, *Robertson;* Portland, *Allitt.*

Tasmania. Port Dalrymple, *R. Brown;* abundant in moist sandy soil, *J. D. Hooker.*

S. Australia. Kangaroo Island, *R. Brown;* towards Rivoli Bay, *F. Mueller.*

31. **M. eleutherostachya,** *F. Muell. Fragm.* iii. 117. A tall bushy shrub with virgate branches, glabrous except the inflorescence. Leaves opposite, linear or lanceolate, mostly erect and recurved, acuminate with a short recurved point, flat or concave, nerveless or 3-nerved, ¼ to ½ in. long. Flowers (white?) in oblong or cylindrical spikes of about 1 in., not very dense, lateral and sessile or shortly pedunculate on the old wood, the axis very rarely growing out into a leafy branch, the rhachis woolly. Calyx-tube nearly glabrous, campanulate, ½ to ¾ line long; lobes as long as the tube, orbicular and much imbricate, rigid and striate, with a narrow scarious minutely ciliate border. Petals about 1 line diameter, not striate. Staminal bundles about 5 lines long, the claws narrow, much longer than the petals, each with 15 or more filaments at the end. Ovules rather numerous, erect on a short placenta; stigma not dilated. Fruiting-spike dense, rarely above 1 in. long; calyxes nearly globular, 1½ to 2 lines diameter, crowned by the persistent inflexed lobes. Cotyledons broad, concave or slightly folded.

W. Australia. Murchison river, *Oldfield;* Sharks' Bay, *Milne.*

Var. *abietina.* More rigid. Leaves spreading, short, very rigid, decussate on the younger branches.—*Drummond* 5*th Coll. n.* 160, *J. S. Roe.*

32. **M. linariifolia,** *Sm. in Trans. Linn. Soc.* iii. 278, *and Exot. Bot. t.* 56. A tall tree, with slender branches, the young shoots and inflorescence usually pubescent, the adult foliage glabrous and often glaucous. Leaves mostly opposite, linear or linear-lanceolate, concave or keeled, rigid, acute, ¾ to 1½ in. long. Flowers in distinct pairs, in rather dense spikes of 1 to 1½ in., at first terminal or in the upper axils, the axis soon growing out into a leafy branch, the rhachis and calyxes more or less pubescent. Calyx-tube ovoid-globular, 1 to 1½ lines long; lobes shorter, broad, obtuse, with scarious or petal-like margins. Petals about twice as long as the calyx-lobes. Sta-

minal bundles often ½ in. long or more, the claws long and narrow, sometimes filiform, each with numerous pinnately-arranged filaments; anthers very small. Ovules very numerous in each cell, covering a peltate placenta; style rather thick, with a broadly capitate stigma. Fruiting-calyx not much enlarged. Seeds minute, cuneate; cotyledons not folded and not much longer than the radicle.—DC. Prod. iii. 214; *Metrosideros hyssopifolia*, Cav. Ic. iv. 20. t. 336.

Queensland. Moreton Bay, *C. Stuart.*

N. S. Wales. Port Jackson, *R. Brown, Fraser*, and others; Hastings river, *Beckler.*

Var. *trichostachya.* Leaves usually smaller. Flowers smaller in looser spikes. Bracts very narrow. Stamens more crowded on a shorter claw. Fruiting-calyx rather more open.—*M. trichostachya*, Lindl. in Mitch. Trop. Austr. 277; Belyando river, *Mitchell;* Burdekin and Gilbert rivers and along the N.E. Coast, *F. Mueller;* Cooper's Creek, *Howitt's Expedition.*

33. **M. radula,** *Lindl. Swan Riv. App.* 8. A tall glabrous shrub with virgate branches. Leaves opposite, linear, acute, concave or with involute margins, ¾ to 1½ or even 2 in. long, nerveless or obscurely 1-nerved. Flowers pink or white, rather large, closely sessile in pairs at the base or below the ends of leafy branchlets, the pairs distant or rarely forming interrupted spikes. Calyx-tube glabrous, attached by the broad base, 1 to 1½ lines long and often broader than long; lobes very short and broad, usually reduced to a narrow scarious rim. Petals 2 to 3 lines diameter. Staminal bundles attaining nearly ½ in., the claws usually shorter than the petals, each with very numerous (50 to 60 or more) filaments on the inner face as well as on the edges; anthers small. Ovules numerous in each cell, covering a peltate placenta; style rather thick, with a broad peltate stigma. Fruiting-calyx when perfect globular, smooth, 3 to 4 lines diameter, but often scarcely enlarged although apparently ripe. Seeds linear-cuneate; cotyledons semiterete or slightly folded.—Schau. in Pl. Preiss. i. 145.

W. Australia. Swan River, *Drummond*, 1*st Coll.*; Canning river, *Preiss, n.* 306, 307; Champion Bay and Murchison river, *Oldfield.* The foliage is that of *M. fulgens*, but the flowers are very different.

34. **M. pulchella,** *R. Br. in Ait. Hort. Kew. ed.* 2. iv. 414. A spreading shrub of 2 to 3 ft., usually glabrous. Leaves scattered, oblong or ovate, obtuse, spreading or recurved, rarely exceeding 2 lines, rigid, nerveless or obscurely 1- or 2-nerved. Flowers rather large, solitary or 2 or 3 together below the ends of the branches. Calyx-tube glabrous, adnate by its broad base, especially after flowering, about 1 line long and often broader than long; lobes scarcely shorter than the tube, ovate-triangular, herbaceous. Petals 2½ lines diameter. Staminal bundles 4 lines to nearly ½ in. long but inflexed so as to appear short, the claws exceeding the petals, each with a few filaments at the end and very numerous shorter ones on the inner face especially near the base. Ovules exceedingly numerous, covering a peltate placenta; style clavate at the end but the stigma scarcely dilated. Fruiting-calyx urceolate-globular, 3 lines long or rather more, crowned by the persistent spreading lobes.—DC. Prod. iii. 214; *M. densa*, Colla, Hort. Ripul. App. 3. t. 4; Lodd. Bot. Cab. t. 200; *M. serpyllifolia*, Dum. Cours. according to DC.

W. Australia. Lucky Bay, *R. Brown;* King George's Sound, and to the eastward,

Drummond, 5*th Coll. n.* 145; gravelly soil in the interior, Phillip's river, Cape Le Grand and Cape Arid, *Maxwell.* In the male flowers the calyx is much less adnate at the base than in the perfect ones.

35. **M. conferta,** *Benth.* A densely-branched glabrous shrub. Leaves scattered, crowded, erect, narrow-linear, concave or semiterete, nerveless, obtuse or almost acute, about 2 lines long. Flowers few, rather large, sessile below the ends of the branches, surrounded by scarious deciduous bracts. Calyx urceolate, the tube fully 1½ lines long; lobes nearly 1 line, ovate with scarious margins. Petals reflexed, nearly 2 lines long. Staminal bundles about 4 lines long, the claws shorter than the petals, each with very numerous filaments on the inner face as well as at the end and on the edges. Ovules very numerous in each cell, on a peltate placenta; style long, with a dilated stigma. Fruiting-calyx globular, truncate, 3 to 4 lines diameter.

W. Australia. In the interior, *J. S. Roe.* Allied to *M. pulchella*, but with a very different foliage.

36. **M. leucadendron,** *Linn. Mant.* 105. A tree often attaining a considerable size, with a thick often spongy bark peeling off in layers, the branches slender and often pendulous, but in some situations remaining a small tree or shrub with rigid erect branches. Leaves alternate, often vertical, elliptical or lanceolate, straight, oblique or falcate, acuminate, acute or obtuse, when broad very rigid and 2 to 4 in. long, when narrow sometimes 6 to 8 in. long, narrowed into a petiole, 3- to 7-nerved with anastomosing veins. Flower-spikes elongated, more or less interrupted, solitary or 2 or 3 together, from under 2 to above 6 in. long, at first terminal but the axis growing out after flowering into a leafy branch, the rhachis and calyxes glabrous pubescent tomentose or woolly. Calyx-tube ovoid, usually about 1½ lines long; lobes short, orbicular, often scarious on the margin. Petals 1 to 1½ lines diameter. Staminal bundles under ½ in. long, the claws sometimes exceedingly short, sometimes exceeding the petals, each with 5 to 8 filaments at the end. Ovules numerous, ascending on an oblong placenta. Fruiting-calyx usually about 2 lines diameter, varying from globular to almost hemispherical. Seeds obovoid or cuneate; cotyledons obovate, thick, much longer than the radicle.—F. Muell. Fragm. iv. 55; *M. leucadendron*, Linn.; *M. minor*, Sm.; and *M. viridiflora*, Gærtn.; DC. Prod. iii. 212, and the same names with the addition of *M. saligna*, Blume, Mus. Bot. i. 66, with the several synonyms quoted by DC. and Blume; *Metrosideros albida*, Sieb. Pl. Exs., referred in Spreng. Syst. Cur. Post. 194 to *M. coriacea* (attributed by mistake to Labill. instead of Salisb. Prod. 352).

N. Australia. Islands of the Gulf of Carpentaria, *R. Brown;* common from the Victoria river to the Gulf of Carpentaria, *F. Mueller*, and others.

Queensland. On the coast at various points from the Burdekin to Moreton Bay, *Banks and Solander, R. Brown, A. Cunningham, F. Mueller*, and others; also in the interior, *Mitchell.*

N. S. Wales. Port Jackson to the Blue Mountains, *R. Brown, Sieb. n.* 319, and others; Hastings and Clarence rivers, *Beckler.*

This species, very widely spread and abundant in the Indian Archipelago and Malayan Peninsula, varies exceedingly in the size, shape and texture of the leaves, in the young shoots very silky or the spikes silky-villous or woolly or the whole quite glabrous, in the short and dense or long and interrupted spikes, in the size of the flower, in the greenish-yellow, whitish, pink or purple stamens, etc., and at first sight it is difficult to believe that they all can be forms of one species, but on examination none of these variations are sufficiently constant or so combined as to allow of the definition of distinct races. In general the name of *M. leucadendron* is given to the glabrous forms, and *M. minor* to the silky or villous-flowered ones, but the indumentum is here the most uncertain of all characters. *M. lancifolia*, Turcz. in Bull. Mosc. 1847, i. 164, and *M. Cumingiana*, Turcz. l. c. from the Philippine Islands, belong to one of the common Archipelago forms with rather thin leaves and small flowers, and I cannot find the auricles of the staminal bundles mentioned as characterizing the former. *M. Cunninghami*, Schau. in Walp. Rep. ii. 927, is a large silky form with large broad thick leaves and large flowers; *M. saligna*, Schau. l. c., from Endeavour river, is more glabrous with long acuminate leaves and long glabrous interrupted spikes; *M. mimosoides*, A. Cunn., Schau. l. c., is very little different from the last. Some specimens from Rockingham Bay, *Dallachy*, and from Endeavour river, *R. Brown*, are remarkable for their dark coloured stamens.

Var. ? *parvifolia.* Leaves mostly ½ to 1 in. long. Flowers small and only very slightly pentadelphous.—*M. lanceolata*, R. Br. Herb.; *Callistemon nervosus*, Lindl. in Mitch. Trop. Austr. 335; *Leptospermum speciosum*, Schau. in Walp. Rep. ii. 923 (described from Cunningham's specimens in bud only). Behind the Government House, Sydney, *R. Brown;* rocks, Balmy Creek, in the interior of Queensland, *Mitchell;* Moreton Bay. *A. Cunningham.* This may perhaps prove to be a distinct species, but I can find no character to distinguish it from the small-leaved specimens of *M. leucadendron.*

M. Sieberi, Schau. in Walp. Rep. ii. 928, from the character given, is most probably to be included among the forms of *M. leucadendron.*

37. **M. lasiandra,** *F. Muell. Fragm.* iii. 115. A small tree, the young foliage silvery-silky, becoming glabrous and glaucous with age. Leaves alternate, often vertical, from elliptical-lanceolate to almost linear, acute or acuminate, narrowed at the base, rigid, thick, 1 to 2 in. long, obscurely 3- or 5-nerved. Flowers small, more or less distant, forming irregularly interrupted cylindrical spikes, at first terminal, but the axis soon growing out into a leafy shoot, the rhachis and calyxes softly pubescent or villous. Calyx-tube ovoid, about 1 line long; lobes ovate, about half as long as the tube. Petals not much longer than the calyx-lobes, often pubescent. Staminal bundles about 3 lines long, the claws short, more or less pubescent outside, irregularly divided, each into 12 to 20 filaments, of which some are often free almost to the base; anthers small. Ovules exceedingly numerous, covering a peltate placenta; style pubescent at the base; stigma small. Fruiting-calyx not much enlarged, crowned by the persistent lobes. Seeds not winged.

N. Australia. Arid country, on the Upper Victoria and Fitzmaurice rivers, *F. Mueller.*

38. **M. genistifolia,** *Sm. in Trans. Linn. Soc.* iii. 277, *and Exot. Bot. t.* 55. A tall shrub or a tree, attaining 30 to 40 ft. or even more, glabrous or more or less pubescent or hirsute. Leaves scattered, lanceolate or linear-lanceolate, rigid, acute and often pungent-pointed, flat, usually about ½ in., but in some specimens longer, in others much shorter, finely striate, with 7 or more nerves, conspicuous on the floral leaves, almost evanescent on others. Flowers in loose oblong or cylindrical spikes, sometimes terminal, but the axis often growing out before the flowers expand, often much interrupted, and many of the bracts developed into leaves like the stem ones or shorter and broader, rarely all small scale-like and deciduous, the rhachis and calyxes glabrous pubescent or hirsute. Calyx-tube ovoid, above 1 line long; lobes triangular, sometimes acute, almost as in *M. styphelioides*, sometimes more ob-

tuse as in *M. Preissiana.* Petals very deciduous. Staminal bundles about 3 lines long, the claws usually shorter but sometimes longer than the petals, each with numerous filaments. Ovules numerous, closely packed on a small bifid placenta. Fruiting-calyx not much enlarged, nearly globular, crowned by the persistent lobes.—DC. Prod. iii. 212; *M. lanceolata*, Otto, from the diagnoses in DC. Prod. iii. 212; *M. bracteata*, F. Muell. Fragm. i. 15; *Metrosideros decora*, Salisb. Prod. 352.

N. Australia. Sturt's Creek, Van Alphen and Upper Gilbert rivers, *F. Mueller*, also *M'Douall Stuart's Expedition.*

Queensland. Brisbane river, *Fraser, A. Cunningham,* and others; Pine river and Mount Elliott, *Fitzalan;* Marlborough, *Bowman.*

N. S. Wales. Port Jackson, *Burton;* Paramatta, *R. Brown, Woolls;* New England, *C. Stuart;* Cox's river, *Fraser.*

39. **M. styphelioides,** *Sm. in Trans. Linn. Soc.* iii. 275. A tall tree, attaining sometimes 80 ft., the young shoots and inflorescence silky-pubescent or villous, otherwise glabrous. Leaves alternate, ovate or ovate-lanceolate, acuminate, pungent-pointed, mostly about ½ in. long, rigid, finely striate, with many nerves. Flowers in rather dense oblong or cylindrical spikes, the axis growing out before the flowering is over, the floral leaves either like the stem ones and persistent or reduced to deciduous bracts. Calyx-tube ovoid, above 1 line long; lobes lanceolate, acuminate, rigid, acute or pungent, as long as or longer than the tube. Petals as long as the calyx lobes, but very deciduous. Staminal bundles about 3 lines long, the claws not much longer than the calyx-lobe, each with several filaments shortly pinnate along the upper portion. Ovules very numerous, closely packed on a small placenta. Fruiting spikes often leafy, the calyxes crowned by the rigid erect lobes.—Colla, Hort. Ripul. App. t. 6.

N. S. Wales. Port Jackson to the Blue Mountains, *R. Brown, Fraser, Woolls, Miss Athinson;* Hastings, Clarence, and Richmond rivers, *Beckler, C. Moore.*

40. **M. Huegelii,** *Endl. in Hueg. Enum.* 48. An erect shrub, of 4 to 6 ft., with virgate or spreading branches, usually glabrous, except the inflorescence. Leaves alternate, spiral, sessile, attached by the broad concave base, ovate or ovate-lanceolate, acutely acuminate, 1 to 3 lines long, more or less prominently 3- or 5-nerved. Flowers in rather dense, but slender cylindrical spikes, of 1 to 3 in., the axis growing out before the flowering is over, the rhachis tomentose. Calyx-tube broad, scarcely 1 line long; lobes short, broad, obscurely striate. Petals almost scarious, about 1 line diameter. Staminal bundles about 4 lines long, the claws exceeding the petals, irregularly divided at the end, each into 7 to 11 filaments. Ovules not very numerous in each cell, erect, on a short placenta; stigma small. Fruiting spikes cylindrical, not very dense, the calyxes about 2 lines diameter, globular, crowned by the persistent lobes.—Schau. in Pl. Preiss. i. 144.

W. Australia. Sands, Swan River, *Huegel, Fraser, Drummond* (*3rd Coll.?*), *n.* 43; *Preiss, n.* 293, *Oldfield.*

41. **M. dissitiflora,** *F. Muell. Fragm.* iii. 153. Very closely allied to *M. linophylla*, and probably a variety, and chiefly distinguished by the flowers twice as large. Young shoots silky-pubescent. Leaves alternate, linear-lanceolate, acutely acuminate, narrowed at the base, 1 to 1½ in. long, flat,

rigid, obscurely 3-nerved or nerveless. Flowers distant at the base of leafy branches. Calyx glabrous, closely sessile, almost urceolate, the tube about 1 line; lobes not half so long. Petals about $\frac{3}{4}$ line. Staminal bundles 3 lines long or rather more, the claws often exceeding the petals, each with 15 to 20 filaments, more or less pinnately arranged along the upper half. Ovules rather numerous, on a peltate placenta; style rather thick, with a broad stigma.

N. Australia. Between the Bonney and Mount Morphett, *M'Douall Stuart's Expedition.* (A single specimen in *Herb. F. Mueller.*)

42. **M. linophylla,** *F. Muell. Fragm.* iii. 115. Glabrous, except the inflorescence. Leaves alternate, linear-lanceolate, acutely acuminate, narrowed at the base, $\frac{3}{4}$ to $1\frac{1}{2}$ in. long, flat, rigid, obscurely 3-nerved. Flowers very small, often distant, forming loosely cylindrical spikes at the base of the leafy branchlets. Calyx-tube glabrous or pubescent, scarcely $\frac{1}{2}$ line long; lobes broad, obtuse. Petals very small. Staminal bundles about $1\frac{1}{2}$ lines long, the claws exceeding the petals, with several filaments more or less pinnately arranged on the upper half. Ovules numerous? but not seen perfect; style short, with a peltate stigma. Fruiting-calyx not above 1 line diameter.

N.W. Australia, *F. Gregory.* (A single specimen in *Herb. F. Mueller.*)

43. **M. Preissiana,** *Schau. in Pl. Preiss.* i. 143. A tall shrub or tree, the young shoots and often the inflorescence more or less pubescent or hirsute, becoming glabrous with age. Leaves scattered, rather crowded, erect, spreading or recurved, lanceolate or oblong-linear, acute or obtuse, flat thick and rigid, obscurely 1- or 3-nerved, rarely exceeding $\frac{1}{2}$ in. in length. Flowers not large, white or yellowish, in loose oblong or cylindrical spikes, 1 to 2 in. long, rarely terminal, the axis growing out very early into a leafy shoot, and sometimes much interrupted, many of the bracts then leafy like the stem-leaves; rhachis and calyxes glabrous or tomentose. Calyx-tube ovoid, thick, above 1 line long; lobes much shorter, ovate, often persistent. Petals about 1 line diameter or smaller. Staminal bundles 3 to nearly 4 lines long, the claws rather exceeding the petals, each with 10 to 12 or more filaments on the upper portion. Ovules very numerous, covering a broad peltate placenta. —*M. pubescens,* Schau. in Walp. Rep. ii. 928; *M. curvifolia,* Schlecht. Linnæa, xx. 654.

Victoria. Port Phillip, *R. Brown, Gunn;* Bacchus Marsh, *F. Mueller;* on the Murray, *Dallachy.*

S. Australia. Port Lincoln, *R. Brown;* Light River, *Behr;* St. Vincent's Gulf, *F. Mueller;* Kangaroo Island, *Waterhouse, F. Mueller.*

W. Australia. Goose Island Bay, *R. Brown;* King George's Sound to Cape Riche, and to Swan River, *Collie, Drummond,* 1*st Coll.; Preiss, n.* 265; *Harvey, Milne.*

Var. *leiostachya.* Inflorescence quite glabrous. Leaves often smaller, narrower and more recurved. Ovules fewer.—*M. parviflora,* Lindl. Swan Riv. App. 8; King George's Sound and to the eastward, *Harvey, Oldfield, Maxwell;* Isle Boniche, *Fraser;* Murchison river, *Drummond,* 6*th Coll. n.* 74; Dirk Hartog's Island, *Milne.*

44. **M. crassifolia,** *Benth.* Quite glabrous, with virgate branches. Leaves scattered, not crowded, erect or spreading, often incurved, otherwise flat, oblong or oblong-linear, obtuse, narrowed at the base, thick, nerveless or obscurely 1- or 3-nerved, mostly about $\frac{1}{2}$ in., but in some specimens nearly $\frac{3}{4}$,

and in others $\frac{1}{4}$ to $\frac{1}{2}$ in. long. Flowers (pale red or white?) in interrupted leafy spikes, forming the base of lateral shoots, the rhachis and calyxes quite glabrous. Calyx-tube attached by the broad base, thick, rather above 1 line long; lobes very short and broad. Petals 1 line diameter or rather more. Staminal bundles about 3 lines long, the claws as long as the petals, each bearing towards the end 11 to 15 filaments. Ovules numerous, on a peltate placenta; style thick, with a truncate stigma. Fruiting-spikes more or less interrupted, the calyxes fully 2 lines diameter, often slightly immersed in the rhachis.

W. Australia, *Drummond, 5th Coll. n.* 142, 154, and a form with smaller narrower leaves, and smaller flowers, *5th Coll. n.* 141, 153.

45. **M. armillaris,** *Sm. in Trans. Linn. Soc.* iii. 277. A tall glabrous shrub or sometimes a small tree, of 20 to 30 ft. Leaves scattered, crowded, narrow-linear, acute and often recurved at the end, mostly $\frac{1}{2}$ in. long or rather more. Flowers almost immersed in the rhachis of dense or interrupted cylindrical spikes, forming the base of the previous year's or of young lateral shoots. Calyx-tube about 1 line long; lobes shorter, almost acute. Petals above 1 line long. Staminal bundles 3 to 4 lines long or rather more, each with numerous filaments pinnately arranged along the upper half. Ovules very numerous in each cell, covering a peltate placenta; stigma broad.—DC. Prod. iii. 213; *M. ericæfolia,* Andr. Bot. Rep. t. 175; Vent. Jard. Malm. t. 76; Wendl. Coll. i. t. 29, not of Sm.; *Metrosideros armillaris,* Gærtn. Fruct. i. 171. t. 34; Cav. Ic. t. 335.

N. S. Wales. Port Jackson, *R. Brown, R. Cunningham,* and others; northward to Richmond river, *C. Moore;* southward to Twofold Bay, *A. Cunningham,* and Towamba river, *F. Mueller.*

Victoria. Common on river-banks at the south-eastern extremity of the colony, *F. Mueller.*

S. Australia. Kangaroo Island, *R. Brown.*

Var. (?) *tenuifolia.* Leaves semiterete, very narrow, under $\frac{1}{2}$ in. long. Flowers smaller. —*M. cylindrica,* R. Br. Herb.—Dunk river, *R. Brown,* perhaps a distinct species.

46. **M. hamulosa,** *Turcz. in Bull. Mosc.* 1847, i. 165. A glabrous shrub, with the virgate branches and foliage of *M. viminea,* from which it chiefly differs in the elongated inflorescence. Leaves scattered, rather crowded, erect, slightly spreading or recurved at the end, linear, with usually a fine recurved point or at length obtuse, $\frac{1}{4}$ to $\frac{1}{2}$ in. long. Flowers white, in rather dense cylindrical spikes, of 1 to 2 in., on short lateral peduncles, the axis however often growing out into a leafy shoot before the flowering is over. Calyx-tube attached by a rather broad base, about 1 line diameter; lobes short. Petals 1 line long. Staminal bundles about 3 lines long in the perfect flowers or 4 lines in the males, the claws exceeding the petals, each with 12 to 15 or more filaments at the end. Ovules numerous, on a peltate often bifid placenta. Fruiting-spikes more or less interrupted; calyxes about $1\frac{1}{2}$ lines diameter.

W. Australia, *Drummond, 3rd Coll. n.* 44, *5th Coll. n.* 149; Phillips Range, *Maxwell.* Notwithstanding the inflorescence, which in an artificial arrangement removes this to a distance from *M. viminea,* it may possibly prove to be a variety only of that species.

47. **M. brachystachya,** *F. Muell. Fragm.* iii. 119. A spreading

bushy shrub, of 4 or 5 ft., glabrous or the young shoots and inflorescence silky-pubescent. Leaves scattered, linear, flat or semiterete, rigid, acute, obscurely 1-nerved, mostly ½ to ¾ in. long. Flowers pink, in oblong or cylindrical, rather dense or interrupted spikes, the axis growing out before the flowers expand, the rhachis and calyxes usually pubescent. Calyx-tube attached by the broad base, about 1 line long; lobes much shorter, acute or obtuse. Petals above 1 line diameter. Staminal bundles fully 4 lines long, the claws rather narrow, usually exceeding the petals, with 11 to 15 or even more filaments at the end. Ovules rather numerous, on a peltate placenta.

W. Australia, *Drummond, 5th Coll. n.* 150; Gardiner river and Middle Mount Barren, *Maxwell.*

48. **M. glaberrima,** *F. Muell. Fragm.* iii. 119. Apparently diffuse, the specimen quite glabrous. Leaves scattered, rather crowded, linear, semiterete, obtuse or with a short straight point, mostly 2 to 3 lines long. Flowers pink, in rather dense oblong or cylindrical spikes, forming the base of lateral branches, the rhachis and calyxes glabrous. Calyx-tube attached by the broad base, ¾ line long, thick, with short obtuse lobes. Petals fully 1 line long. Staminal bundles about 3 lines long, the claws about as long as the petals, with 7 to 11 filaments at the end. Ovules rather numerous, on a peltate placenta; stigma small. Fruits rather dense, about 2 lines diameter.

W. Australia. Middle Mount Barren, *Maxwell.* Described from a single small specimen in Herb. F. Mueller. It is evidently nearly allied to *M. brachystachya,* and very likely a more glabrous small-leaved variety.

49. **M. rhaphiophylla,** *Schau. in Pl. Preiss.* i. 143. A tall shrub or tree, attaining sometimes 40 to 50 ft., glabrous, except sometimes the inflorescence, the bark deciduous in paper-like sheets. Leaves alternate, narrow-linear, terete or slightly flattened, mostly acute and ¾ to 1 in., rarely only ½ in. and occasionally 1½ in. long. Flowers yellowish-white, in oblong or cylindrical, somewhat interrupted spikes, either terminal or the axis grown out before the flowers expand, the rhachis and calyxes glabrous or slightly pubescent. Calyx-tube closely sessile, with a broad base, 1 to 1½ lines long and almost as broad; lobes very short, broad and scarcely scarious. Petals 1 to 1½ lines diameter. Staminal bundles 4 to 5 lines long, the claws usually exceeding the petals, each with about 15 to 20 filaments at the end or on the inner face above the middle. Ovules exceedingly numerous on a peltate placenta. Fruiting-calyxes smooth, nearly globular, 2 to 3 lines diameter, broadly sessile on the somewhat thickened rhachis.

W. Australia, *Drummond, 5th Coll. n.* 143 *and* 150; Cape Naturaliste, *Collie;* Swan River, *Preiss, n.* 264 (also 267, according to Schauer, but that n. in Herb. Sonder, is *M. trichophylla*); Murchison, Blackwood, Tweed and Fitzgerald rivers, *Oldfield;* S. Hutt river, *Gregory;* Gardiner river, *Maxwell;* and a shorter-leaved form, Fitzgerald and Phillips rivers, *Maxwell.*

Series VI. Capitatæ.—Leaves alternate or opposite. Flowers, at least the males, in terminal heads, the perfect ones occasionally in oblong or cylindrical dense spikes, the axis not growing out until after the flowering is over, the rhachis usually woolly-hirsute. Fruiting spikes very dense, globular or oblong, rarely reduced to 2 or 3 fruits.

50. **M. cymbifolia,** *Benth.* Much branched and quite glabrous. Leaves opposite, rather crowded, oblong-linear, very obtuse, thick, concave or flat above, very convex underneath, about 2 or rarely 3 lines long, usually smooth, shining, and nerveless. Flowers few (white?), in small terminal heads, surrounded by decussate imbricate bracts, falling off during flowering. Calyx-tube glabrous, more or less 5-angled, about ¾ line long. Staminal bundles not 2 lines long, the claws shorter than the petals, each with 11 to 15 filaments. Ovules rather numerous in each cell, on an oblong peltate placenta. Fruiting-calyxes few together, truncate, 2 lines diameter or rather more.

W. Australia, *Drummond, 3rd Coll. n.* 51, *5th Coll. n.* 155.

51. **M. cuticularis,** *Labill. Pl. Nov. Holl.* ii. 30. *t.* 171. A tall shrub or tree, quite glabrous, with rigid tortuous branches, the bark deciduous in paper-like layers. Leaves opposite, linear oblong or narrow-lanceolate, obtuse, thick, flat or concave above, convex underneath, ¼ to nearly ½ in. long. Flowers solitary or 2 or 3 together at the ends of the branches, surrounded by scale-like decussately imbricate bracts. Calyx-tube glabrous, campanulate, about 1½ lines long; lobes nearly as long, erect, lanceolate or triangular. Petals concave, reflexed. Staminal bundles 4 to 5 lines long, the claws about as long as the calyx-lobes, with a dense tuft of above 20 filaments at the end; anthers rather small. Ovules numerous in each cell, on a peltate placenta. Fruiting-calyx thick, campanulate, about 3 lines diameter, with thick more or less persistent lobes.—DC. Prod. iii. 214; Schau. in Pl. Preiss. i. 145; *M. abietina,* Sm. in Rees Cycl. xxiii.; DC. Prod. iii. 214.

W. Australia. King George's Sound and Swan River, *Menzies, Drummond, 1st Coll. and 5th Coll. n.* 155; *Preiss, n.* 303 *and* 304, and others.

52. **M. sparsiflora,** *Turcz. in Bull. Mosc.* 1847, i. 167. A bushy shrub, the young shoots and inflorescence more or less pubescent. Leaves opposite, decussate on the smaller branchlets, oblong, very obtuse, thick, flat or concave above, very convex and nerveless underneath, 1½ to 3 lines long. Flowers solitary or 2 or 3 together at the ends of the branches, surrounded by numerous decussately imbricate bracts, pubescent or tomentose as well as the calyxes. Calyx-tube ovoid, about 1 line long; lobes scarcely shorter. Petals above 1 line long. Staminal bundles fully 3 lines long, the claws narrow, as long as the petals, with 15 or more filaments at the end. Ovules numerous in each cell, on a peltate bifid placenta. Fruiting-calyx usually solitary, urceolate, above 2 lines diameter.

W. Australia, *Drummond, 3rd Coll. n.* 50 *and* 68.

53. **M. calycina,** *R. Br. in Ait. Hort. Kew. ed.* 2. iv. 416. A tall rigid shrub, glabrous or the young shoots slightly pubescent. Leaves opposite, cordate-ovate or ovate-lanceolate, acute, flat or concave, rigid, 3- or 5-nerved in cultivated specimens, thicker and almost nerveless in the wild ones, rarely exceeding ½ in. Flowers 2 or few together, in terminal heads or clusters, surrounded by rather numerous decussately imbricate bracts, the axis growing out soon after flowering. Calyx-tube glabrous, turbinate, about 1½ lines long; lobes herbaceous, acute, fully 2 lines long in cultivated specimens, smaller in the wild ones. Petals almost boat-shaped, scarcely exceeding the calyx-lobes.

Staminal bundles 4 to 5 lines long, the claws scarcely exceeding the petals, with 20 or more filaments at the end. Ovary exceptionally glabrous on the top; ovules numerous in each cell, on a peltate bifid placenta.—DC. Prod. iii. 215.

W. Australia. Lucky Bay, *R. Brown;* also *Drummond, 5th Coll. n.* 165.

54. **M. cordata,** *Benth.* Rigid and glabrous, except the inflorescence. Leaves numerous, alternate, very spreading, ovate or orbicular, cordate or almost stem-clasping, acute or obtuse, rigid, more or less distinctly 3- or 5-nerved, ¼ to ⅓ in. long. Flowers red, rather small, in dense, terminal, globular heads, the axis not growing out till after flowering, the rhachis and calyxes densely tomentose-villous. Bracts deciduous. Calyx-tube ovoid, about 1 line long; lobes very short and broad or scarcely prominent. Petals about 1 line diameter, usually with a deep-coloured centre. Staminal bundles 4 to 5 lines long, the claws usually exceeding the petals, with 7 to 11 filaments at the end. Ovules not very numerous, erect, on a bifid placenta; stigma small. Fruiting-calyx smooth, nearly 2 lines diameter.

W. Australia, *Drummond, 5th Coll. n.* 156, *and Suppl. 5th Coll. n.* 31. There are two forms, one with the leaves 3 to 4 lines diameter, and obscurely nerved, the other with the leaves nearly twice as large and distinctly 5-nerved, but they do not otherwise differ.

55. **M. globifera,** *R. Br. in Ait. Hort. Kew. ed.* 2. iv. 411. A tall shrub or tree, attaining 30 ft., glabrous, or the young shoots and inflorescence slightly silky-hairy, the bark deciduous in paper-like layers. Leaves alternate, from almost obovate to narrow-oblong, flat, obtuse or mucronulate, narrowed at the base, distinctly 5- or rarely 7-nerved, mostly 1½ to 2½ in. long. Flowers in dense globular terminal sessile heads, of ¾ in. or more diameter. Bracts broad, scale-like, imbricate, exceeding the calyx, but very deciduous. Calyx-tube nearly globular, nearly 1½ lines diameter; lobes very short and broad, obtuse, more or less petal-like. Petals about 1 line diameter. Staminal bundles 4 to 5 lines long, the claws often longer than the petals, and more or less confluent at the base, very irregularly divided each into 5 to 9 filaments, some of them almost free. Ovules not numerous, erect, on a short placenta. Fruiting calyxes more or less concrete, forming dense globular masses often 1 in. diameter.—DC. Prod. iii. 212.

W. Australia. King George's Sound or to the eastward, *Baxter;* Cape Arid and Middle Island, *Maxwell.*

56. **M. megacephala,** *F. Muell. Fragm.* iii. 117. A very bushy, rigid shrub, attaining 8 to 10 ft., the young shoots more or less villous with loose spreading hairs, or rarely glabrous from the first. Leaves alternate, from broadly obovate-orbicular and under ½ in. to obovate-oblong and nearly 1 in. long, very obtuse or scarcely mucronate, narrowed at the base, coriaceous, prominently 3- or 5-nerved. Flowers yellowish-white, in dense terminal globular heads, the axis not growing out until after flowering, the rhachis usually tomentose. Bracts large, broad, scale-like, imbricate on the young head, but soon falling off. Calyx-tube glabrous and membranous, or tomentose and more rigid, about 1½ lines long; lobes very thin and scarious, more or less ciliate, persistent. Petals scarious, 1½ to 2 lines diameter. Staminal bundles 5 to 6 lines long, the claws petal-like, rather broad, each with 10 to

20 filaments at the end. Fruiting-calyx often 3 lines diameter, globular, villous; the capsule separable into 3 cocci. Seeds short, thick, cuneate; cotyledons very broad, folded over each other.

W. Australia. Champion Bay and Murchison river, *Oldfield, Walcott, Drummond, 6th Coll. n.* 72: Sharks' Bay, *Martin.*

57. **M. nesophila,** *F. Muell. Fragm.* iii. 113. A shrub of 6 to 8 ft., glabrous, or the young shoots very slightly silky. Leaves alternate, obovate-oblong, obtuse or rarely mucronate, thickly coriaceous, flat and often vertical, obscurely 3-nerved, ½ to nearly 1 in. long. Flowers pink, in dense terminal heads, the rhachis and calyxes glabrous or slightly villous. Bracts short. Calyx-tube villous, broad, above 1 line long; lobes short, broad, scarious. Petals scarious, scarcely 1 line diameter. Staminal bundles about 4 lines long, the claws broad, not much exceeding the petals, and sometimes very short, with 10 to 15 filaments at the end. Ovules not very numerous, erect on a small placenta; stigma small. Fruiting spikes very dense, the calyxes truncate, often 3 lines diameter.

W. Australia. Doubtful Island, *Oldfield;* also *Drummond, 5th Coll. n.* 157 (with rather smaller flowers and fruits), *3rd Coll. n.* 54 (with rather longer and more acute leaves, and the rhachis and calyxes more villous).

58. **M. Oldfieldii,** *F. Muell. Fragm.* iii. 118. A shrub of 3 or 4 ft., with slender branches, glabrous except the inflorescence. Leaves alternate, oblong-lanceolate, with a fine almost pungent point, narrowed into a rather long petiole, rigid, obscurely 3- or 5-nerved, ½ to 1 in. long. Flowers not seen. Fruiting-heads dense, globular, ½ to ¾ in. diameter, the rhachis and calyx-tubes tomentose, and at length more or less concrete, the lobes scarious, and at length wearing away. Seeds not numerous in each cell, erect on a small placenta, but not seen perfect.

W. Australia. Murchison river, *Oldfield.*

59. **M. uncinata,** *R. Br. in Ait. Hort. Kew. ed.* 2. iv. 414. A tall shrub, the young shoots more or less silky pubescent. Leaves alternate, linear-subulate, terete or rarely slightly compressed, smooth, sulcate or almost angular, 1 to 2 in. long, with a fine recurved point, or rarely obtuse. Flowers small, numerous, in very dense terminal ovoid-oblong or almost globular heads, the axis often growing out before the flowering is over; the rhachis and calyxes woolly, hirsute, or rarely quite glabrous. Calyx-tube not 1 line long; lobes exceedingly small and short. Staminal bundles about 2 lines long, the claws about as long as the petals, each with 5 to 7 filaments. Fruiting-spikes very dense and compact, rarely above ½ in. long; the calyxes turbinate, truncate, about 1½ lines long.—DC. Prod. iii. 213; Schau. in Pl. Preiss. i. 138; *M. hamata*, Field. and Gardn. Sert. Pl. t. 74; *M. Drummondii*, Schau. in Pl. Preiss. i. 138 (rather shorter-leaved specimens); *M. semiteres*, Schau. l. c. 143 (leaves longer, not hooked).

N. S. Wales. Barren branches, Lachlan river, *Fraser.*

Victoria. Wimmera, *Dallachy;* N.W. part of the colony, *L. Morton.*

S. Australia. Port Lincoln, *R. Brown;* Boston Point and Lake Victoria, *F. Mueller;* Kangaroo Island, *F. Mueller, Waterhouse.*

W. Australia. From the south coast to Vasse, Swan, and Murchison rivers, *Drummond, 1st Coll. n.* 114 *and* 116, *3rd Coll. n.* 43; *Preiss, n.* 270, 278; *Baxter; Oldfield.*

60. **M. concreta,** *F. Muell. Fragm.* iii. 118. An erect shrub of several ft., with rather slender branches, the young shoots silky-silvery, otherwise glabrous. Leaves alternate, linear or lanceolate, obtuse or acute, flat but thick, and sometimes very narrow, 1-nerved, mostly 2 to 3 in. long. Flowers yellowish-white, in globular terminal and axillary heads, the axis not growing out until after flowering, the rhachis usually tomentose. Calyx-tube broad, often hirsute at the base, not $\frac{3}{4}$ line long; lobes exceedingly short and broad. Petals $\frac{3}{4}$ line diameter. Staminal bundles nearly 3 lines long, the claws scarcely exceeding the petals, each with about 7 filaments at the end. Stigma small. Fruiting-spikes ovoid, very compact, about $\frac{1}{2}$ in. long; the calyxes about $1\frac{1}{2}$ lines long, very closely packed and angular, but not really connate, truncate at the top. Seeds narrow-cuneate; cotyledons not folded.

W. Australia. Murchison river, *Oldfield.*

Var. *brevifolia.* Leaves more acute, 1 to $1\frac{1}{2}$ in. long.—Murchison river, *Oldfield.*—Only seen in fruit, and therefore doubtful.

61. **M. filifolia,** *F. Muell. Fragm.* iii. 119. Erect, attaining several ft., glabrous except the inflorescence. Leaves alternate, linear-subulate, terete, obtuse or acute, rigid but not pungent, $\frac{3}{4}$ to $1\frac{1}{2}$ in. long in some specimens, 1 to 2 in. or even more in others. Flowers (yellow or white?) in ovoid-oblong or almost globular terminal spikes, the axis growing out into a leafy branch after flowering; the rhachis and calyxes more or less tomentose or woolly. Calyx-tube scarcely 1 line diameter; lobes orbicular, scarious. Petals small. Staminal bundles about 3 lines long, the claws short, each with 7 to 9 filaments at the end. Ovules few, erect, on a short oblong placenta. Fruiting-calyxes about 2 lines diameter, either closely packed and more or less concrete in ovoid or globular spikes, or looser by abortion and quite free.

W. Australia. Arid rocky places, Murchison river, *Oldfield.*

M. nematophylla, F. Muell. Fragm. iii. 119, with longer leaves, does not appear to me to be otherwise distinguishable from *M. filifolia.* None of the specimens of either species have good flowers, and the colour of those of *M. nematophylla* appears to me to be rather whitish-yellow than purple.

62. **M. hakeoides,** *F. Muell. Herb.* A tall shrub, the young shoots softly silky-pubescent and somewhat silvery, the older foliage glabrous. Leaves alternate, linear-subulate, terete or slightly compressed, usually sulcate, obtuse or acute, 1 to 2 in. long, the point straight. Flowers small, in dense globular or rarely ovoid terminal heads, the rhachis and calyxes usually villous. Calyx-tube about $\frac{1}{2}$ line long; lobes very small. Petals scarcely 1 line diameter. Staminal bundles not 2 lines long, the claws short, each with 3 to 7 filaments. Ovules few in each cell, erect; stigma slightly dilated. Fruiting-spikes very dense, globular or ovoid, the calyxes truncate, about 1 line diameter.

N. S. Wales. Mount Goningberi, near Cooper's Creek, *Victorian Expedition.*

63 ?. **M. glomerata,** *F. Muell. Rep. Babb. Exped.* 8. Softly pubescent or glabrous with age. Leaves alternate, linear, thick, but more or less flattened, narrowed at the base, nerveless or obscurely 1- or 3-nerved, obtuse or acute, 1 to 2 in. long. Flowers unknown. Fruiting-heads small, dense, globular; calyxes nearly globular, truncate, 1 to $1\frac{1}{2}$ lines diameter.

N. Australia. Upper Victoria river, *F. Mueller.*
S. Australia. N.W. interior, *M'Douall Stuart's Expedition.*
W. Australia. Murchison river, *Oldfield.*

The specimens being in fruit only, the species is doubtful, and may possibly include *M. hakeoides.*

64. **M. pentagona,** *Labill. Pl. Nov. Holl.* ii. 27. *t.* 166. A tall shrub with virgate branches, the young shoots often silky-downy, otherwise glabrous except the inflorescence. Leaves alternate, lanceolate or linear, mucronate or pungent, flat, but thick and rigid, obscurely veined, mostly ½ to ¾ in. long. Flowers small, pink or purple, in very dense globular heads, terminal axillary or lateral, the axis not growing out till after flowering, the rhachis tomentose or woolly. Calyx-tube campanulate or turbinate, about ½ line long; lobes exceedingly short and broad, scarious and often confluent. Petals under 1 line diameter. Stamens not above 2 lines long, shortly united in bundles of 3 to 7. Ovules not numerous, erect from a small placenta. Fruiting-calyxes about 2 lines diameter, often few only in an ovoid head, when more numerous the head very compact and globular.—DC. Prod. iii. 213; Schau. in Pl. Preiss. i. 136.

W. Australia, *Labillardière; Drummond,* 5*th Coll. n.* 152; sandy soil near salt lagoons, Espérance Bay, *Maxwell.*

Var. *subulifolia,* Schau. Leaves linear-subulate, terete, rigid, spreading, furrowed underneath.—King George's Sound, *R. Brown; A. Cunningham; Preiss, n.* 309; *Drummond,* 3*rd Coll. n.* 52.—This form seems almost to pass into *M. nodosa.*

Var. (?) *latifolia.* Leaves oblong-cuneate, rigid, obtuse or mucronate, *Drummond,* 3*rd Coll. n.* 57; granite hills, Cape Paisley, *Maxwell.* Possibly a distinct species, but our specimens of the different forms of *M. pentagona* are not sufficiently good to judge of their limits. The species sometimes approaches *M. striata* in foliage, but is readily distinguished by the small globular flower-heads.

65. **M. ciliosa,** *Turcz. in Bull. Mosc.* 1862, ii. 326. Branches rigid, pubescent. Leaves scattered, obovate ovate-orbicular or very broadly oblong, obtuse, 2 to 3 lines long, flat, concave, thick, rigid, 1-nerved, the somewhat thickened margin ciliate with deciduous hairs. Flowers (yellow-red?) in dense globular terminal heads, the rhachis tomentose-hirsute. Bracts rigid, striate, but very deciduous. Calyx-tube membranous, about 1 line long; lobes united in a scarious continuous border. Petals about 1 line diameter, so deciduous as to be only seen in the bud. Staminal bundles about 5 lines long, divided to about the middle or lower down into 9 to 11 filaments; anthers small. Ovules rather numerous in each cell, on a short placenta.

W. Australia. Between Moore and Murchison rivers, *Drummond,* 6*th Coll. n.* 76.

M. leptospermoides, Schau. in Pl. Preiss. i. 139, from Quangen Plains, *Preiss, n.* 312, may possibly be the same species, and if so this name should be preferred to Turczaninow's. In one of the very imperfect specimens seen, the leaves are broader than in Drummond's; in the other they are mostly narrower.

66. **M. polycephala,** *Benth.* Rigid, with divaricate branches, the young shoots slightly hoary. Leaves alternate, spreading, ovate, ovate-lanceolate or oblong, acute, flat, rigid, 1-nerved, mostly 3 to 4 lines long. Flowers small, pink, in small dense terminal globular heads, the axis not growing out until after flowering, the rhachis and calyxes villous. Calyx-

tube membranous, broadly campanulate, scarcely above ½ line long; lobes short, broad, scarious, occasionally confluent. Petals ½ to ¾ line diameter. Stamens about 2 lines long, shortly united in bundles of 3 or rarely more. Ovules not numerous, erect on a small placenta. Fruiting calyxes about 1½ lines diameter, very densely packed in globular heads.

W. Australia, *Drummond,* 5*th Coll. n.* 175.

67. **M. spathulata,** *Schau. in Pl. Preiss.* i. 134. A small spreading shrub, the young shoots silky-pubescent or softly hirsute, soon becoming glabrous. Leaves scattered, spreading or recurved, obovate-spathulate, obtuse or scarcely mucronate, much narrowed at the base, 2 to 3 or rarely 4 lines long, flat, thick, smooth, obscurely 1-nerved. Flowers small, red, in dense terminal globular heads, the axis not growing out till after flowering, the rhachis and sometimes the calyxes hirsute. Calyx-tube thin, about ½ line long; lobes half as long as the tube, rounded. Petals about ½ line diameter. Stamens 2 to 2½ lines long, very shortly united in bundles of 2 to 5 (usually 3); anthers small. Ovules not numerous in each cell, erect on a small placenta; stigma small. Fruiting-heads globular, very dense, about 4 lines diameter.

W. Australia, *Drummond,* 5*th Coll. n.* 177; and gravelly sides of Konkoberup hills near Cape Riche, *Preiss, n.* 301; Phillips range and Gordon river, *Maxwell.*

Some specimens of *Drummond's n.* 109, in fruit only, appear to be a variety very densely branched with very numerous globular heads, and the calyx-lobes more persistent.

68. **M. eriantha,** *Benth.* A bushy shrub, with the aspect nearly of *M. seriata*, but readily distinguished by its hoary foliage and white-tomentose young shoots and inflorescence. Leaves linear-oblong or more or less cuneate, obtuse, thick, obscurely 1-nerved, narrowed at the base, 2 to 4 lines long. Flowers pink, in terminal globular heads, not so dense as in most of the allied species, the rhachis and calyxes covered with a close but dense white tomentum. Calyx-tube ovoid, almost urceolate, about 1 line long; lobes truncate, tomentose at the base, otherwise scarious and often confluent. Petals about 1 line diameter. Staminal bundles 3 to 4 lines long, the claws about as long as the petals, each with 7 to 11 filaments at the end. Ovules not very numerous, erect on a small placenta; stigma small. Fruiting-heads globular, the calyxes not numerous, smooth, about 2 lines long, narrower and more distinct than in the allied species.

W. Australia, *Drummond,* 5*th Coll. Suppl. n.* 30.

69. **M. subtrigona,** *Schau. in Pl. Preiss.* i. 139. A densely-branched bushy shrub, either low and diffuse, or erect and 3 or 4 ft. high, the young shoots hoary-pubescent or hirsute, the full-grown foliage glabrous or nearly so. Leaves scattered, usually crowded, linear, thick, very obtuse or rarely mucronate, semiterete, mostly 2 to 3 lines long. Flowers small, pink, in small dense terminal globular heads, the axis not growing out until after flowering; the rhachis woolly-tomentose. Calyx-tube glabrous or slightly villous, about 1 line diameter; lobes broad, truncate, scarious, often confluent. Petals about 1 line diameter. Stamens about 3 lines long, shortly united in bundles of 3 to 7. Ovules not very numerous in each cell, erect on a small placenta. Fruiting-calyxes smooth, 1½ to 2 lines diameter, very closely packed in small globular heads.—*M. tuberculata,* Schau. in Pl. Preiss. i. 139.

W. Australia, *Drummond* (*3rd Coll.*), *n.* 57, *5th Coll. n.* 152, 167, 170, 172; King George's Sound to York, *Preiss, n.* 261; low places, Gordon river, *Oldfield.*

70. **M. seriata,** *Lindl. Swan Riv. App.* 8. Branches villous pubescent or glabrous. Leaves scattered, erect or recurved, linear or linear-cuneate, obtuse, narrowed at the base, mostly 3 to 4 lines long, thick but flat, obscurely 1-nerved. Flowers small, red or purple, in dense terminal globular heads, the rhachis tomentose or woolly. Bracts very deciduous. Calyx-tube pubescent at the base, about $\frac{3}{4}$ line long; lobes sometimes short, broad, scarious, and more or less confluent, sometimes more ovate and nearly as long as the tube. Petals under 1 line diameter. Staminal bundles 3 to 4 lines long, the claws exceeding the petals, each with 5 to 9 filaments at the end. Ovules not numerous, on a small peltate placenta; stigma small.—*M. Endlicheriana,* Schau. in Pl. Preiss. i. 134; *M. seriata,* Lindl., and *M. ornata,* Schau. l. c. 135.

W. Australia. Swan River and adjoining districts, *Drummond, 1st Coll. n.* 113; *Preiss, n.* 298, 299, 302, 308; Tone and Gordon rivers, *Oldfield.*

M. Wæberi, Reichb., Schau. in Otto and Dietr. Allgem. Gartnz. iii. 167, at least as to Preiss's specimens, n. 317, from the Konkoberup hills towards Cape Riche, quoted Pl. Preiss. i. 137, appears to be the same plant, although the staminal claws are said to be shorter with only 3 to 5 filaments, as in *M. subtrigona.*

Metrosideros sororia, Endl. in Hueg. Enum. 49, is, according to Schauer, his *M. Endlicheriana,* of which Preiss's specimens do not differ from *Melaleuca seriata,* but Endlicher's character agrees much better with *Kunzea recurva.*

71. **M. scabra,** *R. Br. in Ait. Hort. Kew. ed.* 2. iv. 414. A bushy shrub, either low and spreading or attaining several ft., with virgate branches, glabrous or hirsute when young. Leaves scattered, usually crowded under the flower-heads, erect, incurved or rarely recurved, linear-terete, semiterete or channelled above, obtuse or acute, thick and nerveless, smooth or tuberculate, $\frac{1}{4}$ to $\frac{1}{2}$ in. long. Flowers red, in dense terminal globular heads, varying considerably in size, the rhachis tomentose or hirsute. Bracts striate, very deciduous. Calyx-tube more or less hirsute, $\frac{1}{2}$ to 1 line long; lobes broad, scarious, distinct or confluent. Petals not large, very deciduous. Staminal bundles $\frac{1}{4}$ to nearly $\frac{1}{2}$ in. long, the claws short, each with 7 to 11 or rarely more filaments at the end. Ovules not very numerous, erect on a small placenta; stigma small. Fruiting-heads globular, dense; calyxes truncate, more or less urceolate.—DC. Prod. iii. 213; Sweet, Fl. Austral. t. 10; *M. parviceps,* Lindl. Swan Riv. App. 8; Schau. in Pl. Preiss. i. 136; *M. Manglesii,* Schau. l. c. i. 135.

W. Australia. Lucky Bay, *R. Brown;* from Swan River to the south coast, *Drummond, 1st Coll. n.* 112; *Preiss, n.* 260, 297, 310, 320, 326; *Oldfield,* and others; and eastward to Cape Le Grand, *Maxwell.* Drummond's 3rd Coll. n. 58, has more numerous stamens and long hairs to the calyx; n. 176 has flatter leaves, approaching those of *M. seriata.* In general, several of the foregoing and following species appear almost to pass into this one. In all, the smaller more globular heads have often none but male flowers, with a very small, abortive ovary at the base of the calyx.

72. **M. urceolaris,** *F. Muell. Herb.* A tall bushy shrub, more or less hoary, and often hirsute with spreading hairs, becoming glabrous with age. Leaves scattered, often crowded, linear, semiterete, obtuse or mucronulate, nerveless, mostly about $\frac{1}{2}$ in. long. Flowers yellowish, in dense terminal

globular heads, the axis not growing out till after flowering, the rhachis and calyxes pubescent or villous. Bracts more persistent than in *M. scabra.* Calyx-tube membranous, rather above 1 line long; lobes broad, scarious, often confluent. Petals above 1 line diameter, often ciliate. Staminal bundles 4 to 5 lines long, the claws short and broad, each with 10 to 15 filaments; anthers ovate. Ovules not numerous in each cell, erect on a small peltate placenta. Fruiting-calyxes urceolate-globular, very smooth, 2 to nearly 3 lines diameter, not numerous, but closely packed in a globular head. Seeds obovoid-cuneate; cotyledons broad, more or less folded.

W. Australia. Murchison river, *Oldfield; Drummond, 6th Coll. n.* 73.

Var. *virgata.* Scarcely villous. Branches elongated with less crowded leaves. Flowers rather smaller.—*Drummond, 6th Coll., n.* 71.

The species differs from *M. scabra* chiefly in the colour of the flowers.

73. **M. trichophylla,** *Lindl. Swan Riv. App.* 8. Very spreading or sometimes prostrate, often extending to several ft., either hirsute with fine spreading hairs, especially the young shoots, or quite glabrous. Leaves scattered, crowded, especially under the flower-heads, linear-terete, slender, obtuse or almost acute, $\frac{1}{4}$ to $\frac{1}{2}$ in., or in other specimens $\frac{1}{2}$ to 1 in. long, smooth or tuberculate. Flowers usually but not always larger than in *M. scabra*, pink, in dense terminal globular heads, the axis not growing out till after flowering, the rhachis and calyxes tomentose or villous. Bracts very deciduous or rarely persistent at the time of flowering. Calyx-tube under 1 line long; lobes scarious, ovate or oblong, usually longer than the tube, but very irregular in size, and often more or less confluent. Petals not much longer than the calyx-lobes. Staminal bundles about $\frac{1}{2}$ in. long, claws narrow, exceeding the petals, with 3 to 9 filaments at the end; anthers small. Ovules not very numerous, on a small peltate placenta; stigma small. Fruiting-calyxes either capitate, or two or three together, smooth, truncate, 2 to 3 lines diameter.—Schau. in Pl. Preiss. i. 136; *M. eremæa,* F. Muell. Fragm. iii. 114.

W. Australia. Swan River, *Drummond, 1st Coll.;* Murchison river and Champion Bay, *Oldfield.*—Very near *M. scabra,* differing chiefly in the more slender leaves and longer calyx-lobes.

74. **M. holosericea,** *Schau. in Pl. Preiss.* i. 139. A bushy shrub of 2 to 3 ft., the branches and foliage hoary-tomentose or sometimes silky-villous. Leaves scattered, crowded, linear, terete or semiterete, acute or obtuse, $\frac{1}{4}$ to $\frac{1}{2}$ in. long or more on the principal branches. Flowers (pink?) in dense terminal globular heads, the axis not growing out till after flowering; rhachis and calyxes tomentose-villous. Calyx-tube campanulate, scarcely 1 line long; lobes broad, scarious, often confluent. Petals about 1 line diameter. Staminal bundles about 4 lines long, the claws often exceeding the petals, with 5 to 11 filaments at the end. Ovules not very numerous, erect on a small peltate placenta.

W. Australia. Swan River, *Drummond, 1st Coll. n.* 111; sandy plains near Quangen, *Preiss, n.* 315; Dirk Hartog's Island, *Marten.*

75. **M. squamea,** *Labill. Pl. Nov. Holl.* ii. 28. *t.* 168. A shrub (or tree?), the young shoots more or less villous, with soft loose hairs, at length glabrous. Leaves scattered, numerous, usually spreading, from ovate-lanceo-

late to almost linear, acute or acuminate, flat or concave, distinctly 3-nerved, mostly about ¼ in. and rarely ½ in. long. Flowers rather small, reddish-purple white or yellowish, in small globular terminal heads, the axis not growing out until after flowering, the rhachis and calyxes villous. Bracts acuminate, deciduous, or a few external ones more persistent. Calyx-tube ovoid, about 1 line long; lobes much shorter, very obtuse. Petals under 1 line long. Stamens about 3 lines long, very shortly united in bundles of 5 to 9. Ovules rather numerous in each cell, erect on a small placenta. Fruiting calyxes often considerably enlarged, more or less urceolate, in a globular head.—DC. Prod. iii. 213; Hook. f. Fl. Tasm. i. 128; Bot. Reg. t. 477; Lodd. Bot. Cab. t. 412.

N. S. Wales. Near Appin, *Backhouse.*

Victoria. In the Grampians and on the Glenelg, *F. Mueller.*

Tasmania. Huon river and Port Dalrymple, *R. Brown;* very common in peaty soil in many parts of the colony, ascending to 4500 ft., *J. D. Hooker.*

S. Australia. Mount Gambier, at the S.E. extremity, *F. Mueller.*

76. **M. densa,** *R. Br. in Ait. Hort. Kew. ed.* 2. iv. 416. A bushy rigid shrub, usually glabrous, except the inflorescence. Leaves scattered or in irregular whorls of three, ovate, concave, spreading or recurved, obtuse, 2 or rarely 3 lines long, rigid and more or less prominently 1- or 3-nerved. Flowers small, in ovoid oblong or rarely globular terminal heads, the axis soon growing out into a leafy shoot; the rhachis woolly or nearly glabrous. Bracts broad, often herbaceous. Calyx-tube broadly campanulate; lobes rounded, more or less scarious, nearly as long as the tube. Petals about 1 line diameter. Staminal bundles scarcely above 2 lines long, the claws rarely exceeding the petals, each with 3 to 7 filaments at the end; anthers small. Ovules not very numerous in each cell, erect on a small placenta; style long with a small stigma. Fruiting spike ovoid or cylindrical, dense, ½ to 1 in. long.—DC. Prod. iii. 215; Schau. in Pl. Preiss. i. 144; *M. propinqua,* Schau. l. c.; *M. epacridioides,* Turcz. in Bull. Mosc. 1847, i. 165.

W. Australia. King George's Sound and adjoining districts, *R. Brown, Menzies, A. Cunningham, Baxter, Drummond,* 3*rd Coll. n.* 46, 5*th Coll. n.* 149, *Preiss, n.* 285, 286, 288, *Oldfield, Maxwell.*

In luxuriant shoots the leaves are occasionally longer and broader, attaining even 4 lines. In some garden specimens they are longer and more acute. In Preiss's specimens of *M. propinqua* they are rather smaller but not otherwise different. In Drummond's 5th Coll. n. 167, they are all narrow and regularly in 6 rows, in n. 166 of the same Coll. all very spreading, rigid, and squarrose, almost like those of *M. elachophylla.*

77. **M. thymoides,** *Labill. Pl. Nov. Holl.* ii. 27. *t.* 167. A tall shrub, usually glabrous, the branchlets rigid but slender, virgate or divaricate, and often spinescent. Leaves scattered, from linear-lanceolate to oblong-elliptical or almost ovate, rigid, usually acute, narrowed at the base, flat, 3- or 5-nerved, ¼ to ½ in. long. Flowers yellowish-white, in dense terminal globular heads, the axis rarely growing out until after flowering, the rhachis usually tomentose or woolly. Bracts deciduous. Calyx-tube pubescent, broadly campanulate, about ½ line long; lobes broad, membranous, but more or less distinctly 3-nerved in the centre. Petals striate in the centre. Staminal bundles 3 to 4 lines long, the claws as long as or exceeding the petals, with 5 to 9 filaments at the end; anthers very small. Ovules not

very numerous in each cell, and sometimes very few, erect on a small placenta.—DC. Prod. iii. 213; Schau. in Pl. Preiss. i. 140; *M. spinosa*, Lindl. Swan Riv. App. 8; Schau. in Pl. Preiss. i. 140.

W. Australia. King George's Sound, *R. Brown* and others, and thence to Swan River, *Drummond*, 1*st Coll.*, 3*rd Coll. n.* 44; *Preiss, n.* 280, 281, 282, 283, and others, and eastward to West Mount Barren, *Maxwell*, Vasse and Tone rivers, *Oldfield*. The spinescent branches are a very uncertain character, both in the Swan River and in the King George's Sound specimens.

78. **M. striata,** *Labill. Pl. Nov. Holl.* ii. 26. *t.* 165. A tall and bushy or low and straggling shrub, the young shoots silky, at length glabrous. Leaves alternate, lanceolate or linear, acute and often pungent-pointed, flat, 3- or 5-nerved, sometimes all under ½ in. long and very thick and rigid, sometimes narrower and above ½ in. Flowers pink (or sometimes white?), in dense oblong or cylindrical terminal spikes, the axis not growing out until after flowering, the rhachis and calyxes woolly. Calyx-tube broad, about 1 line diameter; lobes very short and broad. Petals about 1 line diameter. Staminal bundles 5 to 6 lines long, the claws longer than the petals, each with 7 to 11 filaments at the end; anthers small. Ovules not very numerous, on a broad shortly bifid placenta. Fruiting spikes cylindrical, very dense, rarely above 1 in. long, the calyxes 1½ to 2 lines long.—DC. Prod. iii. 212; *M. Fraseri*, Hook. Bot. Mag. t. 3210.

W. Australia. King George's Sound and adjoining districts, *Labillardière, R. Brown, Fraser, Drummond*, 3*rd Coll. n.* 53, and others, and eastward to Young river, and Orleans Bay, *Maxwell*. This species sometimes comes near to *M. pentagona*, but differs in its oblong or cylindrical spikes.

79. **M. polygaloides,** *Schau. in Pl. Preiss.* i. 142. Glabrous or the young shoots pubescent or silky. Leaves scattered or in whorls of 3, those on the main branches oblong or lanceolate, acute, flat, 1- or 3-nerved, and often ½ in. long, on the smaller branches linear or linear-lanceolate, about ¼ in. long, or all narrow and ¼ to ½ in., flat or concave and almost nerveless. Flowers small, white, in dense ovoid or cylindrical terminal spikes, rarely exceeding ½ in., the axis not growing out in any of our specimens, the rhachis and calyxes pubescent. Bracts often persistent. Calyx-tube campanulate, scarcely 1 line diameter; lobes short, obtuse. Petals about ½ line diameter. Stamens not above 2 lines long, shortly united in bundles of 7 to 11. Ovules not numerous, erect on a small placenta; style long, with a small stigma. Fruiting spikes dense, cylindrical, ½ to ¾ in. long, but not seen quite ripe.

W. Australia. Swan River, *Collie, Drummond*, 1*st Coll. and* 3*rd Coll. n.* 45; Kalgan river, *Oldfield*. Preiss's specimens in bud only, n. 327, certainly are conspecific with Drummond's, but the fruiting ones, n. 328, though in a very imperfect state, appear to be different.

80. **M. incana,** *R. Br. in Bot. Reg. t.* 410. A tall shrub, the branches often tortuous or spreading, more or less hoary-tomentose or pubescent. Leaves scattered or irregularly opposite or in whorls of three, very spreading, often crowded, linear or lanceolate, mostly acute, rather rigid but often incurved, ¼ to ½ in. long, 1- or 3-nerved, becoming sometimes glabrous with age, but usually hoary, especially underneath. Flowers rather small, yellowish-white, in dense terminal ovoid or oblong spikes, the axis rarely grow-

ing out until after flowering, the rhachis pubescent. Calyx-tube glabrous or pubescent, broadly campanulate, about 1 line diameter; lobes broad, about half as long as the tube. Petals about 1 line diameter. Stamens 3 to 4 lines long, shortly united in bundles of 3 to 9. Ovules not very numerous, erect on a short placenta. Fruiting spikes dense, cylindrical, ¾ to 1 in. long.—DC. Prod. iii. 215; Schau. in Pl. Preiss. i. 141; *M. canescens*, Link and Otto, Pl. Sel. Hort. Berol. 81. t. 37; *M. tomentosa*, Colla, Hort. Ripul. 87. t. 37; *M. hypochondriaca*, Dehnh. according to Schau.

W. Australia. King George's Sound, *R. Brown, Wakefield;* towards Cape Riche and Canning river, *Preiss, n.* 262, 266, 329, *Drummond, 3rd Coll. n.* 46, and *2nd Coll. n.* 63; Vasse river, *Oldfield.*

81. **M. nodosa,** *Sm. in Trans. Linn. Soc.* iii. 276, *and Exot. Bot. t.* 35. A tall shrub or small tree. Leaves alternate, linear or subulate, rigid, pungent-pointed, mostly ½ to ¾ in., or on luxuriant shoots nearly 1 in. long. Flowers in small dense globular or rarely ovoid axillary or terminal heads, the axis not growing out until after flowering, the rhachis tomentose. Calyx-tube broadly campanulate, about ¾ line long; lobes much shorter, obtuse, and petal-like. Petals about as long as the calyx-lobes. Staminal bundles about 3 lines long, the claws about as long as the petals, with 3 to 6 or rarely more filaments at the end. Ovules not very numerous, erect on a small placenta. Fruiting-heads very dense, globular, 3 to 4 lines diameter, the calyxes truncate.—DC. Prod. iii. 213; Vent. Jard. Malm. t. 112; *Metrosideros nodosa*, Gærtn. Fruct. i. 172. t. 34; Cav. Ic. iv. t. 334; *Melaleuca juniperina*, Sieb.; Reichb. Iconogr. Exot. ii. 4. t. 112; *M. juniperoides*, DC. Prod. iii. 213; *Metrosideros juniperina* and *M. pungens*, Reichb. in Spreng. Syst. Cur. Post. 194.

Queensland. Moreton Bay, *W. Hill.*
N. S. Wales. Port Jackson to the Blue Mountains, *R. Brown, Sieber, n.* 316, 317, and others; Clarence river, *Beckler.*

Of the numerous small flower-heads in this species the great majority appear to consist of deciduous flowers, male by abortion.

M. tenuifolia, DC. Prod. iii. 213, described from fruiting specimens which I have not seen, is probably, from the characters given, a variety of *M. nodosa* with longer leaves.

82. **M. pungens,** *Schau. in Pl. Preiss.* i. 138. A shrub of several ft., the young shoots more or less pubescent. Leaves alternate, spreading, linear-subulate, terete, rigid, with a straight, often pungent point ½ to 1 in. long. Flowers yellowish-white, rather small, in dense ovoid oblong or rarely globular terminal spikes, the axis not growing out until after flowering, the rhachis tomentose or woolly. Calyx-tube pubescent or hirsute, campanulate, under 1 line long; lobes short, broad, scarious, usually ciliate. Petals scarcely 1 line long. Staminal bundles about 3 lines long or rather more, the claws short, each with about 7 filaments at the end; anthers not larger than in *Beaufortia*, but distinctly versatile. Ovules very few in each cell, erect on a small placenta; stigma small. Fruiting spikes dense, oblong or cylindrical, the calyxes not attaining 2 lines.

W. Australia, *Drummond, 1st Coll. n.* 115, *5th Coll. n.* 146; barren gravelly places near Mount Barrow, *Preiss, n.* 316; Kalgan river ranges and East Mount Barren, *Maxwell* (with shorter leaves, rounder flower-heads, the rhachis less woolly); Fitzgerald flats, *Maxwell* (with thicker leaves).

Var. *obtusifolia.* Leaves more crowded, mostly obtuse. Flowers longer, yellow, in very dense and oblong spikes. *Drummond, 5th Coll. n.* 148; West Mount Barren, *Maxwell.*

The species often approaches *M. pentagona,* on the one hand, or *M. nodosa* on the other.

83. **M. ericifolia,** *Sm. in Trans. Linn. Soc.* iii. 276. *and Exot. Bot. t.* 34. A shrub or tree, attaining sometimes a considerable height, usually glabrous and often glaucous, with virgate branchlets. Leaves scattered, numerous, often recurved, narrow-linear, semiterete or convex underneath, obtuse or scarcely acute, rarely above ½ in. long. Flowers yellowish-white or rarely red, not large, (the males?) in ovoid or nearly globular terminal heads, or the perfect ones in oblong and cylindrical spikes of ½ to 1 in., with the axis soon growing out into a leafy branch, the rhachis tomentose. Calyx glabrous or nearly so, short, broad, with short, broad, obtuse, herbaceous lobes. Petals above 1 line long. Staminal bundles 3 to 4 lines long, the claws exceeding the petals, each with about 7 filaments at the end. Ovules rather numerous in each cell, on a short peltate placenta. Fruiting spikes compact; calyxes truncate.—DC. Prod. iii. 213; Hook. f. Fl. Tasm. i. 129; *M. nodosa,* Sieb. Pl. Exs., not of Sm.; *M. Gunniana,* Schau. in Walp. Rep. ii. 928; *M. heliophila,* F. Muell.; Miq. in Ned. Kruidk. Arch. iv. 120 (from the character given).

N. S. Wales. Port Jackson to the Blue Mountains, *R. Brown, Sieber, n.* 318, and others; Lord Howe's Island, rare, on rocks facing the sea, *Milne.*

Victoria. On the Yarra and Goulburn rivers and Dandenong mountains, *F. Mueller.*

Tasmania. Port Dalrymple and islands of Bass's Straits, *R. Brown.* Abundant, especially in swampy ground in the northern parts of the colony, the *Swamp Tea-tree* of the colonists, *J. D. Hooker.*

Var. *erubescens.* Flowers red. Stamens usually more numerous.—*M. erubescens,* Otto, Hort. Berol. 37, according to DC. Prod. i. 214. *M. diosmifolia,* Dum. Cours. according to DC. l. c.—Port Jackson, Paramatta, *Woolls.*

84. **M. viminea,** *Lindl. Swan Riv. App.* 8. A tall glabrous shrub, with virgate branches. Leaves scattered, erect or recurved, narrow linear, flat or convex underneath, with a recurved point or almost obtuse, ¼ to ½ in. long. Flowers small (white?), in terminal globular heads and mostly males, or the perfect ones in oblong-cylindrical spikes, the axis not growing out till after flowering, the rhachis and calyxes glabrous. Calyx-tube broadly campanulate, rather thin, under 1 line diameter; lobes short. Petals under 1 line long. Staminal bundles 2 to 3 lines long, the claws exceeding the petals, each with 7 to 11 or rather more filaments at the end; anthers small. Ovules rather numerous, on a small peltate placenta. Fruiting spikes cylindrical, rather dense or interrupted, the calyxes about 1½ lines diameter.—Schau. in Pl. Preiss. i. 142; *M. Lehmanni,* Schau. l. c.

W. Australia. Swan River, *Drummond, 1st Coll. and n.* 109; *Preiss, n.* 291, 292.

Var. *major.* Leaves rather broader. Flowers larger.—Swan River, Port Gregory, and Champion Bay, *Oldfield.*

M. hamulosa, Turcz, above n. 46, closely resembles this species, but the spikes are all cylindrical and above 1 in. long.

85. **M. microphylla,** *Sm. in Rees Cyclop.* xxiii. A glabrous spreading shrub, very closely allied to *M. viminea* and *M. hamulosa,* but with the foliage of *M. blæriæfolia.* Leaves scattered, spreading or recurved, rather crowded and all nearly of the same size, linear, semiterete or rather thick, obtuse,

about 3 lines long. Flowers white in cylindrical spikes, terminating short lateral branches or the males in ovoid heads. Calyx-tube broad, about 1 line diameter; lobes short. Petals 1 line long. Staminal bundles 3 to 4 lines long, the claws exceeding the petals, each with 11 to 15 or more filaments at the end. Ovules numerous, on a peltate placenta.—*M. brachyphylla*, Schau. in Pl. Preiss. i. 141; *M. tenuissima*, Tausch. in Flora, 1846, 411, according to Schau.; *M. brevifolia*, F. Muell. Fragm. i. 116.

W. Australia. King George's Sound and adjoining districts, *Menzies, Preiss, n.* 255, *Maxwell, Oldfield.*

A fruiting specimen from Menzies, in Herb. R. Br., has the calyxes very much enlarged with thickened obtuse warted lobes, the whole spike very dense and above ½ in. diameter; but this may be abnormal.

86. **M. tenella,** *Benth.* A shrub of about 4 ft., with slender branches, glabrous or minutely pubescent. Leaves scattered or in whorls of 3, narrow-linear, incurved or spreading, flat or concave, acute or almost obtuse, obscurely 1-nerved, 2 to 3 lines long. Flowers small (white?), in globular, oblong, or shortly cylindrical terminal spikes, the axis not growing out until after flowering, the rhachis pubescent or nearly glabrous. Calyx-tube campanulate, scarcely 1 line diameter. Stamens about 2 lines long, very shortly united in bundles of 3, 4 or rarely 5. Ovules not very numerous, erect on a small peltate placenta; style rather thick; stigma not dilated.

W. Australia. Moist soil, tributaries of Phillips River, *Maxwell.*

87? **M. leiopyxis,** *F. Muell. Herb.* A spreading shrub of 2 to 3 ft., glabrous except the inflorescence. Leaves linear-oblong, obtuse, flat, thick, rigid, nerveless, ¼ to ½ in. long. Flowers not seen. Fruiting-calyxes 2 to 4 together on a short pubescent or woolly rhachis, thick and smooth, attached by the broad base, fully 3 lines diameter, truncate at the top.

W. Australia. Limestone Hills, Murchison river, *Oldfield.*

88. **M. pustulata,** *Hook. f. in Hook. Lond. Journ.* vi. 476, and *Fl. Tasm.* i. 129. A glabrous bushy shrub, varying from 2 or 3 ft. to twice that height. Leaves scattered, often crowded, from elliptical-oblong or lanceolate to linear or linear-cuneate, obtuse, thick, concave, nerveless, mostly 2 to 3 lines long, but in some specimens all under 2 lines. Flowers small, not numerous, in small terminal leafy heads, the axis soon growing out into a leafy shoot, the rhachis and calyx glabrous. Calyx-tube ovoid, about 1 line long; lobes nearly as long, lanceolate. Petals about 1 line. Staminal bundles not exceeding 3 lines, the claws longer than the petals, each with 15 to 20 filaments at the end. Ovules rather numerous in each cell, erect on a short peltate or bifid placenta. Fruiting-calyxes very few in the spike, not much enlarged, crowned by the persistent lobes.—*M. halmaturorum*, F. Muell.; Miq. in Ned. Kruidk. Arch. iv. 122.

N. S. Wales. Darling river, *Neilson.*

Victoria. Wimmera, *Dallachy* (with much shorter stamens).

Tasmania. Oyster Bay and on a tributary of the South-Esk river, *Gunn.*

S. Australia. St. Vincent's Gulf, *Whittaker, F. Mueller;* Kangaroo Island, *Waterhouse.*

89. **M. acerosa,** *Schau. in Pl. Preiss.* i. 137. An erect bushy shrub of several ft., loosely silky-hairy or at length glabrous. Leaves scattered,

crowded, spreading or incurved, narrow-linear, flat or concave, rather thick, acute or obtuse, under ½ in. long. Flowers (yellowish-white?) rather small in terminal globular heads, especially the males, the perfect ones fewer in the head and often reduced to 5 or 6, the axis not growing out till after flowering, the rhachis woolly-tomentose. Calyx-tube ovoid, about 1 line long; lobes scarious, broad, truncate and often confluent. Petals scarcely 1 line diameter, so deciduous as rarely to be seen except in the young bud. Stamens 3 to 4 lines long, very irregularly united in bundles of about 7 to 11; anthers ovoid. Ovules not numerous, erect on a small placenta. Fruiting-calyxes usually 2 or 3 only together, very smooth, about 2 lines diameter.

W. Australia. Swan River, *Fraser*, *Drummond*, 1*st Coll.* and *n.* 52, *Preiss*, *n.* 263.
Var. *bracteata.* Bracts larger and more persistent.—*M. bisulcata*, F. Muell. Fragm. iii. 118. Murchison river, *Oldfield.*

90. **M. pauperiflora,** *F. Muell. Fragm.* iii. 116. A shrub of 4 to 6 ft., the young shoots pubescent. Leaves scattered, linear, rather thick, semi-terete or nearly terete, acute or almost obtuse, nerveless, mostly 3 to 4 lines long. Flowers (pink or white?) in small terminal heads of about 6, the axis not growing out until after flowering, the rhachis pubescent. Bracts short, imbricate. Calyx-tube glabrous, ovoid, above 1 line long; lobes small, distinct, with scarious margins. Petals scarcely 1 line long. Stamens 2 to 3 lines, very shortly united in bundles of 7 to 12. Ovules rather numerous in each cell, erect on a small peltate placenta; stigma small. Fruiting-calyxes globular, smooth, about 2 lines diameter.

W. Australia, *Drummond*, 5*th Coll. n.* 154, 158; Phillips Range, *Maxwell.*

91. **M. aspalathoides,** *Schau. in Pl. Preiss.* i. 140. A rigid shrub of about 2 ft., the young branches and foliage white with a close, dense, silky tomentum. Leaves scattered, crowded, subulate, terete, not rigid, about ½ in. long. Flowers few together. Calyx-tube white-tomentose, about 1 line long; lobes rather longer, narrow and apparently persistent. Staminal bundles (3 or 4 lines long?) the claws as long as the calyx-lobes, each with 7 to 9 filaments.

W. Australia. York district and Quangen Plains, *Preiss*, *n.* 2425 and 2426. The only specimens seen (in Herb. Sonder) are very incomplete, but the species appears to be quite distinct from *M. incana* or *M. holosericea*, both of which have white-tomentose foliage.

SERIES VII. PELTATÆ.—Leaves opposite or alternate, very small, often scale-like, more or less peltately attached and closely appressed to the branch or the upper end spreading. Flowers small, in dense heads or spikes.

92. **M. deltoidea,** *Benth.* Glabrous. Leaves opposite, ovate-triangular acute or obtuse, spreading, but thick and more or less peltately attached in the middle of their broad dilated base, mostly 1 to 1½ lines long. Flowers rather small (pink or white), in globular or ovoid terminal heads, the axis soon growing out into a leafy shoot, the rhachis and calyxes glabrous. Calyx-tube ovoid, about ¾ line diameter; lobes very short and broad and rather thick. Petals about ½ line diameter. Staminal bundles 2 to 3 lines long, the claws shorter or rather longer than the petals, each with 5 to 9 filaments at the end. Ovules few in each cell, erect, on a small placenta; stigma small. Fruiting-calyxes in globular clusters, each about 2 lines diameter.

W. Australia, *Drummond*, 5*th Coll. n.* 151; Phillips Range, *Maxwell, in Herb. Oldfield.*

93. **M. minutifolia,** *F. Muell. in Trans. Phil. Soc. Vict.* iii. 45. Nearly allied to *M. tamariscina*, but the branchlets are much more slender and not excavated. Leaves opposite, scale-like, appressed and imbricate, almost stem-clasping and peltately attached near the base, ovate or ovate-lanceolate, acutely acuminate, ½ to nearly 1 line long. Flowers small, in small ovoid terminal heads, the rhachis woolly. Calyx-tube broad, about ½ line long; lobes about as long, broad, striate. Petals nearly 1 line long. Staminal bundles about 3 lines lóng, the claws narrow, nearly twice as long as the petals, each with 7 to 11 filaments at the end; anthers very small. Fruiting-spikes short, the calyxes globular, truncate, about 1½ lines long.

N. Australia. Victoria river, *F. Mueller.*
Queensland (?). Flinders river, *Bowman.*

94. **M. foliolosa,** *A. Cunn. Herb.* Branchlets very numerous, erect and slender, excavated for the leaves, the margins of the excavations forming a fringe round them. Leaves opposite, scale-like, broad, thick, obtuse, triquetrous, peltately attached, closely appressed and imbedded in the excavations, scarcely 1 line long. Flowers only seen in very young bud, few, in terminal heads. Calyx campanulate, with short broad striate lobes. Petals striate. Stamens in bundles of 15 to 20, the claws already as long as the petals. Stigma rather broad. Fruiting-calyxes few in the head or solitary, globular, about 2 lines diameter.

Queensland. Cape Flinders, *A. Cunningham.*

95. **M. micromera,** *Schau. in Pl. Preiss.* i. 146. A tall shrub, with very numerous short slender branches, covered with a short close white tomentum, often concealed by the minute leaves. Leaves mostly in whorls of 3, closely appressed, ovate, scale-like, but thick, peltately attached, rarely above ½ line long. Flowers, at least the males, small, in globular terminal heads, the axis soon growing out into a leafy shoot. Calyx-tube campanulate, membranous, nearly 1 line long, the lobes very short. Petals about ½ line long. Stamens 1 line long or rather more, in bundles of 5 to 9. Ovary not seen. Fruiting-heads dense, globular, 4 to 6 lines diameter, the calyxes open, 1½ to 2 lines diameter, the disk much thickened opposite the persistent lobes; capsule convex, on a level with the calyx; style persistent, with a peltate stigma. Seeds rather numerous in each cell, erect.

W. Australia, *Drummond*, 3*rd Coll. n.* 49; gravelly places, Wariup hills, Goodrich district, rare, *Preiss, n.* 183 *a.*

96. **M. thyoides,** *Turcz. in Bull. Mosc.* 1847, i. 167. A tall shrub, with numerous small slender branchlets, usually whitish, but glabrous or nearly so. Leaves spirally arranged, scale-like, closely appressed and imbricate, thick, peltate and concave, very obtuse and scarcely ½ line long on the smaller branchlets, more distant, acuminate, and nearly 1 line long on the longer branches. Flowers whitish, in ovoid globular or oblong heads, terminal or the axis very soon growing out into a leafy shoot. Calyx-tube ovoid-campanulate, about ¾ line long or shorter and broader in the males, with very short and broad lobes. Petals ½ to ¾ line diameter. Staminal bundles 2 to

nearly 3 lines long, the claws exceeding the petals, each with 5 to 9 filaments at the end. Stigma dilated. Fruiting calyxes truncate, in some specimens about 1½ lines diameter, in not very compact globular heads, in others 2 lines diameter, in very dense oblong spikes, in others again still larger and only 2 or 3 together. Cotyledons very broad and folded.—*M. cupressina*, F. Muell. Fragm. iii. 114.

W. Australia, *Drummond, 3rd Coll. n.* 48, *also n.* 57, *and n.* 169; Phillips Ranges, *Maxwell;* seashore, Sharks' Bay, *Milne* (doubtful, the specimens in leaf only).

97. **M. tamariscina,** *Hook. in Mitch. Trop. Austr.* 262. Branchlets numerous, slender and excavated for each leaf as in *M. foliolosa*, but in a rather less degree. Leaves scarcely opposite, scale-like, peltate and half stem-clasping, closely appressed and half immersed in the excavations, ovate, concave, rarely above ½ line long, the lower ones of each branchlet very obtuse, the upper ones often acuminate. Flowers not seen. Fruiting-spikes oblong or cylindrical, ½ to 1 in. long, the calyxes often densely packed, globular, about 1½ lines diameter.

Queensland. Belyando river, *Mitchell.*

23. CONOTHAMNUS, Lindl.

Calyx-tube campanulate, adnate to the ovary at the base, the free part broad; lobes 5, short, imbricate or open. Petals 5, orbicular, spreading or none. Stamens indefinite, collected in clusters or united in bundles opposite the petals or alternating with the calyx-lobes; anthers versatile, the cells parallel, opening longitudinally. Ovary semi-inferior, enclosed in the calyx-tube, villous, 3-celled, with one ovule in each cell; style filiform, with a small stigma. Capsule enclosed in the hardened and somewhat enlarged calyx, but often nearly free, within or separable from it, opening loculicidally in 3 valves. Seeds ovate or obovate (not seen ripe).—Shrubs, with the habit of *Melaleuca.* Leaves opposite, small, 1- or 3-nerved, rigid. Flowers in terminal globular heads.

The genus is limited to Western Australia, and only differs from *Melaleuca* in the ovules solitary in each cell, as in *Beaufortia*, whilst the anthers are those of *Melaleuca.*

Leaves ½ to 1 in. long, 3-nerved. Petals 5. Stamens united in 5 bundles . . . . 1. *C. trinervis.*
Leaves 2 to 4 lines long, 1-nerved. Petals 0. Stamens collected in 5 clusters, but not united . . . . 2. *C. divaricatus.*

1. **C. trinervis,** *Lindl. Swan Riv. App.* 9. A very rigid spreading shrub, of 2 to 3 ft., the young shoots and inflorescence softly villous. Leaves lanceolate, oblong or oblanceolate, very rigid, pungent-pointed, narrowed at the base, prominently 3-nerved, ½ to 1 in long. Flowers rather large (white?), in dense globular or ovoid terminal heads, the axis growing out after flowering into a leafy shoot, the rhachis and calyxes villous or woolly. Bracts rather large, broadly ovate, very deciduous. Calyx-tube above 1 line diameter, the lobes obtuse, scarious only at the margins. Petals above 1 line diameter, obscurely striate in the centre, with scarious margins. Stamens about 4 lines long, united to about the middle in bundles of 7 to 11 each. Ovary almost free within the calyx-tube. Capsule also entirely enclosed

within the tube. Seeds solitary in each cell, oblong-triquetrous, erect, but not seen quite ripe.—Schau. in Pl. Preiss. i. 147 ; *Melaleuca cuspidata*, Turcz. in Bull. Mosc. 1862, ii. 327.

W. Australia. Swan River, *Drummond, 1st Coll.; Preiss, n.* 2639; sand plains near Cabingong, Murchison river, *Oldfield.*

2. **C. divaricatus,** *Benth.* A low shrub, with opposite, rigid, divaricate, almost spinescent branches, hoary or silky-pubescent, as well as the foliage. Leaves ovate or ovate-lanceolate, obtuse, rigid, 1-nerved and transversely veined, 2 to 3 or rarely 4 lines long. Flowers small, in small dense globular heads terminating the smaller branchlets, the axis not growing out until after flowering, the rhachis and calyxes villous. Bracts rather large, broad, imbricate, but very deciduous. Calyx-tube membranous, above ½ line long; lobes short, scarious. Petals none. Stamens about 1½ lines long, quite distinct, but collected in clusters of about 3, alternating with the calyx-lobes. Ovules solitary in each cell and peltate. Fruiting-calyxes about 1½ lines diameter, in dense globular heads; capsule enclosed in the calyx-tube, but readily separable from it. Seeds ovoid, but not seen perfect.

W. Australia, *Drummond, 5th Coll. n.* 147.

SUBTRIBE III. BEAUFORTIEÆ.—Leaves opposite or scattered, small or narrow and coriaceous, 1- or several-nerved. Flowers closely sessile and solitary in the axils of the floral leaves and bracts. Stamens indefinite, united in bundles opposite the petals or rarely free; anthers erect, attached at the base, the dehiscence various. Ovules 1 or more in each cell of the ovary. Embryo straight or slightly curved, the cotyledons longer than the radicle.

This subtribe differs from the *Euleptospermeæ* chiefly in the anthers.

24. BEAUFORTIA, R. Br.

(Schizopleura, *Lindl.*)

Calyx-tube ovoid or campanulate, adnate to the ovary at the base, the free part erect, contracted or rarely dilated; lobes 5, herbaceous or with scarious margins. Petals 5, spreading. Stamens indefinite, longer than the petals, united in 5 distinct bundles opposite the petals, the filaments or free parts filiform; anthers very small, erect, the cells parallel, opening at the top in 2 valves, the outer valve of each cell usually larger and often deciduous. Ovary enclosed in the calyx-tube, inferior or half-superior, the convex summit villous, with a central depression round the style, 3-celled, with 1 perfect ovule in each cell, peltately attached to the centre of a peltate placenta, with the addition sometimes of 2 imperfect ovules, erect at the top of the placenta, and concealed under the perfect one; style filiform, with a small stigma. Capsule enclosed in the enlarged and hardened calyx-tube, opening loculicidally in 3 valves. Seeds, where known, solitary in each cell, attached by their inner face, with a thin testa; embryo straight; cotyledons flat or plano-convex, much longer than the radicle.—Rigid, often heath-like shrubs, glabrous or pubescent. Leaves opposite, or in one species scattered, small, rigid, 1- or several nerved. Flowers usually red, closely sessile, solitary within each bract, in dense heads or short spikes, either terminal or at the base of the new branch formed by the growing out of the central axis. Bracts membranous, usually very deciduous ; bracteoles small.

This genus is confined to Western Australia. It is closely allied to *Melaleuca*, with the habit and foliage of the smaller-leaved species of that genus, and only differs in the anthers and ovules. As in *Melaleuca*, the flowers are often more or less unisexual, the males usually in smaller more globular heads, the female or hermaphrodite heads more oblong.

Staminal bundles above ½ in. long, the claw much longer than the free part of the filaments.
 Leaves scattered . . . . 1. *B. sparsa.*
 Leaves all opposite.
  Leaves ovate or orbicular, usually recurved or spreading.
   Petals not exceeding the narrow calyx-lobes. Outer valve of the anther-cells conical . . . . 2. *B. decussata.*
   Petals narrow, twice as long as the calyx-lobes. Anther-valves orbicular, ciliate . . . . 3. *B. squarrosa.*
   Petals broad, shortly exceeding the calyx-lobes, unequal. Anther-valves orbicular.
    Leaves small, orbicular. Staminal bundles not very unequal . . . . 4. *B. orbifolia.*
    Leaves ovate or ovate-lanceolate. Inner staminal bundles half the size of the outer ones . . . . 5. *B. anisandra.*
  Leaves narrow-lanceolate or linear.
   Calyx-lobes narrow, as long as or longer than the tube . . 6. *B. macrostemon.*
   Calyx-lobes triangular, much shorter than the tube . . . . 7. *B. cyrtodonta.*
Staminal bundles under ½ in. long, the claw shorter or scarcely longer than the free part of the filaments.
 Leaves lanceolate linear or triquetrous, mostly 3 lines long or more.
  Leaves mostly linear. Calyx-lobes triangular, shorter than the tube . . . . 8. *B. Schaueri.*
  Leaves mostly lanceolate. Calyx-lobes narrow, much longer than the tube and exceeding the petals . . . . 9. *B. purpurea.*
 Leaves ovate obovate or orbicular, under 3 lines long.
  Leaves mostly obovate or orbicular, spreading. Fruits 2 to 3 lines diameter, smooth and only 2 or 3 together . . . . . 10. *B. Dampieri.*
  Leaves recurved or spreading, 1 to 3 lines long. Fruits about 1 line diameter, in dense heads or spikes . . . . 11. *B. elegans.*
  Leaves erect, appressed, under 1 line long. Fruits of *B. elegans.* 12. *B. micrantha.*

1. **B. sparsa,** *R. Br. in Ait. Hort. Kew. ed.* 2. iv. 419. Glabrous, except the inflorescence. Leaves scattered, rather crowded, ovate-elliptical or ovate-lanceolate, obtuse, erect or recurved, many-nerved, flat or concave, but not keeled, ¼ to nearly ½ in. long. Flower-spikes very dense and oblong, the axis already growing out before flowering, the rhachis and calyx glabrous or slightly pubescent, the flowers rather small without the stamens. Calyx-tube about 1 line long; lobes scarcely so long, broad and very obtuse. Petals orbicular, twice as long as the calyx-lobes. Staminal bundles scarlet, fully 1 in. long, the claws slender, each with about 5 filaments at the end, scarcely ¼ as long as the claw; anther-valves small, orbicular. Staminal disk glabrous. Ovules solitary in each cell, the imperfect ones wanting or inconspicuous. Fruit-spikes about 1 in. long, the calyxes but slightly enlarged, the short lobes persistent or at length wearing off. Seeds obovoid; cotyledons plano-convex.—DC. Prod. iii. 211 ; Schau. in Nov. Act. Nat. Cur. xxi. 14 (by misprint 18), and in Pl. Preiss. i. 149 ; *B. splendens*, Paxt. Brit. Fl. Gard. xiii. 145, with a fig.

W. Australia. King George's Sound and adjoining districts, *Menzies, R. Brown, Drummond, 3rd Coll. n.* 42, *4th Coll. n.* 59; *Preiss, n.* 319, 355, and others.

2. **B. decussata,** *R. Br. in Ait. Hort. Kew. ed.* 2. iv. 418. An erect shrub, of 3 or 4 ft., with rigid virgate branches, glabrous or loosely pubescent. Leaves opposite, decussate, ovate, obtuse or scarcely acute, recurved towards the end, concave, keeled, many-nerved, rarely ½ in. long, the floral ones in the spike larger and often persistent. Flower-spikes ovoid or oblong, usually on the old wood below the year's branches, the rhachis and calyxes glabrous or nearly so. Calyx-tube nearly 2 lines long; lobes linear-lanceolate, about the length of the tube. Petals orbicular, not exceeding the calyx-lobes. Staminal bundles of a rich red, often 1 in. long, the claws linear, each with 7 to 11 filaments at the end, about half as long as the claw; outer valve of each anther-cell much larger than the other, conical and deciduous. Staminal disk densely fringed within the stamens with woolly hairs. Ovary with 2 imperfect ovules in each cell, often of a considerable size, although concealed under the perfect one. Seeds oblong; cotyledons ovate, flat or plano-convex.—DC. Prod. iii. 211; Schau. in Nov. Act. Nat. Cur. xxi. 13, and in Pl. Preiss, i. 148; Bot. Mag. t. 1733; Bot. Reg. t. 18; Colla, Hort. Ripul. t. 22.

W. Australia. King George's Sound and adjoining districts, *R. Brown, Baxter, Drummond, 3rd Coll. n.* 59, *4th Coll. n.* 61; *Preiss, n.* 356, and others.

3. **B. squarrosa,** *Schau. in Nov. Act. Nat. Cur.* xxi. 15 (*by misprint* 19), *and in Pl. Preiss.* i. 149. A straggling shrub, of 3 or 4 ft., glabrous, except the inflorescence, and sometimes the young shoots. Leaves opposite, crowded, decussate, ovate or obovate, recurved, concave, 5- or 7-nerved, rarely exceeding 2 lines, the floral ones or bracts nearly orbicular and flatter. Flowers in dense, terminal, globular heads, the axis only growing out after flowering, the rhachis usually pubescent. Calyx-tube 1 to nearly 2 lines long, rather narrow; lobes about 1 line long, prominently 3-nerved. Petals oblong, at least twice as long as the calyx-lobes. Staminal bundles crimson, at least ¾ in. long, the claws slender, each with 3 to 7 filaments at the end, shorter than the claw; anther-valves orbicular, ciliate. Ovary with 2 imperfect ovules in each cell under the perfect one, often enlarged and hardened after flowering, but without any embryo. Fruits, in some specimens, nearly solitary, ovoid, smooth, and 3 to 4 lines long, in others scarcely half so large and many together in little heads; the calyx-lobes always deciduous.

W. Australia. Canning river, *Preiss, n.* 358; between Moore and Murchison rivers, *Drummond, 6th Coll. n.* 75; Port Gregory and Murchison rivers, *Oldfield.*

4. **B. orbifolia,** *F. Muell. Fragm.* iii. 110. A spreading straggling shrub, of 2 or 3 ft., glabrous, except the inflorescence, or the young shoots slightly pubescent. Leaves opposite, crowded and decussate on the smaller branches, orbicular or broadly ovate, obtuse, flat or slightly concave, rigid, 5- or 7-nerved, 1½ to 3 lines long. Flowers in dense globular heads, larger and more numerous than in *B. anisandra*, the axis occasionally growing out before the flowering is over, the rhachis and calyxes pubescent or villous. Calyx-tube about 1½ lines long; lobes rigid, 1- or 3-nerved, about half as long as the tube or the inner ones smaller. Petals broad, about as long as the calyx-tube, scarcely ciliate. Staminal bundles red, ¾ in. long, the claws narrow,

tapering at the end, each with 3 to 5 rigid divaricate filaments, about $\frac{1}{3}$ as long as the claw; anther-valves orbicular, the outer one of each cell much larger than the inner one. Ovary with 2 rudimentary ovules in each cell under the perfect one. Fruit-spikes globular or oblong, the calyx-lobes at length wearing off.

W. Australia, *Drummond, 5th Coll. n.* 178; E. Mount Barren, *Maxwell.* The petals are those of *B. anisandra,* but the flowers are more regular and the leaves broader.

5. **B. anisandra,** *Schau. in Nov. Act. Nat. Cur.* xxi. 17. *t.* 1 A, *and in Pl. Preiss.* i. 149. Glabrous, except the inflorescence, or the young shoots slightly pubescent. Leaves opposite, erect or spreading, ovate to ovate-lanceolate, obtuse or scarcely acute, 1½ to 3 lines long, rigid, concave, keeled, with 1 or 2 faint lateral nerves on each side of the keel. Flowers in globular terminal heads, the long stamens of a rich purple when dry, of a dark bluish-purple when fresh or rarely red, the rhachis and calyxes pubescent or hirsute. Calyx-tube about 1 line long, the lobes not longer than the tube, and the inner ones (next the axis of the spike) often much smaller. Petals broad, slightly ciliate, the external one of each flower often 1½ lines diameter, the inner ones much smaller. Staminal bundles very unequal in the same flower, the claws of the larger ones $\frac{1}{2}$ to $\frac{3}{4}$ in. long, slender, but rigid, bearded inside at the base, each with 3 to 7 rigid digitate filaments much shorter than the claw, the inner bundles very much shorter with fewer filaments; outer valves of each anther-cell orbicular, twice as large as the inner one. Ovary with one large ovule in each cell, the 2 rudimentary ones minute or wanting.

W. Australia. King George's Sound and adjoining districts, *A. Cunningham, Drummond, 3rd Coll. n.* 57; *Preiss, n.* 362, and others, and a var. with redder stamens, *Drummond* (*3rd Coll.?*), *n.* 46; Kojonerup ranges, *Maxwell.*

6. **B. macrostemon,** *Lindl. Swan Riv. App.* 10. A small shrub, often not above 1 ft. high, more or less pubescent or hirsute. Leaves opposite, often broadly lanceolate and 3- or 5-nerved on the main stem, linear or linear-lanceolate, crowded and 1-nerved on the smaller branches, rather rigid, but scarcely acute, mostly 3 to 5 lines long. Flowers in dense globular heads, the rhachis and calyxes villous. Calyx-tube oblique, 1 to 1½ lines long; lobes narrow-lanceolate or linear, as long as the tube or the outer ones longer. Petals ovate, ciliate, rarely exceeding the calyx-lobes. Staminal bundles unequal, the longer ones $\frac{3}{4}$ in. long, the claws narrow, more or less hairy inside at the base, tapering at the top, each with about 3 spreading filaments, much shorter than the claw; inner valve of each anther-cell scarcely conspicuous. Ovary with one large ovule in each cell, the rudimentary ones scarcely conspicuous or wanting.—Schau. in Nov. Act. Nat. Cur. xxi. 16, and in Pl. Preiss. i. 149.

W. Australia. Swan River, *Drummond, 1st Coll.; Preiss, n.* 357; Hampden, *Clarke.*

Var. *incana.* Leaves white-tomentose on both sides, the nerves inconspicuous. Swan River, *Drummond.*

7. **B. cyrtodonta,** *Benth.* This has the linear, decussate, crowded foliage and inflorescence of *B. Schaueri,* and may possibly be a variety, but the staminal bundles are longer, more unequal and rigid, of a rich red colour, and the claws three times as long as the filaments, as in *B. anisandra,* the

longest claw of each flower above ½ in., and often ¾ in. long, with usually 3 filaments to each claw. The calyx-lobes lanceolate, not longer than the tube, 1-nerved, with the petals twice as long, readily distinguish it from *B. macrostemon.*—*Melaleuca cyrtodonta*, Turcz. in Bull. Mosc. 1849, ii. 24.

W. Australia, *Drummond, 4th Coll. n.* 65, *5th Coll. n.* 174; Upper Kalgan river, *Oldfield.*

8. **B. Schaueri,** *Preiss, according to Schau. in Nov. Act. Nat. Cur.* xxi. 18 (*by misprint* 14), *and in Pl. Preiss.* i. 150. A handsome bushy shrub, of 2 to 6 ft., glabrous, except the inflorescence, or the young shoots slightly pubescent. Leaves opposite, sometimes broadly lanceolate, and 3- or 5-nerved on the larger branches, but mostly linear, crowded and decussate, obtuse, keeled, triquetrous or semiterete, 2 or 3 lines long in some specimens, twice as much in others. Flowers in dense globular heads, the rhachis usually tomentose-villous. Calyx-tube glabrous or pubescent, scarcely 1 line long; lobes triangular, 1-nerved, shorter than the tube. Petals rather broad, twice as long as the calyx-lobes. Staminal bundles pink, unequal, the longest 4 to 5 lines long, each with 3 to 7, but usually 5 filaments not shorter and often longer than the claw; inner valve of the anther-cells often scarcely conspicuous. Ovary with 1 ovule in each cell, without any or with 2 minute rudimentary ones. Fruit-heads globular or ovoid, about ½ in. diameter.

W. Australia. King George's Sound and adjoining districts, *Baxter, Drummond, n.* 151, *and 5th Coll. n.* 171; *Preiss, n.* 305, *Oldfield;* Phillips Ranges and Salt river, *Maxwell.*

Var. (?) *atrorubens.* Stamens dark red, longer and more rigid. In *Maxwell's* collection in Herb. F. Mueller.

9. **B. purpurea,** *Lindl. Swan Riv. App.* 10. *t.* 3 A. (Manglesia, *in the plate.*) Branches virgate, glabrous or slightly hoary. Leaves opposite, erect or spreading, ovate-lanceolate to lanceolate-linear on the main branches, linear and decussate on the smaller ones, keeled, rigid, 3- or 5-nerved, obtuse or scarcely acute, 2 to 4 lines long, the floral ones or bracts ovate-cordate, striate with 7 or 9 nerves, the lower ones exceeding the calyx. Flowers rather small in dense globular heads, the axis soon growing out, the rhachis tomentose-villous. Calyx-tube pubescent, under 1 line long; lobes subulate, erect, often twice as long as the tube but variable in length. Petals ovate, shorter than the calyx-lobes. Stamens purple, 3 to 4 lines long, in bundles of 3 to 7, usually 5, the narrow claw about as long as the filaments. Ovary with 1 perfect ovule in each cell without any rudimentary ones. Fruiting-spikes ovoid, under ½ in. long.—Schau. in Nov. Act. Nat. Cur. xxi. 18 (by misprint 14), and in Pl. Preiss. i. 150.

W. Australia. Swan River, *Drummond, 1st Coll. n.* 129; *Preiss, n.* 258.

10. **B. Dampieri,** *A. Cunn. in Bot. Mag. t.* 3272. A rigid straggling tortuous glabrous shrub. Leaves opposite, recurved or very spreading, usually crowded and decussate, orbicular or broadly obovate, obtuse or mucronulate, 1 to 1½ lines long, rigid, prominently 3-nerved besides the nerve-like margins. Flowers small in dense globular heads, the axis soon growing out, the rhachis glabrous or scarcely tomentose. Calyx-tube rather above ½ line long, the lobes triangular, about as long as the tube. Petals orbicu-

lar, exceeding the calyx-lobes. Stamens pale pink, 3 to 4 lines long, in bundles of 9 to 15, the claw shorter than the filaments; outer valve of each anther-cell large and orbicular. Ovary not seen, the flowers examined all males. Fruiting-calyxes usually few in the head or occasionally solitary, ovoid, thick, smooth, attaining 3 lines diameter or even more. Seeds (not seen quite ripe) one large perfect one in each cell with 2 small hard erect sterile ones under it.—Schau. in Nov. Act. Nat. Cur. 19 (by misprint 15).

W. Australia. Sands of Sharks' Bay, *Baudin's Expedition, Denham*; Dirk Hartog's Island, *A. Cunningham, Milne.*

Melaleuca sprengelioides, DC. Prod. iii. 215; Mem. Myrt. t. 3, appears to me to be referable rather to this plant than to the *Regelia ciliata.*

11. **B. elegans,** *Schau. in Nov. Act. Nat. Cur.* xxi. 20, *and in Pl. Preiss.* i. 150. A rigid bushy shrub, usually glabrous except the inflorescence. Leaves opposite, erect, recurved at the end, ovate obovate or broadly oblong, acute or rather obtuse, concave, rigid, 3-nerved, 1½ to 3 lines long. Flowers (yellowish-white?) in dense ovoid or oblong spikes, the axis soon growing out, the rhachis woolly. Calyx-tube hirsute at the base, ovoid, about ¾ line long; lobes lanceolate or triangular, nearly as long as the tube. Petals oval-oblong, ciliate, often not much exceeding the calyx-lobes. Staminal bundles about 4 lines long, the claws narrow, exceeding the petals, each with 5 to 7 filaments about as long as the claw. Ovary with 1 large perfect ovule in each cell and 2 minute rudimentary ones under it, sometimes quite wanting. Fruiting-calyxes small in dense ovoid or cylindrical spikes of about ½ in. or less.

N. Australia. N. W. Cape, *Martin.*

W. Australia. Swan River, *Drummond, 1st Coll., 2nd Coll. n.* 62; Granger plain, *Preiss, n.* 284.

Var. *minor.* Leaves mostly about 1 line long. Flowers smaller, of a deep pink, mostly in globular heads. Calyx-lobes short.—*B. microphylla*, Turcz. in Bull. Mosc. 1849, ii. 24; *Drummond, n.* 130, *4th Coll. n.* 64; Point Henry, *Oldfield*, also the above-mentioned specimen from N.W. Cape. Some specimens are quite intermediate between the original form and the small variety, and the latter again approaches in many respects the *B. micrantha.*

12. **B. micrantha,** *Schau. in Nov. Act. Nat. Cur.* xxi. 22, *and in Pl. Preiss.* i. 151. A small much-branched shrub, the branches often woolly-tomentose but concealed by the more glabrous foliage. Leaves opposite, erect, appressed, decussate and imbricate on the smaller branches, more distant on the larger ones, ovate-triangular, rather obtuse, thick, keeled, ½ to 1 line long, attached by the broad base, but not really peltate. Flowers small, pink, the males in small globular heads, the hermaphrodite in ovoid or oblong spikes, the rhachis tomentose. Calyx-tube pubescent, not ½ line long; lobes ovate, obtuse, concave, shorter than the tube. Petals rather longer than the calyx-lobes, ciliolate. Stamens about 2 lines long, in bundles of 3, the claws much shorter than the filaments, and often hairy at the base. Ovary with 1 perfect ovule in each cell and 2 small rudimentary ones behind it. Fruiting-spikes 3 to 6 lines long, the calyxes not 1 line diameter, with the lobes usually inflexed and persistent the first year, but at length falling off as in other species.—*Regelia adpressa*, Turcz. in Bull. Mosc. 1849, ii. 25.

W. Australia. Stirling ranges and Konkoberup hills towards Cape Riche, *Drummond, 4th. Coll. n.* 63; *Preiss, n.* 256; *Maxwell.*

Var. *puberula.* Leaves rather larger, imbricate, minutely pubescent, the keel less prominent. Flowers rather larger, with usually 5 stamens to each bundle.—*Drummond, 4th Coll. n.* 151; *5th Coll. n.* 173.

Var. *empetrifolia.* Leaves rather narrower and more spreading, ⅓ to nearly 3 lines long.—*Melaleuca empetrifolia*, Reichb. Icon. Exot. ii. 1. t. 102; *Beaufortia empetrifolia*, Schau. in Nov. Act. Nat. Cur. xxi. 21; Lucky Bay, *R. Brown.*

Melaleuca Regelii, Planch. in Hort. Donat. 88. t. 4, may be the same species, but it has only been described from the drawing, which gives no analysis.

25. REGELIA, Schau.

Calyx-tube ovoid or nearly globular, adnate to the ovary at the base, the free part usually contracted; lobes 5, usually deciduous. Petals 5, spreading. Stamens indefinite, united in 5 bundles opposite the petals, the filaments or free parts filiform; anthers erect, the cells placed back to back and opening outwards in longitudinal slits or terminal pores. Ovary inferior, the convex summit villous, with a slight central depression round the style, 3-celled, with 4 ovules in each cell, peltately attached in pairs to a peltate placenta; style filiform, with a small stigma. Capsule enclosed in the enlarged and hardened truncate calyx-tube, opening loculicidally in 3 valves. Seeds 1 or 2 perfect in each cell. Embryo . . .—Rigid shrubs, usually pubescent or villous, with the habit of *Beaufortia.* Leaves small, opposite, mostly 3- or more-nerved. Flowers closely sessile and solitary within each bract in dense heads at first terminal, but the central axis soon growing out into a leafy branch, and often polygamous as in *Melaleuca* and *Beaufortia.*

The genus is limited to West Australia, and only differs from *Beaufortia* in the anthers and the number of ovules.

Staminal bundles ¾ in. long, the claw much longer than the filaments.
Leaves ovate-lanceolate, ½ in. long . . . . 1. *R. grandiflora.*
Staminal bundles 2 to 4 lines long, the claw shorter or not longer than the filaments.
Leaves broad, spreading, 2 to 3 lines long . . . . 2. *R. ciliata.*
Leaves erect, peltately attached, ½ to 2 lines long . . . . 3. *R. inops.*

1. **R. grandiflora,** *Benth.* Branches stout, rigid, long and virgate, hoary-villous. Leaves erect or spreading, decussate, ovate-lanceolate, obtuse or almost acute, keeled and about 5-nerved, mostly about ½ in. long, silky-hoary on both sides. Flower-spikes large, dense, globular or ovoid, the rhachis and calyxes very silky-villous. Lower bracts often leaf-like and exceeding the calyx. Calyx-tube oblique, a little above 1 line long; lobes linear, fully 2 lines long. Petals narrow-oblong, about as long as the calyx-lobes, fringed with fine cilia. Staminal bundles of a rich red, nearly ¾ in. long, the linear claws at least twice as long as the petals, each with 7 to 11 filaments shorter than the claw; anthers conical, the cells opening outwards in longitudinal slits. Fruits not seen.

W. Australia, *Drummond, 5th Coll. n.* 179.

2. **R. ciliata,** *Schau. in Nov. Act. Nat. Cur.* xxi. 11, *and in Pl. Preiss.* i. 148. A spreading shrub of 3 to 5 ft., more or less pubescent or hirsute. Leaves erect, spreading or recurved, broadly ovate obovate or almost orbicu-

lar, obtuse, flat or concave, rigid, prominently 3- or rarely 5-nerved, 2 to 3 lines long. Flower-heads small, dense, globular, the rhachis woolly. Calyx-tube woolly tomentose or hirsute, nearly globular, above 1 line long; lobes erect, lanceolate or almost linear, shorter than the tube. Petals scarcely exceeding the calyx-lobes, rather broad and entire. Staminal bundles scarcely 4 lines long, the claws linear, each with 7 to 11 filaments almost as long as the claw; anther-cells opening in small terminal pores. Fruiting-calyxes not very numerous, in a globular head, and when perfect concrete nearly to the top, each one about 2 lines diameter, with a broad open truncate orifice.

W. Australia. Swan River, *Drummond*, 1*st Coll. and* 2*nd Coll. n.* 51; Vasse river, *Preiss, n.* 287.

Schauer refers here *Melaleuca sprengelioides*, DC., but the plate, Mem. Myrt. t. 3, appears to me rather to represent *Beaufortia Dampieri*, especially in the venation of the leaves and in the anthers.

3. **R. inops,** *Schau. in Pl. Preiss.* ii. 224. A low straggling densely-branched shrub, or sometimes taller with virgate branchlets, more or less pubescent. Leaves appressed, ovate or triangular, peltately attached near the base as in the small-leaved *Melaleucas*, obtuse, thick and obtusely keeled, ½ to 1 line long and decussately imbricate on the smaller branchlets, more distant and often 2 lines long on the more luxuriant branches, the floral ones, especially under the hermaphrodite flowers, often much broader. Flowers pink, in small globular heads. Calyx-tube in the males scarcely ½ line long, glabrous or pubescent; lobes ovate, rather shorter than the tube, in the perfect flowers the tube is villous, 1 line long, but the lobes not longer than in the males. Petals broad, longer than the calyx-lobes. Stamens nearly 3 lines long, in bundles of 7 to 11, the linear claw about as long as the filaments; anthers opening in oblong pores or short slits, longitudinal or somewhat oblique. Fruiting-calyxes 3 or 4 together or sometimes solitary, ovoid-globular, attaining 3 lines, with a broad truncate orifice.—*Beaufortia inops*, Schau. in Nov. Act. Nat. Cur. xxi. 21, and in Pl. Preiss. i. 150; *Regelia globosa*, Turcz. in Bull. Mosc. 1847, i. 168.

W. Australia. King George's Sound to Swan River, *Maclean; Drummond*, 3*rd Coll. n.* 55; *Preiss, n.* 257.

26. PHYMATOCARPUS, F. Muell.

Calyx-tube nearly globular, adnate to the ovary at the base, the free part somewhat contracted; lobes 5, persistent. Petals 5, spreading. Stamens indefinite, very shortly united in a ring at the base, and higher up into 5 bundles opposite the petals, the filaments or free parts filiform; anthers erect, obovoid, the cells placed back to back and opening outwards towards the top in transverse slits. Ovary free, inferior, the truncate summit villous, with a slight central depression round the style, 3-celled, with 2 to 4 ovules in each cell erect on a short basal almost peltate placenta; style filiform, with a small stigma. Capsule enclosed in the enlarged and hardened calyx-tube, opening loculicidally in 3 valves. Seeds few, erect, with a thin testa; embryo straight; cotyledons plano-convex, longer than the radicle.—Shrub, with the small opposite leaves, globular flower-heads and habit of *Regelia* and *Beaufortia*.

This genus is limited to the single Australian species, only differing from *Beaufortia* and *Regelia* in the anthers and ovules.

1. **P. porphyrocephalus,** *F. Muell. Fragm.* iii. 121. A shrub of 2 to 3 ft., usually glabrous except the inflorescence. Leaves erect and imbricate or spreading, orbicular or broadly ovate, obtuse, flat or concave, thick, 1-nerved, 1 to 2 lines long. Flowers small (varying in colour purple or white on the same bush, according to Oldfield) in dense globular heads, the rhachis and calyxes woolly. Calyx-tube $\frac{3}{4}$ line long; lobes ovate-triangular, erect, shorter than the tube. Petals nearly 1 line diameter, ciliolate. Stamens 2 to 3 lines long, shortly united in clusters of 11 to 15, and all connected at the base into a complete ring. Ovules usually 3 in each cell. Fruiting-calyx when old depressed-globular, 2 to $2\frac{1}{2}$ lines diameter, coarsely warted outside, the orifice very open, crowned by the short thick lobes. Seeds few and erect as in *Melaleuca.*

W. Australia. Sand plains, Murchison river, *Oldfield,* and apparently the same species, in fruit only, *Drummond, n.* 71.

27. CALOTHAMNUS, Labill.

(Billottia, *Colla.*)

Calyx-tube campanulate, adnate to the ovary at the base, the free part erect or dilated; lobes 4 or 5, persistent or deciduous. Petals 4 or 5, spreading, usually scarious. Stamens indefinite, much longer than the petals, united in 4 or 5 bundles opposite to them, the filaments or free parts filiform, the lower bundles of each flower sometimes reduced to a single stamen, or without any anther; anthers oblong or linear, erect, attached by the base, the cells parallel and opening inwards in longitudinal slits. Ovary enclosed in the calyx-tube, half inferior or almost free, the convex summit villous with a central depression round the style, 3- or very rarely 4-celled, with several ovules in each cell, erect or ascending on a small placenta; style filiform with a small stigma. Capsule enclosed in the hardened and enlarged calyx-tube, opening loculicidally in 3 or rarely 4 valves. Perfect seeds few, angular and often ciliate or winged at the angles; testa thin; embryo straight; cotyledons plano-convex, longer than the radicle.—Shrubs usually stout, glabrous or hirsute. Leaves scattered, narrow, rigid, terete or flat, 1-nerved or nerveless. Flowers showy, usually red, in lateral clusters or spikes usually turned to one side, immersed in the rhachis when young, and either protruding and free from the time of flowering or remaining immersed till the maturity of the seed. Bracts none. Flowers often polygamous as in the allied genera.

The genus is confined to Western Australia.

A. *Flowers 4-merous. Two upper staminal bundles broad and flat, two lower ones narrow and without anthers (except in* C. torulosa). *Calyx-lobes as long as the tube.*

Calyx entirely immersed in the thick swollen rhachis.
- Leaves flat, 4 to 6 in. long. Rhachis of the spike densely clothed with long hairs . . . . 1. *C. pachystachyus.*
- Leaves terete, 6 to 12 in. long. Rhachis of the spike softly but shortly pubescent . . . . 2. *C. longissimus.*

Calyx not immersed in the rhachis.
  Leaves mostly ½ to 1 in. long. Lower staminal claws simple without anthers.
    Leaves more or less flattened . . . . 3. *C. blepharantherus.*
    Leaves terete . . . . 4. *C. sanguineus.*
  Leaves mostly 1 to 2 in. or more. Lower staminal claws much narrower than the upper ones, but with several stamens. Fruits very large . . . . 5. *C. torulosus.*

B. *Flowers* 5-*merous. Staminal bundles nearly equal. Perfect seeds (where known) ciliate on the angles.*

Flowering calyx-tube not exceeding 1 line, immersed or half-immersed in the rhachis. Leaves terete, rigid.
  Calyx entirely immersed in the corky rhachis . . . . . . 6. *C. gibbosus.*
  Calyx only half immersed when in flower . . . . 7. *C. gracilis.*
Flowering calyx-tube 2 to 3 lines long, almost or quite free.
  Leaves flat, oblanceolate, 2 to 3 in. long . . . . 8. *C. blepharospermus.*
  Leaves terete, rigid, 2 to 4 in. long . . . . 9. *C. chrysantherus.*
  Leaves terete or slightly flattened, mostly under 1 in. . . . 10. *C. Oldfieldii.*

C. *Flowers* 4-*merous* (*except in* C. villosus). *Staminal bundles nearly equal or the lower ones rather smaller or very rarely reduced to a single stamen.*

Flowering calyx-tube more or less immersed. Fruiting-calyx depressed, globular.
  Fruiting-calyx nearly or wholly immersed in the swollen rhachis.
    Leaves flat, oblanceolate or cuneate, 1 to 2 in. long . . . 11. *C. planifolius.*
    Leaves terete, 2 to 12 in. long . . . . 12. *C. lateralis.*
  Fruiting-calyx almost entirely exserted. Leaves terete.
    Stamens above ½ in. long.
      Leaves rigid, often slightly flattened. Staminal bundles nearly equal . . . . 13. *C. microcarpus.*
      Leaves slender. Staminal bundles very unequal . . . 14. *C. Preissii.*
    Stamens 3 to 4 lines long.
      Leaves long. Staminal bundles nearly equal . . . . 15. *C. Schaueri.*
      Leaves short. Lower staminal bundles reduced to a single stamen . . . . 16. *C. Lehmanni.*
Flowering-calyxes exserted from the first and free.
  Calyx very hirsute. Leaves short, terete.
    Flowers mostly 5-merous. Calyx-lobes usually deciduous . 17. *C. villosus.*
    Flowers mostly 4-merous. Calyx-lobes usually persistent.
      Calyx-tube 2 lines, lobes 1 line long . . . . 18. *C. pinifolius.*
      Calyx-tube 3 to 3½ lines, lobes 2 lines long. Fruits large 19. *C. rupestris.*
  Calyx glabrous or closely hoary-pubescent. Flowers 4-merous.
    Leaves terete or slightly flattened. Fruiting-calyx 2-lobed . 20. *C. quadrifidus.*
    Leaves 1 line broad. Fruiting-calyx equally 4-lobed . . . 21. *C. asper.*
    Leaves flat, 1½ to 3 lines broad. Fruiting-calyx unequally 4-lobed . . . . 22. *C. homalophyllus.*

1. **C. pachystachyus,** *Benth.* Branches very thick, densely clothed with long loose hairs. Leaves rather crowded, linear, flat, acuminate-acute, much-narrowed at the base, thick, 1-nerved, 4 to 6 in. long. Flowers large, 4-merous, in dense ovoid or oblong unilateral spikes. Calyx-tube almost entirely immersed in the swollen densely hairy rhachis, about 2 lines broad but much shorter; lobes orbicular, spreading, as long as the tube. Petals obovoid-orbicular, fully 3 lines long. Staminal bundles unequal, the 2 upper claws broadly petal-like, fully 3 in. long, each with 15 to 20 or more short

filaments crowded at the end, the anthers villous in the bud, the 2 lower claws narrower, undivided, acute and without anthers. Ovules rather numerous in each cell. Fruiting-calyxes more or less immersed, 3 to 4 lines diameter, with 2 thick hard inflexed lobes, the 2 others worn away, the dilatations of the rhachis enclosing the fruits corky at the base, thin round the calyx and densely clothed with long hairs.

W. Australia, *Drummond, 2nd Coll. n.* 70, 71; *3rd Coll. n.* 53.

2. **C. longissimus,** *F. Muell. Fragm.* iii. 112. A low shrub, the thick more or less corky branches softly but shortly pubescent. Leaves terete, slender but rigid, acute, 6 in. to above 1 ft. long, glabrous but scabrous. Flowers large, 4-merous, few in globular or ovoid more or less unilateral spikes. Calyx-tube villous, immersed in the swollen pubescent corky rhachis; lobes 1 to 1½ lines long. Petals fully 3 lines long. Staminal bundles unequal, the 2 upper claws often nearly 1 in. long, broad and petal-like, with 15 to 30 short filaments, not so crowded at the end as in *C. pachystachyus*, the 2 lower claws narrow, undivided, acute, without anthers. Fruiting-calyx entirely immersed or nearly so, 2 to 3 lines long, with 2 thick connivent lobes, the 2 others obliterated.

W. Australia, *Drummond, 2nd Coll. n.* 74; *3rd Coll. n.* 54. Sandy plains near Cujong, *Oldfield.*

3. **C. blepharantherus,** *F. Muell. Fragm.* iii. 112. Very closely allied to *C. sanguineus*, differing chiefly in the short leaves, not terete, but more or less flattened; they are linear, rather thick, ½ in. or rarely ¾ in. long. Flowers and fruits as in *C. sanguineus*, the two lower staminal claws as in that species, simple and without anthers, or rarely bearing a very few filaments with perfect anthers.

W. Australia. Murchison river, *Oldfield.*

4. **C. sanguineus,** *Labill. Pl. Nov. Holl.* ii. 25. *t.* 164. A tall shrub, either hirsute with long spreading hairs especially on the young shoots, or glabrous from the first. Leaves subulate, terete, acute, slender, sometimes all from ½ to 1 in., sometimes 1 to 1½ in. long. Flowers 4-merous, rather large, unilateral, few together or in short spikes, not at all immersed in the rhachis. Calyx-tube villous, broad, about 1½ lines long; lobes ovate, as long as the tube, with scarious margins. Petals ovate, 2 to 3 lines long, the 2 upper ones often larger than the lower. Staminal bundles of a rich red, about 1 in. long, the 2 upper claws usually united into one, but readily separating, broad, with very numerous filaments, the 2 lower ones narrow-linear, undivided, without anthers. Fruiting-calyxes ovoid or almost globular, very thick and woody, quite smooth, 4 to 6 lines long, including the thick connivent lobes, of which 2, opposite to each other, are usually larger than the 2 others.—DC. Prod. iii. 211; Schau. in Nov. Act. Nat. Cur. xxi. 24, and in Pl. Preiss. i. 151; *C. eriocarpus*, Lindl. Swan Riv. App. 9.

W. Australia. Géographe Bay, *Labillardière;* King George's Sound to Vasse and Swan rivers, *Baxter; Collie; Drummond, 1st Coll. and n.* 127, 128; *Preiss, n.* 214, 216, 219, 220, 221; and others.—The anthers in this and some of the allied species are more or less ciliate, and the division between the cells is often so narrow as to make the anther appear 1-celled.

5. **C. torulosus,** *Schau. in Nov. Act. Nat. Cur.* xxi. 25, and in Pl. Preiss. i. 152. Closely resembles the longer-leaved forms of *C. sanguineus*, and may be a variety. Leaves slender, erect, terete, about 1½ in. long. Inflorescence of *C. sanguineus*. Calyx rather more open, with a short tube and longer lobes, the 2 upper staminal bundles more separate, each with very numerous filaments, the 2 lower claws, although very narrow, bear a few filaments. Fruiting-calyxes larger than in *C. sanguineus*, often 6 lines long without the lobes, very thick, with thick connivent lobes, of which two larger than the others, as in *C. sanguineus*.

W. Australia. Cape Naturaliste, *Oldfield;* Swan River, *Drummond, 1st Coll.;* foot of Darling Range, *Preiss, n.* 212.

Var. (?) *leptophylla.* Leaves slender, 2 to 4 in. long. Only seen in fruit. *Drummond, n.* 32 *and* 39.

6. **C. gibbosus,** *Benth.* Glabrous except the young shoots, more or less plumose-villous, the branches usually thick and corky. Leaves terete, rigid, mucronate, 1 to 2 or rarely 3 in. long. Flowers 5-merous, in short ovoid or oblong spikes, forming unilateral distinct corky excrescences. Calyx-tube glabrous, buried in the corky rhachis, about 1 line long; lobes rather shorter, very deciduous. Petals ovate, twice as long as the calyx-lobes. Staminal bundles red, about ¾ in. long, all nearly equal, the claws rather narrow, linear, each with 7 to 11 filaments. Fruiting-calyxes depressed-globular, about 1½ lines diameter, almost completely enclosed in the cork, the orifice rather broad and truncate.

W. Australia, *Drummond, 5th Coll. n.* 180; Gardner river, *Maxwell.*

7. **C. gracilis,** *R. Br. in Ait. Hort. Kew. ed.* 2. iv. 418. A low shrub, quite glabrous in all our specimens. Leaves numerous, terete, rigid, mostly 1½ to 2½ but sometimes 3 to 4 in. long. Flowers 5-merous, of a rich red, in short scarcely prominent unilateral clusters of 2 to 4 each. Calyx-tubes under 1 line long, more or less immersed in the slightly prominent rhachis; lobes short. Petals about 1 line long. Staminal bundles about 1 in. long, all nearly equal, the claws narrow, each with 3 to 7 filaments. Fruiting-calyxes globular or slightly depressed, 3 to 4 lines diameter, adnate by their broad base to a slight cavity of the rhachis, but not immersed; lobes inflexed and usually persistent, at least the first year. Seeds when perfect more or less ciliate on the angles.—DC. Prod. iii. 211; Schau. in Nov. Act. Nat. Cur. xxi. 33, and in Pl. Preiss. i. 155.

W. Australia. Lucky Bay, *R. Brown;* King George's Sound and adjoining districts to Cape Riche, *Baxter, Harvey, Maxwell, Drummond, 5th Coll. n.* 180; Middle Mount Barren, *Maxwell* (with the fruiting-calyx breaking out into corky excrescences).—Drummond's n. 55 is perhaps a variety of this species, with the fruits less depressed, and very thick calyx-lobes.

8. **C. blepharospermus,** *F. Muell. Fragm.* iii. 111. A rigid straggling or diffuse shrub, hirsute with spreading hairs, the foliage scarcely becoming glabrous with age. Leaves flat, oblanceolate, obtuse or mucronate, 2 to 3 in. long and often 3 lines broad, very rigid and scabrous along the midrib and margins. Flowers 5-merous, few, large, not strictly unilateral, and sometimes all round the stem in clusters or irregular spikes, the rhachis tomentose or hirsute, somewhat corky, and slightly dilated round the calyx.

Calyx-tube nearly 3 lines long, densely hirsute, adnate by the broad base but not immersed; lobes half as long as the tube. Petals 3 to 4 lines long. Staminal bundles at least 1¼ in. long, the claws narrow, each with numerous rather short filaments from the middle upwards. Ovary almost free in the bottom of the calyx-tube. Fruiting-calyxes nearly cylindrical, 5 to 6 lines long, very thick, more or less verrucose, the 5 erect or spreading lobes persistent at least the first year. Perfect seeds usually 1 in each cell, hirsute especially at the angles with thick transparent cilia, the remaining seeds imperfect, linear-cuneate and glabrous.

W. Australia. Murchison river, *Oldfield.*

Var. (?) *glaber.* Quite glabrous. Leaves rather longer and narrower. Flowers more unilateral. With the typical form, *Oldfield.*

9. **C. chrysantherus,** *F. Muell. Fragm.* iii. 112. An erect shrub, with thick more or less corky branches, the young shoots villous with spreading hairs, the older foliage glabrous. Leaves terete, thick, mucronate-acute, 2 to 4 in. long. Flowers mostly 5-merous, large, few and unilateral, the rhachis slightly swollen and excavated. Calyx-tube sessile by the broad base, but otherwise free, hirsute, very thick, nearly 2 lines long; the lobes 1 to 1½ lines. Petals obovate-oblong, often ciliate, 2 to 3 lines long. Staminal bundles of a rich red, above 1 in. long, nearly equal, each with numerous filaments. Fruits thick, ovoid or cylindrical, smooth or verrucose, 4 to 6 lines long, the thick erect connivent lobes usually persistent. Seeds ciliate on the angles as in *C. blepharospermus.*

W. Australia. Murchison river, *Oldfield.*

10. **C. Oldfieldii,** *F. Muell. Fragm.* iii. 113. A shrub of 1 to 4 ft., with thick often corky branches, usually glabrous except the calyxes. Leaves terete or slightly flattened, thick, obtuse or mucronate, sometimes all under 1 in., rarely 1 to 2 in. long. Flowers mostly 5-merous, few, large, in unilateral clusters of 3 or 4, the rhachis slightly swollen, sometimes corky, and adhering irregularly to the calyx-tubes. Calyx otherwise free, adnate or very shortly immersed at the base, the tube thick, about 2 lines long; lobes broad, obtuse, about 1 line long. Petals about 2 lines. Staminal bundles ¾ to 1 in. long, nearly equal, each with numerous filaments. Fruiting-calyxes ovoid or nearly globular, very thick, 4 to 6 lines diameter, smooth, warted or breaking out into corky excrescences.

W. Australia. Murchison river, *Oldfield.*

11. **C. planifolius,** *Schau. in Nov. Act. Nat. Cur.* xxi. 35. *t.* 1 B, *and in Pl. Preiss.* i. 155. A rigid erect shrub of 2 to 3 ft., with villous branches. Leaves oblanceolate or linear-cuneate, flat, thick, acute when young, obtuse and more or less mucronate when old, 1 to 2 in. long. Flowers 4-merous, unilateral. Calyx-tube almost or entirely immersed in the swollen rhachis, glabrous, about 1 line long; lobes broad, the 2 larger ones shorter than the tube. Petals broad, above 1 line diameter. Staminal bundles red, ¾ to 1 in. long, the claws narrow-linear but flat, each with 3 to 7 pinnately distant filaments. Fruiting-calyx depressed, nearly 2 lines diameter, the tube from one-half to wholly immersed, the 2 larger opposite lobes connivent; the 2 smaller ones soon wearing away.

W. Australia. King George's Sound and adjoining districts, *Drummond, 3rd Coll. n.* 58; *Preiss, n.* 205, 206.

Var. *pallidifolius.* Branches glabrous. Leaves not so thick, more obtuse, 2 to 3 in. long, and often 3 to 4 lines broad, more or less distinctly penniveined when dry. Stamens apparently greenish-yellow, *Drummond, n.* 40, *and 2nd Coll. n.* 72.

12. **C. lateralis,** *Lindl. Swan Riv. App.* 9. Quite glabrous. Leaves terete, usually slender, often many in. long, obtuse or mucronate. Flowers 4-merous, in unilateral spikes of 1 to 3 in., the rhachis usually glabrous and smooth until the flowers break out, swelling to 3 or 4 lines diameter. Calyx entirely immersed, the tube about 1 line long; lobes obtuse, nearly as long as the tube. Petals nearly twice as long as the calyx-lobes. Staminal bundles red, often above 1 in. long, but slightly unequal, the claws narrow-linear, each with about 5 filaments. Fruiting-calyx entirely immersed, the lobes wearing away, or only 2 remaining persistent, the thick rhachis after flowering appearing to be excavated in a number of holes, at the bottom of which are the young capsules.

W. Australia, *Drummond, Preiss,* and others.

The following forms are sometimes very distinct, but often pass into each other:—

1. *longifolius.* Leaves slender, ½ to 1 ft. long or even more.—*C. longifolius,* Lehm. Del. Sem. Hort. Hamb. 1842, 7, according to Schau. in Nov. Act. Nat. Cur. xxi. 34, and in Pl. Preiss. i. 155.—King George's Sound, *R. Brown;* wet or sandy places from King George's Sound to Swan River, *M'Lean; Drummond, 1st Coll. n.* 125; *Preiss, n.* 200, 203, 204; *Oldfield; Turner.*

2. *rigidus.* Leaves more rigid, 2 to 4 in. long.—*C. nodosus,* Turcz. in Bull. Mosc. 1847, i. 168. *C. Huegelii,* Schau. in Nov. Act. Nat. Cur. xxi. 34 (from the description given).—*Drummond, 3rd Coll. n.* 60.

3. *crassus.* Leaves still more rigid. Rhachis of the spike 3 in. long and 6 to 9 lines thick, with very numerous flowers.—*Drummond, n.* 37, *and 2nd Coll. n.* 73.

13. **C. microcarpus,** *F. Muell. Fragm.* iii. 113. A spreading shrub of 2 to 4 ft., glabrous, or the young shoots closely pubescent. Leaves linear, thick, terete or flattened and more or less marked with 2 longitudinal furrows, obtuse or mucronate, 1 to 2 in. long. Flowers 4-merous, in short prominent clusters of 2 to 4, forming an interrupted unilateral spike of 1 to 2 in. Calyx-tube nearly buried in the rhachis, glabrous, ¾ line long; lobes much shorter. Petals as long as the calyx-tube. Staminal bundles about ¾ in. long, the claws very narrow, each with 3 to 9 rather long and slender filaments. Fruiting-calyxes depressed-globular, about 2 lines diameter, immersed at the base only in the prominent rhachis; lobes persistent, inflexed, nearly equal.

W. Australia. Between King George's Sound and Swan River, *Drummond;* Kalgan river, *Oldfield.*

Var. *teres.* Leaves all terete. Spikes more continuous. Flowers rather less immersed. —*Drummond, 5th Coll. n.* 182.

14. **C. Preissii,** *Schau. in Nov. Act. Nat. Cur.* xxi. 31, *and in Pl. Preiss.* i. 154. Young shoots silky-hairy, otherwise glabrous. Leaves terete, very slender, mostly mucronate and often curved or hooked at the end, 3 to 5 in. long. Flowers 4-merous, rather small, in prominent clusters, forming a continuous or interrupted unilateral spike of 2 to 3 in., the rhachis more or less enclosing the calyxes, or forming a torn thin border round them

at the time of flowering. Calyx-tube glabrous, about 1 line long; lobes rather shorter. Petals 1½ lines long. Staminal bundles slender, very unequal, the largest one in each flower ¾ in. long, with 5 to 9 filaments, the smaller with 1 or 2 filaments, or sometimes undivided and without any anthers, the 2 others usually intermediate. Fruiting-calyxes in closely packed clusters, not immersed in the rhachis, but not seen full grown.—*C. laxus*, Kunze in Linnæa, xx. 58 (from the description given).

W. Australia, *Drummond, n.* 234; Gordon river, *Preiss, n.* 209.—The species is nearly allied to *C. Schaueri*, but the staminal bundles are longer and more unequal.

15. **C. Schaueri,** *Lehm. Sem. Hort. Hamb.* 1842, 7, *according to Schau. in Nov. Act. Nat. Cur.* xxi. 32, *and in Pl. Preiss.* i. 154. Glabrous, or the young shoots more or less hairy. Leaves filiform, slender, terete, acuminate, acute, 4 to 8 in. long. Flowers 4-merous, small, in dense unilateral spikes of ½ to 1 in. Calyx wholly immersed at the time of flowering. Staminal bundles nearly equal, 3 to 4 lines long, each with 2 to 4 filaments. Fruiting-calyxes 1 to 1½ lines diameter, forming dense spikes, but less than half immersed in the rhachis.—*C. schœnophyllus*, Schau. in Nov. Act. Nat. Cur. xxi. 33, and in Pl. Preiss. i. 154.

W. Australia. King George's Sound, *R. Brown;* and adjoining districts, *Preiss, n.* 201, 202; *Drummond; Oldfield; Baxter.*

16. **C. Lehmanni,** *Schau. in Nov. Act. Nat. Cur.* xxi. 31, *and in Pl. Preiss.* i. 153. A low branching shrub, the young shoots usually plumose-hirsute, the foliage rarely quite glabrous till the second year. Leaves crowded, terete, mostly acute, about 1 in. long. Flowers 4-merous, small, forming dense slightly prominent unilateral spikes of ½ to 1 in. Calyx-tube immersed in the rhachis at the time of flowering, about ½ line long; lobes nearly as long as the tube. Petals broad, nearly 1 line diameter. Staminal bundles 3 to 4 lines long, the claws narrow, one with 5 to 7 filaments, one with 3 to 5, the two others tapering into a single filament with one anther. Fruiting-calyxes depressed-globular, about 1½ lines diameter, collected into dense spikes, but immersed at the base only; lobes connivent, all equal, or 2 small or obliterated.—*C. plumosus*, Turcz. in Bull. Mosc. 1847, i. 168.

W. Australia. Between King George's Sound and Swan River, *Drummond*, 3rd *Coll. n.* 59; *Preiss, n.* 218.

17. **C. villosus,** *R. Br. in Ait. Hort. Kew. ed.* 2. iv. 418. A low bushy shrub, more or less hirsute with spreading hairs, the branches thick and often corky. Leaves crowded, linear, terete or slightly flattened, mostly incurved, mucronate-acute, ½ to 1 in. long. Flowers 5-merous, or rarely here and there 4-merous, in dense unilateral clusters of 4 to 8, closely sessile, but not immersed in the rhachis at the time of flowering. Calyx-tube globular, very hispid, about 2 lines diameter; lobes erect, not 1 line long. Petals glabrous, 2 lines long or rather more, often very deciduous. Staminal bundles ¾ to 1 in. long, nearly equal, the claws narrow, each with about 9 to 15 filaments. Fruiting-calyxes 3 to 4 lines diameter, globular, usually truncate, although occasionally 1 or 2 of the lobes are thickened and persistent as in *C. quadrifidus.* When many ripen in the same cluster, they are very closely packed or even connate at the base.—DC. Prod. iii. 211; Bot. Reg. t. 1099; Lodd. Bot. Cab. t. 92; Schau. in Nov. Act. Nat. Cur. xxi.

27; Colla, Hort. Ripul. App. t. 15; *C. robustus*, Schau. l. c. 26, and in Pl. Preiss. i. 152.

W. Australia. Lucky Bay, *R. Brown*; King George's Sound, *Harvey*, *Baxter*; and thence to Swan River, *Drummond*, *n.* 55 and *n.* 87; *4th Coll. n.* 61; Hampden, *Clarke*; foot of Konkoberup hills towards Cape Riche, *Preiss*, *n.* 213.—In describing his *C. robustus*, Schauer appears to have examined a flower accidentally 4-merous; in the specimen of Preiss's, which I have seen, they are mostly at least 5-merous.

Var. *ericifolius*. Leaves all under ½ in. long. Flowers smaller. Swan River, *Drummond*, *1st Coll.*

18. **C. pinifolius,** *F. Muell. Fragm.* iii. 153. A shrub of about 2 ft., more or less hirsute with spreading hairs, or at length nearly glabrous, the branches thick and sometimes corky. Leaves very densely crowded, linear, terete, usually straight, rigid, acute and pungent-pointed, ½ to 1 in. long. Flowers 4-merous, few together, in dense unilateral clusters, closely sessile, but not immersed in the rhachis. Calyx-tube globular, densely hirsute, about 2 lines diameter; lobes not 1 line long. Petals about 2 lines. Staminal bundles as in *C. villosus*, ¾ to 1 in. long, each with about 9 to 15 filaments. Fruiting-calyxes 3 to 4 lines diameter, adnate by their broad bases, with 2 thick opposite connivent lobes of 1½ lines, the 2 others much smaller.

W. Australia, *Drummond*, *4th Coll. n.* 62; Phillips Range, *Maxwell.*

19. **C. rupestris,** *Schau. in Nov. Act. Nat. Cur.* xxi. 26, *and in Pl. Preiss.* i. 152. Branches thick, with short crowded leaves, as in *C. villosus*, but the spreading hairs appear to be entirely wanting. Leaves linear-terete, mucronate-pungent, mostly incurved, rigid, rather thick, ½ to above 1 in. long. Flowers 4-merous, 2 to 6 together in unilateral clusters or spikes, closely sessile, but not immersed in the rhachis. Calyx twice as large as in *C. villosus*, densely villous, the tube thick, ovoid-campanulate, fully 3 lines, the lobes 2 lines long. Staminal bundles as in *C. villosus*, ¾ to 1 in. long. Fruiting-calyxes very thick and hard, and often becoming glabrous, ½ to ¾ in. long, including the 2 opposite thick hard connivent lobes.

W. Australia. Swan River, *Drummond*; *Preiss*, *n.* 211.

20. **C. quadrifidus,** *R. Br. in Ait. Hort. Kew. ed.* 2. iv. 418. An erect shrub, attaining 6 to 8 ft., glabrous or more or less hirsute with spreading hairs, as in *C. villosus*, but the branches not so thick. Leaves crowded, linear, from slender terete and mucronate-acute to flat clavate and very obtuse, ½ to 1 in. long. Flowers 4-merous, clustered and closely sessile, but not immersed in the rhachis, forming usually dense unilateral spikes of 1 to 2 in. Calyx-tube ovoid-campanulate, scarcely 2 lines long, glabrous or closely and minutely pubescent; lobes short and broad, with scarious margins. Petals about 2 lines diameter, exceedingly deciduous. Staminal bundles of a rich crimson, ¾ to 1 in. long, nearly equal, the claws narrow, each with 15 to 20 or more filaments at or near the end. Fruiting-calyxes 2 to 3 lines diameter, with 2 opposite thick hard connivent lobes, the 2 others becoming obliterated.—DC. Prod. iii. 211; Bot. Mag. t. 1506; Lodd. Bot. Cab. t. 737; Reichb. Ic. et Descr. Pl. t. 9; Schau. in Nov. Act. Nat. Cur. xxi. 29, and in Pl. Preiss. i. 153; *C. lævigatus*, Schau. ll. cc. xxi. 30, and i. 153.

W. Australia. King George's Sound to Swan and Murchison rivers, and eastward to Cape Le Grand.

The specimens show three rather distinct forms:—

1. *acerosus.* Leaves slender, terete, sometimes above 1 in. long.—*Billottia acerosa*, Colla, Hort. Ripul. 20. t. 23; *C. purpureus*, Endl. in Hueg. Enum. 48.—Swan River, *Drummond*, 1*st Coll.*; *Preiss*, *n.* 207, 215, 217.

2. *normalis.* Leaves under 1 in., more or less flattened, mostly acute.—*C. clavatus*, Lodd. Bot. Cab. t. 1447; Schau. in Nov. Act. Nat. Cur. xxi. 28, and in Pl. Preiss. i. 152.—Lucky Bay, *R. Brown*; King George's Sound and to the eastward, *Preiss*, *n.* 210, and others; also *Drummond*, 1*st Coll. n.* 126.

3. *obtusus.* Leaves flat but thick, linear-clavate, obtuse, sometimes 1 line broad.—Murchison river, *Oldfield*.

21. **C. asper,** *Turcz. in Bull. Mosc.* 1849, ii. 25. Bushy and more or less hirsute with spreading hairs, the branches rather thick. Leaves crowded, linear, flat, obtuse or mucronate-acute, narrowed at the base, ½ to 1 in. long, and mostly about 1 line broad, rigid and scabrous. Flowers 4-merous, in short dense clusters, closely sessile but distinct, the rhachis slightly excavated and dilated round their base. Calyx-tube broadly campanulate, glabrous or nearly so, about 1½ lines, the lobes about 1 line long. Petals about 2 lines. Staminal bundles about ¾ to 1 in. long with many filaments, as in *C. quadrifidus*. Fruiting-calyxes ovoid-globular, thick, 3 to 4 lines diameter, usually crowned by 4 thick erect or connivent lobes, 2 opposite ones rather smaller than the 2 others.

W. Australia, *Drummond*, 3*rd Coll. n.* 52, 4*th Coll. n.* 60.—Near *C. quadrifidus*, but different in foliage and in the large fruits.

22. **C. homalophyllus,** *F. Muell. Fragm.* iii. 111. An erect shrub of 4 to 6 ft., our specimens quite glabrous. Leaves from oblong-cuneate to oblanceolate or almost linear, very obtuse, flat but thick, in some specimens all narrow and ½ to 1 in. long, in others broad and ¾ to 1½ in. long, in others again narrow and 1 to 2 in. long. Flowers 4-merous, sessile, but not immersed in the rhachis, forming unilateral spikes. Calyx-tube ovoid, nearly 2 lines long, rounded at the base; lobes very short and broad, spreading. Petals above 2 lines long, very thin and scarious, falling off as soon as the flower expands. Staminal bundles of a rich crimson, above 1 in. long, all nearly equal, the claws narrow, pinnately divided from the middle upwards; anthers not ciliate. Fruiting-calyxes nearly globular, thick, hard, and smooth, 3 to 4 lines diameter, with 2 of the lobes very thick, connivent and persistent, smooth, or breaking out into warty excrescences.

W. Australia. Murchison river, *Oldfield*; Champion Bay, *Walcott*.—Like the last, this is very closely allied to *C. quadrifidus*.

28. EREMÆA, Lindl.

Calyx-tube campanulate, adnate to the ovary at the base, the free part dilated or erect; lobes 5, triangular or acuminate. Petals 5, obovate or orbicular, spreading, usually scarious. Stamens indefinite, longer than the petals, more or less united in bundles opposite the petals, the filaments or free parts filiform; anthers obovoid, erect on a short connective continuous with the filament, the cells placed back to back and opening outwards in longitudinal slits. Ovary inferior, the summit flat or convex, villous, with a short depression round the style, 3-celled, with several ovules in each cell, erect on a small basal placenta; style filiform, with a small stigma. Capsule enclosed in the hardened and enlarged usually smooth calyx-tube, entirely or more than half inferior, opening loculicidally in 3 cells. Perfect seeds (only observed in *E. ebracteata*) 1 or 2 in each cell, obovoid or cuneate, winged on the angles; testa thin; embryo straight; cotyledons broad and folded over each other, longer than the radicle.—Bushy shrubs, usually more or less hirsute. Leaves alternate, flat and short or narrow and heath-like, often crowded on the young shoots. Flowers solitary or few, sessile, more or less surrounded by imbricate scale-like bracts, rarely entirely wanting.

The genus is limited to West Australia, differing from *Calothamnus* chiefly in inflorescence and in the anthers.

Flowers solitary, surrounded by numerous bracts.
- Leaves flat, elliptical or lanceolate, rigid, recurved. Stamens distinctly 5-adelphous . . . . 1. *E. fimbriata.*
- Leaves narrow, linear. Stamens very shortly and irregularly 5-adelphous.
  - Leaves pungent-pointed . . . . 2. *E. acutifolia.*
  - Leaves not pungent. Flowers small . . . . 3. *E. pilosa.*

Flowers 1 to 3, the bracts few and small or none.
- Leaves linear. Stamens distinctly 5-adelphous . . . . . . 4. *E. ebracteata.*
- Leaves ovate. Stamens irregularly 5-adelphous . . . . . . 5. *E. beaufortioides.*

1. **E. fimbriata,** *Lindl. Swan Riv. App.* 11. A coarse rigid shrub, attaining 3 or 4 ft., more or less hirsute with soft spreading hairs, especially on the young branches, margins of the leaves, and calyxes. Leaves from broadly elliptical to oblong-lanceolate, $\frac{1}{4}$ to $\frac{1}{2}$ in. long, rigid, erect, spreading or recurved, striate, flat or concave, obtuse or almost acute. Flowers solitary, sessile within the last leaves and surrounded by a considerable number of imbricate ovate scale-like bracts about as long as the calyx. Calyx-tube broad and open, about 2 lines diameter, silky-villous; lobes about as long as the tube. Petals about $2\frac{1}{2}$ lines long. Stamens very numerous, united into 5 flat bundles, the claws short and broad; slightly connected at the base. Ovary convex on the top. Fruiting-calyx very thick and hard, about $\frac{1}{2}$ in. diameter, crowned by the persistent lobes surrounding the somewhat prominent summit of the capsule.—Schau. in Pl. Preiss. i. 156.

W. Australia. Swan River and adjoining districts, *Drummond,* 1*st Coll., also n.* 47, *and* 64; *Preiss, n.* 254. Sandy woods near Mouger's Lake, Hill river, *Oldfield.*

Var. *brevifolia.* Leaves short, very broad, concave, squarrose, 7-nerved.—Valley of the Hutt river, *Oldfield.*

2. **E. acutifolia,** *F. Muell. Fragm.* ii. 30. Branches stout and rigid, more or less hirsute. Leaves crowded, spreading, linear, rigid and pungent-pointed, under $\frac{1}{2}$ in. long, ciliate or hirsute with long spreading hairs or glabrous when old. Flowers not so large as in *E. fimbriata,* solitary and sessile within the last leaves, surrounded by numerous imbricate linear or oblong-spathulate bracts shorter than the calyx. Calyx-tube broad and open, about 2 lines diameter, lobes herbaceous, about as long as the tube. Petals scarcely exceeding the calyx-lobes. Stamens very numerous, shortly and irregularly united into 5 bundles, a few filaments often almost free. Ovary flat-topped. Capsule convex, but not exceeding the persistent lobes of the fruiting calyx.

W. Australia. Champion Bay, *Walcott, Oldfield.*

3. **E. pilosa,** *Lindl. Swan Riv. App.* 11. An erect or spreading heath-like shrub of 3 or 4 ft., more or less pubescent or hirsute, the branches much more slender than in *E. fimbriata.* Leaves linear, flat, semiterete or triquetrous, obtuse or nearly so, 2 to 3 or rarely 4 lines long, glabrous or hairy. Flowers much smaller than in *E. fimbriata*, solitary and sessile within the last leaves, surrounded by imbricate scale-like bracts, the inner ones nearly as long as the calyx-tube, the outer ones much smaller. Calyx-tube rather narrow, about 1½ lines long, silky-pubescent; lobes triangular, much shorter than the tube. Petals about 1½ lines long. Stamens very numerous, slightly united in 5 bundles, but with several filaments often free between them. Ovary slightly convex; stigma almost capitate. Fruiting-calyxes globular, very smooth and shining, above ¼ in. diameter, the capsule not prominent.—Schau. in Pl. Preiss. i. 157; *E. ericifolia,* Lindl. l. c., Schau. l. c.; *Metrosideros pauciflora,* Endl. in Hueg. Enum. 50, according to Schau.

W. Australia. Swan River, *Drummond, 1st Coll., and n.* 117; *Preiss, n.* 294, 295, 296; Tone and Gordon rivers, *Oldfield.*

4. **E. ebracteata,** *F. Muell. Fragm.* ii. 29. A low bushy shrub, the branches and young leaves more or less hirsute. Leaves linear, obtuse or scarcely acute, not pungent, 3 to 4 lines long. Flowers rather large, 2 or 3 together, sessile at the ends of the branches, becoming lateral by the elongation of the shoot. Bracts scale-like but few, small, and very deciduous, so as to be rarely seen. Calyx-tube villous, rather narrow, nearly 2 lines long; lobes broadly-triangular or shortly acuminate, scarcely more than ½ line long. Petals about 2 lines diameter. Stamens very numerous, united into 5 flat bundles, the claws nearly as long as the filaments. Ovary convex; stigma acute. Fruiting-calyx smooth, fully 4 lines long, the capsule filling the tube. Seeds when perfect bordered by 2 to 5 longitudinal very transparent wings; cotyledons very broad and folded lengthwise so as to embrace each other.

W. Australia. Murchison river, *Oldfield, Drummond, 6th Coll. n.* 78.

5. **E. beaufortoides,** *Benth.* Apparently a straggling shrub, the young shoots more or less hirsute, at length nearly glabrous. Leaves recurved or spreading, broadly ovate, acute, rigid, 3- or 5-nerved, 2 to 3 lines long. Flowers rather large, usually 2 to 4 together at the ends of the branches, the imbricate scale-like bracts much shorter than the calyx and very deciduous. Calyx narrow, much wrinkled and apparently viscid, about 3 lines long, lobes triangular, acute, about 1 line long. Petals fully 2 lines diameter. Stamens very numerous, more or less united at the base, but scarcely forming regular bundles. Ovary convex on the top. Fruiting-calyxes ovoid or cylindrical, truncate, thick and very smooth, 6 lines long and about 4 lines diameter. Capsule much shorter than the calyx.

W. Australia. Between Moore and Murchison rivers, *Drummond, 6th Coll. n.* 79.

SUBTRIBE 4. EUCALYPTEÆ.—Leaves opposite or alternate, coriaceous, usually large. Flowers usually 3 or more, in umbels, sometimes reduced to heads or very rarely in cymes or solitary, the common peduncles axillary or in a terminal corymb or panicle. Calyx truncate, entire or remotely toothed.

Petals attached by a broad base, distinct or consolidated into an operculum. Stamens indefinite, in several series, free or obscurely united into 4 bundles; anthers various. Ovules indefinite in each cell. Embryo with the cotyledons longer than the radicle and often folded over it.

29. ANGOPHORA, Cav.

Calyx-tube turbinate-campanulate, adnate to the ovary at the base, the free part broad and open, 5-angled, truncate, with 5 small distinct teeth. Petals 5, attached by their broad base, herbaceous and aristate, with coloured margins, much imbricate in the bud, spreading and separately deciduous. Stamens numerous, free, in several series, filaments filiform; anthers versatile, the cells parallel, opening longitudinally. Ovary inferior, the flat summit glabrous, 3- or 4-celled, with many ovules in each cell, ascending on a peltate placenta; style subulate, with a capitate stigma. Capsule enclosed in and adnate to the hardened truncate persistent calyx-tube, opening loculicidally in 3 or 4 valves. Perfect seeds (where known) 1 in each cell, large, broad, very flat, peltately attached on the inner face; testa thin; embryo straight; cotyledons thin, flat, or folded over each other at the edge, deeply cordate, the radicle slightly clavate, scarcely protruding beyond the lobes of the cotyledons.—Trees or shrubs, usually glaucous, pubescent or hispid with bristly hairs. Leaves opposite or here and there alternate, coriaceous, penniveined. Flowers in umbel-like cymes arranged in terminal corymbs. Bracts exceedingly deciduous.

The genus is limited to Eastern Australia. It is very nearly allied to *Eucalyptus*, the petals similarly truncate at the base, but not connate, and the calyx-teeth although small are more prominent than in any *Eucalyptus*.

Leaves mostly or all sessile and cordate at the base.
  Bark smooth and deciduous. Flowers rather large, not numerous. Calyx-tube about 3 lines . . . . 1. *A. cordifolia.*
  Bark rough and persistent. Flowers small and numerous. Calyx-tube about 2 lines long . . . . 2. *A. subvelutina.*
Leaves petiolate, lanceolate, not cordate.
  Bark rough and persistent. Flowers small and numerous. Calyx-tube about 2 lines long. . . . . 3. *A. intermedia.*
  Bark smooth and deciduous. Flowers rather large, not very numerous. Calyx-tube about 3 lines long . . . . 4. *A. lanceolata.*

1. **A. cordifolia,** *Cav. Ic.* iv. 21. *t.* 338. A tall shrub or small tree, more or less pubescent with minute rigid hairs or glaucous, the smaller branchlets and inflorescence hispid with bristly often reddish hairs, the older bark smooth and falling off in large flakes. Leaves ovate or oblong, mostly obtuse, nearly sessile and deeply cordate with rounded auricles, 2 to 4 in. long, glabrous and shining above, glaucous or pubescent underneath. Flowers rather large, 4 to 6 in each umbel, forming a rather dense terminal corymb. Calyx-tube 3 lines long and opening out nearly flat to $\frac{1}{2}$ in. diameter. Petals acutely acuminate, 3 to 4 lines diameter. Fruiting-calyx very hard, often $\frac{3}{4}$ in. broad at the top, and as much in length.—DC. Prod. iii. 222; *Metrosideros hispida*, Sm. in Trans. Linn. Soc. iii. 267, and Exot. Bot. t. 42; Bot. Mag. t. 1960; Lodd. Bot. Cab. t. 106; *M. hirsuta*, Andr. Bot. Rep. t. 281;

M. anomala, Vent. Jard. Malm. t. 5; *M. cordifolia*, Pers. Syn. Pl. ii. 25; *Eucalyptus hirsuta*, Link, Enum. Hort. Berol. ii. 31.

N. S. Wales. Port Jackson, *R. Brown*, *Sieber*, *n.* 971, and others.

2. **A. subvelutina,** *F. Muell. Fragm.* i. 31. A tree attaining a considerable size with a rough persistent bark as in *A. intermedia*, of which F. Mueller now thinks it may be a variety. Foliage and young shoots glaucous or minutely pubescent, with often a few bristles on the flowering branches and inflorescence. Leaves sessile or nearly so, ovate or ovate-lanceolate, mostly acute, all (excepting rarely the upper ones) cordate at the base with rounded auricles as in *A. cordifolia*, 2 to 4 in. long, the veins numerous but not usually so much so nor so fine as in *A. intermedia*. Flowers small, in loose corymbs, precisely as in *A. intermedia*. Fruiting calyxes 3 to 4 lines diameter.—*A. velutina*, F. Muell. Fragm. iv. 170.

Queensland. Brisbane, Burnett, and Boyd rivers, *F. Mueller*.

N. S. Wales. Grose river, *R. Brown;* Paramatta, *Cayley*, *Woolls* (the inflorescence more bristly than usual); Clarence and Macleay rivers, *Beckler*.

3. **A. intermedia,** *DC. Prod.* iii. 222. A tree attaining a considerable size with a rough persistent fibrous bark, quite glabrous or slightly pubescent, or rarely with a few bristles on the inflorescence Leaves distinctly petiolate, lanceolate or sometimes ovate-lanceolate, acutely acuminate, 2 to 4 in. long, or even more in some specimens. Flowers rather small, in loose corymbs or trichotomous panicles. Calyx usually about 2 lines long and 3 lines diameter at the top, but sometimes rather larger, the 5 ribs very prominent and the secondary ones also conspicuous; the teeth shortly subulate, rarely half as long as the tube. Fruiting calyx 3 to 4 lines diameter at the top and about as long.—*Metrosideros floribunda*, Sm. in Trans. Linn. Soc. iii. 267 (not of Ventenat).

Queensland. In the interior, *Mitchell*.

N. S. Wales. Grose river, *R. Brown;* Port Jackson, *Cayley*, *Woolls*, and others; northward to Clarence river, *Beckler*, and New England, *C. Stuart;* southward to Twofold bay, *F. Mueller*.

Victoria. Mouth of the Genoa river, *F. Mueller*.

4. **A. lanceolata,** *Cav. Ic.* iv. 22. *t.* 339. A tree of considerable size, the bark deciduous in large smooth flakes as in *A. cordifolia;* branches and foliage glabrous and scarcely glaucous, or rarely a few bristles on the inflorescence. Leaves distinctly petiolate, lanceolate, acuminate, mostly 3 to 5 in. long, coriaceous, with numerous fine parallel pinnate veins. Flowers in rather dense terminal corymbs or short panicles, larger and more dense than in *A. intermedia*, rather smaller than in *A. cordifolia*. Calyx usually about 3 lines long and 4 lines broad at the top, the teeth very minute or at any rate shorter and thicker than in *A. intermedia*, and the secondary ribs often very short or quite inconspicuous. Fruiting calyx usually thick and very smooth.—DC. Prod. iii. 222; *Metrosideros costata*, Gærtn. Fruct. i. 171. t. 34. f. 2; *M. lanceolata*, Pers. Syn. Pl. ii. 25 (not the sp. with the same name l. c. 26); *M. apocynifolia*, Salisb. Prod. 351.

Queensland. Burnett river, *F. Mueller;* Boyd river, *Herb. F. Mueller;* Moreton bay, *C. Stuart*.

N. S. Wales. Port Jackson to the Blue Mountains, *Burton*, *A. and R. Cunningham*, and others, and in the interior north of Bathurst, *A. Cunningham*.

30. EUCALYPTUS, Lhér.

(Eudesmia, *R. Br.* Symphyomyrtus, *Schau.*)

Calyx-tube obconical campanulate or oblong, adnate to the ovary at the base or rarely to the top, truncate and entire after the falling off of the operculum or with 4 minute teeth; the orifice closed by a hemispherical conical or elongated operculum covering the stamens in the bud and falling off entire when the stamens expand, this operculum usually simple (formed of the concrete petals?), thin or more frequently thick, fleshy or woody, the veins longitudinal, numerous and parallel or rarely anastomosing, the separation from the calyx-tube usually but not always marked in the bud by a distinct line; there is also frequently in the very young bud a very thin membranous external operculum more continuous with the calyx-tube and very rarely this external one persists nearly as long as the internal one and is as thick or nearly so. Stamens numerous, in several series, free or very rarely very shortly united at the base into 4 clusters; anthers versatile or attached at or close to the base, the cells parallel and distinct or divergent and confluent at the apex, opening in longitudinal slits or rarely in terminal pores, the connective often thickened into a small gland either separating the cells or behind them when they are contiguous. Ovary inferior, the summit glabrous, flat, convex or conical, 3- to 6-celled, with numerous ovules in each cell, in 2 to 4 rows, on an adnate or oblong and peltate axile placenta; style subulate or rarely almost clavate, with a small truncate capitate or rarely peltate stigma. Fruit consisting of the more or less enlarged truncate calyx-tube enclosing the capsule, usually of a hard and woody texture and interspersed with resinous receptacles, the persistent disk usually thin and lining the orifice of the calyx-tube when the capsule is deeply sunk; concave, horizontal, convex, or conically projecting, and more or less contracting the orifice when the capsule is not much shorter than, as long as, or longer than the calyx-tube; the capsule always adnate to the calyx-tube although often readily separable from it when quite ripe and dry, very rarely protruding from the orifice left by the disk before maturity, but opening at the apex in as many valves as there are cells, which often protrude, especially when acuminate by the persistent and split base of the style. Seeds for the greater part abortive but more or less enlarged, variously shaped and of a hard apparently uniform texture, one or very few in each cell perfect, usually ovoid or flattened and ovate when solitary, variously shaped and angular when more than one ripen; testa black, dark coloured, or rarely pale, smooth or granular, not hard, in a few species expanded into a variously-shaped wing; hilum ventral or lateral. Embryo with broad cordate 2-lobed or bipartite cotyledons, folded over the straight radicle but otherwise flat.—Shrubs or trees, attaining sometimes a gigantic size, secreting more or less of resinous gums, whence their common appellation of *Gum-trees.* Leaves in the young saplings of many species, and perhaps all in some species, horizontal, opposite, sessile, and cordate, in the adult shrub or tree of most species vertical (or sometimes horizontal), alternate, petiolate and passing more or less from broadly ovate to lanceolate acuminate and falcate, always rigid whether thick or thin, penniveined, the midrib conspicuous; the primary veins often scarcely perceptible when the leaves are thick; in some species few, irregular, oblique, and anastomosing and passing through every gradation from that to numerous parallel diverging or transverse veins, always converging into an intramarginal vein, either close to or more or less distant from the edge, the intermediate reticulate veinlets rarely very prominent, and scarcely any when the primary veins are closely parallel. Flowers large or small, in umbels or heads, usually pedunculate, rarely reduced to a single sessile flower, the peduncles in most species solitary and axillary or lateral (by the abortion of the floral leaves) either at the base of the year's shoot below the leaves or at the end of the older shoot above them. Bracts and bracteoles when present so early deciduous as only to have been observed in a very few species.

With the exception of two species extending to Timor, and two or three or perhaps one single somewhat doubtful species from the Indian Archipelago, the Eucalypti are all Australian, where they constitute a large portion of the forest vegetation. Their size and abundance, as well as the great value of their timber and other products, cause them to be well known to colonists under their local appellations of *Gum-*, *Mahogany-*, and *Box-trees*, *Stringy-barks*, *Iron-barks*, etc., but to the botanist who is unable to compare them in a living state, the due limitation and classification of their species present almost insuperable obstacles. The extraordinary differences in the foliage of many species at different periods of their growth add much to the ordinary difficulties arising from the gradual transition of varieties, races, or species one into the other; moreover, a considerable portion of our herbarium specimens have been gathered to illustrate collections of woods by persons little acquainted with botany, and are but too frequently not in a state to supply the most essential characters. The old division of the genus according to the opposite or alternate leaves is now found to be quite fallacious, so many species having them opposite at an early stage and alternate when full-grown, the second character generally made use of in books, the comparative length of the operculum and calyx-tube is too indefinite for practical use. F. Mueller has proposed sections founded on the nature of the bark, of the value of which I am totally unable to judge, nor have I any means of availing myself of them, for the specimens themselves never show the character and a large proportion of them are either unaccompanied by any notes of it, or the collectors' notes are from various causes indefinite, unreliable, or even contradictory. I have thus been compelled to establish groups upon such characters as appeared to me the most constant among those which are supplied by the specimens; in the first place upon the form of the anthers and secondly upon that of the fruit, and in some cases on the inflorescence or the calyx. It must be admitted, indeed, that these groups, distinct as they may be in the typical species, pass very gradually into each other through intermediate forms, but I have endeavoured to supply cross-references to facilitate the determination of dried specimens in doubtful cases. It is to be hoped that, in the elaborate monograph of the genus with plates representing all the species promised by Dr. Mueller in his 'Fragmenta,' he, from his knowledge of the Gum-trees in a living state, will be able to give us a truly natural arrangement founded upon the proposed cortical or any other system which experience may induce him to adopt. In the meantime, as far as I can gather from the information supplied, it appears to me that among large trees the majority of the "Stringy-barks" are to be found in my first series with reniform anthers, and of the "Iron-barks" and "Box-trees" in the following three series with very small globular or truncate anthers, that other marked peculiarities in the bark are typical rather of species than of groups, and that, among shrubs or small stunted or scrubby trees, the cortical character is of very little avail, even for the discrimination of species.

A few notes may be required on some of the minor characters which I have made use of or neglected in the specific diagnoses and descriptions.

I have thought it generally useless to describe the branchlets terete or angular, for in those species such as *E. pruinosa*, *E. tetragona*, *E. tetraptera*, etc., where the angles are often so prominent as to be almost transformed into wings, there occur branches, often on the same specimen, quite terete.

The form, size, and venation of the leaves described has always been taken from those of the flowering branches of what have been supposed to be adult trees or shrubs; when not stated to the contrary, they are always alternate and petiolate. A great majority of the species are now known to have on the young sapling, or even on adventitious barren branches of older trees, opposite sessile broad or cordate leaves passing gradually into the ordinary alternate petiolate narrower ones. It appeared quite useless in any manner to describe these sapling leaves in the several species where they have been observed, for they present at once the greatest similarity in the corresponding leaves of different species and the greatest dissimilarity in the different leaves of the same species or specimen. Where in the following pages the leaves are described as opposite and sessile, it is meant that they retain that form on the flowering branches. So also in the venation, characteristic as it often is in the lanceolate leaves, the specific modifications disappear in a great measure as the leaf gets broader, and it is only very rarely that there are any appreciable specific differences in the venation of the sapling leaves. A very few at that age, especially in the *Corymbose* series, appear to be already alternate, but to have the lamina peltately inserted on the petiole above the base, but our data on that point are but very scanty.

Diagnostic characters are sometimes taken from the position of the leaves, horizontal or vertical, and the comparative colour of their surfaces, dark above and pale underneath or similar on both sides, but this can rarely be ascertained from dried specimens. In general, it would appear that the horizontal leaves have the two surfaces different and the veins very divergent or transverse, and the vertical leaves have the surfaces similar and the veins oblique; so that where the leaves of the adult tree are alternate lanceolate and foliate with *oblique* veins they are usually vertical, whilst the opposite ones of the sapling of the same species are horizontal.

The inflorescence is often characteristic of species or even of groups, but cannot always be taken absolutely in single specimens. The umbels are as a rule universal, but are always in a very few large-flowered species, and occasionally in others, reduced to a single flower. The length of the peduncle supporting it, either absolute or compared to that of the petiole, to which importance is given in old diagnoses, appears to be rarely available as a specific character. Rarely above 1 in., generally varying from ¼ to ¾ in. and sometimes entirely disappearing, it is only in the few cases where it is constantly long or short as compared to these dimensions that I have referred to it. These peduncles with their umbels are, however, in their general arrangement, of some importance, constituting three types:—1. axillary or lateral, that is, solitary in the axils of the leaves or along the branchlet above or below the leaves; 2. several together in short simple panicles at the end of the branchlet or in the axils of the leaves, and, 3. in a compound terminal corymbose panicle. But these forms appear to pass into each other very much in imperfect specimens. In the first and simplest form, the floral leaves of the uppermost umbels or of very short axillary flowering branches are sometimes quite abortive, converting the inflorescence into the second form; in this again the lower axillary panicles may be occasionally reduced to single umbels as in the first, and even in the terminal corymb, characteristic of the *Corymbosæ*, a single specimen may here and there show an axillary umbel, or after flowering the branches of the corymb may occasionally though rarely grow out into leafy shoots, leaving the fruiting umbels lateral below the new leaves.

The form and dimensions of the calyx-tube (*hypanthium* of Schauer, *cupula* of De Candolle) are taken when the stamens are expanded but still adhering; after they fall, it often alters so much that it neither indicates the form it had in flower nor yet that which it will assume in fruit.

The operculum described is always the single one, probably representing the petals, as it appears when ready to fall off for the expansion of the stamens. The outer one, of whose nature there is still much doubt, exists probably in nearly all species at an early stage, but it is usually thin and falls off too soon to be worth mentioning in descriptions. Where, as in *E. platyphylla*, it persists rather longer, it appears to do so in a very variable degree in the same species. It is only, as far as hitherto observed, in *E. variegata* and *E. eximia* that it is more constantly persistent till nearly the time of expansion of the flower, and equals or exceeds in thickness and consistency the inner one.

The dimensions given for the stamens refer to the outer ones; the inner ones are almost universally gradually shorter.

The style is omitted in the specific descriptions, because I have been unable to ascertain the constancy of the few differences observed. It is certainly longer in some species, thickein others, the stigma a mere point or more or less dilated, but these differences appear to be almost as frequently individual as specific.

The number of cells of the ovary is also very rarely a guide to the species. They generally vary from 3 to 4 or from 4 to 5, very rarely 6, and not constantly so in any species I have seen. In *E. phœnicea* I have only seen two; but as the specimens known are but few, and all probably gathered from one tree, we have no means of judging whether the character is constant.

For similar reasons I have very seldom mentioned the seeds; for great as are the differences observed, we have very seldom means of judging whether they are individual or specific. The fruiting specimens in our herbaria and museums have generally shed their seeds, at least the perfect ones. The abortive seeds are usually numerous in the capsule, unimpregnated and of a hard granular uniform texture, but enlarged, especially those near the top of the capsule, and variously shaped according to the degree of mutual pressure, the several seeds of the same specimen often differing more from each other than the corresponding ones of different species. Of perfect seeds there generally only ripen either 2 or 3 or a single one in each cell, and their shape is accordingly modified. They are, moreover, always near the orifice of the capsule and the first to be shed, and are thus unknown in a large portion of the species. The most remarkable are those of the majority of the *Corymbosæ*, which are large and more or less expanded into a membranous wing; but even that character would appear to be of little value if we consider that species so closely allied in every other respect as *E. calophylla* and *E. ficifolia*, or *E. citriodora* and *E. corymbosa*, only differ from each other in their wingless or winged seeds; that even this difference is proved only by the examination of seeds most probably derived from a single tree of each, and that the wing, when it exists, varies remarkably in size and shape in different seeds from the same specimen.

The embryo in *Eucalyptus* appears always to have the cotyledons folded over the radicle, but varies much in the shape of the cotyledons, very broad or rather narrow, entire, cordate, 2-lobed or 2-partite, and in the comparative length of the radicle, and these differences are very likely of specific constancy; but there are but two or three species in which I have been able to examine the embryo taken from several specimens, and not many where I have had perfect seeds enough to spare more than one or even a single one for dissection. I have therefore thought it very unsafe to rely upon any of the modifications observed for specific distinction.

As some further guide to the determination of species, though often a fallacious one, I have taken the following index of colonial names from the collectors' notes.

"Black Butt" = *E. pilularis, hæmastoma, patens, ficifolia.*
"Bloodwood" = *E. marginata, corymbosa, eximia.*
"Box" = *E. amygdalina, hemiphloia, brachypoda, viminalis.*
"Bastard-Box" = *E. polyanthemos.*
"White Box" = *E. albens.*
"Yellow Box" = *E. melliodora, bicolor.*
"Cider-tree" = *E. Gunnii.*
"Flintwood" = *E. pilularis.*
"Blue Gum" = *E. hæmastoma, globulus, botryoides, megacarpa, viminalis, tereticornis, diversicolor.*
"Drooping Gum" = *E. Risdoni, viminalis.*
"Flooded or Swamp Gum" = *E. coriacea, decipiens, botryoides, rostrata, rudis.*
"Green Gum" = *E. stellulata.*
"Grey Gum" = *E. saligna, resinifera.*
"Lead Gum" = *E. stellulata.*
"Red Gum" = *E. amygdalina, melliodora, odorata, rostrata, tereticornis, resinifera, Stuartiana, calophylla.*
"Risdon Gum" = *E. Risdoni.*
"Rusty Gum" = *E. eximia.*
"Spotted Gum" = *E. hæmastoma, goniocalyx, (citriodora ?), maculata.*
"Turpentine Gum" = *E. Stuartiana.*
"Weeping Gum" = *E. coriacea, viminalis.*

"White Gum" = *E. stellulata, coriacea, amygdalina, paniculata, hæmastoma, albens, botryoides, saligna, goniocalyx, rostrata, Stuartiana, redunca.*
"York Gum" = *E. loxophleba.*
"Gum-top" = *E. virgata.*
"Hiccory" = *E. Stuartiana, resinifera.*
"Iron-bark" = *E. (macrorhyncha?), leucoxylon, hemiphloia, siderophloia, melanophloia, drepanophylla, crebra.*
"Leather-Jacket" = *E. resinifera.*
"Lignum-Vitæ" = *E. polyanthemos.*
"Mahogany (Bastard)" = *E. marginata, botryoides;* (Forest) *E. resinifera, var.;* (Red) *E. resinifera;* (Swamp) *E. robusta, botryoides;* (White) *E. pilularis, var. acmenioides, robusta.*
"Messmate" = *E. obliqua.*
"Mountain Ash" = *E. virgata, hæmastoma;* (Black) *E. leucoxylon.*
"Peppermint" = *E. coriacea, amygdalina, capitellata, piperita, odorata, viminalis.*
"Spearwood" = *E. doratoxylon.*
"Stringy-bark" = *E. amygdalina, obliqua, capitellata, macrorhyncha, piperita, pilularis, tetradonta.*
"Woolly Butt" = *E. longifolia.*

SERIES I. **Renantheræ.**—*Stamens all perfect or very rarely* (*especially in* E. virgata) *some of the outer ones with abortive anthers; anthers reniform or broad and flat, the cells divergent or at length divaricate, contiguous and usually confluent at the apex.*

The species are all Eastern except *E. marginata* and *E. buprestina,* and all extratropical except *E. piperita* and *E. pilularis.*
(The truncate anthers of 17. *E. leucoxylon* and a few others among the *Heterostemones* sometimes open out when old so as to assume almost the appearance of the *Renantheræ.*)

Leaves of the flowering branches sessile, opposite, cordate or connate.
  Calyx-tube contracted at the top after flowering. Leaves mostly connate . . . . 5. *E. Risdoni.*
  Calyx-tube expanded at the top after flowering. Leaves rather large, cordate and broad . . . . 8. *E. dives.*
Leaves of the flowering branches alternate and petiolate.
  Leaf-veins not numerous, very oblique, a few almost parallel to the midrib, giving the leaf a 3- to 7-nerved appearance, sometimes inconspicuous when the leaf is narrow and thick.
    Flowers small, in almost sessile umbels. Buds narrow-oblong. Leaves acute at both ends, usually narrow . . . . . . 1. *E. stellulata.*
    Flowers in distinctly pedunculate umbels. Buds clavate.
      Leaves usually rather broad. Peduncles terete or slightly flattened. Fruit obovoid-globose, the rim flat or scarcely concave . . . . 2. *E. coriacea.*
      Leaves long and falcate. Peduncles much flattened. Fruit narrow-obovoid, contracted at the base, the rim usually concave . . . . 3. *E. virgata.*
  Leaf-veins not close, often very oblique but all inserted along the midrib.
    Fruit pear-shaped obovoid or subglobose, more or less contracted at the orifice, the rim concave, the capsule sunk. Flowers pedicellate.
      Buds clavate. Eastern species.
        Leaves very thick, mostly straight or nearly so and under 3 in. Fruit obovoid-cylindrical, above ½ in. long . . 9. *E. obtusiflora.*
        Leaves mostly very oblique and rather broad. Fruit obovoid or pear-shaped, much under ½ in. Peduncles terete or nearly so . . . . 7. *E. obliqua.*

Leaves long, falcate, the veins inconspicuous or very oblique. Fruit narrow-obovoid, under ½ in. Peduncles much flattened . . . . 3. *E. virgata.*
Buds obovoid. Western species. Leaves usually under 3 in. and prominently veined. Fruits about 1 in. diameter 10. *E. buprestium.*
Fruit subglobose-truncate, the rim usually flat. Buds clavate.
Leaves narrow and thin, or broader thick and smooth.
Operculum convex or hemispherical, smooth or papillose.
Lower leaves and often those of the flowering branches opposite and connate. Bark smooth and deciduous . 5. *E. Risdoni.*
Leaves of the flowering branches all alternate and lanceolate, oblong or linear; lower ones when opposite not connate. Bark of the trunk usually rough and persistent . . . . 4. *E. amygdalina.*
Operculum depressed and rugose. Leaves under 3 in. long 6. *E. coccifera.*
Fruit subglobose, much contracted at the orifice, the rim thin, the capsule sunk. Buds ovoid. Operculum as long as the calyx-tube. Leaf-veins fine . . . . 14. *E. piperita.*
Fruit subglobose or depressed-globose, the rim very convex or prominent. Buds ovoid or obovoid. Operculum usually as long as or longer than the calyx-tube.
Flowers and fruits sessile. Leaves of *E. obliqua.* Operculum very obtuse . . . . 11. *E. capitellata.*
Flowers and fruit pedicellate. Leaves of *E. obliqua.* Operculum acuminate or conical. Calyx-border prominent in the bud . . . . 13. *E. macrorhyncha.*
Flowers and fruit pedicellate. Leaves very thick and rigid, and often nearly straight . . . . 12. *E. santalifolia.*
Leaf-veins numerous, fine and parallel (not very close). Buds ovoid-acuminate or oblong. Operculum as long as or longer than the calyx-tube.
Stamens much inflected in the bud. Leaf-veins oblique or diverging, often scarcely visible on the upper surface. Fruit under ½ in. diameter. Eastern species.
Operculum about as long as the calyx-tube. Fruit contracted at the orifice. Peduncles terete or nearly so . . . 14. *E. piperita, var.*
Operculum longer than the calyx-tube. Fruit straight or scarcely contracted at the orifice.
Peduncles terete or nearly so. Umbels mostly in a terminal panicle. Fruit-rim usually thin and concave 15. *E. pilularis, var. acmenioides.*
Peduncles more or less flattened. Umbels mostly axillary. Fruit-rim usually broad and flat . . . . 15. *E. pilularis.*
Stamens flexuose, but not at all or scarcely inflected in the bud. Leaf-veins more transverse. Fruit above ½ in. diameter. Western species . . . . 16. *E. marginata.*

SERIES II. **Heterostemones.**—*Outer stamens anantherous or with small abortive anthers; anthers of the perfect ones small, globular, or truncate, the cells contiguous, opening in pores or in oblong slits, sometimes at length confluent.*

The species are all Eastern, one only (*E. gracilis*) extending also into Western Australia, and all extratropical, two only (*E. hæmastoma* and *E. microcorys*) also tropical or subtropical.

(The outer stamens appear also to be anantherous or with abortive anthers only in 3. *E. virgata*, and perhaps occasionally but only in a slight degree in some others of the *Renantheræ*, and sometimes but rarely in 27. *E. bicolor*, amongst *Porantheræ*, but I have never found them so in any of the other species.)

Umbels all axillary or lateral. Buds ovoid or rarely obovoid.
Peduncles terete or nearly so.
Flowers large, usually 3 to 5 in the umbel. Leaves very coriaceous . . . . 17. *E. leucoxylon.*
Flowers small, 4 to 8 in the umbel.
Leaves rather thin. Buds ovoid, often acuminate. Flowers distinctly pedicellate . . . . 18. *E. melliodora.*
Leaves thick, narrow, black-dotted, mostly under 3 in. long. Buds obovoid, tapering into short pedicels or nearly sessile . . . . 19. *E. gracilis.*
Peduncles flattened. Leaves thick, long, falcate, not dotted. Buds ovoid, often acuminate . . . . 3. *E. virgata.*
Upper umbels forming a terminal corymb or panicle. Buds obovoid.
Leaves thick, narrow, black-dotted, veins scarcely visible. Flowers small. Pedicels short . . . . 19. *E. gracilis.*
Leaves rather thin, veins very oblique, often distinct. Flowers small. (Anthers usually all perfect) . . . . 27. *E. bicolor.*
Leaves rather thick, not dotted, veins numerous and more regular but fine, sometimes indistinct. Flowers moderate-sized or rather large, distinctly pedicellate . . . . 20. *E. paniculata.*
Upper umbels forming a terminal corymbose panicle. Buds clavate, tapering into a rather long pedicel.
Buds very angular. Operculum often conical . . . . 20. *E. paniculata.*
Buds not angular. Operculum very short and obtuse.
Leaves very coriaceous, with oblique veins (as in *E. obliqua*). Fruit pear-shaped with a broad flat rim . . . . 21. *E. hæmastoma.*
Leaves rather thin, with almost transverse veins. Fruit oblong . . . . 22. *E. microcorys.*

SERIES III. **Porantheræ.**—*Stamens all perfect (except rarely in* E. bicolor, *and perhaps in* E. polyanthemos) ; *anthers small and globular, or broader than long, the cells distinct, opening in small circular pores, sometimes extending at length into oblong slits.*

The species are all Eastern or tropical, including most of the "Box-trees," *E. uncinata* alone extending also into West Australia. The leaves when narrow have always an oblique irregular venation. The operculum is short, and the capsule sunk in the fruit.

This series passes through *E. bicolor* into the *Heterostemones*, and when fully out the anthers sometimes are very nearly those of the *Microntheræ*, whilst amongst *Micrantheræ* there are several species, especially 36. *E. albens*, 38. *E. siderophloia*, 32. *E. stricta*, and 34. *E. decipiens*, in which the anther-cells are so short that their slits are at first little more than pores.

Umbels few-flowered, usually several together, in short leafless axillary or terminal panicles, or in terminal corymbs. Operculum short, obtuse.
Leaves sessile, opposite, cordate or ovate. Flowers in terminal corymbose panicles . . . . 23. *E. pruinosa.*
Leaves orbicular or ovate, obtuse, alternate on long petioles.
Flowers rather large, in a terminal corymbose panicle. Calyx about 3 lines diameter . . . . 24. *E. oligantha.*
Flowers small, mostly in short dense axillary or terminal panicles. Calyx not above 2 lines diameter . . . . . . 25. *E. polyanthemos.*
Leaves ovate, acute or broadly lanceolate, very coriaceous . . 26. *E. Behriana.*
Leaves oblong or lanceolate, rather thin . . . . 27. *E. bicolor.*
Umbels several-flowered, often solitary in the axils, the upper ones in terminal corymbose panicles. Operculum usually conical, about as long as the calyx-tube. Pedicels short.

Leaves linear, thick, the oblique veins seldom visible. Umbels all axillary . . . . 32. *E. stricta.*
Leaves (of *E. obliqua*) broad or falcate, very coriaceous . . . 30. *E. hemiphloia.*
Leaves coriaceous, with numerous fine diverging veins, sometimes inconspicuous . . . . 38. *E. siderophloia.*
Umbels mostly axillary, or lateral and solitary. Leaves ovate-lanceolate to linear.
Leaves large, broad, very glaucous. Buds long and narrow. Operculum acuminate . . . . 36. *E. albens.*
Leaves lanceolate, rigid, without black dots. Operculum short, obtuse. Filaments flexuose . . . . 28. *E. odorata.*
Leaves usually narrow, rigid, copiously black-dotted. Operculum as long as or longer than the calyx-tube. Outer filaments slender, folded in with an acute angle . . . . 29. *E. uncinata.*

(See also 35. *E. corynocalyx*, with the filaments of *E. uncinata*, but a long calyx and longer openings to the anthers.)

SERIES IV. **Micrantheræ.**—*Anthers very small, globular, or broader than long, with globular distinct cells opening in lateral slits.*

The species, with the exception of the Western *E. micranthera* and *E. decipiens*, are all Eastern or tropical, and include most of the "Iron-barks,"—one species, *E. brachypoda*, extending also into the west. The series, which closely connects the *Porantheræ* with the *Normales*, is by no means a distinctly marked one. The anthers have at first sight, in their shape and small size, the appearance of the former, whilst their dehiscence is almost or quite that of the *Normales*. As in *Porantheræ*, the operculum is short, rarely slightly longer than calyx-tube, and the capsule more or less sunk, although the points of the valves often protrude.

Leaves all or mostly sessile, opposite and cordate . . . . . . 39. *E. melanophloia.*
(See also 23. *E. pruinosa*, amongst *Porantheræ*.)
Leaves of the flowering branches ovate-lanceolate to linear-coriaceous, with oblique or inconspicuous veins. Umbels axillary or lateral (except sometimes *E. Bowmani*). Eastern species.
Flowers sessile in the head. Leaves narrow, thick, the veins inconspicuous. Operculum short, obtuse . . . . 31. *E. cneorifolia.*
Calyx-tube tapering into a very short, sometimes scarcely distinct pedicel.
Leaves narrow, thick, the veins inconspicuous. Operculum short. Peduncles very short, lateral . . . . 32. *E. stricta.*
Leaves mostly lanceolate-falcate or ovate-lanceolate, obliquely veined.
Calyx-tube much longer than broad. Operculum conical or acuminate. Fruit cylindrical. Foliage glaucous or white . . . . 36. *E. albens.*
Calyx-tube turbinate. Operculum very obtuse. Foliage not white . . . . 37. *E. Bowmani.*
Leaves mostly lanceolate-falcate, the veins inconspicuous. Upper umbels usually paniculate . . . . 38. *E. siderophloia.*
Flowers distinctly pedicellate.
Leaves narrow, thick, apparently veinless. Calyx-tube not 2 lines diameter, turbinate . . . . 32. *E. stricta.*
Leaves lanceolate, falcate. Calyx-tube above 2 lines diameter, turbinate . . . . 37. *E. Bowmani.*
Leaves ovate-lanceolate or lanceolate. Calyx-tube long, cylindrical. Operculum very flat . . . . 35. *E. corynocalyx.*

(See also 103. *E. oleosa*, and its allies amongst *Normales Inclusæ*, which have the anthers almost globular.)

Leaves of the flowering-branches mostly lanceolate or falcate (sometimes broad in *E. decipiens*), coriaceous with numerous fine diverging veins, often scarcely conspicuous.
  Flowers sessile or nearly so. Umbels all axillary or lateral. Western species.
    Operculum obtuse, shorter than the calyx-tube . . . . . 33. *E. micranthera.*
    Operculum acuminate, longer than the calyx-tube . . . . 34. *E. decipiens.*
  Flowers on distinct but short and angular pedicels. Upper umbels usually in a short terminal panicle. Eastern species 38. *E. siderophloia.*
Leaves of the flowering-branches from ovate and obtuse to narrow-lanceolate and acuminate, with numerous fine parallel diverging or transverse veins, usually visible. Flowers small, the upper umbels often in a short panicle. Operculum about as long as the calyx-tube.
  Leaves usually long and narrow.
    Fruit subglobose, truncate, about 3 lines diameter, scarcely contracted at the orifice . . . . 40. *E. drepanophylla.*
    Fruit ovoid, truncate, contracted above the middle, about 3 lines long . . . . 41. *E. trachyphloia.*
    Fruit from nearly globose to narrow-ovoid, contracted at the orifice, not above 2 lines diameter . . . . 43. *E. crebra.*
    Fruit of *E. crebra*, but 4 lines diameter . . . . 42. *E. leptophleba.*
    Fruit hemispherical, very open, not above 2 lines diameter, the valves protruding . . . . 44. *E. brachypoda.*
  Leaves ovate or oblong, very obtuse. Fruit scarcely above 1 line diameter . . . . 45. *E. brachyandra.*

SERIES V. **Normales.**—*Stamens all perfect; anthers oblong-ovate or nearly globose, the cells perfectly distinct, parallel (either contiguous with the connective-gland behind them, or back to back, with the connective between them), and opening longitudinally.*

(In 103. *E. oleosa*, and its allies, the anthers are smaller, almost globular, and passing into those of the *Micrantheræ*.)

SUBSERIES I. **Subsessiles.**—*Flowers axillary or lateral, usually large,* 1 *to* 3, *sessile or nearly so on the stem, or on an exceedingly short terete or angular peduncle.*

Leaves all opposite, sessile, cordate orbicular or ovate.
  Leaves large. Flowers solitary, the calyx about 1½ in. diameter 46. *E. macrocarpa.*
  Leaves mostly under 3 in. Flowers 3 together on a very short common peduncle. Calyx under ½ in. diameter.
    Leaves crenate. Calyx obtuse at the base . . . . 47. *E. cordata.*
    Leaves quite entire. Calyx tapering at the base . . . . 48. *E. pulverulenta.*
Leaves, at least on the flowering-branches, alternate and petiolate.
  Flowers sessile on the stem. Buds usually very rugose or ovate. Calyx about ½ in. diameter.
    Leaves falcate-acuminate, mostly above 6 in. long . . . . 49. *E. globulus.*
    Leaves broad, obtuse, rarely above 3 in. . . . . 50. *E. alpina.*
  Flowers 3 together, on a very short common peduncle. Buds smooth, about ½ in. diameter. Leaves ovate or lanceolate, acuminate . . . . 51. *E. cosmophylla.*

(In a few specimens of 63. *E. incrassata,* and perhaps some others of the *Robustæ*, in 34. *E. decipiens,* 90. *E. platyphylla,* and a few others, the peduncles are occasionally so much reduced as to bring them almost into the *Subsessiles.*)

SUBSERIES II. **Recurvæ.**—*Flowers axillary or lateral, often large, usually* 3 *or rarely* 5 *together, pedicellate on a recurved terete peduncle. Calyx-tube turbinate or urceolate. Leaves alternate, thick.*

Calyx-tube turbinate, above 1 in diameter. Disk forming a raised ring within the stamens . . . . 52. *E. pyriformis.*

Calyx-tube turbinate, not above ⅓ in. diameter. Disk flat or concave, but often raised above the calyx-border.
Operculum as long as or scarcely longer than the calyx-tube. Eastern species . . . . 53. *E. longifolia.*
Operculum much longer than the calyx-tube. Western species . 54. *E. conoidea.*
Calyx-tube urceolate, about ½ in. long. Ovary and capsule deeply sunk . . . . 55. *E. urnigera.*
Calyx-tube ovoid? Fruit ovoid, ¾ in. diameter, the disk or rim very concave . . . . 56. *E. cæsia.*

(104. *E. decurva*, 105. *E. doratoxylon*, and a few others among the *Inclusæ*, have loose reflexed umbels, but more numerous and smaller flowers. 17. *E. leucoxylon*, often resembles *E. longifolia*, but with very different stamens, and some specimens of 74. *E. occidentalis*, have loose recurved 3-flowered umbels, but with the long operculum and straight stamens of the *Cornutæ*.)

SUBSERIES III. **Robustæ.**—*Peduncles axillary or lateral, or very rarely the upper ones in a terminal corymb, usually flattened, each with several* (*rarely only* 1) *large or moderate-sized flowers, sessile or tapering into thick pedicels. Leaves usually thick and alternate, or in* E. Preissiana *often opposite. Rim of the fruit concave, with a sunk capsule, except in the last four species.*

(Some varieties of *E. dumosa* and *E. incrassata* have pedicellate flowers and more terete peduncles connecting them with the *Inclusæ*, and *E. robusta* and *E. botroyides* are near *E. resinifera* and its allies in foliage, but very different in fruit. *E. vernicosa* connects the series with *E. viminalis* amongst *Exsertæ*, as *E. grossa* does with the *Cornutæ*.)

Leaves with numerous close parallel very diverging or transverse veins. Fruit ovoid-oblong or urceolate, the capsule deeply sunk.
Calyx ribbed or winged, 1 to 2 in. long.
Peduncles 1-flowered. Fruit 4-winged. Western species . 57. *E. tetraptera.*
Peduncles 3- to 5-flowered. Fruit several-ribbed. Tropical species . . . . 58. *E. miniata.*
Calyx not ribbed. Fruit rarely above ½ in. long. Eastern species.
Buds narrow, acuminate, ½ in. long or more. Operculum as long as or longer than the calyx-tube . . . . . . 56. *E. robusta.*
Buds ovoid-oblong, not above 4 lines long. Operculum not so long as the calyx-tube . . . . 60. *E. botryoides.*

(See also 95. *E. saligna*, and its allies, which have the foliage and almost the inflorescence of *E. botryoides*, but a very different fruit; 99. *E. concolor*, from W. Australia, with a hard globular fruit; and 90. *E. platyphylla*, from N. Australia, with large broad leaves.)

Leaves with more oblique and irregular veins often inconspicuous. Fruit ovoid-truncate, the capsule sunk.
Operculum obtuse or umbonate, much shorter than the calyx-tube. Leaf-veins inconspicuous. Peduncles not much flattened . . . . 62. *E. dumosa.*
Operculum umbonate or conical, shorter than the calyx-tube. Peduncles much flattened . . . . 61. *E. goniocalyx.*
Operculum rostrate, often longer than the calyx-tube . . . . 63. *E. incrassata.*
Operculum oblong-obtuse, thin and smooth, as in many *Cornutæ* 65. . *grossa.*

(See also 100. *concolor* and 101. *E. gonianthа*, allied to *E. oleosa* amongst *Inclusæ*.)

Leaves thick, with irregular or inconspicuous veins. Rim of the fruit or disk flat or slightly convex or concave.
Flowers small. Fruit flat-topped, under ½ in. diameter. Leaves ovate or orbicular, rarely much above 1 in. long. Tasmanian species . . . . 66. *E. vernicosa.*

(See also the sessile-flowered variety of 86. *E. viminalis.*)

Flowers rather large. Operculum globular, broader than the calyx-tube. Fruit turbinate, ¾ in. long, the broad rim flat, the valves protruding and erect. Western species . . . 64. *E. gomphocephala.*

Flowers large. Fruit hemispherical, ¾ to 1 in. diameter. Western species.

Leaves acuminate. Fruit with prominent thick incurved capsule-valves . . . . 67. *E. megacarpa.*

Leaves obtuse, often opposite. Fruit flat or concave at the top, the valves not prominent . . . . 68. *E. Preissiana.*

(See also the subseries *Eudesmieæ*, which, when the calyx-teeth become obliterated, resemble the *Robustæ* in their inflorescence and fruit, but have generally more or less opposite leaves and broad short opercula.)

SUBSERIES IV. **Cornutæ.**—*Peduncles axillary or lateral, several- often many-flowered, flattened (except in* E. cornuta). *Flowers sessile or shortly pedicellate. Operculum long, smooth, and not thick. Stamens erect or flexuose in the bud, but not inflected. Fruit turbinate, urceolate or obovoid, the capsule not much sunk. Leaves thick, with irregular oblique veins often inconspicuous.*

The habit of this subseries is that of the *Robustæ*, from which it is readily distinguished by the stamens not inflected in the bud. The species are all Western.

Calyx-tube and fruit more or less immersed in the large thick receptacle. Capsule valves exserted, acuminate, connivent . . . 69. *E. Lehmanni.*

Flowers and fruit sessile. Leaves ovate-lanceolate or lanceolate. Peduncles scarcely flattened. Ovary flat-topped, the style slightly thickened. Capsule slightly convex before opening; valves, when open, very prominent, with long points, often connivent . 70. *E. cornuta.*

Flowers and fruit sessile. Leaves narrow-lanceolate. Peduncles very short broad and flat. Ovary and capsule with a conical top, surrounded by a free annular disk. Valves, when open, prominent and acuminate . . . . 71. *E. annulata.*

Flowers and fruit nearly sessile. Leaves mostly orbicular. Peduncles very flat, broad and often long. Ovary and capsule with a conical top, rim of the fruit scarcely prominent. Valves acuminate, usually prominent . . . . 72. *E. platypus.*

Flowers and fruit nearly sessile. Leaves of *E. cornuta*. Peduncles flattened. Ovary flat-topped, the style not thickened. Fruit truncate, the valves not acuminate nor protruding . . . . . 73. *E. macrandra.*

Flowers and fruit distinctly pedicellate. Leaves of *E. cornuta*. Peduncles flattened. Ovary and capsule convex or conical at the top. Fruit urceolate; valves acuminate, protruding when open . . . . 74. *E. occidentalis.*

Flowers and fruit shortly pedicellate. Leaves narrow. Peduncles flat. Ovary flat-topped, the style not thickened. Fruit obovoid, much contracted at the orifice. Points of the valves often protruding . . . . 75. *E. spathulata.*

(See also 65. *E. grossa*, which has the operculum of the *Cornutæ*, but the stamens distinctly inflected in the bud.)

(In 16. *E. marginata*, the stamens are not inflected, but the anthers are very different. In 89. *E. tereticornis*, 78. *E. Oldfieldii*, and perhaps some others of the *Exsertæ*, the stamens are only slightly inflected at the ends.)

SUBSERIES V. **Exsertæ.**—*Peduncles axillary or lateral, or rarely also the upper ones in a short terminal corymb, terete or scarcely flattened, each with several, often many flowers, usually pedunculate. Fruit globose or depressed, usually more or less contracted at the orifice, the rim convex or prominent, rarely flat, the capsule valves protruding beyond it.*

The general shape of the fruit, with its peculiar broad prominent rim or disk, is like that of 12. *E. capitellata*, and its allies among *Renantheræ*, and is not found in other groups, although 103. *E. oleosa* and its allies, which I have placed among *Inclusæ*, often show some approach to it, and, on the other hand, *E. viminalis*, *dealbata*, and *cinerea*, are frequently exceptional in the present subseries by the rim, and more rarely the valves remaining quite flat.

Leaves opposite, cordate-ovate. Operculum conical, shorter than the calyx-tube. Eastern species . . . . 84. *E. cinerea.*

(See also 47. *E. pulverulenta*, among the *Subsessiles*, differing from *E. cinerea* in its larger less pedicellate flowers, always 3, and in the fruit.)

Leaves alternate, from very broadly ovate to ovate-lanceolate, usually obtuse, whitish, veined. Operculum obtuse, longer than the calyx-tube. Eastern species . . . . 85. *E. dealbata.*

(See also the tropical 91. *E. alba* among the *Subexsertæ*.)

Leaves orbicular ovate or lanceolate, very thick, the veins (fine and parallel or irregular and oblique) scarcely conspicuous.
- Calyx not 3 lines diameter. Operculum short. Tropical species . 76. *E. pallidifolia.*
- Calyx 4 to 6 lines diameter or more. Operculum as long as or longer than the calyx-tube.
  - Disk forming a raised ring or prominent rim round the somewhat depressed capsule.
    - Calyx angular. Tropical species . . . . 77. *E. pachyphylla.*
    - Calyx terete. Stamens dark, scarcely inflected. Western species . . . . 78. *E. Oldfieldii.*
  - Disk concave in the flower, very convex in the fruit, the capsule not depressed. Western species.
    - Ovary slightly convex, shorter than the calyx . . . . 79. *E. pachyloma.*
    - Ovary conical in the centre, rather longer than the calyx-tube. Leaves ovate-oblong or lanceolate . . . . . 80. *E. Drummondii.*
    - Ovary with a large conical summit, entirely exserted. Leaves orbicular . . . . 81. *E. orbifolia.*

Leaves long-lanceolate or linear, not very thick, the oblique veins fine, but often conspicuous. Calyx rarely above 3 lines diameter.
- Leaf-veins inconspicuous. Western species.
  - Leaves narrow-linear. Umbels 2- to 4-flowered. Operculum obtuse . . . . 82. *E. angustissima.*
  - Leaves linear-lanceolate. Umbels with many small flowers, on slender pedicels. Operculum conical . . . . . . 83. *E. leptopoda.*
- Leaf-veins usually conspicuous. Eastern species, except *E. exserta.*
  - Pedicels short. Operculum obtuse or conical, not much longer than the calyx-tube. Fruit-rim not very convex, and often flat 86. *E. viminalis.*
  - Pedicels slender. Operculum more or less rostrate. Fruit-rim very convex or conical.
    - Eastern or tropical species, with a smooth white deciduous bark . . . . 87. *E. rostrata.*
    - Western or tropical species, with a dark rough bark, persistent or falling off in fragments . . . . 88. *E. exserta.*
  - Operculum 2 to 4 times as long as the calyx-tube, obtusely conical. Fruit-rim very convex or conical . . . . . 89. *E. tereticornis.*

SUBSERIES VI. **Subexsertæ.**—*Peduncles axillary or lateral, or also the upper ones more or less paniculate, terete or flattened, several-flowered. Calyx-tube broad at the orifice. Fruit turbinate, the orifice not contracted, the capsule level or slightly sunk, the valves often protruding when open.*

This subseries differs from the *Exsertæ* and the *Inclusæ*, chiefly in the fruit.

Leaves broad, with very diverging veins and distinctly reticulate. Tropical or subtropical species.
Flowers nearly sessile or on short thick pedicels. Operculum hemispherical, short . . . . 90. *E. platyphylla.*
Flowers small, distinctly pedicellate. Operculum conical . . 91. *E. alba.*
Leaves mostly lanceolate, rather thick, the veins fine or obscure, oblique and irregular. Operculum conical. Eastern species . 92. *E. Stuartiana.*
Leaves thick and shining, the veins inconspicuous. Operculum globular, broader than the calyx-tube. Peduncle flattened. Western species. . . . . 64. *E. gomphocephala.*
Leaves long-lanceolate, with numerous, rather regular oblique veins, and more or less reticulate. Fruit rather large.
Fruit with a broad flat dilated rim, the valves protruding. Tropical species . . . . 93. *E. patellaris.*
Fruit with a narrow rim, the capsule somewhat sunk, the points of the valves protruding. Western species . . . . 94. *E. rudis.*

(See also 38. *E. siderophloia,* which has nearly the fruit of the *Subexsertæ*, but with very different anthers; and some varieties of 86. *E. viminalis,* which, when the fruit-rim is less prominent, come near the *Subexsertæ.*)

Leaves ovate-lanceolate to lanceolate, with very numerous, fine, close, parallel veins.
Operculum conical, about as long as the calyx-tube. Leaves usually narrow. Flowers nearly sessile . . . . 95. *E. saligna.*
Operculum much longer than the calyx-tube. Leaves usually broad-lanceolate. Flowers distinctly pedicellate.
Calyx-tube under ½ in. diameter . . . . 96. *E. resinifera.*
Calyx-tube 6–8 lines diameter . . . . 97. *E. pellita.*

SUBSERIES VII. **Inclusæ.**—*Umbels usually several-flowered, axillary or lateral and solitary or several together, in lateral clusters or very short panicles, and then sometimes reduced to one or two flowers each, the peduncles terete or rarely flattened. Fruit more or less contracted at the orifice, the capsule sunk, the valves not protruding, excepting their points when acuminate by the split base of the style.*

Umbels solitary and simple, axillary or the upper ones almost paniculate.
Fruit hard, depressed-globose or subglobose, the rim usually flat and the capsule often scarcely sunk; valves often acuminate by the split base of the style. Leaves alternate, the veins scarcely conspicuous. Western species, the *E. oleosa* also in more eastern deserts.
Operculum hemispherical, not so long as the calyx-tube. Calyx much dilated above the ovary. Umbels few-flowered . . . . 99. *E. patens.*
Operculum longer than the calyx-tube, conical or acuminate.
Peduncles much flattened.
Flowers sessile. Calyx broad at the base or shortly tapering, smooth as well as the operculum . . . . 100. *E. concolor.*
Calyx tapering into a short pedicel, deeply furrowed as well as the operculum . . . . 101. *E. goniantha.*
Peduncles terete or nearly so. Flowers pedicellate.
Calyx depressed-globose, deeply furrowed. Flowers small. Operculum very long . . . . 102. *E. falcata.*
Calyx subglobose or obovoid, smooth or slightly furrowed.
Stamens inflected, flexuose. Capsule slightly sunk . 103. *E. oleosa.*
Stamens slender, inflected with an acute angle. Capsule much sunk. Umbels usually reflexed . . . 104. *E. decurva.*

Fruit pear-shaped, the points of the valves sometimes protruding. Leaves alternate, the veins scarcely conspicuous. Peduncles with 3 almost sessile flowers. Tasmanian species . 98. *E. Gunnii.*
Fruit obovoid or ovoid-oblong, the rim thin, the capsule deeply sunk.
Leaves mostly opposite.
Leaves cordate ovate. Branches scabrous. Operculum short, obtuse. Tropical species . . . . 106. *E. aspera.*
Leaves mostly lanceolate. Operculum beaked. Umbels recurved. Western species . . . . 105. *E. doratoxylon.*
Leaves alternate.
Operculum at least twice as long as the calyx-tube. Western species . . . . 114. *E. redunca.*
Operculum hemispherical or flat.
Peduncles with 3 almost sessile flowers. Tasmanian species . . . . 98. *E. Gunnii.*
Peduncles with several pedicellate flowers.
Calyx ½ in. long, narrow. Stamens long and red. Fruit long, with a distinct neck. Tropical species . 110. *E. phœnicea.*
Calyx 3 to 4 lines long. Leaf-veins parallel, fine, and nearly transverse. Western species . . . . . 111. *E. diversicolor.*

(See also the large-flowered varieties of 62. *E. dumosa.*)

Calyx 2 to 3 lines long. Western species.
Leaves (often above 4 in. long) with irregular, distant, usually prominent veins, the intramarginal one distant from the edge . . . . 112. *E. loxophleba.*
Leaves (usually under 3 in. and often narrow) with fine parallel veins, scarcely conspicuous, the intramarginal one near the edge . . . . 113. *E. fœcunda.*

(See also 135. *E. tetradonta*, and 134. *E. odontocarpa*, amongst *Eudesmieæ*, in which the calyx-teeth occasionally disappear.)

Umbels several together, on very short lateral peduncles, forming short panicles or clusters; operculum very short and flat. Tropical or subtropical species.
Leaves mostly opposite, large, broad, thick, and rigid. Umbels irregular, each often reduced to 1 or 2 flowers. Calyx 4 lines diameter or more . . . . 107. *E. grandifolia.*
Leaves from broadly cordate and opposite to broadly lanceolate, irregularly and conspicuously veined. Umbels many-flowered. Pedicels long. Calyx under 3 lines diameter . . . . . 108. *E. clavigera.*
Leaves all narrow-lanceolate, with more regular veins. Pedicels shorter. Calyx small (the whole inflorescence sometimes reduced to an apparently simple cluster) . . . . 109. *E. tesselaris.*

(See also 43. *E. crebra*, and its allies amongst *Micrantheræ*, which have frequently a compound inflorescence, and a similar fruit, but a conical operculum and very small anthers.)

SUBSERIES VIII. **Corymbosæ.**—*Flowers usually large, (the umbels or very rarely heads) all in a terminal corymbose panicle, or rarely a few of the lower ones axillary. Fruit often large, more or less urceolate, the capsule deeply sunk. Seeds usually large, flat, with acute edges, often more or less expanded in a variously-shaped wing.*

Leaves opposite, connate, large . . . . 115. *E. perfoliata.*
Leaves opposite, sessile, cordate.
Branchlets rusty-pubescent. Leaves large. Fruit above 1 in. long . . . . 116. *E. ferruginea.*
Branchlets and calyx bristly. Leaves small. Fruit ½ to ¾ in. long . . . . 117. *E. setosa.*

Leaves alternate, peltately attached to the petiole above the base.
  Leaves oblong or lanceolate . . . . 118. *E. melissiodora.*
  Leaves broadly ovate . . . . 119. *E. peltata.*
Leaves alternate, petiolate, broadly ovate. Fruit globose, truncate or with a very short neck . . . . 120. *E. latifolia.*
Leaves ovate-lanceolate or lanceolate, acuminate, with numerous fine, close, almost transverse veins.
  Tropical species. Fruit 1 to 1½ in. long, prominently ribbed . 121. *E. ptychocarpa.*
  Western species. Fruit 1½ to 2 in. long, not ribbed.
    Seeds large, not winged . . . . 122. *E. calophylla.*
    Seeds (very irregularly) winged . . . . 123. *E. ficifolia.*
  Eastern species. Fruit ½ to ¾ in. long, not ribbed.
    Seeds more or less winged . . . . 124. *E. corymbosa.*
    Seeds not winged . . . . 125. *E. citriodora.*
Leaves long-lanceolate, thick and smooth, the very fine close almost transverse veins scarcely conspicuous.
  Fruit oblong. Operculum depressed, continuous with the calyx till the moment of separation . . . . 126. *E. terminalis.*
  Fruit globular or ovoid, contracted at the mouth. Operculum of *E. terminalis.* Flowers large . . . . 128. *E. pyrophora.*
  Fruit nearly globular, with a short neck. Operculum depressed (Flowers smaller than in the other species ?) . . . . 127. *E. dichromophloia.*
Leaves narrow-lanceolate, rigid, with more oblique veins. Operculum double.
  Flowers pedicellate in 3-flowered umbels . . . . 129. *E. maculata.*
  Flowers sessile, in heads . . . . 130. *E. eximia.*

SUBSERIES IX. **Eudesmieæ.**—*Leaves, including the petiolate ones, mostly opposite or nearly so. Peduncles usually 3-flowered. Calyx with 4 minute teeth, more or less conspicuous below the globular hemispherical or flattened operculum. Stamens sometimes in 4 clusters.*

The habit of the subseries is generally that of the *Robustæ* or of the *Inclusæ*.

(The calyx-teeth are sometimes scarcely conspicuous in *E. tetragona*, and occasionally flowering branches with alternate leaves may be observed in most species.)

Stamens more or less distinctly in 4 clusters, usually very shortly united at the base (or inserted on 4 lobes of the disk). Western species.
  Flowers very large. Operculum red, very broad, depressed, with crossed raised ribs. Fruit very large, hemispherical, with a broad raised hat-like disk . . . . 131. *E. erythrocorys.*
  Flowers not very large. Operculum hemispherical, smooth. peduncles flattened. Fruit ovoid or globular; rim narrow; capsule sunk. Leaves usually broad . . . . 132. *E. tetragona.*
  Flowers small. Operculum hemispherical, smooth. Peduncles short, not flattened. Fruit ovoid or oblong; rim narrow; capsule sunk. Leaves mostly narrow . . . . 133. *E. eudesmioides.*
Stamens very numerous, not separated into clusters, the disk not lobed. Tropical species. Fruit oblong-cylindrical; rim narrow; capsule sunk.
  Fruit above ½ in. long. Leaves long-lanceolate, the veins usually conspicuous. Flowers rather large. . . . . 135. *E. tetradonta.*
  Fruit under ½ in. long. Leaves short lanceolate, the veins scarcely prominent. Flowers small . . . . 134. *E. odontocarpa.*

(The calyx-teeth are also sometimes distinguishable in 40. *E. globulus*, with large nearly globular sessile and very rugose buds, and in 57. *E. tetraptera*, with angular buds 1½ in. long.)

The following enumerated or more or less described species, all unknown to me, are all most probably synonyms to some of those here described :—

E. ovata, Labill. Pl. Nov. Holl. ii. 13. t. 153, from West Australia, does not occur in the distributed sets of Labillardière's plants which I have seen. From the figure, it appears probable that the specimen represented was from an adventitious branch, with much broader leaves than the ordinary flowering ones. It is very likely, therefore, a form of some one of the described Western species, possibly *E. brachypoda*.

E. multiflora, Poir. Dict. Suppl. ii. 594, appears to me very probably to be the same as *E. resinifera*.

E. oblonga, DC. Prod. iii. 217, is described from Sieber's specimens, n. 583, which are in young bud and fruit with a few leaves, and in that state may equally well be referred to *E. piperita*, *E. hæmastoma* or others.

E. pallens, DC. Prod. iii. 219, is described from Sieber's specimens, n. 606, which I have not seen. The short diagnosis agrees in many respects both with *E. albens* and *E. dealbata*, and especially with the latter, but the operculum described is rather different from either.

E. curvula, Sieb. in Spreng. Syst. Cur. Post. 195, is described from specimens of Sieber's without reference to his number. I have not met with the name in the Sieberian collections I have seen, and the short diagnosis is equally applicable to several species.

E. triantha, Link, Enum. Hort. Berol. ii. 30, *E. salicifolia* and *E. racemosa*, Cav. Ic. iv. 24, *E. linearis* and *E. procera*, Dehnh. in Walp. Rep. ii. 164, although evidently seen by the authors in bud or in flower as well as in leaf, are far too imperfectly described to render their identification possible.

The following, which have been shortly described from young plants in leaf only, could most probably not be identified if we had the origiual specimens before us, and should be entirely discarded :—

E. microphylla, Willd. Enum. Hort. Berol. 515, probably not an *Eucalyptus* at all.

E. reticulata, *media*, *mucronata*, *elongata*, *myrtifolia*, *angustifolia*, *stenophylla*, and *purpurascens* (*oppositifolia*, Desf.), Link, Enum. Hort. Berol. ii. 29 to 31.

E. hypericifolia, *umbellata*, and *connata*, Dum. Cours., quoted by DC. Prod. iii. 221, 222.

E. glaucophylla, *androsemæfolia*, and *rigida*, Hoffm. Verz. 1826, quoted by DC. Prod. iii. 216, 218, 221.

E. perfoliata, *discolor*, *glandulosa*, *rubricaulis*, *pulchella*, and *populifolia*, Desf. Cat. Hort. Par. 1829, 408.

E. tuberculata, Parm. in DC. Prod. iii. 221, and *E. glauca*, DC. l. c.

The following are names only entered in Steudel's Nomenclator from garden catalogues: —*E. alata*, *albicaulis*, *undulata*, and *verrucosa*, Loud.; *E. cotinifolia*, *orbicularis*, and *phillyreoides*, Lodd.

E. serratifolia, Desf. in Steud. Nomencl., is a mistake of Steudel's in copying *Eucalyptus* for *Cratægus*.

SERIES I. RENANTHERÆ.—Stamens all perfect or very rarely some of the outer ones with abortive anthers; anthers reniform or broad and flat, the cells divergent or at length divaricate, contiguous and usually confluent at the apex.

1. **E. stellulata,** *Sieb. in DC. Prod.* iii. 217. A small tree, the furrowed bark coming off at length in layers (*F. Mueller*), rugose below, very smooth above and of a lead colour (*Woolls*). Leaves elliptical, lanceolate or the lower ones ovate, rarely much above 3 in. long, usually straight or nearly so, acuminate and much narrowed towards the base, the veins very oblique and anastomosing, a few of the principal ones prominent, starting from near the base, and almost parallel to the midrib as in *E. coriacea*. Flowers very small and numerous, nearly sessile on very short lateral or axillary peduncles, the buds very narrow. Calyx-tube narrow-turbinate, about 1½ lines long. Operculum conical, about as long as the calyx-tube. Stamens not above 2 lines long, all perfect, inflected in the bud; anthers small,

reniform, the cells divergent and confluent at the apex. Ovary flat-topped. Fruit globular-truncate or pear-shaped, rarely exceeding 2 lines in diameter, often contracted at the orifice, the rim flat or slightly concave, the capsule slightly or scarcely sunk.—DC. Mem. Myrt. t. 6 ; F. Muell. Fragm. ii. 45.

N. S. Wales. Port Jackson to the Blue Mountains, *Sieber, n.* 478; *Fraser,* and others; Goulburn plains, Argyle county, "White Gum," *A. Cunningham;* Mudgee, *Woolls;* New England, "Olive Green Gum," *Leichhardt;* Low Flats, Berrima, "Lead Gum," *Woolls.*

Victoria. Stony hills towards Lake Omeo, subalpine ranges near Mount Barkley, ranges on the Macalister river and on the Upper Genoa river, *F. Mueller.*

Var. *angustifolia.* Leaves narrow, very thick and smooth, scarcely showing the venation. —*E. microphylla,* A. Cunn. in Field, N. S. Wales, 350 (partly); *E. Cunninghamii,* G. Don, Gen. Syst. ii. 821 (partly); Blue Mountains, *A. Cunningham* (partly). Some specimens belong to this species and some to *E. stricta,* both having a similar foliage when the leaves are too narrow and thick to show the venation. The shape of the buds, however, distinguishes *E. stellulata* from all others with thick linear apparently veinless leaves.

2. **E. coriacea,** *A. Cunn.; Schau. in Walp. Rep.* ii. 925. A tree attaining sometimes a considerable height, the exterior bark deciduous, the inner smooth and whitish (*Herb. F. Mueller*). Leaves mostly ovate-lanceolate or lanceolate, acuminate and falcate, from 3 or 4 in. to twice that length, very thick, smooth and shining, the veins not numerous, very oblique, a few starting from below the middle and almost parallel to the midrib giving the leaf a several-nerved appearance. Peduncles axillary or lateral, rather thick, terete or slightly compressed, each with about 5 to 10 flowers, the buds clavate and tapering into a short thick pedicel. Calyx-tube very open, nearly 3 lines diameter. Operculum hemispherical, obtuse or with a small point or shortly conical, shorter than the calyx-tube. Stamens 2 to 3 lines long, all perfect or perhaps occasionally a few of the outer ones with abortive anthers; anthers small, reniform, with short divergent cells confluent at the apex. Ovary flat-topped. Fruits often nearly sessile, smooth, pear-shaped, truncate, 3 or rarely nearly 4 lines diameter, more or less contracted at the orifice, as long as broad or rather longer and slightly tapering at the base, the rim flat or concave, the capsule somewhat sunk or nearly on a level with the border, the valves horizontal or scarcely protruding.—Hook. f. Fl. Tasm. i. 136; F. Muell. Fragm. ii. 52; *E. pauciflora,* Sieb. in Spreng. Syst. Cur. Post. 195; *E. phlebophylla,* F. Muell.; Miq. in Ned. Kruidk. Arch. iv. 140.

N. S. Wales. Port Jackson to the Blue Mountains, *Sieber, n.* 470, and others; Argyle county, *A. Cunningham;* Berrima, "White Gum," *Woolls;* New England, "White Gum," *Leichhardt.*

Victoria. Mountain or marshy forests, ascending in the Alps to 4000 or 5000 ft., where it forms rather extensive woods; "Mountain White Gum" and "Flooded Gum," *F. Mueller;* Creswick, "Peppermint," *Whan.*

Tasmania. Abundant in most parts of the colony, "Weeping Gum," *J. D. Hooker.*

Var. *alpina,* F. Muell. Leaves short and nearly straight. Flowers rather smaller and peduncles shorter.—Mountains on the Macalister river, *F. Mueller.*

E. submultiplinervis, Miq. in Ned. Kruidk. Arch. iv. 138, or *E. sylvicultrix,* F. Muell. in Herb. Sond., is a narrow straight-leaved variety with the flowers of the ordinary size. In a specimen from the Blue Mountains in Herb. F. Muell. the leaves are long and almost linear-lanceolate, but very thick with the longitudinal veins of *E. coriacea,* of which it has also the flowers.

3. **E. virgata,** *Sieb. in DC. Prod.* iii. 217. A tree of considerable size with a furrowed persistent fibrous bark (*Oldfield*), apparently very near to *E. coriacea* and *E. obliqua*, and perhaps a variety of the former. Leaves lanceolate, usually narrow falcate and acuminate, 4 to 6 in. long or sometimes longer, thick and shining with the veins more oblique than in *E. obliqua*, less so than in *E. coriacea* and often very indistinct. Peduncles more or less flattened, with several flowers, on rather long pedicels. Calyx-tube nearly hemispherical, about 2 lines diameter. Operculum hemispherical and short, or more frequently conical and as long as or rather longer than the calyx-tube. Stamens about 3 lines long, inflected in the bud, the outer filaments with small abortive anthers or rarely quite anantherous, the perfect anthers small, reniform, with divergent confluent cells. Fruit narrow pear-shaped, 4 to 6 lines long, scarcely contracted at the orifice, the rim broad and at first concave, but generally flat when quite ripe, the capsule somewhat sunk.

N. S. Wales. Port Jackson or Blue Mountains, *Sieber*, *n.* 467; Twofold Bay, "Mountain Ash," *Oldfield, F. Mueller;* Berrima, "Mountain Ash," *Woolls.*

Victoria. Sealers' Cove, "Gum-top," *Walters.*

S. Australia. A shrub of 10 to 15 ft. with a white and grey bark, in the stunted Stringy-bark forests 15 miles N.W. of Mount Gambier, (*Wilhelmi ?*) *in Herb. F. Mueller.*

I have described this species chiefly from Oldfield's, Woolls's, and F. Mueller's specimens. Sieber's appear to be the same, but they are only in young bud and therefore uncertain. It differs from both *E. coriacea* and *E. obliqua* in the outer stamens bearing only abortive anthers, and in that respect approaches *E. hæmastoma*, from which it differs as well in foliage and in fruit as in these imperfect stamens being much fewer and rarely if ever quite without anthers.

4. **E. amygdalina,** *Labill. Pl. Nov. Holl.* ii. 14. *t.* 154. A tree, usually small or moderate-sized but sometimes attaining a considerable height, the bark sometimes described as persistent and fibrous, sometimes as more or less deciduous in large flakes, the branches slender. Leaves from linear to broadly lanceolate, straight or falcate, mostly acuminate and 2 to 4 in. long, when narrow rather thin, when broad thicker, the veins few and oblique but often inconspicuous, the intra-marginal one at a distance from the edge or rarely near to it. Peduncles axillary or lateral, terete or nearly so, with about 4 to 8 rather small flowers. Buds clavate, often glandular and rough. Calyx-tube turbinate, about 2 lines diameter, tapering into a pedicel often as long as itself. Operculum hemispherical, shorter than the calyx-tube, very obtuse or slightly umbonate. Stamens under 2 lines long, inflected in the bud, all perfect; anthers small, with diverging more or less confluent cells. Ovary flat-topped. Fruit subglobose-truncate, usually under 3 lines diameter, but larger in some varieties, slightly contracted at the orifice, the rim flat or slightly concave and rather broad, the capsule not at all or only slightly sunk, the valves flat or slightly protruding.—DC. Prod. iii. 219; Bot. Mag. t. 3260; Hook. f. Fl. Tasm. i. 135; F. Muell. Fragm. ii. 53; *E. longifolia*, Lindl. Bot. Reg. t. 947; *E. Lindleyana*, DC. Prod. iii. 219; *E. tenuiramis*, Miq. in Ned. Kruidk. Arch. iv. 128.

N. S. Wales. Port Jackson to the Blue Mountains, *R. Brown*, and others; Argyle County, *A. Cunningham;* "N. S. Wales Stringy-bark," *Backhouse*, and others; and southward to Twofold Bay, *F. Mueller.*

Victoria. Low sterile hills from near Ballarat to Gipps' Land, "Box" and "Peppermint-tree," *F. Mueller;* Creswick, "Red Gum," *Whan.*

Tasmania, *R. Brown.* Abundant throughout the island, "Peppermint tree," *J. D. Hooker*, and others.

E. ligustrina, DC. Prod. iii. 219, described from Sieber's specimens n. 617, which I have not seen, is probably this species.

Var. *radiata.* Leaves rather broader, 3 to 4 in. long. Flowers usually more numerous, sometimes near 20 in the umbels. Fruit almost pear-shaped.—*E. radiata,* Sieb. in DC. Prod. iii. 218; DC. Mem. Myrt. t. 7.—Chiefly in N. S. Wales, *Sieber, n.* 475, and others; Beut's Basin and Nepean rivers, "White Gum," with a smooth bark, *Woolls;* South of Argyle, *A. Cunningham,* but also in Victoria and Tasmania passing into the ordinary form.

Var. *nitida.* Leaves broader and more rigid. Peduncles and pedicels shorter. Flowers rather longer.—*E. ambigua,* DC. Prod. iii. 219? from the diagnosis taken from Labillardière's specimen. *E. nitida,* Hook. f. Fl. Tasm. i. 137. t. 29.—In the dried specimens this variety appears to pass into the variety *elata* of *E. Risdoni.*

Var ? *hypericifolia.* Leaves of the fruiting branches all opposite, oblong-lanceolate, rounded or cordate at the base, and sessile or nearly so.—Risdon Cove, *R. Brown.*—*E. hypericifolia,* R. Br. Herb.—The specimens are large and good but in fruit only. To this form may belong also some of the garden plants described from the foliage only under the same name.

5. **E. Risdoni,** *Hook. f. in Hook. Lond. Journ.* vi. 477 *and Fl. Tasm.* i. 133. *t.* 24. A small or moderate-sized tree, rarely attaining 50 to 60 ft., the bark smooth, coming off in irregular patches, the branches usually pendulous (*Oldfield*), dark brown or ashy-white or a mixture of both (*R. Brown*), the flowering shoots often glaucous or nearly white. Leaves sometimes all, even on the flowering branches, opposite ovate-cordate and more or less connate, or sometimes those of the flowering branches alternate, broadly lanceolate and falcate, rather thick with oblique veins scarcely conspicuous, the intramarginal one at a distance from the edge. Peduncles axillary or lateral, terete or angular, bearing each an umbel of 4 to 8 or even more. Flowers larger than those of *E. amygdalina.* Buds obovoid-clavate. Calyx-tube very open, attaining 3 lines diameter. Operculum hemispherical, obtuse, shorter than the calyx-tube. Stamens nearly 3 lines long, inflected in the bud, all perfect; anthers with divergent confluent cells. Ovary flat-topped. Fruit subglobose-truncate, attaining 4 lines diameter, somewhat contracted at the orifice, the rim rather broad, flat, or slightly convex, the capsule not sunk, the valves flat or slightly protruding.

Tasmania. In the southern parts of the island rather abundant, "Risdon" or "Drooping Gum," *R. Brown, J. D. Hooker, Oldfield.*

Var. *elata.* A beautiful tree of the largest size, the bark of the trunk grey and deciduous, that of the extremities of the branches purplish-red or reddish-brown (*Gunn*). Leaves broadly lanceolate-falcate, 2 to 4 in. long, rather thick, sometimes almost as in *E. obliqua.* Flowers of *E. Risdoni.* Fruit pear-shaped, 4 lines diameter, with a broad convex rim.—Lake St. Clair, *Gunn.* This variety in the dried specimens appears to connect *E. amygdalina* with *E. obliqua,* but without doubt belongs to *E. Risdoni* as observed by Oldfield, although the dried specimens were included by J. D. Hooker among the varieties of *E. radiata,* Sieb., now united to *E. amygdalina.*

F. Mueller unites *E. Risdoni* altogether with *E. amygdalina.* J. D. Hooker and Oldfield, both of them from observations made on the spot, have assured me that the two are quite distinct, in habit as well as in the bark. The sessile opposite leaves occupy frequently the flowering branches of *E. Risdoni,* and are only on the saplings and adventitious flowerless branches of *E. amygdalina*; they are, moreover, broad, frequently connate, and usually glaucous or nearly white in the former, always as far as known narrow-ovate or oblong-lanceolate in *E. amygdalina.* When the leaves are alternate they appear to be broader in *E. Risdoni* than in *E. amygdalina,* the pedicels thicker and more angular, the flowers and fruits larger, differences, however, of degree only, to which our dried specimens do not admit of our fixing any precise limits, and in that state it is sometimes scarcely possible to decide to which species they should be referred.

6. **E. coccifera,** *Hook. f. in Hook. Lond. Journ.* vi. 477, *and Fl. Tasm.* i. 133. *t.* 25. A small tree generally very glaucous. Leaves lanceolate, acuminate or obtuse, mostly 2 to 3 in. long, thick and shining, the veins oblique, not numerous nor very conspicuous. Peduncles axillary or lateral, short, thick and much flattened upwards, each with 3 to 6 flowers, sessile or nearly so. Calyx-tube narrow-turbinate, tapering at the base, prominently angled, fully 3 lines long and not above 2 diameter. Operculum exceedingly short, broad, flat or depressed and rugose. Stamens about 3 lines long, inflected in the bud; anthers reniform with diverging or divaricate cells, confluent at the apex. Ovary short, flat-topped. Fruit obovoid-truncate, scarcely contracted at the orifice and often losing the angles of the calyx, 4 to 5 or even 6 lines diameter, the rim flat and rather broad, the capsule scarcely depressed, with short valves.—Bot. Mag. t. 4637 ; *E. daphnoides*, Miq. in Ned. Kruidk. Arch. iv. 133.

Tasmania. Summits of the mountains at an elevation of 3000 to 4000 ft., *J. D. Hooker.*

Var. *parviflora.* Flowers much smaller, the peduncles exceedingly short.—Mount Fatigue, *Gunn.*

The species has much the aspect of some thick-leaved forms of *E. amygdalina,* but is readily known by the depressed operculum and longer calyx.

7. **E. obliqua,** *Lher. Sert. Angl.* 18. *t.* 20. An immense tree attaining from 150 to 250 ft., although flowering already when young and small, with a very tenacious rugged fibrous bark. Leaves in the usual form mostly ovate-lanceolate, falcate and very oblique at the base, more or less acuminate, 4 to 6 in. long, thick with very oblique distant anatomosing veins, the intramarginal one at some distance from the edge. Peduncles axillary or lateral, terete or slightly compressed, bearing each an umbel of about 4 to 8 flowers. Buds shortly clavate. Calyx-tube fully 3 lines diameter, rather short and tapering into a pedicel usually as long or longer. Operculum hemispherical or flattened, very obtuse, shorter than the calyx-tube. Stamens fully 3 lines long, all perfect; anther-cells diverging or at length divaricate and confluent at the apex. Ovary flat-topped. Fruit more or less pear-shaped, truncate at the top, 3 to 5 lines diameter, slightly contracted at the orifice, the rim rather broad and concave, the capsule more or less sunk.—DC. Prod. iii. 219; F. Muell. Fragm. ii. 172; *E. gigantea*, Hook. f. in Hook. Lond. Journ. vi. 479; Fl. Tasm. i. 136. t. 28; F. Muell. Fragm. ii. 44, 171; *E. falcifolia*, Miq. in Ned. Kruidk. Arch. iv. 136, as to the S. Australian specimens; *E. nervosa*, F. Muell.; Miq. in Ned. Kruidk. Arch. iv. 139.

N. S. Wales. Probably in the southern districts adjoining Gipps' Land, but I have only seen specimens doubtfully referable to this species.

Victoria. Constitutes vast "Stringy-bark" forests, covering many hilly parts of the country and extending to the Grampians, *F. Mueller*, and others.

Tasmania. Abundant in most parts of the island, forming a great part of the hill-forests and ascending to 4000 ft., "Stringy-bark," *J. D. Hooker.*

S. Australia. Forming "Stringy-bark" forests at the mouth of the Glenelg, on Mount Gambier, etc., *F. Mueller, Robertson*, and others.

E. fabrorum, Schlecht. Linnæa, xx. 656, is referred by F. Mueller to *E. obliqua*, owing to the author's stating it to be the "Stringy-bark" of the colonists, and very possibly some of Behr's specimens may be of that species, but the only authentic one I have seen in a perfect state is evidently *E. viminalis.* The "Messmate," from Dandenong and other parts of

Victoria is, according to F. Mueller's specimens, also referable to *E. obliqua*, although it has the leaves rather thinner with the veins more conspicuous.

Decaisne, in Herb. Tim. Descr. 126, enumerates *E. obliqua* among Timor plants, a very unlikely station, and Blume may be right in considering the Timor species (which I have not seen) as distinct (*E. Decaisneana*, Blume, Mus. bot. 83), although in his diagnosis, evidently taken from Decaisne's, there is no character incompatible with the true *E. obliqua*.

E. heterophylla, Miq. in Ned. Kruidk. Arch. iv. 141, from Tasmania, described from barren leafy branches, appears to be one of the forms assumed by the saplings or by the adventitious shoots of *E. obliqua*.

8. **E. dives,** *Schau. in Walp. Rep.* ii. 926. A small tree of 12 ft. Leaves sessile, opposite, cordate or ovate, acute or acuminate, rather large, on one branch the upper ones tending to become alternate and oblique. Peduncles mostly on the stem below the leaves, bearing each a dense umbel of 8 to 12 or even more flowers. Buds clavate. Calyx-tube short and broad, about 2 lines diameter, tapering into a rather thick pedicel longer than the calyx. Operculum short obtuse and hemispherical. Anther-cells divergent and confluent at the apex. Fruit unknown.

N. S. Wales. Forest land north of Bathurst, *A. Cunningham*. Probably an opposite-leaved state of some species very nearly allied to or even identical with *E. obliqua*, of which it has the flowers. I have, however, seen no specimen of the true *E. obliqua* from so far north.

9. **E. obtusiflora,** *DC. Prod.* iii. 220, *and Mem. Myrt. t.* 10. Leaves mostly straight, oblong elliptical or almost lanceolate, acuminate, often all under 3 in. long, but in some luxuriant specimens more falcate, acuminate and attaining 5 in., very thick and rigid, the veins oblique and parallel, but not close, the intramarginal one at a distance from the edge. Peduncles lateral or axillary, somewhat compressed, rigid, with an umbel of 4 to 8 rather large flowers. Buds clavate. Pedicels much thickened upwards. Calyx-tube short and broad, fully 3 lines diameter. Operculum broadly hemispherical, obtuse or umbonate, thick, shorter than the calyx-tube. Stamens 2 to 3 lines long, all perfect; anthers reniform, with divergent cells usually confluent at the apex. Fruit very hard and woody, ovoid-truncate, above ½ in. long, the orifice scarcely contracted, the rim rather broad and concave, the capsule depressed.—*E. rigida*, Sieb. Pl. Exs.

N. S. Wales. Port Jackson, *Sieber, n.* 473; *F. Mueller*; Bargo Brush, *Backhouse*.—Allied to *E. obliqua*, but with much more rigid straighter leaves, the flowers larger, and the fruit much larger and differently shaped. I have not seen De Candolle's specimens, and his figure represents parallel-celled anthers, but that is probably the fault of the artist. In other respects it agrees well with our plant.

10. **E. buprestium,** *F. Muell. Fragm.* iii. 57. A shrub of 8 to 10 ft. (*Maxwell*). Leaves lanceolate or rarely oblong, usually narrow, acute or mucronate, mostly under 3 in., rigid but not very thick, with the oblique reticulate veins usually prominent, the intramarginal one at a distance from the edge. Peduncles terete or slightly flattened, mostly lateral below the leaves, each usually with about 6 to 10 flowers, on short but not thick pedicels. Buds obovoid. Calyx-tube about 2 lines long, dilated above the ovary. Operculum hemispherical, obtuse, shorter than the calyx-tube. Stamens inflected in the bud, 2 or 3 lines long; anthers broad and flat, opening in short divergent slits confluent at the apex. Fruit nearly globular, about

1 in. diameter when full grown, but sometimes apparently ripe when much smaller, thick and hard, the orifice much contracted, the rim narrow, the capsule sunk. Perfect seeds very few, large, very irregularly shaped, the acute edge sometimes expanded into a narrow wing.

W. Australia, *Drummond, 3rd Coll. Suppl. n.* 12; sandy plains S. of Kojonerup, Helen's Peak, Salt River, etc., the flowers always swarming with a species of *Buprestis*, *Maxwell.*

11. **E. capitellata,** *Sm. Bot. Nov. Holl.* 42, *and in Trans. Linn. Soc.* iii. 285. A moderate-sized or large tree, with a dark-grey furrowed fibrous bark (*F. Mueller*). Leaves from ovate-lanceolate to long-lanceolate, generally very oblique and falcate, and about 3 to 6 in. long, very thick and shining, with the oblique venation of *E. obliqua.* Peduncles axillary or lateral, usually thick and angular, with about 5 to 10 sessile flowers. Buds oblong-clavate or almost ovoid. Calyx-tube turbinate, usually about 3 lines diameter, and rather more in length. Operculum thick, very obtuse, and about as long as the calyx-tube, or rather longer and obtusely conical. Stamens 2 to 3 lines long, all perfect; anthers with divergent cells, confluent at the apex. Ovary flat-topped. Fruit depressed-globose, 4 to 5 lines diameter, the broad rim convex and often very prominent, the valves of the capsule usually protruding beyond it.—DC. Prod. iii. 218; *E. piperita*, Sm. in White's Voy. 226, with a fig. of leaves and fruit, but not the one described in Trans. Linn. Soc.; *E. piperita*, Reichb. Ic. et Descr. Pl. t. 42 (from the figure and description).

N. S. Wales. Port Jackson, *R. Brown;* "Peppermint-tree" (partly), *White;* North shore, *Woolls* (with large flowers and a rather longer operculum); N. S. Wales, "Stringy-Bark," *C. Moore, Leichhardt;* Blue Mountains, *Wilhelmi.*

S. Australia. Rocky hills, Mount Gambier to St. Vincent's Gulf, *F. Mueller.*

Although nearly allied to *E. obliqua*, this species appears to differ slightly in the thicker leaves with rather less oblique veins, and more essentially in the sessile flowers and fruits, and in the shape of the fruit. When the fruit is not well ripened, the rim is scarcely prominent, yet not quite flat.

Var. (?) *latifolia.* Leaves short, obliquely ovate, very thick and much more straight, the bark deciduous (*Robertson*).

Victoria. Heath near Portland, *Robertson.* Possibly a sessile-flowered form of *E. santalifolia*, but the form of the calyx is more that of *E. capitellata*, and quite different from that of *E. santalifolia*, var. *Baxteri.*

12. **E. santalifolia,** *F. Muell. in Trans. Vict. Inst.* i. 35. A large shrub or tree, attaining sometimes a considerable size (*F. Mueller*). Leaves oblong or lanceolate, acute or acuminate, mostly under 3 in., and often nearly straight, very thick and rigid, scarcely showing the oblique veins. Peduncles axillary or lateral, short, terete or nearly so, each bearing about 6 to 8 rather large flowers on very short pedicels. Calyx-tube short and open, above 2 lines diameter. Operculum not seen. Stamens at least 3 lines long; anthers reniform with diverging confluent cells. Fruit nearly globular, about ½ in. diameter, slightly contracted at the orifice, the rim broad, convex, and prominent, the capsule on a level with it, the valves usually horizontal.—*E. santalifolia*, var. *firma*, Miq. in Ned. Kruidk. Arch. iv. 133.

S. Australia. Hills near Guichen Bay, Marble range and Venus Bay, *F. Mueller.* (*Herb. F. Mueller* and *Herb. Sonder.*)—This is now reduced by F. Mueller to a form of

E. obliqua, but besides the foliage the shape of the fruit is different, being nearly that of *E. macrorhyncha* or *E. capitellata.*

Var. (?) *Baxteri.* Leaves ovate or ovate-oblong, obtuse, usually very oblique, under 3 in. long, very thick, with oblique scarcely conspicuous veins. Peduncles thick and angular, mostly very short. Flowers closely sessile in a dense head. Calyx-tube nearly 3 lines diameter, and shorter than broad. Operculum thick and hemispherical, the buds nearly globular. Ovary flat-topped.—*E. Baxteri*, R. Br. Herb.—S. coast, probably Kangaroo Island, *Baxter* (*Herb. R. Br.*). The heads of flowers are very much like those of *E. dumosa*, var. *conglobata*, but the operculum and especially the anthers are quite different. Fruit not seen.

13. **E. macrorhyncha,** *F. Muell. Herb.* A tall tree with a dark dull-grey furrowed and fibrous bark (*F. Mueller*). Leaves mostly falcate, rather narrow and acuminate, 3 to 5 in. long, the lower ones broader, thick and coriaceous, the very oblique rather distant veins prominent. Peduncles axillary or lateral, terete or scarcely compressed, bearing each about 6 to 8 flowers on pedicels longer than the calyx-tube. Calyx-tube turbinate, smooth, the edge forming a prominent ring round the bud. Operculum conical or acuminate, longer than the calyx-tube. Stamens all perfect; anthers reniform, the cells divergent, confluent at the apex. Fruit depressed-globose, 4 to 6 lines diameter, the rim very broad, convex, and prominent, the valves projecting beyond.—*E. acervula*, Miq. in Ned. Kruidk. Arch. iv. 137, not of Sieb.

Victoria. Mountains on the Macalister river "Iron-bark," Mount Ligar, between Goulburn and Ovens rivers, Avon river "Stringy-bark," *F. Mueller;* near Melbourne, "Stringy-bark," *Adamson.*

Although allied to *E. obliqua*, this is readily distinguished by the buds, and especially by the shape of the fruit.

Var. (?) *brachycorys.* Operculum short and obtuse. Fruit of *E. macrorhyncha.* Expanded flowers not seen, and therefore affinities uncertain.

N. S. Wales. New England, "Stringy-bark," *C. Stuart.*

14. **E. piperita,** *Sm. in Trans. Linn. Soc.* iii. 286 (partly). A tree attaining a considerable height, with a persistent fibrous bark at least on the trunk. Leaves from ovate-lanceolate and very oblique to lanceolate and nearly straight, rarely above 1 in. long, rather thick and rigid, the veins very oblique, almost as in *E. obliqua*, but usually fine and less conspicuous, and more numerous, especially in the narrower leaves. Peduncles axillary or lateral, usually slightly angular, bearing each about 6 to 12 flowers on short thick pedicels. Buds ovoid, acuminate, very narrow when young. Calyx-tube about 2 lines long and almost as much diameter. Operculum conical or acuminate, rarely very obtuse, about as long as the calyx-tube. Stamens all perfect, about 2 lines long; anther-cells diverging or divaricate, usually confluent at the apex. Ovary flat-topped. Fruit obovoid-globular, 2 to 3 lines diameter, always contracted at the orifice, the rim concave or rarely nearly flat, the capsule sunk, the very small valves not at all or scarcely protruding. —*E. acervula*, Sieb. in DC. Prod. iii. 217; F. Muell. Fragm. ii. 64.

N. S. Wales. Port Jackson, *R. Brown*, "Peppermint-tree" (partly), *White, Sieber, n.* 469; Macleay and Clarence rivers, *Beckler;* Twofold Bay, "Stringy-bark," *F. Mueller.*

Victoria. Entrance of the Genoa river, *F. Mueller.*

Var. *laxiflora.* Pedicels rather long. Fruit more obovoid, the rim more depressed. —Manly Beach, "Peppermint-tree," *Woolls;* Twofold Bay, "Stringy-bark," *F. Mueller;* Camden, "Stringy-bark," *Backhouse;* Macleay and Clarence rivers, *Beckler.*

Var. (?) *brachycorys*. Operculum short, hemispherical. One of the "Stringy-barks" from Twofold Bay.

Var. *eugenioides*. Leaf-veins rather more regular and divergent. Pedicels rather longer, and buds broader. Fruit less contracted at the orifice, with a thinner rim.—*E. eugenioides*, Sieb. Pl. Exs. n. 479, and Fl. Mixt. n. 603, and consequently *E. scabra*, Dum. Cours., according to DC. Prod. iii. 218.—N. S. Wales, "Stringy-bark," *Caley*, *Woolls*; Twofold Bay, "Stringy-bark," *F. Mueller*.

The species is sometimes difficult to distinguish in the dried state from some forms of *E. obliqua*, and on the other hand it approaches *E. pilularis*, differing from both of them generally but not strictly, as well in foliage as in the bud and operculum, but more readily in the fruit. The variety *eugenioides* is, however, in some respects almost intermediate between *E. piperita* and *E. pilularis*, var. *acmenioides*.

Some specimens of a "Blue Gum," a tree of 100 ft., from Bathurst plains, *Fraser* (*in Herb. R. Brown*), appear to belong to *E. piperita*, but are only in flower with the operculum fallen off and no fruit.

15. **E. pilularis,** *Sm. in Trans. Linn. Soc.* iii. 284. A moderate-sized or large tree, with a dark coloured rough and somewhat furrowed persistent bark. Leaves mostly lanceolate, falcate or nearly straight, acuminate, 3 to 6 in. long, rather thick and smooth, the veins rather oblique, but much less so and more numerous and parallel than in *E. obliqua* and *E. piperita*; they are also finer and often scarcely conspicuous or slightly impressed on the upper side. Peduncles axillary or lateral, or the upper ones forming more or less of a terminal panicle, distinctly flattened in the typical form, bearing each about 6 to 12 flowers, the pedicels often thick and angular, but sometimes rather long and more slender. Buds acuminate. Calyx-tube about 2 lines long and as much in diameter. Operculum conical or acuminate, longer than the calyx-tube. Stamens 2 to 3 lines long, all perfect, inflected in the bud; anthers reniform or broad, the cells diverging or divaricate, confluent at the apex. Ovary flat-topped. Fruit semiglobose or subglobose, truncate, 4 to 5 lines diameter, straight or slightly contracted at the orifice, the rim rather broad, flat or slightly convex or concave, the capsule somewhat sunk or nearly level, the valves usually horizontal.—*E. persicifolia*, DC. Prod. iii. 217, and F. Muell. Fragm. ii. 61 (in part only), not of Lodd.; *E. semicorticata*, F. Muell. in Journ. Linn. Soc. iii. 86; *E. ornata* and *E. incrassata*, Sieb. Pl. Exs.

Queensland. Brisbane river, "Black-butt," *F. Mueller*.

N. S. Wales. Port Jackson, *R. Brown*, *Sieber*, *n.* 477, and others; northward to Hastings river, *Beckler*; and southward to Twofold Bay, *Oldfield*.—"Black-butt" of numerous collectors, both as "Black-butt" and "Manly-Beach Stringy-bark," *Woolls*, and as "Black-butt" and "Flint-wood" in M'Arthur's wood collection for the London Exhibition, 1862. In some specimens designated as the "Great black-butted Gum," the leaves are thicker, and the flowers larger, with dark-coloured stamens.

Victoria. Macalister river, *F. Mueller*.

Var. (?) *acmenioides*. Leaf-veins finer and more distinct; peduncles less flattened and often terete or nearly so; podicels more slender, sometimes 2 to 3 lines long; operculum rather shorter; fruit smaller, with a much thinner rim.—*E. acmenioides*, Schau. in Walp. Rep. ii. 924.—Rockingham Bay, "Stringy-bark," *Dallachy*; summit of Mount Archer, *Thozet*; Pine River, *Fitzalan*; Hastings river, *Tozer*; Paramatta, *A. Cunningham*; Baulkham hills, "White Mahogany," *Woolls*.

I have much doubt whether this might not be adopted as a distinct species, although it seems sometimes to pass into the typical *E. pilularis*. In bud, the specimens bear some resemblance to those of *E. Stuartiana*, but the stamens and fruit are different. In the typical *E. pilularis* the leaves are thickish, resembling those of *E. siderophloia* (confounded with it by De Candolle under the name of *E. persicifolia*), but the veins are more impressed above and the anthers quite different. The fruit of the typical form is nearly that of *E. hæmastoma*, but the stamens very different. In the var. *acmenioides* the general shape of the fruit is nearly the same, but the thinner rim and more sunk capsule give it sometimes a very different appearance.

16. **E. marginata,** *Sm. in Trans. Linn. Soc.* vi. 302. Usually a large shrub or small tree with a smooth or roughish bark, but sometimes a tree of 12 to 50 ft., with a persistent rough bark (*Oldfield*), or a large forest-tree (*Fraser*). Leaves ovate-lanceolate or lanceolate, acuminate, often falcate, mostly 3 to 5 in. long, with rather numerous very diverging veins, conspicuous especially underneath when the leaf is not very thick, much less so when it is thickly coriaceous, the intramarginal vein at some distance from the edge, the upper surface said to be dark green, and the under one whitish, but the difference scarcely perceptible in dried specimens. Peduncles axillary, or the upper ones without floral leaves, terete or flattened, especially in coarser specimens, each with about 4 to 8 or sometimes more, rarely only 3 flowers, on pedicels of about 2 or 3 lines. Calyx-tube short and very open, 2 to 3 lines diameter. Operculum oblong-conical, from a little longer than to more than twice as long as the calyx-tube, obtuse or acuminate. Stamens 3 to 4 lines long, the filaments very flexuose but not inflected in the bud; anthers reniform, the cells diverging, confluent at the apex. Ovary flat or convex in the centre. Fruit obovoid or subglobose, ½ in. diameter or larger, thick, hard, and smooth, contracted at the orifice, the rim usually flat and not very broad, with the capsule scarcely depressed, but sometimes the rim is still thinner with a sunk capsule; valves small, not protruding.—DC. Prod. iii. 217; *E. floribunda*, Hueg. Enum. 49; Schau. in Pl. Preiss. i. 128; F. Muell. Fragm. ii. 40; *E. hypoleuca*, Schau. in Pl. Preiss. i. 131; *E. Mahagoni*, F. Muell. Fragm. ii. 41.

W. Australia. Dry rocky hills about King George's Sound, *Menzies, R. Brown*, and others, and thence to Swan River, *Fraser; also Drummond, n.* 85, (*5th Coll.?*) *n.* 185, *Suppl. to 3rd Coll. n.* 15; *Preiss, n.* 229, 242, 244, 251. "Bastard Mahogany" or "Mahogany," *Oldfield*, and others.

The species has something of the foliage of *E. pilularis*, var. *acmenioides*, but is readily distinguished amongst *Renantheræ* by the longer operculum and the arrangement of the stamens in the bud. A specimen with numerous umbels of few flowers, each forming almost a leafy panicle, and resembling at first sight *E. patens*, is marked in F. Mueller's herbarium as "M. Gregory's Bloodwood," but the anthers and all other characters are those of *E. marginata*. Our specimens of the species vary much in the consistency of the leaves and the size of the flowers. In some of the Southern ones the leaves are very thick and coarse, and the flowers almost like those of *E. robusta*. The Swan River ones have generally, but not always, thinner and more veined leaves. The species was originally described by Smith from specimens raised at Kew from seeds brought by Menzies from King George's Sound.

According to Fraser's notes this forms the chief forest vegetation about Swan River, but there may be some mistake, as I find in other collections the same memorandum attached to *E. calophylla*.

SERIES II. HETEROSTEMONES.—Outer stamens (usually longer than the others) anantherous or with small abortive anthers; anthers of the perfect ones small, globose or truncate; the cells contiguous, opening in pores or oblong slits, sometimes at length confluent.

17. **E. leucoxylon,** *F. Muell. in Trans. Vict. Inst.* i. 33, *and Fragm.*

ii. 60. A middle-sized or tall tree, with a persistent rough dark iron-grey bark (*F. Mueller*), dark grey and spongy on the trunk, soft and white on the branches (*Oldfield*). Leaves lanceolate, acuminate, often falcate, mostly 3 to 6 in. long, thicker and more coriaceous than in *E. melliodora*, the veins very oblique and irregular, sometimes scarcely conspicuous, the intramarginal one usually more prominent, not far from the edge, except when the leaf is broad. Peduncles axillary, terete or slightly flattened, with 3 or sometimes more rather large flowers, on pedicels often as long as or longer than the calyx-tube. Buds ovoid, acuminate. Calyx-tube turbinate, usually about 3 lines long and as much in diameter, but sometimes longer. Operculum conical or acuminate, about as long as the calyx-tube. Stamens usually very unequal, red or white, the outer ones often ½ in. long or more, and usually anantherous, the inner much shorter; anthers very small, truncate, with contiguous cells opening in terminal pores or short oblong slits, sometimes at length confluent. Ovary flat-topped. Fruit obovoid or subglobular, truncate, not contracted at the orifice, 3 or sometimes 4 lines diameter, the rim thick, flat, or slightly convex, the capsule slightly depressed.—Miq. in Ned. Kruidk. Arch. iv. 126; *E. sideroxylon*, A. Cunn. in Mitch. Trop. Austr. 339 (name only).

N. S. Wales. George's River, *Caley*.—Common "Iron-bark" of the interior, *Fraser*, *A. Cunningham*; red "Iron-bark" of Liverpool, also Paramatta and Mudgee, *Woolls*; "Iron-bark" of New England, *C. Stuart*; "Iron-bark" or "Black Mountain-Ash" of Twofold Bay, *Oldfield*.

Victoria. Avoca river, *F. Mueller*; Murray river, *Dallachy*.

S. Australia. From the Murray to St. Vincent's Gulf, *F. Mueller*, and others.

Var. *angulata*. Flowers large, the calyx distinctly angled.—Devil's Country, S. Australia, *F. Mueller*.

Var. *pallens*. Leaves not so coriaceous and whitish.—New England, *C. Stuart*.

Var. *minor*. Flowers rather smaller and often more numerous at the ends of the branches.—Paramatta, *Woolls*; also several of the S. Australian specimens, "White Gum," *Behr*.—This variety seems almost to pass into *E. melliodora*.

18. **E. melliodora,** *A. Cunn. Herb.; Schau. in Walp. Rep.* ii. 924. A moderate-sized tree of irregular growth, with a smooth bark of a pale lead colour (*A. Cunningham*), scaling off in flakes in the upper part of the tree (*C. Moore*), furrowed and persistent (*F. Mueller*). Leaves lanceolate, usually narrow, acuminate and often falcate, mostly 3 to 4 in. long, rather thick, with very fine and rather numerous but oblique veins, the intramarginal one at a distance from the edge. Peduncles axillary or lateral, somewhat angular but not thick, usually short, each with an umbel of 4 to 8 rather small flowers on pedicels of 1 to 2 lines. Calyx-tube campanulate, about 2 lines long and diameter. Operculum hemispherical or shortly conical, with a small point, varying from a little shorter to rather longer than the calyx-tube. Stamens about 2 lines long, the outer ones rather longer and anantherous, anthers of the others small, with contiguous cells opening in terminal pores, sometimes at length confluent. Ovary short, flat-topped; stigma dilated. Capsule subglobose, truncate, not contracted at the orifice, or rarely ovoid and somewhat contracted; the rim rather broad, flat or nearly so, the capsule more or less depressed, but the valves sometimes prominent when open.—*E. patentiflora*, Miq. in Ned. Kruidk. Arch. iv. 125.

N. S. Wales. A frequent gum about Bathurst and to the north and west, *A. Cun-*

ningham, C. Moore; Rocky Creek, head of the Gwydir and Severn rivers, "Yellow Box," *Leichhardt;* New England, *C. Stuart,* with a smooth white bark, the specimen in bud only.

Victoria. Hills on the Yarra and Snowy rivers, Wanganatta, near Mount Ligar, etc., "Yellow Box," *F. Mueller;* "Red Gum," *Adamson.*

19. **E. gracilis,** *F. Muell. in Trans. Vict. Inst.* i. 35, *and Fragm.* ii. 55. A tall shrub or small tree, with a silvery-grey smooth bark (*Beckler*). Leaves narrow-lanceolate or oblong-linear, mostly mucronate, and under 3 in. long, thick and densely dotted, the numerous very oblique veins scarcely visible. Peduncles short, axillary or the upper ones in a short terminal panicle, terete or slightly angular, each with about 4 to 8 rather small flowers. Calyx-tube obconical, usually rather narrow and prominently 4-angled, about 2 lines long, tapering into a very short pedicel, or almost sessile. Operculum shorter than the calyx-tube, hemispherical conical or shortly acuminate. Stamens inflected and flexuose, the outer ones anantherous and nearly 3 lines long, the perfect ones shorter; anthers small, globular, the cells distinct, opening in circular or oblong pores. Ovary short. Fruit oblong or narrow-urceolate, about 3 lines long, the rim narrow, the capsule deeply sunk.—Miq. in Ned. Kruidk. Arch. iv. 124; *E. fruticetorum,* F. Muell. Fragm. ii. 57 (partly).

N. S. Wales. Desert of the Darling and Murray, *Victorian Expedition.*

Victoria. Desert of the Murray and N.W. portion of the colony, *Dallachy, L. Morton.*

S. Australia. Near Spencer's Gulf, *F. Mueller.*

W. Australia. Phillips Ranges, Fitzgerald and Salt rivers, *Maxwell;* also *Drummond,* 5*th Coll. n.* 184 and *suppl. n.* 36; *n.* 34 *of the* 5*th Coll.* is also the same, with a shorter fruit more distinctly pedicellate.

Var. *breviflora.* Calyx-tube scarcely angled, 1½ to nearly 2 lines long. Fruit about 2 lines only, but the deeply sunk capsule and the stamens entirely as in the ordinary form.—Darling and Murray Desert, also *F. Mueller's* Spencer's-Gulf specimens, which being in fruit only are somewhat doubtful.

The Western specimens have the umbels almost all solitary and axillary, but do not appear otherwise to differ from the Eastern ones.

20. **E. paniculata,** *Sm. in Trans. Linn. Soc.* iii. 287. A large shrub or small or moderate-sized tree, the notes on the bark uncertain. Leaves lanceolate, falcate, acuminate, usually rather broad, 3 to 5 in. long, coriaceous and smooth with numerous fine but oblique veins usually concealed in the thick texture. Peduncles short, angular, usually in a short terminal corymbose panicle or a few solitary in the upper axils, each with about 3 to 6 or sometimes more flowers. Calyx-tube broadly turbinate, 2 to 3 lines diameter, often angular, tapering into a short pedicel. Operculum from obtuse and short to conical and as long as the calyx-tube. Stamens 2 to 3 lines long or sometimes more, inflected in the bud, the outer ones anantherous, anthers of the perfect ones small, at first truncate, the cells opening in terminal pores or at length spreading out, divaricate and confluent. Ovary short, flat-topped. Fruit from subglobose to obovoid-oblong, truncate, and often slightly contracted at the orifice, varying from 2 to 4 lines diameter, the rim narrow, the capsule more or less sunk.—DC. Prod. iii. 220; *E. terminalis,* Sieb. Pl. Exs.

N. S. Wales. Port Jackson, *R. Brown, Sieber, n.* 468, and many others; a "She-Iron-bark," *Woolls.*

Var. *fasciculosa.* Flowers rather smaller, operculum usually short.—*E. fasciculosa,* F. Muell. in Trans. Vict. Inst. 34.

S. Australia. Lofty, Bugle, and other ranges along St. Vincent's Gulf, *F. Mueller;* Banks of the Three-Well river, *Waterhouse;* "White Gum," *Behr.*

Var. *angustifolia.* Leaves narrow and thin, as in some varieties of *E. crebra.* Umbels loose, paniculate. Operculum conical. Outer stamens anantherous.—N. S. Wales, "Narrow-leaved Iron-bark," *Woolls.*

Var ? *conferta.* Flowers still smaller, like those of *E. gracilis,* Leaves rather short and broad.

W. Australia, *Drummond, (3rd Coll. ?) Suppl. n.* 9.

The species is allied to *E. gracilis,* with which F. Mueller (Fragm. ii. 67) proposes to unite his *E. fasciculosa,* but both the foliage and the flowers appear to me to be distinct. When large, the flowers almost assume the aspect of the smaller forms of *E. corymbosa.*

21. **E. hæmastoma,** *Sm. in Trans. Linn. Soc.* iii. 285. A large timber tree, with a smooth deciduous bark, leaving a spotted or variegated trunk (*F. Mueller*) or the bark sometimes smooth and sometimes half-barked, like *Black-butt* (*Woolls*). Leaves usually oblique or falcate, lanceolate, about 4 to 6 in. long, thickly coriaceous, the veins very oblique not close and often anastomosing as in *E. obliqua,* the lower ones sometimes broader and more reticulate. Peduncles more or less angular or compressed, axillary, lateral or a few in a short terminal oblong panicle, each with about 4 to 8 flowers. Buds clavate. Calyx short and broad, scarcely 2 lines diameter, shortly tapering into a rather long, thick or rather slender pedicel. Operculum very short, hemispherical, obtuse. Stamens 2 to 3 lines long, inflected, the outer ones longer and anantherous; anthers of the perfect ones small, the cells opening in short oblong divergent at length confluent slits. Fruit globular-truncate or pear-shaped, 3 to 4 lines diameter, the rim broad, flat or nearly so, usually deeply coloured; the capsule slightly depressed, the valves often protruding when open but very soon falling away.—DC. Prod. iii. 219; F. Muell. Fragm. ii. 51; *E. signata,* F. Muell. in Journ. Linn. Soc. iii. 85; *E. falcifolia,* n. 22 and 23, from N. S. Wales, Miq. in Ned. Kruidk. Arch. iv. 137.

Queensland. Wide Bay, *C. Moore;* Brisbane river, Moreton Bay, "Spotted Gum," *F. Mueller.*

N. S. Wales. Port Jackson, "Blue or White Gum," *Woolls;* Illawarra, "Black-butt," *A. Cunningham;* in the interior, "Mountain Ash" and "Spotted Gum," *Macarthur.*

Var. *micrantha.* Leaves often 6 to 8 in. long or even more, the veins less conspicuous. Flowers and fruit much smaller, but not otherwise different.—*E. micrantha,* D.C. Prod. iii. 217, and Mem. Myrt. t. 5; Port Jackson, *R. Brown, Sieber, n.* 497; Paramatta, *Woolls.*

22. **E. microcorys,** *F. Muell. Fragm.* ii. 50. A tall tree with a persistent furrowed fibrous bark (*F. Mueller*). Leaves mostly ovate-lanceolate or broad lanceolate, acuminate, straight or very unequal at the base, about 3 to 4 in. long, not very thick, the veins very divergent and fine but prominent and not close. Peduncles axillary or in short terminal corymbs, terete or somewhat angular, compressed, ½ to 1 in. long, each with about 4 to 8 flowers. Buds clavate, short but tapering into thick pedicels of 2 to 3 lines. Calyx-tube short, with the free part much dilated, about 2 lines diameter. Operculum much shorter than the calyx, broad, flat, very obtuse or slightly umbonate. Stamens inflected in the bud, the outer ones about 3 lines long, anantherous or with small abortive anthers, the inner ones much shorter and

perfect; anthers small with diverging at length confluent cells. Ovary flat-topped. Fruit obovoid-oblong, contracted at the orifice, tapering at the base, about 3 lines long and scarcely 2 lines diameter, the rim narrow, the capsule sunk.

Queensland. Brisbane river, *F. Mueller;* Sandy-mount Range, towards Brisbane, *Leichhardt.*

N. S. Wales. N.W. interior, *Fraser;* Hastings river, *Beckler.*

The species has the flowers nearly of *E. hæmastoma,* with a differently-shaped fruit, and the foliage almost of *E. marginata.*

SERIES III. PORANTHERÆ.—Stamens all perfect, except rarely in *E. bicolor* and perhaps in *E. polyanthemos;* anthers small and globular or broader than long, the cells distinct, opening in terminal or more or less lateral circular pores, sometimes extending at length into oblong slits.

23. **E. pruinosa,** *Schau. in Walp. Rep.* ii. 926. A tree with a persistent whitish-grey rough and fissured bark (*F. Mueller*), the foliage often glaucous or mealy-white. Leaves sessile, opposite or nearly so, very rigid, orbicular-cordate, ovate or oblong, obtuse or rarely almost acute, mostly 2 to 4 in. long. Umbels 3- to 6-flowered, on short peduncles in a terminal corymb or rarely in the upper axils. Pedicels terete, nearly or quite as long as the calyx-tube. Calyx-tube 2 to 3 lines diameter, not angled, more or less tapering into the pedicel. Operculum hemispherical or shortly conical, more or less acuminate, rarely as long as the calyx. Stamens 2 to nearly 3 lines long, inflected in the bud; anthers very small and globular, with distinct parallel cells, opening in very short slits or circular pores. Ovary slightly convex in the centre. Fruit from ovoid-truncate to almost cylindrical, 3 to 5 lines diameter, scarcely or not at all contracted at the orifice, the rim narrow, the capsule slightly sunk, the valves sometimes protruding.—F. Muell. Fragm. iii. 132; *E. spodophylla,* F. Muell. Fragm. ii. 71.

N. Australia. Islands of the Gulf of Carpentaria, *R. Brown, Henne;* dry ridges on the Victoria river and near Sea Range, *F. Mueller.*

Like many other species, this varies with the young branches acutely 4-angled, almost winged, or even on the same specimen quite terete, and very much in the size of the flowers and fruit.

24. **E. oligantha,** *Schau. in Walp. Rep.* ii. 926. Leaves all petiolate but very broad, orbicular or ovate, obtuse or shortly acuminate, 3 to 4 in. long, rigidly coriaceous with prominent diverging veins, parallel but rather distant. Umbels 3- to 6-flowered, collected in a short terminal panicle. Peduncles terete. Calyx-tube campanulate, about 3 lines long and as much in diameter, tapering into a short pedicel. Operculum rather thick, conical, shorter than the calyx. Stamens 2 to 3 lines long, all perfect, inflected in the bud; anthers very small and globular, with distinct parallel cells opening in circular pores or very short slits. Fruit unknown.

N. Australia. Copeland Island, N. coast, *A. Cunningham.* Until the fruit is known, the precise affinities of this species cannot be determined. It is very unlike any other one I have seen.

25. **E. polyanthemos,** *Schau. in Walp. Rep.* ii. 924. A tree sometimes small, sometimes attaining 40 to 50 ft., with an ash-grey persistent rough and furrowed bark (*F. Mueller*). Leaves on rather long petioles,

broadly ovate-orbicular or rhomboidal, obtuse or rarely shortly acuminate, mostly under 3 in. long, passing in older trees into ovate-lanceolate obtuse and 3 in. long or more, rather rigid with fine diverging anastomosing veins, the intramarginal ones distant from the edge. Umbels of 3 to 6 small flowers, shortly pedunculate and usually several together in short oblong or corymbose panicles in the upper axils or at the ends of the branches. Pedicels rarely longer than the calyx-tube and sometimes very short. Calyx-tube 1 to 1½ lines long. Operculum obtusely conical or almost hemispherical, nearly as long as the calyx-tube. Stamens 1 to 2 lines long, all perfect or rarely a few of the outer ones anantherous; anthers small, with globular distinct cells, opening in round pores. Ovary flat-topped. Fruit turbinate, in some specimens 3 lines diameter, in others not 2 lines, not contracted at the orifice, the rim narrow, the capsule sunk.—*E. populifolia*, Hook. Ic. Pl. t. 879; *E. populnea*, F. Muell. in Journ. Linn. Soc. iii. 93.

N. Australia. Islands of the Gulf of Carpentaria, *R. Brown.*

Queensland. Burdekin, Mackenzie, and Dawson rivers, *F. Mueller;* Wide Bay, *Bidwill.*

N. S. Wales. George River, *R. Brown;* Goulburn Plains and flat country near Bathurst, and the Lachlan river, "Bastard Box," *A. Cunningham.* Forest-land of the interior, *Caley;* Nepean river and Mudgee Road, "Bastard Box," and "Lignum Vitæ," *Woolls;* also in *Leichhardt's* collection.

Victoria. Ovens river, *F. Mueller.*

The tropical specimens to which, from the character given, belongs *E. Baueriana*, Schau. in Walp. Rep. ii. 924, have generally smaller flowers and fruits than the southern ones, but do not otherwise differ.

26. **E. Behriana,** *F. Muell.; Miq. in Ned. Kruidk. Arch.* iv. 139. A tall shrub or small tree (*F. Mueller*). Leaves from ovate to ovate-lanceolate, rarely lanceolate, mostly acute or acuminate, rarely above 3 in. long, thick and smooth, the fine very oblique veins scarcely conspicuous, the intramarginal one at some distance from the edge. Peduncles short, terete or slightly angular, with few rather small sessile flowers, the umbels generally several together forming short oblong or thyrsoid panicles terminal or in the upper axils or several of these together in a compound terminal panicle. Buds obovoid. Calyx not 2 lines long, more or less attenuate at the base. Operculum short, hemispherical, obtuse or scarcely umbonate, the outer membranous one often still persistent in the advanced bud. Stamens all perfect, not 2 lines long, anther-cells small, globular, opening in circular pores, rarely at length confluent. Ovary flat-topped. Fruit obovoid-globular, truncate, about 2 lines diameter, the rim flat, the capsule slightly sunk.

Victoria. Bacchus Marsh, Avoca river, and Pine Forest, *F. Mueller.*

Var. *purpurascens*, F. Muell. Flowers larger. Peduncles and calyx angular, the latter fully 2 lines long. Operculum obtusely conical, but shorter than the calyx-lobe. Stamens purplish.—Lake Wangaroo, *Wilhelmi.*

27. **E. bicolor,** *A. Cunn. Herb.; Hook. in Mitch. Trop. Austr.* 390. A large shrub or sometimes a tree of 30 to 40 ft., with a persistent ash-grey or blackish bark (*F. Mueller, A. Cunningham*), or a tall tree with a smooth white bark (*Dallachy*). Leaves lanceolate, narrow or rarely passing into ovate-lanceolate, mostly 3 to 4 in. but sometimes 5 or 6 in. long, not very thick, often glaucous or pale coloured, the veins fine, oblique, not close, the

marginal one at a distance from the edge and sometimes very prominent towards the base of the leaf. Flowers small, about 3 to 8 together on short peduncles, the umbels forming usually axillary or terminal panicles shorter than the leaves. Pedicels shorter than the calyx. Calyx-tube turbinate, nearly 1½ lines long. Operculum rather thin, hemispherical, obtuse or umbonate, shorter than the calyx-tube. Stamens 1 to 2 lines long, all perfect or occasionally a few of the outer ones without anthers; anthers small, with 2 small globular cells opening in round pores or short oblong slits. Ovary flat-topped. Fruit globular-truncate or pear-shaped, about 2 lines diameter or rarely nearly 3, contracted at the orifice, the rim rather broad, flat or depressed; the capsule somewhat depressed.—F. Muell. in Journ. Linn. Soc. iii. 90; *E. pendula*, A. Cunn. in Steud. Nom. Bot. ed. 2; *E. largiflorens*, F. Muell. in Trans. Vict. Inst. i. 34 and Fragm. ii. 58; *E. hæmastoma*, Miq. in Ned. Kruidk. Arch. iv. 130, as to the Murray specimens, not of Sm.

Queensland. Port Denison, *Dallachy.*

N. S. Wales. Port Jackson and Williams river, *R. Brown;* Baulkham hills, "Ironbark," *Caley;* on the Maranoa, S. of St. George's Bridge, "Bastard Box," *Mitchell;* in the interior, *A. Cunningham;* from the Darling to the Barrier range, *Victorian Expedition.*

Victoria. Mallee scrub, near the Avoca and generally in the N.W. portion of the colony, "Bastard Box," *F. Mueller*, and others.

S. Australia. Scrub near the Murray and thence to St. Vincent's Gulf, *F. Mueller*, and others; Three-Well river, *Waterhouse*, with rather larger flowers.

Var. *parviflora*, F. Muell. Flowers much smaller. Stamens not 1 line long.—Burdekin river, *F. Mueller.*

The southern and desert specimens have rather thicker leaves than those from Queensland, but I can find no other difference. In all there are occasionally 2 or 3 flowers on the specimen twice the size of the others, with the stamens elongated and anantherous, perhaps owing to some insect. The species differs from *E. polyanthemos* in its narrow leaves and from *E. odorata* in inflorescence.

28. **E. odorata,** *Behr in Linnæa.*, xx. 657. A small or moderate-sized tree with a dark grey rough persistent bark (*F. Mueller*). Leaves lanceolate, usually narrow, but sometimes broad, rarely above 4 in. long, rather rigid, the veins oblique and sometimes very much so, and not close, the intramarginal one at some distance from the edge. Peduncles mostly axillary, rather thick and short but scarcely angular. Pedicels sometimes scarcely any and rarely as long as the calyx-tube. Calyx-tube campanulate, about 2½ lines long and as much in diameter. Operculum hemispherical or obtusely conical, shorter than the calyx-tube. Stamens 2 to 3 lines long, all perfect, very flexuose and slightly inflected in the bud; anthers very small, with globular distinct cells, opening in pores or short oblong slits. Ovary flat-topped. Fruit obovoid-truncate, about 2 lines diameter, slightly contracted at the orifice or almost urceolate, tapering at the base, the rim not broad; the capsule deeply sunk.—F. Muell. Fragm. ii. 66, and Pl. Vict. Suppl. t. 17; Miq. in Ned. Kruidk. Arch. iv. 129; *E. porosa*, Miq. in. Ned. Kruidk. Arch. iv. 132; *E. cajuputea*, Miq. l. c. 126.

S. Australia. Hills chiefly calcareous near Flinders Ranges and towards Spencer's and St. Vincent's gulfs, "Peppermint tree" and "Red Gum," *F. Mueller.*

Some specimens bear much resemblance to *E. melliodora*, but have the thicker leaves and the anthers all perfect and nearly globular of *E. odorata*, they are not in fruit.

Var. *floribunda.* Inflorescence occasionally compound, connecting the species in some measure with *E. bicolor*, but the foliage and shape of the flowers and fruits are rather those of *E. odorata.*

Victoria. On the Yarra, *F. Mueller.*

29. **E. uncinata,** *Turcz. in Bull. Mosc.* 1849, ii. 23. A tall shrub, with a smooth red or ash-grey bark, coming off in coriaceous plates (*Oldfield*). Leaves narrow-lanceolate or linear, usually under 3 in. thick, the very fine veins scarcely visible, distant and rather oblique, but not so much so as in *E. gracilis*, always conspicuously black-dotted, especially underneath. Peduncles axillary, rather short, terete or scarcely flattened, bearing each an umbel or head of about 6 to 8 small flowers. Buds ovoid or oblong. Calyx-tube about 1½ lines long, sessile or tapering into a short pedicel. Operculum obtusely conical or acuminate, as long as or rather longer than the calyx-tube. Stamens about 2 lines long, all perfect, the filaments slender and inflected with an acute angle, as in *E. corynocalyx* and *E. decurva;* anthers very small, nearly globular, with contiguous cells opening in terminal pores. Ovary flat-topped. Capsule globular-truncate or pyriform, 2 to nearly 3 lines diameter, contracted at the orifice, the rim concave or at length nearly flat, the capsule sunk, but the valves often acuminate by the split base of the style, and then the subulate tips protruding.—*E. leptophylla*, Miq. in Ned. Kruidk. Arch. iv. 123; *E. oleosa*, F. Muell., Miq. l. c. 127; F. Muell. Fragm. ii. 56 (partly).

N. S. Wales. In the Euryalean scrub of the interior, *Fraser;* desert of the Murray and Darling, *Herb. F. Mueller.*

Victoria. Wimmera and desert of the Murray, *Dallachy, F. Mueller.*

S. Australia. Gawler Town, *Behr;* Murray desert, *F. Mueller.*

W. Australia. Plantagenet and Stirling ranges and eastward to Cape Riche, *Maxwell, Harvey, Drummond, 3rd Coll. n.* 66; Murchison river, *Oldfield* (mostly with very narrow leaves).

Var. *latifolia*, W. Australia, *Drummond, 4th Coll. n.* 76.

Var. (?) *major.* Flowers larger, contracted into very short thick pedicels, the peduncles more flattened. Fruit rather larger, scarcely contracted at the orifice, the rim broader and flatter, the valves not acuminate.—Murchison river, *Oldfield.*

Var. *rostrata.* Flowers more distinctly pedicellate, the operculum acuminate and longer than the calyx.—Phillips Range, *Maxwell;* Murchison river, *Oldfield, also Drummond, 5th Coll. n.* 186.

The species has much the habit of *E. gracilis*, but is very different in stamens and fruit. It is also sometimes very near *E. micranthera*, but differs in the stamens. The young plant has sometimes sessile ovate opposite leaves.

30. **E. hemiphloia,** *F. Muell. Fragm.* ii. 62. A tall tree, sometimes reduced to a shrub. Leaves ovate-lanceolate or lanceolate, falcate or nearly straight, about 3 to 5 in. long, thick and rigid, with very oblique distant veins, almost as in *E. obliqua* and *E. hæmastoma.* Peduncles slightly angular, about 4- to 8-flowered, the umbels mostly forming short terminal panicles, although the fruiting ones are usually lateral below the leaves. Calyx-tube 2 to 2½ lines long and scarcely so much in diameter, tapering into a short thick pedicel or almost sessile. Operculum conical, acuminate, as long as the calyx-tube or rarely shorter and more obtuse. Stamens pale-coloured, about 2 lines long or rather more, all perfect, inflected in the bud; anthers very small, globular, the cells distinct, but opening in pores rather than in slits. Ovary rather deep, slightly conical or convex in the centre.

Fruit ovoid-oblong, about 3 to 4 lines long, truncate and slightly contracted at the orifice, very smooth, the rim narrow, the capsule deeply sunk.

Queensland. Moreton Bay, "Box-tree," *F. Mueller.*
N. S. Wales. Paramatta, "Box-tree," *Woolls.*
S. Australia. Memory Cave and Kangaroo Island, *R. Brown*, Port Lincoln, *Wilhelmi.*

This species has the foliage of *E. obliqua* and of *E. hæmastoma*, but the anthers and fruit are quite different. In Brown's S. Australian specimens the leaves are smaller, but in Wilhelmi's they are the same as in the northern ones, and I can find no character to distinguish them. Both R. Brown and F. Mueller had given them the MS. name of *E. purpurascens.*

Var. (?) *parviflora.* Flowers considerably smaller, Mount Elliott, "Iron-bark," *Fitzalan.* Specimens in Leichhardt's collection, marked "Box," from the range behind the Condamine, appear to be the same with rather longer very angular flowers.

SERIES IV. MICRANTHERÆ.—Anthers very small and globular or broader than long, almost as in the *Porantheræ*, but opening in more oblong or longitudinal slits, almost as in the *Normales*, the cells more distinct than in the *Porantheræ*, less so than in the *Normales.*

31. **E. cneorifolia,** *DC. Prod.* iii. 220, *and Mem. Myrt. t.* 9, *from the char. and fig.* A shrub or small tree, of 6 to 10 ft. (*F. Mueller*). Leaves from narrow-linear to oblong-lanceolate, straight or rarely falcate, mostly under 4 in. long, thick, with the fine diverging veins scarcely ever visible. Peduncles short, terete or scarcely angular, each with a head of 4 to 8 flowers, closely sessile or obscurely pedicellate. Calyx 2 to 2½ lines long, rather thick but not angular. Operculum hemispherical, much shorter than the calyx-tube. Stamens about 2 lines long, inflected in the bud; anthers very small, nearly globular, with distinct parallel cells. Fruit pear-shaped or nearly globular, about 3 lines diameter, contracted at the orifice, the rim rather thick, flat or slightly convex, the capsule more or less sunk, but the valves often slightly protruding.—*E. santalifolia*, F. Muell. in Trans. Vict. Inst. 35, partly.

Victoria? Dense scrub on Mount Useful, *F. Mueller*, specimens in fruit only, and therefore doubtful.
S. Australia. Beyond Salt's Creek, and near Port Lincoln, *F. Mueller;* Kangaroo (or Decres) Island, *R. Brown, Baudin's Expedition, Waterhouse.* This comes near to some narrow-leaved forms of *E. dumosa*, but the fruit is quite different, nearer to that of *E. oleosa*, and the anthers are very much smaller. The large-fruited specimens, originally sent by F. Mueller and described by Miquel as *L. santalifolia*, belong to a distinct species of the *Renantheræ*, for which I have retained the name.

32. **E. stricta,** *Sieb. in DC. Prod.* iii. 218. A shrub or small tree, the bark stringy (*Woolls*), Leaves linear-lanceolate or linear, straight or falcate, obtuse or acuminate, mostly 2 to 4 in. long, very thick and shining, the apparently oblique and distant veins very rarely visible. Peduncles short, slightly angular or terete, each with about 4 to 8 shortly pedicellate small flowers. Buds ovoid. Calyx not 2 lines diameter; operculum hemispherical and mucronate or conical, not longer than the calyx-tube. Stamens not above 2 lines long, inflected in the bud; anthers very small and globular, with distinct parallel cells, opening at first in round pores which extend into oblong slits. Fruit globose-truncate, smooth, 3 to 4 lines diameter, con-

tracted at the orifice, the rim narrow, the capsule sunk, the valves not protruding.—DC. Mem. Myrt. t. 8 (the anthers incorrect); *E. microphylla*, A. Cunn. in Field, N. S. Wales, 350 (partly); *E. Cunninghamii*, G. Don, Gen. Syst. ii. 821 (partly).

N. S. Wales. Port Jackson or Blue Mountains, *Sieber*, *n.* 472; forms brushes in the elevated parts of the Blue Mountains, *A. Cunningham*, *Woolls*. Some specimens, confounded with it by A. Cunningham, belong to the narrow-leaved form of *E. stellulata*, in which the veins are sometimes inconspicuous, but which is readily distinguished by the shape of the buds, the reniform anthers, etc.

33. **E. micranthera,** *F. Muell. Herb.* A shrub, of 6 to 10 ft., with a smooth bark (*Maxwell*). Leaves oblong-lanceolate, acuminate or almost obtuse, 2 to nearly 4 in. long, very thick and smooth so as wholly to conceal the veins. Peduncles very short, often flattened, with 3 to 6 flowers like those of *E. uncinata* or *E. oleosa*, but larger. Calyx-tube turbinate, 2 to nearly 3 lines long, tapering into a very short thick pedicel or almost sessile. Operculum very obtuse and shorter than the calyx-tube. Stamens inflected, sometimes almost as acutely so as in *E. corynocalyx* and *E. uncinata*, but the filaments not so fine and the anthers very minute, with parallel contiguous cells. Ovary flat-topped. Fruit globose-truncate, 4 to 5 lines diameter, somewhat contracted at the orifice, the rim broad, flat or slightly concave, the capsule very slightly sunk.

W. Australia. Sandy hummocks, from Israelite Bay to Eyre's Relief, *Maxwell*. Possibly a form of *E. uncinata*, but both the operculum and the stamens appear different.

34. **E. decipiens,** *Endl. in Hueg. Enum.* 49. Varies from a shrub of 6 to 8 ft., to a small or even a large tree, attaining 60 to 70 ft., with the bark rough and persistent (*Oldfield*), fragile, soft and spongy (*Maxwell*). Leaves ovate, ovate-lanceolate or lanceolate, acuminate, rarely exceeding 4 in. and often under 3, rather thick, the fine diverging veins scarcely conspicuous; the intramarginal one usually at a distance from the edge. Peduncles short, mostly axillary, terete or slightly flattened, each with a head of 6 to 12 sessile flowers. Calyx-tube turbinate, about 2 lines long, the border usually prominent in the bud. Operculum conical or acuminate, from a little longer to nearly twice as long as the calyx-tube. Stamens inflected in the bud; anthers very small, globular, but with distinct cells, parallel or nearly so, opening at first in round pores which become at length longitudinal slits. Ovary conical in the centre. Fruit broadly turbinate, pear-shaped or globose, truncate, 3 lines diameter or rather more, contracted at the orifice, the rim rather broad, flat or scarcely convex, the capsule more or less sunk, but the points of the valves usually protruding.—Schau. in Pl. Preiss. i. 129.

W. Australia. Sand plains, Kalgan river, *Oldfield*, and eastward towards Cape Riche, *Harvey*, *Drummond*, *3rd Coll. Suppl. n.* 14, *Preiss*, *n.* 241, all apparently the shrubby form; the arborescent one, Limestone hills, Swan River, and Banestee river, on the road to King George's Sound, "Flooded Gum," *Oldfield*; swamps about Tulbrinup lake, *Maxwell*. The species is allied in its fruit to *E. uncinata* and *E. oleosa*, and almost intermediate between them as to stamens, differing from both in foliage and in the shape of its sessile flowers.

35. **E. corynocalyx,** *F. Muell. Fragm.* ii. 43. A tall elegant shrub. Leaves usually rather broad, ovate-lanceolate or lanceolate, obtuse or acumi-

nate, mostly 3 to 5 in. long, thick and coriaceous, the veins rather numerous, oblique and often prominent, the intramarginal one at some distance from the edge. Peduncles usually lateral below the leaves, ½ to 1 in. long, terete or slightly angular, erect or spreading, each with 6 to 12 or more distinctly pedicellate flowers. Calyx narrow-urceolate or almost cylindrical, 3 to 5 lines long and rarely 2 lines diameter, smooth or ribbed. Operculum broad and very short, flat or slightly umbonate. Stamens 2 to 3 lines long, the filaments slender and acutely inflected in the bud as in *E. uncinata;* anthers very small, globular, with distinct parallel cells. Ovary flat-topped. Fruit ovoid, often strongly ribbed, nearly ½ in. long, contracted at the orifice, the rim narrow, the capsule deeply sunk.—*E. cladocalyx,* F. Muell. in Linnæa, xxv. 388; Miq. in Ned. Kruidk. Arch. iv. 135.

S. Australia. Marble Range, *Wilhelmi.*

36. **E. albens,** *Miq. in Ned. Kruidk. Arch.* iv. 138. A tree, attaining 60 to 80 ft., with a dull green persistent bark (*F. Mueller*), separating in smooth laminæ or strips (*C. Stuart*), the foliage usually very glaucous or almost mealy-white. Leaves usually large, broad, ovate-lanceolate or lanceolate, often 6 in. long or more, rigid, with oblique veins, the intramarginal one at a distance from the edge. Peduncles lateral, rigid, scarcely flattened, sometimes ¾ in. long, but often much shorter, bearing 4 to 8 rather large flowers. Buds long and acuminate, apparently sessile, but really tapering into short thick angular pedicels. Calyx-tube 3 to 4 lines long and scarcely 2 lines diameter, 2-angled or nearly terete. Operculum conical, acuminate, as long as or rather shorter than the calyx-tube. Stamens 3 to 4 lines long, all perfect, inflected; anthers very small and globular, with distinct parallel cells, opening at length to the base or nearly so. Ovary short, slightly conical in the centre. Fruit obovoid-oblong, truncate, nearly ½ in. long, the rim narrow, the capsule deeply sunk.

N. S. Wales. Macquarrie river, *A. Cunningham;* New England, "White Gum," *C. Stuart;* between Alford's and the Range, "Box," *Leichhardt.*

Victoria. Poor plains, between Ten-mile Creek and Broken River, "White Box," *F. Mueller.*

A very distinct species with something of the habit of the *Robustæ,* but with the anthers of the *Micranthera.* F. Mueller refers it to *E. pallens,* DC., which I have not seen. De Candolle's character agrees rather better with *E. dealbata* than with *E. albens,* but the short hemispherical operculum he describes occurs in neither.

37. **E. Bowmani,** *F. Muell. Herb.* Stature and bark unknown. Leaves ovate-lanceolate or broadly lanceolate, mostly 4 to 6 in. long, straight or falcate, obtuse or acuminate, rigid, with oblique veins, the marginal one at a distance from the edge, like those of *E. albens,* but not glaucous. Peduncles axillary or lateral, more or less flattened, bearing 4 to 8 rather large flowers. Buds obtuse, tapering into a short very thick pedicel or nearly sessile. Calyx-tube obovoid or turbinate, thick, about 2 lines long and as much diameter. Operculum thick, obtuse, longer than the calyx-tube. Stamens 3 to 4 lines long, the filaments slender, inflected in the bud, but not showing the acute angle of *E. corynocalyx;* anthers very small and globular, but with distinct parallel cells, opening longitudinally. Ovary conical in the centre. Fruit unknown.

Queensland, *Bowman.* I have some hesitation in describing the species without having seen the fruit, but it appears quite distinct from any other one known to me. It seems to be allied to *E. albens* and *E. corynocalyx*, but differs from both in the shape of the flowers.

Specimens of two other trees or shrubs, in F. Mueller's collection, are probably closely allied to, if not varieties of the same; one from the head of the Gwydir, *Leichhardt*, in bud only, is glaucous like *E. albens*, and has the calyx-tube shorter and the operculum longer than in *E. Bowmani*, which it agrees with in other respects. The other from Mount Elliot, *Fitzalan*, in flower, only differs from *E. Bowmani* in the upper umbels almost paniculate, in the more distinct pedicels and in the operculum rather shorter and broader.

38. **E. siderophloia,** *Benth.* A tall tree, with a hard, persistent, rough, and furrowed bark (*F. Mueller* and others). Leaves ovate-lanceolate or lanceolate, much acuminate, straight or more frequently falcate, about 3 to 6 in. long, often rather thick, with numerous fine diverging veins, the intramarginal one close to the edge, Peduncles axillary or in terminal corymbose panicles, more or less angular, each with about 6 to 12 flowers, on distinct angular pedicels. Calyx-tube shortly turbinate, about 2 lines diameter. Operculum conical or acuminate, rather longer than the calyx-tube in the ordinary form. Stamens 2 to 3 lines long, all perfect, inflected in the bud; anthers very small and nearly globular, the cells very short, opening at first in oblong slits, extending at length to the base or sometimes almost confluent. Ovary convex or conical in the centre. Fruit globular-truncate or obovoid, 3 to 4 lines diameter, not at all or scarcely contracted at the orifice, the rim slightly prominent, the capsule not much or sometimes scarcely sunk, the valves often protruding.—*E. persicifolia*, DC. Prod. iii. 217, and F. Muell. Fragm. ii. 61 (in part only), not of Lodd.

Queensland. Moreton Bay, "Iron-bark," *A. Cunningham, Leichhardt*, and others.

N. S. Wales. Port Jackson, "Iron-bark," *R. Brown*, and others; "Iron-bark" and "She Iron-bark," *Woolls;* Hastings and Richmond rivers, "Iron-bark," *Beckler, C. Moore.*

Var. (?) *rostrata.* Operculum ¼ to ½ in. long; capsule-valves more prominent.—Port Jackson, "Iron-bark," *R. Brown, Caley;* "Greater Iron-bark," *Backhouse;* "Large-leaved Iron-bark," *Woolls.*

This species is evidently allied on the one hand to *E. albens*, and on the other to *E. crebra* and other Iron-barks. When the operculum is short, specimens in bud only are much like those of the Black-butt, *E. pilularis*, with which they appear to have been confounded both by De Candolle and F. Mueller, although distinguished by all collectors; when the flowers are open the anthers give a ready character, and the venation of the leaves is somewhat different. The rostrate variety, when in young bud, resembles *E. resinifera*, and even *E. tereticornis*, but the venation, and still more the anthers, distinguish it.

E. fibrosa, F. Muell. in Journ. Linn. Soc. iii. 87, from the Brisbane, is only known from specimens in young bud, in which state I am unable to distinguish them from the var. *rostrata* of *E. siderophloia*. F. Mueller, however, designates it as a Stringy-bark. It may therefore prove to be distinct.

39. **E. melanophloia,** *F. Muell. in Journ. Linn. Soc.* iii. 93. A small tree with a blackish persistent deeply-furrowed bark (*F. Mueller*), the foliage more or less glaucous or mealy-white. Leaves sessile, opposite, from cordate-ovate or orbicular to ovate-lanceolate, obtuse or acute. Peduncles short, terete or nearly so, 3- to 6-flowered, axillary or several in a short terminal corymb. Buds tapering into a pedicel shorter than the calyx-tube or almost sessile. Calyx-tube slightly angular, about 2 lines long or rather more, and as much in diameter. Operculum obtusely conical, shorter than the calyx-tube. Stamens 2 to 3 lines long, inflected in the bud; anthers very small and globular, but the cells parallel and distinct. Fruit pear-shaped or glo-

bular-truncate, 2 to nearly 3 lines diameter, more or less contracted at the orifice, the rim thin, the capsule nearly on a level with it and the valves slightly protruding, or more sunk with the valves included.

Queensland. Dawson, Gilbert, and Burnett rivers, *F. Mueller;* Moreton Bay, "Silver-leaved Iron-bark," *C. Moore;* summit of the Leichhardt Range, "Iron-bark," *Bowman.*

N. S. Wales. On the Narran, *Mitchell;* also n. 6 of the N. S. Wales woods of the Paris Exhibition, 1855, *C. Moore;* Cassilis, "Iron bark," *Leichhardt.*

The species is very nearly allied to *E. crebra* and may prove to be an opposite-leaved state of the form described as the "Mackenzie river Box-tree." It sometimes resembles *E. cinerea*, but differs in the bark, the stamens, and the fruit.

40. **E. drepanophylla,** *F. Muell. Herb.* A tree, usually low and stunted, the bark dark-grey and ribbed (*Dallachy*). Leaves long-lanceolate, often exceeding 6 in. and usually falcate, acuminate, with numerous fine parallel and very diverging veins, often scarcely conspicuous, the intra-marginal one close to or very near the edge. Umbels 3- to 6-flowered, usually 3 or 4 together in short axillary or terminal panicles or the lower ones solitary, the peduncles short and terete or nearly so. Calyx-tube ob-conical, nearly 2 lines long, tapering into a short thick pedicel. Operculum conical or obtuse, usually about as long as the calyx-tube. Stamens about 2 lines long, inflected in the bud; anthers very small, nearly globular, with distinct parallel cells. Fruit subglobose-truncate, about 4 lines diameter, slightly contracted at the orifice, the rim rather thin, the capsule somewhat sunk, but convex, so that the valves often slightly protrude.

N. Australia. N.W. coast, *A. Cunningham.*

Queensland. E. coast, *A. Cunningham;* Keppel Bay and Shoalwater Bay, *R. Brown;* Burdekin Expedition, *Fitzalan;* Port Denison, "Iron-bark tree," *Fitzalan, Dallachy;* Bowen river, "Ironbark," *Bowman.*

The species differs from *E. crebra* chiefly in the large flowers and in the larger, harder, and more globular fruit. From *E. leptophleba* it is chiefly distinguished by the leaves not so thick with more oblique veins. It is not impossible, however, that *E. melanophloia, drepanophylla, trachyphloia, leptophleba,* and *crebra,* all of them Iron-barks, may be but forms of one species.

41. **E. trachyphloia,** *F. Muell. in Journ. Linn. Soc.* iii. 90. A moderate-sized tree, with a dark grey rugged bark, persistent. Leaves long-lanceolate, often falcate, 4 to 6 in. long, with very numerous fine parallel almost transverse veins, the marginal one close to or very near the edge. Flowers not seen. Fruiting-umbels several together in terminal panicles or in the upper axils, each with 3 to 6 pedicellate fruits. Fruit ovoid-truncate, contracted towards the orifice, about 3 lines long, the rim thin, the capsule deeply sunk.

Queensland. Burnett river, *F. Mueller.* The specimens are in fruit only, and the affinities of the species are therefore very doubtful.

42. **E. leptophleba,** *F. Muell. in Journ. Linn. Soc.* iii. 86. A moderate-sized or large tree, with a dark persistent rugged bark, of which only fragmentary fruiting specimens have been preserved. These appear to me to differ but slightly from *E. crebra*, in the leaves rather thicker and broader, and in the fruits much larger, attaining 4 lines diameter or rather more.

Queensland. Gilbert river, *F. Mueller.*

43. **E. crebra,** *F. Muell. in Journ. Linn. Soc.* iii. 87. A small middle-

sized or sometimes a large tree, with a hard blackish rough persistent bark (*F. Mueller* and others). Leaves oblong-lanceolate or linear, straight or more frequently falcate, obtuse, mucronate-acute or acuminate, attaining 4 to 6 in. long, rather thick and glaucous or yellowish when dry in the northern specimens, thinner in the subtropical ones, with numerous very diverging fine parallel veins, the intramarginal one very near or close to the edge. Peduncles short, terete or nearly so, each with about 3 to 6 small flowers on short but distinct pedicels; umbels usually 3 or 4 together in short panicles either terminal or axillary, or rarely the lower ones solitary in the axils. Calyx-tube turbinate, about 1 line diameter. Operculum conical or hemispherical, about as long as the calyx-tube. Stamens 1 to 2 lines long, all perfect, inflected in the bud; anthers very small and globular, like those of the *Porantheræ*, but the cells distinct and opening longitudinally to the base. Ovary flat-topped or slightly convex in the centre. Fruit obovoid-truncate, not 2 lines in diameter, somewhat contracted at the orifice and often shortly attenuate at the base, the rim narrow, the capsule more or less sunk but the tips of the valves often protruding when open.—*Metrosideros salicifolia*, *β*, Soland. in Gærtn. Fruct. i. 171. t. 34.

N. Australia. Between the Flinders and Lynd rivers, Gulf of Carpentaria, "Iron-bark tree," *F. Mueller*, including the fruiting specimens of *E. parviflora*, F. Muell., referred to in Journ. Linn. Soc. iii. 90.

Queensland. From the Burdekin to Moreton Bay, often forming large forests, *F. Mueller;* Rockhampton, *Dallachy*, all under the name of "Iron-bark."

N. S. Wales. "Iron-bark," from Smithfield, *Woolls;* Hastings river, *Beckler;* New England, *C. Stuart;* also in *Leichhardt's* collection, all under the name of "Iron-bark."

In flower, this species, especially in the thicker-leaved specimens, is sometimes difficult to distinguish from *E. brachypoda;* the leaves are generally but not always thinner with more oblique veins, and the flowers not so glaucous with the calyx less open; the fruit is, however, very differently shaped. It is very possible, however, that *E. melanophloia, drepanophylla, trachyphloia*, and *leptophleba*, which all differ only in the size of the flowers and fruit, and very slightly in the shape of the fruit, may when more fully known prove to be varieties of *E. crebra*, as well as the following forms:—

1. "Box-tree" of the Mackenzie river, *Leichhardt*, also on the Suttor river, *Bowman*, described by both as having the bark persistent and fissured. The specimens are somewhat glaucous, the leaves rather thin and broad and often obtuse. The flowers quite those of *E. crebra*, the fruit not seen. This is very probably an alternate-leaved state of *E. melanophloia*.

2. "Gum-topped Box," from Suttor river, *Bowman*, described as having the bark furrowed and persistent on the trunk, coming off in layers on the branches. Flowers of *E. crebra*. Fruits of the same shape but rather larger, much smaller, however, than in *E. drepanophylla*.

3. Specimens from New England, *C. Stuart*, described as having the bark white, separating in thin strips, the colour of the specimens not at all glaucous, and the inflorescence rather less compound, but the shape of the leaves, their venation and the flowers and fruits precisely those of *E. crebra*. To this form appear to belong also Sieber's specimens of *E. gracilis*, Pl. Exs. n. 476, referred by De Candolle to *E. hæmastoma*, but very different from Smith's plant of that name. They are in young bud and in fruit.

4. "Gum-tree," from the Brisbane, *Leichhardt*, with small globular fruits much contracted at the orifice, but no flowers; the leaves those of the common Moreton-Bay *E. crebra*.

5. A specimen from the dividing range towards the Gloucester, *Leichhardt*, with the same foliage, with young buds like those of *E. crebra*, but with very small globular-truncate fruits, scarcely contracted at the orifice.

From *E. amygdalina*, with which fruiting specimens have sometimes been confounded, as well as from *E. bicolor* and its allies, *E. crebra* is readily distinguished both by the venation and the anthers.

44. **E. brachypoda,** *Turcz. in Bull. Mosc.* 1849, ii. 21. A tall shrub or small or moderate-sized tree, the bark varying from smooth and whitish to dark and rugged, persistent or shed in large patches (*Oldfield*), dark and rough on the trunk, smooth whitish and deciduous on the branches (*F. Mueller*). Leaves from ovate obtuse and under 2 in. to long-lanceolate obtuse acute or acuminate and attaining 6 to 8 in., more or less pale or glaucous, with numerous very fine parallel almost transverse veins, scarcely conspicuous when the leaf is thick, the marginal one near or close to the edge. Peduncles short, terete or nearly so, each with about 3 to 6 or sometimes more small flowers; umbels usually 3 or 4 together in short panicles either terminal or in the upper axils, or rarely the lower ones solitary and axillary. Calyx short broad and open, 1 to 1¼ lines diameter. Operculum conical or obtuse, not longer than the calyx-tube. Stamens 1 to 2 lines long, inflected in the bud; anthers very small, globular, with distinct parallel cells. Ovary convex in the centre. Fruit almost hemispherical, rarely 2 lines diameter, the orifice open or almost dilated, the rim narrow, the capsule slightly sunk, but very convex in the centre, the valves protruding when open.—*E. brevifolia*, F. Muell. in Journ. Linn. Soc. iii. 84; *E. microtheca*, F. Muell. in Journ. Linn. Soc. iii. 87.

N. Australia. N.W. coast, *A. Cunningham;* table land of the upper Victoria river, "Box-tree," also in the scrub between Flinders and Albert rivers, Gulf of Carpentaria, *F. Mueller.* Macdonnell Ranges, *M'Douall Stuart's Expedition.*

N. S. Wales. Between the Darling river and Barrier range, *Victorian Expedition.*

S. Australia. Cooper's Creek, *Howitt's Expedition.*

W. Australia, *Drummond, 4th Coll. n.* 73. Wet places near the Murchison river, among flooded gums, called "Colaille," *Oldfield,* who remarks on the variability of the bark, but there appears to be some confusion in his notes.

With the habit and inflorescence of *E. crebra,* this species differs from all others of the group in the very open fruit with exserted valves.

45. **E. brachyandra,** *F. Muell. in Journ. Linn. Soc.* iii. 97. A tall shrub or small tree. Leaves ovate or oblong, on long petioles, very obtuse, 2 to 4 in. long, thick with numerous parallel very diverging veins, fine but not very close. Flowers not seen. Umbels several together in a short panicle. Calyx after flowering very small, ovoid-globose, with a few very short stamens with minute globose anthers remaining about the orifice. Fruit urceolate-globose, scarcely more than 1 line long, the rim thin, the capsule sunk.

N. Australia. Rocky declivities of the Upper Victoria river, *F. Mueller.* The specimens preserved are very fragmentary.

Series V. Normales.—Stamens all perfect; anthers oblong-ovate or nearly globose, the cells perfectly distinct, parallel, and opening longitudinally, either contiguous with the connective-gland behind them or back to back with the connective between them.

Subseries I. Subsessiles.—Flowers axillary or lateral, usually large, solitary or 2 or 3 together, sessile or nearly so on the stem, or on an exceedingly short terete or angular peduncle.

46. **E. macrocarpa,** *Hook. Ic. Pl. t.* 405 *to* 407. A stout shrub of 6 to 10 ft., usually more or less mealy-white. Leaves opposite, sessile, broadly cordate-ovate, acute or obtuse, often 6 in. long, or even more, very thick and rigid. Flowers very large, solitary on very short thick axillary peduncles. Calyx-tube broadly hemispherical, hard and woody, smooth or obscurely ribbed, about 1½ in. diameter. Operculum thick and hard, broadly conical or slightly acuminate, about twice as long as the calyx-tube. Stamens about 1 in. long, connivent and inflected at the end, their insertion raised to about 2 lines above the edge of the calyx by the thick edge of the disk, which is also often slightly raised within the stamens in a ring round the ovary; anthers ovate or oblong with parallel distinct cells. Fruit depressed-hemispherical, 2 to 3 in. diameter or even more, the very broad disk forming a raised rim, and the capsule or at least the broad valves protruding still farther in the centre.—Bot. Mag. t. 4333; Paxt. Mag. Bot. xv. 29. with a fig.; Schau. in Pl. Preiss. i. 132; F. Muell. Fragm. ii. 41.

W. Australia. Forest bordering the Quangen plains, S. of Swan River, *Drummond, n.* 13, *Preiss, n.* 235. A specimen of *Labillardière's,* without flower or fruit, from the Maria Island, on the S. coast, appears to be the same species.

47. **E. cordata,** *Labill. Pl. Nov. Holl.* ii. 13. *t.* 152. A small tree, the bark not described, the foliage usually glaucous or mealy-white. Leaves opposite, sessile, cordate, broadly ovate or orbicular, more or less distinctly crenate, mostly under 3 in. long. Peduncles axillary, very short and thick, terete or angular, each with 3 rather large sessile flowers. Calyx broadly campanulate, obtuse at the base, smooth, usually about 4 lines diameter. Operculum depressed-hemispherical, obtuse or umbonate, much shorter than the calyx-tube. Stamens 3 to 4 lines long, inflected in the bud; anthers ovate-oblong, with distinct parallel cells. Fruit globular-truncate, thick and hard, often ½ in. diameter, not contracted at the orifice, the rim slightly projecting, the capsule somewhat sunk, the valves rarely protruding when open. —DC. Prod. iii. 221 (in part); Hook. f. Fl. Tasm. i. 132.

Tasmania. Recherche Bay and Huon River, *Labillardière, J. D. Hooker.* Oldfield expressed an opinion that this might be the young tree of *E. obliqua,* the flowers, however, as well as the fruit, and especially the anthers, are far too dissimilar to admit of the approximation of the two species without more conclusive evidence. Like *E. pulverulenta,* it appears to be much more nearly allied to *E. cosmophylla.*

48. **E. pulverulenta,** *Sims, Bot. Mag. t.* 2087. A small tree, the bark not described, the foliage more or less glaucous or mealy-white. Leaves sessile, opposite, cordate, orbicular or broadly ovate, obtuse or almost acute, always quite entire. Peduncles axillary, very short terete or angular, each with 3 flowers not large and sessile or nearly so. Calyx-tube broadly turbinate, tapering at the base, about 3 lines diameter. Operculum obtusely conical or shortly acuminate, about as long as the calyx-tube. Stamens 3 to 4 lines long, inflected in the bud; anthers small but ovate, with distinct parallel cells. Fruit subglobose-truncate, not contracted at the orifice, usually about 4 lines diameter, the rim thick and convex, the capsule scarcely depressed, the valves slightly protruding.—DC. Prod. iii. 221; Colla, Hort. Ripul. App. t. 1; *E. pulvigera,* A. Cunn. in Field, N. S. Wales, 350; *E. cordata,* Lodd. Bot. Cab. t. 328, not of Labill.

N. S. Wales. Near Cox's River, *A. Cunningham;* Argyle county, *Backhouse;* Berrima, "Argyle Apple," *Woolls.* F. Mueller (Fragm. ii. 70) considers this to be the same as his *E. cinerea,* but, as far as our specimens go, it appears to differ in the foliage, in the larger sessile flowers, and in the larger thicker fruit with a very prominent thick rim.

49. **E. globulus,** *Labill. Voy.* i. 153. *t.* 13, *and Pl. Nov. Holl.* ii. 121. A lofty tree, sometimes exceeding 200 ft., but in many situations flowering when not above 10 ft. high, the young shoots and foliage often glaucous-white, the bark somewhat fibrous but deciduous, leaving the inner bark on the trunk smooth (*F. Mueller*). Leaves of the young tree opposite sessile and cordate, of the full-grown tree lanceolate or ovate-lanceolate, acuminate, falcate, often ½ to 1 ft. long, the veins rather conspicuous, oblique and anastomosing, the intramarginal one at a distance from the edge. Flowers large, axillary, solitary or 2 or 3 together closely sessile on the stem or on a peduncle not longer than thick. Calyx-tube broadly turbinate, thick, woody, and replete with oil-receptacles, more or less ribbed and rugose or warty or rarely smooth, ½ to ¾ in. diameter, the border prominent and the 4 teeth sometimes conspicuous. Operculum thick, hard and warty, depressed-hemispherical with an umbonate or conical centre, shorter than the calyx-tube. Stamens above ½ in. long, inflected in the bud, raised above the calyx by the thick edge of the disk; anthers ovate, with parallel cells. Ovary as long as the calyx, slightly convex. Fruit semiglobular, ¾ to 1 in. diameter, the broad flat-topped disk or rim projecting above the calyx, the capsule nearly level with it, the valves flat, not protruding.—DC. Prod. iii. 220; Hook. f. Fl. Tasm. i. 133; F. Muell. Fragm. ii. 68; Pl. Vict. Suppl. t. 16.

Victoria. Valleys and moist declivities of wooded mountains from Apollo Bay to beyond Wilson's Promontory, extending here and there gregariously to the Buffalo range, *F. Mueller.*

Tasmania. S. parts of the island from 40 miles N. of Hobarton to the extreme south, "Blue Gum," *J. D. Hooker.*

Most of the Victorian specimens have smaller fruits and flowers, and the fruit more convex than those from Tasmania.

50. **E. alpina,** *Lindl. in Mitch. Three Exped.* ii. 175. A rigid scrubby bush of several feet, the young shoots viscid. Leaves mostly broad and very obtuse, orbicular ovate or oblong, straight or oblique at the base, 2 to 3 in. long, very thick, the veins not numerous and oblique. Flowers rather large, solitary or 2 or 3 together, axillary or lateral, closely sessile on the stem by their broad base, but not seen open. Buds irregularly globular, hard woody and rugose as in *E. globulus,* 4 to 5 lines diameter. Operculum very thick, hemispherical, nearly as long as the calyx-tube. Stamens much inflected in the bud; anthers ovate, with distinct parallel cells. Ovary shorter than the calyx, convex in the centre. Fruit very hard and woody, depressed-globular, ¾ to 1 in. diameter, the broad rim convex, the capsule not at all or scarcely sunk, the short valves protruding.

Victoria. Summit of Mount William, *Mitchell, F. Mueller.*

51. **E. cosmophylla,** *F. Muell. in Trans. Vict. Inst.* 32. A tall shrub or small tree, with a smooth ash-coloured bark (*F. Mueller*). Leaves ovate ovate-lanceolate or lanceolate, acute or acuminate, 3 to 5 in. long, very thick and rigid, the veins diverging and much reticulate, the intramarginal one at

a distance from the edge. Peduncles axillary or lateral, short and thick, sometimes scarcely any, each with 3 rather large flowers, sessile or the central one shortly pedicellate. Calyx-tube broad and short, thick and hard, about 5 lines diameter, obtuse at the base. Operculum hard, acuminate, scarcely shorter than the calyx-tube. Stamens 4 to 6 lines long, inflected in the bud; anthers rather small but ovate, with distinct parallel cells. Ovary flat-topped or slightly convex in the centre. Fruit subglobose-truncate, not contracted at the orifice, hard and smooth, 7 to 8 lines diameter, the rim thick and slightly convex, the capsule sunk, the valves not protruding.—Miq. in Ned. Kruidk. Arch. iv. 134.

S. Australia. Bugle and Lofty Ranges, *F. Mueller;* Kangaroo island, *Waterhouse;* Encounter Bay, *Whittaker.*

SUBSERIES II. RECURVÆ.—Flowers axillary or lateral, often large, usually 3 or rarely 5 together, pedicellate on a recurved terete peduncle. Calyx-tube turbinate or urceolate. Leaves alternate, thick.

52. **E. pyriformis,** *Turcz. in Bull. Mosc.* 1849, ii. 22. A shrub attaining 8 to 12 ft. Leaves ovate-lanceolate or lanceolate, acute or acuminate, rarely exceeding 3 in., very thick, the numerous fine oblique parallel veins rarely conspicuous, the intramarginal one at a distance from the edge. Flowers very large, red when fresh (*Oldfield*), 1 to 3 together on thick reflexed peduncles, sometimes 2 to 3 in. long sometimes very short, the pedicels from ½ to 1 in. long. Calyx-tube turbinate or obconical, more or less prominently 2- to 4-ribbed or almost winged, ¾ to nearly 1 in. long and as much in diameter at the top, tapering into the thick pedicel. Operculum conical or hemispherical, usually shortly mucronate, about as long as the calyx-tube. Stamens often ¾ in. long or rather more, inflected in the bud; anthers ovate, with distinct parallel cells. Disk very broad, forming within the stamens a thick prominent ring round the depressed top of the ovary. Fruit almost hemispherical, very hard and woody, about 2 in. diameter, the ring formed by the disk remaining very prominent round the somewhat sunk convex-topped capsule, the valves not protruding beyond the ring.—*E. pruinosa*, Turcz. in Bull. Mosc. 1849, ii. 23, not of Schau.; *E. erythrocalyx*, F. Muell. Fragm. ii. 32.

W. Australia, *Drummond, n.* 58, 61, 4*th Coll. n.* 69, 70; *Gilbert, n.* 256. Sandy plains between Port Gregory and Murchison river, *Oldfield.*

53. **E. longifolia,** *Link and Otto, Ic. Pl. Sel.* 97. *t.* 45. A tree with a rough fibrous persistent or partially deciduous bark (*F. Mueller*), somewhat smooth or fibrous and wrinkled according to the age of the tree (*Woolls*). Leaves lanceolate, usually long and falcate, often exceeding 6 in., the veins fine and divergent but rather distant, the intramarginal one not far from the edge. Peduncles axillary or lateral, usually recurved with 3 or very rarely 4 rather large pedicellate flowers. Calyx-tube turbinate, thick and hard, sometimes slightly angular, 4 to 5 lines long and as much in diameter. Operculum thick and hard, conical, about as long as the calyx-tube or sometimes longer. Stamens fully ½ in. long, inflected in the bud; anthers ovate-oblong, with distinct parallel cells. Ovary rather shorter than the calyx, convex in the

centre. Fruit somewhat pear-shaped, truncate, nearly ¾ in. long, straight or scarcely contracted at the orifice, the broad rim prominent, the capsule slightly sunk but the valves sometimes protruding, or the whole fruit is shorter with a flat rim.—DC. Prod. iii. 216; *E. Woollsii*, F. Muell. Fragm. ii. 50.

N. S. Wales. Port Jackson, "Bastard Box," *R. Brown, Caley;* near Paramatta, "Woolly-butt," *Woolls;* Twofold Bay, *F. Mueller.*

Victoria. Eastern extremity of Gipps' Land, *F. Mueller.*

54. **E. conoidea,** *Benth.* Leaves narrow-oblong or lanceolate, mostly obtuse and under 3 in. long, thick and shining, the very oblique veins scarcely conspicuous, the intramarginal one at a distance from the edge. Peduncles axillary or lateral, usually recurved, terete or slightly angular, each with 3 to 5 rather large pedicellate flowers. Calyx-tube obconical, more or less distinctly ribbed, 3 lines long or rather more, tapering into the pedicel. Operculum broad and conical, smooth or ribbed, not thick, nearly twice as long as the calyx-tube. Stamens nearly ½ in. long, inflected in the bud, raised by the thick disk ½ to 1 line above the border of the calyx; anthers oblong, with parallel distinct cells. Fruit turbinate-truncate, 4 to 6 lines long and as much in diameter on the top, the rim raised above the calyx-border, broad and flat or concave, the capsule level with it or more or less depressed, the short broad valves often protruding when open.

W. Australia, *Drummond, 5th Coll. n.* 37.

Var. *marginata.* Border of the calyx expanded into a prominent horizontal or reflexed ring.—*Drummond, 3rd Coll. n.* 56.

55. **E. urnigera,** *Hook. f. in Hook. Lond. Journ.* vi. 477, *and Fl. Tasm.* i. 134. *t.* 56. A tree usually small, with spreading branches and drooping branchlets often glaucous (*Hooker*), attaining sometimes 50 ft. (*Oldfield*), with a pale brown smooth bark (*R. Brown*). Leaves ovate oval-oblong or lanceolate, obtuse, 2 to 4 in. long, straight or rarely oblique, very thick so as to conceal the oblique rather regular veins. Flowers 3 together or rarely solitary, pedicellate on rather long usually recurved terete pedicels. Calyx-tube more or less urceolate, contracted under the somewhat dilated orifice, rather above 3 lines long and as much in diameter. Operculum short, obtuse or shortly conical. Stamens fully 3 lines long, inflected in the bud; anthers with distinct parallel cells. Fruit hard, oblong-ovoid or nearly globose but always more or less urceolate, the rim narrow, slightly prominent, the capsule much sunk.

Tasmania. Table mountain, *R. Brown;* alpine districts, not uncommon, *J. D. Hooker.*

56. **E. cæsia,** *Benth.* Branches rather slender, pale glaucous or nearly white as well as the foliage and fruits. Leaves ovate-lanceolate or lanceolate, acuminate, rarely above 3 in. long, rather thick, the veins fine and very oblique but numerous. Flowers unknown. Fruiting peduncles axillary or lateral, recurved, terete, above 1 in. long with scars of 3 flowers. Pedicels terete, ½ in. long or more. Fruit ovoid, truncate, ¾ in. diameter and nearly 1 in. long, slightly contracted towards the orifice, the rim very broad and concave but with a thin edge, the capsule deeply sunk, the points of the valves protruding from the centre of the disk but shorter than the border of the fruit.

W. Australia, *Drummond,* 5*th Coll. Suppl. n.* 36.

SUBSERIES III. ROBUSTÆ.—Peduncles axillary or lateral or very rarely the upper ones in a terminal corymb, usually flattened, each with several or rarely only one large or moderate-sized flowers, sessile or tapering into thick pedicels. Leaves usually thick and alternate or in *E. Preissiana* often opposite. Rim of the fruit often concave with a sunk capsule except in the last four species.

57. **E. tetraptera,** *Turcz. in Bull. Mosc.* 1849, ii. 22. A shrub or small tree (rarely above 10 ft., *Maxwell*), the branches nearly terete or very prominently 4-angled almost winged. Leaves oblong-lanceolate, more or less falcate, mostly under 6 in. long, but in luxuriant branches attaining 10 in. or even more, very thick and rigid, shining above, the veins divergent and parallel but scarcely prominent, the intramarginal one not close to the edge. Peduncles axillary or lateral, recurved, very broad and flat but thick and undulate, bearing each a single very large sessile flower. Calyx at least 1½ in. long and 1¼ in. diameter, very prominently 4-angled, the prominent margin forming a border or cup round the operculum and the 4 teeth sometimes prominent. Operculum pyramidal, hard, 4-angled, not half so long as the calyx. Stamens white or red, inflected in the bud, not above ½ in. long; anthers oblong with parallel distinct cells. Ovary short, somewhat convex in the centre. Fruit prominently angled or 4-winged like the calyx or 2-winged at the base like the peduncle and often not much enlarged, but sometimes attaining 2 or even 3 in.; the rim concave, the capsule rather deeply sunk.—F. Muell. Fragm. ii. 34.

W. Australia. Between Swan River and King George's Sound, *Drummond,* 4*th Coll. n.* 71, 5*th Coll. n.* 189; *Harvey;* Fitzgerald river and Granite hills N. of Cape Le Grand, *Maxwell.*

58. **E. miniata,** *A. Cunn.; Schau. in Walp. Rep.* ii. 925. A moderate-sized or large tree, the bark fibrous and persistent but readily separable in flakes (*F. Mueller*), the young shoots sometimes glaucous or mealy white. Leaves ovate-lanceolate or lanceolate, acuminate, mostly 4 to 6 in. long, the veins diverging and parallel but not very close, the intramarginal one very near the edge. Peduncles axillary or lateral, very thick and broad, more or less flattened, ½ to 1 in. long, with about 5 to 7 rather large closely sessile flowers. Calyx-tube thick, turbinate or almost urceolate, about 6 lines long, more or less prominently 8-angled. Operculum hemispherical, obtuse, thick, shorter than the calyx-tube. Stamens richly coloured, nearly ½ in. long, inflected in the bud; anthers oblong with distinct parallel cells. Ovary short, flat-topped. Fruit ovoid or urceolate, very thick and hard, more or less prominently ribbed, 1 to nearly 2 in. long, the rim rather thick, the capsule deeply sunk.—*E. aurantiaca,* F. Muell. in Journ. Linn. Soc. iii. 91.

N. Australia. Hunter's River, York Sound, and Greville island, N.W. Coast, *A. Cunningham;* islands of the Gulf of Carpentaria, *R. Brown;* sandy plains and rocky tablelands round the Gulf of Carpentaria, *F. Mueller;* between the Lynd and Port Essington, *Leichhardt.*

59. **E. robusta,** *Sm. Bot. Nov. Holl.* 40. *t.* 13, *and in Trans. Linn. Soc.* iii. 283. A moderate-sized tree, with a rough furrowed bark. Leaves ovate-

lanceolate, nearly straight or the upper ones narrower and falcate, 4 to 6 in. long or sometimes more, with numerous fine but prominent parallel veins almost transverse, the intramarginal one very near or close to the edge. Peduncles axillary or lateral, stout, angular or flattened, often 1 in. long, each with about 4 to 12 rather large flowers, on thick angular pedicels. Calyx-tube narrow-turbinate or slightly urceolate, 3 to 4 lines long, tapering into the pedicel. Operculum thick, obtusely acuminate, usually rather longer than the calyx-tube. Stamens 4 to 6 lines long, somewhat raised above the calyx-border by the annular margin of the disk; anthers ovoid-oblong, with distinct parallel cells. Ovary flat-topped or slightly conical in the centre. Fruit ovoid-oblong, truncate, smooth, contracted above the middle, about ½ in. long or rather more, the rim thin and slightly prominent, the capsule much sunk.—DC. Prod. iii. 216; F. Muell. Fragm. ii. 43; *E. rostrata*, Cav. Ic. iv. 23. t. 342.

N. S. Wales. Port Jackson to the Blue Mountains, *R. Brown, Sieber, n.* 480, *and Fl. Maurit.* ii. 318, and others; "Swamp Mahogany" and "White Mahogany," *Woolls.*

60. **E. botryoides,** *Sm. in Trans. Linn. Soc.* iii. 286. A tall handsome tree, with a rough furrowed persistent bark. Leaves ovate-lanceolate or lanceolate, acuminate, straight or rarely falcate, 4 to 6 in. long or sometimes more, with numerous fine very diverging parallel veins, the intramarginal one very near or close to the edge. Peduncles axillary or lateral, thick, angular or flat, bearing each about 4 to 10 rather large flowers, sessile or nearly so. Calyx-tube ovoid-turbinate, 2 to nearly 3 lines long. Operculum from very obtuse and much shorter than the calyx-tube to broadly conical and nearly as long as the calyx-tube. Stamens about 3 lines long, or rather more, inflected in the bud; anthers ovoid-oblong, with distinct parallel cells. Ovary convex in the centre. Fruit obovoid-oblong, 4 to 5 lines long when fully ripe, somewhat contracted at the orifice, the rim narrow, the capsule more or less sunk, flat or slightly convex in the centre, the valves not protruding.—DC. Prod. iii. 219; *C. platypodos*, Cav. Ic. iv. 23. t. 341.

Queensland. Brisbane, "Blue Gum," *M'Arthur, n.* 91, *of Paris Exhibition woods.*
N. S. Wales. Port Jackson, *R. Brown,* and others; Manly Beach, "Bastard Mahogany," and Baulkham Hills, "Blue Gum" (the latter not seen in fruit, but apparently the same species), *Woolls.*
Victoria. Snowy River, Cabbage-tree river, and towards the mouth of Broadrip river, "Bastard or Swamp Mahogany," *F. Mueller.*

Var. with the ovary more conical in the centre and the operculum shortly beaked, Paterson's River, "Blue Gum," *Herb. R. Brown.*

61. **E. goniocalyx,** *F. Muell. Fragm.* ii. 48. A tree of moderate size, with the bark rough and persistent on the trunk, at least when the tree is large, deciduous in the upper part (*Oldfield*), usually deciduous, but sometimes persistent (*F. Mueller*). Leaves ovate-lanceolate to lanceolate, usually falcate and often above 6 in. long, usually pale coloured, with the pellucid dots rather conspicuous, the veins oblique and numerous, but not close, the intramarginal one at a distance from the edge. Peduncles short, thick and flat, each with 3 to 7 flowers, sessile or tapering into short thick pedicels. Calyx-tube 3 to 4 lines long, about 2½ lines diameter, with 2 to 4 prominent angles or almost terete. Operculum conical or hemispherical, much shorter

than the calyx-tube. Stamens about 3 lines long, inflected in the bud; anthers ovate, with distinct parallel cells. Ovary conical in the centre and tapering into the style. Fruit ovoid-truncate, about 4 lines long and rather less in diameter, the rim rather thin, the capsule more or less sunk, but the points of the valves, when open, sometimes on a level with the rim, or when the fruit is not so well ripened more or less protruding.—Miq. in Ned. Kruidk. Arch. iv. 134; *E. elæophora*, F. Muell. Fragm. iv. 52.

N. S. Wales. Twofold Bay, "Spotted Gum" and "White Gum," *Oldfield.*

Victoria. Scrubby stony hills of the Buffalo Range, sources of the Yarra and Barwan rivers, grassy hills on the Macalister river, especially near Mount Ligar, Sealer's Cove, etc., "Spotted Gum," *F. Mueller.*—Very near in flowers to some forms of *E. dumosa*, but with a very different foliage.

Var. *acuminata.* Flowers more distinctly pedicellate, the bud narrow, the operculum longer and more acuminate. Gipps' Land, *F. Mueller.*

Var. *pallens.* Specimens glaucous-white, as in *E. dealbata.* Mountains on Snowy River, *F. Mueller.*

62. **E. dumosa,** *A. Cunn.; Schau. in Walp. Rep.* ii. 925. A shrub or small tree, with a smooth whitish persistent bark. Leaves from oblong or almost ovate and obtuse, to lanceolate falcate and acuminate, under 4 in. and rarely above 3 in. long, very thick and smooth, the oblique parallel veins scarcely conspicuous. Peduncles axillary or lateral, terete or flattened, usually short, with 4 to 8 flowers closely sessile or on very short thick pedicels. Calyx-tube ovoid, almost cylindrical, thick and sometimes slightly angular, 2 to nearly 3 lines long. Operculum hemispherical and very obtuse or umbonate or shortly conical, shorter than the calyx-tube. Stamens about 2 lines long, inflected in the bud; anthers ovate, with distinct parallel cells. Ovary flat-topped. Fruit obovoid-truncate or almost oblong, usually about 3 lines long, not contracted at the orifice or very slightly so, the rim not very thick, the capsule more or less sunk.—F. Muell. Fragm. ii. 59 (partly); *E. lamprocarpa*, F. Muell.; Miq. in Ned. Kruidk. Arch. iv. 129; *E. fruticetorum*, F. Muell.; Miq. in Ned. Kruidk. Arch. iv. 131 (partly); *E. santalifolia*, Miq. l. c. 133 (except the var. *firma*), not of F. Muell.

N. S. Wales. Blue Mountains, *Backhouse;* Euryalean scrub in the interior, *A. Cunningham;* Darling desert, *Victorian Expedition.*

Victoria. Mallee scrub, near Lake Baga, *F. Mueller.*

S. Australia. Heath, W. of Glenelg river, *Robertson;* Murray river, *Herrgolt;* Gawler river, *Behr.*

Var. *conglobata*, R. Brown. Peduncles shorter than broad. Flowers closely sessile, the calyx-tube shorter than broad, angular, and operculum conical, as in *E. goniocalyx*, but leaves of *E. dumosa.*—Port Lincoln, *Wilhelmi;* S. coast, *R. Brown.*

Var. *scyphocalyx*, F. Muell. Leaves narrow. Flowers large. Operculum very obtuse, broader than the calyx. Peduncles very short and thick. This approaches in some measure *E. gomphocephala.*

W. Australia. Eyre's Relief, *Maxwell.* Another form, very much like this one, but with longer, not much flattened peduncles, and the fruit nearly ½ in. long, in Herb. R. Br., gathered in Baudin's Expedition on the Ile des Amiraux.

Var. *puncticulata.* Leaves copiously black-dotted. Flowers small.—W. Australia, from Gordon river, *Oldfield*, to Mount Barren Ranges, *Maxwell.*

Var. (?) *rhodophloia.* Bark salmon-coloured. Leaves black-dotted. Flowers rather small, the operculum conical or almost acuminate. Capsule on a level with the rim of the fruit. Possibly a distinct species.—W. Australia, Phillips Bluffs, near Eyre's Relief, *Maxwell.*

63. **E. incrassata,** *Labill. Pl. Nov. Holl.* ii. 12. *t.* 150. A shrub or small tree, attaining sometimes 25 ft., with a smooth bark, persistent or shedding in large patches (*Oldfield, Maxwell*). Leaves ovate ovate-lanceolate or lanceolate, obtuse or rarely acuminate, mostly under 4 in. long, very thick, with oblique usually inconspicuous veins. Peduncles axillary or lateral, short, thick, usually flat or much dilated upwards, bearing each 3 to 8 rather large sessile or shortly pedicellate flowers. Calyx-tube obovoid or turbinate, from under 4 to above 5 lines long, smooth in the original form, but ribbed in the more common varieties. Operculum thick, obtusely acuminate or rostrate, as long as or longer than the calyx-tube. Stamens often ½ in. long, inflected in the bud; anthers ovate-oblong, with distinct parallel cells. Fruit thick, ovoid-cylindrical, from under ½ in. to nearly 1 in. long, not at all or but slightly contracted at the orifice, the rim not very thick when the flowers are small, very broad and flat in some large-flowered forms, the capsule deeply sunk, but sometimes the valves terminating in long protruding points formed by the split base of the style.—DC. Prod. iii. 217.

W. Australia, *Labillardière, Drummond,* 3*rd Coll. n.* 65; scrubby undulating country N. of Stirling Range, *Maxwell.*

Var. *angulosa.* Calyx-tube and operculum more or less prominently angled or several-ribbed, but varying much in this respect as well as in the size of the flowers and fruits.—*E. angulosa,* Schau. in Walp. Rep. ii. 925; *E. cuspidata,* Turcz. in Bull. Mosc. 1849, ii. 21; *E. costata,* F. Muell. in Trans. Vict. Inst. 33; Miq. in Ned. Kruidk. Arch. iv. 136; *E. Muelleri,* Miq. l. c. 130 (a small-flowered form). The locality given by Miquel, "Madam Pepper-weath," is a mistaken reading for "in modum Pepper-menth," or *like Peppermint.*

N. S. Wales. Mallee scrub of the Murray desert to the Barrier Range, *Victorian Expedition.*

S. Australia. Various points of the S. coast, *R. Brown;* Kangaroo island, *Labillardière;* from the Murray to Spencer's and St. Vincent's Gulfs, *F. Mueller, Behr,* and others.

W. Australia. Sandy plains N. of Stirling Range, *Maxwell;* near the sea, King George's Sound, *R. Brown, A. Cunningham, Drummond, n.* 230, 4*th Coll. n.* 75, and others; and eastward to Espérance Bay, Phillips Ranges, Moir's Inlet, Cape Le Grand, *Maxwell.*

F. Mueller, Fragm. ii. 59, is disposed to reduce this variety, and perhaps the whole species, to *E. dumosa.*

64. **E. gomphocephala,** *DC. Prod.* iii. 220, *and Mem. Myrt. t.* 11. A tree, of 40 to 50 ft., with a smooth or rough persistent bark, very dark on the Swan River, iron-grey on the Vasse (*Oldfield*). Leaves ovate-lanceolate to narrow-lanceolate, mostly falcate and acuminate, often exceeding 6 in., thick and shining, the fine rather numerous oblique veins scarcely conspicuous, except on old leaves. Peduncles axillary or lateral, thick and hard, broad and flat, ½ to 1 in. long, each with 3 flowers, either sessile or on very short thick flat pedicels. Calyx-tube obovoid or somewhat urceolate, 4 to 5 lines long and nearly 4 lines diameter. Operculum globular, very thick and hard, broader than the calyx-tube, usually nearly ½ in. diameter. Stamens exceedingly numerous, nearly 4 lines long, inflected in the bud; anthers oblong, with distinct parallel cells. Ovary convex in the centre. Fruit ¾ in. long, somewhat dilated at the orifice, the rim broad and convex, the capsule scarcely sunk, conical in the centre, the open valves protruding.—F. Muell. Fragm. ii. 36.

W. Australia. Géographe Bay, *Leschenault;* Swan River, *Oldfield, Harvey;* Vasse river, and perhaps Murchison river (specimen imperfect), *Oldfield*, towards Cape Leeuwin, *Gregory*.

65. **E. grossa,** *F. Muell. Herb.* A stunted shrub (*Maxwell*). Leaves from ovate and obtuse to lanceolate and acute, very thick and shining, under 3 in. long, the veins oblique, rarely conspicuous, the intramarginal one at a distance from the edge. Peduncles axillary or lateral, often recurved, thick and much flattened, with usually 3 large sessile flowers. Calyx-tube turbinate, prominently ribbed, 4 to 5 lines long. Operculum oblong, very obtuse, thin and smooth as in the *Cornutæ*, as long as or rather shorter, perhaps sometimes longer than the calyx-tube. Stamens about ½ in. long, inflected in the bud; anthers ovate-oblong, with parallel distinct cells. Ovary short, convex in the centre. Fruit not seen.

W. Australia. Phillips river and its tributaries, *Maxwell*. I feel uncertain as to the affinities of this species, the smooth cylindrical obtuse operculum is like that of some of the *Cornutæ*, but the stamens are much inflected in the bud, and the flowers are otherwise quite those of the larger forms of *E. incrassata*.

66. **E. vernicosa,** *Hook. f. in Hook. Lond. Journ.* vi. 478, *and Fl. Tasm.* i. 135. A low bushy shrub, not exceeding 4 ft. in exposed situations, growing perhaps to a small tree where more sheltered (*J. D. Hooker*). Leaves mostly ovate or almost orbicular or the upper ones oblong, obtuse or mucronate, rarely above 1 in. long, thick, smooth and shining so as to conceal the veins, which are rather oblique and distant. Peduncles exceedingly short, thick, and more or less flattened, each with 1 to 3 closely sessile flowers. Calyx-tube thick, about 2 lines long and as much in diameter. Operculum shorter than the calyx-tube and usually shortly acuminate. Stamens inflected in the bud (not seen fully expanded); anthers ovate, the cells closely contiguous, but parallel and distinct, at least in the bud. Ovary not much shorter than the calyx-tube. Fruit hard, ovoid-truncate or almost urceolate, about 3 lines diameter, slightly dilated at the orifice, the rim flat or slightly convex, the capsule somewhat sunk, but the valves protruding when open.

Tasmania. Summit of Mount Fatigue, *Milligan*, and of Mount Lapeyrouse, *Oldham*. The species is readily known by its small leaves. It is in some respects nearly allied to *E. viminalis*, in others to *E. dumosa*.

67. **E. megacarpa,** *F. Muell. Fragm.* ii. 70. A tree, the bark not described. Leaves lanceolate, falcate, mostly 4 to 6 in. long, thick and smooth, the veins irregular, oblique, fine, and scarcely conspicuous. Peduncles axillary or lateral, thick and flat, rather short, each bearing usually 2 sessile flowers. Calyx-tube broadly turbinate, smooth, under 6 lines long, the margin acutely prominent in the bud. Operculum shortly conical. Stamens about ½ in. long, inflected in the bud; anthers ovate-oblong, with distinct parallel cells. Fruit depressed globular, thick and hard, ¾ to 1 in. diameter, the rim very convex and prominent, continuous with the thick, conical, obtuse, incurved and prominent valves of the capsule.

W. Australia. King George's Sound and to the eastward, *R. Brown*, *Drummond*, 3*rd Coll. Suppl. n.* 18; margin of Wilson's Inlet, *Maxwell*; near Augusta, *Gilbert*, *n.* 257. Specimens are also in F. Mueller's herbarium from a tree cultivated at Sydney as a "Blue Gum."

68. **E. Preissiana,** *Schau. in Pl. Preiss.* i. 131. A stout rigid shrub,

of 8 to 12 ft. Leaves mostly opposite, although petiolate, from broadly ovate to ovate-lanceolate, very obtuse or rarely acute, 3 to 5 in. long, very thick and rigid, the veins diverging and parallel but not close, the marginal one at a distance from the edge. Peduncles axillary or lateral, very thick and much dilated, sometimes almost winged, under 1 in. long, each with 3 large flowers, either sessile or tapering into a very short, thick, flattened pedicel. Calyx-tube broadly turbinate or almost hemispherical, very thick and smooth, 7 to 8 lines diameter. Operculum only slightly convex, not broader than the calyx-tube. Stamens 6 to 8 lines long, inflected in the bud; anthers ovate, with parallel distinct cells. Disk broad and concave, the ovary with as many protuberances in the centre as valves. Fruit very hard and shining, broadly turbinate or hemispherical, 1 to 1½ in. diameter, the top flat or concave, the rim fully 3 lines diameter, the capsule slightly depressed, the valves (4, 5, or rarely 6) usually flat.—Hook. Bot. Mag. t. 4266; F. Muell. Fragm. ii. 38; *E. plurilocularis*, F. Muell. Fragm. ii. 70.

W. Australia. From the Kalgan river to Cape Riche, *Preiss, n.* 239, *Drummond, 3rd Coll. n.* 63, *Oldfield.*

The fruit may be seen occasionally, apparently when not well grown, much less widened at the top, thus losing the characteristic form of the species. In some specimens from Salt River, *Maxwell*, the leaves are more acute and the capsule (not perfect) almost contracted at the orifice, but they appear to belong to the same species, having the broad flowering-calyx and flattened peduncle of the obtuse-leaved specimens. In a specimen sent by F. Mueller from a tree grown in the Melbourne Botanic Garden from W. Australian seeds, and named by him *E. pachypoda*, the leaves are acute as in Maxwell's specimen, but the peduncle is very thick and scarcely flattened, bearing more than 3 flowers, with ovoid calyxes. The tree had not yet fruited, but it will probably not prove specifically distinct from *E. Preissiana.*

SUBSERIES IV. CORNUTÆ.—Peduncles axillary or lateral, several-, often many-flowered, flattened (except in *E. cornuta*). Flowers sessile or shortly pedicellate. Operculum long, smooth, and not thick. Stamens erect or flexuose in the bud, not inflected. Fruit turbinate, urceolate or obovoid, the capsule not much sunk. Leaves thick, with irregular oblique veins, often inconspicuous.

69. **E. Lehmanni,** *Preiss, Herb. according to Schau. in Pl. Preiss.* i. 127. A tall shrub or small tree, with a roughish reddish bark coming off in irregular sheets (*Oldfield*). Leaves from ovate to oblong or almost lanceolate, obtuse, under 3 in. long, very thick, the veins very oblique and rather distant, the intramarginal one at a distance from the edge. Flowers several, often 20 or more together in dense heads upon thick recurved peduncles 1 to 3 in. long, and sometimes much flattened, the receptacle forming a globose mass of ½ in. or more diameter, in which the calyx-tubes (usually 2 to 3 lines diameter) are more or less immersed. Operculum cylindrical, dilated at the base, obtuse, often 1½ in. long. Stamens 1½ to 2 in. long, erect in the bud as in *E. cornuta;* anthers oblong, parallel-celled. Ovary convex at the top. Fruits half immersed in the receptacle, about ½ in. diameter, the rim very narrow, the capsule not depressed, the exserted valves connivent into a cone, tapering into the persistent base of the style.—*Symphyomyrtus Lehmanni*, Schau. in Pl. Preiss. i. 127; *E. macrocera*, Turcz. in Bull. Mosc. 1849, ii. 20 (described apparently from an imperfect specimen).

W. Australia. S. coast to the east of King George's Sound, *R. Brown;* stony hills from Bald Island and Stirling range eastward to Cape Arid, *Oldfield; Maxwell; Preiss, n.* 227; *Drummond, 4th Coll. n.* 67, *in most sets.*

70. **E. cornuta,** *Labill. Voy.* i. 403. *t.* 20. A tall shrub or small tree with a smooth bark (*Oldfield*), or more frequently a moderate-sized or tall tree with a bushy head, the bark brown or black, hard, rugged, persistent, half-fibrous or fibrous (*Oldfield*), or the bark smooth and falling off in pieces (*Maxwell*). Leaves lanceolate or ovate-lanceolate, mostly under 4 in. long, rather thick, the veins irregularly oblique, often more conspicuous than in the adjoining species, the intramarginal one at a distance from the edge. Peduncles axillary, terete or slightly compressed, each bearing 6 to 12 or even more flowers, sessile but not immersed in the receptacle. Calyx oblong-turbinate, about 3 lines long and rather less in diameter. Operculum 1 to 1¼ in. long, more or less tapering upwards, but obtuse. Stamens erect or slightly flexuose in the bud, but not inflected; the outer ones often above 1 in. long; the inner ones much shorter; anthers oblong, parallel-celled. Ovary almost on a level with the calyx-rim, the top flat or at length slightly convex, the style thickened at the base. Fruit obovoid-truncate or shortly cylindrical, about 4 lines long, not contracted at the orifice, the rim narrow and scarcely distinct from the slightly convex summit of the fruit; valves when open raised and acuminate by the long often connivent points formed by the split and persistent base of the style. Cotyledons of the seeds very deeply lobed, almost 2-partite.—DC. Prod. iii. 216; Schau. in Pl. Preiss. i. 127; F. Muell. Fragm. ii. 39 (partly).

W. Australia. King George's Sound, *R. Brown,* and eastward to Cape Riche, *Labillardière; A. Cunningham; Drummond, 2nd Coll. n.* 83, *4th Coll. n.* 68; *and in some sets n.* 67; *Preiss, n.* 238, and others; Vasse river, *Gilbert, n.* 270; "Yeit" of the colonists, *Oldfield.*—F. Mueller proposes to unite with this the *E. Lehmanni* as well as several of the following species. I have not ventured to do so at present, as amongst the numerous specimens examined from various sources, I have not yet met with intermediates connecting the different forms, especially as to the summit of the ovary and the fruit.

71. **E. annulata,** *Benth.* A tall shrub with a smooth bark (*Maxwell*). Leaves narrow-lanceolate, acuminate, mostly under 4 in. long, thick and smooth with oblique veins usually very indistinct, the intramarginal one near the edge. Peduncles axillary or lateral, short, thick, flat, and almost as broad as long, each with about 6 to 12 sessile flowers. Calyx-tube turbinate-campanulate, about 3 lines diameter. Operculum 6 to 8 lines long, usually incurved and very obtuse or almost clavate at the end. Stamens straight as in *E. cornuta*, but apparently of a yellowish-white colour as in *E. macrandra,* the margin of the disk that bears them forming a raised inflexed ring about ¾ line broad; anthers oblong with parallel cells. Ovary conical at the top, tapering into the style. Fruit depressed-globose, 4 to 5 lines diameter, the convex rim protruding into a thick ring, quite distinct from the valves, which project much, tapering into long erect or connivent points formed by the persistent base of the style.

W. Australia. Salt River, *Maxwell.*

72. **E. platypus,** *Hook. Ic. Pl. t.* 849. A tree attaining 30 ft., with a smooth bark (*Maxwell*). Leaves very broadly ovate or orbicular, often

coarsely crenate, mostly under 2 in. long, very thick, smooth, and shining, the oblique veins scarcely visible. Peduncles axillary, thick and hard but flat, and often ¼ to ½ in. broad, erect or recurved, mostly above 1 in. long, each bearing about 3 to 7 flowers. Calyx-tube usually 3 to 4 lines long, thick but narrow-turbinate, smooth and nearly terete, or with 2, 3, or sometimes 4 more or less prominent ribs or angles, and generally tapering into a very short, thick, angular or flattened pedicel. Operculum tapering upwards, longer and oftener narrower than the calyx-tube. Stamens erect in the bud as in *E. cornuta*, the outer ones attaining 7 to 8 lines; anthers ovate-oblong, with parallel cells opening longitudinally. Ovary conical in the centre, with as many raised lines as cells. Fruit obovoid-truncate or turbinate, ½ to nearly ¾ in. long and 4 to 7 lines diameter, slightly contracted at the orifice, the rim rather broad, convex; the capsule somewhat sunk, but the valves often acuminate by the split base of the style, with the points protruding.

W. Australia. From about 6 miles N. of the W. end of Stirling range, extending far away eastward beyond Phillips ranges, forming dense impenetrable thickets, "Maalok" of the natives, *Maxwell; also Drummond, 5th Coll. n.* 183 (given by *Maxwell*).

Var. *nutans.* Flowers and fruits larger, the ribs more prominent, one or two sometimes expanded into thick wings.—*E. nutans,* F. Muell. Fragm. iii. 152.—In the interior from Bremer's Inlet, forming dense thickets, *Maxwell.*

In R. Brown's collections are some specimens in very young bud and fruit from Goose Island Bay, apparently of a variety of this species, with leaves from ovate to ovate-lanceolate, but obtuse and under 2 in. long, as in the broad-leaved form. I have not seen the stamens.

73. **E. macrandra,** *F. Muell. Herb.* A shrub or small tree with a smooth bark (*Maxwell*): Leaves from ovate-lanceolate to narrow-lanceolate, rarely exceeding 4 in., very thick and smooth, the veins more numerous and more diverging than in *E. cornuta*, and the intramarginal one usually nearer the edge, but generally scarcely visible. Peduncles rigid and flattened, mostly ½ to 1 in. long, with 8 to 16 or even more flowers, sessile or on very short pedicels. Calyx-tube obovoid-campanulate, usually 2½ to 3 lines long and rather less in diameter, but in some specimens smaller. Operculum usually above 1 in. long. Stamens when dry yellowish, erect in the bud as in *E. cornuta*, the edge of the disk inflected; anthers oblong, with parallel cells. Ovary flat-topped, the style not thickened at the base. Fruit semi-ovoid, truncate, 3 to 4 lines diameter, or in some specimens rather smaller, the rim narrow, on a level with the calyx as well as the flat-topped capsule, the small valves not protruding.

W. Australia. From the valleys S. of Stirling range to Salt River and Phillips range, *Maxwell.*

74. **E. occidentalis,** *Endl. in Hueg. Enum.* 49. A tall shrub or tree, attaining sometimes 80 ft. (*Oldfield*). Leaves from oval-oblong and under 2 in. to narrow-lanceolate, falcate and above 4 in. long, very thick, with oblique veins scarcely conspicuous or rarely prominent underneath, the intramarginal one a little distant from the edge. Peduncles axillary or lateral, more or less flattened and often recurved, with 3 to 5 flowers on rather thick pedicels of 2 to 3 lines. Calyx-tube urceolate-oblong, 3 to 4 lines long at the time of flowering, smooth or obscurely ribbed, usually somewhat dilated at the orifice. Operculum ½ to ¾ in. long, very obtuse or rarely almost acute.

Stamens ½ to ¾ in. long, erect as in *E. cornuta;* anthers oblong, with parallel cells. Ovary very convex or conical at the top. Fruit urceolate, 6 to 8 lines long when full grown, about 5 lines diameter at the top and narrower below, the rim narrow, not prominent, the capsule somewhat sunk but conical in the centre, and the valves protruding when open.—Schau. in Pl. Preiss. i. 128; F. Muell. Fragm. ii. 39.

W. Australia. From Kalgan river and the W. end of Stirling range eastward to Cape Riche and Cape Le Grand, *Maxwell; Oldfield; Harvey; Preiss, n.* 228 *and* 240; *Drummond, 4th Coll. n.* 74, *also n.* 152.

In some of Drummond's and Oldfield's specimens the leaves are smaller and narrower, the calyx and fruit smaller, the orifice slightly contracted, and the very small valves scarcely protrude.

75. **E. spathulata,** *Hook. Ic. Pl. t.* 611. A shrub of 6 to 8 ft. or rather more. Leaves linear, linear-lanceolate or rarely oblong-lanceolate, straight or slightly falcate, under 3 in. long, thick and rigid so as wholly to conceal the veins. Peduncles short, axillary or lateral, flattened but usually not very broad, each with about 4 to 6 flowers. Calyx-tube obovoid, thick, about 2 lines long, tapering into a short thick pedicel. Operculum cylindrical, obtuse, often narrower than the calyx and about twice as long. Stamens erect, slightly flexuose, about 4 lines long, the border of the staminal disk inflected over the sunk ovary; anthers oblong, parallel-celled. Style slightly thickened at the base. Fruit obovoid, 3 lines or rather more in length and nearly as much in diameter, contracted at the orifice, which is further closed by the rather broad flat rim; capsule sunk, but the points of the valves sometimes slightly protruding.

W. Australia. Between Perth and King George's Sound, *Harvey; Drummond, 3rd Coll. n.* 68.

Var. *grandiflora.* Leaves rather broader. Flowers and fruits larger.—Phillips range, *Maxwell.*

The species has much of the aspect of the narrow-leaved forms of *E. redunca*, but in that the operculum is acuminate, and the stamens more or less inflected in the bud.

SUBSERIES V. EXSERTÆ.—Peduncles axillary or lateral, or rarely the upper ones in a short terminal corymb, terete or slightly flattened, each with several, often many, flowers usually pedicellate. Fruit globose or depressed, usually more or less contracted at the orifice, the rim convex or prominent or rarely flat, the capsule-valves protruding beyond it.

76. **E. pallidifolia,** *F. Muell. Fragm.* iii. 131. A small tree with an ash-coloured smooth bark (*F. Mueller*). Leaves ovate-oblong or lanceolate, very obtuse and rarely 3 in. long, thick and smooth, the fine parallel very diverging veins scarcely visible, the intramarginal one close to the edge. Peduncles axillary or lateral, short, nearly terete, with 4 to 6 nearly sessile or shortly pedicellate flowers. Calyx-tube short, about 2 lines diameter. Operculum hemispherical or obtusely conical, shorter than the calyx-tube. Stamens about 2 lines long, inflected in the bud; anthers ovate with parallel distinct cells. Ovary flat-topped. Fruit obovoid-globose, 3 to 4 lines diameter, slightly contracted at the orifice, the rim broad, convex, and prominent, the capsule not sunk, the valves protruding and sometimes acuminate by the persistent split base of the style.

N. Australia. Sandstone table-land on the Upper Victoria river and Sturt's Creek, *F. Mueller.*—As observed by F. Mueller, this resembles in some respects *E. oleosa*, but the venation of the leaves and the fruit are different.

77 ? **E. pachyphylla,** *F. Muell. in Journ. Linn. Soc.* iii. 98. A tall shrub. Leaves ovate or ovate-lanceolate, abruptly acuminate, under 4 in. long, very thick and smooth, the fine diverging parallel veins scarcely conspicuous. Flowers not seen. Fruiting-umbels nearly sessile; fruits on thick terete pedicels, nearly hemispherical, 4-ribbed, very hard and woody, $\frac{3}{4}$ to 1 in. diameter, the rim very broad and conically exserted, the capsule depressed below the rim, the valves scarcely protruding. Seeds broad and flat, bordered by a narrow wing.

N. Australia. Sandy desert at Hooker's Creek, *F. Mueller.*—The specimens are insufficient to determine the affinities of this species. In some respects they resemble *E. cosmophylla* and its allies, but the fruit, the seeds, and perhaps the inflorescence are different.

78. **E. Oldfieldii,** *F. Muell. Fragm.* ii. 37. A shrub of 8 to 10 ft., with a smooth ash-grey bark coming off in layers (*Oldfield*). Leaves ovate-lanceolate or lanceolate-acuminate, often falcate, mostly under 4 but sometimes above 6 in. long, very thick, the veins numerous and rather oblique but scarcely conspicuous, the intramarginal one near the edge, or when the leaf is broad, distant from it. Peduncles axillary or lateral, very short or scarcely any, each with 2 or 3 rather large flowers, sessile or on very short pedicels. Calyx-tube broadly hemispherical, hard and smooth, about $\frac{1}{2}$ in. diameter. Operculum hemispherical, as long as or rather longer than the calyx-tube, usually umbonate or with a small point. Stamens dark-coloured, connivent in the bud, but only slightly inflected, showing their anthers; anthers oblong, with distinct parallel cells. Disk forming a more or less raised ring within the stamens round the flat-topped ovary. Fruit depressed-globose, 7 to 8 lines diameter, the rim very broad, at length convex and much raised, the capsule somewhat depressed in the centre, with the valves slightly prominent.

W. Australia, *Drummond, 5th Coll. Suppl. n.* 35; Murchison river, *Oldfield.*

79. **E. pachyloma,** *Benth.* A shrub of 5 ft. (*Maxwell*). Leaves mostly lanceolate or linear-lanceolate, acuminate, under 3 in. long, thick and rigid, the very oblique veins scarcely conspicuous, the intramarginal one at a distance from the edge. Peduncles axillary or lateral, short and thick, terete or slightly angular, each with 2 to 4 rather large flowers. Calyx-tube broadly turbinate or almost hemispherical, about 4 lines diameter, smooth and tapering into the very short thick pedicel. Stamens pale-coloured, $\frac{1}{2}$ in. long or more, slender and inflected in the bud; anthers ovate with distinct parallel cells. Disk concave. Fruit sessile, depressed-globose, 7 to 8 lines diameter, with the very thick broad convex and raised rim of *E. Oldfieldii*, but without any depressed centre, the capsule not sunk, and the small valves protruding as in *E. rostrata.*

W. Australia, *Drummond, 4th Coll. n.* 64; sand plains, Kalgan river, *Oldfield;* valleys of the Stirling range, *Maxwell.*

80. **E. Drummondii,** *Benth.* Leaves from ovate-oblong to lanceolate, obtuse or acuminate, under 3 in. long, very thick, with very fine close parallel veins, very diverging or almost transverse, but scarcely conspicuous,

the intramarginal one close to the edge. Peduncles axillary or lateral, ½ to 1½ in. long, terete or nearly so, each bearing an umbel of 3 to 6 rather large flowers on terete pedicels often ½ in. long. Calyx-tube broadly hemispherical, hard and smooth, 4 to 5 lines diameter. Operculum conical, rather broader and considerably longer than the calyx-tube. Stamens about ½ in. long, inflected in the bud; anthers rather small, ovate, with distinct parallel cells. Disk very broad, nearly flat, forming a prominent ring round the ovary, of which the obtusely conical centre protrudes about 1 or 1½ lines above the disk at the time of flowering. Fruit unknown.

W. Australia. Between Swan River and King George's Sound, *Drummond, 2nd Coll. n.* 86; *also* 5*th Coll.*

81. **E. orbifolia,** *F. Muell. Fragm.* v. 50. A shrub of 5 ft. (*C. Harper*), the foliage nearly white or yellowish in the single small specimen seen. Leaves nearly orbicular, very obtuse, under 2 in. diameter, very thick and smooth, the veins irregular and distant but scarcely conspicuous. Peduncle axillary, terete, not ½ in. long, with the scars of 5 flowers. Pedicels short and terete. Calyx-tube broadly hemispherical, smooth, about ½ in. diameter. Operculum thick, conical, nearly twice as long as the calyx-tube. Stamens very numerous, inflected in the bud; anthers ovate, with distinct parallel cells. Disk narrow round the conical summit of the ovary, which protrudes 3 or 4 lines above the border of the calyx, tapering into the short thick style. Fruit unknown.

W. Australia. Granite hills in the interior to the north of Swan river, *C. Harper.* Although evidently allied to *E. Drummondi*, this appears to be specifically distinct both in the leaves and the parts of the flowers.

82. **E. angustissima,** *F. Muell. Fragm.* iv. 25. A bushy shrub of 5 ft. (*Maxwell*). Leaves narrow-linear, acuminate or almost aristate, 2 to 3 in. long, the veins inconspicuous. Peduncles axillary, very short, terete, each with 2 to 4 small flowers, only seen in bud. Calyx-tube depressed-hemispherical, not 2 lines diameter. Operculum very obtuse, rather longer than the calyx-tube. Stamens inflected in the bud; anthers with parallel distinct cells. Fruit depressed-globular, about 3 lines diameter, contracted at the orifice, the rim convex, the capsule on a level with it, the valves worn away in the specimens seen.

W. Australia. Point Malcolm and eighty miles away to the eastward, *Maxwell.*

83. **E. leptopoda,** *Benth.* Branchlets slender. Leaves linear-lanceolate, acuminate, often above 4 in. long, not very thick but the veins inconspicuous. Peduncles axillary or lateral, slender, terete or slightly flattened, bearing each a loose umbel of 10 to 15 small flowers on slender pedicels much longer than the buds. Calyx-tube broadly turbinate or almost hemispherical, about 1½ lines diameter. Operculum conical, from a little shorter to a little longer than the calyx-tube and not so broad. Stamens inflected in the bud, flexuose, not 2 lines long; anthers ovate or almost globose, with parallel distinct cells. Fruit depressed-globular, nearly 3 lines diameter, the rim broad, flat or slightly convex, the capsule not sunk, the valves protruding when open.

W. Australia, *Drummond,* 5*th Coll. Suppl. n.* 33 *and* 36, *also n.* 151 *and* 188 *of other sets.*

In the specimens n. 188 the buds are rather larger than in the others, the peduncles and pedicels shorter and the fruits smaller, scarcely 2 lines diameter, with long prominent points to the valves.

84. **E. cinerea,** *F. Muell. Herb.* A moderate-sized tree, with a whitish-brown persistent bark, somewhat fibrous, the foliage more or less glaucous or mealy white. Leaves opposite, sessile, cordate, ovate- or ovate-lanceolate, obtuse or acute, mostly 2 to 4 in. long. Peduncles axillary or in short terminal corymbs, terete or nearly so, each with 3 to 7 pedicellate flowers. Calyx broadly turbinate, about 2 lines diameter or rather more. Operculum conical, shorter than the calyx-tube. Stamens 2 to 3 lines long, inflected in the bud; anthers small but ovate, with distinct parallel cells. Ovary convex in the centre. Fruit semiglobose or subglobose-truncate, about 3 lines diameter, often slightly contracted at the orifice, the rim thin, the capsule very slightly sunk but the valves protruding.

N. S. Wales. Lachlan river, near Bathurst, *A. Cunningham;* also Lake George, *Herb. F. Mueller.*

F. Mueller (Fragm. ii. 70) unites this with *E. pulverulenta,* of which it may be a variety, but, as far as the specimens go, the differences in the leaf, in the size of the flower, and in the shape of the fruit appear to be constant. It may, however, be an opposite-leaved state of *E. dealbata,* and possibly, as well as that species, a form of *E. viminalis.*

85. **E. dealbata,** *A. Cunn.; Schau. in Walp. Rep.* ii. 924. A small stunted tree, the foliage often glaucous-white, the bark rugose or separating in scales, leaving the inner bark white and smooth (*C. Stuart*). Leaves from ovate to ovate-lanceolate and under 4 in. long or sometimes lanceolate and longer, obtuse or acute, the veins oblique and irregular, the intramarginal one at a distance from the edge, all usually conspicuous. Peduncles axillary or lateral, very short and scarcely flattened, bearing each 3 to 6 flowers on short pedicels. Calyx-tube very open, about 2 lines diameter and not so long. Operculum broad, rather thin, hemispherical or conical, longer than the calyx-tube. Stamens about 3 lines long, inflected in the bud; anthers ovate, with parallel distinct cells. Ovary more or less conical in the centre, tapering into the style. Fruit almost hemispherical, about 3 lines diameter, the rim flat, the valves protruding even before they open.

Queensland. In the interior, *Mitchell.*

N. S. Wales. Rocky situations in the interior, *A. Cunningham;* New England, *C. Stuart,* also probably a specimen in young bud of a "Box," *Leichhardt;* Mudgee, "River Gum," *C. Moore.* It is possible that this may prove to be the true *E. pallens,* DC. F. Mueller thinks it may be reducible to a variety of *E. viminalis.*

86. **E. viminalis,** *Labill. Pl. Nov. Holl.* ii. 12. *t.* 151. A tree usually of moderate size, but sometimes attaining a great height with a rough persistent bark, at least on the trunk and main branches, that of the smaller branches often smooth and deciduous, and sometimes the whole described as deciduous. Leaves lanceolate and more or less falcate and acuminate, 3 to 6 in. long, the veins rather regular, numerous and diverging, the intramarginal one near the edge. Peduncles short, axillary or lateral, bearing in some specimens especially southern ones always 3 flowers on short pedicels, in others 6 to 8 flowers more distinctly pedicellate, but always much less so than in *E. rostrata.* Calyx-tube turbinate or hemispherical, 2 lines or rather more in diameter. Operculum somewhat conical and about as long as the calyx-

tube or rarely rather longer and acuminate. Stamens about 3 lines long, inflected in the bud; anthers ovate with parallel distinct cells. Ovary short, flat-topped. Fruit subglobose-truncate, from 3 to 4 or 5 lines diameter, the rim rather broad, at first flat but if well ripened usually prominent above the border of the calyx, the capsule not sunk, the valves short horizontal or protruding when open.—DC. Prod. iii. 218; Hook. f. Fl. Tasm. i. 134; Miq. in Ned. Kruidk. Arch. iv. 125; F. Muell. Fragm. ii. 64; *E. diversifolia*, Bonpl. Pl. Malm. 35. t. 13; DC. Prod. iii. 220; *E. elata*, Dehnh.; Walp. Rep. ii. 163; *E. mannifera*, A. Cunn., and perhaps also Moodie; Walp. Rep. ii. 163, although incorrectly described; *E. persicifolia*, Lodd. Bot. Cab. t. 501 (from the fig.), not of DC.; *E. granularis*, Sieb. Pl. Exs.; *E. pilularis*, DC. Prod. iii. 218, not of Sm.

N. S. Wales. Port Jackson or Blue Mountains, *Caley*, *Sieber*, *n.* 474, *and Fl. Mixt. n.* 604, and others; very generally dispersed through the country bordering on Bathurst Downs, "Blue Gum," *A. Cunningham;* Argyle county, *Backhouse;* Exhibition woods, n. 108, *Macarthur;* near Duck river, "Drooping Gum," *Woolls;* New England, *C. Stuart;* and Camden, "Woolly butt," *Woolls;* also *Caley* (specimens with a hemispherical calyx-tube and broad almost globular operculum).

Victoria. Port Phillip, *R. Brown;* in fertile districts in plains as well as in the hills, "Box-tree" and "Peppermint Gum" of Ovens river, *F. Mueller;* "Weeping Gum," with red and with white timber, *Robertson.*

Tasmania. Port Dalrymple and Derwent river, *R. Brown;* abundant throughout the island, *J. D. Hooker.*

S. Australia. Memory Cove, *R. Brown;* Mount Gambier to Rivoly Bay, Lofty and Bugle ranges, *F. Mueller;* Kangaroo Island, *R. Brown*, *Waterhouse*, these specimens precisely agreeing with those of *E. diversifolia* from French gardens, originally raised from Kangaroo Island seeds.

The species varies very much in the size and number of the flowers, and the shape of the operculum. In the original Tasmanian form, common also in Victoria, the peduncles are mostly 3-flowered, although occasionally many-flowered specimens occur. In the S. Australian *E. diversifolia*, the flowers are rather numerous in the umbel, and the fruit large. In the N. S. Wales specimens the flowers and fruits are usually small, the buds very smooth and shining, and the bark sometimes said to be quite smooth, probably when the rough bark has shed.

E. patentiflora, F. Muell., is referred here in F. Muell. Fragm. ii. 64. The specimens described under that name by Miq. in Ned. Kruidk. Arch. iv. 125 belong to *E. melliodora.*

E. fabrorum, Schlecht. in Linnæa, xx. 656, was supposed by F. Mueller to refer to *E. obliqua*, owing to his stating it to be the "Stringy-bark" of the colonists, but Behr's specimen in Herb. Sonder, communicated by Schlechtendahl, is evidently the large-fruited form of *E. viminalis.*

E. Gunnii, Miq. in Ned. Kruidk. Arch. iv. 126 (not of Hook. f.), from Streleczky range, Victoria, appears to be *E. viminalis.*

87. **E. rostrata,** *Schlecht. Linnæa*, xx. 655. A tall tree with a greyish-white bark, smooth and separating in thin layers (*F. Mueller*, and others), rarely persistent and rough? (*F. Mueller*). Leaves lanceolate, mostly falcate and acuminate, 3 to 6 in. long or even more, the lower ones sometimes ovate or ovate-lanceolate and straight, not thick, the veins rather regular, numerous and oblique, the intramarginal one not close to the edge, or in some desert specimens thick with the veins much less conspicuous. Peduncles rather short, terete or scarcely compressed, bearing each about 4 to 8 flowers on rather long pedicels. Calyx-tube hemispherical, 2 to 2½ lines diameter. Operculum more hemispherical than in *E. viminalis* and about as long as or shorter

than the calyx without the point or beak, which is almost always prominent and sometimes rather long, or very rarely the whole operculum is elongated and obtuse without any beak, but much shorter than in *E. tereticornis*. Stamens about 2 lines long, inflected in the bud; anthers small, ovate, with parallel distinct cells. Ovary short, convex or conical in the centre. Fruit nearly globular, rarely above 3 lines diameter, the rim broad and very prominent, almost conical, the capsule not sunk and the valves entirely protruding even before they open.—F. Muell. in Journ. Linn. Soc. iii. 83; *E. longirostris*, F. Muell.; Miq. in Ned. Kruidk. Arch. iv. 125.

N. S. Wales. Lachlan and Darling rivers to the Barrier range and Cooper's Creek, *Victorian and other Expeditions;* New England, *C. Stuart;* "Flooded Gum" of the colonists, *F. Mueller*.

Victoria. From the Yarra to the Murray, *F. Mueller*.

S. Australia. Banks of streams, "White Gum," *Behr;* from the Murray to St. Vincent's Gulf, "Red Gum," *F. Mueller*, and others; Three-Well River, *Waterhouse;* W. of Lake Torrens, *Babbage*, in Herb. R. Br.

This species, designated as "Red Gum" and "White Gum" by several collectors, is, as observed by F. Mueller, very closely allied to *E. viminalis* and *E. tereticornis*. From the former it differs in the longer pedicels, in the operculum, and in the shape of the fruit, the rim and capsule always much more exserted. From *E. tereticornis* it is chiefly distinguished by the operculum. It has also usually smaller flowers and fruits. In one specimen from the granite hills between Nine-mile Creek and Broken River, Victoria, F. Mueller has appended the note that the bark is persistent like that of "Box."

E. acuminata, Hook. in Mitch. Trop. Austr. 390, from the interior of Queensland, appears to be a variety of *E. rostrata*, with the operculum more conical and less rostrate, approaching the var. *brevirostris* of *E. tereticornis*.

88? **E. exserta,** *F. Muell. in Journ. Linn. Soc.* iii. 85. A moderate-sized or small tree, the bark ash-brown, rough and fissured outside and falling in fragments, somewhat fibrous inside (*F. Mueller*), dark iron-grey and roughish (*Oldfield*). Leaves lanceolate, mostly falcate and acuminate, 3 to 6 in. long or sometimes much more, the lower ones often ovate, rather thick, the veins rather regular, numerous and oblique, the intramarginal one not close to the edge. Peduncles axillary or lateral, terete or scarcely compressed, bearing each 3 to 8 flowers on distinct often rather long pedicels. Calyx-tube hemispherical, about 2 lines diameter (or sometimes nearly 3?). Operculum hemispherical or broadly conical, more or less beaked, acuminate and rather longer than the calyx-tube. Stamens about 2 lines long or rather more, inflected in the bud; anthers ovate with parallel distinct cells. Fruit nearly globular, 3 to 4 lines diameter, the rim broad and very prominent, almost conical, the capsule not sunk, and the valves entirely protruding even before they open.

Queensland. Burnett river, *F. Mueller*.

W. Australia. Murchison river, *Oldfield*.

This is probably the same as *E. rostrata*, notwithstanding the differences described in the bark. There may be also some confusion in Oldfield's specimens, the larger-flowered ones may belong to *E. rudis*, which differs in its large flowers, shorter pedicels, and in the much larger fruit with a flat rim.

89. **E. tereticornis,** *Sm. Bot. Nov. Holl.* 41, *and in Trans. Linn. Soc.* iii. 284. A tall tree, with a smooth whitish or ash-coloured bark shedding in thin layers (*F. Mueller* and others). Leaves lanceolate, mostly falcate and acuminate, often exceeding 6 in. long, the veins rather regular and numerous and oblique as in *E. rostrata*, but often rather coarser, the intramarginal

one rather distant from the edge. Peduncles axillary or lateral, not very short, terete or angular, the upper ones sometimes forming a short panicle, each bearing about 4 to 8 flowers on pedicels of 1 to 3 lines. Calyx tube turbinate, 2 to nearly 3 lines diameter. Operculum conical, acuminate, usually about ½ in. long, always much longer than the calyx-tube and usually broader, of a rather thin texture and smooth. Stamens often ½ in. long, more or less inflected in the bud, but sometimes only very shortly so at the ends; anthers small, ovate, with parallel distinct cells. Ovary nearly as long as the calyx-tube and convex or conical in the centre. Fruit obovoid or almost globular, 3 to 4 lines diameter, the rim broad and very prominent, the capsule not sunk, the valves protruding beyond the rim.—DC. Prod. iii. 216; F. Muell. in Journ. Linn. Soc. iii. 83, and Fragm. ii. 65; *Leptospermum umbellatum*, Gærtn. Fruct. i. 174. t. 35; *E. subulata*, A. Cunn.; Schau. in Walp. Rep. ii. 924.

Queensland. Bay of Inlets, *Banks and Solander;* Broad Sound, Shoalwater, and Keppel Bay, *R. Brown;* Percy island, *A. Cunningham;* Brisbane river, Moreton Bay, *A. Cunningham;* Port Denison, *Fitzalan, Dallachy;* Rockingham Bay, "Red Gum" and "Blue Gum," *Dallachy.*

N. S. Wales. Port Jackson, *Woolls*, and others; Clarence river, *Wilcox;* Macleay and Hastings rivers, *Beckler;* Richmond river, *C. Moore;* "Bastard Box," *Woolls.*

Victoria. Snowy River, Mitchell river, and Providence ponds, *F. Mueller.*

Var. *latifolia.* Leaves ovate to lanceolate. Flowers with a strong cimicine smell.—Shoalwater passage, *R. Brown.*

Var. *brachycorys.* Operculum more obtuse, 3 to 4 lines long.—With the other specimens from Brisbane Macleay and Hastings rivers, from Paramatta, and from the Blue Mountains. To this also probably belong the Mitchell river specimens, in which, however, the buds are not full grown.—*E. punctata*, DC. Prod. iii. 217, founded on Sieber's specimens, n. 623, which I have not seen, appears from his diagnoses and from the figure Mem. Myrt. t. 4, to be the same variety with a short operculum, also described in a state of young bud.

Var. *brevifolia.* Leaves mostly ovate or oblong, obtuse.—New England, in very exposed situations in the mountains, *C. Stuart.*

The common form with a long operculum, when in very young bud, requires some caution in distinguishing it from the rostrate varieties of *E. siderophloia* and *E. resinifera.* The venation of the leaf is then the best guide.

Subseries VI. Subexsertæ.—Peduncles axillary or lateral or also the upper ones more or less paniculate, terete or flattened, several-flowered. Calyx-tube broad at the orifice. Fruit turbinate, the orifice not contracted, the capsule level or slightly sunk, the valves often protruding when open.

90. **E. platyphylla,** *F. Muell. in Journ. Linn. Soc.* iii. 93. A handsome tree, with a light green foliage and smooth white deciduous bark (*F. Mueller*). Leaves ovate or rhomboid, acuminate or obtuse, the larger ones sometimes 8 to 10 in. long and broad and almost cordate, but mostly much smaller and sometimes passing into ovate-lanceolate, rather rigid, the veins prominent, diverging, and anastomosing. Peduncles axillary or lateral, very short and rather thick, each with 3 to 6 or rarely more flowers on short thick angular pedicels. Calyx-tube turbinate or nearly hemispherical, about 3 lines diameter, the margin prominent in the bud after the outer operculum has fallen. Operculum not thick, hemispherical, shorter than the calyx-tube. Stamens 3 to 4 lines long, all perfect, inflected in the bud; anthers oblong, with parallel distinct cells. Ovary flat-topped. Fruit obconical, 4 to 5 lines

diameter, not contracted at the orifice, the rim thick, convex and prominent, the capsule nearly on a level with it, and the valves shortly protruding.

N. Australia. Islands of the Gulf of Carpentaria, *R. Brown.*

Queensland. Shoalwater Bay, *R. Brown;* fertile pastures on the Burdekin, *F. Mueller;* Percy Island, *A. Cunningham;* Endeavour river, *W. Hill;* common about Rockhampton, *Dallachy;* Broad Sound, *Fitzroy;* Bowen river, *Bowman.*

E. populifolia, Hook. in Mitch. Trop. Austr. 204, from near Mount Owen, *Mitchell;* without flowers or fruits, but with remarkably-shaped galls on the branches, belongs more probably to this species than to *E. polyanthemos.*

E. bigalerita, F. Muell. in Journ. Linn. Soc. iii. 96, from the upper Roper river, appears to me to be the same species, with the outer operculum persisting till the bud has nearly attained its full size, whilst in the majority of specimens it falls off at a very early stage.

91. **E. alba,** *Reinw. in Blume, Bijdr.* 1101. A tall tree with a pale ash-coloured rough persistent bark (*F. Mueller*), the foliage of a pale glaucous hue. Leaves from ovate-oblong and 2 to 3 in. long, to ovate-lanceolate or broadly lanceolate, obtuse or scarcely acuminate and 5 to 6 in. long, with diverging veins and very much reticulate, the intramarginal vein very near the edge. Peduncles axillary, terete or nearly so, short, with few pedicellate flowers, not seen expanded. Buds small, ovoid, the operculum obtusely conical, as long as the calyx-tube. Fruit turbinate or obconical, about 3 lines diameter, the rim somewhat convex and rather broad, the capsule slightly depressed, the valves exserted.—Dcne. Herb. Tim. Descr. 126; *E. tectifica,* F. Muell. in Journ. Linn. Soc. iii. 92.

N. Australia, *Baudin's Expedition* (Herb. R. Brown, from Herb. Mus. Par. marked "Côte occidentale," but as in other plants from the same expedition probably in error); grassy valleys, Macarthur river, Gulf of Carpentaria, *F. Mueller.* The Timor specimens from the Herb. Mus. Par. in Herb. R. Brown are in the same state of fruit only as Baudin's Australian one, so also is a Timor specimen of Zippelius's, communicated by Miquel to the Hookerian Herbarium. The *E. moluccana,* Roxb. Fl. Ind. ii. 498, referred here by Miquel, Fl. Ind. Bat. i. part i. 398, must, from Roxburgh's short description, be very different. No specimens of it have been transmitted, and the tree is probably lost from the Calcutta Gardens. That was probably the best evidence as yet obtained of the genus existing in the Indian Archipelago beyond Timor, for *E. deglupta* is described by Blume, and *E. multiflora,* by A. Gray, from specimens without flowers or fruit, and the others are only taken up from Rumphius's very incomplete descriptions and figures of the trunk and foliage, also without flowers or fruit.

Mitchell's specimens, referred by Black in Journ. Linn. Soc. iii. 92, to *E. tectifica,* belong to *E. dealbata,* the leaves of which sometimes assume the form of those of *E. alba,* but with a different venation.

92. **E. Stuartiana,** *F. Muell.; Miq. in Ned. Kruidk. Arch.* iv. 131. A tree attaining a considerable elevation, the bark of the branches smooth and deciduous, that of the trunk rough and rigid and somewhat stringy (*F. Mueller, Oldfield*). Leaves from broadly ovate-lanceolate to narrow lanceolate, mostly 3 to 6 in. long, much narrowed at the base, usually equal or nearly so, but sometimes oblique, thick, the nerves rather regular and diverging but scarcely conspicuous. Peduncles axillary or lateral, terete or slightly angular, with about 4 to 8 flowers on rather short thick pedicels. Calyx-tube smooth, often shining, turbinate, about 2 lines diameter, the border usually prominent in the bud. Operculum conical, sometimes acuminate, from rather shorter to rather longer than the calyx-tube. Stamens dark-coloured, 2 to nearly 3 lines long, inflected in the bud; anthers ovate-oblong,

with parallel distinct cells. Ovary short, flat-topped. Fruit almost turbinate, usually about 3 lines but varying from 2 to 4 lines diameter, not contracted at the orifice, the rim not thick, slightly prominent, the capsule level with it or slightly sunk, the valves horizontal or protruding when open.—*E. acervula*, Hook. f. Fl. Tasm. i. 135, not of Sieb.; *E. Gunnii*, F. Muell. Fragm. ii. 62, not of Hook.; *E. persicifolia*, Miq. in Ned. Kruidk. Arch. iv. 137, not of Lodd.; *E. Baueriana*, Miq. l. c. not of Schauer; *E. falcifolia*, Miq. l. c. 136 (one specimen).

Queensland? A specimen with the Queensland woods of the Exhibition, 1862, *W. Hill*, appears to be this species but is in bud only.

N. S. Wales. Bathurst plains, *Fraser*.

Victoria. In plains and moist valleys, ascending the wooded moist mountains of the Australian Alps, extending to the western frontier, "White Gum," *F. Mueller*, also "Apple-tree" of the colonists from a label in Herb. F. Mueller.

Tasmania. Abundant in many parts of the colony, "Red Gum," *J. D. Hooker*.

S. Australia. From the Glenelg to Guichen Bay, *F. Mueller*.

This species is, as observed by F. Mueller, well distinguished from *E. piperita* (*E. acervula*, Sieb.) by the anthers; he unites it with *E. Gunnii*, but it appears to differ from that species in the more numerous, more pedicellate flowers, the shape of the fruit, etc. It is perhaps nearest to *E. viminalis*, differing, however, in foliage and in the shape of the fruit.

Var. *longifolia*. Leaves very long (4 to 8 in.) and acuminate, more or less falcate, but thick, with the veins scarcely conspicuous, the intramarginal one often near the edge. Umbels several-flowered. Operculum short.

N. S. Wales. "Yellow or Grey Gum and Bastard Box," *Woolls?* in Herb. F. Mueller; Twofold Bay, "Turpentine Gum" or "Hiccory," *Oldfield, F. Mueller*. In foliage and inflorescence this resembles in some measure *E. virgata*, but the buds, anthers, and fruit are quite different.

93? **E. patellaris,** *F. Muell. in Journ. Linn. Soc.* iii. 84. A tall tree with a rough furrowed persistent dull whitish bark (*F. Mueller*). Leaves lanceolate, falcate, acuminate, about 4 to 6 in. long, the veins rather numerous and regular, oblique, the intramarginal one rather distant from the edge. Perfect flowers unknown. Inflorescence perhaps compound. Calyx-tube (only seen in a diseased persistent bud) hard, hemispherical, about 5 lines diameter, the border prominent. Operculum much depressed, umbonate. Fruit pedicellate, broadly urceolate, about 5 lines diameter, the orifice dilated, the rim broad and flat, the valves protruding.

N. Australia. Dry banks of the Roper river, *F. Mueller*. Described from specimens far too imperfect to determine the affinities.

94. **E. rudis,** *Endl. in Hueg. Enum.* 49. A moderate-sized tree, the bark hard, rough, iron-grey, and persistent (*Oldfield*), corky (*Maxwell*). Leaves from ovate-lanceolate to narrow-lanceolate, often falcate, much acuminate, the longest exceeding 6 in., not very thick, the fine diverging veins not close nor yet very prominent, the intramarginal one not close to the edge. Peduncles axillary or lateral, terete or slightly flattened, each with about 4 to 8 flowers on pedicels from rather shorter to rather longer than the calyx-tube. Calyx-tube smooth, broadly turbinate, rather above 3 lines diameter. Operculum conical, as long as or rather longer than the calyx-tube. Stamens 3 to 4 lines long, inflected in the bud; anthers ovate or oblong with parallel distinct cells. Ovary shorter than the calyx, conical in the centre. Fruit broadly turbinate or almost hemispherical, 4 to 5 lines diameter, more or less

dilated at the orifice, the rim narrow, the capsule somewhat sunk but very convex or conical in the centre and the valves protruding when open.—Schau. in Pl. Preiss. i. 130.

W. Australia. Sandy woods, Swan River, *Preiss, n.* 252, *Drummond, n.* 58; Vasse river, "Flooded Gum," *Oldfield;* Gardner river and grassy flats near Salt River, "Swamp Gum," *Maxwell.*

I have not seen Huegel's specimens, but quote them on Schauer's authority, who has compared them. The fruiting specimens distributed by Preiss (not described by Schauer) belong to *E. patens*, which has much resemblance with *E. rudis* in foliage, but differs in inflorescence, flowers, and fruit.

95. **E. saligna,** *Sm. in Trans. Linn. Soc.* iii. 285. A tall tree with a smooth silver-grey shining bark, shedding in thin longitudinal strips (*Beckler*). Leaves from ovate-lanceolate to long-lanceolate, but usually narrow, acuminate, 4 to 6 in. long, with very numerous fine close transverse parallel veins, the intramarginal one close to the edge. Peduncles short, mostly flattened, each with 4 to 8 flowers. Calyx-tube narrow-turbinate, 2 to nearly 3 lines long, sessile or tapering into a short thick pedicel, the border of the calyx prominent in the bud and the orifice usually expanding after flowering. Operculum conical, about as long as the calyx-tube. Stamens 2 to 3 lines long, inflected in the bud, anthers ovate, with distinct parallel cells. Ovary conical in the centre. Fruit subglobose-truncate, not contracted at the orifice, the rim narrow, slightly raised above the calyx-border, the capsule somewhat or scarcely sunk, the valves more or less protruding.—DC. Prod. iii. 218.

N. S. Wales, *White;* Cox's river and Glendon, *Leichhardt;* Paramatta, "White Gum," *Woolls;* "Grey Gum," *Herb. F. Mueller,* without the collector's name; Richmond and Clarence rivers, *Beckler.*

96. **E. resinifera,** *Sm. in White. Voy.* 231, *in Trans. Linn. Soc.* iii. 284, *and Exot. Bot. t.* 84. A tall tree with a rough persistent bark on the trunk but more or less deciduous on the branches (*Woolls* and others). Leaves ovate-lanceolate to lanceolate, acuminate, straight or falcate, mostly 4 to 6 in. long, rather thick, with numerous fine close parallel and almost transverse veins, sometimes scarcely conspicuous, the intramarginal one close to the edge. Peduncles axillary or lateral, more or less flattened, each with about 6 to 8 or sometimes more flowers on pedicels usually short but sometimes longer than the calyx-tube. Calyx-tube broadly turbinate, 2½ to 3 or rarely 4 lines diameter. Operculum conical or acuminate, much longer than the calyx-tube and often broader at the base as in *E. tereticornis.* Stamens 4 to 6 lines long, raised above the calyx-border by the disk, inflected in the bud; anthers small, ovate, with parallel distinct cells. Ovary not much shorter than the calyx, conical in the centre. Fruit obconical, subglobose-truncate or almost hemispherical, not contracted at the orifice, the rim not broad, convex or prominent, the capsule somewhat sunk or nearly level with it, the valves protruding.—DC. Prod. iii. 216.

Queensland. Valleys of the Upper Brisbane (with a very long operculum), *F. Mueller;* Head of the Cape, *Bowman.*

N. S. Wales. Port Jackson, *R. Brown;* "Red Gum," *White;* Cumberland and Paramatta, "Red Mahogany," "Red Gum," "Grey Gum," "Leather-Jacket," and "Hickory," *Woolls.*

This species is allied in the fruit and foliage to *E. saligna*, differing chiefly in the pedicellate flowers and large operculum, and in the fruit to *E. Stuartiana*, from which it is readily distinguished by the venation of the leaves as well as by the operculum. When the operculum is long, the buds resemble those of *E. siderophloia*, var. *rostrata*, and of *E. tereticornis*, but the venation of the foliage and other characters are quite different. It varies much in the size of the flowers, the length of the pedicel, and in the operculum from under twice to four times the length of the calyx-tube. Smith's specimen is a garden one, with the operculum about twice the calyx-tube, but a native one in the Banksian herbarium, probably seen by Smith, has it three times the calyx-tube. Gærtner's figure and description of the fruit of *Metrosideros gummifera*, quoted by Smith as belonging to *E. resinifera*, and which has thus prevented the recognising the species, was taken from a specimen in the Banksian herbarium of *E. corymbosa*.

Var. *grandiflora*. Buds ovoid, about 4 lines diameter, the operculum broad and thick at the base, with a rather long beak or gradually tapering. Fruit about 4 to 6 lines diameter, with a raised rim and exserted valves.—Andr. Bot. Rep. t. 400; *E. hemilampra*, F. Muell. Herb.—Mauly Beach, "Forest Mahogany," *Woolls*; "Swamp Mahogany," *Caley*. Very near and possibly referable to *E. pellita*.

97. **E. pellita,** *F. Muell. Fragm.* iv. 159. A tree of 40 to 50 ft., with a rough dark grey bark (*Dallachy*). Leaves ovate-lanceolate or almost ovate, acuminate, nearly straight, 5 to 6 in. long or more, rigid, with numerous parallel almost transverse veins, the intramarginal one near the edge. Peduncles axillary or lateral, stout and much flattened, often 1 in. long, each with about 4 to 8 rather large flowers on thick angular pedicels often as long as the calyx-tube. Calyx-tube much broader and shorter than in *E. botryoides*, 5 to nearly 6 lines diameter and more or less angular. Operculum thick, hemispherical, broader than the calyx-tube, with a short obtuse beak. Stamens about ½ in. long, somewhat raised above the calyx-border by the disk, inflected in the bud; anthers ovate-oblong, with parallel distinct cells. Ovary very conical in the centre. Fruit subglobose-truncate or nearly hemispherical, 6 to 8 lines diameter, not contracted at the orifice, the rim raised above the calyx-border, slightly convex and rather broad, the capsule scarcely sunk, the valves much projecting.—*E. spectabilis*, F. Muell. Fragm. v. 45.

Queensland. Rockhampton, *Dallachy*. The species, as observed by F. Mueller, resembles *E. botryoides*, but differs in the larger especially broader flowers, in the conical ovary, and in the shape of the fruit. It is, however, very closely allied to *E. saligna* and *E. resinifera*, differing chiefly in the size of its leaves, flowers, and fruit, and should perhaps include the var. *grandiflora*, which I have referred to the latter.

Subseries VII. Inclusæ.—Umbels usually several flowered, axillary or lateral and solitary or several together in lateral clusters or very short panicles and then sometimes reduced to one or two flowers each, the peduncles terete or rarely flattened. Fruit more or less contracted at the orifice, the capsule sunk, the valves not protruding, except their points when acuminate by the split base of the style.

98. **E. Gunnii,** *Hook. f. in Hook. Lond. Journ.* iii. 499, *and Fl. Tasm.* i. 134. *t.* 27. A small often scrubby tree but attaining sometimes 30 ft., with a smooth bark (*J. D. Hooker*), the young foliage glaucous. Leaves from ovate-lanceolate or elliptical and obtuse to lanceolate-acute, under 3 in long, usually much narrowed at the base and rarely oblique, thick, with the veins not numerous and scarcely conspicuous. Peduncles axillary, very short, each with 3 rather large almost sessile flowers. Calyx-tube turbinate,

tapering at the base, 2 to nearly 3 lines long, and not so much in diameter. Operculum hemispherical and umbonate or conical, much shorter than the calyx-tube. Stamens 2 to nearly 3 lines long, inflected in the bud; anthers ovate, with parallel distinct cells. Ovary flat-topped or nearly so. Fruit pear-shaped, truncate, somewhat contracted at the orifice, 2 to nearly 4 lines diameter, the rim rather thin and scarcely protruding, the capsule more or less sunk, the points of the valves sometimes slightly protruding.—*E. ligustrina*, Miq. in Ned. Kruidk. Arch. iv. 134, but probably not of DC.

Victoria. Summit of the Baw-Baw mountains, *F. Mueller.*

Tasmania. Abundant in Alpine districts at an elevation of 3 to 4000 ft., often forming small forests, "Cider-tree," *J. D. Hooker.*

In some old small fruits the valves are more exserted, but the shape is always different from that of *E. Stuartiana*, and in well-formed fruits the capsule is distinctly sunk. In other respects the species is as nearly allied to *E. viminalis* as to *E. Stuartiana.*

99. **E. patens,** *Benth.* A tree attaining a great height, with a rough, half-fibrous, persistent bark (*Oldfield*), or a shrub of 6 to 10 ft., with a smooth bark (*Maxwell*). Leaves mostly falcate, from ovate-lanceolate and under 3 in. to lanceolate-acuminate and 4 to 6 in. long, not very thick, with fine diverging veins rather numerous, the intramarginal one more or less distant from the edge. Peduncles axillary or lateral or forming short panicles, short, terete or slightly angular, each with 3 to 6 flowers on short pedicels. Calyx-tube turbinate at the base, very broad and open above the ovary, about 3 lines diameter. Operculum hemispherical and umbonate or broadly conical, not so long as the calyx-tube. Stamens pale-coloured, about 3 lines long, inflected in the bud; anthers ovate, with contiguous cells parallel and distinct. Ovary flat-topped. Fruit globular-truncate, nearly ½ in. diameter, more or less contracted at the orifice, the rim narrow, the capsule sunk but not deep, flat-topped before opening, the valves not protruding.

W. Australia. Harvey river, "Black-butt," *Oldfield;* Tone river and granite rocks near Cape Arid, *Maxwell;* also *Drummond, 4th Coll. n.* 72; *Gilbert; J. S. Roe;* and in *Preiss's* collection in fruit distributed with the flowering specimens of *E. rudis*, but apparently not seen by Schauer.

100. **E. concolor,** *Schau. in Pl. Preiss.* i. 129. A tree of 30 to 40 ft., with a smooth bark (*Oldfield*), a small tree of 8 to 12 ft. (*Preiss*), with much of the aspect of *E. decipiens*, but larger and more rigid in all its parts. Leaves ovate-lanceolate to lanceolate-acuminate, often 4 to 5 in. long, thick and rigid, the fine diverging veins numerous and parallel but scarcely conspicuous, the intramarginal one nearer the edge than in *E. decipiens.* Peduncles short, axillary, broad and flat but thick, each with a head of 6 to 12 or more sessile flowers. Calyx-tube turbinate, thick and often angled, but otherwise smooth, about 3 lines long. Operculum conical or acuminate, rather longer than the calyx-tube. Stamens inflected; anthers globular, small, but not so small as in *E. decipiens*, with distinct parallel cells. Ovary conical or convex in the centre. Fruit globose-truncate, about 4 lines diameter, contracted at the orifice, the rim broad, flat or slightly convex, the capsule sunk, but the points of the valves usually protruding.

W. Australia. Doubtful-Island Bay and shady ravines, Point Irwin, *Oldfield;* near Freemantle, *Preiss, n.* 225; also *Drummond, 4th Coll. n.* 77.

Var. With the calyx tapering into a very short pedicel as in *E. goniantha,* but smooth as in *E. concolor.*—Doubtful Island, Peninsula, and Cape Arid, *Maxwell.*

101. **E. goniantha,** *Turcz. in Bull. Mosc.* 1847, i. 163. Leaves ovate-lanceolate or lanceolate-acuminate, mostly falcate, rarely under 3 in. and sometimes above 4 in. long, thick and rigid, the very fine rather oblique veins numerous and parallel but scarcely conspicuous, the intramarginal one close to or very near the edge. Peduncles axillary or lateral, short, rather thick and flattened, mostly recurved, each with 4 to 8 flowers on short thick angular pedicels. Calyx-tube very broadly turbinate, thick and very prominently ribbed, 3 to 4 lines diameter. Operculum strongly ribbed, nearly hemispherical at the base, with a thick obtuse beak as long as or rather longer than the calyx-tube. Stamens 4 to 5 lines long, inflected in the bud; anthers small, ovate, with parallel distinct cells. Fruit depressed-globular or subglobular, truncate, hard, more or less ribbed, or sometimes almost smooth, 4 to 5 lines diameter, somewhat contracted at the orifice, the rim rather broad and nearly flat, the capsule somewhat sunk, but the valves occasionally protruding.

W. Australia. King George's Sound or to the eastward, *Collie; Baxter; Drummond,* 3*rd Coll. n.* 71; Franklin river, *Maxwell* (in fruit only with rather broad leaves).

102. **E. falcata,** *Turcz. in Bull. Mosc.* 1847, i. 163. A shrub of 10 to 12 ft. (*Maxwell*). Leaves lanceolate, acuminate, often falcate, mostly under 4 in. long, thick and smooth, the very fine oblique veins scarcely visible. Peduncles axillary or lateral, terete or slightly angular, each with about 6 to 12 flowers on slender pedicels of 3 to 4 lines. Calyx-tube short, depressed, about 2 lines diameter, thick, and more or less distinctly furrowed, but not so much so as in *E. goniantha.* Operculum conical, acuminate, fully twice as long as and much narrower than the calyx-tube. Stamens 2 to 3 lines long, or rather more, inflected in the bud; anthers ovate, with parallel distinct cells. Fruit depressed-globular, 3 to 4 lines diameter, much contracted at the orifice, the rim narrow and flat, but the disk within the staminal margin forming a protruding ring over the capsule, which is sunk, but the long points of the valves, formed by the split base of the style, usually protrude.

W. Australia, *Drummond,* 3*rd Coll. n.* 70; plains to the north and south of Stirling range, *Maxwell.*

103. **E. oleosa,** *F. Muell. Fragm.* ii. 56 (partly). A shrub or small tree, the bark of the trunk rough and persistent, that of the branches smooth (*F. Mueller*). Leaves mostly lanceolate, obtuse or acuminate, under 4 in. long, thick and smooth, the oblique and rather numerous veins scarcely conspicuous. Peduncles axillary or lateral, terete or slightly angular, each with about 4 to 8 more or less pedicellate flowers. Calyx-tube obovoid, more or less contracted at the base, and sometimes at the top, 2 to 2½ lines long. Operculum obtusely conical or shortly acuminate, usually exceeding the calyx-tube, and sometimes much longer and not very thick. Stamens 2 to 3 lines long, inflected in the bud, but without the acute angle of *E. uncinata;* anthers small, ovate, with parallel distinct cells. Ovary short, convex or conical in the centre. Fruit ovoid or globose, truncate, contracted at the orifice, about 3 lines long, the rim flat or concave, the capsule sunk, but the

slender points of the valves formed by the split base of the style often protruding.—*E. socialis*, F.Muell.; Miq. in Ned. Kruidk. Arch. iv. 132; *E. turbinata*, Behr. and Muell.; Miq. in Ned. Kruidk. Arch. iv. 137.

N. S. Wales. Mallee scrub of the Murray desert, *Beckler.*

Victoria. Murray desert, *F. Mueller, Dallachy.*

S. Australia. Port Lincoln, *Wilhelmi.*

W. Australia. Gravelly places near Moir's Inlet, *Maxwell.*—These specimens, as well as a few of those from the Murray desert, are distinguished by the long beak to the operculum.

The foliage of the species is that of *E. dumosa*, but it is well distinguished by the longer pedicels, the shape of the calyx, the thinner operculum, and the shape of the fruit.

104. **E. decurva,** *F. Muell. Fragm.* iii. 130. A large shrub of 10 to 12 ft., or a small tree of 10 to 30 ft., with a smooth bark (*Oldfield, Maxwell*). Leaves lanceolate, usually narrow, rarely ovate-lanceolate, acuminate, rarely exceeding 4 in. and often under 3 in. long, thick or rather thin, the veins diverging, but not close and scarcely visible, the intramarginal one more or less distant from the edge. Peduncles axillary or lateral, terete or somewhat flattened, each bearing an umbel of 3 to 7 flowers usually recurved and on rather long pedicels, but sometimes erect. Calyx-tube ovoid or almost cylindrical, 2 to 2½ lines long and nearly 2 lines diameter, abruptly contracted or obtuse at the base, not ribbed. Operculum hemispherical and broad at the base, with a central beak sometimes very short, sometimes above 2 lines long. Stamens about 3 lines long, the filaments slender and acutely inflected as in *E. uncinata* and *E. corynocalyx;* anthers very small, globular, with distinct parallel cells. Ovary short, convex or conical in the centre. Fruit ovoid, contracted at the orifice, 3 to 4 lines long and rather less in diameter, the rim narrow, the capsule deeply sunk.

W. Australia. Low flats and rich soil to the east of Kojonerup from the Stirling Range to East Mount Barren, *Maxwell*, also *Drummond, 5th Coll. n.* 186, all with narrow not very thick leaves; from Kalgan river and King George's Sound to the eastward, *Harvey, Oldfield, Maxwell,* with broader and thicker leaves; Vasse river, *Gilbert, n.* 266, with thick but narrow leaves.

A specimen in fruit only from Murchison river, *Oldfield,* looks like the same species. The *E. doratoxylon,* which in many respects resembles this species, differs in the leaves mostly opposite as well as in the stamens. The *E. decurva* itself is very closely allied to *E. oleosa,* but the shape of the calyx and fruit and the arrangement of the stamens are somewhat different. Both species, from the smallness of their anthers, come near to the *Micrantheræ.*

105. **E. doratoxylon,** *F. Muell. Fragm.* ii. 55. A large shrub. Leaves all opposite or nearly so, lanceolate, acuminate, nearly straight, under 3 in. long, the veins fine, oblique, and rather numerous, but scarcely conspicuous, the intramarginal one at a distance from the edge. Peduncles axillary or lateral, terete or nearly so, recurved, each bearing about 4 to 7 flowers on rather slender pedicels. Calyx-tube ovoid or almost cylindrical, about 2 lines long. Operculum hemispherical or shortly conical, with a rather long beak. Stamens 2 to 3 lines long, inflected in the bud, but not acutely so as in *E. decurva;* anthers ovate-oblong, with parallel distinct cells. Fruit ovoid, much contracted at the orifice, about 3 lines long and rather less in diameter, the rim narrow, the capsule deeply sunk.

W. Australia. Lucky Bay, *R. Brown;* Sullinup ranges, "Spearwood," and Russell

Range, *Maxwell; Baxter;* also *Drummond, 3rd Coll. n.* 69, *4th Coll. n.* 97.—Allied in many respects, especially in the inflorescence and shape of the flowers, to *E. decurva;* this species is readily distinguished by the leaves mostly opposite, and by the stamens.

106. **E. aspera,** *F. Muell. in Journ. Linn. Soc.* iii. 95. A small tree, with a smooth ashy-white bark (*F. Mueller*), the branchlets and often the leaf-veins scabrous or hispid, the foliage often glaucous. Leaves sessile, opposite, cordate, ovate or oblong, obtuse, mostly under 2 in. long. Peduncles axillary or lateral, very short, each bearing 2 to 6 flowers, on pedicels either very short or longer than the calyx. Calyx-tube short and broad, 2 to nearly 3 lines diameter. Operculum hemispherical, obtuse, shorter than the calyx-tube. Stamens 2 to 3 lines long, inflected in the bud; anthers oval-oblong, with parallel distinct cells. Fruit ovoid-truncate, slightly contracted or straight at the orifice, 3 to 4 lines long, the rim thin, the capsule deeply sunk.

N. Australia. Sandstone table-land, upper Victoria river, *F. Mueller.*—The specimens are not in good state, but the species is evidently different from any others known to me.

107. **E. grandifolia,** *R. Br. Herb.* A small tree, with the outer bark brown and deciduous, the inner whitish and very smooth (*R. Brown*). Leaves opposite or nearly so, petiolate, from ovate to ovate-lanceolate, 4 to 6 in. long in the specimens, but probably often larger, rigid, with rather fine diverging veins, the intramarginal one remote from the edge. Flowers rather large, on pedicels of ½ to ¾ in., 3 to 10 together, rather clustered than umbellate on a very short lateral peduncle, reduced sometimes to a tubercle (probably the inflorescence consists of several umbels reduced to 1 or 2 flowers each). Calyx-tube very short, broad, and open, 4 to nearly 5 lines diameter. Operculum convex or almost hemispherical, obtuse or umbonate, much shorter than the calyx-tube. Stamens 4 to 5 lines long or rather more, inflected in the bud; anthers oblong, with parallel distinct cells. Ovary flat-topped. Fruit unknown.

N. Australia. Islands of the Gulf of Carpentaria, *R. Brown* (*Herb. R. Brown*).

108. **E. clavigera,** *A. Cunn. in Walp. Rep.* ii. 926. A large shrub or small tree (*R. Brown*), with an ash-coloured bark (*F. Mueller*). Leaves from opposite, sessile or nearly so, and broadly ovate-cordate or almost orbicular, to alternate and broadly ovate or ovate-lanceolate, rarely above 4 in. long, rather rigid, the veins prominent, diverging or almost transverse, but not close. Peduncles short, two or more together on a short leafless branch forming lateral clusters or very short panicles, each peduncle bearing an umbel of several rather small flowers on slender pedicels often ½ in. long. Calyx-tube turbinate, about 2 lines long and as much in diameter. Operculum very flat or convex, rarely almost hemispherical but much shorter than the calyx-tube. Stamens about 3 lines long, inflected in the bud; anthers ovate or oblong, with parallel distinct cells. Ovary flat-topped. Fruit from nearly globular to ovoid-oblong, 4 to 5 lines long, more or less contracted at the orifice, the rim thin, the capsule deeply sunk.—*E. polysciadia,* F. Muell. in Journ. Linn. Soc. iii. 98.

N. Australia. Careening Bay, N.W. coast, *A. Cunningham*; Islands of the Gulf of

Carpentaria, *R. Brown;* arid rocky hills near Macadam range, *F. Mueller;* Albert river, *Henne.*

109. **E. tesselaris,** *F. Muell. in Journ. Linn. Soc.* iii. 88. A middle-sized or large tree, the bark dark-brown, smooth and deciduous, the inner whitish and very smooth (*R. Brown*), the bark persistent on the trunk, dull ash-coloured, marked with longitudinal and transverse furrows forming separable pieces (*F. Mueller*), casts its bark in small angular pieces (*Mitchell*). Leaves lanceolate to almost linear, straight or falcate, 3 to 6 in. long, with numerous fine parallel diverging or almost transverse veins and more or less reticulate, the intramarginal vein close to the edge. Peduncles very short, usually several together in lateral clusters or very short panicles, often so reduced as to appear like a single compact irregular umbel, each peduncle with 3 to 6 (or when the inflorescence is compact 1 or 2) flowers on short or slender pedicels. Calyx-tube short, much widened above the ovary, 2 to 2½ or rarely nearly 3 lines diameter. Operculum very short and only slightly convex. Stamens 2 to 3 lines long, inflected in the bud; anthers ovate-oblong, with parallel distinct cells. Ovary flat-topped. Fruit ovoid or oblong, 3 to 4 lines long, slightly contracted at the orifice, the rim thin, the capsule deeply sunk.—*E. viminalis,* Hook. in Mitch. Trop. Austr. 157, not of Labill.; *E. Hookeri,* F. Muell. in Journ. Linn. Soc. iii. 90.

N. Australia. Careening and Vansittart's bays, N.W. coast, *A. Cunningham;* islands of the Gulf of Carpentaria, *R. Brown;* S.E. coast of the Gulf of Carpentaria, *F. Mueller.*

Queensland, *Bowman;* Fitzroy Downs, *Mitchell;* Port Denison, *Fitzalan.*

Var. *Dallachiana.* Veins of the leaves more oblique, the intramarginal one not so close to the edge, the cluster of umbels so dense as to be reduced almost to a sessile head.—Queensland, *Bowman;* Rockhampton, *Dallachy.*

110. **E. phœnicea,** *F. Muell. in Journ. Linn. Soc.* iii. 91. A middle-sized or small tree, the bark persistent or tardily falling off from the upper branches, and readily separable in flakes (*F. Mueller*). Leaves lanceolate, 4 to 6 in. long or even more, with fine diverging veins, numerous but somewhat reticulate, the intramarginal one close to the edge. Peduncles lateral, terete or nearly so, bearing each a dense umbel of numerous large flowers remarkable for their long narrow shape. Pedicels 2 to 3 lines long. Calyx-tube 5 to 6 lines long, obscurely ribbed, about 3 lines diameter at the orifice, and tapering downwards. Operculum hemispherical or conical, shorter than broad and much shorter than the calyx-tube. Stamens about ½ in. long, orange or scarlet, much inflected in the bud; anthers ovate, with parallel distinct cells. Ovary in the flower examined 2-celled. Fruit oblong, ¾ to 1 in. long, crowned by a narrow neck of about 2 or 3 lines, with a thin rim, the capsule sunk to the base of the neck.

N. Australia. Sandstone table-land on the Victoria and Upper Roper rivers, *F. Mueller.*

111. **E. diversicolor,** *F. Muell. Fragm.* iii. 131. A tree attaining 80 to 100 ft., the trunk decorticating by hard layers of ½ to ⅔ in. thick, the limbs and branches by chartaceous laminæ (*Oldfield*). Leaves ovate-lanceolate or lanceolate, acuminate, often falcate, 3 to 6 in. long, rather thick, with numerous fine very diverging veins, often scarcely conspicuous, the intramar-

ginal one at some distance from the edge, dark above, pale underneath when fresh (*Oldfield*). Peduncles axillary or lateral, terete or scarcely angular, each with 3 to 6 rather large flowers, not seen however fully expanded. Calyx-tube turbinate when in bud, about 3 lines long, tapering into a pedicel nearly or quite as long. Operculum hemispherical or obtusely conical, rather shorter than the calyx-tube. Stamens inflected in the bud; anthers ovate with parallel distinct cells. Ovary conical in the centre. Fruit ovoid-truncate, about ½ in. long, 4 to 5 lines diameter, contracted at the orifice or almost urceolate, the rim rather thick, the capsule deeply sunk with a conical top, yet the valves much shorter than the border of the fruit.

W. Australia. King George's Sound, *R. Brown;* on small elevations in swamps near rivers beyond the reach of the water, Blackwood and Hay rivers, Wilson's Inlet and Porongerup ranges, "Blue Gum," *Oldfield.*

112? **E. loxophleba,** *Benth.* A tree from 10 to 30 ft. high, with a rough ash-grey fibrous bark (*Oldfield*), 40 to 45 ft., the bark separable in layers (*Preiss*). Leaves lanceolate, acuminate, narrow and often 4 to 5 in. long or the lower ones shorter and broader, all rather rigid with very oblique rather distant and prominent veins, the intramarginal one distant from the edge. Peduncles axillary or lateral, terete or slightly flattened, each with a dense umbel of 6 to 12 flowers. Calyx-tube obconical, 2 to 2½ or rarely nearly 3 lines long, tapering into a short pedicel. Operculum hemispherical or obtusely conical, shorter than the calyx-tube. Stamens scarcely exceeding 2 lines, inflected in the bud, the filaments usually dark-coloured in the dry specimens; anthers small, with parallel distinct cells. Fruit narrow-obovoid, truncate, straight or slightly contracted at the orifice, rarely above 3 lines long and 2 lines diameter, the rim narrow, the capsule deeply sunk. —*E. amygdalina*, Schau. in Pl. Preiss. i. 130 (from the description given), not of Labill.; *E. fruticetorum*, F. Muell. Fragm. ii. 57 (as to the W. Australian specimens).

W. Australia. Swan River and Darling range, *Collie; Drummond, 2nd Coll. n.* 82; York district, *Preiss, n.* 246 (and 248 ?); Murchison river and Champion Bay, "York Gum," *Oldfield.*

The "Yandee," a tree of 40 to 45 ft., with a nearly black persistent furrowed bark consisting of strap-like pieces, from the Murchison river, *Oldfield,* appears to be otherwise precisely the same.

Var. *fruticosa.* A shrub branching from the ground, the leaves rather broader, the flowers rather larger, the peduncles more flattened.—Murchison river, *Oldfield;* Salt river, *Maxwell.*

113. **E. fœcunda,** *Schau. in Pl. Preiss.* i. 130. A tall shrub with a dark smooth bark (*Oldfield*). Leaves lanceolate, acuminate, rarely exceeding 3 in., thick, with fine veins scarcely conspicuous and much more numerous and less oblique than in *E. loxophleba*, the intramarginal one very near the edge. Peduncles axillary or lateral, rather short, terete or slightly flattened, each with a dense umbel of 4 to 8 flowers. Calyx-tube ovoid-turbinate, 2 to 2½ lines long, obtuse at the base or shortly tapering into the short pedicel. Operculum hemispherical, much shorter than the calyx-tube. Stamens 2 to 3 lines long, inflected in the bud; anthers ovate with parallel distinct cells. Ovary flat-topped. Fruit ovoid oblong or almost cylindrical, slightly contracted at the orifice, about 2 lines diameter and varying in

length from under 3 to about 4 lines, the rim thin, the capsule deeply sunk, but sometimes the base of the style splits into long points to the valves protruding beyond the border of the fruit.

S. Australia? Specimens in young bud and in fruit from the S. coast, *R. Brown*, appear to belong to this species.

W. Australia. Swan River, *Drummond, 2nd Coll. n.* 87; limestone hills near Freemantle, *Preiss, n.* 231; Yenert, *Gilbert, n.* 263; Port Gregory, Murchison and South Hutt rivers, *Oldfield;* Sharks' Bay and Dirk Hartog's Island, *Milne*, also in the collection of Baudin's Expedition.—Different as the long and the short fruits appear, there are numerous intermediate forms, and the specimens do not otherwise differ.

114. **E. redunca,** *Schau. in Pl. Preiss.* i. 127. In the original form, a shrub or small tree with a smooth white bark (*Oldfield, Maxwell*). Leaves ovate-lanceolate or lanceolate-acuminate, under 3 in. long, thick, with fine oblique not close veins, often scarcely visible, the intramarginal one at a distance from the edge. Peduncles axillary or lateral, flattened or rarely terete, each with a dense umbel of 6 to 12 flowers. Calyx-tube narrow, 2½ to nearly 3 lines long, tapering into a short thick or flattened pedicel. Operculum conical, acuminate, at least twice as long as the calyx-tube. Stamens 3 to 4 lines long, more or less inflected in the bud; anthers oblong, with parallel distinct cells. Ovary convex or shortly conical in the centre. Fruit obovoid or obovoid-oblong, 4 to 5 lines long and about 3 diameter, contracted at the orifice, the rim narrow, the capsule considerably sunk, the points of the valves rarely protruding.

W. Australia. King George's Sound and adjoining districts to Swan River and eastward to Cape Riche, *Preiss, n.* 232, 234, 245, 247; *Drummond, 2nd Coll. n.* 81 *and* 84; *Gilbert, n.* 271, and others.

Var. *melanophloia.* Leaves larger, more prominently veined.—Murchison and South Hutt rivers, a small tree with a smooth black bark, *Oldfield.*

Var. *angustifolia.* Leaves linear or linear-lanceolate.—*E. xanthonema*, Turcz. in Bull. Mosc. 1847, i. 163; W. Australia, *Drummond, 3rd Coll. n.* 67, *5th Coll. n.* 187; S. side of Stirling ranges and eastward to Phillips ranges, *Maxwell.*

Var. *elata.* A large tree, the trunk generally swelling out suddenly near the ground, forming a kind of pedestal, the bark smooth, white, decorticating in long chartaceous pieces (*Oldfield*). Operculum rather shorter and the fruit less contracted at the orifice, but not differing otherwise from the normal form.—Kalgan river, "White Gum," *Oldfield.*

The species, especially in the narrow-leaved forms, has much resemblance on the one hand to *E. fœcunda*, on the other to *E. spathulata*, but is readily distinguished from the former by the operculum, from the latter by the stamens and the acuminate operculum.

SUBSERIES VIII. CORYMBOSÆ.—Flowers usually large, the umbels (or very rarely heads) all in a terminal corymbose panicle or rarely a few of the lower ones axillary. Fruit often large, more or less urceolate, the capsule deeply sunk. Seeds usually large, flat, with acute edges, often more or less expanded in a variously-shaped wing.

115. **E. perfoliata,** *R. Brown, Herb.* A large shrub of 10 ft. or more (*A. Cunningham*). Leaves opposite, connate, 6 to 8 in. long and 3 to 4 in. broad, very obtuse, glaucous with numerous parallel transverse veins. Flowers large, sessile in heads of 4 to 6, on terete peduncles forming a corymbose terminal panicle. Calyx-tube thick, broadly turbinate, smooth or nearly so, 7 to 8 lines long and as much in diameter. Operculum not seen. Stamens

above ½ in. long, inflected in the bud; anthers small, ovate-oblong, with parallel distinct cells. Fruit urceolate, 1½ in. long and above 1 in. diameter, smooth, the rim concave, the capsule sunk. Seeds not seen.

N. Australia. Barren hills, Rae's River, N.W. Coast, *A. Cunningham;* N.W. Coast, *Bynoe.*

116. **E. ferruginea,** *Schau. in Walp. Rep.* ii. 926. A moderate-sized tree, with a rough persistent dark grey bark (*F. Mueller*), the young branches and often the foliage more or less rusty-pubescent, or the branches hispid with a few stiff hairs or bristles, but sometimes quite glabrous. Leaves large, often 4 to 5 in. diameter, sessile, opposite, cordate orbicular or oblong, mostly obtuse and sometimes undulate. Flowers rather large, the umbels in a dense terminal corymbose panicle, or in one specimen a single umbel axillary. Peduncles and pedicels short, terete. Calyx-tube very broadly campanulate, 6 to 8 lines diameter. Operculum broadly conical, shorter than the calyx-tube. Fruit ovoid, when perfect about 1 in. long and ¾ in. diameter, contracted towards the orifice, the rim narrow, the capsule deeply sunk. Seeds winged.—F. Muell. in Journ. Linn. Soc. iii. 95; *E. confertiflora,* F. Muell. l. c. 96.

N. Australia. Copeland island, N.W. coast, *A. Cunningham;* Victoria river and Arnhem's Land, *F. Mueller.*

117. **E. setosa,** *Schau. in Walp. Rep.* ii. 926. A small or moderate-sized tree, with a smooth ash-grey bark (*R. Brown*), the branchlets and inflorescence more or less hispid with rust-coloured bristles. Leaves opposite, sessile, cordate orbicular and obtuse or ovate and almost acute, rarely above 2 in. long. Umbels shortly pedunculate, several-flowered, forming short, terminal, rather loose corymbose panicles. Pedicels often longer than the calyx. Calyx-tube obovoid, often slightly 8-ribbed, about 3 lines long, more or less covered with bristles. Operculum conical, shorter than the calyx-tube, often bearing a few bristles. Anthers ovate, parallel-celled. Ovary flat-topped, the style not dilated. Fruit urceolate-globular, much contracted at the top, hard and woody, ½ to ¾ in. diameter, the rim narrow, the capsule sunk. Perfect seeds large, broadly winged.—F. Muell. Fragm. iii. 132.

N. Australia. Islands of the Gulf of Carpentaria, *R. Brown;* Sweers Island, *Henne.*
Queensland. Mount Elliott, *Fitzalan, Dallachy,* with fewer setæ on the buds.

118? **E. melissiodora,** *Lindl. in Mitch. Trop. Austr.* 235. A shrub, exhaling a powerful odour of balm, and covered with a rusty resinous pubescence, short and scabrous on the foliage, almost bristly on the branchlets. Leaves oblong-lanceolate, obtuse, more or less peltately inserted on the petiole above their base, the veins transverse but not close. Flowers and fruit unknown.

Queensland. Sandstone rocks, Balmy Creek, *Mitchell.* Possibly a barren state of *E. citriodora,* or some allied species, in which the leaves of the flowering branches are not peltate.

119? **E. peltata,** *Benth.* A tree with a dark shining brittle and flaky but persistent bark (*F. Mueller*). Leaves from nearly orbicular to oblong-ovate, obtuse, rather large, peltately inserted on the petiole above their base, rusty-scabrous or glabrous and somewhat glaucous, with diverging but not close veins. Flowers rather large, nearly sessile in the umbels, which are arranged in oblong (or corymbose?) terminal panicles, but not seen expanded. Calyx-tube obconical in the bud, about 3 lines long, smooth and shining. Operculum much shorter, obtusely conical or hemispherical. Anthers ovate-oblong, with parallel cells. Fruit urceolate-globose, about 4 lines diameter, contracted above the deeply sunk capsule, the rim thin. Seeds (which I have not seen) smooth and not winged according to F. Mueller.—*E. melissiodora*, F. Muell. in Linn. Journ. Soc. iii. 95, not of Lindl.

Queensland. Porphyritic mountains, Newcastle range, *F. Mueller.*—Possibly a variety or state of some species allied to *E. latifolia* without the peltate leaves. The specimens are very imperfect.

120. **E. latifolia,** *F. Muell. in Journ. Linn. Soc.* iii. 94. A small or middle-sized tree, with a smooth ash-grey bark, tardily separating from the inner brownish bark also smooth (*F. Mueller*). Leaves alternate or here and there almost opposite, petiolate, ovate, obtuse, with transverse parallel veins, rather more prominent and not so close as in the allied narrow-leaved species. Flowers rather large, 4 to 6 in each umbel, in a large terminal corymbose panicle. Peduncles terete; pedicels terete, shorter than the calyx-tube. Calyx-tube broadly turbinate, 4 to 5 lines diameter, rather thick. Operculum very short, slightly convex. Anthers ovate-oblong, with parallel distinct cells. Fruits globose-truncate or urceolate-globose with a very short neck, smooth and not ribbed, 3 to 4 lines diameter, the rim thin; the capsule deeply sunk. Seeds winged.

N. Australia. Islands of the Gulf of Carpentaria, *R. Brown;* upper part of the Roper river, *F. Mueller.*

121. **E. ptychocarpa,** *F. Muell. in Journ. Linn. Soc.* iii. 90. A middle-sized or tall tree, with a persistent bark intermediate between that of the Stringy-barks and the Box-trees (*F. Mueller*). Leaves large, from broadly ovate to ovate-lanceolate, sometimes above a foot long, straight or falcate, with numerous fine closely parallel almost transverse veins. Flowers large, in umbels forming a terminal panicle, peduncles terete, $\frac{1}{2}$ to 2 in. long, pedicels sometimes very short, sometimes 1 to 2 in. long. Calyx-tube turbinate, $\frac{1}{2}$ to $\frac{3}{4}$ in. long, hard, with about 8 longitudinal ribs. Operculum not seen. Stamens above $\frac{1}{2}$ in. long; filaments rigid, inflected in the bud; anthers small, ovate, with distinct parallel cells. Fruits ovoid or slightly urceolate, very thick and hard, 1 to 2 in. long, with about 8 prominent ribs, the rim thick, the capsule sunk. Seeds winged.

N. Australia. Dry river-beds and rocky streams at the sources of the Wentworth, Wickham, and Limmen Bight rivers, *F. Mueller;* Melville Island, *Fraser;* Port Essington, *Gilbert.*—The fruit somewhat resembles that of *E. miniata*, but the venation of the leaves and the inflorescence are quite different.

122. **E. calophylla,** *R. Br. in Journ. Geogr. Soc.* 1831, 20 (*name only*); *Schau. in Pl. Preiss.* i. 131. A beautiful tree, with a more dense foliage than usual in the genus, the rough corky bark coming off in irregular masses (*Oldfield*). Leaves ovate ovate-lanceolate or lanceolate, obtuse or mucronate-acute, rather rigid, with very numerous tranverse parallel veins, the intramarginal one scarcely distant from the edge. Umbels loose, with

rather large flowers, in a terminal corymbose panicle, with one or two sometimes in the upper axils. Peduncles flattened or nearly terete, pedicels longer than the calyx-tube. Calyx-tube turbinate and often ribbed on the adnate part, the free part much dilated, often ½ in. diameter. Operculum hemispherical, obtuse or umbonate, shorter than the calyx-tube and continuous with it till the flower expands. Stamens ½ to ¾ in. long; anthers ovate, with parallel distinct cells opening longitudinally. Ovary flat or slightly convex on the top. Fruit when perfect ovoid-urceolate, 2 in. long and above 1 in. diameter, very thick and hard, with a thick neck contracted at the orifice, but sometimes the fruit is smaller, the neck less distinct and less contracted. Capsule deeply sunk. Seeds large, ovate, black, flat or with a raised angle on one face, the edges acute but scarcely winged, the hilum large on the inner face.—F. Muell. Fragm. ii. 35; *E. splachnicarpa*, Hook. Bot. Mag. t. 4036.

W. Australia. Common about King George's Sound, *R. Brown*, *Fraser*, *Oldfield*, and others; and thence to Swan River, *Fraser*, *Drummond*, *n.* 150; *Preiss*, *n.* 250, and others; rare towards Port Gregory, *Oldfield*; "Red Gum," *Oldfield*.

123. **E. ficifolia,** *F. Muell. Fragm.* ii. 85. Only known from imperfect specimens in fruit, which differ in no respect from *E. calophylla*, except that the seeds are of a pale colour and the testa expanded at one end, or round one side into a broad variously-shaped wing. Further specimens may prove these differences not to be constant.

W. Australia. Broke's Inlet, "Black-butt," *Maxwell*. From the Hay, Gordon, and Tone rivers in the same neighbourhood are flowering specimens undistinguishable from *E. calophylla*, which may possibly belong to this species.

124. **E. corymbosa,** *Sm. Bot. Nov. Holl.* 43, *and in Trans. Linn. Soc.* iii. 287. Usually a small or middle-sized tree, but sometimes attaining a great height, with a persistent furrowed bark (*F. Mueller*). Leaves ovate-lanceolate or lanceolate, acuminate, about 3 to 6 in. long, with numerous fine transverse parallel veins, often scarcely visible. Umbels loose, several-flowered, mostly in a terminal corymbose panicle, the peduncles slightly compressed or angular. Flowers rather large, on pedicels of 2 to 4 lines. Calyx-tube, when open, broadly turbinate, 3 to 4 lines diameter, often dilated at the margin. Operculum short, hemispherical, umbonate or shortly acuminate. Stamens attaining 5 or 6 lines; anthers very small but ovate, with distinct parallel cells opening longitudinally. Ovary short, flat-topped. Fruit more or less urceolate, ½ to ¾ in. long, usually contracted above the capsule and often expanded at the orifice, the rim narrow, the capsule sunk. Seeds large, ovate, more or less bordered by a wing, usually narrow.—DC. Prod. iii. 220; F. Muell. Fragm. ii. 46; *Metrosideros gummifera*, Soland. in Gærtn. Fruct. i. 170. t. 34. f. 1.

Queensland. E. coast, *A. Cunningham*; Rockhampton, *Dallachy*; dry ridges, Brisbane river, Moreton Bay, *F. Mueller*, *W. Hill*, *Fitzalan*.

N. S. Wales. Port Jackson, "Blood tree," *R. Brown*; open forests, Clarence and Richmond rivers, *C. Moore*; Paramatta, "Blood-wood," *Woolls*; Twofold Bay, *F. Mueller*.

It is possible that some of the specimens here referred may belong to *E. citriodora*, or the northern ones to *E. terminalis*, both of which it is often very difficult to distinguish from *E corymbosa*. The figure usually quoted of *E. corymbosa*, Cav. Ic. iv. t. 340, is a very indifferent one, and looks much more like *E. paniculata*.

125. **E. citriodora,** *Hook. in Mitch. Trop. Austr.* 235. A tree with a smooth bark (*F. Mueller*), the foliage emitting a strong odour of citron when rubbed (*Mitchell*), evidently very closely allied to *E. corymbosa.* In the imperfect state of our specimens (in leaf only, with loose fruits or in young bud), it can only be distinguished from that species by the veins of the leaves rather more distinct, the pedicels shorter, the fruit scarcely so large, contracted at the orifice, but without so distinct a neck, and by the seeds almost equally large, but very obscurely or not at all winged.—F. Muell. Fragm. ii. 47.

Queensland. Balmy Creek, *Mitchell;* Wide Bay, *C. Moore.* It is possible also that some of the Brisbane specimens may be referable rather to this than to *E. corymbosa.*

Woolls's "Spotted Gum," from Paramatta, is very much like *E. citriodora.*

126. **E. terminalis,** *F. Muell. in Journ. Linn. Soc.* iii. 89. A tree, very closely allied to *E. corymbosa*, and often scarcely to be distinguished from it in the dried specimens. It is generally of a paler or more glaucous colour, the leaves usually narrower with less conspicuous veins, the operculum very obtuse, hemispherical and not showing the junction with the calyx-tube till just as it is detached, the fruit narrower, more oblong and less urceolate, that is, contracted at the orifice without so distinct a neck; it varies in size from about 7 lines to nearly 1 in. long. Seeds with a rather long wing. —*E. polycarpa*, F. Muell. in Journ. Linn. Soc. iii. 88.

N. Australia. Arnhem's Land and Gulf of Carpentaria, *F. Mueller.*

Queensland. Albany Island, *W. Hill;* Curtis and Gloucester islands, *Henne;* Edgecombe Bay and Rockhampton, *Dallachy*, also *Bowman;* Endeavour river, *Banks and Solander.*

A specimen in fruit only from Careening Bay, on the N.W. coast, *A. Cunningham*, resembles this rather than *E. pyrophòra.*

127. **E. dichromophloia,** *F. Muell. in Journ. Linn. Soc.* iii. 89. A moderate-sized or large tree, the bark smooth, ash-grey, at length separating from the inner reddish bark (*F. Mueller*). Leaves in the imperfect specimens very long lanceolate, narrow, thick, with numerous, very fine, close, parallel veins, the intramarginal one scarcely distant from the edge. Umbels several-flowered, forming loose, terminal, corymbose panicles. Young buds obovoid, with a very short obtuse operculum; perfect flowers unknown. Anthers of *E. corymbosa.* Fruit urceolate-globose, with a contracted neck, smooth, attaining sometimes ½ in. diameter, but mostly much smaller; the rim thin, the capsule sunk. Perfect seeds broadly winged on one side.

N. Australia. Islands of the Gulf of Carpentaria, *R. Brown;* Abel Tasman, M'Arthur and Roper rivers, *F. Mueller.* It appears to differ but slightly from *E. terminalis* in the size and shape of the fruits, and perhaps in the bark.

128. **E. pyrophora,** *Benth.* Nearly allied to the preceding four species, but apparently to be distinguished, unless all be considered as forms of *E. corymbosa.* Leaves long, narrow, and thicker than in any of them. Inflorescence the same. Buds obovoid-pear-shaped, the very obtuse operculum undistinguishable from the calyx-tube till it separates, and then often tearing off irregularly. Flowers larger than in *E. terminalis*, the calyx-tube very broad and open, varying from 4 to 6 lines diameter. Stamens of the allied species. Fruit globose or slightly ovoid, contracted at the orifice, without a

distinct neck, the rim thin, the capsule sunk. Seeds apparently winged, but not seen perfect.

N. Australia. Nichol Bay, *Gregory's Expedition;* Upper Victoria river and Depôt Creek, *F. Mueller*, also with rather smaller flowers, Depuech Island, *Bynoe.*

129. **E. maculata,** *Hook. Ic. Pl. t.* 619. A lofty tree with a smooth bark falling off in patches so as to give the trunk a spotted appearance. Leaves ovate-lanceolate or lanceolate, straight or falcate, acuminate, mostly 4 to 6 in. long or even more, with numerous parallel but rather oblique veins, not so close as in the preceding species, and rather coarse, the intramarginal one close to the edge. Umbels 3-flowered, usually several together, on short leafless branches, forming a panicle or corymb. Peduncles and pedicels short and thick, scarcely angular. Calyx-tube, in the young bud shortly cylindrical, when open broadly turbinate, 3 to 4 lines diameter. Operculum hemispherical, much shorter than the calyx-tube, the outer one much thicker and more persistent than in most species where it has been observed, and usually umbonate or shortly acuminate, the inner one (corresponding to the single one of most species) thin, obtuse, smooth, and shining. Stamens attaining 4 or 5 lines; anthers ovate with parallel distinct cells opening longitudinally. Ovary flat-topped. Fruit ovoid-urceolate, usually about ½ in. long, and nearly as much in diameter, the rim narrow, the capsule deeply sunk.—F. Muell. Fragm. ii. 47; *E. variegata*, F. Muell. in Journ. Linn. Soc. iii. 88.

Queensland. Brisbane river, *F. Mueller.*

N. S. Wales. "Spotted Gum" of Maitland, *Backhouse;* common in the Liverpool district, *Woolls;* above Paramatta, *Cayley.*

130. **E. eximia,** *Schau. in Walp. Rep.* ii. 925. A large tree. Leaves falcate-lanceolate, acuminate, mostly 4 to 6 in. long, with numerous veins, fine and parallel, but scarcely visible owing to the thick coriaceous texture. Flowers several together, closely sessile in heads, which are usually arranged on thick angular or flattened peduncles, in terminal corymbs or panicles. Calyx-tube thick, obconical, somewhat angular, much tapering at the base, 3 to 4 lines long. Operculum broadly conical or shortly acuminate, always much shorter than the calyx-tube, and double, as in *E. maculata*, but the inner one not readily separable in the dried specimens till the flower is ready to open. Stamens 3 to 4 lines long; anthers ovate-oblong, the cells parallel, opening longitudinally. Ovary short, flat-topped. Fruit urceolate, ¾ to 1 in. long, the rim thin, the capsule deeply sunk.

N. S. Wales. Banks of the river Grose, *R. Brown;* "Bloodwood" of the Blue Mountains, *Miss Atkinson, Woolls.* This is evidently a very distinct species, more nearly allied to *E. maculata*, than to the Port Jackson "Blood-wood" (*E. corymbosa*), but different from both. I have not been able quite to satisfy myself of the structure of the operculum, which would require the examination of living specimens.

SUBSERIES IX. EUDESMIEÆ.—Leaves mostly opposite or nearly so. Peduncles usually 3-flowered. Calyx with 4 teeth, more or less conspicuous below the globular hemispherical or flattened operculum. Stamens sometimes very shortly united in 4 clusters, alternating with the calyx-teeth.

131. **E. erythrocorys,** *F. Muell. Fragm.* ii. 33. A shrub of 8 to 10 (*Oldfield*), or a tree of 20 to 30 ft. (*Drummond*). Leaves mostly opposite

or nearly so, or the upper ones alternate, all petiolate, long-lanceolate or broadly linear, often above 6 in. long, rigid, but with the oblique rather irregular veins conspicuous on both sides, the intramarginal one near the edge. Peduncles axillary or lateral, very thick, flat, and broad, ½ to 1 in. long, bearing each 3 large flowers, nearly sessile or tapering into short, thick, flattened pedicels. Calyx-tube turbinate, very thick, irregularly ribbed, ½ to ¾ in. long, and nearly ¾ in. diameter at the top, with 4 more or less prominent angles, terminating in exceedingly short, obtuse, scarcely prominent teeth. Operculum red, thick and fleshy, depressed and flat-topped, broader and shorter than the calyx-tube, obtusely square or almost 4-lobed, divided into 4 quarters by raised ribs, forming a cross on the top, each quarter transversely wrinkled, with a raised rib along the centre, opposite to the calyx-teeth. Stamens very numerous, inflected, forming 4 bundles alternating with the calyx-teeth, the claw or entire part very short and broad, or 4 clusters if the claw be considered as a mere dilatation or lobe of the margin of the staminal disk. Ovary much depressed, flat-topped. Fruit nearly hemispherical, ribbed, 1 to 1½ in. diameter, the margin of the calyx horizontally dilated, the disk very broad and obtusely prominent, giving it the shape of an old-fashioned hat, the capsule depressed in the centre, the valves not raised.

W. Australia. Stony plains, Murchison river, "Illyarie" of the natives, *Oldfield;* limestone hills, west of the Valley of the Lake, *Drummond, 6th Coll. n.* 70, who describes it in Hook. Kew Journ. v. 121, as one of the finest of the genus, with its scarlet cups and fine yellow flowers (*i.e.* stamens).

132. **E. tetragona,** *F. Muell. Fragm.* iv. 51. Varying from a low scrubby shrub, densely covered with a white meal, to a small tree, of 20 to 25 ft., the specimens often entirely deprived of the whiteness; branches mostly 4-angled or almost 4-winged, rarely terete. Leaves mostly opposite or nearly so or the upper ones alternate, from broadly ovate and very obtuse to lanceolate-falcate and almost acute, rarely above 4 in. long, thick and rigid, with diverging but rather distant veins, the intramarginal one at a distance from the edge. Peduncles axillary, short, thick, angular or flattened, with 3 or very rarely 4 or 5 rather large flowers, on thick angular or flattened pedicels. Calyx-tube campanulate, about 3 or rarely nearly 4 lines long and broad, with 4 minutely prominent teeth, sometimes very conspicuous, sometimes scarcely perceptible. Operculum depressed-hemispherical, shorter than the calyx-tube, smooth. Stamens 3 to 4 lines long, more or less distinctly arranged in 4 clusters or bundles, alternating with the calyx-tube, but the claws or dilatations of the disk very short or scarcely perceptible; anthers small, with parallel cells opening longitudinally. Fruit ovoid or nearly globular, truncate, contracted at the orifice, smooth or more or less ribbed, ½ to ¾ in. diameter, the rim scarcely distinct; capsule sunk, usually 4-celled.—*Eudesmia tetragona,* R.Br. App. Flind. Voy. ii. 599. t. 3; Sweet, Fl. Austral. t. 21; *Eucalyptus pleurocarpa,* Schau. in Pl. Preiss. i. 132; F. Muell. Fragm. ii. 37.

W. Australia. In exposed barren places, near the shore, Lucky Bay, *R. Brown;* from the Stirling Range, eastward to Cape Arid, *Maxwell, Preiss, n.* 253, *Baxter; Drummond, 4th Coll. n.* 78, and others. Oldfield observes that from the abundance of essential oil this species contains, it is killed down to the ground by the periodical fires, when other plants are only a little scorched, and is thus generally to be found only in an untidy, ragged, scrubby form, but he had seen dead stems of 25 feet.

133. **E. eudesmioides,** *F. Muell. Fragm.* ii. 35. A shrub, attaining 10 ft., with a smooth bark (*Oldfield*). Leaves from broad-lanceolate and 4 to 5 in. long, to narrow-lanceolate and shorter, mostly mucronate-acute and often falcate, rigid, the veins rather numerous but oblique and anastomosing, very conspicuous in the narrow leaves, much less so in the larger ones, the intramarginal one usually distant from the edge. Peduncles axillary, very short, nearly terete, mostly 3-flowered. Peduncles short. Calyx-tube narrow-turbinate, 2½ to nearly 3 lines long, with 4 minute teeth, sometimes prominent, sometimes scarcely conspicuous. Operculum short, depressed hemispherical, very obtuse and rather thick. Stamens 2 to 3 lines long, distinctly arranged in 4 clusters or bundles alternating with the calyx-teeth; anthers very small, nearly globular, with distinct parallel cells. Fruit ovoid or oblong, usually ½ to nearly ¾ in. long, in some specimens (perhaps not perfect), contracted at the orifice, but usually cylindrical, the rim concave, not broad, the capsule slightly sunk, usually 3-celled.

W. Australia. Sandy plains and limestone hills, Murchison river, *Oldfield.* Very near *E. tetragona* in characters, but the narrow leaves, small flowers, and narrow fruits, give it a very different aspect.

134. **E. odontocarpa,** *F. Muell. in Journ. Linn. Soc.* iii. 98. A shrub of 8 to 10 ft., with slender branches (*F. Mueller*). Leaves opposite or alternate, linear-lanceolate, mostly 3 to 5 in. long, with oblique anastomosing veins, inconspicuous at first, more prominent in the fruiting specimens, the intramarginal one near the edge. Peduncles axillary, short, each with 3 small flowers on short pedicels, but not seen expanded. Calyx-tube in the bud narrow-turbinate, about 2 lines long, with 4 small but prominent spreading teeth. Operculum hemispherical, very obtuse. Stamens apparently not in clusters; anthers small, with parallel cells. Fruit oblong-cylindrical, 4 to 5 lines long,.not contracted at the orifice when fully ripe; rim narrow, concave, the capsule slightly sunk, 3- or 4-celled.

N. Australia. Sturt's Creek Desert, *F. Mueller.* Very much like some specimens of *E. eudesmioides*, but the stamens do not appear to be arranged in clusters, and at once distinguished from the following species by the very much smaller flowers.

135. **E. tetrodonta,** *F. Muell. in Journ. Linn. Soc.* iii. 97. A tree, with a whitish, fibrous, persistent bark (*F. Mueller*). Leaves opposite or alternate, long-lanceolate, acuminate, often falcate and above 6 in. long, coriaceous, but the numerous somewhat oblique veins prominent, the intramarginal one near the edge. Peduncles axillary or 2 or 3 together at the ends of the branches, short and thick but not dilated, each bearing 3 or very rarely 5 rather large flowers, on thick angular or flattened pedicels of 2 to 4 lines. Calyx-tube obconical or turbinate, 3 to 4 lines long, with 4 rounded very obtuse teeth, slightly prominent on the bud. Operculum hemispherical or nearly globular, smooth. Stamens very numerous, the longest attaining 5 or 6 lines, not distinctly arranged in clusters; anthers oblong, with parallel cells opening longitudinally. Ovary flat-topped. Fruit oblong-cylindrical, ½ to ¾ in. long, 4 to 6 lines diameter, not contracted at the orifice, the rim narrow but forming an acutely prominent ring, the capsule sunk, usually 3 celled.

N. Australia. Entrance to Victoria river and elevated sterile districts of Arnhem's

Land, "Stringy-bark," *F. Mueller*; N. coast, *A. Cunningham*; Port Essington, *Armstrong*.

SUBTRIBE V. METROSIDEREÆ.—Leaves opposite or rarely alternate, myrtle-like or large, penniveined. Flowers usually in little cymes corymbs or short racemes, axillary or in terminal panicles, rarely solitary in the axils and then pedicellate. Stamens numerous, free or rarely united in bundles opposite the petals; anthers versatile, the cells parallel, opening longitudinally. Ovules few or many in each cell of the ovary, in 2 or more rows. Embryo straight or slightly curved, the cotyledons longer than the radicle.

This subtribe has nearly the flowers and embryo of *Euleptospermeæ*, but a different inflorescence and a habit approaching that of *Myrteæ*.

31. TRISTANIA, R. Br.

(Lophostemon, *Schott*; Tristaniopsis, *Brongn. and Gris.*)

Calyx-tube turbinate-campanulate or open, adnate to the ovary at the base, the free part broad; lobes 5, short. Petals 5, broad, much imbricate. Stamens indefinite, more or less united in bundles opposite the petals, the filaments or free parts filiform, inflected or rarely erect; anthers versatile, the cells parallel, opening longitudinally. Ovary inferior half superior or free except the broad base, but included in the calyx-tube, flat or convex on the top and very rarely depressed in the centre round the style, 3-celled, with several horizontal or recurved ovules in each cell; style filiform, with a more or less capitate stigma. Capsule adnate or almost free, enclosed in or protruding from the persistent calyx, opening loculicidally in 3 valves. Perfect seeds where known, few in each cell, linear-cuneate or expanded at the end into a flat wing; testa thin, embryo straight; cotyledons broad and folded over each other, as long as or longer than the radicle.—Tall shrubs or trees, Leaves alternate or irregularly verticillate at the ends of the branches, or in one species opposite, penniveined. Flowers small, yellow or white, in pedunculate axillary cymes. Bracts very deciduous or entirely wanting.

Besides the Australian species, the genus comprises at least two from New Caledonia and about four from the Indian Archipelago.

SECTION I. **Neriophyllum.**—*Leaves opposite. Stamens erect, shortly and irregularly 5-adelphous. Ovary inferior, flat-topped, with very numerous ovules in each cell. Seeds unknown* . . . . 1. *T. neriifolia.*

SECTION II. **Lophostemon.**—*Leaves alternate. Stamens inflexed, 5-adelphous, with long claws. Ovary inferior, flat-topped, with very numerous horizontal or recurved ovules in each cell. Seeds linear-cuneate.*

Staminal claws half as long as the petals. Flowers usually small (yellow?).
 Calyx-lobes short and very obtuse . . . . 2. *T. suaveolens.*
Staminal claws as long as the petals. Flowers few in the cyme, rather
 large. Calyx-lobes lanceolate, acute . . . . 3. *T. conferta.*

SECTION III. **Eutristania.**—*Leaves alternate. Stamens inflexed, 5-adelphous, with very short claws. Ovary adnate or half superior, ovules all reflexed. Seeds very flat or expanded at the end into a flat wing, the embryo in the thickened base.*

Ovary adnate. Flowers small, white, and numerous . . . . 4. *T. lactiflua.*
Ovary half superior.
 Stamens scarcely exceeding the petals. Seeds winged. Flowers yellow.
  Flowers small and numerous. Calyx not 1 line diameter . . . . 5. *T. exiliflora.*

Flowers few. Calyx 1½ to 2 lines diameter . . . . 6. *T. laurina.*
Stamens much longer than the petals, clustered but almost free. Seeds not winged . . . . 7. *T. psidioides.*
(Doubtful species. Leaves opposite. Fruit of *T. psidioides* . . . 8. *T. umbrosa.*)

SECTION I. NERIOPHYLLUM.—Leaves opposite. Stamens erect, shortly and irregularly 5-adelphous. Ovary inferior, flat-topped, with very numerous ovules in each cell. Seeds unknown.

1. **T. neriifolia,** *R. Br. in Ait. Hort. Kew. ed.* 2. iv. 417. A tall slender shrub or small tree, glabrous or the young shoots and under side of the leaves minutely glaucous-pubescent. Leaves opposite, lanceolate, acute, narrowed into a short petiole, nerveless except the prominent midrib, 1½ to 3 in. long. Flowers yellow, in opposite axillary cymes, but forming usually a terminal corymb, the central shoot not growing out till after the flowering. Calyx-tube turbinate, 5-angled, 1 to 1½ lines long, lobes ovate, as long as the tube. Petals often above 2 lines long. Stamens erect, longer than the petals, almost 1-seriate, more or less distinctly but irregularly united in clusters of 3 to 5 each, opposite the petals. Ovary completely adnate, glabrous and concave on the top, with a deep central depression round the style. Ovules exceedingly numerous in each cell, covering a peltate placenta. Fruiting-calyx about 2 lines long, the capsule much shorter. Seeds not seen.—DC. Prod. iii. 210; Bonpl. Pl. Malm. t. 30; F. Muell. Fragm. iv. 56; Lodd. Bot. Cab. t. 157; *Melaleuca neriifolia*, Bot. Mag. t. 1058; *M. salicifolia*, Andr. Bot. Rep. t. 485; *Tristdnia salicina*, A. Cunn. in Bot. Reg. under n. 1839.

N. S. Wales. Port Jackson to the Blue Mountains, *R. Brown, Sieber, n.* 219, and others; southward to Illawarra, *Shepherd.*

SECTION II. LOPHOSTEMON.—Leaves alternate. Stamens inflexed, 5-adelphous, with long claws. Ovary inferior, flat-topped, with very numerous ovules in each cell. Seeds linear-cuneate, not expanded at the end.

2. **T. suaveolens,** *Sm. in Rees Cycl.* xxxvi. A shrub or tree, more or less glaucous or hoary, or the young shoots hirsute, rarely quite glabrous. Leaves alternate, petiolate, ovate-elliptical, ovate-lanceolate or elliptical-oblong, obtuse or acuminate, more or less distinctly penniveined and reticulate, in some specimens 1½ to 3 in., in others 3 to 6 in. long. Flowers usually small, in axillary cymes, the common peduncle ¼ to ½ in. long, more or less flattened. Calyx-tube campanulate, usually hoary-pubescent, 1 to 1½ lines long; lobes very short and broad. Petals 1½ lines diameter. Staminal bundles about as long as the petals, the claws half as long as the petals, rather broad, with numerous inflexed filaments. Ovary wholly adnate, flat or concave at the top and glabrous, not depressed round the style; ovules very numerous in each cell, on an oblong reflexed placenta. Fruiting-calyx very open, 2 to nearly 4 lines diameter, the capsule not exceeding the tube. Seeds linear-cuneate, not winged; cotyledons rather broad and folded.—DC. Prod. iii. 210; *Melaleuca suaveolens*, Gærtn. Fruct. t. 173. t. 35; *Tristania depressa*, A. Cunn. in Bot. Reg. under n. 1839; DC. Prod. iii. 210; *T. rhytiphloia*, F. Muell. Fragm. i. 81.

N. Australia. Victoria river and Sea range, *F. Mueller;* Gulf of Carpentaria, *R. Brown.*
Queensland. Cape York, *M'Gillivray;* Endeavour river, *Banks and Solander;* Re-

pulse Bay, *A. Cunningham;* Rockingham Bay, Rockhampton, *Dallachy;* Burnett river, *F. Mueller;* Mount Elliott, *Fitzalan;* Brisbane river, Moreton Bay, etc., *Backhouse, W. Hill,* and others.

N. S. Wales. Richmond and Clarence rivers, *Beckler.*

Var. ? *grandiflora.* Very hoary-tomentose. Flowers nearly twice as large. Petioles of the leaves very short.—Attack Creek, *M'Douall Stuart.* Perhaps a distinct species, but there is but a single specimen.

3. **T. conferta,** *R. Br. in Ait. Hort. Kew. ed.* 2. iv. 417. A tall tree, with a smooth brown deciduous bark and dense foliage, the young shoots often clothed with spreading hairs, otherwise glabrous except the inflorescence, the buds of the succeeding year covered with large imbricate coloured scales. Leaves alternate, crowded at the ends of the branches so as to appear verticillate, petiolate, ovate or ovate-lanceolate, acuminate or rarely almost obtuse, usually 3 to 6 in. long, penniveined and minutely reticulate underneath. Flowers in cymes of 3 to 7, usually on the young wood below the cluster of leaves, the floral leaves mostly abortive, the peduncle flattened, $\frac{1}{4}$ to $\frac{1}{2}$ in. long, or rarely elongated. Calyx-tube more or less pubescent or hirsute, turbinate, $1\frac{1}{2}$ to near 3 lines long; lobes narrow, acute, nearly as long as the tube. Petals undulate, often 3 lines diameter. Staminal bundles often $\frac{1}{2}$ in. long, inflexed, the claws long and linear, with numerous short slender filaments nearly along their whole length; anthers very small. Ovary wholly adnate, flat-topped without any central depression; ovules exceedingly numerous in each cell, covering an oblong reflexed placenta. Fruiting-calyx 3 to 4 lines diameter, hemispherical or cup-shaped, truncate, smooth, the capsule level with the orifice or shortly exceeding it. Seeds linear-cuneate, not winged; cotyledons folded.—DC. Prod. iii. 210; F. Muell. Fragm. iv. 57; *T. subverticillata,* Wendl. in Ott. Dietr. Allg. Gartenz. i. 186; *T. macrophylla,* A. Cunn. in Bot. Reg. t. 1839; F. Muell. Fragm. i. 82; *Lophostemon arborescens,* Schott in Wien. Zeitschr. iii. (1830) 772.

N. Australia. Port Essington, *Armstrong.*

Queensland. Sandy Cape and Keppel Bay, *R. Brown;* mouths of the Burdekin river, *F. Mueller;* Rockhampton, Edgecombe Bay, etc., *Dallachy, Henne;* Brisbane river, Moreton Bay, *A. Cunningham* and others.

N. S. Wales. Hastings river, *Beckler.*

SECTION III. EUTRISTANIA.—Leaves alternate. Stamens inflexed, 5-adelphous with very short claws, or clustered only. Ovary adnate or half superior, ovules all reflexed. Seeds very flat or expanded at the end into a flat wing.

To this section belong the Asiatic species, as well as the New Caledonian ones forming Brongniart and Gris' genus *Tristaniopsis.*

4. **T. lactiflua,** *F. Muell. Fragm.* i. 82. A tree attaining 30 ft., glabrous or the young shoots, under side of the leaves, and inflorescence glaucous-pubescent. Leaves alternate, often almost verticillate at the ends of the branches, ovate or broadly ovate-lanceolate, obtuse or acute, penniveined, 3 to 6 in. long, on a petiole often of 1 in. Flowers white, small and very numerous, in axillary cymes, the common peduncle often 1 to 2 in. long. Calyx-tube broad, scarcely 1 line long, with very short rounded lobes as in *T. suaveolens.* Petals about $1\frac{1}{2}$ lines diameter. Staminal bundles about as long as the petals, inflexed, the claws short and broad, each with 15 to 20

filaments. Ovary wholly adnate, concave at the top without any central depression. Ovules reflexed, not so numerous as in *T. suaveolens*. Fruit not seen.

N. Australia. Foot of M'Adam Range, *F. Mueller;* Port Essington, *Armstrong.*

5. **T. exiliflora,** *F. Muell. Fragm.* v. 11. Glabrous or the inflorescence minutely hoary-pubescent. Leaves alternate, lanceolate or elliptical, almost acute, much narrowed into the petiole, penniveined, 2 to 4 in. long. Flowers yellow, small, rather numerous, in small axillary shortly pedunculate cymes, the pedicels at length longer than the calyx. Calyx-tube turbinate or almost hemispherical, rather above ½ line long, lobes ovate, rather shorter. Petals about ¾ line diameter. Staminal bundles inflexed, not exceeding the petals, the claws short, each with 2 to 4 filaments, often hairy at the base. Ovary half adnate, the summit very convex, pubescent, not depressed round the style, with 3 to 6 pendulous ovules in each cell. Fruit obovoid-globular, about 2 lines diameter, adnate at the base only, filling the calyx-tube and protruding considerably beyond it. Seeds obovoid, not much flattened, the testa sometimes with a short appendage or quite wingless, often lined with a granular substance. Cotyledons broad, deeply lobed and closely folded over the radicle.

Queensland. Rockingham Bay, *Dallachy.* The species is very closely allied to *T. laurina,* with the same foliage, floral characters, and fruit, but with the flowers as small as in *T. lactiflua,* and the seeds, as far as known, not winged.

6. **T. laurina,** *R. Br. in Ait. Hort. Kew. ed.* 2. iv. 417. A somewhat scrubby shrub in exposed localities, becoming in moist situations a tree, often of great height, the young shoots more or less glaucous or silky-pubescent, especially the under side of the leaves, the older foliage glabrous. Leaves alternate, lanceolate, elliptical or obovate-lanceolate, acuminate, penniveined, 2 to 4 in. long, narrowed into a petiole. Flowers yellow, in short axillary cymes, on a very short common peduncle, the pedicels rarely longer than the calyx. Calyx-tube broadly campanulate, 1½ to 2½ lines diameter, lobes small, triangular, distant at the time of flowering although imbricate in the young bud. Petals 1½ to 2 lines long, usually undulate. Staminal bundles inflexed, scarcely exceeding the petals, the claws very short, each with 15 to 20 filaments. Ovary half-adnate, the summit very convex, hirsute, not depressed round the style, with several (about 10) reflexed ovules in each cell. Capsule obovoid or almost globular, 3 to 5 lines diameter, adnate at the base only, filling the calyx-tube and protruding considerably beyond it. Seeds oblong, flat, laterally attached near the top, the upper part thin and winglike, embryo in the lower thickened portion; cotyledons deeply cordate and folded over each other; radicle superior, rather long.—DC. Prod. iii. 210; F. Muell. Fragm. i. 81; *Melaleuca laurina,* Sm. in Trans. Linn. Soc. iii. 275.

Queensland. Brisbane river, Moreton Bay, *A. Cunningham, Fraser,* and others.

N. S. Wales. Port Jackson to the Blue Mountains, *R. Brown, Sieber, n.* 220, and others; northward to Hastings, Macleay, and Clarence rivers, *Beckler;* southward to Illawarra, *M'Arthur* and others, and Twofold Bay, *F. Mueller.*

Victoria. Banks of rivers, Gipps' Land, *F. Mueller.*

7. **T. psidioides,** *A. Cunn. in Bot. Reg. under n.* 1839. A small tree, the branchlets and inflorescence hoary-tomentose. Leaves alternate, petio-

late, oval-elliptical, 1½ to 2½ in. long, obtuse or almost acute, narrowed at the base, penniveined, glabrous above, white underneath with a close tomentum. Flowers in dense corymbose cymes in the upper axils or almost terminal. Calyx broadly turbinate, tapering into a short pedicel, the tube about 1½ lines long, the lobes lanceolate, about as long as the tube. Petals short and broad, pubescent outside. Stamens free or nearly so but in 5 clusters, about ½ in. long, the filaments slender. Ovary almost inferior with a prominent convex summit, with about 6 closely-packed flat ovules laterally attached but pendulous from a short placenta. Fruit nearly globular, about 3 lines diameter, free except the broad base, adnate to and resting on the flattened calyx-tube, the lobes spreading or deciduous. Seeds very flat, but not winged.

N. Australia. Brunswick Bay, N.W. coast, *A. Cunningham;* islands of the N. coast, *R. Brown.*

8. **T. (?) umbrosa,** *A. Cunn. in Bot. Reg. under n.* 1839. The specimens are in fruit only and much resemble those of *T. psidioides*, except that they are nearly glabrous, the leaves are not white underneath and all opposite. Fruits and seeds of *T. psidioides.*

N. Australia. Hunter's River, York Sound, N.W. coast, *A. Cunningham.*

32. SYNCARPIA, Ten.

(Kamptzia, *Nees.*)

Calyx-tube turbinate or campanulate, adnate to the ovary at the base, the free part erect or dilated; lobes 4 or rarely 5, persistent. Petals 4 or rarely 5, spreading. Stamens indefinite, free, in 1 or 2 series, sometimes interrupted between the petals, filaments filiform; anthers versatile, cells parallel, opening longitudinally. Ovary inferior, flat-topped or convex, scarcely depressed round the style, 2- or 3-celled, with 1 or several ovules in each cell, erect on a basal placenta; style filiform with a small stigma. Capsule included in and adnate to the calyx-tube, opening loculicidally in 2 or 3 valves. Seeds linear-cuneate, testa thin, embryo straight, cotyledons plano-convex, longer than the radicle.—Trees. Leaves opposite, penniveined. Flowers in dense globular heads, either solitary on axillary peduncles or forming terminal panicles.

The genus consists of two species exclusively Australian, and differing perhaps as much from each other as either one does from *Metrosideros.*

Calyxes connate. Petals broad. Ovary 3-celled, with several ovules in each cell . . . . 1. *S. laurifolia.*
Calyxes free. Petals narrow. Ovary 2-celled, with 1 ovule in each cell 2. *S. leptopetala.*

1. **S. laurifolia,** *Ten. in Mem. Soc. Ital. Sc. Moden.* xxii. *t.* 1. A slender tree, the young shoots and under side of the leaves more or less hoary-pubescent or glaucous. Leaves appearing sometimes in whorls of 4 from 2 pairs being close together, from broadly ovate to elliptical-oblong, obtuse or obtusely acuminate, glabrous above, 2 to 3 in. long, on petioles of ¼ to ½ in. Flowers white, united, 6 to 10 together in globular heads, on peduncles of ¾ to 1 in. at the base of the new shoots, with 2 to 4 bracts close under the head, either short and scale-like or leaf-like and exceeding the flowers.

Calyxes connate at the base, the free parts broadly campanulate, softly hoary-pubescent, 1 to 1½ lines long, lobes short, broad and obtuse. Petals broadly ovate or orbicular, about 1½ lines long. Stamens 3 to 4 lines long, in about 2 rows round a flat disk fully 3 lines diameter. Ovary flat-topped, tomentose, 3-celled, with rather numerous ovules in each cell, erect on an oblong placenta. Fruiting-heads about ½ in. diameter, the calyxes connate to about the middle.—F. Muell. Fragm. i. 79; *Metrosideros glomulifera*, Sm. in Trans. Linn. Soc. iii. 269; DC. Prod. iii. 225; *Tristania albens*, A. Cunn. in Bot. Reg. under n. 1839; DC. Prod. iii. 210; *Kamptzia albens*, Nees in Nov. Act. Nat. Cur. xviii. Suppl. Præf. 9. t. 1; *Metrosideros procera* and *M. propinqua*, Salisb. Prod. 351?

Queensland. Shoalwater Bay Passage, *R. Brown;* Moreton Bay, *F. Mueller.*
N. S. Wales. Port Jackson to the Blue Mountains, *Burton, A. Cunningham, Miss Atkinson;* northward to Hastings river, *Beckler;* southward to Illawarra, *A. Cunningham.*

Var. *glabra.* Quite glabrous, even the calyx. Flowers rather small.—Hastings river, *Beckler.*

2. **S. leptopetala,** *F. Muell. Fragm.* i. 79. A tree of 50 to 60 ft., the young shoots, under side of the leaves, and inflorescence minutely and closely tomentose or almost scurfy, or at length glabrous, the young branches angular. Leaves ovate-elliptical or ovate-lanceolate, acutely acuminate, penniveined, glabrous above, 2 to 4 in. long, tapering into rather short petioles. Flowers small and numerous, in dense globular heads but quite free from each other, the common peduncles slender, 1 to 1½ in. long, in terminal clusters or panicles. Bracts very small, linear or lanceolate. Calyx-tube pubescent or nearly glabrous, membranous, turbinate-campanulate, 1 to 1½ lines long; lobes short, rounded. Petals narrow, ¾ line long. Stamens in a single row round the margin of the calyx-tube but interrupted between the petals, 3 to 4 lines long. Ovary convex, pubescent, 2-celled, with 1 erect ovule in each cell.

Queensland. Brisbane river, Moreton Bay, *F. Mueller, C. Moore,* and others.
N. S. Wales. In the interior, *A. Cunningham* (in Oxley's 2nd Expedition).

The great difference in the ovary, the free flowers, and the habit, which is that of a *Calycanthus* or an *Adina*, might perhaps justify the considering this as generically distinct from *S. laurifolia*, but the fruit is unknown, and perhaps both are too nearly allied to *Metrosideros.*

33. LYSICARPUS, F. Muell.

Calyx-tube campanulate, adnate to the ovary at the base; lobes 5, small, almost valvate. Petals 5, spreading. Stamens indefinite, free or nearly so, in 2 or more series interrupted opposite the sepals, the inner ones shorter, a few of the outer ones with reniform indehiscent anthers, the others with versatile anthers, the cells parallel, opening longitudinally. Ovary enclosed in the calyx-tube, but free except the broad base, tapering above, but with a distinct depression round the style, 3-celled with numerous ovules in each cell, erect on a basal placenta; style filiform, with a capitate almost 3-lobed stigma. Capsule oblong, protruding from the persistent calyx, opening loculicidally in 3 valves. Seeds . . .—Tree. Leaves opposite or whorled, narrow. Flowers polygamous, the males in irregular cymes, the hermaphrodites often solitary.

The genus is limited to the single Australian species. It is very nearly allied to *Metrosideros*.

1. **L. ternifolius,** *F. Muell. in Trans. Phil. Inst.* ii. 68. A tree attaining about 30 ft., with a soft thick fibrous bark, the young branchlets and inflorescence softly tomentose-pubescent. Leaves opposite or in whorls of 3, narrow-linear, mucronate-acute or rarely obtuse, 1½ to 3 in. long, with closely revolute margins, shining above, whitish-pubescent or at length glabrous underneath. Male flowers in irregular terminal or almost terminal leafy cymes, the hermaphrodite often solitary on opposite pedicels below the ends of the branches. Calyx-tube softly tomentose, about 1½ lines long, broader in the hermaphrodite than in the male flowers. Petals above 1 line diameter, orbicular, pubescent or ciliolate. Stamens exceeding the petals. Ovary pubescent. Capsule often twice as long as the calyx.—*Tristania angustifolia*, Hook. in Mitch. Trop. Austr. 198.

Queensland. On the Maranoa, *Mitchell;* Darling Downs and between the Mackenzie and Dawson rivers, *F. Mueller.*

34. METROSIDEROS, Banks.

(Nania, *Miq.*)

Calyx-tube (in the Australian species) campanulate, adnate to the ovary at the base, lobes 5, rarely 4, slightly imbricate. Petals 5, rarely 4, spreading. Stamens indefinite, free, in 1 or more series, exceeding the petals, filaments filiform; anthers versatile, the cells parallel, opening longitudinally. Ovary included in the calyx-tube, inferior or half superior, slightly depressed round the style, 3-celled, with numerous ovules in each cell closely packed in several series, on a peltate or oblong adnate placenta; style filiform, with a small stigma. Capsule inferior, half superior, or almost free, but surrounded by or enclosed in the persistent calyx-tube, opening loculicidally in 3 valves or rarely irregularly dehiscent. Seeds usually numerous, flat, cuneate or linear, erect; embryo straight, the cotyledons flat or folded, longer than the radicle.—Shrubs or trees, rarely climbing. Leaves opposite, penniveined. Flowers often showy, in dense terminal trichotomous cymes, or rarely axillary.

The genus comprises several very variable species dispersed over the islands of the Pacific and Indian Archipelago from New Zealand to the Sandwich Islands, with one somewhat anomalous species from South Africa. The single Australian species belongs to a group represented only by one other one from the Archipelago and generically distinguished by Miquel under the name of *Nania*, chiefly on account of its flat broad seeds.

1. **M. eucalyptoides,** *F. Muell. Fragm.* i. 243. A moderate-sized tree, glabrous or the young shoots glaucous or slightly tomentose. Leaves closely sessile and somewhat cordate, broadly elliptical-oblong, obtuse, mostly 4 to 8 in. long, thinly coriaceous. Flowers without the stamens rather small, in rather dense cymes in the upper axils. Bracts very small and narrow. Pedicels slender, 2 to 4 lines long. Calyx very open, about 2 lines diameter, lobes distant, narrow. Petals linear or oblong, narrowed into a distinct claw. Stamens numerous, about ½ in. long. Ovary half superior. Capsule nearly globular, attached only by the broad base to the persistent calyx. Ovules very numerous, flat, amphitropous, erect, densely imbricate

and completely covering the scarcely prominent placenta. Seeds very flat, obovate-falcate, but not seen quite ripe.—*Xanthostemon eucalyptoides*, F. Muell. Fragm. i. 81.

N. Australia. Arid banks of the Fitzmaurice river, *F. Mueller.*

M. aromatica, Salisb. Prod. 351, from Port Jackson, *Burton*, is evidently not a *Metrosideros*, as the genus is now constituted, but I have not met with any specimens corresponding to the imperfect diagnoses given.

35. XANTHOSTEMON, F. Muell.

(Fremya, *Brongn. and Gris.*)

Calyx-tube broadly campanulate or open, adnate to the ovary at the base; lobes 4 or 5, slightly imbricate, often unequal. Petals 4 or 5, spreading. Stamens indefinite, free or slightly united at the base, in one or more series much exceeding the petals; filaments often rigid; anthers versatile or, from a dilatation of the connective round the filament, apparently attached by the base, the cells parallel, opening longitudinally. Ovary enclosed in the calyx-tube, half-inferior or free except the broad base, 2- to 6-celled, with numerous ovules in each cell, closely packed in a single ring round a clavate or peltate placenta; style filiform with a small stigma. Capsule free except the broad base, seated on the expanded calyx, or half enclosed in the cup-shaped calyx-tube, opening loculicidally in 2 to 6 valves. Seeds flat or angular; testa thin; cotyledons broad, flat or folded over each other, longer than the straight or incurved radicle.—Trees or shrubs. Leaves alternate, penniveined. Flowers in dense cymes on terminal or axillary peduncles, or (in species not Australian) solitary or nearly so. Bracts and bracteoles usually very small or none.

Besides the two Australian species there are a considerable number in New Caledonia. The genus has since been reunited by F. Mueller with *Metrosideros*, which it closely resembles. The constantly alternate leaves, with the insertion and arrangement of the ovules, seem, however, to justify the maintaining it either as a genus or section at least as distinct as *Syncarpia*, *Lysicarpus*, and the non-Australian *Cloezia*, *Tepualia*, and *Spermolepis*.

Leaves acuminate, 4 to 6 in. long. Calyx above 3 lines diameter, half enclosing the capsule . . . . 1. *X. chrysanthus.*
Leaves obtuse, 2 to 3 in. long. Calyx about 2 lines diameter, opening flat under the capsule . . . . 2. *X. paradoxus.*

1. **X. chrysanthus,** *F. Muell. Herb.* A tall handsome tree, the specimens quite glabrous. Leaves lanceolate or elliptical, acuminate or almost acute, 4 to 6 in. long, narrowed into a short petiole. Peduncles in the upper axils about ½ in. long, bearing each a dense cyme of 5 to 10 rather large flowers of a golden-yellow. Calyx-tube broadly campanulate, 3 to 3½ lines diameter, somewhat enlarged and half enclosing the fruit; lobes ovate-triangular, shorter than the tube, and half as long as the orbicular petals. Stamens 20 to 25, in a single series, the longest nearly 1 in. long; anthers oblong, versatile, the connective scarcely thickened. Ovary more than half-superior, usually 3-celled, with numerous flat ovules closely packed in a single whorl round a peltate somewhat clavate placenta. Style very long, not at all immersed. Capsule about 5 lines diameter. Seeds few perfect, flat, with a thin testa; cotyledons broad, 2-lobed, conduplicate, more or less en-

closing the incurved radicle; sterile seeds numerous, of the same shape, but hard and homogeneous.—*Metrosideros chrysantha*, F. Muell. Fragm. iv. 159.

Queensland. Along streams, Rockingham Bay, *Dallachy*.

2. **X. paradoxus,** *F. Muell. Fragm.* i. 80. A tall shrub or small tree, the inflorescence and under side of the leaves tomentose pubescent or glaucous, at length glabrous, the upper leaves assuming a yellowish hue in the dried state. Leaves obovate-oblong or elliptical, obtuse, mostly 2 to 3 in. long, narrowed into a short petiole. Cymes dense, terminal or in the upper axils. Calyx-tube broadly campanulate, about 2 lines diameter, opening flat under the fruit; lobes ovate-triangular, varying from ½ to 1 line long. Petals ovate or orbicular, ciliate, 1½ to 2 lines long. Stamens yellow, rigid, nearly ¾ in. long; anthers really attached at the back, but the thick fleshy connective enclosing the summits of the filament so as to make them appear attached by the base. Ovary nearly superior, 2- or 3-celled, the style quite terminal; ovules in a ring round the clavate placenta. Capsule globular or almost ovoid. Seeds as in *X. chrysanthus*, the broad cotyledons folded over the incumbent radicle.—*Metrosideros paradoxa*, F. Muell. Fragm. i. 243.

N. Australia. Montague Sound, N.W. coast, *A. Cunningham*; rocky hills on the Victoria river and Arnhem's Land, *F. Mueller*.

36. BACKHOUSIA, Hook. and Harv.

Calyx-tube turbinate or broadly campanulate, adnate to the ovary at the base; lobes 4, almost petal-like or scarious, persistent. Petals 4, shorter than or scarcely exceeding the calyx-lobes, usually persistent. Stamens indefinite, free, in several series; anthers versatile, the cells parallel, opening longitudinally. Ovary in the bottom of the calyx-tube, inferior or half-superior, 2-celled, with several ovules in each cell, recurved or pendulous, attached either in 2 rows to an axile placenta, or to a placenta pendulous from the apex of the cell; style filiform, with a small stigma. Capsule enclosed in the persistent calyx-tube or protruding from it, apparently indehiscent or separating into 2 cocci. Seeds obovoid or cuneate; embryo straight, cotyledons (where known) conduplicate and longer than the radicle. —Trees or shrubs. Leaves opposite, penniveined. Flowers in cymes sometimes reduced to heads or in umbels, on axillary peduncles often forming terminal leafy panicles. Bracts very deciduous.

The genus is confined to Australia, and may be considered in some measure as connecting the true *Myrteæ* with the *Leptospermeæ*, but is readily known by the calyx, ovary, and fruit.

Cymes dense, corymbose. Pedicels shorter than the calyx-tube.
- Leaves ovate-acuminate. Calyx-lobes nearly equal . . . . . . 1. *B. myrtifolia.*
- Leaves lanceolate or oblong-linear. Inner calyx-lobes large and petal-like . . . . 2. *B. angustifolia.*

Cymes umbel-like. Pedicels filiform, many times longer than the calyx-tube.
- Leaves ovate-obtuse. Placentas axile . . . . 3. *B. sciadophora.*
- Leaves ovate-lanceolate, acuminate. Placentas pendulous from the summit of the cell . . . . 4. *B. citriodora.*

1. **B. myrtifolia,** *Hook. and Harv. in Bot. Mag. t.* 4133. A tall shrub or small tree, the young shoots and under side of the leaves and the

inflorescence more or less pubescent or softly hirsute, the older foliage glabrous. Leaves ovate, acutely acuminate, penniveined, 1 to 2 in. long, narrowed into a petiole of 1 to 2 lines. Flowers white, in small cymes sometimes contracted into heads, on peduncles of ¾ to 1 in. at the base of the new shoots, forming terminal leafy panicles. Bracts narrow, falling off long before flowering. Calyx-tube turbinate, softly pubescent or rarely glabrous, nearly 1½ lines long; lobes from ovate-oblong to lanceolate, petal-like but rigid, 2 lines or in large-flowered forms 3 lines long. Petals not half so long. Ovary inferior, slightly convex and villous on the top; ovules 8 to 10 in each cell, campylotropous, attached in two rows to a somewhat thickened placenta adnate to the axis. Fruit enclosed in the calyx-tube, but not seen ripe.—F. Muell. Fragm. i. 78; *B. riparia*, Hook. in Bot. Mag. under n. 4133.

Queensland. Moreton Bay, *W. Hill;* Pine river, *Fitzalan.*
N. S. Wales. Hawkesbury river, *R. Brown;* Port Jackson, *Burton;* Paramatta, *Woolls;* Hastings river, *A. Cunningham, Beckler;* Macleay river, *Beckler;* Clarence river, *Wilcox.*

2. **B. angustifolia,** *F. Muell. Fragm.* i. 79. A tall shrub, the young shoots and inflorescence minutely hoary-pubescent or tomentose, the adult foliage glabrous. Leaves lanceolate or narrow-oblong, obtuse or mucronate, very obliquely penniveined, 1 to 1½ in. long. Flowers rather small, in cymes or heads of 3 to 9 each, on peduncles of ½ in. or less in the upper axils, forming a divaricate leafy panicle. Calyx-tube turbinate, ribbed, pubescent, about 1 line long, outer lobes orbicular, as long as the tube, inner ones much larger and petal-like. Petals shorter than the inner calyx-lobes. Outer stamens above 2 lines long. Ovary inferior, convex and pubescent on the top; ovules about 6 in each cell, campylotropous, and attached in two rows to an axile placenta as in *R. myrtifolia.*

Queensland. Dawson river, *F. Mueller.*

3. **B. sciadophora,** *F. Muell. Fragm.* ii. 26, 171. A tree, either glabrous or the young shoots minutely pubescent. Leaves broadly ovate, obtuse, 1 to 2½ in. long, on short petioles. Flowers small, numerous, in umbel-like cymes or clusters on a common peduncle of ½ to ¾ in. in the upper axils, the slender pedicels often ½ in. long. Calyx-tube glabrous, broadly campanulate, about 1 line long; outer lobes rounded and rather shorter, the inner ones rather longer than the tube. Petals broad, slightly exceeding the calyx-lobes. Stamens about 2 lines long. Ovary semiadnate to the bottom of the calyx, the convex top slightly pubescent; ovules 4 to 6 in each cell, campylotropous, attached in 2 rows to an axile placenta. Capsule filling the calyx-tube, flat-topped, apparently indehiscent but readily separable into 2 cocci.

Queensland. Rockhampton, *Thozet,* with small leaves.
N. S. Wales. Hastings and Macleay rivers, *Beckler.*

4. **B. citriodora,** *F. Muell. Fragm.* i. 78. A tall shrub or small tree, the young shoots under side of the leaves and inflorescence hoary-tomentose or at length glabrous. Leaves ovate or ovate-lanceolate, acuminate, coriaceous, glabrous above, 3 to 5 in. long, on petioles of ¼ to ½ in. or more. Flowers

small, numerous in umbel-like clusters on peduncles of 1 in. or more in the upper axils or at the ends of the branches, the slender pedicels above $\frac{1}{2}$ in. long. Calyx-tube pubescent, broadly campanulate, about 1 line long, outer lobes broad and scarcely longer than the tube, inner ones longer and narrowed at the base. Petals shorter than the calyx, but not seen expanded. Ovary in the bottom of the calyx, semiadnate with a conical top, very rarely with a third cell; ovules about 6 to 8 in each cell, pendulous from a cuneate placenta suspended from the summit of the cell. After flowering the summit of the ovary protrudes much from the calyx, and shows no sign of splitting, but the ripe fruit not seen.

Queensland. Woods near Moreton Bay, *W. Hill, F. Mueller.*—Notwithstanding the difference in the placentation, this species cannot well be generically separated from the preceding one.

37. OSBORNIA, F. Muell.

Calyx-tube turbinate, not produced above the ovary; lobes 8, nearly equal, persistent. Petals none. Stamens indefinite, free, in 2 or 3 series, scarcely exceeding the calyx-lobes; filaments filiform; anthers small, versatile, the cells parallel, opening longitudinally. Ovary inferior, imperfectly 2-celled, with several ovules attached to a basal placenta or short dissepiment; style subulate, rather thick, with a small stigma. Fruit adnate to and included in the scarcely enlarged calyx-tube, and crowned by the persistent lobes, apparently dry and indehiscent. Seeds 1 or 2, obovoid, with a thin testa; embryo straight, with thick flattened or hemispherical cotyledons longer than the radicle.—Shrub. Leaves opposite, penniveined. Flowers small, sessile, solitary in the axils or terminal and three together. Bracteoles deciduous.

The genus is limited to the single Australian species, and shows no immediate affinity to any other one, except in some measure to *Backhousia.*

1. **O. octodonta,** *F. Muell. Fragm.* iii. 31. A bushy shrub, glabrous except the flowers. Leaves obovate-oblong, very obtuse, $\frac{3}{4}$ to $1\frac{1}{2}$ in. long, much narrowed into a very short petiole, thickened at the base, and leaving a contraction at the nodes when they fall off. Flowers sessile, solitary in the axils between 2 concave deciduous tomentose bracteoles, or 3 together at the ends of the branches. Calyx white with a close tomentum or short down, tube narrow, 2 to $2\frac{1}{2}$ lines long; lobes shorter, oblong, very obtuse, much imbricate in the bud. Fruit apparently dry, but not hard.

N. Australia. Islands of the Gulf of Carpentaria and Arnhem N. and S. Bays, *R. Brown;* Port Essington, *Armstrong;* Trinity Bay, *Henne.*

TRIBE II. MYRTEÆ.—Ovary divided into 2 or more cells, or if 1-celled with 2 placentas. Fruit an indehiscent berry or a drupe. Leaves opposite, dotted.

38. RHODOMYRTUS, DC.

Calyx-tube turbinate, oblong or nearly globular, scarcely or not at all produced above the ovary; lobes 4 or 5, herbaceous, persistent. Petals 4 or 5, spreading. Stamens numerous in several series, free; filaments filiform; anthers versatile or attached near the base, with parallel cells opening longitudinally. Ovary really 1- 2- or 3-celled, with several ovules in 2 rows in each cell, but owing to spurious dissepiments interposed between the ovules, appearing either 2-, 4- or 6-celled or divided into numerous 1-ovulated cells superposed in 2, 4, or 6 rows; style filiform, with the stigma usually peltate. Fruit a berry or almost a drupe, globular, ovoid, or cylindrical, divided into 1-seeded cells or nuts superposed in 2 to 6 or almost in a single row. Seeds compressed, reniform, or nearly orbicular, with a hard testa; embryo horse-shoe-shaped or ring-shaped, with a long radicle and very small cotyledons.—Trees or shrubs more or less tomentose or villous. Leaves opposite, penniveined or triplinerved. Peduncles axillary, bearing 1 or 3 or rarely a raceme or cyme of 5 or more flowers, pink or white. Bracts small, or when the peduncles are several-flowered the lowest sometimes leaf-like. Bracteoles small and deciduous.

Besides the Australian species, there is one which is widely distributed over the Indian Archipelago, extending to S. China, but which has not yet been detected in Australia. The genus is nearly allied to *Myrtus*, and still nearer to *Psidium*, but appears to be sufficiently characterized by the ovary and fruit to be distinguished from both.

Leaves penniveined. Flowers rather large (3, 5, or 7). Ovules and seeds in 6 rows . . . . 1. *R. psidioides.*
Leaves prominently triplinerved. Flowers small (usually 3). Ovules and seeds in 4 rows . . . . 2. *R. trineura.*
Leaves penniveined, but with an intramarginal vein often prominent, so as to be almost triplinerved.
Flowers small in a loose dichotomous cyme. Ovules and seeds in 4 or 6 rows . . . . 3. *R. cymiflora.*
Flowers rather large (1, 3, or 5). Ovules in 2 rows. Fruit long, cylindrical, with the seeds in 1 or 2 rows . . . . 4. *R. macrocarpa.*

1. **R. psidioides,** *Benth.* A tree attaining sometimes a great size, the young shoots more or less hoary-pubescent; the older foliage glabrous. Leaves petiolate, from oval-elliptical to ovate-lanceolate or oblong, shortly and obtusely acuminate, mostly 3 to 4 in. long, shining above, penniveined and prominently reticulate on both sides, the margins usually recurved. Peduncles axillary, rarely 1-flowered, mostly with 1, 2, or 3 pairs of pedicels besides the terminal one, the lowest often again 3-flowered, the pedicels all articulate below the calyx. Calyx-tube hoary-tomentose, thick, fully 2 lines long; lobes 5 or rarely 4, shorter than the tube, ovate, the inner ones rather larger and thinner than the outer. Petals about 3 lines long. Stigma broadly peltate. Berry ovoid-globular. Ovules and seeds superposed in 6 rows.—*Nelitris psidioides*, G. Don, Gen. Syst. ii. 829; *Myrtus Tozerii*, F. Muell. Fragm. ii. 86. t. 13.

Queensland. Brisbane river, *Hill, F. Mueller.*
N. S. Wales. Hunter's river, *R. Brown, Scott*; Hastings River, *Fraser, Beckler*; Clarence river, *Beckler.*

2. **R. trineura,** *F. Muell.* A shrub, the young shoots more or less velvety-tomentose. Leaves petiolate, ovate-lanceolate, acuminate, $1\frac{1}{2}$ to $2\frac{1}{2}$ in. long, triplinerved, much reticulate, glabrous above, loosely pubescent or tomentose underneath. Flowers usually 3 together, sessile in the axils, or borne on a short common peduncle. Calyx-tube tomentose-villous, above 1 line long; lobes 5, as long as the tube. Petals twice as long as the calyx-lobes, minutely pubescent or glabrous. Berry globular, villous, about 3 lines

diameter. Ovules and seeds superposed in 4 (or sometimes 6 ?) rows.—*Myrtus trineura,* F. Muell. Fragm. iv. 117.

Queensland. Wooded shores of Rockingham Bay, *W. Hill, Dallachy.*

3. **R. cymiflora,** *F. Muell.* Shrubby and glabrous. Leaves ovate-elliptical, shortly and obtusely acuminate, narrowed into a short petiole, finely and rather distantly penniveined, the veins united in a nerve much within the margin, and more prominent towards the base of the leaf, which thus appears almost triplinerved. Flowers several but not numerous, in loose dichotomous cymes, axillary, pedunculate, and sometimes exceeding the leaves. Calyx-tube turbinate or almost globular, above 1 line diameter; lobes 5, broad, shorter than the tube. Petals fully 2 lines diameter. Ovules superposed in 6 or rarely 4 rows. Fruit only seen young.—*Myrtus cymiflora,* F. Muell. Fragm. v. 12.

Queensland. Seaview Range, Rockingham Bay, *Dallachy.*

4. **R. macrocarpa,** *Benth.* A tall shrub, the young branches and inflorescence hoary with a close tomentum. Leaves petiolate, oval-elliptical or obovate, obtuse or shortly acuminate, often 6 to 10 in. long, penniveined and reticulate, glabrous or minutely pubescent underneath. Peduncles in the upper axils short, bearing either 1 or 3 flowers, or a short compact leafy raceme. Calyx-tube cylindrical; lobes 5, unequal. Petals tardily expanding. Style large, peltate. Ovules usually superposed in 2 rows on a parietal placenta protruding between the rows (the ovary reduced to a single cell). Fruit cylindrical, $\frac{3}{4}$ to $1\frac{1}{4}$ in. long, almost torulose. Seeds large, superposed usually in a single row, or very rarely the 2 rows perfect, and separated by firm partitions, the fruit then shorter and broader.

Queensland. Albany Island, *W. Hill;* Rockingham Bay, *Dallachy.*

39. MYRTUS, Linn.

Calyx-tube turbinate, scarcely or not at all produced above the ovary; lobes 4 or 5, small, usually persistent. Petals 4 or 5, spreading. Stamens numerous, in several series, free; filaments filiform; anthers versatile, or attached near the base, with parallel cells opening longitudinally. Ovary completely 2- or 3-celled, or imperfectly so, the dissepiments not quite reaching to the summit, with several ovules in each cell attached without order, or in 2 rows to an axile placenta either scarcely prominent or divided into 2 lamella; style filiform, with a small or rarely capitate stigma. Fruit a berry, globular or rarely ovoid, with few or rather numerous seeds not distinctly superposed in rows. Seeds more or less reniform, or almost circular, the testa hard or crustaceous, rarely membranous; embryo curved, horse-shoe-shaped, circular or spirally involute, with a long radicle; cotyledons very small, or rarely larger and folded.—Shrubs or rarely trees, glabrous or rarely pubescent or silky. Leaves opposite, penniveined, Peduncles axillary, usually slender, 1-flowered, or with several flowers in a centrifugal cyme, assuming, in the Australian several-flowered species, the form of a 5- or 7-flowered raceme, with a terminal flower sessile or on a shorter pedicel than the others. Bracteoles small and usually deciduous.

The genus is rather numerous in extratropical S. America and the Andes, extending more sparingly to other parts of S. America, to Mexico, and the W. Indies. There are also 4 New Zealand species, and one widely spread over S. Europe and W. Asia, besides the Australian ones, which are all endemic. There is no positive character to separate it from *Eugenia*, except the embryo, and the 1-flowered species of the two genera are not very dissimilar in foliage. Generally speaking, however, the Myrtles have smaller leaves, a more simple inflorescence, and more generally 5-merous flowers than the *Eugenias* of the Old World.

Peduncles axillary, solitary, slender, 1-flowered.
  Calyx-limb shortly and broadly sinuate-lobed. Ovary 3-celled . 1. *M. rhytisperma.*
  Calyx-limb divided to the base into 5 lobes.
    Leaves linear or lanceolate, hoary underneath. Ovary 2-celled 2. *M. tenuifolia.*
    Leaves ovate or ovate-lanceolate, acuminate, glabrous.
      Branchlets angular. Calyx slightly pubescent. Ovary 2-celled . . . . 3. *M. gonoclada.*
      Branchlets terete. Calyx hoary-pubescent. Ovary 2-celled, with many ovules . . . . 4. *M. Hillii.*
      Branchlets terete. Calyx glabrous. Ovary 3-celled, with few ovules in each cell . . . . 5. *M. Beckleri.*
Peduncles clustered in each axil, or bearing 3 or more flowers.
    Ovary 2-celled.
  Calyx 5-lobed, glabrous.
    Leaves very shining, usually acuminate. Flowers numerous. Pedicels usually in pairs in the racemes. Ovules few . . 6. *M. Bidwillii.*
    Leaves scarcely shining. Veins oblique and irregular. Pedicels slender, solitary along the raceme. Ovules numerous . 7. *M. racemulosa.*
    Leaves scarcely shining, acuminate. Veins diverging and regular. Pedicels short, clustered on a very short common peduncle. 8. *M. acmenioides.*
  Calyx 4-lobed, pubescent . . . . 9. *M. fragrantissima.*

1. **M. rhytisperma,** *F. Muell. Fragm.* i. 77. A shrub or small tree, with the habit of the common European Myrtle, the young shoots slightly pubescent, the older foliage glabrous. Leaves oblong-elliptical or oval-oblong, obtuse, $\frac{3}{4}$ to $1\frac{1}{2}$ in. long, narrowed or rounded at the base, finely penniveined, green on both sides. Peduncles axillary, 1-flowered, slender, nearly as long as the leaves, with minute bracteoles under the calyx. Calyx glabrous or nearly so; tube turbinate, $1\frac{1}{2}$ to 2 lines long; lobes 5, short, broad, rounded, connate into a broad sinuate limb. Petals 5. Ovary imperfectly 3-celled, the dissepiments not reaching the axis in the upper part; ovules 5 or 6 in each cell; stigma peltate. Berry 4 to 5 lines diameter. Seeds few, above 2 lines broad; testa not hard, slightly granular-rugose. Embryo long, more or less involute, with very short cotyledons.

Queensland. Wide Bay, *C. Moore*; Moreton Bay, *W. Hill*; in the interior, *Leichhardt.*

Var. *grandifolia.* Leaves ovate, shortly acuminate, $1\frac{1}{2}$ in. long. Flowers larger.

N. S. Wales. Clarence river, *C. Moore.*

2. **M. tenuifolia,** *Sm. in Trans. Linn. Soc.* iii. 280. A small elegant spreading shrub, the young shoots more or less silky. Leaves from linear-lanceolate to ovate-lanceolate, obscurely penniveined, flat or with recurved margins, rarely exceeding 1 in., glabrous above, hoary or silky-white underneath. Peduncles axillary, 1-flowered, slender, shorter than the leaves. Bracteoles small, close under the calyx. Calyx-tube tomentose, rather broad, about $\frac{3}{4}$ line long; lobes 5, broad, obtuse, nearly equal, rather longer than

the tube. Petals 5, ovate-orbicular, about 2 lines long. Ovary 2-celled; ovules rather numerous in each cell on a 2-lobed placenta. Seeds not numerous, testa hard, embryo semicircular, narrow, with 2 small cotyledons.

Queensland. Moreton Island, *Backhouse.*
N. S. Wales. Port Jackson to the Blue Mountains, *R. Brown; A. and R. Cunningham,* and others; Clarence and Richmond rivers, *C. Moore.* The latter specimens, as well as the Moreton Island ones, are broad-leaved, the Blue Mountain ones have generally narrow leaves.

3. **M. gonoclada,** *F. Muell. Herb.* A tree, attaining about 25 ft., quite glabrous, excepting sometimes the calyx, the young branches often marked with raised lines decurrent from the leaves. Leaves ovate, obtuse or obtusely acuminate, narrowed at the base, smooth and shining, with an intramarginal vein as in *M. acmenioides,* but the veins less numerous. Pedicels solitary, 1-flowered, axillary or below the leaves on the young shoot, slightly thickened at the end, articulate, with a pair of minute bracts under the calyx. Calyx-tube turbinate, glabrous or minutely hoary; lobes 5, nearly equal, much shorter than the tube. Petals 5, about 1½ lines diameter, minutely pubescent-ciliate. Ovary pubescent at the top, 5-celled; ovules rather numerous, on a peltate 2-lobed placenta. Fruit not seen.

Queensland. Moreton Bay, *C. Stuart.* This is very much like the European *M. communis,* but at once distinguished by the 2-celled ovary.

4. **M. Hillii,** *Benth.* A shrub or small tree, glabrous except the flowers, the branchlets terete. Leaves ovate, acuminate, narrowed into a short petiole, 1 to 2 in. long, very smooth and shining, penniveined, with the veins irregularly confluent into an intramarginal one. Pedicels axillary, slender, ½ to ¾ in. long, solitary or 2 or 3 together on a very short common peduncle. Calyx tomentose-pubescent; tube nearly globular, under 1 line long; lobes 5, broad, rounded, slightly unequal and rather longer than the tube. Petals 5, 2½ lines long, pubescent and ciliate. Ovary pubescent on the top, very fleshy, 2-celled, with about 16 to 20 ovules in each cell. Fruit nearly globular, crowned by the spreading or reflexed calyx-lobes, but not seen ripe. Seeds several.

Queensland, *W. Hill;* Pine river, Moreton Bay, *Fitzalan.*

5. **M. Becklerii,** *F. Muell. Fragm.* ii. 85. A tall shrub, quite glabrous. Leaves ovate or ovate-lanceolate, acuminate, cuneate at the base, 1 to 2 in. long, rather thick, penniveined or obscurely triplinerved, the lateral nerves scarcely conspicuous. Peduncles solitary, axillary, filiform, rarely above ½ in. long, with very minute bracteoles a short distance from the flowers. Calyx glabrous; lobes 5, short and broad. Petals not seen. Ovary 3-celled, with 8 to 10 ovules in each cell in 2 rows; stigma slightly peltate. Fruit globular, about 2 lines diameter. Seeds several, flat, nearly orbicular, the testa minutely granulate-reticulate.

N. S. Wales. Mountain woods, Cloud's Creek, Clarence river, *Beckler.*

6. **M. Bidwillii,** *Benth.* A shrub or small tree, quite glabrous. Leaves broadly ovate but usually contracted into a long lanceolate obtuse point, cuneate at the base, on a short broad petiole, 2 to 3 in. long, finely and distantly penniveined, coriaceous and very smooth and shining. Flowers much

more numerous than in *M. racemulosa,* in short loose racemes, clustered in the axils, the pedicels generally in opposite pairs along the rhachis, with a cluster of 5 at the end. Calyx-tube short; lobes 5 or rarely 4, spreading to a little more than 1 line diameter. Petals usually 5, sometimes 4 or 6, 1½ lines diameter, minutely ciliolate. Stamens much more numerous than in *M. racemulosa,* and covering half the radius of the flat disk. Ovary completely 2-celled, with a small cluster of ovules in each cell.

Queensland. Wide Bay, *Bidwill.* Some specimens of Dallachy's, from Port Denison, with less acuminate leaves, appear to belong to the same species, but are in bud only.

7. **M. racemulosa,** *Benth.* A small tree, quite glabrous, the branchlets terete or slightly flattened. Leaves ovate, obtuse or shortly acuminate, rounded or scarcely cuneate at the base, 1½ to 2½ in. long, penniveined, with a few of the veins more prominent, the lower ones very oblique, and the lowest pair sometimes forming an intramarginal one nearly to the end. Pedicels slender, usually 5 or 7 in a loose axillary raceme, not exceeding the leaves, the terminal one short, the lateral ones longer, solitary and opposite, and sometimes 2 racemes in each axil. Bracteoles minute, close under the flower. Calyx glabrous; tube somewhat turbinate, under 1 line long; lobes 5, broad, about as long as the tube. Petals 5, fully twice as long as the calyx-lobes. Stamens numerous, as in all *Myrti,* but occupying only the margin of the disk. Ovary 2-celled, with 12 to 16 ovules in each cell, on a broad placenta, the dissepiments scarcely complete to the top. Fruit globular, about 2 lines diameter, crowned by the calyx-limb. Seeds 1 or 2, nearly globular or reniform; testa hard. Embryo very long, irregularly twisted or doubly folded or involute, the radicular end thickened, the cotyledons very small.

Queensland. Broad Sound, *R. Brown;* Port Denison, *Fitzalan;* Edgecombe and Rockingham Bays, *Dallachy.*

Var. *conferta.* Racemes short, almost reduced to the clusters of *M. acmenioides,* but the venation of the leaves as in *M. racemulosa.*—Port Denison, *Fitzalan.*

8. **M. acmenioides,** *F. Muell. Fragm.* i. 77. A tree, of 20 to 40 ft., quite glabrous, with a reddish bark. Leaves ovate, acuminate, narrowed into a short petiole, 1½ to nearly 3 in. long, scarcely shining, finely penniveined, with the veins much more regular and diverging than in *M. racemulosa,* confluent in a fine intramarginal one. Pedicels rather firm, 3 to 4 lines long, usually several together in the axils or at the old nodes, in a cluster or short raceme, on a very short common peduncle. Bracteoles minute, deciduous, close under the flower. Calyx-tube broad, about 1 line long; lobes 5, broad, obtuse, shorter than the tube, all equal or the inner one larger with petal-like margins. Petals 5, more or less ciliate, the outermost about 2 lines diameter, the others rather smaller. Ovary 2-celled, with about 12 to 16 ovules in each cell on a 2-lobed placenta. Fruit about 2 lines diameter, usually crowned by the calyx-lobes. Seeds few and sometimes only one, globular, reniform or hemispherical; testa hard, smooth and shining. Embryo long, spirally involute, the radicular end thickened; cotyledons very small.

Queensland. Moreton Bay and Wide Bay, *W. Hill, C. Moore.*
N. S. Wales. Hastings and Clarence rivers, *Beckler, Wilcox.*

9. **M. fragrantissima,** *F. Muell. Herb.* A shrub or tree, the young shoots slightly hoary. Leaves very shortly petiolate, broadly ovate, 1 to 2 in. long, glabrous, penniveined, without any intramarginal vein. Flowers small, few, in short pedunculate axillary racemes, with the terminal one sessile, or the pedicels solitary and 1-flowered at the base of the shoots. Flowers smaller than in the other species and apparently all 4-merous. Calyx pubescent, the tube nearly globular, about 1 line diameter; lobes 4, rather shorter than the tube. Petals 4, twice as long as the calyx-lobes. Ovary 2-celled, with rather numerous ovules crowded on the small placenta; stigma small. Fruit not seen.

Queensland. Moreton Bay, *Herb. F. Mueller.*
N. S. Wales. Richmond river, *C. Moore? in Herb. F. Mueller.*

The seed being unknown, the genus of this plant must be uncertain, but, notwithstanding its 4-merous flowers, it has in other respects much more the aspect of a *Myrtus* than of a *Eugenia.*

40. RHODAMNIA, Jack.

(Monoxora, *Wight.*)

Calyx-tube ovoid or nearly globular, not produced above the ovary; lobes 4, usually persistent. Petals 4, spreading. Stamens numerous, in several series, free; filaments filiform; anthers versatile, with parallel cells, opening longitudinally. Ovary 1-celled, with 2 parietal placentas, each with several ovules; style filiform; stigma usually peltate. Berry globular, usually crowned by the calyx-limb. Seeds usually few, reniform-globular or variously compressed; testa hard; embryo horseshoe-shaped, with a long radicle and very small cotyledons.—Shrubs or small trees. Leaves opposite, 3-nerved or triplinerved. Flowers usually small, the pedicels clustered in the axils or forming very short racemes. Bracteoles small, deciduous.

The genus is spread over tropical Asia, and comprises about a dozen published species, some of which however will probably be reduced on a careful scrutiny. The three Australian ones appear to be endemic, although it is possible, when better known, that two of them may prove to be extreme forms of the most widely spread among the Asiatic ones. The 1-celled ovary, with parietal placenta, readily distinguishes the genus from all other *Myrteæ*, and the 3-nerved leaves are only in this genus and in *Rhodomyrtus.*

Flowers sessile in the axils. Leaves acuminate, mostly above 3 in. long. 1. *R. sessiliflora.*
Flowers in pedunculate cymes. Leaves mostly under 3 in. long.
Leaves acuminate, 3-nerved, pubescent underneath but not white.
Calyx glabrous or pubescent . . . . 2. *R. trinervia.*
Leaves obtuse, triplinerved, shining above, white underneath. Calyx very tomentose . . . . 3. *R. argentea.*

1. **R. sessiliflora,** *Benth.* Branches tomentose-pubescent. Leaves ovate, acuminate, mostly 3 to 5 in. long, glabrous above, more or less tomentose-pubescent underneath, especially on the nerves, triplinerved and reticulate. Flowers small, usually 3 together, sessile in the axils. Bracteoles small, linear, deciduous. Calyx densely tomentose-pubescent, about 1 line long; lobes orbicular or ovate, obtuse, unequal, the largest about 1 line diameter. Petals 1½ lines diameter. Stamens rather longer. Ovules numerous, in 3 or 4 irregular rows on each placenta. Berry small, globular, pubescent, with 1 to 4 seeds, the calyx-lobes deciduous.

Queensland. Rockingham Bay, *Dallachy.* Evidently nearly allied to the common *R. spectabilis*, Blume, but at once distinguished by the sessile flowers and fruits.

2. **R. trinervia,** *Blume, Mus. Bot.* i. 79. A tall shrub or small tree, the young shoots, under side of the leaves, and inflorescence, more or less velvety-pubescent, but not white. Leaves ovate-oblong or ovate-lanceolate, acuminate, glabrous and much reticulate above, prominently 3-nerved from the base. Peduncles slender, axillary, 3 together in a cluster or on a short common peduncle, each with 1 or rarely 3 flowers, with minute bracteoles under the calyx. Calyx pubescent or nearly glabrous; tube about 1 line long; lobes nearly as long. Petals twice as long as the calyx-lobes. Stamens shorter than the petals. Stigma small. Berry globular, about 3 lines diameter or rather more, with few or with rather numerous seeds.—*Myrtus trinervia*, Sm. in Trans. Linn. Soc. iii. 280; *Eugenia (?) trinervia*, DC. Prod. iii. 279; Bot. Mag. t. 3223; *Monoxora rubescens*, Benth. in Hook. Lond. Journ. ii. 219; *Myrtus melastomoides*, F. Muell. Fragm. i. 76.

Queensland. Damp woods, Moreton Bay, and in the interior, *A. Cunningham, Fraser, W. Hill.*

N. S. Wales. Port Jackson to the Blue Mountains, *R. Brown, Woolls, Miss Atkinson;* northward to Clarence river, *C. Moore;* southward to Illawarra, *A. Cunningham, Shepherd, Ralston.*

3. **R. argentea,** *Benth.* A tall tree, the young shoots, under side of the leaves, and inflorescence more or less silvery-white with a close minute tomentum. Leaves oval or elliptical, obtuse, narrowed at the base, triplinerved, with transverse veins and scarcely reticulate, 2 to 3 in. long, smooth and shining above. Peduncles axillary, solitary or 2 or 3 together, 2 to 4 lines long, each bearing either 3 or a trichotomous cyme of 5 to 9 flowers on very short pedicels. Calyx tomentose; tube about 1 line diameter; lobes about as long as the tube but rather unequal. Petals slightly tomentose, fully twice as long as the calyx-lobes. Stamens shorter than the petals. Ovules rather numerous to each placenta.

Queensland. Moreton Bay, *A. Cunningham* (a doubtful form, with acuminate leaves, longer than as above described, perhaps distinct, but the specimens insufficient). Also among Queensland woods, Exhibition, 1862, *W. Hill.*

N. S. Wales. Clarence river, *C. Moore, Wilcox.*

The species is very near *R. cinerea*, Jack, from which *R. spectabilis*, Blume, and several others may prove not to be specifically distinct.

41. **FENZLIA,** Endl.

Calyx-tube ovoid, not produced above the ovary; lobes 5, acute, persistent. Petals 5, spreading. Stamens numerous, in several series, free; filaments filiform; anthers versatile, with parallel cells opening longitudinally. Ovary 1-celled with a parietal placenta, or 2-celled with the placentas attached to the dissepiment, with 2 or 3 superposed ovules in each cell; style filiform, with a small stigma. Drupe ovoid or globular, crowned by the spreading or reflexed calyx-lobes, the epicarp thin, the endocarp thick and bony. Seeds 1 or 2, separately enclosed in the endocarp; testa thin; embryo very long, spirally involute, the outer radicular end somewhat thickened, the cotyledons linear, in the centre of the coil.—Shrubs more or less hoary-tomentose. Leaves opposite, penniveined. Flowers pink, solitary and pedicellate in the axils, with a pair of bracteoles under the calyx.

This genus is limited to the two species endemic in Australia.

Leaves usually glabrous above. Calyx-tube and fruit at length glabrous or moderately tomentose, ovoid . . . . 1. *F. obtusa.*
Leaves tomentose on both sides, usually small. Calyx-tube and fruit very tomentose, globular . . . . 2. *F. retusa.*

1. **F. obtusa,** *Endl. Atakta,* 19. *t.* 17. A low bushy shrub, the young shoots, inflorescence, and under side of the leaves hoary-tomentose. Leaves petiolate, obovate or oblong, very obtuse, mostly ¾ to 1 in. long, coriaceous, finely penniveined, smooth and shining above. Pedicels sometimes very short, sometimes 3 to 4 lines long, with a pair of subulate bracteoles under the calyx. Flowers pink. Calyx tomentose, the tube ovoid-oblong, about 1 line long; lobes narrow lanceolate-subulate, usually longer than the tube and united at the base in a short open limb. Petals obovate, 2 to 3 lines long, pubescent or nearly glabrous. Stamens shorter than the petals. Fruit very hard, ovoid, 2 to 3 lines long, glabrous or tomentose. Seeds usually 2 or 3.

Queensland. Shoalwater Bay Passage, Broad Sound, etc., *R. Brown;* Cape York, *M'Gillivray, W. Hill;* Islands of Torres Straits, *Hutchinson, C. Moore;* Rockingham Bay, *Dallachy.*

Var. *microphylla.* Leaves 3 to 4 lines long.—Dividing ranges between Thomson and Burdekin rivers, *S. Sutherland* (a small fragment and another in *Bowman's* collection in Herb. F. Mueller).

2. **F. retusa,** *Endl. Atakta,* 20. *t.* 18. Very near *F. obtusa,* but much more stellate-tomentose. Leaves usually but not always smaller, mostly under ¾ in. long, in the original specimens narrow and notched at the end, scarcely losing their tomentum on the upper side. Pedicels short. Flowers small. Calyx-tube more globular than in *F. obtusa* and densely tomentose, the lobes shorter than the tube. Petals tomentose outside, not so much contracted at the base in our specimens as represented in the plate. Fruit usually almost globular, much smaller than in *F. obtusa,* more or less tomentose.

N. Australia. Islands of the Gulf of Carpentaria, *R. Brown;* Victoria river, *F. Mueller.*

42. NELITRIS, Gærtn.

Calyx-tube campanulate, not at all or scarcely produced above the ovary; lobes 4 or 5. Petals 4 or 5, spreading. Stamens numerous, in several series, free; anthers versatile, with parallel cells opening longitudinally. Ovary 4- or 5-celled, with 2 or very few ovules in each cell, and sometimes each cell divided into 2 by a spurious dissepiment; style filiform, the stigma in the perfect flowers peltate. Berry globular, crowned by the calyx-lobes. Seeds few, reniform-globose; testa hard; embryo horseshoe-shaped or circular, with a long radicle and short linear cotyledons.—Shrubs or small trees. Leaves opposite, penniveined. Flowers small, pedicellate in axillary racemes, often forming terminal leafy panicles.

The genus is dispersed over tropical Asia, especially the Indian Archipelago and the Pacific islands, the Australian species apparently identical with the commonest Asiatic one. It is nearly allied to *Myrtus,* but readily distinguished by the number of cells to the ovary.

1. **N. paniculata,** *Lindl. Collect. under n.* 16. A shrub or small tree, the young shoots and inflorescence silky-pubescent. Leaves ovate-lanceo-

late, acutely acuminate, narrowed at the base, 1 to 2 in. long, glabrous above, with fine scarcely conspicuous nearly transverse veins, silky-pubescent underneath or at length glabrous. Flowers smaller than in other Australian *Myrtles*, the racemes usually shorter or scarcely longer than the leaves, but often forming an elegant leafy panicle. Calyx very silky-pubescent, the tube about ½ line long, and the lobes about the same length. Petals twice as long as the calyx-lobes, more or less silky-pubescent. Anthers small, nearly globular. Berry about 2 lines diameter. Seeds few, with a hard tubercular-rugose almost bony testa; cotyledons nearly one-third the length of the embryo.—DC. Prod. iii. 231; Wight, Ic. t. 521; *Myrtus elachantha*, F. Muell. Fragm. iv. 56.

Queensland. Moreton Bay, *W. Hill;* Pine woods, Wide Bay, *Bidwill.* Common in the Indian Archipelago up to the Philippine Islands and in the eastern provinces of India to Khasia.

Var. *laxiflora.* Leaves longer, the veins more or less transverse (only visible in the old leaves). Flowers more numerous, in looser racemes and rather larger, the calyx glabrous or very slightly pubescent. Ovary 5-celled with 5 to 7 ovules in each cell (usually 2 or 3 in the common form). Fruit not seen. Perhaps a distinct species.—Rockingham Bay, *Dallachy.*

43. EUGENIA, *Linn.*

(Jossinia, *Comm.;* Jambosa, *DC.;* Syzygium, *Gærtn.;* Acmena, *DC.*)

Calyx-tube from globular to narrow-turbinate, not at all or more or less produced above the ovary; lobes 4, very rarely 5, from large and imbricate to very short and scarcely prominent above the truncate margin. Petals 4, very rarely 5, either free and spreading, or more or less connivent, or connate and falling off in a single calyptra. Stamens numerous, in several series, free or obscurely collected in 4 bundles; anthers versatile, usually small, the cells parallel or very rarely divaricate, opening longitudinally. Ovary 2-celled or very rarely (in species not Australian) 3-celled, with several ovules in each cell, or only 2 in an American section. Fruit a berry or sometimes almost a drupe, or nearly dry with a fibrous rind. Seeds either solitary and globose, or few and variously-shaped by compression; testa membranous or cartilaginous; embryo thick and fleshy, with a very short radicle, the cotyledons either united in an apparently homogeneous mass or more or less separable.—Trees or shrubs. Leaves opposite, penniveined. Flowers (in the Australian species) either solitary in the axils, or in lateral or terminal trichotomous cymes or panicles.

A most numerous genus, spread over the tropical and subtropical regions both of the New and the Old World. Of the 16 Australian species 12 or 13 are endemic, 3 or perhaps 4 common to East India and the Archipelago. The genus has been variously subdivided into sections or genera by different botanists according to whether they have worked chiefly upon American or upon Asiatic species. The most convenient course, however, appears to be that proposed by Wight, A. Gray, and others, to retain under the genus all *Myrteæ* with fleshy fruits and thick fleshy cotyledons with a very short radicle, except, perhaps, a very few American species with very different floral characters.

SECT. I. **Eueugenia.**—*Pedicels short, 1-flowered, solitary or 2 together in the axils or at the old nodes. Calyx-tube not at all or scarcely produced above the ovary. Petals free and spreading* . . . . 1. *E. carissoides.*

SECT. II. **Syzygium.**—*Flowers in trichotomous panicles or cymes. Calyx-tube more* *or less produced above the ovary, the border entire or very shortly sinuately-lobed, or with more prominent but very deciduous lobes. Petals more or less cohering in a calyptra, or rarely spreading and separately deciduous.*

Flowers in loose panicles, terminal or in the upper axils.
  Panicles corymbose. Petals cohering. Anther-cells divaricate . 2. *E. Smithii.*
  Panicles oblong or pyramidal. Petals often more or less distinct. Anther-cells parallel . . . . 3. *E. Ventenatii.*
Flowers in dense or trichotomous panicles, lateral on the old wood.
  Panicles reduced to a short dense corymb or head. Buds long, slender and clavate. Stamens very short. Leaves narrow . 4. *E. leptantha.*
  Panicles trichotomous, divaricate. Buds nearly globular. Leaves large, broad, rigid, shining and reticulate . . . . 5. *E. Jambolana.*

(See also 8. *E. grandis*, with dense panicles mostly terminal, which has almost the calyx of *Syzygium.*)

Sect. III. **Jambosa.**—*Flowers in trichotomous panicles or cymes. Calyx-tube more or less produced above the ovary, prominently lobed, the lobes usually persistent. Petals free and spreading.*

Flowers in divaricate trichotomous cymes or panicles, lateral on the old wood.
  Calyx-tube urceolate, 4 to 5 lines long. Outer stamens above 1 in. long. Fruit large, ovoid . . . . 6. *E. cormiflora.*
  Calyx-tube turbinate, about 2 lines long. Stamens scarcely above ½ in. long. Fruit nearly globular . . . . 7. *E. Tierneyana.*
Flowers large, in a large trichotomous terminal panicle. Leaves broad, obtuse, coriaceous. Calyx-tube turbinate.
  Calyx-tube sessile, about 3 lines long; lobes very short . . . 8. *E. grandis.*
  Calyx-tube nearly 5 lines long, tapering into a thick pedicel; lobes 3 to 5 lines . . . . 9. *E. suborbicularis.*
Flowers in a dense terminal sessile cyme. Calyx-tube narrow-clavate. Stamens purple, ¾ to 1 in. long. Leaves long . . 10. *E. Wilsonii.*
Flowers rather large, few in a terminal cyme. Calyx-tube turbinate; lobes as long as the tube.
  Leaves long, narrow, very obliquely and irregularly veined . . 11. *E. eucalyptoides.*
  Leaves ovate or elliptical, under 3 in. long, with fine irregular very diverging veins . . . . 12. *E. myrtifolia.*
Flowers rather small, in a corymbose terminal panicle. Calyx turbinate-campanulate, under 2 lines long; lobes small.
  Flowers tapering at the base, sessile or nearly so. Calyx usually 5-lobed . . . . 13. *E. angophoroides.*
  Flowers distinctly pedicellate. Calyx 4-lobed.
    Calyx-lobes very deciduous, leaving a truncate margin at the time of flowering . . . . 3. *E. Ventenatii.*
    Calyx-lobes persistent at the time of flowering . . . . . 14. *E. Armstrongii.*
Flowers in slender trichotomous cymes, opposite on young shoots or in terminal pairs. Calyx-lobes very small . . . . 15. *E. oleosa.*
Flowers 3 or few in axillary cymes. Calyx-lobes rather large.
  Leaves 2 to 3 in. long, finely and transversely penniveined . . 12. *E. myrtifolia.*
  Leaves broad, 3 to 5 in. long, almost triplinerved . . . . . 16. *E. Dallachiana.*

Specimens are before me of two other species, probably *Eugenias*, but insufficient for definition. One, a shrub, evidently allied to *E. myrtifolia*, but with larger more coriaceous leaves, and a looser more divaricate inflorescence, from Albany island, *W. Hill.* The other, with the foliage nearly of the E. Indian *E. nervosa*, but the calyx quite different; leaves only and unripe loose fruits without perfect seeds (*E. jucunda*, F. Muell.); Rockingham Bay, *Dallachy.*

Section 1. Eueugenia.—Pedicels short, 1-flowered, solitary or 2 together in the axils or at the old nodes. Calyx-tube (in the Old World species) not at all or scarcely produced above the ovary. Petals free and spreading.

This section, more definitely characterized by the inflorescence than by the calyx, comprises only a few of the Old World species, but very numerous American ones, and, according to the views of those who have studied chiefly American *Myrtaceæ*, should, with other species having a racemose or clustered (not trichotomous or cymose) axillary inflorescence, constitute the whole genus *Eugenia*, to the exclusion of *Syzygium* and *Jambosa*.

1. **E. carissoides,** *F. Muell. Fragm.* iii. 130. A shrub, with short divaricate glabrous branches. Leaves shortly petiolate, ovate orbicular or almost rhomboidal, very obtuse, $\frac{3}{4}$ to $1\frac{1}{2}$ in. long, coriaceous, irregularly penniveined and loosely reticulate. Flowers solitary or 2 together at the old nodes, nearly sessile or on pedicels rarely 2 lines long. Calyx glabrous or minutely pubescent; tube campanulate, about 1 line long, not produced above the ovary; lobes 4, nearly orbicular, persistent, about as long as the tube. Petals 4, spreading and falling off separately. Anthers short. Ovules rather numerous. Berry globular, 3 to 4 lines diameter, and 1-seeded, or oblong with 2 superposed seeds, or broader than long and somewhat didymous with 2 collateral seeds, crowned by the calyx-lobes.—*E. hypospodia*, F. Muell. Fragm. v. 15.

Queensland. Northumberland Islands, *R. Brown;* Cape York, *M'Gillivray;* common on rocks at Port Denison and Rockingham Bay, *Dallachy.*

The species is very nearly allied to, and perhaps not really distinct from, *E. rariflora*, Benth. in Hook. Lond. Journ. ii. 221; A. Gray, Bot. U. S. Expl. Exped. i. 514, t. 60, a species widely spread over the S. Pacific islands, and differing chiefly, as far as known, in its much larger fruit.

SECTION 2. SYZYGIUM.—Flowers in trichotomous panicles or cymes. Calyx-tube more or less produced above the ovary, the border entire or very shortly sinuately-lobed, or with more prominent but very deciduous lobes. Petals more or less cohering in a calyptra and falling off together, or rarely spreading and separately deciduous.

These species are all natives of the Old World, although a very few have in some measure become naturalized in some parts of tropical America. The section is often considered as a genus, but there are too many species in which the character derived from the calyx and petals is doubtful or variable, to allow of its being distinctly separable from *Jambosa.*

2. **E. Smithii,** *Poir. Dict. Suppl.* iii. 126. A tree, sometimes small and slender, but attaining in some places a considerable height, quite glabrous. Leaves petiolate, from ovate to ovate-oblong or ovate-lanceolate, obtuse or more or less acuminate, narrowed at the base, mostly 2 to 3 in. long, smooth and finely penniveined. Flowers small and numerous, in a terminal trichotomous panicle, sometimes corymbose and shorter than the leaves, sometimes longer and more pyramidal. Bracts minute and deciduous. Calyx-tube turbinate, about 1 line long, the free part very much broader; lobes either all very short broad and scarcely prominent, or 1 or 2 rather larger almost petal-like and deciduous. Real petals 4, united in a small flat very deciduous calyptra. Stamens scarcely 1 line long; anthers small, with distinct globular divaricate cells. Ovules rather numerous. Fruit white or purple, globular, $\frac{1}{4}$ to $\frac{1}{2}$ in. diameter, crowned by the circular prominent calyx-rim; endocarp thick and hard. Cotyledons closely combined.—*E. elliptica*, Sm. in Trans. Linn. Soc. iii. 281, not of Lam.; Bot. Mag. t. 1872;

Myrtus Smithii, Spreng. Syst. ii. 487; *Acmena floribunda, var. β*, DC. Prod. iii. 262; Bot. Mag. t. 5480 (wrong as to the petals); *Syzygium brachynemum*, F. Muell. Fragm. iv. 59 and Pl. Vict. Suppl. t. 18 (the petals not quite correct); probably also *Acmena Kingii*, G. Don, Gard. Dict. ii. 851.

N. Australia. Port Essington, *Armstrong.*

Queensland. Cape York, *W. Hill;* Rockingham Bay, *Dallachy;* Brisbane river, Moreton Bay, *F. Mueller.*

N. S. Wales. Port Jackson to the Blue Mountains, *R. Brown*, and others; northward to Hastings, Clarence, and Macleay rivers, *Beckler, Wilcox;* New England, *C. Stuart;* southward to Illawarra, *A. Cunningham;* Twofold Bay, *F. Mueller.*

Victoria. Snowy River, Lake King, Sealers' Cove, Cape Wilson, etc., known as "Lilly Pillies," *F. Mueller.*

The anthers with divaricate cells are, so far as hitherto observed, exceptional in the genus.

3. **E. Ventenatii,** *Benth.* A tall tree, quite glabrous. Leaves petiolate, oblong-lanceolate or rarely ovate-lanceolate, acuminate, narrowed at the base, mostly 3 to 5 in. long, finely penniveined as in *E. Smithii.* Flowers larger than in that species, in compound thyrsoid or oblong panicles, the pedicels short but slender and distinct. Buds nearly globular. Calyx-tube broadly turbinate-campanulate, about 1½ lines long, the adnate portion very short, the margin truncate with 4 lobes or teeth very short, or if larger and petal-like falling off as the flower expands. Petals 4, ovate, concave, under 1 line long, usually distinct and very deciduous, but according to F. Mueller sometimes cohering, and occasionally there is an inner series of smaller ones. Stamens attaining about 2 lines; anther-cells parallel. Ovules about 10 in each cell. Fruit not seen.—*Metrosideros floribunda*, Vent. Jard. Malm. t. 75, not of Sm.; *Syzygium floribundum*, F. Muell. Fragm. iv. 58.

Queensland. Rockingham Bay, *Dallachy;* Brisbane river, Moreton Bay, *F. Mueller, W. Hill, C. Stuart;* Ipswich, *Vernet;* also in *R. Brown's* collection without a label.

N. S. Wales. Clarence river, *Beckler.*

4. **E. leptantha,** *Wight, Illustr.* ii. 15, *and Ic. t.* 528. A tree? glabrous but pale, or the inflorescence hoary-pubescent. Leaves from oval-elliptical to oblong-lanceolate, obtusely acuminate, narrowed into a very short petiole, 4 to 5 in. long, finely penniveined. Flowers in short dense raceme-like cymes, almost reduced to heads, on the previous year's wood, either in the axils of the old leaves or at the nodes of the denuded branches, the peduncles and pedicels very short. Calyx-tube 5 to 6 lines long, very narrow, clavate, glabrous or powdery-pubescent, the free part short, slightly dilated, obscurely sinuate-toothed. Petals cohering and falling off together in a small calyptra. Stamens short. Ovules 12 to 20 in each cell. Fruiting cymes looser, the calyx cylindrical, but not seen ripe.—*Syzygium longiflorum*, Wall. Cat. Herb. Ind. n. 3572.

Queensland. Rockingham Bay, *Dallachy.*—The species is also found in the Malayan Peninsula.

5. **E. Jambolana,** *Lam. Dict.* iii. 198. A tall shrub or tree, attaining sometimes a considerable size, quite glabrous. Leaves oval-oblong, obtuse or shortly acuminate, usually 4 to 6 in. long and 2 to 3 in. broad, but sometimes longer, very firm, shining, with numerous fine pinnate veins and reticulate between them, the principal ones confluent but not forming a regular intramarginal vein. Flowers not large, numerous, in broad trichotomous

panicles lateral on the old wood below the leaves, the ultimate cymes dense. Calyx sessile, turbinate-campanulate, the lobes very short and broad at the margin, almost entire when the flower is fully out. Petals cohering and falling off together in a calyptra. Berry roundish, from the size of a cherry to that of a pigeon's egg, usually with a single seed (*Roxburgh*).—Wight, Illustr. ii. 16, and Ic. t. 535, 624; *Syzygium Jambolanum*, DC. Prod. iii. 259; Wight and Arn. Prod. 329, with the synonyms adduced; *Eugenia Moorei*, F. Muell. Fragm. v. 33.

Queensland. Albany island, *W. Hill.*
N. S. Wales. Tweed river, *C. Moore.*
Very common in East India and the Archipelago, where the fruit is much eaten.

SECTION III. JAMBOSA.—Flowers in trichotomous panicles or cymes, calyx-tube more or less produced above the ovary, prominently lobed, the lobes usually persistent. Petals free and spreading.

This section, like *Syzygium*, is limited to the Old World, excepting where naturalized from cultivation.

6. **E. cormiflora,** *F. Muell. Fragm.* v. 32. A tree of 30 to 40 ft., with a fine head (*Dallachy*). Leaves ovate-elliptical to almost oblong, obtuse or shortly acuminate, 3 to 5 in. long, narrowed into a petiole often ½ in. long, not very thick, the principal veins rather distant and uniting irregularly far within the margin. Flowers large, in short trichotomous cymes, clustered on the trunk not above 3 ft. from the ground, the peduncles and pedicels very short. Calyx-tube urceolate, nearly ½ in. long, very thick, the free part short, dilated at the top; lobes 4, very unequal, the largest nearly orbicular, 4 to 5 lines broad. Petals 4, free, broad, unequal, the largest above ½ in. long. Stamens erect and more rigid than in most species, the outer ones above 1 in. long; anthers oblong. Ovary very thick and fleshy, with 2 small cells, each with about 8 ovules. Fruit ovoid-urceolate, crowned by the calyx-lobes, nearly 2 in. long.

Queensland. Rockingham Bay, *Dallachy;* Maryborough, *W. Hill.* The species appear to be very nearly allied to *E. Malaccensis*, Linn., common in India and the Archipelago.

7. **E. Tierneyana,** *F. Muell. Fragm.* v. 14. A tree of 60 to 70 ft., with an ashy bark and spreading branches (*Dallachy*), quite glabrous. Leaves elliptical-oblong to almost obovate, shortly and obtusely acuminate, 3 to 6 in. long, narrowed into a short petiole, not very thick, the primary nerves rather distant and uniting far within the margin. Flowers rather large, not numerous, in loose trichotomous cymes on the old wood, in the axils of the old leaves or at the nodes of denuded branches, not exceeding the leaves and often several from the same node. Calyx-tube turbinate, about 3 lines long, rapidly contracted into a short pedicel; lobes 4, orbicular, distinct, unequal, the largest nearly 2 lines, the smallest scarcely above 1 line diameter. Petals nearly 4 lines diameter, spreading and separately deciduous. Stamens half as long again as the petals. Ovary in the narrow base of the calyx, with numerous ovules in each cell. Fruit globular, red, about ½ in. diameter.

Queensland. Rockingham Bay, the red fruit produced in large quanties and making very good jam, *Dallachy.* The species is very nearly allied to the E. Indian *E. laurifolia*, Roxb., differing chiefly in the leaves narrowed at the base.

8. **E. grandis,** *Wight, Illustr.* ii. 17, *and Ic. t.* 614. A large and handsome tree, quite glabrous. Leaves from broadly oval to oval-oblong, obtuse or obtusely acuminate, 4 to 6 in. long, very firm and shining as in *E. Jambolana*, but thicker, and the veins more distant, forming a continuous intramarginal nerve. Flowers rather large and numerous, in dense trichotomous cymes, either terminal or in the upper axils. Calyx-tube thick, turbinate, shortly produced above the ovary, about 3 lines long; lobes 4, broad and short but unequal, wearing off after flowering. Petals usually spreading and falling off separately. Fruit globular, white, above 1 in. diameter, with 1 or 2 seeds, or smaller with 1 seed.—*E. cymosa*, Roxb. Fl. Ind. ii. 492, not of Lam.; *E. firma*, Wall. Cat. Herb. Ind. n. 3603; *Syzygium grande*, Walp. Rep. ii. 180; *Jambosa grandis* and *J. firma*, Blume, Mus. Bot. i. 108; *Eugenia fortis*, F. Muell. Fragm. v. 13.

Queensland. Lizard islands, *Banks and Solander;* Albany island, *W. Hill;* Rockingham Bay, *Dallachy.* The species is widely spread over the eastern provinces of India and the Archipelago. It is placed by Wight in the section *Syzygium* and by Blume in *Jambosa*, and is in some respects intermediate between the two.

9. **E. suborbicularis,** *Benth.* A tree attaining a considerable size, quite glabrous. Leaves broadly obovate or almost orbicular, very obtuse, 4 to 6 in. long, on a rather long petiole, coriaceous but not so thick and shining as in *E. grandis*, with numerous parallel diverging veins, confluent within the margin, and finely reticulate between them. Flowers large, in a short terminal trichotomous cyme. Calyx-tube narrow-turbinate, 7 to 8 lines long, broad and campanulate above the ovary; lobes 4, broad, the inner ones nearly ½ in. diameter, with scarious margins, the outer ones rather smaller. Petals spreading and separately deciduous, the larger outer one nearly ¾ in. diameter. Stamens exceedingly numerous, readily separable in the bud into 4 parcels. Ovules ascending.

Queensland. Cape York and Endeavour river, *W. Hill*, N.E. coast, *A. Cunningham.*

10. **E. Wilsonii,** *F. Muell. Fragm.* v. 12. Glabrous. Leaves broadly lanceolate, acuminate, 5 to 6 in. long, rounded at the base, with a short petiole, finely and transversely penniveined. Flowers large, in a short dense terminal cyme almost contracted into a head. Calyx-tube very narrow-turbinate, about 4 lines long; lobes 4, rounded, about 1 line diameter and nearly equal. Petals about 1½ lines diameter, separately deciduous. Stamens reddish-purple, the longer ones nearly 1 in. long. Anthers small. Ovary about half the length of the calyx-tube, concave at the top and scarcely fleshy. Ovules numerous in each cell, in 2 rows, ascending from a pendulous placenta. Fruit ovoid, about ½ in. long, narrowed at the top and crowned by the small calyx-lobes. Seeds usually 2 or 3; cotyledons thick and fleshy but separate.

Queensland. Rockingham Bay, *Dallachy.*

11. **E. eucalyptoides,** *F. Muell. Fragm.* iv. 55. A tall shrub or small tree, glabrous and somewhat glaucous, with pendulous branches. Leaves lanceolate, often falcate, 4 to 6 in. long or more, narrowed into a very short petiole, remotely and irregularly penniveined and reticulate, the principal veins more or less confluent at some distance from the edge. Flowers rather

large, few, in compact terminal cymes. Calyx-tube broadly turbinate, about 2 lines long, the free part broad; lobes 4, broadly orbicular, the inner larger ones almost as long as the tube. Petals orbicular, the larger outer ones fully 3 lines diameter, all separately deciduous. Ovary about half the length of the calyx; ovules incurved, acuminate. Fruit globular, 1-seeded, crowned by the calyx limb, but only seen young —*Jambosa eucalyptoides*, F. Muell. Fragm. i. 226.

N. Australia. Gravelly places on the Victoria river, *F. Mueller*. From the appearance of the leaves, their shape and venation, they are probably vertical as in many species of *Eucalyptus*.

12. **E. myrtifolia,** *Sims, Bot. Mag. t.* 2230. An evergreen glabrous shrub. Leaves petiolate, varying from oval-oblong or almost obovate to oblong-elliptical or almost lanceolate, obtuse or acuminate, 2 to 3 in. long, cuneate or narrowed at the base, finely and almost transversely penniveined. Peduncles axillary, lateral or terminating short leafy shoots, bearing usually 3 or 5 flowers but sometimes more, in a loose trichotomous panicle. Calyx-tube turbinate, 1½ to nearly 2 lines diameter; lobes very unequal, the largest nearly as long as the tube. Petals nearly 3 lines diameter, spreading and separately deciduous. Outer stamens nearly ½ in. long. Ovary about ½ the length of the calyx-tube, with a cluster of 8 to 10 ovules in each cell. Fruit red, ovoid or nearly globular, crowned by the calyx-limb.—Bot. Reg. t. 627; Lodd. Bot. Cab. t. 625; *E. australis*, Wendl. in Link, Enum. Hort. Berol. ii. 28; Colla in Hort. Ripul. App. t. 8; *Jambosa australis*, DC. Prod. iii. 287; *J. Thozetiana*, F. Muell. Fragm. i. 225.

Queensland. Moreton Bay, *F. Mueller, C. Stuart;* Wide Bay, *Bidwill;* Rockhampton, *Dallachy, Thozet;* Ipswich, *Nernst.*
N. S. Wales. Botany Bay, *Banks and Solander;* Hunter's River, *R. Brown, Macleay;* Hastings and Clarence rivers, *Beckler*; Illawarra, *A. Cunningham, Shepherd.*

13. **E. angophoroides,** *F. Muell. Fragm.* v. 33. A glabrous tree. Leaves petiolate, oblong-lanceolate or elliptical, acuminate, mostly 2 to 3 in. long, narrowed at the base, finely penniveined as in *E. Ventenatii*, but the veins more prominent, and not so much reticulate as in *E. Armstrongii*. Flowers in a compound terminal corymbose panicle, shorter than the leaves. Buds obovoid, nearly sessile or tapering into a very short pedicel. Calyx-tube turbinate, scarcely more than 1 line long, and about 1½ lines diameter; lobes or teeth either 5, all small and triangular, or one larger and more petal-like. Petals broad, about 1 line diameter, separately deciduous. Stamens about 2 lines long. Ovules several in each cell of the ovary. Fruit unknown.

Queensland. Rockingham Bay, *Dallachy*. With the habit and aspect of *E. Ventenatii*, but readily distinguished by the more sessile flowers as well as by the calyx and petals.

14. **E. Armstrongii,** *Benth.* A glabrous tree. Leaves petiolate, oblong-lanceolate, obtusely acuminate, narrowed at the base, mostly 3 to 5 in. long, more coriaceous than in *E. Ventenatii*, the intramarginal vein more distant from the edge. Flowers in a rather dense corymbose terminal panicle, much shorter than the leaves. Calyx-tube turbinate-campanulate, about 1½ lines long; lobes short and very broad, the inner ones larger and often petal-

like on the margin. Petals quite separate, 1½ to 2 lines diameter. Stamens rather longer than the petals. Ovary very short in the bottom of the calyx, with about 6 recurved ovules in each cell.

N. Australia. Port Essington, *Armstrong;* N. coast, *A. Cunningham.*

15. **E. oleosa,** *F. Muell. Fragm.* v. 15. A small handsome tree of 15 to 20 ft. (*Dallachy*), quite glabrous. Leaves from elliptical to lanceolate, acuminate, narrowed at the base, 2 to 3 or rarely 4 in. long, not very thick, the veins oblique and prominent underneath. Flowers white, remarkable for their long slender stamens, in trichotomous pedunculate cymes, either opposite at the base of the new shoots, or terminal in pairs, the peduncles, branches and pedicels slender. Calyx narrow-turbinate, nearly 3 lines long, tapering into a pedicel, sometimes short, sometimes as long as the calyx; lobes 4, ovate or broad, about ½ line long. Petals quite separate, about 1½ lines diameter. Filaments very numerous and fine, ½ in. long or more. Ovary not half so long as the calyx-tube, with about 8 ovules in each cell; style long and slender. Fruit globular.

Queensland. Rockingham Bay, *Dallachy.* Very near *E. rivularis*, Seem., from the Fiji Islands, but the veins of the leaves are not nearly so numerous or close, and more oblique, and the stamens nearly twice as long as in that species.

16. **E. (?) Dallachiana,** *F. Muell. Herb.* Leaves broadly ovate, 3 to 5 in. long, of a thinner consistence than in most *Eugenias*, and the one or two lower pairs of veins more prominent than the others, and continued almost to the apex of the leaf, so as to make it appear almost triplinerved or quintuplinerved like some *Rhodomyrti.* Cymes axillary, pedunculate, rather loose, and apparently only few-flowered, but the specimens seen are only in young fruit. Calyx-tube in that state nearly globular, about 3 lines diameter, not produced above the ovary; lobes 4, broad, spreading, unequal, all shorter than the tube. Petals not seen. Remains of stamens those of *Eugenia.* Ovary 2-celled, with rather numerous ovules in each cell, but only one or two from the same cell enlarged. Young seed apparently that of *Eugenia*, but not far enough advanced to determine.

Queensland. Rockingham Bay, *Dallachy.* The aspect of this plant is very different from that of any *Eugenia* known to me, yet, as far as the specimens go, they supply no character to separate it from the genus.

TRIBE IV. LECYTHIDEÆ.—Ovary divided more or less completely into 2 or more cells. Fruit woody fibrous or fleshy, indehiscent or opening in an operculum at the top. Leaves alternate, not dotted.

SUBTRIBE I. BARRINGTONIEÆ.—Stamens inserted on a regular broad or cup-shaped disk (not unequally produced on one side as in the American *Eulecythideæ*). Fruit fibrous or fleshy but not woody. Calyx usually almost, but not quite valvate.

44. BARRINGTONIA, Forst.

(Stravadium, *Juss.*)

Calyx-tube ovoid or turbinate, not at all or scarcely produced above the ovary, the limb either closed in the bud and splitting into 2 to 4 valvate segments or rarely with 3 or 4 lobes, imbricate in the bud. Petals 4 or 5, adhering at the base to the staminal cup. Stamens numerous, in several series, shortly united at the base into a ring or cup; anthers small, with parallel cells opening longitudinally. Ovary inferior, with an annular disk on the top within the stamens, 2- to 4-celled, with 2 to 8 ovules in each cell, horizontal or pendulous, in 2 rows; style filiform with a small stigma. Fruit pyramidal ovoid or oblong, hard and fibrous, indehiscent. Seed usually solitary, with a thick testa; embryo undivided, consisting of a thick woody stratum, and a more or less distinct pith in the centre.—Trees. Leaves alternate, usually crowded at the ends of the branches, penniveined and not dotted. Flowers in terminal or lateral spikes or racemes. Bracts small and deciduous.

The genus is confined to the tropical regions of the Old World. The Australian species are both widely dispersed over the Indian Archipelago, and one is also common in East India.

Leaves often above 1 ft. long. Flowers large, in short racemes. Stamens 2 to 4. in. long. Fruit large . . . . 1. *B. speciosa.*
Leaves under 6 in. long. Flowers small, in long racemes. Stamens 3 to 4 lines long. Fruit small . . . . 2. *B. acutangula.*

1. **B. speciosa,** *Linn. f.; DC. Prod.* iii. 288. A large handsome tree. Leaves sessile, obovate, entire, attaining more than 1 ft. in length. Flowers very large, in short terminal racemes, the rhachis thick, the pedicels 1 to 2 in. long. Calyx deeply divided into 2 or 3 oval-oblong concave almost leaf-like segments, above 1 in. long. Petals from half as long again to twice as long as the calyx-segments. Stamens very numerous, red, 2 to 4 in. long. Ovary imperfectly 4-celled, with about 6 ovules in each cell. Fruit large, pyramidal, 4-angled, crowned by the persistent calyx-lobes.—Wight, Ic. t. 547.

Queensland. Cape York and Dayman's Island, Endeavour Straits, *W. Hill.*—Widely dispersed over the Indian Archipelago and Pacific Islands. The Australian specimens are imperfect, but there is very little doubt of their identity with the Archipelago plant, from which the above description is taken.

Some specimens, named *B. calyptrata*, by R. Br., from the Lizard Islands, *Banks and Solander*, have the foliage nearly of *B. speciosa*, but the flowers in long racemes like those of *B. acutangula*. None are fully out, but they appear to be intermediate in size between those of *B. speciosa* and *B. acutangula.*

2. **B. acutangula,** *Gærtn. Fruct.* ii. 97. *t.* 101. A large handsome tree. Leaves from obovate or oblong-cuneate to almost elliptical, obtuse or shortly acuminate, rarely much above 4 in. long, serrulate or entire, narrowed into a short petiole. Flowers red, rather small, in very long slender pendulous racemes. Bracts oblong, very deciduous. Pedicels 2 to 4 lines long. Calyx-tube ovoid-globose, about 1 line long; lobes 4, rather longer than the tube, orbicular. Petals about twice as long as the calyx-tube. Stamens not much longer than the petals. Ovary 2-celled, with 2 pendulous ovules in each cell. Fruit oblong, 4-angled, 1 in. long or rather more.—Wight and Arn. Prod. 333; *Stravadium rubrum*, DC. Prod. iii. 289.

N. Australia. Rivulets of M'Adam range and Fitzmaurice river, *F. Mueller.*—Common in most parts of India as well as in the Archipelago. I find but 2 cells to the ovary both in the Indian and the Australian specimens.

45. CAREYA, Roxb.

Calyx-tube thick, turbinate or ovoid, not produced above the ovary, the limb deeply 4-lobed. Petals 4, spreading. Stamens numerous in several series, quite free, the outermost longer ones or the innermost shorter ones or both without anthers, the intermediate ones or nearly all perfect; anthers small, with parallel cells opening longitudinally. Ovary inferior, 4- or rarely 5-celled, with several small ovules in 2 rows in each cell; style elongated, with a somewhat capitate or slightly 4-lobed stigma. Fruit globular, fleshy, with a hard rind, crowned by the calyx-limb. Seeds several, enveloped in a fleshy pulp, and usually irregularly scattered; testa thick; embryo undivided. —Trees, or in one instance an undershrub. Leaves alternate, usually crowded at the ends of the branches, penniveined and not dotted. Flowers large, in racemes or interrupted spikes, usually short.

The genus comprises three E. Indian species, one of which is supposed to be the same as the Australian one.

1. **C. arborea,** *Roxb. Pl. Corom.* iii. 14. *t.* 218, *and Fl. Ind.* ii. 638. —Var. (?) *australis.* A tree attaining a large size. Leaves from ovate and shortly acuminate to obovate and very obtuse, minutely crenulate or entire, not above 4 in. long in the Australian specimens seen, but much larger in Indian ones. Flowers large, pedicellate, few together in very short cymes, terminating short leafy shoots. Calyx-lobes 4, orbicular, unequal, the larger ones minutely ciliolate. Petals when fully out obovate-oblong, in some specimens 2 in. long, in others much smaller. Perfect stamens as long as the petals, without any barren filaments outside, but a few short ones inside without anthers. Ovary 4-celled, with 10 to 12 ovules in each cell. Fruit broadly ovoid, 1½ in. long or more, not at all angled, crowned by the persistent calyx-lobes.

N. Australia. Brunswick Bay, N.W. coast, *A. Cunningham;* plains at the mouth of the Victoria river, *F. Mueller;* Islands of the Gulf of Carpentaria, *R. Brown, Henne.*

Queensland. Cape Grafton, *Banks and Solander;* Estuary of the Burdekin, *Fitzalan;* Rockhampton, *Dallachy.*

I have some doubts whether this be really identical with Roxburgh's *C. arborea* from the Coromandel coast, figured also in Wight, Illustr. ii. t. 99, 100, which has usually much larger leaves, and is said to have the flowers closely sessile, very numerous ovules in each cell of the ovary, and the fruit globular. Our specimens of the Indian plant are very imperfect, and those of the Australian one in F. Mueller's as well as in the Hookerian herbarium, although numerous from various localities, are for the most part fragmentary. R. Brown's alone are in fruit as well as in flower.

F. Mueller proposes to unite *Careya* with *Barringtonia* on account of the very few anantherous stamens in *C. arborea* (perhaps sometimes none), showing a connecting link between the two genera. But as far as known, there appears to be also a marked difference in the shape and structure of the fruit in the two cases, besides some minor differences, of the constancy of which we have at present no means of judging.

ORDER XLIX. MELASTOMACEÆ.

Calyx-tube enclosing the ovary, and either cohering with its angles, leaving intermediate cavities, or entirely free or more or less adnate to it, the limb entire or with 3 to 5 or rarely 6 lobes or teeth, usually imbricate in the bud. Petals as many as calyx-lobes, inserted at the orifice of the calyx-tube, imbricate (usually contorted) in the bud. Stamens usually twice as many, sometimes only as many as petals and inserted with them, the filaments curved down in the bud; anthers 2-celled, opening in 1 or 2 pores at the top or very rarely in longitudinal slits, and before flowering their tips are usually contained in the cavities between the ovary and calyx, the connective often variously extended or thickened. Ovary enclosed in the calyx-tube and adnate to it, or more or less free, with 2 to 6 or rarely more cells, with the placenta in the axis, or rarely 1-celled by the abortion of the partitions. Style simple, with a minute or capitate or peltate stigma. Ovules several, rarely 2 only to each placenta, anatropous. Fruit enclosed in the calyx or combined with it, either succulent and indehiscent, or bursting irregularly, or capsular and opening in as many valves as there are cells. Seeds usually numerous and small, straight or *cochleate* (*i. e.* curved somewhat like an univalve shell), without albumen; testa coriaceous, crustaceous or membranous. Embryo straight or curved; cotyledons plano-convex or thick and variously folded; radicle short.—Herbs shrubs or rarely trees. Leaves opposite, simple, petiolate, 3- to 11-nerved, or in *Memecyleæ* 1-nerved and penniveined, entire or rarely serrulate. Stipules none. Flowers usually in terminal panicles or clusters, rarely axillary or solitary.

A large Order, chiefly American, and most abundant within the tropics, a considerable number also in tropical and subtropical Asia, especially in the Eastern Archipelago, and a few in tropical and southern Africa. The four Australian genera are all Asiatic and three of them also African.

TRIBE I. **Osbeckieæ.**—*Leaves with* 3, 5 *or more ribs. Anthers opening in a single terminal pore. Ovary more or less adherent, except the convex or conical summit,* 2- *to* 6-*celled. Fruit capsular or rarely pulpy. Seeds cochleate.*

Anthers all similar and equal or nearly so. Fruit capsular, opening in valves.
Calyx-lobes 4, rarely 5, with bristle-like appendages between them. Anthers without any or scarcely any appendage at the base . . . 1. OSBECKIA.
Calyx-lobes 5 or 6, without appendages between them. Anthers with a short 2-lobed inflected appendage at the base . . . . 2. OTANTHERA.
Anthers alternately smaller or dissimilar. Fruit succulent or pulpy, bursting irregularly . . . . 3. MELASTOMA.

TRIBE II. **Memecyleæ.**—*Leaves with the midrib prominent, the veins pinnate or inconspicuous. Anther-cells adnate to a much thickened connective, and opening in separate slits or pores. Ovary adnate,* 1-*celled, with a central placenta. Fruit a berry* . . . . 4. MEMECYLON.

1. OSBECKIA, Linn.

Calyx-tube ovoid globular or urceolate; lobes or teeth 4 or 5, deciduous, with appendages between them, which are usually bristle-like, terminating in a tuft of hairs. Petals obovate. Stamens twice as many as petals, all equal and similar or nearly so; anthers opening in a single pore at the summit, and without any or scarcely any appendage at the base of the connective. Ovary 4- or 5-celled, crowned with bristles. Fruiting-calyx usually truncate after the fall of the lobes; capsule opening at the top in as many valves as there were cells to the ovary. Seeds cochleate.—Herbs undershrubs or rarely shrubs. Leaves sessile or petiolate, 3-, 5- or 7-nerved. Flowers usually terminal, in clusters or short racemes, often forming leafy panicles, rarely solitary. Calyx-tube often more or less covered with bristles or ciliate scales.

The genus comprises a few African species besides a considerable number from tropical Asia and the Archipelago, including one of the Australian ones, the other Australian species is endemic.

Flowers 4-merous. Scales of the calyx with long bristles, 5 alternating with the lobes and sometimes a few below the middle of the tube. Anthers with long slender beaks . . . . 1. *O. chinensis.*
Flowers 5-merous. Scales of the calyx with short bristles, very numerous and completely covering the tube. Anthers with short beaks . 2. *O. australiana.*

1. **O. chinensis,** *Linn. Spec. Pl.* 490 (*not Bot. Reg. nor Bot. Mag.*). A herb undershrub or shrub, from 1½ to 3 ft. high, glabrous or with a few short stiff hairs. Leaves very shortly petiolate, linear linear-oblong or almost lanceolate, 1 to 2 in. long. Flowers several together, sometimes very few, forming sessile terminal clusters, almost condensed into heads. Calyx-tube about 3 lines long or rather more; lobes 4, not quite so long as the tube, broad or narrow, acute, ciliate, but without any terminal tuft of hairs, with 4 accessory ciliate scales inserted between and a little below them on the outside, and occasionally a few ciliate scales on the tube below the middle. Petals 4. Anthers produced into a slender beak. Capsule 4-celled.—Benth. Fl. Hongk. 115, with the synonymes adduced; F. Muell. Fragm. iv. 160; *O. angustifolia*, Don; Wall. Pl. As. Rar. iii. t. 251.

Queensland. Rockingham Bay, *Dallachy*. The species extends over the Indian Archipelago and the eastern provinces of India to S. China and Formosa.

2. **O. australiana,** *Naud. in Ann. Sc. Nat. ser.* 3. xiv. 59. A shrub, attaining several feet, more or less scabrous-pubescent. Leaves linear or lanceolate, 3-nerved, mostly 1 to 2 in. long or smaller on the side-branches. Flowers usually 3 to 7, at the ends of the branches, in a cyme, sessile, but not so dense as in *O. chinensis*. Calyx-tube nearly globular, about 3 lines diameter, densely covered with tufts of rather short bristles (bristly scales); lobes 5, ovate or ovate-lanceolate, much shorter than the tube, ciliate but without a terminal tuft of hairs. Petals twice as long as the calyx-tube. Anthers with a short broad beak. Capsule 5-celled.

N. Australia. Melville Island, *Fraser;* M'Adam Range and Arnhem's Land, *F. Mueller;* Port Essington, *Armstrong*.

2. OTANTHERA, Blume.

(Lachnopodium, *Blume.*)

Calyx-tube ovoid; lobes 5 or 6, deciduous, alternating with as many short bristly scales or appendages. Petals obovate. Stamens twice as many as petals, all equal and similar; anthers opening in a single pore on the summit, the connective produced at the base into a short 2-lobed appendage turned up on the inner face. Ovary 5- or 6-celled, crowned with bristles. Fruiting-calyx truncate after the fall of the lobes; capsule (in the Australian species) opening at the top in as many valves as there were cells to the ovary, in other species more pulpy and less regularly dehiscent. Seeds cochleate, small and very numerous.—Shrubs more or less strigose, with the habit of the smaller-flowered *Melastomas*. Leaves 5- or 7-nerved. Flowers in terminal trichotomous cymes or panicles.

The genus consists of very few species natives of the Indian Archipelago, one of which, the same as the Australian one, differs slightly from the others in the fruit drier and more capsular, and was therefore distinguished by Blume under the name of *Lachnopodium*.

1. **O. bracteata,** *Korth. Verh. Nat. Gesch. Bot.* 235. *t.* 51. A shrub of several feet, the branches more or less covered with pale-coloured or rusty hairs or bristles. Leaves petiolate, ovate or ovate-elliptical, mostly 3 to 5 in. long, membranous, rough with short strigose hairs. Flowers few, in short terminal trichotomous cymes, the peduncles and pedicels with a few small leaves at the base of the cyme, and a short, broad, concave, almost cordate bract at the base of each branch or pedicel. Calyx-tube about 2 lines long, densely covered with small scales, divided each into 3 to 5 long erect cilia or bristles; lobes 5 or 6, linear, scarcely so long as the tube, ciliate with a few long bristles, the intermediate bristly scales short and obtuse. Petals white or pink, 5 to 6 lines long, each with a bristle at the end. Ovary adnate to about half the calyx-tube, the convex summit very bristly. Fruit nearly globular, crowned by the scars of the calyx-lobes. Capsule apparently dry, the placentas projecting far into the cells.—Naud. in Ann. Sc. Nat. ser. 3. xiii. 354; *Lachnopodium bracteatum*, Blume, Mus. Bot. i. 56.

Queensland. Dalrymple Creek, Rockingham Bay, *Dallachy*. Also in Sumatra.
Korthals figures the calyx-lobes rather broad; I find them narrow, as described by Blume, both in the Sumatran and the Australian specimens. *Lachnopodium rubro-limbatum*, Blume, Mus. Bot. i. 56, taken up from the *Melastoma rubro-limbatum*, a garden plant, figured in Link and Otto, Ic. Pl. Sel. 89. t. 41, appears to be the same species.

3. MELASTOMA, Linn.

Calyx-tube campanulate or ovoid; lobes or teeth 5 or rarely 6, deciduous, with or without small alternate accessory lobes or appendages. Petals obcordate or obovate. Stamens twice as many as petals; anthers elongated, opening at the top in a single pore, very unequal, 5 larger, with the connective produced below into a long appendage incurved and 2-lobed or 2-pointed at the lower end, 5 smaller, with the appendage shorter or wanting. Ovary 5- or rarely 6-celled, crowned with a few stiff hairs or bristles. Fruit truncate after the fall of the calyx-lobes, the capsule or berry more or less succulent or pulpy and bursting irregularly. Seeds cochleate.—Shrubs, more or less strigose or hairy. Leaves usually ovate, 3- or more-nerved. Flowers terminal, solitary or few together in cymes, often large and showy; the calyx usually covered with bristles or scales.

A considerable genus, extending over tropical Asia and the Pacific islands. The only Australian species is a common one in India and the Archipelago.

1. **M. malabathricum,** *Linn. Spec. Pl.* 559, var. *polyanthum*. A shrub of a few feet in height, more or less clothed with hairs or bristles, often very rigid and scale-like on the branches, rigid and strigose on the upper side of the leaves, longer and softer on the under side, but sometimes nearly all rigid and scale-like, or nearly all long and soft. Leaves petiolate, from ovate almost cordate and 6 in. long, to oblong-lanceolate and 3 in. long, with 3 or 5 nerves besides a fine intramarginal one. Flowers usually about 5 to 11 in terminal almost sessile cymes. Bracts very deciduous, from large and broadly ovate to small and narrow-lanceolate. Calyx-tube ovoid-globular, 2 to 3 lines long,

densely covered with appressed chaffy scales or bristles; lobes usually 5, from ovate to lanceolate, more or less acuminate, longer and sometimes much longer than the tube or rarely rather shorter, alternating with 5 small subulate or short chaffy scales or accessory lobes. Petals large, pale purple or white. Fruit nearly globular, 3 to nearly 4 lines diameter. Seeds imbedded in a purple pulp.—*M. polyanthum*, Blume, Mus. Bot. i. 52. t. 6; *M. denticulatum*, Labill. Sert. Austr. Caled. i. 65. t. 64; *M. Novæ-Hollandiæ*, Naud. in Ann. Sc. Nat. ser. 3. xiii. 290.

N. Australia. Between Providence Hill and M'Adam Range, and Adelaide river, *F. Mueller*; Port Essington, *Armstrong*.

Queensland. Endeavour river, *Banks and Solander, A. Cunningham*; Brisbane river, Moreton Bay, *A. Cunningham*, and others; Mount Elliott, *Dallachy*.

N. S. Wales. Clarence river, *Wilcox*.

The typical *M. malabathricum* is usually distinguished by its larger flowers, with the bracts and calyx-lobes larger in proportion, but some of the Moreton Island specimens have them nearly as large as the Indian ones. Many Australian specimens correspond exactly either with those of *M. polyanthum* from the Archipelago, or with those of *M. denticulatum*, from New Caledonia, and it is probable that the species should include the whole of the twenty-four adopted or proposed by Naudin, Ann. Sc. Nat. ser. 3. xiii. 283 to 293, as "Species magis ad *M. malabathricum* vergentes ideoque difficilius distinguendæ," besides several of the "Species addendæ," p. 294, not seen by him. The characters are generally most trifling.

4. MEMECYLON, Linn.

Calyx-tube hemispherical or campanulate, the limb entire or obtusely 4-lobed, rarely 5-lobed. Petals 4 or rarely 5, ovate or orbicular. Stamens twice as many as petals, all equal and similar; anthers short, with a thick connective, forming a conical spur at the base, the cells opening in longitudinal slits. Ovary entirely adnate to the calyx-tube, 1-celled, with 6 to 12 ovules, verticillate round a short central placenta; style filiform, with a small stigma. Fruit a berry, crowned by the calyx-teeth or border, or by a circular scar only. Seeds solitary or rarely 2 or 3; testa somewhat crustaceous; cotyledons very much convolute or variously folded, usually enclosing the radicle.—Trees or shrubs. Leaves coriaceous, with 1 prominent midrib and pinnate veins often scarcely perceptible. Flowers usually small, in axillary clusters or cymes.

The genus is spread over the tropical regions of the Old World, the species especially numerous in Ceylon and the Indian Archipelago. The only Australian one is also in Ceylon and the Indian Peninsula.

1. **M. umbellatum,** *Burm. Fl. Ind.* 87. A bushy or divaricately-branched shrub, quite glabrous. Leaves shortly petiolate, from broadly ovate to ovate-lanceolate, obscurely and obtusely acuminate, 1 to 2 in. long, of a dark green and shining above, paler or sometimes yellowish underneath, the veins usually quite inconspicuous. Peduncles axillary, very short, bearing an umbel-like or shortly racemose cluster of small flowers, on slender pedicels of 1 line or rather more. Adnate part of the calyx-tube very short, the free part broadly campanulate, less than 1 line diameter, broadly and shortly 4-lobed. Petals ovate, acute, about 1 line long. Stamens exceeding the petals. Fruit green, smooth, nearly globular, about 3 or rarely 4 lines diameter, crowned by the small persistent calyx-limb; pericarp slightly fleshy. Seed

solitary, globular; cotyledons fleshy and very much contortuplicate.—Thwaites, Enum. Ceyl. Pl. 111; *M. ramiflorum*, Lam. Dict. iv. 88, DC. Prod. iii. 6 (at least as to the Indian plant); Wight, Illustr. i. 214. t. 93 (*M. tinctorium*, Kœn. on the plate); *Myrcia? Australasiæ*, F. Muell. Rep. Burd. Exped. 7.

N. Australia. North-west coast, *A. Cunningham.*

Queensland. Estuary of the Burdekin, *Fitzalan;* Mount Elliott, Edgecombe and Rockingham Bays, *Dallachy*, Cleveland Bay, *Bowman.*

The species is common in Ceylon and the Indian Peninsula, and perhaps also in the Mauritius.

Specimens of a tree from the Clarence river, *Beckler and Wilcox*, and from Richmond river, *C. Moore*, in fruit only and bearing F. Mueller's MSS. name of *Nelitris (?) ingens*, may possibly belong to a *Memecylon*, although unlike any species known to me. The leaves are penniveined, not unlike those of *Eugenia myrtifolia*, but rather larger and not dotted. The fruits are in cymes, either terminal or in the upper axils, globose, very hard, about 1 in. diameter, marked with the scar of the calyx-limb. Seed 1 only, globular. Embryo thick and hard, the cotyledons very complicately folded and contortuplicate as in *Memecylon.*

ORDER L. LYTHRARIEÆ.

Calyx-tube free, but usually enclosing the ovary; lobes or primary teeth 4, 5, or sometimes more, very rarely 3, valvate in the bud, the sinus sometimes produced externally into as many accessory teeth. Petals as many as primary calyx-teeth or lobes, rarely deficient, inserted at the top of the calyx-tube, usually crumpled in the bud. Stamens as many or twice as many as petals or fewer, or rarely indefinite, inserted in the calyx-tube at various heights; filaments inflected in the bud; anthers versatile, with parallel cells opening longitudinally. Ovary free from the calyx, but usually enclosed in its tube, 2- or more-celled, or rarely 1-celled by the abortion of the partitions; style simple, the stigma capitate or rarely 2-lobed. Ovules usually numerous, anatropous, attached to the axis, or very rarely parietal. Fruit a membranous coriaceous or hard capsule, variously dehiscent, enclosed in or surrounded by the persistent calyx, the valves usually detaching themselves from the central persistent placentiferous column. Seeds without albumen; testa coriaceous, membranous or rarely thick; embryo straight; cotyledons oblong or orbicular-cordate; radicle short, or rarely cotyledons small and radicle long.—Herbs shrubs or trees. Leaves opposite, verticillate or sometimes alternate, entire, without stipules. Flowers in axillary or terminal panicles cymes or clusters, rarely solitary.

A considerable Order, some of the herbaceous genera spread over the greater part of the globe, the larger woody-stemmed ones confined to the tropics in the Old or the New World. The five Australian genera are all Asiatic, three of them at least are also African, and the two herbaceous genera extend to America and Europe.

Annual or perennial herbs, very rarely becoming woody at the base.
  Calyx short, membranous, the ribs inconspicuous or only as many as primary teeth; accessory teeth minute or none. Petals very small or none . . . . 1. AMMANNIA.
  Calyx narrow, with twice as many ribs as primary teeth; accessory teeth prominent. Petals usually conspicuous . . . . 2. LYTHRUM.
Shrubs or trees.
  Stamens twice as many as petals,

Calyx-lobes 6, with accessory teeth. Capsule enclosed in the calyx. Maritime shrub, with solitary flowers in the upper axils . . . 3. PEMPHIS.
Calyx-lobes 4, without accessory teeth. Capsule exserted. Flowers in leafy panicles . . . . 4. LAWSONIA.
Stamens indefinite. Calyx-lobes 4 to 8, without accessory teeth. Fruit large, fleshy. Flowers large, 1 to 3 in the upper axils . . . . 5. SONNERATIA.

1. AMMANNIA, Linn.

(Rotala, *Linn.*; Ameletia, *DC.*)

Calyx membranous, short, the ribs not at all, or the primary ones only, prominent, with 4 or 5, rarely 3 or 6 primary teeth, without any or with very small external accessory ones. Petals small and fugacious or none. Stamens as many as primary calyx-teeth, or twice as many or fewer, inserted towards the middle of the tube or lower down. Ovary 2- to 5-celled, or 1-celled by the abortion of the partitions. Style often short, with a capitate stigma. Capsule included in the persistent calyx or protruding from it, opening in septicidal valves or bursting irregularly. Seeds very small.—Annual herbs, chiefly frequenting wet situations, usually glabrous, with a 4-angled stem. Leaves opposite or verticillate. Flowers very small, subsessile or pedicellate, solitary or in trichotomous cymes or clusters, with a pair of small bracteoles under the calyx, sometimes very minute or scarcely conspicuous.

A considerable genus, chiefly tropical and Asiatic or African, with a few species from tropical or Northern America, or from more temperate Asia. Of the seven Australian species two are endemic, the others widely spread over tropical Asia, and at least four extend into Africa.

Flowers sessile, solitary in the axils. Capsule opening in as many valves as cells. (**Rotala.**)
Leaves narrow, in whorls of 3 to 8. Capsule 3- or 4-valved . . . 1. *A. Rotala.*
Leaves ovate-lanceolate or oblong, opposite or rarely in threes. Capsule 3- or 4-valved . . . . 2. *A. pentandra.*
Leaves orbicular, opposite. Capsule 2-valved . . . . 3. *A. diandra.*
Flowers pedicellate, solitary or in cymes. Capsule bursting irregularly or transversely.
Flowers solitary, on long filiform pedicels. Leaves oblong-linear. Petals present . . . . 4. *A. crinipes.*
Flowers 1 to 3, on short axillary peduncles. Leaves broadly oblong, petiolate. No petals . . . . 5. *A. triflora.*
Flowers several in axillary cymes.
Leaves narrowed at the base. Calyx-lobes triangular. No petals. 6. *A. indica.*
Leaves dilated or cordate, auriculate at the base. Calyx-teeth very short. Petals present.
Capsule under 1 line diameter. Stamens 4 or fewer . . . . 7. *A. multiflora.*
Capsule about 1½ lines diameter. Stamens above 4, usually 6 to 8 . . . . 8. *A. auriculata.*

1. **A. Rotala,** *F. Muell. Fragm.* iii. 108. A slender annual, simple or slightly branched, often creeping at the base, and not above 3 in. long in the Australian specimens, twice as much in some Indian ones. Leaves in whorls of 3 to 6 or sometimes more, linear, not exceeding 3 or 4 lines. Flowers minute, nearly sessile and solitary in the axils. Calyx smooth and membranous, not above ½ line diameter, with 5 or sometimes 4 or 3 acute teeth without accessory ones. Petals none or minute and fugacious in the Austra-

lian specimens, nearly as long as the calyx-teeth in some Indian ones. Stamens 3 (4 or 5?), inserted near the base of the calyx and not exceeding it. Ovary 1-celled or more or less divided into 3 by very thin evanescent partitions. Style short. Capsule 3-valved.—*Rotala verticillaris*, Linn. Mant. 195; DC. Prod. iii. 76; Wight, Ic. t. 260; *Rotala apetala*, F. Muell. Fragm. iii. 108.

N. Australia. Beds of streams periodically inundated, Sturt's Creek, *F. Mueller.* Spread over E. India and the Archipelago.

2. **A. pentandra,** *Roxb. Fl. Ind.* i. 427. Annual or perhaps a perennial of short duration, often shortly creeping at the base, with ascending or erect stems, 6 to 8 in. high and scarcely branched when luxuriant, but often only 2 or 3 in. and much branched. Leaves opposite or very rarely the floral ones in threes, from ovate-cordate to oblong and almost cuneate, acute or obtuse, the larger ones ½ in. long, but usually not above ¼ in., the floral ones always exceeding the flowers. Flowers solitary in the axils, sessile or nearly so. Calyx scarcely above ½ line diameter, with 5 or rarely 4 or 3 short lobes, without accessory teeth. Petals very small or none. Stamens 5, or sometimes 4 or 3 inserted near the base of the calyx and not exceeding its lobes. Capsule opening in 3 or rarely 4 valves.—DC. Prod. iii. 79; W. and Arn. Prod. 305, with the synonyms adduced; Blume, Mus. Bot. ii. t. 46; *Rotala Roxburghiana*, Wight, Ic. t. 260.

Queensland. Endeavour river, *R. Brown;* water-holes, Moreton Bay, *C. Stuart.*

Var. *decussata.* Smaller and more branched. Petals usually none.—*Rotala decussata*, DC. Prod. iii. 76; *Ortegioides decussata*, Soland. in Herb. Banks; *Entelia ammannioides*, R. Br. Herb.; *Ammannia illecebroides*, Arn. in Wight, Cat. n. 2317.

N. Australia. Islands of the Gulf of Carpentaria, *R. Brown;* Victoria river, *F. Mueller;* Port Essington, *Armstrong.*

Queensland. Shoalwater Bay, *R. Brown;* E. coast, *Banks and Solander.*

3. **A. diandra,** *F. Muell. Fragm.* iii. 108 (under *Ameletia*). Erect or creeping at the base, branched or nearly simple, not exceeding 6 in. but not so slender as the preceding species. Stem leaves sessile, orbicular, very obtuse, cordate at the base, 2 to 3 lines diameter, the floral ones scarcely smaller, orbicular or ovate, and often very close, forming imbricate decussate spikes. Flowers solitary in the axils, sessile or nearly so. Calyx small, very thin and membranous, somewhat 4-angled, with 4 acute lobes shorter than the tube, without accessory teeth. Petals rudimentary. Stamens usually 2, inserted below the middle of the tube. Ovary 1-celled or imperfectly 2-celled. Capsule opening in 2 valves.

N. Australia. Around the lagoons and moist banks of the Upper Victoria and Fitzmaurice rivers, *F. Mueller.* The specimens are all in fruit; in some the calyx is scarcely 1 line long and shorter than the oblong capsule; in others the calyx is nearly 2 lines long, with a very much shorter globular capsule. In all I have found either small rudimentary petals or their scars, and the stamens adherent to about ⅓ of the calyx.

4. **A. crinipes,** *F. Muell. in Trans. Phil. Soc. Vict.* iii. 49. A slender branching annual of 3 or 4 in. Leaves linear-oblong, obtuse, 2 to 3 lines long, narrowed at the base. Pedicels solitary in the axils, filiform, 1-flowered, often exceeding ½ in. Calyx about 1 line long, narrowed at the base, broader and somewhat folded at the orifice, truncate, slightly sinuate and readily splitting into 4 retuse lobes. Petals 4, small, white, orbicular. Stamens 4 or

fewer, shorter than the calyx and inserted about the middle. Ovary 2-celled; stigma large. Fruiting-calyx broad; capsule ovoid or nearly globular, as long as the calyx, bursting irregularly.

N. Australia. About pools and lagoons, between the Victoria and Fitzmaurice rivers, *F. Mueller*. This has the solitary flowers of the section *Rotala*, with the capsule of the many-flowered true *Ammannias*, and differs from the whole genus in the long filiform peduncles.

5. **A. triflora,** *R. Br. Herb.* A diffuse much-branched annual, with slender ascending stems of ½ to 1 ft., minutely hoary-pubescent or glabrous. Leaves distinctly petiolate, oval-oblong, narrowed or rounded at the base, mostly under ½ in. long. Peduncles short, with 1 to 3 sessile or very shortly pedicellate flowers, much larger than in *A. indica.* Calyx-tube broadly campanulate, with 4 broad triangular lobes, and the sinuses produced into as many short horizontally spreading accessory lobes. Petals none. Stamens 4, inserted in the middle of the calyx-tube. Capsule 2-celled, depressed, irregularly circumsciss.

N. Australia. Islands of the Gulf of Carpentaria, *R. Brown.*

6. **A. indica,** *Lam. Illustr. n.* 1555 ?; *DC. Prod.* iii. 77. Erect, more or less branched, and often exceeding 2 ft. in height. Leaves lanceolate or oblong-linear, acute, narrowed at the base, mostly ½ to 1 in. long, but luxuriant ones sometimes longer, and those of the side branches smaller Flowers very small, in little axillary cymes or clusters, the pedicels slender, but rarely 1 line long, and the common peduncle very short or scarcely any. Calyx broadly campanulate, usually about ¾ line diameter, with 4 short broad triangular lobes, without accessory teeth. Petals none. Stamens 2 to 4. Ovary 2-celled. Capsule depressed-globular, usually exceeding the calyx, and bursting irregularly.—W. and Arn. Prod. 305; Blume, Mus. Bot. ii. 133. t. 46; *A. vesicatoria*, Roxb. Fl. Ind. i. 426; DC. Prod. iii. 78; W. and Arn. Prod. 305.

N. Australia. Careening and Brunswick bays, N.W. coast, *A. Cunningham;* Nichol Bay, *Gregory's Expedition;* Victoria river and Sturt's Creek, *F. Mueller.*

Queensland, *Bowman;* Endeavour river, *Banks and Solander;* Shoalwater Bay, *R. Brown.*

S. Australia. Cooper's Creek, *Howitt's Expedition.*—Common in tropical and subtropical Africa and Asia. Lamarck describes the leaves as decurrent, but this is undoubtedly the species to which his plant has been referred by De Candolle and others.

7. **A. auriculata,** *Willd.; DC. Prod.* iii. 80. Erect and not much branched, usually 6 in. to 1 ft. high, and coarser than the other Australian species, with larger flowers. Leaves lanceolate or oblong-linear, mostly ½ to 1 in. long, sessile and dilated at the base, and more or less cordate-auriculate. Flowers in little axillary cymes, shorter than the floral leaves, the peduncles, branches, and pedicels all short. Calyx at first narrow at the base, with the upper part broader and folded, with 4 short teeth, above 1 line diameter when fully out, with the border truncate, the teeth scarcely prominent. Petals 4, orbicular. Stamens usually, but perhaps not always, 6 to 8. Ovary 2-celled; style rather longer than in the preceding species. Capsule depressed-globular, scarcely exceeding the calyx, about 1½ lines diameter, bursting irregularly and transversely.

Queensland. Point Look-out, *Banks and Solander;* Wide Bay, *Bidwill.*—Abundant in tropical and subtropical Africa, perhaps rather less so in Asia, where it is commonly replaced by the following species or variety.

8. **A. multiflora,** *Roxb. Fl. Ind.* i. 426. Erect and branched, but usually smaller than *A. indica* or *A. auriculata*, and often only 3 to 4 in. high. Leaves linear or lanceolate, often above ½ in. long, and narrowed below the middle, but always more or less dilated and cordate-auriculate at the base, as in *A. auriculata.* Flowers minute, in little axillary dichotomous cymes shorter than the floral leaves, the peduncles, branches, and pedicels short but filiform. Calyx about ¾ line long, at first narrow at the base with the upper part folded, with 4 very short teeth, afterwards truncate, with the teeth scarcely conspicuous. Petals 4, minute. Stamens 4, or fewer. Ovary 2-celled; style rather long. Capsule depressed-globular, under 1 line diameter, scarcely exceeding the calyx, bursting irregularly and transversely.—DC. Prod. iii. 79; W. and Arn. Prod. 305; *A. australasica,* F. Muell. Trans. Phil. Soc. Vict. i. 41.

Queensland. Keppel Bay, *R. Brown.*
N. S. Wales. Darling river, *Victorian Expedition.*
Victoria. Lagoons on the Murray river, *F. Mueller.*
Widely spread over tropical Asia and Africa. I have great doubts whether it be not a small-flowered variety of *A. auriculata.*—*A. microcarpa,* DC. Prod. iii. 78; Dcne. Herb. Tim. Descr. 125, from Timor, and perhaps from the N. coast of Australia, appears to be a form of *A. multiflora,* with a narrower capsule.

2. LYTHRUM, Linn.

Calyx tubular, 8- to 12-ribbed, with 4 to 6 triangular often very short primary lobes or teeth, the sinus produced into as many external accessory ones, either short and spreading, or erect and longer than the primary ones. Petals 4 to 6. Stamens twice as many as petals or fewer, inserted below the middle of the calyx. Ovary 2-celled (or very rarely 3-celled ?), with several ovules in each cell; style filiform, with a minute or capitate stigma. Capsule included in the persistent calyx, oblong or globular, opening in septicidal valves at the top or bursting irregularly. Seeds numerous, small.—Herbs or rarely undershrubs, glabrous or villous. Leaves opposite verticillate or the upper ones alternate, usually narrow. Flowers solitary, or 3 to 5 together in the axils, sessile or pedunculate, but not forming a head as in most *Nesæas.*

The genus is spread over most parts of the globe. Of the three Australian species one is endemic, the other two have a geographical range nearly as wide as that of the genus.

Calyx outer-lobes erect, longer than the inner ones. Capsule oblong, hard, septicidally dehiscent.
  Tall perennial. Leaves opposite or verticillate. Flowers nearly sessile, several in each axil, forming showy terminal spikes more or less leafy . . . . 1. *L. Salicaria.*
  Decumbent annual. Upper leaves alternate. Flowers small, solitary, sessile or shortly pedicellate . . . . 2. *L. hyssopifolium.*
Calyx outer-lobes very small, spreading. Capsule membranous, irregularly dehiscent. Erect annual. Peduncles filiform, 1- to 3-flowered . . . . 3. *L. arnhemicum.*

1. **L. Salicaria,** *Linn.; DC. Prod.* iii. 82. Rootstock perennial, with stout

annual erect stems, 2 to 3 ft. high, slightly branched, glabrous or pubescent. Leaves opposite or sometimes in threes, sessile and stem-clasping, lanceolate, entire, 2 to 3 in. long. Flowers reddish-purple or pink, 3 to 5 together, nearly sessile in the axils, forming handsome terminal spikes, more or less leafy at the base, the upper floral leaves reduced to bracts scarcely longer or even shorter than the flowers. Calyx about 3 lines long, with 6 (rarely 5) short triangular primary lobes or teeth, the sinuses produced into as many subulate erect outer lobes much longer than the primary ones. Stamens usually 12, 6 longer than the calyx and 6 shorter. Capsule oblong, rather hard, enclosed in the calyx, splitting septicidally into 2 carpels opening in their inner face.—Hook. f. Fl. Tasm. i. 126.

Queensland. Along watercourses, Brisbane river, Moreton Bay, *Fitzalan, Leichhardt.*

N. S. Wales. Port Jackson to the Blue Mountains, *R. Brown* and others; northward to Clarence river, *Beckler;* and inland to Lachlan and Macquarrie rivers, etc., *A. Cunningham* and others.

Victoria. Banks of streams, Yarra, etc., *F. Mueller;* in the Grampians, *Wilhelmi.*

Tasmania. Common in wet places, *J. D. Hooker.*

S. Australia. From the Murray to St. Vincent's Gulf, *F. Mueller* and others.

The species is common in northern and subtropical Asia, in Europe, and N. America. For curious details on the fertilization of three different sexual forms, see Darwin in Journ. Linn. Soc. viii. 169.

2. **L. hyssopifolium,** *Linn.; DC. Prod.* iii. 81. A glabrous annual, rarely more than 6 or 8 in. high, the stems slightly branched and decumbent at the base, or, in starved specimens, erect and simple. Leaves sessile, narrow, entire, scarcely ½ in. long, the lower ones opposite, the upper ones alternate. Flowers small, solitary in the upper axils, sessile or nearly so. Calyx 1 to 2 lines long, very slender, the inner primary lobes or teeth very minute and membranous, the outer ones longer, erect, lanceolate-triangular and green. Petals 4 to 6, from rather shorter than the calyx-tube to rather longer. Stamens about as many as petals. Capsule included in the calyx, rather hard, opening septicidally at the top.—Hook. f. Fl. Tasm. i. 126; *L. thymifolium,* Linn.; DC. Prod. iii. 81.

Queensland. On the Burdekin, *F. Mueller.*

N. S. Wales. Port Jackson, *Herb. Hooker,* and others; swamps on the Lachlan and other parts of the interior, *A. Cunningham, Victorian Expedition,* etc.

Victoria. In swamps and wet places, *F. Mueller.*

Tasmania. Port Dalrymple, *R. Brown;* northern parts of the island, *J. D. Hooker.*

S. Australia. St. Vincent's Gulf, etc., *F. Mueller* and others.

The species is found in most parts of the world, especially in maritime districts.

3. **L. arnhemicum,** *F. Muell. Fragm.* iii. 109. An erect glabrous annual of ½ to 1 ft. Leaves opposite, linear, narrowed at the base, often exceeding 1 in. Peduncles axillary, slender, 2 to 6 lines long, either 1-flowered with a pair of bracts above the middle, or bearing 3 shortly pedicellate flowers with small bracts at the base of the pedicels, and a pair of bracteoles on each of the lateral ones. Calyx at first ovate-campanulate, but soon broad, 2 lines long, with 12 prominent herbaceous ribs, membranous between them; lobes 6, erect, triangular, about ¼ as long as the tube, each tipped with a dark spot, the folds of the sinuses forming as many horizontal accessory teeth. Petals 6, purple, much longer than the calyx, but very fugacious. Stamens

12 or fewer, longer than the calyx. Ovary 2- or rarely 3-celled, but the dissepiments very soon disappearing; style slender, with a small stigma. Capsule globular, about the length of the calyx, membranous and bursting irregularly.

N. Australia. Moist sandy plains and banks of Victoria river and Sturt's Creek, *F. Mueller.*—The species is remarkable for the inflorescence, more lax than in any other *Lythrum.* The short calyx shows an approach to *Nesæa,* but the ovary is usually 2-celled only, and the inflorescence is not capitate. The nearest approach among *Lythrums,* both in inflorescence and capsule, is shown in the S. African *L. rigidulum,* Sond.

3. PEMPHIS, Forst.

(Maclellandia, *Wight.*)

Calyx campanulate, slightly striate, with 6 short erect triangular primary lobes or teeth, the sinuses produced into as many small accessory spreading ones. Petals 6, oval. Stamens 12, shorter than the calyx, and attached rather above the middle of the tube. Ovary small, 3-celled at the base only, with several ovules in each cell; style rather thick, with a broad capitate stigma. Capsule globular, enclosed in the calyx, transversely circumsciss. Seeds angular or compressed, the testa thick with the angles often expanded into narrow thick wings.—Shrub. Leaves opposite. Flowers solitary in the axils.

The genus is limited to a single species.

1. **P. acidula,** *Forst; DC. Prod.* iii. 89. A small and bushy, or tall and spreading shrub or small tree, more or less hoary with a minute tomentum. Leaves oblong, obtuse, narrowed into a short petiole, rather thick, 1-nerved, about ½ in. long. Flowers in the upper axils, on pedicels shorter or rarely rather longer than the leaves. Bracteoles none. Calyx about 2 lines long, the accessory lobes much shorter than the primary ones. Petals 3 to 4 lines long. Fruiting-calyx not much enlarged.—Blume, Mus. Bot. ii. t. 43; *Maclellandia Griffithiana,* Wight, Ic. t. 1996.

N. Australia. North coast, *A. Cunningham;* Port Essington, *Leichhardt.*

Queensland. Tropical seacoasts and adjoining islands, *R. Brown, A. Cunningham, F. Mueller, M'Gillivray, Leichhardt, W. Hill.*

The species is widely spread over the seacoasts of tropical Asia and the Pacific Islands.

4. LAWSONIA, Linn.

Calyx-tube broadly turbinate; lobes 4, ovate-triangular, the sinuses acute without accessory lobes. Petals 4, broad, sessile. Stamens 8, inserted round an annular disk at the top of the calyx-tube, almost in the same row as the petals, and alternating with them in pairs; filaments rather thick. Ovary filling the calyx-tube, flat-topped or depressed in the centre, 4-celled with very thin dissepiments and many ovules in each cell; style filiform, with a small capitate stigma. Capsule nearly globular, surrounded at the base by the persistent calyx, bursting irregularly. Seeds numerous, cuneate, angular; testa thick.—Shrub. Leaves opposite. Flowers in loose racemes forming leafy panicles.

The genus is limited to a single species.

1. **L. alba,** *Lam.; DC. Prod.* iii. 91. A glabrous, much-branched shrub of several feet, with divaricate branches, the smaller ones often spinescent. Leaves from obovate and obtuse to ovate or lanceolate and acute, narrowed into a short petiole, rarely above 1 in. long, thin and penniveined. Flowers white, numerous, in little loose racemes forming usually a large terminal pyramidal leafy panicle, the ultimate branches usually leafless, and the bracts very minute or none. Pedicels 1 to 2 lines long. Buds globular. Calyx-tube not 1 line diameter, the lobes spreading to a diameter of 3 lines. Petals not twice as long as the calyx-lobes. Capsule about 3 lines diameter.—Wight, Illustr. i. t. 87.

N. Australia. Melville Island, *Fraser.*—Dispersed over tropical and subtropical Africa and Asia, but frequently cultivated under the name of *Henné,* for a yellow dye used especially in Africa for dyeing the nails of ladies' fingers. It appears to be abundant in Timor.

5. SONNERATIA, Linn. f.

Calyx thick, the tube broadly campanulate, adnate to the ovary at the base; lobes 4 to 8, lanceolate or triangular, the sinuses acute without accessory lobes or teeth. Petals 4 to 8, narrow, or none. Stamens numerous, inserted at the top of the calyx-tube, inflected in the bud. Ovary enclosed in and partially adnate to the calyx-tube, depressed-globular, 10- to 15-celled; style elongated with a small capitate stigma. Fruit large, depressed, fleshy, and indehiscent, surrounded by the persistent calyx, and adnate to it at the base. Seeds immersed in pulp, angular, with a thick testa. Embryo curved.—Glabrous trees or shrubs. Leaves opposite, petiolate, rather thick. Flowers large, solitary or 3 together in the upper axils or at the ends of the branches.

Besides the Australian species which is spread over E. India and the Archipelago and extends to eastern Africa, the genus contains one or two others from the same region.

1. **S. acida,** *Linn. f.; DC. Prod.* iii. 231. A tree. Leaves petiolate, broadly obovate or ovate, 2 to 3 in. long. Calyx broad, about 1 in. long, divided to below the middle into about 6 thick valvate lobes. Petals linear, scarcely exceeding the calyx. Fruit (in Indian specimens) above 1½ in. diameter.—Wight, Ic. t. 340.

N. Australia. Frequent in bogs on the N. and N.W. coasts, *A. Cunningham.* The specimens are imperfect, in leaf with a single flower, but as far as they go they are exactly like some of our Malayan ones.

Some specimens in Herb. R. Brown from Arnhem N. Bay have no petals in the expanded flowers, yet they look more like *S. acida* than *S. apetala*, and the petals may have fallen away.

ORDER LI. ONAGRARIEÆ.

Calyx-tube adnate to the ovary, entirely so or produced above it; lobes 2 to 4, rarely 5 or 6, valvate in the bud. Petals as many as calyx-lobes, inserted at the top of the calyx-tube, rarely wanting. Stamens as many or twice as many as petals, or fewer, inserted at the top of the calyx-tube, free (except in a Mexican genus); anthers from ovate to linear, versatile, with parallel cells opening longitudinally. Ovary inferior, more or less completely divided into as many cells as calyx-lobes, or rarely 1-celled; style filiform, or sometimes very short or scarcely any; stigma entire or divided into as many lobes as cells to the ovary. Ovules usually numerous, in 1 or 2 rows in each cell, anatropous, rarely, in genera not Australian, solitary. Fruit various, in the Australian genera capsular and elongated, opening from the apex downwards in as many valves as cells, or splitting laterally between the ribs of the calyx. Seeds usually small; testa membranous, coriaceous or rarely spongy. Albumen none or exceedingly thin. Embryo usually ovoid; cotyledons plano-convex (except in *Trapa*), with a very short radicle.—Herbs, annual or perennial, or, in a few genera not Australian, shrubs or even trees. Leaves opposite or alternate, without stipules, entire serrate or very rarely divided. Flowers usually solitary in the axils, sometimes forming leafy racemes or spikes at the ends of the branches, often with 2 small bracteoles under the calyx.

The Order is dispersed over nearly the whole surface of the globe. Of the 4 Australian genera, one, *Epilobium*, has nearly as extensive a range as the whole Order; two, *Jussiæa* and *Ludwigia*, belong chiefly to the warmer regions, *Ludwigia* extending into temperate climates; the fourth, *Œnothera*, is almost entirely American.

Calyx-tube produced above the ovary. Capsule opening from the summit downwards. Seeds naked. Stamens twice as many as calyx-lobes or petals . . . . 1. ŒNOTHERA.
Calyx-tube not produced above the ovary.
Capsule opening from the summit downwards in 4 valves. Seeds with a tuft of hairs. Stamens 8. Petals 4 . . . . 2. EPILOBIUM.
Capsule opening laterally between the ribs of the calyx or at the summit inside the calyx. Seeds naked.
Stamens twice as many as calyx-lobes or petals . . . . 3. JUSSIÆA.
Stamens of the same number as calyx-lobes or petals . . . . . 4. LUDWIGIA.

1. ŒNOTHERA, Linn.

Calyx-tube more or less produced above the ovary and dilated at the end into a 4-lobed limb, the whole free part deciduous. Petals 4. Stamens 8, inserted at the summit of the calyx-tube; anthers linear. Ovary 4-celled, with many ovules in each cell; style filiform with a capitate clavate or 4-lobed stigma. Capsule usually opening from the summit downwards loculicidally in 4 valves separating from the persistent axis. Seeds without any tuft of hairs.—Herbs or rarely small shrubs. Leaves alternate, or rarely the lower ones opposite, entire or variously toothed or lobed. Flowers axillary, solitary or very rarely in pairs, sometimes forming terminal racemes or spikes, rarely contracted into heads.

A large American genus, chiefly extratropical or Andine, a very few species now naturalized in various parts of the Old World. Of the two Australian species, one is a naturalized one of American origin, the others apparently endemic, but very closely allied to a Chilian species.

Tall erect plant, with large yellow flowers in a terminal spike. Stigma 4-lobed . . . . 1. *Œ. biennis.*
Small prostrate plant. Flowers small in the axils of the stem leaves. Stigma undivided . . . . 2. *Œ. tasmanica.*

* 1. **Œ. biennis,** *Linn.; DC. Prod.* iii. 46. A biennial, 2 or 3 ft. high, the stems almost simple and more or less hairy. Leaves ovate-lanceolate or lanceolate, slightly toothed, hoary or downy. Flowers large, yellow, fragrant,

sessile in a long terminal spike often leafy at the base. Ovary and adnate part of the calyx about 6 to 8 lines long, the free part of the calyx-tube at least 1 in. long. Petals broad and spreading. Stigma divided into 4 linear lobes. Capsules ¾ to 1 in. long, scarcely angular.

A plant of N. American origin, long cultivated in gardens in Europe and other countries, and readily establishing itself in waste places on river banks, etc., and now said to be naturalized in many parts of **N. S. Wales, Victoria,** and **S. Australia.**

2. **Œ. tasmanica,** *Hook. f. Fl. Tasm.* i. 119. A small plant, with slender prostrate or creeping stems of a few inches, glabrous or slightly pubescent. Lower leaves opposite, the others alternate, sessile or nearly so, oblong, obtuse, rarely exceeding ½ in., glabrous, with crisped more or less toothed margins. Flowers small, yellow, sessile or nearly so, solitary in the axils of the leaves. Calyx-tube at the time of flowering not exceeding the leaves, very shortly produced above the ovary; lobes 1 line long or rather more. Petals shortly exceeding the calyx-lobes. Anthers oblong. Stigma entire, clavate, almost capitate. Capsule terete or slightly 4-gonous, ½ in. long or rather more, often curved.

Tasmania. Alpine marshes at Marlborough, *Gunn.*

The specimens have not very good flowers. They very nearly resemble the broad-leaved varieties of *Œ. dentata*, Cav., distinguished by Spach as *Holostigma heterophyllum.* This species ranges from S. Chile to California, and differs from the Tasmanian one chiefly in the rather longer more angular capsule. Further specimens may possibly show the two to be forms only of one species.

2. EPILOBIUM, Linn.

Calyx-tube not at all or scarcely produced above the ovary; lobes 4, deciduous. Petals 4. Stamens 8; anthers linear or oblong. Ovary inferior, 4-celled, with numerous ovules in each cell; style filiform; stigma entire and club-shaped in the Australian species, 4-lobed in some others. Capsule elongated, opening loculicidally in 4 valves from the summit downwards. Seeds small, with a tuft of long hairs at the end.—Herbs, mostly erect, or with a decumbent or creeping base. Leaves opposite or irregularly scattered. Flowers pink or red, rarely white, solitary in the upper axils or forming a terminal raceme.

The genus is diffused over nearly the whole globe, from the extreme Arctic regions of both hemispheres to the tropics. The numerous forms the species assume in every variety of climate make it exceedingly difficult to define them upon any certain principle, and botanists seldom agree as to the number they should admit. The general tendency of late has been to an inordinate multiplication of supposed species. F. Mueller, on the other hand (Veget Chath. Isl. 15), proposes to reduce the whole of the New Zealand and Australian species to the Linnæan *E. tetragonum*, a course which will hardly be concurred in by the majority of botanists. Of the following forms, *E. confertifolium* and *E. pallidiflorum*, at least, appear to me be quite distinct, whatever may be said of the others. The very imperfect state of the majority of the Australian specimens, except those from Tasmania and W. Australia, increases the difficulty of judging of the relative value of the characters observed. The autumnal offshoots, often very useful in distinguishing European species, are not described in any of the Australian ones, and do not appear on the specimens.

Stems short, densely leafy, creeping or shortly ascending at the
tips. Flowers small . . . . 1. *E. confertifolium.*
Stems erect, or decumbent at the base only.

Flowers small. Calyx-lobes under 3 lines long and petals not twice as long.
Stems terete.
Pubescent or hoary. Leaves mostly alternate and narrow . . . 2. *E. junceum.*
Glabrous or slightly hoary in the upper portion. Leaves mostly opposite and oblong . . . . 3. *E. glabellum.*
Stems more or less tetragonous, with raised lines decurrent from the leaves . . . . 4. *E. tetragonum.*
Flowers large. Calyx-lobes 3 lines long or more. Petals twice as long.
Leaves mostly oblong, obtuse and under 1 in. . . . . . . 5. *E. Billardierianum.*
Leaves lanceolate or linear, acute, 1 to 2 in. long . . . . 6. *E. pallidiflorum.*

1. **E. confertifolium,** *Hook. f. Fl. Antarct.* i. 10, *and Handb. N. Zeal. Fl.* 78. Small, almost glabrous, prostrate and creeping, the branches rather stout, 1 to 4 in. long, shortly ascending at the tips. Leaves crowded, all opposite, from ovate-oblong to linear-oblong, obtuse, under ½ in. long, rather thick and shining, obscurely toothed. Flowers small, in the upper axils, on pedicels at first shorter than the leaves, but often much longer when in fruit. Calyx-lobes under 2 lines long, and petals only a little longer. Capsule about ½ in. long.—Hook. Ic. Pl. t. 685; *E. tenuipes*, Hook. f. Fl. Tasm. i. 116.

Victoria. Bogong range, at an elevation of 6000 to 7000 ft., *F. Mueller.*
Tasmania. Summit of Table mountain, Derwent river, *R. Brown;* abundant on the summit of Mount Olympus and Isis river, Middlesex plains, *Gunn.*
The species is also in New Zealand.

2. **E. junceum,** *Forst. in Spreng. Syst.* ii. 233. Stems from a hard decumbent base, erect, terete, hoary-tomentose or softly pubescent, usually 1 to 2 ft. high and rigid, but smaller and slender when starved. Lowest leaves opposite, the upper ones and often nearly all alternate, sessile, linear-oblong, remotely sinuate-toothed, the larger ones often 2 in. long or more, but mostly smaller and the upper floral ones often very much reduced, all hoary or pubescent. Flowers in the upper axils sometimes quite small, but the calyx-lobes usually 2 to nearly 3 lines long, the petals rather longer, the pedicels at first shorter than the floral leaves, but lengthening much after flowering. Capsule slender, usually about 2 in. long.—Hook. f. Fl. Tasm. i. 118, and Handb. N. Zeal. Fl. 80; *E. canescens*, Endl. in Hueg. Enum. 44; Nees in Pl. Preiss. i. 159.

Queensland. Plains of the Condamine, *Leichhardt.*
N. S. Wales. Port Jackson to the Blue Mountains, *R. Brown;* Hastings, Clarence, and Macleay rivers, *Beckler;* in the interior to the N. of Bathurst, *A. Cunningham;* New England, *Leichhardt.*
Victoria. Common in the colony, *F. Mueller, Robertson,* and others.
Tasmania. Port Dalrymple, *R. Brown;* abundant throughout the colony by waysides, in pastures, etc., *J. D. Hooker.*
S. Australia. In grassy plains, *F. Mueller,* and others.
W. Australia. From King George's Sound to Swan and Moore rivers, etc., *Huegel, Drummond, n.* 253, *Preiss, n.* 1946, 1947, 1948, and others.
The species is also in New Zealand. It appears to be the common *Epilobium* of extra-tropical Australia, in situations occupied by *E. montanum* in the northern hemisphere, and generally speaking it is readily distinguished from any of the following, although here and there doubtful specimens are met with.

3. **E. glabellum,** *Forst. in Spreng. Syst.* ii. 233. Nearly allied to *E. junceum*, but usually glabrous or the stem very slightly hoary-pubescent in the upper portion. Leaves all or nearly all except the upper floral ones opposite, sessile or nearly so, from oval-oblong to narrow-oblong, obtuse, always broader and the teeth less prominent than in *E. junceum.* Flowers small, as in that species.—Hook. f. Fl. Tasm. i. 118, and Handb. N. Zeal. Fl. 79.

Victoria. "Australia Felix," *F. Mueller.*
Tasmania. Abundant in many parts of the island, *J. D. Hooker.*
S. Australia. Glenelg river and Bugle Range, *F. Mueller.*
W. Australia. King George's Sound, *R. Brown, Drummond, 2nd Coll. n.* 239; Tweed river, *Oldfield.*

The species is much more common in New Zealand than in Australia, and, although generally distinct, seems sometimes almost to pass into *E. junceum*, on the one hand, and rather more into *E. Billardierianum* on the other.

4. **E. tetragonum,** *Linn. Spec. Pl.* 495. Stems erect, 1 to 2 ft. high, glabrous or slightly hoary-pubescent, and more or less angular, especially in the lower portion, with raised lines decurrent from the leaves. Leaves sessile or nearly so, from ovate-lanceolate to narrow-oblong, the lower ones opposite, usually larger, thinner, with more prominent veins than in *E. glabellum.* Flowers small, the calyx-lobes rarely above 2 lines long and the petals not much longer. Capsule often very long.—Ser. in DC. Prod. iii. 43; Hook. f. Fl. Tasm. i. 117.

N. S. Wales. Macleay river, *Beckler.*
Victoria. Banks of streams near Goulburn river, *F. Mueller.*
Tasmania. Common in moist, especially alpine situations, *J. D. Hooker.*
S. Australia. Gawler river, *F. Mueller.*

In the majority of the Australian specimens, the raised decurrent lines on the stems are less prominent than in the European and Asiatic ones, and I have some doubts whether they may not be luxuriant forms of *E. glabellum*, and whether the true *E. tetragonum* is really Australian, except here and there where introduced from Europe.

5. **E. Billardierianum,** *Ser. in DC. Prod.* iii. 41. Glabrous or minutely hoary-pubescent, especially in the upper portion. Stems usually nearly simple, ½ to 1½ ft. high, terete or rarely with short faint decurrent lines from some of the leaves. Leaves sessile or nearly so, mostly opposite, except the floral ones, from narrow ovate-oblong to linear-oblong, obtuse, more or less toothed, rarely exceeding 1 in. Pedicels shorter than the leaves or the upper ones exceeding them when in fruit. Calyx-lobes about 3 lines long or rather more, the petals usually twice as long. Capsule elongated.—Hook. f. Fl. Tasm. i. 117. t. 21, and Handb. N. Zeal. Fl. 81.

N. S. Wales. Head of the Gwydir, *Leichhardt;* Ben Lomond and Arne river, *Beckler.*
Tasmania. Common in alpine situations, *J. D. Hooker.*

The foliage is nearly that of *E. glabellum*, which by some is included in *E. Billardierianum*, but the leaves are more crowded and the flowers nearly as large as in *E. pallidiflorum.*

6. **E. pallidiflorum,** *Soland.; A. Cunn. in Ann. Nat. Hist.* iii. 34. A handsome plant, readily distinguished by its long acute leaves and large flowers. Stems, from a decumbent base, erect, terete, 2 to 3 ft. high, glabrous or hoary-pubescent in the upper portion. Leaves opposite, except the upper

floral ones, sessile or on very short broad petioles, linear or lanceolate, acute, with a few distant teeth, mostly 1 to 2 in. long. Pedicels usually short, even when in fruit. Calyx-lobes 3 lines long or more; petals twice as long. Capsule long, usually hoary-tomentose.—Hook. f. Fl. Tasm. i. 117; Handb. Fl. N. Zeal. 81; *E. macranthum*, Hook. f. in Hook. Ic. Pl. t. 297.

Victoria. Grampians, *Herb. F. Mueller*.

Tasmania. Common in ditches and marshes, especially in the northern parts of the colony, *J. D. Hooker*.

S. Australia. Meadows near Holdfast Bay, Mount Disappointment, Cox's Creek, *F. Mueller*.

3. JUSSIÆA, Linn.

Calyx-tube not produced above the ovary; lobes 4, 5 or rarely 6, persistent. Petals as many as calyx-lobes. Stamens twice as many as calyx-lobes. Ovary with as many cells as calyx-lobes and numerous ovules in each cell; style short or long or scarcely any; stigma more or less lobed. Capsule terete or with as many or twice as many ribs or angles as calyx-lobes, opening septicidally in valves separating from the persistent ribs or irregularly between the ribs. Seeds usually numerous; testa thin or crustaceous, or thick and spongy.—Herbs, sometimes aquatic, or rarely shrubs. Leaves alternate, entire or very rarely serrate. Flowers yellow or white, solitary in the axils; petals usually broad.

The genus is chiefly American, both tropical and extratropical, a few species also spread over tropical and subtropical Africa and Asia. The Australian species are both of them common in the New as well as the Old World.

Creeping or floating plant. Flowers usually 5-merous, on pedicels longer than the ovary . . . . 1. *J. repens*.
Erect plant. Flowers mostly 4-merous, on very short pedicels or almost sessile . . . . 2. *J. suffruticosa*.

1. **J. repens,** *Linn. Spec. Pl.* 555, *and Mant.* 381. Herbaceous, creeping in mud or floating in water, often sustaining itself by little vesicles round the insertion of the leaves, glabrous or more or less hirsute, with soft spreading hairs. Leaves from obovate or obovate-oblong, to narrow cuneate-oblong or almost lanceolate, acute or rarely obtuse, the upper ones usually 1 to 2 in. long, those about the short creeping branches often very small. Peduncles usually longer than the ovary and fruit, with 2 small bracteoles at the summit. Calyx-tube or ovary cylindrical, rather slender, under $\frac{1}{2}$ in. long when in flower; lobes usually 5, lanceolate, acute, 3 to 4 lines long. Petals broadly obovate, from a little longer to twice as long as the calyx-lobes. Capsule lengthening to about $\frac{3}{4}$ in., and about $1\frac{1}{2}$ lines thick, smooth and shining but usually sprinkled with a few hairs, the 5 primary ribs prominent, the secondary ones less so.—DC. Prod. iii. 54; Wight in Hook. Bot. Misc. iii. 300. t. Suppl. 40; *J. Swartziana*, DC. l. c.

Queensland. Port Curtis, *M'Gillivray;* common in lagoons about Moreton Bay, *C. Stuart*.

N. S. Wales. Richmond and Hunter's rivers, *R. Brown;* Paramatta, *Woolls;* Illawarra, *A. Cunningham;* inland on the Darling, *Victorian Expedition*, to Cooper's Creek, *Howitt's Expedition*.

Victoria. Morasses of Snowy River and bends of Murray river, *F. Mueller*.

S. Australia. Murray river, *F. Mueller*.

2. **J. suffruticosa,** *Linn. Spec. Pl.* 555. An erect branching perennial, attaining 2 or 3 ft., the base of the stem often hard and woody, either softly pubescent or villous in all its parts or rarely almost glabrous, the stem often angular. Leaves lanceolate or almost linear, acute, narrowed at the base, the larger ones 2 to 4 in. long. Pedicels much shorter than the calyx-tube or ovary, the bracteoles reduced to small glands or wanting. Calyx-tube or ovary usually about ½ in. long when in flower, but soon lengthening out; lobes 4 or rarely 5, lanceolate, broad or narrow, 3- to 5-nerved, 4 to 5 lines long. Petals broad, exceeding the calyx-lobes. Capsule 1½ to 2 in. long, usually above 2 lines broad, tapering to the base, nearly terete, the ribs scarcely prominent.—F. Muell. Fragm. iii. 130; *J. villosa* and *J. angustifolia*, Lam. Dict. iii. 331; DC. Prod. iii. 55, 57; *J. villosa*, W. and Arn. Prod. 336, with the synonyms adduced; *J. suffruticosa* and *J. angustifolia*, Griseb. Fl. Brit. W. Ind. 273, with the numerous synonyms adduced.

N. Australia. Victoria river and Macadam Range, *F. Mueller;* Strangways river, *M'Douall Stuart;* Albert river, *Henne.*

Queensland. Broad Sound and Northumberland islands, *R. Brown;* Lizard Island, *M'Gillivray;* Burnett river, *F. Mueller;* Burdekin river, *Bowman;* Rockhampton, *Dallachy;* Brisbane river, Moreton Bay, *A. Cunningham, F. Mueller.*

N. S. Wales. Clarence river, *Beckler;* New England, *C. Stuart.*

The species is common in most tropical countries. The nearly glabrous forms distinguished sometimes as *J. angustifolia,* seem frequently to pass into the villous ones in most localities. In Australia, the two appear to be equally abundant in Queensland, the villous ones more common in N. Australia, and the more glabrous ones in N. S. Wales.

4. LUDWIGIA, Linn.

Calyx-tube not produced above the ovary; lobes 4, 5 or rarely 3, persistent or at length deciduous. Petals as many as calyx-lobes or sometimes none. Stamens as many as calyx-lobes. Ovary with as many cells as calyx-lobes, and numerous ovules in each cell; stigma sessile or nearly so, capitate, furrowed or obscurely lobed. Capsule angular or terete, much longer than broad, opening either in terminal pores or irregularly along the sides between the ribs. Seeds small, numerous, without any tuft of hairs.—Annual or perennial herbs, sometimes somewhat woody at the base. Leaves alternate or the lower ones (in species not Australian) opposite. Flowers axillary, sessile or nearly so, or rarely distinctly pedicellate. Petals usually very small.

The genus is dispersed over the warmer and temperate regions of the globe; the only Australian species is a common Asiatic and African one.

1. **L. parviflora,** *Roxb. Fl. Ind.* i. 419. An erect or diffuse glabrous annual, rarely above 1 ft. high. Leaves alternate, lanceolate, or, in most of the Australian specimens, linear, entire, 1 to 2 or even 3 in. long, narrowed into a short petiole. Flowers very small, solitary in the axils, sessile or very shortly pedicellate. Calyx-tube (or ovary) at the time of flowering, rarely 1½ lines long, but very rapidly enlarging; lobes usually 4 in the Indian specimens, more frequently 5 in the Australian ones, small and very acute. Petals not exceeding the calyx-lobes. Stamens rather shorter. Stigma large, capitate. Capsule 4 to 6 lines long and 1½ lines broad when attaining its full size, but often ripening much smaller.—Wight, Illustr. t. 101.

N. Australia. Victoria river, *F. Mueller;* Port Essington, *Armstrong.*

Queensland. Endeavour river, *A. Cunningham*; Burdekin river, *Bowman.*

The species is widely spread over tropical Asia and Africa. Amongst the synonyms quoted by Wight and Arnott, Prod. 336, are *L. diffusa*, Hamilt. in Trans. Linn. Soc. xiv. 301, and *L. perennis*, Linn. Spec. Pl. 173. These are copied by Miquel into his Fl. Ind. Bat. i. part i. 629, and observing that one of them is an old name of Linnæus's, he, without further inquiry (except perhaps a glance at Rheede's fig. of *Caramba*, Hort. Malab. ii. t. 49, cited by Linnæus, which is the true *L. parviflora*), adopts this name of *L. perennis* for the species, and Miquel's example is followed by F. Mueller, Fragm. iii. 129. But not only is Linnæus's name wholly inapplicable to a plant so constantly and evidently annual, but so is also his specific character "foliis oppositis floribus pedicellatis," and as to the reference to Rheede's *Caramba*, he expressly rejects it in his Mantissa, p. 332, as pointed out in DC. Prod. iii. 59. Although therefore Linnæus may have confounded this plant with some other, it is certainly not the one he had in view in characterizing his *L. perennis*, and Arnott and others are fully justified in adopting Roxburgh's *L. parviflora*. As to *L. diffusa*, Hamilt., although he also thought Rheede's *Caramba* might be the same, it is in fact quite distinct in the long slender ovary and capsule, and in some measure in inflorescence. It is *L. prostrata*, Roxb. Fl. Ind. i. 420; Wight, Ic. t. 762, and includes the three species of *Nematopyxis*, described by Miquel, Fl. Ind. Bat. i. part i. 630. It has not yet been found in Australia.

ORDER LII. SAMYDACEÆ.

Sepals free or united at the base into a 4- or 5-lobed (rarely 2-, 3- or 6- or more-lobed) calyx, free from the ovary or more or less adherent. Petals either as many as the sepals or calyx-lobes, inserted at their base, persistent with them, and resembling them in consistence, or wanting. Stamens perigynous, indefinite or not corresponding in number with the calyx-lobes, or, if equal to them, usually opposite the petals and alternating with small glands or scales. Ovary superior or more or less inferior, with 2, 3 or more parietal placentas and several ovules to each placenta; style entire or more or less divided into as many branches as placentas. Fruit indehiscent or opening in valves between the placentas. Seeds often arillate, with a fleshy albumen. Embryo straight or nearly so, with the radicle next the hilum and flat cotyledons.—Trees or shrubs. Leaves alternate, undivided, usually toothed. Stipules small or none. Flowers hermaphrodite or rarely diœcious.

A considerable Order, if taken with the limits above given, and widely distributed over the New and the Old World, chiefly within the tropics. The two following genera belong to two of those tribes into which it may be divided, and which are considered by some as distinct Orders, viz. *Casearieæ* or *Samydeæ* proper, without petals, the stamens in a single series; and *Homalineæ*, with sepal-like petals, the stamens inserted singly or in clusters opposite the petals.

Petals none. Stamens in a single row, alternating with short ciliate scales. Ovary superior . . . . 1. CASEARIA.
Petals as many as sepals. Stamens opposite them, singly or in clusters. Ovary inferior . . . . 2. HOMALIUM.

1. CASEARIA, Linn.

Calyx-lobes 4 or 5. Petals none. Stamens 6 to 15 or rarely more, alternating with as many short ciliate or hairy scales (staminodia?), all in a single series and united in a perigynous ring at the base. Ovary superior, 1-celled, with 3 or rarely 4 parietal placentas; style entire or shortly 3-lobed. Fruit somewhat succulent, opening in valves or more fleshy and indehiscent. Seeds often with an arillus.—Trees or shrubs. Leaves usually, but not always dotted with a mixture of round and oblong transparent dots. Stipules lateral. Flowers usually small, in axillary clusters.

A considerable genus, chiefly American, with a few African and Asiatic species. The only Australian species appears to be a common Indian one.

Leaves not dotted. Stamens 8 . . . . 1. *C. esculenta.*
Leaves pellucid-dotted. Stamens 10 to 12 . . . . 2. *C. Dallachii.*

1. **C. esculenta,** *Roxb. Fl. Ind.* ii. 422. A large shrub, usually quite glabrous, the branches not angular. Leaves from oval-elliptical to nearly oblong, acuminate, narrowed at the base, 2 to 4 in. long or sometimes rather more, scarcely coriaceous, but not dotted. Flowers very small, in axillary clusters, the pedicels about 1 line long. Calyx glabrous, rather above 1 line diameter when open, 5-lobed. Stamens 8, alternating with as many short truncate staminodia, usually scarcely pubescent. Ovary glabrous, tapering into a short style; stigma entire. Placentas 3, the ovules not numerous.

Queensland. Brisbane river, *F. Mueller.*
The species to which this plant seems referable is widely spread over E. India. It may be the same as *C. ovata*, Willd., and *C. zeylanica*, Thw., as doubtfully suggested by Thwaites, Enum. Ceyl. Pl. 19, but both of those appear to have the ovary hirsute.

2. **C. Dallachii,** *F. Muell. Fragm.* v. 107. Nearly glabrous. Leaves shortly petiolate, ovate, shortly acuminate, 3 to 4 in. long, pellucid-dotted, minutely tomentose near the base or quite glabrous. Flowers densely clustered, hoary-pubescent, the pedicels shorter than the calyx. Calyx-segments 5, orbicular, about 1 line long. Stamens 10 to 12, alternating and more or less united with as many staminodia, which are bearded at the end. Ovary more or less hirsute at the top; style very short and thick, with a large undivided stigma. Placentas 3.

Queensland. Rockingham Bay, *Dallachy.*—Very nearly allied to *C. glabra*, Roxb. (which appears to be a variety only of the common Indian *C. tomentosa*), differing in the rather thicker calyx-lobes, and more numerous stamens.

2. HOMALIUM, Jacq.

(Blackwellia, *Juss.*)

Calyx-tube turbinate or oblong, adherent to the ovary at the base; lobes 4 to 12. Petals as many as calyx-lobes. Stamens 1 or more opposite each petal, with 1 gland opposite each calyx-lobe. Ovary 1-celled, adherent in the lower part, conical and free in the upper portion, crowned with 3 to 5 styles, either free or united into one; placentas as many as styles, in the upper free part of the ovary, with 2 to 6, usually 4 ovules to each. Fruit slightly enlarged, surrounded by the persistent calyx-lobes and petals, and usually opening at the top in short valves between the placentas.—Trees or shrubs. Leaves not dotted. Flowers in axillary spikes or racemes, or in terminal panicles.

A considerable tropical genus, chiefly Asiatic and African, with a few American species. Of the two Australian species, one is also in the islands of the South Pacific, the other is endemic.

Leaves and flower-spikes above 2 in. long. Calyx-segments usually 8 to 10, with as many petals of about the same size. Stamens 2 or 3 opposite each petal . . . . 1. *H. vitiense.*

Leaves and flower-spikes under 2 in. long. Calyx-segments usually 5.
Petals as many but larger. Stamens solitary opposite each petal . 2. *H. brachybotrys*.

1. **H. vitiense,** *Benth. in Journ. Linn. Soc.* iv. 36. A tree, glabrous except the inflorescence, or rarely a few appressed hairs on the under side of the leaves. Leaves broadly ovate, obtuse or very shortly and obtusely acuminate, irregularly and often obscurely sinuate-crenate or undulate, 2 to 4 in. long, on petioles of from $\frac{1}{4}$ to $\frac{1}{2}$ in. or rarely longer. Flowers very nearly sessile, in simple or branched spikes, varying from 2 or 3 in. long and rather dense, to twice that length and interrupted, the rhachis and flowers more or less pubescent. Calyx-tube narrow-turbinate, $\frac{1}{2}$ to $\frac{3}{4}$ line long; lobes 8 to 10 (or rarely 6 or 7?), linear; petals as many, scarcely more cuneate, giving the whole flower the appearance of a 16- to 20-lobed calyx, the enlarged calyx-lobes and petals after flowering about $1\frac{1}{2}$ lines long and ciliate-hirsute. Stamens in pairs or 3 together opposite each petal.—*H. alnifolium*, F. Muell. Fragm. ii. 127.

Queensland. Rockhampton, *Dallachy*. Also in New Caledonia and the Fiji islands. The leaves in the Australian specimens are rather larger and more coriaceous than in those from the Fiji islands, but are precisely as in New Caledonian specimens collected by Deplanche and Vieillard under nos. 23 and 2076, and referred by them to *H. tomentosum*, Benth., from which they differ both in flowers and foliage. *H. vitiense* is much more nearly allied to *H. fœtidum*, Benth.

2. **H. brachybotrys,** *F. Muell. Fragm.* ii. 127. Glabrous or nearly so, except the inflorescence. Leaves oval-elliptical or obovate, obtuse, entire or obscurely sinuate, rarely exceeding 2 in. and mostly about 1 in. long, narrowed into a petiole, drying of a paler colour than most of the genus. Flowers very small, sessile, in simple slender spikes of about 1 in., the rhachis pubescent as well as the flowers. Calyx-tube ovoid, about $\frac{1}{2}$ line long; lobes 5 (or 6?), narrow-linear, rather shorter than the tube. Petals oblong or spathulate, rather longer and much broader than the calyx-lobes. Stamens solitary opposite each petal. Styles and placentas 4 or 5.—*Blackwellia brachybotrya*, F. Muell. in Trans. Vict. Inst. iii. 48.

Queensland. Granite rocks, sources of the Gilbert river, *F. Mueller*.

ORDER LIII. PASSIFLOREÆ.

Calyx-tube short or rarely elongated; lobes 4 or 5, valvate or more or less imbricate in the bud, often coloured inside. Petals as many as calyx-lobes, inserted at their base and alternating with them, often persistent with them and much resembling them, sometimes small or rarely wanting. Stamens usually as many as calyx-lobes, rarely twice as many, inserted at the base of the calyx, but often connate with the ovary-stalk to near the top and appearing to be there inserted. Ovary usually stalked, 1-celled, with 3 or rarely 5 parietal placentas, each with several ovules. Style divided into as many branches as there are placentas, with terminal stigmas. Fruit indehiscent and succulent or opening in valves between the placentas. Seeds often arillate; albumen fleshy. Embryo straight, with leafy cotyledons, the radicle next the hilum.—Climbers, or rarely, in genera not Australian, erect herbs or shrubs. Leaves alternate, entire or divided, with stipules. Flowers hermaphrodite or unisexual, solitary or in cymes or racemes, on axillary peduncles. Tendrils axillary, often accompanying or terminating the peduncles.

The Order is dispersed over the tropical and subtropical regions of the New and the Old World. Of the two Australian genera one is American, with the exception of a few Old World species, the other is African and Asiatic.

Flowers usually large, hermaphrodite. One or several rings of coloured filaments or appendages forming a corona within the petals . . . . 1. PASSIFLORA.
Flowers small, unisexual. Petals small or none. Corona small or none . 2. MODECCA.

1. PASSIFLORA, Linn.

(Disemma, *Labill.*; Murucuja, *Pers.*)

Calyx-tube short. Petals rarely wanting and often like the calyx lobes. One or several rings of coloured filaments or appendages forming a corona within the petals. Stamens as many as calyx-lobes, so united with the ovary-stalk as to appear to be inserted at or near its summit. Styles 3, with large capitate stigmas. Fruit succulent or pulpy, indehiscent, or opening obscurely in 3 valves.—Climbers with axillary tendrils. Leaves entire or palmately-lobed or divided. Flowers usually hermaphrodite, the calyx-lobes coloured inside nearly or quite as much as the petals.

The species are numerous in tropical or subtropical America, with a very few from Africa, Asia, and the Pacific islands. The three Australian ones are supposed to be endemic, and one is probably really so, another is perhaps a variety only of a New Caledonian species, and the third is as yet insufficiently known. They all belong to the section *Disemma*, usually considered as a genus, distinguished by the number of rings of the corona—in *Murucuja* 1, in *Disemma* 2, in other *Passifloras* 3 or more; but this distinction proves too artificial to be taken as of more than sectional value, being unaccompanied by any difference in habit or other characters. In all the Australian species the filaments of the inner corona are united in a crenate or shortly lobed ring or tube.

Pubescent. Leaf-lobes rather acute. Inner ring of the corona connivent or contracted at the top . . . . 1. *P. Herbertiana.*
Glabrous. Leaf-lobes usually obtuse. Inner ring of the corona erect and plicate, scarcely contracted at the top.
Calyx $1\frac{1}{2}$ to 2 in. long . . . . 2. *P. Banksii.*
Calyx under 1 in. long . . . . 3. *P. brachystephana.*

1. **P. Herbertiana,** *Lindl. Bot. Reg. t.* 737. A tall robust climber, more or less pubescent. Leaves broad, truncate or slightly cordate at the base, larger than in *P. Banksii*, often 3 in. long or more, with 3 broad triangular almost acute lobes, pubescent on both sides (sometimes minutely so), the petiole with 2 glands very near the summit. Flowers solitary or in pairs, rather large, on pedicels much shorter than the leaves, with 2 or 3 scattered setaceous bracteoles at or below the middle. Calyx-lobes nearly $1\frac{1}{2}$ in. long, of a greenish-white or pale orange-yellow. Petals narrow, scarcely more than half as long as the calyx-lobes. Inner corona about $\frac{1}{2}$ in. long, broadly tubular but contracted at the orifice, crenate or shortly lobed; outer corona rather shorter, of a single row of filaments. Gynophore rather shorter than the calyx-lobes.—*Disemma Herbertiana*, DC. Prod. iii. 332.

Queensland. Brisbane river, Moreton Bay, *Fraser*.

N. S. Wales. Port Jackson and Newcastle Inlet, *R. Brown*; New England, *C. Stuart*; Illawarra, *A. Cunningham*.

2. **P. Banksii,** *Benth.* Quite glabrous. Leaves broad, usually under 3 in. long, with 3 broad obtuse lobes rarely divided to the middle of the leaf, and each lobe occasionally sinuate or more or less distinctly 2- or 3-lobed, the petiole with 2 glands very near the summit, very rarely obscure or altogether wanting. Flowers rather large, sometimes pale when they first open but soon assuming a brick-red or dull scarlet colour, on pedicels much shorter than the leaves, with 2 or 3 scattered setaceous bracteoles at or below the middle. Calyx-lobes about 1½ in. long or rather more. Petals narrow, scarcely more than half as long as the calyx-lobes. Inner corona broadly tubular, slightly contracted, plicate and shortly lobed at the orifice; outer corona about the same length, of a single row of filaments. Ovary-stalk longer than the petals, shorter than the calyx-lobes.—*P. coccinea*, Soland. in Herb. Banks, not of Aubl.; *Disemma coccinea*, DC. Prod. iii. 333.

Queensland. Endeavour river, *Banks and Solander, A. Cunningham;* Keppel Bay, *R. Brown;* Brisbane river, Moreton Bay, *A. Cunningham* and others. Islands of the coast, *M'Gillivray, Henne.*

The species very closely resembles the original *P. aurantia*, Forst., or *Disemma aurantia*, Labill. Sert. Austr. Caled. t. 79, from New Caledonia, as well as *P. adiantifolia*, Lindl. Bot. Reg. t. 233 (*Murucuja Baueri*, Lindl. Collect. Bot. t. 36, *Disemma adiantifolia*, DC. Prod. iii. 333), from Norfolk Island, the former differing only in the petiolar glands further from the limb and the bracteoles nearer the flower, and the latter in the absence of petiolar glands. But it seems doubtful whether the petiolar glands are constant in Australia, for in a specimen of Macarthur's, communicated by Backhouse, without the precise station, but stated to be from Australia, they are certainly entirely wanting.

3. **P. brachystephana,** *F. Muell.* Glabrous, like *P. Banksii.* Leaves smaller but otherwise precisely the same. Flowers also differing only in size. Calyx under 1 in. long. Petals less than half as long. Corona very short, but otherwise like that of *P. Banksii.*—*Disemma brachystephana*, F. Muell. Fragm. i. 56.

Queensland. Scrub on the Burdekin, *F. Mueller.*

There was but a single expanded flower on the specimens, and in that the petals do not show, but on examining a bud, I found the structure precisely as in *P. Banksii.* The species will require verifying on better specimens.

2. MODECCA, Lam.

Flowers unisexual. Calyx-tube short, campanulate or elongated. Petals small, especially in the females. Stamens as many as calyx-lobes, usually with a small scale opposite to each, free or united at the base, reduced in the females to small staminodia, or wanting. Ovary rudimentary in the males, more or less stalked in the females, with 3 parietal placentas, stigmas 3, sessile or nearly so, or on a 3-fid style. Capsule inflated, coriaceous or thin, more or less dehiscent in 3 valves. Seeds with a small cup-shaped aril.—Tall climbers. Leaves entire or palmately or pinnately lobed or divided; stipules often inconspicuous. Flowers usually very small, white or green, in cymes or racemes, on axillary peduncles, the rhachis produced into a simple tendril.

The genus extends over tropical Africa and Asia. The only Australian species appears to be endemic, although very nearly resembling one from Khasia.

1. **M. australis,** *R. Br. in DC. Prod.* iii. 337. A climber extending greatly amongst underwood (*A. Cunningham*), quite glabrous. Leaves on

long petioles, broadly ovate-cordate, quite entire, scarcely acuminate, 4 in. long or more, membranous, the base of the limb very shortly decurrent on the petiole and expanded into 2 rather large often confluent glands. Peduncles long and slender, terminating in a rather strong tendril, at the base of which are a pair of small opposite pedunculate cymes of very small flowers, very imperfect in our specimens, but according to Bauer's figure, given by Endlicher, presenting all the characters of the genus; the stigmas are on very short distinct styles. Capsule ovoid, inflated, about 2 in. long, very smooth. Seeds ovate, flat, almost muricate.—Endl. Iconogr. t. 114, 115.

N. Australia. Cygnet Bay, *A. Cunningham;* N.W. coast, *Bynoe.*

ORDER LIV. CUCURBITACEÆ.

Flowers usually unisexual. Calyx-tube adherent to the ovary and produced above it into a campanulate or tubular 5-toothed or 5-lobed free portion, which forms the whole calyx in the males. Petals 5, free or united in a lobed corolla, adnate to the free part of the calyx-tube and usually so confluent with it as to appear continuous with it between its teeth or lobes. Stamens 3 or 5, inserted on the calyx-tube below the petals, the filaments free or united; anthers separate or confluent into a waved or curved mass. Ovary usually 1-celled when very young, either with 3 or (rarely 4 or 5) parietal placentas soon thickening and meeting in the axis, dividing into as many or twice as many cells, or with 1 placenta and remaining 1-celled. Style 1, entire or 3-lobed, or rarely 3 almost distinct styles; stigmas 3 (rarely 4 or 5), entire or lobed. Ovules 1 or more to each placenta. Fruit succulent or coriaceous, often with a hard rind, indehiscent or bursting irregularly or rarely opening in 3 valves. Seeds usually flat, often obovate or oblong, without albumen; testa coriaceous or bony. Embryo straight; cotyledons large, usually notched at the base, with a short radicle.—Herbs (except in a few species not Australian) weak, prostrate or climbing by means of tendrils arising from the sides of the stems near the petioles, generally more or less scabrous or hispid. Leaves alternate, without stipules, usually palmately veined and angular, lobed or divided. Flowers unisexual in all the Australian genera, on axillary peduncles, the males usually in racemes or clusters, or sometimes solitary, the females generally solitary.

A considerable Order, dispersed over all but the colder regions of the globe, but most abundant in dry hot countries, especially in Africa. The nine Australian genera are all common to Asia and Africa, five of them are also in America, and one, *Bryonia*, extends to Europe.

TRIBE I. **Cucurbiteæ.**—*Ovules numerous, horizontal.*

Anther-cells very flexuose or conduplicate.
 Calyx-tube elongated. Petals fringed with long cilia . . . . 1. TRICHOSANTHES.
 Calyx-tube broadly campanulate or turbinate. Petals not fringed.
  Tendrils branched.
   Male flowers large, solitary. Fruit large, with a hard rind, dry but not fibrous . . . . 2. LAGENARIA.
   Male flowers in pedunculate racemes. Fruit dry, fibrous . . 3. LUFFA.
   Male flowers small, in clusters or short sessile racemes (in the Australian species). Fruit a small berry . . . . . 6. BRYONIA.

Tendrils simple.
Anthers tipped with an appendage to the connective. Fruit pulpy or fleshy . . . . 4. CUCUMIS.
Anthers without appendage.
Corolla with incurved scales at the insertion of the stamens. Fruit usually pulpy, sometimes dehiscent . . . . . . 5. MOMORDICA.
Corolla without incurved scales. Fruit a small berry . . 6. BRYONIA.
Anther-cells straight, parallel.
Calyx-tube broadly campanulate. Anthers without appendage. Female flowers pedunculate, bearing staminodia . . . . . . 7. MELOTHRIA.
Calyx-tube turbinate. Anthers with a minute appendage. Female flowers sessile, without staminodia . . . . 8. MUKIA.
TRIBE II. **Sicyeæ.**—*Ovules solitary, pendulous.*
Tendrils branched. Flowers small. Fruit small, prickly in the Australian species . . . . 9. SICYOS.

1. TRICHOSANTHES, Linn.

Calyx in the males and free part of it in the females oblong or cylindrical, dilated upwards, 5-lobed. Corolla rotate, deeply divided into 5 oblong or lanceolate lobes, bordered by long hair-like lobes or cilia. Stamens in the males 3, filaments very short, free; anthers 2 with 2 cells, one with 1 cell, the cells conduplicate. Ovary in the females oblong or globular, with 3 placentas; style slender, with 3 linear stigmas, the gynœcium reduced in the males to 3 filiform rudiments. Fruit succulent, often large, with a hard rind. Seeds smooth or with undulate or crenate margins.—Climbing annuals or perennials. Tendrils 2- or 3-branched. Flowers white, large or small, the males in pedunculate racemes, the females solitary.

The genus is dispersed over tropical Asia and America. Of the four Australian species two are common Asiatic ones, the other two are endemic, but as yet insufficiently known.

Leaves palmately or pedately divided into petiolate segments . . . 1. *T. pentaphylla.*
Leaves palmately lobed.
Male racemes without bracts. Fruits acuminate . . . . 2. *T. cucumerina.*
Male racemes with large broad leafy bracts. Fruits not acuminate . 3. *T. palmata.*
Leaves ovate-cordate, not lobed, softly villous. Male racemes with small oblong or lanceolate bracts . . . . 4. *T. Hearnii.*

1. **T. (?) pentaphylla,** *F. Muell. Herb.* Apparently a large climber, the specimens quite glabrous. Leaves palmately or pedately divided into 5 ovate or ovate-lanceolate acuminate entire segments, about 3 to 5 in. long, all petiolulate or the lateral ones rarely united at the base. Tendrils 3-branched. Male flowers unknown. Females solitary, shortly pedicellate. Calyx-tube cylindrical, rather thick, broad and obtuse at the base, produced far above the ovary, rather more than 1 in. long; lobes broadly lanceolate, acuminate, 3 to 4 lines long, entire or with 1 or 2 teeth. Corolla-lobes fringed. Fruit "as round as a ball, beautifully red, the flesh deep yellow, the pulp dark green" (*Dallachy*). Seeds compressed, thick, oblong, the margin entire.

Queensland. Burdekin river, *F. Mueller;* Rockingham Bay, *Dallachy*. The specimens do not admit of the further examination of the flowers, of which there is only one ready to open. The foliage is that of a *Telfairia,* to which it may possibly have to be transferred notwithstanding the narrower seeds, unless the two genera be combined into one.

2. **T. cucumerina,** *Linn. Spec. Pl.* 1432. Stems slender, although sometimes extending to a great length. Leaves nearly orbicular or reniform

in their outline, broadly cordate at the base, mostly 3 to 4 in. diameter, palmately 3- to 7-lobed, the lobes broad, rarely reaching to the middle and irregularly toothed, more or less scabrous-pubescent. Tendrils 3-branched. Male flowers in a short raceme at the end of a long slender peduncle, without bracts. Calyx-tube, in the young bud, short broad and rounded at the base; teeth very short and recurved. Corolla-lobes narrow-oblong, ½ in. long, besides the fringe of long cilia. Female flowers shortly pedicellate. Calyx-tube attenuate above the ovary into a long slender neck. Fruit ovoid-conical, acuminate, not exceeding 2 in., orange-red or yellow when ripe. Seeds about 8 or 10, thick but flattened, with the margin more or less crenate.—Naud. in Ann. Sc. Nat. ser. 4. xviii. 191.

N. Australia. Victoria river, *F. Mueller;* bare rocky hills, Nichol Bay, *Gregory's Expedition.* Common in hedges, etc., in East India.

3. **T. palmata,** *Roxb. Fl. Ind.* iii. 704. A coarse climber. Leaves broad, palmately 3- to 7-lobed, the lobes sometimes broad and short, more frequently especially the central one reaching to below the middle and more or less sinuate-toothed or lobed, pubescent. Male racemes on long stout peduncles, at first short and head-like, at length elongated, with a broadly cuneate or orbicular toothed or jagged bract at least 1 in. diameter under each pedicel. Pedicels very short. Calyx-tube above 1 in. long, attenuate below the middle; lobes ovate or lanceolate, acuminate, 3 to 4 lines long. Petals obovate, fringed with very long cilia. Female flowers shortly pedicellate. Calyx-tube abruptly contracted above the ovary. Fruit nearly globular, not acuminate, 2 to 3 in. diameter.—Wight and Arn. Prod. 350, with the synonyms adduced; Wight, Illustr. t. 104, 105.

Queensland. Brisbane river, *W. Hill;* Rockingham Bay, *Dallachy* (with larger less lobed leaves).

N. S. Wales. Tweed river, *C. Moore.*

The species is common in forests in India, where it climbs to the tops of the loftiest trees (*Roxburgh*).

4. **T. Hearnii,** *F. Muell. Herb.* Of this there are two male fragments in F. Mueller's collection under the name of *T. Hearnii.* Leaves broadly cordate-ovate, denticulate and sometimes obscurely sinuate-lobed, like those of *T. dioica*, Roxb. (now united to *T. nervifolia*) and *T. cordata*, but, instead of being scabrous-pubescent only they are densely and softly villous underneath. Male racemes on long peduncles. Bracts persistent, oblong or lanceolate, entire or toothed, but only 2 or 3 lines long. Calyx-tube slender, attenuate at the base, above ½ in. long; lobes narrow, acute. Petals broadly oblong, densely fringed with long cilia.

Queensland. Rockingham Bay, *Dallachy (Herb. F. Mueller).*

A male specimen in Herb. R. Br., from the islands of the Gulf of Carpentaria, in bud only, may belong to the same species, but some of the leaves are deeply divided into 2 to 5 lobes.

2. LAGENARIA, Ser.

Calyx in the males, and free part of it in the females, campanulate or tubular, with 5 teeth or lobes. Corolla campanulate, deeply 5-lobed. Stamens in the males 3, shorter than the calyx-tube; filaments free; anthers two with 2 cells, one with 1 cell, the cells linear, flexuose, bordering the connective. Ovary in the females from obovoid to cylindrical, with 3 placentas, and numerous horizontal ovules; style short, thick, with 3 bifid stigmas. Fruit large, indehiscent, with a hard rind and fungous flesh. Seeds variously shaped.—Large climber. Tendrils 2-branched. Flowers white, both males and females solitary.

The genus consists only of a single species.

1. **L. vulgaris,** *Ser. in DC. Prod.* iii. 299. A coarse climber, often emitting a musky odour, more or less pubescent or villous. Leaves rather large, broadly orbicular-cordate, angular and denticulate or obscurely or shortly lobed. Tendrils usually 2-branched. Male flowers rather large, white, on peduncles of 2 to 4 in. Calyx-tube turbinate, about ½ in. long; lobes or teeth linear, shorter than the tube. Corolla expanding to 2 or 3 in. diameter. Female flowers rather smaller, on shorter peduncles. Fruit very variable in shape and size.

Queensland. From Broad Sound to Port Denison, *Thozet.*—The species appears to be indigenous in Asia and Africa, but is much cultivated and establishes itself in many tropical and subtropical countries. It includes the *bottle-* and many other *Gourds.*

3. LUFFA, Cav.

Calyx in the males, and free part of it above a narrow tube in the females, campanulate or turbinate, with 5 teeth. Corolla rotate, deeply divided into 5 oblong-obovate or obcordate lobes. Stamens in the males 3 or rarely 5; filaments free, or two connate and the third free; anthers protruding from the calyx-tube, two with 2 cells, one with 1 cell, the cells flexuose, the connective without any appendage. Ovary in the females elongated, with 3 placentas and many horizontal ovules; style columnar, the stigma divided into 3 bifid lobes; rudimentary gynœcium in the males a small gland. Fruit dry, oblong or cylindrical, terete or ribbed, fibrous inside, the small hard conical end (or base of the style) circumsciss and deciduous. Seeds oblong, compressed.—Prostrate or climbing annuals, often large. Leaves palmately 3- or 7-lobed. Tendrils branched. Flowers rather large, yellow or white, the males in pedunculate racemes, the females solitary. Fruits usually rather large.

The genus comprises a few Asiatic and a greater number of African species. The Australian species appear both of them to be included in the Asiatic ones; one of them also abundant in Africa.

Fruit smooth . . . . 1. *L. ægyptiaca.*
Fruit tuberculate or muricate . . . . 2. *L. graveolens.*

1. **L. ægyptiaca,** *Mill. Dict.; Ser. in DC. Prod.* iii. 303. A large climber. Leaves large, broad, the lower ones 5-angled, the upper ones more or less deeply 5-lobed, the lobes, at least the central one, usually acute, often above 6 in. diameter, more or less scabrous. -Tendrils 3-branched. Male racemes elongated, on long peduncles, without bracts. Pedicels short. Calyx broadly turbinate, about ½ in. diameter. Corolla more than 1 in. diameter. Fruit oblong, from 2 or 3 to 8 or 10 in. long, smooth, with 10 deeper coloured streaks when fresh, which in the dry state are often slightly raised ribs, but not acutely prominent as in *L. acutangula.*—*L. pentandra*, Roxb.

Fl. Ind. iii. 712 ; Wight, Ic. t. 499 ; *L. cylindrica*, Rœm. ; Naud. in Ann. Sc. Nat. ser. 4. xii. 119, with the long list of synonyms adduced ; *L. leiocarpa*, F. Muell. Fragm. iii. 107.

Queensland. Gilbert and Burdekin rivers, *F. Mueller*; Fitzroy river, *Thozet* ; Edgecombe Bay, *Dallachy*.

The species is widely spread over tropical and subtropical Africa and Asia. Naudin distinguishes the Australian plant as a variety which F. Mueller raises to a species on account of the fruit said to be not larger than a fowl's egg without longitudinal lines. But in the specimens sent by F. Mueller from the Gilbert river, the fruits are quite as large, and the slightly raised lines quite as conspicuous as in several of the Indian ones.

2. **L. graveolens,** *Roxb. Fl. Ind.* iii. 716? A much more slender and smaller plant than *L. ægyptiaca*, the leaves smaller and less divided, the lobes short and broad, sometimes very obscure, all rounded and slightly sinuate-denticulate, or the central lobe more acute. Flowers smaller than in *L. ægyptiaca*, the males in long racemes, but also a solitary male on a rather long pedicel in the same axil as the female one in all the Australian specimens. Fruits ovoid, 2 to 3 in. long, not ribbed, muricate with scattered rigid tubercles or very short spines. Seeds flat, smooth, about 3 lines long.—Naud. in Ann. Sc. Nat. ser. 4. xii. 124 ; F. Muell. Fragm. iii. 106.

N. Australia. N.W. coast, *Bynoe* ; tributaries of the Victoria river, *F. Mueller*.

The species, if correctly determined, is also on the coast of Coromandel, but the specimens are so imperfect that it is impossible to establish without doubt the identity concluded by Naudin from the fruit. In several of the Australian specimens the leaves are much more acutely lobed than they are represented in Roxburgh's drawing, and the calyx-lobes have a hollow protuberance at the base, which suggested to F. Mueller the specific name of *L. saccata* which he had given to his plant. These protuberances do not appear in the Indian species, nor can I find them in some of the Australian specimens with leaves more like Roxburgh's, but the few flowers are too ill-dried to ascertain the point. Naudin says the fruit is scarcely bigger than a pigeon's egg. Some of those in the Kew herbaria are nearly 3 in. long.

4. CUCUMIS, Linn.

Calyx in the males, and free part of it in the females turbinate or campanulate, with 5 teeth or lobes. Corolla campanulate, deeply 5-lobed or divided to the calyx. Stamens 3 ; filaments short, free ; anthers two with 2 cells, one with 1 cell ; cells linear, flexuose, connective produced into a crest-like appendage beyond the cells. Ovary in the female with 3 (rarely 5) placentas and numerous horizontal ovules ; style short, with 3 (rarely 5) obtuse stigmas. Fruit variously shaped, fleshy with a hard rind, indehiscent or rarely tardily opening in 3 valves. Seeds oblong, compressed, the margin not thickened.—Climbers either annual or with a perennial rhizome, more or less hispid. Tendrils simple. Flowers yellow, the males in axillary clusters or rarely solitary, the females solitary, usually sessile or shortly pedicellate.

The genus extends over the tropical and subtropical regions of the New and the Old World. The only Australian species is a common one in Asia.

1. **C. trigonus,** *Roxb. Fl. Ind.* iii. 722. A rather slender creeper or climber, sometimes rigidly hispid, almost aculeolate, sometimes scabrous-pubescent. Leaves not large, usually broadly ovate-cordate in their outline, either nearly entire or more or less 3- 5- or 7-lobed, the lobes slightly or sometimes more deeply toothed, usually scabrous. Flowers small, on short slender pedicels. Calyx in the males from a little more than 1 line to nearly

2 lines long, pubescent-hirsute or densely woolly; lobes short and narrow. Corolla about ½ in. diameter, the lobes acute. Female flowers usually rather larger, the adnate tube ovoid or oblong, 3 to 4 lines long, tomentose-pubescent or densely woolly. Fruit globular or ovoid, often quite glabrous, but sometimes retaining a few scattered hairs, from under 1 in. diameter to more than twice that size.—Wight, Ic. t. 497; Naud. in Ann. Sc. Nat. ser. 4. xi. 30; *C. pubescens*, Hook. in Mitch. Trop. Austr. 110; *C. jucundus* and *C. picrocarpus*, F. Muell. in Trans. Phil. Inst. Vict. iii. 46.

N. Australia. Oakover river, Nichol Bay, *Gregory's Expedition;* Victoria river, *F. Mueller;* Port Essington, *Armstrong;* Albert river, *Henne;* in the interior, *M'Douall Stuart's Expedition.*

Queensland. Suttor and Bogan rivers, *Bowman;* Fort Cooper, *Thozet.*

N. S. Wales. Narran and Balonne rivers, *Mitchell;* Darling river to Cooper's Creek, *Victorian and other Expeditions.*

The only absolute difference to be gathered from Naudin's investigations between *C. trigonus*, and what he concludes to be the wild Melon (*C. Melo*, var. *agrestis*, Naud. in Ann. Sc. Nat. ser. 4. xi. 73; *C. pubescens*, of Indian botanists, Wight, Ic. t. 496, and probably of Willd.), is, that the former has a perennial root, or rather rhizome, and roots very readily at the joints, whilst the Melon is strictly annual. As, however, the stems are always annual, the existence of the perennial rhizome is rarely ascertained except in cultivation, and no collectors of Australian specimens allude to it. Some of these look very much like Indian specimens of the wild Melon, others have more the appearance of the Indian *C. trigona*, and some are not to be distinguished from the New Caledonian *C. Pancherianus*, Naud. in Ann. Sc. Nat. ser. 4. xii. 112. t. 8. Most probably all are forms only of *C. Melo.*

C. myriocarpus, Naud. l. c. xi. 22, with leaves deeply divided into rounded ciliate lobes, nearly glabrous above, rigidly hispid underneath, and with small globular densely prickly fruits on filiform pedicels, commonly known in gardens as *C. prophetarum*, but not the true Linnæan species of that name, is in F. Mueller's collection from the banks of the Torrens river in S. Australia, as an introduced plant.

5. MOMORDICA, Linn.

Calyx in the males, and free part of it in the females, short, campanulate, with 5 lobes. Corolla rotate or broadly campanulate, usually divided to the calyx into 5 lobes. Stamens in the males 2 or 3; filaments short, free; anthers at first coherent, at length free, one or two 2-celled, the others 1-celled, the cells flexuose, the connective without any appendage. Two (or three?) connivent scales on the tube of the calyx and corolla at the insertion of the stamens. Ovary in the females fusiform or oblong, with 3 placentas and several horizontal ovules; style slender, with 3 stigmas. Fruit oblong, fusiform or cylindric, not fibrous, indehiscent or opening more or less in 3 valves. Seeds imbedded in pulp, flattened or convex, smooth or variously sculptured.—Climbers usually slender. Leaves entire, lobed or 3- to 7-foliolate. Tendrils simple. Peduncles axillary, either all 1-flowered, with a broad bract under the flower, or the males paniculate.

The genus is dispersed over the tropical and subtropical regions of both the New and the Old World; most of the species, however, are African. The only Australian one is common in Asia and Africa.

1. **M. Balsamina,** *Linn.; Ser. in DC. Prod.* iii. 311. A slender annual climber. Leaves thin and glabrous, orbicular in their circumscription, mostly under 2 in. diameter, palmately and deeply 5-lobed, the lobes more or less rhomboidal, deeply and acutely toothed or lobed. Peduncles all slender and

1-flowered, the males usually longer than the leaf, with a reniform or broadly cordate bract a little below the flower, the females shorter, with the bract below the middle. Calyx fully ½ in. diameter, with very thin broad acute lobes longer than the tube. Corolla yellow, nearly twice as long as the calyx. Female flowers rather smaller. Ovary fusiform, attenuate under the free part of the calyx. Fruit ovoid-globular, more or less attenuate at the end, about 1 in. diameter, bursting irregularly. Seeds 5 or 6, rather large, each one enveloped in a red pulp.

Queensland. Rockhampton, *Dallachy.*—Widely spread over Asia and Africa, and now introduced into America.

6. BRYONIA, Linn.

(Bryonopsis, *Blume.*)

Calyx in the males, and free part of it in the females, broadly campanulate, 5-toothed. Corolla campanulate, deeply 5-lobed. Stamens in the males 3; filaments free; anthers two with 2 cells, one with 1 cell, the cells flexuose. Ovary in the females fusiform, ovoid or globular, contracted at the top, with 3 placentas and few horizontal ovules; style slender, with 3 reniform or bifid stigmas. Fruit a globular or ovoid-conical berry. Seeds few, compressed, or with convex faces and a thickened margin enveloped in pulp.—Climbing herbs with simple or 2-branched tendrils. Leaves palmately lobed. Flowers greenish-yellow, small as well as the fruits, in axillary racemes sometimes reduced to clusters.

The genus, taken in the above extended sense given to it by most botanists, although not numerous in species, ranges over the warmer and temperate regions both of the New and the Old World. The Australian species, however, belongs to the section *Bryonopsis*, now adopted by Naudin as a distinct genus, limited to 2 or perhaps 3 Asiatic and African species, of which the Australian is one.

1. **B. laciniosa,** *Linn.; Ser. in DC. Prod.* iii. 308. Stems rather slender, but extending to a great length. Leaves broad, very deeply palmatifid or almost pedatifid, the lobes ovate ovate-lanceolate or sometimes linear-lanceolate, often 3 to 4 in. long, and more or less angular or sinuate-toothed. Tendrils usually 2-branched, but one branch sometimes small or quite wanting. Flowers small, in very short axillary racemes usually reduced to clusters, the males and females often in the same axil, the rhachis rarely 3 to 4 lines long. Pedicels slender, from 1 to 5 or 6 lines long. Calyx 1½ to 2 lines diameter. Corolla scarcely twice the size of the calyx. Berry globular, yellow or red, about 1 in. diameter. Seeds with a very thick transversely-furrowed border, the faces convex or conical within the border.—Wight, l. c. t. 500; Naud. in Ann. Sc. Nat. ser. 4. xii. 139, with the synonyms adduced; *Zehneria erythrocarpa*, F. Muell. in Hook. Kew Journ. viii. 51 (from the character given).

N. Australia. Sir Charles Hardy's Island, *Henne;* Port Essington, *Armstrong.*

Queensland. Broad Sound, *R. Brown;* N.E. coast, *A. Cunningham;* Burdekin river, *F. Mueller;* Suttor river, *Bowman;* Rockhampton, *Thozet, Dallachy;* Brisbane river, Moreton Bay, *F. Mueller.*

N. S. Wales. Macleay river, *Beckler;* Clarence river, *Wilcox.*

The species is dispersed over tropical Asia and Africa. Naudin, Ann. Sc. Nat. ser. 4. xii. 140, and xviii. 193, distinguishes this species, with 1 or 2 closely allied ones (or perhaps varieties) as the above-mentioned genus *Bryonopsis*. This name was originally proposed by Blume for several old *Bryonias* now referred to *Zehneria* and other groups, and is now limited by Naudin to *B. laciniosa* and its allies, characterized especially by the seed, but also by monœcious not diœcious flowers, the clustered not racemose inflorescence, and branched not simple tendrils. But one of our European true *Bryonias* is monœcious, the clusters of *B. laciniosa* are nothing but short racemes, and the branched tendrils, although general, are not constant, and the genus rests solely on the seed, which appears to me to be a much better sectional than generic character.

7. MELOTHRIA, Linn.

Calyx in the males, and upper free part of it in the females, campanulate, shortly 5-toothed. Corolla rotate, deeply 5-lobed, with narrow lobes. Stamens in the males 3; filaments short, free; anthers often slightly cohering, two with 2 cells, one with 1 cell, the cells straight and parallel, 3 small staminodia in the females. Ovary in the females with 3 placentas and several horizontal ovules; style short, with 3 capitate, dilated or bifid stigmas. Fruit a small globular ovoid or fusiform berry. Seeds flat, oval or oblong, enveloped in pulp.—Slender climbing or prostrate herbs. Leaves triangular or palmately lobed. Tendrils simple. Flowers very small, yellow, the males in short racemes almost reduced to pedunculate umbels or sessile clusters, the females on slender axillary pedicels, solitary or clustered.

The genus is dispersed over the tropical and subtropical regions of the New and the Old World, most abundant in Africa. The Australian species are both endemic.

Leaves broadly triangular or hastate. Male flowers in a pedunculate umbel-like raceme. Females on long filiform pedicels . . . . 1. *M. Cunninghamii.*
Leaves palmately 5- or 7-lobed. Male and female flowers minute, clustered in the same axils on filiform but rather short pedicels . 2. *M. Muelleri.*

1. **M. Cunninghamii,** *F. Muell.* (as *Zehneria*). Stems very slender, often filiform. Leaves broadly triangular or hastate, irregularly but not deeply toothed, or rarely obscurely 3- or 5-lobed, thin and somewhat scabrous, the larger ones nearly 3 in. long, but mostly smaller. Tendrils simple, filiform. Male peduncles slender, bearing at the end a short corymbose raceme almost reduced to an umbel of about 6 small yellow flowers. Female flowers usually solitary in the axils, on filiform pedicels of 1 to 2 in., with rarely a male flower in the same axil. Calyx about 1 line diameter. Corolla about 2 lines diameter. Ovary or calyx-tube of the females attenuate into a slender neck. Stigmas capitate. Berry globular, 3 to 4 lines diameter.—*Zehneria Cunninghamii*, F. Muell. in Hook. Kew Journ. viii. 51.

N. Australia. Arnhem N. Bay, *R. Brown.*
Queensland. Brisbane river, Moreton Bay, *F. Mueller;* Breakfast Creek, *Bowman;* Rockhampton, *Dallachy.*
N. S. Wales. Paramatta, *Woolls;* Clarence river, *Beckler.*

This species is nearly allied to the African *M. triangularis*, Benth. The northern specimens in Herb. R. Brown, have the leaves broadly cordate, the flowers rather longer and the fruits rather larger, almost ovoid, but they appear to belong to the same species.

2. **M. Muelleri,** *Benth.* Small and rather slender, very scabrous but not hispid. Leaves on long petioles, deeply cordate, nearly orbicular, 1 to 2 in. diameter, shortly and palmately 5- to 7-lobed, the lobes mostly obtuse, coarsely toothed or lobed. Tendrils small, filiform, simple. Flowers minute, on filiform pedicels of 2 to 3 lines, the males and females clustered in the

same axils. Calyx not 1 line diameter, with minute teeth. Corolla about 2 lines diameter, divided to the calyx into obtuse lobes. Ovary or calyx-tube in the females ovoid, contracted into a short neck, the corolla smaller than in the males. Staminodia 3, very small. Stigmas reniform or shortly 2-lobed. Berry globular, about ½ in. diameter.—*Cucurbita micrantha*, F. Muell. in Trans. Phil. Soc. Vict. i. 17; *Cucumis? Muelleri*, Naud. in Ann. Sc. Nat. Ser. 4, xi. 84; *Zehneria micrantha*, F. Muell. Fragm. i. 182, and Pl. Vict. t. 18; *Mukia micrantha*, F. Muell. Fragm. ii. 180; iii. 107.

N. S. Wales. Hunter's River, *Bauer* (*in Herb. R. Br.*); Darling desert, *Dallachy and Goodwin.*

Victoria. Murray river, *F. Mueller.*

8. MUKIA, Arn.

Calyx in the males, and free part of it in the females, turbinate-campanulate, 5-toothed. Corolla rotate, divided to the calyx into 5 acute lobes. Stamens in the males 3, filaments short, free; anthers two with 2 cells, one with 1 cell, the cells parallel and straight, the connective produced into a short point beyond them; the females without staminodia. Ovary in the females with 2 or 3 placentas and several horizontal ovules; style clavate, with a thick 2- or 3-lobed stigma. Fruit a globular berry. Seeds few, compressed, scrobiculate.—Scabrous-hispid annuals, with the habit of *Cucumis.* Leaves angular or rarely lobed. Flowers small, yellow, the males clustered and pedicellate, the females solitary and sessile or nearly so.

Besides the Australian species, which is widely spread over tropical Asia and Africa, there may be a second African one.

1. **M. scabrella,** *Arn. in Hook. Journ. Bot.* iii. 276. Rather slender, but very scabrous-hispid. Leaves shortly petiolate, deeply cordate, from broadly triangular to ovate-lanceolate, and more or less hastate with broad rounded or angular lobes, usually obtuse, obscurely crenate or rarely shortly 3- or 5-lobed, mostly under 2 in. long. Male flowers clustered in the axils, the pedicels 2 to 3 lines long; females almost sessile. Calyx hirsute, above 1 line long, with small linear teeth. Corolla lobes about 1 line long. Adnate part of the calyx-tube or ovary in the females about 2 lines long, densely hirsute with long white hairs. Style surrounded by a cup-shaped disk. Berry globular, sometimes attaining ½ in. diameter.—Wight, Ic. t. 501; Naud. in Ann. Sc. Nat. ser. 4. xii. 142.

N. Australia. N.W. coast, *Bynoe;* Nichol Bay, *Gregory's Expedition* (with deeply lobed narrow leaves); Upper Victoria river and Gulf of Carpentaria, *F. Mueller;* Port Essington, *Armstrong.*

Queensland. Keppel and Shoalwater Bays and Northumberland islands, *R. Brown;* Burdekin and Gilbert rivers, *F. Mueller;* Port Curtis and Lizard Island, *M'Gillivray;* N.E. coast, *A. Cunningham;* Rockingham Bay, *Dallachy, Thozet.*

9. SICYOS, Linn.

Calyx in the males and free part of it above the narrow tube in the females campanulate, with 5 small subulate teeth. Corolla rotate, divided to the calyx into 5 ovate lobes. Stamens in the males united in a column clavate at the top and more or less lobed, with 3 to 5 linear curved and flexuose anther-cells. Ovary in the females 1-celled with 1 pendulous ovule. Fruit small, dry, ovoid or oblong, acute or beaked, usually covered with prickles.—Prostrate or climbing herbs. Leaves angular or 3- or 5-lobed. Tendrils 3-branched. Flowers small, the males in racemes sometimes reduced to corymbs or clusters; the females pedicellate in the axils or sometimes in the same raceme with the males.

The genus is spread over the warmer regions of the New and the Old World. The only Australian species is a common American one.

1. **S. angulata,** *Linn.; DC. Prod.* iii. 309. Stems rather slender, but extending sometimes to a great length, glabrous or sparingly scabrous. Leaves on long petioles, from broadly ovate-cordate to almost reniform, usually acutely 3-angled or palmately lobed, the central angle or lobe the longest, of a thin texture and often 3 to 4 in. long or more. Male and female flowers often in the same axil, the males in a short raceme on a long peduncle, the females in a small dense cluster on a very short peduncle. Calyx in the males scarcely above 1 line diameter and the corolla rarely 3 lines, the females still smaller. Fruits ovoid, rarely ½ in. long, densely covered with barbed prickles.—Hook. f. Fl. Tasm. i. 143; *S. fretensis*, Hook. f. in Hook. Loud. Journ. vi. 473; *S. australis*, Endl. Prod. Fl. Norf. 67; A Gray, Bot. Amer. Expl. Exped. i. 648.

Queensland. Moreton Bay, *F. Mueller.*
N. S. Wales. Port Jackson and Blue Mountains, *R. Brown, Woolls.*
Victoria. Banks of the Tambo and Gipps' Land, *F. Mueller.*
Tasmania. Islands of Bass's Strait, *Gunn.*

A common weed in tropical and N. America, widely dispersed over the Pacific isles and New Zealand, but not recorded from Asia or Africa.

A. Gray distinguishes *S. australis* from the common American form chiefly by its smaller flowers. It is not easy to judge of this from dried specimens without soaking, and the size appears variable, but certainly in some Australian specimens quite as large as in the common American forms.

Order LV. FICOIDEÆ.

Calyx persistent, free or adnate to the ovary at the base, divided to the middle or to the base into 5 or 4 rarely more or only 3 lobes or segments, imbricate in the bud or very rarely valvate. Petals none or indefinite and narrow, very rarely equal in number to the calyx-segments, inserted at their base. Stamens few or many, usually indefinite, or not corresponding in number to the calyx-lobes, or rarely equal in number to them, inserted on the calyx-tube, or hypogynous when the calyx is divided to the base; filaments free or united in a cup at the base; anthers with parallel cells opening longitudinally. Ovary inferior, half superior or superior, 3- to 5- or more-celled, rarely 2-celled or reduced to a single carpel; styles as many as cells, free or united at the base, usually filiform and stigmatic along the inner side, or rarely with terminal stigmas or very short; ovules 1, 2 or more in each cell usually inserted on a basal placenta more or less adnate to the axis or inner angle of the cell. Fruit a capsule or rarely fleshy or drupaceous, opening loculicidally septicidally or both, in as many or twice as many valves as cells, or transversely circumsciss or indehiscent. Seeds with a crustaceous or rarely membranous or thick testa, usually compressed. Embryo curved round a mealy albumen.—Herbs or rarely undershrubs or almost shrubby. Leaves alternate or more rarely opposite, entire, often succulent. Flowers either solitary, terminal, leaf-opposed or in the forks of the stems, or in axillary cymes or clusters.

The Order is widely dispersed over the globe, although not extending to very cold regions, the majority of species inhabiting sandy or rocky seacoasts or dry wastes or spreading as weeds of cultivation, and particularly abundant in S. Africa. Of the eight Australian genera, three are generally distributed over the warmer regions of the globe, three are especially South African, with a few of their numerous species dispersed over a wider range, and two small ones are endemic in Australia.

TRIBE I. **Mesembryeæ.**—*Calyx-tube adnate to the ovary.*

Petals numerous, linear . . . . 1. MESEMBRYANTHEMUM.
Petals none . . . . 2. TETRAGONIA.

TRIBE II. **Aizoideæ.**—*Calyx free, but with a distinct tube bearing the stamens. Petals none.*

Capsule opening in valves.
  Stamens indefinite . . . . 3. AIZOON.
  Stamens 4 . . . . 4. GUNNIA.
Capsule circumsciss.
  Styles and ovary-cells 3 to 5 . . . . 5. SESUVIUM.
  Styles and ovary-cells 2 or 1 . . . . 6. TRIANTHEMA.

TRIBE III. (or SUBORDER). **Mollugineæ.**—*Calyx divided to the base. Petals* 5 *or fewer or none.*

Stamens 8, united in a cup at the base. Ovules 1, 2 or rarely 3 in each cell . . . . 7. MACARTHURIA.
Stamens few or many, free or rarely slightly united when very numerous. Ovules many or rarely 3 or 4 in each cell 8. MOLLUGO.

TRIBE I. MESEMBRYEÆ.—Calyx-tube adnate to the ovary, either entirely so, or produced above it.

1. MESEMBRYANTHEMUM, Linn.

Calyx-tube adnate to the ovary; lobes 5 or rarely more or fewer. Petals numerous, linear, in one or more series. Stamens numerous, in several series. Ovary inferior, with 5 or more, rarely 4, cells, each with numerous ovules; styles as many as cells of the ovary, free or connate at the base, stigmatic along the inner side. Capsule surrounded by the persistent calyx, the summit flat and loculicidally dehiscent. Seeds minute, with a crustaceous testa. —Herbs or undershrubs, more or less succulent. Leaves opposite or alternate, fleshy, entire or rigidly ciliate, without stipules. Flowers showy, terminal or in the forks of the branches, or leaf-opposed.

The species are very numerous in S. Africa, a few spreading along the seashore to various parts of the world. Of the four Australian ones here enumerated, one is introduced only, one is included amongst the widely diffused maritime ones, found also in S. Africa, and the remaining two belong probably to the same category, although they have not been absolutely identified with any S. African species.

Leaves opposite, triquetrous, linear or oblong.
  Leaves mostly above 1 in. Flowers about 1½ in. diameter on rather long pedicels . . . . 1. *M. æquilaterale.*
  Leaves mostly under 1 in. Flowers not above 1 in. diameter, sessile or shortly pedicellate in tufts of leaves at the nodes . 2. *M. australe.*

Leaves alternate, flat.
Flowers rather small, sessile or shortly pedicellate, terminal or leaf-opposed, white or pink. Leaves undulate, covered with transparent vesicles . . . . 3. *M. crystallinum.*
Flowers rather large, yellow, on long pedicels. Leaves lanceolate or spathulate . . . . 4. *M. pomeridianum.*

1. **M. æquilaterale,** *Haw.; Hook. f. Fl. Tasm.* i. 146. Perennial, with robust prostrate stems, extending sometimes to a considerable length, with short ascending flowering branches, or sometimes more ascending from the base. Leaves opposite, stem-clasping, thickly linear-triquetrous, equal-sided or laterally compressed, attaining 2 to 3 in. Flowers rather large, red, pedicellate or nearly sessile within the last small pair of leaves. Calyx-tube turbinate, ½ in. long or rather more; lobes unequal, the 2 larger ones often as long as the tube, with prominent angles decurrent on the calyx and pedicel, or the calyx quite terete. Petals spreading to about 1½ in. diameter. Styles and ovary-cells varying from 6 to 10. Fruit said to be about the size of a good gooseberry.—*M. æquilaterale, M. glaucescens, M. Rossi,* and *M. nigrescens,* Haw.; DC. Prod. iii. 429; Salm-Dyck, Monogr. § 19. f. 1, 2, 3; *M. præcox,* F. Muell. in Linnæa, xxv. 384.

Queensland. Plains of the Condamine, *Leichhardt.*
Victoria. Along the seacoast in various places, *F. Mueller, Robertson.*
Tasmania. Abundant on the seacoast and ascending the rivers as far as their waters are salt, called "Pig's faces," *J. D. Hooker.*
S. Australia. Murray river, Holdfast Bay, Salt plains on the W. side of Flinders Range, *F. Mueller.*
W. Australia. Swan River, *Drummond, Oldfield;* Murchison river, *Oldfield.*

The same species is also found on the coasts of Chile and California, and scarcely differs from the S. African *M. acinaciforme,* Linn., except in the leaves not so thick and the flowers smaller.

2. **M. australe,** *Soland. in Ait. Hort. Kew. ed.* 1. ii. 187. Perennial, with prostrate stems rooting at the nodes, the flowering branches very short, or reduced to clusters of leaves surrounding the peduncle. Leaves opposite, triquetrous or somewhat flattened and oblong, obtuse or rarely almost acute, ½ to ¾ in. long. Flowers reddish, solitary in the axillary clusters of leaves, or terminating very short leafy branches, the pedicels from rather shorter than the leaves to twice their length. Calyx-tube turbinate, 2 to 3 lines long; lobes unequal, the 2 larger ones as long as or rarely longer than the tube, and rarely forming slightly prominent lines decurrent on it. Petals spreading to about 1 in. diameter. Styles and ovary-cells usually 5. —DC. Prod. iii. 428; Salm-Dyck, Monogr. § 18. f. 2; Hook. f. Fl. Tasm. i. 147; *M. clavellatum,* Haw.; DC. Prod. iii. 428; Salm-Dyck, Monogr. § 18. f. 1; *M. demissum,* Willd. Enum. Suppl. 36 (name only, referred to *M. australe* in Link, Enum. Hort. Berol. ii. 51).

N. S. Wales. Darling river, *Victorian Expedition* (a bad doubtful specimen).
Victoria. Phillips island, *F. Mueller.*
Tasmania. Woolnorth, Circular Head and banks of the Tamar, *Gunn, J. D. Hooker;* King's Island, *F. Mueller.*
S. Australia. Seacoast, *F. Mueller;* Murray river, *Blandowski.*
W. Australia. Bald Island, *Oldfield, Maxwell;* Swan River, *Drummond, 3rd Coll. n.* 59 (leaves ½ to 1½ in. and pedicels 1 to 2 in. long.

The species is found also on the seacoasts of New Zealand and the islands of the South Pacific, and is probably not really distinct from the S. African *M. crassifolium*, Linn.

3. **M. crystallinum,** *Linn.; DC. Prod.* iii. 448. Annual, prostrate and much-branched, the thick stems under 1 ft. long and covered, as well as the foliage, with transparent vesicles, to which it owes the name of "Ice Plant." Radical leaves broadly cordate-ovate; stem-leaves alternate, flat but succulent, from broadly obovate to obovate-oblong, much undulate, obtuse, narrowed into a stem-clasping petiole. Flowers not large, on very short terminal or leaf-opposed pedicels, the upper ones forming a loose leafy cyme. Calyx 3 to 4 lines diameter, the lobes short and obtuse or rarely longer and lanceolate. Petals white or pink, spreading to about ½ in. diameter. Styles and ovary-cells 5.—Sibth. Fl. Græc. t. 481; DC. Pl. Grass. t. 128.

S. Australia. Holdfast Bay and Port Adelaide, *F. Mueller.*

W. Australia. Swan River, *Drummond.*

A common seacoast plant in S. Africa, found also on the coasts of the Canary Islands, southern Europe, and California.

* 4. **M. pomeridianum,** *Linn.; DC. Prod.* iii. 450. Annual, decumbent or ascending, under a foot high, pubescent with a few soft hairs especially on the inflorescence and margins of the leaves. Leaves alternate, flat but succulent, lanceolate or spathulate, often 2 to 3 in. long. Flowers rather large, pale yellow, on long peduncles, terminal or in the forks. Calyx-tube above ½ in. diameter; lobes very unequal, the longest ¾ in. long. Petals very numerous in several rows. Styles and ovary-cells 12 or more.—Bot. Mag. t. 540; Salm-Dyck, Monogr. § 65. f. 1.

A South African species, naturalized on the Darling river, *Dallachy.*

2. TETRAGONIA, Linn.

(Tetragonella, *Miq.*)

Calyx-tube adnate to the ovary at the base and usually produced above it; lobes 4 or 5, or rarely 3. Petals none. Stamens indefinite, few or many, inserted at the top of the calyx-tube, free but usually in clusters alternating with the lobes. Ovary inferior, 2- to 8-celled, with 1 pendulous ovule in each cell, the summit convex or conical, and rarely containing a second erect ovule. Styles as many as cells, linear, stigmatic along the inner side. Fruit indehiscent with a hard almost bony endocarp, the herbaceous or almost fleshy epicarp (or persistent calyx) often variously horned or tubercular.—Herbs or undershrubs. Leaves alternate, flat but rather thick, without stipules. Flowers solitary or few together in the axils, sessile or pedicellate, usually of a yellowish or reddish-green.

The species comprises several S. African species, besides a few dispersed over the seacoasts of New Zealand, the Pacific islands, and some parts of Asia and America. Of the two Australian species, one is also in New Zealand and extratropical S. America and Asia, the other is endemic.

Flowers usually hermaphrodite, with 3 or usually more styles and ovary-cells. Fruit often horned . . . . 1. *T. expansa.*
Flowers mostly unisexual, with 2 or very rarely 3 styles and ovary-cells. Fruit never horned . . . . 2. *T. implexicoma.*

1. **T. expansa,** *Murr.; DC. Prod.* iii. 452. Decumbent or prostrate,

often extending to several feet. Leaves petiolate, the larger ones ovate, triangular or broadly hastate, 2 to 4 in. long, entire, obtuse or acute, the smaller ones narrower. Flowers small, yellow, on very short pedicels or almost sessile in the axils, solitary or 2 together. Calyx-tube broadly turbinate, a little above 1 line diameter; lobes broad and obtuse, about as long as the tube. Stamens in clusters of 3 or 4 opposite each sinus of the calyx. Ovary half-inferior, the free portion depressed-hemispherical, with 3 to 8 external furrows and as many cells. Fruit hard, $\frac{1}{4}$ to $\frac{1}{2}$ in. diameter, from nearly globular and almost without protuberances to turbinate, angular, with 2, 3 or more hard prominent horns, the endocarp woody.—Hook. f. Fl. Tasm. i. 147; Bot. Mag. t. 2362; Payer in Ann. Sc. Nat. ser. 3, xviii. t. 13; *T. inermis*, F. Muell. in Linnæa, xxv. 384.

Queensland. Brisbane river, Moreton Bay, *C. Stuart.*

N. S. Wales. Port Jackson, *J. D. Hooker;* Hastings river, *Beckler;* in the interior at the camp at Meninville, *Victorian Expedition.*

Victoria. Port Phillip, *R. Brown;* Sealers' Cove, *F. Mueller.*

Tasmania. Northern shores, *J. D. Hooker.*

S. Australia. Elders Range, Lake Torrens, *F. Mueller.*

The species is also on the coasts of New Zealand, extratropical S. America, and Japan, and has been cultivated in Europe as "New Zealand Spinach."

2. **T. implexicoma,** *Hook. f. Fl. Tasm.* i. 148. Decumbent or almost climbing. Leaves petiolate, ovate or rhomboidal, usually smaller than in *T. expansa*, and often much narrower or quite oblong, usually covered with transparent vesicles like the Ice-plant. Pedicels filiform, solitary or 2 together in the axils on young leafy shoots. Flowers polygamo-diœcious, the males with a narrow calyx-lobe, the lobes nearly 2 lines long and valvate; stamens in clusters of 3 or 4 opposite the sinuses. Perfect flowers with a broader calyx-tube. Ovary 2-celled or rarely 3-celled, the lower part adnate, with 1 pendulous ovule in each cell, the free part conical, sometimes with 1 erect ovule in each cell, and circumscissly deciduous after flowering. Fruit smaller than in *T. expansa*, 3 to 4 lines long including the red succulent calyx, with irregularly prominent ribs or tubercles, but not horned. Seeds 1 or rarely 2.—F. Muell. Pl. Vict. t. 13; *Tetragonella implexicoma*, Miq. in Pl. Preiss. i. 245.

Victoria. Port Phillip, *F. Mueller*, *Harvey*, and others; mouth of the Glenelg, *F. Mueller.*

Tasmania. Abundant on all the coasts, sometimes festooning the bushes on the shore, *J. D. Hooker.*

S. Australia. Lower Murray river, St. Vincent's Gulf, etc., *F. Mueller.*

W. Australia. Rottenest Island, *Preiss*, *n.* 2393.

TRIBE II. AIZOIDEÆ.—Calyx free, but with a distinct turbinate tube, bearing the stamens at or below the top.

3. AIZOON, Linn.

Calyx free, deeply 4- or 5-lobed. Petals none. Stamens indefinite, usually about 20, inserted at the top of the calyx-tube, free, but more or less in clusters alternating with the lobes. Ovary superior, enclosed in the calyx-tube, 5-celled or in the Australian species 4-celled, with 2 or more ovules in each cell; styles as many as cells, filiform, stigmatic along the inner side. Capsule surrounded by the persistent calyx, depressed, opening loculicidally in as many valves as cells, or in the Australian species the valves split septicidally.—Herbs or undershrubs. Leaves alternate or rarely opposite, without stipules. Flowers solitary or divaricately cymose.

The genus is chiefly African, and especially S. African, but extending to N. Africa and S. Europe. The only Australian species is endemic, and differs from the African ones in habit, in its 4-merous flowers, in the calyx valvate, not imbricate in the bud, and in the dehiscence of the capsule. In this respect it is more nearly allied to *Gunnia*, to which it ought perhaps to be referred, but the stamens are numerous as in *Aizoon*. The opposite leaves occur also in one of the S. African species.

1. **A. quadrifidum,** *F. Muell. Fragm.* ii. 148. A rigid shrub, probably small, with divaricate opposite or dichotomous branches, covered as well as the leaves with a dense but close almost scurfy tomentum. Leaves opposite, narrow-linear, obtuse, rather thick and soft, not above ½ in. long in the specimens. Flowers shortly pedicellate, terminal or in the forks, resembling in shape and size those of *A. hispanicum*. Calyx tomentose; tube short, broadly turbinate, the ribs not prominent; lobes 4, valvate, ovate-acuminate, about 3 lines long. Stamens numerous, densely crowded opposite the sinus of the calyx, more distant opposite the lobes; filaments slender, more or less covered like the ovary with transparent vesicular cells, about as long as the calyx-lobes. Ovary truncate on the top, 4-angled, 4-celled, with 4 styles stigmatic along their inner side. Ovules numerous. Capsule almost free, obpyramidal-truncate, septicidally dehiscent in 4 truncate valves, loculicidally divided almost to the base.—*Sesuvium quadrifidum*, F. Muell. Rep. Babb. Exped. 9.

N. S. Wales. Towards the Barrier Range, between Duroodoo and Nangavera, *Victorian Expedition.*
S. Australia. Desert at Stuart's Creek, *Hergolt.*

4. GUNNIA, F. Muell.

Calyx free, deeply divided into 4 lobes, valvate in the bud. Petals none. Stamens 4, inserted on the calyx-tube and alternating with its lobes. Ovary superior, enclosed in the calyx, 4-celled, with several ovules in each cell; styles 4, filiform, stigmatic along the inner side. Capsule enclosed in the persistent calyx, opening loculicidally, each valve splitting septicidally. Seeds numerous, small; testa thin and smooth; embryo curved round a mealy albumen.—Small diffuse annuals. Leaves opposite. Flowers terminal or in the forks, nearly sessile.

The genus is limited to Australia.
Leaves linear . . . . 1. *G. septifraga.*
Leaves ovate . . . . 2. *G. Drummondii.*

1. **G. septifraga,** *F. Muell. Rep. Babb. Exped.* 9. The specimens described from one small fragment, but are evidently allied to *G. Drummondii*. The branches appear to be more slender, the single pair of floral leaves preserved are linear. Flowers nearly sessile in the forks or terminal. Calyx-lobes acute, about 1½ lines long when in flower, 2 lines long when in fruit.

S. Australia. Stuart's Creek, *Hergolt.*

2. **G. Drummondii,** *Benth.* A diffuse annual, with opposite branches,

the whole plant in our specimens scarcely exceeding 2 in. Leaves opposite, petiolate, ovate or ovate-oblong, 2 to 4 lines long. Flowers large for the plant, sessile or nearly so in the forks and at the ends of the branches. Calyx-tube short, turbinate; lobes 4, valvate, broadly ovate, acute, nearly 3 lines long when closed over the fruit. Stamens 4, inserted below the middle of the calyx-tube and about as long as its lobes. Capsule contracted and very shortly adnate at the base, broad in the centre, pyramidal and 4-angled at the top, opening both loculicidally and septicidally. Seeds numerous.

W. Australia, *Drummond, n.* 241.

5. SESUVIUM, Linn.

Calyx free, deeply 5-lobed. Petals none. Stamens 5, alternating with the calyx-lobes or indefinite, often very numerous, inserted at the top of the tube. Ovary free, enclosed in the calyx, 3- to 5-celled, with numerous ovules in each cell; styles as many as cells, filiform, stigmatic along the inner side, at least towards the end. Capsule surrounded by the persistent calyx, membranous, more or less completely divided by very thin dissepiments, transversely circumsciss about the middle. Seeds several; testa coriaceous, smooth. —Herbs or undershrubs. Leaves opposite, fleshy, without stipules, but sometimes with scarious dilatations of the petiole. Flowers solitary or clustered in the axils or almost cymose, the calyx-lobes usually coloured inside, with more or less scarious margins.

The genus contains about four species, spread over the seacoasts of the tropical and subtropical regions of the globe, the Australian species being the commonest and the most generally diffused both in the New and the Old World.

1. **S. portulacastrum,** *Linn.; DC. Prod.* iii. 453. A succulent herb, procumbent or creeping and rooting at the joints. Leaves linear or linear-oblong, contracted below the middle, broader and stem-clasping at the base, mostly 1 to 2 in. long, rather thick, flat above, convex underneath. Pedicels from very short to rather longer than the calyx. Calyx 3 to 4 lines long or sometimes more, the tube turbinate, ½ to 1 line long; lobes ovate-lanceolate, green on the back, scarious on the margins and pink or purple inside, often shortly mucronate below the end. Stamens very numerous, inserted at the top of the calyx-tube and shorter than its lobes, the filaments sometimes shortly united at the base. Ovary 3- or rarely 4-celled. Capsule ovoid, not exceeding the calyx, circumsciss below the middle.—Bot. Mag. t. 1701.

N. Australia. Albert river, *Henne.*
Queensland. E. coast, *R. Brown;* Port Curtis, *M'Gillivray;* Howicks Group and sandy shores of the islands of Moreton Bay, *F. Mueller;* Fitzroy river, *Thozet.*
N. S. Wales. Clarence river, *Wilcox.*

The styles are free to the base in all the flowers I have examined, but are occasionally 4 in number, as in *Psammanthe marina,* Hance in Walp. Ann. ii. 660, from the Chinese coasts, which appears to be a variety only of *S. portulacastrum,* notwithstanding that the styles are really, as described by Hance, shortly united at the base.

S. repens, Roth, to which the Indian specimens are referred in Wight and Arn. Prod. 361, appears to be a variety or rather a state only of *S. portulacoides,* with smaller flowers and shorter and broader leaves, owing, as suggested by Arnott, to want of luxuriance.

6. TRIANTHEMA, Linn.

(Ancistrostigma, *Fenzl.*)

Calyx free, more or less deeply 5-lobed. Petals none. Stamens inserted at the top of the calyx-tube, either 5 alternating with its lobes or indefinite. Ovary free or nearly so, enclosed in the calyx, either 2-celled with 2 styles, or 1-celled (reduced to 1 carpel) with 1 excentrical or lateral style; ovules 2 or more in each cell, attached to a basal placenta, free or shortly adnate to the partition. Capsule membranous or hard, transversely circumsciss, and when 2-celled the upper portion sometimes separating septicidally into 2 cocci, and in some species, not Australian, divided inside by a transverse partition under the uppermost seed. Seeds orbicular or reniform, the testa often granular.—Prostrate or diffuse herbs, rarely woody at the base. Leaves opposite, the two of each pair unequal in size, the petioles often with a scarious dilatation at the base, but no real stipules. Flowers axillary, solitary or in cymes or clusters. Bracts and bracteoles often somewhat scarious.

The genus is dispersed over the tropical and subtropical regions of the New as well as the Old World. Of the 6 Australian species, 2 are widely distributed over the warmer regions of Asia and Africa, the 4 others are endemic.

Ovary and fruit 2-celled, truncate or concave at the top. Styles 2.
  Flowers clustered . . . . 1. *T. decandra.*
Ovary and fruit 1-celled, with 1 style.
  Ovary and fruit truncate or concave at the top. Flowers clustered.
    Glabrous or slightly pubescent. Stamens 5. Capsule short and broad . . . . 2. *T. crystallina.*
    Hirsute with long hairs. Stamens about 20. Capsule with a narrow beak . . . . 3. *T. pilosa.*
  Ovary and fruit acute or tapering into the style, or rounded at the top.
    Glabrous or sparingly pubescent. Flowers solitary, pedicellate, not very small . . . . 4. *T. oxycalyptra.*
    Hirsute, small and densely tufted. Flowers solitary in the axils, but crowded on the plant, small . . . . 5. *T. rhynchocalyptra.*
    Glabrous. Flowers small, in loose pedunculate cymes . . 6. *T. cypseleoides.*

1. **T. decandra,** *Linn.; DC. Prod.* iii. 352. Procumbent and glabrous, said to be annual, but the specimens sometimes show a hard woody base; branches dichotomous, rarely attaining 1 ft. Leaves from broadly obovate to oblong, ½ to 1½ in. long, narrowed into a rather long petiole. Flowers several together in a cluster, not exceeding the petiole, although sometimes very shortly pedunculate. Bracts and bracteoles small and scarious. Calyx about 1 line long when in flower, somewhat enlarged when in fruit, the lobes longer than the tube, scarious on the margin and mucronate close to the end. Stamens 10 to 12 or sometimes a few more. Ovary ovoid, truncate, with a few prominent tubercles, 2-celled; ovules 2 in each cell, collaterally ascending from a basal placenta, shortly adnate to the dissepiment; styles 2. Capsule about 2 lines long, the seeds superposed in each cell, the upper one ascending the lower one pendulous; when ripe the upper portion separating into 2 cocci, opening on the inner face, the lower portion circumsciss below the insertion of the seeds. Seeds black, rugose.—Wight, Ic. t. 296; F. Muell. Fragm. i. 172.

N. Australia. N.W. coast, *A. Cunningham*, *Bynoe*.
Queensland. Dawson and Burnett rivers, and Peak Downs, *F. Mueller;* Funnel Creek, *Bowman*.
N. S. Wales. Darling desert to Cooper's Creek, *Victorian Expedition*, *Howitt's Expedition*, etc.

2. **T. crystallina,** *Vahl, DC. Prod.* iii. 352. Glabrous or slightly pubescent or covered with little transparent vesicles, prostrate or diffuse, the wiry dichotomous stems sometimes extending to 1 or 2 ft., sometimes short and compact. Leaves from oval-oblong to linear. Flowers small, in axillary cymes or clusters, much shorter than the leaves. Calyx about 1½ lines long, the lobes narrow, obtuse, spreading, rather longer than the tube. Stamens 5. Ovary of 1 carpel, truncate, with 2 erect ovules; style excentrical. Capsule short and broad, the top concave, forming a short broad cup round the style. Seeds 2, granular, flat, obliquely superposed.—Wight and Arn. Prod. 355; F. Muell. Fragm. i. 171; *T. glaucifolia*, F. Muell, Fragm. i. 172.

N. Australia. Lower Victoria river, *F. Mueller;* in the interior, *M'Douall Stuart*.
Queensland. Broad Sound, *R. Brown;* Burdekin and Dawson rivers, *F. Mueller;* Cape river, *Bowman*.
The species is common in tropical Asia and Africa.

3. **T. pilosa,** *F. Muell. Fragm.* i. 174. Procumbent, from a few inches to above 2 ft. in length, hirsute, with spreading hairs, particularly long and dense about the inflorescence. Leaves obovate, narrowed into a rather long petiole, the largest attaining about 1 in. Flowers in axillary sessile clusters. Calyx when full grown about 3 lines long; lobes ovate-lanceolate, very open, as long as the tube. Stamens about 20. Ovary of 1 carpel, the style lateral, below the end, which soon closes round it; ovules 2. Capsule included in the calyx, produced into a cylinder concave or cup-shaped at the top round the style, circumsciss about the middle of the basal seed-bearing portion.

N. Australia. N.W. coast, *Bynoe;* Cygnet Bay, *A. Cunningham;* Nichol Bay and De Grey river, *Ridley's Expedition;* desert between Hooker's and Sturt's Creeks, *F. Mueller;* islands of the Gulf of Carpentaria, *R. Brown*.

4. **T. oxycalyptra,** *F. Muell. Fragm.* i. 173. Prostrate, rather slender, glabrous or sparingly pubescent. Leaves obovate ovate or spathulate, obtuse or almost acute, mostly under ½ in. long and narrowed into a long petiole. Flowers solitary, more or less pedicellate. Calyx 3 to 4 lines long; lobes rather longer than the tube. Stamens about 15 to 20. Ovary of 1 carpel, the style terminal; ovules about 4 or 5. Capsule rather broad, acute, circumsciss below the middle. Seeds 2 to 5, superposed, compressed, elegantly marked with radiating rows of papillæ.

N. Australia. Sturt's Creek, *F. Mueller*.

5. **T. rhynchocalyptra,** *F. Muell. Fragm.* i. 174. Perennial, forming dense prostrate tufts, sometimes only 2 or 3 in. diameter, sometimes woody at the base, the branches extending to 6 in., more or less hirsute, with rigid bristly or soft spreading hairs. Leaves oval or oblong, obtuse, rarely above ½ in. long, narrowed into a short petiole. Flowers sessile and crowded, although solitary in each axil. Calyx thin and membranous, rather narrow, about 2 lines long or rather more when in fruit, the lobes scarcely so long as the tube. Stamens about 10. Ovary of 1 carpel obliquely tapering into

the style, which is rather long and recurved. Ovules about 5 or 6, on panicles of various lengths on a short placenta. Capsule acute, circumsciss below the middle. Seeds 2 to 5, minutely granulose.

N. Australia. Sandy stony hills and plains, Victoria river, *F. Mueller;* islands of the Gulf of Carpentaria, *R. Brown.*

6. **T. cypseleoides,** *Benth.* Glabrous, prostrate, slender and very small. Leaves obovate or orbicular, scarcely above ¼ in. long in the specimens seen, on slender petioles dilated at the base into scarious stipules. Flowers small, in loose axillary pedunculate cymes exceeding the leaves, with scarious bracts under the forks and pedicels. Pedicels slender, about 1 line long. Calyx about 1 line long, the lobes broad, obtuse, rather longer than the tube. Stamens 7 to 10. Ovary of 1 carpel short and rounded, with 6 to 12 ovules on funicles of various lengths; style terminal, but slightly excentrical, linear and recurved. Capsule globular, circumsciss. Seeds smooth. —*Ancistrostigma cypseleoides,* Fenzl, Nov. Stirp. Decad. 85.

N. S. Wales. Hawkesbury river, *R. Brown;* also in *Leichhardt's* collection.

Tribe III (or Suborder). Mollugineæ.—Calyx free, divided to the base or nearly so.

When the calyx is divided quite to the base, the stamens, inserted as in the rest of the Order below the lobes, are necessarily hypogynous. In a few species the base of the calyx is slightly developed and then the stamens are somewhat perigynous. The group has been frequently referred to *Caryophylleæ* or to *Portulaceæ,* with both of which as with *Ficoideæ, Phytolaccaceæ, Chenopodiaceæ, Amarantaceæ,* etc., it agrees in the seeds and embryo. It differs however both from *Caryophylleæ* and *Portulaceæ* in the divided ovary as well as in habit, and although certainly allied to those two as well as to *Phytolaccaceæ,* it appears to me to be much more closely connected with the tribe *Aizoideæ* of *Ficoideæ* through *Trianthema.* Like all the *Ficoideæ* it is remarkable for the general want of symmetry between the stamens and the other parts of the flower.

7. MACARTHURIA, Hueg.

Calyx-segments 5, persistent. Petals 5 or none. Stamens 8, slightly perigynous, the filaments united in a cup at the base. Ovary free, enclosed in the calyx, 3-celled, with 1, 2, or 3 ovules in each cell, attached to a basal placenta; styles 3, with small terminal stigmas. Capsule enclosed in the persistent calyx, opening loculicidally in 3 valves. Seeds reniform or subglobose, the funicle expanded into a small cup-shaped white arillus.—Rigid wiry or rush-like herbs or undershrubs. Leaves few, alternate, narrow, often all reduced to scales. Flowers small, in lateral or terminal short irregular cymes, or forming a spreading dichotomous cyme with opposite bracts.

The genus is confined to Australia.

Flowers in compact lateral or terminal cymes (sometimes reduced to 1 or 2 flowers).
  Flowers ½ to ¾ line long, without petals. Ovules solitary in each cell of the ovary . . . . 1. *M. apetala.*
  Flowers 1½ lines long, with petals. Ovules 2 or 3 in each cell of the ovary . . . . 2. *M. australis.*
Flowers in loose dichotomous terminal cymes, with petals. Floral leaves or bracts often opposite. Ovules solitary in each cell of the ovary . . . . 3. *M. neocambrica*

1. **M. apetala,** *Harv. in Hook. Kew Journ.* vii. 55. Perennial, with the erect rush-like or wiry stems of *M. australis*, but much smaller and more slender, usually forming dense tufts of 6 to 8 in. Leaves linear, few and small, many of them reduced to small scales. Flowers much smaller than in *M. australis*, in little cymes, either lateral or in the forks of the branches. Calyx-segments obtuse, scarcely above ½ line long, shortly united at the base. Petals none. Stamens much more perigynous than in *M. australis*. Ovules 1 in each cell of the ovary; styles exceedingly short. Fruiting-calyx slightly enlarged, but not exceeding ¾ line.

W. Australia. Swan River and to the northward, *Drummond, n.* 10 *and* 677.

2. **M. australis,** *Hueg. Enum.* 11. Stems from a perennial stock, erect, virgate or rush-like, slightly branched, 1 to 2 ft. high. Leaves few and distant, long and linear or the lower ones shorter and broader, or all reduced to small scales, the stems then appearing quite leafless. Flowers in short cymes or clusters, often intermixed with a few small floral leaves or scale-like bracts, nearly sessile along the branches or terminating short leafy branchlets, or loose, irregular and few-flowered. Pedicels 1 to 2 lines long. Calyx-segments about 1½ lines long, the outer ones green, the inner more scarious. Petals scarcely exceeding the sepals, oblong. Ovules 2 or 3 in each cell of the ovary.—Hook. Ic. Pl. t. 408; Steud. in Pl. Preiss. i. 229; Steetz in Pl. Preiss. ii. 359; *M. foliosa*, Steud. l. c. 230; Steetz, l. c. 360 (from the description).

W. Australia. King George's Sound to Swan River, *Huegel, Drummond,* 1*st Coll., Preiss, n.* 1671, 1672, and others; Murchison river, *Oldfield.*

3. **M. neocambrica,** *F. Muell. Fragm.* v. 28. A diffuse plant of 6 to 8 in. (*F. Mueller*). Stem-leaves few, alternate, oblong-linear or linear-cuneate, rather thick. Flowers in a broad loose spreading dichotomous cyme or panicle, each one pedicellate in the forks or terminal. Floral leaves or bracts very small, mostly opposite or nearly so. Calyx about 1 line long. Petals about as long as the calyx, perhaps sometimes wanting. Ovules 1 in each cell of the ovary; styles rather short.

N. S. Wales. Tweed river, *C. Moore.* The specimens seen not perfect.

8. MOLLUGO, Linn.

(Glinus, *Linn.;* Trigastrotheca, *F. Muell.*)

Calyx-segments 5, persistent. Petals none. Stamens few or many, free, sometimes with the addition of a few staminodia, of which 1 to 5 external ones represent petals, alternating with the calyx-segments. Ovary 3- to 5-celled, with several ovules in each cell; styles as many as cells, linear or clavate. Capsule membranous, enclosed in the persistent calyx, opening loculicidally in as many valves as cells. Seeds with a smooth or granulate testa, the funicle sometimes thickened into a small white arillus or strophiole, with or without a filiform appendage.—Erect or diffuse herbs, mostly annual. Radical leaves rosulate, but often disappearing before the flowering. Stem-leaves alternate, but often clustered in the axils so as to appear verticillate. Stipules very small and fugacious. Flowers small, the pedicels usually clustered in the axils, sometimes forming cymes, umbels, or racemes.

The genus is abundantly diffused over the warmer regions of the globe, extending into Europe and North America. Of the five Australian species, three are very common in Asia and Africa, two of them extending also to tropical America, the other two are endemic.

SECTION I. **Glinus.**—*Seeds strophiolate, with a filiform appendage (resembling a funicle) more or less encircling them.*

Softly tomentose. Flowers rather large, in axillary clusters. Stamens about 10 to 15 . . . . 1. *M. Glinus.*
Quite glabrous. Flowers rather large, in terminal clusters. Stamens about 15 . . . . 2. *M. orygioides.*
Glabrous or slightly pubescent. Flowers rather small, in axillary clusters. Stamens under 10 . . . . 3. *M. Spergula.*

SECTION II. **Mollugo.**—*Seeds without any strophiola.*

Glabrous, decumbent. Flowers rather large in loose axillary cymes or racemes. Stamens 5, the filaments much dilated at the base . 4. *M. trigastrotheca.*
Glabrous, filiform, and small. Flowers small, on filiform pedicels. Stamens 5, the filaments not dilated . . . . 5. *M. Cerviana.*

1. **M. Glinus,** *A. Rich. Fl. Abyss.* i. 48. A rather coarse species, softly tomentose all over, sometimes small and erect, but usually diffuse, procumbent or ascending and spreading to above 1 ft. Leaves from obovate-orbicular to oblong-spathulate, sometimes above 1 in. long, but usually much smaller. Flowers clustered at the nodes on short pedicels rarely as long as the calyx. Calyx like the rest of the plant, very tomentose, segments $2\frac{1}{2}$ to 4 lines long. Stamens about 10 to 15, with 5 or fewer external staminodia, flat, very thin and transparent, often forked. Styles usually 5, united at the base. Capsule enclosed in the calyx, 5-lobed. Seeds numerous; testa tuberculate, funicle thickened into a short strophiole or arillus, with a long filiform hair-like white process more or less encircling the seed.—*Glinus lotoides*, Linn. Spec. Pl. 663; Fenzl, in Ann. Wien. Mus. i. 357, with the synonyms adduced; F. Muell. Pl. Vict. i. 202.

N. Australia. Victoria river, *F. Mueller.*
Queensland. Rockhampton, *Thozet.*
N. S. Wales. Darling river, *Dallachy and Goodwin.*
Victoria. Sandy occasionally inundated banks of the Murray river, *F. Mueller.*

Widely dispersed over the tropical and subtropical regions of the Old World, extending to Europe, and found also in various parts of tropical America. There is a less tomentose variety with smaller flowers, approaching *M. Spergula*, which however has not yet been found in Australia.

2. **M. orygioides,** *F. Muell. Herb.* Stout and rigid, apparently perennial, dichotomously branched, quite glabrous. Leaves obovate or oblong, all under $\frac{1}{2}$ in. in our specimens. Flowers rather large, in terminal clusters, on very short pedicels. Outer calyx-segments about 2 lines long, with a narrow scarious border, inner ones rather larger at first, with a broader border, at length 3 lines long, broadly-ovate, white and scarious, with a greenish centre. Stamens about 15, with a few staminodia, either all subulate or 2 or 3 flat thin and transparent. Styles 3 or rarely 4, quite free. Seeds not numerous, larger than those of *M. Glinus*, and the hair-like appendage to the funicle not so long, only half encircling the seed.—*Glinus orygioides*, F. Muell. Pl. Vict. i. 203.

N. S. Wales. Desert plains, east of Grey Range, *Beckler.*
S. Australia. Cooper's Creek, *Wright.*

3. **M. Spergula,** *Linn. Spec.* 131. Glabrous or somewhat pubescent when young, much and dichotomously branched, procumbent and spreading to 1 ft. or more, or nearly erect when small. Leaves from obovate-oblong to almost linear, sometimes almost 1 in. long, much smaller on the flowering branches. Flowers in small clusters at the nodes, the pedicels as long as or longer than the calyx. Calyx-segments glabrous, from 1 to nearly 1½ lines long. Stamens not above 10 and usually much fewer, with occasionally a few staminodia amongst them. Styles or style-branches and capsule-valves 3. Seeds rather numerous, the funicle thickened into a small arillus, with a filiform process more or less encircling the seed as in *M. Glinus.*—*Glinus Mollugo*, Fenzl in Ann. Wien. Mus. i. 359, with the synonyms adduced; F. Muell. Pl. Vict. i. 203; *M. Novæ-Hollandiæ*, F. Muell. in Trans. Phil. Soc. Vict. i. 14; *M. glinoides*, A. Rich. Fl. Abyss. i. 48, not of Cambess.

N. Australia. Victoria river and towards M'Adam Range, *F. Mueller.*

Victoria. Sandy periodically inundated banks of the Murray and its backwaters, *F. Mueller.*

W. Australia, *Drummond, 4th Coll. n.* 166, *5th Coll. n.* 276.

The species is common in tropical Asia and Africa.

4. **M. trigastrotheca,** *F. Muell. Pl. Vict.* i. 201. Glabrous, decumbent or diffuse, dichotomously branched, under 1 ft. long. Leaves linear, clustered so as to appear verticillate, acute, often exceeding 1 in. Flowers rather large, in loose cymes or branched racemes often as long as the leaves. Calyx-segments about 2 lines long, white and petal-like, with a green centre. Stamens usually 5, the filaments much dilated below the middle. Ovary membranous, soon inflated, 3-celled, with 3 or 4 ovules in each cell; styles rather long, with small terminal stigmas. Capsule depressed-globular, membranous, 3-furrowed, 3-valved. Seeds few, tuberculate, without any appendage to the funicle.—*Trigastrotheca molluginea*, F. Muell. in Hook. Kew Journ. ix. 16.

N. Australia. Depuech island, N.W. coast, *Bynoe;* Hearson island, Nichol Bay, *Walcott;* Stuart's Creek and Arnhem's Land, *F. Mueller.*

5. **M. Cerviana,** *Ser. in DC. Prod.* i. 392. A little slender glabrous annual of a few in., with filiform branches. Leaves in distant clusters, linear, mostly under ½ in. long, the radical ones sometimes shorter and oblong. Pedicels filiform, longer than the leaves. Calyx-segments about ½ line long when in flower, lengthening to nearly 1 line. Stamens usually 5, the filaments filiform from the base. Styles 3, short, distinct, stigmatic towards the end. Seeds numerous, small, without any appendage to the funicle.—Fenzl in Ann. Wien. Mus. i. 379; F. Muell. Fragm. ii. 148.

N. S. Wales. Towards the Barrier Range, *Victorian Expedition.*

S. Australia. Near Lake Gillies, *Burkitt.*

The species is diffused over tropical and subtropical Asia and Africa and southern Europe.

ORDER LVI. UMBELLIFERÆ.

Calyx-tube adnate to the ovary; limb forming a slightly raised line round the summit, or 5-toothed or lobed, or quite inconspicuous. Petals 5, usually inflexed at the tip, more or less imbricate or very rarely valvate in the bud.

Stamens 5, alternating with the petals and inserted with them round the epigynous disk at the summit of the adnate calyx-tube; anthers versatile, with parallel cells opening longitudinally. Epigynous disk within the stamens usually 2-lobed, variously shaped, free from the styles or confluent with their thickened base and therefore the disk-lobes often called *stylopodes*. Ovary inferior, 2-celled or very rarely 1-celled by abortion, with 1 anatropous ovule in each cell, pendulous from the summit. Styles 2, with small terminal stigmas. Fruit usually separating into 2 indehiscent 1-seeded nuts or carpels, called *mericarps*, often leaving a persistent filiform central axis called a *carpophore*, either entire or splitting into two. Each carpel is marked with longitudinal *ribs*, of which the primary ones (corresponding with the calyx-teeth and intervening sinuses) are normally 5 to each carpel, *i.e.* 2, *lateral* (one on each side) at the *commissure* or junction of the 2 carpels, 1 *dorsal* on the back of the carpel, and 2 *intermediate* between the dorsal and lateral one on each side, but some of these are occasionally inconspicuous, and in some genera 4 *secondary* ribs to each carpel, between the primary ones, are as conspicuous or more prominent than the primary ones. In many genera there are longitudinal linear oil-vessels called *vittæ*, within or under the pericarp. Seed often adhering to the pericarp; testa very thin; albumen horny, filling the seed or furrowed or excavated on the inner face (next the commissure). Embryo minute near the apex of the seed, with the radicle superior. —Herbs or very rarely shrubs, with alternate leaves, often much cut or divided, the petiole usually dilated into a sheathing base, but without distinct stipules, except in *Hydrocotyle*. Flowers usually small, in terminal or lateral (leaf-opposed) umbels, which are either compound, each ray of the general umbel bearing a partial umbel, or simple or reduced to a globular head. Bracts at the base of the general umbel, either one under each ray or fewer, termed the general *involucre*, and one or three or more under the partial umbel termed the *involucel* or *partial involucre*, or one or both involucres wanting. Flowers frequently more or less polygamous, some, in the same or different umbels from the perfect ones, being males by the constant abortion of the ovary, and occasionally one or a few in the centre of the umbel females without stamens.

A numerous Order, more or less represented nearly all over the globe, especially in the temperate region of the northern hemisphere, where the delimitation of the very numerous genera presents the greatest difficulties. The Australian genera are much more marked, two or three only being rather uncertain in their connection with northern ones. Of the 14 genera here enumerated four have a very wide range within as well as without the tropics in both hemispheres, one, *Seseli*, is more specially characteristic of the temperate regions of the Old World, three range over Andine and Antarctic America and New Zealand, one, *Aciphyllum*, extends to New Zealand and the Antarctic islands only, another, *Trachymene*, extends only to New Caledonia and Borneo, the remaining four are endemic.

Fruits laterally compressed, without vittæ. Seeds laterally compressed.
- Umbels simple.
  - Creeping perennials or slender annuals with scarious stipules . . 1. HYDROCOTYLE.
  - Annuals or perennials, with dissected or toothed leaves without stipules. Fruit very flat . . . . 2. TRACHYMENE.
  - Leafless plant with rush-like stems . . . . 3. SIEBERA.
- Umbels compound, sometimes reduced to 1 or 2 flowers with bracts under the flower as well as under the pedicel.
  - Calyx-teeth small or inconspicuous. Perennials or shrubs, usually

glabrous. Leaves entire or ternately divided into small narrow lobes. Involucral bracts small . . . . 3. SIEBERA.
Calyx-lobes peltate, cordate or usually attached by the whole of their broad base. Herbs villous or glabrous. Leaves toothed, lobed or divided, or rarely entire. Involucral bracts conspicuous . . . . 4. XANTHOSIA.
Fruits scarcely compressed or compressed dorsally, without vittæ, usually furrowed at the commissure. Seed terete or dorsally compressed.
Umbels simple or rarely irregularly compound and few-flowered. Ovary 2-celled.
Carpels nearly terete or angular. Tufted perennials with radical or imbricate leaves, or rarely slender and creeping stems . . 5. AZORELLA.
Carpels much compressed dorsally. Tufted perennials with radical leaves and peduncles . . . . 6. DIPLASPIS.
Umbels simple. Ovary and fruit of a single ovule and seed . . . 7. ACTINOTUS.
Heads of flowers simple, dense. Leaf-lobes and involucral bracts rigid and pungent-pointed . . . . 8. ERYNGIUM.
Fruit slightly or not compressed. Carpels with 5 prominent ribs and usually 1 vitta under each furrow.
Umbels simple.
Small creeping glabrous plant with linear tufted entire leaves. Albumen terete . . . . 11. CRANTZIA.
Tufted pubescent perennial. Leaves much dissected. Albumen concave towards the commissure . . . . 14. OREOMYRRHIS.
Umbels compound.
Fruit-ribs obtuse.
Commissure of the fruit narrow. Seeds terete . . . . . 9. APIUM.
Commissure of the fruit broad. Seeds semiterete . . . . 10. SESELI.
Fruit-ribs very acutely prominent, the lateral ones often almost winged . . . . 12. ACIPHYLLUM.
Fruit scarcely compressed, densely covered with bristles proceeding from 4 prominent secondary ribs on each carpel, with single vittæ under the ribs. Primary ribs inconspicuous . . . . 13. DAUCUS.

Besides the above genera, the following *Umbelliferæ*, introduced from Europe, have more or less established themselves in some of the settled colonies, all with compound umbels.

Petroselinum sativum, Hoffm. (*Parsley*). An erect glabrous plant with dissected leaves, allied to *Apium*, but the umbels all pedunculate with a few involucral bracts, the flowers of a pale greenish-yellow, and the carpophore bipartite.—About Adelaide.

Ammi majus, Linn. With dissected leaves, pedunculate umbels, the flowers and fruit nearly of *Apium*, but the general involucre of a few dissected bracts.—Paramatta, *Woolls*.

Sium latifolium, Linn., and *S. angustifolium*, Linn. Perennials decumbent or sometimes creeping at the base, and erect or ascending stems. Leaves simply pinnate. Fruits nearly those of *Apium*, but the calyx-teeth usually prominent and several vittæ under each furrow. Umbels with general and partial involucres. *S. latifolium*, a large species with the umbels all terminal.—Cape Wilson and Lofty Range, *F. Mueller*. *S. angustifolium*, smaller, with the umbels leaf-opposed or lateral.—Paramatta, *Woolls*.

Pastinaca sativa, Linn. (*Parsnip*). Erect with pinnate leaves. Umbels without involucres. Fruits dorsally compressed, very flat, oval, with scarcely prominent ribs and very conspicuous vittæ.—Near Adelaide, *F. Mueller*.

Caucalis infesta, Curt. Erect, tall but slender, with pinnatifid or pinnate leaves. General involucre none or of one linear bract. Fruit small, bristly as in *Daucus*, but the bristles scattered, the secondary ribs not prominent, the primary ones alone conspicuous, with single vittæ under the furrows.—Near Port Macquarrie in Tasmania, *Milligan*.

Coriandrum sativum, Linn. (*Coriander*). An erect rather slender annual with finely dissected leaves. Umbels without general involucre. Fruits globular, not readily separating into two carpels, and without vittæ.—Near Adelaide.

We have also from W. Australia, *Drummond*, *2nd Coll. n.* 195, specimens in leaf only of an *Umbellifer* which I have been unable to match precisely, although they much resemble the above-mentioned *Sium angustifolium*, Linn.

1. HYDROCOTYLE, Linn.

Calyx-teeth minute or inconspicuous. Petals entire, acute, valvate or imbricate. Disk flat, with a raised annular or cup-shaped margin. Fruit laterally compressed, without vittæ, often didymous, carpophore deciduous with the carpels or persistent; carpels with the dorsal rib prominent, the lateral ribs concealed in the commissure, or distinct and prominent or rarely combined in one prominent rib, the intermediate ribs usually prominent, straight, curved or short and semicircular; secondary ribs very rarely conspicuous. Seed straight, laterally compressed.—Herbs either prostrate and rooting at the nodes or erect and annual. Leaves either orbicular, peltate or deeply cordate and entire or divided, or cuneate at the base and divided. Stipules scarious, often toothed or jagged, especially in the annual species. Flowers small, sometimes unisexual, in simple umbels or also verticillate on the peduncle below the terminal umbel, white or rarely purplish.

The genus is dispersed over the warmer and temperate regions of the globe, most frequent in moist situations or floating in water. Of the 26 Australian species two have a wide range over the New as well as the Old World, a third is also in New Zealand and perhaps in some other countries, the remainder are endemic.

SECTION I. **Euhydrocotyle.**—*Leaves (except in* H. alata) *orbicular and peltate or deeply cordate, or divided to the base into* 3, 5 *or more segments. Petals valvate. Carpels with the intermediate ribs alone prominent on each side or rarely with the lateral ones also distinct and prominent.*

Stems creeping and rooting, at least at the lower joints. Carpophore deciduous with the carpels.
  Fruits more or less didymous, the carpels convex on the sides, the dorsal edge obtuse.
    Leaves orbicular, peltately attached by the centre . . . . 1. *H. vulgaris.*
    Leaves rounded or reniform-cordate with a deep sinus at the insertion of the petiole, crenate or lobed but not divided to the base.
      Fruits nearly sessile in the head.
        Small densely-matted plant. Stipules imbricated on the short flowering branches . . . . 2. *H. peduncularis.*
        Diffuse plants with distinct flowering nodes. Stipules not imbricate . . . . 3. *H. hirta*
      Fruits distinctly pedicellate in the umbel.
        Flowering-stems ascending or erect. Leaves more or less hirsute and lobed. Fruit 1 line broad . . . . 4. *H. laxiflora.*
        Flowering-stems slender and diffuse. Leaves glabrous, scarcely lobed. Fruit ¾ line broad on long slender pedicels . . . . 5. *H. pedicellosa.*
    Leaves divided to the base into 3 or 5 lobes . . . . . . 6. *H. tripartita.*
Fruits flat, striate, the dorsal edge of the carpels acute, the secondary ribs slightly prominent. Plant glabrous . . . . 7. *H. plebeia.*
Fruits very flat, the dorsal edge of the carpels acute or winged. Plants usually glabrous.
  Leaves rounded or reniform, crenate or broadly and obtusely lobed.
    Carpels with the dorsal edge acute but not winged . . . 7. *H. plebeia.*
    Carpels with the dorsal edge expanded into a wing . . . 8. *H. pterocarpa.*

Leaves divided to the base or nearly so into lanceolate segments. Carpels not winged . . . . 9. *H. geraniifolia.*
Small or filiform, erect or diffuse annuals, not rooting at the joints. Carpophore persistent or deciduous.
Fruits broader than long, more or less didymous with 2 (intermediate) ribs on each side, the lateral ribs not distinct from the narrow commissure.
Fruits with the ribs very prominent, forming 6 thick obtuse wings to the fruit, the intermediate ones not much curved . 10. *H. medicaginoides.*
Fruits with the intermediate ribs semicircular, enclosing a pit or inner disk.
Dorsal rib winged . . . . 11. *H. muriculata.*
Fruit not winged.
Fruit smooth or scarcely granular outside the rib.
Leaves nearly divided to the base. Semicircular ribs of the fruit very obtuse, enclosing a small pit . . 12. *H. callicarpa.*
Leaves not divided to the middle. Semicircular ribs of the fruit acute, enclosing a broad disk . . . 13. *H. scutellifera.*
Fruit granular or muricate along the ribs.
Fruit about ½ line broad . . . . 14. *H. hispidula.*
Fruit nearly 1 line broad, very didymous . . . . 15. *H. trachycarpa.*
Fruit transversely rugose, forming 1 or 2 rows of little pits outside the semicircular ribs.
Pits several in 1 or 2 rows.
Plant of 3 to 6 in., often hirsute. Fruit 1 line broad . . . . 16. *H. pilifera.*
Plant of 1 to 2 in., glabrous. Fruit ¾ line broad . 17. *H. capillaris.*
Pits 2, 3 or rarely 4 to each carpel . . . . . . 18. *H. rugulosa.*
Fruits broader than long, very flat, with 4 or apparently 6 ribs on each side (the lateral ribs distinct from the commissure and the dorsal rib very thick, so as to be prominent on each side) . 19. *H. diantha.*
Fruits ovate or not broader than long, striate with 4 nearly parallel ribs on each side (the lateral ones distinct from the commissure).
Leaves orbicular-cordate or reniform.
Stems long and slender. Umbels 6- to 10-flowered . . 20. *H. grammatocarpa.*
Stems floating, leafy part very short. Umbels 3- to 6-flowered . . . . 21. *H. lemnoides.*
Leaves triangular hastate or acutely 5-lobed. Stems erect or shortly diffuse . . . . 22. *H. alata.*
Fruits acutely tetragonous (the lateral ribs forming one prominent angle at the broad commissure, the dorsal rib prominent, the intermediate ones inconspicuous) . . . . 23. *H. tetragonocarpa.*

SECTION II. **Centella.**—*Leaves (except in* H. asiatica) *cuneate at the base or narrow. Petals imbricate.*

Small annuals. Leaves cuneate at the base, toothed or 3-partite.
Leaves deeply toothed. Flowers in dense heads. Fruits didymous, glochidiate. Plant of 1 to 1½ in. . . . . 24. *H. glochidiata.*
Leaves lobed or partite. Flowers in loose umbels. Fruits very flat. Plant of 2 to 6 in. . . . . 25. *H. verticillata.*
Perennial, creeping and rooting at the nodes. Leaves broadly cordate. Carpophore deciduous with the fruit. Fruits obscurely several-ribbed (the secondary ribs sometimes conspicuous) . . 26. *H. asiatica.*

(*Azorella Muelleri* has the habit and stipules of a *Hydrocotyle,* but the fruit is scarcely compressed, furrowed at the commissure, and the calyx-teeth very prominent.)

SECTION I. EUHYDROCOTYLE, *DC.*—Leaves (except in *H. alata*) orbicular and peltate or deeply cordate, or divided to the base into 5 or more segments. Petals valvate. Carpels with the intermediate ribs alone prominent on each side, or rarely with the lateral ones also distinct and prominent.

1. **H. vulgaris,** *Linn.; DC. Prod.* iv. 59. Stems slender, creeping in mud or floating in water, rooting at every node and emitting from the same point tufts of leaves and peduncles. Leaves orbicular, ½ to 1 in. diameter or when very luxuriant twice that size, crenate or slightly lobed, peltately attached by the centre to a rather long petiole, with about 9 or when luxuriant 11 nerves radiating from the same point. Stipules broad and entire but very soon worn away from the rooting nodes. Peduncles shorter than the petioles, either with a single terminal head or umbel or with the addition of 2 or 3 whorls below it of minute white flowers on exceedingly short pedicels. Bracts small, scarious. Petals valvate, slightly induplicate. Fruit 1¼ to 1½ lines broad, not above 1 line long, 2-ribbed on each side (the intermediate ribs alone prominent, the lateral ones concealed in the commissure), the dorsal edge of the carpels obtuse.—Reichb. Ic. Fl. Germ. t. 1842; *H. interrupta*, Muehl.; DC. Prod. iv. 59; *H. verticillata*, Thunb.; Harv. and Sond. Fl. Cap. ii. 527.

Queensland. Moreton Bay, *Leichhardt.*
N. S. Wales. Paramatta, *Woolls;* Hunter's River, *Oldfield.*
Victoria. Lake Wellington, Gipps' Land, and Murray river, *F. Mueller;* near Melbourne, *Adamson.*
S. Australia. St. Vincent's Gulf, Gawler river, *F. Mueller.*
The species is widely distributed over the temperate regions of both hemispheres.

2. **H. peduncularis,** *R. Br. in A. Rich. Hydroc.* 62. *t.* 61. *f.* 26. Stems perennial and creeping to some length, with numerous very short flowering branches covered with imbricate stipules, forming usually little densely matted tufts. Leaves orbicular-cordate or reniform, rarely above 2 lines diameter, shortly broadly and obtusely 5-lobed. Peduncles filiform, usually but not always exceeding the leaves, bearing a head of 3 to 6 small flowers. Petals valvate. Fruits about ¾ line broad, carpels with convex sides, each with a prominent curved rib. Styles short.—DC. Prod. iv. 66; Hook. f. Fl. Tasm. i. 152. t. 32B.

Tasmania. Marshes at Circular Head, summits of the Western Mountains, Hampshire Hills, etc., *J. D. Hooker* and others.

Var. *gracilenta,* Hook. f. More slender and lax.—*H. gracilenta,* Hook. f. in Hook. Lond. Journ. vi. 467. This is much like the most slender and smallest forms of *H. hirta,* but the stipules are much larger, the habit different, the flowers much fewer, the ribs of the fruit more curved and prominent.

3. **H. hirta,** *R. Br. in A. Rich. Hydroc.* 64. Prostrate or creeping and rooting at least at the lower nodes, rather slender and more or less hirsute. Leaves orbicular-cordate or reniform, divided to about ¼ or ¾, rarely deeper, into about 7 broad crenate lobes, usually from ½ to 1½ in. diameter. Stipules usually ciliate or fringed. Peduncles solitary or 2 together, each with a small head of numerous (10 to 40) minute flowers, sessile or very shortly pedicellate. Bracts small. Petals valvate. Fruits not more than 1 line diameter and often smaller, very closely packed in a small globular head,

quite smooth or granular, each with 2 prominent ribs on each side (the intermediate ones), the dorsal edge of the carpels obtuse.—Hook. f. Fl. Tasm. i. 152.

Queensland. Brisbane river, *F. Mueller.*
N. S. Wales. Port Jackson, *R. Brown*, and others.
Victoria. From Gipps' Land to the western frontiers, *F. Mueller*, and others.
Tasmania. Port Dalrymple, *R. Brown;* abundant in wet places throughout the colony, *J. D. Hooker.*
S. Australia, *R. Brown*, Lofty Range, Mount Disappointment, etc., *F. Mueller.*
W. Australia, *Drummond, 3rd Coll. Suppl. n.* 32.

Var. (?) *acutiloba*, F. Muell. Leaves divided to about the middle into triangular more or less acute lobes. Plant rather large. Fruits very small.—Queensland, from Dawson river, *F. Mueller*, Rockhampton, *Dallachy*, to Brisbane river, Moreton Bay, *F. Mueller*, and others.

Var. ? *pusilla.* Smaller and less hirsute, sometimes nearly glabrous.—*H. pulchella*, R. Br. in A. Rich. Hydroc. 59; DC. Prod. iv. 66; *H. elegans*, A. Rich. Hydroc. 58; DC. Prod. iv. 66 (with deeper lobed leaves); *H. tasmanica*, Hook. f. in Hook. Lond. Journ. vi. 467, and Fl. Tasm. i. 152. t. 32A; *H. vagans*, Hook. f. in Hook. Lond. Journ. vi. 468, and Fl. Tasm. i. 153. t. 33A.—From Queensland to Tasmania, including some of Sieber's specimens, n. 14. The more glabrous forms from more aquatic situations.

The species may not be distinct from a common tropical and subtropical one, which includes *H. rotundifolia*, Roxb.; Wight, Ic. t. 564, from tropical Asia; *H. sibthorpioides*, Lam.; A. Rich. Hydroc. t. 54. f. 8. from the Mauritius; *H. Mannii*, Hook. f. in Journ. Linn. Soc. vii. 194, from tropical Africa; *H. marchantioides*, Clos in Gay, Fl. Chil. iii. 67, from Chile, and some forms usually referred to *H. Bonplandi*, A. Rich, from the Andes. The common New Zealand *H. moschata*, Forst., is rather more distinct, but even that might perhaps be included in the same series, and, if so, Forster's name appears to have the right of priority for the collective species.

H. intertexta, R. Br.; A. Rich, Hydroc. 63; DC. Prod. iv. 66, from the single small specimen I have seen, would appear to be one of the forms of *H. hirta*, with rather longer styles. A. Richard describes the carpophore as persistent, a character which the specimen I saw did not show. *H. Gaudichaudiana*, DC. Prod. iv. 67, only known to me by the diagnosis given, is again probably one of the forms of *H. hirta.*

4. **H. laxiflora,** *DC. Prod.* iv. 61. Stems creeping and rooting like the allied species, but the flowering branches often ascending or erect to the length of 6 in. or even more, hirsute as well as the leaves with spreading hairs. Leaves orbicular-cordate, shortly and broadly 5- to 11-lobed and crenate, rarely above 1 in. diameter. Stipules entire or slightly fringed-ciliate. Peduncles short or long, each with a globular umbel of 30 to 40 or even more flowers. Bracts numerous, small and narrow. Pedicels varying from ½ line to 3 lines long, when long usually with infertile flowers. Petals valvate. Styles long. Fruit about 1 line broad or rather more, smooth or granular, with 2 prominent ribs (the intermediate ones) on each side, the dorsal edge of the carpels obtuse.—F. Muell. Fragm. iv. 179; *H. densiflora*, DC. Prod. iv. 67; F. Muell. Fragm. iv. 180.

Queensland. Dawson and Burnett rivers, *F. Mueller.*
N. S. Wales. Blue Mountains and in the interior to Bathurst and Argyle county, *A. Cunningham, Woolls*, and others; New England, *Beckler.*
Victoria. Port Phillip, *Gunn;* Glenelg river, *Allitt.*
S. Australia. Torrens river, St. Vincent's Gulf, etc., *Behr, F. Mueller.*

Var. ? *minor.* Flowers and fruits very much smaller.—Moreton Bay, *C. Stuart.*

De Candolle had already observed the great similarity between his *H. laxiflora* and *H. densiflora*, which appear to be always found growing together, and F. Mueller, Fragm. iv. 180, seems to have suspected dimorphism. An observation of numerous specimens seems to show that the differences are those of semisexual dimorphism. In the long-pedicelled umbels the petals are more expanded, the stamens longer, and the ovary enlarges but little, and ultimately withers without forming good seed. I have only found ripe fruits in the dense umbels, in which the pedicels rarely attain 1 line. In these the petals open less freely, sometimes cohering till they fall, and the stamens are shorter. I have found both kinds of umbels on the same specimen.

5. **H. pedicellosa,** *F. Muell. Fragm.* iv. 182. Stems slender, slightly pubescent or nearly glabrous. Leaves orbicular-cordate or reniform, crenate, scarcely lobed, thin, glabrous or sparingly pubescent, often above 1 in. diameter. Stipules broad, entire. Peduncles filiform, with a loose umbel of 10 to 30 very small flowers, on filiform pedicels varying from 1 to 3 lines in length. Bracts all broad, short, and scarious. Ovary at the time of flowering not ¼ line long and broad. Petals valvate, glandular. Styles rather long. Fruits about ¾ line broad, smooth or granular, with 2 scarcely prominent ribs on each side.

N. S. Wales. Richmond river, *Beckler.*—The species requires further investigation. Some of the apparent characters may be due to the circumstances under which the specimens were growing.

6. **H. tripartita,** *R. Br. in A. Rich. Hydroc.* 69. *t.* 61. *f.* 25. Small and very slender, sometimes densely matted, or the filiform stems extending to several in., glabrous or sprinkled with a few hairs. Leaves divided to the petiole into 3 to 5 cuneate entire or 2- or 3-toothed segments, rarely above 3 lines long, and sometimes not 1½ lines. Stipules entire. Peduncles filiform, shorter than the leaves, each with an umbel or head of 3 to 6 or rarely more small flowers nearly sessile. Fruits ¾ line diameter, with 2 slightly prominent ribs on each side, smooth or granular; carpels convex on the sides, the dorsal edge obtuse.—DC. Prod. iv. 65.

Queensland. Burnett river, *F. Mueller.*
N. S. Wales. Port Jackson, *R. Brown, Sieber, n.* 411 (partly), *Woolls.*

Var. *muscosa.* Very small and densely matted; stems often under 1 in. long; leaves not above ¼ in. diameter.—*H. muscosa,* R. Br. in A. Rich, Hydroc. 68. t. 61. f. 27; DC. Prod. iv. 64; Hook. f. Fl. Tasm. i. 154.
Victoria. Broken River, *F. Mueller.*
Tasmania, *R. Brown;* Circular Head, forming large patches in moist places, *Gunn.*

7. **H. plebeia,** *R. Br. in A. Rich. Monogr.* 46. *t.* 60. *f.* 23. A rather large species, creeping and rooting at the lower nodes, quite glabrous. Leaves orbicular-cordate or reniform, crenate or shortly 9- to 11-lobed, ½ to 1 or even 2 in. diameter, the sinus deep and narrow. Stipules broad, entire. Peduncles filiform, with a small globular head of numerous minute almost sessile flowers. Bracts small, narrow. Petals valvate. Fruits about 1 line diameter, very flat, with 2 prominent ribs on each side, carpels much compressed laterally, with the dorsal edge acute but not winged.

W. Australia. King George's Sound, *R. Brown;* also *Drummond, 2nd Coll. n.* 196. The species is allied to *H. americana,* Linn., which, however, has the fruit much smaller and not so flat, connecting the species with the larger forms of *H. hirta.*

H. striata, Benth. in Hueg. Enum. 53, was described from a small specimen which I now think is a variety of *H. plebeia,* in which the lateral ribs of the carpels are slightly prominent between the primary ones, giving the whole fruit a striate appearance.

8. **H. pterocarpa,** *F. Muell. in Trans. Vict. Inst.* 1855, 126, *and in Hook. Kew Journ.* viii. 69. Very near *H. plebeia*, differing only in the winged fruits, glabrous, creeping, and rooting at the base. Leaves orbicular-cordate or reniform, doubly crenate or obscurely lobed, about ½ to 1 in. diameter, on long petioles. Stipules entire. Peduncles shorter than the petioles, with a small globular head of rather numerous flowers almost sessile. Bracts small. Petals valvate. Fruits rather broader than long, notched above and below, very flat, with 2 prominent ribs on each side, attaining sometimes fully 2 lines diameter, the acute dorsal edge of the carpels being expanded into a wing.—Hook. f. Fl. Tasm. i. 159. t. 33.

Victoria. Rivulets towards Mount Disappointment, *F. Mueller;* Glenelg river, *Robertson.*

Tasmania. Formosa, *Gunn;* South Port, *C. Stuart.*

The species is also in New Zealand. The dilatation of the fruit appears to be somewhat variable, and it is possible that it may prove to be a variety of *H. plebeia.*

9. **H. geraniifolia,** *F. Muell. in Trans. Vict. Inst.* 1855, 126, *and in Hook. Kew Journ.* viii. 70; *Fragm.* iv. 181. Glabrous or rarely sprinkled with a few hairs; stems lax, diffuse, often rooting at the lower joints, 1 to 2 ft. long or more. Leaves deeply divided into 3 to 7 lanceolate-acute toothed or lobed segments, the larger central one often above 1 in. long; the lower leaves sometimes slightly peltate. Stipules fringed. Peduncles slender, with an umbel of rather numerous flowers, on short filiform pedicels. Bracts minute, except two larger scarious very deciduous outer ones. Petals valvate. Fruits nearly 2 lines broad, very flat, deeply notched at the base, the dorsal edge of the carpels acute and expanded into a narrow wing, with 2 (intermediate) prominent ribs on each side, short and semicircular, the cavity inside smooth or with one row of granules, the carpels outside the rib more or less granular-tuberculate.

N. S. Wales. Port Jackson, *R. Brown;* Blue Mountains, *Miss Atkinson;* Hastings and Macleay rivers, *Beckler.*

Victoria. Moist valleys, from the Dandenong range to Gipps' Land, *F. Mueller.*

When large and luxuriant, the species bears some outward resemblance to the Peruvian *H. quinqueloba,* Ruiz and Pav., but the fruit is quite different. In the latter respect as well as in the habit it comes near to *Trachymene procumbens,* but is at once known by the stipules and minute scarious bracts.

10. **H. medicaginoides,** *Turcz. in Bull. Mosc.* 1849, ii. 27. A little slender annual of ½ to 3 in., glabrous or sprinkled with a few white hairs. Leaves few, small, petiolate, orbicular-cordate, divided very deeply or quite to the base into 3 entire or 2- or 3-lobed segments. Stipules fringed-ciliate. Peduncles filiform, ¼ to ½ in. long, with 6 to 12 minute flowers, almost sessile in a head scarcely ½ line diameter. Petals valvate. Fruits nearly sessile, ¾ line broad and about half as long, the intermediate ribs on each side and the dorsal edge very prominent, forming 3 thick wings to each carpel, and very rugose between them —*H. lobocarpa,* F. Muell. Fragm. iv. 178.

Victoria. Towards Lake Hindmarsh, *Dallachy.*

W. Australia, *Drummond, 4th Coll. n.* 144.

11. **H. muriculata,** *Turcz. in Bull. Mosc.* 1849, ii. 28. A little glabrous annual of 1 to 2 in. Leaves deeply 3- or 5-lobed. Stipules

fringed-ciliate. Peduncles filiform, longer than the leaves, but not exceeding 4 lines, bearing each a little head of rather numerous minute flowers. Petals acute, valvate, thin and rather larger than in the allied species. Fruit very flat, 1½ lines broad, and much shorter, the dorsal edges of the carpels expanded into a rather broad thin wing, the intermediate ribs raised with a row of tubercles on each side. Carpophore apparently persistent.

W. Australia, *Drummond, 4th Coll. n.* 143.

12. **H. callicarpa,** *Bunge in Pl. Preiss.* i. 283. A little annual with a tuft of procumbent or ascending stems of 1 to 3 in., usually glabrous. Leaves ¼ to ½ in. diameter, the lower ones sometimes lobed only, but mostly divided into 3 to 5 broadly cuneate toothed or lobed segments, glabrous or sprinkled with a few hairs. Stipules often fringed or jagged. Umbels sessile or pedunculate, each with about 6 to 10 minute flowers, on slender pedicels rarely above ½ line long. Petals valvate. Fruits ½ to ¾ line broad, didymous, smooth, not winged, the intermediate ribs obtusely prominent, semicircular, enclosing a well-defined pit. Carpophore usually persistent.—*H. tripartita*, Hook. f. in Hook. Ic. Pl. t. 312, and Fl. Tasm. i. 153, not of R. Br.

Victoria. Yarra-Yarra and Port Phillip, *F. Mueller;* Creswick, Skipton, etc., *Whan;* Glenelg river, *Robertson.*

Tasmania. Near Launceston, *Gunn* and others.

S. Australia. Murray river, Lofty and Barossa ranges, St. Vincent's Gulf, *F. Mueller.*

W. Australia. King George's Sound, *Preiss, n.* 2074; Swan River, *Drummond, n.* 5.

13. **H. scutellifera,** *Benth.* A diffuse glabrous annual of 2 to 4 in. Leaves orbicular-cordate or reniform, with 3, 5, or 7 broad obtuse lobes, entire or crenate, not reaching to the middle. Stipules broad, jagged. Peduncles usually exceeding the leaves, bearing a globular head or umbel of 10 to 20 very small flowers, sessile or nearly so. Petals valvate. Styles very short. Fruits about ¾ line broad, ½ line long, very flat, the dorsal edge of the carpels rather thick, the surface smooth, the intermediate ribs acutely prominent, semicircular, enclosing a little shield on each side of the fruit. Carpophore usually deciduous with the carpels.

W. Australia. Swan River, *Drummond, 1st Coll. and 2nd Coll. n.* 4.

14. **H. hispidula,** *Bunge in Pl. Preiss.* i. 283. A slender annual of 3 to 6 in., usually glabrous. Leaves orbicular-cordate, rather deeply divided into 5 broad lobes again toothed or lobed, more or less hispid on both sides, the radical leaves about ½ in. diameter on long petioles, the stem leaves few and small. Stipules fringed or jagged. Flowers exceedingly small, 6 to 12 in the umbels or heads, the pedicels scarcely attaining ½ line when in fruit. Petals valvate. Fruits ½ line broad and not so long, didymous, the intermediate ribs semicircular, obtusely prominent, and slightly granular-tuberculate. Carpophore short, persistent.

W. Australia. Sands of Rottenest Island, *Preiss, n.* 2086.

Var. *tenella.* More slender and diffuse. Leaves less lobed, on shorter petioles.—Warren river, *Herb. F. Mueller.*

15. **H. trachycarpa,** *F. Muell. in Linnæa,* xxv. 394. A glabrous

annual, with very slender diffuse stems, elongated but not rooting at the joints. Leaves divided to below the middle into 5 or 3 broadly cuneate lobes, each with 2 or 3 deep teeth or lobes, the larger ones above ½ in. diameter, all very thin. Stipules minute, slightly ciliate. Peduncles filiform, each with an umbel of 3 to 6 very minute flowers. Pedicels at first very short, at length nearly as long as the fruit. Petals valvate. Fruit nearly 1 line broad, but not nearly so long, didymous, the intermediate ribs semicircular, obtusely prominent, usually with a few tubercles in the enclosed pit, and a single row on the outside. Carpophore short, persistent.

N. S. Wales. Darling river, *Victorian Expedition*.—Perhaps a variety only of *H. hispidula*.

16. **H. pilifera,** *Turcz. in Bull. Mosc.* 1849, ii. 26. An erect annual, often coarser than the allied species, and attaining 6 in. to 1 ft., although sometimes small, more or less hirsute with spreading hairs or rarely quite glabrous. Leaves ½ in. diameter or more, deeply divided into about 5 cuneate lobes, usually again toothed or lobed. Stipules scarious, but more adnate to the petiole than in most species. Umbels pedunculate, with about 10 to 20 minute flowers on very short pedicels. Petals valvate. Fruits about 1 line broad, didymous, the intermediate ribs semicircular, obtusely prominent, the enclosed pit smooth, with numerous transverse reticulations outside. Carpophore persistent.

W. Australia, *Drummond*, 1*st Coll.* (2*nd Coll.?*) *n.* 21 *and* 198; Pinjarra, Murray river, *Oldfield*.

Var. *glabrata*. Nearly or quite glabrous. Swan River, *Drummond*, 1*st Coll.*

17. **H. capillaris,** *F. Muell. Fragm.* iv. 178. A minute slender glabrous annual, rarely exceeding 2 in., and often under 1 in. Leaves very small, deeply divided into 3 lobes or segments, obtuse and entire, or shortly 2- or 3-lobed. Stipules broad. Peduncles short, with a head or umbel of 3 to 6 minute flowers on very short pedicels. Petals valvate. Fruits about ¾ line broad, didymous, the intermediate ribs semicircular, very prominent, the enclosed pit smooth, the outside remarkably pitted and rugose.

Victoria. Muddy places often dried up, Port Phillip, Hopkins river, *F. Mueller*.

S. Australia. Mount Gambier and Kangaroo Island, *F. Mueller*.

W. Australia. Geographe Bay, *Oldfield*.

The species differs from *H. pilifera*, var. *glabrata*, chiefly in its minute size and smaller less-divided leaves.

18. **H. rugulosa,** *Turcz. in Bull. Mosc.* 1849, ii. 27. A small slender glabrous annual, with erect or diffuse stems of 2 to 4 in. Leaves not above ½ in. diameter, divided to the base into 3 to 5 broadly-cuneate toothed or lobed segments. Stipules broad, slightly jagged. Peduncles short, slender, with an umbel of 2 to 4 minute flowers, on very short pedicels. Petals valvate. Fruits ½ to ¾ line broad, didymous, the intermediate ribs very prominent, semicircular, connected with the outer margin by 2 or 3 raised transverse lines, thus forming 1 or rarely 2 pits within the rib, and 3 or 4 outside of it.

W. Australia, *Drummond*, 4*th Coll. n.* 146.

19. **H. grammatocarpa,** *F. Muell. Fragm.* ii. 128. A glabrous

annual with filiform stems, diffuse or prostrate, and often elongated but not rooting at the nodes. Leaves cordate-orbicular or reniform, often above ½ in. diameter, crenate or obscurely 5- or 7-lobed, very thin and membranous. Stipules fringed. Peduncles filiform, with a globular head or umbel of 6 to 10 or rather more minute flowers, at first nearly sessile, at length shortly pedicellate. Petals thin, coloured, but acute and valvate. Fruits broadly ovate, about ¾ line broad and rather longer, slightly compressed, not indented at the commissure, with about 6 equal ribs on each side (the secondary ones prominent). Carpophore deciduous with the carpels.

N. Australia. Gulf of Carpentaria, *F. Mueller.*

20. **H. diantha,** *DC. Prod.* iv. 63. A very slender diffuse glabrous annual, from 1 or 2 in. to twice that length. Leaves shortly petiolate, orbicular-crenate or reniform, shortly and obtusely 4- or 7-lobed. Stipules broad, entire. Umbels sessile or shortly pedunculate with 3 to 6 or rarely only 2 pedicellate flowers. Petals acute, valvate. Styles exceedingly short. Fruits on pedicels of ½ to 1 line, very flat, fully 1 line broad and not so long; at first appearing 4-ribbed on each side, when quite ripe the dorsal rib thickens so as to appear almost double, and each carpel is semiorbicular, very flat, each side bordered by a thick margin with the intermediate rib not so prominent.—F. Muell. Fragm. iv. 179.

W. Australia. Swan River, *Huegel;* Blackwood and Tone rivers, *Oldfield.*

21. **H. lemnoides,** *Benth.* A very small Lemna-like plant evidently floating in water, with long matted filiform roots or submerged stems, the leaf-bearing part not above ½ in., and often not ¼ in. long. Leaves petiolate, orbicular-cordate or reniform, obscurely crenate, 1 to 2 lines diameter. Stipules broad and rather large. Flowers apparently unisexual, in umbels of 3 to 6. Male umbels nearly sessile. Petals rather thick, valvate. Styles present, but the ovary abortive. Females (only seen in fruit) shortly pedunculate. Fruits about ½ line broad and long, notched at the base, laterally compressed, the carpels convex and faintly nerved on each side.

W. Australia, *Drummond, n.* 202.

22. **H. alata,** *R. Br. in A. Rich. Hydroc.* 73. *t.* 61. *f.* 28. A little erect glabrous annual of 1 to 2 in., with slender divaricate branches. Leaves shortly petiolate, triangular-hastate and deeply 3-lobed, rather thick, not above 2 lines broad in our specimens. Stipules small, scarious. Peduncles 1 to 2 lines long, with a head or umbel of 6 to 12 almost sessile flowers. Petals not seen. Fruits about ¾ lines long and broad, compressed, with 4 to 6 nearly equal ribs on each side, otherwise smooth or minutely rugose.—*H. cymbalaria,* Benth. in Hueg. Enum. 53.

W. Australia. King George's Sound, *R. Brown, Huegel;* Tweed and Kalgan rivers, *Oldfield.*

23. **H. tetragonocarpa,** *Bunge, in Pl. Preiss.* i. 284. A slender diffuse glabrous annual, from 1 or 2 in. to ½ ft. long or rather more. Leaves cordate-orbicular or reniform, shortly and broadly 5- or 7-lobed, the lobes entire or slightly toothed, the lower ones rarely slightly peltate. Stipules broad, mostly jagged. Peduncles filiform, mostly shorter than the leaves,

each with about 6 to 10 very small almost sessile flowers. Petals valvate. Fruits about $\frac{3}{4}$ line long, not compressed, acutely 4-angled with 2 styles, or very rarely 3-angled with 3 styles, obscurely striate, but not seen quite ripe.

W. Australia. Wet sands, Rottenest Island, *Preiss, n.* 2085; wet places, Swan River, *Oldfield.*

I have not succeeded in finding either in Preiss's or in Oldfield's specimens perfectly ripe fruit with good seed, but those that have apparently attained their full size are remarkable for the acutely-prominent lateral angles of the carpels, the commissure as broad as the opposite diameter of the fruit; it is probable, however, that when quite ripe the dorsal edge of the carpels may be more dilated.

SECTION II. CENTELLA.—Leaves (except in *H. asiatica*) cuneate at the base or narrow. Petals imbricate.

24. **H. glochidiata,** *Benth.* A little erect or diffuse branching annual, $\frac{1}{2}$ to $1\frac{1}{2}$ in. high, the stem and leaves rather thick for the size. Leaves few, obovate or cuneate, with 2 or 3 coarse teeth or lobes. Stipules broad, entire. Flowers numerous, sessile in dense globular or ovoid heads of 1 to 2 lines. Bracts linear-spathulate. Calyx-teeth inconspicuous. Petals broad obtuse, slightly imbricate. Styles very short. Fruits about $\frac{1}{2}$ line broad and long, didymous, hispid with short glochidiate bristles, the intermediate and dorsal ribs very prominent, the commissure very narrow. Carpophore not persistent.

W. Australia. *Drummond, n.* 104 (105?) *and 4th Coll. n.* 247.

25. **H. verticillata,** *Turcz. in Bull. Mosc.* 1849. ii. 28. An erect or diffuse glabrous annual, from 2 or 3 in. to twice that height. Leaves with a cuneate base tapering into the petiole, deeply divided into 3 cuneate lobes or segments which are again usually acutely 2- or 3-lobed. Stipules very minute or none. Umbels sessile, 6- to 12-flowered, the setaceous or almost spathulate bracts very minute or inconspicuous. Petals minute, broad, obtuse, slightly imbricate in the bud. Stamens and styles very short. Fruits on pedicels of about 1 line, very flat, about 1 line broad and not quite so long, smooth, the lateral ribs concealed in the narrow commissure, the intermediate ribs curved and very near the dorsal edge of the carpels. Carpophore persistent, more or less deeply divided, or rarely remaining entire.—*H. homalocarpa*, F. Muell. Fragm. ii. 129.

W. Australia. King George's Sound, *R. Br.*; Swan River, *Drummond*, 1*st Coll. also* 4*th Coll. n.* 145; wet places, Tweed, Murray and Blackwood rivers and Mount Barker, *Oldfield.*

This is a very anomalous species, differing from the other *Centellas*, and, indeed, from the whole genus, in its narrow dissected leaves and usually split carpophore. The fruit is otherwise quite that of *Euhydrocotyle.*

26. **H. asiatica,** *Linn.; DC. Prod.* iv. 62. A creeping perennial, rooting at the nodes, and sometimes half-floating. Leaves broadly cordate, orbicular or almost reniform, entire crenate or sinuate toothed, 1 to $1\frac{1}{2}$ in. diameter, glabrous or pubescent, on petioles varying very much in length. Stipules broad, usually entire. Flowers 3 or 4 in little heads or umbels, on peduncles varying much in length or almost sessile. Two outer bracts under the umbel broad and scarious like the stipules, the inner ones small and narrow. Petals broad and thin, much imbricated in the bud. Fruit nearly 2 lines diameter, laterally compressed, but the dorsal edges obtuse,

showing when young the secondary as well as the primary ribs, when ripe obscurely 4- to 6-ribbed on each side and somewhat reticulate.—Bunge in Pl. Preiss. i. 283; Hook. f. Fl. Tasm. i. 152; Wight, Ic. t. 565; *H. repanda*, Pers.; DC. Prod. iv. 62; *H. cordifolia*, Hook. f. in Hook. Ic. t. 303.

Queensland. Burdekin river, *F. Mueller;* Moreton Bay, *C. Stuart.*

N. S. Wales. Port Jackson to the Blue Mountains, *R. Brown, Sieber, n.* 531, and others.

Victoria. Yarra and Ovens rivers, etc., *F. Mueller.*

Tasmania. Marshes, Arthur's Lake, Circular Head, Launceston, *J. D. Hooker.*

S. Australia. S. coast, *R. Brown;* Torrens river, *F. Mueller.*

W. Australia. Swan River, *Preiss, n.* 2065; also *Drummond, 4th Coll. n.* 147; Murchison river, *Oldfield.*

The species is also in New Zealand, and generally distributed over tropical and subtropical Asia, Africa, and America.

2. TRACHYMENE, Rudge.

(Didiscus, *DC.;* Dimetopia, *DC.;* Pritzelia, *Walp.;* Huegelia, *Reich.;* Cesatia, *Endl.;* Hemicarpus, *F. Muell.*)

Calyx-teeth minute or inconspicuous or rarely 1 or 2 rather longer and subulate. Petals entire, obtuse or nearly so, much imbricate. Disk flat or with slightly prominent margins or scarcely any. Fruit laterally compressed, usually flat, notched at the base, without vittæ; carpophore persistent, undivided; carpels laterally compressed, the dorsal rib prominent, rarely winged, lateral ones concealed in the narrow commissure, intermediate ribs semicircular, shorter than the fruit, enclosing as it were an inner disk. Seed straight, laterally compressed.—Herbs either annual, biennial or with a perennial stock, more or less hirsute or rarely glabrous. Leaves ternately divided or rarely toothed only, without stipules. Flowers white or blue, in simple umbels, on terminal or leaf-opposed peduncles. Involucre of linear bracts usually shortly united at the base. Fruits usually tubercular muricate or villous, one carpel often differently or less muricate than the other or abortive.

Besides the Australian species which are endemic, there is one from New Caledonia and one from Borneo.

Small annuals. Leaves divided. Flowers few in the umbels.
- Fruits tubercular or muricate, one carpel differently or less so than the other, rarely both equally so or one quite smooth . . . . . 1. *T. pilosa.*
- Fruits equally covered with long ciliate bristles . . . . 2. *T. cyanopetala.*
- Fruits densely covered with a white cottony or spongy wool . . . 3. *T. eriocarpa.*

Coarse erect annuals or biennials. Leaves divided or lobed. Flowers numerous in the umbel.
- One carpel winged, the other usually abortive . . . . 4. *T. villosa.*
- Fruit not winged.
  - Involucral bracts about as long as the pedicels. Flowers rather large. Carpels both perfect . . . . 5. *T. cærulea.*
  - Involucral bracts much shorter than the pedicels. Flowers small. Leaves divided. Carpels both perfect or one abortive.
    - More or less hirsute, not glaucous . . . . 6. *T. australis.*
    - Very glabrous and glaucous . . . . 7. *T. glaucifolia.*
  - Involucral bracts short. Leaves deeply 3-lobed, with oblong-cuneate lobes. One carpel abortive . . . . 8. *T. glandulosa.*

Rootstock perennial, with elongated branching stems.
  Leaves deeply divided, both carpels usually perfect.
    Stems erect, rigid. Leaves mostly radical from the base . . . 9. *T. incisa.*
    Stems weak, procumbent, leafy . . . . 10. *T. procumbens.*
  Leaves toothed or lobed, mostly from the base of the stem. One carpel abortive . . . . 11. *T. hemicarpa.*
Stock perennial, densely tufted, bearing toothed or lobed leaves and simple peduncles. Both carpels usually perfect . . . . 12. *T. humilis.*

1. **T. pilosa,** *Sm. in Rees Cycl. Suppl.* An erect or diffuse annual, rarely above 6 in. high and usually only 3 or 4 in., more or less hirsute or nearly glabrous. Leaves shortly petiolate, deeply 3- or almost 5-lobed, with linear or cuneate entire or 3-lobed divisions. Peduncles terminal or leaf-opposed, bearing each a small umbel of about 8 to 12 flowers on very short pedicels. Bracts 6 to 10, nearly as long as the flowers, slightly united at the base. Margins of the disk prominent. Fruits 1½ to 3 lines broad, the carpels usually unequally muricate, one with acute or shortly aristate points, the other with obtuse tubercles or quite smooth.—*Dimetopia pusilla*, DC. Prod. iv. 71; *D. hirta*, Benth. in Hueg. Enum. 54; Bunge in Pl. Preiss. i. 284; *D. Walpersii*, Bunge, Del. Sem. Hort. Dorp. 1846, in Linnæa, xxiv. 156; *D. homocarpa*, Bunge in Bot. Zeit. 1847, 136; *D. isocarpa*, Bartl. according to Walp. Rep. v. 840; *Pritzelia didiscoides*, Walp. Rep. ii. 428.

N. S. Wales. Port Jackson, N. shore, *C. Moore*.
Victoria. Murray desert, *F. Mueller*.
S. Australia. Pine forest, *Behr;* near Adelaide, *F. Mueller;* Port Lincoln, *Wilhelmi*.
W. Australia. King George's Sound, *R. Brown*, and adjoining districts, *Huegel, Oldfield*, and others; thence to Swan River, *Preiss, n.* 2072, *Oldfield*, and Champion Bay, *Walcott*, and eastward to Cape Arid, *Maxwell*.

Var. *Preissiana.* Radical leaves larger, forming a dense tuft and exceeding the short stems. Involucral bracts rather longer.—*Dimetopia Preissii*, Bunge in Pl. Preiss. i. 284. Sandy banks of the lake in Rottenest island, *Preiss, n.* 2089, also *Drummond, n.* 32.

2. **T. cyanopetala,** *Benth.* Usually more slender and rather taller than *T. pusilla*, and glabrous or nearly so. Leaves as in that species, deeply 3- or 5-lobed, with linear or cuneate entire or 2- or 3-lobed divisions. Peduncles short. Involucre of 4 or 5 bracts, rather broader than in *T. pusilla*. Flowers in the umbel 3 to 6, on very short pedicels. Petals blue. Fruits densely covered with soft ciliolate bristles, much longer than those of *T. pusilla* and usually equally dense on both carpels, rarely one carpel almost bare.—*Dimetopia cyanopetala*, F. Muell. Fragm. i. 231.

N. S. Wales. Between the Upper Bogan and Lachlan rivers, *L. Morton*.
Victoria. Murray river, *F. Mueller*.
W. Australia. Swan River, *Drummond, 1st Coll.; 2nd Coll. n.* 30; Murchison river, *Oldfield*.

3. **T. eriocarpa,** *Benth.* An annual, closely resembling the taller and more simple specimens of *T. pusilla* in everything except the fruit. Stems often 6 in. high or more. Leaves deeply 3- or 5-lobed, with linear or cuneate entire or 2- or 3-lobed divisions. Involucral bracts acute, about as long as the pedicels. Flowers 6 to 12 in the umbel. Fruits on pedicels of 1 to 2 lines, scarcely larger than in *T. pusilla*, and of the same shape, but so densely covered with a white cottony almost spongy wool as to make them appear much

larger and nearly globular.—*Dimetopia eriocarpa*, F. Muell. in Trans. Vict. Inst. 1855, 127, and in Hook. Kew Journ. viii. 70; *Cesatia ornata*, Endl. in Ann. Wien. Mus. ii. 200.

S. Australia. Near Cudnaka, *F. Mueller.*

W. Australia. Swan River, *Drummond, 1st Coll. n.* 424; *2nd Coll. n.* 29; Murchison river, *Oldfield;* South-west Bay and Oldfield river, *Maxwell.*

4. **T. villosa,** *Benth.* Erect, apparently annual or biennial, with the habit of *T. cærulea*, but more hirsute with long spreading hairs. Leaves tripartite, the segments again deeply divided into 2 or 3 oblong-cuneate coarsely toothed lobes. Peduncles rigid, glabrous, bearing an umbel about ¾ in. diameter when in flower. Involucral bracts subulate, shortly united, rather shorter than the pedicels. Flowers very small. Calyx-teeth inconspicuous. Disk scarcely any. Fruit usually reduced by abortion to a single carpel, tuberculate or muricate on the surface, the dorsal rib expanded into a broad thin smooth wing.—*Didiscus villosus*, F. Muell. in Proc. Roy. Soc. Tasm. iii. 238; *Hemicarpus villosus*, F. Muell. in Hook. Kew Journ. iv. 18.

N. Australia. Tableland between the upper Victoria river and Hooker's and Sturt's Creeks, *F. Mueller.*

5. **T. cærulea,** *Grah. in Edinb. New. Phil. Journ.* v. 380. A rather coarse erect annual or biennial of 1 to 2 ft., more or less hirsute. Leaves once or twice tripartite, with linear-cuneate 3-fid or incised acute lobes, the upper floral leaves small and simple or 3-fid. Peduncles long, bearing each an umbel of very numerous flowers, 1 to 2 in. diameter. Involucral bracts numerous, linear, nearly as long as the pedicels, shortly united into a turbinate base, the centre of the umbel occupied by a flat disk. Calyx-teeth obsolete. Petals usually blue, unequal, the external rather longer than the inner ones. Disk annular. Fruit usually ripening both carpels, from 2 to 3 lines broad, the surface granular-rugose. Styles rather long.—Bot. Reg. t. 1225; *Huegelia cærulea*, Reichb. Iconogr. Exot. t. 201; *Didiscus cyaneus*, DC. Mem. Ombell. 28. t. 4; *D. cæruleus*, DC. in Hook. Bot. Mag. t. 2875, and Prod. iv. 72; Bunge in Pl. Preiss. i. 285.

W. Australia. Swan River, *Fraser, Drummond, 1st Coll., Preiss, n.* 2055.

In the wild specimens the fruits are scarcely above 2 lines broad, in the cultivated ones about 3 lines.

Var. *leucopetala*, F. Muell. Flowers (in the dried specimens) white. Fruits densely muricate.—Murchison river, *Oldfield.*

6. **T. australis,** *Benth.* Very near *T. cærulea*, and perhaps only a variety. In the original western specimens the leaves are few and mostly radical or at the base of the stem, hispid with long hairs. Peduncles long and distant. Umbels rather smaller than in *T. cærulea*, the involucral bracts linear-subulate and much shorter than the pedicels. In the eastern plant the stems are coarser and more leafy, the hairs fewer and sometimes the whole plant nearly glabrous. Peduncles on the main stem sometimes several together so as to form a large irregular compound umbel, the involucral bracts very short. In both series the fruits have in some specimens or in some umbels both carpels perfect, in others one is constantly abortive, and all vary from quite smooth to more or less tuberculate.—*Didiscus pilosus*, Benth. in Hueg. Enum. 54; Hook. f. Fl. Tasm. i. 154; Hook. Ic. t. 307; *Dimetopia*

anisocarpa and *D. grandis*, Turcz. in Bull. Mosc. 1849, ii. 29; *Didiscus anisocarpus* and *D. grandis*, F. Muell. in Proc. Roy. Soc. Tasm. iii. 238.

Queensland. Newcastle range, *F. Mueller*.

N. S. Wales. North of Bathurst, *A. Cunningham*; Mooni Creek, *Mitchell*; New England, *Beckler*.

Victoria. Sandy hills, Port Phillip, and near Sandridge, *F. Mueller*; "Native Parsnip," *Robertson*.

Tasmania. Coasts between Circular Head and Woolnoth, *J. D. Hooker*, and others.

W. Australia. King George's Sound to Swan River, *Huegel, Drummond, 4th Coll. n.* 132 *and* 133; Perongerup range, *Maxwell*.

7. **T. glaucifolia,** *Benth.* Apparently an annual or biennial, resembling in every respect the more glabrous forms of *T. pilosa*, except that the whole plant is perfectly glabrous, and, according to F. Mueller, glaucous-pruinose when fresh, and the involucral bracts are usually but not always broader.—*Didiscus glaucifolius*, F. Muell. in Linnæa, xxv. 395.

N. S. Wales. Near Duroodoo towards the Barrier Range, *Victorian Expedition*.

S. Australia. Near Elders Range, *F. Mueller*; Flinders Range, *Howitt's Expedition*; Finke river, *M'Douall Stuart*.

8. **T. glandulosa,** *Benth.* Erect and apparently annual or biennial, with the habit of *T. pilosa*, hirsute with short glandular hairs. Leaves shortly petiolate, deeply divided (but not to the base) into 3 oblong-cuneate rather broad coarsely-toothed or incised lobes. Peduncles long, glandular-hirsute. Flowers small, in umbels of about ½ in. diameter. Involucral bracts shorter than the pedicels, united at the base. Calyx-teeth obsolete. Disk broad, rather thick. Fruit reduced by abortion to a single carpel, about 2 lines long and almost as broad, granular-tuberculate, not winged. —*Didiscus glandulosus*, F. Muell. in Proc. Roy. Soc. Tasm. iii. 238.

N. Australia. Nicolson river, Gulf of Carpentaria, *F. Mueller*.

9. **T. incisa,** *Rudge in Trans. Linn. Soc.* x. 300. *t.* 21. Glabrous or rarely with a few long hairs on the radical leaves. Stems from a thick perennial root-stock erect, thin but rigid, 1 to 2 ft. high. Leaves chiefly radical or on the lower part of the stem, on long petioles, not large, 3- or 5-partite, the segments often again twice trifid with narrow acute lobes, the upper leaves few small and less divided. Peduncles long. Umbels smaller and often much smaller than in *T. cærulea*. Bracts much shorter than the pedicels. Flowers small. Calyx-teeth more distinct than in the preceding species. Disk very prominent. Fruit ripening both carpels, about 2 lines broad, obtusely muricate.—*Didiscus albiflorus*, DC. Prod. iv. 72.

Queensland. Near Brisbane, *Mrs. Dietrich*.

N. S. Wales. Port Jackson to the Blue Mountains, *R. Brown, Sieber, n.* 120, and others; Hastings and Clarence rivers, *Beckler*; New England, *C. Stuart*; Gwydir river, *Leichhardt*.

Var. *pilosa*. Sprinkled with a few long hairs; umbels rather larger,—N. coast, *R. Brown*; Clarence river, *Beckler*.

10. **T. procumbens,** *Benth.* Stems from a perennial rootstock elongated, procumbent or ascending, slender, quite glabrous. Leaves of the stem as well as the lower ones on slender petioles, tripartite, each segment again deeply divided into narrow-cuneate or lanceolate toothed or incised

lobes, mostly acute, with a few glandular hairs at the base and at the summit of the petiole, otherwise usually glabrous. Peduncles long and slender. Umbels rarely above ½ in. diameter with numerous small flowers on filiform pedicels of above 2 lines. Involucral bracts setaceous, much shorter than the pedicels. Calyx-teeth minutely prominent. Disk shortly cup-shaped. Fruits usually ripening both carpels, about 1½ lines broad, smooth, or tubercular muricate.—*Didiscus procumbens*, F. Muell. in Proc. Roy. Soc. Tasm. iii. 237.

Queensland. Brisbane river, Moreton Bay, *F. Mueller*, *Leichhardt*, *Mrs. Dietrich*; Lizard Island, *M'Gillivray*.

11. **T. hemicarpa,** *Benth.* Stems from a perennial stock, erect, glabrous or sprinkled with a few long hairs. Leaves on the lower part of the stem rather small, on rigid petioles, broadly cuneate or deeply 3-lobed, coarsely and acutely toothed or lobed, upper leaves small and linear. Peduncles slender. Umbels dense, scarcely above ¼ in. diameter, but rather numerous, in a loose terminal panicle. Flowers very small, calyx with 1 or 2 teeth usually prominent and shortly subulate. Fruit reduced by abortion to 1 carpel about 1½ lines long, granular or tuberculate.—*Didiscus hemicarpus*, F. Muell. in Trans. Bot. Soc. Edinb. vii. 491; *D. setulosus*, F. Muell. in Proc. Roy. Soc. Tasm. iii. 238; *Hemicarpus didiscoides*, F. Muell. in Hook. Kew Journ. vi. 18.

N. Australia. Barren plains from Point Pearce to the mouth of the Victoria river, *F. Mueller*.

Var. *major*. Taller, umbels larger, with more numerous flowers.—Lacrosse Island, Cambridge Gulf and Vansittart Bay, *A. Cunningham*; elevated land, Cape Lambert, Nichol Bay, *Gregory's Expedition*; Glenelg district, *Martin*.

Var.? *rotundifolia*. Leaves nearly orbicular, toothed only, not lobed.—Port Essington, *Armstrong*.

12. **T. humilis,** *Benth.* Stock perennial and densely tufted. Leaves radical, on long petioles, ovate or oblong, entire and obtuse or obtusely 3- or 5-lobed, mostly ½ to 1 in. long, rather thick, glabrous or sprinkled with a few hairs. Peduncles or scapes usually longer than the leaves and sometimes 6 in. to 1 ft. high, with a single terminal umbel of about ½ in. diameter. Flowers numerous. Involucral bracts linear, shorter than the pedicels. Calyx-teeth obsolete. Disk annular. Fruit ripening both carpels, 1½ to 2 lines broad, smooth, the inner circle formed by the intermediate ribs often very small.—*Didiscus humilis*, Hook. f. in Hook. Ic. t. 304. and Fl. Tasm. i. 154.

Victoria. Alpine and subalpine pastures in the Australian Alps, *F. Mueller*.
Tasmania. Abundant in subalpine situations, *J. D. Hooker*.

3. SIEBERA, Reichb.

(Trachymene, *DC.*, not of *Rudge*; Fischera, *Spreng.* (partly), *Sm.*; Platysace, *Bunge*; Platycarpidium, *F. Mueller*.)

Calyx-teeth small but usually conspicuous. Petals entire, induplicate-valvate or slightly imbricate, concave, with the end inflexed, the midrib prominent inside, the bud prominently 5-angled. Disk flat and thick, or scarcely any besides the thick conical base of the styles. Fruit laterally compressed, slightly notched at the base, without vittæ; carpophore persistent; carpels more or less turgid, but flattened at the commissure, the dorsal rib usually prominent, the lateral ones concealed in or slightly prominent at the narrow commissure, the intermediate ones usually faint. Seed more or less compressed but often not filling the cavity.—Rigid herbs with a perennial almost woody stock and virgate branches, or heath-like shrubs, glabrous or slightly glandular-pubescent. Leaves all entire or the lower ones divided or all reduced to small scales, without stipules. Umbels compound or rarely simple, terminal. Involucral bracts small. Flowers small, white. Fruit small.

The genus is confined to Australia.

Perennials or undershrubs. Stems leafless or the lower leaves divided, the upper ones linear-subulate.
- Stems flattened or 2-winged, nearly leafless. Fruits broader than long . . . . . 1. *S. compressa.*
- Stems terete angular or scarcely flattened.
  - Stems nearly leafless except a few divided small leaves at the base. Fruits as long as broad.
    - Umbels compound . . . . . 2. *S. juncea.*
    - Umbels all simple . . . . . 3. *S. haplosciadia.*
  - Stems leafless, twining. Fruits very flat, with acute edges broader than long . . . . . 4. *S. cirrosa.*
  - Stems diffuse, leafy.
    - Lower leaves divided, upper ones subulate. Fruits broader than long . . . . . 4. *S. heterophylla.*
    - Leaves all or nearly all divided. Fruits as long as broad.
      - Peduncles short. Carpels turgid at thedorsal edge, broadly flat towards the commissure . . . . . 5. *S. tenuissima.*
  - Stems rigid, erect, branches divaricate. Peduncles long, rigid.
    - Carpels turgid in the centre, flat at the dorsal edge and commissure . . . . . 6. *S. dissecta.*

Shrubs with numerous small narrow leaves. Fruits nearly flat or with a broad flat furrow at the commissure. Western species.
- Leaves erect or spreading.
  - Umbels nearly sessile. Carpels turgid and very obtuse at the dorsal rib . . . . . 7. *S. commutata.*
  - Umbels on slender peduncles. Carpels slightly turgid in the middle, the dorsal edge acute . . . . . 8. *S. effusa.*
- Leaves closely reflexed. Fruit of *S. effusa* . . . . . 9. *S. deflexa.*

Large leafy shrub. Fruits very flat, with acute edges. Eastern species 10. *S. valida.*

Leafy shrubs. Carpels wholly turgid except a narrow furrow at the commissure. Eastern species.
- Leaves narrow-linear or subulate, all entire.
  - Leaves short. Stems short and diffuse, usually glandular-pubescent . . . . . 11. *S. ericoides.*
  - Leaves mostly $\frac{1}{2}$ in. or more. Stems ascending or erect, usually quite glabrous . . . . . 12. *S. linearifolia.*
- Leaves orbicular, obovate, ovate or lanceolate, all entire . . . . . 13. *S. Billardierii.*
- Lower leaves or nearly all deeply 3- or 5-lobed, rigid, very acute, almost pungent . . . . . 14. *S. Stephensonii.*

1. **S. compressa,** *Benth.* Stems from a perennial rootstalk herbaceous but rigid, $\frac{1}{2}$ to 2 ft. high, very flat and striate or more or less distinctly bordered by 2 opposite herbaceous wings, the whole sometimes very narrow, sometimes attaining a breadth of 2 or 3 lines; branches either few and straight or flexuose, or more numerous and divaricate. Leaves few, small,

setaceous, entire, or the lower ones once or twice 2-fid, or all reduced to minute scales. Umbels terminal or on short lateral branches, ½ to 1 in. diameter, more or less compound, with slender divaricate rays, 2 or 3 of the longer ones bearing an umbellule sometimes again compound, whilst some of the rays of the general umbel are reduced to simple pedicels. Involucral bracts few and small. Fruits about 1 line broad and ¾ line long, the carpels turgid, leaving a rather narrow commissural furrow, the dorsal edge obtuse.—*Azorella compressa*, Labill. Pl. Nov. Holl. i. 75. t. 101; *Trachymene compressa*, Spreng.; DC. Prod. iv. 73; *T. anceps*, DC. l. c.; Bunge in Pl. Preiss. i. 288; *T. platyptera*, Bunge, l. c. 287; *T. stricta*, Bunge, l. c. 288.

W. Australia. King George's Sound and adjoining districts and thence to Swan River, *Labillardière, R. Brown* and others, *Drummond, n.* 93; *Preiss, n.* 2059, 2060, 2061, 2062, 2063.

Var. *filiformis.* Stems very slender, scarcely flattened, but the fruit of *S. compressa*, not of *S. juncea.*—*T. filiformis*, Bunge in Pl. Preiss. i. 289.—Swan River, *Preiss, n.* 2058.

I have been quite unable to sort the specimens into distinct groups, according to the breadth of the stem, which is often variable on the same specimen, and unaccompanied by any corresponding character. In general those gathered on the seacoast appear to have the stems the most dilated.

2. **S. juncea,** *Benth.* Stems erect, almost leafless, terete, angular, or slightly compressed, either rushlike and 1 to 2 ft. high, or shorter more branched and flexuose, or slender and intricately branched. Leaves reduced to a few filiform scales, or a few, at the base of the stem or on very short barren shoots, ½ to 1 in. long and once or twice trifid. Umbels in the taller varieties terminal, compound, larger and more dense than in *S. compressa*, smaller on the slender branching specimens. Involucral bracts few, small, linear, and reflexed or rarely elongated. Fruit rather more than 1 line long and broad, the carpels somewhat turgid, granular-tuberculate, leaving a very narrow smooth furrow on each side between them at the commissure, the dorsal edge almost acute.—*Trachymene juncea* and *T. teres*, Bunge in Pl. Preiss. i. 286.

W. Australia. Swan River and thence to King George's Sound, *Drummond*, 1*st Coll. and n.* 16, 91; *Preiss, n.* 2069, 2083, and some specimens under 2082.

Var. *ramosissima.* Stems shorter, more branched and slender than in the common form.—*Trachymene ramosissima*, Benth. in Hueg. Enum. 54; *T. candelabrum*, Bunge in Pl. Preiss. i. 287; *Platysace flexuosa*, Turcz. in Bull. Mosc. 1849, ii. 29.—Swan River and King George's Sound, *Drummond, 4th Coll. n.* 138; *Preiss, n.* 2057.

Var. *pendula.* Stems short, slender, much branched, the ends reflexed with small, less compound umbels.—*Trachymene pendula*, Benth. in Hueg. Enum. 54; *T. scabriuscula*, Bunge in Pl. Preiss. i. 287.—King George's Sound and neighbouring districts, *R. Brown*; *Drummond, 3rd Coll. n.* 228; *Preiss, n.* 2075.

3. **S. haplosciadia,** *Benth.* Stems from a perennial rootstock apparently leafless, terete or angular, erect and rushlike, but the upper branches often flexuose or recurved as in the var. *pendula* of *S. juncea.* Leaves few, minute, entire or trifid. Umbels terminal, all simple, with very numerous flowers, on short slender pedicels. Involucral bracts linear, reflexed, broader than in *S. juncea.* Fruit about as long as broad, as in *S. juncea*, but smoother and flatter, the dorsal edge more acute, almost winged, the lateral ribs at the commissure thickened and almost as prominent as the somewhat turgid centres of the carpels, but separate from them on each side by a narrow furrow. Styles with a thick conical base, without the broad disk of *S. compressa.*

W. Australia, *Drummond, 2nd Coll. n.* 19; Gales Brook, *Maxwell.*

4. **S. cirrosa,** *Benth.* Rhizome said to be tuberous. Stems terete, rush-like, slender, often twining and spreading over bushes, leafless or with a very few small subulate scale-like leaves (the base of the stem however not seen). Umbels compound, terminal, the rays rather long and slender. Involucral bracts few, small, and narrow. Petals obtuse and slightly imbricate. Fruits very flat, not tuberculate, about 3 lines broad and 2½ lines long, the dorsal edge very acute, the carpels convex on each side in the centre with the intermediate ribs slightly prominent.—*Platysace cirrosa*, Bunge in Pl. Preiss. i. 285; F. Muell. Fragm. i. 231.

W. Australia. Swan River, *Drummond* (*2nd Coll.?*), *n.* 15, *Preiss, n.* 2064; S. Hutt river, *Oldfield.*

4. **S. heterophylla,** *Benth.* Stems from a hard woody base numerous, erect on the young plant, diffuse or decumbent when older, usually 6 to 8 in. long, but sometimes 1 ft., and more slender. Leaves narrow-linear, the lower ones once or almost twice trifid, the upper ones and sometimes nearly all entire. Umbels small, compound but of few rays, with 1 or 2 small involucral bracts. Calyx-teeth prominent. Disk broad. Fruits about 1 line broad and not so long, didymous, more or less granular-tuberculate, the carpels turgid and rounded on the back, but with the dorsal rib prominent, shortly tapering towards the narrow commissure.—*Trachymene heterophylla*, F. Muell. 1st Gen. Rep., erroneously referred to *T. ramosissima*, Benth., by Klatt in Linnæa, xxix. 708.

Victoria. Sandy hills, chiefly near the sea, from the Glenelg to Gipps' Land, *F. Mueller, Robertson,* and others.

S. Australia. Mount Barker district, *F. Mueller;* Marble ranges, *Wilhelmi.*

When the lower leaves are wanting, this species often resembles some specimens of *S. linearifolia* or of *S. ericoides*, but is readily known by the shape of the fruit.

5. **S. tenuissima,** *Benth.* Stems diffuse, elongated sometimes to above 1 ft., branched, almost filiform. Leaves mostly once or twice tripartite, with narrow-linear very acute lobes or segments, a few of the upper ones small and entire. Umbels compound, on short slender peduncles. Bracts linear. Buds prominently 5-angled as in the rest of the genus, but the obtuse petals often slightly imbricate. Fruits rather above 1 line long and broad; carpels turgid at the dorsal edge, the flat part of the fruit in the centre broad, with a raised rib on each carpel near the narrow turgid part.

W. Australia, *Drummond, 2nd Coll. n.* 18.

6. **S. dissecta,** *Benth.* Stems erect, rigid, under 1 ft. high, with rigid divaricate branches. Leaves twice or thrice ternately divided into short but fine subulate lobes. Umbels compact, on thick rigid peduncles; rays numerous, the central one very short; all with many flowers, and one or two occasionally again compound. Involucral bracts few, those of the general umbel sometimes divided into a few subulate divaricate lobes. Calyx-teeth inconspicuous. Fruits about as long as broad, smooth or nearly so, the carpels slightly turgid in the centre, very flat at the dorsal edge and in the commissure, the intermediate ribs very fine or scarcely conspicuous.

W. Australia. Between Moore and Murchison rivers, *Drummond, 6th Coll. n.* 119.

7. **S. commutata,** *Benth.* Shrubby and heath-like, with virgate branches. Leaves often somewhat crowded, linear, rather thick, obtuse or almost acute, 2 to 4 lines long, smooth and nerveless, often slightly twisted. Umbels compound, but compact and not large, on a peduncle not exceeding the last leaves. General involucre of 2 or 3 bracts resembling the stem-leaves, partial ones very small. Calyx-teeth prominent and connected by a narrow border. Fruits nearly orbicular, not quite 1 line broad, the carpels turgid towards the very obtuse dorsal edge, gradually flattened to the commissure, the lateral ribs joined in a single slightly prominent rib at the commissure, the intermediate ones inconspicuous or very slightly prominent.—*Trachymene commutata*, Turcz. in Bull. Mosc. 1849, ii. 30.

W. Australia, *Drummond, 3rd Coll. n.* 229, *4th Coll. n.* 136; towards Cape Riche, *Harvey;* also in *Maxwell's* collection.

8. **S. effusa,** *Benth.* Shrubby and heath-like with slender virgate branches. Leaves linear or linear-cuneate, mostly obtuse, narrowed to the base, 2 to 4 lines long, often crowded on the barren branches. Umbels compound, on slender peduncles much longer than the leaves, each with 4 to 8 rays all bearing partial umbels. Involucral bracts very small and slender or none. Calyx-teeth prominent. Styles on a broadly conical base. Fruits flat, about 1 line broad and scarcely so long, notched at the base, smooth or slightly rugose, the carpels scarcely turgid, with the dorsal edge acute, the lateral ribs contiguous or somewhat distinct at the commissure, the intermediate ribs sometimes inconspicuous, sometimes slightly prominent and curved almost as in *Trachymene.*—*Trachymene effusa*, Turcz. in Bull. Mosc. 1849. ii. 31; *Platysace trachymenioides*, F. Muell. Fragm. i. 232.

W. Australia. Between Swan River and King George's Sound, *Harvey, Drummond, n.* 33, *and 4th Coll. n.* 135; various places eastward from W. Mount Barren to Point Malcolm, *Maxwell;* also Champion Bay, *Wallcott*, with slightly rugose fruits. The species often much resembles *S. commutata*, but is readily distinguished by the longer peduncles and the fruit.

9. **S. deflexa,** *Benth.* Shrubby and virgate or much branched. Leaves usually crowded and closely reflexed on the branches, from broadly lanceolate to linear, obtuse, rather thick, rarely 2 lines long and often only 1 line, the upper ones sometimes almost orbicular. Umbels compound, on rather long slender peduncles, with 3 to 6 slender rays, each bearing a partial umbel. Involucral bracts few and small or none. Calyx-teeth prominent. Fruits as in *S. effusa*, very flat, about 1 line broad and scarcely so long, notched at the base, smooth, the carpels slightly turgid about the centre, the dorsal edge acute, the lateral ribs contiguous at the commissure, the intermediate ones curved, often inconspicuous.—*Trachymene deflexa*, Turcz. in Bull. Mosc. 1849, ii. 31.

W. Australia. King George's Sound to Cape Riche, *Harvey, Drummond, 4th Coll. n.* 137; eastward to Israelite Bay and Eagle Hawk Camp, *Maxwell.*

10. **S. valida,** *Benth.* A tall shrub. Leaves linear or linear-lanceolate, narrowed at both ends, 1 to 2 in. long, coriaceous, faintly 3-nerved. Umbels compound, numerous, forming a broad terminal panicle; rays usually 3 or 4, the central one very short, all bearing partial umbels, sometimes again compound. Involucral bracts few and small. Calyx-teeth shortly prominent.

Petals sometimes slightly imbricate. Fruits very flat, about 3 lines broad and 2 lines long, carpels not turgid, with the dorsal edge acute, the intermediate curved ribs slightly raised.—*Platycarpidium validum*, F. Muell. in Hook. Kew Journ. ix. 310; *Platysace valida*, F. Muell. Fragm. i. 232.

Queensland. Burdekin river, *F. Mueller;* Bowen river, *Bowman;* Rockingham Bay, *Dallachy.*

11. **S. ericoides,** *Benth.* A small, much-branched, divaricate or diffuse shrub, glabrous or more frequently glandular-pubescent towards the ends of the branches. Leaves all entire, linear or subulate, acute, more spreading than in *S. linearifolia*, rarely exceeding ½ in. and mostly shorter. Umbels compound, but small and compact, very shortly pedunculate, with few rays. Involucral bracts short, linear. Fruit nearly as in *S. linearifolia*, but less rugose or quite smooth, the carpels usually more turgid and often furrowed at the intermediate rib, besides the commissural furrow.—*Trachymene ericoides*, Sieb. in DC. Prod. iv. 738; *T. tenuis* and *T. subvelutina*, DC. l. c.

Queensland. Moreton Island, *M'Gillivray.*

N. S. Wales. Port Jackson to the Blue Mountains, *R. Brown, Sieber, n.* 121, and others; northward to Hastings and Clarence rivers, *Beckler*, and southward to Illawarra, *A. Cunningham.*

Victoria. Avon river, in Gipps' Land, *F. Mueller.*

Var. *thymifolia*, A. Cunn. Leaves small (not above 2 lines long), oblong-linear or lanceolate, acute, with recurved margins.—Barren spots, forest land, Moreton Bay, *A. Cunningham.*

The species is often scarcely to be distinguished from *S. linearifolia.*

12. **S. linearifolia,** *Benth.* Shrubby and glabrous with slender branches, decumbent, ascending or erect. Leaves all entire, narrow-linear or subulate, acute, mostly ½ to 1 in. long. Umbels compound, on slender peduncles, usually exceeding the last leaves, with 3 or 4 or rarely more slender rays. Involucral bracts small, linear. Calyx-teeth shortly prominent. Disk broad with a thickened margin. Fruit about 1 line long and broad, more or less rugose, the ribs scarcely conspicuous; carpels turgid with a broad obtuse back, leaving only a narrow groove at the commissure.—*Azorella linearifolia*, Cav. Ic. v. 57. t. 485; *Trachymene linearis*, Spreng.; DC. Prod. iv. 73; *Fischera linearis*, Sm. in Rees Cycl. Suppl.

N. S. Wales. Port Jackson to the Blue Mountains, *R. Brown, Sieber, n.* 126, and others; near Richmond, *Wilhelmi;* New England, *C. Stuart.*

The species passes almost into *S. ericifolia* on the one hand, and the narrow-leaved varieties of *S. Billardieri* on the other.

13. **S. Billardieri,** *Benth.* A shrub, either low and diffuse or erect and attaining 2 or 3 ft., glabrous or with minutely pubescent branches. Leaves orbicular, obovate, ovate, elliptical, cuneate, or broadly or narrow-lanceolate, acute or obtuse, narrowed at the base and almost petiolate or closely sessile and rounded at the base, mostly under ½ in. long when broad and obtuse, often above 1 in. when narrow and acute. Umbels compound, sessile or pedunculate, but the peduncles rarely long; rays often numerous but sometimes few. Involucral bracts linear, small, or rarely as long as the rays. Fruit about 1 line long and broad, more or less tubercular or rugose, the ribs scarcely conspicuous, or both the dorsal and intermediate ones pro-

minent, or the latter depressed. Carpels turgid, leaving a narrow furrow between them at the commissure.

N. S. Wales. Port Jackson to the Blue Mountains, *R. Brown*, *Sieber*, *n.* 122, 124, 125, *and Fl. Mixt. n.* 542, 617, and southward to Illawarra, *Shepherd*, and Twofold Bay, *A. Cunningham*.

Victoria. Gipps' Land, barren ranges beyond Snowy River, Buffalo Ranges, Grampians, *F. Mueller*.

A most variable species as to the form of the leaves, and might indeed include the preceding two species as well as the following six varieties, which appear generally to be found in the same situations, and most of them, however different their extremes, to pass into each other by the finest gradations, or even to show leaves of very different forms on the same plant. Even *S. heterophylla* and *S. Stephensonii*, with the lower leaves divided, may not be quite constant in the characters separating them from *S. Billardieri*.

a. conferta. Leaves crowded, orbicular, about ⅓ in. diameter. Umbels dense, sessile.—*Trachymene conferta*, Gaud.; Benth. in Hueg. Enum. 54; *T. ovalis*, var. *conferta*, DC. Prod. iv. 73.

b. ovata. Leaves ovate, more or less acute, rounded at the base.—*Azorella ovata*, Labill. Pl. Nov. Holl. i. 74. t. 100; *Trachymene ovata*, Spreng.; DC. Prod. iv. 73; *T. ovalis*, DC. l. c.; *T. buxifolia*, Sieb. Pl. Exs.; *Fischera ovata*, Sm. in Rees Cycl. Suppl.

c. myrtifolia. Leaves ovate-lanceolate or oblong, including every shape between the forms *b* and *d*.—*Trachymene myrtifolia*, Sieb. in DC. Prod. iv. 73.

d. lanceolata. Leaves lanceolate, acute, narrowed at the base, mostly above ½ in. long.—*Azorella lanceolata*, Labill. Pl. Nov. Holl. i. 74. t. 99; *Trachymene lanceolata*, Spreng.; DC. Prod. iv. 73; Bot. Mag. t. 3334; *Fischera lanceolata*, Sm. in Rees Cycl. Suppl.

e. cuneata. Leaves more or less cuneate obcordate obovate or oblanceolate, narrowed at the base and almost petiolate. Umbels on peduncles longer than the leaves.—Gipps' Land, *F. Mueller*. This variety, rather more distinct than the preceding ones, may, by some, be considered as a species.

f. crassifolia. Leaves crowded, orbicular obovate or oblong, very small and thick.—*Trachymene crassifolia*, Benth. in. Hueg. Enum. 54. Apparently rare, the only specimens I have seen are in Brown's and in Fraser's collections, and at first sight it appears to be a well-marked species, but the differences may possibly be due to a seacoast station.

14. **S. Stephensonii,** *Benth.* Shrubby with virgate branches, glabrous or nearly so, very nearly allied to the var. *ovata* of *S. Billardieri*, but the leaves linear-lanceolate or lanceolate, rigid, very acute, almost pungent, ¼ to ½ in. long, and the lower ones or nearly all deeply divided into 3 spreading segments, the outer ones sometimes again 2-lobed. Umbels compound, sessile within the last leaves. Fruit of *S. Billardieri*, very tubercular-rugose.—*Trachymene Stephensonii*, Turcz. in Bull. Mosc. 1847, i. 170.

N. S. Wales, *Stephenson, n.* 284, "within 125 miles of Sydney" on the printed labels.

4. XANTHOSIA, Rudge.

(Leucolæna, *R. Br.*; Schœnolæna, *Bunge*; Pentapeltis, *Bunge*.)

Calyx-lobes orbicular ovate or lanceolate, peltate cordate or not attached by the whole of the base. Petals with an induplicate point and reduplicate margins, slightly imbricate or almost valvate. Disk of 2 prominent lobes or glands at the back of the styles or rarely almost flat. Fruit laterally compressed, notched at the base with rounded auricles, without vittæ, the carpophore persistent, the dorsal edge of the carpels obtuse, the primary and often some of the secondary ribs prominent and curved at the base. Seed somewhat compressed.—Herbs or small shrubs, diffuse or decumbent at the base or erect, often clothed with soft long hairs mixed with a stellate tomentum. Leaves toothed, lobed or ternately divided. Umbels usually compound, the partial ones with two or three bracts and several almost sessile flowers, the general one of 3 or 4 rays and as many bracts, but sometimes the whole umbel reduced to very few or to a single flower.

The genus is confined to Australia.

Umbels dense or few-flowered. Bracts of the involucres narrow or small or herbaceous or rigid and chaff-like.
  Leaves slender, nearly terete, chiefly radical. Stems long and rigid. Umbels compact, compound, with rigid chaff-like bracts.
    Calyx-lobes orbicular peltate . . . . 1. *X. juncea.*
    Calyx-lobes cordate-auriculate, acute . . . . 2. *X. tenuior.*
  Leaves orbicular-cordate or reniform, coriaceous, sinuate-toothed or shortly lobed, glabrous or tomentose.
    Calyx-lobes orbicular-peltate. Umbels compound. Bracts narrow, rigid . . . . 3. *X. peltigera.*
    Calyx-lobes acute, neither cordate nor peltate. Umbels few-flowered. Bracts small, coloured . . . . 4. *X. hederifolia.*
  Leaves thin, ovate obovate or oblong, entire or toothed, white underneath. Umbels few-flowered. Bracts small, coloured. 5. *X. candida.*
  Leaves cuneate, equally 3-toothed or entire, mostly white-tomentose underneath.
    Umbels pedunculate, 1- to 4-flowered. Calyx-lobes shortly peltate . . . . 6. *X. tridentata.*
    Umbels nearly sessile, 1-flowered. Calyx-lobes not peltate . . 7. *X. singuliflora.*
  Leaves ovate to lanceolate, lobed. Plant usually pilose and tomentose. Umbels nearly sessile or rarely pedunculate, 1- to 4-flowered . . . . 8. *X. pilosa.*
  Leaves narrow, entire, 3-partite or ternately divided. Low, diffuse or much-branched plants.
    Leaves simple. Umbels 3- to 9-flowered. Calyx-lobes auriculate . . . . 9. *X. ciliata.*
    Leaves 3-partite or ternately divided.
      Umbels 1- to 4-flowered.
        Leaves 3-partite, with entire or 2-lobed narrow segments. Umbels mostly sessile . . . . 10. *X. pusilla.*
        Leaves 3-partite, with 2- or 3-lobed cuneate segments. Umbels on slender peduncles . . . . 11. *X. fruticulosa.*
      Umbels irregularly compound, with several flowers in each umbellule.
        Leaves 3-partite, with entire or 2-lobed segments . . . 12. *X. Huegelii.*
        Leaves twice 3-partite or more divided . . . . 13. *X. dissecta.*
Umbels compound, of 3 to 5 distinct rays of several flowers each, besides a central cluster of flowers. Partial involucres of 3 broad petal-like coloured bracts, exceeding the flowers.
  Flowering-branches leafy. Leaves toothed or lobed.
    Glabrous or woolly-tomentose. Leaves orbicular, toothed . . 17. *X. rotundifolia.*
    Hirsute with long hairs and stellate-tomentose. Leaves ovate, lobed . . . . 15. *X. vestita.*
  Leaves chiefly radical or at the base of the stems, divided.
    Stems elongated and branched. Leaves 3-partite, with cuneate-toothed or lobed segments . . . . 16. *X. Atkinsoniana.*
    Stems short, not much branched. Leaves 3-partite, with linear divided segments.
      Peduncles much longer than the leaves . . . . 14. *X. peduncularis.*
      Peduncles shorter than the leaves . . . . 13. *X. dissecta.*

1. **X. juncea,** *Benth.* Quite glabrous. Stems from a thick rhizome, ascending or erect, slender but rigid, 1 to 2 ft. long. Leaves very few and chiefly radical, linear-terete, those on the stem very small and distant. Peduncles long and slender, bearing a compact irregularly compound umbel of 3 or 4 very short rays, almost contracted into a head. Involucral bracts, both general and partial, exceeding the flowers, rigid, glume-like, with white scarious margins, 3 to 4 lines long. Flowers nearly sessile, 3 or 4 in each partial umbel. Calyx-lobes short, orbicular, peltately attached by the centre. Petals narrow with an inflexed point. Disk-lobes glabrous. Fruit with the primary ribs only slightly prominent.—*Schœnolæna juncea,* Bunge in Pl. Preiss. i. 289.

W. Australia. Swan River, *Preiss, n.* 2082; Vasse river, *Mrs. Molloy.*

2. **X. tenuior,** *Benth.* Very near *X. juncea,* with the same habit, linear-terete, almost subulate radical leaves, nearly leafless stems and compact inflorescence, but smaller and more slender in all its parts, and the calyx-lobes ovate acute, slightly cordate-auriculate at the base, but not at all peltate.—*Schœnolæna tenuior,* Bunge in Pl. Preiss. i. 290.

W. Australia. Near Albany, *Preiss, n.* 2080; King George's Sound, *Maclean, Harvey.*

3. **X. peltigera,** *Benth.* Quite glabrous. Stems from a perennial stock, numerous, ascending to 1 or 2 ft., more or less angular. Leaves on long petioles, broadly orbicular-cordate reniform or ovate-rhomboidal, often 1 to 2 in. broad, coriaceous, several-nerved, sinuate-toothed. Peduncles long, terminal or leaf-opposed. Umbels compound, usually of 4 to 6 many-flowered rays, with a few flowers in the centre. Involucral bracts linear-acute, rigid, striate, those of the general involucre about as long as the rays, the partial ones exceeding the flowers. Calyx-lobes broadly oblong, very obtuse, peltately attached by the centre. Petals shortly unguiculate, not inflexed at the tips. Disk scarcely prominent. Styles erect, straight, thickened in the middle so as to be almost fusiform. Fruit with the secondary ribs as well as the primary ones usually prominent.—*Leucolæna peltigera,* Hook. Ic. Pl. t. 45; *Pentapeltis peltigera,* Bunge in Pl. Preiss. i. 292.

W. Australia. Swan River, *Drummond, n.* 92; Darling Range, *Preiss, n.* 2081; King George's Sound, *Fraser.*

4. **X. hederifolia,** *Benth.* Stems elongated, weak, branching, diffuse, glabrous or tomentose when young. Leaves petiolate, orbicular-cordate, shortly and acutely 5- or 7-lobed, or coarsely toothed, coriaceous, with revolute margins, glabrous above, densely woolly-tomentose underneath, rarely above $\frac{3}{4}$ in. diameter. Peduncles usually exceeding the leaves, with a compact irregularly compound umbel, most frequently of 2 or 3 short 1-flowered rays with 2 or 3 single flowers between them. Bracts of the general involucre narrow, partial ones about as long as the flowers, (purple?) coloured. Calyx-lobes purple, scarcely acute, neither peltate nor cordate. Disk-lobes large, glabrous. Petals narrow. Fruits rather large, the secondary ribs usually prominent as well as the primary ones.

W. Australia, *Drummond,* 5*th Coll. n.* 294.

5. **X. candida,** *Steud.; Bunge in Pl. Preiss.* i. 291. Stems long, slender,

diffuse, tomentose-pubescent or glabrous. Leaves from broadly ovate or orbicular to oblong-cuneate, coarsely and irregularly toothed or lobed or rarely entire, white-tomentose especially underneath when young, at length glabrous above, narrowed into a long petiole, the largest ½ to 1 in. long, but mostly smaller. Peduncles slender, bearing a small compound umbel, usually of 4 short rays, each with 1 flower and 3 bracts, the central one narrow, like those of the general involucre, the 2 lateral ovate and more or less coloured, and 1 central flower without bracts. Calyx-lobes ovate, almost obtuse, not cordate. Petals narrow. Disk-lobes thick, depressed on the top, glabrous.—*Leucolæna candida*, Benth. in Hueg. Enum. 55.

W. Australia. King George's Sound, *R. Brown;* Swan River, *Huegel, Fraser, Preiss, n.* 2088; Sussex district, *Preiss, n.* 2077; Vasse, Harvey, Gordon and Blackwood rivers, *Oldfield.*

6. **X. tridentata,** *DC. Prod.* iv. 75. Slender and diffuse, slightly tomentose and hirsute or nearly glabrous. Leaves cuneate, acutely and nearly equally 3-toothed at the end, mostly under ½ in. long, glabrous or white-tomentose underneath. Peduncles slender, leaf-opposed or terminating short leafy branches, reflexed, irregularly 1- to 3-flowered or rarely 4- or 5-flowered, with 1 or 2 bracts to each branch or pedicel, and 3 or rarely 2 lanceolate acute bracts to each flower, about 2 lines long. Calyx-lobes acute, peltately attached a little above their base. Petals very narrow. Disk-lobes glabrous. Fruit with the lateral ribs distinct as well as the intermediate and secondary ones.

N. S. Wales. Port Jackson to the Blue Mountains, *R. Brown, Leichhardt, Miss Atkinson, F. Mueller;* southward to Twofold Bay, *F. Mueller.*
Victoria. Wilson's Promontory, *F. Mueller.*

7. **X. singuliflora,** *F. Muell. Fragm.* iv. 184. Stems slender, diffuse, much intricate, under 1 ft. long, glabrous or slightly tomentose. Leaves oblong-cuneate, acute, entire or 3-toothed at the end, narrowed into rather a long petiole, mostly under ½ in. long, glabrous or white-tomentose underneath. Umbels reduced to a single flower, sessile or shortly pedunculate at the upper nodes, surrounded by 3 or 4 narrow bracts at the base of a very short pedicel, and 2 broader ones close under the flower. Calyx-lobes acute, not cordate nor peltate, with almost scarious margins. Petals inflexed at the points. Fruits with the secondary ribs prominent.

W. Australia. Near Cape Paisley, *Maxwell.*

8. **X. pilosa,** *Rudge in Trans. Linn. Soc.* x. 301. *t.* 22. *f.* 1. An erect or more frequently diffuse or procumbent, much branched, leafy shrub, of 1 to 2 ft., copiously clothed with rather long hairs intermixed with a brown stellate tomentum, or very rarely in mountain situations becoming almost glabrous. Leaves cuneate, obovate, oblong, lanceolate or broadly ovate, coarsely sinuate-toothed, 3- or 5-lobed or rarely 3-partite, the central lobe always longer than the lateral ones, rarely exceeding 1 in., the petiole short, dilated and ciliate at the base, but without real stipules. Peduncles usually 2 together at the nodes, very short or slender and nearly as long as the leaves, each usually with 2 flowers, more rarely 3 or only 1, with 2 or 3 small narrow bracts forming a general involucre at the base of the short pedicels or rays, and 2 or 3 oblong-lanceolate bracts of 2 or 3 lines, forming a partial

involucre under each flower. Calyx-lobes rather thickened at the base, but neither cordate nor peltate. Petals narrow, with a long inflexed point. Disk-lobes thick, pubescent. Fruit rather above 1 line long and broad, the intermediate and secondary ribs prominent, the lateral ones scarcely distinct from the commissure.—*X. montana*, Sieb. in DC. Prod. iv. 74; Hook. f. Fl. Tasm. i. 155; *X. hirsuta*, DC. Prod. iv. 74; *Leucolæna pannosa*, Benth. in Hueg. Enum. 55 (more densely villous, with longer leaves).

Queensland. Moreton Island, *F. Mueller.*

N. S. Wales. Port Jackson to the Blue Mountains, *R. Brown, Sieber, n.* 247, 248 (the latter less hirsute), and others, to the southward, *A. Cunningham;* Twofold Bay, *F. Mueller.*

Victoria. Scrubby moist valleys in Gipps' Land, *F. Mueller.*

Tasmania. Common in many places on the N. coast, *J. D. Hooker.*

One-flowered and two-flowered involucres, on which two species have been frequently distinguished, occur sometimes on the same specimens. The glabrous specimens, from the Blue Mountains, appear, at first sight, very distinct, but I can find no other difference than the want of hairs.

9. **X. ciliata,** *Hook. Ic. Pl. t.* 726. Diffuse and small, but more shrubby than the following species, pubescent or nearly glabrous. Leaves undivided, linear linear-oblong or linear-cuneate, entire or shortly 3-toothed, the petioles dilated and ciliate at the base. Peduncles short at the nodes or terminating short leafy branches, bearing a small compact more or less compound umbel of from 3 or 4 to three times as many flowers. Involucral bracts 3 to each partial umbel, linear-lanceolate, herbaceous, pubescent, not 2 lines long. Calyx-lobes cordate-auriculate, acuminate. Petals narrow. Disk-lobes glabrous. Fruits with the secondary ribs more or less prominent.

W. Australia, *Drummond, 1st Coll., also n.* 237, 721, *and 4th Coll. n.* 140.

10. **X. pusilla,** *Bunge in Pl. Preiss.* i. 291. Small diffuse and densely branched, rarely exceeding 3 or 4 in. in the original form, more or less hirsute with spreading hairs. Leaves 3-partite, the segments from broadly elliptical and about $\frac{1}{4}$ in. to lanceolate and $\frac{1}{2}$ in. long or rather more, entire or the lateral ones 2-lobed, rather thick. Umbels sessile and leaf-opposed, or pedunculate in the axils and often reflexed, with 1 to 3 flowers, surrounded by narrow hirsute involucral bracts. Calyx-lobes very acute, neither cordate nor peltate. Disk-lobes large, concave. Petals very narrow, with inflexed points. Fruit rather large, the prominent ribs variable in number, usually 4 on each side. —Hook. f. Fl. Tasm. 1. 156; *X. villosa,* Turcz. in Bull. Mosc. 1849, ii. 32.

Victoria. Wilson's Promontory, *F. Mueller;* near Portland, *Allitt.*

Tasmania. Sandy soil, north shore, *J. D. Hooker.*

S. Australia. Lofty Range and near Adelaide, *F. Mueller.*

W. Australia. King George's Sound and adjoining districts, *R. Brown, Wakefield, Baxter, Preiss, n.* 2078, *Drummond, 4th Coll. n.* 139.

Var. *glabrata.* Stems elongated, slender and glabrous or nearly so.—*X. glabrata,* Bunge in Pl. Preiss. i. 290. King George's Sound, *Baxter;* shady woods, Canning river, *Preiss, n.* 2076. The differences shown by this form are probably owing to the situations it grows in.

11. **X. fruticulosa,** *Benth.* Diffuse, much-branched and slender, but more shrubby than the allied species, hirsute with soft hairs or at length glabrous. Leaves shortly petiolate, divided into 3 cuneate entire or more frequently 3-lobed segments, under $\frac{1}{2}$ in. long, glabrous or hairy but not tomentose. Peduncles filiform, exceeding the leaves, bearing each a scarcely

compound umbel of 2 to 4 flowers. Involucral bracts 4 to 6, ovate or ovate-lanceolate, scarcely coloured, spreading to a diameter of 2 or 3 lines. Flowers nearly sessile. Calyx-lobes acute, slightly auriculate. Petals much inflexed. Styles very short, not exceeding the disk-lobes. Fruits short, but not seen quite ripe.

W. Australia. Between Moore and Murchison rivers, *Drummond, 6th Coll. n.* 118.

12. **X. Huegelii,** *Steud.; Bunge in Pl. Preiss.* i. 291. Stems from a perennial or woody stock, erect or rarely diffuse, hirsute with spreading hairs, rarely exceeding 6 to 8 in. Leaves divided into 3 lanceolate or linear segments, entire or the lateral ones 2-lobed or 2-partite, usually ½ to 1 in. long, villous or nearly glabrous. Peduncles short or long, with an irregularly compound umbel of 3 or 4 short rays, each with 3 to 6 sessile flowers, and 1 to 3 flowers pedicellate in the centre. Involucral bracts narrow, acute, herbaceous. Calyx-segments acutely acuminate, cordate at the base. Petals narrow. Disk-lobes large, undulate-lobed. Fruit usually with 6 ribs on each side.—*Leucolæna Huegelii*, Benth. in Hueg. Enum. 55.

W. Australia. Swan River, *Huegel; Drummond, 1st Coll. and n.* 712; *Preiss, n.* 2090, and others. Resembles the long narrow-leaved forms of *X. pusilla*, but the flowers are much more numerous, and the calyx-lobes, disk, and fruit-ribs rather different.

13. **X. dissecta,** *Hook. f. in Hook. Ic. Pl. t.* 302, *and Fl. Tasm.* i. 155. Diffuse with a perennial stock, sometimes forming dense tufts and rarely exceeding 6 in., glabrous. Leaves on long petioles, once or twice 3-partite, with narrow cuneate or linear acutely 3-toothed or 3-lobed segments. Peduncles leaf-opposed or terminating short axillary branches, all shorter than the leaves although sometimes near 1 in. long, bearing in the original slender form small compound umbels of 2 or 3 short rays each with 2 or 3 flowers and the bracts small, but in the more luxuriant and stouter forms the umbels larger, the flowers more numerous, and the bracts larger and more coloured. Calyx-lobes acute, not peltate and scarcely cordate. Petals very narrow. Disk-lobes large, glabrous. Fruit very didymous, with only 4 ribs on each side, the secondary ones rarely conspicuous.—*X. leiophylla*, F. Muell.; Klatt in Linnæa, xxix. 710; *X. pinnatisecta*, F. Muell.; Klatt, l. c. 711.

N. S. Wales. Blue Mountains, *A. Cunningham.*

Victoria. From the Glenelg river, *Robertson*, to Gipps' Land, *F. Mueller;* Wimmera, *Dallachy;* Murray desert, *Irvine.*

Tasmania. Northern shores, Rocky Cape, Georgetown, *J. D. Hooker.*

S. Australia. S. coast, *R. Brown;* Rivoli Bay and Kangaroo island, *F. Mueller;* Port Lincoln, *Wilhelmi.*

Var. *floribunda.* More luxuriant. Umbels of 3 or 4 rays with a single flower in the centre and a general involucre of 3 or 4 small narrow bracts; partial umbels of 3 to 6 nearly sessile flowers and an involucre of 2 or 3 ovate-lanceolate or lanceolate coloured bracts.—To this variety belong several of the Victorian and S. Australian specimens. It is connected with the slender few-flowered forms by numerous intermediates, and scarcely differs from *X. peduncularis*, except in the short peduncles and less numerous flowers.

14. **X. peduncularis,** *Benth.* A rather small plant, forming sometimes close tufts, but with the appearance of being almost annual, glabrous or softly hirsute, attaining (with the inflorescence) 6 in. to 1 ft. Leaves

mostly radical or at the base of the stem, on long petioles, once or twice 3-partite with linear or cuneate acutely-toothed or lobed segments like those of *X. dissecta*, the upper ones small, simple and linear. Peduncles much longer than the leaves, with a compound umbel like that of *X. rotundifolia*, but much smaller; rays 3 or 4, with a few central flowers. Bracts of the general involucre narrow, acute, the partial ones 3, broad and coloured, longer than the flowers, the lateral ones very oblique. Calyx-lobes slightly cordate, acute. Fruits with 4 or 6 prominent ribs on each side.

W. Australia, *Drummond, 5th Coll. n.* 292; Bremer Bay river, *Maxwell.*

15. **X. vestita,** *Benth.* Shrubby, with the dense stellate tomentum, long spreading hairs and foliage of *X. pilosa*, but with the umbels nearly of *X. rotundifolia.* Leaves very shortly petiolate, broadly ovate, irregularly and obtusely toothed and lobed, the larger ones above 1 in. long, tomentose and hirsute underneath, hirsute or at length nearly glabrous above. Peduncles longer than the leaves, solitary or 2 together, each with a rather large compound umbel of 3 or 4 rays with 1 or 2 central flowers. Bracts of the general involucre narrow, very hirsute, shorter than the rays, of the partial ones 3, obovate-oblong, obtuse, nearly equal, longer than the flowers, coloured and petal-like but villous outside. Calyx-lobes petal-like, obtuse, not cordate. Disk-lobes villous. Fruit with the secondary ribs usually prominent.

N. S. Wales. Hawkesbury river and Blue Mountains (*C. Moore?*) in Herb. F. Mueller.

16. **X. Atkinsoniana,** *F. Muell. Fragm.* ii. 127. Glabrous, or the young leaves floccose-tomentose. Stems from a perennial woody stock elongated, slender, ascending to 2 feet or more. Leaves chiefly radical or near the base of the stem, on long petioles, in the eastern specimens 3-partite with cuneate acutely toothed or deeply lobed segments, the central ones often longer and more divided, in the western specimens less divided with the lobes more entire and less acute, the upper leaves in both few, small, and scarcely divided. Peduncles long, with a compound umbel like that of *X. rotundifolia*, but smaller, especially in the eastern specimens, the coloured petal-like bracts of the partial involucres sometimes scarcely longer than the flowers. Calyx-lobes shortly and broadly cordate, acute. Fruits 4- or 6-ribbed.

N. S Wales. Grassy open country north of Richmond, *A. Cunningham;* Clyde river, *C. Moore;* Blue Mountains, *A. and R. Cunningham, Miss Atkinson.*

W. Australia. Swan River, *Drummond, 1st Coll.;* Blackwood river, *Oldfield;* Bald Island, Tone river, and Lake Leven, *Maxwell.*

17. **X. rotundifolia,** *DC. Prod.* iv. 75. Stems erect, simple or slightly branched, often woody at the base, 1 to 2 ft. high, glabrous tomentose or rarely hirsute. Leaves not confined to the base of the stem, on short petioles, nearly orbicular, irregularly and acutely toothed, $\frac{3}{4}$ to $1\frac{1}{2}$ in. diameter, coriaceous, glabrous or woolly underneath especially when young. Peduncles long, bearing each a compound umbel larger than in most species; rays usually 4 with a sessile umbellule in the centre. Involucral bracts petal-like and coloured, those of the general involucre usually 4, ovate-lanceolate or rhomboidal, shorter than the rays, the partial ones 3, exceeding the flowers, often $\frac{1}{2}$ in. long, the lateral ones broadly semi-ovate and falcate, the central

one narrower and equal-sided. Flowers rather numerous, on very short pedicels. Calyx-lobes acute, often slightly cordate. Petals rather broad, the induplicate point ciliate. Fruits 4- to 6-ribbed on each side.—Bunge in Pl. Preiss. i. 292; Bot. Mag. t. 3582.

W. Australia. King George's Sound and adjoining districts, *R. Brown*, and others; *Drummond, 3rd Coll. n. 227; Preiss, n.* 2066.

F. Mueller is disposed to unite the three preceding species with this one as varieties, but the foliage and habit are so different that until I have seen intermediate specimens I cannot but consider them as distinct.

5. AZORELLA, Lam.

(Fragosa, *Ruiz and Pav.;* Pozoa, *Lag.;* Microsciadium, *Hook. f.;* Oschatzia, *Walp.;* Dichopetalum, *F. Muell.*)

Calyx-teeth or lobes prominent, either small and acute or large petal-like and deciduous. Petals obtuse or acute, imbricate in the bud. Disk thick, flat, convex or confluent with the styles. Fruit slightly compressed laterally or scarcely broader than thick, the sides furrowed at the commissure (when quite ripe); carpels nearly terete or angular, with 5 more or less prominent nearly equidistant ribs, the lateral ones not close to the rather narrow comsure. Vittæ none. Carpophore short, persistent. Seed straight.—Perennials, the Australian species either tufted with radical leaves and peduncles or more slender with creeping runners. Leaves (in the Australian species) toothed or lobed but undivided, the base of the petiole rarely expanded into distinct scarious stipules. Umbels simple or rarely irregularly compound, the involucral bracts free or united.

The genus, which I have adopted in the extended sense given to it by A. Gray (Bot. Amer. Expl. Exped. i. 697) and Weddell (Chloris Andiana, ii. 190), includes a considerable number of species from Andine and extratropical S. America, New Zealand and the Antarctic regions generally. Most of the Antarctic and Andine ones belong, however, to a group unrepresented in Australia, to which the genus is sometimes confined, in which the densely matted stocks covered with the imbricated remains of old leaves form large compact masses. The floral and carpological characters are very nearly the same throughout the genus. The fruit, however, must be examined quite ripe. Before that it is often almost equally 4-sided, the longitudinal furrows at the commissure scarcely perceptible.

Small Hydrocotyle-like plant, with creeping stolons and short slender ascending stems. Umbels several-flowered, terminal or sessile at the nodes . . . . 1. *A. Muelleri.*
(*Hydrocotyle tetragonocarpa* has the fruits scarcely compressed, but the petals are valvate.)
Tufted plants with radical leaves. Stems or scapes erect, leafless, simple or branched upwards. Flowers solitary or in irregular umbels of 2 to 6. Calyx-lobes not above half as long as the petals.
Leaves cuneate, 3- to 7-toothed or lobed at the end . . . . . 2. *A. cuneifolia.*
Leaves broadly ovate-cordate, 5- to 9-lobed . . . . 3. *A. saxifraga.*
Tufted plant with radical leaves. Scapes erect, leafless, with an irregular umbel of 6 to 12 flowers sometimes compound.
Calyx-lobes petal-like and as long as the petals . . . . 4. *A. dichopetalum.*

1. **A. Muelleri,** *Benth.* A glabrous perennial with the aspect of a *Hydrocotyle*, the small tufted stock apparently emitting creeping stolons, the slender ascending stems rarely exceeding the radical leaves. Leaves orbi-

cular-cordate or reniform, shortly and obtusely 5- to 9-lobed and crenate, ½ to ¾ in. diameter, the radical ones on long petioles. Stipules as in *Hydrocotyle* scarious, scarcely adnate to the petiole and jagged or ciliate. Umbels almost sessile at the nodes, the last one of the stem appearing pedunculate from the smallness of the floral leaf. Involucral bracts united at the base, membranous, lanceolate, about as long as the flowers, mostly with 1 or 2 setaceous teeth or lobes. Calyx-teeth ovate-triangular, petal-like, half as long as the petals. Petals almost acute, but thin aud imbricate in the bud. Disk rather thick, flat or concave, surrounding the styles. Fruit ovoid, thick, slightly compressed laterally, furrowed at the commissure, but not seen ripe.—*Pozoa Fragosa*, F. Muell. in Trans. Phil. Inst. Vict. i. 102, and in Hook. Kew Journ. viii. 70.

Victoria. Under the shade of rocks on the summits of the Munyang mountains at an elevation of 6000 ft., *F. Mueller.*—This species in many respects approaches *Hydrocotyle* in character as well as in habit.

2. **A. cuneifolia,** *F. Muell.* (*as a* Pozoa). Perennial, forming a densely-tufted stock, quite glabrous. Leaves all radical on long petioles, cuneate, irregularly 3- to 7-lobed or toothed at the end, the narrowest sometimes entire. Flowering stems leafless except a small bract under each branch, erect, 6 in. to 1 ft. high, dichotomously-branched in the upper portion, each branch or peduncle bearing a single terminal flower or an irregular umbel of 2 to 9 flowers. Calyx-teeth minute or scarcely conspicuous. Petals broad, acute, much imbricate. Disk at first flat, but thick and soon becoming conical with the short styles terminating the lobes. Fruit ovoid, about 2 lines long, scarcely compressed, and slightly furrowed at the commissure between the lateral ribs; carpels with 5 equidistant prominent ribs.—*Centella cuneifolia*, F. Muell. in 1st Gen. Rep. and in Hook. Kew Journ. vii. t. 12; *Microsciadium cuneifolium*, F. Muell. in Hook. Kew Journ. vii. 379; *Pozoa cuneifolia*, F. Muell. in Trans. Phil. Inst. i. 103.

Victoria. In turf moss on Mount Macedon, the Cobberas mountains, and others of the Australian Alps, *F. Mueller.*

3. **A. saxifraga,** *Benth.* A smally tufted perennial, quite glabrous. Leaves all radical, broadly ovate-cordate, under ½ in. long, divided to about the middle or sometimes more deeply into 5 to 9 acute or obtuse lobes. Peduncles or scapes slender, 3 to 4 in. high, each with 1, 2 or 3 flowers on rather long pedicels and a bract under each pedicel. Calyx-teeth acute, nearly half as long as the petals. Petals long, attaining fully 1 line. Disk thick and conical. Fruit not seen ripe, but apparently the same as in *A. cuneifolia.*—*Microsciadium saxifraga*, Hook. f. in Hook. Lond. Journ. vi. 468, and Fl. Tasm. i. 159; *Oschatzia saxifraga*, Walp. Ann. i. 140.

Tasmania. Wet heathy places, Loddon Plains, and Macquarrie Harbour, *Gunn.*

4. **A. dichopetala,** *Benth.* Perennial, with a densely-tufted stock, the leaves and peduncles often hispid with long rigid scattered hairs. Leaves radical, on long petioles, orbicular-cordate or reniform, ½ to 1 in. diameter, shortly and broadly 5- to 7-lobed and crenate. Peduncles shorter or longer than the leaves, bearing an irregular umbel of 6 to 12 flowers, the pedicels very variable in length, and one sometimes bearing a partial umbel. Bracts

under the umbel very unequal and sometimes united at the base. Calyx-lobes petal-like, as large as the petals and falling off with them. Petals fully 1 line long, acute but slightly imbricate. Disk thick, at first slightly convex, aftewards conical.—*Dichopetalum ranunculaceum*, F. Muell. in Hook. Kew Journ. vii. 378. t. 11, and in Trans. Phil. Inst. Vict. i. 102; Hook. f. Fl. Tasm. i. 157. t. 35.

Victoria. Munyang mountains at an elevation of 5000 to 6000 ft., *F. Mueller*.
Tasmania. Mount Sorrel, Macquarrie Harbour, *Gunn*, *Milligan*; Mount Lapeyrouse, *Oldfield*.

The Victorian specimens are sprinkled with a few of the long hairs which are very abundant on the Tasmanian ones, but there appears to be no other difference.

6. DIPLASPIS, Hook. f.

(Pozoopsis, *Hook. f.*)

Calyx-teeth inconspicuous. Petals ovate, imbricate in the bud. Disk none besides the broad thick conical base of the styles. Fruit compressed from front to back, deeply furrowed on each side at the commissure. Carpels much flattened from front to back, the outer face flat with the dorsal rib in the centre, the lateral ribs bordering the narrow commissure in the centre of the convex inner face, the intermediate ribs forming the edges of the carpels. Vittæ none. Carpophore persistent. Seeds straight, flattened.—Perennials, with creeping rhizomes and dense tufts of radical leaves. Leaves cordate or orbicular, rather thick. Scapes simple with a terminal simple umbel of several often many flowers.

The genus is confined to Australia.

Glabrous. Leaves broadly ovate-cordate, quite entire. Flowers under 20 in the umbels . . . . 1. *D. hydrocotylea*.
More or less hirsute with long hairs. Leaves orbicular-cordate distinctly crenate. Flowers numerous in the umbel . . . . . . 2. *D. cordifolia*.

1. **D. hydrocotylea,** *Hook. f. in Hook. Lond. Journ.* vi. 469, *and Fl. Tasm.* i. 156. *t.* 34. Perennial, with a creeping rhizome. Leaves radical, tufted, on rather long petioles, cordate, orbicular or ovate, thick with revolute margins, under ½ in. and often not ¼ in. diameter, glabrous or sprinkled with a few hairs. Scapes 2 to 4 in. long, each with a single terminal umbel of 12 to 20 flowers on pedicels of 1 to 2 lines. Involucre of a few unequal linear bracts. Petals obtuse.

Victoria. Haidinger range, Bogong and Upper Mitta-Mitta mountains, *F. Mueller*.
Tasmania. Marshes and wet sandy ground about alpine lakes, *J. D. Hooker*.

2. **D. cordifolia,** *Hook. f. Fl. Tasm.* i. 157. Very near *D. hydrocotylea*, but rather larger, more or less hirsute with spreading hairs, the revolute margins of the leaves distinctly crenate, the flowers much more numerous in the umbels, and the fruits considerably narrower.—*Pozoopsis cordifolia,* Hook. f. in Hook. Ic. Pl. t. 859.

Tasmania. Marshy places Mount Sorrel, Macquarrie Harbour, *Gunn*, *Milligan*; Mount Lapeyrouse, *C. Stuart*.

7. ACTINOTUS, Labill.

(Eriocalia, *Sm.*; Holotome, *Endl.*; Hemiphues, *Hook. f.*)

Calyx-limb distinct, campanulate or open, truncate or 5-lobed. Petals 5, unguiculate, spathulate or orbicular-concave, imbricate in the bud, or none. Disk-lobes or glands at the back of the styles, often not quite at the base. Ovary with a single cell and ovule; styles 2, often united at the base. Fruit ovate, of a single carpel, crowned by the calyx-limb, compressed from front to back, 5-ribbed, one face (the broad commissure?) nearly flat with 1 rib, the other (the back?) convex with 2 ribs, and 1 rib at or near each margin. Vittæ none. Seed filling the cavity.—Herbs either annual or with a perennial rootstock or woody or tufted base. Leaves toothed or ternately divided. Umbels simple, surrounded by a radiating involucre of herbaceous or coloured and often very tomentose or woolly bracts exceeding the flowers. Flowers often very numerous, all with apparently perfect stamens and styles, but the outer ones often males by the abortion of the ovary, rarely all perfect.

The genus is confined to Australia. The characters upon which it has been divided appear to me to be too artificial and too little in correlation with each other to serve even for sections.

Stems branched, more or less leafy. Umbels pedunculate.	
Involucres very woolly or hirsute, coloured, 1 to 3 in. diameter.	
Tall erect plants. Leaves once or twice tripartite.	
Densely tomentose. Calyx 5-lobed. Petals none	1. *A. Helianthi.*
Silky-hairy. Calyx truncate. Petals present	2. *A. leucocephalus.*
Involucres under ½ in. diameter. Stems diffuse or ascending.	
Leaves divided, white underneath. Calyx 5-lobed. Petals none	3. *A. minor.*
Leaves 3-toothed, glabrous. Calyx truncate. Petals present .	4. *A. omnifertilis.*
Leaves orbicular or rhomboidal, toothed. Calyx 5-lobed. Petals present	5. *A. rhomboideus.*
Stock densely tufted. Leaves and peduncles radical. Involucre under ½ in. diameter, the bracts united at the base. Calyx 5-lobed. Petals none (or present?)	6. *A. bellidioides.*
Stems leafy. Leaves divided with linear lobes. Umbels small, sessile. Calyx 5-toothed. Petals none	7. *A. glomeratus.*

1. **A. Helianthi,** *Labill. Pl. Nov. Holl.* i. 67. *t.* 92. Erect, apparently perennial, 1 to 2 ft. high, covered with a soft dense almost floccose or woolly tomentum, rarely wearing off from the upper side of the leaves. Leaves twice 3-partite, with linear or oblong-linear mostly obtuse segments, entire or again 2- or 3-lobed. Umbels dense, on long stout peduncles. Involucre radiating to a diameter of 2 or 3 in., consisting of 10 to 18 coloured softly tomentose bracts. Flowers on filiform pedicels of 1½ to 2 lines, but so numerous as to form a dense head of ½ in. to ¾ in. diameter, the outer ones all males, the central ones perfect, both in numerous rows. Calyx-limb hairy, transparent, about ½ line long, deeply divided into obtuse linear lobes. Petals none. Disk-lobes oblong, gland-like, adnate to the entire base of the style. Fruit about 2 lines long, covered with long silky hairs.—Bot. Reg. t. 654; DC. Prod. iv. 83; *Eriocalia major*, Sm. Exot. Bot. ii. 37. t. 38.

Queensland. Moreton Bay, *F. Mueller.*

N. S. Wales. Port Jackson to the Blue Mountains, *R. Brown*, *Sieber*, *n.* 128, and others; New England, *C. Stuart*; on the upper Maranoa, *Mitchell.*

2. **A. leucocephalus,** *Benth. in Hueg. Enum.* 56. Erect, dichotomously branched, more or less clothed with soft hairs, 1 to 2 ft. high. Radical leaves on long petioles, the others sessile, 3-partite, with linear or linear-cuneate segments, entire or again 3-partite or lobed. Umbels dense on long peduncles. Involucre expanding to 1 or 2 in. diameter, consisting of numerous lanceolate bracts, very densely covered with long silky hairs and twice as long as the flowers. Flowers very numerous, densely packed, but on short filiform pedicels, the males in several rows at the circumference, the central ones perfect. Calyx-limb small, campanulate, truncate, transparent, silky-hairy. Petals small, spathulate, on slender claws. Disk-lobes forming glands on the undivided base of the style. Fruits broad, above 1 line long, very silky-hairy.—Hook. Ic. Pl. t. 847; Bunge in Pl. Preiss. i. 292.

W. Australia. Swan River, *Drummond, n.* 28; Canning river, *Preiss, n.* 2056; Stirling ranges, *Maxwell.*

3. **A. minor,** *DC. Prod.* iv. 83. Stems from a perennial base, long and slender, diffuse or ascending, glabrous or slightly tomentose or rarely silky-hairy. Leaves small, on short petioles, 3-partite, with cuneate or linear segments usually 3 to 4 lines long, entire or 2- or 3-lobed, glabrous above, white tomentose underneath, rarely silky-hairy. Umbels small, on long slender peduncles. Involucre radiating to about ½ in. diameter, the bracts lanceolate, acute, densely white-tomentose on the upper side, glabrous on the back, at least along the centre. Flowers very numerous, the males in several rows at the circumference, the perfect ones in the centre, all on short very hairy pedicels. Calyx-limb deeply divided into acute lobes. Petals none. Disk-lobes sessile, with the styles between them distinct from the base. Fruit about 1 line long, hairy.—*Eriocalia minor*, Sm. Exot. Bot. ii. 39. t. 79.

N. S. Wales. Port Jackson to the Blue Mountains, *R. Brown, Sieber, n.* 127, and others; Kiama, *Harvey;* Illawarra, *Shepherd.*

4. **A. omnifertilis,** *F. Muell.* (as a *Holotome*). Annual or with a tufted perennial base; stems ascending or erect, filiform, ½ to 1 ft. high, usually glabrous. Leaves mostly radical, on long slender petioles, cuneate, acutely 3- to 5-toothed or lobed, rarely above ½ in. long, glabrous in all our specimens. Stem-leaves few and less lobed, the upper ones narrow and entire. Umbels small, on filiform peduncles. Involucre not above ¼ in. diameter, the bracts narrow, acute, glabrous on both sides or slightly silky inside at the base. Flowers 10 to 20 or rarely more, slightly silky-hairy, on very short pedicels, all apparently perfect, but a few of the outer ones more slender from the first and probably not ripening. Calyx-limb transparent, truncate. Petals orbicular, concave, unguiculate. Fruit but little more than ½ line long.—*Holotome omnifertilis*, F. Muell. Fragm. ii. 129.

W. Australia. King George's Sound, *Harvey;* Hay river, *Maxwell.*

5. **A. rhomboideus,** *Benth.* Stems diffuse, branched and shrubby at the base, afterwards erect or ascending to the height of 1 ft. or more, hirsute with long soft hairs mixed with a stellate tomentum. Leaves chiefly in the lower part of the stem, petiolate, nearly orbicular or rhomboidal, irregularly and acutely toothed, under 1 in. long, hairy when young, at length glabrous.

Peduncles long, in the upper almost leafless part of the stem, bearing each a small dense umbel, with an involucre of 6 to 10 or more linear-lanceolate hirsute bracts exceeding the flowers. Flowers numerous, all perfect, or the outer ones males, on short pedicels. Calyx-limb acutely 5-lobed, hirsute. Petals unguiculate, scarcely inflexed, the margins recurved in the upper part. Disk-lobes scarcely distinct from the conical base of the styles. Fruit hairy. —*Xanthosia rhomboidea*, Turcz. in Bull. Mosc. 1849, ii. 32.

W. Australia, *Drummond, 4th Coll. n.* 134.—The structure of the flower and fruit, as well as the inflorescence, are entirely those of *Actinotus*, not of *Xanthosia.*

6. **A. bellidioides,** *Benth.* A dwarf perennial, forming dense tufts more or less covered with long soft hairs. Leaves radical, obovate-spathulate or orbicular, entire or coarsely crenate, thick, under $\frac{1}{2}$ in. long, on a petiole usually shorter. Peduncles $\frac{1}{2}$ to $1\frac{1}{2}$ in. long, bearing a small head or umbel. Involucre radiating to about 3 lines diameter, consisting of 6 to 10 bracts, united at the base or sometimes to half their length. Flowers 6 to 10 or rarely more, 1 or 2 of the outer ones sometimes barren. Calyx-limb deeply 5-lobed. Petals none (or sometimes linear?). Fruits about 1 line long.—*Hemiphues bellidioides*, Hook. f. in Hook. Lond. Journ. vi. 470, and Fl. Tasm. i. 158. t. 36; *H. affinis*, *H. tridentata*, and *H. suffocata*, Hook. f. in Hook. Lond. Journ. vi. 471.

Tasmania. Mount Fatigue, Recherche Bay, at an elevation of 4000 ft., *Gunn.*

7. **A. glomeratus,** *Benth.* Erect and more or less clothed with soft hairs, with wiry branches, about 6 to 8 in. high. Leaves solitary, or 2 or 3 from the same node, deeply 3-partite, with linear entire or 2- or 3-lobed segments shorter and scarcely broader than the petiole. Flowers almost sessile, in lateral or terminal clusters, surrounded by a few short linear very hairy bracts, several males with the perfect ones in the same clusters. Calyx-limb with 5 acuminate lobes. Petals none. Disk scarcely any in the males, the abortive style clavate and hairy, in the females 2 short styles on a large conical disk or base. Fruit about $1\frac{1}{2}$ lines long.

W. Australia. Swan River, *Drummond, 1st Coll.*; King George's Sound, *Oldfield.*

8. ERYNGIUM, Linn.

Calyx-lobes rigid, acute or pungent-pointed. Petals erect, with reduplicate or recurved margins and a long induplicate point, scarcely imbricate in the bud. Disk with a thick raised margin encircling the styles. Fruit obovoid or ovoid, scarcely compressed, the ribs inconspicuous, without vittæ. Carpophore deciduous.—Herbs with prickly leaves and involucres. Flowers in compact spikes or heads, with a bract under each flower, the outer ones and sometimes some of the inner ones much longer than the flowers, rigid and pungent-pointed. Calyx-tube covered with transparent, acuminate or obtuse, transparent, flat or vesicular scales.

The genus is spread over the greater part of the warm and temperate regions of the globe, the species most abundant and most varied in S. America. Of the four Australian species one is also in Chili, another extends to New Zealand, the remaining two appear to be endemic.

Leaves pinnately toothed, lobed, or divided, the radical ones narrow.
  Point of the petals jagged or ciliate.
 Flower-heads ovoid or globular.
  Stems erect, or rarely shortly decumbent at the base . . . . 1. *E. rostratum.*
  Stems prostrate, resembling stolons but not rooting . . . . 2. *E. vesiculosum.*
 Flower-heads oblong or cylindrical . . . . 3. *E. plantagineum.*
Radical leaves obovate or oblong, toothed or lobed. Stem leaves opposite, short, divaricately lobed. Stems dichotomous. Point of the petals obtuse, entire . . . . 4. *E. expansum.*

1. **E. rostratum,** *Cav. Ic. Pl.* vi. 35. *t.* 552. Stems erect, 1 to 2 ft. high, the lower branches sometimes alternate, but more frequently the branches 2, 3, or 4 together, with a peduncle in the fork. Radical leaves elongated, usually linear, pinnatifid, with entire or pinnatifid linear-pungent lobes, but sometimes the rhachis broader-linear, and the lobes reduced to teeth, or the rhachis very narrow with very few distant narrow lobes, or in wet places the leaves quite entire, grass-like, 6 in. long, and marked with raised transverse lines so as to appear jointed. Stem-leaves only under the peduncles or branches, short, once or twice pinnatifid, and very rigid and pungent. Flower-heads ovoid-globose. Bracts very rigid and pungent, linear or linear-lanceolate, the outer ones and sometimes a few of the inner ones ½ to 1 in. long, the others smaller, and some not exceeding the flowers. Calyx-tube densely covered with linear obtuse scales or vesicles. Inflected point of the petals ciliate-denticulate or jagged.—DC. Prod. iv. 89; *E. ovinum*, A. Cunn. in Field, N. S. Wales, 358; Schlecht. Linnæa, xx. 622; DC. Prod. iv. 89; *E. angustifolium*, DC. Prod. iv. 95 (from the diagnosis given); *E. pinnatifidum* and *E. tetracephalum*, Bunge in Pl. Preiss. i. 293.

N. S. Wales. Plains of Bathurst and all the grassy lands in the interior, *A. Cunningham*, *Fraser*.

Victoria. Common about Melbourne, *Adamson;* Wendu valley, Glenelg river, *Robertson;* Skipton plains, *Whan.*

S. Australia. Near Bethanie, *Behr;* Guichen Bay, Lofty Range, Torrens river, etc., *F. Mueller* and others.

W. Australia. Swan River and Darling Range, *Drummond*, 1*st Coll. also n.* 8, 25, 26; *Preiss, n.* 2053, 2054, and others; Canning, Vasse, Blackwood, and Tone rivers, *Oldfield.*

The species is found also in extratropical South America. It is exceedingly variable in size, number of heads, and degree of division of the leaves. In some vigorous specimens, the heads are ¾ in. or rather more in diameter, without the involucral bracts, which are 1 to 1½ in. long, and some of them with a few bristly lobes. In others the heads are few and small, and but few of the bracts attain ½ in. In general, in arid situations the leaves are more divided with narrower more rigid lobes, and in wet situations either entire or simply pinnatifid.

Var. *subdecumbens.* Radical leaves 6 in. to 1 ft. long, linear, entire, or with a few linear lobes. Stems short, sometimes decumbent, almost as in *E. vesiculosum.*—W. Australia, *Drummond*, 3*rd Coll. n.* 230; Tweed river, *Oldfield.*

2. **E. vesiculosum,** *Labill. Pl. Nov. Holl.* i. 73. *t.* 98. Radical leaves lanceolate, oblanceolate, oblong or broadly linear, coarsely prickly-toothed, narrowed into a petiole, rarely above 3 or 4 in. long and usually much shorter. Stems elongated, prostrate, having the appearance of stolons, but not rooting. Floral leaves opposite, cuneate or linear, mostly ½ to 1 in. long, with 3 to 5 pungent teeth or lobes. Peduncles radical or from the

nodes, each with a small hemispherical globular or shortly ovoid head. Outer bracts and sometimes a few of the inner ones linear or lanceolate, rigid, pungent and far exceeding the flowers, the others much smaller. Scales or vesicles of the calyx-tube sometimes lanceolate and acute, sometimes oblong and obtuse. Petals with the inflected points slightly jagged.—DC. Prod. iv. 92; Hook. f. Fl. Tasm. i. 159.

N. S. Wales. Port Jackson, *Clowes;* New England, *C. Stuart.*

Victoria. Near Melbourne, *Adamson;* Wendu valley, Glenelg river, *Robertson;* Portland, *Allitt;* Skipton, *Whan.*

Tasmania, *R. Brown;* marshy places in the northern and central parts of the island, *J. D. Hooker.*

S. Australia. Near Bethanie, Gawler river, Encounter Bay, *F. Mueller.*

The species is also in New Zealand.

3. **E. plantagineum,** *F. Muell. in Proc. Roy. Soc. Tasm.* iii. 235. Very closely allied to *E. rostratum*, with the same habit, foliage, and flowers, but the heads are oblong-cylindrical, and the bracts smaller, a few only of the outer ones and very rarely 1 or 2 of the upper ones projecting far beyond the flowers.

Queensland, *R. Brown;* Peak Downs, *F. Mueller;* tributaries of the Upper Darling river (*Leichhardt ?*).

S. Australia. Flooded ground S. of Wills Creek, *Howitt's Expedition.*

4. **E. expansum,** *F. Muell. in Proc. Roy. Soc. Tasm.* iii. 236. Radical leaves oblong or almost obovate, 2 to 3 in. long, narrowed into a petiole, bordered by coarse prickly teeth or lobes. Stems erect or diffuse, dichotomous, extending to 1 or 2 ft. Floral leaves opposite, short and broad, deeply divided into 3 or rarely 5 divaricate cuneate prickly-toothed lobes. Peduncles in the forks very short, each with a small globular head of 6 to 8 or rarely more small flowers. Bracts linear or lanceolate, pungent, all at least the outer ones much longer than the flowers. Flowers scarcely 1 line long, including the ovary. Calyx-lobes shorter than the adnate tube (or ovary), pungent-pointed as in the rest of the genus. Petals very short, the inflected end obtuse, entire.—Klatt in Linnæa, xxix. 712.

Queensland. Dawson and Burnett rivers and Peak Downs, *F. Mueller;* Wide Bay, *Leichhardt;* Brisbane river, Moreton Bay, *Leichhardt, C. Stuart.*

N. S. Wales. Hunter's River, *R. Brown.*

The species has some resemblance to the tropical American *E. fœtidum*, but it is remarkable for the smallness of its flowerheads.

9. APIUM, Linn.

Calyx-teeth inconspicuous. Petals ovate or broad, with a short inflexed tip, the margins not recurved, scarcely imbricate. Disk rather thick, confluent with the conical base of the styles. Fruit short, slightly compressed laterally. Carpels ovoid, with 5 prominent ribs, the lateral ones close to the rather narrow commissure, with 1 vitta under each furrow, and usually 2 at the commissure. Carpophore undivided. Seed nearly terete, straight. —Erect or prostrate herbs. Leaves ternately or pinnately dissected. Umbels compound, leaf-opposed or terminal, without involucral bracts.

The genus, whether limited to three or four species, or further extended to include several species distinguished upon slight grounds by modern botanists, will be found to extend over most of the temperate and warmer regions of the globe. Both the Australian species have a wide range, one chiefly in the southern hemisphere without the tropics, the other in America and tropical Africa.

Leaves once or twice pinnate, with 3 or 5 more or less divided broad or narrow segments . . . . 1. *A. australe.*
Leaves ternately divided into numerous filiform segments or lobes . . 2. *A. leptophyllum.*

1. **A. australe,** *Thou.; Hook, f. Fl. Tasm.* i. 160. Stems usually prostrate or decumbent, rarely erect, from very short to 1 or 2 ft. long, or even more. Leaves once or twice pinnatipartite, very variable in size and shape, the segments 3-partite, with incised lobes, from broadly obovate to narrow-linear, the lower ones on rather long petiolules. Umbels sessile or very shortly pedunculate at the nodes, of from 3 to 6 rays, each with a small umbel of rather numerous white flowers, without involucral bracts. Disk broad and thick, almost flat. Carpels with the primary ribs very prominent, almost corky, and narrow furrows between them; vittæ usually broad, but not very distinct.—*A. prostratum*, Labill. Pl. Nov. Holl. i. 76. t. 103; Vent. Jard. Malm. t. 81; *Petroselinum prostratum*, DC. Prod. iv. 102; Hook. Ic. Pl. t. 305; *Helosciadium australe*, Bunge in Pl. Preiss. i. 294; *H. prostratum*, Bunge, l. c. 295.

Queensland. Port Curtis, *M'Gillivray;* Brisbane river, Moreton Bay, *F. Mueller;* Fitzroy river, *Thozet.*

N. S. Wales. Port Jackson to the Blue Mountains, *R. Brown, Sieber, n.* 119, and others; northward to Hastings river, *Fraser;* Clarence river, *Wilcox;* southward to Twofold Bay, *F. Mueller;* Lord Howe's Island, *M'Gillivray.*

Victoria. From the Glenelg to Gipps' Land, *F. Mueller* and others; Wimmera, *Dallachy.*

Tasmania. Islands of Bass's Straits, *R. Brown;* common, especially on the northern shores of the island, *J. D. Hooker.*

S. Australia. From the Murray river to Spencer's and St. Vincent's Gulfs, *F. Mueller*, and others.

W. Australia. From the S. coast to Swan and Murchison rivers, *Drummond, 1st Coll., also n.* 65, 124, 141, and 293, *Preiss, n.* 2051, 2052, *Oldfield.*

There are two common forms, one with short broad very obtuse leaf-segments, chiefly found near the sea; and some specimens from the seacoast of Tasmania, the islands of Bass's Straits and adjoining coasts of the mainland, have a thick almost woody stem and large thick leaves divided into very numerous small obtuse segments. The other form has numerous long narrow acute linear segments, and often seems too unlike the maritime one to belong to the same species, but the intermediates between the two are very numerous, passing gradually from the one to the other.

The species is also in New Zealand, the S. Pacific islands, Antarctic America. and perhaps in South Africa. It is very near the wild celery of the northern hemisphere (*A. graveolens*, Linn.), but that has generally an erect stem, and the ribs of the fruit appear to be always much more slender, with broad furrows between them.

2. **A. leptophyllum,** *F. Muell. Herb.* An erect or diffuse slender glabrous annual of 1 to 2 ft. Leaves ternately divided into numerous filiform segments, the lower ones petiolate, the upper ones sessile, with fewer segments. Umbels at the nodes sessile or pedunculate, of 2 or 3 slender rays, each with a partial umbel of many flowers on slender pedicels, without involucral bracts. Disk rather broad, convex, scarcely distinct from the very short styles. Ribs of the carpels very prominent and thick, almost corky, separated by very narrow furrows, with one vitta under each furrow.—

Helosciadium leptophyllum, DC. Prod. iv. 105, with the numerous synonyms adduced.

Queensland. Brisbane river, Moreton Bay, *F. Mueller.*

N. S. Wales. Port Jackson, *Sieber, n.* 481, *Backhouse;* Clarence river, *Beckler.*

The species is common in South America, extending to the Southern States of North America, and is also found in tropical Africa.

10. SESELI, Linn.

Calyx-teeth usually prominent. Petals in the Australian species ovate, tapering into an inflexed point, the margins not recurved. Disk thick, surrounding the base of the style or confluent with it. Fruit ovoid or oblong, not compressed. Carpels with 5 prominent ribs, the lateral ones close to the broad commissure, with 1 vitta under each furrow, and usually 2 to the commissure. Carpophore divided or nearly entire. Seed semiterete, straight.—Herbs, in the Australian species, glabrous, with a perennial stock. Leaves chiefly radical, once or twice pinnately dissected. Umbels terminal, compound. Involucral bracts few. Flowers white.

The genus comprises a considerable number of species, inhabitants of the northern hemisphere in the Old World. The Australian species are both endemic, and though differing in some slight particulars will probably prove to be really congeners of the northern ones.

Leaf-segments narrow. Fruit narrow-oblong, 3 to 4 lines long . . 1. *S. Harveyanus.*
Leaf-segments short and broad. Fruit shortly oblong, under 2 lines . 2. *S. algens.*

1. **S. Harveyanus,** *F. Muell. in Hook. Kew Journ.* viii. 71, *and in Trans. Phil. Inst. Vict.* i. 104. Stems from a thick perennial rootstock, erect, glabrous as well as the whole plant, 1 to $1\frac{1}{2}$ ft. high. Leaves chiefly radical, sometimes as long as the stem, on long petioles, pinnately divided, with linear or linear-lanceolate segments often $\frac{1}{2}$ in. long, the lower ones of each leaf often again 3-partite or shortly pinnate; upper leaves few, narrow, simply pinnate or simple. Umbels terminal, of 4 to 7 very unequal rays, each partial umbel with 1 to 4 perfect flowers on long pedicels, and several small males on short pedicels. Involucres, both general and partial, of 2 or 3 very small bracts. Calyx-teeth prominent, unequal. Disk-lobes thick, surrounding the base of the styles. Fruit narrow-oblong, 3 to 4 lines long; carpels with 5 equally prominent ribs, the furrows rather broad, with 1 vitta under each and 2 at the commissure.—Klatt in Linnæa, xxix. 715.

Victoria. Alpine and subalpine pastures, from the Cobberas to the Munyong mountains, *F. Mueller.*

2. **S. algens,** *F. Muell. in Hook. Kew Journ.* viii. 71, *and in Trans. Phil. Inst. Vict.* i. 104. Quite glabrous. Stems from a thick perennial rootstock, short, decumbent or ascending. Leaves chiefly radical, simply pinnate, with broad rhomboidal obovate or cuneate segments deeply and acutely toothed or incised, mostly under $\frac{1}{2}$ in. long; stem leaves few, the sheathing bases of the petioles long and broad. Umbels of 4 to 6 unequal rays. Involucres, both general and partial, of 2 or 3 narrow bracts. Calyx-teeth scarcely conspicuous. Disk-lobes confluent with the conical base of the styles. Fruit (not seen ripe) very shortly oblong, each carpel with 5 very prominent ribs.—Klatt in Linnæa, xxix. 716.

Victoria. Gravelly borders of alpine rivulets and springs, in the Munyong mountains, *F. Mueller.*

11. CRANTZIA, Nutt.

Calyx-teeth shortly prominent. Petals acute, concave, the margins not recurved, imbricate in the bud. Disk scarcely distinct from the conical base of the styles. Fruit broadly ovoid, very slightly laterally compressed. Carpels nearly terete, with 5 corky ribs, the lateral ones forming a thick mass at the rather broad commissure, with 1 vitta under each furrow and 2 at the commissure. Carpophore not separating from the carpels. Albumen of the seed terete.—Small creeping herb. Leaves linear-terete, undivided. Umbels simple, with minute involucral bracts.

The genus is confined to a single species, extending to New Zealand and extratropical and Andine America.

1. **C. lineata,** *Nutt. Gen. Pl. N. Amer.* i. 178. Stems or rhizomes slender creeping and rooting at the nodes. Leaves solitary or tufted at the nodes, slender, fistulose, marked with transverse nodes, from under 1 in. in some specimens to 2 or 3 in. long, or even more, rarely in American specimens broader and flattened at the upper end. Peduncles filiform, solitary at the nodes, each with an umbel of 8 to 12 or sometimes more minute flowers, on pedicels of 1 to 2 lines. Fruits very small.—DC. Prod. iv. 71; Hook. f. Fl. Tasm. i. 160, and Fl. Antarct. 287. t. 100; Wedd. Chlor. And. ii. t. 68.—*C. australica*, F. Muell. 2nd Gen. Rep. according to Klatt, Linnæa, xxix. 714.

Queensland. Brisbane river, *Mrs. Dietrich.*
N. S. Wales. Twofold Bay, *F. Mueller.*
Victoria. Mouth of Snowy River, Yarra river, *F. Mueller;* Barwan river, *Oldfield.*
Tasmania. Islands of Bass's Straits, *R. Brown;* marshes near Launceston, etc., *J. D. Hooker.*
S. Australia. St. Vincent's Gulf, *F. Mueller.*

12. ACIPHYLLA, Forst.

Calyx-teeth more or less prominent. Petals ovate or lanceolate, often inflected at the tip but not acuminate, the margins not recurved, imbricate in the bud. Disk in the perfect flowers continuous with the conical base of the styles, in the males of 2 thick lobes without styles. Fruit (in the normal species) oblong, somewhat dorsally compressed; carpels dorsally compressed, with 5 (rarely 4 or 3) acutely prominent ribs, the lateral ones bordering the broad commissure and expanded into narrow wings, or at least twice as broad as the other 3, of which 1 or 2 are sometimes wanting; vittæ 1 (or 2?) under each furrow, often obscure. Carpophore bipartite.—Glabrous perennials, the true species rigid and erect. Leaves pinnately decompound or reduced to a simple petiole. Umbels compound. Involucres of linear bracts. Flowers polygamo-diœcious, the male plants without intermixture of perfect ones, but some males often intermixed in the umbels of perfect flowers.

The genus extends to New Zealand and the Antarctic islands. F. Mueller proposes to include in it the whole of the species published by J. D. Hooker as *Anisotome*, and to adopt Forster's name of *Gingidium* for the collective genus, or to reduce it altogether to *Ligu-*

ticum. But Hooker has now shown (Handb. N. Zeal. Fl.) that Forster's *Gingidium* belongs rather to *Angelica*, and the true *Aciphyllæ* appear to me to form as natural and well characterized a genus as the majority of those now adopted in *Umbelliferæ*. Whether those *Anisotomes*, now reduced by Hooker to *Ligusticum*, ought really to be regarded as congeners of the northern species can only be determined by a careful comparison of the numerous allied genera in the northern hemisphere.

Rigid erect plants. Fruits oblong, the lateral ribs or wings twice as broad as the others.
  Leaves all reduced to long simple apparently articulate petioles . . 1. *A. simplicifolia.*
  Leaves dissected with linear segments. . . . . 2. *A. glacialis.*
Small densely tufted or procumbent plant. Fruits ovoid, the ribs all equal. Leaves dissected . . . . 3. *A. procumbens.*

1. **A. simplicifolia,** *F. Muell.* (as a *Gingidium*). Stems from a densely tufted stock erect, rigid, 1 to 2 ft. high, with a short linear leaf or bract under each branch. Leaves otherwise radical, narrow-linear, 6 in. to 1 ft. long, obtuse, terete or flattened towards the end, striate and marked with transverse raised lines, giving them an articulate appearance, undivided, with broad sheathing bases. Umbel on the main stem of 8 to 10 or more unequal rays, those on the lateral branches smaller. General involucre of 3 or 4 unequal linear bracts, shorter than the rays, the partial ones of more numerous bracts. Petals rather narrow, scarcely pointed. Fruit oblong, 4 to 5 lines long, the carpels dorsally flattened, the ribs all acutely prominent, but the lateral ones twice as broad as the others and almost winged. Vittæ often very obscure.—*Gingidium simplicifolium*, F. Muell. in Trans. Phil. Inst. Vict. i. 104, Pl. Vict. t. 27, and in Hook. Kew Journ. viii. 72.

Victoria. Moist grassy subalpine pastures from Mount Wellington to the Munyong mountains, *F. Mueller.*

2. **A. glacialis,** *F. Muell.* (as a *Gingidium*). Stems from a densely tufted stock erect, rigid, 1 to 2 ft. high. Leaves chiefly radical, once or twice pinnate, the rhachis thick, striate, marked at the divisions with transverse raised lines giving it an articulate appearance, broadly sheathing at the base; segments linear, entire. Umbels of several often very unequal rays. Bracts of the general involucre linear-lanceolate, shorter than the rays, of the partial involucres small and narrow. Petals ovate. Fruit oblong, 4 to 5 lines long, the carpels dorsally compressed, the ribs all acutely prominent, the lateral ones much more so than the others but less winged than in *A. simplicifolia.* Vittæ often obscure.—*Gingidium glaciale*, F. Muell. in Trans. Phil. Inst. Vict. i. 104, and in Hook. Kew Journ. viii. 71.

Victoria. Higher regions of the Australian Alps, at an elevation of 5 to 7000 ft., *F. Mueller.*

3. **A** (?) **procumbens,** *F. Muell.* (as a *Gingidium*). A small plant forming short dense tufts or emitting prostrate branches of 2 to 3 in., thickly covered with the membranous imbricated sheaths of old leaves. Leaves twice pinnate, with short crowded linear segments, acute and tipped with a hair-like point. Flowering-stems scarcely 1 in. high, with a single compound umbel of many rays. Involucral bracts few, lanceolate or linear, with scarious edges. Calyx-teeth half as long as the petals. Petals ovate or lanceolate. Fruit ovoid, about 1½ lines long, the ribs all very acutely prominent and almost winged, equal or irregularly unequal. Vittæ ob-

scure.—*Gingidium procumbens*, F. Muell. Fragm. i. 15; Hook. f. Fl. Tasm. ii. 363.

Tasmania. Summit of Mount Lapeyrouse, *Oldfield.*

This differs from the other species in its short equally-ribbed fruits, and comes very near to those which J. D. Hooker now reduces to *Ligusticum*, but differs from the northern species of that genus in the want of the numerous vittæ. The habit is that of some northern species of *Meum* or *Gaya.*

13. DAUCUS, Linn.

Calyx-teeth prominent. Petals with inflexed points, the margins not recurved, slightly imbricate in the bud. Disk small, confluent with the conical base of the styles. Fruit ovoid or oblong, scarcely compressed, bristly; carpels dorsally compressed, the primary ribs inconspicuous or not prominent, the 4 secondary ribs very prominent, expanded into rows of glochidiate bristles, with 1 vitta under each secondary rib and 2 at the broad commissure. Carpophore simple or bipartite.—Annuals or biennials, usually hirsute. Leaves decompound, with narrow segments. Umbels compound, the bracts of the general involucre usually dissected.

Besides the Australian species, which extends over New Zealand and Western America, the genus includes the Carrot and a few other species natives of the northern hemisphere.

1. **D. brachiatus,** *Sieb. in DC. Prod.* iv. 214. An erect or decumbent annual, sometimes small and slender, sometimes stout and attaining 1 to 2 ft., more or less sprinkled or hirsute with short stiff hairs. Leaves on slender petioles, twice pinnate, with short narrow incised or pinnate segments, usually minutely mucronulate. Umbels of about 3 to 5 very unequal rays, with 2 or 3 floral leaves or involucral bracts divided into 2 or 3 linear-subulate segments; one of the rays sometimes growing out into a continuation of the stem and bearing another compound umbel. Fruit ovoid, varying very much in size, usually scarcely 2 lines long, with short bristles, sometimes above 3 lines long, the bristles long and very fine, or stout and dilated at the base.—Bunge in Pl. Preiss. i. 295; Hook. f. Fl. Tasm. i. 161, with the synonyms adduced (except *D. pusillus*, Mich.); *Scandix glochidiata*, Labill. Pl. Nov. Holl. i. 75. t. 102.

Queensland, *Bowman;* Moreton Bay, *F. Mueller;* near Warwick, *Beckler;* in the interior, *Mitchell.*

N. S. Wales. Port Jackson to the Blue Mountains, *Sieber, n.* 115, *A. Cunningham,* and others; Macleay river, *Beckler;* New England, *C. Stuart;* Darling and Lachlan rivers, *A. Cunningham,* and southward to Twofold Bay, *F. Mueller.*

Victoria. Murray river, *F. Mueller;* Wimmera, *Dallachy;* Creswick, *Whan;* Portland, *Allitt.*

Tasmania. Common in the northern parts of the island, *J. D. Hooker.*

S. Australia. From the Murray river to St. Vincent's and Spencer's Gulfs, *F. Mueller* and others.

W. Australia. King George's Sound, *R. Brown,* and thence to Swan River, *Drummond,* 1*st Coll.; Preiss, n.* 2071, 2073; *Oldfield; Collie.*

J. D. Hooker includes among the synonyms of this species the *Daucus pusillus,* Mich., of the United States of N. America, and on that authority F. Mueller takes Michaux's name as the oldest for the species. *D. pusillus* appears however always to have the umbel, although small, regular with numerous rays, and more dissected involucral bracts, as in *D. Carota,* and must probably be retained as a distinct species connecting in some measure *D. brachiatus* with *D. Carota.*

D. Carota, Linn., the *wild carrot*, a tall erect plant, the umbels rather large with numerous crowded rays and the bracts of both involucres pinnatifid, is amongst the plants introduced from Europe and more or less established in waste places near settlements in Victoria and S. Australia.

14. OREOMYRRHIS, Endl.

(Caldasia, *Lag.*)

Calyx-teeth inconspicuous. Petals slightly concave, with short inflexed points, the margins not recurved, imbricate in the bud. Disk rather broad, continuous with the base of the styles. Fruit oblong or narrow, usually tapering towards the end, slightly compressed laterally; carpels nearly terete, with 5 obtusely prominent ribs, the lateral ones close to the rather broad commissure, with 1 vitta under each furrow and usually 2 at the commissure. Seed nearly terete, but longitudinally furrowed towards the commissure.—Perennial tufted herbs. Leaves pinnately dissected. Umbels simple, pedunculate. Involucral bracts ovate or lanceolate.

The genus consists apparently of a very few species, of which the Australian one has the widest range, extending over Antarctic and Andine America as well as New Zealand. The furrow of the albumen, although not deep, is distinct in all the seeds I have examined, and the habit appears to me to be very much that of several *Scandicineæ*. The genus is indeed very nearly allied to *Chærophyllum*.

1. **O. andicola,** *Endl.; Hook. f. Fl. Ant.* ii. 288. *t.* 101. A densely tufted perennial, sometimes with only radical leaves and simple scapes, sometimes producing erect or ascending sparingly branched stems of 1 ft. or more, always more or less pubescent or hirsute. Leaves once twice or three times pinnately divided, the segments short and sessile, the ultimate lobes short, from linear-subulate to oblong, mostly acute. Peduncles usually erect from the stock, from 2 or 3 in. to 1 ft. or even near 2 ft. long, bearing each a single simple umbel, or sometimes a tuft of leaves and peduncles forming an irregularly compound leafy umbel. Flowers numerous, at first nearly sessile within the involucre of about 6 to 10 ovate or lanceolate entire or toothed leafy bracts; fruiting pedicels longer than the bracts, often growing out to 2 lines or more. Fruit usually narrow-oblong, 2 to 3 lines long, tapering at the top, but sometimes almost ovoid, and scarcely 2 lines long, and in one Victorian specimen (perhaps in an abnormal state) the unripe fruits are cylindrical and nearly 5 lines long.—Wedd. Chlor. And. ii. 206; *Caldasia andicola*, Lag.; DC. Prod. iv. 229; *Myrrhis andicola*, H. B. and K. Nov. Gen. et Sp. v. 13. t. 419; *Caldasia eriopoda*, DC. Prod. iv. 229; *Oreomyrrhis eriopoda*, Hook. f. Fl. Tasm. i. 162; *O. argentea*, Hook. f. in Hook. Ic. t. 300, and Fl. Tasm. i. 162 (very hirsute with silky silvery hairs); *O. brachycarpa*, Hook. f. in. Hook. Ic. under n. 300. and Fl. Tasm. i. 162 (with short fruits); *O. sessiliflora* and *O. ciliata*, Hook. f. in Hook. Lond. Journ. vi. 471, and Fl. Tasm. i. 162, 163.

N. S. Wales. Clarence river, *Beckler*.

Victoria. Grassy places in the mountain districts, *F. Mueller*, and others.

Tasmania. Port Dalrymple, *R. Brown*. Common in grassy pastures both in the mountains and plains, *J. D. Hooker*.

The species is also in New Zealand and in Antarctic and Andine America. The several forms described by J. D. Hooker, which appeared distinct enough in numerous specimens originally transmitted from Tasmania, are by no means well marked out in more extensive collections; the most striking are the *O. argentea* and *O. brachycarpa*, but the dense silvery hairs of the former occur in a greater or less degree in several Victorian as well as Andine specimens, and the comparative length of the fruit appears to be exceedingly inconstant.

ORDER LVII. ARALIACEÆ.

Calyx-tube adnate to the ovary; limb forming a slightly raised line or short cup round the summit, truncate or toothed, or quite inconspicuous. Petals 5 or more, or rarely 4, usually valvate and shortly inflected at the tip, and often cohering, rarely with a long inflected point, or (in a few species not Australian) obtuse and imbricate, inserted round an epigynous entire disk. Stamens as many as petals or sometimes (in genera not Australian) more, inserted with them round the epigynous disk; anthers versatile, with parallel cells opening longitudinally. Ovary inferior, 2- or more-celled, or very rarely 1-celled by abortion, with 1 anatropous ovule in each cell, pendulous from the summit. Styles as many as cells, either distinct erect and afterwards recurved with small terminal stigmas, or united in a cone, or reduced to a slight protuberance with as many stigmas as cells radiating on the summit and often scarcely conspicuous. Fruit more or less drupaceous and indehiscent, the epicarp succulent, rarely nearly dry and thin, always distinct from the endocarp, which is hardened into as many one-seeded pyrenes as cells of the ovary, usually laterally compressed. Seed pendulous, testa very thin, albumen the shape of the pyrene, with an even surface, or rarely ruminate. Embryo minute, near the apex of the seed, the radical superior.—Trees, shrubs, or woody climbers, very rarely (in a few species not Australian) herbs. Leaves simple, digitate or pinnately compound, sometimes very large, the rhachis often articulate, the petiole dilated at the base or the dilatations united in an intrapetiolar stipule. Flowers small, often greenish or purple, in umbels heads or rarely racemes, which are usually disposed in large terminal racemes or panicles, the umbels rarely solitary or in compound umbels. Bracts usually small and often inconspicuous or none. Flowers frequently polygamous, the ovary entirely abortive in the males, the stamens often smaller or rarely wanting in the females.

With the exception of a very few species in the temperate regions of the northern and southern hemispheres, the Order is confined to the tropics in the New as well as in the Old World. Of the six Australian species, two are widely spread over tropical Asia and Africa, one of them extending also to New Zealand; one extends only to the Indian Archipelago and the islands of the South Pacific; one, *Hedera*, has as yet only one ascertained congener, spread over the temperate regions of the northern hemisphere in the Old World; the remaining two are endemic.

Generally speaking, *Araliaceæ* differ from *Umbelliferæ* by their tall shrubby or arborescent habit, large leaves, paniculate inflorescence, valvate petals, entire disk and drupaceous fruits, but every one of these characters breaks down in some exceptional case, and some have proposed to unite the two Orders. But such connecting links occur in the case of even the most natural Orders, and it appears to me that if *Astrotriche* and *Horsfieldia* are transferred from *Umbelliferæ*, where they have been hitherto placed, into *Araliaceæ*, there is really very little difficulty in drawing the line of demarcation between the two.

Styles 2 (or exceptionally 3) distinct.
  Petals with the tips slightly or not at all inflected. Umbels heads or racemes paniculate, rarely solitary.
    Leaves all simple. Plant densely stellate-tomentose . . . . 1. ASTROTRICHE.
    Leaves mostly or all compound. Plant glabrous or nearly so . 2. PANAX.

Petals narrow, with long induplicate points. Umbels twice or thrice compound. Leaves digitate . . . . 3. MACKINLAYA.
Styles united in a cone or short style, or reduced to a short protuberance in the disk. Ovary several-celled.
Ovary-cells and pyrenes 5 to 7. Flowers pedicellate, without bracteoles.
Albumen ruminate . . . . 4. HEDERA.
Albumen even . . . . 5. HEPTAPLEURUM.
Ovary-cells and pyrenes 7 to 16. Flowers closely sessile within 4 short broad bracts, forming small heads arranged in long racemes . . . . 6. BRASSAIA.

1. ASTROTICHE, DC.

Calyx-teeth minutely prominent. Petals 5, valvate, usually pubescent outside. Stamens 5. Disk broad and not thick, the margin often prominent. Ovary 2-celled. Styles 2, distinct, at first erect, afterwards recurved. Fruit flattened or thick, the endocarp hardened into 2 pyrenes, furrowed on each side of the commissural edge or curved into spurious empty cells, the endocarp quite distinct as in other *Araliaceæ*, but not so succulent. Albumen even.—Shrubs more or less clothed with a stellate tomentum. Leaves petiolate, undivided, entire. Umbels pedunculate, in large terminal panicles. Flowers articulate on the pedicel.

The genus is limited to Australia. It is usually placed in *Umbelliferæ*, but the structure of the flowers and fruit, as well as the habit, are much nearer those of *Panax*, from which the genus differs slightly in the epicarp rather drier, in the foliage and the stellate tomentum.

Fruit thick, with narrow wings on each side. Endocarp curved into spurious cells on each side of the inner angle of the pyrene. Leaves cordate-lanceolate . . . . 1. *A. pterocarpa.*
Fruit flat. Endocarp grooved only on each side of the inner angle of pyrene.
Leaves from ovate-lanceolate to narrow-lanceolate . . . . 2. *A. floccosa.*
Leaves linear-lanceolate, acute, mostly 3 to 6 in. long. Calyx-teeth scarcely conspicuous . . . . 3. *A. longifolia.*
Leaves from oblong-linear to narrow linear, obtuse, 1 to 3 in. long. Calyx-teeth prominent . . . . 4. *A. ledifolia.*

1. **A. pterocarpa,** *Benth.* A slender shrub (*W. Hill*). Leaves on long petioles, cordate-lanceolate, 6 to 10 in. long, densely floccose-tomentose on both sides as well as the leafy branches. Panicle very large, with narrow leafy bracts under the principal branches. Umbels dense, many-flowered, on short peduncles. Pedicels rarely longer than the flowers. Disk scarcely prominent. Fruits without the wings ovoid-oblong, thick, about 3 lines long, slightly furrowed; the endocarp forming 3 collateral cells in each carpel, the 2 lateral ones empty, the central one enclosing the seed; the epicarp expanded at the commissure into a rather broad wing, often not apparent till the fruit is quite ripe.

Queensland. Fitzroy Island, *W. Hill.*

2. **A. floccosa,** *DC. Mem. Ombell.* 30. *t.* 5; *Prod.* iv. 74. A shrub, attaining from 10 to 30 ft., the young branches inflorescence and under side of the leaves clothed with a dense floccose tomentum. Leaves from ovate-lanceolate to lanceolate, tapering into a narrow point, rounded at the base or

slightly cordate, the larger ones sometimes almost peltate, 4 to 8 in. long, glabrous on the upper side, the floral ones small, linear-lanceolate or the upper ones reduced to small bracts. Umbels numerous, many-flowered, in a large terminal panicle. Petals woolly-tomentose outside. Disk with a slightly raised margin. Fruit nearly 2 lines broad, flat, not winged, the endocarp of each carpel sometimes grooved or folded towards the commissure, but not curved into spurious cells.—*Bolax floccipes*, Sieb. Pl. Exs.

Queensland. Moreton Island, *F. Mueller.*

N. S. Wales. Port Jackson to the Blue Mountains, *R. Brown, Sieber, n.* 258, and others; northward to Hastings river, *Beckler*; and New England, *C. Stuart;* southward to Illawarra, *A. Cunningham*, and others.

Var. *subpeltata.* Leaves more coriaceous, shining above, slightly cordate or shortly peltate at the base.—Blue Mountains, Illawarra, etc.

Var. *angustifolia.* Leaves lanceolate, about ½ in broad.—Blue Mountains and to the northward.

Var. *incana.* Tomentum closer and whiter. Panicle more slender and leafless. Flowers fewer and less tomentose.—*A. latifolia*, Benth. in Hueg. Enum. 55.—Port Jackson and Blue Mountains.

3. **A. longifolia,** *Benth. in Hueg. Enum.* 55. Nearly allied to the narrow-leaved varieties of *A. floccosa*, but the leaves much narrower, the panicle more slender, with fewer flowers to the umbel and the calyx-teeth more prominent. Leaves linear-lanceolate, acuminate, 3 to 5 in. long and rarely above $\frac{1}{4}$ in. broad, glabrous above, with a close white or looser and floccose tomentum underneath, or rarely almost glabrous. Fruits rather larger than in *A. floccosa.*

Queensland. Brisbane river, Moreton Bay, *A. Cunningham, F. Mueller*, and others.

N. S. Wales. Port Jackson to the Blue Mountains, *R. Brown, Lownes;* Port Stevens, *M'Arthur.*

Some specimens are very difficult to distinguish from the narrow-leaved ones of *A. floccosa;* others come near to the longer-leaved ones of *A. ledifolia.*

4. **A. ledifolia,** *DC. Mem. Ombell.* 30. *t.* 6, *and Prod.* iv. 74. A more slender shrub than *A. floccosa*, with virgate branches covered as well as the inflorescence and under side of the leaves with a close or floccose tomentum. Leaves oblong-linear or narrow-linear, obtuse or rarely almost acute, 1 to 3 in. long, glabrous above, the margins recurved or revolute. Panicle smaller and narrower than in *A. floccosa.* Flowers and fruit the same as in that species.—*Bolax ledifolius*, Sieb. Pl. Exs.; *A. hoveoides*, A. Cunn.; Benth. in Hueg. Enum. 55 (with short leaves); *A. linearis*, A. Cunn.; Benth. l. c. (with long narrow leaves); *A. asperifolia*, F. Muell.; Klatt in Linnæa, xxix. 709 (with long leaves).

N. S. Wales. Port Jackson to the Blue Mountains, *Sieber, n.* 257, and others; southward to Illawarra, *A. Cunningham*, and others.

Victoria. Buffalo range, Latrobe river, Grampians, etc., *F. Mueller* (usually the short-leaved form).

2. PANAX, Linn.

(Nothopanax, *Miq.*)

Calyx-border usually slightly prominent, truncate or shortly 5-toothed. Petals 5, valvate, often cohering at the tips, especially in female flowers. Stamens 5. Disk broad and not thick, the margin sometimes prominent.

Ovary 2- or rarely 3-celled. Styles 2, rarely 3, at first erect and sometimes cohering, afterwards distinct and recurved. Fruit flattened, the endocarp hardened into 2 distinct pyrenes not furrowed, sometimes 2-ribbed on the dorsal edge, the exocarp more or less succulent. Albumen even.—Trees or shrubs. Leaves pinnately or digitately compound or rarely a few on the same tree or bush undivided. Flowers often polygamous, articulate on the pedicels, in umbels or rarely in heads or racemes, the umbels or racemes paniculate or rarely solitary.

The genus, if limited according to the views of Planchon and Decaisne, is widely distributed over the tropical regions of the Old World and extends to New Zealand, but is not American, and comprises Linnæus's *P. fruticosa* and others. The northern herbaceous species of Linnæus, with imbricate petals, are united by the same authors with *Aralia*, a course sanctioned by A. Gray and others. Miquel, however, reserves the name of *Panax* for these herbaceous species, and proposes the name of *Nothopanax* for Planchon and Decaisne's *Panax*. As the views of the latter authors will probably meet with more general adoption, they are here followed. The seven Australian species, as far as hitherto known, are all endemic, two of them anomalous in their inflorescence.

Leaves digitate or rarely undivided. Umbels solitary . . . . . 1. *P. Gunnii.*
Leaves pinnate or bipinnate. Flowers umbellate, umbels paniculate or racemose.
  Leaflets glabrous, long, obliquely lanceolate. Calyx-teeth scarcely prominent . . . . . 2. *P. Murrayi.*
  Leaflets softly pubescent underneath, large, ovate or oblong, acuminate. Calyx-teeth scarcely prominent . . . . . 3. *P. mollis.*
  Leaflets glabrous, large, ovate-lanceolate or oblong. Umbels few-flowered, numerous. Calyx-limb cup-shaped, truncate . . . 4. *P. Macgillivræi.*
  Leaflets glabrous, mostly under 3 in., ovate-lanceolate or linear, entire toothed or dissected. Calyx-limb very short, sinuate-toothed . . . . . 5. *P. sambucifolius.*
Leaves 3-foliolate. Flowers sessile, capitate; heads paniculate or racemose . . . . . 6. *P. cephalobotrys.*
Leaves pinnate or bipinnate. Flowers pedicellate, racemose; racemes paniculate . . . . . 7. *P. elegans.*

1. **P. Gunnii,** *Hook. f. in Hook. Lond. Journ.* vi. 466, *and Fl. Tasm.* i. 163. *t.* 37. A sparingly-branched shrub of 2 or 3 ft. with slender branches, more or less strigose with appressed hairs, as well as the petioles and sometimes the ribs of the leaflets. Leaves mostly digitate, with 5 or rarely 6 or 7 lanceolate segments of 1 to 2 in., coarsely serrate or pinnatifid, but on some branches the leaves reduced to 3 lanceolate segments or quite simple. Umbels terminal, solitary, on short peduncles. Flowers numerous, on pedicels of 1 or 2 lines. Calyx-teeth prominent. Disk slightly convex. Styles short. Fruit not seen ripe.

Tasmania. Towards Port Macquarrie, Franklin and Gordon rivers, *Gunn, Milligan*; near Mount Lapeyrouse, *Oldfield, C. Stuart.*

2. **P. Murrayi,** *F. Muell. Fragm.* ii. 106. A splendid tree, the trunk simple to the height of 50 or 60 ft., and then almost trichotomously branched (*F. Mueller, Dallachy*). Leaves simply pinnate, often several ft. long; leaflets obliquely lanceolate, entire or slightly denticulate, herbaceous but not thin, 3 to 6 in. long, or when luxuriant 8 to 10 in., quite glabrous. Umbels many-flowered, pedunculate, in racemes or divaricately-branched panicles. Calyx-teeth inconspicuous. Petals and stamens not seen. Fruit about 2

lines broad; the endocarp not very hard.—*Nothopanax Murrayi*, Seem. Fl. Vit. 114.

Queensland. Rockingham Bay, *Dallachy*.
N. S. Wales. Hastings river, *Beckler;* Twofold Bay, *F. Mueller*.

3. **P. mollis,** *Benth.* A tall shrub. Leaves simply (or doubly?) pinnate; leaflets ovate ovate-lanceolate or oblong, acuminate, 6 to 10 in. long, glabrous above, softly pubescent or villous underneath. Umbels many-flowered, numerous, in large divaricately-branched panicles, the rhachis minutely tomentose. Calyx-teeth slightly and irregularly prominent. Styles long and slender. Fruit about 2 lines broad, but not seen quite ripe.

Queensland. Rockingham Bay, *Dallachy*.

4. **P. Macgillivræi,** *Benth.* A small tree of about 20 ft., quite glabrous. Leaves simply (or doubly?) pinnate, the rhachis articulate; leaflets shortly petiolulate, oval-oblong or ovate-lanceolate, shortly acuminate, often oblique at the base, 6 to 10 in. long, thin and membranous in our specimens. Umbels few-flowered, in a large loose compound panicle with slender branches and pedicels. Calyx-limb prominently cup-shaped, truncate or slightly sinuate-toothed. Petals rather long and narrow. Fruits about 3 lines broad, very flat, the carpels often readily separating, each with a thin exocarp, and a flat smooth hard endocarp.—*Nothopanax Macgillivrayi*, Seem. Fl. Vit. 114.

Queensland. Cape York, *M'Gillivray;* Albany Island, *W. Hill*.

5. **P. sambucifolius,** *Sieb. in DC. Prod.* iii. 255. A tall shrub or tree, quite glabrous. Leaves simply or doubly pinnate; leaflets exceedingly variable, most commonly distant, petiolulate or sessile, ovate elliptical or lanceolate, 1½ to 3 in. long, acute, entire, denticulate or lobed, the lowest of the simply pinnate leaf, or the lowest of each pinna often smaller, broader, and close to the base, but sometimes the leaflets divided, or narrow-linear and pinnatifid with divaricate distant lobes, the rhachis sometimes dilated and as broad as the lobes. Umbels many-flowered in a terminal branched corymbose panicle or in a simple raceme. Calyx-limb shortly prominent, dilated, very shortly sinuate-toothed. Petals in the perfect flowers often cohering at the tips, smaller and more spreading in the males. Fruit 2 to 3 lines broad, with a white or lead-coloured succulent exocarp, the endocarps or pyrenes flat with 2 obtuse dorsal ribs.—*P. angustifolius* and *P. dendroides*, F. Muell. in Trans. Phil. Inst. Vict. i. 42, and Pl. Vict. t. 28; *Nothopanax sambucifolius*, Seem. Fl. Vit. 115.

N. S. Wales. Port Jackson to the Blue Mountains, *R. Brown; Sieber, n.* 256, and others; northward to Hastings, Macleay, and Clarence rivers, *Beckler;* and southward to Illawarra, *A. Cunningham;* Berrima, *Woolls;* and Twofold Bay, *F. Mueller*.

Victoria. Mountains from Dandenong and Mount Macedon to the Buffalo range and a great part of Gipps' Land, *F. Mueller*.

Tasmania. Douglas river, E. coast, *Milligan*, according to F. Mueller, but the specimens are in leaf only, and appear to me to be somewhat doubtful.

The specimens with narrow much-dissected leaves are so very unlike the others or even any *Panax*, that A. Cunningham had distributed them under the name of *Trachymene pinnata*. The manner, however, in which the various forms of leaflets are combined, even on the same specimens, show that all belong to one species.

6. **P. cephalobotrys,** *F. Muell. Fragm.* ii. 83. Shrubby, somewhat

climbing, glabrous except the inflorescence, or the young branches and petioles sprinkled with a few appressed hairs. Leaves on long slender petioles, with 3 petiolulate leaflets, oblong or lanceolate, acuminate, 3 to 4 in. long. Flowers sessile, in small pedunculate heads, forming a simple raceme or a slender slightly-branched panicle scarcely exceeding the leaves. Petals and stamens not seen. Young fruit broadly ovate, compressed, crowned by the short cup-shaped, obtusely 5-lobed calyx-limb. Disk with the margin slightly prominent. Styles rather long.

N. S. Wales. Clarence and Richmond rivers, *Beckler*.

This and the following species differ from the rest of the genus in inflorescence, but the flowers and fruits appear to be otherwise entirely those of *Panax*.

7. **P. elegans,** *F. Muell. in Trans. Phil. Inst. Vict.* ii. 68. A large and handsome tree, glabrous except the inflorescence. Leaves large, simply or doubly pinnate, the rhachis articulate. Leaflets petiolate, opposite, ovate, acuminate, entire, coriaceous, shining, often 3 to 4 in. long. Flowers singly pedicellate in little racemes, which are very numerous and arranged in a large terminal divaricately-branched panicle, the rhachis minutely hoary-pubescent. Calyx-border shortly prominent, entire. Petals and styles of the genus. Disk not prominent. Fruits about 3 lines broad, the endocarp or pyrenes hard.—*Nothopanax elegans*, Seem. Fl. Vit. 114.

Queensland, *Burdekin Expedition;* Rockhampton and Edgecombe Bay, *Dallachy;* Brisbane river, Moreton Bay, *A. Cunningham, F. Mueller, C. Moore.*

N. S. Wales. Clarence river, *Beckler;* Richmond river, *C. Moore;* Illawarra, *Ralston* (according to F. Mueller, the specimen in leaf only).

3. MACKINLAYA, F. Muell.

Calyx with 5 prominent lobes. Petals unguiculate, with long induplicate points, valvate in the bud. Stamens 5. Disk broad, the margin undulate. Ovary 2-celled. Styles 2, at first erect, afterwards recurved. Fruit very flat, the endocarp cartilaginous, forming 2 separate pyrenes, the exocarp succulent.—Shrub or tree. Leaves digitately compound. Flowers polygamous, articulate on the pedicels, in a large compound terminal umbel, with general and partial involucres of narrow bracts.

The genus is limited to a single species, endemic in Australia, differing from *Panax* chiefly in inflorescence and in the petals resembling those of many *Umbelliferæ*.

1. **M. macrosciadia,** *F. Muell. Fragm.* iv. 120. A slender shrub or small tree, quite glabrous. Leaves with the common petiole sheathing at the base, but without distinct stipules; leaflets 3 to 7, usually 5, petiolulate, ovate or oblong, shortly acuminate, entire or with a few coarse distant teeth, 4 to 8 in. long, membranous at least at the time of flowering. Umbels 3 times or even 4 times compound, with numerous rays, the primary ones often 4 or 5 in. long, the secondary and tertiary umbels compact. Involucres both general and partial of several linear or linear-lanceolate bracts, much shorter than the rays. Calyx-lobes acute or acuminate. Fruits when perfect about 7 lines broad and 5 lines long, but one carpel often deformed and semiabortive.—*Panax macrosciadia*, F. Muell. Fragm. ii. 108, 176.

Queensland. E. coast, *R. Brown, A. Cunningham;* Dunk Island, *M'Gillivray;*

Fitzroy Island, *M'Gillivray*, *W. Hill*; Port Molle and Cumberland Islands, *Fitzalan*; Rockingham Bay, *Dallachy*.

4. HEDERA, Linn.

(Irvingia, *F. Muell.*; Kissodendron, *Seem.*)

Calyx-border slightly prominent, entire or sinuate-toothed. Petals 5, valvate. Stamens 5. Disk convex, sometimes very prominent. Ovary 5-celled. Styles united into an obtuse cone or very short cylindrical style, with 5 scarcely prominent stigmas. Fruit nearly globular, with 5 1-seeded pyrenes. Seed with a furrowed or ruminated albumen.—Woody climbers or trees. Leaves entire, lobed or pinnately compound. Flowers umbellate, not articulate on the pedicel, the umbels pedunculate in terminal panicles.

The genus, characterized essentially by the ruminated albumen, contains besides the Australian species, which is endemic, one widely dispersed over the northern hemisphere in the Old World, and probably some other Asiatic ones as yet insufficiently investigated.

1. **H. australiana,** *F. Muell. Fragm.* iv. 120. A small tree, quite glabrous. Leaves large, pinnate, the rhachis articulate; leaflets few, ovate, oval-oblong or ovate-lanceolate, shortly acuminate, often above 6 in. long, smooth and shining, but prominently veined almost as in *Heptapleurum venulosum*. Umbels pedunculate, with the peduncles almost verticillate along the elongated branches of a large loose terminal panicle. Calyx-border slightly sinuate-toothed. Disk broadly conical, though not quite so thick as in *H. helix*. Style very shortly cylindrical or reduced to a small boss on the centre of the disk. Drupe above 2 lines diameter, with 5 hard pyrenes, enclosing a seed with a deeply ruminate surface.—*Irvingia australiana*, F. Muell. Fragm. v. 19; *Kissodendron australianum*, Seem. Journ. Bot. iii. 201.

Queensland. Herbert river, *F. Mueller*; Rockingham Bay, *Dallachy*.

The semi-superior appearance of the ovary of *Hedera helix* is due to the thickness of the epigynous disk, and the only character remaining to separate *H. australiana* generically from it is the compound foliage, which can scarcely be admitted in an Order where it is so peculiarly variable.

5. HEPTAPLEURUM, Gærtn.

(Paratropia, *Blume.*)

Calyx-teeth minute or inconspicuous. Petals 5 or 6, or rarely more, valvate. Stamens as many as petals. Disk flat or convex. Ovary with 5 or 6, rarely more cells. Styles united in a short cone, with as many sessile scarcely prominent stigmas as cells. Fruit nearly globular, the endocarp not very hard, forming 5 or 6, rarely more, 1-seeded pyrenes.—Trees or tall shrubs. Leaves digitately compound. Flowers mostly unisexual, not articulate on the pedicel, umbellate, the umbels arranged in terminal panicles or racemes.

A considerable genus dispersed over tropical and eastern temperate Asia, the only Australian species being one which has the widest range in East India.

1. **H. venulosum,** *Seem. Journ. Bot.* iii. 80. A tall shrub or tree, quite glabrous. Leaflets 5 to 7, on long petiolules, mostly elliptical or oval-oblong, acuminate, 4 to 8 in. long, but in some Indian specimens short and obtuse, coriaceous, somewhat shining, the pinnate veins and reticulate veinlets very prominent. Stipules adnate to the petiole at the base only, united

within it into a single obtuse lamina. Umbels in a divaricately-branched panicle shorter than the leaves. Male flowers with exserted stamens, and scarcely any rudiment of the ovary. Females often with more or less perfect stamens. Parts of the flowers 5 or 6. Fruit about 2 lines diameter.—*Paratropia venulosa*, W. and Arn. Prod. 377; Wight, Illustr. t. 118; F. Muell. Fragm. iv. 121; *Aralia Moorei*, F. Muell. Fragm. ii. 108.

Queensland. Wide Bay, *C. Moore.*—The species is widely dispersed over East India.

6. BRASSAIA, Endl.

Calyx-tube broad, adnate to the ovary, without any prominent border. Petals 7 to 18, usually about 12, valvate, usually cohering at the apex. Stamens as many as petals. Disk not thick, broad, with as many radiating furrows as cells, and confluent with the slightly raised styles or base of the radiating stigmas. Ovary with as many cells as parts of the flower and stigmas. Fruit with as many 1-seeded laterally compressed pyrenes as cells of the ovary.—Tree. Leaves digitately compound. Flowers sessile in little dense heads, shortly pedunculate in long racemes, each flower embedded in a cup-shape involucre of 4 small imbricate bracts.

The genus is limited to a single species, endemic in Australia.

1. **B. actinophylla,** *Endl. Nov. Stirp. Dec.* 89. A handsome tree, attaining 40 ft., quite glabrous. Leaflets 7 to 16, petiolulate, oblong or obovate-oblong, very shortly acuminate, coriaceous, entire, 6 in. to 1 ft. long. Stipules united in a single interpetiolar stipule, adnate to the petiole at the base. Flower-heads scarcely above ½ in. diameter, on peduncles, sometimes very short, rarely ½ to 1 in. long, rather numerous along the stout rhachis of the racemes, which attain sometimes several feet, and are often several together at the end of the branch, each one subtended by long acuminate leafless stipules.—F. Muell. Fragm. ii. 108, iv. 121; Seem. Journ. Bot. ii. 213.

Queensland. Endeavour river, *Banks and Solander*, *A. Cunningham*; Cape York, *W. Hill*; Palm Island, *Henne*; Port Molle, *Fitzalan*; Rockingham Bay, *Dallachy*; Boyd river, *C. Moore.*

ORDER LVIII. CORNACEÆ.

Calyx-tube adnate to the ovary; limb forming a raised border, entire or with as many teeth as petals. Petals 4, 5, or rarely more, valvate in the bud, inserted round an epigynous disk or on the calyx-border, rarely wanting. Stamens as many or rarely twice as many as the petals, and inserted with them; anthers with parallel cells opening longitudinally. Ovary inferior, 1- or 2-celled, with one anatropous pendulous ovule in each cell; style simple, with a terminal entire or rarely lobed stigma. Fruit an indehiscent drupe, with a 1- or 2-celled nucleus. Seeds solitary, pendulous, with a fleshy albumen and thin testa; embryo straight, nearly as long as the albumen, the radicle superior and shorter than the flat cotyledons.—Trees shrubs or very rarely herbs. Leaves opposite or rarely alternate, entire or slightly lobed, without stipules. Flowers usually small, in axillary or terminal heads cymes or corymbose panicles.

A small Order, generally scattered over the globe, but most abundant in the temperate regions of the northern hemisphere. It is represented in Australia by a single genus common to tropical Asia and Africa, and belonging to the small section of *Alangieæ*, differing in alternate leaves and in some other respects from the majority of the Order.

1. MARLEA, Roxb.

(Rhytidandra, *A. Gray*; Pseudalangium, *F. Muell.*)

Calyx-limb minutely toothed. Petals narrow-linear. Stamens the same number as petals, the filaments adhering to the petals at the base and connecting them in an apparently tubular corolla; anthers adnate, long and linear. Ovary 1- or 2-celled; style filiform, with a 2- or 4-lobed or capitate stigma. Drupe often reduced to 1 cell and seed.—Trees or shrubs. Leaves alternate. Flowers in axillary cymes.

Besides the Australian species, which is also in the islands of the South Pacific, there are three others in tropical Asia and Africa.

1. **M. vitiensis,** *Benth.* A tree, attaining a considerable height, glabrous or the young branches pubescent or villous. Leaves ovate ovate-lanceolate or oblong, shortly acuminate, more or less oblique and unequal at the base or rarely equal, 3 to 5 in. long, glabrous or slightly pubescent underneath in the normal form. Flowers in short axillary cymes on slender peduncles, rarely much exceeding the petioles. Calyx-limb cup-shaped, about $\frac{2}{3}$ line diameter. Petals 4 to 6, varying in length from 4 to 6 lines, connected by the stamens up to from $\frac{1}{4}$ to $\frac{1}{2}$ their length, revolute at the ends. Filaments villous; anthers about the length of the corolla, the valves involute, dividing each cell into 2 before they open and marked with transverse constrictions, which give them the appearance of being chambered. Disk cup-shaped, enclosing the base of the style. Style divided at the end into 2 linear stigmatic lobes. Ovary 1-celled with 1 ovule. Drupe ovoid, about $\frac{1}{2}$ in. long.—*Rhytidandra vitiensis*, A. Gray, Bot. Amer. Expl. Exped. i. 303. t. 28, and in Proc. Amer. Acad. vi. 55; *Pseudalangium polyosmoides*, F. Muell. Fragm. ii. 84; *Rhytidandra polyosmoides*, F. Muell. Fragm. ii. 176.

Queensland. Rockingham Bay, *Dallachy*.
N. S. Wales. Clarence river, *Beckler*, *C. Moore*; Richmond river, *Beckler*.

Var. *tomentosa*. Softly villous all over, or the upper side of the leaves alone glabrous. Flowers villous, the petals more deeply free than usual.—Fitzroy river, *Thozet*; Rockhampton, *Dallachy*; Moreton Bay, *W. Hill*.

The apparently chambered anthers are not really so, and traces of the constrictions may often be seen in *M. begonifolia*, the latter differs also in the large thick disk, the 2-celled ovary, and shortly 4-lobed style; but *M. barbata* has the thick disk with a 1-celled ovary and 2-lobed style, and an unpublished Malayan species has a small disk, with a 1-celled ovary, and almost entire style, all these distinctions proving thus to be specific, not generic.

ORDER LIX. LORANTHACEÆ.

Calyx-tube adnate to the ovary, the limb with as many lobes or teeth as petals, or forming an entire border, or none. Petals (or segments of the perianth when the calyx is inconspicuous) 4 to 6 or rarely more, free or united in a lobed corolla, inserted round an epigynous disk, valvate in the bud, rarely wanting. Stamens as many as petals, opposite to and usually inserted on them. Ovary inferior, 1-celled, with 1 erect ovule, usually not perceptible till the flowering is past, and adnate to the wall of the cell so as to have been described as pendulous; style or stigma simple. Fruit an indehiscent berry or drupe, with a single seed. Albumen fleshy. Embryo straight, with a superior radicle.—Shrubs, usually much branched, parasitical on the branches of trees and shrubs, sometimes so near their roots as to appear terrestrial, very rarely (only in two Australian species) really terrestrial shrubs or trees. Leaves opposite or rarely alternate, usually thick and leathery or sometimes fleshy, sometimes reduced to minute scales or none. Bracts usually one under each flower or pedicel, with 2 bracteoles close under the flower, concave or united in a little cup, having the appearance of an external calyx, or the bracteoles or very rarely the bracts also wanting.

A considerable Order, chiefly abundant within or near the tropics both in the New and the Old World, with a very few species from more temperate regions in the northern as well as in the southern hemisphere. Of the five Australian genera, two have a very wide range over nearly the whole area of the Order, the other three are endemic. In the arrangement and delimitation of the groups, I have been guided by Professor Oliver's careful study of the Order in Journ. Linn. Soc. vii.

Calyx-border distinct. Anthers oblong or linear, with parallel cells opening longitudinally. Flowers (in Australian species) elongated, hermaphrodite.
- Terrestrial tree. Fruit broadly 3-winged . . . . 1. NUYTSIA.
- Terrestrial shrub. Fruit a drupe, the endocarp with 8 internal ridges corresponding with furrows in the seed . . . . 2. ATKINSONIA.
- Parasitical shrubs. Fruit succulent. Seed not furrowed . . . . 3. LORANTHUS.

Perianth apparently simple. Anthers broad, opening transversely or in pores. Flowers very small, unisexual.
- Anthers sessile on the perianth-segments, opening inwards in several pores. Plants glabrous, leafless or leafy . . . . 4. VISCUM.
- Anthers at the base of the perianth-segments, opening transversely. Plants leafy, hoary or tomentose . . . . 5. NOTOTHIXOS.

1. NUYTSIA, R. Br.

Flowers of *Loranthus*. Petals free. Anthers versatile. Fruit a dry drupe or nut, with 3 broad longitudinal wings.—Terrestrial tree. Leaves alternate. Flowers racemose.

The genus is limited to a single species, endemic in Australia.

1. **N. floribunda,** *R. Br. in Journ. Geogr. Soc.* i. 17; *Bot. Works*, i. 308. A tree of 30 to 35 ft., quite glabrous, with spreading branches. Leaves linear, acute or obtuse, mostly 1½ to 3 in. long, entire, thick, the lower ones of the new shoots reduced to small scales. Flowers orange-yellow, in showy racemes crowded at the ends of the branches. Pedicels ¼ to ½ in. long, with 3 lanceolate obtuse bracts (1 bract and 2 bracteoles) close under the flower, small at the time of flowering, often enlarging afterwards. Calyx-limb unequally 6-toothed. Petals 6, linear, nearly ½ in. long. Fruit about ½ in. long, the wings short and thick. Embryo with 3 or 4 unequal cotyledons.—Lindl. Swan Riv. App. t. 4; Fenzl in Hueg. Enum. 57; Miq. in Pl. Preiss. i. 279; Oliv. in Journ. Linn. Soc. vii. 96; *Loranthus floribundus*, Labill. Pl. Nov. Holl. i. 87. t. 113.

W. Australia. King George's Sound, *R. Brown* and others, and thence to Swan River, *Drummond, 1st Coll., Preiss, n.* 1608, and others, and Murchison river, *Oldfield.*

2. ATKINSONIA, F. Muell.

Flowers of *Loranthus*. Petals free. Anthers versatile. Fruit a drupe, not winged, the endocarp hard, with 8 longitudinal internal ribs protruding into as many deep furrows of the seed.—Terrestrial shrub. Leaves alternate. Flowers in axillary racemes.

The genus is limited to a single species, endemic in Australia.

1. **A. ligustrina,** *F. Muell. Fragm.* v. 34. An erect bushy shrub, attaining 2 or 3 ft. in barren rocky situations, twice that height in other places. Leaves oblong-lanceolate, obtuse, narrowed into a short petiole, mostly 1 to 1½ in. long, not fleshy, the margins often recurved. Flowers on very short pedicels, in axillary racemes much shorter than the leaves. Bracteoles 2, close under the flower, the third or subtending bract often a little lower on the pedicel. Calyx-limb obscurely toothed. Petals usually 6, occasionally 7 or 8, linear, about 3 lines long. Drupe small, ovoid-oblong, the exocarp thin. —*Nuytsia ligustrina,* A. Cunn.; Lindl. Swan Riv. App. 39; F. Muell. Fragm. ii. 130.

N. S. Wales. Blue Mountains, *A. and R. Cunningham* and others.

3. LORANTHUS, Linn.

Calyx-limb short, truncate or toothed. Petals 4 to 8, free or more or less united in a tubular corolla, spreading at the ends. Stamens inserted on the base of the petals; filaments distinct; anthers adnate or versatile, with parallel cells opening longitudinally. Style filiform, with a terminal stigma. Fruit a berry, usually crowned by the limb of the calyx.—Parasitical shrubs. Leaves opposite or rarely alternate. Flowers hermaphrodite, axillary or terminal, in racemes or cymes or solitary, long and brightly coloured, or green at least at the tips, or, in species not Australian, small and green. Bracts in all the Australian species solitary, small and concave, close under each flower, without the 2 bracteoles which are in many extra-Australian species.

A very large genus, almost wholly tropical or subtropical, in America, Africa, and Asia, with one species as far north as the south of Europe. Of the 15 Australian species, one is common in Asia, another extends into Timor and perhaps over several of the islands of the Indian Archipelago; the remaining 13, as far as hitherto ascertained, are endemic.

Several of the Australian species, besides the Asiatic *L. longiflorus*, appear to have two forms of leaves, so different in aspect that it is difficult to fancy that the two belong to one species; the one sessile, broad, and deeply cordate, the other petiolate, narrow, and contracted at the base. Many also, probably, vary in the colours of the flower more or less red or yellow-orange, with or without green tips or the green extending to below the middle. The notes of the collectors on the trees on which the several species grow, are so varied that there seems to be no evidence that particular species affect particular trees. The most commonly noted are *Eucalyptus, Casuarina,* and *Exocarpus*, but *Acacia, Banksia, Melaleuca, Fusanus,* and many others are also mentioned as feeding species of *Loranthus.*

Anthers versatile, oblong. Petals free. Leaves opposite. Inflorescence mostly terminal.
  Leaves petiolate, thick, from short and obovate to long lanceolate and falcate. Cymes several-flowered . . . . 1. *L. celastroides.*
  Leaves small, sessile or nearly so. Peduncles slender, 2-flowered.
    Leaves linear, narrowed at the base. . . . . 2. *L. Bidwillii.*
    Leaves ovate, rounded at the base . . . . 3. *L. myrtifolius.*

Anthers adnate, linear. Petals united to the middle or higher up.
Leaves alternate or opposite. Inflorescence axillary.
Flowers several, in racemes . . . . 4. *L. longiflorus.*
Flowers several, in cymes.
Flowers and inflorescence glabrous. Calyx-limb truncate, much shorter than the adnate tube.
Leaves linear or linear-lanceolate. S. coast plant . . . 5. *L. angustifolius.*
Leaves obovate to oblong. N. S. Wales and Queensland plant.
Cymes rather loose, shortly pedunculate . . . . . 6. *L. dictyophlebus.*
Cymes reduced to a sessile cluster . . . . 7. *L. alyxifolius.*
Flowers and inflorescence hoary-tomentose. Calyx-limb 5-toothed, nearly as long as the adnate tube. Cymes sessile, few-flowered . . . . 8. *L. odontocalyx.*
Flowers solitary or in pairs.
Leaves mostly opposite, coriaceous.
Leaves terete, slender . . . . 9. *L. linearifolius.*
Leaves flat, from narrow-linear to oblong-cuneate . . . 10. *L. Exocarpi.*
Leaves all alternate, thin. Pedicels slender . . . . 11. *L. acacioides.*
Anthers adnate, linear. Petals free. Leaves mostly opposite.
Flowers in clusters of 3, the clusters in axillary racemes . . . 12. *L. signatus.*
Flowers in loose terminal cymes . . . . 13. *L. maytenifolius.*
Flowers in axillary cymes (or umbels), the common peduncle with 2 to 5 diverging or divaricate umbellate branches.
Branches of the peduncle (usually 3 or 4) bearing each a single flower . . . . 14. *L. sanguineus.*
Peduncle twice forked with 1 flower to each branch . . . 15. *L. bifurcatus.*
Branches of the peduncle usually 3 or 4, each bearing 3 flowers.
Leaves terete . . . . 16. *L. linophyllus.*
Leaves flat.
Lateral flowers of the 3, or all 3, pedicellate. Plant glabrous . . . . 17. *L. pendulus.*
Flowers all 3 closely sessile. Plant more or less hoary-tomentose, at least the calyx . . . . 18. *L. Quandang.*
Flowers sessile on the dilated apex of the peduncle, between 2 large bracts or floral leaves . . . . 19. *L. grandibracteus.*

1. **L. celastroides,** *Sieb. in Roem. and Schult. Syst.* vii. 163. Glabrous. Leaves opposite, from obovate or ovate, 1 to 2 in. long to cuneate-oblong lanceolate or almost linear, and 4 in. long or more, and when narrow often falcate, obtuse or very rarely almost acute, narrowed into a petiole. Flowers in loose terminal trichotomous cymes, shortly pedunculate and always shorter than the last leaves, sometimes appearing axillary from the shortness of the flowering-branch. Calyx-border scarcely prominent, obscurely sinuate-toothed. Petals 5 or 6, free, about 1½ in. long. Anthers versatile, oblong. —DC. Prod. iv. 318; A. Gray, Bot. Amer. Expl. Exped. i. 740. t. 100; F. Muell. Rep. Burdek. Exped. 13; *L. eucalyptifolius*, Sieb. in Roem. and Schult. Syst. vii. 163, not of H. B. and K.; *L. eucalyptoides*, DC. Prod. iv. 318; A. Gray, Bot. Amer. Expl. Exped. i. 741; F. Muell. Pl. Vict. t. 30.

Queensland. Brisbane river, Moreton Bay, *F. Mueller*.

N. S. Wales. Port Jackson to the Blue Mountains, *R. Brown*, *Sieber*, *n.* 242 and 244, and others; northward to Clarence river, *Beckler*, and southward to Twofold Bay, *Mossman*.

Victoria. On the Yarra, *F. Mueller*, *Robertson*; Lake King, *F. Mueller*.

The broad-leaved specimens (*L. celastroides*) are specially noted as growing on *Banksia* and *Casuarina*, and the narrow-leaved (*L. eucalyptoides*) on *Eucalyptus* and *Casuarina.* In Beckler's series of specimens from Clarence river, several are quite intermediate as to the shape of the leaf.

2. **L. Bidwillii,** *Benth.* Glabrous, with slender branches. Leaves opposite, narrow-linear, obtuse, shortly contracted at the base, but scarcely petiolate, rarely exceeding 1 in. Peduncles terminal or apparently axillary from the shortness of the flowering-branch, short but filiform, with 2 filiform 1-flowered branches not exceeding the last leaves. Calyx-border scarcely prominent, obscurely toothed. Petals 5, free, very narrow, nearly 1 in. long. Anthers versatile.

Queensland. Wide Bay, *Bidwill.*

3. **L. myrtifolius,** *A. Cunn. Herb,* Glabrous, with slender branches. Leaves opposite, sessile, ovate, obtuse, rounded or almost cordate at the base, not thick, ½ to 1 in. long. Peduncles terminal or apparently axillary from the shortness of the flowering-branch, filiform, with 2 filiform 1-flowered branches, scarcely exceeding the leaves. Calyx-border scarcely prominent, obscurely toothed. Petals 5, free, very narrow, above 1 in. long. Anthers versatile.

N. S. Wales. Logan Vale, *A. Cunningham.*

4. **L. longiflorus,** *Desr. in Lam. Dict.* iv. 598. Glabrous or the inflorescence slightly tomentose. Leaves alternate, in the ordinary form petiolate, from broadly ovate or ovate-lanceolate to narrow-lanceolate, obtuse, narrowed at the base, mostly 2 to 4 in. long, thick, obscurely veined. Flowers large, in short dense axillary racemes, rarely reduced to 2 or 3 flowers, all distinctly and singly pedicellate. Calyx-limb prominent, truncate, often oblique and sometimes obscurely toothed. Petals 5, 1¼ to 1½ in. long, united to about two-thirds of their length into a slightly swollen tube, occasionally splitting as the corolla fades, the upper portion of the petals reflexed at the time of flowering. Anthers narrow-linear, adnate.—DC. Prod. iv. 304; W. and Arn. Prod. 384; Wight, Ic. t. 302; *L. indicus*, Desr. in Lam. Dict. iv. 601, not of DC.; *L. vitellinus*, F. Muell. Rep. Burdek. Exp. 12.

N. Australia. Victoria river, *F. Mueller;* islands of the Gulf of Carpentaria, *R. Brown.*

Queensland. Shoalwater passage, *R. Brown;* Port Curtis, *M'Gillivray;* Burdekin and Gilbert rivers, *F. Mueller;* Port Denison, *Burdekin Expedition;* Edgecombe Bay, *Dallachy;* Gloucester island, *Henne;* Brisbane river, Moreton Bay, *A. Cunningham, F. Mueller*, and others; in the interior, *Mitchell.*

N. S. Wales. Port Jackson to the Blue Mountains, *Mossman, Woolls*, and others; New England, *C. Stuart;* on the Murrumbidgee, *M'Arthur.*

Var.? *amplexifolius,* Thw. Enum. Ceyl. Pl. 134. Leaves sessile, orbicular cordate.—*L. amplexifolius,* DC. Prod. iv. 305; W. and Arn. Prod. 384.—Arnhem's Land, *F. Mueller;* Suttor river, *Dorsay;* Cooper Creek, *Bowman.* The specimens are fragmentary and do not show whether this be a distinct variety or only a form of leaf assumed at a particular age or on some branches only.

5. **L. angustifolius,** *R. Br. Herb.* Glabrous. Leaves mostly opposite, linear or linear-lanceolate, often falcate, 3 to 4 in. long, thick, veinless or 3-nerved, narrowed into a short petiole. Cymes 3- to 5-flowered, sessile in the axils and the branches very short. Calyx-limb small, with scarcely conspi-

cuous teeth. Corolla 1½ in. long, rather slender, the petals united to above ½ their length into a slightly dilated tube, splitting on the upper side when old. Anthers narrow-linear, adnate.

S. Australia. Port Lincoln or Memory Cove, *R. Brown.* Nearly allied both to *L. dictyophlebus* and *L. alyxifolius*, but slightly differing from both in inflorescence and, as far as hitherto known, very much in foliage. The station is also distant.

6. **L. dictyophlebus,** *F. Muell. Rep. Burdek. Exped.* 14. Glabrous. Leaves mostly opposite, from broadly obovate or orbicular to oblong-elliptical, obtuse, narrowed into a distinct petiole, sometimes all under 3 in. long, broad and thick, with the veins scarcely conspicuous, sometimes 4 to 5 in. long, smooth and shining, with the reticulate veinlets as well as the primary veins prominent. Peduncles short, axillary, each with 3 or 4 short 3-flowered branches. Calyx-tube or ovary very narrow; limb small, the teeth scarcely conspicuous. Corolla 1½ in. long, slender, the petals united to about ¾ their length into a slightly dilated tube, often splitting when old. Anthers narrow-linear, adnate.

Queensland. Rockhampton, *Dallachy;* Brisbane river, Moreton Bay, *A. Cunningham, F. Mueller.*

N. S. Wales. Botany Bay, *R. Brown;* Clarence and Hastings rivers, *Beckler;* Kiama, *Harvey;* Illawarra, *Shepherd.*

7. **L. alyxifolius,** *F. Muell. Herb.* Glabrous, or the inflorescence minutely rusty-tomentose. Leaves opposite, from broadly obovate to obovate-oblong, very obtuse, tapering into a very short petiole, mostly from 2 to 3 in. long, thick and obscurely veined. Flowers long and slender, almost sessile, in dense axillary cymes or clusters. Calyx-border short, obscurely sinuate-toothed. Petals 1½ in. long, united to at least ¾ of their length in a scarcely dilated tube, often split on the upper side. Anthers adnate, narrow-linear.—*L. maytenifolius*, F. Muell. Rep. Burdek. Exped. 14, not of A. Gray.

Queensland. Brisbane river, Moreton Bay, *F. Mueller, W. Hill.*

N. S. Wales. Newcastle, *R. Brown;* Hastings, Clarence, and Richmond rivers, *Beckler;* Illawarra, *Shepherd.*

8. **L. odontocalyx,** *F. Muell. Herb.* Glabrous, except the inflorescence, or slightly hoary-tomentose. Leaves mostly alternate, oblong, cuneate-oblong or lanceolate, obtuse, narrowed into a short petiole, under 3 in. long, thick and scarcely veined. Cymes axillary, sessile or very shortly pedunculate, usually 3- to 7-flowered, the short branches and pedicels hoary-tomentose as well as the buds. Calyx tomentose, the limb more prominent than in the allied species, nearly as long as the adnate tube, 5-toothed. Petals 5, not quite 1 in. long, united to above the middle. Anthers adnate, linear. Young fruits crowned by a long neck formed by the calyx-limb.

N. Australia. Towards M'Adam Range and Providence Hill, *F. Mueller.*

A very imperfect specimen with broader leaves, from Howick's Group on the coast of Queensland, *F. Mueller*, appears to belong to the same species.

9. **L. linearifolius,** *Hook. in Mitch. Trop. Austr.* 102. Glabrous. Leaves mostly opposite, terete, slender, sometimes almost filiform, acute or obtuse, usually 2 to 3 in. but sometimes above 4 in. long. Flowers axillary, pedicellate in pairs, the pedicels sessile or borne on a common peduncle of 1

to 2 lines. Calyx-border obscurely toothed. Petals usually 6, about 1 in. long, united to about the middle in a slightly dilated tube, often split on the upper side. Anthers adnate, linear.

Queensland, *Bowman;* Narran river, *Mitchell.*
S. Australia. N.W. interior, *M'Douall Stuart's Expedition.*
W. Australia. Sharks' Bay, *Denham;* between Moore and Murchison rivers, *Drummond,* 6*th Coll. n.* 117; Port Gregory, *Oldfield.*

10. **L. Exocarpi,** *Behr in Linnæa,* xx. 624. Glabrous. Leaves mostly opposite, but here and there alternate, from oblong-cuneate to narrow-linear, but always flat, obtuse, narrowed into a petiole, mostly 1½ to 2½ in. long, rather thick, often triplinerved. Flowers axillary, pedicellate, solitary or in pairs, the thick pedicels sessile or more rarely borne on a very short common peduncle. Calyx-border obscurely toothed. Petals usually 6, above 1 in. long, united to about ⅓ of their length. Anthers adnate, linear. —*L. subfalcatus,* Hook. in Mitch. Trop. Austr. 224.

N. Australia. N.W. coast, *Bynoe;* Victoria river, *F. Mueller;* islands of the Gulf of Carpentaria, *R. Brown, Henne;* M'Donnell range, *M'Douall Stuart's Expedition.*
Queensland. Keppel Bay, *R. Brown;* Port Denison, *Fitzalan;* Mount Archer, *Dallachy;* Brisbane river, Moreton Bay, *F. Mueller, Fitzalan;* Flinders river, *Bowen;* Lake Salvator, *Mitchell.*
N. S. Wales. Upper Darling river, *Bowman;* Murrumbidgee river, *Backhouse.*
Victoria. Murray river, Bacchus marsh, Yarra-Yarra, etc., *F. Mueller.*
S. Australia. Near the Burossa range, *Behr;* from the Murray to Spencer's Gulf, *F. Mueller* and others; Cooper's Creek, *Howitt's Expedition.*
The tropical specimens have usually broader, more cuneate leaves, the southern ones frequently but not always very narrow linear leaves.

11. **L. acacioides,** *A. Cunn. Herb.* Glabrous, with slender branches. Leaves alternate, oblong or lanceolate, obtuse, narrowed into a petiole, rarely above 2 in. long, rather thin and often 3-nerved. Peduncles axillary, solitary or in pairs, short but slender, reflexed, each with 2 flowers, on short slender pedicels. Calyx-limb prominent, truncate or sinuate-toothed. Petals not quite 1 in. long, united to about ¾ of their length into a slightly dilated tube. Anthers adnate, narrow-linear.

N. Australia. N.W. coast, *A. Cunningham;* Victoria and Fitzmaurice rivers, *F. Mueller.*

12. **L. signatus,** *F. Muell. Herb.* Glabrous. Leaves opposite or nearly so, in the ordinary form from obovate to oval-oblong or elliptical obtuse, narrowed into a short petiole, 2 to 4 in. long, rather thick, but more or less prominently veined, but in some specimens all sessile, orbicular-cordate, with large rounded auricles, and in others mostly narrow-lanceolate and 4 or 5 in. long. Flowers in clusters of 3, almost sessile along the rhachis of dense axillary 1-sided racemes of 1 to 2 in., with 1 bract under each flower as in all the Australian species. Calyx-border obscurely sinuate-toothed. Petals 5 or 6, free, narrow, about 1 in. long, not dilated at the end. Anthers adnate, linear, very narrow.—*L. indicus,* DC. Prod. iv. 305, not of Desr.

N. Australia. Arnhem S. Bay and islands of the Gulf of Carpentaria, *R. Brown;* N. coast, *F. Mueller;* Quail island, *Flood* (the latter with broad cordate leaves).
Queensland. Gilbert river, *F. Mueller;* N.E. coast, *A. Cunningham, R. Brown* (both with long narrow leaves).

The variations in the leaf appear to be the same as in the case of *L. longiflorus* and *L. pendulus*, but not having seen the sessile cordate and the narrow petiolate leaves on the same specimen, I am unable to say whether they represent distinct varieties or different ages or parts of the same individual. The species is also in Timor, whence De Candolle had the specimens in which he thought he had identified Desrousseaux' *L. indicus*, above referred to *L. longiflorus*. Cuming's n. 1945, from the Philippine Islands, may also be the same, and if so, the species has a wide range in the Archipelago, but is not, I believe, in Continental India. *L. insularis*, A. Gray, a S. Pacific island plant, to which F. Mueller had referred these specimens, appears to me to be quite distinct.

13. **L. maytenifolius,** *A. Gray, Bot. Amer. Expl. Exped.* i. 739. *t.* 99. Glabrous. Leaves opposite, petiolate, broadly ovate or obovate, not exceeding 2 in. in our specimen, not thick, irregularly veined. Flowers in terminal shortly pedunculate cymes, with 1 bract under each flower. Calyx-border rather deeply cup-shaped, truncate. Petals 6 or occasionally 5, free, narrow. Anthers adnate, linear.

N. S. Wales. Grose river, *R. Brown;* Woolongong, *American Exploring Expedition.*

This species, as shown by our specimen communicated by A. Gray, and well represented in his plate, differs from all other Australian free-petaled species with adnate stamens by the terminal inflorescence. A. Gray states the inflorescence to be also sometimes in the upper axils, but that was owing probably to his having on a hasty inspection referred to it a specimen in the Hookerian Herbarium, labelled, with doubt, "New Holland, Fraser." This however must be a mistake, the specimen is probably not Australian, differing from all others from that country in the presence of 3 bracts (a bract and 2 bracteoles) under each flower.

14. **L. sanguineus,** *F. Muell. Fragm.* i. 177, *and Rep. Burdek. Exped.* 13. Glabrous and more or less glaucous, usually pendulous. Leaves mostly opposite, oblong-linear to linear-lanceolate and falcate, obtuse, 3 to 6 in. long, thick, obscurely veined or veinless. Flowers axillary, the common peduncle bearing 3 or 4 umbellate divaricate branches, as in *L. pendulus*, but each with only 1 flower. Calyx-border very short, truncate. Petals 5 or 6, free, narrow, about 1½ in. long, the slightly dilated apex of the bud remarkably angular. Anthers adnate, linear. Stigma capitate, much larger than in the allied species.

N. Australia. Victoria river, *F. Mueller;* islands of the Gulf of Carpentaria, *R. Brown;* Bentinck's island and Albert river, *Henne.*

15. **L. bifurcatus,** *Benth.* Glabrous and more or less glaucous. Leaves mostly opposite, linear-lanceolate, falcate, obtuse or almost acute, 4 to 8 in. long, thick, often 3-nerved. Flowers axillary, the common peduncle twice forked, each branch bearing a single flower, without any in the forks. Calyx-border reduced to a scarcely conspicuous line. Petals 5 or 6, free, narrow, about 1 in. long, the buds dilated at the base to a diameter greater than that of the calyx-tube, and clavate, but not angular at the end. Stigma not large.

N. Australia. Islands of the Gulf of Carpentaria, *R. Brown.* Allied to *L. sanguineus* and *L. pendulus*, differing from both chiefly in the ramification of the peduncle.

16. **L. linophyllus,** *Fenzl in Hueg. Enum.* 56. Quite glabrous or the young shoots inflorescence and flowers, or the inflorescence only, hoary-tomentose or almost whoolly. Leaves opposite, terete and usually slender, like those of *L. linearifolius*, or sometimes thicker, but never flat, mostly 2 to 3

in. long, but sometimes above 4 or under 1½ in. Flowers in axillary or lateral cymes, the common peduncle very short, bearing an umbel of 3 or 4, rarely 2 rays, each with a partial cyme of 3 or rarely 5 flowers, as in *L. pendulus*. Calyx-border scarcely prominent, truncate. Buds slender, clavate at the end. Petals free, 9 to 10 lines long. Anthers adnate, oblong-linear. —Miq. in Pl. Preiss. i. 279; *L. Preissii*, Miq. l. c. 280, Behr in Linnæa, xx. 625; *L. Casuarinæ*, Miq. l. c. 279; *L. scoparius*, Miq. l. c. 280.

N. Australia. Bay of Rest, N.W. coast, *A. Cunningham*; Sturt's Creek, *F. Mueller*.
Queensland. Brisbane river, Moreton Bay, *F. Mueller*; Warwick, *Beckler*.
N. S. Wales. Port Jackson, *R. Brown* and others; northward to Clarence river, *Beckler*; New England, *C. Stuart*; Liverpool plains, *A. Cunningham*; southward to Illawarra, *Herb. F. Mueller*; in the interior, *Mitchell*; towards the Barrier Range, *Victorian Expedition*.
Victoria. Port Phillip and near Melbourne, *Gunn, Robertson*; on the Murray towards Ovens river, *F. Mueller*.
S. Australia. From the Murray to Spencer's Gulf, *F. Mueller*; Lake Torrens and Cooper's Creek, *Howitt's Expedition*.
W. Australia. Swan River, *Drummond, Preiss, n.* 1611, 1613, 1615, 1618; Murchison river, *Oldfield*; Sharks' Bay, *A. Cunningham*; Dirk Hartog's Isle, *Milne*.

There are three rather distinct forms: *a*. with glabrous flowers in W. and N.W. Australia, S. Australia, Victoria, and the desert interior of N. S. Wales; *b*. with slightly tomentose flowers, from the Brisbane to Port Jackson and Illawarra; *c*. with woolly-tomentose flowers, in New England and W. Australia.

17. **L. pendulus,** *Sieb. in DC. Prod.* iv. 294, *and Mem. Lor. t.* 1. Glabrous. Leaves mostly opposite, from obovate oblong-cuneate and about 2 in. long, to linear-lanceolate attaining sometimes 10 in. or even more, thick and usually 3- or 5-nerved, in a few specimens sessile, broad and cordate. Flowers in axillary cymes, the common peduncle short, bearing an umbel of 3 or 4 rarely 2 rays, each with a partial cyme of 3 rarely 5 flowers, the central one sessile, or rarely all pedicellate. Calyx-border shortly cup-shaped, truncate. Buds slender, clavate at the tips. Petals free, 1 to 1½ in. long. Anthers adnate, from oblong-linear to narrow-linear.—*L. congener*, Sieb. in DC. Prod. iv. 295, and Mem. Lor. t. 2 (leaves short, central flowers sessile); *L. longifolius*, Hook. Ic. Pl. t. 880 (leaves very long, central flowers sessile); *L. aurantiacus*, A. Cunn.; Hook. in Mitch. Trop. Austr. 101 (flowers all pedicellate); *L. Miquelii*, Lehm. in Pl. Preiss. i. 280 (flowers all pedicellate).

N. Australia. Victoria river, *F. Mueller*; Port Essington, *Armstrong*; Gilbert river, *F. Mueller*.
Queensland. Wide Bay, *Bidwill*; Rockingham Bay and Rockhampton, *Dallachy*; Brisbane river, Moreton Bay, *F. Mueller* and others.
N. S. Wales. Port Jackson to the Blue Mountains, *R. Brown, Sieber, n.* 241, 243, *and Fl. Mixt. n.* 622, *A. Cunningham* and others; northward to Clarence, Macleay, and Hastings rivers, *Beckler*; New England, *C. Stuart*.
Victoria. Port Phillip, Yarra and Latrobe rivers, to the Murray, *F. Mueller* and others; N.W. of the colony, *Morton*.
S. Australia. S. coast, *R. Brown*; Lofty and Bugle ranges, Holdfast Bay, etc., *F. Mueller*.
W. Australia. Swan River and neighbourhood, *Drummond, Preiss, n.* 1617, *Oldfield*.

Var. *amplexifolius*. Leaves sessile, broad, cordate. Roebuck Bay, N.W. coast, *Marten*.

Var. *parviflorus*. Leaves small and narrow; flowers small, often 4-merous, the central ones sessile.—*L. Melaleucæ*, Lehm. in Pl. Preiss. i. 281, Miq. in Ned. Kruidk. Arch. iv. 107; *L. miraculosus*, Miq. in Pl. Preiss. i. 281, and in Ned. Kruidk. Arch. iv. 106.—Paramatta, *Woolls*; New England, *C. Stuart*; towards the Barrier Range, *Victorian Ex-*

pedition; Port Lincoln, *Wilhelmi;* interior of S. Australia, *M'Douall Stuart;* Swan River, *Drummond;* Rottenest island, *A. Cunningham, Preiss, n.* 1616.

18. **L. Quandang,** *Lindl. in Mitch. Three Exped.* ii. 69. Foliage and inflorescence more or less hoary-tomentose, or rarely glabrous except the calyx. Leaves opposite, from obovate or oblong-cuneate and 1 to 2 in. long to lanceolate and 3 or 4 in. long, thick, veinless or obscurely 3-nerved. Flowers in axillary cymes, the common peduncle bearing an umbel of 2 or 3 rarely 4 rays, each with 3 closely sessile flowers. Calyx-tube tomentose, more contracted under the limb than in *L. pendulus;* limb tomentose or very rarely glabrous. Flowers otherwise of *L. pendulus;* petals free, under 1 in. long. Anthers adnate, oblong-linear.—*L. nutans*, A. Cunn.; Hook. in Mitch. Trop. Austr. 158; *L. canus*, F. Muell. in Hook. Kew Journ. viii. 145, and in Trans. Vict. Inst. 1855, 128; Miq. in Ned. Kruidk. Arch. iv. 107.

N. Australia. Victoria river and Sturt's Creek, *F. Mueller;* Thomson river, *A. C. Gregory;* islands of the Gulf of Carpentaria, *R. Brown, Henne.*
Queensland, *Bowman;* Keppel Bay, *R. Brown;* Suttor Desert, islands of Howick's group, and Moreton Bay, *F. Mueller;* Belyando river and Fitzroy Downs, *Mitchell.*
N. S. Wales. Port Jackson, *F. Mueller;* Hunter's River and in the western interior, *A. Cunningham, Fraser;* from the Darling to the Barrier Range, *Victorian Expedition, Dallachy and Goodwin* and others.
Victoria. Grampians, *Mitchell, F. Mueller;* Dandenong Creek, *Wilhelmi;* Buffalo Creek, *F. Mueller.*
S. Australia. Cooper's Creek, *Howitt's Expedition.*
W. Australia Murchison river, *Oldfield, Drummond, 6th Coll. n.* 116.

Var.? *amplexifolius.* Leaves broad, sessile, cordate.—Victoria river, *F. Mueller;* the specimen very imperfect and doubtful.

19. **L. grandibracteus,** *F. Muell. Rep. Burdek. Exped.* 14. Glabrous. Leaves opposite, from oblong-cuneate to linear-lanceolate, obtuse, narrowed into a petiole, 2 to 4 in. long or more when narrow, thick, veinless or obscurely 3-nerved. Peduncles at the forks of the branches, each about 1 in. long, very much flattened and dilated at the end into a truncate receptacle, bearing 4 to 6 closely sessile flowers between 2 broadly ovate or ovate-lanceolate obtuse floral leaves or leafy bracts, 1 to 1½ in. long, and obscurely several-nerved. Bracts under each flower minute or none. Calyx nearly 2 lines long, with a very short denticulate-ciliate limb. Petals free, nearly 1 in. long; anthers adnate, linear.

N. Australia. Islands of the Gulf of Carpentaria, *R. Brown;* between Albert and Flinders rivers, *F. Mueller.*
Queensland. Sandy Cape, *R. Brown;* Suttor river, *Bowman.*
S. Australia. Cooper's Creek, *Victorian Expedition;* N. interior, *Capt. Strutt.*

4. VISCUM, Linn.

Flowers unisexual. Calyx-border inconspicuous, or rarely forming a scarcely prominent line. Petals 3 to 5, very short, having the appearance of a simple perianth. Anthers sessile on the petals, short and broad, opening inwards in several pores in transverse rows. Stigma sessile. Fruit a one-seeded berry.—Parasitical glabrous shrubs. Branches opposite or dichotomous. Leaves opposite or none. Flowers very small, green or yellowish, monœcious and clustered at the nodes in all the Australian species, diœcious and in the forks of the branches in the common European one.

The genus is spread over Asia, Africa, and temperate Europe. The three Australian species are all Asiatic also. They are sometimes found growing upon species of *Loranthus*, as well as upon the trees that feed them.

Leaves opposite at the nodes. Petals deciduous . . . . 1. *V. orientale*.
Leaves none.
Branches slender, angular, not flattened. Petals deciduous . . . 2. *V. angulatum*.
Branches flattened. Petals persistent . . . . 3. *V. articulatum*.

1. **V. orientale,** *Willd.; DC. Prod.* iv. 278. Branches elongated, nearly terete, always leafy. Leaves opposite, narrow-oblong or lanceolate, and 1 to 1½ in. long in the Australian specimen, from that to obovate and varying from 1 to 3 in. in Asiatic ones, narrowed at the base, 3- or 5-nerved. Flowers minute, in 1 to 3 sessile or shortly pedicellate clusters in each axil, each cluster consisting of 3 or 5 flowers, the central one or rarely 3 females, each under 1 line long, the 2 lateral ones males and much smaller, all sessile in the clusters within 1 or 2 small bracts. Petals 4 or rarely 3, deciduous. Berry globular, 2 or rarely 3 lines diameter.

Queensland. Rockhampton, *Bowman*; a single small specimen in Herb. F. Muell. Common in India and the Archipelago, extending westward almost to the Mediterranean.

2. **V. angulatum,** *Heyne; W. and Arn. Prod.* 380. Quite leafless. Branchlets opposite or dichotomous, articulate, rather slender, angular, not flattened, the older branches terete. Flowers minute, in sessile clusters of 3 to 6 at the nodes, the males and females in the same clusters, each one half-immersed in a cup-shaped 2-lobed bract. Petals usually 4, very deciduous. Berry small, globular.

Queensland. Gilbert river, *F. Mueller*; Edgecombe Bay, *Dallachy*; Port Denison, *W. Hill*; Brisbane river, Moreton Bay, *A. Cunningham*, and others. Also in the mountains of the Indian Peninsula, but apparently not common there.

3. **V. articulatum,** *Burm.; DC. Prod.* iv. 284. Very much branched, forming tufts from a few in. to 1 or 2 ft. diameter. Branches flattened, articulate, sometimes forked at almost every node, sometimes elongated; the articles thick, mostly ½ to ¾ in. long and 2 to 4 lines broad, but sometimes as broad as long or above 1 in. long and very narrow. Flowers minute, sessile and clustered at the nodes, males and females in the same clusters, the females scarcely ½ line long, nearly globular, half-buried in a cup-shaped bract, the males still smaller. Petals usually 3, very minute, persistent and crowning the very small globular berry.—*V. moniliforme*, Blume; DC. Prod. iv. 284; Wight, Ic. t. 1018 and 1019.

Queensland. Moreton Bay, *F. Mueller*.
N. S. Wales. Liverpool Plains, *A. Cunningham*; Richmond and Clarence rivers, *Beckler*.

5. NOTOTHIXOS, Oliv.

Flowers unisexual. Calyx-border quite inconspicuous. Petals 4, rarely 3 or 5. Anthers almost sessile, at the base of the petals, not adnate, transversely 2-lobed inside, with parallel lobes obscurely locellate. Stigma sessile. Fruit a 1-seeded berry.—Parasitical dichotomous shrubs, more or less covered with a golden or hoary tomentum, rarely at length nearly glabrous. Leaves opposite, flat, 3- or 5-nerved, but the nerves often obscure. Stipules minute, rigid, acute. Flowers minute, sessile in little pedunculate heads, solitary or several on a common terminal peduncle.

The genus is endemic in Australia. The three species distinguished by Oliver are united into one by *F. Mueller* (Fragm. ii. 109, and iv. 173). It is possible that *N. subaureus* may prove to be a remarkable variety of *N. incanus*, which is only known from specimens with imperfectly-developed inflorescence, but as yet intermediate forms have not been observed an *N. cornifolius* appears to me in all states to be quite distinct. As in the case of *Viscums*, they are found sometimes parasites on species of *Loranthus*.

Leaves small, cuneate or spathulate. Flower-heads solitary (or in threes ?). Plant hoary . . . . 1. *N. incanus.*
Leaves ovate. Flower-heads in threes. Plant more or less golden-tomentose . . . . 2. *N. subaureus*
Leaves obovate-oblong or broadly cuneate. Flower-heads in a terminal raceme. Plant hoary or nearly glabrous . . . . 3. *N. cornifolius.*

1. **N. incanus,** *Oliv. in Journ. Linn. Soc.* vii. 104. Densely branched and hoary with a minute tomentum, the branchlets much flattened below the leaves, the older branches terete. Leaves oblong-cuneate or spathulate, narrowed into a petiole, very obtuse and rarely mucronulate, ½ to ¾ in. long or rarely more, the nerves faint or inconspicuous. Flower-heads solitary (or sometimes 3 together ?) on very short terminal peduncles, usually with few flowers, the females about ½ line long, the males considerably smaller. Fruits about 3 lines long.—*Viscum incanum*, Hook. Ic. Pl. t. 73.

Queensland. Brisbane river, Moreton Bay, *Fraser*, *F. Mueller.*

2. **N. subaureus,** *Oliv. in Journ. Linn. Soc.* vii. 103. Divaricately branched, the young parts and under side of the leaves densely covered with a more or less golden tomentum. Leaves ovate, obtuse, narrowed into a short petiole, mostly about ¾ to 1 in. long. Flower-heads 3, the common peduncle very short, each partial one ¼ to ½ in. long, the lateral ones very divaricate or recurved, the flowers minute and sessile.—*Viscum subaureum*, F. Muell. in Herb. Hook.

Queensland. Brisbane river, Moreton Bay, *F. Mueller;* Ipswich, Vernet.
N. S. Wales, *C. Moore.* Blue Mountains, *Miss Atkinson;* Lake Macquarrie, *Backhouse;* Twofold Bay, *F. Mueller.*

3. **N. cornifolius,** *Oliv. in Journ. Linn. Soc.* vii. 103. A larger species than the two preceding ones, the young parts hoary-tomentose, becoming at length nearly glabrous; branches terete. Leaves obovate-oblong or oblong-cuneate, obtuse, narrowed into a short petiole, mostly 1½ to 2 in. long. Flower-heads several, opposite, in pairs, in a terminal raceme nearly as long as the leaves, with minute bracts both under the short peduncles and under the heads. Flowers sessile in the heads, the females not ½ line long, the males still smaller.

Queensland. Brisbane river, Moreton Bay, *A. Cunningham*, *Fraser.*
N. S. Wales. Port Jackson, *J. D. Hooker;* Upper Hunter river and Liverpool Plains, *A. Cunningham;* Richmond and Clarence rivers, *Beckler.*

Order LX. CAPRIFOLIACEÆ.

Calyx-tube adnate to the ovary, the limb short, truncate or of 4 or 5 rarely more lobes or teeth. Corolla gamopetalous, inserted round the epigynous disk; lobes 4 or 5 rarely 3, imbricate in the bud. Stamens as many as lobes of the corolla, alternate with them, inserted in the tube; anthers versatile with parallel cells opening longitudinally. Ovary inferior, 2- to 5-celled or rarely 1-celled, with 1 or more pendulous ovules in each cell. Stigmas as many as cells, or united into one, sessile or on a single filiform style. Fruit an indehiscent berry, or rarely dry, 1- to 5-celled. Seeds 1 or more in each cell. Embryo in the axis of a fleshy albumen; radicle superior, cotyledons oval or oblong.—Trees, shrubs, or climbers, rarely herbs. Leaves opposite, usually without stipules, simple or rarely pinnate.

A rather small Order, chiefly dispersed over the temperate regions of the northern hemisphere, with a very few tropical or southern species, represented in Australia by a single genus having a wide range in the northern hemisphere, and remarkable for its pinnate leaves. Many other genera scarcely differ from *Rubiaceæ* except in the want of stipules.

1. SAMBUCUS, Linn.

Calyx-limb of 3 to 5 small teeth. Corolla with a very short tube, and 3 to 5 lobes, spreading so as to appear rotate. Stamens inserted at the base of the corolla. Ovary 3- to 5-celled with 1 pendulous ovule in each cell, stigma sessile, 3- to 5-lobed. Fruit a berry-like drupe, with 3 to 5 seed-like pyrenes, each containing a single seed.—Trees, shrubs, or tall herbs. Leaves opposite, pinnate. Flowers white or yellow, rather small, in large terminal corymbose cymes.

The genus is widely dispersed over Europe, temperate Asia, and North America. The Australian species are both endemic, but nearly correspond to the two commonest of the northern ones.

Shrub or tree. Leaves without stipule-like lobes. Flowers mostly 3-merous. Berries yellow . . . . 1. *S. xanthocarpa.*
Tall shrub. Lowest leaflets of each leaf close to the stem, short and broad, looking like stipules. Flowers mostly 4-merous. Berries white . . . . 2. *S. Gaudichaudiana.*

1. **S. xanthocarpa,** *F. Muell. in Hook. Kew Journ.* viii. 145, *and in Trans. Phil. Inst. Vict.* i. 42; *Pl. Vict. t.* 29. A tall shrub or small tree, quite glabrous. Leaflets 3 or 5, all petiolulate, the lower pair sometimes again divided into 2 or 3 each, lanceolate or ovate-lanceolate, acuminate, narrowed at the base, acutely but not deeply serrate or almost entire, mostly 2 to 3 in. long. Primary branches of the corymb umbellate, the others cymose. Flowers mostly 3-merous, rarely 4-merous, yellow (*Mitchell*). Berries yellow (*F. Mueller*).—*Tripetelus australasicus*, Lindl. in Mitch. Three Exped. ii. 14.

Queensland. Brisbane river, Moreton Bay, *F. Mueller*, and others.

N. S. Wales, *A. Cunningham.* Blue Mountains, *Miss Atkinson;* northward to Hastings and Clarence rivers, *Beckler;* southward to Illawarra, *F. Mueller*; in the interior to Lachlan river, *Mitchell.*

2. **S. Gaudichaudiana,** *DC. Prod.* iv. 322. Stems from a perennial stock herbaceous, erect, 3 to 5 ft. high, glabrous as well as the rest of the

plant. Leaflets 5 to 11, sessile or petiolulate, ovate or ovate-lanceolate, coarsely and acutely toothed, 2 to 5 in. long, the lowest of each leaf close to the stem, short, broad, and toothed, resembling leafy stipules. Primary branches of the corymb umbellate, the others cymose. Flowers mostly 4-merous. Berries oblong, white.—Hook. f. Fl. Tasm. i. 164.

Queensland. Brisbane river, Moreton Bay, *Fitzalan.*
N. S. Wales. Paramatta, *Woolls.*
Victoria. Yarra Yarra, Cape Otway, Apollo Bay, *F. Mueller;* Portland, *Allitt;* Wendu valley, *Robertson;* Wimmera, *Dallachy.*
Tasmania. Dense shady woody ravines and alluvial flats in the northern parts of the colony, *J. D. Hooker.*
S. Australia. Mount Gambier, *F. Mueller.*

ORDER LXI. RUBIACEÆ.

Calyx-tube adnate to the ovary; the limb entire or with as many teeth lobes or divisions as lobes of the corolla, rarely more, fewer or none. Corolla gamopetalous, inserted round the epigynous disk; lobes 4, 5 or sometimes more, rarely only 3, either imbricate (often contorted) or valvate in the bud. Stamens as many as lobes of the corolla, alternating with them and inserted in the tube; anthers versatile, with parallel cells opening longitudinally. Ovary inferior, 2- or more-celled, with 1 or more ovules in each cell, rarely 1-celled with parietal placentas, or reduced to one 1-ovulate cell; style more or less divided into as many stigmatic lobes as carpels to the gynœcium, or undivided with a thickened entire or notched stigma. Fruit a capsule, drupe, berry or indehiscent nut. Seeds with a fleshy or horny albumen, and rather small straight embryo with flat cotyledons, or rarely with little or no albumen, and cylindrical embryo with semiterete cotyledons.—Trees, shrubs, herbs, or rarely climbers. Leaves opposite or whorled. Stipules interpetiolar, either free or connate with the petioles in a sheath bordered by cilia or leaf-like lobes, or with one or two points on each side, or connate within the petioles in a short sheath or ring round the stem. Inflorescence various, usually more or less cymose, axillary or terminal. Flowers occasionally polygamous or unisexual, especially in *Guettardeæ* and *Anthospermeæ.*

A very large Order, dispersed over every part of the globe; the *Cinchoneæ* and *Coffeeæ,* with few exceptions, tropical or subtropical; the *Stellatæ* chiefly inhabiting the more temperate or cold regions. Of the 29 Australian genera 2, belonging to *Stellatæ,* are those which have the widest range in that tribe; 2, *Nertera* and *Coprosma,* range more or less over the extratropical or mountain regions of the southern hemisphere; 8 extend over the tropical or subtropical regions both of the New and the Old World; 6 are common to tropical Africa and Asia; 7 are limited to tropical Asia or extend into the Pacific Isles or the islands of the W. coast of Africa; 4 only are endemic in Australia, and of those 4, 3 are monotypic.

TRIBE I. **Cinchoneæ.**—*Leaves opposite or rarely whorled, with small or membranous stipules between or inside of them. Ovules several in each cell of the ovary.*

SUBTRIBE I. **Naucleeæ.**—*Flowers very numerous, closely packed in globular heads on a small receptacle. Ovules pendulous or imbricate on a pendulous placenta. Fruits capsular or rarely fleshy, not pulpy.—Trees or shrubs.*

Calyxes concrete, forming in fruit a hard fleshy mass. Petals imbricate 1. SARCOCEPHALUS.

SUBTRIBE II. **Hedyotideæ.**—*Flowers in cymes, clusters, or solitary. Corolla-lobes valvate. Ovules attached to an axile or basal placenta. Fruit capsular or separating into dry cocci.—Herbs, undershrubs, or rarely shrubs.*

Capsule ovoid or globular, or separating into hard cocci.
  Flowers 4-merous. Petals entire . . . . 2. HEDYOTIS.
  Flowers 5-merous. Petals 2- or 3-toothed . . . . 3. DENTELLA.
Capsule broader than long, much compressed . . . . 4. OPHIORRHIZA.

SUBTRIBE III. **Gardenieæ.**—*Flowers in cymes, clusters, or solitary. Corolla-lobes imbricate. Ovules attached to an axile or parietal placenta. Fruit succulent, indehiscent.*

Ovary 1-celled, with 2, 3 or more parietal prominent placentas.
  Stipules connate within the petioles. Inflorescence usually terminal 5. GARDENIA.
Ovary 2-celled (rarely several-celled) with axile placentas. Stipules interpetiolar.
  Flowers 5-merous.
    Inflorescence axillary. Ovules imbedded in a fleshy placenta . 6. RANDIA.
    Corymbs terminal. Ovules not imbedded in theplacenta . . 7. WEBERA.
  Flowers 4-merous. Ovules few (sometimes only 2 perfect) on a peltate placenta. Inflorescence axillary or terminal . . . . 8. DIPLOSPORA.

TRIBE II. **Coffeeæ.**—*Leaves opposite or rarely whorled, with small or membranous stipules between or inside of them. Ovules solitary or very rarely* 2 *collateral in each cell of the ovary, rarely several to each carpel separated by spurious septa. Cells of the fruit or pyrenes always* 1-*seeded.*

SUBTRIBE I. **Ixoreæ.**—*Corolla-lobes imbricate (usually contorted). Ovules laterally attached, usually peltate. Fruit a berry or drupe. Albumen copious.—Trees or shrubs* . . . . 9. IXORA.

SUBTRIBE II. **Guettardeæ.**—*Corolla-lobes imbricate. Ovules suspended from the summit of the normal or spurious cells. Fruit a drupe, the nucleus or pyrenes usually hard. Albumen little or none.—Trees or shrubs. Flowers in forked cymes or solitary, rarely umbellate.*

Uniovulate spurious cells of the ovary and pyrenes of the fruit much more numerous than the style-lobes, superposed in several series . 10. TIMONIUS.
Uniovulate spurious cells of the ovary 2 to each normal cell and style-lobe, superposed, the upper ovule erect, the lower one suspended . 11. SCYPHIPHORA.
Cells 1-ovulate of the same number as the style-lobes.
  Corolla-tube slender, limb spreading.
    Ovary 2-celled . . . . 12. ANTIRRHŒA.
    Ovary several-celled.
      Calyx-limb lobed . . . . 13. GUETTARDELLA.
      Calyx-limb cupular, truncate, or obscurely toothed . . . . 14. GUETTARDA.
  Corolla-tube ovoid, lobes very small. Flowers umbellate . . . 15. HODGKINSONIA.

SUBTRIBE III. **Vanguerieæ.**—*Corolla-lobes valvate. Ovules laterally attached at or near the top. Fruit a berry-like drupe with* 1-*seeded pyrenes. Albumen copious.—Trees or shrubs.*

Flowers in axillary cymes or clusters. Ovary 2-celled . . . . . 16. CANTHIUM.

SUBTRIBE IV. **Psychotrieæ.**—*Corolla-lobes valvate. Ovules erect from the base or laterally attached below the middle. Style-lobes short. Fruit a berry-like drupe with* 1-*seeded pyrenes. Albumen copious.—Trees or shrubs.*

Flowers in globular heads, the calyx-tubes concrete or immersed in the receptacle. Ovules often twice as many as style-lobes. Fruit compound, pulpy, with 1-seeded pyrenes . . . . 17. MORINDA.
Flowers and fruits quite distinct.
  Ovules and 1-seeded pyrenes 4, style-lobes 4 . . . . 18. CŒLOSPERMUM.

Ovules and 1-seeded pyrenes of the same number as the style-lobes.
Ovary-cells, pyrenes, and style-lobes 4 or more. Flowers in axillary sessile clusters . . . . 19. LASIANTHUS.
Ovary-cells, pyrenes, and style-lobes 2. Flowers in terminal cymes or corymbs . . . . 20. PSYCHOTRIA.

SUBTRIBE V. **Anthospermeæ.**—*Corolla-lobes valvate. Ovules erect from the base. Style-lobes very long. Fruit a berry-like drupe. Albumen copious.—Shrubs, rarely trees or herbs. Flowers often unisexual or polygamous.*

Shrubs, erect or creeping. Flowers clustered or solitary, unisexual or polygamous . . . . 21. COPROSMA.
Slender creeping perennial herbs. Flowers solitary, hermaphrodite, or polygamous . . . . 22. NERTERA.

SUBTRIBE VI. **Opercularieæ.**—*Corolla-lobes valvate. Ovules erect from the base, usually one only to each flower. Style-lobes long. Fruit capsular,* 2-*valved. Albumen copious.—Herbs, undershrubs, or rarely small shrubs.*

Flowers in simple or compound heads, the calyx-tubes connate. Outer valves of the fruits of each partial head connate in a persistent cup, inner valves of the same fruits connate in a deciduous operculum.
Flower-heads connate in a compound head or rarely solitary . . 23. OPERCULARIA.
Flower-heads several in an umbel . . . . 24. POMAX.
Flowers in a globular head, but not connate. Fruits separately 2-valved . . . . 25. ELEUTHRANTHES.

SUBTRIBE VII. **Spermacoceæ.**—*Corolla-lobes valvate. Ovules variously laterally attached to the axis, the ovary usually perfectly* 2-*celled. Style-lobes short. Fruit capsular, or indehiscent and dry. Albumen copious.—Herbs, undershrubs, or rarely shrubs.*

Flowers in terminal cymes. Ovules attached at or near the top.
Cocci separating at the base only, falling off together, leaving a persistent subulate axis . . . . 26. KNOXIA.
Flowers in terminal heads or lateral clusters. Ovules attached at or below the middle. Capsule opening septicidally . . . . 27. SPERMACOCE.

TRIBE III. **Stellatæ.**—*Calyx wholly adnate without any visible border (in the Australian genera). Corolla-lobes valvate. Ovary* 2-*celled, with* 1 *ovule in each cell. Fruit small, indehiscent.—Herbs, rarely undershrubs. Stipules similar to the leaves, and connected with them by a short sheath or ring, forming whorls of* 4 *or more, very rarely (only in* 2 *Australian species) reduced to the* 2 *leaves.*

Corolla funnel-shaped, with a distinct tube . . . . 28. ASPERULA.
Corolla nearly rotate, with scarcely any tube . . . . 29. GALIUM.

The above subtribes will be found probably to comprehend the whole of the Order, except the *Eucinchoneæ* and *Rondeletieæ*, which, numerous in tropical America, less so in Africa, Asia, and the Pacific Islands, are, as far as hitherto known, unrepresented in Australia.

TRIBE I. CINCHONEÆ.—Leaves opposite or rarely whorled, with small or membranous stipules between or inside of them. Ovules several in each cell of the ovary.

SUBTRIBE I. NAUCLEEÆ.—Flowers very numerous, closely packed in globular heads on a small receptacle. Ovules pendulous or imbricate on a pendulous placenta. Fruits capsular or rarely fleshy, not pulpy.—Trees or shrubs.

This subtribe is a very natural one, although it includes genera with imbricate and with valvate corollas, and with pedicellate as well as sessile flowers. The only Australian representative belongs to a genus somewhat anomalous in the fruits forming a compound succulent head, but very different from that of *Morinda* or of *Opercularia*, in which the calyx-tubes are connate at the base or immersed in the receptacle; whilst in *Sarcocephalus* it is the upper part of the calyx-tubes and epigynous disks that are fleshy and connate; the lower portion is often dry and never pulpy.

1. SARCOCEPHALUS, Afz.

(Platanocarpus, *Korth.*)

Flowers densely packed in a globular head, the calyxes cohering. Corolla-tube slender; lobes 4 or 5, spreading, slightly imbricate in the bud. Anthers nearly sessile at the mouth of the corolla-tube. Ovary 2-celled, with several linear ovules in each cell inversely imbricate on a linear placenta, pendulous from the top of the cell. Style much exserted, with 2 short stigmatic lobes. Fruits connate in a dense globular mass, fleshy when fresh, hard when dry, but capsular at the base. Seeds 1 or 2 in each cell, not winged. Albumen abundant.—Trees or shrubs. Stipules interpetiolar, membranous, very deciduous. Flower-heads solitary on terminal peduncles.

The genus is spread over tropical Africa and Asia, the Australian species having a wide range in E. India and the Archipelago.

1. **S. cordatus,** *Miq. Fl. Ind. Bat.* ii. 133. A handsome tree, either quite glabrous or the leaves softly pubescent underneath. Leaves broadly ovate, obtuse, rounded, cuneate or (in specimens not Australian) broadly cordate at the base, from 4 or 5 in. to twice that length. Stipules large, broad, obtuse, but so deciduous as to be rarely seen. Flowers (yellow) in dense globular heads above 1 in. diameter without the styles. Calyx-limb shortly campanulate, with 4 or 5 stipitate clavate gland-like lobes. Corolla-tube slender, about 2 lines long; lobes about half that length, obtuse. Style very long, with a thick ovoid shortly 2-lobed stigma. Fruits united in a hard globular mass of above 1 in. diameter, pitted and rough with the remains of the more or less succulent calyxes and disks. Seeds either 1 oblong, or 2 superposed and truncate in each cell.—*Nauclea coadunata*, Sm. in Rees Cycl. xxiv.; DC. Prod. iv. 344; *N. undulata* and *N. cordata*, Roxb. Fl. Ind. i. 508, 509; *Sarcocephalus undulatus*, Miq. Fl. Ind. Bat. ii. 133.

N. Australia. Glenelg river, N.W. coast, *Herb. Hooker;* Victoria river, *F. Mueller.*
Queensland. Rockhampton, *Dallachy;* Port Denison, *Fitzalan.*
The species is also in Ceylon and in the Archipelago.—*N. Bartlingii*, DC. Prod. iv. 344, or *Sarcocephalus Bartlingii,* Miq. Fl. Ind. Bat. ii. 133, is probably the form with pubescent leaves.

SUBTRIBE II. HEDYOTIDEÆ.—Flowers in cymes, clusters, or solitary. Corolla-lobes valvate. Ovules attached to an axile or basal placenta. Fruit capsular or separating into dry cocci. — Herbs, undershrubs, or rarely shrubs.

This subtribe is characterized especially with reference to the genera represented in Australia, and some others nearly related to them. It is possible that it may require some modification to mark more accurately the line of separation from the *Rondeletieæ* and *Eucinchoneæ*.

2. HEDYOTIS, Linn.

(Oldenlandia, *Linn.;* Houstonia, *Linn.;* Metabolos, *Blume.*)

Calyx-limb of 4 (very rarely 5) persistent teeth or lobes. Corolla-tube short or slender, of 4 (very rarely 5) lobes, valvate in the bud. Anthers usually exserted from the tube. Ovary 2-(rarely 3- or 4-)celled, with several ovules in each cell, attached to placentas arising from near the base. Style entire or with 2 (rarely 3 or 4) stigmatic lobes. Capsule globular or ovoid, sometimes more than half-superior, the carpels separating septicidally, and indehiscent or loculicidally 2-valved, or the whole capsule loculicidally 2-valved, the valves remaining entire or splitting septicidally.—Herbs, undershrubs, or rarely climbers. Stipules interpetiolar, united with the petioles in a short sheath or almost free, either truncate or ovate, entire or fringed with bristle-like subulate lobes.

A large genus, widely spread over tropical and subtropical Asia and Africa, with a few American species. Of the nine Australian species, one or perhaps two are common in India and the Archipelago, the others appear to be endemic.

The two Linnæan genera *Hedyotis* and *Oldenlandia* were united by Wight and Arnott, whom Torrey and Gray followed adding a third Linnæan genus *Houstonia,* and many smaller ones established by Blume and others come within the limits thus assigned to *Hedyotis.* I had subsequently thought that good characters might be found conformable to habit for separating *Hedyotis* from *Oldenlandia.* A. Gray has, however, shown that *Houstonia* is still more distinct in the seeds, but that it would require adding to it species of *Anotis* and *Hedyotis,* and that there are intermediate forms between *Hedyotis* and *Oldenlandia,* and going through all the sections it must be admitted that the adoption of the three genera would produce very unnatural groups. As, moreover, among the Australian species the three genera are represented by species having all the true *Oldenlandia* habit, it appears to be better to return to the union as proposed by Wight and Arnott and by Torrey and Gray.

Capsule hard, indehiscent or separable into 2 indehiscent cocci.
- Coarse plant. Flowers numerous, in sessile axillary clusters . 1. *H. auricularia.*

Capsule slightly protruding from the calyx-tube, and opening loculicidally, or both septicidally and loculicidally. Leaves narrow-linear. Pedicels slender, solitary, or 2 or 3 together.
- Slender, erect, rigid, and virgate. Corolla-lobes longer than the tube. Capsule loculicidal only . . . . 2. *H. cærulescens.*
- Very slender, erect. Panicle loose, almost leafless. Corolla-lobes shorter than the tube. Capsule loculicidal and septicidal . . . . 3. *H. mitrasacmoides.*
- Slender and diffuse. Pedicels axillary. Corolla-lobes shorter than the tube. Capsule loculicidal and septicidal . . . . 4. *H. galioides.*

Capsule at least half-superior, opening in 4 valves.
- Corolla divided almost to the base, and often persistent. Seeds angular, or if peltate not concave. Very small much-branched annuals or perennials.
  - Leaves linear, mucronate-acute. Calyx-lobes mucronate-acute as long as the corolla. Annual of 2–3 in.. . . . . . 5. *H. scleranthoides.*
  - Leaves oblong, obtuse, petiolate. Flowers about 1 line long. Calyx-lobes minute, distant. Annual, under 1 in. . . . 6. *H. elatinoides.*
  - Leaves linear, obtuse. Flowers about ½ line long. Calyx-lobes obtuse, shorter than the corolla. Perennial of 2–3 in. 7. *H. tillæacea.*
- Corolla-lobes shorter than the tube. Seeds broad, concave, peltately attached by a prominent rib on the inner face. Perennials with linear leaves.
  - Pedicels slender: Corolla-tube nearly 2 lines long . . . . 8. *H. trachymenioides.*
  - Pedicels very short. Corolla-tube 1 line long . . . . . 9. *H. pterospora.*

There are also specimens of a distinct *Hedyotis* from Sturt's Creek, *F. Mueller*, but too imperfect for description.

1. **H. auricularia,** *Linn.*; *DC. Prod.* iv. 420. A decumbent straggling herb of 1 to 3 ft., the branches hairy or pubescent all round or on the opposite sides only or almost glabrous. Leaves shortly petiolate, ovate-lanceolate to oblong-lanceolate, 1 to 3 in. long, glabrous or pubescent, with very oblique raised veins diverging from the midrib and very prominent underneath. Stipules short, with long bristle-like lobes or teeth. Flowers but little more than 1 line long, in dense axillary sessile clusters. Calyx-lobes subulate and recurved, about as long as the tube; the corolla scarcely exceeding them. Capsules small, crowned by the calyx-lobes, and quite indehiscent or rarely separating into 2 hard indehiscent cocci, each containing 4 to 6 angular seeds. —W. and Arn. Prod. 412; Benth. Fl. Hongk. 150; *H. geniculata*, Roxb.; DC. Prod. iv. 420; *Metabolos venosus*, Blume; DC. Prod. iv. 435.

Queensland. Cape York, *M'Gillivray*.—Widely spread over E. India, the Archipelago, and the islands of the S. Pacific.

2. **H. cærulescens,** *F. Muell. Fragm.* iv. 38. Annual (or with a perennial stock?), with slender but rigid virgate stems of about 1 ft., quite glabrous. Leaves narrow-linear, acute, attaining sometimes 1 in. or even more. Pedicels in the upper axils, solitary or in pairs, or 3 or 4 on a common peduncle, the pedicels unequal in length, forming a very irregular loose terminal slightly-leafy panicle. Calyx-tube when in flower 1 line long, the lobes much shorter. Corolla about 1 line long, deeply 4-cleft. Fruit about 2 lines long and much narrower, laterally compressed and furrowed at the dissepiment, truncate at the top, the capsule very slightly prominent, opening loculicidally in 2 valves, not splitting septicidally. Seeds small and angular.

Queensland. Broad Sound, *R. Brown;* basaltic pastures, Mackenzie and Isaacs rivers, and Peak Downs, *F. Mueller*.—This species, the only Australian one strictly belonging to the section *Oldenlandia*, is very closely allied to the E. Indian *H. maritima*, and further specimens may possibly show it to be a variety only.

3. **H. mitrasacmoides,** *F. Muell. Fragm.* iv. 37. A slender erect dichotomous glabrous annual of 1 ft. or more. Leaves few, narrow-linear. Stipules small, scarious, with 1 or 2 short bristle-like lobes on each side. Flowers very small, in a loose irregular 2–3-chotomous cyme or panicle, the floral leaves reduced to minute bracts. Pedicels solitary or 2 together, long and filiform. Calyx very small, with minute distant teeth. Corolla scarcely exceeding 1 line, the tube somewhat dilated upwards, the lobes rather shorter than the tube. Fruit ovoid or as broad as long, somewhat compressed, deeply furrowed at the dissepiment, $1\frac{1}{2}$ lines long. Capsule as long as the calyx-teeth, opening loculicidally and septicidally in 4 valves.

N. Australia. Depot Creek and Arnhem's Land, *F. Mueller*.

4. **H. galioides,** *F. Muell. Fragm.* iv. 38. A slender diffuse much-branched glabrous annual. Leaves narrow-linear. Stipules very small, with 1 or 2 small bristle-like lobes on each side. Pedicels filiform, axillary, solitary or 2 or 3 together. Calyx-tube globular, contracted at the orifice, the lobes acute, nearly as long as the tube, separated by narrow acute sinuses. Corolla-tube shorter than the calyx-lobes, the lobes scarcely so long as the

tube. Fruit ovoid-globular, 1 line long or rather more, contracted at the top, furrowed at the dissepiment. Capsule scarcely protruding, the carpels separating septicidally at the top, and opening each in 2 valves. Seeds small, angular.

N. Australia. Along rivulets near M'Adam Range, *F. Mueller;* S. Goulburn Island, *A. Cunningham.*

Queensland. Shoal Bay, *R. Brown;* Port Curtis, *M'Gillivray;* Cape Race and Broad Sound, *Bowman.*

The plant is undistinguishable from the slender forms of the common *H. Burmanniana* W. and Arn. (*Oldenlandia herbacea,* DC.), except by the longer connivent calyx-teeth and the capsule, which are rather those of *Hedyotis* than of *Oldenlandia.*

5. **H. scleranthoides,** *F. Muell. Fragm.* iv. 39. A divaricately-branched or diffuse glabrous annual of 2 or 3 in. Leaves linear, mucronate-acute, 3 to 4 lines long. Stipules short, usually with 2 bristle-like lobes on each side. Flowers very small, on very short pedicels, solitary or 2 together in each axil. Calyx-tube exceedingly short, broadly turbinate; lobes mucronate-acute, about $\frac{1}{2}$ line long. Corolla about as long as the calyx-lobes, divided nearly to the base. Fruit globular, scarcely $\frac{3}{4}$ line diameter. Capsule half-superior, divided to the base loculicidally and septicidally into 4 valves. Seeds nearly globular.

N. Australia. Depôt Creek, *F. Mueller.*

6. **H. elatinoides,** *Benth.* A diffuse glabrous annual, not 1 in. long in our specimen. Leaves oblong, not 2 lines long, narrowed into a petiole. Stipules very short, entire or scarcely toothed. Pedicels slender, as long as the leaves or rather longer. Flowers scarcely $\frac{1}{2}$ line long. Calyx-tube turbinate, with small distant teeth. Corolla divided nearly to the base, and persistent as in *H. tillæacea.* Capsule $\frac{3}{4}$ line broad, half-superior, opening loculicidally and septicidally in 4 valves. Seeds apparently peltate.

W. Australia, *Drummond, 4th Coll. n.* 108.

7. **H. tillæacea,** *F. Muell. Fragm.* iv. 39. Perennial, much-branched, erect or diffuse, rarely exceeding 3 or 4 in., glabrous or minutely pubescent. Leaves linear or linear-oblong, mostly obtuse, rarely above $\frac{1}{4}$ in. long. Stipules small, usually with 2 teeth on each side. Pedicels axillary, not exceeding the leaves. Calyx-tube turbinate, not $\frac{1}{2}$ line long; lobes linear, herbaceous, distant, rather longer than the tube. Corolla broadly campanulate, about 1 line long, divided nearly to the base, persistent on the ripe fruit. Capsule more than half-superior, compressed, about 1 line broad, opening loculicidally in 2 valves, not usually splitting. Seeds ovate, attached by the inner face, which is sometimes broad and flat, sometimes narrow and prominent.

Queensland. Between the Mackenzie and Dawson rivers, *F. Mueller;* Suttor river, *Bowman;* plains on the lower Maranoa, *Mitchell.*

N. S. Wales. Between the Darling river and Barrier Range, *Victorian Expedition.*

S. Australia. Wills's Creek and Cooper's Creek, *Howitt's Expedition.*

8. **H. trachymenioides,** *F. Muell. Fragm.* iv. 40. Perennial, erect, with numerous dichotomous branches, slender but rigid. Leaves linear-filiform, the upper ones very short. Stipular sheaths short, entire or with 1 or

2 short bristle-like lobes on each side. Pedicels slender in the upper axils, forming sometimes an irregular panicle. Calyx-tube very short, broad, with distant teeth. Corolla-tube slender, nearly 2 lines long; lobes much shorter. Fruit $1\frac{1}{2}$ lines broad and scarcely 1 line long, compressed, furrowed at the dissepiment, the capsule half exserted, opening loculicidally in 2 valves. Seeds rather numerous, orbicular, concave, thin, peltately attached by a prominent ridge on the concave face.

Queensland. Dawson river, *F. Mueller;* Isaacs river, *Bowman.*

This and the following species agree perfectly with the N. American *Houstonias* in flower, fruit, and seed, although with the habit of *Oldenlandia.*

9. **H. pterospora,** *F. Muell. Fragm.* iv. 40. Perennial, ascending or erect, much-branched, minutely pubescent or glabrous. Leaves narrow-linear. Stipular sheaths short, with 1 or 2 short bristle-like lobes on each side. Flowers mostly nearly sessile or very shortly pedicellate along the upper branchlets. Calyx about $\frac{1}{2}$ line long, with short distant spreading teeth. Corolla-tube slender, about 1 line long; lobes shorter than the tube. Fruit about $1\frac{1}{2}$ lines broad and not nearly so long, compressed, didymous, the capsule about $\frac{2}{3}$ exserted, opening loculicidally in 2 valves. Seeds few, large, orbicular, concave, so thin as to appear like a wing to the prominent ridge on the concave face by which they are attached.

N. Australia. Sturt's Creek, *F. Mueller.*

3. DENTELLA, Forst.

(Lippaya, *Endl.*)

Calyx-limb tubular, 5-lobed, persistent. Corolla-tube somewhat dilated upwards; lobes 5, usually 2- or 3-toothed, induplicate-valvate in the bud. Anthers included in the tube. Ovary 2-celled, with several ovules in each cell, attached to a placenta arising from near the base. Style with 2 linear stigmatic lobes. Capsule globular, crowned by the calyx-limb, 2-celled, scarcely dehiscent. Seeds more or less angular.—Prostrate herb. Stipules interpetiolar, entire or ciliate. Flowers solitary, sessile in the axils or forks.

The genus is limited to a single species.

1. **D. repens,** *Forst.; DC. Prod.* iv. 419. Stems from a perennial stock, prostrate or creeping, sometimes very small, forming dense patches of 1 or 2 in., sometimes extending to 1 or 2 ft., glabrous or hirsute with transparent almost scarious hairs. Leaves from ovate or oblong obtuse and petiolate, to lanceolate or linear and acute, under $\frac{1}{2}$ in. and often under $\frac{1}{4}$ in. long. Stipules short and scarious. Flowers sessile in the axils of the leaves or in the forks of the branches. Calyx-tube nearly globular, $\frac{1}{2}$ to $\frac{3}{4}$ lines diameter, usually very hispid; limb tubular, membranous, nearly 2 lines long, divided to about the middle into linear lobes. Corolla 2 to 3 lines long, the lobes shorter than the tube. Anthers linear. Capsule about $1\frac{1}{2}$ lines diameter, hispid with long transparent hairs.—W. and Arn. Prod. 405; *Lippaya telephioides,* Endl. Atakta, 13. t. 13.

N. Australia. Islands of the Gulf of Carpentaria, *R. Brown;* Fitzmaurice river and Sturt's Creek, *F. Mueller.*

Queensland. Dawson river, *F. Mueller;* Port Curtis, *M'Gillivray;* Wide Bay, *Bidwill;* Moreton Bay, *C. Stuart.*

N. S. Wales. Blue Mountains, *A. Cunningham.*
S. Australia. Between Stokes Range and Cooper's Creek, *M'Douall Stuart's Expedition.*
The species ranges over East India, the Archipelago, and islands of the S. Pacific.

4. OPHIORRHIZA, Linn.

Calyx-limb of 5 persistent lobes or teeth. Corolla-tube slender; lobes 5, valvate in the bud. Anthers included in the corolla-tube. Ovary 2-celled, with several ovules in each cell, attached to a placenta ascending from near the base. Style usually included, with 2 stigmatic lobes. Capsule much flattened and very broad, almost 2-lobed, opening loculicidally in 2 valves. Seeds several, angular.—Herbs or low straggling shrubs. Stipules united with the petioles in a very short truncate sheath, either entire or occasionally with 1 or 2 long hair-like points. Flowers sessile along the branches of terminal or rarely axillary pedunculate cymes.

A considerable genus, extending over tropical and eastern subtropical Asia. The only Australian species is endemic, although nearly allied to an East Indian one.

1. **O. australiana,** *Benth.* A low shrub with weak branches, the younger ones rusty-tomentose with short crisped hairs. Leaves ovate-lanceolate or elliptical, acutely acuminate, narrowed into a rather long petiole, mostly 2 to 3 in. long, but those of the same pair often unequal, sprinkled with appressed hairs on the upper side, pale underneath, with the veins more or less hirsute. Stipules with long hair-like points. Cymes pedunculate, shorter than the leaves. Bracts small, setaceous. Calyx-lobes linear, about as long as the hirsute tube. Corolla nearly glabrous outside, the tube scarcely above 1 line long, the lobes rather shorter, tomentose inside. Fruit about 3 lines broad and scarcely above 1 line long.

Queensland. Rockingham Bay, *Dallachy.* The species is nearly allied to the common Indian *O. Mungos*, which, however, appears to have the leaves always glabrous, and the corolla-lobes much shorter.

SUBTRIBE III. GARDENIEÆ.—Flowers in cymes clusters or solitary, Corolla-lobes imbricate, frequently contorted. Ovules attached to an axile or parietal placenta. Fruit succulent, indehiscent.

5. GARDENIA, Linn.

Calyx-limb tubular, truncate, toothed, lobed or divided to the base into 5 or more lobes. Corolla-tube cylindrical or slightly dilated upwards; lobes 5 or more, imbricate in the bud. Anthers nearly sessile, usually more or less exserted. Ovary 1-celled, incompletely divided by 2, 3 or rarely more projecting parietal placentas, with several ovules to each placenta. Style with 2, 3 or rarely more thick erect stigmatic lobes, or nearly entire. Fruit succulent, indehiscent, usually crowned by the calyx. Seeds numerous, immersed in the fleshy or pulpy placentas.—Shrubs or trees, the young shoots often exuding a resinous gum. Stipules solitary on each side, entire, more or less connate round the stem within the petioles, and often very deciduous. Flowers usually rather large and solitary or 3 together, terminal or axillary by the non-development of the flowering-branch.

The genus is confined to the Old World, spreading over the tropical and subtropical regions of Asia and Africa. As far as hitherto observed, all the Australian species appear to be endemic.

Calyx-limb large, scarcely toothed, but splitting on one side or into 2 lobes. Plant glabrous.
- Leaves under 1 in. long, narrowed at the base. Corolla-tube dilated upwards, longer than the calyx . . . . 1. *G. edulis.*
- Leaves broadly ovate. Corolla-tube cylindrical, scarcely exceeding the calyx-lobes . . . . 2. *G. resinosa.*

Calyx-limb with 4 to 6 linear obtuse lobes, rarely cohering. Young shoots pubescent.
- Leaves obovate or oblong, under 1½ in. Fruits under 1 in. long . 3. *G. pyriformis.*
- Leaves broadly ovate or orbicular, 2 to 3 in. long. Fruits 1½ to 2 in. long . . . . 4. *G. megasperma.*

Calyx-limb campanulate, truncate, with long subulate-acuminate teeth. Corolla-tube long and cylindrical.
- Quite glabrous. Calyx-limb without the teeth at least 4 lines long. 5. *G. Macgillivræi.*
- Foliage and flowers softly pubescent or villous. Calyx-limb without the teeth not 2 lines long.
  - Tall shrub or tree. Leaves broadly ovate . . . . 6. *G. ochreata.*
  - Low shrub or undershrub. Leaves narrow, oblong . . . . 7. *G. suffruticosa.*

Calyx-teeth distinct from the base. Fruit large. Plant glabrous.
- Leaves oblong, 1 to 2 in. long . . . . 8. *G. fucata.*
- Leaves ovate, 3 to 4 in. long . . . . 9. *G. Jardinei.*

Calyx(villous)-limb campanulate, with short teeth. Corolla-tube short, broad; lobes acute. Pericarp thin . . . . 10. *G. chartacea.*

We have also a specimen of a distinct large-leaved species, from Mount Bremer, Cape York, *W. Hill*, but insufficient for description.

1. **G. edulis,** *F. Muell. Fragm.* i. 54. A small tree, apparently glabrous, the young shoots resinous. Leaves small, obovate or oblong, narrowed into a short petiole, very obtuse, rarely above 1 in. long. Flowers rather small, white with a green tube, solitary or 3 together and almost sessile. Calyx-limb scarcely above 2 lines long, irregularly and shortly toothed, usually splitting on one side. Corolla-tube about 4 lines long, much dilated upwards; lobes 5 or 6, oval-oblong, rather shorter than the tube. Ovary with 3 or 4 parietal placentas. Fruit nearly globular, ½ to ¾ in. diameter, crowned by the remains of the calyx-limb.

N. Australia. Gilbert river and between Flinders and Lynd river, the "Breadfruit-tree" of Leichhardt, *F. Mueller.*

2. **G. resinosa,** *F. Muell. Fragm.* i. 54. A small tree, quite glabrous, the young shoots resinous. Leaves shortly petiolate, from broadly ovate to oval-oblong, obtuse at both ends, coriaceous, penninerved and reticulate, not exceeding 3 in. in our specimens. Flowers terminal, solitary or 3 together, shortly pedicellate. Calyx-tube contracted at the top; limb tubular-campanulate, ¾ to nearly 1 in. long, almost cartilaginous, 6-ribbed, scarcely toothed but splitting more or less into 2 lobes, especially on one side. Corolla-tube scarcely so long as the calyx; lobes 6, ovate, obtuse, 6 to 8 lines long. Placentas apparently 3. Style thickly clavate at the end with short connivent stigmatic lobes. Fruit only seen young.

N. Australia. Rocky hills, Victoria river, *F. Mueller, Bynoe.* Allied in some respects to the Indian *G. costata,* Wall., but at once known by the shorter corolla-tube.

3. **G. pyriformis,** *A. Cunn. Herb.* A shrub or tree, the specimens resembling those of *G. edulis*, but the young foliage and shoots hoary-tomentose or pubescent. Leaves obovate or oblong, ¾ to 1½ in. long, obtuse, narrowed into a short petiole, the older ones often glabrous. Flowers terminal, solitary, shortly pedicellate, larger than in *G. edulis.* Calyx-limb very shortly tubular-campanulate, with 4 to 6 linear lobes, very variable in breadth and length, from ¼ to ½ in. long, usually nearly equal in the same flower and quite distinct. Corolla-tube exceeding the calyx-lobes, scarcely dilated upwards; lobes broadly oblong, shorter than the tube, varying in number from 5 to 8 and often not of the same number as the calyx-lobes. Placentas 2 or 3. Fruit small, ovoid-globular or almost pear-shaped, crowned by the remains of the calyx-limb.

N. Australia. York Sound, N.W. coast, *A. Cunningham;* Victoria river, *Bynoe.*

4. **G. megasperma,** *F. Muell. Fragm.* i. 54. A shrub, with thick branches, the young shoots and buds hoary-pubescent, the older leaves glabrous or nearly so. Leaves petiolate or nearly sessile, broadly ovate or almost orbicular, very obtuse, rounded or cordate at the base, coriaceous, marked as in some other species with ciliate pits in the axils of the primary veins, but these may not be constant. Flowers terminal, solitary, nearly sessile, pubescent. Calyx-limb ribbed, 4 to 6 lines long, divided to about the middle into linear obtuse lobes, occasionally cohering; corolla-tube ¾ in. long, slightly dilated upwards; lobes 4 to 7, oblong, rather shorter than the tube. Fruits ovoid-oblong, nearly 2 in. long, crowned by the base of the calyx-limb.

N. Australia. Cambridge Gulf and Vansittart Bay, N.W. coast, *A. Cunningham;* rocky places, Victoria river, *F. Mueller;* islands of the Gulf of Carpentaria, *R. Brown.*

5. **G. Macgillivræi,** *Benth.* A small tree, quite glabrous or the calyx slightly pubescent. Leaves almost sessile, elliptical or obovate-oblong, acuminate, narrowed at the base, membranous and 3 to 4 in. long in our specimens, but still young. Flowers solitary or in threes, terminal but appearing lateral from the growing out of the new shoot, very shortly pedicellate, 6-merous. Calyx-tube ovoid, about 3 lines long; limb campanulate, at least 4 lines long, besides the long unequal subulate teeth. Corolla-tube 1½ in. long; lobes oblong, about 1 in. Placentas (in the ovary examined) 3. Style slightly thickened at the end. Fruit (if rightly matched) hard, ovoid, 1½ in. long.

Queensland. Cape York, *M'Gillivray, W. Hill.* M'Gillivray's specimens are in flower only, Hill's in fruit only, but they appear to belong to the same species.

6. **G. ochreata,** *F. Muell. Fragm.* i. 55, *and Rep. Burdek. Exp.* 11. A large shrub or small tree, the branches, under side of the leaves, inflorescence, flowers and fruit softly pubescent or villous. Leaves shortly petiolate, broadly ovate, obtuse, 2 to 4 in. long on the flowering-branches, the upper surface minutely pubescent or at length glabrous. Flowers 6-merous, terminal, solitary or in threes, very shortly pedunculate or sessile. Calyx-limb shortly campanulate, with subulate lobes much longer than the entire part. Corolla-tube about ¾ in. long; lobes nearly as long when fully out, though much shorter when first expanding. Placentas 3. Style slightly clavate at the end, entire. Fruit ovoid or nearly globular, 1 to 1½ in. long or rather more, said to be eatable when fresh.

Queensland. Grassy barren places, Burdekin river, *F. Mueller;* granite hills, Cape Upstart, *Fitzalan;* Mount Elliot, Edgecombe Bay, *Dallachy.*

7. **G. suffruticosa,** *R. Br. Herb.* Stems erect, under 1 ft. high, simple and leafless at the base, with 2 or 3 very short leafy branches, forming a tuft at the end. Leaves crowded, oblong, 1½ to 3 in. long, narrowed into a very short petiole, resinous and pubescent on both sides or at least underneath. Flowers solitary and nearly sessile in the fork of the branches. Calyx pubescent; limb campanulate, about 2 lines long, truncate, with 5 or 6 linear-subulate lobes longer than the entire part. Corolla not seen but, according to Brown's MSS., the tube is pubescent, 10 to 11 lines long, with 6 lobes, the anthers included in the tube. Placentas 3. Fruit globular, pubescent, about ¾ in. diameter.

N. Australia. Gulf of Carpentaria, *R. Brown.*

8. **G. fucata,** *R. Br. Herb.* An erect branching shrub, quite glabrous. Leaves oblong, obtuse at both ends or narrowed into the very short petiole, 1 to nearly 2 in. long. Flowers solitary, terminal. Calyx-tube slender, attenuated into a short pedicel; limb divided to the base into 5 or 6 linear lobes, about 3 lines long. Corolla-tube slender, about 6 lines long; lobes 5 or 6, broad, shorter than the tube.

N. Australia. Gulf of Carpentaria, *R. Brown.* I have not dissected the flower, but from R. Brown's notes, as well as from the stipules and resinous shoots, there can be no doubt of its belonging to the genus.

9. **G. Jardinei,** *F. Muell. Herb.* A tree, quite glabrous. Leaves very shortly petiolate, ovate, obtuse at both ends, 3 to 4 in. long in the specimens seen. Flowers not seen, the remains of the pedicels lateral from the new shoot having grown out. Fruit ovoid, glabrous, nearly 2 in. long, crowned by the remains of 5 distinct teeth, without any prominent entire limb. Placentas 3.

Queensland. Port Denison and Mount Elliott, *Dallachy.*

10. **G. chartacea,** *F. Muell. Rep. Burdek. Exped.* 12. Shrubby, the branches slender, pubescent, with short appressed hairs. Leaves on very short petioles or nearly sessile, opposite or whorled, from elliptical-oblong to linear-lanceolate, acuminate, 3 to 5 in. long, with prominent very oblique veins, glabrous above, sprinkled with appressed hairs underneath. Stipules of *Gardenia.* Flowers nearly sessile, apparently axillary, but perhaps really terminal at the base of a new shoot. Calyx hirsute, the tube about 2 lines long; limb about the same length, with 5 short teeth. Corolla-tube broad, almost ovoid, about 3 lines long; lobes 5, lanceolate, acuminate, much longer than the tube. Placentas 2. Fruit oblong, about ¾ in. long, the pericarp thin. Seeds enveloped in pulp.

Queensland. Brisbane river, Moreton Bay, *A. Cunningham, F. Mueller.*
N. S. Wales. Clarence river, *Beckler;* Mount Lindsay and Wilson's Creek, *Herb. F. Mueller.*

Var. (?) *latifolia,* F. Muell. Leaves ovate-elliptical, 4 to 6 in. long, the veins much more diverging than in *G. chartacea.* Flowers in terminal clusters, on short 3-fid peduncles, the pedicels long and villous. Calyx glabrous. Placentas 2.—Richmond river, *C. Moore.*—This has every appearance of being a distinct species, but as it is not in fruit, and the flowers, although glabrous, are not very different in shape, I have thought it for the present safer to retain it, as proposed by F. Mueller, under *G. chartacea*.

6. RANDIA, Linn.

(Stylocoryne, *Cav.*, not of others; Griffithia, *W. and Arn.*; Cupia, *DC.*; Gynopachys, *Bl.*)

Calyx-limb tubular campanulate or annular, truncate toothed or lobed. Corolla-tube cylindrical, short or long, rarely dilated at the top; lobes 5, imbricate (usually contorted) in the bud. Anthers nearly sessile, included in the tube or exserted. Ovary 2-celled, with several, usually numerous, ovules in each cell, attached to a fleshy peltate placenta. Style with 2 thick stigmatic lobes or entire. Fruit succulent, indehiscent, often crowned by the calyx-limb. Seeds several, immersed in the fleshy or pulpy placenta.—Shrubs or rarely trees, often, especially in species not Australian, armed with opposite axillary thorns. Stipules interpetiolar, solitary on each side, pointed, with a broad base but not united, often deciduous. Flowers in axillary cymes or clusters, or solitary at the summit of short branches or tufts of leaves.

A considerable genus, dispersed over the tropical regions of the New and the Old World.

Often thorny. Flowers solitary, pedicellate. Corolla-tube cylindrical, longer than the lobes . . . . 1. *R. Moorei.*
Unarmed. Flowers few, in very loose cymes. Corolla-tube oblong, nearly as long as the lobes . . . . 2. *R. Fitzalani.*
Unarmed. Flowers numerous in dense leaf-opposed cymes. Corolla-tube much shorter than the lobes . . . . 3. *R. densiflora.*

There are specimens also in R. Brown's collection of a shrub from Torres Straits, apparently allied to *Randia triflora*, but scarcely sufficient for accurate description.

1. **R. Moorei,** *F. Muell. Herb.* A shrub of 8 ft. (*C. Moore*), quite glabrous, producing axillary thorns, very small and few in the specimen, but probably sometimes large. Leaves ovate, $1\frac{1}{2}$ to 3 in. long, on short petioles. Stipules broad, with acute points. Pedicels axillary, slender, solitary, 1 to 4 lines long, 1-flowered, with a pair of small bracteoles at the end. Calyx about 2 lines long, the limb campanulate, truncate, obscurely toothed. Corolla-tube about 4 lines long, cylindrical, slightly dilated at the orifice; lobes 5, obtuse, not quite so long as the tube. Fruit not seen.

N. S. Wales. Tweed river, *C. Moore*. A single specimen in Herb. F. Mueller, apparently very nearly allied to the E. Indian *R. (Griffithia) fragrans*, W. and Arn.

2. **R. Fitzalani,** *F. Muell. Herb.* An unarmed tree, quite glabrous. Leaves obovate-oblong or elliptical, obtuse, narrowed into a rather long petiole, often above 6 in. long, smooth and shining. Stipules lanceolate, very deciduous. Cymes loose, few-flowered or the fertile flowers almost solitary, axillary at the base of the young shoot, appearing terminal before the branch grows out. Flowers half diœcious, the males with semiabortive ovaries, the females with imperfect stamens. Calyx-limb campanulate, truncate, nearly 3 lines diameter. Corolla-tube oblong, 4 to 6 lines long, slightly contracted at the orifice; lobes oblong, about as long as the tube or rather longer. Anthers long-linear, included in the tube. Style slightly thickened in the middle, with 2 short linear lobes. Fruit hard, globular and $1\frac{1}{2}$ in. diameter, or ovoid and longer.—*Gardenia Fitzalani*, F. Muell. Rep. Burdek. Exped. 12.

Queensland. Cape Upstart, Magnetical Island, etc., *Burdekin Expedition;* Endeavour river, *W. Hill;* Broad Sound, *Bowman;* Rockingham Bay, *Dallachy, W. Hill.*

3. **R. densiflora,** *Benth. Fl. Hongk.* 155. An unarmed shrub, glabrous except sometimes the flowers. Leaves oval-oblong or almost lanceolate, coriaceous, shining, 4 to 5 in. long. Flowers rather crowded in shortly pedunculate or almost sessile cymes, really axillary, although they appear leaf-opposed by the abortion of the subtending leaf. Calyx-limb cup-shaped, obscurely toothed. Corolla-tube about 1 line long, very hairy inside at the orifice; lobes oblong, nearly 3 lines long. Anthers exserted. Style linear, much exserted, the lobes not separating spontaneously. Berries small and globular or larger and ovoid. Seeds ovoid, compressed or angular, more or less immersed in the pulpy placenta.—*Cupia densiflora*, DC. Prod. iv. 394, and other synonyms quoted Fl. Hongk. l. c.; *Ixora Thozetiana*, F. Muell. Fragm. ii. 132.

N. Australia. Gulf of Carpentaria, *R. Brown* (from his MS.).

Queensland. Northumberland Islands, *R. Brown;* Port Denison, *Thozet, Dallachy;* Rockingham Bay and Rockhampton, *Dallachy.*

The species is common in the Eastern Archipelago. It is evidently closely allied to *Stylocoryne racemosa*, Cav., from the Philippine Islands, which I think should include *S. coffeoides*, A. Gray, as originally proposed by Hooker and Arnott. *Griffithia Gardneri*, Thw. Enum. Pl. Ceyl. 158, from Ceylon, is scarcely to be distinguished specifically from *R. densiflora*. *Stylocoryne Harveyi*, A. Gray, from the Fiji islands, is rather more distinct, but yet belonging to the same group of closely allied species.

7. WEBERA, Schreb.

(Stylocoryne, *W. and Arn. and others, not of Cav.*)

Calyx-limb short, truncate or 5-toothed, deciduous. Corolla-tube cylindrical; lobes 5, imbricate (usually contorted) in the bud. Anthers nearly sessile, more or less exserted. Ovary 2-celled with several ovules in each cell, attached to a peltate placenta. Fruit a globular berry. Style long, slightly thickened upwards, undivided. Seeds angular, without any or with a very thin pulp.—Shrubs or trees, without thorns. Stipules solitary on each side, pointed, with a broad base but not united. Flowers not usually numerous, in broad terminal cymes or corymbs.

The species are scattered over tropical Africa, Asia, and the S. Pacific islands. The Australian species is apparently endemic, although very closely allied to a common one in the S. Pacific.

1. **W. Dallachiana,** *F. Muell. Herb.* A tree of 20 to 30 ft. (*Dallachy*). Leaves petiolate, oval elliptical or oblong, acuminate, narrowed at the base, often 6 to 8 in. long. Flowers very numerous in a terminal trichotomous corymb shorter than the leaves. Calyx very small, the limb short, cup-shaped, obscurely toothed. Corolla-tube slender, about 5 lines long; lobes oblong, less than half the length of the tube; anthers linear. Style very long.

Queensland. Albany Island, *W. Hill*, Rockingham Bay, *Dallachy.*—In fruit and foliage the species is undistinguishable from the common *W. sambucina* (*Pavetta*, DC., *Stylocoryne*, A. Gray) of the Fiji and other Pacific islands, and the specimens have a similar tendency to dry black; the only difference I can detect is in the corolla-tube fully twice as long, but this appears to be constant.

8. DIPLOSPORA, DC.

(Discospermum, *Dalz.*)

Calyx-limb short, 4-toothed or truncate. Corolla-tube short, lobes 4, spreading, imbricate in the bud. Anthers exserted. Ovary 2-celled, with 2 or more ovules in each cell, attached to a small peltate placenta. Style with 2 stigmatic lobes. Fruit a globular berry. Seeds solitary or few in each cell.—Trees or shrubs. Stipules interpetiolar, pointed, with a broad base. Flowers in axillary clusters or close cymes, or in pairs of clusters with one terminal one, forming a short terminal raceme.

Besides the Australian species, which is endemic, the genus comprises 3 or 4 from tropical Asia.

1. **D. australis,** *Benth.* A small glabrous tree. Leaves ovate or elliptical, shortly and obtusely acuminate, narrowed into a short petiole, 2 to 3 in. long, coriaceous and shining. Stipules triangular, acute, deciduous. Peduncles 3 or 5 in a short terminal raceme, the lateral ones opposite in pairs, divaricate, each with 3 flowers sessile within a pair of concave bracts. Flowers altogether not 2 lines long. Calyx-limb cup-shaped, with 4 short rounded teeth or lobes. Corolla-tube very short and broad; lobes 4, much longer than the tube. Ovules about 3 to each placenta. Style short with oblong stigmatic lobes.

Queensland. Cape York, *W. Hill, M'Gillivray.*—The structure of the flower and ovary is precisely that of *D. viridiflora*, DC., although the inflorescence is so different. In both species at the time of flowering there are only 2 or 3 comparatively large perfect ovules in each cell, accompanied generally by 2 or 3 minute abortive ones which in some flowers are more perfect. *Discospermum*, Dalz., with the inflorescence and most of the characters of *D. viridiflora*, differs slightly, in both its species, in the ovules, but nevertheless the whole ought, as suggested by Thwaites, to be included in one genus.

TRIBE II. COFFEEÆ.—Leaves opposite or rarely whorled, with small or membranous stipules between or inside of them. Ovules solitary or very rarely 2 collateral in each cell of the ovary, rarely several to each carpel separated by spurious septa. Cells of the fruit or pyrenes always 1-seeded.

SUBTRIBE I. IXOREÆ.—Corolla-lobes imbricate, usually contorted. Ovules laterally attached, usually peltate. Fruit a berry or drupe. Albumen copious. Trees or shrubs.

The æstivation of the corolla is the most important character of this subtribe, which only differs from the smaller-fruited *Gardenieæ* in the solitary ovules. The peltate attachment of the ovules may not be quite constant even in the genus *Ixora.*

9. IXORA, Linn.

(Pavetta, *Linn.*)

Calyx-limb small, 4-toothed or lobed (rarely 5-toothed). Corolla-tube slender; lobes 4 (rarely 5), imbricate in the bud, usually contorted. Anthers usually exserted. Ovary 2-celled, with 1 ovule in each cell, peltately attached to the centre of the partition or rarely near the base. Style exserted, entire or divided at the end into 2 stigmatic lobes. Fruit a small globular berry or drupe, the endocarp not hard, forming 2 1-seeded pyrenes. Seeds broad, with the inner face flat or more frequently very concave.—Shrubs or small trees. Stipules interpetiolar, pointed, their broad bases often connate within the petioles. Flowers in terminal dense or large corymbs or panicles, or, in species not Australian, in smaller axillary or lateral cymes.

A large genus, widely dispersed over tropical Asia and Africa, with a few tropical American species. Of the seven Australian species three are common in E. India and the Archipelago, another extends at least to Timor, and the remaining three appear to be endemic, and very unlike any Asiatic species. The two Linnæan genera *Ixora* and *Pavetta* have been generally distinguished by the style,—2-lobed in *Ixora*, simple in *Pavetta*,—but owing to there being several species where the lobes rarely spread, and the stigma being really compound in all, Roxburgh, Korthals, Miquel, and others have united the two genera, and, as A. Gray appears to think there are good grounds for the union, I have followed their example. There are great differences in the form of the seeds in different species, but, as far as known, these do not coincide with differences in the style. How far the seeds may be made available for sectional distinction remains to be seen when those of more species shall have been observed.

SECTION I. **Pavetta.**—*Flowers* 4-*merous. Style slender, simple, or the lobes not separating.*

Leaves glabrous, usually narrow . . . . 1. *I. Pavetta.*
Leaves pubescent, at least underneath, usually broad . . . . 2. *I. tomentosa.*

SECTION II. **Ixora.**—*Flowers* 4-*merous. Style-lobes usually spreading.*

Cymes dense, sessile. Corolla-tube 1⅓ in. long; lobes acute, ¼ the length of the tube . . . . 3. *I. coccinea.*
Panicles loose. Corolla-tube 3 to 4 lines long; lobes oblong, nearly as as long as the tube . . . . 4. *I. Timorensis.*
Corymb rather dense, sessile. Corolla-tube 1½ lines long; lobes about the same length . . . . 5. *I. Becklerii.*
Peduncles very short, clustered, each with 3 sessile flowers. Corolla not 2 lines long, the tube very short. Leaves small, coriaceous, and shining . . . . 6. *I. triflora.*

SECTION III. **Pentadium.**—*Flowers* 5-*merous.*

Corymb rather dense, sessile. Corolla-tube about 1 line; lobes about 3 lines long . . . . 7. *I. pentamera.*

SECTION I. PAVETTA.—Flowers 4-merous. Style slender, simple, or the lobes not separating, usually very long. Specimens usually drying black. Flowers in sessile corymbs.

1. **I. Pavetta,** *Roxb. Fl. Ind.* i. 385. A tall shrub or small tree, glabrous or slightly pubescent. Leaves petiolate, oval-oblong or almost lanceolate, acute or acuminate, 3 to 4 in. long, narrowed at the base. Stipules very shortly acuminate, connected at the base within the petioles. Corymb loosely trichotomous, sessile above the last leaves. Calyx about 1 line long; limb loosely campanulate with minute teeth. Corolla-tube 4 to 5 or rarely 6 lines long; lobes oblong, much shorter. Fruits 2 to 3 lines diameter. Seeds hemispherical, very concave on the inner face.—*Pavetta indica*, Linn.; W. and Arn. Prod. 431; Wight, Ic. t. 148.

Queensland. East coast, *R. Brown;* Burdekin river, *F. Mueller;* Port Denison, *Fitzalan;* Edgecombe Bay, *Dallachy;* Curtis island, *Henne;* Rockhampton, *Thozet;* Brisbane river, Moreton Bay, *A. Cunningham*, *F. Mueller*, and others. The species is widely spread over E. India and the Archipelago.

2. **I. tomentosa,** *Roxb. Fl. Ind.* i. 386. A tall shrub or tree, closely

allied to *I. Pavetta.* Leaves rather broader and more obtuse, softly pubescent on both sides when young, rarely becoming glabrous above when old. Corymbs more dense than in *I. Pavetta,* the whole inflorescence and calyxes tomentose or hoary pubescent. Calyx-limb small, with minute teeth. Corolla fruit and seeds of *I. Pavetta.*—Wight, Ic. t. 186; *Pavetta tomentosa,* Sm.; W. and Arn. Prod. 431.

N. Australia. Careening Bay, N. W. Coast, *A. Cunningham;* Victoria river, *F. Mueller, Bynoe*; N. Coast, *R. Brown.*—Not so common in India as *I. Pavetta.* The Australian specimens quite agree with the ordinary Indian ones, both wild and from the Calcutta Botanic Garden. The Ceylon plant, designated by the same name in Thwaites's Enumeration, differs, as remarked by him, in the long narrow calyx-lobes, and is probably a distinct species.

SECTION III. IXORA.—Flowers 4-merous. Style with 2 stigmatic lobes usually spreading, rarely remaining coherent in species not Australian.

3. **I. coccinea,** *Linn.; W. and Arn. Prod.* 427. A tall glabrous shrub. Leaves nearly sessile, from oval or oblong with a more or less cordate base to cuneate-obovate, obtuse and mucronulate, or acute or shortly acuminate, rarely exceeding 3 in. on the flowering branches. Stipules with a fine subulate point from a broad base, shortly connate within the petioles. Cymes forming a dense corymb, nearly sessile above the last leaves, almost contracted into a head. Flowers orange-red. Calyx-limb acutely 4-lobed. Corolla-tube at least 1½ in. long; lobes acute, not above 4 lines long. Seeds hemispherical, very concave on the inner face.—Wight, Ic. t. 153 (not Bot. Mag. t. 169); *I. grandiflora,* Ker, Bot. Reg. t. 154; Wight in Hook. Bot. Misc. iii. 294. t. suppl. 35.

N. Australia. Port Essington, *Armstrong.*—Extends from Ceylon and the Indian Peninsula to the Archipelago. Although the specimens were gathered as part of the native vegetation, it is possible that this may have been one of the exotic shrubs planted during the time that Port Essington was colonized.

4. **I. Timorensis,** *Dcne. Herb. Tim. Descr.* 90. A small tree, quite glabrous. Leaves shortly petiolate, oval-oblong or oblong-elliptical, obtuse, acute, or shortly acuminate, or rarely ovate-lanceolate, 4 to 8 in. long, the uppermost floral pair occasionally but rarely sessile and cordate. Stipules shortly connate within the petioles. Flowers white, in large loose terminal panicles more or less pyramidal or rarely almost corymbose, all pedicellate or a few sessile in the last forks. Calyx small, the short limb truncate or obscurely toothed. Corolla-tube 3 to 4 lines long, slender, hairy inside at the orifice; lobes narrow-oblong, nearly or sometimes quite as long as the tube. Fruit globular, 2 to 3 lines diameter. Seeds not seen.—*I. Klanderiana,* F. Muell. Fragm. v. 18.

N. Australia. North coast, *R. Brown;* Port Essington, *Armstrong.*
Queensland. Cape York and neighbouring islands, *M'Gillivray, W. Hill;* Rockingham Bay, *Dallachy.*
The species extends to Timor, and possibly to other islands of the Archipelago. The only one among the E. Indian ones to which it can be compared is *I. undulata,* Roxb., which has a similar foliage and inflorescence as well as corolla, but the calyx-limb in that as in most Indian species is deeply toothed.

5. **I. Becklerii,** *Benth.* A tall shrub or small tree, quite glabrous.

Leaves ovate or elliptical, shortly acuminate, narrowed at the base, 3 to 4 in. long, smooth and shining. Stipules connate within the petioles, with fine subulate points. Corymbs rather dense, sessile and much shorter than the leaves. Calyx-limb short, irregularly toothed. Corolla-tube about 1½ lines long; lobes 4, about as long as the tube, acute. Style-lobes short, linear. Fruit 3 to 4 lines diameter. Seeds (not seen quite ripe) hemispherical, the inner face not concave, the testa wrinkled, but the albumen not ruminate.

N. S. Wales. Richmond and Clarence rivers, *Beckler.*

6. **I. triflora,** *R. Br. Herb.* A glabrous shrub with dichotomous branches. Leaves ovate or elliptical-oblong, obtuse or scarcely acuminate, shortly narrowed into the petiole, 1 to 1½ in. long, very coriaceous, smooth and shining. Stipules membranous, short, broad, and deciduous. Peduncles very short, in little terminal clusters, each with 3 to 5 very small sessile flowers. Calyx-limb short, obscurely toothed. Corolla not 2 lines long, glabrous inside; lobes 4, longer than the tube. Anthers exserted. Style shortly 2-lobed. Fruit small, ovoid, smooth. Seeds hemispherical, rugose, the inner face not concave, but not seen very perfect.

Queensland. East coast, *R. Brown;* Broad Sound, *Bowman;* Rockhampton, *Thozet, Dallachy.*—Very unlike any other *Ixora* known to me, but the ovules and the æstivation of the corolla are those of the genus.

SECTION III. PENTADIUM.—Flowers 5-merous. Style undivided.

7. **I. pentamera,** *Benth.* A shrub of 8 to 10 ft., the branches and inflorescence minutely hoary-pubescent. Leaves petiolate, oval-elliptical, 4 to 6 in. long, coriaceous, smooth and shining. Stipules broad, slightly connate within the petioles, deciduous. Flowers small in a nearly sessile rather dense corymb like that of *I. Becklerii.* Calyx pubescent; limb short, with 5 broad rounded short lobes or teeth. Corolla glabrous, the tube about 1 line, the lobes oblong, about 3 lines long. Anthers long-linear, exserted. Style long, slightly thickened towards the end, entire.—Fruit ovoid-globular, crowned by the calyx-limb, about 3 lines diameter. Pyrenes smooth. Seeds hemispherical, the inner face not concave, but the albumen ruminate as in the section *Grumilia* of *Psychotria.*

N. Australia. Islands of the Gulf of Carpentaria, *R. Brown.*—In the only flower examined, the ovules appeared to be attached near the base, but the æstivation of the corolla is much contorto-imbricated.

SUBTRIBE II. GUETTARDEÆ.—Corolla-lobes imbricate, not contorted. Ovules suspended from the summit of the normal or spurious cell (except in the upper cells of *Scyphiphora*). Fruit a drupe, the nucleus or pyrenes hard. Seeds attached by a thickened funiculus usually closing the orifice of the pyrene or cells of the nucleus. Albumen little or none. Embryo nearly terete, the cotyledons semiterete and scarcely broader than the radicle.—Trees or shrubs. Flowers in forked cymes, rarely in umbels, or when polygamous the females often solitary.

The exceptional characters of this subtribe have been well pointed out by A. Gray ('Notes on some *Rubiaceæ*'), except that it is much to be doubted whether it includes any genus with really valvate corolla-lobes.

10. TIMONIUS, Rumph.

(Polyphragmon, *Desf.*)

Calyx-limb truncate or irregularly toothed. Corolla-tube cylindrical; lobes 4 or more, short, thick, obtuse, with a central rib prominent inside, the margins more or less imbricate in the bud. Anthers included in the tube. Ovary (normally 5- to 10-celled with several ovules to each cell) divided by spurious dissepiments between the ovules into very numerous 1-ovulate cells, superposed in several irregular rows. Style divided into about 5 to 10 linear lobes. Fruit a drupe, with exceedingly numerous oblong-linear 1-seeded pyrenes, closely packed and diverging in many rows from the axis. Seeds of *Guettarda*.—Trees or shrubs. Stipules membranous, so deciduous as to be rarely seen. Flowers polygamo-diœcious, on axillary peduncles, the females (with small or imperfect stamens) usually solitary, the males (with an abortive ovary) 3 or more together, sessile in the forks or along the branches of pedunculate cymes.

The genus consists of a few species, dispersed over the Archipelago and islands of the South Pacific, the Australian species extending to Sumatra and Amboyna. The peculiar seed of this and other *Guettardeæ* has been very accurately described by A. Gray in the above-mentioned notes. I do not find, however, that *Timonius* is so exceptional in the æstivation of the corolla as it appeared to him, but it is difficult to observe. In the bud the lobes adhere so closely as to require much soaking to open them without tearing, and, when open, they are so thick as to convey the idea that they must have been valvate. But, on examining buds just ready to burst, I have found the margins overlapping, both in *T. Rumphii* and *T. Forsteri*.

1. **T. Rumphii,** *DC. Prod.* iv. 461. A tall shrub or small tree, either glabrous except the inflorescence, or the young shoots silky-hairy, and the older leaves sprinkled with long soft hairs. Leaves from ovate-elliptical to oblong-lanceolate, acuminate, narrowed into a petiole, mostly 3 to 5 in. long. Male flowers several in a forked cyme. Calyx-limb tubular, 2 to nearly 3 lines long, truncate or irregularly toothed, the ovary quite abortive. Corolla tomentose, the tube about 4 lines long; lobes 4 to 10, oblong-linear, rather more than half as long as the tube. Style rudimentary. Female flowers solitary, resembling the males, except that the stamens are small and the ovary perfect, the 1-ovulate cells exceedingly numerous. Style with about 5 to 10 linear unequal lobes. Fruit globular, about ½ in. diameter, crowned by the calyx-limb. —*Polyphragmon sericeum*, Desf. in Mem. Mus. vi. 6. t. 2; DC. Prod. iv. 445; *Guettarda polyphragmoides*, F. Muell. Fragm. ii. 134.

N. Australia. Careening Bay, N.W. coast, *A. Cunningham;* Upper Victoria and Fitzmaurice rivers, *F. Mueller;* Port Essington, *Armstrong;* Sweers Island, *Henne;* Upper Lynd river, *Leichhardt.*

Queensland. Cape York, *M'Gillivray;* Percy Islands, *A. Cunningham*; Rockingham Bay, *W. Hill;* Rockhampton, *Dallachy* and others.

The species is also in Timor, Amboyna, Sumatra, and probably in other islands of the Archipelago.

11. SCYPHIPHORA, Gærtn.

(Epithinia, *Jack.*)

Calyx-limb truncate or minutely toothed. Corolla-tube cylindrical or slightly dilated upwards; lobes 4 or rarely 5, imbricate in the bud. Anthers linear-sagittate, exserted. Ovary really 2-celled, but each cell divided by a spurious dissepiment into 2 superposed ones, with one ovule in each, the upper ovule erect, the lower one pendulous. Style filiform, with 2 short linear stigmatic lobes. Fruit a drupe with a hard endocarp scarcely separable into pyrenes, with 4 1-seeded cells superposed in pairs, or fewer by abortion. Seeds of *Guettarda.*—Shrub. Stipules interpetiolar, broad and short. Flowers in small pedunculate axillary cymes.

The genus consists of a single species, extending from Ceylon over the Indian Archipelago.

1. **S. hydrophylacea,** *Gærtn.; DC. Prod.* iv. 577. A shrub of several feet, quite glabrous, the young shoots resinous. Leaves obovate, very obtuse, narrowed into a rather long petiole, 1½ to nearly 3 in. long, coriaceous smooth and shining. Cymes dense, very shortly pedunculate. Corolla-tube 1½ to nearly 2 lines long, hairy inside at the orifice; lobes ovate-oblong, rather obtuse. Drupe oblong, crowned by the calyx-limb, longitudinally ribbed and furrowed, 3 to 4 lines long. Albumen present but very scanty. —*Epithinia Malayana,* Jack.; DC. Prod. iv. 478.

Queensland. Albany Island, Cape York, *M'Gillivray, W. Hill.*

The species appears to be common on the coasts of Ceylon, of the islands of the Archipelago, and of the Malayan Peninsula. The figures and descriptions of Gærtner, Fr. iii. 91. t. 196, and of A. Rich in Mém. Soc. Hist. Nat. v. 159. t. 14, are incorrect in many particulars. The only accurate account of the structure of the ovary and fruit I am aware of is that of A. Gray, Not. Rub. 19.

12. ANTIRRHÆA, Juss.

Calyx-limb 4-lobed. Corolla-tube slender; lobes 4, imbricate in the bud. Anthers included in the tube. Ovary 2-celled, with 1 pendulous ovule in each cell. Style filiform with 2 short linear stigmatic lobes. Fruit a drupe, the nucleus hard, separating into 2 1-seeded pyrenes. Seeds of *Guettarda.* —Shrubs or trees. Stipules interpetiolar, acuminate, deciduous. Flowers often polygamous, sessile on the branches of a forked cyme or the females solitary, on axillary peduncles.

The genus comprises several species from the Mauritius and Madagascar; one is quoted from Sumatra, and it may possibly include the *Chomelia* from the Sandwich Islands, described by A. Gray. The Australian species appears to be endemic, but I have not had the opportunity of comparing it with Korthals' *A. strigosa* from Sumatra. The genus is perhaps too closely allied on the one hand to *Chomelia,* and on the other to *Guettarda* and *Guettardella.*

1. **A. tenuiflora,** *F. Muell. Herb.* A shrub or tree, glabrous or the young parts silky-pubescent. Leaves oval-elliptical, acuminate, narrowed into a petiole, 3 to 6 in. long, membranous, glabrous above, the underside often sparingly pubescent. Peduncles axillary, forked, with several sessile flowers along the branches. Calyx scarcely 1 line long, the lobes short and spreading. Corolla nearly glabrous, the slender tube between 4 and 5 lines long, the lobes ovate, about 1 line. Fruit not seen.

Queensland. Rockingham Bay, *Dallachy.*

13. GUETTARDELLA, Champ.

Calyx-limb deeply 4- to 6-lobed Corolla-tube slender; limb 4- to 6-lobed, imbricate in the bud. Anthers included in the tube. Ovary 4- to

6-celled, with 1 pendulous ovule in each cell. Style divided at the top into as many linear lobes as cells to the ovary. Fruit a drupe, the hard putamen with 4 to 6 1-seeded cells or separating into as many pyrenes.—Seeds of *Guettarda*, the albumen thin or none.—Shrubs usually slender. Stipules interpetiolar, broad, pointed. Flowers rather small, probably polygamous, sessile on the branches of a forked axillary pedunculate cyme or solitary.

A small genus, containing besides the single Australian species which is endemic, one from the Philippine Islands and one from S. China. It differs slightly from *Guettarda* and *Bobea* in the slender habit and in the calyx, and from *Antirrhœa* and *Chomelia* in the parts of the gynœcium more than two.

1. **G. putaminosa,** *Benth.* Apparently shrubby, much-branched, slender, glabrous or the young parts silky-pubescent. Leaves from obovate to oblong, very obtuse, narrowed into a short petiole, rarely above 1 in. long and mostly smaller, smooth and shining. Peduncles slender, axillary, bearing either 1 or a cyme of 3 flowers, only seen in an advanced state. Calyx-lobes small, obtuse. Corolla-tube very slender, about 2 lines long, silky-pubescent; lobes 4 to 6, very obtuse, about ½ line long and broad. Drupe ovoid or oblong, about 3 lines long, glabrous, the putamen hard, 5- or 6-celled, or (when not quite ripe?) separating into as many pyrenes.—*Bobea putaminosa* or *Timonius putaminosus*, F. Muell. Fragm. iv. 92.

Queensland. Rockhampton, *Thozet.*

14. GUETTARDA, Linn.

Calyx-limb truncate or irregularly toothed. Corolla-tube cylindrical; lobes 4 or more, imbricate in the bud. Anthers included in the tube. Ovary 4- or more-celled, with 1 pendulous ovule in each cell; style with as many linear lobes as cells of the ovary. Fruit a drupe, with a hard several-celled nucleus. Seeds solitary in each cell, oblong, cylindrical or curved, funiculus thick, closing the orifice of the cell; testa thin; albumen none or very thin; embryo the shape of the seed, the cotyledons not broader than the superior radicle.—Trees or shrubs often tomentose. Stipules interpetiolar, broad, acuminate, deciduous. Flowers sessile along the branches of a forked cyme, pedunculate in the axils.

The genus is spread over the tropical regions of both the New and the Old World. The only Australian species is common on seacoasts from Eastern Africa to the Pacific.

1. **G. speciosa,** *Linn.; DC. Prod.* iv. 455. A coarse shrub, attaining 5 or 6 ft., the young branches thick, often flattened, gummy and glabrous or tomentose. Leaves shortly petiolate, broadly obovate-orbicular or ovate, very obtuse, rounded or slightly cordate at the base, 6 to 10 in. long or even more, glabrous above, softly pubescent tomentose or nearly glabrous underneath. Flowers large, in rather dense cymes. Calyx-limb truncate, deciduous, 1 to 1½ lines long. Corolla-tube above 1 in., sometimes 1½ in. long; lobes 4 to 9, oblong, obtuse, not ½ in. long. Ovary-cells 4 to 9, usually 5 or 6. Drupe nearly globular, attaining 1 in. diameter, chiefly consisting of the very hard woody endocarp, more or less lobed, the interstices filled with a hard fibrous mesocarp, the cells and seeds small and curved.—Wight, Ic. t. 40.; Bot. Reg. t. 1393.

N. Australia. Port Essington, *Armstrong.*

Queensland. Torres Straits, *R. Brown,* and along the coast and adjoining islands from thence to Edgecombe Bay and Port Denison, *F. Mueller, W. Hill,* and others.

15. HODGKINSONIA, F. Muell.

Calyx-limb minute, 4-toothed. Corolla-tube ovoid; lobes 4, very short, obtuse (slightly imbricate?). Anthers included, the filaments inserted near the base of the corolla-tube. Ovary 2- to 4-celled, with 1 pendulous ovule in each cell; style with as many linear-lobes as ovary-cells. Fruit a drupe, with a thick hard 2- to 4-celled putamen. Seeds of *Guettarda,* the albumen present but scanty.—Tree. Stipules interpetiolar, acuminate, very deciduous. Flowers polygamo-diœcious, umbellate, on slender axillary peduncles.

The genus consists of a single species, endemic in Australia, allied in several respects to *Guettardella,* but with a different inflorescence and corolla.

1. **H. ovatiflora,** *F. Muell. Fragm.* ii. 132. A tree with slender branches, glabrous or the young shoots with a few appressed hairs. Leaves petiolate, elliptical or ovate, obtuse or obtusely acuminate, narrowed at the base, the veins not prominent, 1½ to 3 in. long. Peduncles slender, shorter than the leaves, bearing either a single umbel, or also a pair of lateral branches each with an umbel, each umbel on the male plant with 10 to 12 flowers, on the female with only 3 to 6. Pedicels short. Calyx-limb very small. Corolla in the males ovoid, almost urceolate, fully 2 lines long, the lobes very short, thick, and obtuse, the ovary abortive. Corolla in the females much smaller and nearly globular; anthers small; style short. Drupe small, ovoid or globular.

Queensland. Wide Bay, *Bidwill;* Rockhampton, *Dallachy;* Brisbane river, Moreton Bay, *Fraser, F. Mueller.*

N. S. Wales. Clarence and Richmond rivers, *Beckler.*

SUBTRIBE III. VANGUERIEÆ.—Corolla-lobes valvate. Ovules laterally attached at or near the top. Fruit a berry-like drupe with 1-seeded pyrenes. Albumen copious.—Trees or shrubs.

16. CANTHIUM, Lam.

Calyx-limb short, more or less toothed. Corolla-tube short or cylindrical; lobes 4 or 5, valvate in the bud. Anthers exserted or rarely included in the tube. Ovary 2-celled, with 1 ovule in each cell, laterally attached near or at the top. Style exserted, with a thick ovoid or mitre-shaped entire or 2-lobed stigma. Fruit a globular compressed or didymous drupe, with 1 or 2 one-seeded pyrenes.—Shrubs either unarmed or with axillary thorns. Stipules interpetiolar, pointed, with a broad base. Flowers in axillary cymes or clusters.

A considerable genus, extending over tropical Africa, Asia, and the Pacific Islands. Of the seven Australian species one extends into the Pacific Islands, the others appear to be endemic.

Stigma ovoid or mitre-shaped, entire or very shortly 2-lobed.
  Flowers in pedunculate cymes, the lobes of the corolla longer
  than the tube.

Leaves broadly ovate, prominently penniveined and reticulate . . 1. *C. latifolium.*
Leaves oblong-elliptical to narrow-lanceolate, prominently and obliquely penniveined . . . . 2. *C. attenuatum.*
Leaves ovate to oblong-elliptical (2 to 6 in. long) very smooth and shining, the veins scarcely prominent . . . . 3. *C. lucidum.*
Leaves oblong, rarely 1½ in. long, scarcely shining, the veins rarely conspicuous. Flowers about 3 lines long . . . . . 4. *C. oleifolium.*
Leaves ovate, rarely above 1 in. long, smooth and shining. Flowers not 2 lines long . . . . 5. *C. buxifolium.*
Stigma thick, deeply 2-lobed. Flowers 2 or 3 together. Lobes of the corolla narrow, rather shorter than the tube. Leaves small . . . . 6. *C. vacciniifolium.*
Stigma broad, peltate, entire. Flowers in sessile clusters. Lobes of the corolla much shorter than the tube . . . . 7. *C. coprosmoides.*

1. **C. latifolium,** *F. Muell. Herb.* A glabrous and apparently glaucous shrub, nearly allied to *C. lucidum.* Leaves broadly ovate, very rigid, the pinnate veins and reticulations much more prominent than in *C. lucidum.* Flowers much smaller than in that species, and in looser cymes, otherwise their structure as well as the inflorescence the same.

N. S. Wales. In the interior towards the Barrier Range (*Nielson ?*) in Herb. F. Mueller.

S. Australia. N.W. interior, *M'Douall Stuart.*

2. **C. attenuatum,** *R. Br. ms.* A glabrous shrub, very nearly allied to *C. lucidum.* Leaves narrower, from oblong-elliptical and about 2 in. to narrow-lanceolate and 5 or 6 in. long, the pinnate veins much more oblique and more prominent than in *C. lucidum.* Flowers nearly the same, but in shorter and more dense cymes, and the tube of the corolla longer in proportion to the lobes.

N. Australia. Brunswick Bay, N.W. coast, *A. Cunningham;* Victoria River and Arnhem's Land, *F. Mueller;* N. coast, *R. Brown;* Sweers Island, *Henne.*

Queensland. Burdekin and Burnett rivers, *F. Mueller;* Port Denison, *W. Hill, Bowman;* St. George's Bridge on the Balonne, *Mitchell.*

3. **C. lucidum,** *Hook. and Arn. Bot. Beech.* 65. A tall shrub or small tree, perfectly glabrous. Leaves ovate, obovate or elliptical-oblong, obtuse or scarcely acuminate, narrowed into a short petiole, scarcely exceeding 2 in. in some specimens, 4 to 6 in. long in others, coriaceous, very smooth and shining, with distant very oblique veins scarcely prominent. Cymes axillary, shortly pedunculate, often large and many-flowered but shorter than the leaves. Pedicels short or sometimes the flowers sessile, except those in the forks. Corolla glabrous outside, slightly hairy inside, the tube about 1 line; lobes about 2 lines long. Anthers exserted. Stigma thick, ovoid, more or less mitre-shaped (hollowed at the base round the style). Fruit, when both carpels ripen, somewhat compressed and didymous, 3 to 4 lines broad, but often one-seeded and nearly globular.—*C. lamprophyllum*, F. Muell. Fragm. ii. 133.

N. Australia. Gulf of Carpentaria, *R. Brown.*

Queensland. E. coast, *R. Brown;* Dawson and Burnett rivers, *F. Mueller;* Port Denison, *Fitzalan;* Edgecombe Bay, *Dallachy;* Rockhampton, *Thozet;* Brisbane river, Moreton Bay, *A. Cunningham, F. Mueller*, and others.

N. S. Wales. Clarence river, *Beckler;* Tweed river, *C. Moore.*

Some of the specimens are precisely the same as those from the Sandwich, the Fiji, and other islands of the N. and S. Pacific, though in some of the latter the pedicels are longer, showing an approach to the inflorescence of the common *C. didymum*, Gærtn. The flowers are, however, in the Australian as well as in the Pacific specimens, almost constantly 4-merous, and usually 5-merous in *C. didymum.*

4. **C. oleifolium,** *Hook. in Mitch. Trop. Austr.* 397. A tall glabrous shrub, sometimes glaucous, a few branchlets occasionally degenerating into short spines. Leaves oblong, obtuse, narrowed into a short petiole, rarely above 1½ in. long in the flowering specimens, larger in barren ones, thick and smooth but scarcely shining, the veins usually inconspicuous. Flowers in short almost sessile axillary cymes, rather smaller than in *C. lucidum*, and varying in the number of parts 4 or 5. Corolla-tube nearly as long as the lobes, the flowers otherwise the same as in *C. lucidum.* Fruit also the same, didymous when both carpels ripen.

Queensland. Burdekin river, *F. Mueller;* Suttor river, *Sutherland.*

N. S. Wales. Plains of the Gwydir, *Mitchell;* Castlereagh river, *C. Moore;* Darling river to Cooper's Creek, *Nielson.*

5. **C. buxifolium,** *Benth.* Glabrous and much-branched. Leaves ovate or broadly elliptical, obtuse or obscurely and obtusely acuminate, narrowed into a short petiole, rarely exceeding 1 in. in length, coriaceous, very smooth and shining, the veins few, very oblique and scarcely conspicuous. Flowers 4-merous, very small, rather numerous, in pedunculate cymes about as long as the leaves, the pedicels short except those in the forks. Corolla not 2 lines long, the tube exceedingly short, glabrous inside, the lobes much longer. Stamens exserted. Stigma mitre-shaped. Fruit of *C. lucidum*, or rather smaller.

Queensland. Burnett and Dawson rivers, *F. Mueller;* also in *Leichhardt's* collection.

The preceding five species are certainly very nearly allied to each other, differing chiefly in foliage and in the size of the flowers and length of corolla-tube, but they can scarcely be united into one species without adding them all to the *C. didymum*, Gærtn.

6. **C. vacciniifolium,** *F. Muell. in Trans. Phil. Inst. Vict.* iii. 47. A shrub attaining 16 ft. or more, with very numerous slender divaricate branches, the smaller branchlets sometimes spinescent. Leaves petiolate, from broadly obovate to oblong, obtuse, ¼ to nearly ½ in. long, the veins scarcely conspicuous. Flowers usually 4-merous, 2 or 3 together in little axillary cymes, the common peduncle and pedicels very short and slender. Corolla about 3 lines long, the lobes narrow, acute, rather shorter than the tube. Stamens exserted. Stigma ovoid, divided to the base into 2 thick lobes. Fruit, when ripening both carpels, a little more than 2 lines diameter, the pyrenes not so hard as in *C. lucidum*, and especially as in *C. coprosmoides.*—*C. microphyllum*, F. Muell. Fragm. ii. 134.

Queensland. Cairncross island and Suttor river, *F. Mueller;* Mount Wyatt, *Bowman;* Kent's Lagoon, *Leichhardt;* Brisbane river, Moreton Bay, *F. Mueller, C. Stuart.*

N. S. Wales. Macleay river, *Beckler.*

7. **C. coprosmoides,** *F. Muell. in Trans. Phil. Inst. Vict.* iii. 47. A tall shrub or small tree, quite glabrous. Leaves obovate ovate or broadly elliptical, obtuse, shortly contracted at the base, in some specimens all under 2 in., in others 3 to 4 in. long, coriaceous but scarcely shining, the veins

distant and not prominent. Flowers 4-merous or 5-merous, very shortly pedicellate, in sessile axillary clusters of 3 to 6. Corolla-tube slender, fully 4 lines long, bearded inside at the orifice, the lobes about ½ as long as the tube. Anthers slightly protruding. Style exserted, with a broad thick peltate stigma. Fruit sometimes ½ in. broad, on a pedicel of 2 to 4 lines.

Queensland. Port Denison, *Fitzalan;* Edgecombe and Rockingham Bays, *Dallachy, W. Hill;* Rockhampton, *Thozet;* Dawson and Brisbane rivers, *F. Mueller.*

N. S. Wales. Port Jackson to the Blue Mountains, *R. Brown* and others; Hastings and Clarence rivers, *Beckler.*

The species is very closely allied to *C. barbatum*, Benth., from the Pacific Islands, but the leaves are more coriaceous and obtuse, the pedicels shorter, and the corolla-lobes more obtuse.

SUBTRIBE IV. PSYCHOTRIEÆ.—Corolla-lobes valvate. Ovules erect from the base or laterally attached below the middle. Style-lobes short. Fruit a berry-like drupe with 1-seeded pyrenes. Albumen copious.—Trees or shrubs.

17. MORINDA, Linn.

Flowers usually several together, united at the base into a small head. Calyx-limb short, scarcely toothed. Corolla-tube cylindrical or slightly dilated at the top; lobes 5, rarely 4, valvate in the bud. Anthers included in the tube or rarely exserted. Ovary 2-celled or more or less completely 4-celled, with 1 ovule in each cell, laterally attached at the base or below the middle; style exserted, with 2 stigmatic lobes or rarely entire. Fruits of each flower-head united in a compound succulent berry, including a number of hard 1-seeded pyrenes, usually 2 to 4, proceeding from each flower.—Shrubs or small trees, or sometimes woody climbers. Stipules usually membranous and united within the petioles in a short sheath. Flower-heads on axillary or terminal solitary or clustered peduncles.

A considerable tropical genus, chiefly Asiatic or African, with 2 or 3 American species. Of the 4 Australian species, one is common in tropical Asia, another as widely distributed over the seacoasts of southern Asia and the Pacific, the two others are endemic.

Peduncles solitary, apparently leaf-opposed. Leaves very large . . 1. *M. citrifolia.*
Peduncles 2 together at the ends of the branches . . . . 2. *M. jasminoides.*
Peduncles 4 or more together at the ends of the branches.
  Leaves ovate to oblong-lanceolate, not much veined. Flower-heads without prominent bracts . . . . 3. *M. umbellata.*
  Leaves broadly ovate or orbicular, coriaceous and prominently reticulate. One large coloured leafy bract to each flower-head . . 4. *M. reticulata.*

1. **M. citrifolia,** *Linn.; DC. Prod.* iv. 446. A tall glabrous shrub, with thick more or less 4-angled branches. Leaves large, ovate, broad or narrow, mostly 6 to 10 in. long, on very short petioles. Stipules large, membranous. Flower-heads on very short peduncles, apparently leaf-opposed from the abortion of the subtending leaf, without prominent bracts. Flowers numerous, the calyx-tubes quite connate. Corolla-tube ¼ to ½ in. long; lobes shorter than the tube. Ovary 2-celled, the ovules ascending, attached below or near the middle. Fruit forming a pulpy mass above 1 in. diameter, the pyrenes orbicular, flattened, about 3 lines diameter.

Queensland. Along the coast and adjoining islands, from Albany island and Cape York to Percy islands and Edgecombe Bay, *A. Cunningham, F. Mueller,* and others. Common on the seacoasts of tropical Asia and especially of the Pacific islands. The Australian specimens are in leaf, in fruit, or in very young bud, and the flowers are described from Asiatic specimens. The fruits received from F. Mueller, as those of the "Leichhardt Tree," or *Sarcocephalus Leichhardtii*, Rep. Burdek. Exped. 12, belong to *Morinda citrifolia.*

2. **M. jasminoides,** *A. Cunn.; Hook. Bot. Mag. t.* 3351. A tall glabrous shrub, attaining sometimes 20 ft., with weak straggling branches. Leaves from nearly ovate to oblong-lanceolate, acuminate, narrowed into the petiole, 1½ to 3 in. long. Stipules connate, deciduous. Peduncles slender, 2 together at the ends of the branches, each with a small head of 6 to 12 or even more flowers, the calyx-tubes quite connate or immersed in the receptacle. Corolla 3 to 4 lines long, the tube straight, usually shorter than the lobes. Ovary 4-celled; style 2-lobed. Drupes concrete, forming a globular compound berry about ½ in. diameter.

Queensland. Brisbane river, Moreton Bay, *F. Mueller; C. Stuart.*

N. S. Wales. Port Jackson to the Blue Mountains, *R. Brown, Woolls*, and others; northward to Hastings and Clarence rivers, *Beckler;* southward to Illawarra, *Shepherd.*

Victoria. Snowy and Broadribb rivers, *F. Mueller.*

The species is very nearly allied to *M. umbellata*, differing chiefly, but apparently constantly, in the peduncles, never more than 2 together and the flowers large.

3. **M. umbellata,** *Linn.; W. and Arn. Prod.* i. 420. A trailing diffuse or somewhat climbing shrub, glabrous or the young branches slightly pubescent. Leaves from ovate-oblong or obovate to oblong-lanceolate, 1½ to 3 in. long. Peduncles ¼ to ½ in. long, 4 to 8 together (usually about 6) at the ends of the branches, each with a small head of about 6 to 12 flowers, the calyx-tubes quite connate or immersed in the receptacle. Corolla scarcely 3 lines long, the tube straight, rather shorter than the lobes. Ovary 4-celled; style 2-lobed. Drupes forming a compound globular berry, 4 to 6 lines in diameter.

Queensland. Rockingham Bay, *Dallachy.* The species has a wide range over E. India and the Archipelago.

4. **M. reticulata,** *Benth.* A low struggling shrub, quite glabrous. Leaves petiolate, orbicular or broadly ovate, very shortly and acutely acuminate, about 2 to 3 in. long, coriaceous and prominently reticulate, like those of *Cœlospermum reticulatum.* Stipules triangular, acute. Peduncles 4 together at the ends of the branches, bearing each a head of about 6 to 12 flowers, one large orbicular petiolate coloured bract to each head, adnate to one of the calyxes, like those of *Mussænda.* Calyx-tubes partly immersed in the receptacle. Corolla-tube slender, about 6 lines long; lobes about 2 lines long. Anthers exserted. Ovary (if rightly observed) 2-celled, with 2 collateral ovules in each cell, attached about the middle. Style long, with two short stigmatic lobes.

Queensland. Albany island, *W. Hill;* N.E. coast, *A. Cunningham.*

In the only two flowers I detached from a head to examine, I did not feel quite certain whether there were 2 or 4 ovules to each flower.

18. CŒLOSPERMUM, Blume.

(Pogonolobus, *F. Muell.*)

Calyx-limb truncate or obscurely toothed. Corolla-tube cylindrical or slightly dilated at the top; lobes 4 or 5, valvate in the bud. Anthers exserted, linear. Ovary 2-celled, with 2 ovules in each cell, laterally almost peltately attached on each side of a very prominent placenta; style 2-lobed. Fruit a drupe, with 4 distinct hard 1-seeded pyrenes.—Shrubs with straggling or climbing branches. Stipules interpetiolar, acuminate, separate or shortly connate within the petioles. Flowers in umbels clusters or cymes, terminal and solitary or forming terminal thyrsoid panicles.

The genus, if rightly identified, extends over the Indian Archipelago, the two Australian species, however, being endemic. Although I have not seen the two species described by Blume and Korthals (but unknown to Miquel), I have little doubt of their being congeners. The chief character consists in the ovules and 1-seeded pyrenes being twice the number of the carpels and style-lobes, as in several species of *Morinda,* from which *Cœlospermum* differs in its separate flowers.

Leaves smooth and shining, the veins not very prominent. Cymes rather dense, in oblong panicles. . . . . 1. *C. paniculatum.*
Leaves rigid, prominently veined and reticulate. Umbels solitary . . 2. *C. reticulatum.*

1. **C. paniculatum,** *F. Muell. Fragm.* v. 19. A woody climber, quite glabrous. Leaves petiolate, ovate, ovate-lanceolate, or oblong, shortly acuminate, coriaceous, shining, 2 to 4 in. long, the veins not very prominent. Stipules (or their remains?) forming a short ring. Flowers white, in cymes forming a dense oblong terminal panicle. Corolla-tube nearly 2 lines long; lobes 5, rather longer than the tube. Style deeply divided into 2 linear lobes. Fruit globular, with 4 one-seeded pyrenes.

Queensland. Rockingham Bay, *Dallachy;* Pine river, *Fitzalan.*
N. S. Wales. Clarence river, *Beckler.*

2. **C. reticulatum,** *Benth.* A scrubby shrub (*Dallachy*), the branches often flexuose and perhaps sometimes climbing, usually glabrous. Leaves obovate or oval-oblong, very shortly and acutely acuminate, $1\frac{1}{2}$ to $2\frac{1}{2}$ in. long, rigid, the pinnate veins and reticulate veinlets very prominent. Stipules acuminate, not connate. Flowers pedicellate, in umbels or clusters either terminal and sessile, or axillary by the reduction of the flowering branches to leafless peduncles. Corolla-tube nearly 3 lines long; lobes 4 or 5, shorter than the tube, villous inside. Style with 2 linear stigmatic lobes. Fruit globular, about 4 lines diameter, containing 4 bony pyrenes.—*Pogonolobus reticulatus*, F. Muell. Fragm. i. 56, and Rep. Burdek. Exped. 11.

N. Australia. Arnhem S. Bay, *R. Brown;* Low Island, *Henne;* M'Adam and Newcastle ranges, *F. Mueller.*
Queensland. Bay of Inlets, *Banks and Solander;* Dayman's Island, *W. Hill;* Port Molle and Port Denison, *Fitzalan;* Rockingham Bay, *Dallachy;* Rockhampton, *Thozet;* Belyando river, *Mitchell.*

19. LASIANTHUS, Jack.

(Mephitidia, *Reinw.*)

Calyx-limb obscurely toothed or lobed. Corolla-tube usually dilated at the top, lobes 4 to 6, valvate in the bud. Anthers included in the tube or shortly exserted. Ovary 4- to 9-celled, with 1 erect ovule in each cell; style divided at the top into as many linear stigmatic lobes as ovary-cells. Fruit a drupe crowned by the calyx-limb, with as many 1-seeded pyrenes as ovary-cells, or fewer by abortion.—Shrubs or undershrubs, often smelling disagreeably, the specimens usually drying black. Stipules interpetiolar, pointed, with a broad base. Flowers small, in dense clusters or heads, axillary or, in some species not Australian, terminal.

The genus extends over tropical Asia, but is chiefly abundant in the Archipelago. The only Australian species appears to be the same as one extending from Ceylon over a part at least of the Archipelago.

1. **L. strigosus,** *Wight in Calc. Journ. Nat. Hist.* vi. 512. An erect shrub, the branches and under side of the leaves more or less hirsute with short rigid hairs. Leaves very shortly petiolate, mostly oblong in the Australian specimens, broader in some others, 3 to 5 in. long, with about 6 or 7 very oblique veins prominent underneath on each side of the midrib, and transverse veinlets. Flowers sessile and clustered, about 3 in each axil. Bracts small. Calyx about 1½ lines long, the limb obscurely toothed, longer than the adnate tube. Corolla rather more than 2 lines long; lobes 4 or 5, scarcely so long as the tube. Stamens not exserted. Ovary usually 4-celled, with a very thick fleshy concave epigynous disk. Fruit ovoid-globular, about ¼ in. diameter.—*Mephitidia strigosa*, Thw. Enum. Ceyl. Pl. 146.

Queensland. Rockingham Bay, *Dallachy*. These specimens quite agree with Cingalese ones. Blume's character of *L. sylvestris*, Bl. Bijdr. 999, agrees also well with our plant, but not the specimens so named by Miquel. *L. chinensis*, Benth. Fl. Hongk. 160, may also possibly prove to be the same, but, as observed by Thwaites, the whole genus requires much further investigation before the extent of variation to which the species are liable can be satisfactorily ascertained.

20. PSYCHOTRIA, Linn.

(Grumilia, *Gærtn.*)

Calyx-limb short, truncate, toothed or lobed. Corolla-tube short; lobes 5, or rarely 4, valvate in the bud. Anthers included in the tube or shortly exserted. Ovary 2-celled, with 1 ovule in each cell, erect from the base. Style short, with 2 stigmatic lobes. Fruit a drupe, with 2 hemispherical pyrenes, smooth or with longitudinal ribs and furrows. Seed hemispherical, with furrows corresponding to those of the endocarp, or ruminate, or smooth.—Shrubs or small trees. Stipules interpetiolar, 1 on each side, membranous, and often connate within the petioles, and very deciduous in the Australian species, small, persistent and 1 or 2 on each side in many American ones. Flowers usually small, in terminal cymes, or, in species not Australian, axillary.

A large genus, ranging over the tropical regions both of the New and the Old World. The Australian species appear to be all endemic. The limits of the genus are not well defined. The above character includes *Grumilia*, usually but vaguely distinguished by the seed, and excludes *Chasalia*, *Palicourea*, and others scarcely differing but by the length of the corolla-tube, characters requiring confirmation by the study of very numerous little known or undescribed species.

SECTION I. **Grumilia.**—*Calyx-limb (at least in the Australian species) obscurely sinuate-toothed. Pyrenes not furrowed. Seeds ruminate.*

Leaves obtuse. Cymes divaricate. Corolla 2 to 3 lines long . . . 1. *P. nesophila.*
Leaves acuminate. Cymes paniculate. Corolla 1 line long . . . 2. *P. Dallachiana.*

SECTION II. **Mapourea.**—*Calyx-limb (at least in the Australian species) distinctly* *toothed. Pyrenes and seeds longitudinally furrowed. Stipules membranous,* 1 *on each side, deciduous.*

Flowers capitate on the branches of the cyme. Leaves usually softly pubescent or tomentose . . . . 3. *P. loniceroides.*
Flowers in corymbs, cymose to the last. Leaves usually glabrous.
Corolla about 2 lines long; lobes as long as the cylindrical tube.
Corolla glabrous outside. Fruit ovoid . . . . 4. *P. daphnoides.*
Corolla hoary-tomentose. Fruit globular . . . . 5. *P. poliostemma.*
Corolla about 3 lines long, lobes shorter than the ovoid tube. Leaves thick and coriaceous . . . . 6. *P. Fitzalani.*

There are also specimens from Rockingham Bay, *Dallachy,* of what appear to be two other species of the section *Grumilia,* and from Cape York, *M'Gillivray,* of another of the section *Mapourea,* but all insufficient for definition.

SECTION I. GRUMILIA.—Calyx-limb (at least in the Australian species) obscurely sinuate-toothed. Pyrenes not furrowed. Seeds ruminate. Stipules 1 on each side, very deciduous.

1. **P. nesophila,** *F. Muell. Fragm.* ii. 135. A shrub or tree, quite glabrous. Leaves broadly ovate obovate or almost oblong, obtuse or very shortly and obtusely acuminate, 1 to 3 in. long, narrowed into a rather long petiole, thinly coriaceous. Flowers in shortly pedunculate very divaricate terminal cymes. Calyx-limb truncate or obscurely toothed. Corolla glabrous outside, very hairy inside; tube 1½ lines long; lobes as long as the tube, thickened and hood-shaped at the tips. Ovules broad. Fruit globular, smooth, the pyrenes very obscurely ribbed. Seeds hemispherical, very much ruminate-rugose.

N. Australia. N. coast, *R. Brown;* N. Goulburn island, *A. Cunningham;* Port Essington, *Armstrong.*

Queensland. Albany island, *F. Mueller.*

2. **P. Dallachiana,** *Benth.* A straggling shrub of 6 to 8 ft., quite glabrous. Leaves elliptical or oblong, acuminate, narrowed into a rather long petiole, mostly 3 to 4 in. long, smooth and almost shining. Flowers very small, in an ovate very shortly pedunculate panicle, dense when first coming out. Calyx-limb slightly sinuate-toothed. Corolla glabrous outside, bearded inside at the throat, about 1 line long but perhaps not yet at its full size; anthers exserted. Ovules ovoid. Fruiting panicle loose, divaricately trichotomous. Fruits globular, about 3 lines diameter, the pyrenes not ribbed. Seeds ruminate.

Queensland. Rockingham Bay, *Dallachy.* I do not feel quite sure that the specimens coming into flower are rightly identified with the fruiting ones, but believe them to belong to the same species.

SECTION II. MAPOUREA.—Calyx-limb (at least in the Australian species) distinctly toothed. Pyrenes and seeds longitudinally furrowed. Stipules 1 on each side, membranous, deciduous.

The American *Mapoureas* are distinguished from *Psychotria* proper, chiefly by the stipules. It remains yet to be ascertained whether the Asiatic and Australian species strictly belong in other respects to the same section.

3. **P. loniceroides,** *Sieb. in DC. Prod.* iv. 523. A shrub attaining 12 to 15 ft., the branches, foliage, and inflorescence more or less rusty-

tomentose or softly pubescent, or rarely the leaves at length glabrous above. Leaves ovate elliptical or oblong, acuminate or almost obtuse, narrowed into a petiole, mostly 2 to 3 in. long. Peduncles terminal but often appearing lateral from the elongation of only one branch of the fork, shorter than the leaves, more or less cymosely branched or almost umbellate, each branch bearing a small dense cyme or head of small sessile flowers. Bracts not exceeding the flowers. Calyx hirsute, the lobes or teeth acute, longer than the tube. Corolla-tube broad, about 2 lines long, lobes shorter than the tube, hirsute outside; anthers exserted. Fruits ovoid, pubescent, crowned by the calyx-limb; pyrenes and seeds longitudinally ribbed and furrowed.

Queensland. Wide Bay, *Bidwill;* Brisbane river, Moreton Bay, *A. Cunningham, F. Mueller.*

N. S. Wales. Port Jackson to the Blue Mountains, *R. Brown, Sieber, n.* 263, and others; Port Macquarrie, *Backhouse, M'Arthur.*

Var. *angustifolia.* Leaves narrow, on very short petioles, softly villous. Cymes with very few flower-heads. Rockingham Bay, *Dallachy.*

4. **P. daphnoides,** *A. Cunn.; Hook. Bot. Mag. t.* 3228. Shrubby, glabrous, except the inflorescence. Leaves obovate or oblong, very obtuse or rarely shortly acuminate, in some specimens under 1 in., in others 2 in. long or more, narrowed into a petiole, somewhat coriaceous, almost veinless except the midrib. Cymes terminal, pedunculate, trichotomous or the primary branches almost umbellate, 1 to nearly 2 in. diameter, rather dense, the flowers all sessile but not capitate, the branches and calyxes hoary-tomentose or glabrous. Calyx-limb 5-toothed. Corolla about 2 lines long, glabrous outside, bearded inside at the throat, the lobes as long as the tube. Fruit ovoid, the pyrenes and seeds prominently ribbed and furrowed.

Queensland. Cape York, *M'Gillivray.;* N.E. coast, *A. Cunningham;* Burnett river, *F. Mueller;* Port Denison, *Fitzalan;* Wide Bay, *Bidwill;* Fitzroy river, *Bowman;* Rockhampton, *Thozet, Dallachy;* Brisbane river, Moreton Bay, *A. Cunningham, F. Mueller.*

N. S. Wales. Clarence river, *Beckler.*

The tropical specimens have mostly large leaves; the small-leaved specimens (mostly under 1 in.) are chiefly subtropical.

Var. *angustifolia.* Leaves narrow-oblong. Queensland, *Bowman.*

5. **P. poliostemma,** *Benth.* Nearly allied to the larger forms of *P. daphnoides*, but the leaves are larger, mostly 3 to 4 in. long, the flowers numerous, in more pedunculate cymes, the corollas hoary-tomentose outside, and the fruit globular, not ovoid.

Queensland. Cape York, *M'Gillivray;* Edgecombe Bay, *Dallachy;* Mount Elliott, *Fitzalan.* A. Gray, from a memorandum in Herb. Hook., thinks this may be a variety of his *P. tephrosanthes*, from the Fiji islands. We have no specimens to compare, but he describes the leaves as acuminate, the pedicels slender, and the calyx truncate, none of which points agree with our plant, and which seem to indicate rather a species of the section *Grumilia.*

6. **P. Fitzalani,** *Benth.* Quite glabrous. Leaves broadly ovate or obovate, obtuse, narrowed into a petiole, 2 to 3 in. long, very thick and coriaceous. Cymes pedunculate, terminal, corymbose, shorter than the leaves, glabrous. Calyx-limb short, distinctly toothed. Corolla nearly 3 lines long, the tube almost ovoid, hairy inside; lobes 5, shorter than the tube. An-

thers exserted. Style with 2 short oblong stigmatic lobes. Ovules of *Psychotria.* Fruit not seen.

Queensland. Port Molle, *Fitzalan* (Burdekin Expedition).

SUBTRIBE V. ANTHOSPERMEÆ.—Corolla-lobes valvate. Ovules erect from the base. Style-lobes very long and subulate. Fruit a berry-like drupe. Albumen copious.—Shrubs, rarely trees or herbs. Flowers often unisexual or polygamous.

21. COPROSMA, Forst.

(Marquisia, *A. Rich.*)

Calyx-limb toothed or lobed. Corolla-tube short; lobes 4, 5 or rarely 6, valvate in the bud. Stamens inserted at the base of the tube; filaments long; anthers exserted. Ovary 2-celled, with one erect ovule in each cell; style divided nearly to the base into 2 long filiform lobes, more or less papillose-hirsute. Fruit a berry-like drupe, with 2 1-seeded pyrenes, usually furrowed on the inner face.—Shrubs, erect and bushy or prostrate and creeping. Leaves usually small. Stipules interpetiolar, acuminate or acute, sometimes denticulate. Flowers solitary or clustered, axillary or terminal, usually diœcious, the males with an imperfect or absolutely abortive style and no ovary, the females with small stamens, more or less imperfect. The whole plant often fœtid when fresh.

The genus is most numerous in New Zealand, but extends also to the Pacific islands and to Borneo. Of the five Australian species, one is also in New Zealand, the others are endemic.

Flowers clustered, 2 or 3 together. Leaves mostly above 1 in. long.
  Leaves acutely acuminate, much and conspicuously veined . . . . 1. *C. acutifolia.*
  Leaves very shortly acuminate; veins few and distant . . . . . 2. *C. hirtella.*
Flowers solitary. Leaves small.
  Erect bushy shrubs. Corolla-lobes much longer than the tube.
    Leaves thin, flat . . . . 3. *C. Billardieri.*
    Leaves coriaceous, shining, with recurved margins . . . . . 4. *C. nitida.*
  Prostrate creeping shrub. Corolla-lobes shorter than the tube . . 5. *C. pumila.*

1. **C. (?) acutifolia,** *F. Muell. Herb.* Quite glabrous. Leaves ovate, acutely acuminate, narrowed into a very short petiole, 1½ to 2½ in. long, very coriaceous and shining, but the numerous veins and reticulations very conspicuous on both sides. Flowers small, sessile, in pairs or threes, on very short peduncles, clustered in the axils, all males in the specimens seen. Corolla about 1½ lines long; lobes valvate, longer than the tube. Anthers exserted. Ovary rudimentary, without any cavity or style.

N. S. Wales. Durandoo, in the western interior, *Goodwin and Dallachy.*

2. **C. hirtella,** *Labill. Pl. Nov. Holl.* i. 70. *t.* 95. A rigid shrub, of 3 to 5 ft., quite glabrous or minutely scabrous-pubescent. Leaves from broadly ovate to elliptical or lanceolate, shortly and acutely acuminate, narrowed into a short petiole, the larger ones 2 to 3 in. long, thick and coriaceous, with few scarcely prominent veins, usually scabrous on the upper side. Flowers few together, in very shortly pedunculate terminal heads, becoming axillary by the growth of the shoot, with a pair of lanceolate acuminate con-

nate bracts or floral leaves under the head. Calyx minutely toothed. Corolla nearly 3 lines long in the males, the lobes 4 to 6, longer than the tube, rather smaller in the females. Style-branches fully ½ in. long, very shortly united at the base.—DC. Prod. iv. 578; Hook. f. Fl. Tasm. i. 165; *C. cuspidifolia*, DC. l. c.

N. S. Wales. Damp situations in the western ranges of the Blue Mountains, and near Bathurst, *A. and R. Cunningham;* Argyle County, *M'Arthur.*

Victoria. Rich wet valleys of the Buffalo Range, *F. Mueller;* Portland Ranges, *Robertson;* Port Phillip, *R. Brown.*

Tasmania. Port Dalrymple, *R. Brown;* abundant in rocky places throughout the colony, *J. D. Hooker.*

3. **C. Billardieri,** *Hook. f. in Hook. Lond. Journ.* vi. 465, *and Fl. Tasm.* i. 165. A slender twiggy shrub, of 6 to 12 ft., glabrous or minutely pubescent, the smaller branchlets often spinescent. Leaves elliptical-oblong or lanceolate, obtuse or acute, all under ½ in. long, of a thin texture, glabrous. Flowers solitary, terminating very short axillary shoots, with a pair of bracts under the calyx. Calyx-limb very short, acutely 4-lobed. Corolla about 2 lines long, the tube very short, smaller and narrower in the females than in the males. Style-branches 4 or 5 lines long. Fruits small, red.—*Canthium quadrifidum*, Labill. Pl. Nov. Holl. i. 69. t. 94; *Marquisia Billardieri*, A. Rich. in Mém. Soc. Hist. Nat. Par. v. 192; DC. Prod. iv. 477.

N. S. Wales. Arne river, *Beckler;* Mount Tomah, *Woolls;* Five Islands, *A. Cunningham;* Twofold Bay, *F. Mueller.*

Victoria. Yarra river, Dandenong Ranges, *F. Mueller;* Corner Inlet, *Wilhelmi.*

Tasmania. Common by the banks of streams in rich soils, in shaded ravines and dense forests, "Native Currant," *J. D. Hooker.*

4. **C. nitida,** *Hook. f. in Hook. Lond. Journ.* vi. 465, *and Fl. Tasm.* i. 165. *t.* 39 A. A rigid bushy shrub, of 5 to 6 ft., quite glabrous, the branchlets rarely spinescent. Leaves elliptical lanceolate or oblong, obtuse or acute, all under ½ in. long, or in luxuriant specimens ½ to 1 in., coriaceous and shining, with recurved margins. Flowers of *C. Billardieri*, solitary and terminating very short axillary shoots. Fruit ovoid oblong or nearly globular, larger than in *C. Billardieri.*

Victoria. Mount Baw-Baw, at an elevation of 4 to 5000 ft., *F. Mueller.*

Tasmania, *R. Brown;* common on open mountain-tops, *J. D. Hooker.*

5. **C. pumila,** *Hook. f. in Hook. Lond. Journ.* vi. 465, *and Fl. Tasm.* i. 166. A prostrate densely-matted glabrous shrub, creeping and rooting sometimes to a great extent. Leaves ovate, obtuse, 2 to 3 lines long, coriaceous, shining. Flowers solitary, terminal, sessile between the last leaves. Calyx-lobes ovate, obtuse. Corolla about 2½ lines long, the tube dilated upwards, almost campanulate; lobes much shorter than the tube in the males, rather deeper in the females. Stamens much exserted. Styles not so long as in the other species.

Victoria. Snowy mossy plains, Baw-Baw, Cobberas mountains, and others of the Australian Alps, at an elevation of 4 to 6000 ft., *F. Mueller.*

Tasmania. Middlesex plains and all the mountain-tops, abundant, *J. D. Hooker.*

The species is also in New Zealand.

22. NERTERA, Banks.

(Cunina, *Clos.*)

Calyx-limb inconspicuous or 2-lobed. Corolla-tube campanulate or slender; lobes 4, valvate in the bud. Stamens inserted at or near the base of the corolla-tube; filaments long; anthers exserted. Ovary 2-celled, with 1 erect ovule in each cell; style divided nearly to the base into 2 long filiform lobes, more or less papillose-hirsute. Fruit a berry-like drupe, with 2 1-seeded pyrenes.—Slender prostrate or creeping perennials. Stipules interpetiolar, very small. Flowers solitary, terminal, sessile or very shortly pedicellate within the last leaves, hermaphrodite or polygamo-diœcious.

A small genus, extending over New Zealand, Antarctic and Andine America, the Indian Archipelago, and the Pacific islands. Of the two Australian species, one is the common one over nearly the whole range of the genus, the other is endemic.

Glabrous. Calyx-limb obsolete. Corolla campanulate . . . . 1. *N. depressa.*
Hirsute. Calyx 2-lobed. Corolla slender . . . . 2. *N. reptans.*

1. **N. depressa,** *Banks; DC. Prod.* iv. 451. A slender prostrate perennial, creeping and rooting at the nodes, quite glabrous, forming sometimes dense patches of very few inches in diameter, sometimes extending to a considerable length, usually drying black. Leaves petiolate, from ovate to orbicular, obtuse or acute, rounded or almost cordate at the base, often all under 2 lines long, but in luxuriant specimens 3 or even 4 lines. Flowers solitary, terminal, but the fruits sometimes appearing axillary after the shoot has grown out. Calyx-limb none or scarcely conspicuous. Corolla glabrous, campanulate, scarcely above 1 line long, the lobes shorter than the tube. Drupe ovoid or globular, the pyrenes quite smooth.—Sm. Ic. Ined. t. 28; Hook. f. Fl. Tasm. i. 167; *Cunina Sanfuentes,* Clos in C. Gay, Fl. Chil. iii. 203. t. 34.

Victoria. Wet gravelly places, Snowy River, summits of Baw-Baw mountains, sources of the Yarra, *F. Mueller.*

Tasmania. By springs on the summits of the western mountains, *Gunn.*

The species extends to New Zealand, Andine and Antarctic America and the Pacific islands, but the Javanese plant referred to it appears to be distinct. In the small Australian form of *N. depressa,* so far as can be judged from dried specimens, the exocarp encloses but little pulp and dries close upon the nucleus; in the more luxuriant New Zealand and American specimens, it is much larger and more pulpy, and, when dried under pressure, assumes the false appearance of a broad membrane bordering the nucleus, which gave rise to Clos's genus *Cunina.*

2. **N. reptans,** *F. Muell. Herb.* A prostrate creeping perennial, like *N. depressa,* but not usually drying black, and more or less hirsute, with short scattered rather rigid hairs. Leaves very shortly petiolate, ovate, obtuse or acute, 3 to 4 lines long. Flowers terminal, very shortly pedicellate within the last pair of leaves. Calyx-limb of 2 triangular ciliate lobes. Corolla-tube very slender, 4 lines long; lobes short, ciliate, with a few long bristles. Drupe ovoid, not quite 2 lines long when dry, crowned by the calyx-lobes.—*Diodia reptans,* F. Muell. in Trans. Vict. Inst. 1855, 128.

Victoria. Mountain pastures and plains along the Snowy River, *F. Mueller.*

Subtribe VI. Opercularieæ.—Corolla-lobes valvate. Ovules erect from the base, usually one only to each flower. Style-lobes long and subulate.

Fruit capsular, 2-valved. Albumen copious.—Herbs undershrubs or rarely small shrubs. Flowers in dense heads, the calyxes connate or in one species free.

23. OPERCULARIA, Gærtn.

Flowers connate by the calyx-tubes in a globular compound or rarely simple and solitary head. Calyx-lobes 3 to 5. Corolla-tube short; lobes 3 to 5, valvate. Stamens inserted at the base of the corolla-tube; filaments long; anthers exserted. Ovary 1-celled, with 1 erect ovule (rarely 2-celled with 1 ovule in each cell, but 1 cell and ovule abortive?); style filiform, divided to the middle or nearly to the base into 2 long usually papillose-hirsute branches, one sometimes abortive. Fruit a 2-valved capsule, the capsules of each compound head distributed into partial heads of 3 to 6, the outer valves in each partial or solitary head united in a persistent cup, the inner valves united in a deciduous operculum. Seeds obovate or oblong, often rugose, the inner face often concave or marked with longitudinal raised ridges. Albumen copious.—Herbs undershrubs or rarely twiners, often very fœtid when fresh. Stipules usually forming with the base of the petioles a short sheath, with 1 or 2 entire or ciliate-toothed points on each side. Flower-heads usually in the forks of the stem or terminal, either upon an erect peduncle often longer than the head, or upon a short recurved peduncle, and then often appearing axillary from the development of only one branch of the fork. Flowers often polygamo-diœcious.

The genus is confined to Australia. The species are very difficult to describe with contrasted characters, differing chiefly in the very variable points of foliage and inflorescence. The seeds, as far as known, appear to present more decided specific differences, but it is only in a small portion of the specimens that I have been able to observe them, and their constancy in the same species remains to be proved.

Peduncles erect, terminal or in the forks. Leaves linear. Seed ovoid-oblong, obtusely 3- or 4-angled. Western species, except *O. scabrida*.
- Glabrous or nearly so. Leaves acute.
  - Herb or undershrub. Upper leaves long. Peduncles usually long. Seeds half enclosed on each side in a concave valve-like appendage . . . . 1. *O. vaginata.*
  - Small shrub. Leaves short, clustered in the axils. Peduncles short. Seeds without appendage . . . . 2. *O. spermacocca.*
  - Herb or undershrub. Leaves short or long. Peduncles usually long. Seeds without appendage . . . . 3. *O. scabrida.*
- Hirsute, with long soft hairs. Leaves obtuse . . . . 4. *O. hirsuta.*

Peduncles recurved, all very short, or those of the primary forks only elongated.
- Eastern species.
  - Usually large, glabrous or scabrous-pubescent. Leaves ovate or lanceolate, often above 1 in. long. Heads many-flowered. Seeds with 2 smooth ribs on the inner face . . . . 5. *O. aspera.*
  - Hirsute. Leaves usually under 1 in. Seeds very rugose, the inner face with a prominent centre, without smooth ribs . . 6. *O. hispida.*
  - Glabrous. Leaves linear-lanceolate. Flower-heads small, sessile or nearly so. Seeds of *O. hispida* . . . . 7. *O. diphylla.*
  - Small procumbent plant, usually glabrous. Leaves ovate. Seeds very broad, smooth, with 3 prominent ribs on the inner face . 8. *O. ovata.*

Glabrous or nearly so, small and diffuse or wiry and elongated. Leaves and flower-heads small. Seeds broad, slightly rugose, with 2 prominent ribs on the inner face . . . . 9. *O. varia.*

Western species.

Stout, erect, glabrous or scabrous. Leaves sessile, ovate-lanceolate or lanceolate. Heads many-flowered . . . . 10. *O. rubioides.*

Large, twining, glabrous. Leaves lanceolate, long-acuminate. Heads many-flowered . . . . 11. *O. volubilis.*

Large, hirsute, often flexuous. Leaves ovate or lanceolate. Heads many-flowered . . . . 12. *O. hispidula.*

Erect, much branched, very hispid, not turning black. Leaves ovate or lanceolate. Flower-heads echinate . . . . 13. *O. echinocephala.*

Glabrous. Stems wiry or flexuose. Leaves few, narrow or small. Heads simple, 2- to 5-flowered . . . . 14. *O. apiciflora.*

1. **O. vaginata,** *Labill. Pl. Nov. Holl.* i. 34. *t.* 46. A perennial or undershrub, with erect virgate stems, 6 in. to 1 ft. high or rather more, usually glabrous. Leaves linear or linear-lanceolate, the upper ones often 1 to 2 in. long, the lower ones sometimes very small or abortive. Stipular sheaths long. Heads globular, compound, on rather long erect peduncles, with or without 2 to 4 linear bracts close under the head. Calyx-lobes subulate-acuminate. Corolla rarely exceeding them. Seeds small, obovoid-triquetrous, each with 2 concave appendages one on each side, as long as the seed, fixed at the upper end and falling with the seed, but readily detached.—DC. Prod. iv. 615; Bartl. in Pl. Preiss. i. 369.

W. Australia. From King George's Sound, *R. Brown, Menzies, Labillardière,* to Swan River, *Drummond,* 1*st Coll.*; *Preiss, n.* 2429, 2433; Flinders Bay, *Collie.*

In some old specimens some branches are flexuose and almost leafless like those of *O. apiciflora,* but more rigid, and the fruiting-heads are compound.

O. multicaulis, Bartl. in Pl. Preiss. i. 369, from Princess Royal Harbour, *Preiss, n.* 2432, which I have not seen, does not appear, from the description given, to differ from *O. vaginata.*

2. **O. spermacocea,** *Labill. Pl. Nov. Holl.* i. 35. *t.* 47. A small bushy erect shrub or undershrub, glabrous or slightly pubescent, the short erect branches usually shining. Leaves narrow-linear, often clustered in the axils, mostly short but sometimes above $\frac{1}{2}$ inch long. Stipules often with linear leaf-like points. Flower-heads all pedunculate, small, with few flowers but compound. Calyx-lobes acuminate, rigid, sometimes enlarged and lanceolate when in fruit. Corolla scarcely exceeding the calyx-lobes. Seeds ovoid, obtusely angular, granular-tuberculose, like those of *O. vaginalis,* but without the lateral appendages.—DC. Prod. iv. 615.

W. Australia, *Labillardière*; Port Gregory, Champion Bay, *Oldfield.*

3. **O. scabrida,** *Schlecht. Linnæa,* xx. 604. Erect or ascending, rather slender, scabrous-pubescent or hirsute. Leaves linear-lanceolate or narrow-oblong, usually narrowed at both ends and under $\frac{1}{2}$ in. long. Flower-heads globular, compound, on long erect peduncles in the forks, with 2 or 3 linear floral leaves close under them. Calyx-lobes subulate-acuminate. Seeds small, obovoid, obtusely 3- or 4-angled.

Victoria. Grampians, *F. Mueller;* Wimmera, *Dallachy;* Portland, *Allitt;* Glenelg river, *Robertson.*

S. Australia. Sandy scrub, *Behr;* Lofty Ranges, *F. Mueller.*

4. **O. hirsuta,** *F. Muell. Herb.* Erect, with virgate stems of about 1 ft., covered as well as the foliage with long soft almost silky hairs. Leaves few, linear, obtuse, ½ to 1 in. long. Flower-heads globular, on erect peduncles, with numerous small flowers. Calyx-lobes scarcely so long as the hirsute corollas. Anthers smaller and style shorter than in most species. Seeds oblong-ovoid, obtusely 4-angled, very rugose except the 2 lateral smooth angles.

W. Australia. Lucky Bay, Oldfield and Young rivers, Esperance Bay, *Maxwell.*

5. **O. aspera,** *Gærtn. Fruct.* i. 112. *t.* 24. A rather coarse or slender species, 1 to 2 ft. long or more, glabrous or scabrous-pubescent, varying much in foliage. In the original form leaves shortly petiolate, ovate or ovate-lanceolate, very scabrous on the upper side, glabrous or pubescent underneath, mostly ½ to 1½ in. long. Flower-heads globular, compound, on short recurved peduncles. Calyx-lobes linear-lanceolate. Corolla funnel-shaped, about 2 lines long, the lobes varying as in other species from 3 to 5. Stamens as many as corolla-lobes or fewer. Style-branches occasionally reduced to 1. Seeds ovate, rugose, the inner face much flattened but with 2 longitudinal smooth ribs, one on each side of the prominent rugose centre. —*O. paleata,* Young in Trans. Linn. Soc. iii. 30. t. 5; DC. Prod. iv. 616; *O. ocymifolia,* Juss. in Ann. Mus. Par. iv. 428. t. 71. f. 3; DC. l. c.

Queensland. Burnett river and Moreton Bay, *F. Mueller.*

N. S. Wales. Botany Bay, *Banks and Solander;* Port Jackson, *R. Brown,* and others; Twofold Bay, *F. Mueller.*

Victoria. Entrance of Genoa River and mouth of Snowy River, *F. Mueller;* Wimmera, *Dallachy.*

Var. *ligustrifolia.* Leaves lanceolate, acutely acuminate, thinner and less scabrous than in the original form.—*O. ligustrifolia,* Juss. in Ann. Mus. Par. iii. 428. t. 71. f. 2 (from the char. and fig.); DC. Prod. iv. 616; *O. rubioides,* Sieb. Pl. Exs., but scarcely of Jussieu. —Port Jackson to the Blue Mountains, *Sieber, n.* 251, *Woolls,* and others.

Var. *hyssopifolia.* Leaves smaller, narrow-lanceolate and thin.—*L. hyssopifolia,* Juss. in Ann. Mus. Par. iii. 428. t. 71. f. 1; DC. Prod. iv. 616.—Queensland coast, *Banks and Solander, R. Brown, Bowen, Dallachy;* Port Jackson to the Blue Mountains, *Woolls,* and others. This variety much resembles *O. diphylla* in foliage, but the seeds are those of *O. aspera.*

6. **O. hispida,** *Spreng. Syst. Veg.* i. 385. Usually smaller than *O. aspera,* and more hirsute. Leaves petiolate, ovate or lanceolate, either very hirsute and scarcely scabrous or very scabrous, mostly about ½ in. long. Flower-heads on short recurved peduncles, smaller than in *O. aspera,* and the calyx-lobes rather shorter. Seeds of the shape of those of *O. aspera,* concave and rugose on the inner face with the centre projecting, but (as far as known) without the two prominent smooth ribs of that species.—DC. Prod. iv. 615; *O. aspera,* Juss. in Ann. Mus. Par. iv. 427. t. 70. f. 1; *O. hispidula,* Miq. in Ned. Kruidk. Arch. iv. 108, not of Endl.

N. S. Wales. Botany Bay, *Banks and Solander;* Port Jackson, *R. Brown,* and others; New England, *C. Stuart;* Castlereagh, *C. Moore.*

S. Australia. Fifteen Mile Creek, *F. Mueller.*

O. hirtella, DC. Prod. iv. 616, may be a form either of this species or of *O. hispidula,* but is insufficiently described for identification.

7. **O. diphylla,** *Gærtn. Fruct.* i. 113. Slender and nearly glabrous,

resembling the weaker forms of *O. aspera*, var. *hyssopifolia*. Leaves linear or linear-lanceolate, attenuate at both ends. Flower-heads nearly sessile at the forks but reflexed, much smaller than in *O. aspera*, and the calyx-lobes much shorter. Seeds very much pitted-rugose on the inner face, without the two smooth ribs of *O. aspera*.—*O. sessiliflora*, Juss. in Ann. Mus. Par. iv. 427. t. 70. f. 2; DC. Prod. iv. 615.

Queensland. Bay of Inlets, *Banks and Solander;* Brisbane river, Moreton Bay, *F. Mueller;* Rockhampton, *Dallachy.*

N. S. Wales. Botany Bay, *Banks and Solander.*

8. **O. ovata,** *Hook. f. in Hook. Lond. Journ.* vi. 465, *and Fl. Tasm.* i. 166. *t.* 38. Diffuse, spreading from a few inches to about 1 ft., glabrous or scabrous-pubescent. Leaves petiolate, ovate or lanceolate, mostly about ½ in. but sometimes ¾ in. long. Heads small, nearly sessile in the forks or on short recurved peduncles. Calyx-lobes usually 3, shorter and broader than in *O. aspera*. Seeds broad, smooth outside, and nearly so on the inner face.

Victoria. Wilson's Promontory and near Brighton, *F. Mueller;* Wendu valley, Glenelg river, *Robertson.*

Tasmania. Near Launceston, in stiff clay soil, *Gunn.*

S. Australia. Lofty Ranges, *F. Mueller.*

9. **O. varia,** *Hook. f. in Hook. Lond. Journ.* vi. 466, *and Fl. Tasm.* i. 167. Either small and diffuse or with wiry stems ascending to nearly 1 ft., glabrous, scabrous or hirsute. Leaves very shortly petiolate, oblong lanceolate or almost linear, usually acute, rarely ½ in. long and often much smaller. Flower-heads small, sessile in the forks or nearly so, recurved. Capsules usually only 3 or 4 to each partial head. Seeds broad, not very rugose, the inner face concave with denticulate margins and 2 prominent longitudinal ribs.—*O. ecliptoides*, F. Muell.; Miq. in Ned. Kruidk. Arch. iv. 110.

N. S. Wales? Port Jackson, *R. Brown.*—A specimen which appears to belong to the long slender form of this species, but not in seed.

Victoria. Portland, *Allitt;* Glenelg river, *Robertson;* Latrobe river, *F. Mueller.*

Tasmania. Very common in dry stony places throughout the colony, *J. D. Hooker.*

S. Australia. South coast, *R. Brown;* Rivoli Bay, Mount Disappointment, etc., *F. Mueller.*

Var. *rigidior.* Stems more rigid and virgate. Leaves small, linear.—*O. turpis*, F. Muell.; Miq. in Ned. Kruidk. Arch. iv. 109; *A. hyssopifolia*, Miq. l. c., not of Juss.—About Adelaide, etc., *F. Mueller.*

10. **O. rubioides,** *Juss. in Ann. Mus. Par.* iv. 428 (*from the char. given*). Glabrous, rather stout and rigid, apparently erect and perhaps woody at the base. Leaves sessile, ovate-lanceolate or lanceolate, acute, ½ to 1 in. long or rather more, the margins recurved, the upper surface scabrous. Flowers numerous, in globular heads on short recurved peduncles or nearly sessile. Calyx-lobes lanceolate. Seeds not seen.—DC. Prod. iv. 616.

W. Australia. Towards Cape Riche, *Drummond*, 5*th Coll. n.* 435.—This accords better with Jussieu's character, especially as to the sessile leaves, than any of the forms of *O. aspera*, which have been sometimes supposed to be his plant.

11. **O. volubilis,** *R. Br. Herb.* Glabrous or nearly so, the stems twining and attaining several feet. Leaves petiolate, lanceolate or ovate-lanceo-

late, narrowed into a long point, often above 2 in. long. Heads many-flowered, on short recurved peduncles. Calyx-lobes subulate-acuminate, often 3 lines long. Seeds broad, with a prominent ridge on the inner face, but not seen ripe.

W. Australia, *Drummond, 2nd Coll. n.* 236; King George's Sound, *R. Brown;* Princess Royal Harbour, *Maxwell.*

12. **O. hispidula,** *Endl. in Hueg. Enum.* 58. A rather coarse species, more or less hirsute with scattered hairs, the stems weak or ascending, often several feet long. Leaves petiolate, from ovate to lanceolate, acute or acuminate, ½ to 1½ in. long, sprinkled with rather long hairs on both sides. Flower-heads compound, usually on very short recurved peduncles or almost sessile, rarely on longer peduncles in the primary forks. Calyx-lobes subulate. Seeds ovoid, obtusely 4-angled, very rugose, almost muricate, but only seen in the few-flowered variety.—Bartl. in Pl. Preiss. i. 369.

W. Australia. Swan River, *Drummond, 1st Coll., Preiss, n.* 2431; Australind and Cape Naturaliste, *Oldfield.*

Var. *pauciflora.* Flower-heads smaller with fewer flowers, and the plant usually less hairy.—*O. pauciflora,* Endl. in Hueg. Enum. 57; King George's Sound, *A. Cunningham, Harvey, Oldfield.*

It appears that Huegel's many-flowered form was from King George's Sound, and the few-flowered one from Swan River. Our specimens are the reverse.

O. purpurea, Bartl. in Pl. Preiss. i. 369, from Darling range, *Preiss. n.* 2430, is unknown to me, but the description given applies very well to several of our specimens of *O. hispidula.*

13. **O. echinocephala,** *Benth.* Erect or diffuse, much-branched, very hispid with short spreading hairs, and not drying black like the other species. Leaves ovate-lanceolate or linear, very shortly petiolate, mostly ½ to ¾ in. long, very scabrous-hispid, with recurved margins. Heads numerous, globular, compound, many-flowered, on very short recurved peduncles or nearly sessile, the terminal ones sessile within the last leaves. Calyx-lobes lanceolate-subulate, hispid, rigid, giving the heads a very echinate appearance. Seeds ovate-oblong, scarcely rugose, the inner face nearly flat, with a very prominent smooth longitudinal rib on each side of the central ridge.

W. Australia. Swan River, *Drummond, 1st Coll.;* Harvey river, *Oldfield.*

14. **O. apiciflora,** *Labill. Pl. Nov. Holl.* i. 35. *t.* 48. Stems slender and diffuse, or erect and twiggy, or flexuose, sometimes almost leafless, glabrous or sprinkled with a few small hairs. Leaves linear or linear-lanceolate, very acute, usually few and small, rarely exceeding ½ in. Peduncles very short and recurved, with a simple head of 3 to 5 flowers. Calyx-lobes acuminate. Corolla very short, campanulate. Seeds ovate, the inner face very concave with denticulate margins and a very prominent central sulcate longitudinal rib.—DC. Prod. iv. 615.

W. Australia, *Labillardière.* Swan River, *Drummond, n.* 437; N. of Murchison river, *Oldfield.*—Some almost leafless specimens of *O. vaginata* have some resemblance to this species, but are not so slender, and the seed is very different.

24. POMAX, Soland.

Flowers connate by the calyx-tubes in simple heads, of which several are pedicellate in a terminal umbel. Calyx-lobes about 3. Corolla-tube short, lobes 3 to 5, valvate. Stamens 5 or fewer, inserted at the base of the corolla-tube; filaments long; anthers exserted. Ovary 1-celled, with 1 erect ovule; style filiform, deeply divided into 2 long exserted filiform hispid branches, one sometimes abortive. Fruit a 2-valvate capsule, the outer valves of all the capsules united in a persistent cup crowned by the outer calyx-lobes, the inner valves united in a deciduous operculum.—A small shrub or undershrub. Stipules interpetiolar.

The genus is limited to a single species, endemic in Australia, only differing from *Opercularia* in the simple flower-heads forming an umbel, instead of being united in a compound head.

1. **P. umbellata,** *Soland. in Gærtn. Fruct.* i. 112. Much-branched, diffuse or erect, usually not exceeding 1 ft. in height, more or less hirsute or rarely glabrous. Leaves petiolate, ovate, elliptical or lanceolate, mostly under ½ in. long or rather more when narrow. Umbel terminal, sessile within the last leaves; rays or peduncles 2 to 3 lines long when in flower, longer when in fruit, each bearing a head usually of about 3 or 4 flowers. Corolla about 1½ lines long. Persistent cup (formed by the outer fruit-valves, but often called an involucre) campanulate, 1 to 1½ lines long, bordered by 5 to 8 ovate spreading teeth (the outer calyx-lobes), the inner calyx-lobes much smaller or scarcely conspicuous on the top of the deciduous operculum.—*Opercularia umbellata*, Gærtn. Fruct. i. 112. t. 24; *Pomax hirta* and *P. glabra*, DC. Prod. iv. 615; *P. rupestris*, F. Muell. in Linnæa, xxv. 395.

Queensland. Suttor, Burdekin, and Burnett rivers, *F. Mueller;* Rockhampton, *Dallachy;* Brisbane river, Moreton Bay, *F. Mueller;* on the Maranoa and near Mount Pluto, *Mitchell.*

N. S. Wales. Port Jackson to the Blue Mountains, *R. Brown, Sieber, n.* 250, and others; New England, *C. Stuart;* Arne river, *Beckler;* and in the interior on the Lachlan, *A. Cunningham;* and Darling river, *Victorian Expedition.*

Victoria. Buffalo range, Snowy and Avon rivers, *F. Mueller.*

S. Australia. Near Lake Torrens, *F. Mueller;* Lake Gillies, *Burkitt.*

25. ELEUTHRANTHES, F. Muell.

Flowers in heads but not concrete. Calyx-limb 4- or 5-lobed. Corolla-tube slender; lobes 4 or 5, valvate in the bud. Stamens 5 or fewer, inserted on the base of the corolla-tube; anthers exserted. Ovary 1-celled, with a single erect ovule; style deeply divided into 2 long filiform branches. Fruits not concrete, capsular, opening apparently in 2 valves. Seed obovate, compressed, but not seen ripe.—A small herb. Stipules reduced to a short sheath. Flower-heads dense, terminal, globular.

The genus is limited to a single species, endemic in Australia.

1. **E. opercularina,** *F. Muell. Fragm.* iv. 92. A small procumbent very hirsute annual, spreading to a few inches diameter. Leaves petiolate, ovate, mostly acute, under ½ in. long, hirsute with long soft hairs. Flowers very numerous, in dense globular hirsute heads of ½ in. diameter or rather more, sessile above the last leaves but without any other bracts or involucre. Calyx-lobes linear or subulate, hirsute, 1 to 1½ lines long. Corolla about

1 line long, the lobes very short, slightly hairy. Young seeds white and rugose.—*Opercularia liberiflora*, F. Muell. l. c.

W. Australia. Rocks on the Fitzgerald river, *Maxwell.*

SUBTRIBE VII. SPERMACOCEÆ.—Corolla-lobes valvate. Ovules variously attached to the axis, the ovary usually perfectly 2-celled. Style-lobes short. Fruit capsular or indehiscent and dry. Albumen copious.—Herbs, undershrubs, or rarely shrubs.

26. KNOXIA, Linn.

Calyx-limb of 4 minute persistent teeth. Corolla-tube slender; lobes 4, valvate in the bud. Anthers scarcely exserted. Ovary 2-celled, with 1 pendulous ovule in each cell; style with 2 short stigmatic lobes. Fruit small and dry, the 2 carpels either separating from the base upwards or falling off together, leaving a persistent filiform axis.—Herbs or undershrubs. Stipules 1 on each side, usually fringed with bristle-like teeth. Flowers in terminal cymes or corymbs, the branches often lengthened into one-sided spikes.

A small genus, extending over tropical Asia and Africa, the only Australian species being the most common one in Asia.

1. **K. corymbosa,** *Willd.; W. and Arn. Prod.* i. 439. A perennial usually erect, 1 to 2 ft. high, often almost woody at the base, more or less pubescent and but little branched. Leaves oblong-lanceolate or rarely nearly ovate, 2 to 3 in. long. Flowers 1 to 1½ lines long, numerous, in loose terminal cymes. Capsule ovoid, about 1 line long, usually falling off entire from the filiform persistent axis.—Wight, Illustr. t. 128.

Queensland. E. coast, *R. Brown;* Palm Island, *Henne;* Rockhampton, *Thozet* and others; Pine river, *Herb. F. Mueller.*—Common in tropical Asia, from Ceylon and the Peninsula to the Archipelago. The Australian specimens have smaller flowers than usual, but I have not seen the corollas well opened.

27. SPERMACOCE, Linn.

(Borreria, *G. F. W. Mey.;* Bigelowia, *Spreng.*)

Calyx-limb of 4 or rarely only 2 teeth or small lobes. Corolla-tube short or rarely slender; lobes 4, valvate in the bud. Anthers exserted or rarely included in the tube. Ovary 2-celled, with 1 ovule in each cell, laterally attached or ascending; style entire or with 2 short stigmatic lobes. Capsule small, separating into 2 carpels more or less opening on their inner face, or leaving more or less of the dissepiment free or attached to one of the carpels. Seeds marked on the inner face by a longitudinal furrow, concavity, or broad opaque surface containing the hilum.—Herbs or rarely undershrubs. Stipules shortly sheathing, bordered with bristle-like teeth. Flowers small, clustered in the axils of the leaves or in terminal heads.

A large genus, widely spread over the tropical and subtropical regions both of the New and the Old World, some species being amongst the commonest of the tropical weeds, but none of these, not even the widely-spread Asiatic and African *S. stricta* or *S. hispida*, have as yet been observed in Australia, the whole of the Australian species being as far as known endemic. Varying in the dehiscence of the capsule from that ascribed to *Borreria* to that of *Spermacoce* proper, their inflorescence is that most prevalent in *Borreria;* the heads are terminal, or if axillary only on one side of the stem, showing that they have become lateral only by the elongation of one only of the branches of a normally forked stem. The shape of the corolla and insertion of the stamens appear to be constant characters, and essential to be attended to in the determination of species otherwise similar in aspect. The length of the stamens may vary from dimorphism.

- Stamens inserted at the base or below the middle of the corolla-tube, the anthers always included. Corolla 1 to 1½ lines long.
  - Leaves linear, lanceolate or narrow-elliptical. Stems erect, divaricate or scarcely diffuse. Stamens at the base of the tube.
    - Corolla-lobes much shorter than the tube . . . . 1. *S. brachystema.*
    - Corolla-lobes longer than the tube.
      - Corolla densely bearded at the throat. Anthers small, ovate 2. *S. pogostoma.*
      - Corolla not bearded, the narrow lobes pubescent inside. Anthers oblong . . . . 3. *S. leptoloba.*
  - Leaves ovate or broadly lanceolate, rigid, with callous margins. Stamens near the middle of the tube . . . . 4. *S. marginata.*
- Stamens inserted at the orifice of the corolla-tube, alternating with its lobes.
  - Calyx usually 4-lobed. Corolla-lobes without internal appendages. Annual or perennial herbs.
    - Corolla-lobes longer than the short broad tube.
      - Leaves linear or linear-lanceolate.
        - Corolla about 1 line long. Stamens much shorter than the lobes . . . . 5. *S. multicaulis.*
        - Corolla about 2 lines long. Stamens as long as or longer than the lobes . . . . 6. *S. exserta.*
      - Leaves ovate or elliptical, on long petioles. Stamens exceeding the corolla-lobes . . . . 7. *S. membranacea.*
    - Corolla-lobes shorter than the tube.
      - Stems diffuse. Leaves petiolate, ovate or broadly lanceolate. Corolla nearly 2 lines long . . . . 8. *S. debilis.*
      - Stems erect or ascending. Leaves narrow-linear.
        - Flowers about 4, in terminal and lateral heads. Corolla about 3 lines long. Cocci almost closed . . . . 9. *S. inaperta.*
        - Flowers numerous, in heads chiefly terminal.
          - Corolla about 3 lines long; lobes nearly as long as the tube . . . . 10. *S. stenophylla.*
          - Corolla about 4 lines long; lobes a little more than half as long as the slender tube . . . . 11. *S. lævigata.*
  - Calyx usually 4-lobed. Corolla-lobes with 2 oblique prominent laminæ or auricles on the inner face. Leaves linear or lanceolate. Annual or perennial herbs.
    - Corolla 2½ to 3 lines long, the lobes about as long as the tube . 12. *S. breviflora.*
    - Corolla 5 to 6 lines long, the lobes shorter than the tube . . 13. *S. auriculata.*
  - Calyx 2-lobed. Corolla-lobes without appendages. Undershrub . 14. *S. suffruticosa.*

(The common Asiatic *S. stricta*, Linn. f., may possibly have been found in Queensland. It differs from *S. multicaulis* in the funnel-shaped corolla, with lobes much shorter than the tube.)

1. **S. brachystema,** *R. Br. Herb.* An erect or spreading rather rigid annual, attaining 1 or 2 ft. and not much branched, or sometimes much smaller, more or less pubescent or hirsute, or sometimes nearly glabrous. Leaves sessile, linear-lanceolate or rarely oblong, mostly 1 to 1½ in. long, occasionally clustered in the axils. Bristles of the stipules rather long. Flowers small, in dense terminal or lateral heads or clusters. Calyx-lobes 4,

acute, often unequal; the longer ones as long as the tube. Corolla about 1 line long, bearded inside at the orifice of the tube, the lobes very short. Anthers small, ovoid, on very short filaments at the base of the tube. Capsule about 1½ lines long, more or less of the dissepiment remaining attached to one of the carpels after dehiscence.—*S. stricta,* F. Muell. Fragm. iv. 41, not of Linn.

N. Australia. Arnhem's Land, *F. Mueller;* Port Essington, *Armstrong.*

Queensland. E. coast, *R. Brown;* Port Denison, *Fitzalan;* Burnett river and Brisbane river, Moreton Bay, *F. Mueller;* Rockhampton, *Thozet, Dallachy* (the latter with rather broader leaves and more hirsute).

2. **S. pogostoma,** *Benth.* Annual, erect, and quite glabrous. Leaves linear or linear-lanceolate. Stipular bristles rather long. Flowers small, not very numerous, in terminal and lateral heads. Calyx-lobes lanceolate, very acute, almost pungent, longer than the tube, and almost as long as the corolla. Corolla a little more than 1 line long, deeply coloured when dry, very densely bearded inside at the orifice of the tube; lobes longer than the tube, inflected at the tips. Anthers small, ovate, almost sessile at the base of the tube as in *S. brachystema.*

N. Australia. Sturt's Creek, *F. Mueller;* Amity Creek (*M'Douall Stuart?*) *Herb. F. Muell.*

3. **S. leptoloba,** *Benth.* Annual, divaricately-branched, scabrous-pubescent. Leaves lanceolate or elliptical, 1 to 2 in. long, acuminate and narrowed at the base, with a few very oblique veins prominent underneath. Stipular bristles very fine and not very long. Flowers numerous, in dense terminal or rarely lateral heads, intermixed with numerous cilia. Calyx-lobes lanceolate-subulate, longer than the tube, and at least as long as the corolla. Corolla about 1½ lines long, the tube broad and nearly glabrous inside; lobes very narrow, longer than the tube, pubescent inside, at least above the middle. Anthers oblong, much larger than in the allied species, inserted at the base of the tube. Fruit about 1 line long, smooth and white, the calyx-lobes on the summit assuming a bluish hue in the dried specimens.

N. Australia. Port Essington, *Armstrong.*—The flowers appear to be more or less unisexual, the females often without any traces of stamens, and the males with a slender abortive ovary.

4. **S. marginata,** *Benth.* Prostrate, rigid, glabrous and smooth, or the angles of the stem and margins of the leaves scabrous. Leaves sessile, ovate to broadly lanceolate, ½ to 1½ in. long, mucronate-acute, rigid, undulate, with thickened callous margins. Stipular sheath very short, with rigid bristles. Flowers numerous, in dense terminal or lateral heads. Calyx-lobes lanceolate, rigidly acuminate, longer than the tube, and usually exceeding the corolla. Corolla about 1 line long, the tube very broad, glabrous inside; lobes longer than the tube, inflexed and thickened at the end, bearded inside at the base. Stamens inserted about halfway up the tube and included in it. Carpels opening inside to the base.

Queensland. N.E. coast, *Banks and Solander, A. Cunningham;* Howick's Isles, *Macgillivray, F. Mueller.*

5. **S. multicaulis,** *Benth.* Annual or with a perennial almost woody base and erect rigid stems, glabrous and smooth or minutely scabrous-

pubescent. Leaves sessile, linear, with revolute margins. Stipular sheath very short. Flowers in small but dense terminal or at length lateral clusters or heads. Calyx-lobes thick, obtuse, incurved, as long as the tube and nearly as long as the corolla. Corolla not above 1 line long, usually pubescent outside, the tube slightly hairy inside; lobes longer than the tube, thickened towards the end. Stamens inserted at the orifice of the tube; anthers oblong, not exceeding the lobes. Carpels opening nearly to the base.

N. Australia. Islands of the Gulf of Carpentaria, *R. Brown* (with rather larger flower-heads).

Queensland. Burnett river, *F. Mueller;* Broadsound and Bowen river, *Bowman;* Brisbane river, *A. Cunningham.*

The flowers are nearly those of *S. semierecta,* Roxb., but the foliage is very different. In the latter respect it resembles the narrow-leaved form of *S. striata,* Linn., but differs in the flowers.

6. **S. exserta,** *Benth.* Apparently annual, erect, more or less hirsute. Leaves long-linear or linear-lanceolate. Stipular bristles rather long and fine. Flowers in dense globular terminal or sometimes lateral heads. Calyx-lobes lanceolate, ciliate, much longer than the tube, and often nearly as long as the corolla, the tube hirsute with almost scale-like bristles. Corolla nearly 3 lines long, the tube short and broad, the lobes much longer, lanceolate. Stamens inserted at the orifice of the tube, as long as or much longer than the lobes. Fruit only seen young.

N. Australia. Islands of the Gulf of Carpentaria, *R. Brown;* Croker's Island, *A. Cunningham;* Port Essington, *Armstrong.*

7. **S. membranacea,** *R. Br. Herb.* A slender annual, sprinkled with minute hairs or nearly glabrous. Leaves on rather long petioles, ovate or elliptical, $\frac{3}{4}$ to $1\frac{1}{2}$ in. long, thin and membranous. Stipular sheath very short, with very fine often long bristles. Flowers very small, in small terminal heads. Calyx-lobes lanceolate, shorter than the corolla. Corolla white, scarcely above 1 line long, the lobes longer than the tube. Stamens inserted at the orifice of the tube, exceeding the lobes. Fruit not seen.

N. Australia. Islands of the Gulf of Carpentaria, *R. Brown.*

8. **S. debilis,** *Benth.* Slender and diffuse, probably annual, glabrous or minutely scabrous. Leaves petiolate, ovate or oblong-lanceolate, obtuse or scarcely acute, not rigid, under $\frac{3}{4}$ in. and mostly under $\frac{1}{2}$ in. long. Flowers numerous, in dense heads mostly terminal. Calyx-lobes lanceolate-subulate, longer than the tube. Corolla nearly 2 lines long; tube cylindrical, slightly hairy inside but not at the orifice; lobes pubescent outside, very much shorter than the tube. Stamens inserted at the orifice of the tube, and shorter than the lobes. Fruit small, the carpels opening to the base.

Queensland. Sir Charles Hardy's Island, *Henne.*—The habit is quite that of a Malacca species in Griffith's collection, but the corolla is very much larger and differently shaped.

9. **S. inaperta,** *F. Muell. Fragm.* iv. 43. An annual with ascending or erect, rigid, but slender branches of 1 ft. or more, glabrous and smooth in our specimens. Leaves sessile, narrow-linear, with revolute margins. Stipular sheaths rather large and scarious, the bristles fine and rather short.

Flowers only about 4 together in axillary sessile clusters, numerous but only in one axil of each pair of leaves. Calyx pubescent; lobes linear-lanceolate, acute. Corolla funnel-shaped, about 3 lines long, glabrous inside, the lobes rather shorter than the tube. Stamens inserted at the orifice of the tube, shorter than the lobes. Fruit-carpels separating, and not always opening on their inner face.

N. Australia. Grassy places, Lower Victoria river, *F. Mueller.*

10. **S. stenophylla,** *F. Muell. Fragm.* iv. 43. Apparently perennial, with erect scarcely branched stems, glabrous and smooth. Leaves long, narrow-linear. Stipular sheaths scarious, rather long, the bristles usually short. Flowers in dense globular terminal or occasionally lateral heads. Calyx-lobes lanceolate, acute, rigidly ciliate, longer than the very short tube. Corolla funnel-shaped, 3 lines long, slightly bearded at the throat, the lobes nearly as long as the tube. Stamens inserted at the orifice of the tube, nearly as long as the lobes. Capsule a little more than 1 line long, the carpels opening at the top.

N. Australia. Sweers Island, Gulf of Carpentaria, *Henne.*

11. **S. lævigata,** *F. Muell. Fragm.* iv. 41. Apparently perennial, glabrous and smooth, or hispid with a few scattered rigid hairs, the stems erect, slightly branched, slender but rigid. Leaves long, linear. Stipular bristles remarkably long. Flowers in dense terminal heads. Calyx-lobes lanceolate-subulate, acute, more than twice as long as the tube. Corolla nearly 4 lines long, the tube long and slender, slightly dilated upwards; lobes oblong-linear, rather above half as long as the tube. Stamens inserted at the orifice of the tube, and nearly as long as the lobes. Carpels opening upwards.

N. Australia. Stony and grassy banks of Victoria river, Wickham's Creek, and Depôt Creek, *F. Mueller.*

Var. (?) *hispida.* Whole plant more or less hispid. Leaves very acute, often above 2 in. long.—*S. purpureo-cærulea,* R. Br. Herb. Gulf of Carpentaria, *R. Brown.*

Var. (?) *dilatata.* Calyx-lobes shorter. Floral leaves much dilated and coloured at the base.—Islands of the Gulf of Carpentaria, *R. Brown.*

12. **S. breviflora,** *F. Muell. Herb.* Annual, diffuse, and more or less hirsute. Leaves linear or linear-lanceolate, with recurved margins. Flowers in very dense terminal globular heads, often above ½ in. diameter. Calyx-lobes lanceolate or lanceolate-subulate, much longer than the tube. Corolla from 2½ to about 3 lines long, the lobes about as long as the rather broad tube, with the internal appendages of *S. auriculata,* but more oblique and often united into one, at least at the base, occupying altogether ⅓ of the lobe. Calyx-lobes usually recurved in the fruiting-head.

N. Australia. Victoria river and Depot Creek, *F. Mueller;* Arnhem N. Bay and islands of the Gulf of Carpentaria, *R. Brown;* Attack Creek, *M'Douall Stuart* (with rather larger flowers).

13. **S. auriculata,** *F. Muell. Fragm.* iv. 42. Low and diffuse or tall and slightly branched, more or less hispid, and sometimes very rigidly so. Leaves linear or lanceolate. Stipular bristles rather long. Flowers in dense terminal globular heads, often above ½ in. diameter. Calyx-lobes long and subulate. Corolla 6 lines long, the tube long and slender, bearded inside,

dilated at the top into a campanulate limb, the lobes shorter than the tube, each with a pair of oblique prominent glandular-toothed appendages on the inner face at about half their length. Stamens inserted at the orifice of the tube, shorter than the lobes. Capsule rather thin, the carpels opening at the top, a portion of the dissepiment remaining attached to one of them.

N. Australia. N. coast, *R. Brown;* Upper Victoria river, *F. Mueller;* Port Essington, *Armstrong* (with long linear glabrous or hirsute leaves).

14. **S. suffruticosa,** *R. Br. Herb.* A rigid glabrous undershrub, with erect stems of about 1 ft. Leaves linear, clustered in the axils. Stipular sheaths short, with about 2 bristles on each side dilated at the base, and sometimes united into a single lanceolate point. Flower-heads terminal and lateral, forming a short forked cyme at the ends of the branches. Calyx-lobes 2 only, short, linear. Corolla 2½ or nearly 3 lines long, the lobes spreading, as long as the tube, with a longitudinal line of hairs inside each. Stamens inserted at the orifice of the tube, and rather longer than the lobes. Fruits obovoid, hard, the carpels separating, the dissepiment remaining attached to one of them.

N. Australia. Islands of the Gulf of Carpentaria, *R. Brown.*

TRIBE III. STELLATÆ.—Calyx wholly adnate without any visible border (in the Australian genera). Corolla-lobes valvate. Ovary 2-celled, with 1 ovule in each cell. Fruit small, indehiscent.—Herbs, rarely undershrubs. Stipules similar to the leaves, and connected with them by a short sheath or ring, forming whorls of 4 or more, very rarely (only in two Australian species) reduced to the 2 true leaves.

28. ASPERULA, Linn.

Calyx completely combined with the ovary, without any visible border. Corolla funnel-shaped, with a distinct tube and 4 spreading lobes, valvate in the bud. Anthers exserted. Style 2-lobed. Ovary 2-celled, with 1 ascending or laterally attached ovule in each cell. Fruit small, dry, 2-lobed (when perfect), indehiscent.—Herbs with slender quadrangular stems. Leaves in whorls of 4 to 8, of which 2 are real leaves and the remainder stipules, although precisely similar in shape and size, in one species reduced to the 2 real leaves. Flowers small, solitary, or in axillary or terminal cymes or clusters, occasionally more or less unisexual, the females with a much shorter corolla-tube than the males, but always more prominent than in *Galium.*

The genus extends over the cooler temperate and subtropical regions of the Old World, but is unknown in America or southern Africa. The Australian species are all endemic. It differs from *Galium* only in the shape of the corolla.

Leaves mostly in pairs . . . . 1. *A. geminifolia.*
Leaves all or nearly all in fours . . . . 2. *A. subsimplex.*
Leaves mostly in sixes, a few whorls sometimes of four only.
  Leaves narrow-linear, with fine points. Plant usually scabrous-hispid 3. *A. scoparia.*
  Leaves linear, obtuse or acute, minutely pubescent . . . . . . 4. *A. conferta.*
  Leaves obovate, oblong or oblong-linear, obtuse . . . . 5. *A. Gunnii.*
Leaves mostly in eights, about 1 line long. Stems short, filiform . . 6. *A. minima.*

1. **A. geminifolia,** *F. Muell. Fragm.* v. 147. Stems long, slender,

weak and diffuse, quite glabrous. Leaves 2 or very rarely 3 to each whorl, narrow-linear, acute, above 1 in. long in luxuriant specimens, but usually shorter. Peduncles terminal, elongated, solitary or 3 together, each with about 5 to 7 flowers, almost sessile, in a small cyme. Corolla about $\frac{3}{4}$ line long, the lobes nearly as long as the tube. Fruit small, very rugose, almost fleshy.

Queensland. Burdekin and Brisbane rivers, *F. Mueller;* Cannon's River, *Bowman.*

2. **A. subsimplex,** *Hook. f. in Hook. Lond. Journ.* vi. 463, *and Fl. Tasm.* i. 168. Small, slender, diffuse and glabrous. Leaves in all our specimens constantly four in each whorl, linear, acute or scarcely obtuse, rarely exceeding $\frac{1}{4}$ in., the upper ones very short and rather broader. Flowers very shortly pedicellate, solitary in the upper axils or about 3 together at the ends of the branches. Corolla about $\frac{3}{4}$ line long, the lobes nearly as long as the tube. Young fruit rugose, but not seen ripe.

Tasmania. Circular Head, Formosa, Lake St. Clair, etc., *Lawrence, Gunn.*

3. **A. scoparia,** *Hook. f. in Hook. Lond. Journ.* vi. 463, *and Fl. Tasm.* i. 169. *t.* 40A. Stems short and densely tufted, erect or decumbent, or in the northern specimens 6 in. long or more, and rather rigid, the whole plant more or less scabrous-pubescent. Leaves 6 in a whorl, linear, acute, with a fine point, mostly 3 to 4 lines long. Flowers 3 or 5 together, nearly sessile within the last whorl of leaves, or about 3 together, on short common peduncles, of which 3 to 5 are umbellate within the last whorl. Corolla of the male flowers about 1 line long, with a slender tube, in the females much shorter, the tube scarcely so long as the lobes.—*Rubia syrticola,* Miq. in Ned. Kruidk. Arch. iv. 111.

Queensland. On the Maranoa, *Mitchell.*
N. S. Wales. New England. *C. Stuart;* near Appin, *Backhouse.*
Victoria. Desert of the Murray, *F. Mueller.*
Tasmania. Dry gravelly fields at Laurenny, *J. D. Hooker.*
S. Australia. Gawler town, Lofty Ranges, Mount Gambier, *F. Mueller.*

4. **A. conferta,** *Hook. f. in Hook. Lond. Journ.* vi. 464, *and Fl. Tasm.* i. 169. Diffuse, decumbent or nearly erect, sometimes forming dense prostrate tufts of a few in., sometimes 6 ft. to 1 ft. long, rigid or slender, nearly glabrous or slightly scabrous-pubescent. Leaves in whorls of 6, linear, almost obtuse or shortly pointed, but without the fine point of *A. scoparia.* Flowers in little terminal cymes or clusters, almost sessile within the last whorl of leaves. Corolla above 1 line long in the males with a slender tube, much shorter with a short broad tube in the females. Fruit smooth or rugose by drying.

Queensland. Peak Downs, *F. Mueller;* near Warwick, *Beckler.*
N. S. Wales. Open plains near Bathurst, *A. Cunningham;* Paramatta, *Woolls;* Darling river, *Dallachy.*
Victoria. Dry places in the western and north-western parts of the colony, *F. Mueller* and others.
Tasmania. Abundant in dry places throughout the colony, *J. D. Hooker.*

Var. *elongata.* Stems long and slender. Leaves $\frac{1}{4}$ to $\frac{1}{2}$ in. long, narrow-linear, obtuse. Flowers very small.—Mackenzie and Suttor rivers, *F. Mueller;* New England, *C. Stuart;* Smythe's ranges, *Whan;* Forest Creek, *F. Mueller.*

5. **A. Gunnii,** *Hook. f. in Hook. Lond. Journ.* vi. 463; *Fl. Tasm.* i. 168. Rather small and diffuse, glabrous or slightly pubescent. Leaves mostly in whorls of 6 or here and there of 4, only in slender specimens, oblong-linear or linear-cuneate, usually obtuse, rarely above ¼ in. long. Flowers mostly in threes in the upper axils or in a small terminal cluster, rather smaller than in *A. conferta*, but otherwise similar. Fruits globular or didymous, about 1 line diameter, not rugose.

N. S. Wales. Near Appin, *Backhouse.*

Victoria. Dandenong ranges and Mount Cobberas, *F. Mueller.*

Tasmania. Alpine situations, not unfrequent, *J. D. Hooker.*

Var. *pusilla.* Smaller and more branched.—*A. pusilla*, Hook. f. in Hook. Lond. Journ. vi. 464, and Fl. Tasm. i. 169. t. 40B.—Common in alpine and subalpine situations in Tasmania, *J. D. Hooker.*

6. **A. minima,** *Hook. f. in Hook. Lond. Journ.* vi. 464, *and Fl. Tasm.* i. 170. A little slender diffuse plant, forming dense tufts of a few in. diameter, slightly scabrous-pubescent. Leaves 6 to 8 (usually 8) in a whorl, linear, acute, mostly about 1 line long. Flowers all pedicellate, mostly in threes within the last leaves, or 3 peduncles bearing each 3 pedicellate flowers. Corolla in the males about ¾ line long, the lobes as long as the tube, scarcely above ½ line long in the females.

Tasmania. Common about George Town, *Gunn.*

29. GALIUM, Linn.

Calyx completely combined with the ovary, without any visible border. Corolla rotate, the tube scarcely perceptible, with 4 spreading lobes valvate in the bud. Anthers exserted. Style deeply 2-cleft. Ovary 2-celled, with 1 ascending or laterally attached ovule in each cell. Fruit small, dry, 2-lobed (when perfect), indehiscent.—Herbs, with weak quadrangular stems. Leaves in whorls of 4 to 8, of which 2 are real leaves and the remainder stipules, although precisely similar in size and shape, in one species reduced to the 2 real leaves. Flowers small, in axillary or terminal trichotomous cymes or panicles, rarely solitary.

An extensive genus, spread over the whole of the temperate regions of the globe, especially abundant in Europe and northern Asia, with very few tropical species, and those chiefly limited to mountain regions. Of the five or six Australian species, one is perhaps the same as a New Zealand one, another is probably introduced from Europe, the remainder appear to be endemic. But the proper discrimination and limitation of species in the whole genus is a very difficult and much disputed question.

Fruit glabrous and smooth or rarely slightly tubercular.
- Leaves in pairs . . . . 1. *G. geminifolium.*
- Leaves in whorls of four.
  - Flowers white. Leaves narrow or rarely ovate . . . . 2. *G. Gaudichaudi.*
  - Flowers yellow. Leaves ovate . . . . 3. *G. ciliare.*

Fruit muricate or hispid.
- Leaves in whorls of four, mostly ovate or lanceolate.
  - Plant scabrous or hispid, with short clinging hairs. . . . . 4. *G. australe.*
  - Plant densely clothed with soft hairs not at all clinging . . . 5. *G. albescens.*
- Leaves in whorls of five or six, usually narrow. Hairs of the plant very clinging . . . . 6. *G. Aparine.*

1. **G. geminifolium,** *F. Muell. in Trans. Vict. Inst.* 1855, 147, *and*

in Hook. Kew Journ. viii. 146; *Pl. Vict. t.* 31. Stems elongated, slender, glabrous or scabrous on the angles. Leaves all in pairs, narrow-linear, mostly obtuse. Flowers very small, about 3 together, sessile within the last leaves or on a common peduncle in the upper axils. Fruits rather large, glabrous and smooth.

N. S. Wales. Darling river, *Victorian Expedition.*

Victoria. Murray and Avoca river, *F. Mueller.*

S. Australia, *Wilhelmi.*

The plant closely resembles *Asperula geminifolia*, but is more rigidly divaricate, and the corolla is that of a *Galium.*

2. **G. Gaudichaudi,** *DC. Prod.* iv. 607. A very variable plant, usually hispid, more rarely glabrous except minute asperities. Stems usually numerous, short erect and densely tufted, or diffuse and extending to 1 or 2 ft. Leaves almost always 4 in a whorl, usually sessile, mostly lanceolate or linear, with recurved margins, 2 to 3 lines long, sometimes, and generally the lower ones, small and ovate, or (when grown in shady places?) thinner, ovate, and much narrowed at the base. Flowers very small, about 3 together, on axillary peduncles, which sometimes grow out irregularly into leafy branches, or the flowers and a few leaves clustered on a very short peduncle. Fruit quite glabrous and smooth when fully ripe, though often appearing rugose when shrivelled in dried specimens.—*G. vagans*, Hook. f. in Hook. Lond. Journ. vi. 461, and Fl. Tasm. i. 170; *G. axiflorum*, F. Muell.; Miq. in Ned. Kruidk. Arch. iv. 113.

N. S. Wales. Port Jackson to the Blue Mountains, *R. Brown* and others; northward to New England, *C. Stuart;* Macleay and Clarence rivers, *Beckler;* open plains of the interior, *A. Cunningham* and others.

Victoria. Dry places in the western and north-western parts of the colony, *F. Mueller* and others.

Tasmania. Dry bushy places, not uncommon, *J. D. Hooker.*

S. Australia. Murray river to St. Vincent's Gulf, *F. Mueller* and others.

The species is closely allied to and may not be really distinct from the New Zealand *G. umbrosum*, Forst.

Var. *glabrescens* Stems tufted, erect, glabrous or nearly so. Leaves broad.—New England, *Beckler.*

Var. *muriculatum.* Fruit slightly tubercular, connecting the species with *G. australe.*—New England, Clarence river, Paramatta, Wilson's Promontory, Cudnata, generally single specimens, possibly hybrids.

3. **G. ciliare,** *Hook. f. in Hook. Lond. Journ.* vi. 461, *and Fl. Tasm.* i. 170. *t.* 41. Slender diffuse and more or less hispid, with small ovate leaves in whorls of 4, closely resembling some broad-leaved forms of *G. Gaudichaudi*, and with the same inflorescence, but the flowers are said to be of a bright yellow, a character which proves constant in species of the northern hemisphere. Fruits glabrous and smooth.

Tasmania. Abundant in dry pastures, *J. D. Hooker.*

4. **G. australe,** *DC. Prod.* iv. 608. Slender and diffuse but perhaps perennial, the stems often elongated and much intricate, more or less rough or hispid, with short rigid clinging hairs. Leaves in whorls of 4, from ovate to lanceolate, sessile or shortly contracted at the base. Pedicels rather long, often 3 together, on a short axillary peduncle, with a whorl of leaves

at their base. Flowers of *G. Gaudichaudi.* Fruit small, more or less muricate or echinate, with hooked bristles.—Hook. f. Fl. Tasm. i. 171; *G. densum*, Hook. f. in Hook. Lond. Journ. vi. 461; *G. erythrorhizum*, F. Muell., Miq. in Ned. Kruidk. Arch. iv. 113.

Queensland. Warwick, *Beckler.*
N. S. Wales. Port Jackson (*Woolls?*).
Victoria. Wilson's Promontory, *F. Mueller.*
Tasmania. Common in dry pastures, etc., *J. D. Hooker.*
S. Australia. Near Cudnaka and Kangaroo island, *F. Mueller.*

Var. *piloso-hispidum.* Stems and foliage hispid.—*G. squalidum*, Hook. f. in Hook. Lond. Journ. vi. 462, and Fl. Tasm. i. 171.

5. **G. albescens,** *Hook. f. in Hook. Lond. Journ.* vi. 462, *and Fl. Tasm.* i. 171. A slender diffuse plant, very near *G. australe*, but densely clothed with soft hairs not at all prehensile. Leaves small, ovate, in whorls of four. Flowers of *G. australe.* Fruits rather larger, muricate or hispid with hooked bristles.

Tasmania. Rocky places on Mount Wellington, *Gunn.*

6. **G. Aparine,** *Linn.; DC. Prod.* iv. 608; var. *minor.* A trailing or climbing annual often several ft. long, but in the Australian specimen under 2 ft., clinging by the recurved asperities or small prickles on the angles of the stems and on the edges and midribs of the leaves. Leaves in the Australian specimens 5 or 6 in the whorl, in the larger forms 6 to 8, linear or linear-lanceolate, often above 1 in. long. Peduncles axillary, bearing a loose cyme of 3 or more small greenish-white flowers, with 3 or 4 leaves at the base of the cyme. Fruits covered with hooked bristles or prickles forming very adhesive burrs.

Victoria. Murray river, *F. Mueller;* Wendu valley, Glenelg river, *Robertson.*
S. Australia. Mount Gambier, *F. Mueller.*
W. Australia. Swan River, *Drummond, n.* 727, *Oldfield.*

The species is common in Europe and northern and central Asia, whence it has been carried with cultivation to most parts of the world, and it is therefore probably introduced only into Australia. Some of the specimens are small and bad, and perhaps not correctly distinguished from imperfect specimens of *G. australe.* The western ones, however, appear to me without doubt to be the *G. Aparine.*

Order LXII. COMPOSITÆ.

Flowers or *florets* collected together in a head (rarely reduced to a single floret), surrounded by an involucre of several bracts, either in one row or imbricated in several rows, the whole having the appearance of a single flower. Receptacle on which the florets are inserted either naked or bearing chaffy scales or hairs or bristles between the florets. In each floret the calyx is wanting or converted into a *pappus* or ring of hairs or scales on the top of the ovary. Corollas either all hermaphrodite, tubular, and 5- or rarely 4-toothed (heads *discoid*), or all hermaphrodite and *ligulate*, that is, with a slender tube and a flat strap-shaped lamina, or those of the centre or *disk* tubular and hermaphrodite or male, and those of the circumference either ligulate and female or neuter, forming a *ray* (heads *radiate*), or filiform and female (heads *discoid* but *heterogamous*). Stamens 5, rarely 4, inserted in the tube of the corolla, the anthers linear and united in a sheath round the style (except in *Xanthium* or where more or less imperfect), 2-celled, opening inwards by longitudinal slits, the connective usually produced at the top into a small erect appendage, the anther obtuse or sagittate at the base, the basal lobes sometimes prolonged into short and acute or long very fine and hair-like points or lobes called *tails*. Ovary inferior, with a single erect ovule. Style filiform, usually divided at the top into two short stigmatic branches. Fruit a small dry seed-like nut or *achene*, crowned by the pappus or naked. Seed erect, without albumen. Embryo straight or rarely curved. Radicle inferior.—Herbs shrubs or very rarely small trees, with alternate or opposite leaves, without stipules. Flower-heads terminal or very rarely axillary, solitary or in panicles usually corymbose, sometimes reduced to clusters or compound heads, the general inflorescence often centrifugal, the inflorescence within the head always centripetal.

The most extensive family amongst flowering plants, and represented in every quarter of the globe and in every variety of station, most abundant in America, in Southern Africa, in Australia, and in the Mediterranean region of the northern hemisphere, the species less numerous in proportion in tropical Asia and Africa. After deducting the foreign weeds, which are evidently of recent introduction, the species known as indigenous to Australia, nearly 500 in number, are here arranged under 88 genera. Of these, 9 are widely-spread genera, represented in most countries; 5 belong more specially to the temperate regions of the Old World, and amongst them one (*Leuzea*) almost limited out of Australia to the Mediterranean region; 10 belong to the tropical and subtropical flora of America and Asia, several of them more especially American; 10 extend into the tropical regions of the Old World, but not into America; 3 are represented, out of Australia, only in S. America (chiefly extratropical); 4 only in S. Africa; 8 only in New Zealand or through New Zealand in extratropical and Andine S. America, and 39 (of which 18 are still monotypic) are endemic in Australia.

The subdivision into groups of this vast Order has always presented the greatest difficulty and exercised the ingenuity of the most distinguished botanists, there not being one of the characters adopted which does not occasionally break down. Even that of the florets entirely ligulate, which distinguishes the *Cichoraceæ*, hitherto supposed to be absolute, has been invalidated by the *Catamixis*, recently described by Thompson (Journ. Linn. Soc. ix.), which, although ligulate, appears to be more nearly allied to those genera which connect *Asteroideæ* with *Mutisiaceæ*. Upon the whole, the minute differences in the shape of the style-branches, relied on by Lessing and by De Candolle, notwithstanding numerous exceptions and ambiguities, are, perhaps, the most reliable; and, after a little experience, the difficulties attending their observation diminish considerably. It is always in the style of the hermaphrodite florets that this character must be observed, for in the females it is almost uniform throughout the Order, and due allowance must always be made for occasional exceptions in isolated species or genera. So it is also with another difficult but important character, the anthers with or without tails. These so-called *tails* are in some *Gnaphalieæ* so exceedingly fine and transparent as to be scarcely visible without a strong reflected light on the stage of the mounted lens, whilst in those *Asteroideæ* where they are present they gradually dwindle down to scarcely acute auricles, and in all cases appear to be sometimes wanting in individual species of groups generally characterized by them. With regard to genera, the large ones run into each other so much as to render it a more than usually difficult task to fix their limits, and the number of monotypic or almost monotypic genera published has been most excessive; for the Australian flora alone I have ventured to propose the suppression of more than 80. Too much importance has probably been attached by modern botanists to slight modifications of the pappus in distinguishing these genera, whilst, on the other hand, the involucre may not always have been sufficiently made use of. The shape of the achene is often an excellent character, but very deceptive, for it is rarely ripe in herbarium specimens and the ovary often gives a very incorrect idea of its future shape.

Taking the most important of the tribual characters it will be found that—

The leaves are always alternate in *Cynarocephalæ*, *Vernoniaceæ*, *Senecionidæ*, *Mutisia-*

ceæ, and *Cichoriaceæ*, and with few exceptions also in *Asteroideæ*, *Ambrosieæ*, *Anthemideæ*, and *Gnaphalieæ*; almost always opposite in *Eupatoriaceæ* and *Helimtheæ*. In *Gnaphalieæ* they are always quite entire, in the other tribes they vary, entire, toothed, lobed, or divided.

The flower-heads are homogamous and discoid in *Cynarocephalæ*, *Vernoniaceæ*, and *Eupatoriaceæ*, homogamous and ligulate in *Cichoriaceæ*, strictly unisexual in *Ambrosieæ*, heterogamous, with the florets of the circumference ligulate or filiform in the other tribes, with exceptionally discoid and homogamous species or genera, rare in *Asteroideæ*, *Anthemideæ*, and *Calendulaceæ*, more frequent in *Gnaphalieæ* and *Senecionidæ*.

The scales of the receptacle are characteristic of *Heliantheæ*, and occur only in a few exceptional genera in other tribes.

Tailed anthers characterize with few exceptions the *Cynarocephalæ*, *Gnaphalieæ*, *Calendulaceæ*, and *Mutisiaceæ*, and prevail in a portion of *Asteroideæ*. They never occur in *Eupatoriaceæ*, *Heliantheæ*, *Ambrosieæ*, *Anthemideæ*, or *Senecionidæ*, and are very rare in *Vernoniaceæ*.

The pappus is generally deficient or reduced to a ring or cup in *Anthemideæ*, consists of rigid awns or scales, or is wanting in *Heliantheæ*, and is most frequently capillary in other tribes, but with so little constancy that it is of very little use in tribual distinctions.

TRIBE I. **Cynarocephalæ.**—*Leaves alternate. Flower-heads discoid, the florets all tubular, hermaphrodite, and regular or nearly so, the lobes usually narrow. Anthers usually fringed or tailed at the base. Style usually slightly thickened at the base of the branches, which are narrow and obtuse or slightly pointed and often erect.*

Pappus none.
  Outer involucral bracts with a large leafy appendage, inner ones with a pungent point. (Flowers orange.) . . . . * CARTHAMUS.
  Involucral bracts ending in a rigid spine with smaller ones at its base. (Flowers purple.) . . . . 3. CENTAUREA.
Pappus of numerous unequal bristles not much longer than the achene.
  Outer involucral bracts ending in small palmate points, intermediate ones in a rigid spine. (Flowers yellow.) . . . . . . 3. CENTAUREA.
  Involucral bracts ending in a long lanceolate prickle. Receptacle without bristles. (Flowers purple.) . . . . * ONOPORDON.
Pappus of long capillary bristles. Receptacle with bristles between the florets.
  Pappus-bristles plumose, in a single row, with or without a few simple ones outside. Involucre not prickly . . . . 1. SAUSSUREA.
  Pappus-bristles numerous, in several rows.
    Pappus-bristles very unequal, the longest or nearly all plumose.
      Involucral bracts ending in short scarious jagged appendages. 2. LEUZEA.
      Involucral bracts ending in spreading rigid leafy or spinous appendages . . . . * CYNARA.
    Pappus-bristles nearly equal, all plumose. Involucres usually prickly . . . . * CIRSIUM.
    Pappus-bristles nearly equal, all simple. Involucres usually prickly . . . . * CARDUUS.
  (27. COLEOCOMA, has almost the habit of *Cynarocephalæ*, but with filiform female florets.)

TRIBE II. **Vernoniaceæ.**—*Leaves alternate. Flower-heads discoid, the florets all tubular, hermaphrodite and regular or nearly so. Involucre imbricate. Anthers usually obtuse at the base, without tails. Style branches subulate and acute, not swollen at the base.*

Flower-heads on separate peduncles.
  Involucre broad, with a few outer leafy bracts. Pappus of a single row of rigid flattened scale-like bristles . . . . . . 5. CENTRATHERUM.
  Involucre ovoid, without leafy bracts. Pappus of capillary bristles with a few or a ring of short ones outside . . . . 4. VERNONIA.

Involucre ovoid, consisting of few herbaceous bracts. Anthers almost tailed. Pappus of 2 to 4 exceedingly deciduous short bristles . . . . 6. PLEUROCARPÆA.

(84. GYNURA, has subulate style-lobes, but involucral bracts in a single row.)

Flower-heads small, sessile, in a cluster or compound head. Involucres narrow, flattened. Pappus of a few rigid bristles . . . . 7. ELEPHANTOPUS.

TRIBE III. **Eupatoriaceæ.**—*Leaves usually opposite. Flower-heads discoid, the florets all tubular, hermaphrodite and regular or nearly so. Anthers obtuse at the base, without tails. Style-branches elongated obtuse and usually club-shaped or thickened at the end.*

Pappus of numerous capillary bristles . . . . 8. EUPATORIUM.
Pappus of 5 or 10 chaffy scales or bristles, dilated at the base . . 9. AGERATUM.
Pappus of 3 to 5 short bristles, each tipped by a gland . . . . 10. ADENOSTEMMA.

TRIBE IV. **Asteroideæ.**—*Leaves alternate or very rarely opposite. Flower-heads either heterogamous or diœcious, the female florets ligulate or filiform, the hermaphrodites or males tubular and 4- or 5-toothed, or (in very few exceptional species) the florets all hermaphrodite and tubular. Anthers various. Style-branches in the hermaphrodite florets usually more or less flattened, produced beyond the stigmatic lines into tips or appendages, papillose on the outside.*

Female florets ligulate, forming a ray to the flower-head.
  Pappus of the ray or of all the florets of capillary simple or plumose bristles.
    Achenes terete or slightly flattened. Ray-florets in a single row.
      Anthers obtuse at the base or shortly pointed. Involucral bracts with dry scarious margins.
        Leafy shrubs or undershrubs . . . . 11. OLEARIA.
        Herb with radical leaves and large heads on simple scapes 12. CELMISIA.
      Anthers with fine tails. Involucral bracts dry or the outer ones leaf-like. Ray-florets often irregular . . . . . 26. PTERIGERON.
    Achenes much flattened. Ray-florets numerous, usually in several rows.
      Achenes all fertile, produced into a slender beak bearing a capillary pappus . . . . 14. PODOCOMA.
      Achenes all fertile, not beaked, the capillary pappus sessile.
        Style-branches with subulate tips . . . . 13. VITTADINIA.
        Style-branches with lanceolate or obtuse tips . . . . 15. ERIGERON.
      Achenes of the ray fertile, with a capillary pappus, those of the disk mostly abortive, with a reduced or scaly pappus . 17. MINURIA.
  Pappus of rigid unequal usually divaricate awns or spines, sometimes accompanied by scales . . . . 18. CALOTIS.
  Pappus none or of scales or bristles, much shorter than the achene.
      Achenes abruptly contracted at the top into a short neck or boss. Pappus none . . . . 19. LAGENOPHORA.
      Achenes obtuse or truncate at the top. Pappus short or none 20. BRACHYCOME.
Female florets filiform or irregular.
  Flower-heads small, closely sessile, in dense clusters or compound heads.
    Involucral bracts linear, herbaceous or scarious.
      Pappus none. Anthers without tails or points at the base . 21. SPHÆRANTHUS.
      Pappus of capillary bristles. Anthers with short tails or points at the base . . . . 22. MONENTELES.
    Involucral bracts dry, rigid and acute. Anthers tailed.
      Pappus of few rigid bristles, flat and scale-like at the base . 28. THESPIDIUM.
  Flower-heads separately pedunculate or rarely sessile, but distinct.
    Pappus of simple capillary bristles.
      Anthers without tails . . . . 16. CONYZA.

(A few species of 11. OLEARIA, and 15. ERIGERON, have the ray-florets minutely and imperfectly ligulate.)
Anthers with fine tails.
Involucral bracts narrow-linear, herbaceous or soft. Style of the disk-florets branched . . . . 23. BLUMEA.
Involucral bracts rigid, often broad. Some or all the disk-florets sterile, with a simple style . . . . . 24 PLUCHEA.
Pappus of the female florets none, of the sterile disk-florets small. Anthers and involucre of *Pluchea* . . . . 25. EPALTES.
Pappus of scabrous denticulate or almost plumose bristles. Ray-florets usually irregular. Anthers tailed. Involucre dry or leafy . . . . 26. PTERIGERON.
Pappus a short scaly jagged tube or cup. Anthers tailed. Involucre dry . . . . 27. COLEOCOMA.
Pappus none or of short obtuse scales.
Anthers obtuse at the base. (See VII. ANTHEMIDEÆ.)
Anthers with fine tails. (See VIII. GNAPHALIEÆ.)
(38. NABLONIUM, has the habit of *Calotis*, but no female florets, and the receptacles with scales.)

TRIBE V. **Ambrosieæ.**—*Leaves alternate. Flower-heads absolutely unisexual. Anthers not united.*

Female florets 2 together, consolidated with the involucre into a prickly burr; males numerous, in globular heads, with a very small involucre . . . . 29. XANTHIUM.

TRIBE VI. **Heliantheæ.**—*Leaves opposite or rarely alternate. Flower-heads either heterogamous, with the female florets more or less ligulate, the central ones tubular hermaphrodite or male, or rarely discoid, with all the florets hermaphrodite and tubular. Receptacles with chaffy scales between the florets. Anthers without tails. Style of the* Senecionidæ *or approaching that of* Asteroideæ. *Pappus of stiff awns or of short scales or none.*

Involucre of 2 rows of bracts, the outer narrow leafy and glandular, the inner ones and the receptacle scales enveloping the florets. Pappus none . . . . 30. SIEGESBECKIA.
Involucre of 2 or 3 rows of bracts, nearly equal or the outer row broader and leafy.
Pappus none or of very short awns or fine bristles.
Receptacle flat or slightly convex.
Style-branches obtuse and flattened. Ray-achenes triangular; disk-achenes flat . . . . 31. ECLIPTA.
Style-branches almost acute. Ray- and disk-achenes usually flattened or thick . . . . 32. WEDELIA.
Style of the disk-florets undivided. Ray-achenes flattened; disk-achenes abortive . . . . 33. MOONIA.
Receptacle conical. Style-branches truncate. Ray-achenes triangular; disk-achenes flattened . . . . 34. SPILANTHES.
Pappus of oblong chaffy scales. Receptacle conical . . . . 35. GALINSOGA.
Pappus of 2 to 4 rigid awns.
Style-branches with acute or subulate points.
Ray-florets, when present, neuter . . . . 36. BIDENS.
Ray-florets, when present, female . . . . 37. GLOSSOGYNE.
Style-branches truncate. Small creeping herb, with 1-headed scapes . . . . 38. NABLONIUM.
Involucre of a single row of bracts, united in a toothed cup. Pappus of unequal scales or awns . . . . 39. TAGETES.
Involucre of 4 broad leafy bracts, 2 outer larger than the 2 inner. Ray-florets in several rows, with very small ligules. No pappus . 40. ENHYDRA.

Flower-heads small, narrow, collected in dense clusters or compound heads. No pappus . . . . 41. FLAVERIA.

TRIBE VII. **Anthemideæ.**—*Leaves alternate. Flower-heads heterogamous, the females ligulate or filiform or without corollas, the disk-florets hermaphrodite or male, or very rarely all the florets tubular and hermaphrodite. Receptacle without or rarely with scales. Anthers without tails. Style of* Senecionidæ. *Pappus none or reduced to a raised border or rarely of short scales.*

Florets of the circumference distinctly ligulate.
 Receptacle very conical, with a few scales between the florets . * ANTHEMIS.
 Receptacle flat or convex, without scales . . . . * CHRYSANTHEMUM.
Florets of the circumference tubular or obscurely ligulate, or without corollas. Disk-achenes often abortive. Small annuals or rarely perennials.
 Female florets without any or with a very short, broad, or conical corolla.
  Achenes flattened, obtuse or truncate. Flower-heads pedunculate . . . . 42. COTULA.
  Achenes flattened, crowned by the hardened style, or by 2 prominent divaricate angles. Flower-heads sessile . . . . . 43. SOLIVA.
 Female florets shortly tubular. Achenes 3- or 4-angled. Flower-heads sessile or nearly so . . . . 44. MYRIOGYNE.
 Female florets slender.
  Achenes 4-angled or flattened. Pappus none.
   Style-lobes long and slender. Leaves radical, spreading. Scapes leafless . . . . 19. LAGENOPHORA.
   Style-lobes short, truncate. Leaves crowded, in dense tufts or on very short creeping stems. Flower-heads sessile or shortly pedunculate . . . . 45. ABROTANELLA.
  Achenes large, flat, with herbaceous involute wings. Pappus none. Flower-heads nearly sessile . . . . 46. CERATOGYNE.
  Achenes not flattened. Pappus of lanceolate or oblong scales.
   Flower-heads terminal. Pappus-scales lanceolate . . . . . 47. ELACHANTHUS.
   Flower-heads clustered within the radical leaves. Pappus-scales oblong . . . . 48. ISOETOPSIS.

TRIBE VIII. **Gnaphalieæ.**—*Leaves alternate, quite entire. Flower-heads discoid, with all the florets tubular and hermaphrodite or the central ones male, or the florets of the circumference female and filiform or very rarely ligulate or irregular, or rarely the heads more or less diœcious. Anthers with very fine hair-like tails at the base, sometimes very short (or rarely quite wanting ?). Style-branches usually nearly terete, very obtuse or truncate. Involucral bracts most frequently scarious.*

SUBTRIBE I. **Angiantheæ.**—*Flower-heads small, usually numerous, sessile or nearly so on a common receptacle, in a dense cluster or compound head, usually surrounded by scarious or leafy bracts, forming a general involucre. Florets all hermaphrodite, a few rarely sterile.*

Receptacles (within the partial heads) without scales between the florets.
 General involucre of numerous bracts, in several rows, usually with scarious tips or small radiating laminæ. General receptacle broad and flat.
  Pappus none or various . . . . 49. MYRIOCEPHALUS.
 General involucre none or of few short scarious bracts, with or without a few leafy bracts or floral leaves outside.
  Pappus none or of 1 or more awned or jagged scales.
   Involucres flattened, with 1 outer flat and 2, rarely 3 lateral

conduplicate or concave scarious bracts, with or without 1 or 2 inner flat ones . . . . 50. ANGIANTHUS.
Involucres of several bracts, the inner ones broader than the outer ones, and very deciduous . . . . 51. GNEPHOSIS.
(See also 62. ERIOCHLAMYS, in which the heads are sometimes clustered.)
Pappus of several bristles or scales, plumose-ciliate or with terminal plumose tufts. General receptacle small or branched, and the partial heads often rather distinct.
Floral leaves none or few, and shorter than the heads.
Involucral bracts about as long as the florets, without any or with very short and broad radiating tips. Pappus-bristles usually more than 6 . . . . 52. CALOCEPHALUS.
Inner involucral bracts with long, petal-like, radiating laminæ. Pappus-bristles 3 to 6 . . . . 53. CEPHALIPTERUM.
Floral leaves several, as long as or longer than the clusters.
Pappus-bristles 5, very elastic. Small annuals . . . . 54. GNAPHALODES.
Receptacles (within the partial heads) with scales between the florets.
Pappus of several plumose-ciliate bristles or scales. Stems or peduncles elongated and erect . . . . 55. CRASPEDIA.
Pappus none or of very short scales. Dwarf, diffuse or stemless annuals . . . . 56. CHTHONOCEPHALUS.

SUBTRIBE II. **Helichryseæ.**—*Flower-heads distinct, pedunculate or sessile. Female filiform florets few or none, rarely forming* 1 *or* 2 *complete outer series.*

Receptacle with scales between the florets.
Pappus none. Receptacle-scales enveloping the florets . . . 57. IXODIA.
Pappus a scarious cup. Receptacle-scales broad, flat or concave . 58. AMMOBIUM.
Pappus of distinct capillary bristles. Receptacle-scales narrow, flat or concave . . . . 59. CASSINIA.
Receptacle without scales.
Pappus none.
Involucre small, narrow, of scarious or petal-like bracts, not radiating. Flower-heads paniculate or corymbose . . . 60. HUMEA.
Involucre turbinate, the outer bracts small, imbricate, the inner petal-like, radiating. Flower-heads on long peduncles . . 61. PITHOCARPA.
Involucre globular, enveloped in dense wool, the inner bracts scarious, not radiating. Receptacle conical. Small annual, the heads often clustered . . . . 62. ERIOCHLAMYS.
Involucre broadly hemispherical, with scarious bracts, not radiating. Flower-heads on long peduncles . . . . . . 63. ACOMIS.
Involucre of few linear herbaceous nearly equal bracts. Corolla slender, recurved. Dwarf annuals . . . . 64. TOXANTHUS.
Pappus a small greenish glandular cup. Involucre of few, narrow, nearly equal, herbaceous bracts. Dwarf annual . . . . 65. SCYPHOCORONIS.
Pappus of chaffy scales.
Involucre hemispherical. Bracts scarious. Pappus-scales obtuse or jagged, or divided into bristle-like branches . . . 66. RUTIDOSIS.
Involucre cylindrical, the bracts scarcely scarious. Pappus-scales produced into long fine awns. Dwarf annual . . . 67. QUINETIA.
Pappus of capillary bristles, simple, barbellate or plumose.
Involucral bracts linear, all herbaceous or the inner ones shortly scarious at the tips.
Achenes contracted into a slender beak. Involucral bracts nearly equal . . . . 68. MILLOTIA.
Achenes not beaked. Involucres imbricate.
Florets all hermaphrodite. Inner involucral bracts with short scarious or radiating tips. Pappus simple . . 69. IXIOLÆNA.

Outer florets female, ligulate, irregular or filiform. Involucral bracts subulate-acuminate. Pappus strongly barbellate or plumose . . . . 70. ATHRIXIA.
Involucral bracts green, almost leaf-like or thin, but not scarious. Pappus plumose. Achenes shortly stipitate . . 71. PODOTHECA.
Involucral bracts imbricate, scarious or with petal-like laminæ. Achenes sessile.
Pappus of simple or shortly barbellate bristles, rarely (in sect. *Chrysocephalum* of *Helichrysum*) with a plumose tuft at the end.
Involucral bracts all very thin and scarious. Outer female florets usually large and irregular . . . . 72. PODOLEPIS.
Outer involucral bracts thin and scarious. Achenes more or less distinctly contracted into a beak . . . . 73. LEPTORHYNCHUS.
Involucral bracts almost all or the inner ones or their laminæ opaquely scarious or petal-like. Achenes not beaked. Female florets usually few (except in sect. *Chrysocephalum*). Stems erect, leafy, simple or branched.
Outer achenes broad and flat, all the others abortive . 74. SCHŒNIA.
Achenes angular or compressed, but not very flat, all perfect, or only few abortive in the centre . . . . 75. HELICHRYSUM.
Involucres more or less scarious, usually elastically spreading after flowering. Female florets usually numerous. Dwarf tufted or shortly diffuse mountain perennials, with solitary, sessile or shortly pedunculate flower-heads . . . . 78. RAOULIA.
Pappus plumose from the base, except in some *Waitziæ*.
Involucral bracts scarious or petal-like, or with petal-like laminæ. Achenes sessile.
Achenes with slender beaks . . . . 76. WAITZIA.
Achenes not beaked . . . . 77. HELIPTERUM.
Involucral bracts more or less green, herbaceous or thin, but not scarious. Achenes stipitate . . . . 71. PODOTHECA.

SUBTRIBE III. **Eugnaphalieæ.**—*Flower-heads distinct or in dense clusters or compound heads, usually small. Female filiform florets numerous, in several rows or in separate heads.*

Pappus of capillary bristles.
Female florets in one or two rows. Dwarf tufted or shortly diffuse mountain perennials, with solitary, sessile or shortly pedunculate flower-heads . . . . 78. RAOULIA.
Flower-heads strictly diœcious . . . . 79. ANTENNARIA.
Flower-heads more or less androgynous, usually small and clustered, rarely solitary. Female florets in several rows . . . . . . 80. GNAPHALIUM.
Pappus of 6 serrulate, thick or broad bristles or scales. Habit of *Raoulia* . . . . 81. PTERYGOPAPPUS.
Pappus none . . . . 82. STUARTINA.
(47. ELACHANTHUS, with a scaly pappus, and 46. CERATOGYNE, and 48. ISOETOPSIS, without pappus, have the entire leaves and numerous female florets of *Gnaphalieæ*, but short anthers without tails, and the female florets wider at the base.)

TRIBE IX.—**Senecionidæ.** *Leaves alternate. Flower-heads either heterogamous, with the female florets ligulate or rarely filiform, or sometimes homogamous, with all the florets hermaphrodite and tubular. Receptacle without scales. Anthers obtuse or scarcely pointed at the base, without tails. Style-branches truncate and penicillate, or rarely with pubescent tips or appendages. Pappus of capillary bristles. Involucral bracts in the Australian genera in a single row, with or without a few small outer ones round their base.*

Outer female florets filiform, usually in 2 or 3 rows . . . . . 83. ERECHTHITES.
Florets all tubular and hermaphrodite, or the outer ones ligulate.
Style-branches with subulate tips . . . . 84. GYNURA.
Style-branches truncate.
Inflorescence terminal. Indumentum cottony or simple . . 85. SENECIO.
Inflorescence axillary. Indumentum stellate . . . . . . 86. BEDFORDIA.

TRIBE X. **Calendulaceæ.**—*Leaves alternate. Flower-heads usually heterogamous, the ray-florets ligulate, female or rarely neuter, the disk-florets tubular, hermaphrodite, but sterile or rarely fertile, and very rarely the heads homogamous and discoid. Anthers usually sagittate, but scarcely tailed. Style-branches in the disk-florets more or less concrete, and thickened at the base.*

Disk-achenes usually perfect. Leaves radical or nearly so, pinnatifid, white underneath. Rays long, spreading.
Achenes glabrous. Pappus none . . . . 87. CYMBONOTUS.
Achenes densely woolly. Pappus of short scales concealed in the wool . . . . 88. CRYPTOSTEMMA.
Disk-achenes all abortive. Stems leafy. Leaves undivided. Pappus none.
Rays very small. Achenes with 3 broad scarious wings . . . * TRIPTERIS.
Rays rather long, spreading. Achenes much incurved, muricate on the back, the margins sometimes dilated but not winged . . * CALENDULA.

TRIBE XI. **Mutisiaceæ.**—*Leaves alternate. Flower-heads either heterogamous, with radiating female florets, or homogamous, with the florets all tubular and hermaphrodite, some or all of the outer florets in all cases more or less 2-lipped. Anthers pointed or tailed at the base. Style varying from that of* Calendulaceæ *to nearly that of* Senecionidæ.

Flower-heads radiate. Achenes villous, with a pappus of capillary bristles. Radical leaves, scape, and large flower-heads nearly of *Celmisia* . . . . 89. AMBLYSPERMA.

TRIBE XII. **Cichoriaceæ.**—*Leaves alternate. Flower-heads homogamous, with all the florets ligulate.*

Pappus none or very minute.
Stem rigid, branching, leafy. Florets large, blue . . . . . * CICHORIUM.
Leaves radical. Scapes leafless, simple or branched. Florets small, yellow . . . . * ARNOSERIS.
Pappus of narrow flat scales tapering into simple or plumose bristles 90. MICROSERIS.
Pappus, at least of the central achenes, of plumose capillary bristles.
Receptacle with a few chaffy scales between the florets. Achenes mostly beaked . . . . 91. HYPOCHŒRIS.
Receptacle without scales.
Achenes tapering into a slender beak. Involucral bracts long, nearly equal . . . . * TRAGOPOGON.
Achenes very shortly contracted at the top. Involucre with small outer bracts.
Stems leafy, hispid. Outer involucral bracts numerous . . 92. PICRIS.
Leaves radical. Stems leafless or nearly so. Outer involucral bracts few and small . . . . * LEONTODON.
Pappus of numerous simple capillary bristles or hairs.
Achenes not at all or scarcely flattened.
Stems leafy. Achenes very shortly contracted at the top . . 93. CREPIS.
Leaves radical. Scapes leafless. Achenes contracted into a long slender beak . . . . * TARAXACUM.
Achenes very flat. Stems leafy.
Achenes not beaked . . . . 94. SONCHUS.
Achenes contracted into a slender beak . . . . * LACTUCA.

TRIBE I. CYNAROCEPHALÆ.—Leaves alternate, often prickly. Flower-heads discoid, the florets all tubular, hermaphrodite and equal or nearly so, the lobes usually narrow. Receptacle bristly or rarely naked. Anthers usually fringed or tailed at the base. Style usually slightly swollen at the base of the branches, which are narrow and obtuse or slightly pointed, and often erect or cohering nearly to the end.

This tribe comprises a great majority of the plants popularly known in Europe under the name of *Thistles*. The two species indigenous to Australia are, however, not prickly. The involucre is usually ovoid or globular, hard, with numerous imbricated scales, unlike that of almost any other *Compositæ* except tropical *Vernoniaceæ*, readily known by their style; *Coleocoma* and a few allied plants distinguished by their slender female florets; and a few tropical *Labiatifloræ*. The *Calendulaceæ*, united by some with *Cynarocephalæ* under the name of *Cynareæ* on account of their style, differ widely in their habit and involucre, and generally in their heterogamous radiate flower-heads, and appear to me to be much better placed between the *Senecionidæ* and the *Labiatifloræ*.

1. SAUSSUREA, DC.

(Aplotaxis, *DC.*)

Involucre ovoid or campanulate, not longer than the florets, the bracts numerous, imbricate, not prickly, the inner ones the longest. Receptacle bearing bristles between the florets. Florets all tubular, regular, with 5 narrow lobes. Anthers tailed. Style-branches linear, slightly thickened at the base. Achenes glabrous. Pappus of several plumose bristles united in a ring at the base, with a few outer simple or short bristles or scales, sometimes on one side only or very small, rarely wholly wanting.—Erect herbs. Leaves toothed or lobed or pinnately divided. Flower-heads rather large and solitary, or smaller and paniculate. Florets purplish.

A considerable genus, spread over the hilly regions of the northern hemisphere, chiefly in the Old World. The only Australian species is also in India and China.

1. **S. carthamoides,** *Benth. Fl. Hongk.* 168. An annual, with a rigid erect branching stem of 1 to 2 ft. or sometimes more, sulcate and slightly cottony. Leaves deeply pinnatifid or lyrate, the lower lobes narrow, the terminal one broad, thin, green above, white and cottony underneath, 3 or 4 in. long; the lower leaves sometimes ovate and nearly entire on long petioles, the uppermost few, with narrow lobes. Flower-heads few, on long peduncles. Involucre campanulate, 6 to 8 lines diameter, the numerous bracts linear-lanceolate, often very pointed but not pungent. Achenes striate, slightly curved. Outer pappus usually of several very short scales or bristles united in a minute oblique ring but very variable in size and number, sometimes very deciduous, leaving only the minute oblique ring, or very small from the first or entirely wanting.—*Serratula carthamoides*, Roxb. Fl. Ind. iii. 407; *Aplotaxis carthamoides*, Ham. in DC. Prod. vi. 540; *A. multicaulis*, DC. l. c. and in Deless. Ic. Pl. iv. t. 68; *A. foliosa*, Edgew. in Trans. Linn. Soc. xx. 77; *Haplotaxis australasica*, F. Muell. Fragm. i. 36.

Queensland, *Bowman;* Dawson and Burnett rivers, *F. Mueller;* Keppel Bay, *Thozet.*
N. S. Wales. Hunter's River, *R. Brown;* Clarence river, *Beckler.*

The species is common throughout India, extending to China and Japan. *A. candicans*, DC. Prod vi. 540, may be a large form of it with less divided leaves, and larger more hoary involucres.

2. LEUZEA, DC.

Involucre ovoid or globular, the bracts imbricate, numerous, not prickly, with broad rigidly scarious jagged tips. Receptacle flat, densely bristly between the florets. Florets all tubular, slender, 5-lobed. Anthers with short tails. Style-branches cohering or shortly spreading at the tips. Achenes oblong, compressed, glabrous. Pappus of numerous unequal fine bristles in several rows, all shortly plumose.—Erect herbs. Leaves toothed or pinnately divided. Flower-heads large, solitary, terminal. Florets purplish.

A small genus, confined to the temperate regions of the northern hemisphere in the Old World, with the exception of the single Australian species, which is endemic.

1. **L. australis,** *Gaudich. in Freyc. Voy.* 462. *t.* 92. A rigid erect herb, probably biennial, attaining 2 ft. or rather more, simple or scarcely branched, with a little loose cottony wool. Leaves petiolate, oblong-lanceolate, either toothed only, or more or less deeply pinnatifid or pinnately divided, the lower ones 6 in. long or more on long petioles, the upper ones few, small, and nearly sessile. Flower-heads solitary on a long terminal peduncle. Involucre ovoid, and 1½ to 2 in. long when in flower, more globular when in fruit, the outer bracts short, with a nearly orbicular appendage, the inner ones gradually longer, the innermost as long as the florets, tapering into narrow-linear tips with scarcely any scarious appendage. Style-branches often remaining united to the end. Achenes smooth, slightly striate, crowned by a slightly projecting border under the pappus.—DC. Prod. vi. 665.

Queensland, *Bowman;* Condamine river, Leydley's Creek, and head of the Gwydir, *Leichhardt;* Dawson river, *F. Mueller.*

N. S. Wales. Sandy shore near Kingstown, *R. Brown;* Fish River, *Gaudichaud;* New England, *C. Stuart.*

Victoria. Rocky grassy declivities, Murrandale river and Lake Omeo, *F. Mueller.*

With all the habit and characters of the genus, this is the only species native of the southern hemisphere. Amongst northern species it is the most nearly related to the Spanish *L. rhaponticoides,* Graells.

3. CENTAUREA, Linn.

Involucre globular or ovoid, the bracts imbricate, numerous, ending either in a prickle or in a fringed or toothed appendage. Receptacle bearing numerous bristles between the florets. Florets all tubular and 5-lobed, the outer row often larger and neuter. Anthers tailed. Style-branches linear, often cohering, thickened at the base. Achenes glabrous, usually obliquely or laterally attached at the base. Pappus short, of simple bristles or scales, sometimes very short, or rarely wholly wanting.—Erect or prostrate herbs, usually rigid. Leaves alternate, entire or pinnatifid, rarely prickly. Flower-heads large and solitary, or smaller and paniculate. Florets purple-blue or yellow.

The species are very numerous in the Mediterranean and Caucasian regions of the northern hemisphere, with a very few American species, and some of the common ones spread with civilization over various parts of the globe. Among these must be included those now found in Australia, of which *C. melitensis* alone has at first sight been taken for an indigenous plant.

1. **C. melitensis,** *Linn.; DC. Prod.* vi. 593. An erect rigid annual of 1 to 2 ft., with a little white cottony wool, or nearly glabrous. Radical leaves pinnately divided; stem-leaves narrow, decurrent, entire or slightly toothed. Flower-heads terminal, sessile above the last leaves, solitary or 2 or 3 in a cluster. Involucre above ½ in. long, the bracts rigid, the appendage of the outer ones small with short palmate spines, of the intermediate ones consisting of a rigid spine spreading to from 2 to 4 lines, with short divaricate spines at the base, the inner ones tapering into a very short simple spine. Florets yellow. Pappus of several series of bristles, the outer ones short, the intermediate gradually longer, the innermost row very short.—*C. apula*, Lam.; DC. l. c.

A native of the Mediterranean region, now spread over cultivated and waste places in many of the warmer regions both of the New and the Old World, especially near the sea, and very abundant in various parts of **Queensland, N. S. Wales, Victoria, Tasmania,** and **S. Australia.**

The following European species of this genus, and of other genera of *Cynarocephalæ*, have also been sent from Australia as introduced weeds:—

C. solstitialis, Linn.; DC. Prod. vi. 594. An annual, with the habit, foliage, and yellow florets of *C. melitensis*, but with a much longer and stouter spine to the intermediate involucral bracts, whilst the inner ones have a jagged scarious appendage without any spine.—S. Australia, *Herb. F. Mueller.*

C. calcitrapa, Linn.; DC. Prod. vi. 597. A coarse annual, green or slightly covered with cottony down, seldom rising to a foot in height, but with very spreading or prostrate branches. Leaves pinnatifid, with a few long linear or lanceolate lobes. Flower-heads sessile in the forks or within the last leaves of the branches. Involucral bracts, at least the intermediate ones, ending in stiff spreading spines of ½ to 1 in., with 1 or 2 small ones at their base. Florets purple. Achenes without any pappus.—Tasmania, *Herb. F. Mueller.*

Carthamus tinctorius, Linn.; DC. Prod. vi. 612. An erect rigid glabrous herb of 1 to 2 ft., with alternate ovate leaves, stem-clasping at the base, and bordered by a few small prickly teeth. Flower-heads terminal. Involucres globular, of closely imbricate rigid bracts, the outer ones terminating in leafy appendages like the stem-leaves, but smaller, the inner ones in a rigid pungent point, the florets all tubular, of a rich orange. Receptacle with linear bristle-like scales. Achenes glabrous, without any pappus.—Near Adelaide, in the neighbourhood of gardens.

Onopordon acanthium, Linn.; DC. Prod. vi. 618 (Scotch Thistle). A tall stout Thistle, covered with a loose cottony wool. Leaves coarsely toothed or pinnatifid, waved and very prickly, their broadly-decurrent margins forming prickly wings all down the stems. Flower-heads large, erect, and solitary at the ends of the branches. Involucres globular, of numerous bracts, ending in a long lanceolate prickle. Florets all tubular. Receptacle pitted with raised jagged edges shorter than the achenes. Achenes glabrous. Pappus of serrulate bristles, not plumose, and rather longer than the achenes.—Victoria and South Australia.

Carduus marianus, Linn. (*Silybum marianum*, Gærtn.; DC. Prod. vi. 616.) An erect Thistle of 1 or 2 ft., glabrous or with very little cottony wool. Leaves smooth and shining, variegated with white veins, the lower ones deeply pinnatifid and very prickly, the upper ones clasping the stem by prickly auricles scarcely decurrent. Flower-heads large, drooping, solitary and terminal. Involucre globular, the bracts imbricated with a very broad base, and a stiff spreading leafy appendage ending in a long prickle, bordered with prickles at its base. Receptacle with bristles between the florets. Achenes glabrous. Pappus of simple hairs.—About Melbourne, *Adamson.*

Cirsium lanceolatum, Scop.; DC. Prod. vi. 636. A rather stout Thistle, attaining 3 or 4 ft. Leaves waved, pinnatifid and very prickly, rough on the upper side, white and cottony underneath, decurrent into prickly wings along the stem. Flower-heads not numerous, rather large. Involucres ovoid, above an inch long, the bracts lanceolate, ending in a stiff prickle. Florets purple, all tubular. Receptacle with bristles between the florets. Achenes glabrous. Pappus of plumose bristles.—Victoria and Tasmania.—*C. palustre*, Scop.; DC. Prod. vi. 645. A tall Thistle, with the stems quite covered with the prickly decurrent margins of the leaves. Leaves pinnatifid and very prickly. Flower-heads rather numerous, small, and ovoid, usually collected in clusters. Involucral bracts numerous, with small somewhat prickly points. Florets, achenes, and pappus of *C. lanceolatum.*—Tasmania. —*C. arvense,* Scop.; DC. Prod. vi. 643. A rather tall Thistle with a creeping rhizome. Leaves pinnatifid, very prickly, clasping the stem with prickly auricles, or shortly decurrent. Flower-heads not large, in loose corymbs, diœcious. Involucral bracts numerous, with very small prickly points. Florets, achenes, and pappus of *C. lanceolatum.*—Tasmania.

Cynara cardunculus, Linn.; DC. Prod. vi. 620, var. *Scolymus* (*C. Scolymus,* Linn.; DC. l. c. the *Artichoke*). A tall stout Thistle, the stems and underside of the leaves with more or less of a white cottony wool. Leaves large, deeply pinnatifid, with narrow spinous lobes or teeth. Flower-heads very large, the bracts much imbricated and fleshy at the base, the hard herbaceous tips spreading, prickly-pointed in the typical wild *Cardoon,* almost obtuse or notched in the cultivated *Artichoke.* Receptacle very fleshy, with bristles between the florets. Florets all tubular, purple. Achenes glabrous. Pappus of numerous very unequal bristles, the longer ones or nearly all plumose.—Near Adelaide.

TRIBE II. VERNONIACEÆ.—Leaves alternate. Flower-heads discoid, the florets all tubular, hermaphrodite and regular or nearly so. Involucres imbricate. Anthers usually obtuse at the base, without tails. Style-branches subulate and acute, not swollen at the base.

The subulate style-branches appear to be constant in the tribe. They occur also exceptionally in a few genera of *Asteroideæ, Gnaphalieæ,* or *Senecionidæ,* which are, however, readily to be distinguished either by their heterogamous flower-heads, or by their involucre. The exceptional anthers of *Pleurocarpæa* occur also in the small group of Mascarene *Vernoniaceæ,* distinguished by De Candolle under the name of *Bojerieæ.*

4. VERNONIA, Schreb.

Involucre ovoid-globular or hemispherical, the bracts imbricate, not longer than the florets, the inner bracts the longest. Receptacle without scales. Florets all tubular and equal, regular, with 5 narrow lobes. Anthers obtuse at the base. Style-lobes subulate. Achenes mostly striate or angular, rarely cylindrical. Pappus of numerous capillary bristles, usually surrounded by an outer row of very short often chaff-like bristles, which are rarely entirely wanting.—Herbs, or in species not Australian, shrubs or climbers. Leaves alternate. Flower-heads terminal or in the upper axils, in cymes or panicles or sometimes solitary. Florets usually purple.

A very numerous genus, widely spread over the warmer regions of the globe, but most abundant in America, where it extends beyond the tropics both to the northward and southward. In Australia it is represented by a single species, a weed of tropical Asia, differing slightly from the great mass of the genus in its habit, which is nearly that of a *Conyza.*

1. **V. cinerea,** *Less.; DC. Prod.* v. 24. Annual or forming a perennial rootstock, erect, 1 to 2 feet high, nearly glabrous, scabrous-pubescent, hirsute, hoary-tomentose or woolly. Lower leaves petiolate, ovate-oblong or lanceolate, often irregularly toothed or sinuate, the upper ones few and narrow, or occasionally nearly all ovate or nearly all narrow. Flower-heads small, on slender peduncles, forming a terminal leafless cymose panicle. Involucral bracts very acute. Achenes cylindrical, scarcely striate, hairy. Pappus white, the outer row very short, and sometimes reduced to very few bristles.—*V. erigeroides,* DC. Prod. v. 25; *V. cyanopioides,* Walp. in Linnæa, xiv. 509, and probably nearly the whole of the section *Tephrodes,* DC.

N. Australia. Victoria river and Macadam range, *F. Mueller.*

Queensland. Port Curtis, *M'Gillivray;* Rockingham Bay, Port Denison, Rockhampton, *Dallachy;* Brisbane river, Moreton Bay, and Peak Downs, *F. Mueller;* Keppel Bay, *Thozet.*

N. S. Wales. Port Jackson to the Blue Mountains, *Woolls* and others; northward to Clarence and Hastings rivers, *Beckler;* New England, *C. Stuart;* southward to Twofold Bay, *Mossman.*

There are two principal varieties in Australia, one either nearly glabrous or rarely hirsute, with short rigid hairs, the other very hoary-tomentose or woolly, with softer more obtuse leaves, and usually smaller flower-heads, the corolla-lobes also appear shorter. The flowers very purple or white, but almost always purple in the more glabrous form.

5. CENTRATHERUM, Cass.

Involucre imbricate, a few of the outer bracts long and leaf-like, the others not longer than the florets, the inner ones the longest. Receptacle naked. Florets all tubular and equal, regular, with 5 narrow lobes. Anthers obtuse at the base; style-lobes subulate. Achenes oblong, nearly cylindrical, striate. Pappus of a single series of rigid, flattened, hirsute, very deciduous, almost chaff-like bristles.—Herbs or shrubs. Leaves alternate. Flower-heads on terminal or leaf-opposed peduncles. Involucres hemispherical. Florets usually purple.

A small American genus, of which two species, including the Australian one, have spread over several of the warmer regions of the Old World.

1. **C. muticum,** *Less.; DC. Prod.* v. 70. A rigid divaricately-branched herb, probably annual, although with a hard almost woody base, spreading to 2 or 3 ft., glabrous or pubescent. Leaves petiolate, lanceolate-oblong or almost ovate, 1 to 2 in. long, coarsely and irregularly toothed. Flower-heads often above ½ in. diameter, the outer leafy bracts ½ to 1 in. long, the inner ones numerous, with spreading more or less coloured or scarious denticulate tips. Florets purple, much longer than the involucre. Achenes usually glabrous, with about 10 very obtuse ribs, the pappus falling off even before they are ripe.

Queensland. Brisbane river, Moreton Bay, *F. Mueller;* Rockhampton, *Dallachy.*

N. S. Wales. Hawkesbury river, *R. Brown;* Clarence river, *Beckler, Wilcox;* Blue Mountains, *Miss Atkinson.*

The species is common in tropical America; we have also apparently the same from the Philippine Islands.

6. PLEUROCARPÆA, Benth.

Involucre ovoid, the bracts few, herbaceous, imbricate, not so long as the florets. Receptacle without scales. Florets all tubular and equal, regular, often incurved, with 5 narrow lobes. Anthers sagittate at the base, the auricles produced into short not fine points. Achenes thick, with prominent ribs. Pappus of 2 to 5 rigid short exceedingly deciduous bristles.—Leaves alternate, entire or toothed. Flower-heads on terminal peduncles.

The genus is limited to a single species endemic in Australia, not nearly allied to any one known to me, unless it be to some anomalous species of *Decaneurum.*

1. **P. denticulata,** *Benth.* Herbaceous, with hard divaricate or decum-

bent branches, our specimens above 1 ft. long, and quite glabrous. Leaves ovate or oval-elliptical, mucronate-acute, contracted into a very short petiole, the larger ones above 2 in. long, irregularly bordered by acute teeth, the upper ones smaller and entire. Peduncles terminal, solitary or 2 together, 1 to 2 in. long or longer after flowering, slightly thickened under the head. Involucre about 4 lines long, thickened at the base, the bracts broadly lanceolate, acuminate. Florets about 10 to 20, of a bluish-purple, the tube exceeding the involucre, often incurved, shortly dilated into a deeply 5-lobed limb.

N. Australia. Islands of the Gulf of Carpentaria, *R. Brown*, who had given it the provisional name of *Lipothrix denticulata*, but he afterwards published as *Lipotriche* a very different African plant.

7. ELEPHANTOPUS, Linn.

Flower-heads of 2 to 5 florets, collected together in compound heads. Involucres compressed, of about 8 bracts, dry, stiff, alternately plane and conduplicate. Receptacle naked. Florets with 4 narrow equal lobes, but deeper cleft on one side, so as to be somewhat palmate. Anthers obtuse at the base. Style-lobes subulate. Pappus of a few stiff bristles, somewhat dilated at the base.—Stiff herbs, usually grey with appressed often silky hairs. Leaves alternate.

A genus of about a dozen American species, one of which is also spread over tropical Africa and Asia as well as Australia.

1. **E. scaber,** *Linn.; DC. Prod.* v. 86. Stock perennial. Stems stiff, erect, about a foot high, with a few forked spreading branches, more or less covered, as well as the leaves and involucres, with greyish hairs. Radical leaves 2 to 4 in. long, obovate-oblong, more or less crenate, and usually narrowed into a petiole. Stem leaves few and more sessile. Flower-heads closely clustered into terminal hemispherical compound heads, of nearly 1 in. diameter, surrounded by about 4 broadly cordate sessile leafy bracts. Involucral bracts narrow, very pointed, almost prickly.—Wight, Ic. t. 1086; Benth. Fl. Hongk. 170, with the synonyms adduced.

N. Australia. Port Essington, *Armstrong*. Common in the warmer regions of America, Africa, and Asia.
Queensland. Endeavour river, *Banks and Solander.*

Tribe IV. Eupatoriaceæ.—Leaves usually opposite. Flower-heads discoid, the florets all tubular, hermaphrodite, and regular or nearly so. Anthers obtuse at the base, without tails. Style-branches elongated, obtuse and usually club-shaped or thickened at the end.

This tribe is scarcely Australian, the following three species, perhaps all introduced, are readily distinguished from all other tribes by their opposite leaves and club-shaped styles.

8. EUPATORIUM, Linn.

Involucre hemispherical, campanulate or cylindrical, the bracts imbricate, in 2 or more series. Receptacle flat or slightly convex, without scales.

Florets numerous or few, all tubular, hermaphrodite, 5-toothed. Anthers obtuse at the base. Style-branches elongated, obtuse. Achenes 5-angled, without intermediate striæ. Pappus of a single series of capillary bristles. —Perennial herbs or shrubs or very rarely annuals. Leaves usually opposite. Flower-heads mostly corymbose. Green parts of the plant often sprinkled with resinous dots.

A vast genus, the great majority of species being American, a few ranging over eastern Asia, and one extending to Europe, and now introduced into Australia.

*1. **E. cannabinum,** *Linn.; DC. Prod.* v. 180. A perennial with erect stems of 3 to 4 ft., slightly pubescent. Leaves divided to the base into 3 broadly lanceolate coarsely-toothed segments, often 4 or 5 in. long, those of the upper leaves smaller and sometimes very narrow, the uppermost leaves rarely undivided. Flower-heads numerous in compact terminal heads. Involucres cylindrical, of few unequal bracts, the inner ones often coloured. Florets usually 5, purple or rarely white.—*E. Lindleyanum*, F. Muell. Fragm. v. 62, not of DC.

Very common in the temperate regions of the northern hemisphere in the Old World; it appears to have established itself on the Tweed river in **N. S. Wales.** *C. Stuart.*

9. AGERATUM, Linn.

Involucre florets and style of *Eupatorium*. Achenes angular. Pappus of 5 or 10 chaffy scales or bristles, dilated at the base.—Herbs with the opposite leaves and habit of *Eupatorium*.

A genus of a small number of American species, one of which is spread all over the warmer regions of the globe.

1. **A. conyzoides,** *Linn.; DC. Prod.* v. 108. An erect branching annual, 1 to 2 ft. high, more or less hirsute with spreading hairs. Leaves opposite, petiolate, ovate, crenate. Flower-heads rather small, in dense terminal corymbs. Involucral bracts striate, acute, in about 2 rows. Florets numerous, pale blue or white. Achenes black. Pappus of 5 lanceolate awned chaffy scales, often serrate in the lower part.—Hook. Exot. Fl. t. 15; F. Muell. Fragm. v. 62.

Queensland. Rockhampton, *Dallachy.* A common weed all over the warmer regions of the globe.

10. ADENOSTEMMA, Forst.

Flower-heads, florets, and style of *Eupatorium*. Achenes obovate-oblong, contracted at the base. Pappus of 3, 4, or 5 short stiff bristles, each tipped by a globular or club-shaped gland.—Herbs, either glabrous or glandular, pubescent. Leaves opposite. Flower-heads usually hemispherical, small.

A genus of a very few American species, one of which spreads all round the warmer zone of the globe.

1. **A. viscosum,** *Forst.; DC. Prod.* v. 111. Stem annual, erect or ascending, rooting at the base, and possibly renewed a second year by a creeping rhizome or stolons, glabrous or glandular-pubescent, 1 to 2 ft.

high. Leaves few, opposite, petiolate, from ovate to broadly triangular, usually coarsely toothed, from barely 1 in. long and rather thick and rough, to 3 or 4 in. long, thin and glabrous. Flower-heads hemispherical, 3 to 4 lines diameter, in a loose spreading terminal 2- or 3-chotomous panicle, with very small leaves under the branches. Involucral bracts oblong, in about 2 rows. Florets numerous, often hairy outside. Achenes more or less muricate or rarely quite smooth.—Wight, Ic. t. 1087, 1088.

Queensland. Endeavour river, *Banks and Solander;* Rockhampton and Rockingham Bay, *Bowman;* Lizard Island, *M'Gillivray.*

N. S. Wales. Hawkesbury river, *R. Brown.*

S. Australia. Entrance to the Murray river, *Wilhelmi.* Probably introduced.

This is a common weed in the warmer regions of the globe, especially in the Old World, where it extends northward to Japan. The species should include all those published by De Candolle and others from the Old World, and at least *A. brasiliense* and *A. triangulare* among the American ones.

TRIBE IV. ASTEROIDEÆ.—Leaves alternate or very rarely opposite. Flower-heads either heterogamous or diœcious, the female florets ligulate or filiform, the hermaphrodites or males tubular, and 4- or 5-toothed, or in very few exceptional species all the florets hermaphrodite and tubular. Anthers obtuse acute or tailed at the base. Style-branches in the hermaphrodite florets usually more or less flattened, produced beyond the stigmatic lines into short and obtuse or lanceolate or almost subulate tips, papillose on the outside.

The majority of the genera are easily distinguished from *Anthemideæ* and *Senecionidæ* by the style, a few, however, with the disk-florets sterile, have, like similarly sterile disk-florets in *Anthemideæ* and *Gnaphalieæ*, the style undivided and obtuse or truncate without stigmatic lines. For these cases no positive character can be given, but the genera or species must be classed from general affinity or minor characters. Thus *Minuria* is distinguished from *Anthemideæ* by the pappus, and *Pluchea* and the allied genera from *Gnaphalieæ* by the foliage and involucre. Very rarely the style-branches are almost subulate, but then the female florets prevent any confusion with *Vernoniaceæ.*

11. OLEARIA, Mœnch.

(Eurybia, *Cass.;* Steetzia, *Sond.*)

Involucres from broadly hemispherical to narrow-ovate, the bracts imbricate in several rows, the margins more or less dry or scarious, without herbaceous tips. Receptacle pitted, the borders of the pits often denticulate, but without scales. Florets of the ray female in a single row, or fewer than those of the disk, usually ligulate, spreading, very rarely slender and filiform or deficient. Disk-florets numerous or few, hermaphrodite, tubular, gradually tapering to the base in most species of the first two sections, more abruptly contracted in some others, usually 5-lobed. Anthers often acute at the base or with minute tails, rarely obtuse. Style-lobes flattened with short obtuse or rarely lanceolate appendages, papillose on the back. Achenes striate, terete or slightly compressed. Pappus of numerous, usually unequal, capillary bristles.—Shrubs undershrubs or very rarely herbs. Leaves alternate or rarely opposite. Heads solitary, corymbose or paniculate, terminal but sometimes appearing axillary from the shortness of the flowering branches. Ray-florets white or blue. Disk-florets yellow or rarely purplish or even blue. The indumentum of the underside of the leaves is usually more or less present also on the branches and inflorescence.

The genus is limited to Australasia, there being, besides the 63 Australian species, which are all endemic, only about 20 others, all natives of New Zealand. It is, however, very closely allied to the extensive genus *Aster*, widely diffused over the temperate regions of the northern hemisphere, especially in America, not separated indeed from *Olearia* by any one definite character, and F. Mueller has recently proposed to unite the whole and several others to *Aster* itself. It appears to me, however, that independently of the convenience of retaining *Olearia* for the Australasian species, there is little risk of its being confounded with the northern genus. The habit of most species is very different. Where the achenes are the same or nearly so (as in the North American *Biotias*, which have the achenes as little compressed as in a few Australian species), the foliaceous-tipped involucral bracts of the former are a ready distinction. Some Australian species again are separated from all the American ones by their styles, others by their anthers, and most of them by the indumentum. There appear to be indeed better grounds for maintaining *Olearia* as distinct from *Aster* than for retaining *Erigeron*, which passes so gradually into it, and that again into *Conyza*, and if all these were united into one, we should have a group quite unmanageable without dividing it into sections corresponding to the present genera, which would be in fact retaining the present arrangement, but with all the evils consequent on the nominal change.

That *Olearia* and *Eurybia* are inseparable even as sections has been shown by J. D. Hooker, Archer, and others, and I follow J. D. Hooker in adopting the former as the older name. Schultz-Bipontinus, under the idea that the genus is the *Shawia* of Forster, adopts the latter name, and accordingly, in the 'Pollichia,' gives to all published species, good or bad, new names, as *Shawias*. This is, however, a mistake. Forster considered the true *Olearias* as *Asters*, and founded *Shawia* on a plant characterized specially by solitary florets. It proved afterwards that this was not a constant character, and his genus was suppressed, and the circumstance that this abnormal peculiarity occurs on some specimens of one species of *Olearia*, can surely be no reason for now transferring the name founded on it to a large genus where it has not been observed in any other species.

In the subdivision of *Olearia*, I have adopted the main sections proposed by Archer (Journ. Linn. Soc. v. 17, 20), founded on the indumentum, which, with the exception of two or three species where it almost disappears, seems to be the most available in a genus where so many species pass into each other by almost insensible gradations.

SECTION I. **Dicerotriche.**—*Indumentum of the underside of the leaves (usually either silvery-shining, tomentose or loosely villous) consisting of centrally-attached or divaricately forked* (T-*shaped*) *hairs, otherwise simple.*

Leaves opposite.
  Leaves oblong, flat, 2 to 4 in. long, densely tomentose underneath.
    Flower-heads large. Ray-florets usually 10 to 12 . . . . 1. *O. megalophylla.*
    Flower-heads small. Ray-florets usually 2 or 3 . . . . . 2. *O. chrysophylla.*
  Leaves narrow-oblong, with recurved margins, densely silky underneath . . . . 3. *O. alpicola.*
  Leaves linear, with revolute margins . . . . 4. *O. rosmarinifolia.*
  Leaves lanceolate, viscid, smooth (not rugose) above, white underneath . . . . 5. *O. viscosa.*
Leaves alternate.
  Flower-heads very large, solitary, on long peduncles. Leaves ovate, 2 to 4 in. long.
    Leaves denticulate, closely silky or silvery underneath . . . 6. *O. grandiflora.*
    Leaves mostly entire, densely and softly cottony underneath . 7. *O. pannosa.*
  Flower-heads small, numerous and paniculate, or rather larger and shortly pedunculate. Indumentum close.
    Leaves prominently reticulate on the upper surface.
      Leaves broadly ovate or elliptical, mostly 3 or 4 in. long, entire or slightly toothed.

Indumentum very silvery and close. Involucres narrow-cylindrical, few-flowered . . . . 8. *O. oliganthema.*
Indumentum silvery and close. Involucres turbinate, few-flowered . . . . 9. *O. argophylla.*
Indumentum white and soft. Involucres hemispherical, many-flowered . . . . 10. *O. cydoniæfolia.*
Leaves from obovate to oblong or lanceolate, mostly obtuse, sinuate-toothed and under 2 in. long . . . . 11. *O. myrsinoides.*
Leaves not reticulate on the upper surface.
Leaves obovate or oblong, entire, under 1 in. long . . . 12. *O. persoonioides.*
Leaves small, cuneate, mostly toothed at the end . . . . 13. *O. obcordata.*
Leaves narrow-linear, crowded.
Leaves pungent-pointed, mostly above 1 in. long . . . 14. *O. pinifolia.*
Leaves obtuse, mostly ½ in. long . . . . 15. *O. ledifolia.*
Flower-heads rather large, usually pedunculate and corymbose.
Leaves ovate, loosely tomentose underneath . . . . 16. *O. dentata.*

SECTION II. **Asterotriche.**—*Indumentum of the under side of the leaves consisting of stellate hairs, sometimes very close and almost mealy.*

Leaves alternate, mostly toothed or sinuate.
Leaves ovate-oblong or lanceolate, smooth or scabrous above. Flower-heads rather small. Involucres broadly turbinate. Achenes hairy . . . . 17. *O. stellulata.*
Leaves oblong-linear, sinuate-toothed, obtuse. Flower-heads rather large. Achenes hairy . . . . 18. *O. asterotricha.*
Leaves ovate-lanceolate or oblong, very. rugose. . Flower-heads large. Achenes glabrous . . . . 19. *O. gravis.*
Leaves ovate-lanceolate, acute, rather thin and somewhat glutinous. Flower-heads in pedunculate corymbs . . . . 20. *O. Nernstii.*
Leaves alternate or opposite, narrow, quite entire.
Leaves alternate, narrow-linear . . . . 21. *O. hygrophila.*
Leaves mostly opposite, lanceolate . . . . 22. *O. viscidula.*

SECTION III. **Eriotriche.**—*Indumentum of the under side of the leaves consisting of densely intricate woolly hairs. Leaves alternate, often small.*

Flower-heads (usually small and ovoid) axillary or on very short axillary branchlets, forming long leafy racemes.
Ray-florets slender, tubular, much shorter than the entire part of the style . . . . 23. *O. tubuliflora.*
Ray-florets ligulate, but shorter than their style . . . . 24. *O. axillaris.*
Ray-florets ligulate, scarcely longer than their style. Flower-heads sessile. Western species.
Leaves 3 to 6 lines long, linear to oblong-cuneate, with recurved margins . . . . 25. *O. revoluta.*
Leaves 1 to 2 lines long, linear, with revolute margins . . . 26. *O. exilifolia.*
Ray-florets conspicuously exceeding the style.
Eastern species.
Leaves 1 to 4 lines long, obovate to oblong-linear, the upper surface glabrous or scabrous . . . . 27. *O. ramulosa.*
Leaves under 1 line, narrow, the upper surface glabrous or scabrous. Flower-heads usually very small and numerous . . . . 28. *O. floribunda.*
Leaves mostly under 1 line and often under ½ line, obovate oblong or globular, the upper surface glabrous or tomentose . . . . 29. *O. lepidophylla.*
Leaves narrow-linear, 3 to 6 lines long. Flower-heads forming short, dense, terminal, leafy racemes . . . . 30. *O. subspicata.*

Western species.
Leaves very small, mostly obovate and 3-toothed . . . . 31. *O. exiguifolia.*
Leaves narrow-linear, ¼ to ¾ in. long, quite entire . . . 32. *O. Cassiniæ.*
Flower-heads usually hemispherical, terminal, solitary or corymbose.
Rays conspicuous.
Leaves mosty reflexed, clustered, about 1 line long. Flower-heads solitary . . . . 33. *O. ramosissima.*
Leaves oblong-cuneate, 3 to 4 lines long.
Flower-heads solitary, on long branchlets . . . . 34. *O. pimeleoides.*
Flower-heads corymbose or on short branchlets . . . . . 35. *O. iodochroa.*
Leaves linear, 2 to 4 lines long, with revolute margins. Plant glandular-pubescent and slightly woolly . . . . 36. *O. adenolasia.*
(See also 51. *O. muricata,* which is sometimes very near *O. adenolasia,* 50. *O. heleophila,* which has often a little wool about the involucre, and 62. *O. arguta,* which has the young foliage sometimes woolly.)

Flower-heads cylindrical, with few tubular florets, without any ray . 37. *O. conocephala.*

SECTION IV. **Adenotriche.**—*Plant glabrous, usually glutinous. Involucre ovoid, turbinate or rarely hemispherical, the bracts usually obtuse, rigid, scarious on the margins and often ciliate.*

Flower-heads solitary or in leafy corymbs or panicles.
Leaves flat, obovate cuneate or oblong-linear, mostly toothed. Flower-heads solitary, almost sessile above the last leaf.
Flower-heads very large. Involucre broadly turbinate, above ½ in. long.
Leaves narrow, cuneate . . . . 38. *O. magniflora.*
Leaves small, obovate . . . . 39. *O. calcarea.*
Flower-heads small. Involucre almost hemispherical, 3 to 4 lines long.
Leaves small, obovate . . . . 40. *O. Muelleri.*
Leaves narrow, oblong or lanceolate . . . . 41. *O. Stuartii.*
Leaves flat, linear or linear-cuneate, entire or toothed.
Flower-heads small, solitary or in an oblong panicle . . . . 42. *O. decurrens.*
Flower-heads ovoid, corymbose . . . . 43. *O. glutinosa.*
Leaves linear or small, quite entire, with revolute margins. Flower-heads ovoid (except *O. imbricata*), nearly sessile, solitary or corymbose.
Leaves slender, ½ to in. long or more. Eastern species . . 43. *O. glutinosa.*
Leaves erect or spreading, about ¼ in. long. Western species . 44. *O. passerinoides.*
Leaves erect, under ¼ in. long. Eastern species . . . . . 45. *O. teretifolia.*
Leaves very spreading or recurved, under ¼ in. long. Eastern species . . . . 46. *O. Hookeri.*
Leaves very short, thick, spreading. Involucre more hemispherical. Western species . . . . 47. *O. imbricata.*
(See also *O. tenuifolia,* which has the involucral bracts rather obtuse.)

Flower-heads small, numerous, in a leafless corymbose panicle.
Leaves elliptical-oblong or lanceolate, 1½ to 3 in. long . . . . 48. *O. elliptica.*
Leaves narrow-linear, ½ to 1½ in. long . . . . 49. *O. glandulosa.*

SECTION V. **Merismotriche.**—*Glabrous, glandular-pubescent or hirsute, and often glutinous, the hairs simple rigid, white or transparent and septate. Involucre hemispherical, with narrow, usually acute bracts.*

Flower-heads terminal, the peduncles mostly shorter than the heads.
Leaves usually under ½ in. long. Involucres under ½ in. diameter.
Glabrous or slightly glandular-pubescent. Leaves linear, with

slightly recurved margins, small and distant or long and crowded. Panicles loose, divaricate . . . . 50. *O. heleophila.*
Glabrous glandular-pubescent or muricate. Leaves linear, the margins revolute, small and obtuse. Panicles loose, divaricate 51. *O. muricata.*
Muricate, scabrous or hispid. Panicles loose, divaricate.
Leaves linear, mucronate, erect, the margins revolute . . . 52. *O. strigosa.*
Leaves obovate-cuneate or almost linear, entire or 3-toothed . 53. *O. paucidentata.*
Very glandular-pubescent. Leaves linear. Panicles oblong, narrow . . . . 54. *O. stricta.*
Flower-heads rather large, few, terminal or in the upper axils, on peduncles shorter than or rarely exceeding the leaves.
Leaves linear, ½ to 2 in. long.
Slightly glandular-pubescent. Leaves slender. Peduncles mostly exceeding the leaves. Involucre scarcely ½ in. diameter . . 55. *O. tenuifolia.*
Very viscid-pubescent or hirsute. Corymb dense. Involucre about ½ in. diameter, the bracts unequal . . . . 56. *O. adenophora.*
Very scabrous. Flower-heads few, almost sessile. Involucre nearly ¾ in. diameter, the bracts nearly equal . . . . . . 57. *O. homolepis.*
Leaves oblong-cuneate or almost linear, toothed. Plant glabrous, glutinous. Involucres much imbricate . . . . 41. *O. Stuartii.*
Leaves obovate or oblong, crenate, very viscid. Peduncles longer than the leaves. Involucres scarcely ½ in. diameter . . . . 58. *O. xerophila.*
Flower-heads solitary, on peduncles very much longer than the leaves.
Leaves obovate-oblong to lanceolate.
Glabrous. Peduncles with subulate bracts. Involucre much shorter than the disk . . . . 59. *O. Ferresii.*
Glabrous or hispid. Peduncles without any or with only one bract. Involucre as long as the disk.
Leaves mostly oblong, obtuse, coarsely toothed. Southern and Western species . . . . 60. *O. rudis.*
Leaves lanceolate, acute, entire or with 2 or 3 acute teeth towards the end. Tropical species . . . . 62. *O. arguta.*
Leaves linear.
Leaves hoary-hispid . . . . 61. *O. picridifolia.*
Leaves ciliate, otherwise glabrous . . . . 63. *O. ciliata.*

Eurybia chrysotricha, Ten. Cat. Hort. Neap. 85, is supposed by Lindley, Bot. Reg. Misc. 1841, 19, to be from Australia. Tenore however says that it was raised from seed sent by Bonpland from America. At any rate the character given is wholly insufficient for identifying it.

SECTION I. DICEROTRICHE, *Archer*.—Indumentum of the under side of the leaves (usually either silvery-shining tomentose or loosely villous) consisting of centrally attached or divaricately forked (T-shaped) hairs, otherwise simple.

1. **O. megalophylla,** *F. Muell. Fragm.* v. 70. A tall shrub, the indumentum dense, almost woolly, consisting of rather loose divaricately forked hairs. Leaves opposite, shortly petiolate, broadly or narrow-oblong, obtuse, 2 to 4 in. long, coriaceous, shining above, densely tomentose underneath. Flower-heads rather large, in a terminal corymb. Involucre broadly ovoid or almost hemispherical, tomentose, 3 or 4 lines long, much imbricate. Ray-florets about 10 to 12. Disk-florets more numerous. Anthers with small but distinct points at the base. Style-appendages very short, obtuse. Achenes nearly glabrous. Pappus nearly equal or a few of the outer bristles rather

shorter.—*Eurybia megalophylla*, F. Muell. in Proc. Roy. Soc. Tasm. iii. 228; *Aster megalophyllus*, F. Muell. Fragm. v. 70.

Victoria. Bushy declivities of the Australian Alps, Cobberas mountains, Mount Buller, etc., *F. Mueller*. This and the two following species are very closely allied to each other.

2. **O. chrysophylla,** *Benth.* A shrub, attaining 6 ft. (*A. Cunn.*), the indumentum sometimes close and silvery, but more frequently dense and soft, white or brown, sometimes quite woolly, consisting of centrally-attached or divaricately-forked hairs. Leaves opposite, oblong, obtuse, entire or sinuate-denticulate, 2 to 4 in. long, glabrous on the upper side. Flower-heads numerous, small, in corymbose panicles. Involucres ovoid, pubescent or nearly glabrous. Florets about 6 to 8, of which 2 or 3 are rays. Anthers with minute points at the base. Style-appendages short, obtuse. Pappus nearly equal.—*Eurybia chrysophylla*, DC. Prod. v. 266; *Eurybia oppositifolia*, F. Muell. Fragm. ii. 88.

N. S. Wales. In the interior, N. of Bathurst, *A. Cunningham*; Guy-Fox Peak, Arne river, and Mount Mitchell, *Beckler*; also Macleay river, *Beckler*, with a very dense wool.

3. **O. alpicola,** *F. Muell. Fragm.* v. 70. A tall shrub, the indumentum dense, almost silvery, consisting of intricate stipitate forked or centrally attached hairs. Leaves mostly opposite, shortly petiolate, oblong-lanceolate to almost linear, 2 to 4 in. long, coriaceous, glabrous on the upper side, the margins slightly recurved. Flower-heads numerous, small, in a terminal corymbose panicle. Involucre ovoid, tomentose, with about 4 to 6 ray-florets and rather more numerous disk-florets. Anthers with minute points at the base. Style-appendages short, obtuse. Achenes glabrous (in all the specimens examined). Pappus nearly equal.—*Eurybia alpicola*, F. Muell. in Proc. Roy. Soc. Tasm. iii. 229; *Aster alpicola*, F. Muell. Fragm. v. 70.

Victoria. Along rivulets and springs in the Australian Alps, at an elevation of 4000 to 5000 feet, *F. Mueller*.

4. **O. rosmarinifolia,** *Benth.* An erect shrub with virgate branches, hoary or silvery with a close tomentum, consisting of centrally attached hairs. Leaves opposite, sessile, narrow linear, acute or obtuse, mostly 1½ to 3 in long, glabrous or scabrous above, the margins so closely revolute as to conceal the tomentose under-surface except the midrib. Flower-heads on axillary peduncles, forming usually a terminal leafy panicle. Involucre turbinate, tomentose. Ray-florets 6 to 8; disk-florets more numerous. Anthers attenuate at the base, but without prominent points. Style-appendages obtuse. Achenes glandular-papillose, not hairy. Pappus rather unequal, but no very short bristles.—*Eurybia rosmarinifolia*, DC. Prod. v. 268.

N. S. Wales. In the interior north of Bathurst, *A. Cunningham*; New England, *C. Stuart*; tributaries of Clarence river, *Beckler*. Also in *Leichhardt's* collection.

5. **O. viscosa,** *Benth.* A bushy shrub of 4 to 5 ft., the young shoots often viscid. Leaves opposite, petiolate, oblong-lanceolate, narrowed at both ends but usually obtuse, 2 to 3 in. long, glabrous above, silvery-white underneath with a close tomentum consisting of centrally attached hairs. Flower-heads small, numerous, in corymbose panicles. Involucral bracts few, glabrous or glandular-pubescent. Florets usually 4 to 6, of which 1 or 2

are ligulate. Anthers with distinct points at the base. Style-appendages very short. Achenes glabrous or hairy in the upper part. Pappus equal or with a few rather shorter bristles.—*Aster viscosus*, Labill. Pl. Nov. Holl. ii. 53. t. 203; *Eurybia viscosa*, Cass.; DC. Prod. v. 266; Hook. f. Fl. Tasm. i. 173.

Victoria. On the shores of Lake King, *F. Mueller.*

Tasmania. Derwent river and islands of Bass's Straits, *R. Brown.* Abundant on the sides of mountains, especially in the southern parts of the colony, emitting a strong musky smell, *J. D. Hooker.*

De Candolle refers here *Balbisia Caledoniæ*, Spreng. Syst. iii. 569, said by the latter author to be from New Caledonia, *Forster*, but this is probably a mistake.

6. **O. grandiflora,** *Hook. Ic. Pl. t.* 862. Stature apparently of *O. pannosa.* Branches, peduncles and under side of the leaves densely clothed with a close glossy intricate tomentum, white or reddish and consisting of centrally attached hairs. Leaves alternate, petiolate, ovate or elliptical, acutely denticulate, narrowed at the base, 2 to 5 in. long, quite glabrous and reticulate on the upper side as in *O. pannosa* but of a thinner texture. Flower-heads hemispherical, on long terminal peduncles like those of *O. pannosa*, but still larger. Florets, achenes and pappus of *O. pannosa.*—*Steetzia grandiflora*, Sond. in Linnæa, xxv. 452; *Aster Sonderi*, F. Muell. Fragm. v. 83.

S. Australia. Lofty and Onkaparinga ranges, *F. Mueller;* near Adelaide, *C. Dutton.*

7. **O. pannosa,** *Hook. Ic. Pl. under n.* 862. A shrub of several feet or sometimes an undershrub, the branches, peduncles and under side of the leaves clothed with a soft dense white or reddish tomentum, consisting of centrally attached hairs and sometimes almost woolly. Leaves alternate, petiolate, from broadly ovate-cordate to oblong, obtuse, entire, 2 to 3 in. long, glabrous above, with the reticulations often much depressed. Flower-heads large, hemispherical, terminal or on axillary peduncles, often thickened under the head. Involucres often above 1 in. diamer, the bracts imbricate, acute, the inner ones as long as the disk. Ray-florets ¾ to 1 in. long, Anthers of the disk-florets acute at the base, but without protruding points. Style-lobes long, with short obtuse appendages. Achenes long, hirsute. Pappus very copious, the outer bristles gradually shorter.—*Steetzia pannosa*, Sond. in Linnæa, xxv. 451; *S. ovata*, Sond. l. c. 452, and *S. Muelleri*, Sond. l. c. 453; *Aster pannosus*, F. Muell. Fragm. v. 83; *Eurybia pannosa*, F. Muell. Pl. Vict. t. 32; *E. cardiophylla*, F. Muell. in Linnæa, xxv. 398.

Victoria. Mount Remarkable, Light River, Mount M'Ivor, *F. Mueller.*

S. Australia. Summit of a mountain near Port Lincoln, *R. Brown;* on the Murray, *Whittaker;* Rivoli Bay, *F. Mueller, Robertson;* Tattiara country, *Woods;* Port Lincoln, *Wilhelmi.*

The narrow-leaved forms with more numerous flower-heads are apparently from old shrubs, the large broad-leaved ones with very long peduncles and large heads, probably from luxuriant shoots grown up when old stems have been cut down.

8. **O. oliganthema,** *F. Muell. Herb.* Apparently allied to *O. argophylla.* Leaves in the specimens alternate, petiolate, ovate, about 2 in. long, entire or slightly sinuate-toothed, glabrous and reticulate above, silvery-shining underneath with a very close tomentum consisting of centrally attached hairs. Flower-heads small and numerous in a dense terminal corymb. Involucres

cylindrical, nearly glabrous. Florets 4 to 6, of which 1 or 2 ligulate. Anthers with minute points. Style-lobes long, with short obtuse appendages. Achenes glandular-pubescent. Pappus of unequal bristles, some of the inner ones thickened upwards.

N. S. Wales. Blue Mountains, *C. Moore*, *Woolls*, in both cases single small specimens in Herb. F. Mueller.

9. **O. argophylla,** *F. Muell. Fragm.* v. 68. A tree attaining 20 to 25 ft., emitting a strong musky smell, the indumentum close and silvery-shining, consisting of centrally affixed hairs. Leaves alternate, petiolate, from oval-elliptical to oblong-lanceolate, acute at both ends, more or less callous-denticulate or rarely quite entire, 3 to 5 in. long, very much reticulate and glabrous above or with minute shining hairs, silvery-silky underneath. Flower-heads small, numerous, in large terminal corymbs. Involucre oblong-turbinate. Ray-florets about 3 to 5; disk-florets 6 to 8. Anthers with minute points at the base sometimes scarcely perceptible. Style-appendages short, obtuse. Achenes sparingly hirsute. Pappus copious, nearly equal, except a few short outer bristles, which are rarely wanting.—*Aster argophyllus*, Labill. Pl. Nov. Holl. ii. 52. t. 201; Bot. Mag. t. 1563; F. Muell. Fragm. v. 68; *Eurybia argophylla*, Cass.; DC. Prod. v. 267; Hook. f. Fl. Tasm. i. 172.

N. S. Wales. Port Jackson, *M'Arthur;* Mittagong range, *Woolls.*

Victoria. Genoa river and Sumut river, *F. Mueller;* Corner Inlet, *Wilhelmi;* Bullarock Range, *Whan.*

Tasmania. Derwent river and Port Dalrymple, *R. Brown.*—Common in forests in rich damp soils, *J. D. Hooker.*

10. **O. cydoniæfolia,** *Benth.* A tall shrub, the indumentum soft, very white, not so close as in *O. argophylla*, but similarly consisting of centrally affixed hairs. Leaves alternate, petiolate, ovate or elliptical, obtuse or almost acute, entire or slightly sinuate-toothed, mostly 3 to 4 in. long, the upper surface glabrous, at first smooth but finely reticulate when full grown, white-tomentose underneath. Flower-heads larger than in *O. argophylla*, in terminal panicles. Involucres hemispherical, tomentose, enclosing numerous florets but only seen young.—*Aster cydoniæfolius*, A. Cunn. Herb.; *Eurybia cydoniæfolia*, DC. Prod. v. 267; *E. Beckleri*, F. Muell. Veg. Chath. Isl. 21.

N. S. Wales. In the N.W. interior, *Fraser;* Clarence river, *Beckler;* Apsley Falls, *Leichhardt.* The New Zealand station given by De Candolle arose from some mistake. Cunningham's specimens, as appears from his herbarium, were from Fraser.

11. **O. myrsinoides,** *F. Muell. Fragm.* v. 69. A shrub, usually low and straggling or densely bushy, the branches, peduncles and under side of the leaves closely silvery or silky-tomentose with centrally attached hairs. Leaves alternate, in the original form nearly sessile, obovate to oblong, very obtuse, minutely denticulate, under ½ in. long, the upper surface glabrous, shining and much reticulate. Flower-heads in the original form narrow, 3 to 5 together on axillary peduncles forming a leafy oblong panicle. Involucre obconical. Florets 4 to 8, of which 2 or 3 ligulate. Anthers with more or less distinct minute points at the base or sometimes scarcely acute. Style-appendages short, obtuse. Achenes glabrous. Pappus-bristles nume-

rous and very unequal.—*Aster myrsinoides*, Labill. Pl. Nov. Holl. ii. 53. t. 202; F. Muell. Fragm. v. 69; *Eurybia myrsinoides*, Nees, Gen. et Sp. Ast. 146; DC. Prod. v. 268; Hook. f. Fl. Tasm. i. 174.

Victoria. Port Phillip, *R. Brown;* Delatite river, Loddon plains, Wilson's Promontory, etc., *F. Mueller.*

Tasmania. Port Dalrymple and Derwent river, *R. Brown;* generally in hilly districts, *J. D. Hooker.*

Var. *serrata*, DC. Leaves sessile, obovate to oblong, ½ to 1 in. long, rounded at the end but acutely and rigidly denticulate. Florets 4 to 8 in the head.—Mount Disappointment and Wilson's Promontory, *F. Mueller.*

Var. *erubescens*, F. Muell. Leaves shortly petiolate, obovate, oblong or lanceolate, often acute, 1 to 2 in. long, acutely and rigidly or rarely minutely denticulate. Heads larger with 3 to 5 florets in the ray and 6 to 8 in the disk or even more.—*Aster erubescens*, Sieb. Pl. Exs.; *Eurybia erubescens*, DC. Prod. v. 267; Hook. f. Fl. Tasm. i. 173.

N. S. Wales. Port Jackson to the Blue Mountains and in the interior to Lachlan river, *A. and R. Cunningham, Sieber, n.* 339, and others.

Victoria. Rocky barren scrubby places from the Glenelg to Wilson's Promontory, in the Grampians, and over a great part of Australia Felix. *F. Mueller* and others.

Tasmania. Derwent river, *R. Brown;* not unfrequent throughout the island, *J. D. Hooker.*

All the varieties are said to smell of musk.

12. **O. persoonioides,** *Benth.* A dwarf bushy shrub rarely exceeding 3 ft. Leaves alternate, petiolate, from obovate to oblong, very obtuse, narrowed at the base, ¾ to 1½ in. long, very smooth and shining above, silvery or silky-tomentose underneath with centrally attached hairs. Flower-heads small, rather numerous, forming leafy panicles. Involucres ovoid. Ray-florets 3 or 4, white; disk-florets about 10 to 12. Anthers with minute points at the base. Style-appendages almost acute. Achenes hairy. Pappus with a few outer shorter bristles.—*Eurybia persoonioides*, DC. Prod. v. 267; Hook. f. Fl. Tasm. i. 174.

Tasmania. Table Mountain, Derwent river, *R. Brown;* generally in alpine situations, *J. D. Hooker.*

Var. *lanceolata.* Leaves lanceolate, almost acute. Flower-heads fewer, but scarcely larger. Achenes glabrous.

Var. *alpina.* Flower-heads fewer and larger, with rather more florets. Achenes glabrous. —*Eurybia alpina*, Hook. f. Fl. Tasm. i. 174. t. 42. Only in alpine situations, *J. D. Hooker.*

13. **O. obcordata,** *Benth.* A small bushy shrub of about 2 ft. with a strong musky odour (*J. D. Hooker*). Leaves alternate, cuneate and obtusely 3- or 5-toothed at the end or obcordate, mostly under ½ in. long, the upper surface as in *O. persoonioides*, glabrous and not reticulate, the under surface white with a close tomentum consisting of centrally attached hairs. Flower-heads mostly solitary, pedunculate. Involucres ovoid, with about 3 or 4 ligulate florets, and about as many in the disk. Anthers with very minute points at the base or only acute. Style-appendages rather obtuse. Achenes glabrous. Pappus slightly unequal.—*Eurybia obcordata*, Hook. f. in Hook. Lond. Journ. vi. 108; Fl. Tasm. i. 174. t. 42; *Aster obcordatus*, F. Muell. Fragm. v. 69.

Tasmania. Table Mountain, Derwent river, *R. Brown;* generally on mountains at from 3000 to 4000 feet, *J. D. Hooker.*

14. **O. pinifolia,** *Benth.* A rigid bushy shrub, with stout tomentose

branches. Leaves alternate, crowded, narrow-linear, rigid, pungent-pointed, the margins closely revolute, $\frac{3}{4}$ to $1\frac{1}{2}$ in. long, glabrous and smooth above, the under side silvery-silky with centrally attached hairs, but almost wholly concealed. Flower-heads mostly solitary, pedunculate. Involucre turbinate. Ray-florets about 8 to 10, those of the disk rather more numerous. Anthers with points at the base, but I have not found them so distinctly tailed as figured. Style-appendages short, obtuse. Achenes glabrous. Pappus rather unequal.—*Eurybia pinifolia*, Hook. f. in Hook. Lond. Journ. vi. 108, and in Fl. Tasm. i. 177. t. 45 ; *Aster pinifolius*, F. Muell. Fragm. v. 71.

Tasmania. Table Mountain, Derwent river, *R. Brown ;* Mount Wellington and Vale of Belvoir, *Gunn ;* Mount Lapeyrouse, *C. Stuart.*

15. **O. ledifolia,** *Benth.* A small bushy or diffuse shrub, with thick branches. Leaves alternate, crowded, oblong-linear, obtuse, with closely revolute margins, under $\frac{3}{4}$ in. long, the upper surface glabrous or sprinkled with a few centrally-attached hairs, the under side rusty or silvery-tomentose. Flower-heads solitary on peduncles rarely as long as the leaves. Involucre turbinate, with obtuse bracts. Ray-florets about 8 to 10 ; disk-florets rather more numerous. Anthers with distinct points at the base. Style-appendages short, obtuse. Achenes glabrous. Pappus with a few of the outer bristles shorter than the others.—*Eurybia ledifolia*, DC. Prod. v. 269 ; Hook. f. Fl. Tasm. i. 177 ; *Aster ledifolius*, A. Cunn. Herb.

Tasmania. Table Mountain, Derwent river, *R. Brown ;* summit of Mount Wellington, *Fraser* and others ; Mount Lapeyrouse, *C. Stuart.*

16. **O. dentata,** *Mœnch ; DC. Prod.* v. 271. A tall rather coarse shrub, the branches and under side of the leaves densely tomentose, with divaricate stipitate hairs. Leaves alternate, petiolate, ovate or almost orbicular, obtuse, sinuate coarsely-toothed or entire, mostly 1 to 2 in. long, or when very luxuriant longer and narrower, rather thick, scabrous or loosely hairy on the upper side. Flower-heads large, on peduncles very variable in length, not numerous, but forming usually a terminal corymb. Involucre hemispherical, $\frac{1}{2}$ to $\frac{3}{4}$ in. diameter. Ray-florets rather numerous, white or blue ; disk-florets scarcely exceeding the involucre. Anthers without prominent points. Style-appendages very short and obtuse. Achenes usually very hairy. Pappus-bristles in two distinct series, the outer ones not half so long as the inner ones.—*Aster dentatus*, Andr. Bot. Rep. t. 61 ; *A. tomentosus*, Schrad. in Wendl. Sert. Hann. 8. t. 24 (DC.) ; *A. ferrugineus*, Wendl. in Flora, 1819, 676 ; *Diplopappus rotundifolius*, Less. in Linnæa, vi. 116 ; *Olearia rotundifolia*, DC. Prod. v. 271.

N. S. Wales. Port Jackson to the Blue Mountains, *R. Brown*, *Sieber*, *n.* 341, and others ; southward to Illawarra, *Backhouse ;* and Twofold Bay, *F. Mueller.*

Some botanists distinguish *O. rotundifolia* from *O. dentata* by the longer peduncles, others reverse these characters. Specimens with toothed leaves have usually but not always smaller flower-heads than those with entire leaves, and may have been taken from older plants. I cannot perceive any constant difference in the colour of the pappus except what may be attributed to the process of desiccation.

SECTION II. ASTEROTRICHE, *Archer*.—Indumentum of the under side of the leaves consisting of stellate hairs.

The indumentum is generally very distinct from that of any other section. It is only when it is very close and short, as in *O. viscidula*, that it requires some care to distinguish it from that of the section *Dicerotriche*.

17. **O. stellulata**, *DC. Prod.* v. 272. An erect shrub of 3 to 5 ft. Leaves alternate, oblong or lanceolate, obtuse or acute, more or less sinuate-toothed or rarely almost entire, glabrous scabrous or stellate-hairy above, white or rusty underneath with a dense stellate tomentum, varying in size from all under ½ in. in some specimens to above 2 or even 3 in. in others. Heads in the original form rather small, in leafy panicles. Involucre turbinate. Ray-florets about 8 to 12; disk-florets rather more numerous. Anthers scarcely auriculate. Style-appendages almost acute. Achenes more or less hirsute. Pappus with a distinct external series of short bristles.—*Aster stellulatus*, Labill. Pl. Nov. Holl. ii. 50. t. 196; *Eurybia fulvida*, Cass.; Hook. f. Fl. Tasm. i. 175; *Diplostephium stellulatum*, Nees, Gen. et Sp. Ast. 187; *Aster phlogopappus*, Labill. Pl. Nov. Holl. ii. 49. t. 195; *A. phlogotrichus*, Spreng. Syst. iii. 525; *Eurybia quercifolia*, Cass. (DC.); *Diplostephium phlogotrichum*, Nees, Gen. et Sp. Ast. 186; *Olearia phlogopappa*, DC. Prod. v. 272; *Eurybia Gunniana*, DC. Prod. v. 268; Hook. f. Fl. Tasm. i. 175; *E. subrepanda*, DC. l. c.

Queensland. Port Bowen, *R. Brown*; Mount Hedlow, near Rockhampton, *Dallachy*.

N. S. Wales. Hastings river, *Fraser*; Twofold Bay, *F. Mueller*; Gabo Island, *Maplestone*.

Victoria. Australian Alps and Wilson's Promontory, *F. Mueller*, near Melbourne, *Robertson*.

Tasmania. Port Dalrymple and islands of Bass's Straits, *R. Brown*; very common throughout the island, *J. D. Hooker*.

Var. *canescens*. Leaves narrow-oblong, entire, ¾ to 2 in. long, hoary-tomentose on both sides.—Apsley river, *Fraser*; Severn river, New England, *C. Stuart*.

Var. *lirata*. Leaves lanceolate, 2 to 5 in. long, deeply furrowed by the impressed veins, almost bullate, obscurely sinuate-toothed or almost entire. Flower-heads larger and broader.—*Aster liratus*, Sims, Bot. Mag. t. 1509; *Diplostephium lyratum*, Nees, Gen. et Sp. Ast. 188; *Eurybia lirata*, DC. Prod. v. 267; Hook. f. Fl. Tasm. i. 175. t. 43.—Victoria and Tasmania, *R. Brown*, *J. D. Hooker*, *F. Mueller*, and others.

Var. *quercifolia*. Leaves oblong, obtuse, entire or obtusely toothed, 1 to 3 in. long, very much bullate, very scabrous above, densely stellate-tomentose and often rufous underneath. —*Aster quercifolius*, Sieb. Pl. Ex.; *Olearia quercifolia*, DC. Prod. v. 272; *Eurybia rugosa*, F. Muell.; Arch. in Journ. Linn. Soc. v. 22.—Blue Mountains, *Sieber*, *n.* 340, *A. Cunningham* and others; Victoria, *F. Mueller*; Tasmania, *R. Brown*.

18. **O. asterotricha**, *F. Muell. Fragm.* v. 79. A shrub, rough with a rigid stellate pubescence. Leaves alternate, oblong-linear, very obtuse, with revolute margins, more or less sinuate-toothed, mostly ¾ to 1½ in. long, bullate and very scabrous above, tomentose underneath. Flower-heads rather large, solitary and terminal, or several in a terminal corymb. Involucre almost hemispherical, often above ½ in. diameter. Ray-florets often above 20; disk-florets slightly exceeding the inner bracts. Anthers, style, achenes, and pappus of *O. stellulata*.—*Eurybia asterotricha*, F. Muell. Fragm. i. 111; Pl. Vict. t. 33; *Aster asterotrichus*, F. Muell. Fragm. v. 79.

Victoria. Glenelg river, Grampians, Mount Disappointment, Dandenong ranges, Gipps' Land, *F. Mueller*.

Var. *parvifolia*. Leaves mostly under ½ in. long.

N. S. Wales. Bargo Brush, *Backhouse;* also near Portland in Victoria, *Robertson.* The species only differs from *O. stellulata*, var. *quercifolia*, in the narrow leaves, and perhaps in the purple colour of the disk-florets.

19. **O. gravis,** *F. Muell. Fragm.* v. 82. An erect rigid shrub of a few feet, more or less hoary or rust-coloured, with a soft or scabrous stellate tomentum. Leaves alternate, oblong-lanceolate, obtuse or almost acute, irregularly toothed or nearly entire, 1 to 2 in. long, rugose or almost bullate, usually hoary above and more densely tomentose underneath. Flower-heads rather large, in a simple terminal corymb. Involucre above ½ in. diameter, the bracts shorter than the disk, almost acute, the outer ones nearly as long as the inner. Anthers not auriculate. Style-appendages lanceolate. Achenes glabrous, rather long and somewhat compressed, but with very prominent ribs. Pappus with a distinct external series of short bristles.—*Aster gravis*, F. Muell. l. c.

N. S. Wales. Near Tenterfield, New England, *C. Stuart.*—Very nearly allied to the var. *quercifolia* of *O. stellulata.*

20. **O. Nernstii,** *F. Muell. Fragm.* v. 81. A shrub with elongated branches, loosely stellate-tomentose and apparently somewhat glutinous. Leaves alternate, ovate-lanceolate, acute, remotely and often acutely sinuate-toothed or nearly entire, 1 to 3 in. long, rather thin, glabrous and smooth above, loosely stellate-tomentose underneath. Flower-heads not large, in terminal pedunculate corymbs. Involucre almost hemispherical, 4 to 5 lines diameter, the bracts not very unequal. Ray-florets 15 to 20; disk-florets numerous, not much exceeding the involucre. Anthers not auriculate. Style-appendages shortly lanceolate. Achenes short, glabrous. Pappus with an external series of very short bristles.—*Aster Nernstii*, F. Muell. l. c.

Queensland. Near Ipswich, *Nernst;* towards Moreton Bay, *Leichhardt.*
N. S. Wales. Hastings river, *Beckler;* Richmond river, *C. Moore, Fawcett.*

21. **O. hygrophila,** *Benth.* Shrubby, with slender virgate branches, very sparingly stellate-tomentose. Leaves alternate, linear, mostly acute, entire with recurved margins, ¾ to 1½ in. long, glabrous above, sparingly and minutely stellate-tomentose underneath. Flower-heads rather large, on slender peduncles, solitary or few in a loose corymb. Involucre scarcely ½ in. diameter, the bracts narrow, somewhat unequal. Ray-florets 12 to 20, rather narrow; disk-florets numerous, scarcely exceeding the involucre. Anthers not auriculate. Style-appendages lanceolate. Achenes glabrous. Pappus-bristles slightly unequal, with very few short external ones or sometimes none at all.—*Eurybia hygrophila*, DC. Prod. v. 269.

Queensland. Stradbrooke Island, Moreton Bay, *Fraser.*

22. **O. viscidula,** *Benth.* A tall rather slender shrub, more or less viscid. Leaves opposite, or rarely alternate, on luxuriant lateral shoots, linear-lanceolate, acute at both ends, entire, mostly 2 to 3 in. long, silvery-white underneath, with a close tomentum consisting of minute stellate hairs. Flower-heads small, numerous, in short axillary leafy panicles. Involucres glabrous or viscid-pubescent. Ray-florets 8 to 10; disk-florets rather more numerous. Anthers not auriculate. Style-appendages shortly lanceolate.

Achenes hirsute. Pappus with an external series of very short bristles.—*Eurybia viscidula*, F. Muell. Fragm. i. 50 ; *Aster Siemssenii*, F. Muell. Fragm. v. 71.

N. S. Wales. Near Cook's River, *R. Brown ;* in the interior, *Fraser ;* near Illawarra, *A. Cunningham ;* near Goulburn, *C. Moore.*

SECTION III. ERIOTRICHE, *Archer.*—Indumentum of the under side of the leaves consisting of densely imbricate woolly hairs. Leaves alternate. Anthers not auriculate.

Most of the species of this section are remarkable for their small leaves. In some viscid species the woolly tomentum nearly disappears, but it always remains sufficiently conspicuous on the under side of the leaves to distinguish them from those of the section *Adenotriche.*

23. **O. tubuliflora,** *Benth.* A tall shrub, with numerous erect virgate branches more or less scabrous-pubescent or viscid, with an admixture of woolly tomentum especially on the under side of the leaves. Leaves narrow-linear, often clustered in the axils, mostly $\frac{1}{4}$ to 1 in. long, the margins revolute, glabrous or scabrous-pubescent above. Flower-heads small, sessile or nearly so in the axils, and shorter than the floral leaves. Involucre ovoid, under 2 lines long, the bracts very obtuse. Ray-florets 3 or 4, shorter than the undivided part of their style, tubular or scarcely expanded into a minute ligula; disk-florets 3 to 5, exceeding the involucre. Anthers, style, achenes, and pappus of *O. axillaris,* the bristles, however, usually fewer.—*Eurybia tubuliflora,* Sond. and Muell. in Linnæa, xxv. 455 ; *Aster tubuliflorus,* F. Muell. Fragm. v. 65.

Victoria. Mount M'Ivor, *F. Mueller,*

S. Australia. Lake Alexandrina, St. Vincent's Gulf, Lofty Ranges, etc., *F. Mueller, Hillebrand, Blandowski.*

Cf *Eurybia artemisioides,* Sond. in Linnæa, xxv. 456, or *Aster artemisioides,* F. Muell. Fragm. v. 65, the majority of the specimens in F. Mueller's herbarium appear to me to be the same as *O. tubuliflora,* but Sonder describes the ligula of the ray florets as equal to its style, and suggests that it is a variety of *O. ramulosa,* and that is probably the case with one of F. Mueller's specimens from Lake Alexandrina, where the two species appear to grow together.

24. **O. axillaris,** *F. Muell. Fragm.* v. 64. An erect much-branched shrub of 3 to 6 ft., more or less hoary or white with a close woolly tomentum. Leaves from obovate or oblong-cuneate and $\frac{1}{4}$ to $\frac{1}{2}$ in. long to linear or linear-lanceolate and $\frac{1}{2}$ to $\frac{3}{4}$ in. long, obtuse, entire, with revolute margins, woolly-white on both sides or glabrous and shining above. Flower-heads sessile in the axils, and shorter than the floral leaves, or very rarely 1 or 2 together on short axillary leafy shoots. Involucre ovoid, 2 to 3 lines long, the bracts obtuse. Ray-florets about 4 to 6, shorter than those of the disk, the small ligula not so long as the style, entire or 2- or 3-toothed ; disk-florets about 6 to 10, exceeding the involucre. Style-appendages short. Achenes hairy or rarely glabrous. Pappus-bristles all nearly equal or occasionally a very few short outer ones.—*Eurybia,* sect. *Brachyglossa,* DC. Prod. v. 265 ; *Aster axillaris,* F. Muell. Fragm. v. 64.

N. Australia. Dampier's Archipelago, N.W. coast, *A. Cunningham.*

Victoria. Common on the sandy seacoasts, *F. Mueller* and others.

Tasmania. Sandy hills of the N. coast and islands of Bass's Straits, *R. Brown, J. D. Hooker* and others.

S. Australia. Sandy hills of the coast both on the mainland and Kangaroo Island, *R. Brown, F. Mueller*, and others.

W. Australia. Goose Island Bay, *R. Brown;* sandy hills on the S. coast and on the west coast to Swan River, Murchison river, Shark's Bay, Dirk Hartog's Island, etc., *A. Cunningham; Baudin; Drummond, n.* 126, 129; *Preiss, n.* 89, 90, 91, 92, 93.

Very variable in the degree of woolliness and in the shape of the leaves. The following forms, described as distinct species, appear to pass very much one into the other, and are all uniform in essential characters:—

a. obovata. Leaves obovate to cuneate-oblong, very tomentose on both sides, or glabrous above. Flower-heads small.—*Eurybia oligantha*, DC. Prod. v. 266; *E. brachyglossa*, DC. l. c. 265 (with glabrous achenes); *E. candidissima*, Steetz in Pl. Preiss. i. 418.—Chiefly on the W. coast, from Swan River to Shark's Bay.

b. normalis. Leaves narrow-linear, mostly under ½ in. Florets rarely more than 10 in the head.—*Eurybia axillaris*, DC. Prod. v. 266; Steetz in Pl. Preiss. i. 417; *E. capitellata*, DC. l. c.; *E. Dampieri*, DC. l. c. (with longer very narrow leaves).—The commonest form on the S. coast.

c. linearis. Leaves linear or linear-lanceolate, mostly above ½ in. Florets usually 10 to 15 in the head.—*Eurybia linearifolia*, DC. Prod. v. 266.—Occasionally both on the S. and W. coasts.

25. **O. revoluta,** *F. Muell. Herb.* An erect, much-branched shrub, with the habit and foliage of some of the narrow-leaved forms of *O. axillaris*, or of the long-leaved forms of *O. ramulosa*, the ligula of the ray-florets much more developed than in the former, but less so than in the latter species. Leaves linear to oblong-cuneate, ¼ to ½ in. long or rather more, obtuse, with revolute margins, glabrous or slightly tomentose above, white or hoary with a close woolly tomentum underneath. Flower-heads sessile or on short leafy peduncles in the axils, rarely exceeding the subtending leaf. Involucres broadly ovoid or almost turbinate, the bracts rather numerous. Ray-florets 4 to 8, the ligula exceeding its style. Disk-florets 6 to 10, exceeding the involucre. Style, achene, and pappus of *O. axillaris*.

W. Australia. Murchison and S. Hutt rivers, *Oldfield.*

Var. *minor.* Leaves small and narrow. Flower-heads small, with 2 or 3 florets in the ray, 3 to 5 in the disk.—King George's Sound and adjoining districts, *Harvey, Oldfield, Maxwell.*

26. **O. exilifolia,** *F. Muell. Fragm.* v. 69. A bushy shrub of 3 to 5 ft., closely resembling the common forms of *O. ramulosa*, but with shorter ligulæ. Leaves linear, obtuse, with revolute margins, 1 to 2 lines long, woolly underneath. Flower-heads of *O. axillaris*, sessile or on very short leafy peduncles in the axils and usually exceeding the subtending leaf. Ray-florets 2 or 3, the ligula scarcely exceeding the style. Disk-florets 3 or 4, longer than the involucre. Style, achene, and pappus of *O. axillaris.*—*Aster exilifolius*, F. Muell. l. c.

W. Australia. Calcareous hills towards the Great Bight, *Maxwell.*

27. **O. ramulosa,** *Benth.* A shrub of 3 to 6 ft., much-branched, more or less scabrous-pubescent and sometimes slightly glutinous. Leaves crowded, usually very small and spreading, sometimes reflexed and clustered in the axils, varying from obovate or almost orbicular, about 1 line long, with a petiole of the same length, to oblong-linear and nearly ½ in. long,

usually obtuse, with recurved or revolute margins, glabrous or scabrous-pubescent above, the under surface with more or less of a thin intricate wool, or sometimes the thin loose wool covering the whole plant. Flower-heads small, very numerous, usually sessile or terminating very short lateral leafy peduncles or branchlets, forming long leafy spikes or racemes along the branches. Involucre ovoid, much imbricate, nearly 3 lines long. Ray-florets 6 to 10, the ligula much longer than its style; disk-florets scarcely more numerous. Style branches short, the appendages triangular, almost as in some *Senecionidæ*, but papillose only not hispid. Achenes short, slightly compressed, striate or 4-angled, hairy or rarely glabrous. Pappus with an outer series of short bristles, but sometimes very few only.

There are two principal forms, which are often distinguished as species, but only differ in the shape of the leaves.

a. microphylla. Leaves obovate or oblong, 1 to 2 lines long. Flower-heads small.—*Aster microphyllus*, Vent. Jard. Malm. under n. 83; *Diplostephium microphyllum*, Nees, Gen. et Sp. Ast. 191; *Eurybia microphylla*, DC. Prod. v. 270.

N. S. Wales. Port Jackson to the Blue Mountains, *R. Brown*, *Sieber, n.* 338, *and Fl. Mixt. n.* 514, and others; Lachlan river, *A. Cunningham.*

b. communis. Leaves narrow, 1 to 6 lines long.—*Aster ramulosus* and *A. aculeatus*, Labill. Pl. Nov. Holl. ii. 51, 52, t. 198, 200; *A. exasperatus*, Link, Enum. Hort. Berol. ii. 328 (erroneously said to be from the Cape); *Diplostephium aculeatum*, Nees, Gen. et Sp. 192, and *D. ramulosum*, Nees, l. c. 193; *Eurybia ramulosa*, DC. Prod. v. 270; Hook. f. Fl. Tasm. i. 178; *E. propinqua*, *E. aculeata*, and *E. epileia*, DC. l. c.; *E. ericoides*, Steetz in Pl. Preiss. i. 423.

N. S. Wales. New England, *C. Stuart;* Mudgee, *Woolls* (with glabrous glandular achenes).

Victoria. Port Phillip, *R. Brown;* abundant from the Glenelg to Gipps' Land, *F. Mueller* and others, Wimmera, *Dallachy;* in the Grampians (very rigid, with small flower-heads), *F. Mueller;* Mount Koroug and Mount M'Ivor (with long very scabrous or woolly leaves and larger flower-heads), *F. Mueller.*

Tasmania. Derwent river, Port Dalrymple, islands of Bass's Straits, *R. Brown;* common throughout the island, *J. D. Hooker.*

S. Australia. S. coast, *R. Brown;* Rivoli, Holdfast and Guichen bays, *F. Mueller;* Memory Cove, *R. Brown* (with thicker longer narrow leaves).

28. **O. floribunda,** *Benth.* A much-branched shrub of 4 to 6 ft., with numerous very small clustered leaves and a profusion of small flower-heads on very short leafy branchlets, forming leafy racemes collected into large dense pyramidal panicles; closely allied to *O. ramulosa*, differing in the much smaller leaves, rarely above 1 line long, the involucres not above 1½ lines long. Florets about 6 to 10, of which 3 to 4 ligulate, their structure as well as the achenes and pappus as in *O. ramulosa.*—*Eurybia floribunda*, Hook. f. in Hook. Lond. Journ. vi. 109, and Fl. Tasm. i. 179. t. 45; *Aster florulentus*, F. Muell. Fragm. v. 82.

Victoria. Along torrents in the Australian Alps, *F. Mueller.*

Tasmania. Derwent river, *R. Brown;* common along the banks of rivers, *J. D. Hooker.*

S. Australia. Memory Cove, *R. Brown.*

29. **O. lepidophylla,** *Benth.* A much-branched shrub of 3 to 6 ft., usually white or hoary with a close dense woolly tomentum. Leaves minute, clustered in the axils, obovate-oblong or almost globular, very obtuse, with revolute margins, often all under ½ line long, but those subtending the

clusters or on luxuriant barren shoots sometimes narrow and above 1 line long. Flower-heads terminating short leafy branchlets, smaller than in *O. ramulosa*, but usually rather larger than in *O. floribunda*, and not so racemose. Florets, achenes, and pappus as in *O. ramulosa*.—*Aster microphyllus*, Labill. Pl. Nov. Holl. ii. 51. t. 199; *A. lepidophyllus*, Pers. Syn. Pl. ii. 442; *Diplostephium lepidophyllum*, Nees, Gen. et Sp. Ast. 190; *Eurybia lepidophylla*, DC. Prod. v. 270; Hook. f. Fl. Tasm. i. 178.

N. S. Wales. Darling river, *Victorian Expedition.*
Victoria. Grampians, *Wilhelmi;* Wimmera, *Dallachy;* N.W. part of the colony, *L. Morton.*
Tasmania. Derwent river, Adventure Bay, islands of Bass's Straits, *R. Brown;* sandy hills near the sea on the N. coast, also on the western and other mountains, *Gunn* and others.
S. Australia. From the Murray to St. Vincent's and Spencer's Gulfs, *F. Mueller* and others.

Eurybia brachyphylla, F. Muell.; Sond. in Linnæa, xxv. 455, appears to be founded on specimens with half-starved flower-heads, with the ray-florets sometimes not much developed, but certainly sometimes longer than their styles.

30. **O. subspicata,** *Benth.* Shrubby, with erect virgate branches, more or less woolly-tomentose. Leaves erect, not clustered, linear, obtuse or nearly so, with revolute margins, rarely exceeding ½ in., glabrous above, woolly-tomentose underneath. Flower-heads ovoid or almost cylindrical, shortly pedunculate or nearly sessile, crowded into short terminal racemes or spikes. Involucre much imbricate, 3 to 4 lines long, the bracts obtuse or nearly so. Ray-florets 3 to 6; the ligula much longer than the style; disk-florets about as many, longer than the involucre. Style-appendages lanceolate. Achenes silky-hairy. Pappus bristles rather unequal, but without any distinct series of short ones.—*Eurybia subspicata*, Hook. in Mitch. Trop. Austr. 293.

Queensland. Maranoa and Belyando rivers, *Mitchell.*
N. S. Wales. Darling river, *Victorian Expedition;* Murray desert, *F. Mueller, Dallachy.*

The species differs from *O. ramulosa* chiefly in its longer leaves and larger flower-heads.

31. **O. exiguifolia,** *F. Muell. Fragm.* v. 67. A much-branched shrub of 3 to 4 ft., closely resembling the smaller forms of *O. ramulosa*, differing chiefly in the leaves, which are obovate-orbicular or broadly-cuneate, very obtuse and mostly 3-toothed, 1 to 2 lines long, narrowed into a short petiole. Flower-heads small, terminating short axillary branchlets. Florets, achenes, and pappus of *O. ramulosa*.—*Aster exiguifolius*, F. Muell. l. c.

W. Australia. Sand hummocks, Eyre's Relief, *Maxwell.*

32. **O. Cassiniæ,** *F. Muell. Fragm.* v. 68. Erect, 6 to 8 ft. high, with numerous slender erect branchlets, slightly hoary, with a close fine intricate tomentum. Leaves narrow-linear, with revolute margins, ¼ to ¾ in. long, glabrous or with a minute woolly tomentum on the under side. Flower-heads small, numerous, nearly sessile in the upper axils, forming short leafy racemes arranged in an oblong or pyramidal panicle, or on short axillary branches in a looser panicle. Involucre ovoid, about 2 lines long, the bracts more glabrous and shining than in *O. ramulosa*. Ray-florets 2 or 3, with long

ligulæ; disk-florets 3 or 4, scarcely exceeding the involucre. Achenes hairy. Pappus-bristles not very unequal.—*Aster Cassiniæ*, F. Muell. l. c.

W. Australia. King George's Sound, *R. Brown*; banks of Lake Leven, *Maxwell*.

The preceding ten species, from *O. tubuliflora* to *O. Cassiniæ*, appear sometimes in the dried specimens to pass into each other by almost insensible gradations.

33. **O. ramosissima,** *Benth.* A shrub of 2 or 3 ft., with numerous rather slender branches, scabrous-pubescent, mixed with a little loose wool. Leaves minute, reflexed, clustered in the axils, lanceolate or linear, entire, with revolute margins, all under 1 line long or rarely the larger ones narrow and nearly 2 lines long, glabrous and smooth or scabrous above, with a thin loose wool underneath. Flower-heads solitary at the ends of the branchlets, forming an oblong or rarely corymbose leafy panicle. Involucre broadly turbinate, about 3 lines long, the bracts often coloured and jagged at the edge. Florets all blue (*F. Mueller*), those of the ray 12 to 15, more numerous in the disk and longer than the involucre. Achenes more or less villous. Pappus white, with a few short outer bristles.—*Eurybia ramosissima*, DC. Prod. v. 270; *Aster cyanodiscus* or *Olearia cyanodiscalis*, Muell. Fragm. v. 82.

N. S. Wales. Port Jackson, *Gaudichaud* (the specimens not seen); in the N.W. interior, *A. Cunningham*; near Exmouth, *Fraser*; near Clifton, New England, *C. Stuart*.

34. **O. pimeleoides,** *Benth.* A shrub of 4 or 5 ft., more or less hoary or white with a close woolly tomentum, the branches rigid and virgate or sometimes short and almost spinescent. Leaves mostly oblong-cuneate and 3 to 4 lines long, but passing into obovate and much shorter, or into longer and almost linear, obtuse, with recurved margins, glabrous or hoary above, tomentose underneath. Flower-heads solitary, terminal. Involucre broadly ovoid or almost hemispherical, the bracts much imbricate, almost acute. Ray-florets 10 to 15 or even more; disk-florets numerous, longer than the involucre. Style-appendages short, obtuse. Achenes silky-villous. Pappus-bristles unequal, but none very short.—*Eurybia pimeleoides*, DC. Prod. v. 268.

N. S. Wales. Lachlan and Darling rivers and all the branches in the west, *Fraser*, *A. Cunningham*, *Victorian Expedition*, etc.

Victoria. Mallee scrub on the Avoca and Murray, *F. Mueller*; Wimmera, *Dallachy*; N.W. districts, *L. Morton*.

S. Australia. Crystal Brook, Lake Torrens, *F. Mueller*.

Var. *minor*. Flower-heads smaller, ovoid, with fewer florets.—Murray desert, *F. Mueller*; Rotton Island, *Wilhelmi*.

35. **O. iodochroa,** *F. Muell. Fragm.* v. 81. A bushy or spreading shrub, the branches slightly tomentose or glutinous or nearly glabrous. Leaves linear, cuneate-oblong or almost obovate, entire or 2- or 3-toothed towards the end, mostly 3 to 4 lines long, with recurved margins, rather thick, glabrous above, white or brown underneath with a close woolly tomentum. Flower-heads terminal, solitary or few in a dense terminal corymb. Involucre hemispherical, nearly ½ in. diameter, the inner bracts coloured on the margin. Ray-florets 15 to 20; disk-florets purple (*F. Mueller*). Style-appendages very short. Achenes silky-hairy, rather short. Pappus with an

outer series of short bristles.—*Eurybia iodochroa*, F. Muell. Fragm. ii. 110; *Aster iodochrous*, F. Muell. Fragm. v. 81.

Victoria. Woody declivities of the Australian Alps, Nangatta Creek, Genoa river, Snowy River, etc., *F. Mueller.*

36. **O. adenolasia,** *F. Muell. Fragm.* v. 67. A shrub of several ft., with rigid virgate branches, more or less scabrous with a viscid pubescence, intermixed with a thin loose woolly tomentum. Leaves linear or linear-oblong, obtuse, with closely revolute margins, under ½ in. long, glandular-scabrous or almost muricate above, or on young shoots woolly on both sides, the lower ones occasionally longer and trifid. Flower-heads terminal, rather small. Involucre broadly ovoid or almost hemispherical, much imbricate and greener than in most species. Ray-florets 8 to 10; disk-florets rather more numerous. Style-appendages almost acute. Achenes hairy. Pappus with an outer series of short bristles.—*Diplopappus glandulosus*, Turcz. in Bull. Mosc. 1851, ii. 62; *Aster adenolasius*, F. Muell. Fragm. v. 67.

W. Australia, *Drummond, 4th Coll. n.* 219, *5th Coll. n.* 369; Phillips ranges, *Maxwell.*—This species approaches in many respects *O. muricata*, but has always more or less of the woolly tomentum of the section *Eriotriche* mixed with the glandular pubescence.

37. **O.? conocephala,** *F. Muell. Fragm.* v. 79. A small compact densely-branched shrub, hoary or white with a close woolly tomentum. Leaves alternate, obovate or cuneate-oblong, under ½ in. long, tomentose on both sides. Flower-heads solitary, terminating the short branches. Involucre cylindrical, above ½ in. long, nearly glabrous, the bracts much imbricate and very obtuse. Florets 4 or 5, all tubular, slender, equal, and longer than the involucre, with 5 narrow linear lobes. Anthers very narrow, free in the florets examined, not auriculate. Style-branches long and narrow, but flattened, quite glabrous, stigmatic nearly to the end. Achenes glabrous, terete, striate, nearly 3 lines long, but not seen ripe. Pappus-bristles barbellate, almost plumose, a few of the outer ones shorter.—*Eurybia conocephala*, F. Muell. in Trans. Vict. Inst. i. 36; *Aster conocephalus*, F. Muell. Fragm. v. 67.

N. S. Wales. Desert of the Murumbidgee, Murray, and Darling, *F. Mueller, Victorian Expedition.*
Victoria. N.W. districts, *L. Morton.*
S. Australia. Head of the Great Bight, *Delisser;* Fowler's Bay, *R. Brown.*
The species has not the style of *Olearia*, and there are no female florets. I also found the anthers quite free in all the flowers I examined, but that may not be constantly the case. Notwithstanding these anomalies, as I know of no genus to which it is more nearly allied, I have left it in *Olearia* as described by *F. Mueller.*

SECTION IV. ADENOTRICHE, *Archer.*—Plant glabrous, usually glutinous. Involucre ovoid turbinate or rarely hemispherical, the bracts usually obtuse, rigid, scarious on the margins and often ciliate, more imbricate than in *Merismotriche.*

38. **O. magniflora,** *F. Muell. Fragm.* v. 80. A shrub of 3 or 4 ft., with divaricate branches, the whole plant glabrous and often glutinous. Leaves not numerous, narrow, cuneate, thick, obtuse and often 3-toothed at

the end, the larger ones above ½ in. long, but mostly smaller. Flower-heads large, solitary, terminal. Involucre broadly turbinate, ¾ in. diameter, the bracts numerous, much imbricate, the inner ones often coloured on the margins. Ray-florets 15 to 20, long and narrow; disk-florets numerous. Style-lobes long, with short obtuse appendages. Achenes long, glabrous. Pappus exceeding the involucre, a few of the outer bristles shorter than the others. —*Aster magniflorus*, F. Muell. l. c.

N. S. Wales. Desert of the Murray, Darling, and Murrumbidgee, *F. Mueller* and others; towards Lachlan river, *Burkitt*.

Victoria. Desert of the N.W., *L. Morton, F. Mueller.*

39. **O. calcarea,** *F. Muell. Herb.* A compact much-branched shrub, more or less glutinous, otherwise glabrous. Leaves obovate or broadly cuneate, thick, more or less toothed, not above ¼ in. long in the specimen seen. Flower-heads solitary, terminal, nearly as large as in *O. magniflora*. Involucre broadly turbinate, the inner bracts 7 to 8 lines long, often coloured. Style-appendages long and pointed. Achenes silky-hairy. Pappus with a few of the outer bristles rather shorter than the others.

S. Australia. Towards Cudnaka and banks of the Murray near Moorundi, *F. Mueller*. (*Herb. F. Mueller*). The species appears to be quite distinct, but requires further elucidation from more complete specimens.

40. **O. Muelleri,** *Benth.* A much-branched bushy shrub, more or less glutinous, otherwise glabrous. Leaves obovate or broadly cuneate, thick, very obtuse, entire or toothed, under ½ in. long, narrowed into a petiole. Flower-heads solitary, terminal, much smaller than in *O. magniflora*. Involucre ovoid or turbinate, much imbricate, the inner bracts about 4 lines long. Ray-florets 8 to 10; disk-florets more numerous. Style-appendages short as in *O. magniflora*. Achenes silky-hairy, much shorter than in the last two species, and the pappus with fewer bristles.—*Eurybia Muelleri*, Sond. in Linnæa, xxv. 459.

N. S. Wales. Darling river, *Victorian Expedition*; desert of the Murray and Murrumbidgee, *F. Mueller*.

Victoria. Avoca river, *F. Mueller*; Wimmera, *Dallachy*; N.W. districts, *L. Morton*.

S. Australia. South coast, *R. Brown*; Flinders' Range, *F. Mueller*; Lake Gairdner, *Babbage*; Northern interior, *M'Douall Stuart's Expedition*.

W. Australia, *Drummond*, with rather smaller flower-heads, but apparently not otherwise different.

41. **O. Stuartii,** *F. Muell. Fragm.* v. 76. Apparently a divaricately-branched small shrub, densely glandular-viscid, otherwise glabrous. Leaves cuneate-oblong or almost linear, obtuse and obtusely toothed, ½ to 1 in. long, narrowed into a petiole. Flower-heads terminal, shortly pedunculate, broadly ovoid or almost hemispherical. Involucral bracts much imbricate, more acute than in the other species of this section. Achenes slightly hairy. Pappus rather unequal.—*Eurybia Stuartii*, F. Muell. Fragm. i. 202; *Aster Stuartii*, F. Muell. Fragm. v. 76.

S. Australia. N.W. of Lake Gairdner, *M'Douall Stuart*. The species might perhaps be transferred to the following section, but requires further elucidation from more perfect specimens.

42. **O. decurrens,** *Benth.* A tall shrub, glabrous and viscid. Leaves

usually distant, linear-oblong, obtuse, entire or with a few coarse teeth or on barren shoots cuneate and lobed, mostly ½ to 1 in. long. Flower-heads small, in a loose leafy panicle or almost solitary. Involucre ovoid-turbinate, imbricate, the inner bracts about 3 lines long. Ray-florets 6 to 8; disk-florets slightly exceeding the involucre. Style-branches rather long, but with short lanceolate appendages. Achenes silky-hairy. Pappus-bristles unequal.—*Eurybia decurrens*, DC. Prod. v. 269.

N. S. Wales. Lachlan river, *A. Cunningham;* Darling and Murray desert, *Victorian Expedition* and others.

S. Australia. Port Lincoln, *R. Brown, Wilhelmi;* Spencer's Gulf, *F. Mueller.*

43. **O. glutinosa,** *Benth.* A bushy shrub of 3 to 5 ft., very glutinous, but otherwise glabrous. Leaves narrow-linear, acute or obtuse, ½ to 1½ in. long, flat or the margins slightly recurved. Flower-heads small, in terminal corymbs, not pedunculate although at first prominent above the last leaves, and the lateral leafy branches often growing out much beyond the inflorescence. Involucres larger and more ovoid than in *O. glandulosa*, the inner bracts above 2 lines long. Ray-florets about 6 to 10; disk-florets scarcely exceeding the involucre. Anthers much exserted. Style-appendages short. Achenes glabrous or sparingly hairy. Pappus-bristles unequal.—*Eurybia glutinosa*, Lindl. Bot. Reg. 1839, Misc. 68; *E. linifolia*, Hook. f. in Hook. Lond. Journ. vi. 109, and Fl. Tasm. i. 179; *E. glutescens*, Sond. in Linnæa, xxv. 462; *Aster glutescens* or *Olearia glutescens*, F. Muell. Fragm. v. 77; *Aster orarius* or *Olearia oraria*, F. Muell. Fragm. v. 78 (in an older state).

Victoria. Port Phillip, *R. Brown;* from the mouth of the Glenelg to Wilson's Promontory, *F. Mueller* and others.

Tasmania. Sandy seacoasts in the northern parts of the island and in the islands of Bass's Straits, *J. D. Hooker* and others.

S. Australia. Murray river near Moorundi and Lofty Ranges, *Wilhelmi;* Lake Victoria and Port Gawler, *F. Mueller.*

44. **O. passerinoides,** *Benth.* A bushy shrub of 1 to 2 ft., glabrous and glutinous. Leaves crowded, linear, erect, obtuse, with somewhat revolute margins, but the broad midrib exposed, mostly 2 to 3 lines long. Flower-heads small, sessile within the last leaves. Involucres ovoid, the inner bracts about 2½ lines long. Ray-florets about 6 to 8; disk-florets rather more numerous, not exceeding the involucre. Style-appendages short. Achenes short, hairy. Pappus-bristles unequal.—*Diplopappus passerinoides*, Turcz. in Bull. Mosc. 1851. ii. 63; *Aster vernicosus* or *Olearia vernicosa*, F. Muell. Fragm. v. 67.

W. Australia, *Drummond, 5th Coll. n.* 371; sandy plains, Middle Mount Barren, Eyre and Phillips ranges, *Maxwell.* The species is very near *O. Hookeri*, and chiefly differs in its erect leaves. Some small-leaved scarcely hispid specimens of *O. muricata* come near to *O. passerinoides*, but may be at once known by their leaves with prominent revolute margins, leaving only a narrow furrow between them, and the involucres are broader with more acute bracts.

45. **O. teretifolia,** *F. Muell. Fragm.* v. 77. A bushy shrub of 2 to 5 ft., glabrous and viscid. Leaves linear, erect, closely appressed, obtuse, the margins closely revolute so as to be nearly terete, usually not exceeding 1 line on the flowering branches, looser and often 2 lines long on barren ones. Flower-heads small, terminal, sessile. Involucre ovoid, not 2 lines long.

Ray-florets 3 or 4; disk-florets 4 to 6. Achenes glabrous or hairy. Pappus-bristles unequal, not numerous.—*Eurybia teretifolia*, Sond. in Linnæa, xxv. 464; *Aster teretifolius*, F. Muell. Fragm. v. 77.

S. Australia. Stony declivities between Adelaide and Lofty ranges, *F. Mueller;* Kangaroo Island, *Waterhouse.*

Var. with rather larger flower-heads.

N. S. Wales. Darling desert, *Herb. F. Mueller.*

The species differs slightly from *O. passerinoides* and *O. Hookeri* in its small erect leaves, and may possibly be a variety of the latter.

46. **O. Hookeri,** *Benth.* A much-branched shrub, glabrous and glutinous. Leaves numerous, linear, spreading or recurved, obtuse, with revolute margins, mostly 2 to 3 lines long. Flower-heads small, terminal, sessile. Involucre ovoid, the inner bracts about 2½ lines long. Ray-florets about 6 to 8; disk-florets rather more numerous, not exceeding the involucre. Style-appendages short. Achenes short, hairy. Pappus unequal.—*Eurybia ericoides*, Hook. f. Fl. Tasm. i. 180, not of Steetz; *Eurybia Hookeri*, Sond. in Linnæa, xxv. 463.

Tasmania. Near Hobarton, *Lyall, Gunn,* and others. A specimen from Great Swan Port, referred here by Hooker, has the leaves much shorter and the flower-heads more hemispherical like those of *O. imbricata.*

Var.? *microcephala.* Leaves very small, but spreading or recurved. Flower-heads small.

Victoria. Murray river, *Dallachy.*

47. **O. imbricata,** *Benth.* A shrub of 1 to 2 ft., with erect or divaricate branches, glabrous and sometimes slightly glutinous. Leaves numerous, linear-cuneate, obtuse, thick, 1 to 2 or rarely 3 lines long, erect or spreading. Flower-heads terminal, shortly pedunculate. Involucre hemispherical, the inner bracts almost acute, 2½ lines long. Ray-florets 15 to 20. Achenes short, hairy. Pappus-bristles not numerous, a few outer ones short.—*Eurybia imbricata*, Turcz. in Bull. Mosc. 1851, ii. 61; *Aster Turczaninowii* or *Olearia Turczaninowii*, F. Muell. Fragm. v. 67.

W. Australia, *Drummond, 5th Coll. n* 370; broken country near Oldfield river, Phillips river, valleys near Eyre's Range, *Maxwell.*

48. **O. elliptica,** *DC. Prod.* v. 271. A tall shrub, glabrous and often glutinous, the foliage sprinkled with minute glandular dots. Leaves elliptical, oblong or lanceolate, acute acuminate or almost obtuse, entire or rarely sinuate-toothed, narrowed at the base, 1½ to 3 in. long. Flower-heads small, numerous, in a terminal corymbose, leafless panicle. Involucre broadly ovoid or almost hemispherical, 3 to 4 lines diameter. Ray-florets 6 to 8; disk-florets rather more numerous. Style-appendages rather long. Achenes slightly hairy. Pappus with few or many outer short bristles.—*Eurybia illita*, F. Muell. Fragm. i. 16; *Aster illitus* or *Olearia illita*, F. Muell. Fragm. v. 76.

Queensland. Near Warwick, *Beckler.*

N. S. Wales. New England, *C. Stuart;* Mount Lindsay, *W. Hill;* Illawarra, *A. Cunningham.*

49. **O. glandulosa,** *Benth.* An undershrub or shrub attaining 3 or 4 ft., quite glabrous but more or less glandular-dotted, the branches slender and erect. Leaves very narrow-linear, acute, with closely revolute margins

so as to appear terete, ½ to 1½ in. long. Flower-heads small, in terminal corymbose leafless panicles. Involucre almost hemispherical, about 3 lines diameter, the bracts narrow, in few rows, the inner ones scarcely 1½ lines long. Ray-florets 12 to 15; disk-florets scarcely exceeding the involucre. Style-appendages short and obtuse. Achenes small, silky-hairy. Pappus-bristles unequal, not very numerous.—*Aster glandulosus*, Labill. Pl. Nov. Holl. ii. 50, t. 197; *Galatella glandulosa*, Nees, Gen. et Sp. Ast. 174; *Eurybia glandulosa*, DC. Prod. v. 269; Hook. f. Fl. Tasm. i. 179.

N. S. Wales. Marshy lands south of Lake George, *A. Cunningham.*

Victoria. Marshy places, Dandenong ranges, Delatite river, *F. Mueller;* near Creswick, *Whan.*

Tasmania. Port Dalrymple, *R. Brown;* abundant on marshy river-banks throughout the colony, ascending to 3500 feet, *J. D. Hooker.*

S. Australia. S. coast, *R. Brown, Whittaker.*

SECTION V. MERISMOTRICHE, *Archer.*—Glabrous, glandular-pubescent or hirsute with simple rigid white or transparent hairs, transversely septate and often glutinous. Involucre hemispherical, with narrow usually acute bracts. Leaves alternate. Anthers not auriculate.

The last four species of this section come near to some species of *Vittadinia,* but have much fewer ray-florets, the style appendages are shorter, and the achenes less compressed.

50. **O. heleophila,** *F. Muell. Fragm.* v. 66. A shrub or undershrub with slender virgate branches, glabrous glandular-pubescent or sometimes with a little wool about the involucres. Leaves narrow-linear, acute or obtuse, with revolute margins, in some specimens mostly under ½ in. long and distant, in others the lower ones 1 to 1½ in. and much crowded. Flower-heads small, forming an irregular loose panicle or corymb. Involucre ovoid-turbinate or at length almost hemispherical, the bracts narrow, acute or rather obtuse, the inner ones about 2 lines long. Ray-florets about 12 to 15; disk-florets scarcely longer than the involucre. Style-appendages short. Achenes small, silky-hairy. Pappus-bristles not very numerous, several of the outer ones short.—*Eurybia elæophila*, DC. Prod. v. 269; *Aster heleophilus*, F. Muell. Fragm. v. 66; *Eurybia affinis*, Steetz in Pl. Preiss. i. 421; *E. paniculata*, Steetz, l. c. 422; *Aster Preissii*, F. Muell. Fragm. v. 66.

W. Australia, *Drummond, 2nd Coll. n.* 173; King George's Sound, *A. Cunningham;* near Guildford and Hay river, *Preiss, n.* 80 and 81.

Var. *major.* More woolly; lower leaves rather broader, occasionally toothed; flower-heads larger.—*Eurybia Lehmanniana*, Steetz in Pl. Preiss. i. 422; *Aster Lehmanni,* F. Muell. Fragm. v. 66.—Swan River, *Preiss, n.* 79.

51. **O. muricata,** *Benth.* A divaricately-branched shrub of 1 to 2 ft., glabrous glandular-pubescent or almost muricate with short rigid hairs and occasionally with a little of the wool of *Eriotriche* about the peduncles. Leaves linear, very obtuse, with closely revolute margins, mostly 2 to 4 lines long. Flower-heads shortly pedunculate, rather small. Involucre from narrow turbinate to almost hemispherical, with narrow bracts. Ray-florets about 8 to 10; disk-florets rather more numerous. Style-appendages short but almost acute. Achenes hairy. Pappus with a few outer short bristles. —*Eurybia muricata*, Steetz in Pl. Preiss. i. 423; *Aster muricatus*, F. Muell. Fragm. v. 66.

W. Australia. Near Cape Riche, *Preiss, n.* 82, *Harvey.* Apparently a variable species, difficult to distinguish on the one hand from *O. adenolasia* and on the other from *O. heleophila*, and some glabrous specimens coming even near to *O. imbricata.* Amongst numerous specimens of Drummond's, we should probably refer to *O. muricata*, n. 77, 85, 3rd Coll. n. 127 (which is *Eurybia leptophylla*, Turcz. in Bull. Mosc. 1851, i. n. 171), and perhaps also 5th Coll. n. 384, the latter, however, possibly a form of *O. imbricata.*

52. **O. strigosa,** *Benth.* An erect, sparingly-branched undershrub of 1 to 3 ft., more or less hispid with short septate hairs intermixed with a slight glandular pubescence. Leaves linear, mucronate-acute, with revolute margins, often almost muricate. Flower-heads larger than in the last two species, more or less pedunculate, the upper leaves gradually reduced almost to bracts. Involucre hemispherical, nearly ½ in. diameter, the inner bracts about 3 lines long, rather acute. Ray-florets about 8 to 12; disk-florets scarcely exceeding the involucre. Achenes silky-hairy. Pappus unequal.—*Eurybia strigosa*, Steetz in Pl. Preiss. i. 419; *E. aspera*, Steetz, l. c. 420; *Aster Steetzii* or *Olearia Steetzii*, F. Muell. Fragm. v. 66.

W. Australia. Vasse river, *Preiss, n.* 83, also a single specimen, without the locality given, in Maxwell's collection. The species appears to be very nearly allied to the last two.

53. **O. paucidentata,** *F. Muell. Fragm.* v. 66. An undershrub or shrub of 2 to 3 ft., scabrous-pubescent or hispid with short septate hairs and often somewhat viscid. Leaves oblong-linear or oblanceolate and ½ to 1 in. long or shorter and obovate, obtuse, narrowed into a petiole, with 1 or 2 prominent teeth or lobes on each side or occasionally entire, the margins usually recurved. Flower-heads rather small, on peduncles usually longer than the leaves, forming a terminal leafy panicle. Involucre hemispherical, the bracts narrow, almost acute, the inner ones about 2 lines or in large heads 2½ lines long. Ray-florets 12 to 20; disk-florets more numerous, scarcely exceeding the involucre. Style-appendages short. Achenes pubescent. Pappus rather short, slightly unequal, with sometimes a very few short outer bristles.—*Eurybia paucidentata*, Steetz in Pl. Preiss. i. 420; *Aster paucidentatus*, F. Muell. Fragm. v. 66.

W. Australia. Swan River, *Drummond*, 1*st Coll., also n.* 31, 34, *and* 4*th Coll. n.* 128; *Preiss, n.* 74, 84, *and n.* 80 *in part;* Franklin and Blackwood rivers, *Oldfield.*

Var. *latifolia, Drummond,* 2*nd Coll. n.* 172; Plantagenet and Stirling ranges, *Maxwell.* The species varies much in the breadth of the leaf, the more or less copious indumentum, and in the size of the flower-heads.

54. **O. stricta,** *Benth,* An erect undershrub or shrub, of 1 to 3 ft., not much branched, very glandular-pubescent and hirsute with short septate hairs. Leaves linear, flat or with recurved margins, not exceeding ½ in., often clustered in the axils. Flower-heads terminating short lateral branches, and sessile within the last leaves, forming a long, narrow, leafy panicle. Involucre hemispherical, with narrow, herbaceous, acute bracts, the inner ones nearly 3 lines long. Ray-florets 12 to 15; disk-florets not longer than the involucre. Achenes short, pubescent. Pappus of nearly equal bristles, with an outer row of short ones.

Victoria. Rocks of Mount Aberdeen, Buffalo Range, at an elevation of 4000 ft., *F. Mueller.*

55. **O. tenuifolia,** *Benth.* A shrub of 3 to 4 ft., often scabrous or glutinous, with glandular papillæ or very short rigid hairs, otherwise glabrous. Leaves narrow-linear, acute or almost obtuse, ½ to 1 in. long, the margins usually revolute. Flower-heads much larger and fewer than in *O. glandulosa*, which the species resembles in foliage, all pedunculate, irregularly corymbose. Involucre nearly hemispherical, the outer bracts with herbaceous lanceolate, but appressed lips, the inner ones dry, nearly 3 lines long. Ray-florets 10 to 15; disk-florets not much exceeding the involucre. Style-appendages short. Achenes more or less hairy. Pappus unequal, the outer bristles sometimes quite short.—*Eurybia tenuifolia*, DC. Prod. v. 269.

N. S. Wales. Rocky Callitris Ranges, S. of Liverpool plains, and brushes near Bathurst, and other sterile broken plains in the interior, *A. Cunningham*, *Fraser*.

56. **O. adenophora,** *F. Muell. Fragm.* v. 78. A shrub, attaining several feet, very viscid and scabrous, with a glandular pubescence intermixed with articulate hairs. Leaves rather crowded, linear, obtuse, entire, with revolute margins, mostly above 1 in. long. Flower-heads rather large, on peduncles shorter than the leaves, solitary or few in a short corymb. Involucre hemispherical, the bracts numerous, acute, the inner ones above 3 lines long, the outer ones gradually shorter. Ray-florets above 20; disk-florets numerous, rather longer than the involucre. Style-appendages very short and obtuse. Achenes silky-hairy. Pappus with an outer row of numerous short bristles.—*Eurybia adenophora*, F. Muell. Fragm. i. 111; *Aster adenophorus*, F. Muell. Fragm. v. 78.

Victoria. Mountains on M'Alister river, at an elevation of 2000 to 3000 ft., *F. Mueller*.

57. **O. homolepis,** *F. Muell. Fragm.* v. 65. A shrub, of 3 or 4 ft., with stout, erect, virgate branches, very scabrous with short, rigid hairs. Leaves crowded, linear, mucronate-acute or obtuse, with revolute margins, ½ to 1 in. long, very scabrous or almost muricate. Flower-heads large, solitary or 2 or 3 together at the ends of the branches, on peduncles shorter than the leaves. Involucre hemispherical, the bracts narrow, the inner ones 5 lines long, the outer ones often scarcely shorter. Ray-florets above 20; disk-florets numerous, scarcely exceeding the involucre. Achenes silky-hairy. Pappus bristles nearly equal, with a few outer short ones.—*Aster homolepis*, F. Muell. l. c.

W. Australia. Murchison river, *Oldfield, Drummond*, 6*th Coll. n.* 151.

58. **O. xerophila,** *F. Muell. Fragm.* v. 76. A shrub, very glutinous and hispid with septate hairs. Leaves from obovate to oblong, obtuse, crenately toothed and almost crisped, narrowed into a petiole, ¾ to 1 in. long. Flower-heads not very large, on axillary peduncles longer than the leaves, solitary or few in a loose corymb. Involucre hemispherical, the bracts narrow, acute, in several rows, the inner ones about 2 lines long. Ray-florets 12 to 15; disk-florets numerous, much longer than the involucre. Style-appendages lanceolate. Achenes silky-pubescent. Pappus sometimes very unequal, a few outer bristles quite short, more rarely all nearly equal.—*Eurybia xerophila*, F. Muell. Fragm. i. 51; *Aster xerophilus*, F. Muell. Fragm. v. 76; *A. Heynei*, F. Muell. Fragm. v. 86.

Queensland. Barren ridges, Upper Burdekin river, *F. Mueller;* Cape river, *Bowman.*

59. **O. Ferresii,** *F. Muell. Fragm.* v. 75. Shrubby, glabrous, except a very few loose hairs about the involucre. Leaves broadly lanceolate, acute, slightly and remotely toothed, narrowed into a petiole, 3 to 4 in. long, green on both sides and not scabrous. Flower-heads large, few in a terminal corymb, the peduncle as long as the leaves, with a few subulate bracts. Involucre hemispherical, slightly glandular, the bracts narrow, mostly acute, in several rows, the inner ones about 4 lines long, the outer ones gradually shorter. Ray-florets about 20, narrow; disk-florets very numerous, much longer than the involucre. Achenes silky. Pappus unequal, but with few very short bristles.—*Eurybia Ferresii,* F. Muell. Fragm. iii. 18. t. 18; *Aster Ferresii,* F. Muell. Fragm. v. 75.

N. Australia. Brindley's Bluff, near M'Donnell Ranges, *M'Douall Stuart's Expedition.* A single specimen with the flowers scarcely expanded (*Herb. F. Mueller*).

60. **O. rudis,** *F. Muell. Fragm.* v. 75. An erect shrub or undershrub, more or less scabrous or hirsute with rigid septate hairs or rarely nearly glabrous. Leaves obovate-oblong, oblong-cuneate or broadly lanceolate, obtuse or rarely acute, coarsely and irregularly toothed or nearly entire, mostly 1 to 2 in. long, usually contracted below the middle, but often broader and stem-clasping at the base. Flower-heads rather larger, solitary or few together in a terminal corymb, the peduncles longer than the leaves, thickened under the head, without any or only a single linear bract. Involucre hemispherical, the bracts narrow, mostly acute, the inner ones about 4 lines long, the outer scarcely shorter. Ray-florets narrow, numerous; disk-florets scarcely so long as the involucre. Style-appendages very short. Achenes glabrous or nearly so, slightly compressed, strongly striate. Pappus-bristles nearly equal.—*Eurybia rudis,* Benth. in Hueg. Enum. 58; Steetz in Pl. Preiss. i. 418; *Aster exul,* Lindl. Swan Riv. App. 24; F. Muell. Fragm. v. 75.

W. Australia. Swan River, *Huegel, Drummond, 1st Coll., also n.* 35, 39, 181, 385; *Preiss, n.* 63, and others.

Var. *scabra.* Very scabrous and hispid. Leaves oblong-cuneate, rarely exceeding 1 in. on the flowering branches. Flower-heads rather smaller.—*Eurybia scabra,* Benth. in Hueg. Enum. 58.

N. S. Wales. Murray desert, near the Murrumbidgee. *F. Mueller.*

Victoria. Wimmera, *Dallachy.*

S. Australia. Memory Cove, *R. Brown;* Venus and Streaky Bays, *Babbage, Warburton.*

Var. *glabriuscula.* Leaves usually large, obovate-oblong, scabrous, but appearing glabrous.—Wimmera, *Dallachy;* Gawler Town, Rivoli Bay, etc., in S. Australia, *F. Mueller.*

61. **O. picridifolia,** *Benth.* An erect shrub, with the rigid pubescence, inflorescence and flower-heads of *O. rudis,* of which F. Mueller now thinks it may be a variety, but the leaves are all very narrow-lanceolate or linear, narrowed at the base and quite entire, a form to which I have seen no approach in any of the varieties of *O. rudis.*—*Eurybia picridifolia,* F. Muell. in Linnæa, xxv. 397.

S. Australia. Barren hills towards Lake Torrens, *F. Mueller.*

62. **O. arguta,** *Benth.* A shrub or undershrub, of 1 to 2 ft., more or

less hirsute with septate hairs and somewhat glutinous. Leaves oblong-lanceolate, acute, entire or with a few pointed teeth towards the end, narrowed below the middle, but broad and stem-clasping at the base, 2 to 4 in. long. Flower-heads rather large, solitary, on peduncles longer than the leaves, with 1 or 2 small bracts. Involucre nearly hemispherical, the bracts narrow, acute, the inner ones 4 to 5 lines long, the outer ones shorter. Ray-florets numerous; disk-florets not longer than the involucre. Achenes pubescent. Pappus-bristles nearly equal.—*Aster argutus*, R. Br. Herb.

N. Australia. Islands of the Gulf of Carpentaria, *R. Brown.*

Var. *lanata.* Young leaves clothed with a dense white deciduous wool, and mostly quite entire.—Arnhem, N. and S. bays, *R. Brown.*

This species is certainly nearly allied in essential characters to *O. rudis*, but the aspect of the specimens is very different, and the stations of the two are widely distant.

63. **O. ciliata,** *F. Muell. Fragm.* v. 79. An undershrub or small shrub, not exceeding 1 ft. without the peduncles, the branches glabrous or with a few septate hairs. Leaves crowded, spreading, linear, rigid, mucronate-acute or almost obtuse, mostly ½ to ¾ in., but sometimes above 1 in. long, the margins revolute and usually ciliate, otherwise glabrous or rarely scabrous-pubescent. Flower-heads rather large, on terminal peduncles, often 4 or 5 in. long. Involucre hemispherical, the bracts narrow, acute, glabrous or ciliate, the inner ones nearly 4 lines long, the outer ones gradually shorter. Ray-florets 15 to 20; disk-florets numerous, scarcely exceeding the involucre. Style-appendages short. Achenes glabrous or silky-pubescent. Pappus-bristles nearly equal, with occasionally a few outer very short ones.—*Eurybia ciliata*, Benth. in Hueg. Enum. 58; Steetz in Pl. Preiss. i. 418; Hook. f. Fl. Tasm. i. 180; *Aster Huegelii*, F. Muell. Fragm. v. 79.

Victoria. Murray desert, *F. Mueller;* Wimmera, *Dallachy;* Mount Abrupt, *Wilhelmi;* Wilson's Promontory, *F. Mueller.*

Tasmania. S. Esk river, *Gunn;* near Swanport, *C. Stuart.*

S. Australia. S. coast, *R. Brown;* Lofty Range, Spencer's Gulf, etc., *F. Mueller* and others.

W. Australia. King George's Sound and adjoining districts, *R. Brown, Huegel, Drummond, n.* 19, 5*th Coll. n.* 375; *Preiss, n.* 76, 77, 78; Cape Naturaliste, *Oldfield.*

Var. *hispida.* Leaves very hispid, the longer ones occasionally 3-toothed.—Hake's Place, S. Australia, *F. Mueller.*

Var.? *squamifolia*, F. Muell. Leaves densely crowded, 1 to 2 lines long, entire or 3-lobed. Flower heads much smaller.—Kangaroo Island, *Waterhouse.*

12. CELMISIA, Cass.

Involucre broadly hemispherical, the bracts imbricate, in several rows, the margins dry or scarious, without herbaceous tips. Receptacle pitted, without scales. Florets of the ray female, in a single row, ligulate, spreading. Disk-florets numerous, hermaphrodite, tubular, 5-lobed. Anthers with acute or pointed auricles at the base. Style-branches flattened, with rather long tips or appendages, papillose on the back. Achenes slightly compressed, with 2 or 3 prominent nerves on each side. Pappus of numerous unequal capillary bristles.—Perennial herbs, more or less silvery-silky. Leaves chiefly radical, narrow, entire. Scapes nearly leafless, bearing a single large flower-head.

The genus comprises but few species, natives of the Antarctic regions and New Zealand; the single Australian species being the same as one of the New Zealand ones. It is closely allied to some of the mountain species of *Aster*, differing chiefly in the anthers not obtuse at the base. From *Olearia* it is chiefly distinguished by its habit and longer style-appendages.

1. **C. longifolia,** *Cass.; DC. Prod.* v. 209. A perennial, with a densely tufted stock, forming often large silvery-white patches. Radical leaves linear or rarely linear-lanceolate, with a broad sheathing base, softly mucronate or obtuse, the margins revolute, varying in length from barely 2 in. in some specimens to 8 or 10 in. in others, densely white tomentose underneath, the silvery-silky indumentum of the upper surface often deciduous, leaving the old leaves glabrous and shining above. Scapes always exceeding the leaves, and attaining sometimes 1½ ft., the leaves all reduced to linear or lanceolate bracts, ½ to 1½ in. long. Involucre broadly turbinate or hemispherical, woolly or at length glabrous, the inner bracts ½ in. long. Ray-florets above 30, pink or white; disk-florets about as long as the involucre. Achenes fully 3 lines long, more or less silky-pubescent. Pappus-bristles very unequal, the shortest half as long as the longest.—Gaud. in Freyc. Voy. 470. t. 91; Hook. f. Fl. Tasm. i. 181; Handb. N. Zeal. Fl. 134; *C. asteliæfolia*, Hook. f. Fl. Ant. i. 35; *Aster Celmisia*, F. Muell. Fragm. v. 84.

N. S. Wales. Bogs of the Blue Mountains, *A. Cunningham* and others.

Victoria. Australian Alps, at an elevation of 4000 to 5000 ft., and summit of Mount William, in the Grampians, *F. Mueller*.

Tasmania. Derwent river, *R. Brown;* frequent in bogs on the summits of the mountains, at an elevation of 3000 to 5000 ft., *J. D. Hooker*.

The species is also in New Zealand.

Var. ? *latifolia*, F. Muell. Leaves 8 in. to 1 ft. long, ¾ to 1½ in. wide, narrowed below the middle, the margins not recurved.—High Alpine ranges on the M'Alister river, Haidinger Range and Mount Buller, *F. Mueller*. A specimen from Mount Barkly appears to connect this with the common narrow-leaved form.

Var. ? *saxifraga*. Very small in all its parts. Stock usually branching and elongated to 1 to 3 in., covered with the remains of old leaves. Leaves in a dense tuft, narrow-linear, with revolute margins, under 1 in. long. Scapes 1 to 3 in. long. Flower-heads, florets, and achenes much smaller than in the common *C. longifolia*.—Table Mountain, Derwent river, *R. Brown;* summit of Mount Lapeyrouse, *Oldfield*.

13. VITTADINIA, A. Rich.

(Microgyne, *Less.;* Eurybiopsis, *DC.*)

Involucre hemispherical or campanulate, the bracts imbricate in several rows, with dry or scarious margins, without herbaceous tips, in the Australian species narrow and mostly acute. Receptacle pitted, without scales. Florets of the ray female numerous and crowded, so as to form more than one row, ligulate and spreading in the Australian species. Disk-florets numerous, but often not so many as those of the ray, hermaphrodite, tubular, dilated upwards, usually 5-lobed. Anthers obtuse at the base. Style-lobes somewhat flattened, with subulate tips or appendages papillose on the back. Achenes narrow, compressed or flat, with or without ribs on the faces. Pappus of numerous often unequal capillary bristles.—Perennial herbs or undershrubs, at length woody at the base, or in species not Australian shrubs. Leaves alternate. Flower-heads terminal, solitary or forming loose leafy corymbs. Ray-florets white or blue. Disk-florets yellow.

The genus extends to New Zealand and extratropical S. America, and (in a slightly modified form) to the Sandwich Islands. Of the 4 Australian species one is also in New Zealand, and very closely allied to the S. American one; the others are endemic. As a genus, the group is nearly allied to *Eurybia*, *Aster*, and especially to *Erigeron*, but, as shown by A. Gray (Proc. Amer. Acad. v. 116), it cannot well be united with either. From *Eurybia* it differs in the more numerous ray-florets and the more flattened achenes, from *Erigeron* in habit, and from both in the subulate tips to the styles.

SECTION I. **Vittadinia vera.**—*Achenes with 2 or more ribs on each face.*

Involucre imbricate in several rows. Achenes shorter than the involucre, with 2 or 3 ribs on each face. Pappus not so long . . 1. *V. brachycomoides.*
Involucre of 2 or 3 rows. Achenes nearly as long as the involucre, many-ribbed or finely striate. Pappus as long again . . . . 2. *V. australis.*

SECTION II. **Eurybiopsis.** *Achenes very flat, the margins slightly thickened, without prominent ribs on the faces.*

Scabrous-pubescent or hirsute. Leaves oblong or cuneate. Ray-florets scarcely exceeding the pappus . . . . 3. *V. scabra.*
Glabrous or scabrous-pubescent. Leaves linear or the lower ones linear-cuneate. Ray-florets longer than the pappus . . . . 4. *V. macrorrhiza.*

SECTION I. VITTADINIA VERA, *A. Gray.*—Achenes with 2 or more ribs on each face.

1. **V. brachycomoides,** *F. Muell. Fragm.* v. 86, *as an* Aster. Stems from a thick woody stock, erect or decumbent, not much branched, ½ to 1½ ft. long, with more or less of a loose white woolly deciduous tomentum. Leaves in the original form linear or lanceolate, ½ to 1½ in. long, entire or rarely 3-toothed at the end. Flower-heads on long terminal peduncles. Involucre hemispherical, the bracts in several rows, the inner ones 2½ to 3 lines long, the outer ones gradually shorter. Ray-florets narrow, elongated, spreading. Achenes narrow, much shorter than the involucre, flat, with 2 or 3 prominent ribs on each face. Pappus of fine white rather unequal and not very copious bristles, not so long as the achene.

N. Australia. Basaltic plains, Hooker and Sturt's Creek and Arnhem's Land, *F. Mueller.*

Queensland. Bustard Bay, *Banks and Solander;* Keppel Bay, *R. Brown, Thozet;* Percy Island, *M'Gillivray;* Rockhampton, *Dallachy.*

Var. (?) *latifolia.* Leaves broader, oblong, entire or toothed. Involucral scales broader and fewer.—Endeavour river and Northumberland Islands, *R. Brown* (the former specimens nearly glabrous, the latter very woolly); Albany Island, Cape York, *M'Gillivray;* Rockingham Bay, *Dallachy;* E. coast, *A. Cunningham, Bowman;* also Purdie's River in the interior of N. Australia, *M'Douall Stuart's Expedition.*—Possibly a distinct species.

2. **V. australis,** *A. Rich.; DC. Prod.* v. 280. Herbaceous, either erect and apparently annual (flowering the first year?) or with diffuse or ascending stems from a woody base, rarely above 1 ft. high, more or less tomentose, with soft almost silky or woolly hairs, or scabrous-hispid with rigid hairs arising from a tubercle. Leaves in the typical form from obovate or spathulate to linear-cuneate, entire or coarsely 3-toothed or lobed, narrowed

into a petiole, under ½ in. long when broad, sometimes above 1 in. when narrow. Flower-heads solitary, terminal. Involucre of 2 or 3 rows of narrow bracts 3 to 4 lines long, or the outer row shorter. Ray-florets narrow, about as long as the pappus or rather longer, but usually revolute so as to appear much shorter; disk-florets slender, much longer than the involucre. Achenes narrow, nearly as long as the involucre, tapering at the base, more or less pubescent, striate with 6 to 8 fine ribs on each face. Pappus longer than the achene, of copious rather unequal brownish bristles, a few outer ones much shorter.—*Aster Behrii*, Schlecht. Linnæa, xxi. 446; F. Muell. Fragm. v. 87; *Vittadinia triloba*, *V. cuneata*, and probably also *V. dentata*, DC. Prod. v. 281; *Eurybiopsis scabrida* and *E. gracilis*, Hook. f. in Hook. Lond. Journ. vi. 110; *E. Hookeri*, F. Muell. in Linnæa, xxv. 453; *Vittadinia scabra* and *V. cuneata*, Hook. f. Fl. Tasm. i. 181, 182; *V. triloba*, *V. cuneata*, and *V. scabra*, A. Gray in Proc. Amer. Acad. v. 118; *Diplopappus australasicus*, Turcz. in Bull. Mosc. 1851, i. 171.

Queensland. Bustard Bay and Bay of Inlets, *Banks and Solander;* Keppel Bay, *Thozet;* Port Curtis, *M'Gillivray* (the achenes in these specimens not so prominently striate but very different from those of *V. scabra*).

N. S. Wales. Port Jackson and Hunter's River, *R. Brown;* Clarence river, *Beckler;* Lachlan river, *A. Cunningham;* Darling and Murray desert, and towards the Barrier Range, *Victorian and other Expeditions.*

Victoria. Common in dry open declivities from Lake King, in Gipps' Land, to the western frontier, in the Murray desert, Wimmera, etc., *F. Mueller* and others.

Tasmania. Derwent river, *R. Brown;* in dry stony places, frequent, *J. D. Hooker* and others.

S. Australia. From the Murray river to St. Vincent's and Spencer's gulfs, *F. Mueller* and others.

W. Australia. From the S. coast to Swan River, *Drummond, n.* 35, 36, 87, 386; *4th Coll. n.* 218; *5th Coll. n.* 373; *Preiss, n.* 102 *and* 104; Murchison river, *Oldfield.*

The species is also in New Zealand. The specimens from that country, like some of the Victorian ones, have all the leaves short, obovate or spathulate; from this there is a very gradual passage to the very narrow linear form of some of the Victorian and N. S. Wales specimens. The indumentum is also very variable, sometimes soft, woolly or almost silky, sometimes all scabrous and rigid. The following forms appear to be more distinct, and possibly, when better known, may be regarded as species. On the other hand, *Microgyne trifurcata*, Less., from S. America, differs but very slightly in the more villous achenes and narrow-lobed leaves:

Var. *dissecta.* Leaves often twice 3-lobed.—Port Jackson, *R. Brown;* New England, *C. Stuart;* Bent's Basin, *Woolls;* Upper Bogan and Lachlan rivers, *L. Morton;* Yarra river, *F. Mueller.*

Var. *tenuissima.* Leaves linear-subulate. Flower-heads small.—Port Jackson, *R. Brown, Woolls;* Burnett river, *F. Mueller* (heads very young, and the identity doubtful).

Var. *pterochæta*, F. Muell. Achenes with very fine scarcely conspicuous striæ. Pappus bristles almost plumose.—Castlereagh river, *C. Moore;* Darling desert, *Victorian Expedition.*

Var. *megacephala*, F. Muell. Flower-heads large. Ray-florets longer than the pappus. —Spencer's Gulf, *F. Mueller;* Tasmania, *Gunn.*

Section II: Eurybiopsis, *A. Gray.*—Achenes very flat, the margins slightly thickened, without any ribs on the faces, or very rarely one short obscure rib.

3. **V. scabra,** *DC. Prod.* v. 281. A rigid herb of 1 to 2 ft. with erect branches, or rarely small, somewhat woody at the base and divaricately

branched, scabrous-pubescent or hirsute. Leaves linear-oblong or cuneate, often above 1 in. long, and when broad stem-clasping at the base, entire or with a few coarse obtuse teeth. Flower-heads on peduncles longer than the leaves, usually forming a terminal corymb. Involucre almost hemispherical, the bracts numerous, narrow, acute or almost obtuse, the inner ones scarcely 3 lines long, the outer ones shorter. Ray-florets numerous, in some specimens all exceedingly narrow, almost filiform, and scarcely exceeding the pappus, in others rather broader and longer; disk-florets less numerous, at first but little longer than the involucre, but the ripe pappus much exceeding it. Achenes very flat, with thickened margins and no longitudinal ribs on the faces, sprinkled with appressed hairs. Pappus bristles unequal, scabrous. *V. hispidula*, F. Muell.; A. Gray in Proc. Amer. Acad. v. 118; *Erigeron Vittadinia*, F. Muell. Fragm. v. 87.

Queensland, *R. Brown;* Bustard Bay and Bay of Inlets, *Banks and Solander;* rocky hills, Cleveland and Rodd's Bays, *A. Cunningham;* Wide Bay, *Leichhardt;* Gilbert river and Peak Downs, *F. Mueller;* flats on the Maranoa, *Mitchell;* Curtis Island, *Henne.*

N. S. Wales. Port Jackson and Paterson's River, *R. Brown.*

4. **V. macrorrhiza,** *A. Gray in Proc. Amer. Acad.* v. 118. A perennial with a thick woody stock and slender erect not much branched stems of about 6 in., or 1 ft. when luxuriant, glabrous or scabrous-pubescent. Leaves linear, or the lower ones linear-cuneate, $\frac{1}{2}$ in. long or less, or rarely nearly 1 in., the upper ones small and distant. Flower-heads on long terminal peduncles, solitary or very loosely corymbose. Involucre hemispherical, the bracts narrow and acutely acuminate, the inner ones 3 lines long. Ray-florets not quite so numerous as in *V. scabra*, narrow but longer than the pappus, although usually revolute so as to appear shorter; disk-florets fewer, longer than the involucre. Achenes very flat, with thickened margins, without any or with a single short rib on their faces, sprinkled with appressed hairs. Pappus rather unequal.—*Eurybiopsis macrorrhiza*, DC. Prod. v. 260.

N. Australia. Brunswick Bay and Prince Regent's Harbour, N.W. coast, *A. Cunningham;* Providence Hill, *F. Mueller;* Port Essington, *Armstrong;* islands of the Gulf of Carpentaria, *R. Brown.*

Queensland. Broad Sound, *R. Brown,* apparently the same species, although with rather longer leaves.

14. PODOCOMA, Less.; R. Br.

(Podopappus, *Hook. et Arn.;* Asteropsis, *Less.?;* Ixiochlamys, *F. Muell.*).

Involucre broadly ovoid or hemispherical, the bracts imbricate in several rows, narrow, acute. Receptacle without scales. Florets of the ray female numerous, crowded in several rows, ligulate but very narrow. Disk-florets few, hermaphrodite, tubular but slender, usually 5-lobed. Anthers obtuse at the base. Style-lobes somewhat flattened, with narrow tips or appendages sometimes almost subulate. Achenes short, flat, produced into a long or short slender beak. Pappus of numerous capillary bristles.—Perennial herb. Leaves alternate. Flower-heads large, terminal, solitary, or very loosely corymbose.

Besides the Australian species, which is endemic, there are three from extratropical South America. Notwithstanding the confusion arising from Lessing's having described the ray-florets as 1-seriate, there seems little reason to doubt that *Erigeron hieracifolium*, Poir. (or

Podocoma hieracifolia and *P. primulifolia*, Cass.), is identical with *Podopappus hirsutus*, Hook. et Arn., and *Asteropsis macrocephala*, Less., is most probably the same as *Podopappus tomentosus*, Hook. et Arn. The genus only differs from *Vittadinia* in the beaked achene, the length of the beak varying even in the same species.

1. **P. cuneifolia,** *R. Br. App. Sturt Exped.* 17. Stems much branched and almost woody at the base, with ascending leafy branches rarely above 4 or 5 in. high, without the peduncles. Leaves crowded, oblong-cuneate or almost linear, acutely toothed or lobed at the end, narrowed into a petiole, mostly above 1 in. long, ciliate as well as the petiole and stem with long rigid white hairs. Peduncles much longer than the leaves. Involucral bracts numerous, linear-lanceolate, acute, the inner ones ½ in. long and coloured at the tips, the outer ones shorter, more or less glandular-pubescent as well as the peduncles. Ray-florets almost filiform, scarcely exceeding the pappus; disk-florets about as long as the involucre. Achenes small, obovate, glabrous or silky-hairy, the filiform beak three or four times as long as the achene itself. Pappus fine and white.—*Ixiochlamys cuneifolia*, F. Muell. and Sond. in Linnæa, xxv. 466.

N. Australia. Nichol Bay, N.W. coast, *F. Gregory's Expedition.*
N. S. Wales. Mount Goningberi, *Victorian Expedition.*
S. Australia. Dry river-beds, Cudnaka, Arkaba, etc., *F. Mueller;* in the N. interior, *M'Douall Stuart's Expedition.*

15. ERIGERON, Linn.

Involucre from ovoid to hemispherical, the bracts numerous, narrow, nearly equal or imbricate in several rows. Receptacle flat or slightly convex, without scales. Ray-florets female, numerous, in 2 or more rows either all ligulate or very narrow, or the inner ones shorter and filiform. Disk-florets few or numerous, hermaphrodite, tubular, 5-toothed. Anthers obtuse at the base. Style-branches narrow, somewhat flattened, with lanceolate tips or appendages papillose outside. Achenes flattened, the margins usually thickened. Pappus of copious capillary nearly equal bristles.—Herbs. Leaves alternate or radical. Flower-heads solitary corymbose or paniculate. Ray-florets white pink or purplish.

A large genus, ranging over the greater part of the globe, but chiefly in the temperate regions of the northern hemisphere, or in mountainous tropical regions. Of the five Australian species, two, both probably of American origin, are common tropical weeds also in the Old World, the other three appear to be endemic. The genus is very closely allied to *Aster*, differing chiefly in the more numerous and narrower ray-florets, and even passes into it by almost insensible gradations among the American species, and on the other hand some species of the section *Cœnotus* might almost equally well be placed in *Conyza*.

SECTION I. **Euerigeron.**—*Female ray-florets all ligulate in many rows.*

Small tufted perennial. Leaves mostly radical. Scapes one-headed, with few small bract-like leaves or none . . . . 1. *E. pappochroma.*
Stems erect, branching, leafy, usually annual.
Flower-heads hemispherical, solitary or corymbose.
Leaves oblong or lanceolate. Pappus-bristles numerous, capillary . . . . 2. *E. ambiguus.*
Leaves narrow-linear. Pappus-bristles few, strongly barbellate, very caducous . . . . 3. *E. minurioides.*
Flower-heads very small, ovoid in an oblong panicle . . . . . . 4. *E. canadensis.*

SECTION II. **Cœnotus.**—*Inner rows or nearly all the female ray-florets filiform and tubular.*

Plant nearly glabrous. Flower-heads corymbose. Ligulate florets numerous, rather longer than the involucre . . . . 5. *E. conyzoides.*
Plant pubescent or hirsute. Flower-heads in a short panicle. Ray-florets nearly all filiform, the outer ones with a very minute ligula 6. *E. linifolius.*

SECTION I. EUERIGERON.—Female florets all ligulate, in several rows.

1. **E. pappochroma,** *Labill. Pl. Nov. Holl.* ii. 47. *t.* 193. Stock short, thick, simple or branched and tufted. Radical leaves spreading, from oblong-linear and nearly sessile to broadly obovate or spathulate, and narrowed into a long petiole, entire or remotely toothed, from under ½ in. to above 1 in. long. Stems or scapes simple, usually exceeding the leaves and sometimes above 6 in. long, with a few small narrow linear leaves or bracts, and a single terminal flower-head, the whole plant glabrous or more or less hirsute. Involucre hemispherical, the bracts linear-lanceolate, in about 2 rows, the inner ones 3 to 4 lines long. Ray-florets very numerous, the ligula very narrow, 1 to 1½ lines long; disk-florets not exceeding the involucre.—DC. Prod. v. 288; Hook. f. Fl. Tasm. i. 182; *E. phlogotrichus*, Spreng. Syst. iii. 520.

Victoria. Summits of the Australian Alps, *F. Mueller.*

Tasmania. Summit of Table Mountain, Derwent river, *R. Brown*, and of most of the higher mountains, descending to Recherche Bay, *J. D. Hooker*, etc.

The following varieties, all alpine, appear at first sight to be distinct species, but it is difficult to assign any precise limits to any of them :—

a. stellatus. Glabrous except a few cilia on the margin of the leaves. Stock often elongated. Leaves densely tufted, linear-cuneate, under ½ in. long, scarcely petiolate, coriaceous. Ray-florets fewer and longer than in the other varieties.—*Aplopappus stellatus*, Hook. f. in Hook. Lond. Journ. vi. 112; *Erigeron tasmanicus*, var., Hook. f. Fl. Tasm. t. 46 A (the left-hand figure).—Tasmania.

b. oblongatus. Glabrous or nearly so. Leaves oblong-spathulate or elliptical-oblong, narrowed into a long petiole, entire, coriaceous.—*Aplopappus tasmanicus*, Hook. f. in Hook. Lond. Journ. vi. 110; *Erigeron tasmanicus*, Hook. f. Fl. Tasm. i. 183. t. 46 A (the right-hand figure).—Tasmania.

c. Billardieri. Glabrous or nearly so. Leaves obovate-oblong or spathulate, narrowed into a long petiole, usually toothed, much thinner than in the last variety.—*Aplopappus pappochroma*, Hook. f. in Hook. Lond. Journ. vi. 111.—Tasmania.

d. Gunnii. Softly hirsute. Leaves obovate-oblong or spathulate, entire or toothed, not thick. Scapes long or short.—*Aplopappus Gunnii* and *A. bellidioides*, Hook. f. in Hook. Lond. Journ. vi. 111, 112; *Erigeron Gunnii*, Hook. f. Fl. Tasm. i. 183.—Tasmania and Victoria.

e. setosa. Leaves small, shortly petiolate, oblong or cuneate, entire, thick, hispid with rigid bristly hairs. Scapes very short.—Munyong Mountains, Victoria, *F. Mueller.*

2. **E. ambiguus,** *F. Muell. in Trans. Phil. Inst. Vict.* iii. 58. Stems several, erect or ascending, corymbosely branched, shortly pubescent and somewhat glandular, attaining about 1 ft. in height. Leaves oblong or lanceolate, entire or with a few coarse teeth. Flower-heads small, in a loose terminal corymb. Involucral bracts narrow-linear or subulate, acuminate, the inner ones about 2 lines long. Ray-florets very numerous and slender, but ligulate, slightly exceeding the involucre; disk-florets much less numerous, about as long as the involucre. Style and achenes of the genus.

Queensland. Gilbert river, *F. Mueller.*

3. **E.? minurioides,** *Benth.* Stems apparently spreading or decumbent, branched, leafy, glabrous or glandular-pubescent near the inflorescence. Leaves narrow-linear, mucronate, acute, bordered by a few minute mucronate or almost hair-like teeth, from under ½ in. to about ¾ in. long. Flower-heads solitary, terminating the rather numerous leafy branches. Involucre broadly hemispherical, about 3 lines diameter, the bracts narrow, acuminate, nearly equal, in 2 or 3 rows. Ray-florets numerous and narrow, but the ligulæ at least 2 lines long, the disk-florets very numerous. Achenes flat and obovate, as in the rest of the genus, but the pappus-bristles not numerous, exceedingly fragile and deciduous and strongly barbellate.

Victoria. Port Phillip, *F. Mueller.*—A very distinct plant, of doubtful affinity, with something of the habit of a *Minuria*, but more branched, and, notwithstanding the difference of the pappus, which is nearly that of *Gymnostephium*, appears to be best placed in *Erigeron*. I have thought, indeed, that it might have been some Cape *Gymnostephium*, allied to *G. gracile*, and accidentally introduced, but the disk-florets are certainly fertile, and the involucre and rays are those of *Erigeron*, and not of *Gymnostephium*.

* 4. **E. canadensis,** *Linn.; DC. Prod.* v. 289. An erect not much branched annual, of 1 to 3 ft., glabrous or hispid with short spreading hairs. Leaves linear, 1 to 3 in. long, entire or rarely with a few distant teeth. Flower-heads small and very numerous in a large oblong or rarely corymbose terminal panicle, the peduncles very slender. Involucre ovoid, nearly glabrous, the bracts narrow, acute, about 2 lines long. Ray-florets very numerous, slender, but ligulate, scarcely exceeding the involucre, white; disk-florets not so many.

N. S. Wales. Port Jackson, *Woolls*, probably introduced. The species, of American origin, is now common as a roadside weed in most tropical countries, as well as in a great part of Europe.

SECTION II. CŒNOTUS.—Inner rows of the female florets or nearly all filiform, shorter than the style and not expanded into a ligula.

5. **E. conyzoides,** *F. Muell. in Trans. Phil. Soc. Vict.* i. 105; *in Hook. Kew Journ.* viii. 146; *and Fragm.* v. 87. An erect annual, of 1½ to 2 ft., more corymbosely branched than *E. linifolius*, and quite glabrous. Stem leaves linear or lanceolate, often 2 to 3 in. long, quite entire, the radical and lower leaves longer and broader, entire or remotely toothed, narrowed into a long petiole. Flower-heads larger than in *E. linifolius*, forming a terminal corymbose panicle. Involucre hemispherical, the bracts linear-subulate, pointed, in several rows, the inner ones above 3 lines long. Ray-florets exceedingly numerous, the outer 1 or 2 rows ligulate, but so narrow as to be almost filiform, exceeding the pappus, the inner rows filiform, tubular, and shorter; disk-florets very few. Style-appendages short. Achenes small, flat.

N. S. Wales. Tributaries of the Clarence river, *Herb. F. Mueller.*
Victoria. Sources of the Murray and Snowy rivers, at an elevation of 4000 to 5000 ft., *F. Mueller.*

6. **E. linifolius,** *Willd. Spec. Pl.* iii. 1955. A coarse erect annual, 1 to

2 ft. high or rather more, clothed with long soft hairs, or more shortly scabrous-pubescent. Radical leaves petiolate, oblong, often coarsely toothed or pinnatifid; stem leaves sessile, linear, entire or occasionally remotely toothed, often above 2 in. long. Flower-heads rather small, pedunculate, more or less paniculate. Involucre broadly ovoid or almost hemispherical, the bracts narrow, acute, in 2 or 3 series. Female florets very numerous, filiform, not so long as the pappus, the outer ones usually dilated at the tip into a minute ligula, the others all tubular; disk-florets few. Style-appendages short. Achenes small, flat, pubescent.—*Conyza ambigua*, DC. Prod. v. 381; Sond. in Linnæa, xxv. 481; *Erigeron ambiguus*, Sch. Bip. in Phyt. Canar. ii. 208.

Queensland. Brisbane river, Moreton Bay, *F. Mueller*; Rockhampton, a troublesome weed, *Thozet.*

N. S. Wales. Port Jackson, *Woolls*, *Backhouse;* Clarence river, *Beckler.*

Victoria. About Melbourne, *Robertson.*

S. Australia. Near Adelaide, and other places about St. Vincent's Gulf, *Behr*, *F. Mueller*, and others.

W. Australia, *Drummond*, *n.* 130.

A common tropical weed, found also in Europe. Some of the above enumerated specimens may belong to *E. albidus*, A. Gray in Proc. Amer. Acad. v. 319 (*Conyza albida*, Willd.), another tropical weed, which, together with other modern botanists, I had set down as *E. bonariensis*, Linn., an error first pointed out by A. Gray. This *E. albidus* has rather smaller flower-heads than *E. linifolius*, and the ligules of the ray rather more developed, approaching nearer to *E. canadensis*, but I now find it scarcely possible clearly to distinguish the two, and, on the other hand, some specimens appear to show no ligula at all, passing as it were into the genus *Conyza.*

16. CONYZA, Linn.

Involucral bracts numerous, narrow, nearly equal or imbricate, in several rows. Receptacle flat or slightly convex, without scales. Ray-florets female, numerous, in several rows, all tubular, filiform, shorter than the involucre. Disk-florets few, hermaphrodite, tubular, 5-toothed. Anthers obtuse at the base. Style-branches narrow, somewhat flattened, with lanceolate tips or appendages papillose outside. Achenes small, flattened, the margins usually thickened. Pappus of copious capillary bristles.—Herbs. Leaves alternate, entire, lobed or dissected. Flower-heads usually paniculate.

The genus as above defined, and as understood by De Candolle, comprises a considerable number of species dispersed over the warmer parts of the globe. The Australian species are both of them widely spread over tropical Asia, and one of them is equally abundant in Africa. The genus is closely allied to the section *Cænotus* of *Erigeron*, differing in the total absence of any ligulate expansion of the ray-florets; the softer, more copious pappus, and the larger proportion of female florets to the hermaphrodite ones, give also to the flower-heads a somewhat different aspect. From *Blumea* (to which Schultz-Bipontinus and Miquel propose to transfer the name of *Conyza*) it differs in the want of tails to the anthers.

Tall branching, nearly glabrous, viscid plant. Leaves lanceolate, mostly entire . . . . 1. *C. viscidula.*
Hirsute annuals or biennials.
Leaves narrow, nearly entire . . . . *Erigeron linifolius.*
Leaves obovate or oblong, coarsely toothed or pinnatifid . . . . . . 2. *C. ægyptiaca.*

1. **C. viscidula,** *Wall.; DC. Prod.* v. 383. A tall, erect, branching

herb, more or less viscid-pubescent, especially the inflorescence. Lower leaves ovate, acuminate, often 3 to 4 in. long, upper ones smaller, ovate-lanceolate or lanceolate, all narrowed to the base but scarcely petiolate, slightly toothed or entire. Flower-heads numerous, rather small, clustered and corymbose on the lateral branches of a large terminal panicle. Involucral bracts nearly equal, about 2 lines long, the outer ones linear-lanceolate, acute, the inner narrower, more acuminate, and more scarious. Pappus slightly exceeding the involucre. Ray-florets exceedingly numerous, the style about as long as the pappus, the filiform corollas very much shorter. Disk-florets about 2 to 6.—*C. Wallichii*, DC. Prod. v. 384; *C. polycephala*, Edgew. in Trans. Linn. Soc. xx. 66.

Queensland. Shoalwater Bay, *R. Brown.*

N. S. Wales. Edge of the scrub, Richmond river, (*C. Moore?*) *in Herb. F. Mueller.*

The species is common in India. It has much the aspect of *Blumea balsamifera*, DC., but differs both in involucre and anthers.

2. **C. ægyptiaca,** *Ait.; DC. Prod.* v. 382. A coarse, erect, hirsute annual or biennial, sometimes 2 to 3 ft. high and nearly simple, except the terminal panicle, sometimes divaricately branched below the middle. Leaves lanceolate or oblong, obtuse or rarely almost acute, coarsely toothed in their whole length or at the base only, or pinnatifid with ovate oblong or rarely linear lobes. Flower-heads rather large for the genus, shortly pedicellate, in dense cymes or clusters, forming a terminal corymbose panicle. Involucral bracts narrow, subulate-acuminate, the inner ones above 3 lines long. Florets and pappus not exceeding the involucre. Ray-florets exceedingly numerous, all filiform, but not so short as in *C. viscidula;* disk-florets also numerous, but varying in different heads.—*C. lineariloba*, DC. Prod. v. 385.

Queensland. Northumberland Islands, Lizard Island, Broad Sound, *R. Brown;* Port Molle, *M'Gillivray;* Burnett and Burdekin rivers, *F. Mueller;* Rockingham Bay and Rockhampton, *Dallachy;* Brisbane river, *Fraser;* Keppel Bay, *Thozet;* also from *Leichhardt's Collection.*

The species is common in tropical and subtropical Asia and Africa. Most of the Australian specimens, like some from Amoy (*Hance*), and the majority of the Mauritius ones I have seen, belong to a variety with the leaves more decidedly pinnatifid than they are usually in the Egyptian and Indian specimens; but in some of the Australian specimens the leaves are toothed only as in the Egyptian and Indian ones, and one gathered on the Nile by Speke and Grant has them precisely like the common Australian form.

17. MINURIA, DC.

(Therogeron, *DC.;* Elachothamnus, *DC.;* Kippistia, *F. Muell.*)

Involucre ovoid or hemispherical, the bracts narrow, in few rows, dry or scarious on the margin. Receptacle without scales. Florets of the ray female, numerous, in several rows, ligulate but very narrow and sometimes short. Disk-florets hermaphrodite, but sterile, numerous or few, tubular, dilated upwards, usually 5-toothed. Anthers obtuse at the base. Style-lobes with obtuse acute or somewhat elongated tips, papillose on the back or the style simple and semiabortive. Achenes of the ray flattened, with thickened margins, obovate or narrow, those of the disk slender and abortive. Pappus of the ray of numerous capillary bristles, that of the disk variously reduced or more paleaceous.—Undershrubs or shrubs, glabrous or the young branches woolly or pubescent. Leaves alternate, narrow, entire or toothed. Flower-heads hemispherical or broadly ovoid, pedunculate, solitary or corymbose. Ray-florets usually white.

The genus is confined to Australia. It is allied to *Erigeron*, differing chiefly in the abortive disk-achenes, with a reduced or altered pappus.

Involucral bracts oblong-linear with scarious ciliate margins. Ray-achenes very densely silky-hairy . . . . 1. *M. leptophylla.*
Involucral bracts narrow, acute. Ray-achenes glabrous or slightly pubescent.
Involucre about 3 lines long. Disk-pappus of 5 to 10 bristles, with several short ones . . . . 2. *M. Cunninghamii.*
Involucre not 2 lines long. Disk-pappus of fewer bristles than the ray, with several or scarcely any short ones.
Glaucous and glabrous. Leaves lanceolate or linear, very acute . . . . 3. *M. integerrima.*
Young shoots woolly. Leaves linear, obtuse, entire or toothed 4. *M. denticulata.*
Involucral bracts oblong-linear, ciliate, about 2 lines long. Ray-achenes glabrous. Disk-pappus united in a tube . . . . . 5. *M. suædifolia.*

1. **M. leptophylla,** *DC. Prod.* v. 298. An undershrub or small shrub, with numerous erect or ascending branches, under 1 ft. and often under 6 in. high, glabrous or pubescent in the upper portion. Leaves narrow-linear, almost filiform, mucronate-acute or almost obtuse, $\frac{1}{4}$ to $\frac{1}{2}$ in. long in some specimens, $\frac{1}{2}$ to 1 in. in others, the upper ones small and few. Flower-heads terminal, pedunculate above the last leaves. Involucre hemispherical, the bracts not numerous, in 2 or 3 rows, with scarious ciliate margins, the inner ones 2 to $2\frac{1}{2}$ lines long, a few of the outer ones smaller. Ray-florets 20 to 30, with oblong-linear ligulæ. Disk-florets about as long as the involucre. Achenes of the ray so densely covered with long silky hairs as to conceal their form, with a pappus of numerous nearly equal bristles; achenes of the disk abortive, terete, nearly glabrous, most of the pappus bristles very short and slightly flattened at the base, a few only elongated and usually thickened or plumose at the end.—*M. tenuissima*, DC. Prod. v. 298; *M. asteroidea*, Sond. in Linnæa, xxv. 467.

N. S. Wales. Lachlan river, *A. Cunningham*; Macquarrie river, *Mitchell*; Lachlan, Darling, and Murray rivers, to the Barrier range, *Victorian and other Expeditions.*
Victoria. Wimmera, *Dallachy.*
S. Australia. Dry grassy places and sandy plains, from the Murray to St. Vincent's Gulf, *F. Mueller* and others; Lake Gillies, *Burkitt.*
W. Australia, *Drummond*, 5*th Coll. n.* 372.

Var.? *hispida.* The whole plant hirsute with short spreading hairs.
Queensland. Rockingham Bay, *Dallachy*, a single slender specimen in Herb. F. Mueller, with a single flower-head, insufficient for examination, and therefore the determination doubtful.

2. **M. Cunninghamii,** *Benth.* A bushy shrub or undershrub, with short slender branchlets, quite glabrous. Leaves narrow-linear, entire, the longer ones acute and above $\frac{1}{2}$ in. long, the smaller ones often obtuse, thick and almost fleshy. Peduncles terminal, longer than the leaves. Involucre

ovoid or almost hemispherical, about 3 lines long, the bracts narrow, acute. Ray-florets numerous, narrow, longer than the pappus; disk-florets usually few. Achenes of the ray glabrous or slightly pubescent, narrow, but flat, with a pappus of numerous very fine white capillary bristles; achenes of the disk nearly terete and abortive, the bristles all, except 5 to 10, very short.—*Elachothamnus Cunninghamii*, DC. Prod. v. 398; F. Muell. Pl. Vict. t. 34; *Eurybiopsis intricata*, F. Muell. in Linnæa, xxv. 396; *Therogeron tenuifolius*, Sond. in Linnæa, xxv. 467.

N. S. Wales. Swampy situations, Lachlan river, *A. Cunningham*; Yayinga mountains, Darling river, *Victorian Expedition*; on the Murray, towards the Murrumbidgee, *F. Mueller*.

S. Australia. Stony shady places, Cudnaka, *F. Mueller*.

3. **M. integerrima,** *Benth.* Perfectly glabrous, smooth and somewhat glaucous. Stems, from a perennial often woody base, erect, rigid, often corymbosely branched, rarely above 1 ft. high. Leaves lanceolate or linear, acute, quite entire, mostly ½ to 1 in. or rarely 2 in. long, the upper ones few and small. Flower-heads small, terminal. Involucre broadly hemispherical, 2 to 3 lines diameter, with numerous narrow bracts, scarious on the edges. Ray-florets exceedingly numerous, in many rows, small, narrow, white; disk-florets sometimes only 2 or 3, sometimes numerous. Achenes of the ray small, flat, the pappus-bristles fine, white and not very numerous; achenes of the disk abortive, with still fewer pappus-bristles, occasionally accompanied by a few short ones.—*Therogeron integerrimus*, DC. Prod. v. 283.

N. Australia. Hooker's and Sturt's Creeks, *F. Mueller*.

Queensland. Burdekin river, *F. Mueller*; Condamine river, *Leichhardt*; Cape river, *Bowman*.

N. S. Wales. Lachlan river, *A. Cunningham*; Bogan river, *Mitchell*; Murray and Darling deserts, and thence to the Barrier Range, *Victorian and other Expeditions*.

Victoria. Murray desert, *F. Mueller*.

S. Australia. Murray river, *F. Mueller*; Cooper's Creek, *Herb. F. Mueller*; N.W. interior, *M'Douall Stuart's Expedition*.

4. **M. denticulata,** *Benth.* Allied to *M. integerrima*, but more branching, with diffuse or ascending stems, rarely above 8 or 9 in. high, the young shoots and sometimes the whole plant clothed with a white woolly tomentum. Leaves linear or linear-oblong, obtuse, entire or remotely toothed. Flower-heads, florets, and achenes of *M. integerrima*, but the disk-pappus more generally accompanied by a few very short bristles slightly dilated at the base.—*Therogeron denticulatus*, DC. Prod. v. 283.

N. S. Wales. Arid plains of the interior, *Fraser*, *A. Cunningham*; Darling and Lachlan rivers, *Victorian Expedition*.

Victoria. Murray desert, *Dallachy*.

S. Australia. In the interior, *M'Douall Stuart's Expedition*.

F. Mueller proposes to unite this with *M. integerrima* as one species under the name of *Erigeron Candollei*.

5. **M. suædifolia,** *F. Muell.* under *Kippistia*. A small bushy shrub or undershrub, under 1 ft. and often under 6 in. high, glabrous and glaucous. Leaves narrow-linear, nearly terete, under ½ in. long, mostly recurved at the end. Flower-heads small, pedunculate. Involucre hemispherical, 2 to 3 lines diameter, the bracts oblong-linear, ciliolate, in few rows. Ray-florets

numerous, very little exceeding the involucre; disk-florets numerous, as long as the involucre. Achenes of the ray flat, with a pappus of rather numerous bristles, the outer ones free, the inner ones more or less connate; achenes of the disk abortive, the pappus united in a more or less toothed scarious tube. —*Kippistia suædifolia*, F. Muell. Rep. Babb. Exped. 12, and Pl. Vict. t. 35.

Victoria. Wimmera, *Dallachy.*

S. Australia. Coorong desert, *Irvine;* towards Spencer's Gulf, *Warburton;* Stuart's Creek, *Babbage's Expedition.*

18. CALOTIS, R. Br.

(Huenefeldia, *Walp.;* Goniopogon, *Turcz.;* Cheiroloma, *F. Muell.*)

Involucre usually hemispherical, the bracts in about 2 rows, nearly equal, broad or narrow, with dry or scarious margins and usually a few inner narrow bracts. Receptacle flat or convex, without scales. Florets of the ray female, ligulate, often numerous, but in a single row. Disk-florets numerous, apparently hermaphrodite but sterile (except in *C. hispidula*), tubular, 5-toothed. Anthers obtuse at the base. Style-branches in the disk-florets somewhat flattened or almost filiform, usually obtuse, papillose outside at the end. Fruiting-heads usually globular. Achenes of the ray flat, obovate or oblong. Pappus of 2 or more barbed bristles, sometimes all short, but more frequently 1, 2 or more growing out into rigid divaricate awns or spines, and accompanied sometimes by 2 or more truncate scales. Disk-achenes usually abortive.—Perennial herbs or rarely annuals, with the habit nearly of *Brachycome.* Leaves alternate, entire toothed or pinnately divided. Flower-heads pedunculate, the rays white, rarely blue or purple.

The genus is confined to Australia.

SECTION I. **Eucalotis.**—*Pappus consisting of rigid barbed awns and flat truncate scales. Perennials with leafy stems.*

Leaves linear-lanceolate or oblong, remotely toothed or pinnatifid.
 Scales of the pappus usually united in a cup; awns 1 or 2 . . . 1. *C. dentex.*
Leaves cuneate or spathulate, toothed at the end. Scales of the pappus 2 or 3, alternating with the awns. Plant not glandular . 2. *C. cuneifolia.*
Leaves oblong-cuneate, deeply toothed at the end. Scales of the pappus 3 to 6, alternating with the awns. Plant glandular-hirsute 3. *C. glandulosa.*

SECTION II. **Cymbaria.**—*Pappus consisting of 2 or more rigid awns, dilated and united at the base, without scales. Perennials with leafy stems, sometimes appearing annual.*

Awns of the pappus 2, almost boat-shaped at the base. Plant usually pubescent or hirsute . . . . 4. *C. cymbacantha.*
Awns of the pappus several, united in a glabrous cup at the base. Stems elongated, glabrous or nearly so . . . . 5. *C. erinacea.*

SECTION III. **Acantharia.**—*Pappus consisting of several unequal awns, all distinct and usually hispid at the base, sometimes all short.*

Achenes not winged. Perennials.
 Stock emitting stolons or creeping rhizomes and a tuft of radical leaves. Scapes simple or with very few heads.
  Radical leaves entire, toothed or lobed.
   Rhizome creeping. Plant usually hirsute. Radical and lower leaves mostly toothed. Fruiting heads ½ in. diameter 6. *C. scabiosifolia.*

Plant stoloniferous, usually glabrous. Leaves radical, narrow. Scapes almost leafless. Fruiting heads small . . . . . 7. *C. scapigera.*
Radical leaves pinnate with pinnatifid or linear segments. Involucral bracts few, broad . . . . 8. *C. anthemoides.*
Stems numerous, erect, several-headed, the radical leaves decayed before flowering. Flower-heads small. Involucral bracts narrow.
Upper leaves linear. Pappus longer than the achenes. Fruiting-heads nearly 3 lines diameter . . . . 9. *C. lappulacea.*
Upper leaves small, cuneate. Pappus shorter than the achenes. Fruiting-heads not 2 lines diameter . . . . 10. *C. microcephala.*
Achenes winged. Pappus short or of very fine awns. Perennial with slender branching stems . . . . 11. *C. breviseta.*
Small annuals.
Ray white. Achenes completely covered with long intricate or plumose hairs . . . . 12. *C. plumulifera.*
Ray purple. Achenes shortly hirsute with ciliate wings . . . 13. *C. porphyroglossa.*
Ray white. Achenes nearly glabrous, the wings very shortly ciliate . . . . 14. *C. pterosperma.*

Section IV. **Cheiroloma.**—*Ray-florets scarcely exceeding the pappus. Disk-florets fertile. Pappus consisting of several unequal awns, alternating with very short, entire or lobed bristles or scales.*

Annual. Achenes not winged . . . . 15. *C. hispidula.*

Section I. Eucalotis.—Pappus consisting of rigid barbed awns and flat truncate scales. Perennials with leafy stems.

1. **C. dentex,** *R. Br. in Bot. Reg. under n.* 504. A perennial, with nearly simple or branched erect or decumbent stems, attaining sometimes 2 or 3 ft., scabrous-pubescent or hispid. Leaves linear, lanceolate or oblong, usually acute, acutely and remotely toothed or pinnatifid, mostly 1 to 2 in. long, often dilated into stem-clasping auricles. Flower-heads pedunculate. Involucre hemispherical, scabrous-pubescent; bracts not numerous, ovate, 3- or 5-nerved, almost membranous, the narrow inner ones very few or none. Achenes of the ray flat, obovate, scabrous or almost muricate. Pappus of 2 or 3 broad truncate scales, almost united into a fringed cup, and 1 or more, frequently 2 rigid bristles or awns, barbed at the end only.—DC. Prod. v. 302.

Queensland. Burdekin river, *F. Mueller;* Brisbane river, Moreton Bay, *Leichhardt, W. Hill.*

N. S. Wales. Port Jackson to the Blue Mountains, *R. Brown* and others; New England, *C. Stuart;* Macleay and Clarence rivers, *Beckler;* head of the Gwydir, *Leichhardt.*

2. **C. cuneifolia,** *R. Br. in Bot. Reg. t.* 504. An erect or spreading branching perennial, rarely attaining 1 ft., more or less hoary, scabrous-pubescent or hirsute. Leaves oblong, cuneate or spathulate, sometimes almost orbicular, coarsely toothed, narrowed into a short or long petiole dilated into stem-clasping auricles. Flower-heads hemispherical, becoming globular when in fruit, on terminal peduncles. Involucral bracts oval, oblong or lanceolate, about 2 lines long. Ray-florets long and narrow. Achenes flat, short, obovate. Pappus of 3 rigid barbed bristles or awns, and 2 broad membranous truncate scales, quite distinct, and sometimes a third smaller

one. Disk-achenes abortive. Fruiting-heads forming a globular burr of 3 or 4 lines diameter.—DC. Prod. v. 302; *C. dilatata*, A. Cunn.; DC. l. c.

Queensland. Suttor river, *Thozet*; Burdekin river, *F. Mueller*.

N. S. Wales. Port Jackson to the Blue Mountains, *R. Brown* and others, and in the interior to the Lachlan, Darling and Murray rivers and on to the Barrier Range, *A. Cunningham*, *Victorian Expedition*, and others; New England, *C. Stuart*, said to be one of the worst burrs for sheep.

S. Australia. Murray river, *F. Mueller*.

3. **C. glandulosa,** *F. Muell. in Trans. Vict. Inst.* 1855, 129, *and in Hook. Kew Journ.* viii. 146. A perennial with a thick woody stock and diffuse or procumbent branching stems of 6 in. to 1 ft., the whole plant hirsute with short glandular hairs. Leaves petiolate, oblong-cuneate, deeply toothed towards the end, the upper ones narrower and sessile. Flower-heads rather large, with blue or whitish rays. Involucral bracts herbaceous, ovate-lanceolate or lanceolate, the largest fully 3 lines long. Achenes of the ray obovate, tuberculate or muricate, otherwise glabrous. Pappus of about 3 to 6 very unequal rigid awns, usually barbed at the end only, with as many obovate truncate scales.—*Huenefeldia coronopifolia*, Walp. in Linnæa, xiv. 307.

Victoria. Dry grassy ridges of the Snowy River and its tributaries towards Maneroo, *F. Mueller*, *Lhotzky*.

Section II. Cymbaria.—Pappus consisting of 2 or more rigid awns, dilated and united at the base, without scales. Perennials with leafy stems, sometimes appearing annual.

4. **C. cymbacantha,** *F. Muell. in Linnæa*, xxv. 400; *Pl. Vict. t.* 36. Apparently perennial, with ascending stems of about 1 ft., slightly scabrous-pubescent. Lower leaves linear-cuneate or oblong, coarsely toothed, 1 to 2 in. long, narrowed into a long petiole; upper ones smaller, sessile, linear or lanceolate, entire. Involucral bracts ovate. Ray-florets numerous, narrow. Achenes flat, obovate, crowned by 2 rigid divaricate scales, broad and concave, almost boat-shaped at the base, tapering into short rigid barbed awns, the summit of the achene convex within the scales.—Sond. in Linnæa, xxv. 469.

S. Australia. Sandy hills, Crystal Brook, *F. Mueller*; N.E. of Lake Gairdner, *Herb. F. Mueller*.

Var. *pumila*. Under 6 in. high and flowering the first year so as to appear annual, pubescent or hirsute. Flower-heads small. Scales of the pappus not so broad at the base, and tapering into longer, more densely barbed awns. .

N. S. Wales. Darling river to Cooper's Creek, *Dallachy*.

5. **C. erinacea,** *Steetz in Pl. Preiss.* i. 424. A glabrous often glaucous perennial, with erect or ascending rigid branching stems of 1 to 2 ft. Leaves not numerous, linear or linear-lanceolate, acute, entire, or the lower ones toothed. Flower-heads of *C. lappulacea*. Involucral bracts oblong or linear, slightly scabrous, 1 to $1\frac{1}{2}$ lines long. Ray-florets yellow (*F. Mueller*). Fruiting-heads 3 to 4 lines diameter. Achenes glabrous, smooth or nearly so, with 3 to 5 awns scarcely barbed and united at the base into a broad open almost cartilaginous cup, often as long as the achene, and the border

sinuate between the awns; the summit of the achene within the cup conical with a few minute bristles.

N. S. Wales. Between Darling river and Cooper's Creek, *Neilson*, *Wheeler*.
Victoria. Wimmera, *Dallachy*.
S. Australia. Spencer's Gulf, *R. Brown*; Holdfast Bay, Port Adelaide, Pfeiffer's Flat, Port Lincoln, Boston Point, *F. Mueller*; in the N.W. interior, *M'Douall Stuart's Expedition*.
W. Australia. Swan River, *Drummond*, *Preiss*, *n.* 2427; Gordon and Bowes rivers, *Oldfield*.

Var. *parviflora*. Leaves linear-cuneate, the larger ones acutely toothed from near the base. Flower-heads smaller. Pappus-bristles more numerous, the united base shorter.—Durandoo, *Victorian Expedition*.

Huenefeldia angustifolia, Walp. in Linnæa, xiv. 506, which I have not seen, is, from the description, most probably this species.

SECTION III. ACANTHARIA.—Pappus consisting of several unequal awns, all distinct and usually hispid at the base, sometimes all short.

6. **C. scabiosifolia,** *Sond. and Muell. in Linnæa*, xxv. 471. Hirsute, scabrous or nearly glabrous, the stock densely tufted, with creeping rhizomes or stolons. Radical leaves petiolate, obovate or oblong, coarsely toothed or pinnatifid or sometimes lyrate, often 2 to 3 in. long. Stems ascending or erect, ½ to 1 ft. high, simple or slightly branched, with few smaller more sessile and less divided leaves. Involucral bracts ovate. Ray-florets white or purplish. Achenes softly pubescent. Pappus of about 8 very rigid awns, 3 or 4 larger ones much thickened at the base and divaricate but very unequal, the alternating smaller ones sometimes minute or wanting.—*C. Muelleri*, Sond. in Linnæa, xxv. 470 (more glabrous with narrower leaves).

N. S. Wales. Darling river, *Victorian Expedition*.
Victoria. Avoca river and Geelong, *F. Mueller*; Wimmera, *Dallachy*.
S. Australia. Pastures, Walpena and Cudnaka, *F. Mueller*.

Var. *lasiocarpa*, F. Muell. Leaves more rigid, less toothed. Flower-heads and achenes larger.—Snowy and M'Alister rivers and Maneroo, *F. Mueller*.

Var. *pubescens*, F. Muell. Softly villous. Leaves cuneate, toothed at the end only. Flower-heads as in the last variety.—Mountains on the Mitta-Mitta river, *F. Mueller*.

Var. *integrifolia*, F. Muell. Nearly glabrous.. Radical leaves narrow, mostly linear, rigid, entire; stem leaves entire or toothed. Flower-heads large. Involucral bracts broad.—Blue Mountains, *A. Cunningham* and others; grassy mountains on the Macalister river and Black Forest, *F. Mueller*.

Var. *elongata*. An apparently etiolated form, the radical leaves 6 in. long or more, with a few remote teeth or lobes. Stems almost filiform, with 1 or 2 long 1-headed branches.—Port Phillip, *F. Mueller*.

Var. (?) *cuneata*, *F. Muell.* Radical leaves cuneate-oblong, more or less toothed. Stems elongated with a few oblong leaves and 1 to 3 flower-heads.

Queensland. Rockhampton and Keppel Bay, *Thozet*; Burdekin river and desert on the Suttor, *F. Mueller*.

7. **C. scapigera,** *Hook. in Mitch. Trop. Austr.* 75. A small tufted perennial emitting creeping stolons, glabrous or slightly hairy. Radical leaves linear or linear-lanceolate, entire or rarely remotely toothed, narrowed into a petiole of 1 to 3 in. Flower-stems or scapes simple, longer than the leaves, bearing a few small narrow leaves and a single head very much smaller than in *C. scabiosifolia*. Involucral bracts broadly-oblong. Ray-florets

small, whitish. Fruiting-heads about 4 lines diameter. Achenes flat, tubercular or muricate, with 3 to 5 divaricate straight or hooked awns, obversely hirsute and very hairy at the base, and 2 to 5 small soft erect very hairy awns.

N. Australia. Northern base of Newcastle Range, *F. Mueller.* (Rather uncertain, the specimens not good.)

N. S. Wales. Macquarrie river, *Mitchell;* Darling river, *Victorian and other Expeditions.*

Victoria. Murray river, *F. Mueller;* near Lake Hindmarsh, *Werth.*

S. Australia. Murray river, rare, *Wood.*

8. **C. anthemoides,** *F. Muell. in Trans. Phil. Soc. Vict.* i. 44, *and in Hook. Kew Journ.* viii. 147. A tufted perennial, emitting creeping stolons and quite glabrous. Radical leaves on long petioles, with numerous linear pinnatifid or entire segments, often 3 to 4 in. long. Flowering-stems or scapes slender, usually simple, exceeding the leaves, with a few small distant, sessile, entire, lanceolate, leafy bracts. Involucral bracts few, broadly ovate or orbicular, glabrous except the slightly ciliate margins. Ray-florets whitish. Achenes flat, obovate, nearly glabrous. Pappus of 6 to 8 rather short unequal awns, hispid at the base, but not seen ripe.

Victoria. Muddy localities near Station Peak, *F. Mueller;* Skipton, *Whan.* In the flower-head I examined I found the disk-achenes abortive, as usual in the genus.

9. **C. lappulacea,** *Benth. in Hueg. Enum.* 60. A perennial, sometimes almost woody at the base, with numerous erect or ascending slender branching stems of ½ to 1 ft., more or less hirsute especially in the lower part, rarely entirely glabrous. Lower leaves oblong-cuneate and often toothed or lobed, upper ones linear and entire, sometimes all under ¼ in., sometimes those in the middle of the stems ½ in. long. Flower-heads small, when in fruit scarcely 3 lines diameter. Involucral bracts linear-lanceolate or oblong, sometimes slightly cuneate, ciliate and hispid. Ray-florets yellow (*F. Mueller*), small and narrow. Achenes muricate. Pappus of 1 to 4 rigid barbed awns about 1 line long and 1 or more very short ones, all hirsute at the base, the total number varying from 4 to 8.—Sond. in Linnæa, xxv. 470.

Queensland. Broad Sound, *R. Brown;* Condamine river, *Leichhardt;* Burdekin river and Suttor desert, *F. Mueller;* Rockhampton, *Dallachy;* Maranoa and Belyando rivers, *Mitchell;* Moreton Bay, *C. Stuart.*

N. S. Wales. Port Jackson to the Blue Mountains, *R. Brown;* Lachlan river and Bathurst, *A. Cunningham;* Glendon, *Leichhardt;* New England, *C. Stuart;* Darling and Murray rivers, *Dallachy.*

Victoria. Snowy River, *F. Mueller.*

S. Australia. Flinders Range and Torrens river, *F. Mueller.*

W. Australia. *Burges* in Herb. Hooker.

C. polyseta, Sond. in Linnæa, xxv. 470, from Cudnaka, appears to be a very slight variety of *C. lappulacea,* with rather larger flower-heads.

10. **C. microcephala,** *Benth.* An erect much-branched undershrub not exceeding 6 to 8 in., more or less hoary-hirsute. Leaves oblong-cuneate, obtuse, entire or toothed, under ½ in. long, those of the branchlets much smaller. Flower-heads scarcely 1½ lines diameter. Involucral bracts narrow-oblong, obtuse. Fruiting-heads scarcely more than hemispherical,

and not 2 lines diameter. Achenes hirsute, obovate, flat but not winged. Pappus of 6 to 8 barbed awns, nearly equal, and all shorter than the achene.

N. S. Wales. Murray and Darling rivers, *Herb. F. Mueller.*

11. **C. breviseta,** *Benth. in Hueg. Enum.* 60. An erect perennial of ½ to 1 ft., or sometimes more, diffuse with slender stems of 1 to 2 ft. Leaves linear, obtuse, entire, or some of the lower ones linear-cuneate and coarsely toothed. Flower-heads very small, on slender peduncles. Involucral bracts numerous, linear, mostly acute, scarcely 1 line long. Ray-florets white. Achenes ovate, shortly hispid, not 1 line long, bordered by a narrow flat edge almost expanded into a wing. Pappus of 5 to 10 or even more little rigid barbed awns, usually much shorter than the achene.—*C. tropica*, F. Muell. in Trans. Phil. Inst. Vict. iii. 58.

N. Australia. Upper Victoria river and barren plains, Fitzmaurice river, *F. Mueller;* Albert river, *Henne.*

12. **C. plumulifera,** *F. Muell. in Trans. Phil. Inst. Vict.* iii. 57. An annual with erect or ascending stems, rarely 6 in. high, glabrous or hirsute. Leaves oblong-lanceolate or cuneate, entire or with a few coarse teeth. Involucres 2 to 2½ lines diameter, the bracts oblong, obtuse or almost acute. Ray-florets very numerous, white. Fruiting-heads about 3 lines diameter. Achenes bordered by densely ciliate wings, and so covered by long mostly plumose hairs as to conceal their form. Pappus of several fine barbed awns, unequal, but mostly about the length of the achene.—*Goniopogon multicaule*, Turcz. in Bull. Mosc. 1851, i. 174. t. 2.

N. S. Wales. Plains of the Murray, *F. Mueller;* from the Darling to the Barrier Range, *Victorian Expedition.*
S. Australia. Cooper's Creek, *A. C. Gregory.*
W. Australia, *Drummond, 3rd Coll. n.* 97, *4th Coll. n.* 215; Irvine river, *Oldfield.*

13. **C. porphyroglossa,** *F. Muell. Herb.* A hispid annual, with erect or ascending stems of about 4 to 5 in. Leaves cuneate or the lower ones petiolate and spathulate, deeply toothed or almost lobed, the upper ones smaller and narrow. Involucres 2 to 2½ lines diameter, the bracts narrow and acute. Ray-florets purple, numerous, very narrow. Fruiting-heads about 4 lines diameter. Achenes shortly hirsute, bordered by rather broad wings, densely ciliate on the edge, but without the long plumose hairs of *C. plumulifera.* Pappus of numerous barbed awns, unequal but all shorter than the achene.

S. Australia. Cooper's Creek, *Murray.* Possibly, according to F. Mueller, a variety only of *C. plumulifera.*

14. **C. pterosperma,** *R. Br. ms.* An erect annual of about 1 ft., pubescent or hirsute when young, at length nearly glabrous. Lower leaves on the young plant obovate or petiolate and spathulate, coarsely toothed, on the older plant all linear and entire or rather broader and toothed at the end. Involucral bracts narrow, acute. Ray-florets numerous, white. Fruiting-heads scarcely 3 lines diameter. Achenes flat, sprinkled with a few short hairs, bordered by a very shortly ciliate wing; pappus of about 8 to 10 very short awns.

N. Australia. Islands of the Gulf of Carpentaria, *R. Brown.*

SECTION IV. CHEIROLOMA.—Ray-florets scarcely exceeding the pappus. Disk-florets fertile. Pappus consisting of several unequal awns, alternating with very short entire or lobed bristles or scales.

15. **C. hispidula,** *F. Muell. in Trans. Vict. Inst.* 1855, 130. A hispid annual, with procumbent or rarely erect branching stems of 3 to 6 in. Lower and radical leaves petiolate, obovate spathulate or cuneate, upper ones oblong-lanceolate, toothed towards the end or entire. Peduncle short. Involucral bracts ovate-lanceolate or oblong, hispid or almost muricate. Ray-florets few and very small, the ligula scarcely exceeding the pappus; disk-florets numerous, also small and all fertile. Fruiting-heads 3 to 4 lines diameter. Achenes quite similar in the ray and in the disk, flattened with thick obtuse margins, slightly hispid; pappus of about 4 to 6 rigid divaricate more or less barbed unequal bristles, alternating with as many much shorter bristles or scales either subulate and entire or palmately 3-fid or sometimes spathulate, and all hispid.—*Cheiroloma hispidulum*, F. Muell. in Linnæa, xxv. 401; Sonder in Linnæa, xxv. 473.

N. S. Wales. Molle's Plains, *A. Cunningham;* Upper Bogan and Lachlan rivers, *L. Morton;* Darling river to the Barrier Range, *Victorian Expedition;* between Stokes Range and Cooper's Creek, *Wheeler.*

Victoria. Wimmera, *Dallachy.*

S. Australia. Crystal Brook and Cudnaka, *F. Mueller.*

W. Australia. Swan River, *Drummond,* 1*st Coll. and n.* 375; Champion Bay, *Walcott.*

19. LAGENOPHORA, Cass.

(Ixauchenus, *Cass.;* Solenogyne, *Cass.;* Emphysopus, *Hook. f.*)

Involucre nearly hemispherical, the bracts in about 2 rows, nearly equal, broad or narrow, with dry or scarious margins. Receptacle convex, without scales. Florets of the ray numerous, female, ligulate or short and tubular. Disk-florets numerous, hermaphrodite, tubular, with a more or less dilated limb, 5-toothed. Anthers obtuse at the base. Style-branches in the disk-florets somewhat flattened, but long and slender, papillose outside at least in the upper portion. Achenes compressed, abruptly contracted at the top either very shortly or into a distinct neck. Pappus none.—Small perennial herbs, with a tufted stock, radical leaves and leafless simple scapes, or rarely the scapes growing out into simple stems decumbent and leafy at the base. Flower-heads terminal, the ray white or purplish.

A small genus, chiefly Australian, but extending also to New Zealand and many parts of tropical Asia. Of the four Australian species, one appears to be the same as the Asiatic one, the other three are endemic.

Ray-florets ligulate, linear, longer than the involucre and spreading.
 Flower-heads without the ray not above 4 lines diameter. Involucral bracts narrow . . . . 1. *L. Billardieri.*
 Flower-heads without the ray nearly 6 lines diameter. Involucral bracts broad . . . . 2. *L. Huegelii.*
Ray-florets not exceeding those of the disk.
 Scapes slender, mostly much longer than the leaves. Ray-florets all tubular, 3-toothed . . . . 3. *L. solenogyne.*

Scapes thickened, shorter or rarely longer than the leaves. Ray-florets minute, mostly opening out into a short concave ligula . . . . 4. *L. emphysopus.*

1. **L. Billardieri,** *Cass., DC. Prod.* v. 307. A perennial with a short thick stock and slender creeping rhizomes, otherwise stemless with radical leaves and scapes, or the stems shortly decumbent and leafy at the base. Leaves from obovate to cuneate-oblong, obtuse, irregularly toothed or shortly lobed, narrowed into a petiole, usually all under 2 in. long, rarely above 3 in. Scapes slender, simple, from 2 or 3 in. to nearly a foot long. Involucre from under 3 to nearly 4 lines diameter, the bracts rather numerous, linear or oblong, acute or rather obtuse. Ray-florets blue, ligulate, exceeding the involucre. Achenes at least as long as the involucre, the margins usually glandular-pubescent, contracted at the base and abruptly contracted at the top into a neck sometimes as long as the breadth of the achene, sometimes very short.—Hook. f. Fl. Tasm. i. 188; *Bellis stipitata*, Labill. Pl. Nov. Holl. ii. 55. t. 205; *Ixauchenus sublyratus*, Cass.; DC. Prod. v. 308; *Brachycome pumila*, Walp. Rep. ii. 584, according to Steetz in Pl. Preiss. i. 428; *Lagenophora gracilis*, Steetz in Pl. Preiss. i. 431.

Queensland. Brisbane river, Moreton Bay, *F. Mueller, Leichhardt.*

N. S. Wales. Port Jackson to the Blue Mountains, *R. Brown, Sieber, n.* 505, and others; Clarence, Macleay, and Hastings river, *Beckler.*

Victoria. In marshy places and common along streams in subalpine situations throughout Gipps' Land, and thence to the Grampians, the Yarra and Glenelg rivers, *F. Mueller, Robertson,* and others.

Tasmania. Port Dalrymple, *R. Brown;* abundant throughout the island, *J. D. Hooker.*

S. Australia. Mount Gambier, Rivoli Bay, *F. Mueller.*

W. Australia. King George's Sound, *Menzies;* Swan River, *Drummond, 1st Coll.;* Gordon river, *Oldfield.*

The species is also in Ceylon, Khasya, the Indian Archipelago, and S. China. There are generally two distinguishable varieties:—1. *microcephala,* glabrous or hirsute, the flower-heads without the rays scarcely 3 lines diameter and the rays short, the most common tropical and subtropical form, and, 2. *normalis,* usually hirsute, the flower-heads without the rays about 4 lines diameter and the rays rather longer; this is most abundant in the southern districts. Labillardière's own specimens are almost intermediate between the two.

Of *L. montana,* Hook. f. in Hook. Lond. Journ. vi. 113, and Fl. Tasm. i. 189, the original specimens, with small narrow nearly glabrous leaves, are intermediate in the size of the flower-heads, but many specimens, especially from Victoria, pass gradually into the normal form. *L. latifolia,* Hook. f. in Hook. Lond. Journ. vi. 113, and Fl. Tasm. i. 189. t. 49 A, with small obovate hirsute leaves, passes also gradually into the normal form. In R. Brown's herbarium is a remarkable variety from Port Phillip, with elongated leafy stems and the involucral bracts very narrow, almost setaceous.

2. **L. Huegelii,** *Benth. in Hueg. Enum.* 59. Very near *L. Billardieri,* but a larger and coarser plant, hirsute or sometimes nearly glabrous. Leaves oblong or obovate-oblong, sinuate-toothed or almost pinnatifid, 2 to 4 in. long. Scapes ½ to 1½ ft. long, often with a few leaves near the base. Flower-heads nearly ½ in. diameter without the rays, the bracts, especially the inner ones, broader and more obtuse than in *L. Billardieri.* Achenes glabrous or sprinkled with short hairs both in the eastern and western specimens.—Steetz in Pl. Preiss. i. 430; *L. Gunniana,* Steetz l. c. 431; Hook. f. Fl. Tasm. i. 189. t. 49 B.

Victoria. Yarra river, *F. Mueller,* Wendu vale, *Robertson.*

Tasmania. Abundant in pastures, Launceston, Macquarrie plains, etc., *J. D. Hooker, Gunn.*

S. Australia. Lofty Ranges, *F. Mueller.*

W. Australia. Swan River to King George's Sound, *Drummond, n.* 60, 377, *Huegel, Preiss, n.* 118, and others.

3. **L. solenogyne,** *F. Muell. Fragm.* v. 62. Glabrous or hirsute. Radical leaves from obovate to oblong-cuneate, obtuse, 1 to 2 in. long, toothed at the end or above the middle, narrowed into a distinct petiole. Scapes filiform, much longer than the leaves, usually with a few distant small narrow-linear leaves. Flower-heads scarcely above 2 lines diameter when in flower, nearly 3 lines when in fruit. Involucral bracts oblong, obtuse, with scarious often denticulate or ciliate margins. Ray-florets numerous, all apparently erect, tubular, and 3-toothed, not longer than the disk. Achenes narrow, flat, with thickened margins, contracted at the base, terminating in a very short conical but obtuse and callous point.—*Solenogyne bellioides,* Cass.; DC. Prod. v. 367 (from the character given); *S. brachycomoides,* F. Muell. Fragm. v. 62.

Queensland. Brisbane river, *F. Mueller* (a single specimen in fruit only, and therefore doubtful).

N. S. Wales. Port Jackson, *R. Brown;* New England, *C. Stuart.*

4. **L. emphysopus,** *Hook. f. Fl. Tasm.* i. 189. Very hirsute or nearly glabrous. Leaves all radical, densely tufted, oblong, obtuse, narrowed at the base, 1½ to 3 in. long. Scapes very little exceeding the leaves, rather thick and often constricted under the head, leafless or with 1 or 2 very small leafy bracts. Involucre about 2 lines diameter when in flower, 3 lines when in fruit, the bracts oblong, obtuse, with scarious margins. Ray-florets very numerous and short, apparently tubular when in bud, but opening out into a short, concave, 2- or 3-toothed ligula. Achenes of the disk abortive, those of the ray as long as the involucre, narrow, flat, contracted at the base and very shortly so at the top, but without the distinct neck of *L. Billardieri.*—*Emphysopus Gunnii,* Hook. f. in Hook. Lond. Journ. vi. 113; *Solenogyne bellioides,* Sond. in Linnæa, xxv. 480, F. Muell. Pl. Vict. t. 37, but scarcely of *Cassini.*

N. S. Wales. Port Jackson to the Blue Mountains, *R. Brown, Woolls,* and others; New England, *C. Stuart;* Clarence river, *Beckler.*

Victoria. Snowy plains on Limestone river and pastures, Bugle Range, *F. Mueller;* Wendu valley, *Robertson.*

Tasmania. Common in various parts of the colony, *Gunn.*

20. BRACHYCOME, Cass.

(Brachystephium, *Less.;* Paquerina, *Cass.;* Steiroglossa, *DC.;* Silphiospermum, *Steetz.*)

Involucre usually hemispherical, the bracts in about 2 rows, nearly equal, broad or narrow, with dry or scarious margins. Receptacle convex or conical, without scales. Florets of the ray female, ligulate, numerous, but usually in a single row. Disk-florets numerous, hermaphrodite, tubular, with a more or less dilated limb, 5-toothed. Anthers obtuse at the base. Style-branches in the disk-florets somewhat flattened, with lanceolate or triangular tips or appendages, papillose outside. Achenes usually compressed when young, when ripe either flat with obtuse or acute or winged margins, or thick and obtusely 4-angled. Pappus consisting of a ring of short scale-like bristles, or scarcely perceptible or none at all.—Herbs either tufted with 1-headed scapes, or annuals or perennials with erect or ascending branching stems. Leaves alternate, entire toothed or divided. Flower-heads terminal, the ray white blue or purplish, very rarely yellow.

Besides the Australian species, which are all endemic, there are only three from New Zealand. The genus is, however, nearly allied to *Bellis*, a group confined to the northern hemisphere, to which F. Mueller proposes to reunite it. The majority of the Australian species, however, differ in habit; they are all distinguished by the dry or scarious margins of the involucral bracts, and most of them by other more trifling characters.

SECTION I. **Brachystephium.**—*Ray inconspicuous. Achenes often compressed, especially when young, but with thickened margins, never winged, and sometimes at length as thick as broad. Pappus conspicuous, stellately spreading. Involucral bracts usually broadly scarious.*

Perennial, with stout erect or ascending stems nearly simple. Lower leaves pinnatifid. Flower-heads large . . . . 1. *B. diversifolia.*
Small erect branching perennial. Leaves cuneate, mostly toothed or lobed. Flower-heads small . . . . 2. *B. melanocarpa.*
Small glabrous perennial, with creeping rhizomes. Leaves radical, linear, entire or pinnate. Scapes simple. Flower-heads small . 3. *B. radicans.*
Small annuals. Flower-heads small, with few very small ray-florets.
Stem branched and leafy. Achenes glabrous, with 4 obtuse transversely crenate angles . . . . 4. *B. goniocarpa.*
Leaves radical, scape simple. Achenes smooth, with 2 thick corky angles, obtuse but bordered by a line of long woolly hairs, and 2 angles narrow and glabrous . . . . 5. *B. pachyptera.*

SECTION II. **Paquerina.**—*Ray conspicuous. Achenes often compressed, especially when young, but with thickened margins, never winged, and sometimes at length as thick as broad. Pappus minute or none.*

Glabrous annuals, with narrow leaves. Ray-achenes different from those of the disk. Western species.
Stems erect and branching. Leaves mostly pinnate . . . . . 6. *B. iberidifolia.*
Stems simple or slightly branched at the base. Leaves entire or rarely toothed at the end . . . . 7. *B. pusilla.*
Slender, decumbent, glandular-pubescent, branching annual (or rarely perennial ?). Leaves mostly obovate, toothed or lobed. Eastern species . . . . 8. *B. microcarpa.*
Perennials, glabrous (except *B. angustifolia*). Leaves narrow, entire or pinnate.
Stock tufted.
Leaves all radical. Scapes simple, leafless or nearly so.
Leaves pinnate . . . . 9. *B. Stuartii.*
Leaves linear or lanceolate, entire . . . . 10. *B. scapigera.*
Stems decumbent or ascending, simple or with few divaricate branches, leafy at the base.
Leaves long, narrow, deeply pinnatifid . . . . 11. *B. Muelleri.*
Leaves small, linear, obtuse, entire . . . . 12. *B. parvula.*
Stems decumbent or creeping at the base. Leaves entire.
Achenes rather broad, with thick edges and concave sides.
Glabrous. Leaves thin. Pappus none . . . . 13. *B. graminea.*
Usually slightly glandular-pubescent. Leaves rigid. Pappus minute . . . . 14. *B. angustifolia.*
Achenes narrow, with thick margins and flat sides.

Plant glabrous. Leaves thin. Pappus minute . . . . . . 15. *B. linearifolia.*
Stems erect, rigid. Leaves entire, mostly lanceolate . . . . . . 17. *B. basaltica.*
Stems slender, erect or slightly decumbent at the base, branched.
Leaves narrow, entire or pinnate. Achenes narrow . . . . . 18. *B. trachycarpa.*
Stems decumbent, branching, often woody at the base. Leaves toothed, pinnatifid or pinnate.
Pappus minute or scarcely any. Leaves all petiolate, acutely toothed or lobed.
Western species. (Disk-achenes with acute margins?) . . . 28. *B. Billardieri.*
Eastern species. Achenes all with obtuse edges . . . . . . 16. *B. heterophylla.*
Pappus conspicuous. Eastern species. (See the last three species of Sect. III.)

SECTION III. **Brachycome.**—*Ray conspicuous. Achenes flat, the margins obtuse, acute or winged. Pappus conspicuous.*

Small annuals, branching at the base. Leaves often lobed or pinnate.
Glabrous or pubescent. Achenes narrow, not winged . . . . . 19. *B. exilis.*
Glabrous. Achenes rather broad, but not winged . . . . 20. *B. ptychocarpa.*
Pubescent or hirsute. Achenes broadly winged . . . . 21. *B. debilis.*
(See also *B. ciliaris*, which the first year has the appearance of an erect corymbose annual.)
Perennials. Leaves all or mostly radical. Scapes leafless or with few leaves, 1-headed.
Achenes not winged. Leaves all radical, obovate or oblong, entire or crenate . . . . 22. *B. decipiens.*
Achenes winged. Scapes sometimes with a few leaves.
Leaves linear, entire . . . . 23. *B. cardiocarpa.*
Leaves once or twice pinnate, with pinnatifid segments . . . 24. *B. nivalis.*
Leaves cuneate-oblong, entire or crenate.
Glabrous or nearly so . . . . 25. *B. scapiformis.*
Hirsute. Stems, when full-grown, several-headed.
Leaves obtusely toothed or lobed . . . . 26. *B. stricta.*
Leaves acutely toothed or lobed . . . . 27. *B. heterodonta.*
Branching perennials with more or less leafy stems. Achenes (at least those of the disk) winged or acutely bordered.
Leaves mostly pinnate, with linear segments. Stems, the first year, erect and corymbose, afterwards ascending . . . . 29. *B. ciliaris.*
Leaves oblong or cuneate, toothed or lobed.
Glabrous, glandular-pubescent or hirsute.
Lobes or teeth of the leaves very acute.
Eastern species. Achenes winged . . . . 27. *B. heterodonta.*
Western species. Achenes acutely bordered, but not winged . . . . 28. *B. Billardieri.*
Lobes or teeth of the leaves obtuse or scarcely acute . . . 26. *B. stricta.*
Stems woolly, at least when young. Leaves narrow . . . . 30. *B. calocarpa.*
Leaves mostly linear, entire. Ray-florets yellow . . . . . . 31. *B. marginata.*
Branching decumbent perennials, with more or less leafy stems. Achenes very flat, but with obtuse edges, not winged.
Leaves thin, obovate or cuneate-oblong, entire or toothed at the end, sessile, except the lower ones . . . . 32. *B. Sieberi.*
Leaves thick, obovate or oblong, all petiolate, coarsely crenate or pinnatifid . . . . 33. *B. discolor.*
Leaves once or twice pinnatisect or deeply pinnatifid, with narrow or rarely cuneate lobes . . . . 34. *B. multifida.*

SECTION IV. **Silphiosperma.**—*Ray-florets not exceeding the involucre. Achenes flat, with an entire or toothed wing. Pappus none. Small annuals.*

Glandular-pubescent, 3 to 6 in. high. Involucral bracts narrow.

Achene-wings entire or slightly toothed . . . . 35. *B. glandulosa.*
Glabrous, ½ to 2 in. high. Involucral bracts broad. Achene-wings conspicuously toothed . . . . 36. *B. collina.*

B. dentata, Gaud. in Freyc. Voy. Bot. 468, too imperfectly described for identification, is referred by DC. with doubt to *Vittadinia*.

SECTION I. BRACHYSTEPHIUM.—Achenes often compressed, especially when young, but with thickened margins, never winged, and sometimes at length as thick as broad. Pappus conspicuous, stellately spreading. Involucral bracts usually broadly scarious.

This section differs from *Paquerina* chiefly in the pappus; the achenes are also more generally thick when ripe, and the outer ones especially often thickened unequally at the top, turning inwards, the pappus becoming quite excentrical.

1. **B. diversifolia,** *Fisch. and Mey. Ind.* ii. *Sem. Hort. Petrop.* 31. A tufted perennial, flowering however the first year so as to appear annual, glabrous or hirsute with septate hairs. Stems usually simple, 1 to 2 ft. high or rarely reduced to a short scape. Lower leaves often crowded, obovate spathulate or oblong, coarsely toothed or pinnatifid, the lobes broad or narrow, sometimes again divided, the upper leaves smaller and less divided, and the upper part of the stem a long almost leafless peduncle. Flower-heads larger than in any other species. Involucre above ½ in. diameter, the scarious margins of the bracts broad. Ray-florets white, long and numerous. Style-branches with a lanceolate papillose appendage, but without the longer hairs at the base figured by Lessing (Syn. Comp. f. 16). Achenes oblong, thickened upwards, marked with longitudinal furrows, the outer ones obliquely incurved at the top, with a dense tuft of short capillary bristles.—Hook. f. Fl. Tasm. i. 187; *Pyrethrum diversifolium*, Grah. in Hook. Exot. Fl. iii. t. 215; Bot. Reg. t. 1025; *Brachystephium leucanthemoides*, Less. Syn. Comp. 389; DC. Prod. vi. 304.

N. S. Wales. Port Jackson, *R. Brown;* Clarence river, *Beckler* (with smooth and with tuberculate achenes).
Victoria. On the Yarra, *F. Mueller, Robertson;* in the Australian Alps, *F. Mueller;* Wimmera, *Dallachy*.
Tasmania. Port Dalrymple, *R. Brown;* common in grassy pastures throughout the island, *J. D. Hooker*.
S. Australia. Near Adelaide, *Blandowski*.

Var. *humilis*. Very small in all its parts, but not otherwise different, probably a starved state.—*Steiroglossa humilis*, DC. Prod. vi. 39. Lachlan river, *A. Cunningham*.
Var. *maritima*. Leaves mostly twice pinnatifid, rather thick. Scapes short.—Islands of Bass's Straits, *R. Brown*.

2. **B. melanocarpa,** *Sond. and F. Muell. in Linnæa*, xxv. 476. A small perennial, but evidently flowering the first year so as to appear annual, erect, slightly branched, rarely much above 6 in. high. Leaves mostly oblong-cuneate, obtuse and coarsely toothed at the end or shortly lobed above the middle, narrowed into a rather long petiole. Flower-heads small, on slender peduncles. Involucral bracts broad, scarious at the end. Ray-florets rather numerous. Achenes narrow-obovate, somewhat compressed, the edges obtuse, the sides often tuberculate, usually very black. Pappus of short bristles stellately spreading.

N. S. Wales. Darling river, *Victorian Expedition*.
Victoria. Murray river, *F. Mueller*.

3. **B. radicans,** *Steetz in Pl. Preiss.* i. 429. A glabrous slender perennial, with a tufted stock, emitting slender creeping rhizomes. Leaves radical, narrow-linear, entire or with a very few linear lobes, rarely under 1 in. and often 3 or 4 in. long. Scapes simple, slender, from 3 or 4 in. to nearly 1 ft. long, leafless or with 1 or 2 small leaves, rarely decumbent and leafy at the base. Flower-heads not very small, the involucral bracts rather broad, with broad scarious margins. Achenes striate, bordered by a broad thick obtusely crenate margin. Pappus of very short stellately spreading bristles.—Hook. f. Fl. Tasm. i. 184.

Victoria. Wet grassy places, near Omeo, *F. Mueller.*

Tasmania. Marshy places, Marlborough, South Esk river, etc., *Gunn.*

4. **B. goniocarpa,** *Sond. and F. Muell. in Linnæa,* xxv. 474. A small, diffuse, much-branched annual, rarely exceeding 3 in., glabrous or sprinkled with septate hairs. Leaves linear or linear-cuneate, the lower ones usually pinnatifid, the upper ones shortly lobed or toothed at the end or entire. Flower-heads not 3 lines diameter, very convex after flowering. Involucral bracts obovate, concave. Ray-florets few and very small; disk-florets very numerous, the corolla minute. Achenes thick, oblong-cuneate, more or less angular, with obtuse rugose angles; the outer achenes incurved at the end. Pappus of the outer florets very oblique, that of the disk stellately spreading.

Victoria. Wimmera, *Dallachy.*

S. Australia. Murray river, Burra-Burra mines, *F. Mueller.*

W. Australia, *Drummond, 5th Coll. n.* 391.

5. **B. pachyptera,** *Turcz. in Bull. Mosc.* 1851, i. 175. A small, apparently stemless, but densely tufted annual. Leaves radical, linear, 3-lobed or shortly pinnate above the middle, with linear segments. Scapes leafless, 2 to 4 in. high. Involucres 3 to 4 lines diameter, the bracts very obtuse and broadly scarious. Receptacle small. Achenes numerous, slightly compressed, with thick almost corky edges, bordered by a longitudinal line of long woolly hairs, and with a narrow acutely prominent glabrous ridge on each side. Pappus of short scaly bristles,

N. S. Wales. Molle's Plains, *Fraser;* Darling river, *Victorian Expedition;* Murray river, *Dallachy.*

S. Australia. Murray river, *F. Mueller.*

W. Australia. Between Swan River and King George's Sound, *Drummond, 4th Coll. n.* 205.

Section II. Paquerina.—Achenes often compressed, especially when young, but with thickened margins, never winged, and sometimes as thick as broad. Pappus minute or none.

6. **B. iberidifolia,** *Benth. in Hueg. Enum.* 59. An erect, glabrous, corymbosely branched annual, about 1 ft. high. Leaves pinnately divided into narrow-linear segments. Involucral scales oblong, with narrow scarious margins. Ray-florets blue or rarely white, rather large. Achenes of the disk more or less angular or furrowed, sometimes slightly flattened, without wings, those of the ray usually much larger, thicker, sometimes almost corky. Pappus very minute or none.—Bot. Reg. 1841. t. 9; Bot. Mag. t. 3876; Steetz in Pl. Preiss. i. 425; *B. capillacea,* Walp. Rep. ii. 584 (according to Steetz); *Steiroglossa chamæmillifolia,* DC. Prod. vi. 39.

W. Australia. King George's Sound, *A. Cunningham*, and thence to Vasse and Swan rivers, *Huegel*, *Drummond*, 1*st Coll.*, 5*th Coll. n.* 371, *Preiss, n.* 94, 95, 96, 97, and others.

Var. *diffusa.* More branching from the base. Achenes apparently flatter, but not seen ripe.—Murchison river, *Oldfield.*

7. **B. pusilla,** *Steetz in Pl. Preiss.* i. 427. A slender glabrous annual, with erect or ascending stems, simple or slightly branched, not exceeding 6 in., including the peduncle. Leaves radical or in the lower part of the stem, narrow-linear, obtuse, entire or very rarely with 1 or 2 teeth, or very short lobes, the upper part of the stem a long leafy peduncle. Flower-heads of *B. iberidifolia*, but usually smaller. Achenes smaller than in *B. iberidifolia*, those of the disk more or less flattened with thickened edges, those of the ray less flattened, the margins much less prominent. Pappus none or very minute, sometimes rather more conspicuous on the ray-achenes.—*B. bellidioides*, Steetz in Pl. Preiss. i. 426.

W. Australia. Swan River, *Drummond*, 5*th Coll. n.* 373; *Preiss, n.* 86, 88, 89; *Oldfield*; Murray and Murchison rivers, *Oldfield.*

This has the foliage and stature of *B. parvula*, but appears to be constantly annual, and the achenes are all narrower. It may prove to be a reduced variety of *B. iberidifolia*, with undivided leaves. The distinction established by Steetz between his two species was owing to his having described the achenes of the ray in his *B. bellidioides*, and those of the disk in his . *pusilla.*

8. **B. microcarpa,** *F. Muell. Fragm.* i. 50. A more or less pubescent annual, sometimes very small and branched, like *B. exilis*, the ascending or decumbent stems sometimes lengthening out to near a foot. Leaves petiolate, obovate or almost orbicular, obtusely and coarsely toothed or lobed, under 1 in. long, or when luxuriant lyrate and nearly 2 in. Flower-heads small, like those of *B. exilis*, on long slender peduncles. Achenes obovate, very flat, with thickened margins, sprinkled with a few hairs or tuberculate on the sides. Pappus very small.

Queensland. Burnett river, *F. Mueller*; Wide Bay to Moreton Bay, *Leichhardt*; Brisbane river, Moreton Bay, *Fraser*, *F. Mueller*; Cape river, *Bowman.*

N. S. Wales. New England, *C. Stuart*; Clarence river, *Beckler*; Richmond river, *Fawcett.*—The achenes are sometimes very much like those of *B. melanocarpa*, but flatter, with a much smaller pappus.

9. **B. Stuartii,** *Benth.* A small glabrous perennial, with a thick tufted or shortly creeping rootstock. Leaves radical, pinnate, with narrow, cuneate, entire toothed or lobed segments, the lower ones small and distant. Scapes leafless, slender, 3 to 6 in. high. Flower-head rather small. Involucral bracts broadly oblong. Receptacle very conical. Achenes small, black, rather narrow, slightly flattened, but thick, with obtuse edges, not winged, more or less tuberculate on the sides. Pappus minute.

N. S. Wales. New England, *C. Stuart.*

10. **B. scapigera,** *DC. Prod.* vii. 277. A glabrous perennial, with a densely tufted stock. Leaves radical, often surrounded at the base by the fibre-like remains of old leaves, oblong or linear-lanceolate, narrowed into a long petiole of a rather thick texture, quite entire. Scapes leafless or with 1 or 2 small leaves, 3 to 6 in. high or when luxuriant rather more. Flower-

head not large. Involucral bracts obtuse and rather broadly scarious. Achenes compressed, but with obtuse edges, not at all winged. Pappus very small.—*Senecio scapiger*, Sieb. in Spreng. Syst. iii. 559; *Brachystephium scapigerum*, DC. Prod. vi. 304.

N. S. Wales. Port Jackson or Blue Mountains, *Sieber, n.* 332.
Victoria. Grassy valleys, Buffalo Range, Delatite river, Cobberas Mountains, *F. Mueller.*

11. **B. Muelleri,** *Sond. in Linnæa,* xxv. 475. Glabrous, tufted at the base, and probably perennial. Stems decumbent, simple and single-headed or slightly branched, not exceeding 6 in. in our specimens. Leaves in the lower part narrow, 1 to 2 in. long, deeply pinnatifid with short, ovate, mucronate, entire or toothed lobes. Flower-heads not small. Involucral bracts oblong, acuminate. Achenes rather narrow, still young, but apparently with thick obtuse margins and narrow sides, as in *B. pachyptera*, but quite glabrous. Pappus minute or none.

S. Australia. Near Gawler Town, *F. Mueller.*

12. **B. parvula,** *Hook. f. Fl. Tasm.* i. 185. A small glabrous perennial, with a densely tufted stock and diffuse divaricately branched stems, rarely attaining 6 in. Leaves linear or linear-cuneate, obtuse, rather thick, the radical and lower ones occasionally 2- or 3-toothed at the end, but often all entire. Flower-heads rather small. Involucral bracts oblong-cuneate. Achenes flat, broad, with thickened margins, not winged. Pappus minute or none.

Victoria. Sandy banks of the Yarra, *F. Mueller, Harvey, Adamson.*
Tasmania. Flinders' Island, *Milligan.*

This species resembles *B. pusilla* in aspect, but besides the evidently perennial duration, the achenes appear to be the same in the ray as in the disk, and all broader than in *B. pusilla.*

13. **B. graminea,** *F. Muell. Fragm.* i. 49. Slender and usually glabrous; stock perennial, creeping or sometimes tufted. Stems occasionally short and numerous, more frequently elongated, decumbent or erect, leafy at the base only or above the middle. Leaves linear or linear-lanceolate, mostly acute, narrowed into a long petiole. Involucre about 3 lines diameter, the bracts narrow-oblong, obtuse. Ray-florets numerous, narrow. Style-appendages narrow. Achenes as long as the involucre, obovate, compressed, with very thick obtuse callous margins, the somewhat concave sides often tuberculate, and the whole achene sometimes glandular. Pappus none.—*Bellis graminea*, Labill. Pl. Nov. Holl. ii. 54. t. 204; *Paquerina graminea*, Cass.; DC. Prod. v. 307; Hook. f. Fl. Tasm. i. 188.

Victoria. Wet places, Merriman's Creek, Hall's Creek, Goulburn river, Dandenong Mountains, Station Peak, etc., *F. Mueller.*
Tasmania. Port Dalrymple, *R. Brown*; marshy places, not uncommon, *J. D. Hooker.*
S. Australia. From the Murray to St. Vincent's Gulf, *F. Mueller* and others.

14. **B. angustifolia,** *A. Cunn. in DC. Prod.* v. 306. A perennial with a creeping rhizome and ascending or erect stems, sometimes short, sometimes 8 to 10 in. long, leafy at the base, more or less glandular-pubescent especially under the flower-head. Leaves linear or linear-lanceolate, mostly acute, narrowed into a petiole, 1½ to above 3 in. long, rather firm, and the midrib pro-

minent underneath. Flower-heads not large. Involucre very glandular, the scarious margins of the bracts very narrow. Achenes obovate-oblong, flattened but thick, often glandular, with thick callous margins. Pappus very minute, and sometimes quite inconspicuous.—Hook. f. Fl. Tasm. i. 186.

N. S. Wales. Open downs, Goulburn Plains, *A. Cunningham.*
Victoria. Omeo, *F. Mueller.*
Tasmania, *Gunn.*

Although very near *B. graminea*, this species appears to be distinct in the shape of the achenes, as well as in the glandular pubescence and narrow rigid leaves.

15. **B. linearifolia,** *DC. Prod.* v. 306. Glabrous, probably perennial, with slender ascending stems and petiolate leaves, either all linear or the lower ones oblong-lanceolate, all entire, and the specimens closely resembling the slender varieties of *B. graminea*, but the achenes are flat, rather narrow, with thickened margins, and crowned by a very small but distinct pappus.

N. S. Wales. Port Jackson, *A. Cunningham, J. D. Hooker,* and others.

16. **B. heterophylla,** *Benth. in Hueg. Enum.* 60. A glabrous perennial with slender decumbent or ascending branching stems, like those of *B. linearifolia* and *B. Sieberi.* Leaves all petiolate, obovate-cuneate or oblong, with a few very acute teeth or lobes. Flower-heads rather small, on slender peduncles. Achenes flat, rather narrow, with thickened margins, and crowned by a very small but distinct pappus.

N. S. Wales. Port Jackson, *R. Brown.*—The specimens are numerous, and show two distinct varieties in foliage, one with broad thin leaves with broad but very acute lobes, the other with narrower, smaller, almost pinnatifid lobes. Both may be varieties of *B. linearifolia*, but the leaves are all toothed and lobed. From *B. Sieberi* they differ in the petioles and acute lobes of the leaves, and the very small pappus.

17. **B. basaltica,** *F. Muell. Fragm.* i. 50. A glabrous and somewhat glaucous perennial, with rigid erect branching stems of 1½ to 3 ft. Lower leaves petiolate, obovate-oblong, the others oblong lanceolate or linear, mostly acute and quite entire, the upper ones few, small, and distant. Peduncles long, somewhat corymbose. Involucral bracts narrow, almost acute. Achenes narrow, at first flat with thickened margins, at length thick, tuberculate, not winged. Pappus very minute.

Queensland, *Bowman;* basaltic plains from Peak range to Darling Downs, *F. Mueller.*

Var. *gracilis.* More slender, 1 to 2 ft. high. Leaves all linear or linear-lanceolate.
Queensland. Rockhampton and Keppel Bay, *Thozet.*
N. S. Wales. Macquarrie marshes, *Mitchell;* Murray and Darling rivers, *Dallachy.*
S. Australia. Tamunda on the Gawler river, *F. Mueller.*

18. **B. trachycarpa,** *F. Muell. in Linnæa,* xxv. 339. A glabrous perennial, erect and corymbosely branched the first year, afterwards with numerous nearly simple or divaricately branched, slender but rigid stems from a more or less woody base. Leaves linear, the lowest often rather broad, entire or pinnately lobed, the stem-leaves mostly narrow and entire. Flower-heads small. Involucral bracts cuneate-oblong, rather broadly scarious at the end. Ray short, lilac. Achenes narrow, with thickened margins, the sides more or less tuberculate, and sometimes hirsute with short hooked hairs. Pappus minute or almost none.

Queensland. Broad Sound, *R. Brown;* Keppel Bay, *Thozet.*
N. S. Wales. Darling river, *Dallachy.*
Victoria. Victoria ranges and Mount Sturgeon, *F. Mueller;* sandy heaths, Glenelg river, *Robertson.*
S. Australia. Lofty Ranges, Cudnaka, Crystal Brook, *F. Mueller.*

SECTION III. BRACHYCOME.—Achenes flat, the margins acute or winged, or in a few species obtuse. Pappus conspicuous.

19. **B. exilis,** *Sond. in Linnæa,* xxv. 473. A slender annual, with erect or ascending branched stems, usually only 2 or 3 in. high without the peduncles, glabrous in the original form. Leaves mostly pinnatifid, with few short linear acute lobes. Peduncles slender, erect, 2 to 4 in. long. Flower-heads small. Involucral bracts few, oblong, scarious. Achenes narrow, with thickened margins, not winged. Pappus at first conspicuous, but when the achene is ripe not longer than its breadth, and sometimes much shorter.—*B. leptocarpa,* F. Muell. in Trans. Phil. Soc. Vict. i. 43, and in Hook. Kew Journ. viii. 147.

N. S. Wales. In the interior, *C. Moore;* Darling river, *Victorian Expedition.*
Victoria. Axe river, Darelin Creek, Esau river, etc., *F. Mueller.*
S. Australia. Lofty and Barossa ranges, Port Lincoln, *F. Mueller.*

Var. (?) *scabrida,* Sonder. Larger, more branched, with broader leaves and smaller flower-heads.—Murray river and Kensington, in both cases old and imperfect specimens.

20. **B. ptychocarpa,** *F. Muell. in Trans. Phil. Soc. Vict.* i. 43, *and in Hook. Kew Journ.* viii. 148. A small glabrous annual, very much like the common glabrous form of *B. exilis,* but the involucral bracts appear to be much broader, the receptacle more conical, and the achenes much broader and flatter, intermediate between those of *B. exilis* and of *B. debilis.* The specimens are, however, few and unsatisfactory.—*Steiroglossa peariloba,* DC. Prod. vi. 39.

N. S. Wales. Lachlan river, *A. Cunningham.*
Victoria. Buffalo range, *F. Mueller.*

21. **B. debilis,** *Sond. in Linnæa,* xxv. 477. A small annual, more or less hirsute, exactly like the scabrous form of *B. exilis,* except that the involucral bracts are rather narrower, and the achenes are bordered by a rather broad wing. Pappus rather long for the genus.

Victoria. Glenelg river, *F. Mueller.*
S. Australia. Kensington, St. Vincent's Gulf, Port Lincoln, *F. Mueller.*
Some very young specimens from the Darling river, *Victorian Expedition,* may possibly belong to this species.

22. **B. decipiens,** *Hook. f. in Hook. Lond. Journ.* vi. 114, *and Fl. Tasm.* i. 184. *t.* 48. An almost stemless perennial, usually glabrous, with precisely the aspect but not the involucre of the European Daisy (*Bellis perennis*). Stock short and tufted. Leaves radical, obovate-oblong, entire or remotely and shortly toothed or crenate, narrowed into a short broad petiole, 1 to 3 in. long. Scapes longer than the leaves, simple and leafless, or with a single small leaf. Heads rather large, the scarious margins of the involucral bracts usually but not always dark coloured. Achenes very flat with a thickened margin, sprinkled with a few short hairs. Pappus very short.

N. S. Wales. Argyle county, *Backhouse, M'Arthur.*
Victoria. Grassy valleys near Mount William, Dandenong range, Plenty Creek, Upper Genoa river, *F. Mueller;* Wendu valley, *Robertson.*
Tasmania. Port Dalrymple, *R. Brown;* abundant in grassy meadows throughout the island, *J. D. Hooker.*
S. Australia. Near Mount Gambier, *F. Mueller.*

Var. *pubescens.* Leaves pubescent, flower-heads smaller.—New England, *C. Stuart.*

23. **B. cardiocarpa,** *F. Muell. Herb.* A glabrous perennial with a tufted stock. Leaves radical, linear, quite entire in all our specimens, and often very narrow and many inches long, but sometimes short and resembling those of *B. scapigera.* Scapes usually erect, often above 1 ft. high, bearing 3 or 4 distant leaves, rarely decumbent and leafy at the base. Flower-heads rather large. Involucral bracts broadly scarious. Achenes flat, bordered by broad thin wings, entire, undulate, or rarely crenate or lobed as in *B. marginata,* the pappus as long as the breadth of the achene or rather more.—*B. linearifolia,* Hook. f. Fl. Tasm. i. 185, not of DC.

Victoria. Swamps of Gipps' Land, *F. Mueller;* Heaths, Glenelg river, *Robertson;* Portland, *Allitt.*
Tasmania. Mount Wellington, Formosa, etc., generally growing in water, *J. D. Hooker* and others.
S. Australia. Rivoli Bay, *C. Mueller.*

Var. *alpina.* Smaller, with shorter leaves.—Baw-Baw, Munyong and Cobra Mountains, Victoria, at an elevation of 4000 to 6000 ft., *F. Mueller.*

24. **B. nivalis,** *F. Muell. in Trans. Phil. Soc. Vict.* i. 43, *and in Hook. Kew Journ.* viii. 148. A glabrous perennial, with a thick tufted or shortly-creeping stock. Leaves radical, pinnate, with once or even twice pinnatifid, or rarely entire linear segments, the whole leaf usually 3 or 4 in. long. Scapes longer than the leaves, simple, leafless. Flower-heads rather large. Involucral bracts narrow. Receptacle at length very conical. Achenes flat, with a broad wing. Pappus conspicuous, the bristles usually united in a ring.

Victoria. Summits of the Australian Alps, Mount Buller and Cobberas mountains (with the leaves mostly simply pinnate), Munyong mountains and Mount Wellington (with the leaves mostly twice pinnate), *F. Mueller.*

25. **B. scapiformis,** *DC. Prod.* v. 306. A perennial with a tufted or slightly-creeping stock, glabrous or nearly so. Radical leaves obovate or oblong-cuneate, coarsely toothed towards the end or rarely nearly entire, narrowed into a petiole, of a thicker consistence than in *B. decipiens.* Flowering-stems 6 in. to 1½ ft. high, sometimes reduced to almost leafless scapes, more frequently leafy below the middle, simple or rarely with a single accessory branch, the leaves smaller and narrower than the radical ones. Flower-heads rather large. Involucral bracts narrow and acute. Achenes flat, the acute edges more or less expanded into a wing. Pappus conspicuous, of short bristles almost dilated into scales.—Hook. f. Fl. Tasm. i. 185.

N. S. Wales. Lachlan river, *A. Cunningham;* Argyle County, *M'Arthur, Backhouse;* Head of the Gwydir, *Leichhardt.*
Victoria. Moist places, summit of the Victoria ranges, *Wilhelmi;* Mount Barkly, Mount Timbertop, Merriman's Creek, etc., *F. Mueller.*
Tasmania. Derwent river, *R. Brown;* abundant in good soil and in marshy places throughout the colony, *J. D. Hooker.*

Var. *tenuiscapa.* A rather small and slender variety with narrow leaves, broader, more obtuse involucral bracts, and in some specimens the immature achenes show no wings; in others they are certainly winged, and some large broad-leaved specimens of *B. scapiformis* have the broad involucral bracts of the *tenuiscapa.*—*B. tenuiscapa,* Hook. f. in Hook. Lond. Journ. vi. 114, and in Fl. Tasm. i. 184. t. 48.—Emu plains, Victoria, *F. Mueller;* Arthur's Lakes and Middlesex plains, Tasmania, *Gunn.*

26. **B. stricta,** *DC. Prod.* v. 305. A perennial with ascending or erect more or less branching stems, attaining 1½ ft. when full-grown, more or less hirsute with short septate hairs or glandular-pubescent, especially under the flower-head, very rarely reduced to short simple scapes. Lower leaves cuneate-oblong, coarsely toothed or almost pinnatifid; upper ones narrow and more entire. Flower-heads rather large, on long peduncles. Involucral bracts narrow as in *B. scapiformis*, but usually more scarious and broader at the end. Achenes very flat, with a broad thin entire or lobed wing. Pappus much smaller than in *B. scapiformis.*—Hook. f. Fl. Tasm. i. 186; *B. leucanthemifolia* and *B. oblongifolia*, Benth. in Hueg. Enum. 60; *B. glauca*, Walp. in Linnæa, xiv. 315.

N. S. Wales. Port Jackson to the Blue Mountains, *R. Brown* and others, and inland to Bathurst, *A. Cunningham;* Mount Mitchell, *Beckler;* Maneroo plains and Argyle county, *Lhotzky;* Twofold Bay, *F. Mueller.*

Victoria. Mount Buller, Mount Timbertop, King's Parrot's Creek, Delatite river, Bacchus Marsh, etc., *F. Mueller;* shady banks, Port Phillip, *Herb. Lemann.*

Tasmania. Port Dalrymple, *R. Brown;* abundant in rocky places throughout the colony, *J. D. Hooker.*

27. **B. heterodonta,** *DC. Prod.* v. 305. A perennial with a tufted stock, emitting creeping stolons and erect or ascending branched stems of 1 to 2 ft. Radical leaves on long petioles, oblong, toothed or pinnatifid, all the teeth or lobes mucronate-acute; stem leaves few, lanceolate or oblong, acute, entire or acutely toothed, all sprinkled, especially underneath, with rather rigid appressed hairs. Flower-heads rather large, on long peduncles. Involucral bracts apparently broad. Achenes (according to DC.) bordered by a toothed wing and tuberculate on the disk.

N. S. Wales. Wet places on the Lachlan river, *A. Cunningham;* N.W. interior, *Fraser.* The specimens are imperfect, but have a different aspect from any other species; the mucronate-acute teeth of the leaves are peculiar.

28? **B. Billardieri,** *Benth.* A perennial, glabrous or hirsute with septate hairs. Stems branched, leafy, apparently decumbent. Leaves cuneate-oblong, pinnatifid with short mucronate-acute lobes. Peduncles 1 to 3 in. long, the flower-heads not large. Involucral bracts oblong. Achenes very flat, oblong, those of the ray apparently with thickened obtuse margins, those of the disk with the margins more acute and slightly ciliate or denticulate, but not seen perfect. Pappus minute or almost none.

W. Australia, *Drummond, 5th Coll. n.* 374.—*Bellis aculeata,* Labill. Pl. Nov. Holl. ii. 55. t. 206, from Van Leeuwin's Land, appears from the figure and description to agree much better with this species than with *B. stricta,* to which it is commonly referred. Our specimens are not very good, and there may still be some doubt about the species.

29. **B. ciliaris,** *Less. Syn. Comp.* 172. Glabrous or more or less clothed, especially when young, with a white wool, probably always perennial, but often flowering the first year so as to appear annual; stems erect, slender,

under 1 ft. high, somewhat corymbosely branched. Leaves pinnate, with narrow-linear usually very divaricate segments. Flower-heads rather small. Involucral scales very variable in breadth. Achenes of the disk flattened, bordered by a broad entire denticulate or ciliate wing, the faces smooth or tuberculate, those of the ray often more tuberculate and without any wing. Pappus usually conspicuous.—DC. Prod. v. 306; Hook. f. Fl. Tasm. i. 187; *Bellis ciliaris*, Labill. Pl. Nov. Holl. ii. 56. t. 209; *Brachycome Drummondii*, Walp. Rep. ii. 584; Steetz in Pl. Preiss. i. 428.

N. S. Wales. Field's Plains, *Fraser;* Lake George, Argyle county and Illawarra, *A. Cunningham;* Darling desert, *Dallachy.*

Victoria. Mount Buller, *F. Mueller;* Murray desert and Wimmera, *Dallachy;*

Tasmania. Derwent river, *R. Brown;* chiefly in the central districts of the island, *J. D. Hooker.*

S. Australia. South coast, *R. Brown;* from Lake Alexandrina to Spencer's and St. Vincent's Gulf, *F. Mueller.*

W. Australia. Chiefly on the South coast to the east of King George's Sound, *R. Brown*, *Harvey*, *Drummond*, 5*th Coll. n.* 387, *Preiss*, *n.* 87, *Maxwell*, and others.

Some specimens, not showing the perennial stock, are very difficult to distinguish from *B. iberidifolia* without the achenes. The flower-heads are, however, usually smaller.

Var. *lanuginosa.* Stems more or less woolly when young.—*B. lanuginosa*, Steetz in Pl. Preiss. i. 427.—Darling desert, *Dallachy*, and W. Australia, *Drummond*, *n.* 86, 95, *and* 4*th Coll. n.* 211; *Preiss*, *n.* 85.

Var. *glandulosa.* More or less glandular-pubescent. Flower-heads small.—S. Australia and W. Australia, *Drummond*, 4*th Coll. n.* 210.

Var. *grandiflora.* Flower-heads as large as in *B. iberidifolia* but achenes of *B. ciliaris.*—W. Australia, *Drummond.*

Var. *subdissecta.* Glabrous. Segments of the lower leaves sometimes again lobed.—Victoria and S. Australia.

Var. *robusta.* More rigid, the stems more leafy, about ½ ft. high. Flower-heads rather large, with the involucral bracts often but not always broader and more scarious.—*B. squalida*, Hook. f. in Hook. Lond. Journ. vi. 115; *B. strongylospermoides*, Walp. in Linnæa, xiv. 305; *B. multicaulis*, F. Muell. in Trans. Phil. Soc. Vict. i. 43, and in Hook. Kew Journ. viii. 148; *Steiroglossa rigidula*, DC. Prod. vi. 39.—N. S. Wales, Victoria and Tasmania.

30. **B. calocarpa,** *F. Muell. in Linnæa*, xxv. 399. A perennial with short branching stems, more or less clothed with a white wool, especially when young, and more robust than *B. marginata.* Leaves linear-lanceolate or cuneate, acute, entire or with a few acute lobes or teeth towards the end or above the middle, of a rather thick texture. Flower-heads rather large, on very long peduncles. Involucral bracts broadly scarious. Rays rather small, white or pink (*F. Mueller*). Achenes flattened, tuberculate or muricate, bordered by a rather broad wing either entire or broken up into lobes.

N. S. Wales. Hawkesbury river, *R. Brown* (nearly glabrous).

Victoria. Wimmera, *Dallachy.*

S. Australia. Murray river, Cudnaka, *F. Mueller.*

The species appears to differ from *B. marginata* chiefly in the colour of the ray; the notes, however, of collectors are in this respect somewhat vague, and it is possible that the two may be varieties only.

31. **B. marginata,** *Benth. in Hueg. Enum.* 60. A perennial with slender branching glabrous stems, almost woody at the base. Leaves linear or linear-oblong, acute, entire or rarely with 1 or 2 teeth, narrowed at the base. Peduncles very long and slender. Ray-florets yellow (*F. Mueller*), white

(*R. Brown*). Involucral bracts rather broadly scarious. Achenes flattened, usually tuberculate, bordered by a rather broad wing, either entire or more frequently broken up into lobes.—*B. chrysoglossa*, F. Muell. in Trans. Phil. Soc. Vict. i. 44, and in Hook. Kew Journ. viii. 148.

Queensland. Burnett river and Peak Downs, *F. Mueller;* Bogan river, *Mitchell.*
N. S. Wales. Paterson's and Hunter's rivers, *R. Brown.*

32. **B. Sieberi,** *DC. Prod.* v. 306. A glabrous perennial, with weak decumbent branching and leafy stems. Leaves from obovate to cuneate-oblong, sessile except the lower ones and often stem-clasping, obtuse and usually with a few small teeth or lobes towards the end or the narrower ones minutely 3-toothed at the end. Flower-heads small, on slender peduncles. Involucral bracts narrow. Achenes narrow-obovate, very flat, with thickened margins, not winged. Pappus conspicuous.

N. S. Wales. Port Jackson, *R. Brown; Sieber, n.* 485; *Harvey.*

33. **B. discolor,** *C. Stuart in Herb. Hook.* A glabrous perennial, stems slightly branched and decumbent at the base, ascending to a foot in length or more. Leaves obovate or cuneate-oblong, obtuse, coarsely crenate or pinnatifid, of a firm texture, purple underneath when fresh (*C. Stuart*), all petiolate. Flower-heads small, on very long peduncles. Involucral bracts narrow. Achenes obovate, flat, often tuberculate, with thickened margins. Pappus conspicuous.

N. S. Wales. New England, *C. Stuart;* Clarence river, *Beckler.*

34. **B. multifida,** *DC. Prod.* v. 306. A branching, erect or diffuse perennial or undershrub, usually glabrous. Leaves pinnate with linear segments, sometimes entire but more frequently lobed or pinnatifid, sometimes long and narrow, sometimes short and slightly dilated. Flower-heads rather small, on long slender peduncles. Achenes black, narrow, slightly compressed, the sides tubercular, the smooth margins often prominent but not winged. Pappus small.—*B. glabra*, Benth. in Hueg. Enum. 59.

N. S. Wales. Port Jackson, *Harvey, Woolls;* Head of the Gwydir, *Leichhardt;* New England, *C. Stuart;* Peel's Range, *A. Cunningham.*

Victoria. From Lake Wellington in Gipps' Land to the Grampians and Murray river, *F. Mueller* and others.

Var. *dilatata.* Leaves thin, the lobes often broadly linear or cuneate.—*B. tenera,* Benth. in Hueg. Enum. 59. Hunter's River, *R. Brown.*

SECTION IV. SILPHIOSPERMA.—Ray-florets not exceeding the involucre. Achenes flat, bordered by an entire or toothed wing. Pappus none. Small annuals.

35. **B. glandulosa,** *Benth.* Very closely allied to *B. collina* but larger, the erect or ascending branching stems attaining 3 to 6 in., and more or less glandular-pubescent. Leaves linear, acutely toothed or pinnatifid. Flower-heads about 2 lines diameter, with the florets of *B. collina*, but the involucral bracts much narrower, almost linear, and usually hirsute with a few glandular hairs. Wings of the achenes sometimes quite entire, sometimes bordered by a few teeth.—*Silphiosperma glandulosum*, Steetz in Pl. Preiss. i. 433.

W. Australia, *Drummond, n.* 15, 99, *5th Coll. n.* 378; *Preiss, n.* 103.

36. **B. collina,** *Benth.* A small erect branching annual, rarely exceeding 2 in., glabrous or slightly pubescent. Leaves small, linear, pinnatifid with short mucronate lobes, the lower lobes reduced sometimes to short cilia. Flower-heads about 2 lines diameter. Involucral bracts few, ovate, the scarious margins very narrow, the inner bracts more oblong. Florets not longer than the involucre, those of the ray with a very small ligula, scarcely so long as the style. Achenes obovate, as long as the involucre, very flat, bordered by a wing divided into linear lobes hooked at the end; the disk sprinkled with a few hairs. Style-lobes quite those of *Brachycome.—Silphiosperma collinum,* Sond. in Linnæa, xxv. 483.

Victoria. Hopkins river, Mount Emu, *F. Mueller;* Wimmera, *Dallachy;* Skipton, *Whan.*

S. Australia. Holdfast Bay, near Gawler Town, foot of Mount Alexander, *F. Mueller.*

Var. *perpusilla.* Stems ½ to 1 in. high. Leaves mostly entire. Achenes rather less toothed.—*Silphiosperma perpusillum,* Steetz in Pl. Preiss. i. 434; *Brachycome tenella,* Turcz. in Bull. Mosc. 1851, i. 176.

W. Australia, *Drummond, 4th Coll. n.* 208; *Preiss, n.* 2416.

21. SPHÆRANTHUS, Willd.

Flower-heads small, sessile in dense globular clusters or compound heads. Involucres ovoid, the bracts linear, imbricate in several rows. Florets not numerous, those of the circumference female filiform and minutely 2- or 3-toothed, hardened at the base; disk-florets very few, hermaphrodite but sometimes sterile, tubular, 5-toothed, thickened at the base. Anthers without tails or points at the base. Styles of the disk-florets bulbous at the base, simple or with 2 slender branches, papillose outside towards the end. Achenes oblong, somewhat flattened, without any pappus.—Coarse erect herbs. Leaves alternate, toothed, decurrent on the stem. Compound flower-heads terminal.

The genus comprises very few species, ranging over tropical Asia and Africa, the two Australian ones being the two most common over the whole area.

Pubescent or hirsute. Involucral bracts ending in a subulate ciliate point. Disk-florets (always?) sterile . . . . 1. *S. hirtus.*
Glabrous. Involucral bracts scarious, often jagged at the end. Disk-florets (always?) fertile . . . . 2. *S. microcephalus.*

1. **S. hirtus,** *Willd.; DC. Prod.* v. 369. Erect with few divaricate branches, more or less hirsute, 1 to 2 ft. high. Leaves obovate, oblong or lanceolate, irregularly and acutely toothed or almost lobed, decurrent along the stem into interrupted toothed wings. Flower-heads numerous, in globular clusters or compound heads of 4 to 5 lines diameter when in flower, ½ in. when in fruit. Bracts linear, scale-like at the base, tapering into subulate hirsute or ciliate points, those surrounding the partial heads rather broader. Female florets 6 to 8 or rather more, not exceeding the involucre; disk-florets 2 or 3, with simple styles. Achenes of the ray slightly hairy, those of the disk abortive.—Wight, Ic. t. 1094; F. Muell. Fragm. iii. 138.

N. Australia. Victoria river and Gulf of Carpentaria, *F. Mueller;* Albert river and Bentinck's Island, *Henne.*

Queensland. Sellheim river, *Bowman;* Maranoa and Belyando rivers, *Mitchell.*

The species is common in tropical Asia, extending into tropical Africa.

2. **S. microcephalus,** *Willd.; DC. Prod.* v. 369. Erect, 1 to 2 ft. high, quite glabrous or rarely minutely glandular-pubescent. Leaves elliptical, oblong or lanceolate, acute, with small acute teeth, decurrent along the stem into continuous entire or slightly toothed wings. Flower-heads in globular clusters or compound heads, rather smaller than in *S. hirtus.* Involucral bracts oblong-linear or cuneate, often jagged at the end. Florets and achenes of *S. hirtus*, but those of the disk fertile as well as those of the circumference.—*S. glaber*, DC. Prod. v. 370; F. Muell. Fragm. iii. 138.

N. Australia. Low flats, Alligator river, and Van Diemen's Gulf, N.W. coast, *A. Cunningham;* islands of the Gulf of Carpentaria, *R. Brown;* Albert river, *F. Mueller.* Also in Ceylon and in the Indian Archipelago.

22. MONENTELES, Labill.

Flower-heads small, sessile, in dense globular clusters or compound heads, interspersed with woolly bracts. Partial involucres ovoid, the bracts linear, usually glabrous and more or less scarious. Florets of the circumference numerous, female, filiform, minutely toothed, not exceeding the disk. Disk-florets solitary or rarely 2 or 3, tubular, hermaphrodite but usually sterile, 4- or 5-toothed. Anthers with more or less prominent tails or points at the base. Style-lobes flattened or almost subulate, papillose outside towards the end. Achenes small, terete or slightly compressed, those of the disk usually abortive. Pappus of capillary simple bristles united in a ring at the base.—Herbs or undershrubs, usually glandular-pubescent and strongly scented, often woolly. Leaves alternate, crenate or toothed, decurrent on the stem. Clusters or compound flower-heads solitary and terminal, or small and numerous in a terminal spike with one bract, usually persistent on the common receptacle under each partial head, the partial involucres often very deciduous.

The species are all Australian, but two of them extend also into New Caledonia and the Eastern Archipelago. With the compound inflorescence and decurrent leaves of *Sphæranthus* the genus is readily distinguished by the pappus and the tailed anthers. The inflorescence is nearly that of the *Angianthеæ*, but there are numerous female filiform florets, and the style-branches of the disk are not truncate. From the American genus *Pterocaulon* it scarcely differs in the hermaphrodite flowers usually reduced to a single one.

Clusters of flower-heads small but numerous, forming a terminal dense or interrupted spike.
- Involucres rigid, about 4 lines long. Leaves ovate or obovate, often above 3 in. long, the wool very dense, almost floccose . 1. *M. verbascifolius.*
- Involucres not 2 lines long. Leaves obovate or oblong, under 2 in long, tomentose or shortly woolly . . . . 2. *M. spicatus.*

Clusters of flower-heads solitary, globular or ovoid-oblong.
- Plant tomentose or woolly. Decurrent wings entire. Clusters globular . . . . 3. *M. sphacelatus.*
- Plant glandular-pubescent, not tomentose. Decurrent wings toothed.
  - Clusters ovoid or oblong, large. Disk-florets solitary . 4. *M. glandulosus.*
  - Clusters globular, rather small. Disk-florets usually 2 . 5. *M. sphæranthoides.*

1. **M. verbascifolius,** *F. Muell. Herb.* A tall erect perennial or undershrub, the foliage densely woolly, often floccose, resembling that of *Verbascum Thapsus.* Leaves ovate or obovate, 2 in. long or more, obtuse, crenulate and sometimes sinuate, very thick and soft, decurrent into entire wings. Clusters of flower-heads numerous, sessile in dense continuous oblong or cylindrical terminal spikes much larger than in *M. spicatus.* Bracts of the common receptacle short, linear, woolly-hairy as in the other species but less spathulate, those of the partial involucres 3 to 4 lines long, linear, rigidly scarious. Ray-florets very numerous; disk-florets solitary.

N. Australia. Glenelg river, N.W. coast, *Marten;* between Victoria and Fitzmaurice rivers, *F. Mueller.*

Queensland, *Bowman.*

2. **M. spicatus,** *Labill. Sert. Austr. Caled.* 43. *t.* 43. An erect perennial of 1 to 3 ft., softly tomentose, pubescent or woolly all over. Leaves obovate or oblong, obtuse, crenate, 1 to 2 in. long or smaller on the flowering branches, soft, rugose, decurrent into narrow entire wings. Clusters of flower-heads small, globular, sessile, forming terminal cylindrical and continuous or interrupted spikes, each cluster rarely above 3 or 4 lines diameter, and containing from 10 to 20 heads. Bracts of the common receptacle linear, acute, woolly, not 2 lines long, those of the partial involucres narrow-linear, acuminate, glabrous. Disk-florets solitary. Achenes sprinkled with a few hairs.—DC. Prod. v. 455.

Queensland. Flinders river, *Bowman;* Fitzroy river, *Fitzalan;* Rockhampton, *Dallachy;* Keppel Bay, *Thozet;* Brisbane river, Moreton Bay, *Fraser, F. Mueller.* The species is also in New Caledonia, Burmah and the Philippine Islands.

3. **M. sphacelatus,** *Labill. Sert. Austr. Caled.* 43. *t.* 44. A perennial or undershrub of 1 to 2 ft., softly tomentose-pubescent or woolly all over. Leaves oblong or lanceolate, obtuse or rarely acute, the larger ones 1 to 2 in. long, those on the flowering stems often smaller, entire or crenulate, soft, rugose, decurrent into narrow wings usually entire. Clusters of flower-heads globular or nearly so, solitary, terminal and pedunculate or rarely becoming lateral and sessile by the elongation of the shoot. Bracts of the common receptacle linear with dark spathulate tips, and densely clothed with long woolly hairs, those of the partial involucres linear, scarious, acute, glabrous or slightly ciliate. Female-florets numerous; disk-florets usually solitary. Style-lobes almost subulate. Achenes glabrous.—DC. Prod. v. 456; *M. globiferus* and *M. intermedius,* DC. l. c. 455.

N. Australia. Nichol Bay, N.W. coast. *Herb. F. Mueller;* Victoria river and M'Adam Range, *F. Mueller;* Albert river, *Henne;* Gulf of Carpentaria, *Landsborough;* Attack Creek and M'Donnell ranges, *M'Douall Stuart's Expedition.*

Queensland. Common along the sandy shores and in the adjacent islands, *R. Brown* and others; Brisbane river, Moreton Bay, *A. Cunningham;* Pine river, *Fitzalan.*

N. S. Wales. Darling river, *Dallachy and Goodwin.*

S. Australia. Towards Spencer's Gulf, *Warburton.*

The species is also in New Caledonia and New Guinea.

4. **M. glandulosus,** *F. Muell. Herb.* A tall erect strongly-scented perennial or undershrub, more or less glandular-pubescent but not woolly. Leaves ovate-lanceolate or lanceolate, acute and acutely toothed, bullate-

rugose, 1 to 2 in. long, decurrent into irregularly and acutely-toothed wings. Clusters of flower-heads terminal and solitary, ovoid or oblong, ½ to 1 in. long. Bracts of the common receptacle linear or linear-cuneate, ciliate with a few long woolly hairs, but not densely woolly as in the other species, those of the partial involucres about 2 lines long, linear, acute, often ciliate with a few hairs. Female-florets numerous; disk-florets solitary.

Queensland. Keppel and Shoalwater bays, Thirsty Sound and Broad Sound, *R. Brown;* rocky hills, Cape Cleveland, *A. Cunningham;* Gilbert river, *F. Mueller;* Flinders and Fitzroy rivers, *Bowman;* Rockhampton and Keppel Bay, *Thozet.*

5. **M. sphæranthoides,** *DC. Prod.* v. 456. A strong-scented shrub or undershrub of 2 to 3 ft., more or less glandular-pubescent or at length glabrous. Leaves linear-lanceolate, obtuse, 1 to 2 in. long on the barren shoots, smaller on the flowering-stems, very rugose or bullate, with revolute toothed margins decurrent into narrow toothed wings. Clusters of flower-heads terminal, solitary, globose or slightly ovoid, fully ½ in. diameter. Bracts of the common receptacle densely crowded, linear, very woolly-hairy, the spathulate tips glandular. Partial involucres very numerous, the bracts linear, scarious, white, glabrous or slightly ciliate. Female-florets numerous; disk-florets usually 2, very rarely 1 or 3. Style-lobes linear-lanceolate. Achenes small, glabrous.

N. Australia. Enderby Island, Dampier's Archipelago, N.W. coast, *A. Cunningham* (the specimens all past flower, showing the persistent bracts of the common receptacle with a few remains of the partial heads); N.W. coast, *Bynoe;*, granite hill, Nichol Bay, *Gregory's Expedition.*

23. BLUMEA, DC.

Involucre ovoid or campanulate, the bracts imbricate in several rows, narrow-linear, herbaceous or soft. Receptacle nearly flat, without scales. Florets all tubular, those of the circumference female, filiform, usually very numerous, those of the disk hermaphrodite, broader, usually few. Anthers with short fine tails or points at the base. Style-branches filiform, papillose outside at the end. Achenes small, usually somewhat compressed, striate or ribbed. Pappus of numerous capillary bristles.—Annual or perennial herbs, usually villous woolly or glandular-pubescent. Leaves alternate, toothed, lobed or rarely entire. Flower-heads in terminal pyramidal or oblong panicles, sometimes contracted into spikes or clusters, or rarely the heads solitary.

A considerable genus, confined to the warmer regions of the Old World, some of them ubiquitous and variable weeds, upon which a large number of spurious species have been fabricated, the real limits of the more distinct forms often very difficult to establish. Of the seven Australian species three appear to be endemic, the other four belong to some of the more common Asiatic forms. The genus is very closely allied to *Conyza* and to *Pluchea,* having very nearly the same florets and achenes, with the involucres and habit of the former and the tailed anthers of the latter.

Flower-heads all distinctly pedunculate.
  Glandular-pubescent, not tomentose.
    Leaves all or almost all narrowed into a petiole. Usually above
      1 ft. high . . . . 1. *B. glandulosa.*

Stem-leaves sessile and stem-clasping. Radical leaves only petiolate. Rarely exceeding 1 ft. in height . . . . . 2. *B. amplectens.*
Usually not exceeding 1 ft., pubescent or villous, not viscid. Stem-leaves sessile and stem-clasping, mostly lanceolate. Peduncles rather long . . . . 3. *B. integrifolia.*
Small. Leaves chiefly radical. Stem-leaves few, small, ovate. Peduncles long and filiform . . . . 4. *B. diffusa.*
Flower-heads, at least the upper ones, sessile and clustered.
Leaves lanceolate, oblong or obovate-oblong, the upper ones sessile.
Involucre about 4 lines long . . . . 5. *B. hieracifolia.*
Involucre under 3 lines long . . . . 6. *B. Cunninghamii.*
Almost all the leaves petiolate and broadly obovate. Involucre 2 to 3 lines long . . . . 7. *B. lacera.*

1. **B. glandulosa,** *DC. Prod.* v. 438. An erect annual, often 2 to 3 ft. high, more or less covered with a glandular-viscid pubescence, with sometimes a little wool at the base of the leaves. Leaves obovate ovate or lanceolate, acutely or coarsely toothed, green on both sides, the lower ones often 3 to 5 in. long and almost lobed at the base, on long stalks, the upper ones smaller and narrower but all petiolate. Flower-heads about 4 lines long, all pedunculate or rarely in some Indian specimens a few of the upper ones almost sessile, forming a large pyramidal panicle.

Queensland. Endeavour river, *Banks and Solander;* Broad Sound and Shoalwater Bay, *R. Brown;* Port Molle, *M'Gillivray.*—The species is widely spread over E. India and the Archipelago, extending to S. China.

Var. *minor.* Not exceeding 1½ ft. in height. Leaves and flower-heads smaller. Port Molle, *M'Gillivray;* Rockhampton, *Thozet.*

2. **B. amplectens,** *DC. Prod.* v. 433. An annual with erect or ascending stems, usually under 1 ft. high, more or less covered with a glandular-viscid pubescence. Radical leaves obovate and petiolate, the stem leaves all sessile and stem-clasping, from ovate to lanceolate, acutely toothed and usually small. Flower-heads all pedunculate, not numerous, in a loose panicle, very divaricate in Indian specimens, less so in the Australian ones. Involucres nearly 3 lines long. Style-branches of the disk-florets long and slender.

Queensland. Keppel Bay, *Thozet;* Connor river, *Bowman.* The species is common in E. India.

3. **B. integrifolia,** *DC. Prod.* v. 433. A rather slender annual, erect or branching from the base, rarely above 1 ft. high, pubescent or hirsute, and perhaps sometimes viscid, the young shoots occasionally silky-woolly. Leaves oblong-lanceolate or rarely the lower ones almost obovate, acute and bordered by distant acute teeth, rather rigid and often scabrous, the lower ones 1 to 3 in. long, the upper ones smaller, narrowed towards the base or broad and stem-clasping at the base. Flower-heads not clustered, the peduncles at first short but generally long and slender when the flowers are fully out. Involucres 2 to 3 lines long.

N. Australia. Port Keatts, N.W. coast, *A. Cunningham;* Victoria river, M'Adam Range, Roper river, *F. Mueller.*

4. **B. diffusa,** *R. Br. Herb.* A small slender and diffuse annual, slightly viscid, with a very little wool at the base of the stems, rarely exceed-

ing 6 in. including the peduncles. Leaves chiefly radical and rosulate, petiolate, obovate-oblong, ½ to 1 or rarely 2 in. long; those on the stem few, small, ovate, stem-clasping. Peduncles long and very slender. Flower-heads small, glabrous or slightly pubescent.

N. Australia. Islands of the Gulf of Carpentaria, *R. Brown;* towards M'Adam Range, *F. Mueller*.

5. **B. hieracifolia,** *DC. Prod.* v. 442. An erect stiff almost simple annual, 1 to 2 ft. high, more or less tomentose or densely villous, but not viscid. Leaves oblong, irregularly and sharply toothed, the lowest nearly obovate, 2 to 3 in. long, and petiolate, the others sessile or nearly so but narrowed at the base, the uppermost almost lanceolate. Flower-heads about 4 lines long, mostly sessile and clustered, the lower clusters distant, the upper ones forming a terminal leafy spike, or more rarely branching into an oblong terminal panicle. Involucres always tomentose, the bracts linear acute and soft as in the rest of the genus, but rather broader than in the following 2 species.—Wight, Ic. t. 1099 (representing a dwarf form).

N. Australia. Between Providence Hill and M'Adam Range, *F. Mueller*.

Queensland. Broad Sound, *R. Brown;* Keppel Bay, *Thozet;* Moreton Bay, *Leichhardt;* also a doubtful fragment in *Bowman's* collection.

The species is one of the common ones in tropical Asia from Ceylon and the Peninsula to the Archipelago, and northwards to S. China and Formosa. R. Brown's and F. Mueller's specimens belong to the very villous var. *holosericea*, Benth. Fl. Hongk. 178, or *B. holosericea*, DC. Prod. v. 442. Leichhardt's and Thozet's are less villous, the leaves of the former nearly glabrous and on longer petioles.

6. **B. Cunninghamii,** *DC. Prod.* v. 435. An erect stiff scarcely branched annual, pubescent or silky-villous. Leaves as in *B. hieracifolia*, the lower ones petiolate and nearly obovate, the uppermost lanceolate often stem-clasping, all irregularly and acutely toothed. Flower-heads 2 to 3 lines long, mostly sessile in small clusters, forming an oblong leafy panicle. Involucres of *B. lacera*.

N. Australia. Careening and Brunswick bays, N.W. coast, *A. Cunningham;* Point Pearce, Victoria river, *F. Mueller;* Gulf of Carpentaria, *R. Brown*.—This plant does not appear to be exactly matched in any of our Indian specimens, although closely allied both to the preceding and the following species. It differs from *B. hieracifolia* chiefly in the smaller flower-heads, from *B. lacera* in the more sessile leaves.

7. **B. lacera,** *DC. Prod.* v. 436. An erect annual, 1 to 2 ft. high, simple or branched, not usually so stiff as *B. hieracifolia*, more or less clothed with soft whitish hairs or pubescence. Leaves all petiolate, obovate ovate or rarely oblong, coarsely toothed or almost lyrate. Flower-heads seldom above 3 lines long and often not above 2 lines, very numerous in narrow oblong and dense or looser and more spreading panicles, leafy at the base, the upper heads sessile and clustered on the branches. Involucral bracts usually very narrow.

N. Australia. Victoria river, *F. Mueller*.

Queensland. Shoalwater Bay, *R. Brown;* Burdekin Expedition, *F. Mueller;* Rockhampton, *Dallachy;* Brisbane river, Moreton Bay, *Fraser*, *F. Mueller*.

This is one of the commonest weeds in tropical Asia, extending from tropical and subtropical Africa to the Archipelago and S. China. It is also extremely variable and sometimes difficult to recognize in the bad specimens often preserved in herbaria. The Australian ones I have seen belong chiefly to, a slender loosely-branched variety with long-petioled thin leaves, often distinguished as a species under the name of *B. Wightiana*, DC. Prod. v. 435; Benth. Fl. Hongk. 178, but which Thwaites appears to be right in reducing to *B. lacera*. The Brisbane river specimens are on the other hand nearly as tall and rigid as *B. hieracifolia*, and the flower-heads are rather larger than usual in *B. lacera*, but the leaves are mostly lyrate and all petiolate, and they appear to represent only a luxuriant state of *B. lacera*.

24. PLUCHEA, Less.

(Spiropodium *and* Eyrea, *F. Muell.*)

Involucre either ovoid with the bracts imbricate in several rows, usually broader, more rigid and less acuminate than in *Blumea*, or hemispherical with narrow bracts. Florets all tubular, those of the circumference female, filiform, usually very numerous, those of the disk broader, hermaphrodite, but usually sterile, few, or the heads almost diœcious. Anthers with short fine tails or points at the base. Style (of the disk-florets) simple, papillose towards the end, or very rarely branched as in *Blumea*. Achenes small, usually somewhat compressed. Pappus of numerous capillary bristles.—Shrubs, undershrubs, or perennial herbs (rarely if ever annuals). Leaves alternate, entire toothed or rarely almost pinnatifid. Flower-heads in terminal corymbs, sometimes contracted into clusters or rarely solitary.

The genus comprises several N. and S. American species, a very few from Africa and tropical and subtropical Asia, besides the Australian ones, five of which are endemic, the sixth a common Asiatic one. They differ generally from *Blumea*, in habit as well as in the involucre and style, but none of these characters are quite constant. They have some affinity also with *Pterigeron*, but the female florets are never ligulate, and the style is scarcely or not at all bulbous at the base.

SECTION I. **Pluchea.**—*Flower-heads ovoid. Involucral bracts lanceolate or the outer ones ovate.*

Shrub of 3 to 4 ft. Leaves obovate. Flower-heads in dense terminal corymbs 2 to 3 in. diameter . . . . 1. *P. indica.*
Herbs or undershrubs of 1 to 2 ft. Flower-heads in loose leafy corymbose panicles, solitary or in small clusters on the branches.
Leaves obovate or oblong-lanceolate . . . . 2. *P. tetranthera.*
Leaves linear . . . . 3. *P. baccharoides.*

SECTION II. **Eyrea.**—*Flower-heads broad or hemispherical. Involucral bracts narrow.*

Flower-heads about ¼ in. diameter, the bracts all dry . . . . . . 4. *P. Eyrea.*
Flower-heads nearly ½ in. diameter, the outer bracts with reflexed tips often leafy.
Inner bracts with acuminate coloured tips. Western species . . 5. *P. squarrosa.*
Nearly all the bracts with fine hair-like tips. Eastern tropical species . . . . 6. *P. dentex.*

SECTION I. PLUCHEA.—Flower-heads ovoid. Involucral bracts lanceolate or the outer ones ovate.

1. **P. indica,** *Less.;—DC. Prod.* v. 451. An erect branching shrub, attaining 3 to 4 ft., glabrous or covered with a minute glandular pubescence. Leaves shortly petiolate, obovate, oblong or rarely ovate, 1 to 2 in. long, with a few small acute teeth or almost entire. Flower-heads in dense terminal

corymbs of about 3 in. diameter and sessile above the last leaves. Involucres ovoid, about 3 lines long, the outer bracts short and obtuse, passing into the innermost acute ones. Female florets very numerous, those of the disk rarely above 6. Anther-tails rather long.—Wight, Illustr. t. 131.

N. Australia. Islands of the Gulf of Carpentaria and the opposite mainland, *R. Brown;* Port Essington, *Armstrong.* Extends over E. India and the Archipelago to S. China.

2. **P. tetranthera,** *F. Muell. Rep. Babb. Exped.* 12. A rigid perennial or undershrub, glabrous or glandular-pubescent in the ordinary forms. Leaves petiolate, from small and obovate to oblong-lanceolate and 1 in. long or rather more, the upper ones sometimes linear, acutely and irregularly toothed or almost entire, those of the barren shoots usually obovate. Flower-heads small, often sessile in clusters of 2 or 3, forming a terminal corymbose panicle. Involucre narrow-ovoid, scarcely 3 lines long, the outer bracts short and obtuse, the inner acute rigid and dry. Florets rather shorter than the involucre, in some heads the female filiform ones very numerous, with only 2 or 3 disk-florets, in others those of the disk at least as numerous as the female ones. Corollas of the disk 4-toothed. Style with 2 filiform papillose branches.

N. Australia. Victoria, Flinders, and Van Alphen rivers, *F. Mueller.*

Queensland. Broad Sound, *R. Brown;* heads of Isaacs river, *Bowman;* Rockhampton, *Thozet.*

Var. *tomentosa.* Taller, leaves larger, and the whole plant closely but softly tomentose. —Arnhem's Land, *F. Mueller.*

3. **P. baccharoides,** *F. Muell. Herb.* A much-branched erect shrub of a few feet, glabrous and somewhat glutinous. Leaves narrow linear, mostly acute, ½ to 1 in. long, rather thick, quite entire. Flower-heads small, sessile or shortly pedunculate in terminal leafy panicles, almost unisexual, the males with ovoid involucres about 2 lines long, the florets chiefly hermaphrodite but sterile, the females narrower and longer, the florets chiefly filiform and shorter than the pappus. Involucral bracts in both dry, much imbricate, minutely ciliolate. Style in the males undivided. Pappus bristles not numerous.—*Spiropodium baccharoides*, F. Muell. Fragm. i. 34.

Queensland. Suttor river, *F. Mueller;* Belyando river, *Mitchell.* This species at first sight much resembles some of the common S. American *Baccharises*, allied to *B. paniculata*, DC., but differs in its terete not angular branchlets, in the shorter florets, in the tailed anthers, and in the heads not perfectly diœcious. I have always found a few central hermaphrodite sterile florets in the female heads and a few filiform female florets in the male heads.

SECTION II. EYREA.—Flower-heads broad or hemispherical. Involucral bracts very narrow.

4. **P. Eyrea,** *F. Muell. Rep. Babb. Exped.* 11 and 12. A glabrous perennial or undershrub, often under 1 ft. high, with erect virgate branches paniculate at the top. Leaves mostly linear, erect, bordered by a few small teeth, under ½ in. long, more or less decurrent along the stem, a few on the main stem lanceolate and broadly decurrent. Flower-heads small, usually numerous, in little corymbs of 3 to 5 each. Involucres broadly ovoid or

hemispherical, the bracts much imbricate, narrower than in the preceding species, the inner ones 2 lines long with subulate points. Female florets very numerous, scarcely thicker than the bristles of the pappus; disk-florets also numerous, with an undivided exserted style. Anther-tails minute. Achenes terete or nearly so.—*Eyrea rubelliflora*, F. Muell. in Linnæa, xxv. 403.

N. Australia. Islands of the Gulf of Carpentaria, *R. Brown;* Arnhem's Land, *F. Mueller;* Nichol Bay, *F. Gregory's Expedition* (a slender form with smaller flower-heads).
Queensland. Keppel Bay and Broad Sound, *R. Brown;* in the interior, *Mitchell.*
W. Australia. Murchison river, *Oldfield, Drummond, 6th Coll. n.* 163.

Var.? *major.* Taller, more divaricately branched. Flower-heads larger and less corymbose. Inner involucral bracts about 3 lines long. Style sometimes branched.
N. S. Wales. Darling river, *Victorian Expedition.*
S. Australia. Dried river-beds, Crystal Brook, Arkaba, *F. Mueller.*

5. **P. squarrosa,** *Benth.* Erect, glabrous or minutely glandular-pubescent, with virgate corymbose branches, 1 to 2 ft. high. Leaves linear linear-cuneate or lanceolate, bordered by a few distant teeth or the larger ones almost pinnatifid, narrowed at the base, often above 1 in. long. Flower-heads much larger than in *P. Eyrea.* Involucres hemispherical, fully ½ in. diameter, the bracts numerous, with herbaceous reflexed tips, the innermost with acuminate, erect, coloured tips. Female florets very numerous, slender, but scarcely filiform, rather longer than the involucre, the style-branches protruding. Disk-florets several, with an undivided style. Achenes pubescent.

W. Australia. Murchison river, *Oldfield, Drummond, 6th Coll. n.* 150.

6. **P. dentex,** *R. Br. Herb.* Erect, much branched, with a hard almost woody base, 1 to 2 ft. high, viscid and minutely pubescent. Leaves linear, mostly above 1 in. long, remotely toothed or pinnatifid with short distant lobes, the upper ones small and entire. Flower-heads larger than in *P. Eyrea.* Involucres hemispherical, the bracts very narrow, the inner ones with a fine hair-like point, a few of the outer ones greener and more recurved. Female florets very numerous, those of the disk usually few. Achenes pubescent.

Queensland. Broad Sound and Thirsty Sound, *R. Brown;* sources of Gilbert river, *F. Mueller;* Port Denison, *Fitzalan.*

25. EPALTES, Less.

(Sphæromorphæa, *DC.* (*partly*); Ethuliopsis, *F. Muell.*)

Involucre ovoid-globular or hemispherical, the bracts imbricate, usually broad, dry and rigid. Receptacle nearly flat, without scales. Florets all tubular, those of the circumference numerous, female, very slender, minutely toothed, those of the disk hermaphrodite, but usually sterile, broader, 3- to 5-toothed, either few or the flower-heads almost diœcious. Anthers with minute tails or points at the base. Style of the disk-flowers undivided or with short obtuse branches, papillose outside. Achenes nearly terete, striate, those of the ray without any pappus, those of the disk usually abortive and with a pappus of 2 or 3 very deciduous bristles.—Herbs, either erect and dichotomous or diffuse. Leaves alternate, entire, toothed or lobed. Flower-heads small in dichotomous cymes or lateral and sessile. Involucral bracts very obtuse in the Australian species, more acute in the Indian *E. divaricata*.

A small genus, comprising, besides the two Australian species (one of which is endemic, and the other only recently found in Formosa), at least one more, a common one in tropical Asia and some parts of Africa. The genus is closely allied to *Pluchea*, differing chiefly in the absence of any pappus to the fertile achenes.

Erect, dichotomously branched. Flower-heads ovoid-globular, terminal, and loosely clustered, often diœcious . . . . 1. *E. Cunninghamii.*
Diffuse annual. Flower-heads hemispherical, lateral, sessile or nearly so . . . . 2. *E. australis.*

1. **E. Cunninghamii,** *Benth.* Glabrous, erect, ½ to 1½ ft. high, dichotomously branched. Leaves sessile and half stem-clasping or sometimes slightly decurrent on the stem, oblong-lanceolate, obtuse, irregularly toothed, the upper ones small, narrow, and sometimes entire. Flower-heads small, clustered at the ends of the branches of loose dichotomous cymes. Involucres ovoid-globular, varying from 1 to 2 lines diameter, the bracts very obtuse. Female florets not exceeding the involucre, in some specimens very numerous with only 1 or 2 hermaphrodite sterile florets in the disk or none at all, in others the female florets few with rather numerous sterile ones. Style of the disk-florets undivided. Fertile achenes without any pappus, sterile ones with a few long deciduous bristles.—*Epaltes australis*, DC. Prod. v. 462, not of Less.; *Ethulia Cunninghamii*, Hook. in Mitch. Trop. Austr. 62; *Ethuliopsis dioica*, F. Muell. Fragm. ii. 155; Pl. Vict. t. 38.

Queensland. Keppel and Shoalwater bays, *R. Brown.*

N. S. Wales. Barren marshes and inundated tracts near the Lachlan and Macquarrie rivers, *Fraser*, *A. Cunningham*, *Mitchell;* Darling and Murray rivers, and thence to the Barrier Range, *Victorian and other Expeditions.*

Victoria. Murray desert, Lake Lalbert, *F. Mueller.*

2. **E. australis,** *Less. in Linnæa*, v. 148, *and Syn. Comp.* 206. Annual (or sometimes perennial?), branching at the base and diffuse or prostrate, small or rarely exceeding 1 ft., glabrous scabrous or hirsute with transparent hairs. Leaves petiolate, obovate or cuneate-oblong, entire toothed or almost lyrate, ½ to 1½ in. long. Flower-heads lateral, sessile or shortly pedunculate. Involucre depressed-hemispherical, 2 to 3 lines diameter, the bracts orbicular concave and very obtuse. Female florets not exceeding the involucre, short, and not so slender as in *E. Cunninghamii*, and very numerous; disk-florets fewer, 3- to 5-lobed. Style usually branched with short narrow obtuse lobes. Achenes all without any pappus, those of the disk mostly, but not always sterile.—*Sphæromorphæa petiolaris*, DC. Prod. vi. 140.

N. Australia. Victoria river to Arnhem's Land, *F. Mueller;* in the interior, *M'Douall Stuart's Expedition;* islands of the Gulf of Carpentaria, *R. Brown.*

Queensland. Keppel Bay, *R. Brown*, *Thozet;* E. coast, *A. Cunningham;* Wide Bay, *Bidwill;* Walloon, *Bowman;* Brisbane river, Moreton Bay, *F. Mueller.*

N. S. Wales. Port Jackson, *R. Brown*, *Woolls*, and others.

S. Australia. Murray river, *F. Mueller;* Cooper's Creek, *D. Murray.*—Recently received also from Formosa (*R. Oldham*). The species has, at first sight, some resemblance with *Myriogyne minuta*, especially in inflorescence, but besides the distinctly tailed anthers and other floral characters, it is readily distinguished by the involucre.

26. PTERIGERON, DC.

(Streptoglossa, *Steetz*; Oliganthemum, *F. Muell.*)

Involucre hemispherical or ovoid, the bracts imbricate in several rows, usually dry and rigid or the outer ones herbaceous, the innermost narrow, acute, often coloured at the tips. Receptacle without scales. Florets of the circumference numerous or few, female, ligulate or, if tubular, less regularly or more deeply lobed than those of the disk; disk-florets numerous or few, hermaphrodite, fertile or sterile, usually 5-lobed. Anthers with fine tails. Style bulbous at the base, the lobes subulate or slightly flattened, sometimes united to the top, papillose outside towards the end. Achenes nearly terete, silky-hairy. Pappus of scabrous denticulate or almost plumose capillary bristles.—Herbs, usually rigid and glandular-pubescent or hirsute. Leaves alternate, entire or toothed, sometimes decurrent. Flower-heads large or small, terminal, and usually forming terminal leafy corymbs.

The genus is exclusively Australian. Originally established by De Candolle as a section of *Erigeron*, it has been shown by A. Gray to form a very distinct genus, allied in many respects to *Pluchea*, but differing in the prominently bulbous base of the style, and usually in the irregularity of the ray-florets, which connect it with *Dicoma*, amongst *Mutisiaceæ*. It ought perhaps to include *Coleocoma* and *Thespidium*, notwithstanding their anomalous pappus.

- Flower-heads large, almost hemispherical.
  - Ray-florets ligulate, exceeding the involucre.
    - Leaves ovate, decurrent. Ray-florets not ¼ line broad . . 1. *P. decurrens.*
    - Leaves narrow or obovate, not decurrent. Ray-florets fully ½ line broad . . . . 2. *P. liatroides.*
  - Ray-florets very slender, about as long as the involucre. Leaves oblong, stem-clasping or slightly decurrent . . . . . . 4. *P. macrocephalus.*
- Flower-heads ovoid.
  - Leaves decurrent. Ray-florets about as long as the involucre . 3. *P. odorus.*
  - Leaves not decurrent. Ray-florets very slender and shorter than the involucre.
    - Leaves mostly oblong. Flower-heads broadly ovoid, 6 to 8 lines long . . . . 5. *P. microglossus.*
    - Leaves mostly linear. Flower-heads numerous, many-flowered, about 4 lines long . . . . 6. *P. adscendens.*
    - Leaves filiform. Flower-heads narrow, 3- or 4-flowered, about 5 lines long . . . . 7. *P. filifolius.*

1. **P. decurrens,** *DC. Prod.* v. 293 (as an *Erigeron*). A rigid branching herb, glandular-pubescent and hirsute. Leaves ovate or oblong, coarsely and irregularly toothed or almost entire, the upper ones broadly decurrent. Flower-heads pedunculate, in a terminal leafy panicle, irregularly corymbose. Involucre ovoid when young, nearly hemispherical and ½ in. diameter when fully out, the outer bracts herbaceous, surrounded by a few floral leaves, the inner ones lanceolate, scarious, very acute, 5 lines long. Ray-florets ligulate but very narrow, longer than the involucre; disk-florets numerous. Style undivided. Achenes clothed with long silky hairs. Pappus almost plumose.—*Streptoglossa Steetzii*, F. Muell. in Trans. Bot. Soc. Edinb. vii. 491 (partly?).

N. Australia. N.W. coast, *Bynoe*; Nichol Bay, *F. Gregory's Expedition.*

2. **P. liatroides,** *Benth.* Erect, rigid and scabrous-pubescent, hirsute or at length nearly glabrous, attaining sometimes 1½ ft., but often flowering when a few in. high. Lower leaves obovate, narrowed into a rather long petiole, the upper ones oblong-cuneate or linear, obtuse, sinuate-toothed or nearly entire, not decurrent. Flower-heads large, not numerous, in a terminal corymb. Involucre hemispherical, nearly 1 in. diameter, the bracts broadly lanceolate, acute or acuminate, the inner ones 7 to 8 lines long and coloured at the apex. Ray-florets ligulate, exceeding the involucre, the ligula fully ½ line broad and 2 lines long, usually 3-lobed. Style of the disk-florets 2-lobed. Achenes silky-villous. Pappus-bristles not numerous, almost plumose.—*Pluchea ligulata,* F. Muell. Rep. Babb. Exped. 12; *Streptoglossa Steetzii,* F. Muell. in Trans. Bot. Soc. Edinb. vii. 491, partly; *Erigeron liatroides,* Turcz. in Bull. Mosc. 1851, i. 172.

W. Australia, *Drummond, n.* 100 and 222; Irvine river, *Oldfield.*

Var. *humilis.* Smaller and more hirsute, the stems in our specimens branching from the base, not above 6 in. high, each usually with a single flower-head.

N. or **S. Australia.** In the interior, Lake Gregory, *G. Hawker;* Strangways river, *M'Douall Stuart's Expedition.*

3. **P. odorus,** *Benth.* Apparently perennial and in the normal form erect, rigid, more or less hirsute, with virgate branches, under 1 ft. high. Leaves sessile and more or less decurrent, the lower ones oblong or lanceolate, obtuse, irregularly toothed, above 1 in. long, the upper ones smaller and linear. Flower-heads mostly terminating very short branches, sessile within 2 or 3 floral leaves. Involucre narrow-ovoid, the bracts imbricate and rigid, the outer ones short, acute or almost obtuse, the inner ones 4 lines long and very acute. Ray-florets about 8 to 10, about as long as the involucre, slender, with 2 or 3, rarely 4 or 5 narrow lobes, more or less distinctly arranged in 2 lips. Style of the disk-florets 2-lobed, the branches slightly flattened and obtuse. Achenes silky-villous. Pappus-bristles denticulate, almost plumose.—*Pluchea odora,* F. Muell. Rep. Babb. Exped. 12.

N. Australia. Victoria river, *F. Mueller.*

Var. ? *major.* Flower-heads larger with more numerous florets. Both the florets and pappus seem to be rather those of *P. odorus* than of *P. adscendens.* The specimens are however very incomplete.—Albert river, *F. Mueller;* Suttor river, *Bowman.*

4. **P. macrocephalus,** *Benth.* Glandular-pubescent or hirsute. Leaves oblong or linear-cuneate, irregularly undulate and toothed, stem-clasping, but only rarely and very slightly decurrent. Flower-heads large, almost hemispherical when fully out, ¾ in. long. Ray-florets not numerous, very slender, about as long as the involucre, 2- or 3-lobed at the end, but not usually ligulate; disk-florets very numerous. Achenes densely silky-hairy.—*Pluchea macrocephala,* F. Muell. Rep. Babb. Exped. 12.

N. Australia. Gulf of Carpentaria, *F. Mueller;* in the interior, lat. 19° 30′, *M'Douall Stuart's Expedition.*

5. **P. microglossus,** *Benth.* Glabrous or slightly glandular-pubescent, the stems rigid, erect, nearly simple or somewhat corymbose, often under 6 in. and none of our specimens above 8 in. high. Leaves from cuneate-oblong to linear-cuneate, entire or denticulate, under 1 inch long, rather rigid,

narrowed towards the base, not decurrent. Flower-heads broadly ovoid, about ½ in. long, the inner involucral bracts very acute. Florets numerous, not exceeding the involucre, those of the ray scarcely so long, with a small, narrow, entire or bifid concave lamina. Achenes less hairy than in *P. ligulatus*. Pappus almost plumose.

N. Australia. Sturt's Creek and Fitzmaurice rivers, *F. Mueller*.

6. **P. adscendens,** *Benth.* A diffuse, ascending or erect, very much branched herb, with a hard almost woody base, glabrous or pubescent. Leaves linear or linear-oblong, narrowed into a petiole, entire or slightly toothed, not decurrent. Flower-heads numerous, rather smaller than in *P. odorus*. Involucre ovoid, the bracts acute, usually but not always surrounded by a few floral leaves. Florets more numerous than in *P. odorus*, those of the ray filiform, scarcely so long as the involucre, with 2 or 3 short linear lobes. Pappus-bristles very minutely denticulate.

Queensland. Suttor, Roper, and Flinders rivers, *F. Mueller*; Cape river, *Bowman*; Belyando river, *Mitchell*.

7. **P. filifolius,** *Benth.* A slender, erect, glabrous annual with filiform dichotomous branches, attaining about 6 in. Leaves filiform, all entire in our specimens, mostly ½ to 1 in. long. Flower-heads solitary or in clusters of 2 or 3. Involucre narrow, 4 to 5 lines long, the bracts rigid, the inner ones very acute. Florets usually 3 or 4, of which 2 or 3 female, slender, shorter than the involucre, 2- or 3-lobed at the top, and apparently one or sometimes none, hermaphrodite or male, but with the corolla already fallen off in all the heads I have examined. Achenes very silky-villous. Pappus rather shorter than the involucre, of numerous unequal bristles, some almost plumose, some entire.—*Oliganthemum filifolium*, F. Muell. Herb.; *Pluchea filifolia*, F. Muell. in Trans. Phil. Inst. Vict. iii. 56.

Queensland. Salsola plains, near the Roper river, *F. Mueller*.—The specimens are old, and I have not succeeded in finding any of the stamen-bearing corollas.

27. COLEOCOMA, F. Muell.

Involucre ovoid, the bracts imbricate in several rows, dry, with slightly scarious tips. Receptacle flat, without scales. Florets all tubular, those of the circumference slender, female, 3- to 5-toothed; disk-florets several, hermaphrodite, sterile, 5-toothed. Anthers tailed. Style of the disk-florets usually undivided. Achenes striate, somewhat compressed, those of the disk abortive. Pappus of linear rigid scale-like bristles, those of the female florets united in a long tube, jagged at the end, those of the disk-florets free almost to the base.—Low rigid herb. Leaves alternate, usually toothed. Flower-heads terminal or lateral.

The genus consists but of a single species, endemic in Australia, differing from *Pterigeron* only in the pappus.

1. **C. centaurea,** *F. Muell. in Hook. Kew Journ.* ix. 19. A low, rigid, erect branching herb, perhaps annual, although almost woody at the base, our specimens quite glabrous. Leaves linear or lanceolate, acute, with a few small acute teeth, contracted at the base, but the upper ones sessile or slightly

decurrent. Flower-heads terminal and sessile within the last leaves or at the base of the lateral branches. Involucre 4 to 5 lines long, the bracts very broad, the inner ones with short broad scarious tips, jagged almost as in *Centaurea.* Florets yellow (F. Muell.). Achenes rather long, but much shorter than the involucre in our specimens, the tips of the pappus of the sterile florets slightly protruding.

N. Australia. Sturt's Creek, *F. Mueller.*

28. THESPIDIUM, F. Muell.

Involucre ovoid, the bracts imbricate in several rows, dry, rigid, and acute. Florets of the circumference female, filiform, shorter than the involucre, 2- or 3-toothed; disk-florets few or only one, hermaphrodite, fertile, 4-toothed. Anthers with minute tails. Style bulbous at the base, the lobes nearly terete, papillose outside towards the end. Achenes cylindrical, striate. Pappus of 10 to 12 bristles, broad and chaff-like at the base, intermixed with a few small setaceous ones.—Low rigid herb. Leaves alternate, entire or toothed. Flower-heads small, often densely clustered in the axils of the leaves and at the base of the stem.

The genus consists of a single species endemic in Australia. It is closely allied to *Pterigeron,* differing chiefly in habit and in the almost scale-like pappus.

1. **T. basiflorum,** *F. Muell. Herb.* A densely-tufted hirsute perennial, with numerous diffuse or ascending stems all under 2 ft. long, crowded on the stock with clusters of flower-heads like some species of *Lepidagathis.* Leaves linear or lanceolate, acute, entire or with a few acute teeth, rarely much above 1 in. long. Flower-heads sessile, solitary or more frequently clustered in the axils and very densely so at the base of the stem. Involucre narrow, about 3 lines long. Female florets rather numerous, those of the disk few or sometimes only one. Achenes of the ray and of the disk slightly contracted at the top and again expanded into the pappus.—*Pluchea basiflora,* F. Muell. Rep. Babb. Exped. 12.

N. Australia. Islands of the Gulf of Carpentaria, *R. Brown;* Port Essington, *Armstrong;* Upper Gilbert river, *F. Mueller.*

TRIBE V. AMBROSIEÆ.—Leaves alternate. Flower-heads strictly unisexual. Anthers not united.

29. XANTHIUM, Linn.

Flower-heads monœcious. Males globular. Involucral bracts small, in a single row. Receptacle cylindrical, with chaffy scales between the florets. Florets numerous, tubular, 5-toothed, without styles. Female heads ovoid. Involucral bracts in 2 or 3 rows, the outer small, the 2 innermost large, consolidated into a hard ovoid 2-celled mass very prickly outside and terminating in 2 tubercles or cones. Florets 2, without corollas. Ovaries each immersed in one of the cells of the involucral mass, the styles protruding with filiform branches. Achenes obovoid, inclosed in the burr-like involucre.

Pappus none. Coarse annuals. Leaves alternate. Flower-heads in axillary or terminal clusters or short racemes.

A genus of two or perhaps three species, natives of the Mediterranean region and the Levant, or one perhaps of Chilian origin, all now spread as weeds over many parts of the world.

*1. **X. spinosum,** *Linn.; DC. Prod.* v. 523. A rigid much-branched annual of 1 to 2 ft., studded with numerous strong trifid spines usually opposite in pairs or rarely solitary, divaricate, placed at the base of the leaves but rather within the petiole. Leaves lanceolate, trifid, with the central lobe much longer than the lateral ones, slightly scabrous above, white-tomentose underneath. Flower-heads in axillary clusters or almost solitary, the upper ones male, the lower female; these are sessile, forming when in fruit oblong burrs about ½ in. long, covered with hooked prickles, the terminal conical beaks exceedingly short, often both reduced to mere tubercles or one only slightly elongated.

An exceedingly troublesome weed, in warm dry situations, supposed to be of Chilian origin, now infesting southern Europe and many warm countries, first observed in Australia about 1852 and now said to be very abundant in many pastures of the interior of **Queensland, N. S. Wales,** and **Victoria,** to the great detriment of the wool.

TRIBE VI. HELIANTHEÆ.—Leaves opposite or rarely alternate. Flower-heads usually heterogamous, the florets of the circumference female or neuter and ligulate, or rarely irregular or wanting, the disk-florets hermaphrodite or male, tubular, 4- or 5-toothed. Receptacle bearing chaffy often rigid scales between the florets. Anthers obtuse at the base. Style of the *Senecionidæ* or approaching that of the *Asteroideæ*. Pappus of stiff awns or short scales or rarely none.

30. SIEGESBECKIA, Linn.

Involucral bracts in about 2 rows, the outer ones linear-spathulate, spreading, glandular-hispid, the inner ones ovate or oblong, half enclosing the achenes, glandular-hispid on the back. Receptacle chaffy, the scales half enclosing the achenes. Ray-florets female, shortly ligulate or irregularly 2- or 3-lobed. Disk-florets hermaphrodite, fertile, tubular, 5-toothed. Anthers obtuse at the base. Style-branches short, somewhat flattened, very obtuse or rarely tipped with a small glandular cone. Achenes somewhat turgid, usually curved. Pappus none.—Herbs with opposite leaves. Flower-heads rather small, pedunculate.

A genus of very few species widely dispersed over the warmer regions of the globe, the only Australian one being the most common, especially in the Old World.

1. **S. orientalis,** *Linn.; DC. Prod.* v. 495. A pubescent branching rather stiff annual, 1 to 2 ft. high. Leaves from broadly ovate-triangular to lanceolate, 1½ to 2 in. long or the lower ones larger, the petioles variable in length, usually dilated upwards but not at the base. Flower-heads 3 to 6 lines broad, in a dichotomous leafy panicle. Outer involucral bracts often 4 or 5 lines long and covered with gland-bearing hairs, but sometimes shorter

than the inner ones and less glandular. Florets small, the rays very short.—Wight, Ic. t. 1103; *S. microcephala* and *S. gracilis*, DC. Prod. v. 496.

Queensland. Wide Bay, *Leichhardt*; Rockhampton, *Thozet*; Brisbane river, Moreton Bay, *F. Mueller*.

N. S. Wales. Paramatta, *Woolls*; Blue Mountains, *Miss Atkinson*; New England, *C. Stuart*; Clarence river, *Beckler*; in the interior on the Darling and Murray and towards the Barrier range, *Victorian and other Expeditions*.

Victoria. Tambo and Murray rivers, *F. Mueller*; Portland, *Allitt*; in the Pyrenees, *Wilhelmi*.

S. Australia. Spencer's Gulf, *R. Brown*; near Adelaide, *Blandowski*; Torrens river, *F. Mueller*.

Most of the southern specimens belong to the variety with short outer involucral bracts.

31. ECLIPTA, Linn.

Involucre of about 2 rows of nearly equal herbaceous bracts. Receptacle chaffy. Florets of the ray female, shortly ligulate, narrow; disk-florets hermaphrodite, usually fertile, tubular, 4-toothed. Anthers obtuse at the base. Style-branches linear, flattened, obtuse. Achenes of the ray triangular, those of the disk flattened. Pappus none or reduced to a border of minute teeth.—Herbs with opposite leaves. Flower-heads small, on axillary or terminal peduncles.

Besides the subjoined species, of which one is endemic in Australia and the other a common weed in warm countries, some Brazilian perennials are included in the genus, but perhaps not correctly so.

Ray white. Involucral bracts ovate . . . . 1. *E. alba*.
Ray yellow. Involucral bracts narrow-lanceolate . . . . 2. *E. platyglossa*.

1. **E. alba,** *Hassk.; Miq. Fl. Ned. Ind.* ii. 65. A branching annual, usually prostrate or creeping, sometimes ascending or erect, 1 ft. long or more, sprinkled with closely appressed short stiff hairs. Leaves shortly petiolate, from nearly ovate to oblong-lanceolate or almost linear, 1 to 2 in. long, coarsely toothed or nearly entire. Peduncles in the upper axils solitary or 2 together, very variable in length, bearing a single flower-head about 3 lines diameter. Involucral bracts usually broadly ovate, obtuse. Scales of the receptacle narrow-linear. Ray-florets small, white. Achenes of the disk with thick almost corky margins, the pappus either quite abortive or reduced to a border of 4 minute obtuse teeth, conspicuous chiefly at the time of flowering.—*E. erecta* and *E. prostrata*, Linn., and the whole section *Eueclipta*, DC. Prod. v. 490.

Queensland. Burdekin river, *F. Mueller*; Brisbane river, Moreton Bay, *C. Stuart*, *F. Mueller*; Rockhampton, *Herb. F. Mueller*. A common weed throughout the warmer regions of the globe.

2. **E. platyglossa,** *F. Muell. Fragm.* ii. 135; *Pl. Vict. t.* 39. Very near to *E. alba*, but the flower-heads are smaller, the involucral bracts narrow-lanceolate, and the ray-florets yellow. Leaves almost or quite sessile, lanceolate, strigose. Achenes quite those of *E. alba*.—*Wollastonia* or *Wedelia ecliptoides*, F. Muell. Pl. Vict. t. 39.

N. Australia. Islands of the Gulf of Carpentaria, *R. Brown*, *Henne*; Albert and Roper rivers, *F. Mueller*.

Queensland. Broad Sound, *R. Brown;* Rockhampton, *Dallachy;* Moreton Bay, *C. Stuart;* also in *Leichhardt's* collection.

N. S. Wales. Cabramatta and Mudgee, *Woolls;* Clarence river, *Beckler;* Darling and Murray rivers, *Dallachy and Goodwin, Herrgolt.*

Victoria. Avoca and Murray rivers, *F. Mueller.*

S. Australia. Torrens river, St. Vincent's Gulf, *F. Mueller.*

A specimen from Facing Island, Port Curtis, *R. Brown*, has the involucral bracts nearly as narrow as in *E. platyglossa*, but according to R. Brown's notes, the ray-florets are white and it seems almost to connect the two species.

32. WEDELIA, Jacq.

(Wollastonia, *DC.*)

Involucre of about 2 rows of herbaceous bracts either all nearly equal or the outer ones larger and more leaf-like. Receptacle chaffy. Florets of the ray female, ligulate; disk-florets hermaphrodite, tubular, 5-toothed, mostly fertile. Anthers obtuse at the base. Style-branches with rather acute tips, usually hirsute. Achenes more or less flattened, or rarely 3-angled, with obtuse or acute rarely winged edges. Pappus none or more frequently consisting of minute scales united in a little cup and occasionally produced into 1 or 2 short bristles.—Herbs with opposite leaves. Flower-heads pedunculate or rarely almost sessile, terminal or in the forks of the branches or axils of the upper leaves. Ray yellow.

A considerable American genus with a few African and Asiatic species. Of the six Australian ones, three are widely spread over India and the Archipelago, one or perhaps two of them extending into Africa, a fourth is also found in Timor, the remaining two are endemic. The *W. biflora* and some other species without any pappus have been separated under the name of *Wollastonia*, but in *W. spilanthoides* the pappus is often so small as to be quite inconspicuous when the achene is ripe, and in every other respect the species are all too closely allied to be generically separated. It is probable that the several genera alluded to by A. Gray under *Lipochæta* will have to be united also with *Wedelia.*

Five or six of the outer involucral bracts more leaf-like and longer than the others. Pappus cup-shaped.
  Leaves oblong or lanceolate, narrowed into a short petiole or nearly sessile . . . . 1. *W. calendulacea.*
  Leaves ovate or ovate-lanceolate, distinctly petiolate. Peduncles mostly longer than the leaves . . . . 2. *W. urticifolia.*
Outer involucral bracts not longer than the inner ones.
  Pappus small and cup-shaped in the centre of the achene, and sometimes 1 or 2 teeth or small bristles from the angles.
    Leaves lanceolate or linear, acuminate. Achenes compressed, not angled . . . . 3. *W. spilanthoides.*
    Leaves ovate, oblong or lanceolate, obtuse. Achenes angled or winged . . . . 4. *W. verbesinoides.*
  Pappus none or of 1 or 2 deciduous bristles.
    Straggling perennial. Leaves ovate or broadly ovate-lanceolate 5. *W. biflora.*
    Erect coarse annual. Leaves lanceolate . . . . 6. *W. asperrima.*

1. **W. calendulacea,** *Less.; DC. Prod.* v. 539. A low decumbent prostrate or creeping perennial, attaining sometimes 2 ft., sprinkled with short appressed hairs. Leaves oblong-lanceolate, 1 to 3 in. long, acute or obtuse, coarsely toothed or nearly entire, narrowed at the base but scarcely petiolate. Flower-heads nearly $\frac{3}{4}$ in. diameter, solitary, on long axillary or

terminal peduncles. Outer bracts of the involucre lanceolate or oblong, 4 or 5 lines long, inner ones smaller. Ray-florets about 10 to 12, rather broad, half as long again as the involucre. Achenes flattened, with a small denticulate cup-shaped pappus.—Wight, Ic. t. 1107.

Queensland. East coast, *A. Cunningham*, a single specimen with a single flower-head, the achenes of which I have been unable to examine and therefore the identity is in some measure doubtful. The species is frequent in India from Ceylon and the Peninsula to the Archipelago and northward to South China.

2. **W. urticifolia,** *DC. in Wight, Contrib.* 18 *and Prod.* v. 539. Erect or decumbent at the base, often 2 to 3 ft. high, more or less hirsute, the hairs not very rigid. Leaves petiolate, ovate or ovate-lanceolate, acuminate, slightly serrate, rather thin, often 3 to 4 in. long, narrowed or rounded at the base, the petiole rather short. Flower-heads small, on rather slender peduncles, the lower ones shorter than the leaves, the upper ones longer but scarcely paniculate. Involucre ovoid, the outer bracts (about 5) acute or acuminate, 3 to 6 lines long, the inner ones shorter. Ray-florets about 8, rather large; disk-florets not very numerous. Achenes somewhat compressed with acute edges, hairy at the top, with a small denticulate cup-shaped pappus, one of the teeth rarely produced into a bristle.—Wight, Ic. t. 1106; *W. Cunninghamii*, DC. Prod. v. 540.

N. Australia. Grassy rocky places, Goulburn island, *A. Cunningham;* islands of the Gulf of Carpentaria, *R. Brown.* The species is frequent in India.

3. **W. spilanthoides,** *F. Muell. Fragm.* v. 64. Very scabrous. Leaves very shortly petiolate, from broadly lanceolate to almost linear, mostly acuminate, with a few coarse irregular teeth especially near the base, or linear and entire, 2 to 4 in. long. Peduncles rigid, long and solitary or with a second shorter one. Involucres hemispherical, smaller than in *W. biflora*, the bracts rather numerous, nearly equal, the outer ones ovate or ovate-lanceolate, scarcely acuminate, the inner ones narrower, all shorter than the florets. Ray-florets 10 to 12 or even more, rather large. Achenes more or less flattened, often pubescent at the top. Pappus a minute denticulate cup, occasionally emitting a short deciduous bristle, but the whole often inconspicuous when the achene is ripe.

Queensland. Port Curtis, *M'Gillivray;* Rodd's Bay, *A. Cunningham;* Wide Bay, *Bidwill;* Burdekin, Burnett, and Brisbane rivers and Newcastle range, *F. Mueller;* Rockingham Bay and Rockhampton, *Dallachy;* Fitzroy river, *Bowman.*

N. S. Wales. Macleay river, *Herb. F. Mueller.*

In some specimens the leaves are few and all narrow-linear and the flower-heads smaller, these are probably old branches or from plants grown in a dry season or locality. In others from Keppel Bay and Broad Sound, *R. Brown*, the petioles are more distinct.

4. **W. verbesinoides,** *F. Muell. Herb.* Erect, rigid, and very scabrous, especially the foliage, with appressed rigid hairs. Leaves petiolate, oblong-lanceolate or ovate-lanceolate, obtuse, slightly toothed or entire, narrowed at the base, 1 to 3 in. long. Flower-heads shortly pedunculate, in irregular terminal panicles. Involucre ovoid-globular, the bracts nearly equal, ovate-oblong, obtuse. Scales of the receptacle obtuse. Ray-florets few and small. Achenes flattened or 3-angled, the edges acute or almost winged, tuberculate, with 1, 2, or 3 unequal short bristles thickened at the

base proceeding from the angles, and in the centre a small cup-shaped pappus occasionally emitting 1 or 2 small bristles but sometimes quite inconspicuous.

N. Australia, *F. Mueller;* Arnhem S. Bay, *R. Brown;* Finke river, *M'Douall Stuart's Expedition.*

5. **W. biflora,** *DC. in Wight, Contrib. Bot. Ind.* 18. A straggling half-scandent branching perennial, sometimes nearly glabrous, but more frequently slightly hoary or even quite white with closely appressed rigid hairs, especially on the under side of the leaves. Leaves petiolate, from broadly ovate to ovate-lanceolate, the lowest sometimes 3 or 4 in. long and very broadly cordate, the others usually smaller and often cuneate at the base, all acute or acuminate, slightly toothed, 3-nerved. Flower-heads most frequently 3 on short or long peduncles or sometimes more in a loose corymb. Involucral bracts nearly equal in about 2 rows. Ray-florets 10 to 12 or sometimes more, the ligules oblong, entire or minutely 3-toothed, 2 to 3 lines long. Achenes obtuse at the edges, without any pappus or with 1 to 3 small slender deciduous awns.—*Wollastonia biflora*, DC. Prod. v. 546; Wight, Ic. t. 1108; Benth. Fl. Hongk. 183, with the synonyms there adduced; *W. insularis* and *W. Forsteriana*, DC. Prod. v. 548.

N. Australia. Islands of the Gulf of Carpentaria, *R. Brown;* South Goulburn island, *A. Cunningham;* Keppel Bay, *Thozet.*

Queensland. Islands of Torres' Straits, *Henne;* Port Curtis and Lizard Islands, *M'Gillivray;* Bay of Inlets, *Banks and Solander;* Rockingham Bay, *Dallachy;* Brisbane river, Moreton Bay, *F. Mueller.*

N. S. Wales. Port Jackson, *R. Brown* and others; New England, *C. Stuart;* Clarence and Macleay rivers, *Beckler;* Lord Howe's Island, *Milne.*

This species is widely spread over East India, extending westward to east tropical Africa, and eastward to the Archipelago and south China.

6. **W. asperrima,** *Benth.* Coarse and erect, 1 to 2 ft. high, but apparently annual, very scabrous, especially the foliage, with rigid appressed hairs. Leaves petiolate, from oblong-lanceolate to linear-lanceolate, mostly acuminate, 2 to 3 in. long, irregularly toothed, narrowed at the base. Flower-heads several, in a loose terminal leafy panicle on rigid peduncles longer than the leaves. Involucre nearly hemispherical, the bracts rather numerous, acuminate, the outer ones 4 to 5 lines long, the inner ones scarcely shorter. Scales of the receptacle very acute and rigid. Ray-florets 10 to 12 or even more, oblong. Achenes with obtuse edges, not at all winged and without any pappus.—*Wollastonia asperrima*, Dcne. Herb. Timor, Descr. 86, and DC. Prod. v. 547 (?), from the character given.

N. Australia. Victoria river and Sturt's Creek, *F. Mueller.* If the identification with Decaisne's plant is correct it is also in Timor.

33. MOONIA, Arn.

(Pentalepis, *F. Muell.*)

Involucre of few (in the Australian species about 5) ovate or lanceolate nearly equal bracts. Receptacle chaffy. Ray-florets about as many as involucral bracts, female, ligulate. Disk-florets more numerous, hermaphrodite but sterile, tubular, 5-toothed. Anthers obtuse at the base. Style undivided.

Achenes of the ray flattened, the margins acute or winged. Pappus minute, cup-shaped, often produced into 2 short bristles. Achenes of the disk abortive.—Herbs with opposite leaves. Flower-heads pedunculate in terminal panicles or in the forks of the branches. Ray yellow.

A small genus containing, besides the Australian species, which are endemic, one or two from Ceylon and the Indian Archipelago. It differs from *Wedelia*, of which it has the habit, chiefly in the abortion of the disk-achenes. The original Ceylon species has more numerous involucral bracts and ray-florets, but does not otherwise differ in essential characters.

Leaves entire. Involucre 5 to 6 lines long . . . . 1. *M. trichodesmoides.*
Leaves coarsely toothed. Involucre about 3 lines long.
  Leaves mostly ovate-lanceolate or lanceolate, nearly sessile . . 2. *M. ecliptoides.*
  Leaves mostly ovate, petiolate . . . . 3. *M. procumbens.*

1. **M. trichodesmoides,** *Benth.* A rigid erect herb attaining 4 ft. in height, scabrous with short hairs or tubercles. Leaves nearly sessile, lanceolate, entire in the specimens seen. Flower-heads narrow-ovoid, in a terminal panicle. Involucral bracts usually 5, lanceolate, acuminate, about 5 or 6 lines long. Ray-florets 5 or 6, ligulate, entire or shortly 2-toothed. Achenes of the ray only seen young, then resembling those of *M. ecliptoides* but pubescent upwards and the 2 bristles of the pappus shorter and softer. Disk-florets as long as the involucre, with a simple style and abortive achenes as in other species.—*Pentalepis trichodesmoides*, F. Muell. in Trans. Bot. Soc. Edinb. vii. 496.

N. Australia. Nichol Bay, N.W. coast, *F. Gregory's Expedition.*

2. **M. ecliptoides,** *Benth.* A stout erect rigid herb, apparently several feet high, scabrous with appressed hairs, the smaller branches often slender and almost leafless. Leaves sessile or nearly so, the lower ones in our specimens ovate-lanceolate, acuminate, coarsely toothed, narrowed at the base, about 3 in. long, the upper ones smaller and narrow-lanceolate. Flower-heads small and numerous in a terminal irregularly corymbose panicle. Involucre ovoid-turbinate, about 3 lines long, of 4, 5 or rarely 6 leafy bracts, acute, and exceeding the disk-florets. Ray-florets as many as involucral bracts, the ligula obovate-oblong, usually 2-lobed. Scales of the receptacle narrow. Style of the disk-florets often long and hirsute. Achenes of the ray with broad incurved almost winged margins, notched at the end, the small cup-shaped pappus usually produced into 2 short bristles.—*Pentalepis ecliptoides*, F. Muell. in Trans. Bot. Soc. Edinb. vii. 496.

N. Australia. N.W. coast, *Bynoe;* Camden Harbour (*Marten?*), Hooker's and Sturt's Creeks, *F. Mueller;* Port Essington, *Armstrong.*

3. **M. procumbens,** *Benth.* Stems branched, diffuse procumbent or ascending, 1 to 2 ft. long. Leaves shortly petiolate, from ovate to lanceolate, coarsely toothed, 1 to 2 in. long, very scabrous with scattered rigid hairs or tubercles like many *Boragineæ*. Peduncles terminal or in the forks, the lower ones short, but mostly much longer than the leaves. Involucral bracts about 5 or 6, herbaceous, as long as or longer than the disk-florets. Ray-florets ovate, 2 to 3 lines long. Scales of the receptacle narrow, acute. Achenes of the ray thick, but compressed with acute margins, not winged.

Pappus small, cup-shaped, usually without bristles. Disk-achenes abortive. —*Wollastonia procumbens*, DC. Prod. v. 548.

N. Australia. Palm Bay, Croker's island, *A. Cunningham;* Port Essington, *Armstrong.*

34. SPILANTHES, Linn.

Involucral bracts in about two rows nearly equal, usually broad and thin. Receptacle chaffy, very conical. Ray-florets female, ligulate, or sometimes none. Disk-florets small, hermaphrodite, tubular, 5-toothed. Anthers obtuse at the base. Style branches truncate. Achenes of the ray 3-angled, those of the disk flattened, the angles or margins usually ciliate. Pappus of 2 or 3 short fine awns or bristles proceeding from the angles, sometimes wanting.—Herbs with opposite leaves. Flower-heads usually on long peduncles, the ray yellow or white.

The genus is widely dispersed over the tropical regions both of the New and the Old World. The two Australian species extend into the Indian Archipelago.

Flower-heads with yellow rays . . . . 1. *S. grandiflora.*
Flower-heads discoid, without rays . . . . 2. *S. anactina.*

1. **S. grandiflora,** *Turcz. in Bull. Mosc.* 1351. i. 185. Decumbent, loosely branched, glabrous or strigose-pubescent, attaining 1 to 2 ft. in length. Leaves shortly petiolate, ovate-lanceolate lanceolate or rarely linear, entire or with a few coarse teeth below the middle, 1 to 2 in. long, 3-nerved, glabrous or sprinkled with a few hairs. Involucral bracts ovate or ovate-lanceolate, scarcely 2 lines long. Ray-florets ligulate, yellow, twice as long as the involucre in the normal form but variable in size. Scales of the receptacle broad and concave, as long as the disk-florets. Bristles of the pappus scarcely thicker and but little longer than the cilia of the angles of the achenes.—*S. macroglossa*, F. Muell. Fragm. v. 63.

N. Australia. Sturt's Creek, *F. Mueller.*

Queensland. Thirsty Sound and Broad Sound, *R. Brown;* Burdekin river, *F. Mueller;* Rockhampton, *Dallachy;* Keppel Bay, *Thozet;* Brisbane river, Moreton Bay, *Mossman, F. Mueller.*

N. S. Wales. New England, *C. Stuart;* Clarence river, *Beckler;* Richmond river, *Fawcett.*

Var. *calva.* More glabrous; achenes smaller with few or no cilia on the angles, and the bristles of the pappus very minute or entirely wanting. To this belong the N. S. Wales and Moreton Bay specimens, but in others also, the achenes, when examined at the time of flowering, have often but few and small cilia and inconspicuous bristles.

Var. *brachyglossa.* Ray-florets very short.—*S. Acmella*, F. Muell. Fragm. v. 63, but scarcely of Linnæus. To this belong the Sturt's Creek specimens. The species, and especially this variety, is very nearly allied to the true *S. Acmella*, Linn., which should include S. *caulorrhiza* and *S. Africana*, DC., and is widely spread over tropical Asia and Africa, but appears distinct in habit, narrower leaves, and more conical usually larger flower-heads. We have received it also from the Indian Archipelago and the Philippine Islands.

2. **S. anactina,** *F. Muell. Fragm.* v. 63. Stems from a procumbent or decumbent base ascending to about 1 ft. high, simple or branched. Leaves linear or oblong-lanceolate, obtuse, entire, rather thick, 1 to 2 or rarely 3 in. long. Involucral bracts broad and thin, the margins often scarious and jagged. Scales of the receptacle broad. Ray-florets entirely wanting. Disk-

florets short and broad. Style-branches very long, flat, truncate. Achenes acutely edged or winged without cilia, the awns rather long for the genus.

N. Australia. Islands of the Gulf of Carpentaria, *R. Brown*, *Henne*. Also in Borneo, *Barber*.

35. GALINSOGA, Cav.

(Vargasia, *DC.*)

Involucre of few (about 5) broad nearly equal bracts. Receptacle conical, chaffy, the scales narrow. Ray-florets few, female, ligulate. Disk-florets hermaphrodite, tubular, 5-toothed. Anthers obtuse at the base. Style-branches acute. Achenes angular, slightly flattened. Pappus of several linear or oblong chaffy scales often plumose-ciliate or wanting, especially on the achenes of the ray.—Herbs with opposite leaves. Flower-heads small, pedunculate. Ray white.

A small genus from tropical and subtropical America, of which one species has spread as a weed of cultivation over many warm and temperate districts of the Old World, and has become introduced as such into Australia.

*1. **G. parviflora,** *Cav.*; *DC. Prod.* v. 677. An erect annual of 1 to 2 ft., glabrous or slightly hairy. Leaves petiolate, ovate or ovate-lanceolate. Peduncles slender. Involucre nearly hemispherical, about 2 lines diameter. Ray-florets about 5, with a small white ligula. Disk-florets short. Achenes small, slightly hairy. Pappus of the ray reduced to a few minute bristles or entirely wanting, that of the disk-achenes consisting of from 12 to 20 chaffy scales, more or less plumose-ciliate.

N. S. Wales. Port Jackson, *Woolls*; Clarence river, *Beckler*. A South American species, introduced with cultivation.

36. BIDENS, Linn.

Involucral bracts few, in about 2 or 3 rows, the outer ones herbaceous, the inner ones usually bordered with a thin whitish margin. Receptacle chaffy. Ray-florets neuter, sterile, ligulate or sometimes wanting; disk-florets tubular, hermaphrodite, 5-toothed. Anthers obtuse at the base. Style-branches with an acute or subulate point. Achenes broad and flattened or slender and 4-angled, often produced into a short beak. Pappus of 2 to 4 rigid retrorsely hispid persistent awns.—Herbs with opposite leaves. Flower-heads on terminal peduncles, the ray yellow or white.

Of the two sections of the genus one, *Platycarpæa*, belongs chiefly to the temperate regions of the northern hemisphere, the Australian species being identical with one of the common northern ones; the other, *Psilocarpæa*, is entirely American with the exception of the two Australian species, which are common weeds in all warm countries.

SECTION I. **Platycarpæa.**—*Achenes flat and rather broad.*

Leaves divided into 3 lanceolate serrate segments . . . . 1. *B. tripartita.*

SECTION II. **Psilocarpæa.**—*Achenes slender, 4-angled.*

Leaves mostly pinnate, with 3 or 5 segments. Ray white . . . . 2. *B. pilosa.*
Leaves mostly bipinnate. Ray yellow . . . . 3. *B. bipinnata.*

SECTION I. PLATYCARPÆA.—Achenes flat and rather broad.

1. **B. tripartita,** *Linn.; DC. Prod.* v. 594. A rather stout erect glabrous annual 1 to 2 ft. high. Leaves deeply cut into 3 lanceolate serrate segments. Flower-heads on terminal peduncles, erect or somewhat drooping. Involucre hemispherical, ½ to 1 in. diameter, the outer bracts sometimes nearly 1 in. long, leaf-like and spreading, the inner ones short broad, often shining and yellow on their edge. Florets either all tubular without any ray or with a few outer ligulate yellow florets. Achenes flat, crowned by 2 or 3, very rarely 4 awns.—*B. repens*, Don; DC. Prod. v. 595.

Victoria. Plains inundated in winter on the Snowy and Mitchell rivers, *F. Mueller*. The species is widely spread over the temperate regions of the northern hemisphere where, however, it is not so common as the closely allied *B. cernua*, which has undivided leaves and more drooping flower-heads.

SECTION II. PSILOCARPÆA.—Achenes slender, 4-angled.

2. **B. pilosa,** *Linn.; DC. Prod.* v. 597. An erect glabrous or slightly hairy annual, 1 to 2 ft. high. Branches angular. Leaves thin, pinnately divided, or the lower ones sometimes simple; segments 3 or sometimes 5, petiolulate, ovate or ovate-lanceolate, 1 to 2 in. long, serrate or rarely lobed. Flower-heads few, terminal, rather small, on slender peduncles. Involucral bracts 2 to 3 lines long. Ray-florets white, few and short or sometimes wholly wanting. Achenes slender, 4-angled, the inner ones often 6 to 7 lines long, the outer ones shorter.—Benth. Fl. Hongk. 183, with the synonyms adduced.

Queensland. Brisbane river, Moreton Bay, *F. Mueller*.

N. S. Wales. Newcastle, *R. Brown*; Port Jackson, *Woolls*; Blue Mountains, *Miss Atkinson*; Clarence river, *Beckler*.

Victoria. Yarra-Yarra, *F. Mueller*.

The species is very common as a weed over most warm countries both in the New and the Old World, and may therefore have been introduced into Australia by cultivation.

3. **B. bipinnata,** *Linn.; DC. Prod.* v. 603. A glabrous annual, resembling *B. pilosa*, but the leaf-segments are usually again divided into small deeply-toothed or lobed segments, the flower-heads are smaller, the involucral bracts less bordered, and the ray-florets small and yellow.

N. Australia. Victoria river, *F. Mueller*; islands of the Gulf of Carpentaria, *R. Brown*.

Queensland. Common on the sandy seacoasts, *R. Brown* and others; islands of Moreton Bay, *F. Mueller*.

The species is common in most warm countries both in the New and the Old World, and apparently truly indigenous in Australia.

37. GLOSSOGYNE, Cass.

(Diodontium, *F. Muell.*)

Involucral bracts few, in about 2 rows, narrow and nearly equal. Receptacle chaffy. Ray-florets female, ligulate, fertile or sometimes wanting; disk-florets tubular, hermaphrodite, 4- or 5-toothed. Anthers obtuse at the base. Style-branches ending in subulate points. Achenes narrow, usually flattened. Pappus of 2 to 4 rigid retrorsely hispid or smooth persistent awns.—Glabrous perennials, with alternate or opposite, pinnate or undivided narrow leaves. Flower-heads small on long terminal peduncles.

A small genus, extending over tropical Asia, to which have also been referred one or two Brazilian species. Of the three Australian species one is also in the Indian Archipelago, the other two are endemic. It is very closely allied to *Bidens*, and further investigation of the allied American plants may induce its reduction to a section of that genus.

Leaves alternate, pinnate. Ray-florets few. Disk-florets 4-toothed.
 Style-branches of the disk-florets very long. Achenes longitudinally striate, the awns erect or slightly spreading . . . . 1. *G. tenuifolia.*
 Style-branches rather short. Achenes transversely rugose, the awns very spreading or reflexed . . . . 2. *G. retroflexa.*
Leaves opposite, entire, slender. Ray-florets wanting. Disk-florets 5-toothed . . . . 3. *G. filifolia.*

1. **G. tenuifolia,** *Cass.; DC. Prod.* v. 632. Stock perennial, tufted, sometimes almost woody, with erect dichotomous stems, 6 in. to 1 ft. high, often almost leafless, or sometimes elongated decumbent and leafy at the base. Leaves chiefly radical or nearly so, the lowest sometimes cuneate and 3-lobed, all the others pinnately divided into 5 or 7 stiff linear segments either entire or 2- or 3-lobed. Flower-heads small, on long slender terminal peduncles. Involucre campanulate, not 2 lines long. Ray-florets small, yellow, spreading. Achenes linear, flattened, about 4 lines long, striate, with 3 or more numerous ribs on each face, crowned by 2 erect or slightly diverging awns. —*Bidens tenuifolia*, Labill. Sert. Austr. Caled. 44. t. 45; *Glossogyne pedunculosa*, DC. Prod. v. 632; *G. bidentidea*, F. Muell. in Linnæa, xxv. 402; *Bidens denudata*, Turcz. in Bull. Mosc. 1851, i. 183.

N. Australia. Goulburn islands, *A. Cunningham.*

Queensland. Northumberland islands and Shoalwater Bay, *R. Brown*; Cape Cleveland and Rodd's Bay, *A. Cunningham;* Cape York and Port Curtis, *M'Gillivray;* Rockingham Bay, *Dallachy;* Albany island and Brisbane river, Moreton Bay, *F. Mueller.*

N. S. Wales. Port Jackson, *R. Brown* and others; New England, *C. Stuart;* Clarence, Hastings and Macleay rivers, *Beckler;* Kiama, *Harvey*, and in the interior on Lachlan river, *A. Cunningham.*

S. Australia. Cudnaka and between Spencer's Gulf and Flinders range, *F. Mueller.*

The species is also in New Caledonia and in the Indian Archipelago, and differs but very little from the East Indian *G. pinnatifida.*

2. **G. retroflexa,** *F. Muell. Fragm.* i. 51. A tufted perennial, with precisely the habit and foliage of the ordinary form of *G. tenuifolia*, and the same inflorescence, involucre and florets, but the style-branches although filiform are shorter and glabrous, the achenes are thicker, transversely rugose, and the awns, usually 3, are very spreading or reflexed on the achene.

Queensand. Basaltic plains between Peak Range and Darling Downs, *F. Mueller.*

3. **G. filifolia,** *F. Muell. Herb.* Glabrous, erect, with 2- or 3-chotomous branches. Leaves opposite, linear-terete, slender, 1 to 2 in. long or even longer, undivided but clustered at the base of the branches. Flower-heads small, rather numerous in a loose corymbose terminal panicle. Involucre of 4 to 6 unequal lanceolate or linear bracts, the longest inner ones rarely exceeding 2 lines and shorter than the florets. Ray-florets wanting; disk-florets tubular, 5-toothed. Style thickened upwards, with 2 filiform papillose almost hairy branches. Achenes about as long as the involucre, ovate, flattened, with thin almost winged margins; awns almost divaricate, not

so long as the breadth of the achene, smooth or with a few reversed prickles. —*Diodontium filifolium*, F. Muell. in Hook. Kew Journ. ix. 19.

N. Australia. Sources of Hooker's Creek, *F. Mueller.*

38. NABLONIUM, Cass.

Involucre turbinate, with few membranous nearly equal bracts, the outer ones broad, the inner ones narrower, passing into the scales of the receptacle. Florets all hermaphrodite, tubular, 5-toothed. Anthers obtuse at the base. Style-branches somewhat flattened, truncate. Achenes flattened, with 2 rigid pungent divergent awns thickened at the base and continuous with the achene.—Small creeping herb, with radical leaves and 1-headed scapes.

The genus is limited to the single Australian species.

1. **N. calyceroides,** *Cass.; DC. Prod.* vi. 37. A small plant with creeping stolons, having something of the aspect of *Calotis scapigera*, with more or less of a white deciduous wool on the scape and foliage or quite glabrous. Leaves radical, linear or lanceolate, acute and often bearing a hair-like point, entire, narrowed at the base, rarely 1 in. long. Scape 1-headed, exceeding the leaves, leafless or rarely with 1 or 2 small linear bracts. Involucre about 2 lines long, the bracts with more or less transparent margins, the narrow inner ones jagged or lobed at the end. Florets not numerous. Anthers linear. Achenes glabrous, obovate or cuneate, the rigid thick divaricate awns very prominent in the fruiting-head. Testa of the seed cohering to the pericarp.—Hook. f. Fl. Tasm. i. 190. t. 48 A.

Tasmania. King's and other islands of Bass's Straits, *R. Brown* and others; Macquarrie Harbour, *Milligan*, *Gunn.*

39. TAGETES, Linn.

Involucral bracts in a single row, united in a toothed cup or tube. Receptacle flat, without scales. Florets of the ray female, ligulate; disk-florets tubular, 5-toothed. Anthers obtuse at the base. Style-branches flattened, obtuse or truncate, usually hirsute. Achenes linear, flattened. Pappus of several narrow very unequal scales or bristles.—Herbs, usually glabrous, the foliage and involucres bearing oblong or round transparent glands or vesicles filled with a strongly-scented oil. Leaves opposite, entire or pinnate. Flower-heads large and solitary or small and corymbose or paniculate. Ray yellow or orange-red.

An American genus, of which a few species have spread as weeds into the Old World.

*1. **T. glandulifera,** *Schranck; DC. Prod.* v. 644. A tall glabrous erect annual, attaining sometimes 6 to 8 ft., with numerous erect branches. Leaves pinnate with linear-lanceolate serrate segments. Flower-heads small and numerous in dense terminal panicles. Involucre tubular, about $\frac{1}{2}$ in. long. Florets 6 to 12, scarcely exceeding the involucre, about 3 bearing each a small yellow ligula, the others tubular. Achenes linear, black, with a pappus of 5 or 6 chaffy bristles, one much longer than the rest.

Queensland. Brisbane river, *F. Mueller;* Neerkool Creek, *Bowman.*
N. S. Wales. Now much spreading on the Hunter and Nepean river, *F. Mueller.*
The species is of S. American origin, introduced with cultivation into Australia as into several of the warmer districts of the Old World.

40. ENHYDRA, Lour.

(Tetraotis, *Reinw.*)

Involucre of 4 broad leafy bracts closely enveloping the florets, the two outer ones larger than the inner ones. Receptacle conical, chaffy, the scales enclosing the florets and achenes. Ray-florets in several rows, female, with very short 3-toothed ligulas; disk-florets hermaphrodite but usually sterile, tubular, 5-toothed. Anthers obtuse at the base. Style-branches flattened, scarcely truncate. Achenes of the ray flattened, with obtuse edges, without any pappus, those of the disk usually abortive.—Herbs with opposite leaves. Flower-heads sessile in the forks of the stem or in the axils of the leaves.

The genus is limited to a very few species from tropical Asia, which may indeed be all varieties of one, and a few others from S. America. The Australian plant is evidently the same as the common Indian one.

1. **E. paludosa,** *DC. Prod.* v. 637. Glabrous or slightly scabrous-pubescent. Stems elongated, creeping and rooting in the mud, the flowering branches ascending, simple or forked. Leaves shortly petiolate, oblong or lanceolate, coarsely toothed or nearly entire, narrowed at the base, or sometimes, especially those under the forks, slightly hastate, the petiole often dilated at the base and stem-clasping. Involucral bracts broadly ovate, the 2 outer larger ones 3 to 6 lines long, exceeding the florets.—*Tetraotis paludosa,* Reinw. in Blume, Bijdr. 892; *Enhydra longifolia* and *E. Heloncha,* DC. Prod. v. 637; *E. Woollsii,* F. Muell. Fragm. iii. 139.

N. S. Wales. In marshes, sometimes quite under water, South Head, *Bauer;* Manly beach, *Woolls.* Also in *Leichhardt's* collection.
The species is common in E. India and the Archipelago, and may not be really distinct from the original *E. fluctuans,* Lour. The Australian specimens have the leaves more narrowed at the base than is usual in Asia, but precisely the same form occurs also in India.

41. FLAVERIA, Juss.

Flower-heads collected in dense clusters or compound heads surrounded by a few leafy bracts or floral leaves. Involucres cylindrical or compressed, with few conduplicate dry bracts. Florets few, the female ones often solitary in the involucre with a small ligula, hermaphrodite florets tubular, 5-toothed. Anthers obtuse at the base. Style-branches truncate. Achenes somewhat compressed. Pappus none.—Herbs with opposite leaves. Clusters of flower-heads terminal or sessile in the forks. Florets yellow.

Besides the Australian species, which is endemic, there are a few American ones, one of which, from S. America, is closely allied to the Australian one.

1. **F. australasica,** *Hook. in Mitch. Trop. Austr.* 118. An erect rigid perfectly glabrous pale green annual, usually about 1 ft. but attaining sometimes 2 or 3 ft. in height, with opposite or dichotomous divaricate branches. Leaves linear or linear-lanceolate, 1 to 2 in. long, entire or with small remote

teeth, the lower ones narrowed towards the base but stem-clasping, the upper ones much dilated at the base. Flower-heads numerous, in dense globular or hemispherical sessile clusters often ½ in. diameter, surrounded by a few broad ovate-acuminate or lanceolate floral leaves longer than the clusters. Involucres 2 to 3 lines long, the outer ones of each cluster usually consisting of 2 or 3 obtuse narrow bracts and including a single ligulate floret, the others containing 2 to 6 disk-florets, the corollas slightly dilated over the achene and glandular at the base. Achenes prominently ribbed.—F. Muell. Fragm. i. 183.

N. Australia. Nichol Bay, N.W. coast, *F. Gregory's Expedition ;* Victoria river and Hooker's Creek, *F. Mueller ;* islands of the Gulf of Carpentaria, *R. Brown ;* in the interior, *M'Douall Stuart's Expedition ;* Albert river, *Henne.*

Queensland. Balonne river, *Mitchell.*

There is very little to distinguish this from the common S. American *F. Contrayerva*, except the narrow leaves and the more sessile and compact clusters of flower-heads with broader floral leaves.

TRIBE VII. ANTHEMIDEÆ.—Leaves alternate, often much cut, rarely quite entire. Flower-heads heterogamous, the florets of the circumference female, ligulate, tubular or without corollas, those of the disk hermaphrodite or male, tubular, or very rarely the heads homogamous. Anthers obtuse or scarcely pointed at the base, usually short. Style of *Senecionidæ.* Pappus none or reduced to a small cup, ring or auricle, or rarely of flat chaffy scales.

A few of the genera have nearly the aspect of the *Asteroid* genera allied to *Brachycome*, but are readily known by the short truncate style-branches, a very few with entire linear leaves approach some of the smaller *Gnaphalieæ*, but have very different anthers. The Australian genera belong all to the group of *Cotuleæ*, which are never radiate ; the following three species, however, of the typical European *Anthemideæ*, with radiate flower-heads, have been more or less introduced as weeds.

Anthemis Cotula, Linn. ; (*Maruta Cotula*, DC. Prod. vi. 13.) A glabrous annual emitting a disagreeable smell when rubbed, with alternate once twice or thrice pinnate leaves with very narrow lobes. Flower-heads loosely corymbose with white sterile rays, a very conical receptacle with small chaffy scales between the florets, and small achenes without any pappus. —N. S. Wales, S. Australia, and W. Australia.

Chrysanthemum segetum, Linn. ; DC. Prod. vi. 64. A glabrous annual, with alternate leaves, the lower ones obovate and petiolate, the upper ones narrow and stem-clasping, generally deeply-toothed at the end. Flower-heads large, on terminal peduncles, with golden-yellow female rays, the receptacle nearly flat, without scales, the achenes small without any pappus.—N. S. Wales and Victoria.

Chrysanthemum Parthenium, Pers. ; (*Pyrethrum parthenium*, DC. Prod. vi. 58.) A perennial with a short stock and erect flowering-stems. Leaves pinnate, with ovate or oblong, toothed or pinnatifid segments. Flower-heads numerous in a terminal corymb, with white female rays, the receptacle nearly flat, without scales, the achenes small, crowned by a minute toothed border.—Port Jackson.

42. COTULA, Linn.

(Gymnogyne, *Steetz ;* Strongylospermum, *Less. ;* Pleiogyne, *C. Koch ;* Symphyomera *and* Ctenosperma, *Hook. f. ;* Leptinella, *Cass.*)

Involucre hemispherical or campanulate, with few nearly equal bracts, in about 2 rows. Receptacle flat, convex or conical, without scales. Florets of the circumference in 1 or several rows, female, without any or with a short broad or conical corolla. Disk-florets numerous, tubular, hermaphrodite, sometimes sterile, 4- or 5-toothed. Anthers obtuse at the base. Style-branches obtuse or truncate, or the style sometimes undivided. Achenes flattened, sometimes winged, without any pappus.—Herbs, usually small or decumbent, with alternate, entire lobed or dissected leaves. Flower-heads small, pedunculate.

A considerable genus, dispersed over the warmer and temperate regions of the Old World, with a few American species. Of the nine Australian species, one is common to the extratropical regions both of the northern and southern hemispheres, another is found in S. Africa, and a third in New Zealand, the remaining six appear to be endemic.

SECT. I. **Cotula.**—*Receptacle flat or convex. Achenes of the female florets on long stalks, in a single row. Female florets without any corolla.*

Involucral bracts very broad. Disk-achenes not winged. Leaves not sheathing at the base, filiform, entire . . . . 1. *C. filifolia.*
Involucral bracts ovate or oblong. Disk-achenes mostly winged. Leaves sheathing at the base.
Plant of 1 to 2 in. with filiform entire leaves . . . . 2. *C. integrifolia.*
Plant often attaining 6 in. or more. Leaves mostly lanceolate or oblong and toothed . . . . 3. *C. coronopifolia.*

SECT. II. **Strongylosperma.**—*Receptacle flat or convex. Achenes of the female florets numerous, in several rows, sessile or stipitate. Female florets without any corolla.*

Achenes of the female florets with broad thin wings. Leaves linear or filiform, entire . . . . 4. *C. gymnogyne.*
Achenes of the female florets with narrow thick wings or obtuse edges. Leaves dissected.
Stems slender. Peduncles filiform, mostly longer than the leaves . 5. *C. australis.*
Stems short, stout. Peduncles thick, mostly shorter than the leaves . . . . 6. *C. alpina.*

SECT. III. **Leptinella.**—*Receptacle conical. Achenes of the ray in several rows, sessile. Female florets with a short corolla.*

Glabrous or nearly so. Female florets inflated . . . . 8. *C. reptans.*
Stems woolly-hairy.
Leaves simply pinnate. Female florets conical . . . . 9. *C. filicula.*
Leaves twice pinnate. Female florets inflated . . . . 7. *C. Drummondii.*

SECTION I. COTULA.—Receptacle flat or convex. Achenes of the female florets in a single row, on long stalks. Female florets without any corolla.

At the time of flowering the female florets sometimes appear nearly sessile, but the stalks grow out considerably as the achene ripens, and usually persist after it has fallen off.

1. **C. filifolia,** *Thunb.; DC. Prod.* vi. 77. A small much-branched annual, glabrous or sprinkled with a few hairs, with slender ascending stems, rarely above 6 in. long. Leaves narrow-linear or filiform, entire, slightly dilated and stem-clasping at the base, mostly ½ to 1 in. long. Peduncles long and slender, sometimes dilated under the flower-head, as in the section or genus *Cenia*. Flower-heads smaller than in *C. coronopifolia*. Involucral bracts few, broad, very obtuse, with scarious margins. Receptacle flat or slightly convex. Female florets in a single row, without any corolla; disk-florets scarcely dilated at the base. Achenes of the female florets, when ripe

supported on a stipes often as long as the involucre and persistent, bordered by a broad thin wing. Achenes of the disk flat and smooth, the margins scarcely thickened but not at all winged.

Victoria. Port Phillip and inundated meadows near Brighton, *F. Mueller.*
Tasmania. Islands of Kent's group, Bass's Straits, *R. Brown.*
S. Australia. Murray river, Holdfast Bay, Gawler river, and Crystal Brook, *F Mueller.*
W. Australia. King George's Sound, *R. Brown.*
The species is also in S. Africa. It has precisely the aspect of *C. gymnogyne,* but with a very different floral structure.

2. **C. integrifolia,** *Hook. f. Fl. Tasm.* i. 192. *t.* 50 B. A little, glabrous scarcely branched annual, of 1 to 2 in., possibly a reduced form of *C. coronopifolia.* Leaves entire, linear, obtuse, rather thick, dilated at the base into a short sheath. Peduncles terminal, slender. Flower-heads not 2 lines diameter. Involucral bracts ovate or oblong, obtuse, with scarious margins. Florets and achenes of *C. coronopifolia.*

Tasmania. Moist ground near George Town, *Gunn.*

3. **C. coronopifolia,** *Linn.; DC. Prod.* vi. 78. Glabrous and diffuse, the stems rooting at the base and ascending from ½ to nearly 1 ft. Leaves lanceolate oblong or almost linear, coarsely toothed, pinnatifid or almost entire, 1 to 2 in. long, dilated at the base into a short sheath round the stem. Flower-heads 3 to 5 lines diameter, on peduncles longer than the leaves. Involucral bracts oblong-linear. Receptacle flat or slightly convex. Female florets in a single row, on flattened pedicels, half as long as the involucre, the ovary bordered by a transparent wing notched at both ends, the style very short in the terminal notch, without any corolla; disk-florets exceedingly numerous, on much shorter persistent pedicels, the corolla tubular, more or less dilated above the ovary, 4-toothed. Achenes of ths female florets nearly ¾ line long, including the thickish spongy wing surrounding them, those of the disk smaller, with a narrower wing.—Steetz in Pl. Preiss. i. 434; Hook. f. Fl. Tasm. i. 191.

N. S. Wales. Port Jackson, *R. Brown;* Twofold Bay, *F. Mueller.*
Victoria. Common in wet pastures, *F. Mueller* and others.
Tasmania. Common in wet pastures, even in brackish water, *J. D. Hooker.*
S. Australia. Near Adelaide and Lofty Range, *F. Mueller;* Kangaroo Island, *Waterhouse.*
W. Australia. Swan River, *Preiss, n.* 128; *Drummond, n.* 33, 62, 379; Hay river and Cape Naturaliste, *Oldfield.*
The species extends over New Zealand, extratropical S. America, S. Africa, and some parts of Europe, especially near the sea.

Section II. Strongylosperma.—Receptacle flat or convex. Achenes of the ray numerous, in several rows, sessile or stipitate. Ray-florets (in the Australian species) without any corolla.

4. **C. gymnogyne,** *F. Muell. Herb.* A slender, glabrous, simple or branched annual, rarely 6 in. high, closely resembling *C. filifolia* in aspect. Leaves linear, almost filiform, often above 1 in. long, dilated and sheathing at the base. Flower-heads 2 to 3 lines diameter or rather larger when in

fruit, on filiform peduncles. Receptacle thick, convex, almost hemispherical. Female florets numerous, in several series, without any corollas; disk-florets much less numerous, the corollas attenuate at the base, the styles apparently perfect or more or less abortive. Achenes of the female florets sessile, broadly winged, with a broad truncate base; those of the disk abortive.—*Gymnogyne cotuloides*, Steetz in Pl. Preiss. i. 432.

W. Australia. Swan River, *Drummond, n.* 380; *Preiss, n.* 101; Kalgan and Gordon rivers, Port Gregory, Champion Bay, Murchison river, *Oldfield*.

5. **C. australis,** *Hook. f. Fl. N. Z.* i. 128. Slender and diffuse, with an apparently perennial creeping rhizome, more or less clothed with long soft hairs or nearly glabrous. Leaves pinnate, with small pinnæ, entire or deeply 3-lobed or pinnatifid, the segments mostly mucronate-acute. Flower-heads small, on slender peduncles. Involucral bracts linear-oblong. Receptacle nearly flat. Female florets numerous, in several rows, without any corollas; disk-florets slightly dilated at the base, 4-toothed, fertile. Achenes of the female florets bordered by a narrow wing, not cordate at the base, on pedicels nearly as long as themselves. Achenes of the disk not winged, on very short pedicels.—Hook. f. Fl. Tasm. i. 191. t. 50 A; *Anacyclus australis*, Sieb. Pl. Exs.; *Strongylospermum australe*, Less. Syn. Comp. 261; DC. Prod. vi. 82; *Pleiogyne australis*, C. Koch in Bot. Zeit. 1843, 40; Sond. in Linnæa, xxv. 484.

N. S. Wales. Port Jackson, to the Blue Mountains, *R. Brown*, *Sieber*, and others, invariably following sheep, *C. Moore*; Clarence river, *Beckler*.
Victoria. Yarra river, *F. Mueller* and others; Creswick diggings, *Whan*.
Tasmania. Moist banks near Hobarton, *J. D. Hooker*.
S. Australia. Near Adelaide, *F. Mueller*.
W. Australia. Between Swan River and King George's Sound, *Drummond*.

The species is also in New Zealand and in the island of Tristan d'Acunha; it is very near the common Asiatic and African *C. anthemoides*, Linn., but in that species the peduncles are usually much shorter, the achenes less stipitate, and the female florets have almost always a short corolla, which I never find in *C. australis* even on Sieber's specimens, although described by De Candolle, who copied from Lessing.

6. **C. alpina,** *Hook. f. Fl. Tasm.* i. 192. *t.* 51 A. A small but stout glabrous perennial, shortly creeping or tufted. Leaves deeply pinnatifid or pinnate, with oblong-linear acute entire or 2- or 3-toothed segments. Flower-heads, when in fruit, fully 3 lines diameter, on thick hollow peduncles, often shorter than the leaves. Involucral bracts ovate-oblong. Receptacle flat or slightly convex. Female florets in several rows, without any corollas. Achenes of the female florets sessile, bordered by rather thick wings, those of the disk usually abortive.—*Ctenosperma alpinum*, Hook. f. in. Hook. Lond. Journ. vi. 115.

Victoria. Cobberas mountains, *F. Mueller*.
Tasmania. Subalpine and alpine situations near Marlborough, *Gunn*.

SECTION III. LEPTINELLA.—Receptacle conical. Achenes of the ray in several rows, sessile. Female florets with a short corolla.

7. **C. Drummondii,** *Benth.* A perennial, with prostrate or creeping

stems, densely clothed with long woolly hairs. Leaves twice pinnate, with oblong-linear or cuneate, toothed or pinnatifid segments. Flower-heads like those of *C. reptans*, on peduncles usually shorter than the leaves. Involucral bracts not so broad as in *C. reptans*, the corolla of the female florets short and inflated, larger in proportion to the achenes than in that species. Achenes not seen ripe, but apparently like those of *C. reptans*.

W. Australia, *Drummond, 3rd Coll. n.* 113; Don river, *Oldfield*.

The species is evidently near *C. reptans*, but much larger than the cut-leaved forms of that species, with the leaves much more cut.

8. **C. reptans,** *Benth.* A slender creeping perennial, glabrous or sprinkled with a few soft hairs, especially on the peduncles. Leaves pinnate, with ovate segments, toothed or pinnately divided into short linear lobes. Flower-heads rarely above 2 lines diameter, on peduncles usually longer than the leaves. Involucral bracts nearly orbicular. Receptacle conical. Female florets in several rows, the corollas very short and broad, inflated, contracted at the orifice, obliquely 2- or 3-toothed, the style shortly exserted. Achenes of the female florets sessile, flattened, with thickened obtuse margins, scarcely forming distinct wings, obtusely notched at the top, those of the disk usually abortive.—*Strongylosperma reptans*, Benth. in Hueg. Enum. 60; DC. Prod. vi. 82; *Pleiogyne reptans*, C. Koch in Bot. Zeit. 1843, 40; Sond. in Linnæa, xxv. 484; *Leptinella intricata*, Hook. f. in Hook. Lond. Journ. vi. 117, and Fl. Tasm. i. 193. t. 52 B; *Leptinella multifida*, Hook. f. in Hook. Lond. Journ. vi. 118; *Pleiogyne multifida*, Sond. in Linnæa, xxv. 484.

N. S. Wales. Twofold Bay, *F. Mueller*.

Victoria. Wet places, from Gipps' Land to the Yarra, *F. Mueller*.

Tasmania. Port Dalrymple, *R. Brown;* marshes in various parts of the island, *J. D. Hooker* and others.

S. Australia. Between Rivoli Bay and Mount Gambier, *F. Mueller*.

Var. *major*. Rather coarser, usually glabrous; petioles and peduncles longer; leaf-segments obovate and almost succulent; flower-heads larger.—*Leptinella longipes*, Hook. f. in Hook. Lond. Journ. vi. 117, and Fl. Tasm. i. 193. t. 52 A.—Victoria and Tasmania.

9. **C. filicula,** *Hook. f. Herb.* A small and stout perennial, tufted, diffuse or shortly creeping, more or less clothed with long soft or woolly hairs. Leaves pinnate, with linear-oblong or cuneate, entire or toothed segments. Flower-heads about 3 lines diameter when in fruit, on stout peduncles rarely exceeding the leaves. Involucral bracts ovate. Receptacle conical. Female florets in many rows, with small conical corollas, oblique and 2-toothed at the orifice, the style shortly exserted. Achenes of the female florets sessile, bordered by thick wings, scarcely distinguishable from the achene itself; those of the disk abortive.—*Symphyomera filicula*. Hook. f. in Hook. Lond. Journ. vi. 116; *Leptinella filicula*, Hook. f. Fl. Tasm. i. 194. t. 51 B.

Victoria. Mounts Wellington and Useful, Buffalo ranges, Baw-Baw mountains, at an elevation of 4000 to 5000 ft,, *F. Mueller*.

Tasmania. Hampshire hills, Mount Wellington, Christmas Rock, *Gunn*.

43. SOLIVA, Ruiz and Pav.

(Gymnostyles, *Juss.*)

Involucral bracts in about 2 rows nearly equal, with scarious margins. Receptacle flat, without scales. Florets of the circumference in several rows, female, without any corolla; disk-florets tubular, tapering at the base, 2- or 3-toothed. Anthers obtuse at the base. Style-lobes short, truncate. Achenes of the female florets flattened, bordered by a thick wing, without any pappus but the wings tapering into the rigid persistent style or produced into two divaricate points or prominent angles, those of the disk usually abortive. —Small diffuse herbs. Leaves alternate, usually finely dissected. Flower-heads sessile.

A small genus, apparently limited to the warmer regions of America, except where introduced with cultivation.

1. **S. anthemifolia,** *R. Br. in Trans. Linn. Soc.* xii. 102. Stems very much shorter than the leaves, forming a dense tuft. Leaves petiolate, 2 to 4 in. long, twice or even thrice pinnate with linear or linear-acute, entire or 3-fid segments, the primary ones often distant along the petiole, clothed with long soft hairs or nearly glabrous. Flower-heads closely sessile and clustered on the short stems, nearly globular when in fruit, $\frac{1}{4}$ to $\frac{1}{2}$ in. diameter. Involucral bracts oblong or lanceolate. Achenes numerous, bordered by a thick transversely rugose wing, which tapers into a rigid style longer than the achene itself, without any lateral angles or points.—DC. Prod. vi. 142; *Gymnostyles anthemifolia*, Juss., according to Br. l. c.

Queensland. Brisbane river, Moreton Bay, *F. Mueller.*
N. S. Wales. Port Jackson, *Woolls,* but only in cultivated ground, and probably introduced from S. America, *R. Brown.*

The species is also in Brazil, Gardner's specimens from Rio Janeiro, n. 225, being precisely the same as the Australian ones. It is also very near *S. acaulis*, Hook. and Arn., from Buenos Ayres, but in that species the wings of the achenes are broader, thinner, and produced into prominent angles or short horns. The pedunculate species referred to the genus, *S. pygmæa*, H. B. and K. (including *S. Mexicana*, DC., and perhaps *S. pedicellata*, Ruiz and Pav.), does not appear to differ from a true *Cotula*.

44. MYRIOGYNE, Less.

(Centipeda, *Lour.*; Sphæromorphæa, *DC. partly.*)

Involucral bracts in about 2 rows, nearly equal, scarious at the edges. Receptacle flat or slightly convex, without scales. Florets of the circumference in many rows, female, with short tubular corollas; disk-florets hermaphrodite, fertile, broadly campanulate, 4-lobed. Anthers short, obtuse at the base. Style-lobes very short, obtuse or truncate. Achenes not at all or scarcely compressed, with 3 or 4 very prominent obtuse ribs or angles, without any pappus.—Herbs with alternate usually toothed leaves. Flower-heads small, sessile, lateral or in a short terminal raceme.

A small genus, chiefly S. Asiatic, with one species from extratropical S. America. Of the two Australian species, one is the common Indian one, the other is endemic. The genus has been united by J. D. Hooker with *Cotula*, but the peculiar habit and the shape of the achenes appear to me to be sufficient characters to retain it as distinct.

Flower-heads all or nearly all sessile and lateral . . . . 1. *M. minuta.*
Flower-heads in terminal leafless racemes . . . . 2. *M. racemosa.*

1. **M. minuta,** *Less. in Linnæa*, vi. 219. A prostrate branching annual or perhaps sometimes a perennial of short duration, the slender stems 2 to 3 in. or rarely 6 in. long in tropical regions, glabrous or clothed with short white woolly intricate hairs, in most of the extratropical Australian specimens more robust, glabrous, attaining sometimes 1 ft. Leaves oblong, $\frac{1}{4}$ to $\frac{1}{2}$ in. long, narrowed at the base or almost petiolate, toothed or almost pinnatifid. Flower-heads $1\frac{1}{2}$ to 2 lines diameter, solitary, at first terminal, but soon becoming leaf-opposed, closely sessile or rarely accompanied by a second pedunculate one (*i. e.* by a short 1-headed flowering branch with the floral leaf abortive); florets very minute, the female corollas scarcely above a third of the length of their ovaries. Styles of the disk-florets with very short truncate lobes. Achenes slightly hairy.—DC. Prod. vi. 139, with the numerous synonyms adduced; Hook. f. Fl. Tasm. i. 194; *M. Cunninghamii*, DC. Prod. vi. 139; F. Muell. Pl. Vict. t. 41; *Centipeda orbicularis*, Lour.; Miq. Fl. Ned. Ind. ii. 89; *Sphæromorphæa centipeda*, DC. Prod. vi. 140; *Sphæromorphæa Russeliana*, DC. l. c.; Deless. Ic. Sel. iv. t. 49; *Cotula minuta*, Forst. Prod. 57; Hook. f. Handb. N. Zeal. Fl. 144.

N. Australia. Arnhem's Land, *F. Mueller.*

Queensland. Broad Sound and Shoalwater Bay, *R. Brown;* Port Curtis, *M'Gillivray;* Suttor river, *Thozet;* Rockhampton, *Dallachy* and others.

N. S. Wales. Port Jackson to the Blue Mountains, *R. Brown, Sieber, n.* 491, and others; Clarence river, *Beckler;* in the interior, Lachlan and Darling rivers, etc., *A. Cunningham, Victorian and other Expeditions* (these specimens generally very robust with larger flower-heads).

Victoria. Very common in marshy places, *F. Mueller* and others.

Tasmania. Waste places, cultivated ground, etc., *J. D. Hooker.*

S. Australia. Common on the Murray and thence to St. Vincent's Gulf, *F. Mueller* and others; Kangaroo island, *Waterhouse.*

W. Australia. Swan River, *Drummond, n.* 89, 179, 183; Capel and Murchison rivers, *Oldfield.*

2. **M. racemosa,** *Hook. in Mitch. Trop. Austr.* 353. Glabrous, with ascending or erect stems, $\frac{1}{2}$ to $1\frac{1}{2}$ ft. high. Leaves linear or linear-lanceolate, acutely toothed, slightly narrowed towards the base or the lower ones petiolate. Flower-heads small, in short terminal leafless racemes. Involucral bracts oblong. Ray-florets minute, almost globular. Styles of the disk-florets with very short obtuse lobes. Achenes very prominently 4- or 5-angled, the angles ciliate.

Queensland, *Bowman;* Maranoa river, *Mitchell;* Burdekin and Gilbert rivers and Newcastle range, *F. Mueller.* The inflorescence is that of *Dichrocephala*, but the achenes and florets are very differently shaped.

45. ABROTANELLA, Cass.

(Scleroleima *and* Trineuron, *Hook. f.*)

Involucral bracts few, in about 2 rows, nearly equal or the outer ones shorter. Receptacle nearly flat, without scales. Florets all tubular, those of the circumference female, 3- or 4-toothed, those of the disk hermaphrodite or male, 4-toothed. Anthers obtuse or slightly pointed at the base. Style-lobes short, truncate. Achenes 4-angled or flattened, not winged, without any pappus.—Dwarf perennials usually densely tufted with closely imbricate small leaves, more rarely with short decumbent or ascending stems and alternate entire leaves. Flower-heads solitary, or few in small corymbs, sessile or shortly pedunculate.

The genus ranges over New Zealand and the Antarctic regions generally; the three Australian species appear to be endemic.

Leaves closely imbricate, acuminate, about ¼ in. long. Florets about 3 or 4 in each head . . . . 1. *A. forsterioides.*
Leaves spreading. Florets numerous in each head.
Leaves ½ in. long. Peduncles not exceeding the leaves . . . . 2. *A. nivigena.*
Leaves ¾ to 1½ in. long. Flowering-stems 2 to 4 in. high, with more than one flower-head . . . . 3. *A. scapigera.*

1. **A. forsterioides,** *Hook. f. Handb. N. Zeal. Fl.* 139. A densely tufted plant forming often large patches. Leaves persistent and densely imbricate on the short closely-packed stems, oblong with a broad sheathing base, subulate-acuminate, 2 to 3 lines long, with an obscurely denticulate cartilaginous margin. Flower-heads small and solitary on each branch, on a peduncle at first exceedingly short, at length nearly as long as the leaves. Involucre of 3 to 5 linear or oblong obtuse unequal bracts, usually shorter than the leaves. Florets 3 or 4, of which 1 male and sterile and 2 or 3 female. Achenes 4-angled.—*Scleroleima forsterioides*, Hook. f. in Hook. Lond. Journ. v. 444. t. 14, and Fl. Tasm. i. 195.

Tasmania. Summit of the Table Mountain, *R. Brown*, and of all the mountains above 4000 ft., forming green velvety cushions, *J. D. Hooker*.

The habit is nearly that of *Pterygopappus*, but it may be at once recognized by the longer narrower leaves, independently of the generic characters.

2. **A. nivigena,** *F. Muell. Herb.* A small glabrous plant forming dense tufts or the stems shortly diffuse but not exceeding 2 or 3 in. Leaves linear, obtuse, about ½ in. long, sheathing at the base, thick and smooth as in the other species. Flower-heads solitary, on peduncles shorter than the leaves. Involucre not 2 lines long, the bracts very obtuse or truncate, a few of the outer ones shorter. Florets and achenes apparently like those of *A. scapigera*, but only seen very young.—*Trineuron nivigenum*, F. Muell. in Trans. Phil. Soc. Vict. i. 105, in Hook. Kew Journ. viii. 149, and Pl. Vict. t. 40.

Victoria. Snowy summits of the Munyong mountains, at an elevation of 5000 to 6000 ft., *F. Mueller*.

3. **A. scapigera,** *F. Muell. Herb.* Stems creeping or tufted at the base, ascending to from 2 to 4 in., glabrous. Leaves oblong-linear, narrowed below the middle, sheathing at the base, ¾ to 1½ in. long, crowded at the base of the flowering stems with a few smaller ones under the peduncles. Flower-heads few, in a small terminal leafy corymb. Involucre nearly 2 lines long, the bracts oblong, nearly equal. Florets of the ray in 2 or 3 rows, those of the disk about the same number. Anthers shortly pointed at the base. Achenes of the female florets oblong, compressed, those of

the disk apparently smaller, but not seen perfect.—*Trineuron scapigerum*, F. Muell. in Trans. Phil. Inst. Vict. ii. 70, and in Hook. Kew Journ. ix. 301, Hook. f. Fl. Tasm. ii. 364.

Tasmania. Crevices of rocks, summit of Mount Lapeyronse, *Oldfield.*

46. CERATOGYNE, Turcz.

(Diotosperma, *A. Gray.*)

Involucre cylindrical, of few oblong bracts, green with scarious edges. Receptacle without scales. Florets of the circumference few, female, filiform, 2- or 3-toothed or shortly ligulate; disk-florets few, hermaphrodite, sterile, tubular, 3- or 4-toothed. Anthers short, thin (obtuse at the base?). Style-lobes slender, acute, hispid. Achenes of the ray large, flat, bordered by herbaceous wings, involute on the margins and produced at the top into incurved auricles; those of the disk abortive.—Small annual. Leaves alternate. Flower-heads small, terminal or axillary.

The genus is limited to a single species endemic in Australia.

1. **C. obionoides,** *Turcz. in Bull. Mosc.* 1851. ii. 69. An annual with erect slender branching stems not exceeding 6 in., and often much smaller, more or less hirsute with soft simple hairs. Leaves petiolate, obovate or oblong, under ½ in. long, the upper ones small, narrow and sessile. Flower-heads nearly sessile or on very short axillary leafy peduncles. Involucre a little more than 1 line long, of 4 to 6 oblong bracts. Florets about as long as the involucre, of which 3 or 4 females and about as many sterile central ones. Achenes of the female florets very soon growing out to at least twice the length of the involucre, remarkable for their broad herbaceous wings the incurved ciliate auricles at the top as long as the corolla.—*Diotosperma Drummondii,* A. Gray in Hook. Kew Journ. iv. 275.

W. Australia, *Drummond,* 5*th Coll. Suppl. n.* 56.

47. ELACHANTHUS, F. Muell.

Involucre of few oblong bracts with scarious edges, nearly equal or with 1 or 2 outer short ones. Receptacle small, without scales. Florets of the circumference several, female, tubular, slender but broader towards the base, minutely 2- or 3-toothed or almost entire. Disk-florets few, hermaphrodite but sterile, slender, 3- or 4-toothed. Anthers short, obtuse at the base. Style undivided or with linear hispid branches. Achenes of the female florets nearly terete, with a pappus of several narrow-lanceolate chaffy scales; those of the disk slender, abortive, with a reduced pappus.—Small annual. Leaves alternate, entire. Flower-heads terminal.

The genus is limited to a single species endemic in Australia. With a habit approaching that of *Millotia, Quinetia,* and other small *Gnaphalieæ,* the florets and anthers, as in *Isoetopsis,* indicate a near affinity with *Cotula* and *Myriogyne,* and although pappus scales are rare in the tribe, they occur also in a few of the subtribe *Tanaceteæ.*

1. **E. pusillus,** *F. Muell. in Linnæa,* xxv. 411. A slender erect branching annual attaining 3 or 4 in. but sometimes much smaller, glabrous or

slightly pubescent especially under the inflorescence. Leaves small, narrow-linear. Involucres at first cylindrical, at length turbinate, 2 to 2½ lines long. Female florets about 5 to 9, about as long as the involucre at the time of flowering but the achene and pappus soon growing out so as to exceed it by 1 line or more. Sterile florets fewer and rather shorter. Achenes of the ray densely silky-villous. Pappus scales 12 to 20, narrow-lanceolate, acute, entire or slightly denticulate, about as long as the achene.

N. S. Wales. Darling river, *Victorian Expedition.*
S. Australia. Dry hills, Akaba and Cudnaka, *F. Mueller.*

48. **ISOETOPSIS,** Turcz.

Involucres of few broad scarious bracts, the outer ones with linear leaf-like tips. Receptacle small, without scales but with a few long hairs. Florets of the circumference in several rows, female, tubular, slender but broader towards the base, 2- or 3-toothed. Disk-florets hermaphrodite but sterile, the tube exceedingly slender, expanded into a campanulate 4-lobed limb. Anthers short, obtuse at the base. Style undivided. Achenes of the female florets nearly terete, with a pappus of obtuse chaffy scales. Achenes and pappus of the disk entirely abortive or rudimentary.—Dwarf tufted herb, the flower-heads small and densely tufted within the grass-like radical leaves.

The genus is limited to a single species, endemic in Australia. Notwithstanding the foliage, the affinity is evident with *Cotula* and *Myriogyne.*

1. **I. graminifolia,** *Turcz. in Bull. Mosc.* 1851, i. 175. *t.* 3. A dwarf, almost stemless densely-tufted plant, the numerous small flower-heads sessile and densely crowded within the radical leaves, which are linear, grass-like, 1 to 2 or even 3 in. long, the inner ones dilated and more or less scarious at the base, passing into the outer involucral bracts, which have short green linear tips, the inner bracts broad and scarious, without tips, about 2 lines long. Florets about as long as the involucre. Achenes of the female florets rather thick, silky-hairy.

N. S. Wales. Between the Lachlan and Darling rivers, *Burkitt*, *Victorian Expedition.*
Victoria. Avoca river, *F. Mueller.*
S. Australia. Murray scrub, *F. Mueller.*
W. Australia, *Drummond*, 4*th Coll. n.* 207, 5*th Coll. n.* 382, 390, *and Suppl. n.* 70.

Tribe VIII. Gnaphalieæ.—Leaves alternate or very rarely irregularly opposite, quite entire. Flower-heads homogamous, with all the florets hermaphrodite, tubular, 4–5- or rarely 3-toothed, or heterogamous, with filiform or very rarely ligulate or irregular female florets. Anthers with fine points or tails at their base. Style of *Senecionidæ.* Pappus of capillary simple or plumose bristles or rarely of short scales or none.—The majority of species have more or less of a loose cottony wool.

Subtribe I. Angiantheæ.—Flower-heads small, sessile or nearly so on a common receptacle in dense clusters or compound heads, often closely surrounded by imbricate bracts or by a few floral leaves forming a general involucre. Florets all tubular and hermaphrodite.

49. MYRIOCEPHALUS, Benth.

(Hyalolepis, *DC.*; Antheidosorus, *A. Gray*; Gilberta, *Turcz.*; Lamprochlæna, Elachopappus, *and* Polycalymma, *F. Muell.*)

Flower-heads exceedingly numerous and sessile on a broad very flat receptacle, in a dense cluster or compound head, surrounded by a general involucre of numerous narrow bracts in many rows, each usually with a scarious tip or radiating appendage. Partial heads 1- or few-flowered. Involucre of few bracts (the outer ones including one on the general receptacle subtending each head), usually contracted into a stalk-like base and scarious at the tip, the inner ones scarious and transparent from the base or rarely all narrow and rigid. Receptacle without scales. Florets hermaphrodite, tubular, slender, 3- to 5-toothed. Anthers with more or less conspicuous points or tails at the base. Style-branches nearly terete, truncate. Achenes more or less compressed. Pappus none or of 1 or more awns or bristle-like scales, simple or more or less plumose.—Herbs, either annual or with a perennial or woody base, often hoary or white, especially when young, with woolly or cottony hairs. Leaves alternate, entire. Clusters or compound heads terminal, usually globose or hemispherical; the flat receptacle sometimes so broadly dilated that the outer flower-heads are reflexed.

The genus is limited to Australia. It differs from *Angianthus* in the more perfect general involucre and more developed common receptacle, the partial involucres sometimes reduced to 2 bracts with a single floret, so as in some species to bring the compound head of *Myriocephalus* in close analogy to the simple head of *Helichrysum*.

Appendages of the general involucral bracts under 1 line long or inconspicuous.
  Dwarf plant. Leaves much longer than the flower-heads. Partial heads 1-flowered. Pappus of 1 very fine awn or none . . . 1. *M. rhizocephalus.*
  Plants of ½ to 1 ft. Leaves under 1 in. long. Partial heads 4- to 6-flowered. Pappus none or of microscopic scales.
    Appendages of the general involucral bracts minute and yellowish or inconspicuous . . . . 2. *M. nudus.*
    Appendages of the general involucral bracts white and nearly 1 line long . . . . 3. *M. appendiculatus.*
  Decumbent or ascending plant, under 6 in. Leaves small, mostly cuneate or spathulate. Partial heads 4-flowered. Pappus of 1 to 4 bristle-like scales . . . . 4. *M. Rudallii.*
Appendages of the general involucral bracts 1 to 2 lines long, broad and very conspicuous.
  Appendages yellow. Partial-heads 1-flowered. Achenes glabrous. Pappus of 5 to 7 bristles, plumose from the base or at the end . . . . 5. *M. gracilis.*
  Appendages white.
    Herbaceous and flaccid. Partial heads 2- or 3-flowered. Achenes hairy. Pappus of 1 to 4 bristles . . . . . 6. *M. helichrysoides.*
    Shrubby at the base with rigid erect branches. Partial heads 2-flowered. Achenes glabrous. Pappus of several bristles tipped wlth transparent globules . . . . 7. *M. suffruticosus.*
    Herbaceous and tall. Partial heads 5- to 8-flowered. Achenes woolly. Pappus of numerous ciliate bristles . . . . . 8. *M. Stuartii.*

1. **M. rhizocephalus,** *Benth.* A small tufted annual, the stem from ½ to 2 or 3 in. long, covered with the broad sheathing bases of the leaves.

Leaves above the broad base linear, grass-like, often 3 to 4 in. long. Clusters of flower-heads globose or hemispherical, ½ to ¾ in. diameter, sessile amidst the upper leaves which form an outer involucre. General involucre of very numerous bracts in several rows, transparent with short green midribs, ciliate with long woolly hairs, except the small glabrous tips which form a ray round the cluster. General receptacle very broad and flat. Partial heads very numerous, shorter than the general involucre, 1-flowered. Partial involucres of 3 bracts, one subtending and two lateral, all linear, hard and more or less concrete at the base, transparent but narrow above. Florets very numerous, 3- or 4-toothed. Achenes narrow, sparingly hairy especially near the top. Pappus of a single hair-like awn, slightly dilated at the base or occasionally none.—*Hyalolepis rhizocephala*, DC. Prod. vi. 149; *H. occidentalis*, F. Muell. Fragm. iii. 155.

N. S. Wales. Molle's Plains, *A. Cunningham;* between the Lachlan and Darling rivers, *Burkitt.*

Victoria. Wet grassy places near the Victoria range, *F. Mueller;* Skipton, *Whan;* Wimmera, *Dallachy.*

S. Australia. Murray river, *F. Mueller;* Cygnet Bay, Kangaroo Island, *Waterhouse.*

W. Australia, *Drummond;* Gordon river, *Oldfield.*

The hair-like bristle of the pappus occurs on the Western as well as on the Eastern specimens, but is often so fine as readily to escape observation.

2. **M. nudus,** *A. Gray in Hook. Kew Journ.* iii. 174. Erect, either annual or with a small perennial rhizome, nearly glabrous except towards the inflorescence, scarcely branched, ½ to 1 ft. high. Leaves linear, with a somewhat dilated base, the lower ones above ½ in. long, the upper ones smaller. Clusters of flower-heads when full-sized ½ in. diameter, without floral leaves. General involucre of several rows of numerous very small scarious woolly-ciliate bracts, shorter than the full-grown partial heads, with very small pale yellow tips forming a ray to the young cluster. Partial heads about 4-flowered, the involucres of 5 or 6 narrow scarious bracts, the subtending and outer ones with a rigid midrib and slightly woolly-ciliate, the inner ones quite glabrous with a scarcely conspicuous midrib. Achenes slightly pubescent. Pappus none.

W. Australia. Swan River, *Drummond, 1st Coll. and n.* 53; Murchison river, *Oldfield.*

Var. *Oldfieldii.* Weaker and more branched, sometimes loosely woolly, the radiating tips of the general involucre sometimes more conspicuous, sometimes less so than in the original form.—*Lamprochlæna Oldfieldii*, F. Muell. Fragm. iii. 157.—Murchison river and Champion Bay, *Oldfield; Drummond, 6th Coll. n.* 162.

3. **M. appendiculatus,** *Benth. in Hueg. Enum.* 61. Stems erect, sparingly branched, with a hard base and possibly perennial, ½ to 1 ft. high. Leaves linear, rarely above ½ in. long, with a slightly dilated base. Clusters of flower-heads depressed-globular, above ½ in. diameter when full grown. General involucre of many rows of very numerous scarious bracts woolly-ciliate at the base, with white ovate-oblong spreading tips forming a ray round the cluster. Partial heads 4- to 6-flowered, the involucres of about 6 cuneate-oblong scarious woolly-ciliate bracts, the subtending and outer ones with a rigid midrib, the inner ones transparent almost throughout. Florets

5-toothed. Achenes minutely hairy. Pappus none or of 1 or 2 microscopic scales.—DC. Prod. vi. 150; A. Gray in Hook. Kew Journ. iii. 174.

W. Australia. Swan River, *Huegel, Drummond, 1st Coll. n.* 59.

4. **M. Rudallii,** *Benth.* An annual, branching at the base, with decumbent or ascending stems under 6 in. long, loosely woolly. Leaves oblong-linear, cuneate or spathulate, very obtuse, often dilated at the base, not above ½ in. long. Clusters of flower-heads when full-grown about ½ in. diameter. General involucre of very numerous woolly-ciliate bracts, those of the outer rows shorter, with green midribs and no laminæ, those of the inner rows with small white petal-like spreading tips, forming a ray to the cluster. Partial heads about 4-flowered, the involucre of about 5 very narrow bracts, the subtending and outer ones little more than the midrib dilated and exceedingly thin and transparent at the end, all ciliate with fine long woolly hairs. Achenes with 1 to 4 unequal bristle-like scales.—*Elachopappus Rudallii,* F. Muell. Fragm. iii. 157.

S. Australia. Cooper's Creek, *Howitt's Expedition.*

5. **M. gracilis,** *Benth.* An annual, with erect slender branches not exceeding 6 in. Leaves narrow-linear. Clusters of flower-heads numerous, hemispherical, about ¼ in. diameter. General involucre of numerous dry appressed bracts, those of several outer rows with a broad bright-yellow scarious spreading lamina of nearly 1 line, forming a shining ray to the cluster, the inner bracts ovate, more scarious, without spreading tips and passing into the subtending bracts within the cluster. Partial heads 1-flowered, the involucres of 3 to 5 narrow scarious bracts usually shorter than the subtending one, especially in the centre of the cluster where all are smaller. Florets 4- or 5-toothed. Achenes glabrous. Pappus of 5 or 6 slender bristles of which 1 or 2 are very fine and hair-like with a short clavate plume at the end, the others not quite so long, more rigid, plumose-ciliate from the base.—*Antheidosorus gracilis,* A. Gray in Hook. Kew Journ. iii. 173; *Gilberta tenuifolia,* Turcz. in Bull. Mosc. 1851, i. 193.

W. Australia. Swan River, *Drummond;* Murchison river, *Oldfield.*

6. **M. helichrysoides,** *A. Gray in Hook. Kew Journ.* iii. 175. Glabrous or nearly so. Stems usually weak and decumbent or ascending, somewhat branched, ½ to 1 ft. long. Leaves linear, obtuse, often above 1 in. long. Clusters of flower-heads when full-sized about ½ in. diameter. General involucre of very numerous bracts in several rows, each one dark in the centre with white margins and white spreading petal-like laminas, sometimes above 1 line long, forming a conspicuous ray to the cluster. Partial heads 2- or 3-flowered, the involucral bracts few, oblong, not ciliate, the subtending and outer ones with a prominent rigid midrib, less conspicuous in the inner ones. Achenes hairy, especially at the top. Pappus of 1, 2 or very rarely 3 bristles, very fine at the time of flowering, thickened and hardened at the base on the ripe achene.

W. Australia. Swan River, *Drummond, 1st Coll.;* very wet places, Cape Naturaliste, *Oldfield.*

7. **M. suffruticosus,** *Benth.* Shrubby at the base, the branches white

with a close cotton, the flowering ones simple, erect, above 1 ft. long. Leaves linear or linear-lanceolate, half stem-clasping, somewhat coriaceous with revolute margins, the larger ones above 1 in. long, the upper ones small and distinct. Clusters of flower-heads hemispherical, ½ to ¾ in. diameter. General involucre of very numerous bracts in many rows, a few small narrow herbaceous and woolly, all the others with white obovate spreading laminæ 1 to 1½ lines long, forming a very conspicuous ray to the clusters. Partial heads mostly 2-flowered, the involucral bracts about 5 or 6 besides the subtending one, usually very narrow without scarious margins. Achenes glabrous or nearly so, but as well as the whole partial flower-heads, apparently glutinous. Pappus of several exceedingly fine bristles, nearly as long as the corolla, simple but bearing usually at their tips 1 to 4 little globular transparent bodies (glands ?).

W. Australia. Between Moore and Murchison rivers, *Drummond, 6th Coll. n.* 153.

8. **M. Stuartii,** *Benth.* Erect, apparently annual, not much branched, 1 to 2 ft. high, pubescent or woolly-white. Leaves linear or lanceolate, half stem-clasping, 1 to 2 in. long, the smaller distant upper ones often with white scarious tips passing into the involucral bracts. Clusters of flower-heads hemispherical, attaining 1 in. diameter or even more. General involucre of numerous bracts in several rows, herbaceous at the base, with white ovate spreading laminæ fully 2 lines long, forming a very conspicuous ray to the cluster. Partial heads 5- to 8-flowered or the central one with more numerous florets; the involucral bracts exceedingly thin and transparent, fringed at the end, about 2 of the outermost of each involucre as well as the subtending one contracted into a short rigid midrib, the others transparent from the base and very broad. Achenes densely clothed with very long woolly hairs. Pappus of 15 to 20 unequal bristles, slightly dilated and ciliate or plumose with short hairs.—*Polycalymma Stuartii*, F. Muell. and Sond. in Linnæa, xxv. 494 ; F. Muell. Pl. Vict. t. 42.

N. S. Wales. Lachlan and Darling rivers to the Barrier range and Cooper's Creek, *Victorian and other Expeditions ;* on the Murray and Murrumbidgee, *F. Mueller.*

Victoria. Murray river, *F. Mueller.*

S. Australia. Murray river, *F. Mueller ;* between Lake Gairdner and Devonport Range, *Babbage's Expedition.*

50. **ANGIANTHUS,** Wendl.

(Siloxerus, *Labill. ;* Ogcerostylus, *Cass. ;* Styloncerus, *Spreng. ;* Cylindrosorus and Phyllocalymma, *Benth. ;* Skirrophorus, *DC. ;* Chrysocoryne, *Endl. ;* Eriocladium, *Lindl. ;* Pogonolepis, *Steetz ;* Piptostemma, Epitriche *and* Gamozygis, *Turcz. ;* Cephalosorus, Hyalochlamys *and* Dithyrostegia, *A. Gray ;* Pleuropappus, *F. Muell.*)

Flower-heads numerous and sessile on a cylindrical conical convex or flat receptacle, in a dense cluster spike or compound head, surrounded by a general involucre of large and leaf-like or of small and herbaceous or scarious bracts, or of both in few rows without radiating tips or sometimes very few or none. Partial heads 1- or few-flowered, very rarely many-flowered. Involucre compressed, of few scarious transparent bracts, the subtending one flat, two lateral ones conduplicate and keeled or concave, and sometimes 2 or more inner ones flat or slightly concave. Receptacle without scales. Florets hermaphrodite, tubular, slender, 4- or 5-toothed, often hardened at the base. Anthers more or less distinctly pointed or tailed at the base. Style-branches nearly terete, truncate. Achenes usually compressed. Pappus none or of 1, 2 or more jagged or awned scales, often united in a ring or cup at the base. Annual or rarely perennial herbs, or in one species shrubby, glabrous or more or less cottony or woolly-white. Leaves alternate or very rarely irregularly opposite, entire. Clusters or spikes of flower-heads terminal, sessile or pedunculate, cylindrical, oblong-ovoid, globular or hemispherical, the partial involucres usually very deciduous with the achenes, or rarely the subtending bract persistent.

The genus is limited to Australia. The numerous genera, mostly monotypic or nearly so, which it is here proposed to unite, have been established chiefly upon minute distinctions in the pappus, which appear to me to afford a much better specific than generic character. In the genus the general involucre is less perfect and the general receptacle less developed than in *Myriocephalus*, more so than in *Gnephosis* and *Calocephalus*, and the partial involucre more reduced and flattened than in either of the two latter genera. For the common name, *Siloxerus* of Labillardière has undoubtedly the right of priority, but it has by common consent been rejected as being at complete variance with the etymology given by the author. Cassini's emendation (rejected as barbarous) and Sprengel's generally adopted one are both more recent than Wendland's name, which typically represents the tribe, and has been applied to several of the species, whilst Labillardière's has only been given to a single one. The general rules of the science appear therefore to be best observed by applying the name of *Angianthus* to the whole genus.

Clusters or spikes of flower-heads cylindrical, oblong-ovoid or rarely globose, the receptacle a cylindrical rhachis. Annuals or rarely herbaceous perennials.

Pappus conspicuous.
Spikes cylindrical without floral leaves. Pappus of 2 or 3 scales ending in bristles plumose at the end . . . . . 1. *A. tomentosus.*
Spikes cylindrical, surrounded by floral leaves. Pappus of 1 very oblique fringed boat-shaped scale . . . . 2. *A. pleuropappus.*
Spikes oblong, often surrounded by floral leaves. Pappus a short irregularly-fringed cup . . . . 3. *A. brachypappus.*
Spikes ovoid, surrounded by floral leaves. Involucral bracts with white appendages. Pappus of several lanceolate fringed scales Stems very short. . . . . 4. *A. humifusus.*
Pappus none or a minute ring.
Bracts of the general involucre all scarious.
Spikes slender, cylindrical, ½ to 1 in. long, attenuate at the base, of a shining brown . . . . 5. *A. myosuroides.*
Spikes short, cylindrical, obtuse at both ends, brown . . 6. *A. tenellus.*
Spikes oblong, attenuate at the base . . . . 7. *A. pusillus.*
Outer bracts of the general involucre leafy or tomentose, although short. Spikes ovoid-oblong . . . . 8. *A. Milnei.*

Clusters of flower-heads ovoid, the receptacle conical. Plant shrubby at the base.

Pappus none or a minute ring. General involucre shorter than the heads . . . . 9. *A. Cunninghamii.*

Clusters of flower-heads ovoid globular or hemispherical, the receptacle flat, convex or rarely conical (almost oblong in A. globifer). *Annuals or rarely herbaceous perennials.*

Pappus conspicuous.

Tall plant. Clusters of flower-heads surrounded by broadly ovate floral leaves. Pappus cup-shaped, deeply jagged . . 10. *A. phyllocephalus.*
Small plant. Clusters of flower-heads surrounded by long linear bracts. Pappus of 5 ovate awned scales . . . . 11. *A. micropoides.*
Small plant. Clusters of flower-heads surrounded by 3 or 4 floral leaves not exceeding the heads. Pappus of 3 ovate awned scales . . . . 12. *A. microcephalus.*
Pappus none or a minute ring.
Involucral bracts usually 4 enclosing 2 or rarely 1 floret.
Stems 2 to 3 in. long or more. Florets 5-merous.
Floral leaves broadly ovate . . . . 13. *A. platycephalus.*
Floral leaves ovate-lanceolate . . . . 14. *A. Drummondii.*
Stems 2 to 3 in. long. Florets 4-merous.
Clusters of flower-heads hemispherical. Florets thickened at the base . . . . 15. *A. Preissianus.*
Clusters of flower-heads ovoid or globular. Florets not thickened at the base . . . . 16. *A. eriocephalus.*
Stems scarcely any. Clusters of flower-heads depressed-globular, almost radical. Florets 5-merous.
Heads 2-flowered . . , ... 17. *A. pygmæus.*
Heads 1-flowered . . . . 18. *A. globifer.*
Involucral bracts 2, enclosing 1 floret.
Dwarf plant, the clusters of flower-heads almost radical . . 19. *A. demissus.*
Erect or ascending, 1 to 6 in. high with terminal clusters.
Leaves narrow.
Stems rigid, 2 to 6 in. high. Floral leaves longer than the heads . . . . 20. *A. strictus.*
Stems slender, 1 to 3 in. high. Floral leaves not longer than the heads. . . . . 21. *A. plumiger.*
Leaves ovate, stem-clasping. Stem slender, 1 to 3 in. high. Floral leaves 2, broad, often connate . . . 22. *A. amplexicaulis.*

1. **A. tomentosus,** *Wendl. Coll. Pl.* ii. 31. *t.* 48. An annual, more or less clothed with white wool, the stems numerous, erect or decumbent, rarely exceeding 6 in. in the Eastern specimens, often above 1 ft. in the Western ones. Leaves from oblong-cuneate to linear, obtuse, narrowed at the base and slightly decurrent, the upper ones small. Clusters of flower-heads oblong or cylindrical, yellowish or pale straw-coloured, those terminating the stem sometimes ¾ or even 1 inch long, those on the branches smaller and more ovoid, all obtuse at the base and usually not close to the last leaves, the rhachis filiform, with few outer empty bracts. Partial involucres about 1½ lines long, the bracts transparent with an opaque midrib in the lower part, 2 conduplicate and 2 or 3 flat. Florets 2, or sometimes 1 or 3, 5-toothed, the corolla not at all or scarcely thickened at the base. Pappus of 2 or 3 ovate denticulate scales, each terminating in a rather long bristle slightly plumose at the end.—DC. Prod. vi. 150; Sond. in Linnæa, xxv. 487; *Cylindrosorus flavescens*, Benth. in Hueg. Enum. 627; DC. Prod. vi. 152; *Angianthus flavescens*, Steetz in Pl. Preiss. i. 438.

Victoria. Murray river, *F. Mueller, Dallachy.*

S. Australia. Nuyt's Archipelago, *R. Brown;* scrub on the Murray river and thence to St. Vincent's Gulf, *Behr, F. Mueller* and others.

W. Australia. King George's Sound to Swan River, *Drummond, 4th Coll. n.* 217, *5th Coll. n.* 352, 353; *Preiss, n.* 62, and others; Fitzgerald and Phillips ranges, *Maxwell;* Bowes and Murchison rivers, *Oldfield.*

2. **A. pleuropappus,** *Benth.* An erect slender branching annual, woolly-tomentose or at length glabrous. Leaves linear. Clusters of flower-heads cylindrical, ½ in. long or more, of a golden-yellow and shining, sessile among a few floral leaves like the stem ones or more lanceolate and shorter than the spike. A few short broad scarious bracts within the floral leaves forming a general involucre and passing into the subtending bracts. Receptacle cylindrical, slender. Partial involucres of 2 keeled bracts and 2 flat stipitate ones expanded into an orbicular erect lamina. Florets 5-merous, thickened and bulb-like at the base. Pappus annular at the base, very oblique, the inner side very short and jagged, the outer side boat-shaped, fringed and terminating in a point nearly as long as the floret.—*Pleuropappus phyllocalymmeus,* F. Muell. in Trans. Vict. Inst. 1855, 37.

S. Australia. Sterile plains, Port Lincoln, *Wilhelmi.*

3. **A. brachypappus,** *F. Muell. in Trans. Phil. Soc. Vict.* i. 44, *and in Hook. Kew Journ.* viii. 149. An annual clothed with white wool, resembling *A. tomentosus,* but the stems more diffuse. Leaves linear or linear-cuneate, the uppermost short ones close under the inflorescence. Clusters of flower-heads oblong or cylindrical, attaining ½ to ¾ in., not so obtuse at the base as in *A. tomentosus,* closely sessile above the last leaves. Receptacle cylindrical. Partial involucres as in *A. tomentosus,* 1½ lines long, with 2 keeled and 2 or 3 flat bracts. Florets usually 2. Pappus a short irregularly-fringed or ciliate cup, not divided into distinct scales, and without any long bristles.

N. S. Wales. Darling and Lachlan rivers, *Burkitt, Victorian Expedition;* Murray river, *Dallachy.*

S. Australia. N.W. interior, *M'Douall Stuart's Expedition.*

4. **A. humifusus,** *Benth.* A dwarf diffuse annual, the slender stems often shorter than the inflorescence, rarely 3 or 4 in. long. Leaves very narrow-linear. Clusters of flower-heads ovoid or globose, of a brown-reddish or nearly white colour, ½ to ¾ in. long, surrounded by floral leaves resembling the stem ones, the inner ones bordered at the base by scarious margins. Receptacle cylindrical, slender, the subtending bracts oblong, ciliate. Partial involucres of 4 to 6 bracts, broader than the subtending one, the tips expanded into short broad plicate white appendages. Florets 2 or 3, 5-merous. Pappus of 5 or 6 lanceolate shortly aristate jagged scales shorter than the floret.—*Siloxerus humifusus,* Labill. Pl. Nov. Holl. ii. 58. t. 209; *Styloncerus humifusus,* Spreng. Syst. iii. 451; DC. Prod. vi. 149; Steetz in Pl. Preiss. i. 435; *S. cylindraceus,* Steetz, l. c. i. 436.

W. Australia. King George's Sound, *R. Brown, Preiss, n.* 41, and others, thence to Swan River, *Drummond,* 1*st Coll.,* 5*th Coll. n.* 361; *Preiss, n.* 40, and others; Gordon river, *Preiss, n.* 1361, *Oldfield,* and Murchison river, *Oldfield,* and eastward to Esperance Bay, *Maxwell.*

Var. *minor.* Clusters of flower-heads and the flowers themselves much smaller, the pappus nearly as long as the floret.—Kalgan river, *Oldfield.*

Var. *grandiflorus.* More robust, 2 to 4 in. high. Clusters of flower-heads and flowers larger. Involucral bracts with whiter tips.—*Styloncerus suberectus,* Steetz in Pl. Preiss. i. 436; *Preiss, n.* 42.

5. **A. myosuroides,** *Benth.* Very closely allied to and perhaps a variety of *A. tenellus,* with the same stature and foliage, but the spike or

cluster of flower-heads is long and slender, sometimes exceeding 1 in., more tapering at the base, the involucral bracts not so broad and less ciliate, and the florets mostly 5-merous and solitary with a very slight thickening at the base.—*Chrysocoryne myosuroides*, A. Gray in Hook. Kew Journ. iii. 152; *C. uniflora*, Turcz. in Bull. Mosc. 1851, i. 188.

W. Australia, *Drummond, 3rd Coll. n.* 116.

6. **A. tenellus,** *Benth.* A slender annual of 1 to 2 in., tomentose or at length glabrous. Leaves small, oblong-linear, thick, obtuse. Spikes or clusters of flower-heads cylindrical, obtuse and scarcely tapering at the base, the longest about ½ in. long but mostly shorter, brown and shining. Receptacle cylindrical, slender. Subtending bracts very broad and transparent with a rather more opaque oblong part at the base, very few of the lower ones empty and no floral leaves round the spike. Partial involucres of 2 complicated rather broad transparent bracts ciliate on the margin. Florets 2, 3-merous, not thickened at the base. Pappus none.—*Crossolepis pusilla*, Hook. Ic. Pl. t. 413, not of Benth.; *Chrysocoryne Drummondii*, A. Gray in Hook. Kew Journ. iii. 152; *C. tenella*, F. Muell. in Trans. Vict. Inst. 1855, 130, and in Hook. Kew Journ. viii. 149.

S. Australia. Between the Fountain and Long Lake, Spencer's Gulf, *Wilhelmi*.

W. Australia. King George's Sound, *R. Brown*; Swan River, *Drummond, 1st Coll.*

A. Gray's specific name has the priority over F. Mueller's, but has been applied to another species by Turczaninow.

7. **A. pusillus,** *Benth.* A slender erect corymbosely-branched annual of 2 to 6 in., slightly woolly when young but soon becoming glabrous. Leaves thick, linear or linear-cuneate, or the upper ones small and ovate. Clusters of flower-heads oblong-clavate, 3 to 4 lines long, from a pale yellow to a rich brown, shining. Receptacle cylindrical, slender. Subtending bracts very broad and transparent, with an opaque truncate base, very few of the lower ones smaller and empty, and no floral leaves round them. Partial involucres of 2 keeled bracts and 2 or 4 flat ones. Florets 2 or 3 in the upper heads or sometimes only 1 in the lower ones, 5-merous, scarcely thickened at the base. Pappus annular and fringed but usually very minute, adhering to the corolla and falling off with it.—*Crossolepis pusilla*, Benth. in Hueg. Enum. 61; *Chrysocoryne pusilla*, Endl. in Bot. Zeit. 1843, 458; Steetz in Pl. Preiss. i. 441; *C. Huegelii*, A. Gray in Hook. Kew Journ. iii. 151; *C. angianthoides*, F. Muell. in Linnæa, xxv. 404, 488.

N. S. Wales. Murray and Darling desert, *Victorian Expedition*.

S. Australia. Murray river and near Cudnaka, *F. Mueller*.

W. Australia. Swan River, *Huegel, Drummond, n.* 355, *Preiss, n.* 45; Murchison river, *Oldfield*.

Var. *polyanthus*. Spikes or clusters of flower-heads, when full grown ½ in. long, the upper partial involucres often containing 3 to 6 florets.—Murray and Darling desert.

8. **A. Milnei,** *Benth.* Annual or possibly perennial, covered with white wool, corymbosely branched, about 6 in. high. Leaves linear. Clusters of flower-heads ovoid, under ½ in. long, surrounded by a few herbaceous or tomentose floral leaves or bracts, not exceeding the partial heads. Receptacle cylindrical. Involucral bracts transparent, with a short opaque midrib, 2

keeled and 2 flat besides the subtending one. Florets 2, 5-toothed, much thickened, bulb-like and truncate at the base. Pappus none.

W. Australia. Sharks' Bay and Dirk Hartog's Island, abundant, *Milne.*

9. **A. Cunninghamii,** *Benth.* A low much-branched bushy shrub or undershrub usually very white, with a close wool and very different in habit from the rest of the genus. Leaves spreading or recurved, from oblong-cuneate and under ¼ in. long to linear and above ½ in., all obtuse. Clusters of flower-heads ovoid-globose, rather small, in terminal corymbs, the upper leaves passing into oblong imbricate floral leaves, forming a general involucre but not exceeding the flower-heads. Receptacle conical. Partial involucres of 2 keeled and 2 flat bracts. Florets usually 2, 5-merous, thickened and bulb-like at the base. Pappus a minute ring falling off with the corolla.—*Skirrophorus Cunninghamii,* DC. Prod. vi. 150, and in Deless. Ic. Sel. iv. t. 51; Steetz in Pl. Preiss. i. 438; *Eriocladium pyramidatum,* Lindl. Swan Riv. App. 24.

W. Australia. Swan River, very abundant, *Fraser, Drummond, 3rd Coll. n.* 125, *Preiss, n.* 30; Murchison river, *Oldfield, Drummond, 6th Coll. n.* 150 or 159; Sharks' Bay and Dirk Hartog's Island, very common, *A. Cunningham, Milne.*

10. **A. phyllocephalus,** *Benth.* Erect, not much branched, stout and rigid, 1 to 2 ft. high, with a loose deciduous wool. Leaves few, oblong, spathulate, the lower ones petiolate and often above 1 in. long, the upper smaller and sessile. Clusters of flower-heads at first hemispherical at length globose, attaining ¾ in. diameter, surrounded by broadly ovate floral leaves, closely imbricate at the base, spreading but herbaceous at the tips. Receptacle convex. Partial involucres exceedingly deciduous, flattened, of about 3 inner complicated bracts often woolly towards the top and 1 or perhaps 2 or 3 outer narrower flat ones. Florets solitary. Pappus cup-shaped, deeply jagged and often oblique, usually coming off with the corolla.—*Cephalosorus phyllocephalus,* A. Gray in Hook. Kew Journ. iii. 152; *Piptostemma carpesioides,* Turcz. in Bull. Mosc. 1851, i. 192; *Cephalosorus brevipapposus,* F. Muell. Fragm. iii. 159.

W. Australia, *Drummond, 4th Coll. n.* 200; Murchison river, *Oldfield.*

11. **A. micropoides,** *Benth.* An erect or decumbent white-woolly annual of a few inches or when very luxuriant above ½ foot. Leaves linear. Clusters of flower-heads nearly globular, closely surrounded by linear leaves often lanceolate at the base and much longer than the cluster, with a few scarious empty bracts within them. Receptacle conical. Partial involucres of 2 concave and about 4 flat narrow-spathulate transparent bracts. Florets 2, 5-merous, slightly thickened at the base. Pappus of about 5 ovate jagged scales terminating in a simple awn not quite so long as the floret.—*Phyllocalymma micropoides,* Benth. in Hueg. Enum. 62; Steetz in Pl. Preiss. i. 436; *P. filaginoides,* Steetz, l. c. 437.

W. Australia. Swan River, *Huegel, Preiss, n.* 36 *and* 37.

The habit is that of *A. strictus,* it is also very near *A. Drummondii,* but is readily distinguished by the pappus. In Preiss's specimens, n. 36, the pappus scales without the awn are nearly half as long as the corolla and scarcely jagged, in n. 37 they are much shorter and more fringed and the awn is shorter. Huegel's specimens are intermediate.

12. **A. microcephalus,** *Benth.* A small diffuse much-branched annual, nearly glabrous. Leaves linear or linear-cuneate, scarcely exceeding ¼ in. Clusters of flower-heads numerous, depressed-globular, attaining nearly 3 lines diameter, surrounded by 3 or 4 floral leaves scarcely so long as the clusters. Receptacle very convex. Partial involucres of 2 keeled bracts and rarely a third flat one besides the subtending one. Florets solitary, 5-merous. Pappus of 3 ovate scales quite distinct or more or less united in a cup, each one slightly jagged and with a fine awn shorter than the floret.—*Cephalosorus microcephalus*, F. Muell. Fragm. iii. 158.

W. Australia. Salt swamp at the estuary of the Murchison river, *Oldfield.*

13. **A. platycephalus,** *Benth.* A small slender annual, none of our specimens 3 in. long. Leaves linear or linear-cuneate. Clusters of flower-heads depressed, hemispherical, 4 to 5 lines diameter when fully out, surrounded by ovate-acuminate floral leaves longer than the cluster, and a few inner scarious broadly-obovate bracts. Receptacle slightly convex. Partial involucres of 2 keeled and 2 flat bracts besides the subtending one. Florets 2, 5-merous, not thickened at the base. Pappus a short jagged ring readily falling off with the corolla.

W. Australia. Tone river, *Oldfield.*

14. **A. Drummondii,** *Benth.* A little slender annual of 2 or 3 in. Leaves linear. Clusters of flower-heads hemispherical, about 3 lines diameter, surrounded by ovate-lanceolate floral leaves longer than the cluster, and a few inner scarious bracts as long as the involucres. Receptacle convex. Partial involucres of 2 keeled and 2 flat bracts rather broad. Florets usually 2, 5-merous, not thickened at the base. Pappus none.—*Skirrophorus Drummondii*, Turcz. in Bull. Mosc. 1851, i. 188.

W. Australia, *Drummond, 3rd Coll. n.* 123. No. 178 may be a variety of the same with longer and narrower floral leaves, but the specimens are too young for accurate determination.

15. **A. Preissianus,** *Benth.* An erect annual of 2 to 4 in., sometimes slender and nearly simple, sometimes much branched from the base, more or less woolly-white. Leaves linear, mostly alternate. Clusters of flower-heads depressed-globular, or hemispherical when fully out, 4 lines diameter or more, surrounded by ovate-lanceolate or linear-lanceolate closely appressed floral leaves longer than the cluster. Receptacle broad, convex. Partial involucres of 2 keeled or concave and 2 flat bracts, a little more than 1 line long. Florets usually 2, 4-merous, thickened and bulb-like at the base. Pappus a minute denticulate ring.—*Skirrophorus Preissianus*, Steetz in Pl. Preiss. i. 439.

Victoria. Plain near Stratham, *Whan;* near Melbourne, *Adamson.*

Tasmania. Islands of Kent's group, Bass's Straits, *R. Brown.*

S. Australia. Holdfast Bay, Spencer's Gulf, *F. Mueller.*

W. Australia, *Drummond, n.* 122; Woodman's Point, *Preiss, n.* 38; Moir's Inlet, *Maxwell.*

There may be some doubt whether the characters separating this and *A. eriocephalus* are constant. In Drummond's specimens, all very much a like, I find sometimes the slender 4-merous corollas bulb-like at the base of this species, sometimes 5-merous corollas not thickened at the base and apparently sterile. No. 206 of Drummond may be the same in a very young state. The species requires however further investigation.

16. **A. eriocephalus,** *Benth.* A little slender annual of 1 to 2 in. Leaves narrow-linear, often opposite. Clusters of flower-heads ovoid or at length globose, but rarely exceeding 2 lines diameter and surrounded by floral leaves not much longer than the cluster. Receptacle smaller and more convex than in *A. Preissianus.* Partial involucres the same as in that species. Florets usually 2, slender, 4-merous, but not at all thickened at the base. Pappus a minute ring or quite inconspicuous.—*Skirrophorus eriocephalus,* Hook. f. Fl. Tasm. i. 198. t. 53 A.

Victoria. Victoria range, Yarra river, sandy and salt plain near Brighton, Hobson's Bay, *F. Mueller.*

Tasmania. Seacoast near George Town, *Gunn.*

W. Australia. King George's Sound, *R. Brown.*

In most specimens the opposite leaves, narrow clusters of flower-heads and florets not thickened at the base appear to separate this from *A. Preissianus,* but sometimes one or all these characters are inconstant, and this may be a variety only of that species.

17. **A. pygmæus,** *Benth.* A little diffuse annual, forming tufts of about 1 in. diameter, white, with a close tomentum. Leaves linear. Clusters of flower-heads depressed-globular, 3 to 4 lines diameter when full-grown, surrounded by a few ovate or lanceolate floral leaves, scarcely exceeding the cluster, and a few inner broad scarious bracts. Receptacle convex. Partial involucres of 2 folded and 2 flat bracts, besides the subtending one. Florets 2, 5-merous, not thickened at the base at the time of flowering. Pappus none.—*Skirrophorus pygmæus,* A. Gray in Hook. Kew Journ. iii. 148; *S. mucronulatus,* Turcz. in Bull. Mosc. 1851, ii. 72.

W. Australia, *Drummond, 5th Coll. n.* 59.

18. **A. globifer,** *Benth.* A dwarf plant, apparently annual, the stems not ½ in. long or scarcely any, forming little clustered tufts ½ to 1 in. diameter. Leaves linear, few and short. Clusters of flower-heads globular, 3 to 4 lines diameter, more or less woolly, surrounded by a few bracts either all broad and scarious or the outer ones narrower and more woolly with short leafy tips, the subtending bracts in the clusters with an opaque midrib, scarious margins, and often produced into a small petal-like pale pink lamina. Receptacle oblong. Involucres of 2 keeled and 2 inner flat bracts. Florets solitary slender, 5-merous, becoming somewhat enlarged and hardened at the base so as to cover the top of the achene. Pappus none.—*Hyalochlamys globifera,* A. Gray in Hook. Kew Journ. iii. 101.

W. Australia, *Drummond, 4th Coll. n.* 204. The subtending bracts within the clusters of flower-heads are more persistent than in most species of the genus, but the involucres are much more those of *Angianthus* than of *Gnephosis.*

19. **A. demissus,** *Benth.* A little diffuse annual, forming loose tufts of 1 to 2 in. diameter. Leaves linear. Clusters of flower-heads obovoid-globose, densely enveloped in long wool, surrounded by a few reflexed floral leaves, longer than the involucres and scarcely any scarious bracts. Receptacle small. Partial involucres of 2 conduplicate bracts besides the subtending flat one, narrow and woolly in the upper half. Florets solitary, 5-merous, dilated and bulb-like at the base; bulb fringed with a ring of long wool, more adherent to the achene than to the rest of the corolla, but readily detached

from both and carrying off the top of the achene. Pappus none.—*Skirrophorus demissus*, A. Gray in Hook. Kew Journ. iii. 149; *Epitriche cuspidata*, Turcz. in Bull. Mosc. 1851, ii. 75.

W. Australia, *Drummond, 5th Coll. Suppl. n.* 58.

20. **A. strictus,** *Benth.* Probably annual, with ascending or erect rigid stems of 2 to 6 in. Leaves very narrow. Clusters of flower-heads obovoid-turbinate, surrounded by rather numerous narrow and rigid recurved floral leaves, longer than the involucres, often acute and almost pungent, with a few inner oblong scarious bracts passing into the subtending ones within the cluster. Receptacle convex. Partial involucre of 2 narrow-oblong concave bracts. Florets solitary, 5-merous, not thickened at the time of flowering, more decidedly bulb-like as the fruit ripens. Pappus none.—*Pogonolepis stricta*, Steetz in Pl. Preiss. i. 440; *Skirrophorus strictus*, A. Gray in Hook. Kew Journ. iii. 149; *S. Muellerianus*, Sond. in Linnæa, xxv. 486.

N. S. Wales. Between the Murray and Darling rivers, *Victorian Expedition.*
Victoria. Plains on Avoca river, *F. Mueller;* Wimmera, *Dallachy.*
S. Australia. Crystal Brook and St. Vincent's Gulf, *F. Mueller.*
W. Australia, *Drummond, 5th Coll. n.* 357; Vasse river, *Preiss, n.* 39, *Oldfield;* Murchison river and Champion Bay, *Oldfield.*

21. **A. plumiger,** *Benth.* A slender annual of 1 to 3 in. Leaves narrow-linear. Clusters of flower-heads obovoid, about 1½ lines diameter, surrounded by herbaceous floral leaves, shorter than or scarcely exceeding the involucres, with rather numerous inner scarious fringed or jagged bracts, as long as the involucres. Partial involucres of 2 conduplicate bracts, besides the flat subtending one which is much shorter than the involucres, but usually with the midrib produced into a jagged almost plumose appendage almost as long as the florets. Florets solitary, 5-merous, not thickened at the base. Pappus none.

W. Australia. Swan and Murchison rivers, *Oldfield.*

22. **A. amplexicaulis,** *Benth.* A slender, erect. glabrous annual of 1 to 3 in., simple or slightly branched. Leaves ovate, concave, stem-clasping, the upper ones often shortly sheathing at the base, obtuse, under ¼ in. long, the 2 uppermost floral ones larger, broader, closely embracing each other, or shortly connate, forming an involucre round the cluster of flower-heads. Receptacle small, slightly branched. Subtending and involucral bracts generally reduced to tufts of a very long white wool, filling the general involucre, but usually shortly united into exceedingly thin, transparent, membranes at the base, 2 such bracts to each floret, rather more prominent, but excessively thin and often united into a single membrane enclosing the achene, which is densely silky-hairy. Pappus of very short scales or bristles dilated and united at the base.—*Dithyrostegia amplexicaulis*, A. Gray in Hook. Kew Journ. iii. 100; *Gamozygis flexuosa*, Turcz. in Bull. Mosc. 1851, ii. 76. t. 1.

W. Australia, *Drummond, Suppl. to 5th Coll. n.* 57.

51. GNEPHOSIS, Cass.

(Cephalosorus (*partly*); Nematopus *and* Crossolepis, *A. Gray*; Leptotriche, *Turcz.*; Trichanthodium *and* Cyathopappus, *F. Muell.*)

Flower-heads numerous and usually more or less stipitate, on a convex or rarely cylindrical receptacle, in an ovoid or globular dense cluster or compound head, without any general involucre or surrounded by a few leafy or scarious bracts rarely exceeding the florets. Partial heads 1- or few-flowered, very rarely many-flowered. Involucre of several bracts, the outer ones like the subtending ones, narrow and often more persistent, the inner ones broader. transparent and very deciduous. Receptacle without scales. Florets hermaphrodite, tubular, slender, 4- or 5-toothed, sometimes hardened at the base. Anthers more or less distinctly pointed or tailed at the base. Style-branches nearly terete, truncate. Achenes usually compressed. Pappus none or forming a jagged ring or cup, or rarely of several short distinct scales.—Annual or rarely perennial herbs, glabrous or more or less cottony. Leaves alternate, entire. Cluster of flower-heads terminal, sessile or pedunculate.

The genus is limited to Australia. It is closely allied to *Angianthus*, but the general involucre and receptacle are less developed than in that genus and the partial involucres much more so, consisting of much more numerous bracts and not flattened.

Partial heads 1- or 2-flowered.
  Minute almost stemless plant. Pappus of several short plumose scales . . . . 1. *G. Burkittii.*
  Stems ascending. Heads 1- or 2-flowered. Involucral bracts appendiculate. Pappus none. Achenes woolly (glabrous or or nearly so in the rest of the genus) . . . . 2. *G. eriocarpa.*
  Stems erect, virgate. Heads 1-flowered. Pappus cup-shaped.
    Clusters of flower-heads ovoid. Involucral bracts as long as the florets. Pappus-cup open . . . . 3. *G. macrocephala.*
    Clusters of flower-heads globular. Florets protruding beyond the involucres. Pappus-cup tubular . . . . 4. *G. skirrophora.*
  Stems erect, corymbose. Heads 2-flowered. Clusters turbinate-globular.
    Involucral bracts appendiculate. Pappus cup-shaped.
      Stem hard. Leaves under ½ in. long . . . . 5. *G. cyathopappa.*
      Stems very slender. Leaves almost subulate, above ½ in. long. 6. *G. leptoclada.*
    Involucral bracts not appendiculate. Pappus none . . . . 7. *G. arachnoidea.*
Partial heads 3- to 6-flowered. Erect plants with filiform branches. Clusters of flower-heads very shining. Pappus annular, jagged or none.
  Branches divaricate, dichotomous. Clusters of flower-heads globular . . . . 8. *G. tenuissima.*
  Branches flexuose. Clusters of flower-heads turbinate at the base 9. *G. acicularis.*
Partial heads 3- to 20-flowered. Dwarf or shortly diffuse annuals with very woolly clusters of flower-heads. Pappus none.
  Leaves linear-subulate. Heads 3- to 5-flowered. Plant under 1 in. . . . . 10. *G. pygmæa.*
  Leaves linear-oblong or oblong. Plant of 1 to 3 in.
    Floral leaves ovate or broadly oblong, woolly only inside. Heads 10- to 15-flowered . . . . 11. *G. brevifolia.*
    Floral leaves oblong, concealed in the abundant wool. Heads 5- to 10-flowered . . . . 12. *G. eriocephala.*

Hirnellia cotuloides, Cass., and *Crossolepis linifolia*, Less., from the short character given, Less. Syn. Comp. 269, 270, are probably species of *Gnephosis*, but insufficiently described for identification.

1. **G. (?) Burkittii,** *Benth.* A dwarf plant, our specimens scarcely above ½ in. high. Leaves narrow-linear, closely surrounding the cluster of flower-heads and exceeding it, the cluster itself globular, 2 to 3 lines diameter, densely enveloped in white wool. Receptacle small. Partial heads 1-flowered, with numerous linear bracts, each with a green centre and scarious margins bordered by long wool, 2 or 3 inner ones almost entirely scarious. Corolla slender, minutely 4- or 5-toothed. Pappus of several short but fine and plumose scales.

S. Australia. Lake Gillies, *Burkitt.* The habit of this minute plant is that of the smallest species of *Angianthus,* but the numerous involucral bracts are rather those of *Gnephosis.*

2. **G. (?) eriocarpa,** *Benth.* Described by F. Mueller from a single fragment, a slender slightly branched stem or branch of about 4 in., with much of white wool about the inflorescence. Leaves linear-cuneate or linear, about ½ in. long. Clusters of flower-heads globose, very young. Bracts narrow, scarious, woolly, with glabrous pink appendages, very numerous but most of them apparently belong to the partial involucres, a very few forming a general involucre to the cluster. Partial heads 1- or 2-flowered. Corolla slender, 5-toothed. Achenes densely covered with long wool, without any pappus.—*Skirrophorus eriocarpus,* F. Muell. Fragm. iii. 156.

N. S. Wales. Between Stoke's Range and Cooper's Creek, *Wheeler.* The inflorescence is much too young to be certain of the floral characters.

3. **G. macrocephala,** *Turcz. in Bull. Mosc.* 1851, i. 190. Stems erect, slender but rigid, 1 to 1½ ft. high. Leaves linear, almost filiform, the upper ones short. Clusters of flower-heads ovoid when fully out, about ½ in. long, with a very few short empty woolly outer bracts, the subtending bracts within the cluster narrow and shorter than the involucres. Receptacle cylindrical. Partial involucres nearly sessile, of about 4 outer persistent bracts, and about 3 or 4 inner deciduous ones much broader, rather longer and folded, all scarious, the tips of the same colour, erect and persistent. Florets solitary. Pappus broadly cup-shaped, crenulate.—*Cephalosorus gymnocephalus,* A. Gray in Hook. Kew Journ. iii. 153.

W. Australia, *Drummond, 4th Coll. n.* 202; Murchison river, *Oldfield.* Oldfield's specimens are more branched, the outer persistent bracts are narrower and more numerous, and the pappus shorter than in Drummond's, but they probably all belong to one species.

4. **G. skirrophora,** *Benth.* An annual, more or less woolly-white, with numerous stems, from 2 or 3 in. to nearly 1 ft. high, hard at the base. Leaves narrow-linear, almost filiform. Clusters of flower-heads globular, 3 to 4 lines diameter, with prominent florets, surrounded by 2 or 3 rows of small scarious woolly bracts, forming a general involucre, but much shorter than the partial heads, and without the radiating tips of *Myriocephalus.* Receptacle small, convex, hirsute, with long fine hairs, the persistent bases of the partial heads prominent. Involucre slightly compressed, the bracts scarious, 2 or 3 outer ones (one subtending?) slightly woolly at the top with a small spreading ovate coloured lamina, 3 or 4 inner ones broader, glabrous, without spreading tips, all exceedingly deciduous. Florets solitary, 4- or 5-merous.

Pappus a cylindrical cup toothed or jagged and nearly half as long as the corolla, usually adhering to and falling off with the thickened base of the corolla.—*Trichanthodium skirrophorum*, Sond. and Muell. in Linnæa, xxv. 490.

N. S. Wales. Dry Lake, near Menindie, *Victorian Expedition.*
Victoria. Murray river, *F. Mueller;* Wimmera, *Dallachy.*
S. Australia. Cudnaka, *F. Mueller*; towards Spencer's Gulf, *Waterhouse.*
W. Australia. Dirk Hartog's Island, *Milne.*

This has fewer and more deciduous outer bracts than most species, but it is much nearer allied to *Gnephosis* than to *Angianthus* in habit as well as in most characters.

5. **G. cyathopappa,** *Benth.* Rigid, erect, and corymbosely branched, some specimens appearing annual and simple at the base, others with a hard and woody base, with numerous stems, but none above 6 in. high, all nearly glabrous or woolly tomentose, especially about the inflorescence. Leaves linear, rather short. Clusters of flower-heads turbinate, very numerous in a dense corymb, surrounded by a few small outer empty bracts like the subtending ones, but without the yellow appendage. Receptacle small, convex or obovoid. Partial heads 2-flowered, very shortly stipitate and usually 2 within the same subtending bract. Bracts scarcely woolly, narrow, with a short broad yellow deciduous appendage, 4 or 5 outer ones of each involucre and the subtending one narrower and more persistent than the 4 or 5 inner very deciduous ones. Pappus cup-shaped, as long as the slender part of the floret, slightly toothed or jagged, falling off with the corolla.—*Cyathopappus gnephosioides* or *Cephalosorus gnephosioides*, F. Muell. Fragm. ii. 158.

N. S. Wales. Near Menindie, Darling river, *Victorian Expedition.*

6. **G. leptoclada,** *Benth.* A slender, erect, corymbosely branched annual, nearly glabrous. Leaves linear, almost filiform. Clusters of flower-heads depressed-globular, 3 to 4 lines diameter, surrounded by a few narrow-oblong leafy bracts or floral leaves shorter than the florets. Receptacle rather broad, convex. Partial heads 2-flowered, very shortly stipitate. Involucral bracts all scarious and deciduous, with a small yellow deciduous appendage, 4 or 5 outer ones (1 subtending?) narrow and about 4 inner rather broader ones. Pappus cylindrical, cup-shaped, slightly jagged.—*Cephalosorus leptocladus*, F. Muell. Fragm. iii. 158.

W. Australia. Stony places, Barrel Well, Murchison river, *Oldfield*, a single specimen in Herb. F. Mueller, the inflorescence very rotten. When examined in a better state it may prove to be a variety of *G. cyathopappa.*

7. **G. arachnoidea,** *Turcz. in Bull. Mosc.* 1851, i. 189. A slender, erect, corymbosely-branched annual, usually 6 in. to 1 ft. high, glabrous, except the inflorescence or more or less woolly-tomentose. Leaves very narrow, linear or subulate. Clusters of flower-heads nearly globose, about 2 lines diameter, surrounded by a few small empty scarious bracts, without any appendage, the subtending ones within the cluster cuneate-oblong, all except sometimes the lower ones with a small broad deciduous lamina. Partial heads 2-flowered, very shortly stipitate and sometimes 2 or 3 within 1 subtending bract. Involucral bracts oblong-cuneate, with a short broad deciduous yellow lamina, about 6 outer ones apparently more persistent than the inner ones, which have a rather larger lamina. Pappus none apparent at the time of flowering.—*Nematopus effusus*, A. Gray in Hook. Kew Journ. iii. 150.

W. Australia, *Drummond, 3rd Coll. n.* 120. The flower-heads have not been seen in an advanced state.

Var. *foliata.* Rather more woolly and the bracts apparently more woolly and more numerous, but the specimens all in very young bud.—*Nematopus foliatus,* Sond. in Linnæa, xxv. 486.

S. Australia. Near Cudnaka, *F. Mueller.*

8. **G. tenuissima,** *Cass.; DC. Prod.* vi. 151. A very slender corymbose annual, at length glabrous, scarcely exceeding 6 in. in height. Leaves few, linear. Clusters of flower-heads obovoid-globular, very shining, the subtending bracts within the cluster and perhaps a few small outer empty ones with a broadly cuneate opaque base, and a very broad scarious lamina. Partial heads stipitate, 3- or 4-flowered. Involucral bracts rather numerous, the outer ones smaller, 4 to 6 or more inner ones oblong-cuneate, scarious, with a broad deciduous lamina. Pappus annular, jagged.

W. Australia. Sharks' Bay, *Gaudichaud;* Dirk Hartog's Island, *A. Cunningham.* All our specimens, from both sources, are far advanced, the florets and achenes almost all fallen away. The species will require further examination when seen in a better state.

9. **G. acicularis,** *Benth.* An erect glabrous annual, very slender although rigid, above 1 ft. high, the upper filiform branches bent in zigzag at every node, the subulate leaves usually in a line with the previous internode. Clusters of flower-heads in young buds nearly globular, shining, tapering at the base, surrounded by small empty scarious bracts. Partial heads about 6-flowered, stipitate. Involucral bracts numerous, oblong-cuneate, fringed except at the top with long woolly hairs, several of the inner ones with a readily detached lamina. Florets in the specimens too young to ascertain their structure and pappus.

W. Australia, *Drummond, 6th Coll. n.* 201. Of this I have seen a considerable number of specimens, showing a species allied to but evidently quite distinct from *G. tenuissima,* but unfortunately too young for a full description.

10. **G. pygmæa,** *Benth.* A dwarf annual, all our specimens under 1 in. high, at length glabrous except the inflorescence. Leaves narrow-linear, scarcely exceeding the flower-heads. Clusters of flower-heads globular, woolly, about 3 lines diameter, surrounded by a few almost leafy bracts, with scarious woolly-ciliate margins, or the inner 2 or 3 as well as the subtending bracts within the clusters wholly scarious and woolly-ciliate. Partial flower-heads 3- to 5-flowered, more or less stipitate. Involucral bracts about 5, outer ones narrow with rigid midribs and scarious woolly-ciliate margins, and 1, 2 or 3 inner ones rather smaller and completely transparent, all without appendages. Florets slender, 3- or 4-toothed. Pappus none, but the achenes have a few hairs, especially at the summit, where they are longer and almost paleaceous, assuming the appearance of a pappus.—*Crossolepis pygmæa,* A. Gray in Hook. Kew Journ. iii. 176; *Leptotriche perpusilla,* Turcz. in Bull. Mosc. 1851, ii. 73.

W. Australia, *Drummond, 5th Coll. Suppl. n.* 60.

11. **G. brevifolia,** *Benth.* A slender branching plant of 1 to 3 in., glabrous except the inflorescence. Leaves small, oblong. Clusters of flower-heads depressed-globular or hemispherical, 4 to 5 lines diameter, surrounded by a few oblong leafy bracts not exceeding the cluster, and very woolly inside.

Partial flower-heads 10- to 15-flowered. Involucral bracts several outer ones herbaceous and bordered by very long intricate woolly hairs, and about 5 inner ones membranous, transparent and ciliate with long wool. Florets 5-merous. Pappus none.—*Crossolepis brevifolia*, A. Gray in Hook. Kew Journ. iii. 175; *Myriocephalus cotuloides*, Turcz. in Bull. Mosc. 1851, ii. 73.

W. Australia, *Drummond, 5th Coll. Suppl. n.* 61.

12. **G. eriocephala,** *Benth.* A slender decumbent branching annual of 2 to 3 in., very loosely and copiously woolly or the lower part glabrous. Leaves small, linear-oblong. Clusters of flower-heads 3 to 5 lines diameter, surrounded by a few oblong herbaceous bracts densely enveloped in wool. Partial heads 6- to 10-flowered. Involucres of several outer herbaceous bracts and about 4 inner transparent ones, all very densely woolly-ciliate. Pappus none.—*Crossolepis eriocephala*, A. Gray in Hook. Kew Journ. iii. 176; *Myriocephalus villosissimus*, Turcz. in Bull. Mosc. 151. ii. 74.

W. Australia, *Drummond, 5th Coll. Suppl. n.* 62.

52. CALOCEPHALUS, R. Br.

(Leucophyta, *R. Br.;* Pachysurus, *Steetz;* Blennospora *and* Achrysum, *A. Gray.*)

Flower-heads numerous and usually more or less stipitate on a small and branching or globose or conical receptacle in an ovoid or globular dense cluster or compound head without any general involucre or surrounded by a few leafy or scarious bracts rarely exceeding the florets. Partial heads 2- or more-flowered. Involucre of several bracts, the outer ones like the subtending ones narrow and often more persistent, the inner ones broader transparent and very deciduous. Receptacle without scales. Florets hermaphrodite, tubular, 5-toothed, not at all or scarcely hardened at the base. Anthers more or less distinctly tailed. Style-branches nearly terete, truncate. Achenes usually compressed. Pappus of several narrow linear scales or bristles plumose-ciliate from the base or at the end only, all free or united in a ring at the base.—Annual or perennial herbs or rarely undershrubs or small shrubs, more or less cottony or woolly, white or rarely glabrous. Leaves alternate or in two species opposite, entire. Clusters of flower-heads terminal, sessile or pedunculate.

The genus is limited to Australia. It is very near to *Gnephosis*, differing chiefly in the pappus. The general receptacle is also sometimes broken up or slightly branched, the partial heads are thus more distinct and having often more florets connect the genus through *Cephalipterum* and *Gnaphalodes* with *Helipterum*.

Partial heads 2- or 3-flowered.
  Leaves alternate.
    Pappus plumose from the base. Involucral bracts without appendages.
      Annual of 2 to 3 inches . . . . 1. *C. Drummondii.*
      Branching shrub of about a foot . . . . 2. *C. Brownii.*
    Pappus plumose at the end. Involucral bracts with small petal-like appendages . . . . 3. *C. Sonderi.*
  Leaves mostly opposite. Pappus plumose chiefly at the end.
    Leaves mostly obtuse. Clusters of flower-heads white . . . 4. *C. lacteus.*
    Leaves mostly acute. Clusters of flower-heads yellow . . . 5. *C. citreus.*

Partial heads 6- or more-flowered.
Clusters of flower-heads dense and globular. Appendages of the involucral bracts radiating round the partial heads.
Leaves linear-cuneate. Pappus plumose at the end . . . . 6. *C. angianthoides.*
Leaves narrow-linear. Pappus plumose from the base . . . 7. *C. Francisii.*
Clusters of flower-heads loose, the heads distinctly stipitate. Involucral bracts without any or with very small appendages. Pappus of woolly-plumose hair-like bristles.
Leaves narrow-linear. Clusters of flower-heads terminal.
Perennial or undershrub. Pappus-bristles distinct . . . . 8. *C. platycephalus.*
Annual. Pappus-bristles united at the base in a broad paleaceous ring . . . . 9. *C. multiflorus.*
Leaves ovate on long petioles. Clusters of flower-heads very densely woolly, clustered along the branches of the panicle . 10. *C. æruoides.*

(See also *Gnephosis Burkittii,* which has a pappus of plumose-ciliate scales, but they are very short, and *Angianthus tomentosus* in which the pappus-scales, only 3 in number, end in plumose bristles.)

1. **C. Drummondii,** *Benth.* A loosely woolly annual with several erect stems of 2 to 3 in. Leaves alternate, narrow-linear. Clusters of flower-heads ovoid-globular, 3 to 4 lines diameter or even more, without any or with a single floral leaf. Receptacle small or branching, subtending bracts within the cluster and a few outer empty ones narrow-oblong, scarious, woolly. Partial heads stipitate, 2-flowered, the outer persistent involucral bracts oblong-cuneate, very woolly near the top of the stipes; inner bracts about 4, longer than the others, ovate, concave, deciduous. Pappus very transparent, oblique, of about 8 to 10 unequal slender woolly-plumose bristles, united in a short cup at the base.—*Blennospora Drummondii,* A. Gray in Hook. Kew Journ. iii. 173.

W. Australia, *Drummond, 5th Coll. n.* 359, *and Suppl. n.* 68.

2. **C. Brownii,** *F. Muell. Rep. Babb. Exped.* 13. A much-branched rigid shrub rarely exceeding 1 ft. in height, white with a close woolly tomentum. Leaves alternate, linear, obtuse, mostly under 1 line and rarely 2 lines long. Clusters of flower-heads globular, 4 to 6 lines diameter, surrounded by a few floral leaves much shorter than the involucres. Receptacle nearly globular. Partial heads 3- or sometimes 2-flowered, raised on very short stipes or protuberances of the receptacle. Bracts all very deciduous, cuneate-oblong or linear-oblong, very woolly above the middle, the subtending ones rather broader, those of the involucre about 10, the inner scarcely broader than the outer, all without appendages. Pappus of about 8 to 10 linear plumose-ciliate scales as long as the corolla, slightly united at the base.—*Leucophyta Brownei,* Cass.; Less. Syn. Comp. 271; DC. Prod. vi. 152; Steetz in Pl. Press. i. 442; Hook. f. Fl. Tasm. i. 196.

Victoria. Seacoasts from Glenelg river to Wilson's Promontory, *F. Mueller, Robertson,* and others.

Tasmania. Port Dalrymple, *R. Brown;* seacoast of the north shores of the island, *J. D. Hooker.*

S. Australia. Common on the sandy seacoasts both of the mainland and Kangaroo Island, *F. Mueller* and others.

W. Australia, *Drummond, 4th Coll. n.* 124; Goose Island Bay, *R. Brown;* King George's Sound, *Fraser* and others; Swan River aud Rottenest Island, *Preiss, n.* 31, 32; Point Irwin, *Oldfield;* Sharks' Bay, *Denham.* The Western specimens are generally more vigorous than the others, with longer leaves and larger heads.

3. **C. Sonderi,** *F. Muell. Rep. Babb. Exped.* 13. An erect branching loosely woolly annual, with a hard stem, attaining sometimes above 1 ft. in height. Leaves alternate, linear, the upper ones short and decurrent. Clusters of flower-heads yellow, ovoid or globose, rarely 5 lines long. Receptacle cylindrical, branched. Partial heads stipitate, 2- or 3-flowered. Subtending and outer involucral bracts 2 or 3, in the lateral heads more numerous, in the terminal one narrow but scarious and woolly-ciliate with a small lamina or appendage, a few inner ones broader, less ciliate, with a more prominent yellow lamina. Pappus of several very unequal scales united in a ring or cup at the base, and more or less produced into awns or bristles of which the longer ones are plumose at the end.

Victoria. Avoca and Murray rivers, *F. Mueller.*
S. Australia. Murray scrub, *F. Mueller.*

4. **C. lacteus,** *Less. Syn. Comp.* 271. A perennial with rather slender but hard ascending or erect stems of 1 to 2 ft., hoary or white with a close tomentum. Leaves linear, obtuse, the lower ones or nearly all opposite. Clusters of flower-heads oblong-ovoid or nearly globular, white, $\frac{1}{4}$ to $\frac{1}{2}$ in. long, without any or only 1 or 2 very small leafy bracts under them. Receptacle cylindrical. Partial heads shortly stipitate, usually 3-flowered. Subtending bracts within the cluster short, scarious. Involucres of about 10 bracts, the outer ones narrow-cuneate, slightly or not at all ciliate, without appendages, the inner ones more deciduous, oblong, with a white broad lamina. Pappus of 6 to 10 very narrow linear scales united in a ring at the base and plumose at the end.—DC. Prod. vi. 151; A. Brongn. Voy. Coq. t. 60; Hook. f. Fl. Tasm. i. 196.

Victoria. Glenelg river, Port Phillip, Melbourne, Ballarat, Bacchus Marsh, etc., *F. Mueller* and others.
Tasmania. Risdon Cove, Derwent river, *R. Brown;* salt-marshes and seacoast, northern parts of the island, *J. D. Hooker.*
S. Australia. Bethanie, Onkaparinga river, Gawler river, *F. Mueller.*
W. Australia, *Drummond, 3rd Coll. n.* 118.

5. **C. citreus,** *Less. Syn. Comp.* 271. A perennial with a woody base, closely resembling *C. lacteus* and perhaps a variety, but the leaves are narrower, less obtuse or almost acute and the clusters of flower-heads yellow. Partial heads more sessile than in *C. lacteus,* the subtending bracts often very small, the outer involucral bracts more woolly at the base, the lamina of the inner ones bright yellow. Pappus-scales more plumose than in *C. lacteus* but variable, sometimes few only and slender, sometimes 8 to 10 and broader. —DC. Prod. vi. 151; Brongn. Voy. Coq. t. 60.

N. S. Wales. Lachlan river, *Fraser;* New England, *C. Stuart.*
Tasmania. Risdon Cove, Derwent river, *R. Brown.*
S. Australia. Guichen Bay, Gawler river, Mount Lofty, Mount Barker, *F. Mueller.*

6. **C. angianthoides,** *Benth.* A slender branching annual. Leaves linear. Clusters of flower-heads from globose and about 3 lines diameter to ovoid and fully 5 lines long, surrounded by a few scarious woolly bracts shorter than the florets. Receptacle small. Partial heads on a stalk-like rhachis, 6- or more-flowered. Subtending bracts within the cluster rather narrow, con-

cave, stipitate. Involucral bracts numerous, the outer ones linear-spathulate, ciliate with long woolly hairs, the broad scarious end glabrous but brown and not spreading, about 6 inner ones oblong scarious, glabrous, with a scarious yellow or white radiating reniform lamina. Pappus annular, jagged, with several bristles as long as the corolla and plumose towards the end with long cilia.—*Pachysurus angianthoides*, Steetz in Pl. Preiss. i. 442.

W. Australia, *Drummond*, 5*th Coll. n.* 354; foot of Mount Eliza, *Preiss, n.* 44.

7. **C. Francisii,** *Benth.* An annual with numerous slender stems of 3 to 6 in., glabrous or nearly so. Leaves very narrow-linear, short. Clusters of flower-heads globose or ovoid, about ½ in. diameter, without any or very few outer empty bracts. Receptacle small. Partial heads shortly stipitate, 6- to 20-flowered. Involucral bracts numerous, scarious, woolly-ciliate, the outer and subtending ones narrow with a small lamina, 12 or more inner ones with a short broad almost reniform spreading lamina, forming a ray round the partial head. Pappus of several exceedingly fine hair-like plumose bristles.—*Pachysurus Francisii*, F. Muell. Fragm. iii. 155.

W. Australia. Murchison river and Champion Bay, *Oldfield*; *Drummond*, 6*th Coll. n.* 161.

9. **C. platycephalus,** *Benth.* A perennial or undershrub, more or less woolly-white, with simple or branched stems of ½ to 1 ft. Leaves linear. Clusters of flower-heads globose and about ½ in. diameter or at length larger and irregularly lobed, without outer empty bracts. Receptacle small. Partial heads shortly stipitate, many-flowered. Involucral bracts very numerous, the outer ones narrow and woolly, the inner broader transparent and glabrous, all with a reniform folded lamina, radiating but very small and scarcely conspicuous when the flowers are fully out. Pappus of hair-like woolly-plumose bristles.—*Pachysurus platycephalus*, F. Muell. Fragm. iii. 154.

N. S. Wales. Darling river to the Barrier range, *Victorian Expedition, Dallachy* and *Goodwin.*

S. Australia. Towards Spencer's Gulf (a fragment only), *F. Mueller*; in the interior, *M'Douall Stuart's Expedition.*

9. **C. multiflorus,** *Benth.* An erect branching annual, of 1 to 6 in., the white wool usually persistent. Leaves narrow-linear. Clusters of flower-heads often ½ in. diameter on the main stem, with the partial heads almost distinct, more compact and ¼ in. diameter on the side branches, all with a few outer leafy bracts or floral leaves, very woolly, and not exceeding the heads. Receptacle small. Partial heads stipitate, 10- to 15-flowered. Involucral bracts numerous, the outer ones persistent, with linear opaque centres and scarious woolly-ciliate margins; inner ones very deciduous, transparent, and slightly ciliate, with small yellow deciduous tips. Pappus of several very fine woolly-plumose bristles, united at the base in a broad paleaceous ring.—*Pachysurus multiflorus*, Turcz. in Bull. Mosc. 1851, i. 192; *Achrysum glomeratum*, A. Gray in Hook. Kew Journ. iv. 229.

W. Australia. *Drummond*, 3*rd Coll. n.* 117, 5*th Coll. n.* 389.

10. **C. æruoides,** *Benth.* An annual, with a hollow, ascending, paniculately-branched stem of above 1 ft., the wool floccose and deciduous.

Leaves few, on long petioles, ovate or rhomboidal, the larger ones above 1 in. long. Clusters of flower-heads ovoid or globular, sessile and irregularly clustered on the branches of the panicle, mostly about $\frac{1}{4}$ in. diameter and very densely woolly. Partial heads not very numerous in the cluster, 10- to 15-flowered. Involucral bracts scarious, shorter than the florets, the outer and subtending ones woolly, with a persistent, rather rigid midrib; the inner ones entirely scarious. Pappus of several long, hair-like, woolly-plumose bristles.—*Pachysurus æruoides*, F. Muell. Fragm. iii. 154.

W. Australia. Moist places, Port Gregory, *Oldfield*. The foliage and aspect of this species are very different from those of any other *Angianthea*.

53. CEPHALIPTERUM, A. Gray.

Flower-heads several together, sessile in a large terminal nearly globular cluster or compound head. Involucres ovoid-turbinate, the bracts numerous, scarious, the inner ones with long, spreading, petal-like laminæ. Receptacle without scales. Florets numerous, all tubular, hermaphrodite, the innermost sterile, with undivided styles. Anthers tailed. Style-branches in the perfect florets truncate. Achenes densely woolly. Pappus of 3 to 6 shortly ciliate bristles, terminating in a plumose tuft, accompanied by an exterior oblique or cup-shaped scale.—Annual. Leaves alternate, entire. Stem simple, with a terminal cluster of flower-heads.

The genus is limited to a single species, endemic in Australia. It is closely allied to *Calocephalus;* but the more distinct flower-heads, with more numerous florets and large petal-like laminæ of the inner involucral bracts, connect it with *Helipterum,* and the external scale of the pappus is peculiar.

1. **C. Drummondii,** *A. Gray in Hook. Kew Journ.* iv. 272. An erect annual of 1 ft. or more, sprinkled with short transparent hairs. Radical and lower leaves oblong-spathulate, narrowed into a long petiole; upper ones linear or lanceolate. Flower-heads rather numerous, in a single terminal nearly globular cluster of $\frac{3}{4}$ to above 1 in. diameter, without any or with only a very few small linear floral leaves. Involucres about 3 lines diameter, the outer bracts very thin and broad, the spreading white or pale yellow laminæ of the inner ones 3 to 4 lines long. Florets about 12 to 15. Anther-tails very short. Pappus-bristles 3 to 6, as long as the corolla, shortly ciliate the whole length, the terminal tuft very prominent, the small outer scale varying from nearly regularly cup-shaped to very oblique and one-sided, entire or fringed by long hairs.

S. Australia. In the interior from the head of the Great Bight, *Delisser*.
W. Australia, *Drummond;* Bowes river and Champion Bay, *Oldfield*.

54. GNAPHALODES, A. Gray.

Flower-heads several, sessile in a dense cluster or compound head, surrounded by a few leafy bracts, each head many-flowered. Involucre ovoid, the bracts scarious, imbricate, very woolly, the inner ones with small radiating tips. Receptacle without scales. Florets all slender, tubular, hermaphrodite, 5-toothed. Anthers with fine tails. Style-branches truncate. Achenes glabrous. Pappus of about 5 elastically-spreading, rather broad, plumose bristles.—Dwarf branching annuals, more or less white-tomentose. Leaves alternate, entire. Clusters of flower-heads terminal.

The genus is limited to Australia. The habit is almost that of *Gnaphalium*, but there are no female florets. Like *Cephalipterum*, it connects the *Angianthææ* with *Helichryseæ*.

Floral leaves broadly ovate. Pappus-bristles without any terminal tuft . . . . 1. *G. uliginosum.*
Floral leaves oblong. Pappus-bristles with a stipitate terminal tuft . 2. *G. condensatum.*
Floral leaves linear-filiform. Pappus-bristles without any terminal tuft . . . . 3. *G. filifolium.*

1. **G. uliginosum,** *A. Gray in Hook. Kew Journ.* iv. 228. A dwarf, diffuse, white tomentose annual, the central stems exceedingly short, the lateral ones often 2 to 3 in. long. Leaves small, petiolate, obovate. Clusters of flower-heads nearly ½ in. diameter, sessile amid broadly ovate or oblong woolly floral-leaves of 2 to 4 lines. Bracts of the involucre closely connected by an intricate wool. Florets numerous. Pappus-bristles plumose from the base, without any terminal tuft, spreading out elastically the moment they are released from the wool of the involucre.—*G. evacinum*, Sond. in Linnæa, xxv. 520.

N. S. Wales. Darling river, *Victorian Expedition, Bowman.*
Victoria. Murray river and Wimmera, *Dallachy.*
S. Australia. From the Murray to the head of Spencer's Gulf, *F. Mueller.*
W. Australia, *Drummond,* 1*st Coll. also n.* 69, *and* 5*th Coll. n.* 360; Bowes and Murchison rivers, *Oldfield;* Gardiner river, *Maxwell;* Swan River, *Preiss, n.* 2415.

2. **G. condensatum,** *A. Gray in Hook. Kew Journ.* iv. 228. Very near *G. uliginosum,* but the flower-heads are larger, in more compact clusters, almost sessile on the ground; the outer floral-leaves oblong, much longer than the cluster, and narrowed into a broad petiole, the inner ones shorter and ovate. Pappus as in *G. uliginosum,* but the bristles terminating in little stipitate tufted appendages.

W Australia, *Drummond,* 5*th Coll. n.* 363; Sharks' Bay, *M. Brown.*

3. **G. filifolium,** *Benth.* A dwarf branching annual, with very few woolly hairs, our specimen scarcely 1 in. high. Leaves linear-filiform, the floral ones closely surrounding the clusters of flower-heads and exceeding them. Clusters dense, nearly globular, 2 to 3 lines diameter. Involucres loosely woolly, the bracts rather numerous, mostly with very short broad and obtuse white radiating tips. Florets very few. Pappus of 5 plumose-ciliate narrow scales or bristles.

W. Australia. Murray river, *Oldfield.* I have only seen a single specimen with the florets still in bud, but there seems to be no doubt of its belonging to this genus.

55. CRASPEDIA, Forst.

(Richea, *Labill.;* Pycnosorus, *Benth.*)

Flower-heads numerous, sessile or nearly so on a convex oblong or cylindrical receptacle, in a globular or ovoid dense cluster or compound head, surrounded by several more or less scarious bracts, forming a general involucre not exceeding the florets. Partial heads 3- to 8-flowered. Involucre of several scarious bracts, the tips sometimes coloured but not radiating, and similar bracts or scales on the partial receptacle under each floret. Florets hermaphrodite, tubular, 5-toothed. Anthers more or less distinctly tailed. Style-branches nearly terete, truncate. Achenes usually compresed, silky-hairy. Pappus of several narrow-linear scales or bristles, plumose-ciliate from the base or towards the end only, all free or slightly united at the base. —Herbs, more or less woolly or silvery-silky, or rarely nearly glabrous. Leaves radical or alternate, entire. Clusters of flower-heads terminal, often rather large.

The genus extends to New Zealand. Of the four Australian species, one is probably the same as the common New Zealand one, the others are endemic. The genus is nearly allied to *Angianthus* and *Calocephalus,* differing from both in the scales of the receptacle within the partial heads.

Clusters of flower-heads pale-coloured, depressed-globular. Outer bracts ovate, with broad brown margins . . . . 1. *C. Richea.*
Clusters of flower-heads bright yellow, globular or ovoid. Outer bracts lanceolate, with brown margins . . . . 2. *C. pleiocephala.*
Clusters of flower-heads yellow, globular or slightly ovoid. Outer bracts very small, and concealed by the reflexed partial heads.
Clusters about ½ in. diameter. Leaves at length glabrous above, loosely woolly underneath . . . . 3. *C. chrysantha.*
Clusters about 1 in. diameter. Leaves silvery-silky on both sides . 4. *C. globosa.*

1. **C. Richea,** *Cass.; DC. Prod.* vi. 152. A perennial, more or less woolly- or silky-white, or nearly glabrous, the tufted stock emitting thick fleshy fibrous roots and simple erect stems. Radical leaves from obovate-oblong to lanceolate, often several inches long and narrowed into a long petiole; stem-leaves narrow, stem-clasping, the upper ones small and distant. Cluster of flower-heads solitary, depressed-globular, ½ to 1 in. diameter, surrounded by about 6 to 10 ovate bracts shorter than the heads, with broad brown scarious margins, the subtending bracts within the cluster also ovate, with brown margins, the inner ones smaller and more scarious. Receptacle globular. Partial heads 6- to 8-flowered. Involucral bracts thin and transparent, ovate, or the inner ones oblong, shorter than the florets, passing into the similar scales of the receptacle. Florets with a slender tube and campanulate limb. Pappus of 10 to 15 filiform plumose scales or bristles, as long as the floret. —Hook. f. Fl. Tasm. i. 197; *C. uniflora,* Forst. Prod. 58; *Richea glauca,* Labill. Voy. t. 16, and Pl. Nov. Holl. ii. 123; *Podospermum pedunculare,* Sieb. Pl. Exs.; *Craspedia glauca* and *C. pilosa,* Spreng. Syst. iii. 441; Lindl. Bot. Reg. t. 1908; *C. fimbriata,* DC. Prod. vi. 152; *C. gracilis,* Hook. f. in Hook. Lond. Journ. vi. 118.

N. S. Wales. Port Jackson, *R. Brown, Sieber, n.* 334, and others; northward to Clarence river, *Beckler,* and New England, *C. Stuart;* southward to Illawarra, *A. Cunningham,* and Twofold Bay, *F. Mueller;* in the interior, to Molle's Plains, *Fraser.*

Victoria. Australian Alps, Munyong mountains, *F. Mueller;* Glenelg river, *Robertson;* Portland, *Allitt;* Wimmera, *Dallachy.*

Tasmania. Derwent river, *R. Brown;* abundant throughout the island, ascending to 4000 ft., *J. D. Hooker,*

S. Australia. From the Murray river to St. Vincent's and Spencer's Gulfs, *F. Mueller,* and others.

W. Australia. From the South coast to Swan River, *Drummond, n.* 47, *Oldfield,* and others.

Var. *macrocephala.* Tall, nearly glabrous, with larger clusters of flower-heads.—*C. macrocephala*, Hook. Bot. Mag. t. 3415; DC. Prod. vi. 153; Hook. f. Fl. Tasm. i. 197.—Tasmania.

Var. *alpina.* Wool very dense and white.—*C. alpina*, Backh. in Hook. Lond. Journ. Bot. vi. 119; Hook. f. Fl. Tasm. i. 198.—Tasmanian Alps and Mount Buller in Victoria.

Three forms, one nearly glabrous, another sprinkled with articulate spreading hairs, and the third white, wtth long silky wool, appear at first sight very distinct, but the indumentum is sometimes mixed, and has no relation to the shape of the leaves or other differences.

2. **C. pleiocephala,** *F. Muell. in Linnæa*, xxv. 404. Apparently annual, but rather robust, branching at the base into erect or ascending stems, mostly simple, of about 1 ft., more or less woolly-hairy. Leaves lanceolate or linear, the lower ones petiolate, often 2 in. long, the upper ones sessile stem-clasping or decurrent. Cluster of flower-heads either solitary terminal and globular, 4 or 5 lines diameter, or ovoid and ½ to ¾ in. long, and then often 2 smaller sessile ones close under it, each surrounded by lanceolate bracts with brown scarious margins, shorter than the florets, with a few inner ones quite scarious, and similar subtending bracts within the cluster. Receptacle cylindrical. Partial heads 3- to 6-flowered. Involucral bracts and scales of the partial receptacle scarious, with bright yellow tips, but scarcely dilated into distinct laminæ. Achenes silky-hairy. Pappus of several plumose bristles.

N. S. Wales. Darling river, *Victorian Expedition;* between Stokes Range and Cooper's Creek, *Neilson.*

S. Australia. Murray river, and east side of Flinders Range, *F. Mueller.*

The habit is that of *Calocephalus Sonderi*, but there appears to be always a scale under each floret.

3. **C. chrysantha,** *Benth.* A perennial, branching at the base, with ascending or erect flowering stems, not above 1 ft. long, and more slender than in *C. globosa.* Leaves linear, generally becoming glabrous above, woolly-white underneath, the upper ones few, small, and distant. Clusters of flower-heads terminal, globular or ovoid, not above ½ in. diameter, surrounded by few very small outer bracts, the subtending bracts within the clusters narrow, brown but scarcely scarious, shorter than the florets, woolly at the base. Receptacle oblong. Partial heads 4- to 6-flowered. Involucral bracts and scales of the receptacle thin and transparent, obovate-oblong, shorter than the florets. Pappus-bristles much more paleaceous than in *C. globosa*, and often here and there connate, scarcely plumose below the middle, very much so towards the end, and bright yellow.—*Calocephalus (?) chrysanthus*, Schlecht. Linnæa, xx. 592; *Pycnosorus chrysanthus*, Sond. in Linnæa, xxv. 492; *P. globosus*, Mitch. Trop. Afr. 323; not of Benth.

Queensland. Maranoa river, *Mitchell;* Peak Downs, *F. Mueller.*

N. S. Wales. Head of the Gwydir, *Leichhardt;* Lachlan river, *A. Cunningham;* Darling river, *Victorian Expedition.*

Victoria. Geelong, *F. Mueller;* Wimmera, *Dallachy.*

S. Australia. Near Bethanie, *Behr;* Light river, *F. Mueller;* Cooper's Creek, *Howitt's Expedition* (a small specimen, with more oblong clusters of flower-heads. Flowers smaller. Pappus more slender and more plumose).

4. **C. globosa,** *Benth.* Apparently perennial, with erect, simple rigid

stems of 1 to 3 ft., silvery-white with a close wool. Lower leaves oblong, linear, or lanceolate, often several inches long; upper ones few, small, and distant, all silvery-white on both sides. Clusters of flower-heads solitary and terminal, globular, bright yellow, $\frac{3}{4}$ to 1 in. diameter, without any general involucre. Receptacle convex, hollow. Subtending bracts within the cluster narrow and short, woolly-ciliate as well as the receptacle. Partial heads 6- to 8-flowered. Involucral bracts and scales of the receptacle ovate or oblong, concave, very thin and transparent, with yellowish tips, but not spreading. Pappus of 12 to 15 plumose bristles.—*Pycnosorus globosus*, Benth. in Hueg. Enum. 67; DC. Prod. vi. 153.

Victoria. Wimmera, *Dallachy.*

S. Australia. South coast, *R. Brown;* Cudnaka, and near Lake Torrens, *F. Mueller;* Mount Searle, *Warburton.*

56. CHTHONOCEPHALUS, Steetz.

(Chamæsphærion, *A. Gray;* Gyrostephium, *Turcz.;* Lachnothalamus, *F. Muell.*)

Flower-heads numerous and sessile on a small receptacle, in a globular or depressed cluster or compound head, surrounded by a few more or less scarious or leafy bracts, forming an irregular general involucre. Partial heads few- or many-flowered. Involucre of several scarious bracts, the outer ones rarely with leafy tips, and similar bracts or scales on the partial receptacle under each floret. Florets hermaphrodite, tubular, 4- or 5-toothed. Anthers more or less distinctly tailed. Style-branches nearly terete, truncate. Achenes usually somewhat compressed, glabrous. Pappus none or short, annular, fringed and very deciduous.—Very dwarf branching or tufted annuals. Leaves radical or alternate, entire. Clusters of flower-heads sessile.

The genus is endemic in Australia. It connects the *Angianthea* with the following genera, the flower-heads being more closely clustered than in the latter, but yet with a few floral leaves occasionally intervening between them.

Stems 1 to 2 in. long. Involucral bracts with broad obtuse tips. No pappus . . . . 1. *C. tomentellus.*
Stemless plants, with the sessile cluster of flower-heads surrounded by the radiating radical leaves.
Involucral bracts with broad obtuse tips. No pappus . . . . . 2. *C. pseudoevax.*
Involucral bracts with rigid acute tips. Pappus a very deciduous fringed ring . . . . 3. *C. pygmæus.*

1. **C. tomentellus,** *Benth.* A small annual, branching from the base with ascending or erect stems of 1 to 3 in., more or less woolly. Leaves obovate or the upper ones oblong, under $\frac{1}{2}$ in. long. Clusters of flower-heads terminal, globose, 3 to 4 lines diameter, surrounded by 3 or 4 broad scarious bracts, and sometimes by 1 or 2 small leafy ones. Heads globular, many-flowered. Involucre of about 5 linear-spathulate or oblong herbaceous bracts with densely woolly margins and scarious tips. Receptacle with ovate transparent woolly scales under each floret. Corollas 5-toothed or rarely 4-toothed. Achenes glabrous, without any pappus.—*Lachnothalamus tomentellus*, F. Muell. Fragm. iii. 156.

W. Australia. Murchison river, *Oldfield;* Useless Harbour, Sharks' Bay, *M. Brown* (with larger clusters of flower-heads and smaller involucral bracts).

2. **C. pseudoevax,** *Steetz in Pl. Preiss.* i. 445. A small annual with scarcely any stem, consisting of numerous sessile flower-heads, forming patches of ½ to 1 in. diameter, surrounded by a few oblong spreading leaves, ¼ to ½ in. long, forming as it were a general involucre, with rarely a leaf or two protruding between the heads, there are also immediately under most of the heads or at least the outer ones 2 or 3 linear leafy bracts more or less woolly with scarious tips not exceeding the heads. Heads ovoid-globular, nearly 2 lines diameter, consisting of numerous broad imbricate scarious bracts or scales, very few or even only one of the outer ones empty, each of the others with a floret in its axil not exceeding the bract. Corolla very slender, 4-toothed. Pappus none.—*C. Drummondii,* A. Gray in Hook. Kew Journ. iii. 178.

N. S. Wales. Between Darling river and Cooper's Creek, *Neilson.*
Victoria. Murray river, *Dallachy.*
S. Australia. Lake Gillies, *Burkitt.*
W. Australia. Swan River, *Drummond, 1st Coll.;* Avon river, *Preiss, n.* 2414 *b;* Murchison river, *Oldfield.*

Preiss's and Drummond's specimens appear to me to have precisely the same structure; the scales of the receptacle become stiffer as the fruit ripens and slightly impressed at the base by the thickening of the achene. There appear however to be 2 forms, the one with larger browner scales, the other with smaller white ones; the two are sent as one by Oldfield, separately in other collections, but as the former are always in flower, the latter in fruit, the difference may be owing to age.

3. **C. pygmæus,** *Benth.* A dwarf stemless plant, the narrow-linear radical leaves forming an involucre round a sessile depressed-globular cluster of flower-heads of 3 to 5 lines diameter. Partial heads about 5- to 7-flowered. Involucral bracts 1 or 2 and the scales of the receptacle with lanceolate acute nearly white rigid tips projecting beyond the florets, a few scales in the centre of the head usually empty. Corollas slender, 3- or rarely 4-toothed. Achenes glabrous. Pappus a fringed scaly ring, very deciduous.—*Chamæsphærion pygmæum,* A. Gray in Hook. Kew Journ. iii. 177; *Gyrostephium rhizocephalum,* Turcz. in Bull. Mosc. 1851, ii. 77.

W. Australia, *Drummond, 5th Coll. n.* 55.

Subtribe II. Helichryseæ.—Flower-heads pedunculate or rarely almost sessile but distinct and not forming a compound head. Florets either all tubular and hermaphrodite or a few of the outer ones (rarely the whole outer row or even two rows where the florets are numerous) female and either very slender or rarely larger and irregular or ligulate.

57. IXODIA, R. Br.

Involucre ovoid, the bracts imbricate, appressed, dry and glutinous, the inner ones with white spreading laminæ. Receptacle shortly conical, bearing scales closely rolled round each floret. Florets all tubular, hermaphrodite,

5-toothed. Anthers tailed. Style-branches truncate. Achenes without any pappus.—Erect shrub. Leaves alternate, entire. Flower-heads rather small, in terminal corymbs.

The genus is limited to a single species, endemic in Australia. It is very near *Ammobium*, differing only in the involute scales of the receptacle and in the want of any pappus.

1. **I. achilleoides,** *R. Br. in Bot. Mag. t.* 1534. An erect glabrous and glutinous shrub. Leaves linear-lanceolate or slightly spathulate, usually acute and mostly above 1 in. long, more or less decurrent on the stem. Flower-heads in a dense terminal corymb, very much resembling those of an *Achillea*, the white petal-like radiating laminæ of the inner involucral bracts imitating the ray-florets of that genus. Involucre without the ray, 2 to 3 lines long, the outer bracts closely appressed, glutinous, with green centres and often slightly woolly. Scales of the receptacle jagged at the end and the outer ones sometimes produced into a small white appendage. Achenes oblong-cylindrical, slightly papillose-pubescent.—DC. Prod. vi. 154; Sond. in Linnæa, xxv. 495; *I. ptarmicoides*, F. Muell. in Linnæa, xxv. 405; *I. alata*, Schlecht. in Linnæa, xx. 493; Sond. l. c. xxv. 495.

Victoria. Rocky places in the Grampians, *F. Mueller;* common at the mouth of the Glenelg towards Portland, *Robertson*, *Allitt*.

S. Australia. Memory Cove and Port Lincoln, *R. Brown;* rocky places, chiefly near the coast, from the Murray to St. Vincent's and Spencer's gulfs, and Kangaroo Island, *F. Mueller*, *Wilhelmi*, and others.

There are usually two forms, of which the extremes look very different, *I. achilleoides*, with narrow leaves, obtuse or with recurved points, not very decurrent, and with small flower-heads, and *I. alata*, with broader more acute and more decurrent leaves and larger flower-heads, but there are many specimens equally referable to the one or to the other.

58. AMMOBIUM, R. Br.

Involucre hemispherical, the bracts either petal-like and spreading or scarious and more appressed. Receptacle more or less convex with flat or slightly concave scales between the florets. Florets all tubular, hermaphrodite, 5-toothed. Anthers with fine tails. Style-lobes truncate. Achenes 4-angled. Pappus a membranous cup, either truncate and entire or more or less produced into 2 or 4 unequal teeth or short awns.—Herbs more or less white-tomentose. Leaves entire. Flower-heads solitary, terminating the stem or branches.

The genus is limited to Australia.

Involucral bracts white, petal-like, spreading, longer than the florets. Stem winged and branching . . . . 1. *A. alatum.*
Involucral bracts scarious, jagged or undulate, shorter than the florets. Stems simple, single-headed. Leaves shortly decurrent . 2. *A. craspedioides.*

1. **A. alatum,** *R. Br. in Bot. Mag. t.* 2459. An erect, branching, white-tomentose herb, attaining 2 or 3 ft., the stems bordered by wings decurrent from the margins of the leaves. Radical leaves lanceolate, several inches long, narrowed into a long petiole; stem-leaves small and distant. Involucral bracts in many rows, spreading to about $\frac{3}{4}$ in. diameter, the white petal-like laminæ broadly ovate, the inner ones $\frac{1}{4}$ in. long on very short claws,

the outer ones shorter and sessile. Florets numerous. Scales of the receptacle rigid, slightly concave, mucronate. Teeth or awns of the pappus-cup very variable, usually very small.—DC. Prod. vi. 153; Gaudich. in Freyc. Voy. 467. t. 90 (*A. spathulatum* on the plate); Sweet, Brit. Fl. Gard. t. 48.

N. S. Wales. Hunter's River, *R. Brown;* sandy plains near Bathurst and on the Macquarrie, *A. Cunningham, Fraser;* Cardington, *Ramsay;* New England, *Beckler.*

2. **A. craspedioides,** *Benth.* Stock perennial, with simple single-headed stems of 1 to 2 ft., loosely woolly as well as the under side of the leaves. Leaves chiefly radical, oblong-lanceolate, narrowed into a petiole, entire, 3 to 4 in. long, scabrous-hirsute on the upper side; stem-leaves narrow, shortly decurrent, the upper ones small and distant. Involucres about 1 in. diameter, the bracts ovate, scarious, straw-coloured, rigid at the base, the outer ones short and rigid, all much shorter than the florets. Scales of the receptacle very broad, rigidly scarious, truncate and jagged at the end. Florets numerous. Anthers with rather long fine tails. Style-lobes truncate, penicillate. Pappus-cup with awns usually rather longer and more rigid than in *A. alatum.*

N. S. Wales. Near Nangas, *M'Arthur.*

59. CASSINIA, R. Br.

(Achromolæna, *Cass.;* Apalochlamys *and* Rhynea, *DC.*)

Involucre narrow-ovoid or oblong, the bracts imbricate, scarious or coloured, without any or, in species not Australian, with small radiating laminæ. Receptacle with scarious chaffy scales between the florets. Florets few, all hermaphrodite, tubular, 5-toothed or, in species not Australian, a very few of the outer ones slender and female. Anthers very shortly or obscurely tailed. Style-branches nearly terete, truncate. Achenes short, angular or nearly terete, usually papillose. Pappus of several simple entire or minutely denticulate capillary bristles, in a single row and slightly cohering in a ring at the base.—Shrubs or rarely herbs. Leaves alternate, entire. Flower-heads small, numerous, in terminal corymbs or panicles.

Besides the Australian species, which are endemic, there are four from New Zealand and one from S. Africa (*Rhynea,* DC.), all differing slightly from the Australian ones in the small white radiating laminæ of their inner involucral bracts. The genus is closely allied on the one hand to *Humea,* on the other to the small-headed *Helichrysa,* with precisely the same habit, differing from the former in the pappus and from both in the scales of the receptacle. A few scales may indeed be occasionally found among the central florets of a few species of *Helichrysum,* especially when they are sterile, but in *Cassinia* they subtend the fertile florets.

Shrubs with small or narrow rigid leaves. Florets under 10.
  Corymbs dense, sessile and shorter than the upper leaves.
    Involucres linear, sessile and very densely packed, the bracts tipped with yellow. Florets usually 2. Leaves scabrous-hirsute . . . . 1. *C. leptocephala.*
    Involucres oblong, the bracts without coloured tips. Florets usually 4. Leaves smooth above.
      Leaves lanceolate. Involucral bracts thinly scarious, very obtuse, shorter than the florets . . . . 2. *C. compacta.*

Leaves narrow-linear. Involucral bracts narrow, scarcely obtuse, as long as the florets . . . . 3. *C. tenuifolia.*
Corymbs flat or convex, dense, but much exceeding the upper leaves.
Leaves mostly oblong, under ¾ in. long, glabrous and smooth above. Involucral bracts broad, white or straw-colour . . 4. *C. denticulata.*
Leaves lanceolate or linear, acute, mostly 1 to 2 in. long, glabrous and smooth above.
Involucral bracts white or pale straw-colour . . . . 5. *C. longifolia.*
Involucral bracts bright yellow . . . . 6. *C. aurea.*
Leaves narrow-linear, obtuse or with recurved points, tuberculate, muricate or hispid above, very rarely nearly smooth . . 7. *C. aculeata.*
Panicles pyramidal or not longer than broad, loose or rather compact. Leaves narrow-linear with small straight or recurved points.
Branches cottony-white. Leaf-points usually recurved . . . 8. *C. lævis.*
Branches and leaves glabrous or slightly viscid. Leaf-points usually straight . . . . 9. *C. quinquefaria.*
Panicles long and loose.
Leaves narrow-linear, heath-like, under ⅓ in. long . . . . . . 10. *C. arcuata.*
Leaves obovate or oblong, flat, under 1 in. long . . . . . . 11. *C. subtropica.*
Panicles almost reduced to oblong spikes. Leaves very small and erect. Florets solitary or rarely 2 in the head . . . . 12. *C. Theodori.*
Tall herb, with large flaccid leaves and ample loose panicles. Florets about 10 . . . . 13. *C. spectabilis.*

(*C. glossophylla*, Cass.; DC. Prod. vi. 155, described from Sieber's specimen, n. 592, is unknown to me. The character would refer to *C. denticulata*, but that De Candolle places it in the section with radiating laminæ to the involucral bracts, which I have not seen in any Australian species.)

1. **C. leptocephala,** *F. Muell. Fragm.* iii. 138. A shrub, the branches as well as the foliage, densely scabrous-hirsute with short rigid hairs. Leaves crowded, narrow-linear, mostly above 2 in. long, with closely revolute margins. Flower-heads sessile and crowded in compact clusters, forming a very dense sessile corymb, 1 to 2 in. diameter, shorter than the last leaves. Involucre linear, about 3 lines long, the bracts very narrow, appressed, thinly scarious, with yellow tips, not spreading. Florets 2 or rarely 3, with scales between them. Pappus-bristles not numerous.

N. S. Wales. Port Jackson, *F. Mueller.*

2. **C. compacta,** *F. Muell. Fragm.* i. 18. A shrub, the branches and under side of the leaves hoary with a very short close tomentum. Leaves lanceolate, scarcely acute, mucronulate, the margins recurved, 1½ to above 2 in. long, glabrous and smooth above. Flower-heads small, in a very dense rather flat sessile corymb shorter than the last leaves. Involucre straw-coloured or pale brown, narrow, scarcely above 1½ lines long, the bracts obtuse, thinly scarious on the margins, shorter than the florets. Florets 4 or rarely 5 or 6.

N. S. Wales. Mount Lindsay, on the borders of Queensland, at an elevation of 5000 feet, *W. Hill.*

3. **C. tenuifolia,** *Benth.* A low bushy shrub, the branches and under side of the leaves white with a close but dense tomentum. Leaves rather

crowded, very narrow-linear, obtuse or with short recurved points, above 1 in. long, smooth and glabrous above. Flower-heads (like those of *C. lævis*) in compact corymbs much shorter than the surrounding leafy branches in our specimens. Involucre oblong, about 2 lines long, the bracts narrow, the inner ones as long as the florets, white, without spreading tips. Florets usually 4, the scales of the receptacle very prominent.

N. S. Wales. Lord Howe's Island, near the shore, *Milne.* It is possible that this may be an abnormal state of *C. lævis,* but the inflorescence appears to be quite different.

4. **C. denticulata,** *R. Br. in Trans. Linn. Soc.* xii. 127. A shrub with glabrous or hoary-tomentose branches. Leaves oblong or broadly lanceolate, acute, often narrowed below the middle but half stem-clasping at the base, often all under ½ in. and rarely ¾ in. long, coriaceous, the margins recurved and minutely scabrous-denticulate or quite entire, glabrous and smooth above, hoary or rusty underneath with a minute scarcely perceptible tomentum, rarely loose and more copious. Flower-heads in a broad rather loose convex corymb much exceeding the leaves. Involucre broadly ovoid, 2 lines long or rather more, white or straw-coloured, the bracts very obtuse. Florets about 10 to 12, the receptacle-scales as long as the florets.—DC. Prod. vi. 155.

N. S. Wales. Port Jackson, *White, R. Cunningham, Woolls* and others; Illawarra, *A. Cunningham.*

5. **C. longifolia,** *R. Br. in Trans. Linn. Soc.* xii. 127. A shrub, the branches and under side of the leaves more or less hoary or white-tomentose or almost glabrous. Leaves linear or lanceolate, acute, narrowed at the base, mostly 1 to 2 in. long, the margins more or less recurved, glabrous and smooth on the upper side. Flower-heads numerous, in a broad rather dense flat or convex corymb much exceeding the last leaves. Involucres oblong, about 2 lines long, pure white in the original form, the bracts very obtuse, opaque. Florets usually about 6 to 8.

N. S. Wales. Port Jackson, *R. Brown, Woolls;* Illawarra, *Shepherd.*
Victoria. Bacchus Marsh and between Ovens river and Mayday Hills, *F. Mueller;* Portland, *Allitt.*

Var. *straminea.* Involucral bracts straw-coloured or pale brown.—*C. longifolia,* DC. Prod. vi. 156.—N. S. Wales, *F. Mueller;* New England, *C. Stuart;* Macleay and Clarence rivers, *Beckler.*

6. **C. aurea,** *R. Br. in Trans. Linn. Soc.* xii. 127. A shrub with the habit, foliage and inflorescence of *C. longifolia,* and probably a variety only, differing only in being usually rather more glabrous, and the involucres of a bright yellow.—Bot. Reg. t. 764; DC. Prod. vi. 155.

N. S. Wales. Port Jackson to the Blue Mountains, *R. Brown, Caley, A. Cunningham, Woolls,* and others.

7. **C. aculeata,** *R. Br. in Trans. Linn. Soc.* xii. 127. A shrub with the branches and under side of the leaves more or less tomentose-pubescent. Leaves crowded, narrow-linear, obtuse or with small recurved points, the margins revolute, varying from ½ to 1½ in. long, usually scabrous or muricate on the upper side with very short rigid hairs or tubercles, very rarely smooth

or nearly so. Flower-heads very numerous, in terminal corymbs, usually many inches broad, sometimes small and dense but always much longer than the last leaves. Involucres narrow-ovoid, usually white but sometimes passing into pink or pale brown, the bracts obtuse, the small outer ones and the base of the inner ones often browner or more tomentose than in *C. longifolia*. Florets varying from 6 to 10 or even 12. Receptacle-scales as long as the florets.—DC. Prod. vi. 156; Hook. f. Fl. Tasm. i. 200; *Calea aculeata*, Labill. Pl. Nov. Holl. ii. 41. t. 185; *Cassinia affinis*, R. Br. in Trans. Linn. Soc. xii. 127; DC. Prod. vi. 156; *C. adunca*, F. Muell. in Linnæa, xxv. 496.

N. S. Wales. Blue Mountains, *Caley, Miss Atkinson* and others; Gabo island, *Maplestone*.

Victoria. Gipps' Land, Dandenong ranges, *F. Mueller;* Portland, *Allitt*.

Tasmania. Port Dalrymple, *R. Brown;* abundant throughout the colony, *J. D. Hooker*.

S. Australia. Murray scrub, Loddon, *F. Mueller*.

Var. *uncata*. Leaves short, sometimes almost smooth above, corymbs usually small.—*C. uncata*, A. Cunn. in DC. Prod. vi. 156.—Liverpool plains, *A. Cunningham;* Clarence river, *Beckler;* also in *Leichhardt's* collection.

Some specimens with slender leaves from Macalister river and other parts of Victoria, *F. Mueller*, with the inflorescence of the common form, have the involucral bracts more or less distinctly superposed in 4 or 5 rows, but do not otherwise differ from the species, and this character does not appear constant, at least in the dried specimens.

8. **C. lævis,** *R. Br. in Trans. Linn. Soc.* xii. 128. A rather slender shrub, the branches and under side of the leaves white-tomentose. Leaves narrow-linear with closely revolute margins, $\frac{1}{2}$ to $1\frac{1}{2}$ in. long, smooth above without any of the asperities of *C. aculeata*. Flower-heads smaller than in that species, in a shortly pyramidal rather dense panicle of 2 or 3 in., rarely condensed into a very convex corymb, and always looser than in *C. aculeata*. Involucre narrow, $1\frac{1}{2}$ lines long, of a pure white, the bracts obtuse but narrow. Florets usually about 4 or 5.—DC. Prod. vi. 156; *C. rosmarinifolia*, DC. l. c.

Queensland. Dawson river, *Herb. F. Mueller;* Warwick, *Beckler*.

N. S. Wales. Liverpool plains and Lachlan river, *A. Cunningham;* New England, *C. Stuart;* Barrier and Mutanie ranges, *Victorian Expedition*.

S. Australia. Spencer's Gulf, *R. Brown*.

9. **C. quinquefaria,** *R. Br. in Trans. Linn. Soc.* xii. 128. A shrub, glabrous or nearly so, and sometimes appearing somewhat viscid. Leaves narrow-linear, with revolute margins, without asperities, mostly above 1 in. long. Flower-heads numerous, in pyramidal panicles, usually looser and larger than in *C. lævis* but not nearly so long as in the following species. Involucres narrow-oblong, scarcely 2 lines long, of a pale straw-colour, the bracts rather narrow, obtuse, distinctly or sometimes obscurely superposed in 5 rows. Florets about 5.—DC. Prod. vi. 157; *C. hygrophila*, A. Cunn. in DC. Prod. vi. 156; *Achromolæna viscosa*, Cass.; quoted by DC. l. c.

N. S. Wales. Blue Mountains, *Caley;* rocks on the Lachlan to the west of Bathurst and barren forest land, Argyle county, *A. Cunningham;* New England, *C. Stuart;* Darling Downs, *Herb. F. Mueller;* also in *Leichhardt's* collection.

10. **C. arcuata,** *R. Br. in Trans. Linn. Soc.* xii. 128. An erect shrub

of 5 or 6 ft., the branches and under side of the leaves white-tomentose. Leaves narrow-linear, obtuse or with short recurved points, the margins closely revolute. Flower-heads small and numerous in a long loose terminal panicle. Involucres cylindrical, often curved, not 2 lines long, straw-coloured white or brown, the bracts very thin, smooth and shining. Florets 2, 3 or rarely 4.—DC. Prod. vi. 156; *C. paniculata*, Behr and Muell. in Linnæa, xxv. 496; F. Muell. Pl. Vict. t. 43.

N. S. Wales. Darling river, *Dallachy.*
Victoria. Grampians, *F. Mueller;* N.W. portion of the colony, *L. Morton.*
S. Australia. Spencer's Gulf, *R. Brown;* Open places in the Murray scrub, *Behr.*
W. Australia. Middle Mount Barren, *Maxwell* (the specimens very young and therefore somewhat uncertain).

11. **C. subtropica,** *F. Muell. Fragm.* i. 17. Apparently a slender shrub, the branches tomentose-pubescent. Leaves petiolate, from obovate to elliptical-oblong, obtuse or mucronulate, under 1 in. long, glabrous and smooth above, white or rusty-tomentose underneath, the margins scarcely recurved. Flower-heads small and numerous, in a long loose terminal panicle. Involucre cylindrical, about 2 lines long, straw-coloured or reddish-brown, the bracts very thin. Florets usually 3, with receptacle scales between them as in the rest of the genus.

N. S. Wales. Mount Lindsay in the borders of Queensland, *W. Hill.*

12. **C ? Theodori,** *F. Muell. Fragm.* v. 148. An erect heath-like shrub, the branches and under side of the leaves cottony. Leaves almost scale-like, erect, 1 to 2 lines long, linear, obtuse, with revolute margins. Flower-heads sessile in short spikes, forming an oblong compact leafy panicle. Involucre narrow, 1 to $1\frac{1}{2}$ lines long, the bracts few, narrow, shining, of a reddish-brown. Florets solitary, or (according to F. Mueller) sometimes 2 together. Achenes glabrous. Pappus-bristles fine, not thickened upwards.

N. S. Wales. Phonolith at the head of the Gwydir, *Leichhardt.* The specimens appear very different from any that I am acquainted with, but, the florets not being yet expanded, the characters cannot be fully observed. The single floret occupying the centre of the receptacle, there can be no scales besides the involucral bracts, but the inflorescence is very different from any as yet known in *Helichrysum*, and shows more affinity to that of *Cassinia arcuata.*

13. **C. spectabilis,** *R. Br. in Trans. Linn. Soc.* xii. 128. An erect robust herb of 3 to 5 ft., the stem hard, almost woody, clothed with white wool. Lower leaves oblong or obovate-oblong, shortly acuminate, broadly stem-clasping and often shortly decurrent, 4 to 6 in. long, pubescent above, more tomentose underneath, the upper ones smaller and lanceolate. Flower-heads very numerous, in a large loose terminal panicle. Involucre ovoid-turbinate, rather above 2 lines long, straw-coloured or pale brown, the bracts thinly scarious, mostly obtuse. Florets above 10 and often nearly 30, with narrow scales between them. Achenes small, prominently ribbed.—Bot. Reg. t. 678; *Calea spectabilis*, Labill. Pl. Nov. Holl. i. 42. t. 186; *Apalochlamys Billardieri*, DC. Prod. vi. 157; Hook. f. Fl. Tasm. i. 199; *A. Endlicherii* and *A. Kerii*, DC. l. c.

Victoria. Hills near the coast from the Glenelg to Gipps' Land, *F. Mueller* and others.

Tasmania. Northern parts of the island and islands of Bass's Straits, *J. D. Hooker* and others.

S. Australia. Kangaroo Island, *F. Mueller*.

60. HUMEA, Sm.

(Calomeria, *Vent.;* Hæckeria, *F. Muell.*)

Involucre oblong, the bracts imbricate, scarious. Receptacle small, without scales. Florets very few or solitary, hermaphrodite, tubular, 5-toothed. Anthers with fine, sometimes minute tails. Style-branches truncate. Achenes narrow, without any pappus.—Herbs or shrubs. Leaves alternate, quite entire. Flower-heads small and numerous in a loose terminal panicle or in compact corymbs.

The genus is limited to Australia, the habit is that of *Cassinia*, from which it differs in the absence of pappus and of receptacle-scales.

Flower-heads in a large loose terminal drooping panicle. Involucral bracts thin and scarious. Tall herb . . . . 1. *H. elegans.*
Flower-heads in dense corymbs. Involucral bracts rigid or petal-like. Shrubs.
Glabrous or glutinous.
Leaves terete, obtuse. Florets usually 3 . . . . 2. *H. cassiniacea.*
Leaves keeled, acute. Florets usually solitary . . . . . 3. *H. punctulata.*
Pubescent or tomentose. Leaves with revolute margins . . . 4. *H. ozothamnoides.*

1. **H. elegans,** *Sm. Exot. Bot.* i. *t.* 1. A robust erect biennial, attaining 5 or 6 ft. or more, glandular-pubescent or nearly glabrous, strongly scented. Lower leaves ovate-lanceolate or oblong, acuminate, stem-clasping or decurrent at the base, 6 to 10 in. long, rugose and scabrous-pubescent but green on both sides, the upper leaves small. Flower-heads very numerous in a large loose terminal panicle with gracefully pendulous branches. Involucre about 3 lines long, of a brown-red or pink, the bracts very thin and scarious, with small ones continued along the peduncles. Florets 3 or 4. Achenes glandular but otherwise glabrous.—DC. Prod. i. 158; F. Muell. Fragm. i. 17; *Calomeria amaranthoides*, Vent. Jard. Malm. t. 73.

N. S. Wales. Port Jackson, *R. Brown*, *Woolls* and others.

Victoria. Tambo, Monkey Creek, Snowy River, *F. Mueller;* Victoria ranges, *Wilhelmi.*

2. **H. cassiniacea,** *F. Muell. Fragm.* i. 17. An erect shrub of 3 or 4 ft., glandular-viscid and strongly scented, otherwise glabrous. Leaves linear, semiterete, obtuse, in some specimens rarely exceeding ½ in., in others nearly 1 in. long, clustered in the axils. Flower-heads very numerous in a very compact corymb of 2 to 3 in. diameter. Involucres white, about 2 lines long, the bracts narrow but obtuse, the inner ones almost as long as the florets. Florets usually 3. Achenes slightly fusiform, glabrous.—*Hæckeria cassiniæformis*, F. Muell. in Trans. Phil. Soc. Vict. i. 45; in Linnæa, xxv. 406; and in Hook. Kew Journ. viii. 156.

S. Australia. Port Lincoln, frequent, *R. Brown*, *Wilhelmi* and others.

3. **H. punctulata,** *F. Muell. Fragm.* iii. 137. An erect shrub or undershrub, with virgate branches, glutinous but otherwise glabrous. Leaves

linear-subulate, acute, keeled or triquetrous, under 1 in. long. Corymbs of flower-heads usually smaller than in the other species. Involucre straw-coloured, about 1½ lines long. Florets solitary in each head.—*Cassinia quinquefaria*, Sond. in Linnæa, xxv. 496, not of R. Br.

S. Australia. Spencer's Gulf, *R. Brown;* Flinders and Elder's Ranges, *F. Mueller.*

4. **H. ozothamnoides,** *F. Muell. Fragm.* i. 17. Branches erect from a woody base, 1 to 2 ft. high, more or less tomentose with a white deciduous wool or sometimes glutinous with scarcely any wool. Leaves very narrow-linear, mucronate-acute, with revolute margins, ½ to 2 in. long, soon becoming glabrous above, white-tomentose underneath. Flower-heads numerous in a compact corymb. Involucre about 2 lines long, straw-coloured or the inner bracts white towards the end. Florets 3 or 4.—F. Muell. Pl. Vict. t. 44; *Hæckeria ozothamnoides*, F. Muell. in Trans. Phil. Soc. Vict. i. 45, and in Hook. Kew Journ. viii. 150.

Victoria. Dry places on Barker's Creek, Upper Murray and Snowy rivers, *F. Mueller.*

61. PITHOCARPA, Lindl.

Involucre turbinate, the bracts imbricate, the outer ones small and appressed, the inner with coloured radiating laminæ. Receptacle flat, without scales. Florets numerous, all hermaphrodite, tubular, 5-toothed. Anthers with fine tails. Style-branches truncate. Achenes angular, without any pappus.—Erect branching annual. Leaves alternate, entire. Flower-heads in a loose erect panicle.

The genus is limited to a single species, endemic in W. Australia, differing from *Humea* in habit and involucre.

1. **P. corymbulosa,** *Lindl. Swan Riv. App.* 23. An erect annual, with long slender but rigid paniculate stems of 1 to 2 ft., slightly woolly and often nearly leafless. Leaves chiefly in the lower part of the stem, linear, soft, the upper ones very few, small and distant. Flower-heads on elongated peduncles. Involucres in the largest form about 3 lines long without the spreading laminæ, the outer bracts gradually smaller and extending down the summit of the peduncle, woolly, with rigid acute or almost obtuse glabrous tips, the inner ones with gradually increasing white spreading laminæ, the innermost nearly 3 lines long. Florets small. Achenes papillose.—Steetz in Pl. Preiss. i. 445; *P. major*, Steetz, l. c. 446.

W. Australia. King George's Sound and adjoining districts, *R. Brown* and others, and thence to Swan River, *Drummond*, 1*st Coll., also n.* 84, 169, *and* 5*th Coll. n.* 353, *Preiss, n.* 27, and others; Moore river, *Oldfield;* and eastward to the Mount Barren Range, *Maxwell.*

The plant varies much in the size of the flower-heads, which in some of Oldfield's specimens are larger even than in the *P. major* of Steetz, whilst in the slender specimens distinguished as *P. pulchella*, Lindl. Swan Riv. App. 23, the involucres are not 2 lines long without the laminæ, which are themselves not 2 lines long. I find no other difference, and the intermediate sizes are numerous.

62. ERIOCHLAMYS, Sond. and Muell.

Involucre broadly ovoid or almost globular, the bracts imbricate, the outer ones herbaceous with revolute margins, the inner ones with broad scarious margins, without coloured laminæ. Receptacle slightly convex, without scales. Florets numerous, all hermaphrodite, tubular, 5-toothed. Anthers with very short points or tails. Style-branches broadly truncate. Achenes oblong, without any pappus.—Small annual. Leaves alternate, small, entire. Flower-heads woolly, sessile, solitary or clustered.

The genus is limited to a single species, endemic in Australia. Like *Pithocarpa*, it only differs from *Humea* in habit, involucre, and the number of florets. The inflorescence approaches sometimes that of *Angianthea*, but the individual heads are occasionally solitary and always more distinct than in that subtribe.

1. **E. Behrii,** *Sond. and Muell. in Linnæa*, xxv. 488. Very much branched, woolly-tomentose and usually diffuse and under 6 in. high, apparently annual, but sometimes larger with a hard almost woody base. Leaves linear, obtuse, often all under $\frac{1}{4}$ in. long. Flower-heads numerous, small, sessile amongst the last leaves and often crowded at the ends of the branches. Involucre more or less woolly, about $1\frac{1}{2}$ lines long. Florets scarcely exceeding the involucre, the corollas often woolly at the base.

N. S. Wales. From the Lachlan and Darling rivers to the Barrier Range, *Victorian and other Expeditions.*

S. Australia. Between Salt Creek and Pine Forest, *Behr;* Murray river, Crystal Brook, Dombey Bay, *F. Mueller;* Port Lincoln, *Wilhelmi.*

63. ACOMIS, F. Muell.

Involucre broadly hemispherical, the bracts loosely imbricate, scarious, slightly thickened at the base. Receptacle convex, without scales. Florets numerous, all hermaphrodite, tubular, 5-toothed. Anthers with fine tails. Style-branches long, truncate. Achenes oblong or narrow, without any pappus.—Erect slender herbs. Leaves linear, entire. Flower-heads on slender terminal peduncles.

The genus is limited to Australia. It is allied to *Rutidosis*, with nearly the same involucre, differing in the absence of pappus and in the much more distinctly tailed anthers.

Leaves lanceolate. Corollas not dilated at the base. Usually woolly . 1. *A. Rutidosis.*
Leaves linear-subulate. Corollas dilated at the base. Nearly glabrous . 2. *A. macra.*

1. **A. Rutidosis,** *F. Muell. Fragm.* ii. 89. Erect, branching, and apparently annual, 1 to 2 ft. high, more or less woolly. Leaves petiolate, lanceolate or almost ovate-lanceolate, acute, narrowed at the base, 1 to $1\frac{1}{2}$ in. long, losing the wool from the upper side. Involucres about 4 lines diameter, the outer bracts ovate, the inner ones oval-oblong. Florets slightly dilated upwards, but not at the base. Achenes narrow.—*Rutidosis acoma*, F. Muell. l. c.

N. S. Wales. Hastings river, *Beckler.*

2. **A. macra,** *F. Muell. Fragm.* iv. 145. Nearly glabrous, the stems slender, almost filiform. Leaves linear-subulate, acute, under 1 in. long. Involucre about 3 lines diameter, the bracts broadly lanceolate, thickened in the centre but not clawed. Corollas glandular and dilated over the achene at the base as in *Cotula*. Achenes oblong.—*Rutidosis macra*, F. Muell. l. c.

Queensland. Cape river, *Bowman.*

64. **TOXANTHUS,** Turcz.

(Anthocerastes, *A. Gray.*)

Involucral cylindrical, of very few narrow herbaceous nearly equal bracts. Receptacle small, without scales. Florets few, hermaphrodite, the corolla tubular, 4- or 5-toothed, continuous with the ovary, persistent and recurved. Anthers with short fine tails. Style-branches with lanceolate papillose tips. Achenes cylindrical, tapering at the top, without any pappus.—Dwarf annuals. Leaves alternate, entire. Flower-heads small, sessile or shortly pedunculate.

The genus is limited to Australia. The style is rather more that of *Asteroideæ* than of *Gnaphalieæ*, but the general character appears to be quite that of the latter tribe.

Involucral bracts recurved after flowering. Achenes distinctly beaked . 1. *T. perpusillus.*
Involucral bracts erect. Achenes tapering at the top but scarcely beaked 2. *T. Muelleri.*

1. **T. perpusillus,** *Turcz. in Bull. Mosc.* 1851, i. 177. A little slender diffuse annual, either scarcely ½ in. long and densely enveloped in wool, or lengthening out to 1 or 2 in. and becoming glabrous. Leaves linear. Flower-heads small, sessile in the tufts of leaves, or 2 or 3 together clustered at the ends of the branches. Involucral bracts 3 to 5, linear, rigid and recurved as the flowering advances. Corolla 4-toothed, woolly at the base. Achenes slender, striate, tapering into a beak continuous with the rigid persistent recurved corolla.—*Anthocerastes Drummondii,* A. Gray in Hook. Kew Journ. iv. 226; Sond. in Linnæa, xxv. 480.

S. Australia. Burra-Burra river and between Hutt and Broughton rivers, *F. Mueller.*
W. Australia, *Drummond, 4th Coll. n.* 203; Murray river, *F. Mueller.*

2. **T. Muelleri,** *Benth.* A diffuse annual of ½ to 1 in., slightly glandular-pubescent. Leaves linear. Flower-heads small, terminal. Involucre about 1½ lines long, of 4 or 5 bracts, linear-lanceolate with slightly scarious margins. Florets 5 to 10, very slender, 4- or 5-toothed. Achenes terete, shortly hirsute, not distinctly beaked but shortly tapering at the top and continuous with the recurved corolla, which is usually persistent but sometimes at length deciduous.—*Anthocerastes Muelleri,* Sond. in Linnæa, xxv. 480.

S. Australia. Murray river, *F. Mueller.*

65. **SCYPHOCORONIS,** A. Gray.

Involucre cylindrical, of very few herbaceous nearly equal bracts. Receptacle small, without scales. Florets hermaphrodite, the corolla tubular, 5-toothed, deciduous. Anthers with short very fine tails. Style-branches with somewhat lanceolate obtuse papillose tips. Achenes cylindrical, crowned by a short herbaceous persistent cup outside the corolla.—Dwarf annual. Leaves alternate or here and there opposite, entire. Flower-heads shortly pedunculate.

The genus is limited to a single species, endemic in Australia, which Turczaninow was perhaps right in including in *Toxanthus,* to which it is closely allied, but differs in the achene continuous with a herbaceous cup outside the corolla, not with the corolla itself.

1. **S. viscosa,** *A. Gray in Hook. Kew Journ.* iv. 225. A little diffuse

annual of 1 to 2 in., sprinkled with short rigid glandular hairs. Leaves linear. Flower-heads small, terminal. Involucre about 2 lines long, of 4 or 5 linear bracts. Florets about 8 to 12, the corolla slender. Achenes linear, glandular.—Hook. Ic. Pl. t. 854; *Toxanthus major*, Turcz. in Bull. Mosc. 1851, ii. 64.

W. Australia, *Drummond, 5th Coll. n.* 53.

66. RUTIDOSIS, DC.

(Pumilo, *Schlecht.*; Actinopappus, *A. Gray.*)

Involucre hemispherical or ovoid, the bracts loosely imbricate, broad, very scarious, the inner ones with a broad more rigid base. Receptacle convex or small, without scales. Florets all hermaphrodite, tubular, 4- or 5-toothed. Anthers very shortly or not at all tailed. Style-branches truncate. Achenes oblong or obconical, scarcely compressed. Pappus of several chaffy scales entire jagged or divided into bristle-like lobes.—Annual or perennial herbs, usually woolly-tomentose. Leaves alternate, entire. Flower-heads terminal, pedunculate or rarely in small dense cymes.

The genus is limited to Australia. The involucre is much like that of *Podolepis*, in other respects the genus is allied to *Helichrysum*, but differs from both in the scale-like pappus.

Stock tufted with erect 1-headed stems.
  Leaves chiefly radical. Scapes under 6 in. high. Pappus-scales oblong-spathulate, obtuse . . . . 1. *R. leiolepis.*
  Stems leafy, slender, above 6 in. high. Pappus-scales lanceolate, acute, ciliate-plumose . . . . 2. *R. leptorrhynchoides.*
Annuals or perennials, erect, branching and cottony.
  Leaves decurrent. Pappus-scales deeply divided into bristle-like lobes . . . . 3. *R. Brownii.*
  Leaves not decurrent. Pappus-scales undivided.
    Anthers much exserted. Pappus-scales 5 to 7, spathulate, quite entire. Flowers yellow . . . . 4. *R. helichrysoides.*
    Anthers not exserted. Pappus-scales about 10.
      Flowers white. Pappus-scales cuneate obtuse, scarcely denticulate . . . . 5. *R. leucantha.*
      Flowers yellow. Pappus-scales lanceolate, acute, ciliate . 6. *R. Murchisonii.*
Small diffuse annual of 1 to 3 in. Flower-heads numerous and small . . . . 7. *R. Pumilo.*

(See also *Helipterum Pyrethrum* and *H. dimorpholepis*, in which the pappus-bristles are more or less scale-like.)

1. **R. leiolepis,** *F. Muell. in Trans. Vict. Inst.* 1855, 131, *and in Hook. Kew Journ.* viii. 149. Stock densely tufted or shortly branched and woody. Leaves chiefly radical, linear, with revolute margins, $\frac{3}{4}$ to $1\frac{1}{2}$ in. long. Scapes 1-headed, under 6 in. long, with a few small leaves. Involucre broadly hemispherical, $\frac{3}{4}$ to 1 in. diameter, the inner bracts with a dry cuneate base and ovate or ovate-lanceolate scarious lamina, the outer ones short and scarious from the base. Receptacle very convex. Florets very slender. Achenes glandular-papillose. Pappus of 10 to 15 white oblong-spathulate almost stipitate scales, as long as the achenes, minutely ciliolate.

Victoria. Rocks along Snowy River, *F. Mueller.*

2. **R. leptorhynchoides,** *F. Muell. Fragm.* v. 148. Stems from a tufted woody stock erect, leafy but simple and 1-headed, often exceeding

1 ft., usually glabrous, except a little loose wool near the base. Leaves narrow-linear with revolute margins, or the lower ones lanceolate and flat. Involucres broadly hemispherical, ½ to ¾ in. diameter, the bracts not ciliate and their broad claws longer in proportion than in the other species. Achenes papillose. Pappus of 10 to 15 narrow-lanceolate ciliate almost plumose scales.

N. S. Wales. Kingstown, Newcastle, *R. Brown, Leichhardt.*

Victoria. Grassy clumps about Nangeala, *Robertson;* near Station Peak, *Fullager.*

The species is allied in habit to *R. leiolepis*, in the pappus to *R. Murchisonii.*

3. **R. Brownii,** *Benth.* An erect rather slender branching annual of ¾ to 1½ ft., more or less covered with cottony wool. Leaves lanceolate or linear, decurrent on the stem, silky-cottony underneath, becoming glabrous above. Peduncles terminal, long and slender, bearing a single small head. Involucre broadly hemispherical, not above 3 lines diameter, the bracts scarious, the shining straw-coloured tips of the inner ones not exceeding the florets. Achenes papillose. Pappus-scales not longer than the achene, deeply divided into bristle-like lobes.

N. Australia. Islands of the Gulf of Carpentaria, *R. Brown.*

Queensland Endeavour river, *Banks and Solander.*

4. **R. helichrysoides,** *DC. Prod.* vi. 159. A perennial with a hard woody stock and erect branching stems of 1 to 2 ft., cottony-white as well as the foliage. Leaves linear or the lower ones oblong-spathulate, 1 to 2 in. long, or the upper ones smaller, obtuse or with callous points, not decurrent. Flower-heads on terminal peduncles. Involucre broadly hemispherical, in some specimens all under ½ in. diameter, in others larger, the bracts cuneate at the base, with elegantly ciliate scarious transversely-wrinkled laminæ, loose but scarcely radiating. Florets yellow, 5-toothed. Anthers exserted. Achenes glandular-papillose. Pappus of 5 to 7 spathulate scales, about as long as the achene.—*R. auricoma*, F. Muell. in Linnæa, xxv. 408.

N. S. Wales. Wet flats, Molle's Plains, *A. Cunningham;* Macquarrie river, *Mitchell;* from the Lachlan and Darling rivers to the Barrier Range, *Victorian and other Expeditions.*

Victoria. Murray river, *F. Mueller;* Wimmera, *Dallachy.*

S. Australia. Cudnaka and Flinders range, *F. Mueller;* Cooper's Creek and Wills Creek (with narrower and less ciliate involucral bracts), *Howitt's Expedition.*

5. **R. leucantha,** *F. Muell. Fragm.* i. 35. Erect, branching, woolly-tomentose, 1 to 2 ft. high and perhaps annual, although with a hard almost woody base. Leaves lanceolate or the upper ones linear, not decurrent, losing the wool on the upper side. Flower-heads on long peduncles, about the size and shape of those of *R. helichrysoides*, but the involucral bracts not ciliate and the florets white and rather more slender. Achenes glandular. Pappus of about 10 cuneate scales, slightly denticulate at the end.

N. Australia. Mackenzie river, *F. Mueller.*

6. **R. Murchisonii,** *F. Muell. Fragm.* i. 34. Erect and branching, rather slender, slightly woolly-tomentose, the specimens under 1 ft. high. Leaves linear, not decurrent. Flower-heads on long peduncles, smaller than in *R. helichrysoides.* Involucres broadly hemispherical, rarely ½ in. diameter, the laminæ of the bracts ovate, scarcely ciliate. Florets yellow, 5-toothed. Anthers not exserted. Achenes papillose. Pappus of 8 to 12 lanceolate, acute, ciliate scales.

Queensland. Burnett river, *F. Mueller;* Wide Bay, *Bidwill.*

7. **R. Pumilo,** *Benth.* A little-slender branching annual, of ½ to 3 in., glabrous or slightly cottony about the inflorescence. Leaves linear, stem-clasping at the base, the lower ones sometimes opposite. Flower-heads very small, usually several in a dense terminal cyme, each one surrounded by 2 to 4 leafy bracts, rather longer than the head. Involucre hemispherical, 2 to 3 lines diameter, of about 8 to 10 broad very scarious bracts, as in the rest of the genus, but nearly equal, the claws of the inner ones very short. Florets sometimes only 3 or 4 in the lateral heads, usually numerous in the terminal ones, the corolla 4- or rarely 5-toothed. Achenes often curved. Pappus of 6 to 12 obovate or oblong-cuneate scales, obtuse and entire or minutely denticulate.—*Pumilo argyrolepis*, Schlecht. Linnæa, xxi. 448; Sond. in Linnæa, xxv. 487; *Pumilo Preissii*, Sond. l. c.; Hook. f. Fl. Tasm. i. 199, t. 53; *Styloncerus multiflorus*, Nees in Pl. Preiss. ii. 244; *Actinopappus perpusillus*, Hook. f., and *A. Drummondii*, A. Gray in Hook. Kew Journ. iv. 226.

Victoria. Yarra river, *F. Mueller, Harvey.*

Tasmania. Trap rocks, West Head, Tamar river, and Georgetown, *Gunn.*

S. Australia. Near Tonunda, *F. Mueller.*

W. Australia, *Drummond, 5th Coll. n.* 67; Swan River, *Preiss, n.* 127; Kalgan and Vasse rivers, *Oldfield.*

Notwithstanding the dwarf slender habit of this little plant, it appears to me not to be generically separable from *Rutidosis.*

67. QUINETIA, Cass.

Involucre cylindrical, of very few nearly equal narrow bracts. Receptacle small, without scales. Florets very few, hermaphrodite, tubular, 4- (or 5-?) toothed. Anthers shortly or obscurely tailed. Style-branches subulate, almost acute. Achenes cylindrical, not beaked. Pappus of several scales, lanceolate at the base, tapering into a fine awn.—Small slender annual. Leaves alternate, entire. Flower-heads small, terminal or almost axillary.

The genus is limited to a single species, endemic in Australia. The style is somewhat anomalous in *Gnaphalieæ*, but in other respects it appears to be more nearly allied to them than to any other tribe.

1. **Q. Urvillei,** *Cass.; DC. Prod.* vi. 158. A little slender erect annual, of 2 to 4 in., more or less woolly-tomentose. Leaves petiolate, from linear-cuneate to broadly obovate, rarely above ½ in. long. Flower-heads very shortly pedunculate, terminal or appearing axillary from the shortness of the lateral shoots. Involucre about 2 lines long, of 3 or 4 oblong-linear bracts, slightly scarious on the edges. Florets 2 to 4, very slender. Achenes, when ripe, nearly as long as the involucre, silky-hairy at the top, and contracted at the base into a very short hairy stipes. Pappus-scales varying from 3 to 8, the awns as long as the achene itself.

W. Australia. King George's Sound, *D'Urville, Huegel;* Bowes and Blackwood rivers, *Oldfield.*

68. MILLOTIA, Cass.

Involucre cylindrical or almost ovoid, of few nearly equal narrow bracts. Receptacle small, without scales. Florets all hermaphrodite, tubular, 4- or

5-toothed. Anthers with fine, usually ciliate-plumose tails. Style-branches terminating in a very short cone. Achenes cylindrical, contracted into a slender beak. Pappus of fine barbellate or ciliate capillary bristles.—Small annuals. Leaves alternate, linear. Flower-heads small, on terminal peduncles.

The genus is limited to Australia. It differs from *Leptorhynchus* and *Waitzia* in the involucre, from other *Gnaphalieæ* in the slender beak of the achenes.

Pappus as long as the corolla, scarcely barbellate . . . . 1. *M. tenuifolia.*
Pappus shorter than the corolla, the bristles ciliate-plumose with a few long hairs . . . . 2. *M. Greevesii.*

1. **M. tenuifolia,** *Cass.; DC. Prod.* vi. 161. A slender erect annual, of 2 to 6 in. or rather more, simple or branched, hoary or white with close or woolly hairs, or becoming almost glabrous. Leaves narrow-linear, or the lower ones broader and contracted into a long petiole. Involucres varying from under 3 to above 4 lines in length, the bracts linear, herbaceous, hoary, with minute coloured tips. Florets 20 to 30, longer than the involucre. Achenes papillose or almost muricate, the slender beak very variable in length, the pappus of very shortly barbellate capillary bristles.—Steetz in Pl. Preiss. i. 456; Hook. f. Fl. Tasm. i. 209; Sond. in Linnæa, xxv. 503; *Senecio myosotidifolius*, Benth. in Hueg. Enum. 66; *Millotia myositidifolia*, Steetz in Pl. Preiss. i. 457; *M. glabra* and *M. robusta*, Steetz, l. c. 458.

N. S. Wales. Piper's Hill, *Fraser;* Mallee scrub, *Victorian Expedition.*

Victoria. About Melbourne, *Adamson;* Glenelg river and in the Grampians, *F. Mueller;* Wimmera, *Dallachy.*

Tasmania. Port Dalrymple, *R. Brown;* common in dry stony places, *J. D. Hooker.*

S. Australia. Fiedler's Section, *Behr;* from the Murray to St. Vincent's and Spencer's gulfs, *F. Mueller* and others.

W. Australia. King George's Sound, *R. Brown*, and thence to Swan River, *Huegel, Drummond, 5th Coll. n.* 365, *Preiss, n.* 66, 67, 68, 100, and others; Murchison river, *Oldfield;* eastward to the Great Bight, *Maxwell.*

2. **M. Greevesii,** *F. Muell. Fragm.* iii. 18. *t.* 19. A slender erect or diffuse annual, under 6 in. high, more or less woolly-white or at length glabrous, except the inflorescence. Leaves filiform. Involucres about $2\frac{1}{2}$ lines long, the bracts about 8 to 10, linear, with fine points, but densely cottony to the end. Florets usually fewer than in *M. tenuifolia*, longer than the involucre. Achenes slender, with long beaks. Pappus shorter than the corolla, the bristles very fine, ciliate-plumose with fine long hairs.

N. S. Wales. Desert near the Barrier Range, *Beckler.*

69. IXIOLÆNA, Benth.

Involucre campanulate or hemispherical, the bracts numerous, linear or narrow linear-lanceolate, herbaceous or rigid, the inner ones with small scarious or coloured tips, sometimes radiating. Receptacle flat or convex, without scales. Florets all hermaphrodite, tubular, 5-toothed, or rarely a very few outer ones female and filiform. Anthers with long fine tails. Style-branches slender, capitate or truncate at the end. Achenes angular or terete, not beaked. Pappus of fine capillary bristles, entire or shortly barbellate.—

Herbs, either annual or with a perennial base. Leaves alternate, entire. Flower-heads on terminal peduncles.

The genus is limited to Australia. It is nearly allied to *Helichrysum*, differing chiefly in the involucre.

Involucres hemispherical, the bracts very narrow, the inner ones with small scarious tips, not coloured.
 Pappus scarcely barbellate, scarcely half as long as the corolla . . 1. *I. brevicompta.*
 Pappus barbellate, nearly or rarely quite as long as the corolla.
  Leaves mostly lanceolate, acute. Pappus-bristles about 8 to 12 . 2. *I. leptolepis.*
  Leaves oblong-spathulate. Pappus-bristles 20 to 30 . . . . 3. *I. supina.*
  (See also *Athrixia tenella*, which has the female florets few and scarcely larger than the others, but the involucral bracts very fine, with filiform points.)
Involucres campanulate, the bracts linear or linear-lanceolate. Pappus as long as the corolla.
 Involucral bracts woolly, the inner ones with small scarious tips, not coloured . . . . 4. *I. tomentosa.*
 Involucral bracts glandular-pubescent, the inner ones with small white radiating laminæ . . . . 5. *I. viscosa.*
  (See also *Helichrysum podolepideum*, allied to *I. tomentosa*, but with small linear scarious laminæ to all the bracts.)

1. **I. brevicompta,** *F. Muell. Fragm.* i. 53. A branching, erect or decumbent annual, or perhaps perennial, with a hard almost woody base, more or less cottony-white. Leaves linear or lanceolate, the larger ones stem-clasping at the base. Peduncles long, leafless or with a few very small linear leafy bracts. Involucre hemispherical, 4 to 5 lines diameter, the bracts numerous, linear, rather rigid, pubescent, the inner ones with small narrow scarious tips. Florets very numerous, longer than the involucre, all hermaphrodite. Achenes glabrous, angular. Pappus of about 10 bristles, much shorter than the corolla, shortly united at the base, simple or scarcely barbellate.

Queensland. Peak Downs, *F. Mueller;* Narran, Maranoa, and Balonne rivers, *Mitchell.*
N. S. Wales. Head of the Gwydir, *Leichhardt.*

2. **I. leptolepis,** *Benth.* A branching perennial with a hard woody base, with a loose cottony wool, which often disappears by the time of flowering. Leaves linear or lanceolate, very acute, usually rigid and scarcely stem-clasping at the base, flat or with revolute margins. Peduncles longer than in *I. tomentosa,* with fewer smaller leafy bracts. Involucre hemispherical (or, when young, broadly campanulate), with numerous linear rigid bracts, more or less glandular, the inner ones 2 to 2½ lines long, with narrow scarious tips, more prominent than in *I. tomentosa.* Florets very numerous, on a broad flat receptacle, longer than the involucre. Achenes glabrous or very slightly pubescent. Pappus of about 8 to 12 bristles, barbellate from the base, very much shorter than in *I. tomentosa,* although sometimes nearly as long as the corolla.—*Helichrysum leptolepis,* DC. Prod. vi. 194; *I. tomentosa,* var. *glabrata,* Sond. in Linnæa, xxv. 504.

N. Australia. Sturt's Creek, *F. Mueller.*
N. S. Wales. Molle's Plains, *A. Cunningham;* Darling and Murray rivers to the Barrier Range, *Victorian and other Expeditions.*

Victoria. Wimmera, *Dallachy*.
S. Australia. Avoca, Cudnaka, Murray river, *F. Mueller;* in the interior, *M'Douall Stuart's Expedition*.

3. **I. supina,** *F. Muell. Trans. Vict. Inst.* 1855, 37. A low decumbent or divaricately-branched perennial, glabrous or scabrous-pubescent. Leaves mostly oblong-spathulate, rather thick, either all under ½ in. long or the larger ones ¾ in., narrowed into a short petiole. Peduncles not very long, with a few small distant leafy bracts. Involucre as in *I. leptolepis*, broad with narrow rigid scabrous-pubescent bracts. Florets as in that species, but perhaps 1 or 2 of the outer ones may be more slender and female. Pappus of 20 to 30 very fine capillary bristles, shortly barbellate upwards.

S. Australia. Memory Cove, *R. Brown;* Kangaroo Island, *F. Mueller*.
Tasmania. Kent's group, Bass's Straits, *R. Brown*.

4. **I. tomentosa,** *Sond. and Muell. in Linnæa*, xxv. 504. A branching perennial, erect or decumbent at the base, under 1 ft. high, covered with a loose cottony wool which it rarely loses. Leaves mostly lanceolate, mucronulate, rather flaccid, with recurved margins, dilated and stem-clasping at the base, or a few of the lower ones narrowed into a petiole. Peduncles short, leafy. Involucre narrow-campanulate, usually about 4 lines long, the bracts linear or narrowly linear-lanceolate, woolly, the inner ones with small narrow scarious tips, not radiating. Florets numerous, scarcely exceeding the involucre. Achenes angular, rather long, glabrous or slightly hirsute. Pappus of 10 to 20 capillary bristles, barbellate from the base, and as long as the corolla.—*Helichrysum Sonderi*, F. Muell. in Seem. Journ. Bot. iv. 121.

N. S. Wales. Darling river, *Victorian Expedition, Mrs. Ford*.
S. Australia. Port Lincoln, *R. Brown;* Murray Desert, *F. Mueller;* Flinders Range, *Howitt's Expedition;* Head of the Great Bight, *Delisser*.

5. **I. viscosa,** *Benth. in Hueg. Enum.* 66. An erect, much-branched annual, of 1 to 2 ft., with a hard stem, more or less hoary-pubescent with a mixture of glandular hairs. Leaves linear or lanceolate, the lower ones contracted into a long petiole, those under the main branches often stem-clasping or shortly decurrent. Flower-heads solitary, terminating numerous peduncles or leafless branches. Involucre broadly ovoid, 3 to 4 lines long, the bracts narrow-linear, herbaceous, glandular-pubescent, with very narrow scarious margins and small white spreading laminæ, a few of the outer ones shorter, without scarious margins or tips. Florets very numerous, all hermaphrodite. Achenes nearly terete, very shortly hirsute. Pappus of 15 to 20 or more capillary bristles, very shortly and finely barbellate.—DC. Prod. vi. 162; Steetz in Pl. Preiss. i. 458; *Helichrysum asteroides*, DC. Prod. vi. 194.

W. Australia. King George's Sound, and adjoining districts, *R. Brown, A. Cunningham, Huegel, Drummond, n.* 174, *Preiss, n.* 120; eastward to Cape Arid, *Maxwell*.

70. ATHRIXIA, Ker.

(Asteridia, *Lindl.;* Chrysodiscus, *Steetz;* Trichostegia, *Turcz.*)

Involucre broadly campanulate or hemispherical, the bracts numerous, very narrow-linear with subulate tips, or almost filiform. Receptacle flat, without scales. Florets of the circumference in a single row (sometimes very few), female ligulate, irregular, or, if tubular, usually longer than the others; disk-florets numerous, hermaphrodite, tubular, 5-toothed. Anthers with fine tails. Style-branches cylindrical truncate or very obtuse. Pappus of capillary bristles, either minutely barbellate or more frequently plumose in the upper part.—Herbs, either annual or perennial. Leaves alternate, entire. Flower-heads on terminal peduncles, the ray-florets usually white.

Besides the Australian species, which are endemic, there are six from S. Africa, the commonest of which, *A. capensis*, is nearly allied to the Australian *A. australis*. The genus is readily known by its involucre, different from that of all other *Helichryseæ*, although approaching that of *Ixiolæna*. From the latter genus it differs chiefly in the female florets either ligulate or enlarged, as in *Podolepis*.

Florets of the circumference ligulate.
Pappus-bristles barbellate only or denticulate . . . . 1. *A. australis.*
Pappus-bristles elegantly plumose in the upper half.
Involucre ¼ in. diameter, glandular-hirsute . . . . 2. *A. gracilis.*
Involucre ⅓ in. diameter or more, very woolly . . . . 3. *A. multiceps.*
Florets of the circumference few, tubular, irregular or filiform. Pappus-bristles plumose in the upper half.
Stems leafy, cottony-white . . . . 4. *A. stricta.*
Stems very short, pubescent with long slender leafless peduncles . . 5. *A. tenella.*

1. **A. australis,** *Steetz in Pl. Preiss.* i. 482. An erect corymbosely branched annual, attaining 1 ft. or rather more, sprinkled with white septate hairs. Leaves linear or linear-lanceolate, stem-clasping at the base, mostly above 1 in. long. Flower-heads rather large, few, in a loose terminal corymb. Involucre about 4 lines long, glandular-pubescent or hirsute, the bracts very numerous but in few rows, ending in hair-like glabrous points, few of the innermost broader and ciliate at the end. Ray-florets 12 to 20, with white spreading 3-toothed ligulæ of 4 to 5 lines. Achenes short, glabrous or slightly glandular-papillose. Pappus of about 10 to 12 bristles, as long as the corolla, barbellate but scarcely plumose.—*Asteridia pulverulenta*, Lindl. Swan Riv. App. 24.

W. Australia. Princess Royal Harbour, *R. Brown;* Swan River, *Drummond*, 1*st Coll.*, *Fraser*, *Preiss*, *n.* 65; Geographe Bay and Blackwood river, *Oldfield;* inland from Cape Le Grand, *Maxwell.*—The species is nearly allied to the S. African *A. capensis.*

2. **A. gracilis,** *Benth.* A slender annual, slightly woolly-pubescent or nearly glabrous. Leaves linear or linear-lanceolate. Flower-heads much smaller than in the other radiate species. Involucre broadly ovoid rather than hemispherical, glandular-hispid, about 3 lines long, the bracts linear, the outer ones finer and shorter, the innermost with tufts of fine hairs at the end. Ray-florets not numerous, with ligulæ of about 3 lines. Pappus of about 3 to 6 bristles, elegantly plumose in the upper half.—*Asteridia gracilis*, A. Gray in Hook. Kew Journ. iv. 274.

W. Australia, *Drummond.*

3. **A. multiceps** *Benth.* Stems numerous, 3 to 6 in. high, rather stout for their size, with scaly hairs mixed with a white cottony wool. Radical leaves oblong-spathulate, the others linear-oblong or lanceolate, obtuse, mostly above 1 in. long. Flower-heads rather large, few but corymbose. Involucre

hemispherical, $\frac{1}{2}$ to $\frac{3}{4}$ in. diameter, the bracts all subulate and woolly-ciliate. Ray-florets numerous, with ligulæ of 3 to 6 lines. Achenes longer than in *A. australis.* Pappus of 8 to 10 bristles, elegantly plumose in the upper part.—*Asteridia multiceps*, A. Gray in Hook. Kew Journ. iv. 274; *Trichostegia asteroides*, Turcz. in Bull. Mosc. 1851, ii. 81.

W. Australia, *Drummond, 5th Coll. Suppl. n.* 66.

Var. *tenella.* More slender and less branched. Ray-florets narrow and more deeply 3-toothed.—Galesbrook and banks of Salt river, *Maxwell.*

4. **A. stricta,** *Benth.* In the original form apparently annual or biennial with rigid, erect, simple or sparingly branched stems of $\frac{3}{4}$ to $1\frac{1}{2}$ ft., very white with a close cottony wool. Radical leaves petiolate and oblong; stem leaves few, small, narrow-linear, becoming glabrous on the upper side. Flower-heads terminal. Involucre hemispherical, $\frac{1}{2}$ to $\frac{3}{4}$ in. diameter, the bracts very numerous, narrow-linear, woolly ciliate, the outer ones glandular, the inner ones with small coloured tips. Florets of the circumference female, not numerous, scarcely longer than those of the disk, with an oblique limb, but without any distinct ligula. Disk-florets very numerous. Pappus of 3 to 6 white bristles, elegantly plumose towards the end.—*Asteridia stricta*, A. Gray in Hook. Kew Journ. iv. 275; *Chrysodiscus niveus*, Steetz in Pl. Preiss. i. 460.

W. Australia. King George's Sound and adjoining districts, *Drummond. n.* 32, 90, *Preiss, n.* 69, *Roe, Oldfield*, and eastward to Lucky Bay, *Maxwell.*

Var. *suffruticosa.* A decumbent or divaricately branched undershrub, with the white cottony branches of the original form, as well as the other characters, except that the involucres are smaller and less woolly.—Lucky Bay, *R. Brown;* Esperance Bay and far to the eastward, *Maxwell.*

5. **A. tenella,** *Benth.* A small branching annual, scarcely 2 in. without the peduncles, more or less woolly or at length nearly glabrous. Leaves linear or lanceolate. Peduncles filiform, rigid, glabrous and leafless, often 3 to 4 in. long. Involucres hemispherical, 3 to 4 lines diameter, the bracts linear, almost filiform, plumose-ciliate, the inner ones ciliate towards the end only, with small glabrous coloured tips. Florets very numerous, longer than the involucre, a few of the outer ones female, but not ligulate nor longer than the others. Achenes narrow, contracted under the pappus. Pappus of 3 or 4 bristles, shorter than the corolla and plumose towards the end.—*Panætia athrixioides*, Sond. and Muell. in Linnæa, xxv. 506.

Victoria. Wimmera, *Dallachy.*
S. Australia. Murray river near Mooruudee, Crystal Brook, and Lake Lalbert, *F. Mueller;* Port Lincoln, *Wilhelmi;* Venus and Streaky bays, *Babbage.*
W. Australia, *Roe;* Murchison river, *Oldfield.*

71. PODOTHECA, Cass.

(Podospermum, *Labill.*; Phænopoda, *Cass.;* Lophoclinium, *Endl.*)

Involucre cylindrical conical or campanulate, the bracts imbricate, herbaceous, often very thin but not scarious, except the innermost linear ones. Receptacle without scales, but retaining the persistent stalks of the achenes. Florets all hermaphrodite, tubular, 5-toothed. Anthers with very fine tails, sometimes scarcely perceptible. Style-branches filiform, truncate or capitate at the end. Achenes angular or terete, very shortly stipitate. Pappus of rigid sometimes almost scale-like bristles, more or less ciliate or plumose and often united at the base.—Erect annuals, glabrous or scabrous-pubescent, not woolly or rarely the involucre very slightly so. Leaves alternate, entire. Flower-heads rather large, sometimes very long, on terminal peduncles, usually dilated under the involucre.

The genus is limited to Australia, differing from *Helipterum* chiefly in the stipitate achenes, and generally in the involucre, which however is less foliaceous in *P. fuscescens* than in the other species.

Involucre narrow, cylindrical or conical.
  Involucre 1 to $1\frac{1}{2}$ in. long.
    Florets considerably longer than the involucre . . . . . 1. *P. gnaphalioides.*
    Florets scarcely exceeding the very narrow involucre . . . 2. *P. angustifolia.*
  Involucre about $\frac{1}{2}$ in. long . . . . 3. *P. pygmæa.*
Involucre turbinate-campanulate, rather broad.
  Pappus-bristles barbellate or shortly and regularly plumose, much longer than the achenes . . . . 4. *P. chrysantha.*
  Pappus-bristles very plumose at the base, scarcely longer than the densely hairy achenes . . . . 5. *P. fuscescens.*

1. **P. gnaphalioides,** *Grah. in Bot. Mag. t.* 3920. An annual of 1 to $1\frac{1}{2}$ ft., with few erect branches, glabrous or sprinkled with a few short hairs. Leaves linear or lanceolate, the lower ones narrowed below the middle, all stem-clasping and sometimes shortly decurrent. Peduncles long, dilated and hollow under the head. Involucre cylindrical or narrow-conical, 1 to $1\frac{1}{2}$ in. long, the bracts broadly lanceolate and herbaceous, the innermost narrow-linear, paleaceous and very acute. Florets very slender, considerably longer than the involucre and pappus. Achenes silky-hairy. Pappus of 4 to 6, usually 5, flattened plumose bristles, united in a cylinder at the base.—Steetz in Pl. Preiss. i. 449; *Lophoclinium citrinum*, Endl. in Bot. Zeit. 1843, 457.

W. Australia, *Drummond, 5th Coll. Suppl. n.* 63; Swan River, *Collie;* Lake Mathilda, *Preiss, n.* 107; Kalgan and Murchison rivers, *Oldfield;* Salt river, *Maxwell.*

2. **P. angustifolia,** *Cass.; DC. Prod.* vi. 159. An erect scabrous-pubescent or nearly glabrous annual, usually branching from the base, rarely above 1 ft. high and often under 6 in. Leaves narrow-linear. Peduncles short, dilated and hollow under the head. Involucre narrow-cylindrical or slightly conical, 1 to $1\frac{1}{4}$ in. long, the bracts herbaceous, linear or linear-lanceolate, acuminate, the innermost linear and scarious. Florets very slightly exceeding the involucre, filiform and much more slender than in *P. gnaphalioides.* Achenes and pappus as in that species, but more slender.—Steetz in Pl. Preiss. i. 448; *Podosperma angustifolia*, Labill. Pl. Nov. Holl. ii. 35. t. 177; *Phænopoda angustifolia*, Cass. (DC.); *Lophoclinium Manglesii*, Endl. in Bot. Zeit. 1843, 457.

Victoria. Near Melbourne, *Harvey;* Wimmera, *Dallachy;* Portland, *F. Mueller.*

S. Australia. Murray river and Encounter Bay, *F. Mueller;* near Adelaide, *Whittaker.*

W. Australia. King George's Sound, *R. Brown* and others; Rottenest Island, *Preiss, n.* 106; also *Drummond, n.* 48, 49, *5th Coll. n.* 319, 320 (the latter specimens only 2 or 3 in. high, including the involucre).

3. **P. pygmæa,** *A. Gray in Hook. Kew Journ.* iv. 227. A little annual, not 2 in. high in our specimens. Leaves linear, glabrous. Involucre cylindrical or conical, but only about ½ in. long and nearly 3 lines diameter at the base, the outer bracts broadly lanceolate and not so acute as in the preceding species, thus connecting them with the following ones. Florets slender, scarcely exceeding the involucre. Pappus-bristles few, plumose, free or slightly united at the base.

W. Australia, *Drummond, 5th Coll. Suppl. n.* 64.

4. **P. chrysantha,** *Benth.* An erect annual, simple or slightly branched, glabrous or scabrous-pubescent, rarely above 1 ft. high. Leaves linear. Peduncles rather long, dilated and hollow under the head. Involucre broadly turbinate-campanulate, 6 to 8 lines long, the bracts thinly herbaceous, broadly lanceolate or ovate-lanceolate, not numerous, and only 2 or 3 outer ones shorter than the others, the innermost linear and scarious. Florets numerous, exceeding the involucre. Achenes hairy. Pappus of 8 to 10 barbellate, but scarcely plumose bristles.—*Ixiolæna chrysantha*, Steetz in Pl. Preiss. i. 459.

W. Australia. Swan River, *Drummond, 1st Coll., Preiss, n.* 105, *Oldfield;* Murchison river, *Oldfield, Drummond, 6th Coll. n.* 148.

5. **P. fuscescens,** *Benth.* A slender annual, our specimens not above 6. in. high and glabrous or nearly so. Leaves small, the lower ones oblong or lanceolate and opposite, the upper ones alternate lanceolate and broader, passing into a few leafy bracts upon the somewhat dilated peduncle close under the flower-head. Involucre campanulate, about ½ in. diameter, the bracts not numerous, of nearly equal length, the outer ones herbaceous though thin, the inner more scarious, with small scarious tips. Florets slender, all hermaphrodite, but the inner ones often sterile, their stalks rather more prominent than those of the outer florets, and irregularly cohering so as to give the receptacle almost a conical form. Achenes, when perfect, densely villous. Pappus of 10 to 15 rigid bristles, plumose especially at the base, less so towards the end.—*Helipterum fuscescens*, Turcz. in Bull. Mosc. 1851, ii. 80; *Acroclinium phyllocephalum*, A. Gray in Hook. Kew Journ. iv. 271.

W. Australia, *Drummond, 5th Coll. Suppl. n.* 64.

Lophoclinium album, Endl. in Bot. Zeit. 1843, 457, may very probably be this species.

72. PODOLEPIS, Labill.

(Scalia, *Sims;* Panætia, *Cass.;* Scaliopsis, *Walp.;* Siemssenia, *Steetz;* Stylolepis, *Lehm.;* Rutidochlamys, *Sond.*)

Involucre hemispherical or rarely ovoid, the bracts imbricate, in several rows, with very thin scarious laminæ, not radiating, the outer ones sessile, the inner ones on rigid or scarious stalks or claws. Receptacle flat, without scales. Florets of the circumference few or numerous, in a single row, female, either ligulate or irregular or with fewer lobes than the disk-florets, which are hermaphrodite, tubular, with 5 or rarely 4 narrow lobes, sometimes slightly irregular. Anthers with fine tails, sometimes very short. Style-branches filiform, truncate or capitate. Achenes nearly terete or slightly compressed, not beaked, glabrous or papillose. Pappus of capillary bristles, simple or shortly barbellate, often slightly united at the base.—Annual or perennial herbs, the cottony wool usually very deciduous, leaving the stem and leaves glabrous and smooth. Leaves alternate, lanceolate or linear, very frequently stem-clasping. Flower-heads terminal, pedunculate or rarely sessile, the ray-florets yellow pink or purple, the scarious laminæ of the involucral bracts imbricate so as to conceal the claws, in all except *P. longipedata*.

The genus is limited to Australia. It is allied to *Athrixia* through those species where the female florets are ligulate, but differs in the involucre; where the female florets are less developed it passes almost into *Helichrysum*, differing chiefly in the very thinly scarious, not opaque or petal-like laminæ of the involucral bracts. A nearly similar involucre occurs in some species of *Helipterum*, but with a plumose pappus.

Involucres ovoid, almost sessile in clusters . . . . 1. *P. rutidochlamys.*
Involucres hemispherical, pedunculate or rarely here and there almost sessile, ½ to 1½ in. diameter (or only 4 lines in *P. Lessoni*).
Laminæ of the intermediate involucral bracts much shorter than their claws, obtuse or almost acute . . . . 2. *P. longipedata.*
Laminæ of the intermediate involucral bracts as long as or longer than their claws and concealing them.
Laminæ very acute or acuminate. Ray-florets yellow, 3- or 4-lobed.
Perennial. Flower-heads large. Eastern species . . . 3. *P. acuminata.*
Annuals. Involucres under 1 in. diameter.
Eastern species. Involucral bracts acute, smooth or slightly rugose . . . . 4. *P. canescens.*
Western species. Involucral bracts very acuminate, smooth . . . . 5. *P. aristata.*
Western species. Involucral bracts acute, very rugose . 6. *P. pallida.*
Laminæ obtuse or scarcely acute. Ray-florets purple, ligulate, entire.
Claws of the involucral bracts glabrous. Pappus usually thickened upwards. . . . . 7. *P. nutans.*
Claws of the involucral bracts glandular. Pappus not thickened upwards . . . . 8. *P. gracilis.*
Laminæ of the involucral bracts obtuse. Ray-florets 3- or 4-lobed.
Ray-florets yellow. Perennial. Involucre 1 to 1½ in. diameter; bracts very rugose . . . . 9. *P. rugata.*
Ray-florets pink, small. Annual. Involucre under ½ in. diameter; bracts smooth . . . . 10. *P. Lessoni.*
Involucres not 3 lines diameter, turbinate or at length hemispherical. Slender annuals. Ray-florets irregular, not much longer than those of the disk.
Involucres turbinate, rather narrow . . . . 11. *P. Siemssenia.*
Involucres at length hemispherical . . . . 12. *P. microcephala.*

1. **P. rutidochlamys,** *F. Muell. Fragm.* iv. 79. An erect branching annual of 2 or 3 ft., with more or less of a loose deciduous wool. Leaves lanceolate or ovate-lanceolate, stem-clasping with rounded auricles and often slightly decurrent. Flower-heads nearly sessile and more or less clustered at the ends of the slender branches of a loose panicle. Involucre ovoid, 4 to 5 lines long, the laminæ of the bracts imbricate, lanceolate, acuminate, transversely rugose, decurrent along the claws. Female florets very few, not longer than the others, but more slender and only 3-lobed. Pappus-bristles

rather numerous, scabrous or minutely barbellate.—*Rutidosis arachnoidea*, Hook. in Mitch. Trop. Austr. 341; *Rutidochlamys Mitchelli*, Sond. in Linnæa, xxv. 497.

Queensland. Shoalwater Bay, *R. Brown;* Upper Burdekin and Burnett rivers, *F. Mueller;* Port Curtis, *M'Gillivray;* Emu Creek, *Thozet;* near Mount Pluto, *Mitchell.*
N. S. Wales. Liverpool Plains, *C. Moore;* Darling river, *Neilson*; Murray river, *F. Mueller;* also in *Leichhardt's Collection.*

2. **P. longipedata,** *A. Cunn. in DC. Prod.* vi. 163. Either annual or with a perennial stock and annual stem of 1 to 2 ft., simple or divaricately-branched and retaining very little cottony wool. Leaves linear or lanceolate or the radical ones oblong-spathulate. Flower-heads smaller than in *P. acuminata*, pedunculate or rarely 3 or 4 together almost sessile. Involucres hemispherical, about ¾ in. or rarely nearly 1 in. diameter, the scarious laminæ of the bracts smaller than in all the allied species, ovate, obtuse or acute, not rugose, those of the intermediate bracts much shorter than their rather broad claws, a few of the outer bracts entirely scarious. Ray-florets as in *P. acuminata*, yellow, longer than the others, ligulate or irregular, 3- or 4-lobed. Pappus-bristles numerous, scarcely barbellate.—*P. Mitchelli*, Sond. in Linnæa, xxv. 508; *P. hieracioides*, F. Muell. Fragm. i. 112; *Scaleopsis Lucæana*, Walp. in Linnæa, xiv. 318.

N. Australia. Mackenzie river, *F. Mueller.*
Queensland. Shoalwater Bay and Broad Sound, *R. Brown;* sandy shores of Moreton Bay, *A. Cunningham;* Maranoa river, *Mitchell.*
N. S. Wales. Hunter's River, *R. Brown;* near Bathurst, *A. Cunningham;* also in *Leichhardt's Collection.*
Victoria. Grassy valleys, Delatite, Mitta-Mitta, Macalister, Omeo, Snowy rivers, *F. Mueller;* Maneroo, *Lhotzky.*

The Northern specimens generally have an annual appearance, with smaller flower-heads, than the Southern ones, which have sometimes several stems from a hard stock, but Brown's specimens quite connect the two forms.

3. **P. acuminata,** *R. Br. in Ait. Hort. Kew. ed.* 2. v. 82. Stems from a perennial stock erect, 1 to 2 ft. high, simple or corymbosely branched, glabrous or with a few scattered short hairs and sometimes a little deciduous wool about the base. Radical and lower leaves petiolate, oblong or lanceolate, often several in. long, the upper ones few, lanceolate or linear, stem-clasping or rarely slightly decurrent. Flower-heads large, yellow, pedunculate. Involucre hemispherical, 1 to 1½ in. diameter, the scarious laminæ of the bracts acute or acuminate, smooth and not rugose, the inner ones on long linear claws. Florets all longer than the involucre, those of the ray longer than the others, the limb ligulate or irregularly tubular, deeply 3- to 5-lobed. Pappus-bristles numerous, white, shortly barbellate, united at the base.—DC. Prod. vi. 162; Hook. f. Fl. Tasm. i. 209; *Scalia jaceoides*, Sims, Bot. Mag. t. 956.

Queensland. Burnet river and Peak Downs, *F. Mueller;* Rockhampton, *Dallachy;* Maranoa river and Mantuan Downs, *Mitchell.*
N. S. Wales. Port Jackson, *R. Brown;* New England, *C. Stuart;* Hastings river, *Beckler;* Richmond river, *Fawcett;* Nangas, *Backhouse;* Darling river, *Victorian Expedition.*
Victoria. Yarra river, Avoca river, Cobberas mountains, *F. Mueller;* Creswick,

Whan; Wimmera, *Dallachy;* Haidinger range at an elevation of 5000 feet (the laminæ of the involucral bracts almost obtuse), *F. Mueller.*

Tasmania. Port Dalrymple, *R. Brown;* abundant in many parts of the colony, ascending to 4000 feet, *J. D. Hooker.*

S. Australia. Holdfast Bay, Mount Gambier, *F. Mueller.*

The Northern specimens have usually smaller flower-heads than the Southern ones.

4. **P. canescens,** *A. Cunn. in DC. Prod.* vi. 163. An annual with erect or ascending branching stems, rarely much exceeding 1 ft. Leaves linear or lanceolate, chiefly at the base of the stem, 1 to 2 in. long or even more, stem-clasping and sometimes shortly decurrent, the upper ones small and distant. Flower-heads rather small, on slender peduncles. Involucre hemispherical, 6 to 8 lines diameter, the scarious laminæ of the bracts acute or acuminate, smooth or slightly rugose, their claws with broad scarious margins. Florets yellow, the outer ones slightly exceeding the others, irregularly 3- or 4-lobed, ligulate or almost 2-lipped.—*P. inundata,* A. Cunn. in DC. Prod. vi. 163; *P. affinis,* Sond. in Linnæa, xxv. 507.

N. S. Wales. Inundated banks of Lachlan river and exposed rocky situations near Croker's Range, *A. Cunningham;* Darling and Murray rivers and Mount Goningberi, *Victorian Expedition.*

Victoria. Wimmera, *Dallachy.*

S. Australia. From the Murray to St. Vincent's and Spencer's Gulfs, *F. Mueller, Wilhelmi,* and others.

5. **P. aristata,** *Benth. in Hueg. Enum.* 64. An erect annual, often exceeding 1 ft., glabrous or slightly woolly. Leaves linear or lanceolate, stem-clasping and often decurrent. Flower-heads usually corymbose, the peduncles rather short. Involucre hemispherical, ½ to ¾ in. diameter, the bracts numerous, their laminæ smooth, acute, and ending usually in a rigid point or awn, the claws of the inner ones narrow and glandular. Ray-florets yellow, longer than the others, irregularly tubular, 3- or 4-lobed. Pappus-bristles scarcely cohering at the base.—DC. Prod. vi. 163; Steetz in Pl. Preiss. i. 466; *P. chrysantha,* Endl. in Bot. Zeit. 1843, 458; Rev. Hortic. 1857, 263, with a figure.

W. Australia. Swan River, *Fraser, Huegel, Drummond,* 1*st Coll.;* Blackwood river, *Oldfield;* between Moore and Murchison rivers, *Drummond,* 6*th Coll. n.* 154; also 5*th Coll. n.* 328.

Var. *minor.* Flower-heads smaller in all their parts. Ray-florets shortly toothed.—*P. subulata,* Steetz in Pl. Preiss. i. 465.—Vasse river, *Preiss, n.* 54; between Moore and Murchison rivers, *Drummond,* 6*th Coll. n.* 155.

6. **P. pallida,** *Turcz. in Bull. Mosc.* 1851, ii. 78. An erect annual, more or less woolly-pubescent, evidently very nearly allied to *P. aristata,* with the same linear or lanceolate leaves and inflorescence. Flower-heads rather larger, the laminæ of the involucral bracts large, ovate-lanceolate, acute or acuminate, and all very conspicuously transversely wrinkled as in *P. rugata.* Ray-florets very numerous (yellow?), with 3 or 4 narrow lobes.

W. Australia, *Drummond,* 5*th Coll. n.* 387.

7. **P. nutans,** *Steetz in Pl. Preiss.* i. 464. An erect paniculately branched annual, usually under 1 ft., closely resembling *P. gracilis.* It

retains more of the cottony wool, the flower-heads are rather smaller, the laminæ of the involucral bracts larger in proportion and the claws of the inner ones quite smooth, not glandular. Ray-florets as in *P. gracilis* ligulate, entire, and apparently purple or pink. Pappus-bristles usually, but not always, slightly thickened upwards.

W. Australia. King George's Sound, *R. Brown, A. Cunningham;* and thence to Swan River, *Drummond, Preiss, n.* 58.

I do not observe anything in the specimens likely to have given rise to Steetz's name. *P. rosea,* Steetz in Pl. Preiss. i. 463, from Preiss's specimens from Swan River, n. 53 and 61, appears to me to be the same species, which should probably be included as a variety in *R. gracilis.*

8. **P. gracilis,** *Grah. in Edinb. New Phil. Journ.* v. 379. An erect, glabrous, paniculately-branched annual, often exceeding 1 ft. Leaves linear or lanceolate, stem-clasping and often shortly decurrent. Peduncles usually rather long. Involucres hemispherical, when fully out ½ to ¾ in. diameter, the bracts numerous, their laminæ broad, obtuse, smooth, the lower sessile ones often descending on the peduncle, the claws of the inner ones narrow, glandular. Ray-florets numerous, ligulate, rather long, entire and truncate, purple or lilac. Pappus of about 15 to 20 capillary bristles, not thickened upwards, quite free or scarcely cohering at the base.—DC. Prod. vi. 163; Bot. Mag. t. 2904; Sweet, Brit. Fl. Gard. t. 285; Steetz in Pl. Preiss. i. 463; *Stylolepis gracilis,* Lehm. in Linnæa, v. 385.

W. Australia. Swan River, *Preiss, n.* 51, 56; *Drummond,* 5*th Coll. n.* 327; Tone river, *Maxwell;* Murchison river, *Oldfield.*

P. filiformis, Steetz in Pl. Preiss. i. 465, from Woodman's Point, *Preiss, n.* 57, appears to be a slender starved state of *P. gracilis.* *P. auriculata,* DC. Prod. vi. 162, from Sharks' Bay, *Gaudichaud,* which I have not seen, may be a variety of *P. gracilis,* with transversely rugose involucral bracts, or perhaps the same as *P. pallida.*

9. **P. rugata,** *Labill. Pl. Nov. Holl.* ii. 57. *t.* 208. A rather stout perennial, with the habit of *P. acuminata,* but usually retaining more of a white loose or close cottony wool. Leaves oblong, lanceolate or linear, the lower ones petiolate and several inches long, the upper ones small narrow and stem-clasping, degenerating into scarious scales on the peduncles. Flower-heads at least as large as in *P. acuminata,* and the florets as in that species yellow, those of the ray larger, ligulate or irregular, 3- or 4-lobed. Involucres 1 to 1½ in. diameter, the large scarious laminæ of the bracts obtuse and always transversely wrinkled.—DC. Prod. vi. 162; Steetz in Pl. Preiss. i. 462; Regel, Gartenfl. t. 320.

Victoria. Wimmera, *Dallachy;* mouth of the Glenelg river, *F. Mueller, Allitt.*

S. Australia. Grassy plains, etc., around St. Vincent's and Spencer's gulfs, *F. Mueller, Whittaker,* and other.

W. Australia. King George's Sound and neighbouring districts, *R. Brown, Labillardière, Drummond,* 4*th Coll. n.* 179, 5*th Coll. n.* 388, *Preiss, n.* 50, and others; eastward to Cape Arid, *Maxwell.*

10. **P. Lessoni,** *Benth.* A slender erect branching annual of ½ to 1 ft., loosely woolly or at length glabrous. Leaves small, from ovate to lanceolate, stem-clasping and sometimes slightly decurrent. Peduncles filiform. Involucres hemispherical, about 4 lines diameter, the very numerous imbricate thinly scarious bracts broadly ovate, obtuse or almost acute, ciliate, the claws

of the inner ones linear, not at all or scarcely glandular. Florets all nearly equal and longer than the involucre with narrow lobes, the outer female ones more slender, irregularly 3- or 4-lobed. Pappus of few capillary bristles, barbellate towards the end, usually reduced in the achenes of the ray to a single bristle or entirely wanting.—*Panætia Lessonii*, Cass.; DC. Prod. vi. 162; Steetz in Pl. Preiss. i. 461; *Panætia Muelleri*, Sond. in Linnæa, xxv. 505; *Podolepis Gilberti*, Turcz. in Bull. Mosc. 1851, i. 195.

N. S. Wales. Darling river, *Herb. F. Mueller.*

S. Australia. Flinders Range, St. Vincent's Gulf, *F. Mueller.*

W. Australia. King George's Sound, *Menzies, Preiss, n.* 59, and thence to Swan River, *Drummond,* 1*st Coll. and n.* 329, 330; 5*th Coll. n.* 386, *Oldfield,* and others; Champion Bay, *Oldfield.*

11. **P. Siemssenia,** *F. Muell. Herb.* A slender glabrous much-branched annual of ½ to 1 ft. Leaves linear. Flower-heads small, on filiform peduncles. Involucre turbinate, about 3 lines long, the scarious laminæ of the bracts oblong, imbricate, the outer ones short. Ray-florets 3- or 4-lobed, ligulate or irregular, exceeding the longest involucral bracts; disk-florets with 5 narrow lobes but often slit on one side nearly to the base. Pappus-bristles not numerous, exceedingly fine, not perceptibly barbellate, shorter than the corolla.—*Siemssenia capillaris,* Steetz in Pl. Preiss. i. 467.

N. S. Wales. Darling river, *Victorian Expedition*; Stokes's Range to Cooper's Creek, *Wheeler;* Murrumbidgee and Murray rivers, *F. Mueller.*

Victoria. Wimmera, *Dallachy.*

S. Australia. Cudnaka, Spencer's Gulf, *F. Mueller.*

W. Australia, *Preiss, n.* 72; *Drummond, n.* 107; Port Gregory, *Oldfield.*

12. **P. microcephala,** *Benth.* Slender, erect, much-branched and perfectly glabrous, the stem hard but probably annual, our specimens not above 6 in. high. Leaves narrow-linear, obtuse. Peduncles slender, with a few very short scale-like but not scarious bracts. Involucres broadly turbinate, at length hemispherical, scarcely exceeding 2 lines diameter, the bracts less scarious than in the other species, obtuse, glandular, the inner ones with small scarious laminæ. Florets shortly exceeding the involucre, all with 5 narrow lobes, those of the circumference more deeply and irregularly slit than the inner ones and often rather larger. Pappus-bristles few, very fine, shortly barbellate.

W. Australia. Sharks' Bay, *Milne.*

73. LEPTORHYNCHUS, Less.

(Rhytidanthe, *Benth.*)

Involucre broadly turbinate, campanulate or hemispherical, the bracts much imbricate in several rows, the short outer ones and the tips or laminæ of the others very thinly scarious and not spreading. Receptacle flat, without scales. Florets all tubular, a few in the circumference usually female, more slender but not longer than the others, 3- or 4-toothed; disk-florets hermaphrodite, 5-toothed. Anthers with fine tails. Style-branches nearly terete, truncate. Achenes small or narrow, somewhat compressed, glabrous or papillose, contracted at the top or produced into a short beak. Pappus of several capillary bristles, scabrous, shortly barbellate or almost plumose towards the end.—Annual or perennial herbs or undershrubs, more or less cottony or glandular-pubescent. Leaves alternate, entire. Flower-heads pedunculate, the outer scarious involucral bracts often descending along the peduncle. Florets almost always longer than the involucre.

The genus is limited to Australia. It is scarcely distinct from *Helichrysum*, differing in the involucral bracts with more thinly scarious tips, neither spreading nor petal-like nor opaque, and in the contraction of the achene at the top, which can in some species be only seen at maturity, and then not always very decided. From *Waitzia*, *Leptorhynchus* differs in the involucre and usually in the shorter beak to the achene. The florets, as in *Ixiolæna* and *Podolepis*, are usually longer in proportion to the involucre than in *Helichrysum* or *Helipterum*.

Achenes contracted at the top, but not distinctly beaked. Perennials or undershrubs.	
Pappus-bristles equally and minutely denticulate.	
Outer involucral bracts very thinly scarious, passing into the scarious scales of the peduncle	1. *L. squamatus.*
Outer involucral bracts with thinly scarious woolly-ciliate margins and dark centres	2. *L. panætioides.*
Pappus-bristles more barbellate or almost plumose towards the end.	
Outer involucral bracts with dark tips and woolly scarious margins, passing into numerous scales on the peduncle	3. *L. tenuifolius.*
Involucral bracts wholly scarious, with woolly-ciliate margins, abruptly distinct from the few leafy bracts on the peduncle . .	4. *L. ambiguus.*
Achenes more distinctly though sometimes shortly beaked. Annual, except *L. linearis.*	
Involucral bracts narrow, acute.	
Involucral bracts small, very numerous, all ciliate with long hairs. Flower-heads 2 to 3 lines long	5. *L. pulchellus.*
Outer involucral bracts very thin and transparent, inner ones with rigid glandular centres. Flower-heads 5 to 6 lines long .	6. *L. elongatus.*
Involucral bracts broad, obtuse.	
Annual. Involucral bracts ciliate	7. *L. Waitzia.*
Rhizome creeping. Involucral bracts not ciliate, the scarious tips broad and almost spreading	8. *L. linearis.*

1. **L. squamatus,** *Less. Syn. Comp.* 273. A perennial with decumbent or ascending stems from 6 in. to 1 foot long, with a little deciduous cottony wool on the young parts and the under side of the leaves. Leaves lanceolate or linear, mucronate-acute, narrowed at the base, flat or with recurved margins, glabrous above, the longest above 1 in. long, the lowest much shorter. Peduncles long with distant scarious scales passing into the involucral bracts. Involucre broadly turbinate or almost hemispherical, not ½ in. diameter, the bracts imbricate in numerous rows, oblong-lanceolate, scarious with woolly-ciliate margins and small coloured but not spreading glabrous tips, the innermost narrow and 3 lines long, the others gradually shorter. Florets longer than the involucre, a very few of the outer ones more slender, female 3- or 4-toothed. Achenes very shortly contracted at the top. Pappus-bristles cohering in a ring at the base, shortly barbellate, those of the female florets few.—DC. Prod. vi. 160; Hook. f. Fl. Tasm. i. 208; F. Muell. Fragm. i. 52; *Chrysocoma squamata*, Labill. Pl. Nov. Holl. ii. 40. t. 184; *Helichrysum dubium*, Cass. (DC.); *Leptorhynchus hemisphæricus*, DC. Prod. vi. 160; *L. gracilis* and *L. Lhotzkyanus*, Walp. in Linnæa, xiv. 317, and probably also *L. nitidulus*, DC. Prod. vi. 160.

N. S. Wales. Port Jackson to the Blue Mountains, *R. Brown* and others; Bathurst plains, *Fraser;* New England, *C. Stuart.*

Victoria. From the Glenelg to Gipps' Land, very common, *Robertson, F. Mueller*, and others; Wimmera, *Dallachy.*

Tasmania. Abundant throughout the colony, *J. D. Hooker.*

S. Australia. Port Adelaide, *Blandowski;* Kensington, *F. Mueller.*

2. **L. panætioides,** *Benth.* Erect, woolly-tomentose, attaining 1 ft. or even more, with a hard base and sometimes evidently suffruticose. Leaves linear, acute but soft and cottony. Peduncles long, slender, leafless, except a few minute linear bracts. Involucre hemispherical, 3 to 4 lines diameter, the bracts imbricate in many rows, narrow with scarious woolly-ciliate margins and very small not spreading glabrous tips. Florets longer than the involucre, a few of the outer ones female and 3-lobed with exserted styles. Achenes shortly contracted at the top. Pappus-bristles slightly barbellate. *Helichrysum panætioides*, DC. Prod. vi. 194.

N. S. Wales. Lachlan river and foot of Mount Aiton, *Fraser, A. Cunningham;* Murray river, *F. Mueller;* also in *Leichhardt's* collection.

This has something of the habit but not the involucre of *Ixiolæna leptolepis.* It is evidently nearly allied to *Leptorrhynchus squamatus*, but more erect and cottony, the peduncles without scarious scales, and the involucres more hemispherical.

3. **L. tenuifolius,** *F. Muell. Fragm.* i. 52. Stock perennial, tufted and more or less woolly, the stems slender, erect, branching, with a little cottony wool about the axils and under side of the leaves. Leaves narrow-linear or subulate, acute, with revolute margins. Peduncles long and filiform. Flower-heads smaller than in *L. squamatus.* Involucre turbinate or at length hemispherical, about 3 lines diameter, the bracts in numerous rows, linear, woolly-ciliate, with very narrow coloured tips. Florets all longer than the involucre, a few of the outer ones female, slender, and usually 4-toothed. Achenes very slightly contracted at the top. Pappus of 5 or 6 capillary bristles, almost plumose towards the end.

Victoria. Open wet forest land, Glenelg river, *Robertson;* common about Melbourne, *Adamson;* Dandenong ranges, Mount M'Ivor, etc., *F. Mueller.*

Some specimens very closely resemble some of those of *L. ambiguus*, but they can always be known by the outer scale-like involucral bracts much narrower and descending along the peduncle as in *L. squamatus.*

4. **L.? ambiguus,** *Benth.* A perennial with erect stems attaining 1 ft. or more, glandular-pubescent, glabrous or with a little cottony wool. Leaves linear or linear-lanceolate, acute, with revolute margins, stem-clasping at the base, the upper ones small and distant. Peduncles with only very small distant leafy bracts. Involucres almost hemispherical, 4 to 5 lines diameter, the scarious bracts numerous, closely imbricate, rather broad, woolly-ciliate, without glabrous tips, the inner ones shortly stipitate. Florets scarcely exceeding the involucre, a few of the outer ones female. Achenes shortly contracted at the top. Pappus of 12 to 20 capillary bristles, barbellate or almost plumose towards the end, few or scarcely any to the achenes of the female florets.—*Helichrysum ambiguum*, Turcz. in Bull. Mosc. 1851, i. 195.

W. Australia, *Drummond, n.* 41, *3rd Coll. n.* 121, *4th Coll. n.* 220.

Var. *semicalvus.* More woody at the base, the stems more branched and woolly, the lower leaves narrow-oblong, obtuse or mucronate.—*Helichrysum semicalvum*, F. Muell. Fragm. ii. 156.

N. S. Wales. Darling river, *Panton;* Barrier range, *Victorian Expedition.*

S. Australia. Flinders range, *F. Mueller;* M'Donnel ranges, *M'Douall Stuart.*

The involucre and florets are very nearly those of *L. pulchellus*, but the achenes are not beaked and the characters are nearly those of *Helichrysum*, sect. *Chrysocephalum;* the achenes however (of which I have seen but few ripe) appear to be distinctly contracted at the top.

5. **L. pulchellus,** *F. Muell. Fragm.* i. 53. An erect corymbosely-branched annual, rarely above 6 in. high, the lower part of the stem sometimes hard so as to appear woody. Leaves linear, acute. Flower-heads small and rather numerous on filiform peduncles. Involucre broadly turbinate or almost hemispherical, about 3 lines diameter, the bracts imbricate in many rows, with scarious ciliate margins and very small coloured tips, the inner ones stipitate. Florets longer than the involucre, the outer female ones few and slender. Achenes contracted into a short but distinct beak. Pappus-bristles barbellate from the base, 2 or 3 to the achenes of the female florets, 4 or 5 to the others.—Sond. in Linnæa, xxv. 500; *Doratolepis? tetrachæta*, Schlecht. in Linnæa, xx. 593.

N. S. Wales. Near Bathurst, *Herb. F. Mueller.*

Victoria. Avoca river, *F. Mueller;* Wimmera, *Dallachy.*

S. Australia. Dry places from the Murray river to St. Vincent's and Spencer's gulfs, *F. Mueller* and others.

6. **L. elongatus,** *DC. Prod.* vi. 160. Apparently annual, but often with a hard tufted base, the stems ascending or erect, often exceeding 1 ft., simple or sparingly branched, glabrous or sprinkled with short almost scale-like hairs. Leaves linear or lanceolate or the lower ones linear-spathulate. Flower-heads rather large, the peduncles bearing a few transparent scales passing into the outer involucral bracts. Involucre campanulate, 4 to 5 lines long, the outer bracts narrow-lanceolate, very scarious and transparent, passing into the inner ones with linear glandular claws and small lanceolate scarious tips. Florets longer than the involucre, all hermaphrodite. Achenes contracted into a beak scarcely perceptible at first, but rather long when the achene is ripe. Pappus of numerous fine capillary bristles.—Hook. f. Fl. Tasm. i. 208; Sond. in Linnæa, xxv. 502.

N. S. Wales. Blue Mountains, near Bathurst, *A. Cunningham;* Argyle county, *Backhouse.*

Victoria. Near Melbourne, *F. Mueller;* Glenelg river, *Robertson;* near Skipton, *Whan;* Wimmera, *Dallachy.*

Tasmania. Not uncommon in various parts of the colony, *J. D. Hooker.*

S. Australia. Sturt river, Bugle and Barossa ranges, Tonunda, *F. Mueller.*

Var. *peduncularis.* Stems more branched and leafy at the base, the upper branches reduced to long peduncles, leafless except the scale-like bracts.—*L. medius*, A. Cunn. in DC. Prod. vi. 160; *Rhytidanthe scabra*, Benth. in Hueg. Enum. 63; *Leptorhynchus rhytidanthe*, DC. Prod. vi. 679.

W. Australia. King George's Sound and adjoining districts, *R. Brown, A. Cunningham,* and others; *Drummond, 3rd Coll. n.* 36, *4th Coll. n.* 180; eastward to the Great Bight, *Maxwell;* Murchison river, *Oldfield.*

7. **L. Waitzia,** *Sond. in Linnæa,* xxv. 501. A loosely woolly-pubescent annual, the stems usually erect and under 6 in., rarely attaining nearly 1 ft.

Leaves linear or linear-oblong, obtuse or mucronate, rarely above 1 in. long, scabrous-pubescent, the uppermost on the peduncles often passing into scarious scales. Flower-heads campanulate or almost globular, about ½ in. diameter, the bracts scarious but not transparent, loosely imbricate, oblong, obtuse, shortly ciliate, the innermost row rigid, with a very small lamina. Florets longer than the involucre, all hermaphrodite. Achenes contracted into a rather long beak. Pappus of 12 or more capillary bristles barbellate from the base.

N. S. Wales. Darling river, *Victorian Expedition.*

Victoria. Avoca and Murray rivers, *F. Mueller;* Wimmera, *Dallachy.*

S. Australia. Dry hills, Salt Creek, *Behr;* Gawler town, *F. Mueller.*

8. **L. linearis,** *Less. Syn. Comp.* 273. Rhizome slender and creeping, emitting tufts of erect or ascending stems, simple or sparingly branched, often 1 ft. high, thinly cottony or glabrous. Leaves linear, the radical ones broad and 1 to 2 in. long, the upper ones gradually narrower, smaller, and more distant, usually glabrous on both sides. Flower-heads few or solitary. Involucre broadly hemispherical, ½ to ¾ in. diameter, the bracts oblong-ovate or almost orbicular, not ciliate, with broadly ovate thinly scarious laminæ looser than in any other species. Florets (as far as I have seen) all hermaphrodite, longer than the involucre. Achenes contracted into a short but distinct beak. Pappus of numerous fine bristles shortly barbellate.—DC. Prod. vi. 160.

N. S. Wales. Moist pastures between Sydney and Paramatta, *R. Brown.*

Victoria. Open woods and sandy hills, Latrobe and Snowy rivers, and heathy ridges near Mombaya river, *F. Mueller.*

Tasmania. Derwent river, *R. Brown.*

74. SCHŒNIA, Cass.

Involucre turbinate or campanulate, the outer bracts adpressed, scarious, imbricate, the inner row with petal-like spreading laminæ. Receptacle without scales. Florets all tubular, 5-toothed, those of the circumference hermaphrodite, fertile. Anthers with fine tails. Style-branches terminating in a short cone. Disk-florets sterile, with an undivided style. Achenes of the circumference very flat, not beaked, those of the disk abortive. Pappus of numerous barbellate bristles.—Annual. Leaves alternate or the lower ones opposite, entire. Flower-heads in a loose corymb.

The genus is limited to a single species endemic in Australia. It differs from the section *Lawrencella* of *Helichrysum*, chiefly in the flat achenes of the circumference.

1. **S. Cassiniana,** *Steetz in Pl. Preiss.* i. 481. An erect corymbosely-branched annual of 1 to 2 ft., scabrous-pubescent or with more or less of cottony wool. Leaves lanceolate or linear, or the lower ones oblong-spathulate, the longest above 2 in., the upper ones few and small. Flower-heads in a loose terminal corymb. Involucre varying from 3 to 5 lines long without the ray, the outer scarious bracts usually brown, the radiating laminæ of the inner bracts white or pink, oblong, 4 to 6 lines long. Perfect achenes in a single row at the circumference, flat, with the edges ciliate with long hairs. Sterile florets numerous.—*Helichrysum Cassinianum,* Gaud. in Freyc. Voy. Bot.

466. t. 87; *Helipterum Cassinianum*, DC. Prod. vi. 216; *Schœnia oppositifolia*, Steetz in Pl. Preiss. i. 480; Bot. Mag. t. 4560; *Pteropogon Cassinianus* and *P. oppositifolius*, F. Muell. in Linnæa, xxv. 415.

S. Australia. Finke river, *M'Douall Stuart's Expedition;* in the scrub N.E. of Lake Gairdner, *Herb. F. Mueller.*

W. Australia. Sharks' Bay (*Gaudichaud*), Champion Bay, Murchison river, *Oldfield;* also *Drummond, n.* 70, *Roe.*

75. HELICHRYSUM.

(Petalolepis *and* Faustula, *Cass.;* Ozothamnus, *R. Br.;* Swammerdamia, *DC.;* Lawrencella, *Lindl.;* Argyrophanes, *Schlecht.;* Chrysocephalum, *Walp.;* Conanthodium, *A. Gray;* Xanthochrysum *and* Argyroglottis, *Turcz.;* Acanthocladium, *F. Muell.*)

Involucre from broadly hemispherical to narrow-ovoid or cylindrical, the bracts imbricate in several rows, either entirely or their laminæ rigidly or opaquely scarious or petal-like, more or less spreading or rarely appressed. Receptacle flat, convex or almost conical, without scales (or very rarely a few in the centre amongst sterile florets). Florets either all hermaphrodite, tubular, and 5- rarely 4-toothed, or a few in the circumference (very rarely 1 or 2 outer rows) female, slender but not longer than the others, 2- or 3-toothed, a few of the central ones sometimes sterile. Anthers with fine tails. Style-branches nearly terete, truncate or rarely with small conical tips. Achenes angular, terete or slightly compressed, not contracted at the top, glabrous papillose or rarely silky-villous. Pappus of capillary bristles simple or more or less barbellate or plumose at the end, not distinctly plumose from the base, those of the female florets often fewer or rarely wanting.—Herbs undershrubs or shrubs, with leafy stems, usually more or less clothed with cottony wool. Leaves alternate or the lower ones very rarely opposite, entire. Florets yellow, the laminæ of the involucral bracts usually white yellow brown or pink, often varying in all these colours with intermediate shades in the same species.

A large genus represented in most warm and temperate regions of the globe, especially numerous in S. Africa and Australia, but without any cosmopolitan species, the Australian ones being all endemic. The limits to be assigned to the group are very uncertain, as it is connected with so many others by almost insensible gradations. The radiating or irregular female florets which separate *Podolepis* and *Athrixia* are, in *P. Lessoni*, *P. rutidochlamys*, and *A. tenella*, and sometimes in *P. canescens*, but little different from those of *Helichrysum*. *Ixiolæna* passes into *Helichrysum* through *I. tomentosa* and *H. podolepideum*, *Leptorhyncnus* through *L. ambiguus* and *L. tenuifolius*, which are very near the section *Chrysocephalum*, and yet cannot be generically separated from *L. squamatus*. The plumose pappus of *Helipterum* is not very distinct from the strongly barbellate pappus of *Lawrencella*, or the semiplumose one of *Chrysocephalum*. The more numerous female florets of *Chrysocephalum*, and the elastically spreading involucre of *H. collinum* and its allies, connect *Helichrysum* with *Raoulia* and through that genus with the *Eugnaphalieæ*, whilst through *Helipterum* and *Cephalipterum* there is a gradual passage into the *Angiantheæ*. And many other connections with other genera of *Gnaphalieæ* may be traced through South African and northern forms. We are obliged, therefore, as in the case of *Asteroideæ*, to make arbitrary demarcations, in order not to unite the whole tribe into one unmanageable genus. Those here adopted are the best that have suggested themselves after much consideration, although it must be admitted that in some instances they are not altogether satisfactory.

SECTION I. **Lawrencella.**—*Annuals. Involucre hemispherical or campanulate, the outer bracts brown sessile and appresssed, the inner ones with coloured radiating laminæ. Achenes with erect transparent obtuse hairs (or elongated papillæ), or the centre ones sometimes abortive and glabrous.*

Branches and peduncles elongated and slender.
  Involucres pink or white . . . . 1. *H. Lawrencella.*
  Involucres golden-yellow.
    Involucre hemispherical, $\frac{3}{4}$ to 1 inch diameter without the ray . . . . 2. *H. subulifolium.*
    Involucre broadly campanulate, not $\frac{1}{2}$ in. diameter without the ray . . . . 3. *H. filifolium.*
Branches compact. Peduncles very short. Involucre ovoid-campanulate with yellow or white rays . . . . 4. *H. semifertile.*

SECTION II. **Xerochlæna.**—*Perennials, sometimes almost woody at the base, rarely also annual. Involucre broad, hemispherical, the outer sessile broad bracts passing more or less gradually into the intermediate or inner ones with scarious or linear claws and radiating coloured laminæ (scarcely conspicuous in* H. rutidolepis). *Achenes glabrous, papillose or rarely shortly villous.*

Flower-heads singly terminating the branches on long peduncles.
    Pappus-bristles simple at the base, barbellate or almost plumose towards the end.
  Involucral bracts elegantly ciliate, the intermediate (white or pink) more than twice as long as the florets, the innermost small and plumose . . . . 5. *H. Baxteri.*
  Involucral bracts not ciliate, the intermediate (yellow) not twice as long as the florets.
    Stems mostly erect and single headed. Involucre $\frac{3}{4}$ to 1 in. diameter . . . . 6. *H. scorpioides.*
    Stems decumbent, often branched. Involucres not exceeding $\frac{1}{2}$ in. diameter . . . . 7. *H. rutidolepis.*
  Involucral bracts not ciliate, the intermediate (white or pink) more than twice as long as the florets.
    Achenes glabrous or papillose.
      Stems erect, branching . . . . 8. *H. obtusifolium.*
      Stems branching at the base only, the branches erect or ascending, single-headed . . . . 9. *H. dealbatum.*
    Achenes shortly hirsute. Tufted plant with short erect single-headed stems . . . . 10. *H. pumilum.*
Flower-heads singly terminating the branches on long peduncles.
    Pappus-bristles simple or equally denticulate.
  Radiating involucral bracts rigid, mostly obtuse (yellow, brown, red, or white).
    Stock tufted with short single-headed stems . . . . . 11. *H. Milligani.*
    Stem tall, usually branched . . . . 12. *H. bracteatum.*
  Radiating involucral bracts petal-like, mostly acute (white or pink).
    No floral leaves.
      Leaves lanceolate or oblong, cottony underneath or on both sides . . . . 13. *H. elatum.*
      Upper leaves linear, glutinous, without cottony wool . . 14. *H. glutinosum.*
    A few narrow linear floral leaves close under the involucre.
      Leaves linear or linear-lanceolate, dilated and stem-clasping at the base, glutinous, without cottony wool . . 15. *H. adenophorum*
      Leaves linear to oblong-spathulate, scabrous above, cottony-white underneath . . . . 16. *H. leucopsidium.*
Flower-heads corymbose. Outer involucral bracts very woolly,

radiating ones white. Pappus-bristles more barbellate towards the end . . . . 17. *H. Blandowskianum.*

SECTION III. **Oxylepis.**—*Perennials or undershrubs. Involucres hemispherical or broadly campanulate, the bracts all with linear claws and radiating laminæ, narrow, acute, and often revolute. Pappus-bristles simple or shortly barbellate, not thickened nor more barbellate upwards.*

Involucres broadly hemispherical, radiating laminæ subulate-acuminate.
  Leaves linear with revolute margins, glabrous or glandular-pubescent . . . . 18. *H. oxylepis.*
  Leaves oblong or lanceolate, cottony-white underneath.
    Laminæ of the involucral bracts spreading or reflexed, the long ones longer than their loosely woolly or nearly glabrous claws . . . . 19. *H. collinum.*
    Laminæ of the involucral bracts revolute, all shorter than their densely woolly claws . . . . 20. *H. rupicola.*
Involucres campanulate, laminæ small, narrow, scarious. Leaves obovate or spathulate, woolly-white on both sides . . . . 21. *H. podolepideum.*

SECTION IV. **Chrysocephalum.**—*Herbs with corymbose flower-heads. Involucr ovoid-turbinate or almost globose, the bracts coloured in many rows scarcely exceeding th florets, appressed or squarrose, ciliate. Female florets in 1 or 2 rows, but not so numerou as the hermaphrodite ones. Achenes glabrous or papillose. Pappus-bristles few, simpl at the base, plumose at the end.*

Leaves usually flat and cottony. Flower-heads often 6 to 8 lines diameter, loosely corymbose, the bracts scarcely squarrose . . 22. *H. apiculatum.*
Leaves very narrow or not cottony. Flower-heads small, densely corymbose, the bracts often squarrose . . . . 23. *H. semipapposum.*

(The remarkable ciliate bracts and almost the pappus of *Chrysocephalum* occur also in *H. Baxteri*, which has solitary flower-heads and large radiating involucres, in *Leptorhynchus tenuifolius* and *L. ambiguus*, which have solitary flower-heads and the achenes usually contracted under the pappus, and in *Helipterum pterochætum*, which has the pappus plumose from the base.)

SECTION V. **Ozothamnus.**—*Shrubs or rarely undershrubs or herbs. Involucre small oblong-ovoid or turbinate-campanulate (rarely larger and ovoid-conical), the bracts imbricate, not exceeding the florets or the inner ones with small coloured radiating tips not much longer than the florets. Female florets few or none. Achenes glabrous or shortly villous. Pappus-bristles simple, often thickened or more denticulate towards the end.*

Flower-heads rather large, solitary, ovoid or turbinate.
  Cottony-white undershrub with spinescent branches. Involucral bracts scarious, without coloured tips . . . . . 24. *H. Dockerii.*
  Divaricately-branched shrub (or undershrub ?) with little wool.
    Involucral bracts rigid, the inner ones with radiating coloured tips . . . . 25. *H. argyroglottis.*
Flower-heads small, in compound or small panicles.
  Panicles loose, compound, almost leafless. Herbs or undershrubs with flat leaves.
    Leaves nearly sessile, narrow. Involucres scarcely woolly, inner bracts with white spreading tips . . . . . . 26. *H. ramosum.*
    Leaves distinctly petiolate. Involucres woolly.
      Lower leaves cordate. Inner involucral bracts with white spreading tips . . . . 27. *H. cordatum.*
      Leaves narrowed at the base. Tips of the involucral bracts scarcely squarrose . . . . 28. *H. obovatum.*

Panicles small, compact, terminating the leafy branches of shrubs (or undershrubs ?). Involucral bracts scarious without spreading tips.

Leaves ovate on slender petioles. Branches slender, divaricate . . . . 29. *H. Bidwillii.*

Leaves elliptical-oblong or linear, almost sessile. Branches slender, divaricate . . . . 30. *H. Becklerii.*

Leaves very small, erect with recurved margins, sessile with rounded auricles. Branches virgate . . . . 31. *H. diotophyllum.*

Flower-heads small, usually numerous, in terminal corymbs. Erect shrubs.

Leaves linear with revolute margins, rarely almost lanceolate and flat.

Involucral bracts with concave erect or loose but not spreading obtuse tips, all or the inner ones white or pink.

Leaves not decurrent. Florets above 20 . . . . 32. *H. diosmifolium.*

Leaves decurrent in prominent lines. Florets about 10 to 12 . . . . 33. *H. adnatum.*

Involucral bracts erect, few, narrow, almost acute. Florets 3 34. *H. Cunninghamii.*

Involucral bracts more or less scarious, obtuse, without any or with scarcely conspicuous white tips.

Involucre broadly campanulate. Florets above 20. Achenes villous. Leaves coriaceous, obtuse . . . . 35. *H. reticulatum.*

Involucre ovoid-turbinate. Florets above 15. Achenes glabrous or papillose.

Leaves not decurrent . . . . 36. *H. cinereum.*

Leaves shortly decurrent . . . . 37. *H. bracteolatum.*

Involucre narrow. Florets 10 to 15. Achenes glabrous 38. *H. cassinioides.*

Involucral inner bracts with white tips usually spreading. Florets under 15 and usually under 12.

Leaves much revolute, very narrow, obtuse, ¾ to 1 in. long. Corymbs compound, very dense. Involucres nearly 3 lines long . . . . 39. *H. Gunnii.*

Leaves much revolute, thick and very obtuse, under ½ in. long, tomentose above when young. Corymbs small and dense . . . . 40. *H. ledifolium.*

Leaves narrow with revolute margins, mostly obtuse, from under ½ to above 1 in. long. Corymbs usually in leafy panicles . . . . 41. *H. rosmarinifolium.*

Leaves nearly flat, linear or lanceolate, often acute, from under ½ to near 2 in. long. Corymbs compound, mostly flat . . . . 42. *H. ferrugineum.*

Leaves obovate or cuneate, flat or with recurved margins.

Involucral bracts without any or with very small white not spreading tips.

Involucres turbinate-campanulate. Florets above 20 . . 43. *H. antennarium.*

Involucres narrow. Florets about 12 . . . . . . 44. *H. obcordatum.*

Involucral inner bracts with small white spreading tips.

Florets about 12 to 15. Leaves mostly obovate or almost oblong, under ½ in. long . . . . 45. *H. Backhousii.*

Florets about 4 to 6. Leaves mostly cuneate, ½ to 1½ in. long . . . . 46. *H. cuneifolium.*

Leaves almost scale-like with revolute margins, ½ to 1 line long.

Leaves closely erect. Flower-heads capitate. Inner involucral bracts with white spreading tips. Florets 2 to 4 . 47. *H. baccharoides.*

Leaves closely reflexed.

Flower-heads corymbose. Involucral bracts with white concave tips. Florets 8 to 10 . . . . 48. *H. lepidophyllum.*

Flower-heads capitate. Involucral bracts without white tips. Florets 12 to 18 . . . . 49. *H. scutellifolium.*
Leaves small, convex or keeled underneath, concave above.
Leaves erect and closely appressed, scale-like. Flower-heads corymbose . . . . 50. *H. pholidotum.*
Leaves loosely erect or spreading. Flower-heads capitate.
Involucral bracts with concave tips. Florets above 20 . 51. *H. lycopodoides.*
Involucral inner bracts with spreading tips. Florets 8 to 12 . . . . 52. *H. selaginoides.*

SECTION I. LAWRENCELLA.—Annuals. Involucre hemispherical or campanulate, the outer bracts brown, sessile and appressed, the inner ones with linear-scarious claws and radiating coloured laminæ. Receptacle flat or convex. Achenes with erect transparent obtuse hairs (or elongated papillæ), the central ones sometimes abortive and glabrous. Pappus equally and sometimes strongly barbellate. Annuals.

1. **H. Lawrencella,** *F. Muell. Herb.* An erect slender branching annual, often exceeding 1 ft., glabrous or nearly so in the typical form. Leaves linear, very narrow, frequently opposite. Flower-heads on long peduncles. Involucre hemispherical, in the ordinary form about ½ in. diameter without the rays, the outer bracts brown, lanceolate, appressed, with scarious margins and slightly woolly, the inner with a broad linear claw, and oblong obtuse spreading petal-like laminæ, 4 to 6 lines long, pink or white. Receptacle flat. Florets all hermaphrodite, about as long as the claws of the inner bracts. Achenes, at least the perfect ones, more or less hispid with transparent hairs (or elongated papillæ). Pappus of numerous barbellate bristles, not thickened upwards.—*Lawrencella rosea,* Lindl. Swan Riv. App. 23.

W. Australia. Swan River, *Drummond,* 1*st Coll.;* Cape Riche, Salt river and Bremer Bay, *Maxwell;* Murchison river, *Oldfield* (with large flower-heads).

Var. *Davenportii.* Glabrous or papillose-pubescent, the leaves sometimes lanceolate. Flower-heads large, the central florets sterile with elongated abortive achenes and the pappus-bristles almost plumose.—*H. Davenportii,* F. Muell. Fragm. iii. 32.

S. Australia. Neales river, *M'Douall Stuart.*

W. Australia. Near Sapphire Lake, *Harper;* Bowes river, *Oldfield.*

I should have maintained this as a distinct species, but that among the very few specimens seen (in Herb. F. Mueller) some have precisely the foliage of *H. Lawrencella,* and among those we have of the typical form with smaller flower-heads a few of the central florets are occasionally abortive with elongated achenes, and Oldfield's Murchison river specimens have the heads quite as large as in *H. Davenportii,* but all the florets perfect.

2. **H. subulifolium,** *F. Muell. Fragm.* iii. 134. An erect glabrous annual, with the narrow leaves of *H. Lawrencella,* but they are seldom opposite and the stems are simple or not much branched. Flower-heads large, on long peduncles. Involucre hemispherical, ¾ to above 1 in. diameter without the ray, the outer bracts brown and appressed as in *H. Lawrencella,* the inner with linear claws and large oblong spreading laminæ of a bright golden colour. Receptacle very convex. Florets as long as the claws of the inner bracts. Achenes hispid with erect transparent hairs. Pappus-bristles not so numerous as in *H. Lawrencella,* strongly barbellate, not thickened upwards.

W. Australia, *Drummond.* Greenough river and Champion Bay, *Oldfield* (with particularly large flower-heads).

3. **H. filifolium,** *F. Muell. Fragm.* iii. 134. A slender erect annual, glabrous or nearly so, simple or corymbosely branched, 6 in. to 1 ft. high. Leaves very narrow-linear, the lower ones often opposite. Flower-heads nearly of *H. subulifolium,* but much smaller. Involucre campanulate, not ½ in. diameter without the rays, the outer bracts appressed, the inner with radiating laminæ of a golden-yellow about 3 lines long. Achenes hispid with transparent hairs. Pappus-bristles about 15 to 20, strongly barbellate.—*Xanthochrysum filifolium,* Turcz. in Bull. Mosc. 1851, i. 199. t. 4 (the upper hairs of the achene mistaken for an outer pappus); *Helipterum tenellum,* A. Gray in Hook. Kew Journ. iv. 230, not of Turcz.

W. Australia, *Drummond, 3rd Coll. n.* 119 *or* 219.

This and the preceding species (which are very closely allied) have the pappus-bristles more strongly barbellate than usually in the genus, but scarcely enough so to be classed with the plumose pappus of *Helipterum.*

4. **H. semifertile,** *F. Muell. Rep. Babb. Exped.* 14. An erect much-branched annual of 2 to 6 in., very slightly woolly-tomentose or glabrous. Leaves narrow-linear, obtuse. Flower-heads on very short peduncles, forming an irregular leafy corymb. Involucre ovoid-campanulate, about 3 lines long, the outer bracts sessile, obtuse and appressed, brown or pale-coloured, the innermost with oblong whitish or bright yellow radiating laminæ, smaller than in the preceding species. Florets 10 to 15, some of the inner ones frequently sterile. Achenes sprinkled with erect transparent hairs. Pappus-bristles rather numerous, barbellate especially towards the end.—*Pteropogon ramosissimus,* F. Muell. in Linnæa, xxv. 412.

N. S. Wales. Goyinga mountains, *Victorian Expedition;* between Stokes range and Cooper's Creek, *Wheeler.*

S. Australia. Between Flinders range and Spencer's Gulf, near Cudnaka, *F. Mueller;* Wills Creek, *Howitt's Expedition.*

SECTION II. XEROCHLÆNA.—Herbaceous perennials, sometimes woody at the base, rarely flowering also the first year so as to appear annual. Involucre broad, hemispherical, the outer bracts sessile, passing more or less gradually into inner ones with linear or broad scarious claws and petal-like radiating laminæ (scarcely conspicuous in *H. rutidolepis*), the innermost of all often shorter and narrower. Receptacle flat or convex. Achenes glabrous, papillose, or very rarely shortly villous.

5. **H. Baxteri,** *A. Cunn.; DC. Prod.* vi. 193. Stems shortly branching and almost woody at the base, with erect usually simple rather slender but rigid branches, of ½ to 1½ ft., white with a close cottony wool. Leaves narrow-linear, ½ to 1 in. long, with revolute margins, cottony-white when young, at length glabrous above. Flower-heads terminal and solitary. Involucre broadly hemispherical, expanding to above 1 in. diameter, the bracts in many rows, sometimes slightly woolly at the base, all elegantly ciliate, the outer ones short, broadly lanceolate, usually tinged with brown, passing into the intermediate ones with petal-like laminæ, pure white or slightly straw-coloured on slender claws, the inner rows much shorter with small plumose laminæ. Florets very numerous, not half so long as the involucre. Achenes glabrous. Pappus of 6 to 8 bristles, simple at the base, more or less plu-

mose at the end.—*Argyrophanes Behrii*, Schlecht. Linnæa, xx. 596; *Chrysocephalum Behrianum*, Sond. in Linnæa, xxv. 517.

Victoria. In the Grampians, *Wilhelmi;* Mount Sturgeon, *Robertson;* Wimmera, *Dallachy* (with smaller straw-coloured heads).

S. Australia. South coast, *Baxter;* between Gawler and Light rivers, *Behr;* from the Murray to St. Vincent's Gulf, common; also near Spencer's Gulf, *F. Mueller.*

6. **H. scorpioides,** *Labill. Pl. Nov. Holl.* ii. 45. *t.* 191. Stems from a perennial tufted or decumbent and shortly branching base, ascending or erect, usually simple, often exceeding 1 ft. and rather weak, but sometimes shorter and more rigid, clothed with a white deciduous cottony wool. Leaves from oblong-spathulate to linear, mostly acute, glabrous or scabrous above, loosely woolly underneath, the upper ones few and small. Involucre broadly hemispherical, from about $\frac{3}{4}$ to 1 in. diameter, the bracts very numerous and spreading, not ciliate, the outer ones short, often tinged with brown, passing into the intermediate ones, of a bright yellow, usually narrow, but obtuse and gradually contracted into the claw, the innermost smaller. Florets exceedingly numerous, more than half as long as the involucre. Achenes glabrous. Pappus-bristles not very numerous, slender, shortly and often sparingly barbellate towards the end.—DC. Prod. iii. 194; Hook. f. Fl. Tasm. i. 211; *Gnaphalium scorpioides*, Poir. Dict. ii. 808; *Helichrysum buphthalmoides*, Sieb. Pl. Exs.; *H. Gunnii*, Hook. f. in Hook. Ic. Pl. t. 320.

N. S. Wales. Port Jackson to the Blue Mountains, *R. Brown, Sieber, n.* 333, and others; northward to Macleay and Clarence rivers, *Beckler;* in the interior in all the fertile lands towards Bathurst, *Fraser, A. Cunningham;* southward to Twofold Bay, *F. Mueller.*

Victoria. In cool hilly pastures from the Glenelg to Gipps' Land, *F. Mueller, Robertson*, and others; Wimmera, *Dallachy.*

Tasmania. Abundant throughout the island, especially in moist pastures, ascending to 4000 ft., *J. D. Hooker.*

S. Australia. Rivoli Bay, Bugle ranges, *F. Mueller;* Onkaparinga river, *Whittaker.*

7. **H. rutidolepis,** *DC. Prod.* vi. 194. A decumbent or loosely branched perennial, almost woody at the base, with a little loose deciduous wool. Leaves oblong, lanceolate or linear, acute or mucronate, mostly narrowed below the middle, but stem-clasping at the base, $\frac{3}{4}$ to nearly 2 in. long. Flower-heads solitary on almost leafless peduncles. Involucre hemispherical, not exceeding $\frac{1}{2}$ in. diameter and often smaller, the bracts spreading in many rows, the outer short ones tinged with brown and passing into the inner or intermediate ones, which have a more distinct obtuse yellow lamina, rather longer than the florets and often transversely wrinkled. Florets very numerous, a few of the outer ones female and almost or quite without pappus, the others hermaphrodite. Achenes glabrous. Pappus-bristles minutely serrulate or simple at the base, more distinctly but sparingly barbellate towards the end. —*H. erosum*, Schlecht. Linnæa, xx. 595.

N. S. Wales. Pastures about Camden and Argyle county, *A. Cunningham;* Bathurst, *Clowes;* Blue Mountains, *Woolls.*

Victoria. Near Melbourne, *F. Mueller;* Wendu vale, *Robertson.*

S. Australia. Onkaparinga and Torrens river, *F. Mueller.*

F. Mueller now considers this as a variety of *H. scorpioides*, to which it is certainly nearly allied, but the loose branching habit and small flower-heads appear to be constant, as far as shown by our numerous specimens,

8. **H. obtusifolium,** *F. Muell. and Sond. in Linnæa*, xxv. 513. Erect and branching with a hard often woody base, but perhaps flowering sometimes the first year so as to appear annual, very variable in stature, above 1 ft. high in the original form, silvery-white with a close cottony wool or at length nearly glabrous. Leaves linear, obtuse, with revolute margins, sometimes narrow and above ½ in. long, sometimes very small. Flower-heads solitary, terminating the branches. Involucre hemispherical, spreading from about ½ in. to above 1 in. diameter, the outer short brown bracts often woolly at the base, passing gradually or almost abruptly into the inner ones, which have a short claw and a white narrow lamina, ¼ to ½ in. long, rarely tinged with pink. Florets very numerous, much less than half as long as the involucre, a few of the outer ones female with a reduced or no pappus, the others hermaphrodite. Achenes glabrous or papillose. Pappus-bristles 15 to 20, simple at the base, strongly barbellate towards the end.—*Helipterum niveum*, Steetz in Pl. Preiss. i. 475.

N. S. Wales. Twofold Bay, *F. Mueller.*

Victoria. Wilson's Promontory, *F. Mueller;* Glenelg river, *F. Mueller, Robertson;* Grampians, *Wilhelmi;* Wimmera, *Dallachy.*

S. Australia. Encounter Bay, *F. Mueller;* Adelaide, *Whittaker.*

W. Australia. King George's Sound and adjoining districts, *Baxter, Drummond, n.* 84, *Preiss, n.* 11, and others (the leaves usually, but not always, narrower than in the Eastern specimens).

Var. *tephrodes.* Under 6 in. high and dichotomously branched, the branchlets sometimes ending in a short spine. Leaves few and small. Flower-heads small.—*Ozothamnus tephrodes*, Turcz. in Bull. Mosc. 1851, ii. 79.—Wimmera, *Dallachy;* Encounter Bay, *Whittaker;* Kangaroo Island, *Waterhouse;* Phillips river, *Maxwell;* also *Drummond*, 5*th Coll. n.* 385.

Var. *squamiger.* Like the preceding variety, but more slender and the leaves all reduced to minute linear scales.—Near Oldfield river, *Maxwell.*

9. **H. dealbatum,** *Labill. Pl. Nov. Holl.* ii. 45. *t.* 190. Stems from a perennial tufted or shortly creeping and branching base, ascending or erect, simple and single-headed, rarely attaining 1 ft., white with a close cottony wool. Leaves radical or at the base of the stems, oblong-lanceolate or slightly spathulate, acute or obtuse, rarely exceeding 1 in., usually flat and glabrous above, cottony-white underneath, the upper ones few, small and narrow. Involucre broadly hemispherical, spreading to from 1 to nearly 1½ in. diameter, the bracts in many rows, the outer short brownish or reddish ones passing gradually into the intermediate lanceolate petal-like ones, of a pure white and often striate, the innermost smaller. Florets very numerous, much less than half as long as the involucre, a few of the outer ones female. Achenes glabrous or papillose. Pappus-bristles barbellate towards the end, scarcely so at the base.—DC. Prod. vi. 189; Hook. f. Fl. Tasm. i. 213; *Gnaphalium niveum*, Poir. Dict. Suppl. ii. 808.

Tasmania. Derwent river and Port Dalrymple, *R. Brown;* in various parts of the colony, generally in a poor wet soil, *J. D. Hooker.*

10. **H. pumilum,** *Hook. f. Fl. Tasm.* i. 213. *t.* 60. A small perennial, with a tufted stock. Leaves chiefly radical, linear or linear-cuneate, obtuse

with revolute margins, 1 to 2 in. long, glabrous above when full grown, cottony-white underneath. Scapes or flowering stems under 6 in. high, simple and single-headed with 1 or 2 narrow, small, distant leaves. Flower-heads of *H. dealbatum*, but rather smaller and the bracts rather more rigid, the intermediate radiating ones white as in that species. Florets rather longer than in *H. dealbatum*, but much less than half the length of the involucre. Achenes in all the specimens seen more or less silky-hirsute. Pappus-bristles barbellate toward the end, scarcely so at the base.

Tasmania. Around Macquarrie Harbour, *Milligan*, *Gunn*.

11. **H. Milligani,** *Hook. f. Fl. Tasm.* i. 214. *t.* 60. A small perennial with a tufted almost woody stock and simple single-headed erect stems, rarely exceeding 6 in., covered with cottony wool. Leaves chiefly crowded near the base, but also scattered on the stem, the lowest petiolate and almost ovate, passing into oblong-spathulate, and the uppermost lanceolate, all flat, glabrous or nearly so. Flower-heads large, the involucres spreading to 1½ in. diameter, white straw-coloured or tinged with red, the bracts as well as the florets, achenes, and pappus those of *H. bracteatum*.

Tasmania. Summits of Mount Pearse and of Mount Sorrell, *Gunn*, *Milligan*; Mount Lapeyrouse, *C. Stuart*.—Notwithstanding the habit, which is that of *H. pumilum*, this may prove to be an alpine variety of *H. bracteatum*.

12. **H. bracteatum,** *Willd.*; *DC. Prod.* vi. 188. An erect branching or simple perennial, of 1 to 2 ft., often flowering the first year so as to be also annual, glabrous scabrous or sprinkled with a few hairs, without cottony wool. Leaves from linear to oblong-lanceolate or the lower ones obovate-oblong, the longer ones often attaining 3 or 4 in., green and sometimes somewhat glutinous. Flower-heads large, solitary or few together on separate peduncles. Involucre hemispherical, spreading to from a little more than 1 in. to nearly 2 in. diameter, in the original form shining yellow or straw-coloured with more or less of a reddish-brown, the outer bracts short ovate and sessile, the inner gradually longer and more lanceolate on a short broad claw, the innermost narrow and rather shorter, all rigid and usually obtuse. Florets very numerous, very much less than half the length of the involucre, the outer ones female and slender. Achenes glabrous or slightly papillose. Pappus-bristles denticulate, not thickened upwards, slightly cohering at the base.—Hook. f. Fl. Tasm. i. 210; Steetz in Pl. Preiss. i. 471; *Xeranthemum bracteatum*, Vent. Jard. Malm. t. 2; *Helichrysum chrysanthum*, Pers. Syn. Pl. ii. 414; *Elichrysum lucidum*, Henckel, Adumbr. (DC.); *Helichrysum viscosum*, Sieb. Pl. Exs.; *H. Banksii*, A. Cunn. in DC. Prod. vi. 188 (with rather broad leaves); *H. bicolor*, Lindl. Bot. Reg. t. 1844 (with narrow leaves); *H. acuminatum*, DC. Prod. vi. 188 (with less obtuse involucral bracts); *H. macrocephalum*, A. Cunn. in DC. l. c. (with large flower-heads).

N. Australia. Port Essington, *Armstrong*.

Queensland. Abundant along the whole coast, *Banks and Solander*, *R. Brown*, and many others.

N. S. Wales. Port Jackson to the Blue Mountains, *R. Brown*, *Sieber*, *n.* 345, and others; lagoons of the interior, *A. Cunningham*; New England, *C. Stuart*; Clarence and Hastings rivers, *Beckler*.

Victoria. From the Glenelg to Gipps' Land, *F. Mueller, Robertson,* and others; snowy top of Mount Buller, *F. Mueller;* Wimmera, *Dallachy.*

Tasmania. Abundant throughout the island, especially in marshy situations, *J. D. Hooker.*

S. Australia. From the Murray to St. Vincent's and Spencer's Gulfs, *F. Mueller* and others.

W. Australia. Swan River, *Drummond, Preiss, n.* 4.

Var. *albidum,* DC. Prod. vi. 189. Involucral bracts white, passing into straw-colour pale brown or pink.—*H. papillosum,* Labill. Pl. Nov. Holl. ii. 46. t. 192; DC. Prod. vi. 189; Hook. f. Fl. Tasm. i. 212; *Gnaphalium papillosum,* Poir. Dict. Suppl. ii. 808; *H. glabratum,* DC. Prod. vi. 189 (from the character given); *H. macranthum,* Benth. in Hueg. Enum. 65; DC. Prod. vi. 189; Paxt. Mag. v. 247, with a fig.; Steetz in Pl. Preiss. i. 471; Bot. Reg. 1838, t. 58; *H. niveum,* Grah. in Bot. Mag. t. 3857; Steetz in Pl. Preiss. i. 471.—Victoria, Tasmania, and W. Australia, in the latter colony at least as abundant as the yellow (including *Drummond, 3rd Coll. n.* 170, *4th Coll. n.* 114, *also n.* 197, and *Preiss, n.* 1).

13. **H. elatum,** *A. Cunn. in DC. Prod.* vi. 193. A stout erect herb or undershrub, sometimes under 1 ft. high and almost simple, in rich shaded situations branching, attaining 7 or 8 ft. and almost woody at the base, clothed with a loose or close cottony wool. Leaves lanceolate or ovate-lanceolate, contracted into a petiole, stem-clasping at its base, the larger ones attaining 2 to 3 or rarely 4 in., the upper surface becoming at length glabrous. Flower-heads large, solitary or loosely paniculate. Involucre hemispherical, spreading to from 1 to 1½ in. diameter, the bracts narrow, acute, petal-like, white or tinged with pink, the outer ones short and sessile, the intermediate long ones on a narrow claw, the innermost shorter and very narrow. Florets exceedingly numerous, much less than half as long as the involucre, a few of the outer ones slender and female. Achenes glabrous or papillose. Pappus-bristles slender, scarcely denticulate, not thickened upwards, shortly cohering at the base.—*H. lanuginosum,* A. Cunn. in DC. Prod. vi. 193; *H. albicans,* Sieb. Pl. Exs.

Queensland. Keppel Bay and Northumberland islands, *R. Brown;* Dawson and Mackenzie ranges, *F. Mueller;* Rodd's Bay, *A. Cunningham;* Port Denison, *Fitzalan* (the leaves quite glabrous); Rockhampton, *Dallachy.*

N. S. Wales. Port Jackson to the Blue Mountains, *R. Brown, Sieber, n.* 346, and *Fl. Mixt. n.* 516, and others; northward to Hastings, Macleay, and Clarence rivers, *Beckler;* New England, *C. Stuart;* southward to Illawarra, *A. Cunningham;* Twofold Bay, *F. Mueller;* Gabo Island, *Maplestone.*

Victoria. Damp valleys, between Wombaza and Genoa river, *F. Mueller.*

Var. *Fraseri.* More shrubby. Leaves crowded, the wool long and very deciduous or none, and sometimes the branches slightly glutinous.—Rocks of Mount Lindsay, at an elevation of 5000 ft., *Fraser, W. Hill;* Port Curtis, *M'Gillivray.*

14. **H. glutinosum,** *Hook.* (as a *Helipterum*). A tall erect branching herb or undershrub, the lower part of the stem and foliage sometimes woolly-white, the upper portion glutinous and scabrous or pubescent without wool. Lower leaves narrow-lanceolate, sometimes densely woolly underneath, glutinous above, the upper ones crowded, narrow-linear with revolute margins, glutinous on both sides, without wool, not dilated at the base, and passing into a few distant subulate bracts on the peduncle. Flower-heads of *H. elatum* or rather smaller. Florets, achenes, and simple pappus-bristles of that species.—*Helipterum glutinosum,* Hook. in Mitch. Trop. Austr. 361.

Queensland. On the Maranoa, *Mitchell;* ridges of the Suttor, *F. Mueller.*
N. S. Wales. Sources of the Boyd river, *Leichhardt.*
Considered by F. Mueller, and perhaps correctly so, as a variety of *H. elatum.*

15. **H. adenophorum,** *F. Muell. in Trans. Vict. Inst.* 1855, 38. A perennial or undershrub, with erect branching stems of 1 to 3 ft., glandular-scabrous without cottony wool. Leaves linear or linear-lanceolate, with revolute margins, slightly dilated and stem-clasping at the base or very shortly decurrent. Flower-heads large, solitary at the ends of the branchlets. Involucre broadly hemispherical, spreading to from 1 to 1½ in. diameter, closely surrounded by 2 or 3 small linear glandular-scabrous leaves, the bracts white or tinged with pink, thin and petal-like, the outer short sessile ones sometimes very pale brown, the intermediate lanceolate-oblong, on dilated claws, a very few of the innermost smaller. Florets very numerous, rather less than half as long as the involucres. Achenes glabrous. Pappus-bristles slender, finely but very shortly barbellate, not perceptibly thickened upwards in the specimens seen.

S. Australia. Barren elevations, Kangaroo Island, *F. Mueller;* scrub near Wallan's Hut, *Waterhouse.*

In general aspect and foliage it closely resembles some specimens of *Waitzia nivea,* but is readily distinguished by the broader claws of the inner involucral bracts, and the achenes not beaked.

16. **H. leucopsidium,** *DC. Prod.* vi. 193. A perennial, with decumbent ascending or almost erect stems, nearly simple or loosely-branched, bearing a little cottony wool. Leaves from linear to oblong-spathulate, obtuse, with recurved margins, mostly ¾ to 1½ in. long, usually very scabrous above and cottony-white underneath. Flower-heads large, solitary at the ends of the branches. Involucre broadly hemispherical, spreading to from 1 to 1½ in. diameter, closely surrounded by 2 or 3 small linear woolly floral leaves, the bracts white or tinged with pink, thin, petal-like, and usually striate, the outer short sessile ones often pale brown, the intermediate lanceolate on broad claws, the innermost narrower and shorter. Florets very numerous, rather less than half as long as the involucres. Achenes glabrous. Pappus-bristles slender, numerous, very slightly barbellate, not thickened upwards.—Hook. f. Fl. Tasm. i. 213. t. 59.

N. S. Wales. Port Jackson, *R. Brown.*
Victoria. From the Glenelg to Wilson's Promontory, *F. Mueller* and others; Wimmera, *Dallachy.*
Tasmania. Port Dalrymple, *R. Brown;* sand hills, North coast, and Flinders' Island, *J. D. Hooker* and others.
S. Australia. Memory Cove, *R. Brown;* Encounter Bay and near Adelaide, *Whittaker;* Streaky Bay and Spencer's Gulf, *Warburton.*
W. Australia. King George's Sound and adjoining districts, *Baxter, Drummond, n.* 346; eastward to Cape Arid, *Maxwell.*

17. **H. Blandowskianum,** *Steetz; Sond. in Linnæa,* xxv. 512. Stems, from a perennial almost woody stock, erect, branched only at the top, 1 to 1½ ft. high, covered as well as the foliage with a dense cottony wool. Leaves linear or lanceolate, soft and thick, 1 to 2 in. long. Flower-heads rather large, in a terminal corymb. Involucre broadly hemispherical,

spreading to a diameter of $\frac{3}{4}$ to 1 in., the outer bracts few, densely woolly with small laminæ, the intermediate with narrow claws and white petal-like radiating laminæ, the innermost small. Florets very numerous, not half so long as the involucre. Achenes papillose. Pappus-bristles about 12, slender, slightly thickened and barbellate upwards.

Victoria. In the Grampians, *Wilhelmi*; Wimmera, *Dallachy*; Glenelg river, *Robertson, Allitt.*

S. Australia. Near Adelaide, *Blandowsky*; Mount Gambier, Rivoli Bay, Encounter Bay, *F. Mueller.*

SECTION III. OXYLEPIS.—Perennials or undershrubs. Involucre hemispherical or broadly campanulate, bracts all with linear claws and radiating laminæ, narrow-acute and often revolute. Pappus-bristles simple or shortly barbellate, not thickened nor more barbellate upwards.

18. **H. oxylepis,** *F. Muell. Fragm.* i. 35. A perennial, with erect or ascending stems of 1 to 2 ft., branching, hard and almost woody at the base, with a little loose wool. Leaves linear, with revolute margins, or very narrow-lanceolate and flat, $1\frac{1}{2}$ to nearly 3 in. long, becoming nearly or quite glabrous, the upper ones few and very small. Flower-heads rather large, solitary. Involucre broadly hemispherical, the bracts numerous and narrow, the claws nearly glabrous, the laminæ (brownish-yellow when dry) subulate-pointed and spreading like those of *H. collinum*, but shorter, although longer than the florets. Florets exceedingly numerous, all or nearly all hermaphrodite. Achenes and pappus of *H. collinum.*

Queensland. Bustard Bay, *Banks and Solander*; sandy shore, Moreton Island, *F. Mueller.*

N. S. Wales. Grose Head, *R. Brown*; also *Backhouse.*

This may possibly be a variety of *H. collinum.*

19. **H. collinum,** *DC. Prod.* vi. 190. A herb or undershrub (of 1 to 2 ft.?), with the habit of some of the compact varieties of *H. elatum*, dotted with white cottony wool. Leaves oblong-lanceolate, mostly 1 to 2 in. long, woolly-white underneath or on both sides. Flower-heads rather large, solitary, the almost leafless peduncles rarely exceeding 2 in. and sometimes very short. Involucre broadly hemispherical, spreading to a diameter of 1 in. or rather more, the bracts very numerous, all with narrow appressed loosely-woolly or sometimes nearly glabrous claws, and very narrow subulate-pointed spreading or reflexed laminæ, pale brown-yellow when dry, varying in length, but the longest intermediate ones always exceeding the florets. Florets exceedingly numerous, all hermaphrodite or a very few outer ones female. Achenes glabrous. Pappus-bristles numerous, fine, minutely serrulate.

Queensland. Bare ridges of Endeavour river, and around Port Bowen, *A. Cunningham*; Harvey's Bay, Sandy Cape, *R. Brown* (with smaller flower-heads).

N. S. Wales. New England, *C. Stuart*; Bent's Basin, *Woolls.*

20. **H. rupicola,** *DC. Prod.* vi. 190. Apparently an undershrub, with the habit, foliage, and nearly all the characters of *H. collinum*, of which it may be a variety; but the almost leafless woolly peduncles are usually 6 to 9 in. long, the flower-heads rather smaller and flatter, the appressed claws of

the involucral bracts very woolly and their laminæ smaller and revolute, the longest intermediate ones, although exceeding the florets, are yet much shorter than their claws. Florets and pappus of *H. collinum.*

Queensland. Cape Grafton, *Banks and Solander*; rocky shores of Cleveland Bay, *A. Cunningham*; Dunk and Goold islands, *M'Gillivray*; Rockingham Bay, *Dallachy.*

21. **H. podolepideum,** *F. Muell. Rep. Babb. Exped.* 13. An undershrub or perhaps a low shrub, densely clothed with a cottony wool. Leaves petiolate, obovate or oblong, 1 to 2 in. long, soft and thick. Flower-heads on almost leafless peduncles of 2 to 4 in. Involucre broadly campanulate, $\frac{1}{2}$ to nearly $\frac{3}{4}$ in. diameter, the bracts numerous, appressed, narrow, with narrow, acute or slightly jagged scarious laminæ, the longest slightly exceeding the florets, of a pale straw-colour or dirty white. Florets exceedingly numerous, all (or nearly all?) hermaphrodite. Achenes glabrous. Pappus-bristles not very numerous, shortly barbellate.

N. S. Wales. Mount Goningberi, *Victorian Expedition*; near the Barrier Range, *Panton.*

S. Australia. To the N.W. of Lake Torrens, *Herb. F. Mueller.*

This species, as well in habit as in characters, connects the subsection *Oxylepis* with *Ixiolæna tomentosa.*

SECTION IV. CHRYSOCEPHALUM.—Herbs, with corymbose flower-heads. Involucre ovoid-turbinate or almost globose, the bracts coloured, in many rows, scarcely exceeding the florets, ciliate, appressed or squarrose, not radiating. Female florets in 1 or 2 rows, but not so numerous as the hermaphrodite ones. Achenes glabrous or papillose. Pappus-bristles few, simple at the base, plumose at the end.

22. **H. apiculatum,** *DC. Prod.* vi. 195. A perennial or perhaps annual, usually branching and hard at the base, with several erect stems, attaining 1 to 2 ft., clothed with a soft silvery tomentum, which rarely disappears from the older leaves. Radical and lower leaves oblong-cuneate and petiolate, the upper ones lanceolate or linear, or sometimes all narrow, the larger ones 1 to 2 in. long. Flower-heads in more or less dense terminal corymbs. Involucres in the original form broadly turbinate or nearly globose, about $\frac{1}{2}$ in diameter, of a bright golden colour, but sometimes much smaller, and, especially in the Western forms, passing into brown, red, straw-coloured, pure white, or pink. Bracts small and very numerous, the laminæ lanceolate, more or less ciliate, the outer ones sessile, the inner ones on woolly claws, all acute or the innermost obtuse, appressed or more rarely squarrose. Florets often as long as the involucre, those of the circumference in 1 or sometimes 2 rows, female, slender, with a reduced or abortive pappus, those of the disk very numerous. Achenes glabrous. Pappus of the disk of 4 to 10 fine bristles, strongly barbellate or almost plumose towards the end.—Hook f. Fl. Tasm. i. 212; *Gnaphalium apiculatum,* Labill. Pl. Nov. Holl. ii. 43. t. 188; Bot. Reg. t. 240; *G. flavissimum,* Sieb. Pl. Exs.; *Helichrysum flavissimum,* DC. Prod. vi. 195; *H. odorum,* DC. l. c. 196; *Chrysocephalum helichrysoides,* Walp. in Linnæa, xiv. 503; *C. apiculatum,* Steetz in Pl. Preiss. i. 474; *C. vitellinum,* Sond. and Muell. in Linnæa, xxv. 514 (the root apparently annual).

N. Australia. Arnhem's Land, *F. Mueller.*

Queensland. Bustard Bay, *Banks and Solander;* Keppel islands, *M'Gillivray;* on the Maranoa, *Mitchell.* (These specimens appear more shrubby and woolly, with broader leaves and larger flower-heads, than the others, but the smaller forms are also in many Queensland collections.)

N. S. Wales. Port Jackson, *R. Brown, Sieber, n.* 336, and others, and from various parts of the colony in numerous collections.

Victoria. Throughout the colony, *F. Mueller* and others.

Tasmania. Port Dalrymple, *R. Brown ;* abundant throughout the island, *J. D. Hooker.*

S. Australia. Memory Cove, *R. Brown :* from the Murray to St. Vincent's and Spencer's gulfs, *F. Mueller* and others.

Var. *minor.* Leaves narrow but woolly, flower-heads smaller but not numerous, connecting this with *H. semipapposum.—H. ramosissimum,* Hook. in Mitch. Trop. Austr. 83.—Chiefly in dry barren situations in all the eastern colonies.

Var. *occidentale.* Leaves narrow. Flower-heads medium-sized, the bracts appressed or squarrose, as frequently white pink or straw-coloured as yellow.—*Chrysocephalum squarrulosum,* Steetz in Pl. Preiss. i. 472; *C. flavissimum,* Steetz, l. c. 473 ; *C. canescens,* Turcz. in Bull. Mosc. 1851, i. 196; *C. glabratum,* Turcz. l. c. 197.

W. Australia. King George's Sound and adjoining districts, *Drummond, n.* 20, 115, 121, 342, 343, *Preiss, n.* 23, 25, *Oldfield, Maxwell.*

23. **H. semipapposum,** *DC. Prod.* vi. 195. Very closely allied to *H. apiculatum,* and probably a variety, often scarcely to be distinguished from some forms of that species. Stems usually more erect and stiff, though not so stout. Leaves narrow, often entirely deprived of wool, but sometimes quite cottony-white. Flower-heads small, more numerous and corymbose than they usually are in *H. apiculatum,* and the bracts more frequently squarrose, but none of these differences constant.—Hook. f. Fl. Tasm. i. 211 ; *Gnaphalium semipapposum,* Labill. Pl. Nov. Holl. ii. 42. t. 187 ; *Helichrysum ciliatum, H. squarrulosum,* and *H. brevicilium,* DC. Prod. vi. 195, 196 ; *Chrysocephalum asperum* and *C. semipapposum,* Steetz in Pl. Preiss. i. 473, 474 ; *C. squarrulosum,* Sond. in Linnæa, xxv. 515.

Appears to be as common as *H. apiculatum,* or nearly so, in **N. S. Wales, Victoria, Tasmania,** and **S. Australia,** but the respective distribution is uncertain, for, having several hundred specimens before me from very numerous localities of this and the preceding species, I have in vain endeavoured to distribute them satisfactorily into two or any greater number of distinct groups.

Var. *brevifolium,* Sond. in Linnæa, xxv. 515. Nearly glabrous and sometimes glutinous. Leaves numerous, small, and often clustered. Flower-heads small, not numerous, the bracts frequently squarrose.—*H. microlepis,* DC. Prod. vi. 195. Scrub of the interior of N. S. Wales, and on the Murray river in Victoria and S. Australia —I was for some time disposed to retain this as a distinct species, from its very different aspect, but the examinations of some specimens lead me to suspect that it is rather a state of *H. semipapposum,* induced by circumstances of growth, than even a variety as adopted by Sonder.

SECTION V. OZOTHAMNUS.—Shrubs or rarely undershrubs or herbs. Involucre small, oblong, ovoid or turbinate-campanulate (rarely larger and ovoid-conical), the bracts imbricate, not exceeding the florets, or the inner ones with small coloured radiating tips not much longer than the florets. Female florets few or none. Achenes glabrous or shortly villous. Pappus-bristles simple, often thickened or more denticulate towards the end.

Ozothamnus was originally proposed as a genus by R. Brown, without any detailed enumeration of species, but from the character he gave as well as from his MS. notes, it is evident he assigned different limits to it from those which have been subsequently adopted, for, relying chiefly on the shape of the involucre and small number of florets (under 20), he included *H. ramosum* and excluded *H. reticulatum*. Lessing, working chiefly upon S. African *Helichrysa*, found it necessary to reunite *Ozothamnus* with that genus as a section. De Candolle and others have subsequently adopted Brown's genus, for those shrubby species only which have the habit of *Cassinia*, but as I have been unable to assign any tangible character to the group so limited, I have felt obliged to return to Lessing's views, extending however the limits of the section so as to include the herbaceous species contemplated by Brown, as well as a few others which have neither the hemispherical radiating involucre of *Xerochlæna*, nor the peculiar characters of *Chrysocephalum*.

24. **H. Dockerii,** *F. Muell.* (as *Acanthocladium*). A rigid divaricately branched undershrub, very white with a close dense cottony wool, the smaller branchlets often spinescent. Leaves oblong-lanceolate or almost ovate, acute or obtuse, narrowed at the base but scarcely petiolate, rather thick and flat, rarely above ½ in. long. Flower-heads nearly sessile, solitary or perhaps clustered. Involucre rather broadly ovoid, 3 to 4 lines long, the bracts straw-coloured or pale brown, opaquely scarious, the outer ones ovate and woolly-ciliate, the inner narrow, with glabrous tips not spreading. Florets above 20, as long as the involucre. Achenes glabrous or papillose. Pappus-bristles numerous, fine, more or less barbellate at the end.—*Acanthocladium Dockerii*, F. Muell. Fragm. ii. 156.

N. S. Wales. Sand hills, Darling river, *Beckler*.

F. Mueller, in the flowers he examined, found no tails to the anthers; in those I dissected the tails were rather long but exceedingly fine and difficult to find.

25. **H. argyroglottis,** *Benth.* A divaricately-branched shrub, scarcely woolly except the under surface of the leaves. Leaves very shortly petiolate, oblong or linear, obtuse, ½ to 1 in. long, mostly scabrous above and white underneath. Flower-heads solitary and nearly sessile. Involucre rather narrow, turbinate-campanulate, 6 to 7 lines long without the rays, the outer bracts broadly lanceolate, acute, rigid, with narrow scarious edges, closely appressed, gradually lengthening, the innermost long and narrow, with white radiating tips or laminæ 2 to nearly 3 lines long. Receptacle conical. Florets all hermaphrodite, scarcely exceeding the appressed involucral bracts. Achenes glabrous. Pappus of several fine slightly denticulate bristles.—*Argyroglottis turbinata*, Turcz. in Bull. Mosc. 1851, ii. 84. t. 1; *Conanthodium Drummondii*, A. Gray in Hook. Kew Journ. iv. 273.

W. Australia, *Drummond, 5th Coll. Suppl. n.* 63. Although overlooked by Turczaninow, I found prominent tails to the anthers in this species as in *H. Dockerii*.

26. **H. ramosum,** *DC. Prod.* vi. 181. Herbaceous and said to be annual, but some specimens have a thick woody rhizome, the stems erect, branching, slender but attaining 2 or 3 ft., quite glabrous or with a little deciduous cottony wool. Leaves not numerous, linear-lanceolate or elliptical-oblong, flat, narrowed into a short petiole, the longest often 2 in. long, but mostly smaller and sometimes more numerous and all under ½ in., glabrous or white underneath. Flower-heads small, in little dense corymbs forming a large loose almost leafless panicle. Involucre ovoid, at length broadly campanulate, about 2 lines long, slightly woolly, the bracts numerous, the inner ones with small white or pink spreading tips. Florets from about 20 to

nearly 30, very few of them female. Achenes glabrous. Pappus-bristles not numerous, thickened and denticulate towards the end.—Steetz in Pl. Preiss. i. 470; *H. gracile*, DC. Prod. vi. 181.

W. Australia. King George's Sound and adjoining districts, *R. Brown*, *A. Cunningham*, and others, *Preiss*, *n.* 26; Swan River and Flinders Range, *Drummond*, *n.* 112, *Oldfield*, and others, and (with pink involucres and smaller leaves) *Drummond*, *4th Coll. n.* 221, *Clarke*, and others.

27. **H. cordatum,** *DC. Prod.* vi. 180. An undershrub with long flexuose reclining or almost climbing branches, closely covered with white cottony wool. Leaves petiolate, the lower ones cordate-ovate, 1 to 2 in. long, losing their wool on the upper side, densely cottony underneath, the upper ones few and small. Flower-heads small and numerous, in little compact corymbs forming a large loose almost leafless panicle. Involucre turbinate-campanulate, about 2½ lines long, woolly-white, the inner and intermediate bracts with small white glabrous spreading tips. Florets about 20, a very few of the outer ones female. Achenes glabrous or papillose. Pappus-bristles scarcely thickened upwards.—Steetz in Pl. Preiss. i. 469.

W. Australia. King George's Sound, *R. Brown* and others, thence to Swan River, chiefly near the sea, *Drummond*, *n.* 168, *Preiss*, *n.* 29, and many others, covering the hills at the entrance to Swan River, *Fraser*. There are occasionally a few small scales on the receptacle.

28. **H. obovatum,** *DC. Prod.* vi. 180. Stems (from a woody base?) weak, branching, reclining or flexuose, woolly-white when young, at length glabrous. Leaves on slender petioles, obovate, ½ to 1 in. long, thin and flat, glabrous or cottony-white underneath. Flower-heads small, in a loose irregular leafless panicle. Involucre broadly turbinate-campanulate, scarcely 2 lines long, woolly as well as the peduncles, the bracts not very numerous, the inner ones with very small or scarcely any scarious spreading tips. Florets about 20. Achenes glabrous. Pappus-bristles scarcely thickened upwards. —F. Muell. Fragm. ii. 89.

N. S. Wales. Rocky slopes on the Hastings river, *A. Cunningham*, Clarence river, *Beckler*.

29. **H. Bidwillii,** *Benth.* Stems weak, straggling or flexuose, with more or less of a deciduous cottony wool. Leaves on slender petioles, ovate or ovate-elliptical, mostly acute, ½ to 1 in. long, flat and thin, glabrous or cottony-white, especially underneath. Flower-heads small, in small rather compact panicles terminating leafy branches. Involucre broadly campanulate, about 2 lines long, glabrous or very thinly woolly, the bracts rather numerous, scarious, appressed, the inner ones with very small slightly spreading acute or jagged tips. Florets about 20 or rather more, a few outer ones female. Achenes glabrous. Pappus-bristles serrulate, but scarcely thickened upwards.

Queensland. Wide Bay, *Bidwill*.
N. S. Wales. Macleay river, *Beckler*.
With the foliage nearly of *H. obovatum*, this has the inflorescence and involucres of *H. Becklerii*.

30. **H. Becklerii,** *F. Muell. Herb.* A shrub (or undershrub?) with

slender divaricate branches, tomentose or pubescent when young. Leaves almost sessile, oblong-lanceolate or linear, flat or with recurved margins, about 4 to 8 lines long, glabrous above, hoary or white underneath. Flower-heads small, in small compact panicles terminating leafy branches. Involucre broadly campanulate, nearly 3 lines diameter, the bracts rather numerous, scarious, loosely appressed, without spreading tips. Florets above 20, several of the outer ones female. Achenes papillose-pubescent or glabrous. Pappus-bristles serrulate, but scarcely thickened upwards.—*Ozothamnus Becklerii*, F. Muell. Fragm. i. 183.

N. S. Wales, *F. Bauer;* Hastings river, *Fraser, Leichhardt, Beckler.*

31. **H. diotophyllum,** *F. Muell. Fragm.* v. 150. A shrub with virgate rather slender branches, covered with a loose cottony wool. Leaves almost scale-like, erect, linear-lanceolate, 1 to 2 lines long, acute, with revolute margins expanded at the base into thick rounded auricles. Flower-heads small, in small very compact panicles terminating leafy branches. Involucre broadly campanulate or almost globular, about 2 lines diameter, slightly woolly at the base, the bracts numerous, scarious, appressed, without spreading tips. Achenes shortly hirsute. Pappus-bristles few, barbellate towards the end.

N. S. Wales. Dogwood Creek, *Leichhardt.* With the foliage nearly of *H. baccharoides*, this has the inflorence and involucres nearly of *H. Becklerii.*

32. **H. diosmifolium,** *Less. in Steud. Nom. Bot. ed.* 2. An erect shrub, said to attain sometimes 20 ft., the branchlets minutely viscid-pubescent or tomentose. Leaves narrow-linear with minute points or almost obtuse, rarely exceeding ½ in., the margins revolute and not decurrent, glabrous or hoary-tomentose underneath. Flower-heads small and numerous, in dense terminal corymbs. Involucre nearly globular or broadly campanulate, 2 to nearly 3 lines diameter, the bracts broad, obtuse, concave, loose and very white or tinged with pink but without spreading tips. Florets about 20, a few outer ones sometimes female. Achenes glabrous or papillose. Pappus-bristles serrulate but scarcely thickened upwards.—*Metalasia rosmarinifolia*, Sieb. Pl. Exs.; *Ozothamnus diosmifolius*, DC. Prod. vi. 166.

Queensland. Burnett river, *F. Mueller;* Pine river, *Fitzalan.*
N. S. Wales. Port Jackson to the Blue Mountains, *R. Brown, Sieber, n.* 342, *and Fl. Mixt. n.* 515, and others; Lachlan and Macquarrie rivers, *A. Cunningham;* New England, *C. Stuart;* Macleay, Hastings, and Clarence rivers, *Beckler;* Richmond river, *C. Moore.*

33. **H. adnatum,** *Benth.* A tall heath-like shrub, the branches tomentose and marked with raised glabrous lines, decurrent from the leaves and persistent after they have fallen. Leaves linear, obtuse retuse or with recurved points, the margins decurrent, glabrous or scabrous above, white or tomentose underneath. Flower-heads small and numerous, in a dense terminal corymb. Involucre narrow, about 2 lines long, the bracts obtuse, concave, rather loose and white or straw-coloured, but without spreading tips. Florets about 10 to 12, 1 to 3 of the outer ones sometimes female. Achenes glabrous or papillose. Pappus-bristles few, thickened and denticulate upwards, only 1 or 2 or entirely wanting to the female florets.—*Ozo-*

thamnus adnatus, DC. Prod. vi. 166 (from the character); *O. retusus*, Sond. and Muell. in Linnæa, xxv. 510.

N. S. Wales. Darling river, *Dallachy and Goodwin.*
Victoria. Barren hills, Bacchus Marsh, *F. Mueller;* Wimmera, *Dallachy.*
S. Australia. Mountains on Gawler river, *Behr;* Murray scrub and Light river, towards Mount Remarkable, *F. Mueller.*

Var. *scabra.* Leaves scabrous-pubescent or almost muricate above, cottony-white underneath.—*Ozothamnus scaber*, F. Muell. in Linnæa, xxv. 407, Sond. in Linnæa, xxv. 511.

The peculiar decurrent lines distinguish at once this species from several others both of *Helichrysum* and *Cassinia*, which have an otherwise similar foliage and inflorescence.

34. **H. Cunninghamii,** *Benth.* A heath-like shrub with virgate woolly-tomentose branches. Leaves narrow-linear, mucronate-acute, rather rigid, about ½ in. long, with revolute margins not decurrent, smooth or with a few asperities above, tomentose underneath. Flower-heads small and numerous in compact terminal corymbs. Involucre very narrow, glabrous, pale straw-colour, about 2 lines long, the bracts not numerous, narrow, rigid, and almost acute, without coloured tips. Florets usually 3, without any scales between them. Achenes glabrous. Pappus-bristles slender, not thickened upwards.—*Cassinia Cunninghamii*, DC. Prod. vi. 156.

N. S. Wales. Summit of Mount Dangan on Hunter's River, *A. Cunningham.* Although the involucre is more like that of some *Cassiniæ* than of any *Helichrysum*, the want of any scales on the receptacle (besides the involucral bracts surrounding the florets) prevents the retaining the species in the former genus.

35. **H. reticulatum,** *Less. in Steud. Nom. Bot. ed.* 2. An erect stout shrub of several feet, with thick woolly-white branches. Leaves broadly linear, very obtuse, with closely revolute margins, 1 to 2 in. long, glabrous above with transverse indented reticulations, tomentose underneath. Flower-heads not so small as in the allied species, numerous, in a rather dense terminal corymb. Involucre broadly campanulate, about 3 lines diameter, woolly-tomentose, the bracts numerous, not spreading, the innermost often with minute white tips and some others with scarious ones, or sometimes all woolly to the end. Florets about 20, a few of the outer ones female. Achenes densely villous. Pappus-bristles serrulate and slightly thickened towards the end.—*Chrysocoma reticulata*, Labill. Pl. Nov. Holl. ii. 40. t. 183; *Faustula reticulata*, Cass. (DC.); *Gnaphalium reticulatum*, Spreng. Syst. iii. 471; *Ozothamnus reticulatus*, DC. Prod. vi. 164; Hook. f. Fl. Tasm. i. 202.

Tasmania. Adventure Bay, *Nelson* (in *Herb. R. Brown*); on the rocky shores of the island, *J. D. Hooker.*

36. **H. cinereum,** *F. Muell. Herb.* An erect much-branched shrub of several feet, the branchlets tomentose. Leaves linear, obtuse, rarely exceeding ½ in., with revolute margins, not decurrent, glabrous above, tomentose underneath, sometimes very narrow, sometimes thick and rather broad. Flower-heads small and numerous, in rather dense terminal corymbs. Involucre at first ovoid, at length broadly turbinate, about 3 lines long, the bracts rather numerous, appressed, often almost acute, the innermost without any or with minute scarcely spreading white tips. Florets 15 to 20 or rather

more, a very few of the outer ones female. Achenes papillose. Pappus-bristles serrulate, slightly thickened upwards.—*Chrysocoma cinerea*, Labill. Pl. Nov. Holl. ii. 39. t. 182; *Ozothamnus cinereus*, DC. Prod. vi. 165; Hook. f. Fl. Tasm. i. 203; *O. turbinatus*, DC. Prod. vi. 164.

N. S. Wales. 'Blue Mountains' (but probably rather from the seacoast), *R. Cunningham*.

Victoria. Frequent along the seacoast from the Glenelg river, *Robertson*, to Wilson's Promontory, Lake King and Snowy River, *F. Mueller*.

Tasmania. Kent's group, Bass's Straits, *R. Brown*; northern shores of the colony, *J. D. Hooker*.

37. **H. bracteolatum,** *Benth.* A shrub with densely woolly-tomentose branches. Leaves linear or almost lanceolate, obtuse or nearly so, with revolute margins shortly decurrent on the stem, $\frac{3}{4}$ to $1\frac{1}{2}$ in. long, glabrous above and sometimes transversely indented, as in *H. reticulatum*, tomentose underneath. Flower-heads nearly as in *H. cinereum*, but in a looser corymb with bracteolate peduncles. Involucre as in that species ovoid-turbinate, about 3 lines long, most of the bracts with very short scarious tips not white. Florets about 15 to 20. Achenes papillose. Pappus-bristles prominently serrulate but scarcely thickened upwards.—*Ozothamnus bracteolatus*, Hook. f. Fl. Tasm. i. 203.

Tasmania. Flinders' Island, *Gunn*, a single specimen. The species requires further elucidation to ascertain how far the characters separating it from *H. cinereum* are constant.

38. **H. cassinioides,** *Benth.* An erect, branching, heath-like shrub, the branches rather slender, tomentose. Leaves narrow-linear, obtuse or with minute recurved points, above $\frac{1}{2}$ in. long, the margins much revolute, not decurrent, glabrous and smooth above, tomentose underneath. Flower-heads small and numerous, in small dense terminal corymbs. Involucre narrow, about 2 lines long, the bracts scarious, concave, appressed, without spreading tips. Florets about 12 to 15. Achenes glabrous. Pappus-bristles scarcely thickened upwards.

Queensland. Keppel Bay and Broad Sound, *R. Brown*. Closely resembles *Cassinia aculeata*, var. *uncata*, but there are certainly no scales between the florets (*Herb. R. Brown*).

39. **H. Gunnii,** *F. Muell. Herb.* A tall shrub, nearly allied to *H. rosmarinifolium*, the branches densely woolly-tomentose. Leaves narrow-linear, mostly obtuse, $\frac{3}{4}$ to $1\frac{1}{2}$ in. long, with revolute margins, not decurrent, tomentose above when young, at length glabrous, woolly underneath. Flower-heads larger than in *H. rosmarinifolium*, very numerous, in a broad dense terminal compound corymb. Involucre narrow, nearly 3 lines long, the bracts rather numerous, the inner ones with small white radiating tips. Florets, achenes, and pappus of *H. rosmarinifolium*.—*Ozothamnus Gunnii*, Hook. f. Fl. Tasm. i. 205.

Tasmania. Sand-hills by the seashore, *Gunn*. Perhaps a variety of *H. rosmarinifolium*.

40. **H. ledifolium,** *Benth.* Closely allied to *H. rosmarinifolium*, and perhaps a variety, the branches rather stouter, the leaves linear with revolute margins, but more crowded and thicker, rusty-tomentose when young, gla-

brous above and mostly reflexed when old, all under ½ in. long. Flower-heads rather larger than in *H. rosmarinifolia*, but otherwise the inflorescence, involucres, florets, achenes, and pappus quite as in that species.—*Cassinia ledifolia*, DC. Prod. vi. 155; *Ozothamnus ledifolius*, Hook. f. Fl. Tasm. i. 204.

Tasmania. Mountains, at an elevation of 3000 ft., *Fraser* and others; Flinders' Island, *Backhouse*. There are certainly no scales to the receptacle in our specimens.

41. **H. rosmarinifolium,** *Less. in Steud. Nom. Bot. ed.* 2. A handsome shrub, attaining 8 or 9 ft., the branchlets more or less clothed with a short white or rusty tomentum or rarely glabrous. Leaves linear, mostly obtuse, from under ½ in. to above 1 in. long, the margins recurved or revolute, not decurrent, glabrous or scabrous above, more or less tomentose underneath. Flower-heads small and numerous, in dense corymbs, usually terminating numerous small leafy branches forming a large leafy panicle. Involucre narrow, scarcely 2 lines long, the bracts imbricate, obtuse, the innermost with small white radiating tips. Florets varying from 6 to 14, a very few of the outer ones female. Achenes strongly ribbed, papillose. Pappus-bristles denticulate, slightly thickened towards the end.—*Eupatorium rosmarinifolium*, Labill. Pl. Nov. Holl. ii. 38. t. 181; *Petalolepis rosmarinifolia*, Cass. (DC.); *Chrysocoma rosmarinifolia*, Spreng. Syst. iii. 424; *Ozothamnus rosmarinifolius*, DC. Prod. vi. 165; Hook. f. Fl. Tasm. i. 205. t. 54.

Victoria. Australian Alps, Baw-Baw, Mitta-Mitta, Cobberas, Mount Barkly, etc., at an elevation of 4000 to 6000 ft., *F. Mueller*.

Tasmania. Derwent river and Port Dalrymple, *R. Brown*; abundant on the banks of streams in the northern parts of the colony, *J. D. Hooker*.

The following appear to me to be but slight varieties of *H. rosmarinifolium*, which, in a large number of specimens, it becomes impossible distinctly to separate from it:—

Ozothamnus thyrsoideus, DC. Prod. vi. 165, Hook. f. Fl. Tasm. i. 205, with the leaves more glabrous and smooth, and the involucres, especially the radiating tips, rather larger.—Tasmania, with intermediate forms from Victoria.

O. ericæfolius, Hook. f. Fl. Tasm. i. 204, t. 57, with the leaves of the side-branches under ¼ in. long, those on the main branches scarcely ½ in.—Tasmania.

O. purpurascens, DC. Prod. vi. 165, with the involucres, except the white radiating tips, more or less pink or purple.—Tasmania and Victoria.

42. **H. ferrugineum,** *Less. in Steud. Nom. Bot. ed.* 2. A tall shrub, very closely allied to *H. rosmarinifolium*, with the same close white or rust-coloured tomentum on the branches and under side of the leaves. Leaves flatter and broader than in that species, and often acute, linear or linear-lanceolate, varying from ½ to nearly 3 in. long, glabrous above. Flower-heads very numerous and small, in a broad dense terminal compound corymb, not usually so paniculate as in *H. rosmarinifolium*. Involucres, florets, achenes, and pappus as in the fewer-flowered forms of that species, the florets usually under 10 and sometimes only 3 or 4.—*Eupatorium ferrugineum*, Labill. Pl. Nov. Holl. ii. 38. t. 180; *Chrysocoma ferruginea*, Spreng. Syst. iii. 424; *Ozothamnus ferrugineus*, DC. Prod. vi. 165; Hook. f. Fl. Tasm. ii. 206; *Cassinia argyophylla*, DC. Prod. vi. 155.

N. S. Wales. Shaded mountains, Illawarra, *A. Cunningham*; New England, *C. Stuart* (with narrower leaves).

Victoria. Port Phillip, *R. Brown*; Yarra river, *F. Mueller*, *Adamson*; Snowy River, *F. Mueller*; Creswick, *Whan*.

Tasmania. Not uncommon in various parts of the colony, *J. D. Hooker.*

S. Australia. Mount Gambier, *F. Mueller* (with narrow leaves, passing into *H. rosmarinifolium*).

43. **H. antennarium,** *F. Muell. Herb.* An erect, glabrous, much-branched shrub of 3 or 4 ft. Leaves obovate or oblong, very obtuse, ½ to 1 or rarely nearly 1½ in. long, contracted into a short petiole, flat or with slightly recurved margins, coriaceous, pale underneath. Flower-heads small and rather numerous, in dense corymbs. Involucre turbinate-campanulate, about 2 lines long, straw colour, the bracts rather numerous, appressed, the innermost often with very short, scarcely spreading white tips. Florets above 20, a very few outer ones occasionally female. Achenes papillose or hirsute towards the top. Pappus-bristles serrulate and slightly thickened towards the end.—*Swammerdamia antennaria,* DC. Prod. vi. 164; *Ozothamnus antennaria,* Hook. f. Fl. Tasm. i. 203.

Tasmania. Table Mountain, Derwent river, *R. Brown;* Mount Wellington and Western mountains, at an elevation of 3000 to 4000 ft., *J. D. Hooker* and others.

44. **H. obcordatum,** *F. Muell. Herb.* An erect shrub, attaining from 2 to 4 or 5 ft. in height, the branches cottony-white or nearly glabrous. Leaves broadly obovate, very obtuse, narrowed into a very short petiole, under ½ in. and sometimes all under ¼ in. long, flat or with slightly recurved margins, coriaceous, glabrous above, pale or white underneath. Flower-heads small and numerous, in a dense terminal corymb. Involucre narrow, about 2 lines long, straw-coloured, the bracts concave and loosely imbricate, without coloured tips. Florets about 6 to 8 in the original form, a few of the outer ones female. Achenes angular, glabrous or papillose. Pappus-bristles not at all or scarcely thickened upwards.—*Ozothamnus obcordatus,* DC. Prod. vi. 165; Hook. f. Fl. Tasm. i. 202.

Tasmania. Dry hillsides about the Derwent river, *R. Brown, J. D. Hooker.*

Var. *major.* Leaves and flower-heads rather larger, florets about 12.—*Cassinia obovata,* DC. Prod. vi. 155.

N. S. Wales. Twofold Bay, *A. Cunningham;* New England, *C. Stuart.*

Victoria. Mount Sturgeon, *Robertson, F. Mueller;* Grampians, *Wilhelmi;* Nangatta Mountains, Forest Creek, entrance to Genoa river, *F. Mueller.*

45. **H. Backhousii,** *F. Muell. Herb.* A small, erect, much-branched shrub of 1 to 3 ft., the branches slightly tomentose or at length glabrous. Leaves from narrow obovate to oblong-cuneate, more or less narrowed at the base, a few rarely almost oblong-linear, obtuse, under ½ in. long, flat or with slightly recurved margins, coriaceous, glabrous above, pale or white underneath. Flower-heads small, in small dense terminal corymbs. Involucre ovoid, above 2 lines long, pale brown, the bracts woolly-ciliate, appressed, a few of the innermost with small white spreading tips. Florets about 15, a very few of the outer ones female. Achenes glabrous or minutely papillose. Pappus-bristles serrulate and slightly thickened towards the end.—*Ozothamnus Backhousii,* Hook. Fl. f. Tasm. i. 204. t. 54; *Cassinia cuneifolia,* DC. Prod. vi. 155.

Tasmania. Port Arthur, *Backhouse;* Mount Wellington, at an elevation of 3000 ft., *A. Cunningham* and others.

46. **H. cuneifolium,** *F. Muell. Herb.* (as an *Ozothamnus*). A shrub, apparently much taller than *H. obcordatum*, with a denser almost floccose tomentum. Leaves cuneate-oblong, very obtuse, $\frac{1}{2}$ to $1\frac{1}{2}$ in. long, flat or with slightly recurved margins, but thinner than in *H. obcordatum*, and the margins often crisped. Flower-heads small and numerous, in large compact corymbs. Involucre narrow, the bracts rather few, the inner ones with white radiating tips, more conspicuous than in *H. Backhousii*. Florets 4 to 6. Achenes pubescent. Pappus-bristles serrulate, slightly thickened towards the end.

Victoria. Snowy river, Latrobe river, and affluents of Genoa river, *F. Mueller*.

47. **H. baccharoides,** *F. Muell. Herb.* An erect, very much-branched shrub, with the habit of *Olearia lepidophylla*. Leaves small, scale-like, erect and closely appressed, $\frac{1}{2}$ to 1 line long, the upper surface concave, but the margins so closely reflexed as to conceal the woolly under surface. Flower-heads small, sessile in dense clusters at the ends of the short branchlets forming long thyrsoid leafy panicles. Involucre very narrow, not 2 lines long, straw-colour, the bracts few, the inner ones with small white scarcely spreading tips. Florets 2 to 4, all hermaphrodite. Achenes papillose. Pappus-bristles rather numerous, scarcely serrulate, but slightly thickened towards the end.—*Baccharis (?) lepidophylla*, DC. Prod. v. 427; *Ozothamnus lepidophyllus*, Hook. f. in Hook. Lond. Journ. vi. 120, not of Steetz; *O. Hookeri*, Sond. in Linnæa, xxv. 509; Hook. f. Fl. Tasm. i. 201. t. 55.

Victoria. Australian Alps, Baw-Baw, Cobberas, Haidinger's Range, at an elevation of 4000 to 6000 ft., *F. Mueller*.

Tasmania. Table Mountain, Derwent river, *R. Brown;* abundant on the mountains throughout the colony, at an elevation of 3000 to 5000 ft., *J. D. Hooker*.

48. **H. lepidophyllum,** *F. Muell. Herb.* A tall, much-branched shrub, the branchlets slender, slightly tomentose or glabrous. Leaves small and scale-like, reflexed and closely appressed, ovate, about $\frac{1}{4}$ line long, thick, with revolute margins, glabrous above, tomentose underneath. Flower-heads small, in dense terminal corymbs. Involucre obovoid, about 2 lines long, slightly woolly at the base, with broad, concave, loosely imbricate, but not spreading white tips. Florets 8 to 10, all hermaphrodite. Achenes papillose. Pappus-bristles denticulate, thickened towards the end.—*Ozothamnus lepidophyllus*, Steetz in Pl. Preiss. i. 468.

W. Australia. King George's Sound to Swan River, *Drummond*, *Preiss*, *n.* 28; Salt river, and eastward to Cape Arid, *Maxwell*.

49. **H. scutellifolium,** *Benth.* A much-branched shrub, not so slender as *H. lepidophyllum*, and more tomentose. Leaves as in that species, small, scale-like, closely reflexed, ovate, about $\frac{1}{2}$ line long, thick with revolute margins, tomentose on both sides or glabrous above. Flower-heads small, sessile in little clusters of 3 to 5, at the ends of the branches. Involucre obovoid, about 2 lines long, the bracts few, pale brown, without spreading tips. Florets about 15 to 18, of which several outer ones slender and female, all remarkably dilated at the base. Achenes papillose. Pappus-bristles barbellate, slightly thickened towards the end.—*Ozothamnus scutellifolius*, Hook. f. Fl. Tasm. i. 202, t. 56.

Tasmania, *Oldfield;* Port Arthur, *Burnett.*

50. **H. pholidotum,** *F. Muell. Herb.* A tall shrub, with numerous slender virgate branchlets, almost destitute of wool. Leaves small, scale-like, erect and closely appressed, lanceolate, about 1 line long, thick, concave above, convex underneath, without recurved margins. Flower-heads small and numerous, in a dense terminal corymb. Involucral bracts mostly with white concave erect appendages.—*Ozothamnus pholidotus* or *Cassinia pholidota,* F. Muell. Fragm. ii. 131.

Victoria. Desert around Lake Hindmarsh, *Dallachy, L. Morton.* The specimens are in young bud only, and the number and structure of the florets cannot be very accurately ascertained, but I see no trace of receptacle-scales between them.

51. **H. lycopodioides,** *Benth.* A small glabrous shrub, with erect branchlets, mostly arising from under the old clusters of flower-heads. Leaves small and erect, but not closely appressed, oblong, or broadly linear, obtuse, not exceeding 2 lines in length, coriaceous, concave above, convex underneath, without recurved margins. Flower-heads small, in a dense terminal cluster or compound head. Involucre ovoid, about 2 lines long, the bracts rather numerous, several of the inner ones with brown or purplish, broad, concave but not spreading tips. Florets above 20, a few of the outer ones female. Achenes papillose. Pappus-bristles barbellate and thickened towards the end.—*Ozothamnus lycopodiodes,* Hook. f. in Hook. Lond. Journ. vi. 119; and Fl. Tasm. i. 201. t. 57.

Tasmania. Sugar-loaf, *Backhouse,* Apsley river, *Burnett,* both near Great Swanport.

52. **H. selaginoides,** *F. Muell. Herb.* A glabrous, much-branched, spreading herb or undershrub, about 1 ft. high. Leaves small, spreading or almost recurved, decurrent at the base, obtuse, mostly about 1 line long, thick, convex or flat underneath, without recurved margins. Flower-heads small, sessile in terminal clusters. Involucre rather narrow, about 2 lines long, straw-colour, the bracts not numerous, the innermost with short broad white spreading tips. Florets 8 to 10 or rarely 12, all hermaphrodite. Achenes papillose. Pappus-bristles not numerous, very slightly thickened and barbellate towards the end.—*Ozothamnus selaginoides,* Sond. and Muell. in Linnæa, xxv. 510; Hook. f. Fl. Tasm. i. 201. t. 56.

Tasmania. Table Mountain, west of Oatland, at an elevation of 3000 ft., *C. Stuart, Gunn, Oldfield.*

76. WAITZIA, Wendl.

(Viraya, *Gaudich.;* Morna, *Lindl.;* Pterochæte, *Steetz.*)

Involucre broadly turbinate-campanulate, hemispherical or almost globular, the bracts imbricate in many rows, all coloured and petal-like, the inner ones on narrow claws, spreading or appressed, but rarely and only shortly radiating. Receptacle flat, without scales. Florets numerous, all hermaphrodite, tubular, 5-toothed. Anthers with very fine tails. Style-branches nearly terete, truncate or with very short cones, almost capitate. Achenes somewhat compressed, glabrous or papillose, terminating in a slender beak (rarely very short). Pappus of capillary bristles usually cohering at the base, simple barbellate or plumose.—Herbs, usually annual. Leaves alternate, linear. Flower-heads in terminal corymbs, or rarely in oblong leafy racemes. Laminæ of the involucral bracts usually serrate-ciliate at the base.

The genus is limited to Australia. It is closely allied to *Leptorhynchus*, *Helichrysum*, and *Helipterum*, differing from the first in the involucre and habit, in the very long beak to the achene in some species, and in the plumose pappus in others; and from *Helichrysum* and *Helipterum* in the beaked achenes. Steetz found no tails to the anthers; I have found exceedingly fine ones in all the species.

Involucre broadly turbinate-campanulate, the bracts loosely imbricate or spreading, the outer passing gradually into the inner. Beak of the achene very long. Pappus-bristles scabrous or barbellate at the base only.
- Involucral bracts (yellow white or pink) narrow and acutely acuminate, the outer ones descending along the peduncle. Corymbs compact . . . . 1. *W. corymbosa.*
- Involucral bracts obtuse or acute, not acuminate, rarely any on the peduncle, usually with a few floral leaves.
  - Involucres golden-yellow, distinctly exceeding the florets. Corymbs compact . . . . 2. *W. aurea.*
  - Involucres white or pink, scarcely exceeding the florets. Corymbs loose . . . . 3. *W. nivea.*

Involucre broadly hemispherical, about ½ in. diameter. Beak of the achene short or long. Pappus-bristles strongly barbellate . . 4. *W. Steetziana.*

Involucre nearly globular, the inner row of bracts more distinctly radiating than the others. Pappus-bristles plumose.
- Peduncles long. Beak much longer than the achene . . . . . 5. *W. podolepis.*
- Peduncles short. Beak much shorter than the achene . . . . 6. *W. paniculata.*

1. **W. corymbosa,** *Wendl. Coll. Pl.* ii. 13. *t.* 42, *not of Steetz.* An erect annual of 1 to 2 ft., scabrous-pubescent or hoary but scarcely woolly. Leaves linear, the lower ones often 2 to 3 in. long, the margins revolute, stem-clasping at the base. Flower-heads usually numerous in a dense terminal corymb. Involucres about ¾ in. diameter, varying in colour from a pale to a dark yellow, white or bright pink, the intermediate bracts with a very slender claw and lanceolate laminæ very acutely acuminate, sometimes 3 or 4 lines long, the outer with gradually shorter broader claws and smaller laminæ passing gradually into small scales more or less descending on the peduncle, 2 or 3 innermost rows of bracts with linear broader scarious claws without any or with very small laminæ. Florets rather shorter than the involucres. Achenes produced into a slender beak several times longer than the achene itself. Pappus-bristles fine and slightly scabrous, slightly united at the base.—Hook. Bot. Mag. t. 5443; *W. acuminata*, Steetz in Pl. Preiss. i. 453; Regel, Gartenfl. t. 401; *W. discolor*, Turcz. in Bull. Mosc. 1851, i. 194.

Victoria. Wimmera and Murray river, *Dallachy.*

S. Australia. Pine forests and scrub to the N.E. of Lake Gairdner, *F. Mueller;* Lake Gillies, *Burkitt*

W. Australia. Swan River, *Drummond, 1st Coll., 4th Coll., n.* 198, *Preiss*; Kalgan, Murray, and S. Hutt rivers, *Oldfield;* Salt Creek, *Maxwell.*

Wendland's figure is an excellent representation of the pink variety, Regel's of the yellow one. The figure in the 'Botanical Magazine' also appears to represent the true *W. corymbosa*, although the erroneous synonymy is copied from Steetz. The figure of *Morna nitida*, Lindl. Bot. Reg. t. 1941, perhaps also rather represents this species than *W. aurea.*

2. **W. aurea,** *Steetz in Pl. Preiss.* i. 452. An erect annual with the habit and linear scabrous-pubescent or almost glabrous leaves of *W. corymbosa.* Flower-heads rather larger and fewer in a looser corymb. Involucre golden-yellow or tinged with brown, the bracts gradually passing from the sessile outer to the stipitate inner or intermediate ones, but more rigid, obtuse or acute but not acuminate, distinctly exceeding the florets, not descending along the peduncle but usually, especially when young, closely surrounded by 2 or 3 small linear floral leaves. Achenes with the long slender beak and simple pappus of *W. corymbosa,* except that the bristles are sometimes barbellate or ciliate-plumose quite at the base.—*Leptorhynchus aureus,* Benth. in Hueg. Enum. 64; DC. Prod. vi. 161; *Morna nitida,* Lindl. Bot. Reg. t. 1941 (?) (referable perhaps to *W. corymbosa*).

W. Australia. Swan River and adjoining districts, *Huegel, Drummond,* 1*st Coll. n.* 333, *and* 5*th Coll. n.* 381, *Preiss, n.* 2, 4, 5, 8, and others; Blackwood and Bowes rivers and Champion Bay, *Oldfield;* Roebuck Bay, *Marten;* South coast, *Baxter;* S.W. Bay and Middle Mount Barren, *Oldfield.*

3. **W. nivea,** *Benth.* An erect annual with linear scabrous-pubescent or almost glabrous leaves and rather large flower-heads usually few in a loose corymb as in *W. aurea,* the involucres surrounded by a few floral leaves as in that species of which it may be a variety, with the involucral bracts of a pure white or pink or very rarely with a very pale yellowish tinge, they are also not rigid, more frequently acute, though not acuminate as in *W. corymbosa,* the almost herbaceous linear claws are more conspicuous, and the florets are longer in proportion to the involucre. Achenes and pappus as in *W. aurea.* —*Morna nivea,* Lindl. Bot. Reg. 1838, t. 9; *Leptorhynchus suaveolens,* Benth. in Hueg. Enum. 64; DC. Prod. vi. 160; *Waitzia odontolepis,* Turcz. in Bull. Mosc. 1851, ii. 77; *W. corymbosa,* Steetz in Pl. Preiss. i. 450, not of Wendland.

W. Australia. King George's Sound, *R. Brown* and others, and thence to Swan River, *Drummond,* 1*st Coll. n.* 334; 5*th Coll. n.* 382, 383; *Preiss, n.* 12, 13, and others; and Murchison river, *Oldfield, Drummond,* 6*th Coll. n.* 158.

Helichrysum rigidulum, DC. Prod. vi. 193, described from a fragment communicated by Sweet, which I have not seen, is probably this species, of which a single flower-head, without examining the achene and pappus, might be taken almost for that of *Helichrysum adenophorum.*

4. **W. Steetziana,** *Lehm. in Pl. Preiss.* i. 454. An erect annual, more slender than the preceding species and almost always under 1 ft. high, glabrous or with a little loose wool. Flower-heads solitary or in loose corymbs smaller than in the preceding species. Involucres hemispherical, about ½ in. diameter, varying from a pure white to a pale or bright yellow, the bracts rather broad, obtuse or scarcely acute, without external floral leaves, passing gradually from the outer sessile to the inner ones on rather broad claws. Florets much smaller than in *W. nivea.* Achenes very papillose, contracted into a slender beak, usually very short at the time of flowering, more or less lengthened when the achene is ripe, sometimes to twice or three times its own length. Pappus bristles short, strongly barbellate or shortly plumose. —*W. tenella,* Hook. Bot. Mag. t. 5342, and *W. dasycarpa,* Turcz. in Bull. Mosc. 1851, ii. 77 (both with long beaks to the achenes); *Leptorhynchus*

citrinus, Benth. in Hueg. Enum. 64; DC. Prod. vi. 161; *Waitzia citrina*, Steetz in Pl. Preiss. i. 454, and *W. sulphurea*, Steetz, l. c. 453 (with the involucres more or less yellow); *W. brevirostris*, Steetz in Pl. Preiss. i. 451 (with short beaks to the achenes and white involucres).

W. Australia. Swan River and adjoining districts, *Drummond*, 1*st Coll. n.* 336, 337; 5*th Coll. Suppl. n.* 65; *Preiss, n.* 6, 7, 9, 10, 15, and others; Murchison river, *Oldfield*; Stirling range, Gairdner and Salt rivers, *Maxwell.*

The species, in habit, foliage, and general form of the involucres, often so closely resembles *Helipterum Cotula*, that it is only to be distinguished by the beak of the achene and by the outer involucral bracts as petal-like as the inner ones instead of being more rigid and brown. The length of the beak to the achene appears to be very variable and sometimes short, even in the ripe achene, but in general it is short at the time of flowering and lengthens considerably afterwards.

5. **W. podolepis,** *Steetz in Pl. Preiss.* i. 450. An erect simple annual. Leaves linear, 1 to 2 in. long, hoary-tomentose (*Gaudichaud*) or glabrous (*Sonder*). Flower-heads in the upper axils on peduncles longer than the leaves, forming a loose corymb or short raceme. Involucre described as hemispherical, figured rather as broadly campanulate, about ½ in. diameter, the outer bracts white and appressed, the inner row with short spreading laminæ of a brown-yellow (*Gaudichaud*), all pure white (*Sonder*). Achenes with a slender beak much longer than themselves. Pappus-bristles strongly barbellate.—*Viraya podolepis*, Gaudich. in Freyc. Voy. Bot. 466. t. 89; *Leptorhynchus podolepis*, DC. Prod. vi. 160; Sond. in Linnæa, xxv. 501.

W. Australia. Sharks' Bay (*Gaudichaud*). I have seen no plant answering to the description and figure. It seems to partake in some respects of the characters of *W. Steetziana*, in others of *W. paniculata*. Sonder, who saw a specimen at Berlin, says it is a *Leptorhynchus*, not a *Waitzia*, but he does not say on what character he relies for distinguishing the two genera. The involucre, achene and pappus figured are certainly those of *Waitzia* not of a *Leptorhynchus*.

6. **W. paniculata,** *F. Muell. Herb.* An erect woolly-tomentose annual rarely exceeding 6 in. Leaves oblong-spathulate or the upper ones linear, soft. Flower-heads nearly sessile on very short axillary branches, forming an oblong dense leafy panicle or raceme. Involucre broadly ovoid or almost globose, of a pale straw-colour, about 3 lines diameter, the scarious ciliate bracts imbricate, the outer ones appressed, the intermediate ones (about 1 row) with flat claws and small oblong spreading laminæ, the innermost with narrow claws and very small ciliate-fringed laminæ. Florets 5 to 20, small. Achenes strongly papillose, with a short slender beak. Pappus-bristles 12 to 15, plumose.—*Pterochæte paniculata*, Steetz in Pl. Preiss. i. 456.

W. Australia. King George's Sound, *R. Brown* and others, and thence to Swan River, *Drummond*, 5*th Coll. n.* 358, *Preiss, n.* 35, and others; Murchison river, *Oldfield.*

77. HELIPTERUM, DC.

(Pteropogon, *DC.*; Rhodanthe *and* Xyridanthe, *Lindl.*; Anisolepis *and* Hyalosperma, *Steetz*; Triptilodiscus, *Turcz.*; Acroclinium, Monencyanthes *and* Dimorpholepis, *A. Gray*; Duttonia *and* Cassiniola, *F. Muell.*)

Involucre from broadly hemispherical to narrow-ovoid or cylindrical, the bracts imbricate in several rows, either entirely or only their laminæ scarious or petal-like, more or less spreading or appressed .Receptacle flat, convex or conical, without scales, and in the Australian species without bristles or fringed pits. Florets in the Australian species hermaphrodite tubular and 5- rarely 4-toothed, or very rarely a few in the circumference female, slender but not longer than the others, 2- to 4-toothed, several in the centre frequently sterile. Anthers with fine tails. Style-branches nearly terete, truncate. Achenes angular terete or somewhat flattened, very rarely contracted at the top but not distinctly beaked, glabrous, papillose or more frequently densely silky-villous. Pappus of capillary or very rarely dilated and almost scale-like bristles, finely plumose-ciliate from the base, those of the female florets or of the central sterile ones sometimes fewer or wanting.—Herbs frequently annual, sometimes perennial, or very rarely slender divaricate shrubs with leafy stems, clothed with cottony wool or nearly glabrous. Leaves alternate or the lower ones very rarely opposite, entire. Florets yellow, the laminæ of the involucral bracts usually white yellow brown or pink, often varying in all these colours with intermediate shades in the same species.

A considerable genus, but confined to South Africa and Australia, the species of each of the two regions all endemic. It differs from *Helichrysum* solely in the plumose pappus, and, although annual duration, more scarious outer involucral bracts, and some other minor characters are more prevalent in *Helipterum* than in *Helichrysum*, yet there are several species in each genus closely allied to corresponding ones in the other, and the section *Lawrencella* of *Helichrysum* is an approach, both in habit and in the prominently barbellate pappus-bristles, to *Helipterum* ; the species of each genus are, however, numerous both in Australia and South Africa, and the character is not difficult to appreciate, it may therefore be convenient to retain the two as distinct genera. Two species, *H. pyrethrum* and *H. dimorpholepis*, have the pappus-bristles dilated towards the base, tending towards the scales of *Rutidosis* but with a very different involucre. Several S. African species have the pits of the receptacle fringed with short bristles. The genus is also closely connected with *Waitzia* through *W. Steetziana*, and with the *Angianthea* through *Cephalipterum*, which is very nearly allied to *Helipterum condensatum*.

SECTION I. **Euhelipterum.**—*Involucre broadly hemispherical, with or without radiating petal-like laminæ to the inner or intermediate bracts.*

Achenes villous with long silky hairs.	
Receptacle flat. Involucres with radiating laminæ.	
Outer bracts thinly scarious or petal-like, inner petal-like with narrow claws. Leaves broad, stem-clasping . . .	1. *H. Manglesii.*
Outer bracts brown, scarious, inner with coloured laminæ and broad claws. Leaves narrow.	
Pappus with a terminal tuft of compact hairs. Annual. Rays white or pink	2. *H. roseum.*
Pappus equally plumose.	
Perennial. Involucral ray white	3. *H. anthemoides.*
Annual. Involucral ray yellow	4. *H. polygalifolium.*
Receptacle conical. Involucres with radiating laminæ and broad claws.	
Pappus with a terminal tuft of compact hairs.	
Radiating involucral laminæ reddish	5. *H. rubellum.*
Radiating involucral laminæ metallic-green when dry . .	6. *H. chlorocephalum.*
Pappus equally plumose. Radiating involucral laminæ white	7. *H. floribundum.*
Pappus-bristles dilated and scale-like, plumose-ciliate. Radiating involucral laminæ white, Leaves small . . .	8. *H. pyrethrum.*
Receptacle flat or convex, honeycombed. Involucral bracts with scarious tips, scarcely radiating.	

Involucre hemispherical, the outer bracts small and scarious . . 9. *H. heteranthum.*
Involucre broadly campanulate, the outer bracts foliaceous though thin . . . . *Podotheca fuscescens.*
Achenes glabrous or papillose.
Outer bracts few, sessile, intermediate with subulate claws and radiating petal-like laminæ.
Outer bracts subulate . . . . 10. *H. stipitatum.*
Outer bracts broad . . . . 11. *H. incanum.*
Outer bracts brown, scarious or coloured, intermediate with broad claws and radiating petal-like laminæ.
Outer bracts wholly scarious or petal-like . . . . . . 12. *H. Cotula.*
Outer bracts with a lanceolate rigid although coloured centre. Achenes much compressed . . . . 13. *H. hyalospermum.*
(See also the section *Lawrencella* of *Helichrysum*, in which the pappus is almost plumose, and *Rutidosis Brownii*, in which the pappus-scales are almost divided into bristles.)

SECTION II. **Pteropogon.**—*Involucres ovoid, turbinate-campanulate or cylindrical, the outer bracts appressed, scarious or rigid, the intermediate or inner with or rarely without radiating petal-like laminæ or tips. Achenes villous. Florets rarely above* 12.

Flower-heads small with conspicuous rays, in compact terminal corymbs. Achenes shortly hirsute.
Almost all the bracts with white laminæ as long as or longer than the claws . . . . 14. *H. condensatum.*
Laminæ of the inner bracts shorter than their claws, those of the outer ones minute or none.
Involucre ovoid-turbinate, the laminæ white . . . . . 15. *H. polyphyllum.*
Involucre cylindrical, the laminæ yellow . . . . . . 16. *H. Humboldtianum.*
Flower-heads on long peduncles or loosely corymbose, with conspicuous rays. Achenes densely silky-villous.
Outer involucral bracts with linear green tips.
Involucres cylindrical, the rays as long as the outer bracts, linear tips squarrose. Corymb rather compact . . . . 17. *H. involucratum.*
Involucres turbinate-campanulate, the rays as long as the outer bracts, linear tips erect . . . . 18. *H. tenellum.*
Involucres ovoid, the rays not half so long as the outer bracts . . . . 19. *H. gracile.*
Outer involucral bracts with obtusely scarious tips.
Glabrous. Flower-heads on long peduncles. Involucres ovoid . . . . 20. *H. strictum.*
Cottony. Flower-heads loosely corymbose. Involucres broadly turbinate . . . . 21. *H. corymbiflorum.*
Flower-heads nearly sessile. Involucres narrow.
Dwarf branching plants, the flower-heads in an irregular leafy corymb. Involucres with very small white laminæ.
Cottony. Florets 10 to 15 . . . . 21. *H. corymbiflorum* var.
Nearly glabrous. Florets 4 to 6 . . . . 22. *H. pygmæum.*
Stems ½ to 1½ ft. high. Flower-heads very small and numerous in an oblong or globular spike-like leafless panicle. No radiating tips . . . . 23. *H. spicatum.*

SECTION III. **Monencyanthes.**—*Involucres ovoid, cylindrical or campanulate, the bracts scarious or coloured without any or with very minute scarious radiating tips. Achenes glabrous or papillose. Flower-heads often very small.*

Flower-heads in dense corymbose clusters almost contracted into heads.
Erect cottony herb of ½ to 1½ ft. Flower-heads small and numerous. Involucral bracts very woolly inside at the base.

Florets 2 to 4 . . . . 24. *H. moschatum.*
Slender divaricate undershrub or shrub. Flower-heads few.
Involucral bracts ciliate. Florets 15 to 20 . . . . 25. *H. pterochætum.*
Flower-heads in loose leafless corymbs or panicles. Involucres small, narrow, scarious, coloured. Florets 10 to 15.
Leaves linear or lanceolate, acute. Corollas dilated upwards.
Involucres all pedicellate, under 2 lines long. No radiating tips . . . . 26. *H. polycephalum.*
Involucres mostly sessile or nearly so, about 2 lines long, with minute scarious radiating tips . . . . 27. *H. corymbosum.*
Leaves linear obtuse. Corollas very slender. Involucres without radiating tips . . . . 28. *H. læve.*
Flower-heads singly sessile within the floral leaves. Small annuals. Florets above 20.
Plant of 1 to 1½ in. Involucres campanulate. Pappus-bristles about 10, plumose, not dilated . . . . 29. *H. exiguum.*
Plant of 3 to 6 in. Involucres broadly ovoid. Pappus-bristles of the perfect florets about 5, dilated and almost scale-like . . . . 30. *H. dimorpholepis.*

(*Olearia conocephala*, p. 480, has homogamous flower-heads and a plumose pappus, but the involucre is less scarious than in *Helipterum*, and the style different.)

SECTION I. EUHELIPTERUM.—Involucre broadly hemispherical, with or rarely without radiating petal-like laminæ to the inner or intermediate bracts.

1. **H. Manglesii,** *F. Muell. Herb.* An erect glabrous corymbosely-branched annual attaining 1 to 2 ft. Leaves ovate-oblong or broadly lanceolate, clasping the stem with rounded auricles. Flower-heads showy on long peduncles bearing a few scarious scales. Involucre hemispherical, when fully out the outer bracts sessile and scarious, the inner ones with a narrow claw and oblong radiating petal-like lamina ¼ to ½ in. long, varying from a pale to a rich pink and sometimes deep purple at the base. Receptacle flat. Florets short, all hermaphrodite, yellow or purple. Achenes densely woolly-hairy. Pappus of 15 to 20 equally plumose bristles.—*Rhodanthe Manglesii*, Lindl. Bot. Reg. t. 1703; Bot. Mag. t. 3483, 5283, 5290; Sweet, Brit. Fl. Gard. ser. 2. t. 295; Steetz in Pl. Preiss. i. 447.

W. Australia. Swan River, *Drummond, 1st Coll., also 5th Coll. n.* 383; near Woodbridge, *Preiss, n.* 49; Kalgan river and Champion Bay, *Oldfield;* Salt river, *Maxwell.*—The species varies very much in the size of the flower-heads.

2. **H. roseum,** *Benth.* An annual with erect or ascending simple or slightly branched stems from under 1 ft. to nearly 2 ft. long, glabrous or nearly so. Leaves linear, acute or almost obtuse, the upper ones few and small, the lower ones sometimes shorter and more obtuse. Flower-heads large, solitary, terminal. Involucre hemispherical, the outer scarious bracts sessile, short, tinged with brown, passing gradually into the inner ones with broad linear claws and radiating petal-like laminæ often ½ in. long, varying from a bright pink to pure white. Receptacle flat. Florets all hermaphrodite, the inner ones often sterile. Achenes densely woolly-villous. Pappus of 10 to 15 plumose bristles, terminating in a dense brush, formed of cilia closer packed and deeper coloured but not longer than the others.—*Acroclinium roseum*, Hook. Bot. Mag. t. 4801.

W. Australia. Murchison river, *Oldfield, Drummond, 6th Coll. n.* 157.

3. **H. anthemoides,** *DC. Prod.* vi. 216. Rootstock perennial, with numerous erect simple rather slender stems, rarely much above 1 ft. high and often short, glabrous. Leaves linear, often rather crowded, mostly ½ in. long, rarely linear-lanceolate and longer, glabrous and smooth or more frequently marked with impressed dots. Flower-heads solitary. Involucre hemispherical, spreading to about ¾ to 1 in. diameter, including the ray, the outer bracts short, broad, scarious, tinged with brown, the inner with broad scarcely ciliate claws and radiating petal-like laminæ 3 to 4 lines long, of a pure white. Receptacle flat. Florets all hermaphrodite. Achenes densely silky-hairy. Pappus-bristles 15 to 20, equally plumose.—Hook. f. Fl. Tasm. i. 215. t. 61; *Helichrysum anthemoides*, Sieb. in Spreng. Syst. iii. 484; *Helipterum punctatum*, DC. Prod. vi. 216.

Queensland, *Bowman;* near Mount Faraday, *Mitchell;* head of the Gwydir, *Leichhardt.*

N. S. Wales. Port Jackson to the Blue Mountains, *R. Brown, Sieber, n.* 344, and others; New England, *C. Stuart;* Richmond river, *Fawcett;* Macquarrie river, *Fraser, A. Cunningham.*

Victoria. Port Phillip, *Gunn;* Grampians, *Wilhelmi;* Wimmera, *Dallachy.*

Tasmania. Derwent river, *R. Brown;* Formosa, Western Mountains, Launceston, etc., *Gunn* and others.

S. Australia. Near Adelaide, *Whittaker;* Mount Remarkable, *F. Mueller.*

4. **H. polygalifolium,** *DC. Prod.* vi. 216. A glabrous annual branching at or near the base, ascending or erect, ½ to 1 ft. high. Lower leaves oblong-spathulate or linear-cuneate, the upper ones linear-lanceolate and small. Flower-heads solitary. Involucre hemispherical, above 1 in. diameter including the ray, the outer short broad scarious bracts tinged with brown, inner ones with a broad slightly woolly-ciliate claw and spreading petal-like lamina, often ½ in. long, of a pale or bright yellow. Receptacle flat. Florets all hermaphrodite. Achenes densely silky-villous. Pappus-bristles 15 to 20, equally plumose with long fine cilia, not tufted at the end.—*H. diffusum*, DC. Prod. vi. 216.

N. S. Wales. Molle's Plains, Lachlan river, and Peel's Range, *Fraser, A. Cunningham;* Darling and Lachlan rivers, *Victorian and other Expeditions.*

S. Australia. Murray river, *F. Mueller;* Venus and Streaky bays, *Warburton.*

5. **H. rubellum,** *Benth.* An annual with simple rather slender stems of 6 to 9 in., more or less glandular-pubescent. Flower-heads solitary. Involucre hemispherical, about ½ in. diameter without the ray, the outer bracts ovate, scarious, tinged with brown, the inner with rather broad scarious claws and spreading petal-like laminæ, about 3 lines long, of a pale or rather dark pink. Receptacle conical. Florets all hermaphrodite, the inner ones usually sterile. Achenes densely silkv-villous. Pappus of the outer perfect achenes of about 20 very plumose bristles, the upper short cilia condensed into a small tuft. A few inner achenes abortive, with a reduced pappus.—*Acroclinium rubellum*, A. Gray in Hook. Kew Journ. iv. 271.

W. Australia, *Drummond, n.* 347, *Roe.* I have not succeeded in finding any tails to the anthers in this species, but may have overlooked them if exceedingly fine.

6. **H. chlorocephalum,** *Benth.* An annual with numerous mostly simple stems of ½ to 1 ft., glabrous or nearly so. Leaves linear or slightly

spathulate, rarely above ½ in. long, the upper ones few and small. Flower-heads terminal. Involucre hemispherical, about ½ in. diameter without the ray, the outer short scarious bracts tinged with brown, the inner ones with scarious claws and radiating petal-like laminæ, 3 to 4 lines long, of a metallic green or yellowish-brown colour when dry (perhaps yellowish when fresh). Receptacle conical. Florets all hermaphrodite, the inner ones sterile. Achenes densely villous. Pappus-bristles 10 to 15, plumose, the upper short cilia of each bristle condensed in a terminal tuft, usually deeper coloured than the rest of the pappus, the inner achenes abortive with a reduced pappus.—*Schœnia chlorocephala*, Turcz. in Bull. Mosc. 1851, i. 193; *Acroclinium multicaule*, A. Gray in Hook. Kew Journ. iv. 271.

W. Australia, *Drummond, 4th Coll. n.* 199. This may prove to be a variety of *H. rubellum*.

7. **H. floribundum,** *DC. Prod.* vi. 217. Stems erect and nearly simple when flowering the first year, at length diffuse, much-branched and woody at the base, the branches ascending from a few inches to above 1 ft. high, glabrous or loosely woolly. Leaves linear or rarely linear-lanceolate, acute. Flower-heads solitary on each branch, but the upper branches often numerous and paniculate. Involucre hemispherical, rather smaller than in *H. anthemoides*, the bracts all white and petal-like, the outer ones short and sessile, passing into the inner ones with a scarious claw and radiating lanceolate lamina of 3 lines or more. Receptacle hemispherical or conical. Florets all hermaphrodite. Achenes densely silky-villous. Pappus of 7 to 10 rigid equally plumose bristles.—*H. chionolepis*, F. Muell. in Linnæa, xxv. 416.

N. S. Wales. Molle's Plains, *Fraser;* Darling and Lachlan rivers, *Victorian and other Expeditions;* between Stokes Range and Cooper's Creek, *Wheeler*.

S. Australia. Flinders Range, Cudnaka, *F. Mueller;* Wills' Creek, *Howitt's Expedition;* Mount Searl, *Warburton*.

Var. *Stuartianum*. Rather taller; flower-heads larger, the outer bracts assuming a straw-colour.—*H. Stuartianum*, Sond. in Linnæa, xxv. 518.—Murray river, *F. Mueller*.

8. **H. Pyrethrum,** *Benth.* An erect glabrous simple or slightly-branched annual of 6 to 8 in., the stem thickened at the base. Leaves lanceolate or linear but small and almost scale-like. Flower-heads solitary on the branches. Involucre hemispherical, 2 to 3 lines diameter without the ray, the outer bracts few short and scarious, passing into the upper stem-leaves, the inner ones with short brown rigid claws and radiating petal-like laminæ, spreading to ¾ in. diameter, pure white or tinged with pink. Receptacle conical. Florets all hermaphrodite, the inner ones probably sterile. Achenes densely silky-villous. Pappus of 10 to 15 plumose-ciliate bristles, more or less dilated scale-like and united at the base or sometimes to the middle.—*Anisolepis pyrethrum*, Steetz in Pl. Preiss. i. 447.

W. Australia. Swan River, *Preiss, n.* 14, *Drummond, 5th Coll. n.* 351. The pappus is rather variable, approaching that of *Rutidosis*, but the habit and involucre are totally unlike those of any species of that genus.

9. **H. heteranthum,** *Turcz. in Bull. Mosc.* 1851, i. 198. An erect annual of 6 in. to 1 ft., scabrous-pubescent and somewhat viscid, branching

and leafy near the base. Lower leaves petiolate, oblong or lanceolate, upper ones small and linear. Flower-heads solitary on long almost leafless peduncles. Involucre hemispherical, 4 to 6 lines diameter, the outer bracts broad, short and scarious, the inner with broad rigid claws and small scarious scarcely spreading laminæ. Receptacle slightly convex, prominently honeycombed. Florets all hermaphrodite, the central ones sterile. Achenes densely silky-villous. Pappus of about 20 rigid equally plumose bristles.—*H. discoideum*, A. Gray in Hook. Kew Journ. iv. 231; *H. anactinum*, F. Muell. Fragm. iii. 137.

W. Australia, *Drummond, n.* 96, *4th Coll. n.* 214, *5th Coll. n.* 374; Champion Bay, *Oldfield.*

The habit recalls sometimes that of *Podolepis Lessoni*, but it is at once known by the short florets, the villous achenes, and plumose pappus.

Var. *majus.* Stems 1 to 1½ ft. high. Lower leaves 2 to 3 in. long. Flower-heads nearly 1 in. diameter.—Between Moore and Murchison rivers, *Drummond, 6th Coll. n.* 152.

10. **H. stipitatum,** *F. Muell. Herb.* Probably perennial, loosely woolly-hairy, only the branches seen. Leaves linear, stem-clasping, 1 to 2 in. long, flaccid. Flower-heads rather large, on long peduncles, leafless except a few linear bracts, the upper ones with rigid coloured points. Involucre hemispherical, above ½ in. diameter without the ray, a few of the outer bracts subulate, coloured and glabrous, passing into the upper bracts of the peduncle, the inner ones very numerous with very narrow rigid glandular claws of about 3 lines, and ovate radiating petal-like laminæ of about 2 lines long, of a bright yellow, a few of the innermost smaller, with very small laminæ. Receptacle flat. Florets numerous, all hermaphrodite. Style-branches tipped with a prominent cone. Achenes glabrous or papillose. Pappus-bristles rather shortly plumose.—*Helichrysum stipitatum*, F. Muell. Fragm. iii. 133.

S. Australia. Finke river, *M'Douall Stuart's Expedition.* The species is evidently allied to *H. incanum*, but known at once by the subulate outer involucral bracts.

11. **H. incanum,** *DC. Prod.* vi. 215. A densely tufted perennial with ascending or erect simple stems or branches of 6 in. to 1 ft., woolly-white as well as the foliage. Leaves crowded at the base of the stems, from narrow-linear to linear-oblong, often 2 to 4 in. long, the upper ones smaller and distant. Flower-heads large on leafless peduncles. Involucre hemispherical, spreading to 1½ in. diameter, the bracts all petal-like but rigid, the outer ones short and sessile, the inner with linear or subulate claws, glandular or woolly-ciliate, and radiating laminæ 4 to 5 lines long, of a pure white or tinged with pink or brown or passing into a pale or bright yellow. Receptacle flat. Florets numerous, all hermaphrodite. Achenes glabrous. Pappus of 10 to 20 equally plumose bristles.—Hook. Ic. Pl. t. 318; Hook. f. Fl. Tasm. i. 214; *Elichrysum incanum*, Hook. Bot. Mag. t. 2881; *Helichrysum molle*, DC. Prod. vi. 194; *Elichrysum albicans*, A. Cunn. in Field, N. S. Wales, 359; *Helipterum albicans*, DC. Prod. vi. 215; *H. bicolorum*, DC. l. c.; *Waitzia brachyrhyncha*, F. Muell. in Linnæa, xxv. 407, and *Helipterum brachyrhynchum*, Sond. in Linnæa, xxv. 517 (with yellow involucres).

Queensland, *Mitchell.*
N. S. Wales. Port Jackson, *F. Mueller;* Blue Mountains, *A. Cunningham, Woolls*

and others; in the interior to the Lachlan and Molle's Plains, *A. Cunningham, Fraser;* New England, *C. Stuart;* Hastings river, *Beckler.*

Victoria. Snowy River, Forest Creek, Mount Timbertop, Maneroo, *F. Mueller;* Grampians, *Wilhelmi;* Wimmera, *Dallachy;* Skipton, *Whan;* near Woodalich, *Robertson.*

Tasmania. Northern parts of the island at all elevations, *J. D. Hooker.*

S Australia. Murray river, Cudnaka, Flinders Range, *F. Mueller.*

The yellow as well as the white variety sent from most localities in N. S. Wales, Victoria, and S. Australia, but only the white from Tasmania. See also F. Muell. Rep. Babb. Exped. 14, as to the diversity of colour.

12. **H. Cotula,** *DC. Prod.* vi. 215. A slender erect or ascending simple or branching annual, rarely exceeding 1 ft. and often under 6 in., flowering when only 1 or 2 in., with a few loose woolly hairs or at length glabrous. Leaves linear, very narrow, but not long, the first often opposite. Flower-heads solitary. Involucre hemispherical, spreading to from ¾ to 1 in. diameter including the ray, the outer bracts short, broad, obtuse and usually tinged with brown, the inner with short, broad, scarious, slightly woolly claws and radiating petal-like laminæ of 3 to 5 lines, varying from pure white to pale or bright yellow, a few of the innermost sometimes small. Receptacle flat. Florets numerous, all hermaphrodite, a few inner ones sometimes sterile. Achenes glabrous or papillose. Pappus of 10 to 15 plumose bristles, the upper cilia forming a yellow tuft.—Steetz in Pl. Preiss. i. 474; Hook. Bot. Mag. t. 5604; *Helichrysum Cotula,* Benth. in Hueg. Enum. 65; *Helipterum simplex,* Steetz in Pl. Preiss. i. 475; *H. citrinum,* Steetz, l. c. 474 (with yellow involucres); *H. pusillum,* Turcz. in Bull. Mosc. 1851, ii. 80; *H. præcox,* F. Muell. in Trans. Vict. Inst. 1855, 58; *H. semisterile,* F. Muell. Fragm. ii. 157; *Helichrysum Oldfieldii,* F. Muell. Fragm. iii. 134.

Queensland. Maranoa river, *Mitchell.*

N. S. Wales. Darling river, *Beckler.*

Victoria. Avoca river, *F. Mueller.*

W. Australia. King George's Sound, *Menzies,* and thence to Swan River, *Drummond, n.* 29, 169, 338, 339, *5th Coll. n.* 384, *Preiss, n.* 18, 21, *Oldfield;* Geographe Bay and Champion Bay, *Oldfield.*

The white and yellow varieties were generally gathered together by Drummond. Without examining the achene and pappus, this species might be readily confounded with *Waitzia Steetziana.*

13. **H. hyalospermum,** *F. Muell. Herb.* A slender erect glabrous or slightly woolly annual, rarely much above 6 in. high and often smaller. Leaves narrow-linear, almost filiform. Flower-heads on long peduncles, leafless except a few small scarious scales passing into the outer involucral bracts. Involucre hemispherical when fully out, 4 to 5 lines diameter without the ray, the outer bracts short, sessile, brown or yellow with a more rigid lanceolate centre, the inner with a broad scarious brown claw and yellow petal-like radiating lamina, about 2 lines long. Receptacle flat. Florets rather numerous, all hermaphrodite. Achenes glabrous or papillose, rather more compressed than in most *Heliptera.* Pappus-bristles 8 to 12, equally plumose, but yellow at the tips.—*Hyalospermum strictum* and *H. glutinosum,* Steetz in Pl. Preiss. i. 477; *Hyalospermum variabile,* Sond. in Linnæa, xxv. 519.

N. S. Wales. Lachlan river, *A. Cunningham;* Lachlan and Darling rivers, *Victorian and other Expeditions.*

Victoria. Wimmera, *Dallachy.*

S. Australia. Gawler Town, Burra-Burra, St. Vincent's Gulf, *F. Mueller;* Lake Gillies, *Burkitt.*

W. Australia. Swan River, *Drummond, n.* 340; Murchison river, *Oldfield.*

Some of the smaller specimens, where the thickening of the centre of the outer involucral bracts is less marked, are difficult to distinguish from yellow specimens of *H. Cotula.* The achenes appear to be larger than in that species (almost transparent in unripe specimens), and the scales on the peduncles do not occur in *H. Cotula.*

SECTION II. PTEROPOGON.—Involucres ovoid, turbinate-campanulate or cylindrical, the outer bracts appressed, scarious or rigid, the inner or intermediate ones with or rarely without radiating petal-like lamina or tips. Achenes villous. Florets rarely above 10.

14. **H. condensatum,** *F. Muell. Fragm.* iii. 136. Stature probably of *H. polyphyllum* and *H. Humboldtianum,* but only the upper portion of the stems seen, the inflorescence and foliage with loose cottony wool. Leaves linear or linear-oblong, obtuse, flaccid. Flower-heads small, numerous, in a compact corymb. Involucre ovoid, 2 lines long without the ray, the bracts woolly and appressed at the base, but nearly all of them with petal-like snow-white spreading laminæ as long as their claws, which are broad and scarious, the inner ones with a green central line. Florets about 8 to 10, all hermaphrodite. Achenes shortly hirsute, but only seen very young. Pappus of 15 to 20 finely plumose bristles.

W. Australia. Murchison river, *Oldfield.* This species appears to connect the genus with *Cephalipterum,* which scarcely differs except in its much more compact inflorescence.

15. **H. polyphyllum,** *F. Muell. Fragm.* i. 35. An erect annual of about 1 to 1½ ft., not much branched and slightly woolly. Leaves narrow-linear, 1 to 2 in. long. Flower-heads small, rather numerous, in terminal corymbs. Involucre ovoid or narrow-campanulate, nearly 3 lines long, slightly woolly, the outer bracts appressed without laminæ, the inner with broadly scarious claws and white petal-like radiating laminæ, rather shorter than their claws. Florets about 10 to 12, all hermaphrodite. Achenes shortly silky-hirsute. Pappus of 15 to 20 plumose bristles.

Queensland, *Bowman* and others; basaltic plains, from the Brisbane to Peak Range, *F. Mueller;* Kent's Plains, *W. Hill*; plains of the Condamine, *Leichhardt;* Rockhampton, *Thozet.*

16. **H. Humboldtianum,** *DC. Prod.* vi. 216. Erect, probably annual, 1 to 2 ft. high, woolly-white or at length nearly glabrous. Leaves linear or linear-lanceolate, acute. Flower-heads small, numerous, in dense terminal corymbs. Involucre cylindrical, 2 to 2½ lines long without the rays, the outer bracts closely imbricate with very short squarrose scarious tips, the inner with slightly woolly-ciliate claws and radiating rather rigid petal-like laminæ, 1½ to 2 lines long, of a bright yellow passing (when dry) into a metallic-green. Florets about 10 to 12, all hermaphrodite, a few of the inner ones sometimes sterile. Achenes silky-hirsute. Pappus of 15 to 20 plumose bristles.—*Elichrysum Humboldtianum,* Gaudich. in Freyc. Voy. Bot. 465. t. 88; *Schœnia? Humboldtiana,* Steetz in Pl. Preiss. i. 481; *Pteropogon*

Humboldtianus, F. Muell. in Linnæa, xxv. 415; *Helipterum Sandfordii*, Hook. Bot. Mag. t. 5350; *H. largiflorens*, F. Muell. Fragm. iii. 135.

W. Australia. Sharks' Bay (*Gaudichaud*); Murchison river, *Oldfield*, *Drummond*, *6th Coll. n.* 160.

17. **H. involucratum,** *F. Muell. Fragm.* iii. 135. A slender, erect, glabrous annual, corymbosely branched, under 1 ft. high. Leaves linear, almost filiform. Flower-heads numerous in a broad flat loose corymb. Involucre cylindrical, nearly 3 lines long without the ray, the outer bracts shortly woolly or nearly glabrous with narrow-linear herbaceous squarrose tips, often passing into the upper leaves, the inner ones with scarious rather broad glabrous claws and petal-like oblong radiating laminæ, about 3 lines long, of a bright yellow. Florets about 10 to 15. Achenes densely villous. Pappus-bristles about 20, plumose.

W. Australia. Murchison river, *Oldfield*. Nearly allied to *H. tenellum*, with a somewhat different involucre.

18. **H. tenellum,** *Turcz. in. Bull. Mosc.* 1851, i. 198. An erect, slender, corymbosely branched annual, of 6 in. to 1 ft., glabrous or slightly woolly. Leaves narrow-linear or filiform. Flower-heads pedunculate or nearly sessile, in a loose or rather dense broad flat corymb, the small upper leaves passing into the outer involucral bracts. Involucre turbinate, rather more than 2 lines long without the rays, the outer bracts with linear-subulate erect herbaceous tips, the inner with radiating petal-like laminæ, about 1½ lines long, varying from bright yellow to pure white or slightly pink. Florets about 12 to 15, all hermaphrodite. Achenes densely villous. Pappus-bristles 15 to 20, plumose.—*Pteropogon ramosus*, A. Gray in Hook. Kew Journ. iv. 270.

W. Australia, *Drummond*; the white variety between Moore and Murchison rivers, *Drummond*, *6th Coll. n.* 156.

19. **H. gracile,** *Benth.* A slender, erect, nearly simple or corymbosely branched annual, from 6 in. to nearly 1 ft. high, glabrous or very slightly woolly. Leaves linear-filiform, the lower ones occasionally opposite. Flower-heads few, solitary on the branches or in clusters of 2 or 3. Involucre ovoid, nearly 3 lines long, the outer bracts of a shining brown, appressed, a few with linear-subulate erect almost herbaceous tips, the inner ones with narrow acute petal-like radiating laminæ, shorter than their claws, of a bright yellow. Florets 10 to 15. Achenes densely villous. Pappus of about 10 plumose bristles.—*Pteropogon gracilis*, A. Gray in Hook. Kew Journ. iv. 269.

W. Australia. Swan river, *Drummond*, *1st Coll.* Possibly a slender variety of *H. tenellum*.

20. **H. strictum,** *Benth.* An erect glabrous annual, attaining 1 to 2 ft., but often under 6 in. Leaves oblong-lanceolate or linear, mostly stem-clasping, the lower ones petiolate, the uppermost small and narrow. Flower-heads solitary on long peduncles. Involucre ovoid or at length campanulate, ¼ to ½ in. diameter, the bracts rigid, closely imbricate, the inner ones with small oblong white radiating petal-like laminæ. Florets 3 or 4 only in some

heads, 8 to 10 in the larger ones, all hermaphrodite. Achenes densely silky-villous. Pappus of 20 or more equally plumose bristles.—*Xyridanthe stricta*, Lindl. Swan Riv. App. 23; *Pteropogon platyphyllus*, F. Muell. in Linnæa, xxv. 413.

N. S. Wales. Darling river, *Herb. F. Mueller;* Goyinga mountains, *Victorian Expedition;* between Stokes Range and Cooper's Creek, *Wheeler.*
S. Australia. Cudnaka, *F. Mueller.*
W. Australia. Swan River, *Drummond, 1st Coll.*

F. Mueller, in Rep. Babb. Exped. 14, unites this with *H. polygalifolium*, of which it has nearly the foliage, but the involucres and pappus appear to me to be very different.

21. **H. corymbiflorum,** *Schlecht. in Linnæa*, xxi. 448. An erect woolly-white corymbose annual, of ½ to 1 ft. Leaves linear or lanceolate, mostly obtuse, soft, the upper ones few and small. Flower-heads, in the original form, in a rather loose terminal leafless corymb, with a few small scarious bracts on the branches and peduncles. Involucre turbinate, about 3 lines long without the ray, the outer bracts wholly scarious, broad, obtuse, slightly woolly-ciliate, the inner with linear or cuneate claws, woolly towards the top and radiating petal-like white laminæ, about 2½ lines long. Florets all hermaphrodite, but some of the central ones usually sterile. Achenes densely silky-villous. Pappus of about 15 to 20 plumose bristles.—Sond. in Linnæa, xxv. 519.

N. S. Wales. Lachlan, Darling, and Murray rivers, *Victorian and other Expeditions.*
Victoria. Avoca river, *F. Mueller;* Wimmera, *Dallachy.*
S. Australia. Fiedler's Section, *Behr;* near Gawler Town, Holdfast Bay, *F. Mueller.*

Var.? *microglossa*, F. Muell. Herb. Dwarf, much branched, very woolly and only 2 to 4 in. high. Inflorescence compact and leafy, scarcely corymbose. Involucres on very short peduncles or almost sessile, narrow, about 2½ lines long, the radiating laminæ very small.—Goyinga mountains, *Victorian Expedition.* Possibly a distinct species, approaching in habit the *H. pygmæum.*

22. **H. pygmæum,** *Benth.* An annual, branching from the base, diffuse or erect, 2 to 4 in. high, slightly woolly or at length glabrous. Leaves narrow-linear, almost filiform. Flower-heads not very numerous, sessile in dense terminal leafy corymbs or clusters. Involucre narrow, about 4 lines long, the bracts brown, scarious, obtuse and appressed, the inner ones with small white ovate radiating tips. Florets 4 to 6, all hermaphrodite, but about half in the centre sterile, the corolla very slender. Achenes densely silky-villous. Pappus of numerous plumose bristles.—*Pteropogon pygmæus*, DC. Prod. vi. 245; A. Gray in Hook. Kew Journ. iv. 267; *P. australis*, Nees in Linnæa, xvi. 223.

N. S. Wales. Molle's Plains, *A. Cunningham, Fraser;* Darling and Lachlan rivers, *Burkitt.*
Victoria. Wimmera and Murray river, *Dallachy.*
S. Australia. Flinders Range, Cudnaka, towards Lake Torrens, *F. Mueller.*

Var. *occidentale.* Usually, but not always more glabrous, with rather smaller flower-heads, the radiating tips of the bracts sometimes very minute.—*Pteropogon Drummondii*, A. Gray in Hook. Kew Journ. iv. 267.
W. Australia, *Drummond, 4th Coll. n.* 175.

23. **H. spicatum,** *F. Muell. Herb.* Said to be annual, but the base

often hard and almost woody and showing sometimes a creeping rhizome, the stems slender, erect, often exceeding 1 ft., woolly-white or hoary or at length glabrous. Leaves linear, almost filiform, the lower ones sometimes above 2 in. long, the upper ones few, small and distant. Flower-heads very small, in a dense globular or ovoid terminal leafless cluster, sometimes extended into an oblong spike interrupted at the base. Involucre narrow, ovoid-cylindrical, 2 to 2½ lines long, the bracts scarious but rather rigid, usually of a rich shining reddish-brown, rarely pale-coloured, without spreading tips or wool inside. Florets about 5 or 6, all hermaphrodite and fertile or 2 or 3 of the inner ones sterile. Achenes densely silky-villous. Pappus-bristles rigid, plumose.—*Pteropogon spicatus*, Steetz in Pl. Preiss. i. 479; A. Gray in Hook. Kew Journ. iv. 268.

W. Australia, *Drummond, n.* 19, 362; Mount Lehmann, *Preiss, n.* 24; Champion Bay, Toodyay and Murchison rivers, *Oldfield.*

Var. *pallens.* Cluster of flower-heads larger, more compound and not so dense, the involucres very pale brown, all other characters precisely as in the original form.—*Helipterum monencyanthioides*, F. Muell. Fragm. iii. 137.—Tone and Salt rivers, *Maxwell.*

SECTION III. MONENCYANTHES.—Involucres ovoid, cylindrical or campanulate, the bracts scarious or coloured, without any or with very minute scarious radiating tips. Achenes glabrous or papillose. Flower-heads often very small.

24. **H. moschatum,** *Benth.* Annual or perhaps sometimes perennial with a creeping rhizome. Stems erect or decumbent at the base, ½ to 1½ ft. high, densely woolly as well as the foliage. Lower leaves petiolate, obovate or spathulate, the upper ones lanceolate or oblanceolate and stem-clasping. Flower-heads small and numerous, nearly sessile in dense corymbose or almost globular clusters. Involucres ovoid, scarcely 2 lines long, the bracts scarious, varying from a rich brown to pale straw-colour, without spreading tips, densely woolly inside at the base. Florets in all the specimens examined 2 or 3, all hermaphrodite and fertile. Achenes narrow-oblong, contracted at the base, smooth and glabrous, but so closely enveloped in the long intricate surrounding wool that it is difficult to extract them.—*Gnaphalium moschatum*, A. Cunn. in DC. Prod. vi. 236; *Calocephalus gnaphalioides*, Hook. in Mitch. Trop. Austr. 378; *Monencyanthes gnaphalioides*, A. Gray in Hook. Kew Journ. iv. 230.

Queensland. Near the Balonne river, *Mitchell.*

N. S. Wales. Molle's Plains, *A. Cunningham, Fraser;* Darling and Murray rivers, to the Barrier Range, *Victorian and other Expeditions.*

Victoria. Wimmera, *Dallachy.*

S. Australia. Cudnaka, Crystal Brook, Mount Lofty Ranges, *F. Mueller;* Wills' Creek, *Howitt's Expedition.*

The species has something of the aspect of *Gnaphalium luteo-album.*

25. **H. pterochætum,** *Benth.* An undershrub or small shrub, with slender but rigid divaricate woolly-white branches. Leaves narrow-linear, with revolute margins, nearly glabrous. Flower-heads small, nearly sessile in small terminal clusters. Involucre narrow-turbinate, straw-coloured, the bracts appressed, scarious, ciliate, with very small tips, deeper coloured but not radiating. Florets 15 to 20, all apparently hermaphrodite. Achenes

oblong, papillose. Pappus of 8 to 15 very fine plumose bristles.—*Chrysocephalum pterochætum*, F. Muell. in Linnæa, xxv. 416; *Helichrysum pterochætum*, F. Muell. Rep. Babb. Exped. 14.

N. S. Wales. Mount Goningberi, *Victorian Expedition.*
S. Australia. Near Cudnaka, *F. Mueller;* Stuart's Creek, Lake Gregory, *Babbage's Expedition.*
The tube of the corolla is bulbous, not at the base, as occurs more or less in many species, but just above the base in the specimens examined.

26. **H. polycephalum,** *Benth.* A slender erect corymbosely branched annual of 6 in. to 1 ft., at first woolly-white but soon glabrous, except the under side of the leaves. Leaves linear or linear-lanceolate, acute, the larger ones often decurrent. Flower-heads very small, all pedunculate in a large loose terminal corymb or panicle. Involucre narrow, scarcely 2 lines long, the bracts rigidly scarious, of a light shining reddish-brown, without spreading tips. Florets 10 to 15, all hermaphrodite and often all fertile. Achenes glabrous or sprinkled with a few short hairs. Pappus-bristles about 10, plumose.—*Pteropogon polycephalus*, A. Gray in Hook. Kew Journ. iv. 268; *Cassinia cuprea* or *Cassiniola cuprea*, F. Muell. Fragm. iii. 139.

W. Australia, *Drummond, Preiss, n.* 43; Murchison river, *Oldfield.*

27. **H. corymbosum,** *Benth.* An erect corymbosely branched annual, 6 in. to 1 ft. high, woolly-white or at length glabrous. Leaves linear or linear-lanceolate, acute, not decurrent. Flower-heads very small, shortly pedunculate in little clusters of 2 to 5, forming a loose irregular corymb. Involucre narrow, rather above 2 lines long, the bracts shining, of a reddish-brown, appressed, the inner ones usually but not always with minute white spreading tips. Florets achenes and pappus of *H. polycephalum*, of which this may possibly prove a variety.—*Pteropogon corymbosus*, A. Gray in Hook. Kew Journ. iv. 268.

W. Australia, *Drummond, 5th Coll. n.* 364; Darling Range, *Collie.*

28. **H. læve,** *Benth.* A slender erect branching annual under 6 in. high and quite glabrous. Leaves small, linear, rather obtuse, narrowed at the base. Flower-heads very small, all pedunculate, forming an irregularly corymbose panicle. Involucre narrow, about 2 lines long, the bracts of a rich shining brown, without spreading tips. Florets 10 to 12, all hermaphrodite. Achenes nearly glabrous. Pappus-bristles about 10.—*Pteropogon lævis*, A. Gray in Hook. Kew Journ. iv. 269.

W. Australia, *Drummond, 5th Coll. n.* 366. Nearly allied to the two preceding species.

29. **H. exiguum,** *F. Muell. in Trans. Vict. Inst.* 1855, 39. A dwarf, very much branched, nearly glabrous annual, forming little tufts 1 to 1½ in. diameter. Leaves small, linear-filiform. Flower-heads small, rather numerous, sessile. Involucre campanulate, about 1½ lines diameter, the bracts scarious, of a reddish-brown, obtuse, without radiating tips. Florets 20 to 40, all hermaphrodite. Achenes glabrous or papillose. Pappus of about 10 plumose bristles.—*Pteropogon demissus*, A. Gray in Hook. Kew Journ. iv. 269.

Victoria. Sandy stony declivities, Grampians, Serra and Victoria ranges, *F. Mueller;* Wimmera, *Dallachy;* Skipton, *Whan.*

S. Australia. Bugle Ranges and near Gawler Town, *F. Mueller.*

W. Australia, *Drummond, 5th Coll. Suppl. n.* 66.

30. **H. dimorpholepis,** *Benth.* An annual with erect or ascending branching stems, green, with a few long hairs, but scarcely any wool: Leaves linear, rather broad, ½ to 1 in long. Flower-heads small, sessile within a few floral leaves exceeding the head, terminal or sometimes lateral. Involucre broadly ovoid, above 2 lines long, the outer bracts lanceolate, scarious, fringed with long cilia, the inner with rigid glandular claws and small scarious tips not spreading. Receptacle conical. Florets rather numerous, exceeding the involucre, a few of the outer ones female. Achenes glabrous. Pappus of about 3 or 4 plumose-ciliate bristles, more or less flattened and scale-like, the outer achenes, especially those of the female florets, often without any pappus, and the innermost usually abortive.—*Dimorpholepis australis,* A. Gray in Hook. Kew Journ. iv. 227; Hook. Ic. Pl. t. 856; *Triptilodiscus pygmæus,* Turcz. in Bull. Mosc. 1851, ii. 66; *Duttonia sessiliceps,* F. Muell. in Linnæa, xxv. 410.

N. S. Wales. Port Jackson, *F. Mueller;* Blue Mountains, beyond Berrima, *Woolls;* Nangas, *M'Arthur;* New England, *C. Stuart.*

Victoria. Murray river, Yarra-Yarra, Mount M'Ivor, *F. Mueller*; Firy Creek, *Whan.*

S. Australia. Rocky Creek, *F. Mueller.*

W. Australia, *Drummond, 5th Coll. Suppl. n.* 54.

78. RAOULIA, Hook. f.

(Merope, *Wed.*)

Involucre ovoid campanulate or hemispherical, the bracts imbricate in several rows, more or less scarious, often opening out elastically when old, the inner ones rarely with radiating coloured tips. Receptacle flat or convex, without scales. Florets all tubular, the outer ones in 1 or rarely 2 rows, female and very slender, those of the disk hermaphrodite, 5-toothed, sometimes sterile. Anthers with fine tails. Style-branches nearly terete, truncate or capitate, sometimes undivided in sterile florets. Achenes oblong or obovate, glabrous or slightly hirsute, not beaked. Pappus of capillary bristles, simple or denticulate and occasionally thickened towards the end.—Dwarf densely tufted perennials, the short branches closely covered with small imbricate leaves, rarely shortly diffuse with crowded but less imbricate leaves. Flower-heads solitary, sessile or shortly pedunculate at the ends of the branches.

An alpine genus, extending over New Zealand and antarctic and Andine S. America; the two Australian species both endemic. In essential character the genus can scarcely be said to differ from *Helichrysum,* but from the peculiar habit the species have been always more readily connected with *Gnaphalium,* from which it differs in the female florets always much fewer than the hermaphrodites. A. Gray unites the American species with *Lucilia,* in which he may be right, although the habit and shape of the involucre appear to be different. The group requires much further revision, and may be better characterized when we obtain more numerous and more perfect specimens of several species which are as yet but little known.

Leaves imbricate, densely woolly-villous. Scapes leafless, ½ to 1 in. . 1. *R. Planchoni.*
Leaves crowded on decumbent stems, but not imbricate, closely silky-tomentose. Flower-heads sessile . . . . 2. *R. Catipes.*

1. **R. Planchoni,** *Hook. f.* A dwarf perennial, forming broad dense tufts, the branches concealed by the remains of old leaves. Leaves imbricate, obovate, 3 to 4 lines long, thick and densely villous with rust-coloured woolly hairs, becoming hoary when old. Scapes or terminal peduncles ½ to 1 in. high, leafless or with a single small bract, bearing a single flower-head. Involucre ovoid, about 3 lines long, the bracts scarious, narrow-linear, acute, the outer ones slightly woolly. Female florets few, in a single row; disk-florets numerous, mostly fertile. Achenes slightly hirsute. Pappus-bristles numerous, free, not thickened upwards.—*Gnaphalium Planchoni,* Hook. f. Fl. Tasm. i. 217. t. 62 C.

Tasmania. Table mountain, Derwent river, *R. Brown;* summit of Mount Olympus, *Gunn.*

2. **R. Catipes,** *Hook. f. Fl. Tasm.* i. 206. *t.* 58 (*R. tasmanica* on the plate). A low decumbent much-branched perennial, forming large tufts of a silvery-white from the uniform closely-appressed tomentum, the stems ascending to from 2 to nearly 6 in. Leaves crowded but scarcely imbricate, spreading and shortly decurrent, obovate-spathulate or oblanceolate, flat or concave, 2 to 3 or rarely 4 lines long. Flower-heads terminal, solitary and sessile. Involucre hemispherical, 3 to 4 lines diameter, the bracts scarious brown and appressed at the base, all but the outermost with small oblong spreading white tips. Florets numerous, mostly hermaphrodite with 1 (or rarely 2?) rows of female florets in the circumference, the majority of the hermaphrodites apparently sterile in some specimens, nearly all fertile in others, but the species does not appear to be so completely diœcious as are most *Antennariæ.* Style of the hermaphrodite florets always branched in all the specimens examined, though the branches do not always spread. Achenes pubescent. Pappus-bristles not very numerous, slightly cohering at the base, denticulate and thickened towards the end, especially in the hermaphrodite florets, very caducous.—*Gnaphalium Catipes,* DC. Prod. vi. 236; *Antennaria nubigena,* F. Muell. in Trans. Phil. Soc. Vict. i. 45, and in Hook. Kew Journ. viii. 161. Pl. Vict. t. 45.

Victoria. Cobberas mountains, at an elevation of 6000 ft,, *F. Mueller.*
Tasmania. Bare rocks on the summits of the highest mountains, *Gunn.*

Subtribe III. Eugnaphalieæ.—Flower-heads distinct or in dense clusters or compound heads, usually small. Female filiform florets numerous, in several rows or in separate heads.

79. ANTENNARIA, Gærtn.

Involucre ovoid campanulate or hemispherical, the bracts imbricate in several rows, more or less scarious, with or without spreading coloured laminæ. Receptacle without scales. Flower-heads diœcious; florets in the female individuals all filiform, 2- or 3-toothed, those in the male individuals apparently hermaphrodite but sterile, tubular, 4- or 5-toothed. Anthers with fine tails. Style in the males undivided. Achenes oblong, terete or compound, not beaked, abortive in the males. Pappus of capillary bristles, usually thickened and denticulate towards the end in the male florets or in all.—Perennial herbs with a tufted or branching base, the normal species more or less cottony-white. Leaves alternate, entire. Flowering-stems in the true species ascending or erect, bearing a cluster or corymb of small flower-heads. In the Australian species the flower-head exceptionally sessile and solitary.

The genus is spread over the northern hemisphere, chiefly in mountain regions, extending into the Andes of South America. The only Australian species (doubtfully inserted in the genus) is endemic.

1. **A. (?) uniceps,** *F. Muell. in Trans. Phil. Soc. Vict.* i. 105, *and in Hook. Kew Journ.* viii. 161. A small densely-tufted or shortly creeping perennial, quite glabrous. Leaves crowded linear or linear-cuneate, mucronate-acute, rigid, concave, about 3 lines long, narrowed towards the base and again dilated into a purplish stem-clasping sheath. Flower-heads small, solitary, almost sessile within the last leaves and shorter than them. Involucre ovoid, the inner bracts with slightly spreading tips, but without coloured laminæ. Florets rather numerous, all sterile with an undivided style in the specimens seen. Achenes abortive with a pappus of simple bristles in a single row, slightly serrulate but scarcely thickened upwards.

Victoria. Gravelly plains near springs, Munyong mountains, at an elevation of 5000 to 6000 ft., *F. Mueller.* I leave this for the present in the genus in which it was originally placed, for, until the fertile florets shall have been observed, its true affinities cannot be ascertained. The habit is not at all that of *Antennaria.* It will most probably prove to be a *Raoulia.*

80. GNAPHALIUM, Linn.

(Leontopodium, *R. Br.*; Euchiton, *Cass.*)

Involucre ovoid or campanulate (rarely hemispherical?), the bracts imbricate in several rows, more or less scarious, with or without small spreading tips. Receptacle without scales. Florets of the circumference female, filiform, in 2 or more rows, often very numerous, those of the disk fewer, often very few, hermaphrodite, tubular, 5-toothed. Anthers with fine tails. Style-branches in the disk-florets nearly terete, truncate. Achenes oblong or obovate, not striate, glabrous or papillose. Pappus of capillary bristles, in a single row.—Herbs annual or perennial, more or less cottony or woolly. Leaves alternate, entire, usually soft. Flower-heads small, usually clustered, either in the upper axils or in terminal spikes, corymbs or compound heads, rarely solitary.

A considerable genus, distributed over nearly the whole globe, but as yet very imperfectly defined. Of the 8 Australian species, 4 occupy a very wide range in the Old World, and 2 of them also in America. Of the remaining 4, 1 is also in New Zealand, the other 3 are endemic.

Clusters or compact corymbs of flower-heads terminal, leafless, solitary or several in an irregular panicle . . . . 1. *G. luteo-album.*
Clusters forming a terminal globular head, usually surrounded by a few floral leaves.
Annual . . . . 2. *G. japonicum.*
Perennials with a tufted or creeping rhizome.
Floral leaves narrow, glabrous above . . . . 3. *G. collinum.*

Floral leaves oblong, spreading, cottony on both sides. Flower-heads almost monœcious . . . . 4. *G. alpigenum.*
Clusters of flower-heads axillary or forming a terminal spike.
Flower-heads about 2 lines long . . . . 5. *G. purpureum.*
Flower-heads about 1 line long . . . . 6. *G. indicum.*
Flower-heads in little leafy corymbs. Branching annual of 1 to 3 in. 7. *G. indutum.*
Flower-heads solitary on leafy scapes of ½ to 3 in.
Involucre about 2 lines long, surrounded by linear floral leaves . 3. *G. collinum*, var.
Involucre at least 3 lines long, without floral leaves, besides very short woolly outer bracts . . . . 8. *G. Traversii.*

1. **G. luteo-album,** *Linn.; DC. Prod.* vi. 230. An annual or perhaps biennial, densely woolly-white, with ascending or erect stems of 1 to 1½ ft. when full grown. Lower leaves petiolate obovate or oblong-spathulate, obtuse; upper ones sessile linear or lanceolate, acute, all usually soft and retaining the wool on both sides. Flower-heads in loose terminal nearly globose clusters or dense corymbs, without floral leaves, either solitary and terminal or several of the clusters in the forks or on the branches of an irregular corymbose panicle. Involucres about 2 lines diameter, nearly globose, the bracts scarious, pale brown or straw-colour, with obtuse scarious tips not spreading. Female florets exceedingly numerous with a few hermaphrodites in the centre.—Hook. f. Fl. Tasm. i. 216; Steetz in Pl. Preiss. i. 478.

Queensland. Cape Upstart, Barnard Isles, Port Curtis, *M'Gillivray;* Suttor river, *Bowman;* Keppel Bay, *Thozet.*

N. S. Wales. Port Jackson to the Blue Mountains, *R. Brown* and others; northward to Clarence river, *Beckler;* southward to Gabo island, *Maplestone;* Lord Howe's Island, *M'Gillivray.*

Victoria. Murray river, Station Peak, *F. Mueller;* Wimmera, *Dallachy;* Skipton, *Whan;* Portland, *Allitt.*

Tasmania. Abundant in rocky places and wet or dry pastures, *J. D. Hooker.*

S. Australia. Mount Gambier, Lofty Range, Kangaroo Island, *F. Mueller;* Burra Burra, *Hinteraecker.*

W. Australia. King George's Sound and neighbouring districts, *Drummond, Preiss, n.* 33, 34; eastward to Esperance Bay, *Maxwell;* Murchison river, *Oldfield.*

The species is common in almost all the warm and temperate regions of the globe.

2. **G. japonicum,** *Thunb. Fl. Jap.* 311. An erect annual, usually under 1 ft., but when luxuriant 1½ ft. high, more or less cottony-white, the base of the stem often hard and almost woody. Leaves from oblong-spathulate and narrowed into a long petiole to linear and sessile, becoming glabrous above, cottony-white underneath. Flower-heads small, in dense globose clusters or compound heads, surrounded by a few floral leaves, either terminal and ½ to ¾ in. diameter or axillary and smaller. Involucres oblong, imbedded at the base in a dense white wool, the bracts scarious, brown or straw-colour, erect, obtuse or the inner ones acute, without spreading tips. Female florets 20 or more, hermaphrodite ones in the centre, solitary or very few. Achenes slightly compressed. Pappus-bristles very fine, scarcely cohering at the base. —Miq. Prolus. Fl. Jap. 109; F. Muell. Fragm. v. 150; *G. involucratum,* Forst. Prod. 55; DC. Prod. vi. 235; Hook. f. Fl. Tasm. i. 216; Steetz in Pl. Preiss. i. 478; Bot. Mag. t. 2582; *Euchiton Forsteri* and *E. pulchellus,* Cass. (DC.).

Queensland. Northumberland islands, *R. Brown;* Keppel Bay, *Thozet;* Brisbane river, Moreton Bay, *F. Mueller.*

N. S. Wales. Port Jackson, *R. Brown*, *Sieber*, *n.* 343, and others; New England, *C. Stuart;* Clarence river, *Beckler;* Darling river, *Victorian Expedition.*

Victoria. Murray river, *Dallachy;* Creswick and Ballarat, *Whan.*

Tasmania. Common in many parts of the island, *J. D. Hooker.*

S. Australia. Near Adelaide, Torrens river, Kensington, *F. Mueller;* Kangaroo island, *Waterhouse.*

W. Australia. South coast to Swan River, *Drummond, n.* 22, *5th Coll. n.* 370, *Preiss*, *n.* 46, 47, and others; Murchison river, *Oldfield.*

The species is also in New Zealand, and extends over some parts of the Eastern Archipelago, and northwards to Japan, from whence we have many specimens, easily recognized by Thunberg's description, but first identified by Miquel. With regard to the several synonyms quoted by De Candolle and copied by F. Mueller, some must remain doubtful, for De Candolle gives the perennial sign to his species, whilst all the perfect specimens I have seen show an annual root, as described by Thunberg.

3. **G. collinum,** *Labill. Pl. Nov. Holl.* ii. 44. *t.* 189. Very nearly allied to *G. japonicum*, and referred to it as a synonym by *F. Mueller.* It appears, however, to be always perennial, forming a tufted stock or emitting underground creeping rhizomes, the stature is usually smaller, the indumentum closer. Leaves more acute, the radical ones more persistent, usually glabrous above, white underneath. Flower-heads not forming so compact a head, and each involucre broader, usually brown. Florets and achenes the same as in *G. japonicum.*—DC. Prod. vi. 235; Hook. f. Fl. Tasm. i. 216; *Euchiton collinum*, Cass. (DC.).

N. S. Wales. New England, *C. Stuart;* Clarence river, *Beckler.*

Tasmania. Dry pastures, not very common, *J. D. Hooker.*

Var.? *radicans*, F. Muell. Dwarf and tufted. Flower-heads few, much larger than in the ordinary form.

Victoria. Summits of the Australian Alps, *F. Mueller.*

Tasmania. Western mountains, *Archer.*

Var.? *monocephalum*, Hook. f. Fl. Tasm. ii. 364. Very dwarf, with linear leaves and solitary flower-heads.—Australian Alps, *F. Mueller.*

These two varieties may possibly prove to be reduced states of *G. alpigenum*, for the specimens are scarcely in a normal condition.

G. gymnocephalum, DC. Prod. vi. 235, from Port Jackson, *G. cephaloideum*, Willd.; DC. Prod. vi. 236, raised in Continental gardens from Baudin's Australian seed, are probably, as suggested by De Candolle, *G. collinum* and *G. japonicum* (*G. involucratum*) respectively; but without the inspection of authentic specimens, it is impossible to ascertain the point.

4. **G. alpigenum,** *F. Muell.*; *Hook. f. Fl. Tasm.* i. 217. *t.* 62 A. A perennial with a tufted stock, emitting creeping stolons or prostrate barren stems. Radical leaves on long petioles, ovate or oblong, rather thick, cottony-white on both sides or at length nearly glabrous above. Flowering-stems not above 6 in. high, with a few oblong or spathulate leaves, all petiolate. Flower-heads in terminal globose clusters, surrounded by a few oblong cottony floral leaves, and a few small clusters sometimes in the upper axils; the heads nearly monœcious, in some the florets all or nearly all hermaphrodite but sterile, in others nearly all female, with 1 or 2 hermaphrodite in the centre. Involucre ovoid, about 2 lines long, cottony at the base only, the bracts oblong, obtuse, scarious, usually of a pale brown. Pappus-bristles numerous, quite free.

Victoria. Australian Alps, at an elevation of 4000 to 6000 ft., *F. Mueller.*
Tasmania. Table mountain, Derwent river, *R. Brown;* Western mountains, *Lawrence, Gunn.*

The species is allied on the one hand to *G. collinum,* on the other to the European *G. leontopodium.*

5. **G. purpureum,** *Linn.; DC. Prod.* vi. 232. An annual or perennial of short duration. Stems simple or branching from the base, ascending or erect, 6 in. to 1 ft. high. Leaves mostly petiolate and spathulate, or the upper ones rarely linear, cottony-white on both sides as well as the stem. Flower-heads in short dense clusters in the axils of the upper leaves, the lower clusters distant, the upper ones forming a terminal leafy spike. Involucre about 2 lines long, the bracts oblong-linear, scarious but woolly, of a dirty white or pale brown. Female florets very numerous, with 2 or 3 hermaphrodite ones in the centre. Pappus-bristles slightly scabrous, cohering in a ring at the base.

Queensland. Brisbane river, Moreton Bay, *Leichhardt, F. Mueller, Henne.*
N. S. Wales. In cultivated places, Port Jackson, *R. Brown;* common about Sydney, *Woolls.*

This appears to be a N. American species, now spread over many parts of the Old World, and probably introduced into Australia since its settlement.

6. **G. indicum,** *Linn.; DC. Prod.* vi. 231. A decumbent ascending or erect annual, rarely exceeding 6 in., covered with a loose cottony wool, sometimes very abundant. Leaves petiolate, spathulate or linear. Flower-heads small, densely clustered in ovate or oblong terminal leafy spikes, with a few in the upper axils. Involucres ovoid, about 1 line long, the bracts densely imbedded in wool, the tips only usually protruding. Female florets very numerous, with 2 or 3 hermaphrodite ones in the centre. Pappus-bristles numerous, cohering in a ring at the base.—*G. niliacum,* Raddi; DC. l. c.

Queensland. Upper Roper and Alligator rivers, *F. Mueller.*
N. S. Wales. Darling river, *Victorian Expedition.*
Victoria. Near Station Peak, *Herb. F. Mueller.*

The species is common in India, extending westward to the Nile and eastward to the Malayan peninsula and China. F. Mueller, Fragm. v. 149, unites it with *G. purpureum;* they appear to me, however, to be always readily distinguished, the *G. indicum* being of the Old World, the *G. purpureum* of American origin. The Australian specimens are small, with very narrow leaves.

7. **G. indutum,** *Hook. f. in Hook. Lond. Journ.* vi. 121, *and in Fl. Tasm.* i. 217. *t.* 62 B. A little slender, erect, much-branched annual, rarely above 3 in. high and often not above 1 in., densely cottony-white. Leaves linear, soft. Flower-heads small, sessile amongst leafy bracts, at first dense but not in globular heads, and at length looser, forming leafy corymbs. Involucre ovoid, about 1 line long or scarcely more, woolly at the base; the bracts oblong, with erect, scarious, brown or straw-coloured tips. Female florets very numerous, with 2 to 4 hermaphrodite ones in the centre. Pappus-bristles quite free.—*G. sericeum,* Turcz. in Bull. Mosc. 1851, ii. 83.

N. S. Wales. Twofold Bay, *F. Mueller.*
Victoria. Near Melbourne, Brighton, *F. Mueller.*
Tasmania. Circular Head, George Town, *Gunn.*

S. Australia. Crystal Brook, Rivoli Bay, *F. Mueller.*
W. Australia, *Drummond, 5th Coll. n.* 392.

The inflorescence of this species is different from that of any other *Gnaphalium* known to me. F. Mueller, Fragm. vi. 150, compares it with the S. African *G. pauciflorum*, DC., a little-known species, of which we have no specimen, but the inflorescence described is different from that of *G. indutum.*

8. **G. Traversii,** *Hook. f. Handb. N. Zeal. Fl.* 154. A dwarf, tufted perennial. Leaves radical, petiolate, obovate or broadly oblong, white with a close cottony wool on both sides. Scapes erect, 1 to 3 in. high, bearing a few small narrow leaves and a single flower-head. Involucre broadly ovoid, about 3 lines long, the bracts linear, scarious, of a pale brown or tinged with red immediately under the erect tips, slightly woolly at the base. Female florets very numerous, with few hermaphrodite ones in the centre. Pappus-bristles very fine and copious, scarcely uniseriate.

Victoria. Subalpine pastures, Snowy River, *F. Mueller.* Also in New Zealand.

81. PTERYGOPAPPUS, Hook. f.

(Maja, *Wedd.*)

Involucre ovoid, the bracts imbricate, scarious, without spreading tips. Receptacle without scales. Florets of the circumference female, filiform, those of the disk fewer, hermaphrodite, tubular, 5-toothed. Anthers with small fine tails. Style-branches in the hermaphrodite but sterile florets exceedingly short, truncate, scarcely spreading. Achenes oblong, papillose. Pappus of few thick bristles, strongly barbellate or almost plumose.—Tufted perennial. Leaves imbricate. Flower-heads solitary and nearly sessile.

The genus is limited to a single species, endemic in Australia.

1. **P. Lawrencii,** *Hook. f. in Hook. Lond. Journ.* vi. 120, *and Fl. Tasm.* i. 207. *t.* 58. A very densely-tufted, almost moss-like plant, with the habit of many species of *Abrotanella.* Leaves densely imbricated on the short closely-packed stems, ovate, acute, thick, concave, about 1 line long, the upper surface hirsute, especially near the base, with long hairs, nearly glabrous outside. Involucre scarcely above 1 line long, the bracts few, pale brown. Florets about 10, of which about 6 slender and female, and 2 to 4 hermaphrodite but sterile, in the flower-heads examined. Pappus-bristles 6 or fewer.—*Maja compacta,* Wedd. Chlor. And. i. 229. t. 27* (introduced amongst Andine plants, from an error in the Herbarium whence it was received).

Tasmania. On the summits of all the mountains, forming, with *Abrotanella forsterioides,* and others, large pulvinate masses, *J. D. Hooker.*

The truly Andine *Antennaria aretioides,* A. Gray, is evidently nearly allied to this species, but its leaves are very obtuse, the hairs white and more woolly, the flower-heads said to be absolutely diœcious, and the pappus-bristles are much finer and more numerous than in *Pterygopappus.*

82. STUARTINA, Sond.

Involucre ovoid, the bracts imbricate, appressed, without appendages, or the inner ones with recurved horn-like tips. Receptacle without scales. Florets few, those of the circumference female filiform, those of the disk very few, hermaphrodite, 4- or 5-toothed. Anthers with small fine tails. Style-branches terete, truncate. Achenes obovoid-oblong. Pappus none.—Annual, with the habit of *Gnaphalium.*

The genus is limited to a single species, endemic in Australia, differing from *Gnaphalium* in the absence of the pappus.

1. **S. Muelleri,** *Sond. in Linnæa,* xxv. 522. A small diffuse or slender annual, rarely 6 in. high. Leaves on long petioles, nearly orbicular, about ¼ in. diameter, woolly-tomentose or at length glabrous above. Flower-heads very small, in little globular clusters, sessile amongst floral leaves similar to those of the stem, the petioles much longer than the clusters. Involucres narrow, scarcely 1 line long, surrounded by a tuft of long woolly hairs, the bracts appressed, but after flowering 1 to 3 of the inner ones are usually produced into recurved horns. Florets from 5 to 7, of which 1 or 2 in the centre hermaphrodite. Achenes glabrous or papillose.

Victoria. Barossa range, Geelong, *F. Mueller.*
S. Australia. Lofty range, Onkaparinga, Cudnaka, *F. Mueller.*

TRIBE IX. SENECIONIDÆ.—Leaves alternate. Flower-heads either heterogamous, with the female florets ligulate or rarely filiform, or sometimes homogamous, with all the florets hermaphrodite and tubular. Receptacle without scales. Anthers obtuse or scarcely pointed at the base, without tails. Style-branches truncate and penicillate, or rarely with pubescent tips or appendages. Pappus of capillary bristles. Involucral bracts in the Australian genera in a single row, with or without a few small outer ones round their base.

83. ERECHTHITES, Rafin.

(Neoceis, *Cass.*)

Involucre of several nearly equal bracts, apparently in a single row, the margins often scarious and imbricate, with a few small ones round the base. Receptacle without scales. Florets all tubular, those of the circumference in 2 or more rows, female, filiform, 3- or 4-toothed or rarely a few outer ones very slightly dilated at the tips, and deeper cleft on the inner side; disk-florets hermaphrodite, 5-toothed. Anthers obtuse at the base. Style-branches truncate. Achenes striate or angular. Pappus of numerous simple fine capillary bristles.—Herbs, annual or perennial. Leaves alternate, entire toothed lobed or pinnately divided. Flower-heads in terminal corymbs. Florets small, usually yellow.

The genus is dispersed over New Zealand and South America, and extends into Africa and Asia, but belongs chiefly to the southern hemisphere. Of the six Australian species, three are also in New Zealand, the three others are endemic. All are nearly allied to *Senecio,* but differ constantly in their filiform female florets. F. Mueller proposes to unite the two genera, on account of those supposed intermediate species forming De Candolle's section *Plagiotome.* It appears to me, however, that there is here some mistake, owing probably to De Candolle, from imperfect specimens, having confounded his *E. Bathurstiana* (a variety of *E. arguta*) with *Senecio brachyglossus.* In the numerous specimens I have examined, I have always found in *E. Bathurstiana,* DC., and its allies, at least two rows of filiform female florets, the outer ones rarely slightly dilated, but not really ligulate, and in *S. brachyglossus* a single row of minute but distinctly-marked ligulæ, without any filiform female florets. The generic name is more frequently spelt *Erechtites* than *Erechthites*, but the latter appears to be the more correct, for in Greek etymologies the *t* after *ch* is always replaced by *th*. The species are exceedingly difficult to limit by any fixed tangible characters, and require further study, especially of the *ripe* achenes, which are seldom present in herbarium specimens.

Involucres small, mostly of 8 to 10 bracts. Plants without wool, glabrous or nearly so.
- Annual. Leaves toothed or lobed . . . . 1. *E. prenanthoides.*
- Perennial. Leaves once- twice- or thrice-pinnatifid, with linear segments . . . . 2. *E. Atkinsoniæ.*

Involucres of about 12 bracts. Plants usually more or less cottony-woolly or scabrous-hirsute.
- Leaves mostly toothed lobed or divided. Involucres above 3 lines long. Achenes short. Annual, often hard at the base (rarely perennial ?) . . . . 3. *E. arguta.*
- Leaves deeply lobed or divided. Involucres about 4 lines long. Perennial . . . . 4. *E. mixta.*
- Leaves linear, mostly entire. Achenes rather long, sometimes contracted into a short beak. Perennial . . . . 5. *E. quadridentata.*

Involucres of about 15 to 20 bracts. Leaves linear, entire toothed or pinnatifid. Achenes often contracted at the top. Perennial . . 6. *E. hispidula.*

F. Mueller informs me that an *E. Muelleri*, was published by Lange in the seed-catalogue of the Copenhagen garden for 1861. It is entirely unknown to me.

1. **E. prenanthoides,** *DC. Prod.* vi. 296. An erect herb, apparently annual, of 1 to 3 ft., usually glabrous and always without wool. Leaves lanceolate or almost linear, 2 to 5 or even 6 in. long, irregularly or almost regularly denticulate along their whole length, clasping the stem by toothed auricles. Flower-heads numerous in a loose terminal corymb. Involucres narrow, about 3 lines long, of about 8 or very rarely a few of 10 or 11 very narrow bracts. Female florets about 12 to 15, hermaphrodite about half as many. Achenes angular, slightly hirsute or nearly glabrous, much shorter than in *E. quadridentata* and not contracted at the top.—Hook. f. Fl. Tasm. i. 218; *Senecio prenanthoides*, A. Rich. Sert. Astrol. 96; also probably *E. sonchoides*, DC. Prod. vi. 296, and *Senecio flaccidus*, A. Rich. Sert. Astrol. 110, from the characters given.

N. S. Wales. Port Jackson, *R. Brown* and others; Kiama, *Harvey.*
Victoria. Wendu Vale, *Robertson;* Dandenong Ranges and Gipps' Land, *F. Mueller.*
Tasmania. Kent's group, *R. Brown;* margins of streams in cool shady places, *J. D. Hooker.*

The species is also in New Zealand.

Var. *picridioides.* Tall and stout, more or less scabrous with short scattered hairs. Leaves larger and broader, coarsely lobed. Panicles involucres and florets of the typical form.—*E. picridioides*, Turcz. in Bull. Mosc. 1851, i. 200.

W. Australia, *Drummond, 3rd Coll. n.* 132.

2. **E. Atkinsoniæ,** *F. Muell Fragm.* v. 88. An erect glabrous herb several feet high, closely allied to *E. prenanthoides*, with the same small flower-heads in a large terminal panicle, but perennial and the leaves once, twice or three times pinnately divided into narrow-linear segments, the involucral bracts are more frequently 10 and sometimes 11 or 12, and the female florets rather more numerous than in *E. prenanthoides.*

N. S. Wales. Grose river, *R. Brown;* Blue Mountains, *Miss Atkinson.*

3. **E. arguta,** *DC. Prod.* vi. 296. A rather coarse erect herb of 1 to 2 ft. when full grown, usually annual but with a hard base, sometimes probably biennial (or perhaps with a more persistent rhizome), more or less scabrous-hirsute with crisped hairs, and occasionally with white cottony wool on the under side of the leaves and about the inflorescences, rarely nearly or even quite glabrous. Leaves lanceolate, oblong or almost linear, irregularly and often acutely toothed, lobed or divided, sessile or petiolate, but almost always clasping the stem with toothed auricles. Flower-heads small, in a terminal corymb much more dense than in *E. quadridentata.* Involucre in the normal form about 3 lines long, of about 12 narrow bracts, often squarrose at the tips and surrounded by a few minute outer ones. Florets rather numerous, the females in 2 or more rows, with about 6 to 10 hermaphrodites in the centre. Achenes short, angular, shortly hirsute or glabrous.—Hook. f. Fl. Tasm. i. 219; *Senecio argutus,* A. Rich. Fl. Nov. Zel. 258, and Sert. Astrol. 104; *S. multicaulis,* A. Rich. Sert. Astrol. 105; *Erechthites lacerata,* F. Muell. in Linnæa, xxv. 417; *Senecio Lessoni,* F. Muell. Cat. Hort. Melb. 1858, 26.

N. S. Wales. Port Jackson, *R. Brown, Woolls;* New England, *C. Stuart;* Clarence river, *Beckler.*

Victoria. Wendu vale, *Robertson;* Yarra-Yarra, Port Phillip, mouth of the Glenelg, *F. Mueller.*

Tasmania. Not uncommon in waste places, *J. D. Hooker.*

S. Australia. Cudnaka, Onkaparinga, Kangaroo Island, *F. Mueller.*

W. Australia, *Drummond, 5th Coll. n.* 376, also *n.* 43 (with narrower leaves sometimes entire).

The species is also in New Zealand.

Var. *microcephala.* Involucres only about 2 lines long but not narrower than in the typical form.—Flats near the Broadribb river, *F. Mueller* (*Herb. F. Muell.*).

Var. *dissecta.* Taller with the leaves more divided, once or almost twice pinnatifid, a few of the outermost florets sometimes obliquely split at the end but never in the heads examined really ligulate.—*E. Bathurstiana,* DC. Prod. vi. 297.—Rocky hills in the neighbourhood of Bathurst, *A. Cunningham;* Wimmera, *Dallachy.*

Senecio apargiæfolius, Walp. Linn. xiv. 309, or *Erechthites apargiæfolia,* Sond. in Linnæa, xxv. 524, from Lhotzky's specimens from Maneroo, appears to be *E. arguta,* although it is said to have a perennial rhizome. To the same species belong also probably *Senecio glomeratus,* Desf. Cat. Hort. Par. 124, or *E. glomerata* var. *subincisa,* DC. Prod. vi. 297, and *Neoceis microcephala,* Cass. (DC.), *Senecio pumilus,* Poir. Dict. Suppl. v. 130, or *E. pumila,* DC. Prod. vi. 297, and *Senecio pusillus,* A. Rich. Sert. Astrol. 99; DC. Prod. vi. 370, of which I have received from M. Decaisne a flower-head, in which the outer florets are certainly female and 4-toothed.

4. **E. mixta,** *DC. Prod.* vi. 297. Apparently perennial and tall, scabrous-pubescent with the divided leaves often cottony underneath of *E. arguta* var. *dissecta,* but with much longer flower-heads, the involucres exceeding 4 lines like those *E. hispidula* but much more slender, consisting of about 12 bracts, and the panicle loose like that of *E. quadridentata.*—*Senecio mixtus,* A. Rich. Sert. Astrol. 112. t. 36 (the leaves more divided than in our specimens); *E. picridioides,* Sond. and Muell. in Linnæa, xxv. 523.

N. S. Wales. Piper's Hill (in the interior?), *Fraser.*

S. Australia. Memory Cove, *R. Brown;* Murray river, near Moorundee, *F. Mueller;* Spencer's Gulf, *Warburton.*

The species requires further investigation from more perfect specimens.

5. **E. quadridentata,** *DC. Prod.* vi. 295. An erect herb more or less clothed with a white deciduous cottony wool, from 1 to above 2 ft. high, with a perennial rhizome. Leaves linear, linear-lanceolate or very rarely oblong-lanceolate, the radical ones sometimes petiolate, the stem leaves sessile, entire or with a few small distant teeth, the longer ones attaining 3 or 4 in. with or without small stem-clasping auricles. Flower-heads slender in a terminal corymbose panicle usually loose but sometimes more crowded. Involucre narrow, about 4 lines long, of about 12 very narrow bracts. Female filiform florets 30 to 40 or more, the hermaphrodites in the centre few (not above 8). Achenes glabrous or papillose-pubescent, striate, usually slender and contracted at the top but very variable, from scarcely longer than in *E. arguta* to nearly twice as long, the terminal contraction amounting to a distinct beak or scarcely perceptible.—Steetz in Pl. Preiss. i. 483; Hook. f. Fl. Tasm. i. 219; *Senecio quadridentatus,* Labill. Pl. Nov. Holl. ii. 48. t. 194; *S. tenuiflorus,* Sieb. Pl. Exs.; *E. tenuiflora,* DC. Prod. vi. 296; *E. incana,* Turcz. in Bull. Mosc. 1851, ii. 85.

Queensland, *F. Mueller;* Moreton Bay, *Leichhardt;* Keppel Bay, *Thozet.*

N. S. Wales. Port Jackson or Blue Mountains, *Sieber, n.* 435; rocky hills near Bathurst, *A. Cunningham.*

Victoria. Wendu vale, *Robertson;* Wimmera, *Dallachy;* Station Peak, *Herb. F. Mueller.*

Tasmania. Table Mountain, Derwent river and Port Dalrymple, *R. Brown;* Gun Carriage Island, Bass's Straits, *Gunn.*

S. Australia. Spencer's Gulf, Third Creek, *F. Mueller*

W. Australia. Swan River, *Drummond, n.* 332, and 5*th Coll. n.* 379; *Preiss, n.* 73, 126, *Oldfield;* Kalgan river, *Oldfield.*

The species is also in New Zealand.

Var. *glabrescens.* Less tomentose, the lower leaves petiolate, oblong, with a few coarse teeth, upper leaves linear and entire. *E. glabrescens,* DC. Prod. vi. 295.—Plains subject to irrigation south of Lake George, *A. Cunningham;* Lord Howe's Island, *Milne;* near Melbourne, *F. Mueller, Adamson;* Circular Head, *Gunn.*—Nearly the same variety passing into the following one, Wimmera, *Dallachy.*

Var. *Gunnii.* Very woolly-white. Leaves mostly petiolate, oblong, entire with a few remote teeth.—*E. Gunnii,* Hook. f. in Hook. Lond. Journ. vi. 122, and Fl. Tasm. i. 220. t. 63.—Snowy River and Yarra-Yarra, *F. Mueller;* common on the summits of the Western Mountains, *Gunn.*

E. glandulosa, DC. Prod. vi. 295, from Lachlan river, *A. Cunningham,* or *Senecio glandulosus,* A. Cunn., appears to be a luxuriant form of *E. quadridentata.*

6. **E. hispidula,** *DC. Prod.* vi. 296 (?). A stout erect perennial of 1 to 2 ft. or rather more, glabrous or scabrous-pubescent and frequently with a little white cottony wool on the under side of the leaves. Leaves linear or lanceolate, entire, coarsely toothed or remotely pinnatifid, often petiolate, sometimes with stem-clasping auricles. Flower-heads larger than in *E. quadridentata,* in a terminal loose or compact corymbose panicle. Involucre 4 to 5 lines long or even more, thicker than in *E. quadridentata,* usually of 15 to 20 bracts, with several small outer ones. Florets very numerous, the females in several rows. Achenes slender and striate as in *E. quadridentata,* or rather shorter.—Hook. f. Fl. Tasm. i. 220; *Senecio hispidulus,* A. Rich. Sert. Astrol. 92 (from the description but scarcely the figure, t. 34); *S. squarrosus,* A. Rich. l. c. 107. t. 35 (from the description and plate); *E. Richardiana,* DC. Prod. vi. 297.

Victoria. Wendu vale, *Robertson;* Glenelg river, near Melbourne, Darbent Creek, Snowy River, Omeo plains, *F. Mueller;* Skipton, *Whan;* Wimmera, *Dallachy.*
Tasmania. Launceston, Circular Head, etc., *Gunn.*
S. Australia. Lofty Range, *F. Mueller.*
W. Australia, *Drummond.*

A. Richard's figure above mentioned, both in general habit and in the analytical details, seems to present a true *Senecio*, probably a form of *S. odorata*, but the description in the text of the flower-heads and florets, totally at variance with the plate, certainly refers to a true *Erechthites*, and is probably not taken from the specimen figured. De Candolle's character is entirely taken from A. Richard's.

84. GYNURA, Cass.

Involucre of nearly equal bracts in a single row with a few small outer ones round their base. Receptacle without scales. Florets all tubular, hermaphrodite, 5-toothed. Anthers obtuse at the base. Style bulbous at the base, the branches ending in long linear hairy points. Achenes striate. Pappus of numerous capillary bristles.—Herbs, often somewhat succulent. Leaves alternate. Flower-heads terminal, solitary or loosely corymbose. Florets yellow.

A small genus, confined to the tropical and subtropical regions of the Old World, the only Australian species being the same as an Indian one.

1. **G. pseudochina,** *DC. Prod.* vi. 299. Rootstock perennial, thick and fleshy. Stems erect or ascending, 1 to 2 ft. high, somewhat succulent, leafy in the lower part only, ending in a long almost leafless peduncle bearing either a single flower-head or a loose corymb of 2 to 7 or 8 heads. Leaves petiolate, obovate ovate-oblong or lanceolate, coarsely toothed, rather thick, pubescent or nearly glabrous, 2 to 3 or even 4 in. long, the petiole often expanded at the base into 2 auricles or lobes. Flower-heads about 7 lines long. Involucre of about 12 narrow bracts with several short outer ones.—*G. ovalis*, DC. and other synonyms adduced in Benth. Fl. Hongk. 189; *Senecio drymophilus*, F. Muell. in Trans. Phil. Inst. Vict. ii. 69.

Queensland. Bustard Bay, *Banks and Solander;* East coast, *R. Brown;* Wide Bay, *Leichhardt;* Brisbane river, rare, *W. Hill, F. Mueller;* Fort Cooper, near Isaac's River, *Bowman;* Port Denison, *Fitzalan;* Rockingham Bay, *Dallachy.*

A native of China and probably also of the Archipelago and various parts of India, but the specimens received are often cultivated. The Himalayan *G. nepalensis*, DC., ought, however, perhaps to be added to the synonyms as a more pubescent variety, the Australian ones being in this respect intermediate between that and the typical form.

85. SENECIO, Linn.

(Centropappus, *Hook. f.*)

Flower-heads homogamous and discoid or heterogamous and radiate. Involucre of nearly equal bracts apparently in a single row, linear or very rarely ovate, the margins often scarious and imbricate, with or rarely without a few small ones at the base passing into the bracts on the peduncles. Receptacle naked or pitted, the borders of the pits rarely toothed or produced into a few short scales. Florets of the ray when present female or rarely neuter, ligulate. Disk-florets tubular, hermaphrodite, 5-toothed. Anthers obtuse at the base, the upper portion of the filament often thickened. Style-branches truncate, usually bearing a tuft of minute hairs and very rarely a short obtuse appendage. Achenes striate or angular. Pappus of numerous simple scabrous or denticulate bristles.—Herbs or very rarely shrubs, glabrous-pubescent or clothed with cottony wool. Leaves alternate, entire or divided, often rather thick. Flower-heads terminal, solitary, corymbose or paniculate. Florets usually yellow, rarely purple or white.

The largest genus among *Compositæ*, and ranging nearly over the whole world, although the individual species are often very local. Of the 28 Australian species 1 only extends to New Zealand, the others are all endemic. The rays are yellow in all of them except *S. leucoglossus*.

SERIES I. **Radiati.**—*Flower-heads radiate.*

Erect leafy annuals.
  Flower-heads few, large. Involucres broadly campanulate. Ligulæ of the ray longer than the involucre.
    Leaves entire. Involucral bracts united above the middle . 1. *S. Gregorii.*
    Leaves pinnatifid with toothed lobes. Involucral bracts united at the base only . . . . 2. *S. platylepis.*
  Flower-heads small. Involucres cylindrical. Ligulæ of the ray very small and rolled back.
    Ray-florets 3, white (or purplish ?). Disk-florets under 10 . 19. *S. leucoglossus.*
    Ray-florets 6 or more, very small, yellow. Disk-florets above 10 . . . . 20. *S. brachyglossus.*
Tufted or shortly creeping perennials with leaves chiefly radical. Flower-stems or scapes erect with 1 or rarely 3 to 5 heads. Involucres broad. Rays long, spreading.
  Leaves ovate, not cordate, nearly entire, tuberculate or muricate above . . . . 3. *S. papillosus.*
  Leaves ovate, cordate, nearly entire, glabrous or woolly underneath . . . . 4. *S. primulifolius.*
  Leaves linear or oblong, entire crenate distantly toothed or pinnatifid . . . . 5. *S. pectinatus.*
Maritime much-branched spreading undershrub. Leaves mostly toothed. Involucres broad. Rays long spreading . . . . 6. *S. spathulatus.*
Glabrous erect shrubs (or in the first two species undershrubs ?).
  Flower-heads few, large. Involucres broad. Rays long.
    Leaves oblong or oblanceolate, the upper ones stem-clasping. Flower-heads (including the ray) nearly 3 in. diameter . . . . 7. *S. megaglossus.*
    Leaves obovate, the upper ones lanceolate, cordate-auriculate at the base. Flower-heads (including the ray) under 2 in. diameter . . . . 8. *S. magnificus.*
  Flower-heads smaller, corymbose. Involucres campanulate. Ligulæ scarcely longer than the involucre.
    Leaves on long petioles, ovate, deeply-toothed . . . . 9. *S. insularis.*
    Leaves sessile, broadly linear, entire . . . . 10. *S. centropappus.*
Herbaceous erect perennials (sometimes woody at the base). Flower-heads corymbose or paniculate, few or numerous.
  Flower-heads large or middle-sized. Involucres campanulate, 3 to 6 lines long. Ligulæ longer than the involucre. Disk-florets numerous.
    Leaves twice pinnate with filiform segments . . . . . 13. *S. capillifolius.*
    Leaves linear, entire toothed or pinnatifid.
      Flower-heads including the ray above 2 in. diameter . . 11. *S. macranthus.*

Flower-heads, including the ray, rarely exceeding 1 in. diameter. Involucral bracts prominently 2-ribbed . 12. *S. lautus.*
Leaves lanceolate, toothed . . . . 12. *S. lautus* var.
Leaves stem-clasping, the lower ones toothed at the end, the upper ones entire, broadly cordate-auriculate . . . . 16. *S. velleioides.*
Leaves all distinctly petiolate.
Leaves large, deeply pinnatifid or pinnatisect with few lanceolate segments . . . . 14. *S. vagus.*
Leaves ovate-lanceolate, toothed but undivided . . . 15. *S. amygdalifolius.*
Flower-heads small. Involucres cylindrical, 2 to 2½ lines long. Ligulæ not longer than the involucres. Disk-florets 10 to 15.
Stems 1 to 3 ft. high. Leaves linear or lanceolate. Flower-heads numerous in a terminal glabrous corymb . . . 17. *S. australis.*
Stems 6 to 10 in. high. Leaves linear. Flower-heads few, usually hoary . . . . 18. *S. Behrianus.*

SERIES II. **Discoidei.**—*Flower-heads small, discoid. Involucres cylindrical.*

Herbaceous perennials, sometimes woody at the base, or almost shrubby.
Involucral bracts about 12. Florets above 20.
Leaves narrow, mostly petiolate, without auricles. Panicle corymbose. Involucre above 2 lines long . . . . . 21. *S. Georgianus.*
Leaves deeply pinnatifid, very white underneath. Panicle corymbose. Involucre above 2 lines long . . . . . 22. *S. Gilberti.*
Leaves auriculate at the base, much-toothed. Panicles large, subpyramidal. Involucre under 2 lines long . . 23. *S. ramosissimus.*
Involucral bracts usually 8. Florets under 20.
Leaves oblong or lanceolate, sessile or petiolate, with a dilated stem-clasping base . . . . 24. *S. odoratus.*
Leaves linear or lanceolate, narrowed at the base or petiolate, without any or with very small auricles . . . . 25. *S. Cunninghamii.*
Leaves ovate or lanceolate, very white underneath, on long petioles not auriculate . . . . 26. *S. hypoleucus.*
Leaves once or twice divided into long linear, almost filiform segments . . . . 27. *S. anethifolius.*
Annuals.
Glabrous. Leaves linear, entire . . . . 28. *S. Gaudichaudianus.*
Slightly and loosely woolly. Leaves pinnatifid . . . . . * *S. vulgaris.*

SERIES I. RADIATI.—Flower-heads radiate, the ligulæ either long and spreading or small and rolled back, in *S. brachyglossus* sometimes not protruding from the involucre.

1. **S. Gregorii,** *F. Muell. in Pl. Greg., quoted Rep. Babb. Exped.* 14. Apparently an erect annual, under 1 ft. high, slightly branched, glabrous and glaucous. Leaves sessile, linear or linear-lanceolate, entire, 1 to 3 in. long. Peduncles dilated under the solitary rather large flower-head. Involucre campanulate, 3 to 4 lines or at length nearly ½ in. long, of about 10 to 12 rather broad bracts, concrete nearly to the apex, without any small outer ones. Ray-florets about 10 to 12, the ligulas long and spreading; disk-florets numerous, exceeding the involucre. Achenes striate, the more perfect ones 2½ lines long and densely hirsute, but some in the same heads often smaller and glabrous. Pappus at first short, but lengthening out to ¾ in.

Queensland. Maranoa river, *Leichhardt.*

N. S. Wales. From the Lachlan and Darling rivers to the Barrier Range, *Victorian and other Expeditions.*

Victoria. N.W. desert, *L. Morton ;* Wimmera, *Dallachy.*

S. Australia. Cooper's Creek, *A. C. Gregory ;* Wills' Creek, *Howitt's Expedition ;* Fincke river, *M'Douall Stuart's Expedition.*

The species differs from all other *Senecios* known to me, in the involucre with the bracts almost as closely connate as in *Werneria* and *Euriops,* but in other respects the characters are entirely those of *Senecio.*

2. **S. platylepis,** *DC. Prod.* vi. 371. An erect slightly branched annual of 1 to 2 ft., glabrous or with a little loose wool. Leaves narrow, irregularly pinnatifid, with obtuse or acute coarsely-toothed lobes, the petioles often dilated and auriculate at the base. Flower-heads rather large, not numerous, in an irregular terminal leafy corymb, the peduncles dilated at the top. Involucre broadly campanulate, 4 to 5 lines long, the bracts rather broad, united at the base, without any or with 1 or 2 minute outer bracts. Ray-florets 12 to 20, the ligulæ long and spreading; disk-florets numerous. Achenes striate, pubescent or hirsute.

N. S. Wales. Low flat land at the foot of Peel's Range, *A. Cunningham, Fraser.*

3. **S. papillosus,** *F. Muell. in Trans. Phil. Inst. Vict.* ii. 69, *and in Hook. Kew Journ.* ix. 301. A perennial with a short thick nearly glabrous stock. Leaves mostly radical, spreading, petiolate, ovate, obtuse, narrowed at the base, nearly entire, under 1 in. long, thick ; upper surface very scabrous or muricate with tubercles or short rigid bristles, otherwise glabrous or very slightly woolly underneath. Flower-stems or scapes simple, under 1 ft. high, often somewhat woolly, bearing a few small linear leaves or scales and a single rather large flower-head. Involucre of rather numerous narrow bracts with scarcely any small outer ones. Ray-florets 15 to 20, with long spreading ligulæ ; disk-florets numerous, not exceeding the involucre.—Hook. f. Fl. Tasm. ii. 365.

Tasmania. Mount Lapeyrouse, *Oldfield.* This may possibly prove to be a variety of the New Zealand *S. bellidioides,* Hook. f.

4. **S. primulifolius,** *F. Muell. in Trans. Phil. Inst. Vict.* ii. 69. A perennial with a short thick glabrous or slightly woolly stock. Leaves chiefly radical, petiolate, cordate-ovate, obtuse, slightly sinuate-toothed or irregularly crenate, 1 to 3 in. long, rather thick, glabrous or with a little loose deciduous wool, pale underneath but not white. Flower-stems or scapes not exceeding 1 ft., single or nearly so, bearing a few small distant leaves or sometimes one larger one below the middle. Flower-heads 1 or 2 or rarely 3 or 4, rather large. Involucre broadly campanulate, the bracts about 4 lines long with a few small outer ones. Ray-florets about 10 to 12, the ligulæ long and spreading ; disk-florets numerous, scarcely exceeding the involucre. Achenes glabrous, but not seen ripe.—Hook. f. Fl. Tasm. ii. 365.

Tasmania. Mount Lapeyrouse, *Oldfield.*

5. **S. pectinatus,** *DC. Prod.* vi. 372. A perennial with a tufted or shortly creeping stock, glabrous or rarely woolly under the inflorescence. Leaves chiefly radical or nearly so, linear or oblong, mostly obtuse and 1 to 2 in. long, shortly pinnatifid, crenate, remotely toothed or rarely quite entire, green on both sides. Flower-stems in the typical form simple and single-

headed, from a few in. to 1 ft. high, with a few leaves smaller and more acute than the radical ones. Flower heads rather large. Involucre broadly campanulate, the bracts often shortly united at the base, with a few small outer ones. Ray-florets 15 to 20 or even more, long and spreading; disk-florets very numerous, not exceeding the involucre. Achenes glabrous.—Hook. f. Fl. Tasm. i. 222.

Victoria. Mount Cobberas and Baw-Baw mountains, at an elevation of 6000 ft., *F. Mueller.*

Tasmania. Table Mountain, Derwent river, *R. Brown;* Mount Wellington, *Gunn;* Mount Lapeyrouse, *Oldfield.*

Var. *pleiocephalus.* Flower-heads rather smaller, 3 to 5 together in a loose terminal corymb.—*S. leptocarpus,* DC. Prod. vi. 372; Hook. f. Fl. Tasm. i. 222. t. 64 (representing an unusually large-leaved specimen).—Mount Wellington, *Gunn;* Mount Sorrel, *Milligan;* Mount Lapeyrouse, *Oldfield.*

6. **S. spathulatus,** *A. Rich. Sert. Astrol.* 125. Diffuse and much-branched, said to be suffruticose. Leaves from narrow-oblong to almost obovate, irregularly toothed or crenate, the lower ones narrowed into a petiole, the others stem-clasping and often auriculate, all rather thick and fleshy, mostly $\frac{3}{4}$ to $1\frac{1}{2}$ in. long. Flower-heads rather large, in an irregular leafy corymb. Involucre campanulate, the bracts about 4 lines long, with a few very small outer ones. Ray-florets about 12 to 20, the ligulæ long and spreading. Disk-florets numerous, exceeding the involucre. Achenes quite glabrous in some specimens, pubescent in others.—DC. Prod. vi. 373; Hook. f. Fl. Tasm. i. 222.

N. S. Wales. Sandy seashores, Port Jackson, *R. Brown;* Bondee Bay *Leichhardt, R. Cunningham?* (with pubescent achenes).

Victoria. Mouth of Snowy River, *F. Mueller* (with glabrous achenes).

Tasmania. King's Island, *R. Brown;* Woolnorth, *Gunn;* Macquarrie Harbour, *Milligan;* straits of Dentrecastreaux (*Lesson*) (with pubescent achenes).

S. anacampserotis, DC. Prod. vi. 374, from Port Jackson, *Fraser,* appears, from the description, to be the same species.

7. **S. megaglossus,** *F. Muell. in Linnæa,* xxv. 419. An erect much-branched glabrous and glaucous shrub (or undershrub?). Leaves oblong or oblanceolate, entire or with a few minute teeth, often 3 to 4 in. long, the lower ones narrowed at the base, the upper ones sessile and stem-clasping. Flower-heads few, very large, the peduncles thickened upwards. Involucre broadly campanulate, the bracts 8 to 9 lines long, without any or very few small outer ones. Pits of the receptacle with jagged margins, often produced into short points or scales. Ray-florets 15 to 20, the ligulæ spreading to a diameter of nearly 3 in. Disk-florets numerous, not exceeding the involucre. Achenes glabrous, Pappus-bristles minutely serrulate.—Sond. in Linnæa, xxv. 527.

S. Australia. Barren hills near the Burra mines and rocks on the Broughton river, *F. Mueller.*

8. **S. magnificus,** *F. Muell. in Linnæa,* xxv. 418. An erect glabrous and glaucous shrub (or undershrub?). Lower leaves obovate-oblong, coarsely and acutely toothed, narrowed at the base, the upper ones lanceolate or ovate-lanceolate, deeply cordate-auriculate, stem-clasping and sometimes slightly de-

current, entire or nearly so, the larger ones 1½ to 2 in. long. Flower-heads rather large, in a loose terminal leafless corymb. Peduncles scarcely dilated. Involucre broadly campanulate, the bracts 4 to 5 lines long, without any or with very minute outer ones. Receptacle not jagged. Ray-florets about 8 to 12, the ligulæ long and spreading; disk-florets numerous, scarcely exceeding the involucre. Achenes pubescent.—Sond. in Linnæa, xxv. 526.

Victoria. Grampians near Mount Zero, *Fisher;* Wimmera, *Dallachy.*
S Australia. Cudnaka, *F. Mueller;* Flinders Range, *Howitt's Expedition, also in M'Douall Stuart's Collection.*

9. **S. insularis,** *Benth.* A low erect branching, perfectly glabrous shrub. Leaves on long petioles, ovate, acute, deeply toothed, narrowed at the base, 2 to 4 in. long. Flower-heads not large, rather numerous, in a leafless corymb shorter than the leaves. Involucre narrow-campanulate, of about 8 bracts 3 to 4 lines long, with very few minute outer ones. Ray-florets 3 or 4, the ligulæ narrow, spreading but (without the tube) about the length of the involucre; disk-florets about 12, shortly exceeding the involucre, the corolla-lobes narrow. Achenes glabrous.

N. S. Wales. Lord Howe's Island, frequent, *Milne;* on the ascent of Mount Ligbird, *M'Gillivray.* Very unlike any other Australian species, but approaching in several respects the New Zealand *S. glastifolius.*

10. **S. centropappus,** *F. Muell. Cat. Hort. Melb.* 1858, 26. An erect, glabrous, branching shrub, attaining 8 to 10 ft. Leaves rather crowded at the ends of the branches, sessile, broadly linear, mostly obtuse, quite entire, 2 to 3 in. long. Flower-heads not large, in small ovate panicles at the ends of the branchlets of a general somewhat corymbose leafy panicle. Involucre broadly campanulate, of about 8 ovate very obtuse bracts, 2 to 2½ lines long, with more scarious and fringed margins than in any other species, with very few small outer ones. Ray-florets 4 to 6, the ligulæ spreading, although scarcely longer (without the tube) than the involucre; disk-florets about 10 to 12, exceeding the involucre. Achenes glabrous. Pappus-bristles very prominently denticulate and almost plumose at the end.—*Centropappus Brunonis,* Hook. f. in Hook. Lond. Journ. vi. 124; Fl. Tasm. i. 225. t. 65.

Tasmania. Table Mountain, Derwent river, *R. Brown;* above the limits of the forests, on Mount Wellington, at an elevation of 3500 to 4000 ft., *Gunn, Oldfield.* Although very different in habit from other Australian species, it resembles in this respect some of the Andine ones. The involucral bracts are broader and more obtuse than in any other species I know, and the prominently denticulate bristles of the pappus are also more marked perhaps than in any others, but the inflorescence, flowers, and fruit are, in all other respects, quite those of *Senecio,* to which F. Mueller appears to have been right in uniting it.

11. **S. macranthus,** *A. Rich. Sert. Astrol.* 126. An erect glabrous perennial, attaining about 4 ft. Lower leaves not seen, upper ones long-linear or linear-lanceolate, with long points, entire or with a few remote teeth, narrowed into a petiole but with acuminate auricles at its base. Flower-heads large, in a loose terminal corymb. Involucre broadly campanulate, the bracts 4 to 5 lines long, prominently 2-ribbed, as in *S. lautus,* with a few small outer ones. Ray-florets rather numerous, the long ligulæ spreading to

a diameter of nearly 2 in. Achenes apparently glabrous, but not seen ripe.—DC. Prod. vi. 374.

N. S. Wales. In the N.W. interior, *Fraser.* Allied to *S. lautus,* but the flower-heads very much larger.

12. **S. lautus,** *Forst. Prod.* 91. An erect glabrous perennial, from 1 to 3 or even 4 ft. high. Leaves usually linear or linear-lanceolate, entire remotely toothed or deeply pinnatifid, rarely broadly lanceolate, either narrowed into a petiole or, especially when broad, dilated and auriculate or stem-clasping at the base. Flower-heads not very large, several in a loose terminal irregular corymb. Involucre campanulate, the bracts 3 to 4 lines long, more prominently 2-ribbed than in most Australian species, with several very small outer ones. Ray-florets about 10 to 15, the ligulæ spreading to from $\frac{3}{4}$ to 1 in. diameter; disk-florets numerous, scarcely exceeding the involucre. Achenes glabrous or pubescent.—Hook. f. Fl. Tasm. i. 221; *S. tripartitus,* A. Rich. Sert. Astrol. 114; DC. Prod. vi. 372; *S. crithmifolius,* A. Rich. l. c. 116; DC. l. c. 372; Steetz in Pl. Preiss. i. 485; *S. pinnatifolius,* A. Rich. l. c. 117; *S. carnulentus,* DC. l. c. 372; Steetz in Preiss. i. 484; *S. rupicola,* A. Rich. l. c. 119. t. 37; DC. l. c. 372; *S. Macquariensis,* DC. l. c. 372.

Queensland. Hervey's Bay, Sandy Cape, *R. Brown;* Suttor river and Moreton Island, *F. Mueller;* on the upper Maranoa, *Mitchell.*

N. S. Wales. Port Jackson, *R. Brown, Woolls;* northward to Hastings river and Mount Mitchell, *Beckler;* New England, *C. Stuart;* southward to Twofold Bay, *F. Mueller;* in the interior to Lake George, *Herb. Hooker;* Macquarrie river, *Fraser;* Lachlan and Darling rivers, *Victorian and other Expeditions.*

Victoria. From the Glenelg to Gipps' Land, and in the interior to the Grampians and Murray river, *F. Mueller* and others.

Tasmania. Port Dalrymple, *R. Brown;* abundant, especially near the coasts, *J. D. Hooker.*

S. Australia. From the Murray to St. Vincent's Gulf and Kangaroo Island, *F. Mueller* and others; Cooper's Creek, *Wheeler.*

W. Australia. King George's Sound, *R. Brown;* from the South coast to Vasse and Swan rivers, *Drummond, n.* 28 *and* 380, *Preiss, n.* 108, 109, 110, 112, 114, *Oldfield,* and others; and eastward to Cape Legrand and Esperance Bay, *Maxwell.*

S. Endlicheri, DC. Prod. vi. 373, appears from the character given to be the same species, so also is *S. ciliolatus,* DC. Prod. vi. 374, judging from Cunningham's own imperfect specimens, which however show that it is not an annual. The species is also in New Zealand.

Var. *lanceolatus.* Leaves rather broadly lanceolate, deeply and acutely toothed, sessile and stem-clasping.—Port Phillip, *Gunn, Adamson.*

13. **S. capillifolius,** *Hook. f. in Hook. Lond. Journ.* vi. 123, *and Fl. Tasm.* i. 222. *t.* 64. Apparently a perennial, quite glabrous. Leaves as in *S. anethifolius,* crowded, deeply divided into long, almost filiform segments, either entire or again divided into similar segments. Flower-heads and inflorescence of the smaller forms of *S. lautus,* the corymb compact. Involucre campanulate and prominently ribbed as in that species. Florets the same. Achenes pubescent.

Tasmania, *Gunn.* Only known from a single specimen, showing the foliage of *S. anethifolius,* with the inflorescence and flowers of *S. lautus.*

14. **S. vagus,** *F. Muell. in Trans. Phil. Soc. Vict.* i. 46, *and in Hook. Kew Journ.* viii. 162; Pl. Vict. t. 46. A tall, erect, glabrous perennial. Leaves petiolate, broadly lanceolate, deeply pinnatifid or pinnately divided

into few lanceolate entire or slightly toothed lobes or segments, of which the middle ones usually large, the upper and lower ones small, the whole leaf sometimes 5 or 6 in. long, the upper leaves smaller and less divided, and on side-branches often all entire or with only one deep lobe on each side, all of a thinner consistence than in most species, the petioles not auriculate. Flower-heads rather large in an irregular corymb. Involucre campanulate, the bracts 4 to 5 lines long, with few smaller outer ones. Ray-florets about 10 to 15, the ligulæ long and spreading; disk-florets numerous, shortly exceeding the involucre. Achenes glabrous.

N. S. Wales. Grose river, *R. Brown;* Hastings and Clarence rivers and Mount Mitchell, *Beckler;* swamps near Port Jackson, *Lowne, Clowes;* Blue Mountains, *Miss Atkinson;* Bent's Basin and near Camden, *Woolls.*

Victoria. Shady moist valleys of the Dandenong ranges, of Mount Disappointment and on Delatite river, *F. Mueller.*

15. **S. amygdalifolius,** *F. Muell. Fragm.* i. 232. A tall, erect, glabrous perennial. Leaves petiolate, ovate-lanceolate to oblong-lanceolate, acute and acutely and coarsely serrate, 2 to 4 in. long. Flower-heads rather large and few in a loose corymb or smaller and more densely corymbose. Involucre narrow-campanulate, the bracts narrow, 4 to 5 lines long, with few small outer ones. Receptacle pitted, with the edges of the pits occasionally produced into short teeth or very rarely into a small scale. Ray-florets rarely above 6 and often only 3 or 4, rather long and spreading; disk-florets shortly exceeding the involucre. Achenes glabrous.

N. S. Wales. Hastings river, *Beckler;* Port Macquarrie, *Backhouse,* also *Vicary.*

16. **S. velleioides,** *A. Cunn. in DC. Prod.* vi. 374. An erect glabrous perennial, attaining 3 ft. or sometimes more. Leaves ovate-oblong or lanceolate, the lower ones shortly petiolate, the intermediate ones 2 to 4 in. long, coarsely serrate, narrowed below the middle but sessile and stem-clasping with broad rounded auricles, the upper ones cordate-ovate or lanceolate, entire, broadly stem-clasping. Flower-heads not large and often rather numerous, in a terminal corymb. Involucre campanulate, the bracts 3 to nearly 4 lines long, with a very few small outer ones. Ray-florets usually about 10 or rather fewer, spreading and rather long; disk-florets scarcely exceeding the involucre. Achenes glabrous or slightly pubescent.—Hook. f. Fl. Tasm. i. 223.

N. S. Wales. Rocky hills to the north of Bathurst, *A. Cunningham;* Nepean river, *R. Cunningham;* Blue Mountains, *Miss Atkinson;* Twofold Bay, *F. Mueller.*

Victoria. Wooded hills, Wilson's Promontory, Apollo Bay, Upper Barwan river, Dandenong range, etc., *F. Mueller.*

Tasmania. Derwent river, *R. Brown;* damp shaded alpine parts of the colony, growing in very rich soil, *J. D. Hooker.*

17. **S. australis,** *Willd. Spec.* iii. 1981. A tall perennial, either quite glabrous or the under side of the leaves slightly cottony-white. Leaves, in the typical form, linear or lanceolate, sessile, quite entire or the base dilated into small acuminate auricles, the larger leaves 5 to 6 in. long. Flower-heads small and numerous, in a large terminal corymb. Involucre cylindrical, of about 10 to 12 bracts, not 2 lines long, with scarcely any small

outer ones. Ray-florets 4 to 6, the ligula (without the tube) not longer than the involucre; disk-florets about 10 to 12, exceeding the involucre. Achenes glabrous or pubescent.—A. Rich. Sert. Astrol. 131. t. 39; DC. Prod. vi. 374; Hook. f. Fl. Tasm. i. 223; *S. dryadeus*, Sieb. Pl. Exs.; *S. linearifolius*, A. Rich. Sert. Astrol. 129; DC. Prod. vi. 374; *S. cinerarioides*, A. Rich. l. c. 128; *S. Richardianus*, DC. l. c. 374.

N. S. Wales. Port Jackson, *R. Brown*, *Sieber*, *n.* 337.
Victoria. Common about Melbourne, *Adamson* and others; Emu Creek, *Whan;* Snowy River and Lake King, *F. Mueller*.
Tasmania. Port Dalrymple and islands of Bass's Straits, *R. Brown;* common throughout the colony, *J. D. Hooker*.
S. Australia. Loddon river, *F. Mueller*.

The New Zealand *S. angustifolius*, Forst., referred here by Willdenow, and after him by De Candolle and others, is probably the same as *S. lautus*. The *S. australis* of modern authors, has not, as far as I am aware of, been found in New Zealand.

Var. *macrodontus*. Leaves lanceolate, more or less serrate, usually whitish underneath, narrowed into a petiole, most frequently auriculate at the base.—*S. persicifolius*, A. Rich. Sert. Astrol. 123; *S. macrodontus*, DC. Prod. vi. 373.—Port Jackson, *R. Brown;* Blue Mountains, *Miss Atkinson;* New England, *C. Stuart;* Hastings and Macleay rivers, *Beckler*.

This may be a distinct species, but I have reduced it to *S. australis*, on the authority of F. Mueller. *S. pauciligulatus*, A. Rich. Sert. Astrol. 121. t. 38; DC. Prod. vi. 373, is a form of this *S. macrodontus*, with broader leaves not auriculate, and a specimen from the Cobberas mountains in Herb. F. Mueller, has the same broad leaves, but sessile and broadly auriculate at the base.

18. **S. Behrianus,** *Sond. et Muell. in Linnæa*, xxv. 527. A low perennial or almost an undershrub, with a shortly creeping woody rhizome and erect stems, usually 6 to 10 in. high, more or less hoary-tomentose, especially about the inflorescenee. Leaves linear, with revolute margins, entire or remotely toothed, rarely above 1 in. long, and the upper ones small. Flower-heads rather small, few in a loose terminal corymb. Involucre cylindrical, of about 10 prominently 2-ribbed bracts, 2 to 2½ lines long, and a very few small outer ones. Ray-florets about 6 to 8, the ligulæ (without the tube) scarcely so long as the involucre; disk-florets 10 to 15, exceeding the involucre. Achenes pubescent.

N. S. Wales. Darling river, *Dallachy*.
S. Australia. Murray river near Moorundi and Wood's Station, *F. Mueller*.

19. **S. leucoglossus,** *F. Muell. Fragm.* ii. 15. An erect glabrous annual of 1 to 2 ft. Leaves few, ovate or lanceolate, acutely and irregularly toothed and lobed, the lower ones petiolate without auricles, the next also petiolate but the petioles dilated into acutely-toothed stem-clasping auricles, the upper ones sessile with broad stem-clasping auricles. Flower-heads small, in small corymbs forming a large very loose panicle. Involucre cylindrical, of about 8 bracts, scarcely 2 lines long, without any or only 1 or 2 small outer ones. Ray-florets about 3, the ligula white (or pale purple?) short and often rolled back. Disk-florets usually under 10, slightly exceeding the involucre. Achenes glabrous or papillose-pubescent.

W. Australia, *Drummond, n.* 29; Harvey river, *Oldfield*.

20. **S. brachyglossus,** *F. Muell. in Linnæa*, xxv. 525. A slender

annual, 6 in. to a foot or rarely 1½ ft. high, glabrous or sprinkled with a few short white hairs. Leaves linear with a few small distant teeth or irregularly pinnatifid with few distant linear lobes. Flower-heads small, solitary or clustered at the ends of the branches of a loose irregular panicle. Involucre cylindrical, of about 8 bracts, about 2 lines long, with 1 or 2 minute outer ones. Ray-florets about 6, the ligulæ oblong but very short and rolled back. Disk-florets 10 to 12, slender, 5-toothed, scarcely exceeding the involucre. Achenes densely pubescent, those of the ray usually longer than those of the disk.—*Erechthites glossantha*, Sond. in Linnæa, xxv. 524.

N. S. Wales. Darling river, *Victorian Expedition;* between Stokes' Range and Cooper's Creek, *Wheeler.*

Victoria. Near Melbourne, *Adamson;* Wendu valley, *Robertson;* Wimmera, *Dallachy.*

S. Australia. Near Adelaide, *F. Mueller.*

W. Australia. Swan River, *Drummond, n.* 44, *Preiss* (no *number*); Murchison river, *Oldfield.*

Var. (?) *major.* Flower-heads larger. Involucres about 3 lines long with about 12 bracts. Florets also more numerous. Point Nepean and Wilson's Promontory, *F. Mueller*, also in W. Australia, *Drummond, n.* 377. This variety, perhaps a distinct species, is only to be distinguished from the European *S. sylvaticus* by the less divided leaves.

Var. (?) *elatior.* Tall and stout (the lower part of the plant not seen). Leaves pinnatifid, with linear or lanceolate unequal lobes, the larger ones denticulate. Flower-heads numerous in a large terminal corymb, otherwise as in the typical form. Achenes glabrous.—Blue Mountains, *Herb. F. Mueller.*

Series II. Discoidei.—Flower-heads small, discoid. Involucre cylindrical.

21. **S. Georgianus,** *DC. Prod.* vi. 371. An erect rigid perennial of 1 to 2 ft., covered when young with white cottony wool which usually persists on the under side of the leaves, or rarely nearly glabrous. Leaves linear or lanceolate, entire or scarcely toothed, rarely almost pinnatifid, 1 to 3 or even nearly 4 in. long, usually narrowed into a petiole which is sometimes slightly dilated at the base but not auriculate. Flower-heads rather small, in a corymbose panicle, but fewer and larger than in *S. odoratus*, varying in size in different specimens. Involucre cylindrical, of about 12 bracts, from a little more than 2 lines to above 3 lines long, the coloured reflexed tips more prominent than in most species. Florets above 30, all tubular, exceeding the involucre. Achenes pubescent.—*S. barkhausioides*, Turcz. in Bull. Mosc. 1851, ii. 86; *S. helichrysoides*, F. Muell. in Trans. Vict. Inst. 1855, 39.

N. S. Wales. Clarence river, *Beckler* (a drawn-up imperfect specimen); banks of Lake George, *A. Cunningham.*

Victoria. Grassy subalpine ridges, Macalister river, Mitta-Mitta Range, Lake Omeo, *F. Mueller.*

Tasmania. Derwent river, *R. Brown* (appears to be this species).

S. Australia. Salt Gulley, *Behr;* hills about Wheal Barton mines, *F. Mueller.*

W. Australia, *Drummond, 5th Coll. n.* 378.

This species has the foliage and habit nearly of *Erechthites quadridentata,* but the involucres are much larger and the florets all hermaphrodite and 5-toothed.

22. **S. Gilberti,** *Turcz. in Bull. Mosc.* 1851, i. 208. A tall erect perennial, the young stems clothed with a deciduous white wool. Leaves

sessile and stem-clasping, lanceolate, deeply pinnatifid, with lanceolate often toothed lobes, woolly-white when young, the wool persisting on the under side, the larger ones 2 to 4 in. long. Flower-heads small, numerous, in a corymbose panicle. Involucre cylindrical, 2 to 2½ lines long, of about 12 very narrow bracts with a few small outer ones. Florets above 20, all tubular, rather longer than the involucre.

W. Australia. Swan River, *Drummond, n.* 325. This has the aspect of *Erechthites arguta*, but is a true *Senecio*. The flower-heads are much larger than in *S. ramosissimus*, smaller than in *S. Georgianus*.

23. **S. ramosissimus,** *DC. Prod.* vi. 371. A stout erect perennial of 2 to 5 ft., usually glabrous. Leaves broadly lanceolate on the main stem and often 4 to 5 in. long, narrower or narrowed below the middle on the side branches, prominently and acutely toothed, sessile and clasping the stem with acutely toothed auricles. Flower-heads small and numerous, in a large more or less pyramidal panicle. Involucre almost campanulate, of about 12 bracts, nearly 2 lines long, almost without small outer ones. Florets above 20, all tubular, slightly exceeding the involucre. Achenes glabrous, much smaller and with a less prominent terminal ring than in *S. odoratus*.—*S. cygnorum*, Steetz in Pl. Preiss. i. 483.

W. Australia. King George's Sound and adjoining districts, *R. Brown, A. Cunningham, Oldfield;* Swan River, *Fraser, Drummond, n.* 28, 328, *Preiss, n.* 70, *Oldfield.*

24. **S. odoratus,** *Hornem.; DC. Prod.* vi. 371. A stout erect perennial of 2 to 3 ft., glabrous or rarely with a loose white deciduous wool on the under side of the leaves. Leaves oblong or lanceolate, irregularly toothed, often narrowed below the middle and sometimes almost petiolate, but dilated and stem-clasping, with toothed auricles at the base, attaining 2 to 4 in. in length, the upper ones and those of the side branches sometimes almost linear. Flower-heads small and numerous, in dense corymbs at the ends of the branches, forming usually a large corymbose panicle. Involucre cylindrical, rarely exceeding 2 lines, of about 8 bracts, with 1 or 2 small outer ones rarely wanting. Florets about 10 to 12, all tubular, considerably longer than the involucre when fully out. Achenes slightly pubescent, the pappus inserted on a callous ring more prominent than in the allied species.—A. Rich. Sert. Astrol. 109; Hook. f. Fl. Tasm. i. 223; *Cacalia odorata*, Desf. Hort. Par. 165 and 400, according to DC.

Victoria. Wendu vale, *Robertson;* Skipton, *Whan;* Wilson's Promontory, *F. Mueller;* Glenny Islands, *Wilhelmi.*

Tasmania. Port Dalrymple, King's Island, *R. Brown;* Woolnorth, *Gunn;* Macquarrie Harbour, *A. Cunningham.*

S. Australia. Memory Cove, Port Lincoln, *R. Brown;* Mount Kaiserstuhl, Tamunda Creek, *F. Mueller;* Kangaroo Island, *R. Brown, Baudin's Expedition, F. Mueller.*

J. D. Hooker in the Handb. N. Zeal. Fl. 160, refers to this species his *S. Banksii*, Fl. N. Zeal. ii. 146, which has indeed the foliage of luxuriant specimens of *S. odoratus*, but appears to me quite distinct in the radiate flower-heads and in the shape of the involucre and florets, which are more like those of a miniature *S. lautus*.

25. **S. Cunninghamii,** *DC. Prod.* vi. 371. This is now considered by F. Mueller as a variety of *S. odoratus*, of which it has the flower-heads and florets. It is however smaller and more woody at the base, the stems

more branched, ascending or erect, often under 1 ft. and rarely above 2 ft. high. Leaves linear or lanceolate, entire or coarsely toothed, narrowed into a short petiole without any or only with very minute auricles at the base. Flower-heads usually fewer than in *S. odoratus*, in small corymbs.—*S. brachylænus*, DC. Prod. vi. 370.

N. S. Wales. Lake George and Lachlan river, *A. Cunningham;* Macquarrie marshes, Duck Creek and Darling river, *Mitchell;* Lachlan and Darling rivers, *Victorian and other Expeditions.*

Victoria. Murray river, *Hergolt;* Wimmera, *Dallachy.*

S. Australia. Murray river to St. Vincent's Gulf and Flinders range, *F. Mueller;* Cooper's Creek, *Howitt's Expedition;* in the interior, *M'Douall Stuart's Expedition.*

W. Australia. Murchison river, *Oldfield, Drummond,* 6*th Coll. n.* 149.

26. **S. hypoleucus,** *F. Muell. Herb.* (as a var. of *S. odoratus*). A tall and erect perennial, the base of the stems apparently more woody than in *S. odoratus*, of which it has the inflorescence and florets and of which F. Mueller believes it to be a variety. Leaves ovate-lanceolate or lanceolate, 2 to 3 in. long, entire or slightly toothed, abruptly contracted into a long petiole not auriculate at the base, all very white and cottony underneath. Corymbs usually very dense.—*S. odoratus*, var. *petiolatus*, Sond. in Linnæa, xxv. 526.

Victoria. Wimmera, *Dallachy.*

S. Australia. Mount Lofty, *Wilhelmi, Whittaker.*

27. **S. anethifolius,** *A. Cunn. in DC. Prod.* vi. 371. A glabrous branching shrub (or undershrub?) attaining 4 or 5 ft. Leaves crowded, pinnately divided into long narrow linear or almost filiform segments, either entire or again bearing a few equally narrow lobes. Flower-heads small, numerous, in a very compact corymb. Involucre cylindrical, above 2 to nearly 3 lines long, of about 8 bracts, without any or with 1 or 2 very small outer ones. Florets about 10 to 12, all tubular, longer than the involucre. Achenes glabrous or scabrous-pubescent.—*S. angustilobus*, F. Muell. in Linnæa, xxv. 418; *S. angustifolius*, Sond. l. c. 526.

N. S. Wales. Mount Caley and Peel's Range. *A. Cunningham.*

S. Australia. Summits of hills near Cudnaka, *F. Mueller.*

28 (?) **S. Gaudichaudianus,** *A. Rich. Sert. Astrol.* 98. An erect glabrous annual of about 6 in. Leaves linear and entire or the lower ones petiolate oblong and obscurely toothed. Flower-heads few, in a small terminal corymb or almost solitary. Involucre cylindrical, nearly 3 lines long, of about 8 bracts with very few small outer ones. Florets all tubular and hermaphrodite, slightly exceeding the involucre. Achenes pubescent.—DC. Prod. vi. 370.

N. S. Wales. Port Jackson (*Gaudichaud*). This is a very doubtful species, and may possibly prove to be a small state of *S. brachyglossus*, in which the small ligulæ have been overlooked. Decaisne informs me that the authentic specimen in Herb. Mus. Par. had suffered much from sea-water, and was not in a state to ascertain the point. In a small one in Herb. F. Mueller, which otherwise agrees with the description, there appear to be no ligulæ, but the specimen is very imperfect and they may have fallen away.

S. vulgaris, Linn.; DC. Prod. vi. 341 (the common European *Groundsel*). An annual of 6 in. to a foot, bearing a little loose cottony wool, with irregularly pinnatifid and toothed leaves, and small flower-heads with the florets all tubular and hermaphrodite, and involucres of about 20 bracts, has appeared as an introduced weed in some parts of N. S. Wales *Woolls.*

86. BEDFORDIA, DC.

Flower-heads homogamous. and discoid. Involucre of nearly equal bracts apparently in a single row, the margins imbricate and scarious. Receptacle pitted. Florets all tubular, hermaphrodite, 5-toothed. Anthers obtuse at the base. Style branches somewhat flattened, very obtuse but not truncate, papillose from below the middle. Achenes angular or striate. Pappus of numerous denticulate bristles.—Shrubs more or less stellate-tomentose. Leaves alternate, entire or irregularly crenate. Flower-heads axillary, solitary or in clusters or dense panicles shorter than the leaves. Florets yellow.

The genus is limited to the two species endemic in Australia. F. Mueller has proposed to unite it with *Senecio*, but the stellate tomentum and axillary inflorescence are quite unknown in that extensive genus, and the style is almost as much that of some *Astereæ* as of *Senecio*. The same style occurs it is true in the Andine *S. iodopappus*, Sch.-Bip., but that species, as observed by Weddell, is anomalous also in several other respects.

Leaves lanceolate, mostly 3 to 5 in. long. Flower-heads in axillary clusters or panicles . . . . 1. *B. salicina.*
Leaves linear, rarely above 2 in. long. Flower-heads solitary or 2 together 2. *B. linearis.*

1. **B. salicina,** *DC. Prod.* vi. 441. A shrub attaining sometimes 12 to 14 ft., the branches, under side of the leaves, and involucres covered with a stellate tomentum either close or loose and almost floccose. Leaves lanceolate, obtuse, 3 to 5 in. long, entire or irregularly crenate, narrowed into a petiole, glabrous above when full grown, the reticulate veinlets impressed in the upper surface, prominent underneath. Flower-heads in axillary dense panicles much shorter than the leaves. Involucral bracts about 8, 2 to 2½ lines long, obtuse or scarcely acuminate. Achenes glabrous, striate, with 4 or 5 prominent ribs.—Hook. Lett. on Duke Bedf. with a fig.; Hook. f. Fl. Tasm. i. 224; *Cacalia salicina*, Labill. Pl. Nov. Holl. ii. 37. t. 179; Bot. Reg. t. 923; *Culcitium salicinum*, Spreng. Syst. iii. 431; *Senecio Bedfordii*, F. Muell. Cat. Hort. Melb. 1858, 26.

Victoria. Bullaruok forest, *Whan*; Muddy Creek, Corner Inlet, *Wilhelmi.*

Tasmania. Port Dalrymple and Derwent river, *R. Brown*; common on the skirts of forests and in the brush, *J. D. Hooker.*

2. **B. linearis,** *DC. Prod.* vi. 441. A shrub of about 4 to 6 ft., the branches, under side of the leaves, and inflorescence whitish with a stellate or floccose tomentum. Leaves rather crowded, linear, obtuse, entire, with revolute margins, scarcely petiolate, coriaceous, glabrous above, the veins inconspicuous, varying in size from all under ½ in. long in some specimens to mostly about 2 in. in others. Flower-heads solitary or 2 together, nearly sessile and usually larger than in *B. salicina.* Flowers otherwise as in that species.—Hook. f. Fl. Tasm. i. 225; *Cacalia linearis*, Labill. Pl. Nov. Holl. ii. 36. t. 178; *Culcitium lineare*, Spreng. Syst. iii. 431; *Senecio Billardierii*, F. Muell. Cat. Hort. Melb. 1858, 26.

Tasmania. Port Dalrymple, *R. Brown*; common especially in the central mountainous and southern parts of the island in rocky soil, *J. D. Hooker.*

TRIBE X. CALENDULACEÆ.—Leaves alternate or radical. Flower-heads usually heterogamous, the ray-florets ligulate, female or rarely neuter, the disk-florets tubular, hermaphrodite but sterile or rarely fertile, and very rarely in species not Australian, the heads homogamous. Receptacle pitted, the margins of the pits rarely produced into irregular scales or bristles. Anthers usually sagittate at the base but scarcely tailed. Style-branches, in the disk-florets, more or less concrete and thickened at the base.

This tribe, which is chiefly S. African, has the style nearly of *Cynarocephalæ*, and on that account has been associated with them under the general name of *Cynareæ*, but the habit and other characters appear much more to connect it with *Senecionidæ* and some *Mutisiaceæ*.

87. CYMBONOTUS, Cass.

Involucre campanulate, at length hemispherical, the bracts imbricate in several rows, the inner ones broader. Receptacle pitted, the margins of the pits often produced into rigid points or deciduous scales. Florets of the ray ligulate, female; disk-florets tubular, hermaphrodite, sometimes sterile. Anthers shortly sagittate but scarcely tailed. Style-branches rather broad, less concrete than usual in the tribe but erect or scarcely spreading. Achenes oblong, glabrous, smooth on the inner face, with 3 or 5 prominent ribs on the back and sides. Pappus none.—Perennial with toothed or pinnatifid radical leaves and single-headed leafless scapes.

The genus is limited to the single species endemic in Australia.

1. **C. Lawsonianus,** *Gaudich. in Freyc. Voy. Bot.* 462. *t.* 86. A perennial either stemless or with very short tufted stems. Leaves radical, spreading, on long petioles, ovate, coarsely toothed and 2 or 3 in. long or longer and lyrately pinnatifid, thin, green and somewhat scabrous above, cottony-white underneath. Scapes or peduncles shorter than the leaves and sometimes very short, cottony-white as well as the involucres. Involucre at length about ½ in. diameter. Ray-florets yellow, spreading; disk-florets shorter than the involucre.—DC. Prod. vi. 491; Sond. in Linnæa, xxv. 528; Hook. f. Fl. Tasm. i. 226; *C. Preissianus*, Steetz in Pl. Preiss. i. 486.

N. S. Wales. Port Jackson, *R. Brown;* Glendon, *Leichhardt ;* Twofold Bay, *R. Brown.*

Victoria. Common about Melbourne, *Adamson, Robertson, F. Mueller;* Station Peak, *Herb. F. Mueller.*

Tasmania. Abundant throughout the island, *J. D. Hooker.*

S. Australia. Near Adelaide, *F. Mueller.*

W. Australia. Grassy places, Mount Barker, and Upper Kalgan river, *Oldfield;* also *Drummond* and *Preiss, n.* 130.

88. CRYPTOSTEMMA, R. Br.

Involucre broadly hemispherical, the bracts imbricate in several rows, the innermost row with much longer lanceolate membranous tips. Receptacle pitted, without bristles or scales. Florets of the ray ligulate, neuter. Disk-florets short, hermaphrodite, tubular, 5-toothed. Anthers sagittate at the base but scarcely tailed. Style branches thickened and concrete nearly to the end. Achenes of the ray abortive, of the disk oblong, densely enveloped in an intricate wool. Pappus of about 6 to 8 short, lanceolate scales concealed in the wool.—Perennial with radical or alternate leaves, pinnatifid or pinnately divided. Scapes or peduncles single-headed.

The genus consists of but two species, both South African, one of them the same as the Australian one. It is very closely allied to *Arctotis*, scarcely differing except in the pappus.

1. **C. calendulacea,** *R. Br.*; *DC. Prod.* vi. 495. A perennial usually tufted and almost stemless, the leafy prostrate stems sometimes however lengthening out especially under cultivation. Leaves 3 to 6 in. long, deeply pinnatifid or pinnately divided into oblong or lanceolate acutely toothed or lobed segments, glabrous or scabrous above, cottony-white underneath. Scapes or peduncles leafless, 1-headed, glandular, rarely exceeding the radical leaves and often shorter than them. Involucre 7 to 8 lines diameter, glabrous or slightly cottony. Ray-florets yellow, the ligula narrow, ½ to 1 in. long. Disk purple.—Harv. and Sond. Fl. Cap. iii. 467; Steetz in Pl. Preiss. i. 487.

Victoria. Common about Melbourne, the whole pastures strewed with its woolly achenes like pellets in autumn, *Adamson*.

Tasmania. Introduced about Hobarton, *Gunn*.

S. Australia. Very common on roadsides about Adelaide, *F. Mueller*.

W. Australia. Very abundant about Perth, where it occurs sometimes with double flowers, the ray-florets variously divided or much increased in number, *Preiss, n.* 129.

The species is South African, and may have been introduced into Australia from the Cape. The following *Calendulaceæ* have also been sent from Australia as introduced weeds:—

Tripteris clandestina, Less.; Harv. and Sond. Fl. Cap. iii. 428. An erect branching slightly pubescent annual of 1 to 2 ft. Leaves oblong or lanceolate, the lower ones petiolate and sinuate-toothed, the upper ones small distant sessile or stem clasping. Flower-heads all nodding at the ends of the branches. Involucral bracts narrow, all equal in a single row with scabrous centres and scarious margins. Receptacles without scales. Ray-florets ligulate, female, the ligulæ scarcely exceeding the involucre. Disk-florets tubular, hermaphrodite, but sterile. Achenes of the ray with 3 angles produced into broad longitudinal scarious wings. Pappus none. Disk-achenes abortive.—*T. atropurpurea*, Turcz. in Bull. Mosc. 1851, i. 212.—W. Australia, *Drummond, 3rd Coll. n.* 131. Probably introduced from the Cape.

Calendula officinalis, Linn.; DC. Prod. vi. 451. An erect or spreading much-branched annual of about 1 ft., green, pubescent and more or less glandular. Lower leaves spathulate, upper ones lanceolate or oblong, entire or slightly toothed, stem-clasping at the base. Flower-heads terminal. Ray-florets ligulate, female, much longer than the involucre, of an orange-yellow. Disk-florets tubular, hermaphrodite but sterile. Achenes of the ray in 2 or 3 rows, elongated and incurved, the inner face smooth, the margins of the outer ones dilated with a ridge of tubercles or prickles along the centre of the back.—A S. European species introduced about Adelaide, from whence is also, in Herb. F. Mueller, a single specimen of *C. arvensis*, Linn., a species or variety distinguished by its much smaller flower-heads and smaller achenes, the outermost ones narrow and elongated, the others very much shorter and broader.

TRIBE XI. MUTISIACEÆ.—Leaves alternate. Flower-heads either heterogamous with radiating female florets or homogamous with tubular florets, some or all of the outer florets in all cases more or less 2-lipped. Receptacle mostly without scales. Anthers pointed or tailed at the base. Style usually thickened under the branches which are erect, truncate and often short as in the *Calendulaceæ*, more rarely elongated and obtuse, almost as in *Senecionidæ*.

89. AMBLYSPERMA, *Benth.*

Involucre broadly hemispherical, the bracts imbricate in several rows, the outer ones gradually shorter. Receptacle without scales. Ray-florets female, ligulate, with an upper lip divided to the base into 2 filiform segments; disk-florets tubular, hermaphrodite, 5-lobed. Anthers with long tails. Style-branches elongated, obtuse. Achenes short, thick, villous. Pappus of numerous rather rigid capillary bristles. Perennial. Leaves radical. Scapes leafless, single-headed.

The genus is limited to a single species endemic in Australia, and the only representative there of a tribe numerous in species in South America, with a few African and tropical Asiatic ones.

1. **A. scapigera,** *Benth. in Hueg. Enum.* 67. A perennial with a tufted woolly stock. Leaves all radical, from ovate to narrow-oblong, 1 to 3 in. long, besides the long petioles, obtuse, sinuate, shortly and broadly lobed or almost lyrate, cottony-white underneath, surrounded at the base by the persistent woolly remains of old petioles. Scapes simple, 1 to 2 ft. high, bearing a few small bracts and a large terminal flower-head. Involucre $\frac{3}{4}$ to $1\frac{1}{2}$ in. diameter, more or less covered with white wool, the bracts lanceolate acuminate. Florets of the ray 12 to 20, the lower lip or ligula above 1 in. long, the filiform segments of the upper not half so long, the anthers present but imperfect. Disk-florets numerous, as long as the involucre, the lobes linear. Anther-tails slightly bearded at the end.—DC. Prod. vii. 20; Steetz in Pl. Preiss. i. 487; *Celmisia spathulata*, A. Cunn. in DC. Prod. v. 209.

W. Australia. King George's Sound and thence to Cape Riche, *Huegel*, *A. Cunningham*, *Preiss*, *n.* 64, *Drummond*, *n.* 389, and others; Tone river, *Maxwell;* Murray river, *Oldfield.*

TRIBE XII. CICHORIACEÆ.—Leaves alternate. Flower-heads homogamous, with all the flowers ligulate.

90. MICROSERIS, Don.

(Monermios, *Hook. f.;* Phyllopappus, *Walp.*)

Involucre of several nearly equal bracts in about 2 rows with a few short imbricate ones outside. Receptacle without scales. Florets all ligulate. Achenes cylindrical with smooth longitudinal ribs, not beaked. Pappus of linear chaffy flat scales in about 2 rows tapering into simple or shortly plumose bristles.—Herbs with radical usually pinnatifid leaves. Scapes leafless, single-headed. Florets yellow.

Besides the Australian species which extends to New Zealand, there is a closely-allied one from extratropical S. America.

1. **M. Forsteri,** *Hook. f. Fl. Nov. Zel.* i. 151, *and Fl. Tasm.* i. 226. *t.* 66. A glabrous perennial with fleshy roots thickened into tubers and a milky juice. Leaves radical, attaining 8 to 10 in. in luxuriant specimens, but often not half so long, narrow-lanceolate or linear, entire or pinnatifid with short distant lobes. Scapes exceeding the leaves. Involucre 6 to 8 lines long in most Australian specimens. Florets exceeding the involucre. Pappus-

bristles or scales in the commonest form not much dilated in the lower part and only very minutely serrulate.—*Scorzonera scapigera*, Forst. Prod. 91; *Scorzonera (Monermios) Lawrencii*, Hook. f. in Hook. Lond. Journ. vi. 124; *Phyllopappus lanceolatus*, Walp. in Linnæa, xiv. 507; Sond. in Linnæa, xxv. 529.

N. S. Wales. Goulburn plains, *A. Cunningham;* Nangas, *M'Arthur;* New England, *C. Stuart.*

Victoria. Common about Melbourne and to the Glenelg, *Robertson, F. Mueller*, and others; Skipton, *Whan;* Mount Remarkable, Delatite river, *F. Mueller.*

Tasmania. Derwent river, *R. Brown*; abundant in good soil in many parts of the island, *J. D. Hooker.*

S. Australia. Reedy Creek and Guichen Bay, *F. Mueller.*

Also in New Zealand.

Var.? *subplumosa.* Pappus-bristles strongly ciliate, almost plumose. I can perceive no other difference.

W. Australia, *Drummond, 5th Coll. n.* 366; Scott's Brook, near Cape Arid, *Maxwell.*

91. **HYPOCHŒRIS,** Linn.

(Cycnoseris, *Endl.*)

Involucre broad or narrow, often elongated after flowering, the bracts imbricate. Receptacles with a few linear chaffy scales between the florets. Florets all ligulate. Achenes usually striate, all or the inner ones only tapering into a slender beak bearing a pappus of plumose bristles.—Annuals or perennials. Leaves radical. Stems simple or with a few long branches, leafless or nearly so. Florets yellow.

The genus is spread over the northern hemisphere, extending also to extratropical South America. The only Australian species is found nearly over the whole range of the genus.

1. **H. glabra,** *Linn.; DC. Prod.* vii. 90. A glabrous annual. Leaves all radical, narrow, spreading, more or less toothed or pinnately lobed. Stems 6 in. to 1 ft. high, usually divided into a few slender branches, leafless except small scales at the base of the branches. Flower-heads solitary at the ends of the branches. Involucre cylindrical, at first small, but lengthening out to from ½ to ¾ in. when in fruit, of a few imbricated bracts, the outer ones short and appressed. Achenes striate and transversely pitted, the pappus of the outer ones sessile, that of the others borne on a long beak terminating the achene.—Steetz in Pl. Preiss. i. 488; *Cycnoseris australis*, Endl. in Bot. Zeit. 1843, 459.

N. S. Wales. Paramatta, *Woolls.*

Victoria, *Robertson;* near Skipton, *Whan;* Wimmera, *Dallachy.*

Tasmania, *Gunn;* Flinders Island, *Milligan.*

S. Australia. Common, *F. Mueller* and others.

W. Australia. Kalgan river, *Oldfield;* Swan River, *Drummond, n.* 74, *Preiss, n.* 119.

H. radiata, Linn.; DC. Prod. vii. 91. A perennial resembling *H. glabra*, but taller, with larger flower-heads, the leaves hispid, and all the achenes terminating in a slender beak bearing the pappus, a common European species, is amongst the introduced plants in the neighbourhood of Paramatta, *Woolls.*

92. **PICRIS,** Linn.

Involucre of several nearly equal erect inner bracts, with 2 or 3 rows of smaller outer ones, usually spreading. Receptacle without scales. Florets all ligulate. Achenes transversely striate or muricate, not at all or very shortly beaked. Pappus of whitish fine bristles, of which the inner ones at least are plumose.—Coarse hispid annuals. Leaves alternate, toothed. Flower-heads in a loose irregular corymb. Florets yellow.

A genus containing but few species, natives of the temperate and subtropical regions of the northern hemisphere in the Old World, one of which is also the Australian one, and is found as an introduced plant in other parts of the world.

1. **P. hieracioides,** *Linn.; DC. Prod.* vii. 128. A biennial from 1 to 2 or 3 ft. high, covered with short rough hairs, most of which are minutely hooked so as to cling to whatever they come in contact with, but rather less so in some of the Australian than in the northern specimens. Leaves lanceolate or, especially in Australian specimens, linear, the lower ones tapering into a petiole, and often 6 in. long or more, the upper ones few and small. Peduncles rather long and stiff, the upper ones sometimes irregularly umbellate. Involucres from under $\frac{1}{2}$ in. to nearly $\frac{3}{4}$ in. long. Achenes very strongly transversely striate or muricate, usually contracted under the pappus or tapering into a very short beak. Pappus-bristles usually very plumose, except a few of the outer ones of each achene.—Hook. f. Fl. Tasm. i. 227; *P. angustifolia*, DC. Prod. vii. 130; Sond. in Linnæa, xxv. 529; *P. attenuata*, A. Cunn. in Ann. Nat. Hist. ii. 125; *P. barbarorum*, Lindl. in Mitch. Three Exped. ii. 149, and in Bot. Reg. 1838, Misc. 58; *P. asperrima*, Lindl. in Bot. Reg. 1838, Misc. 58; *P. hamulosa*, Wall.; DC. Prod. vii. 129.

Queensland. Shoalwater Bay, *R. Brown;* Moreton Island, *M'Gillivray;* Rockhampton, *Dallachy.*

N. S. Wales. Port Jackson, *R. Brown, Woolls;* Macleay, Clarence, and Hastings rivers, *Beckler;* New England, *C. Stuart;* north of Bathurst, *A. Cunningham;* Lachlan and Darling rivers, *Victorian Expedition.*

Victoria. About Melbourne, *Adamson;* Wimmera, *Dallachy.*

Tasmania. Port Dalrymple, *R. Brown;* common in the northern parts of the island, *J. D. Hooker.*

S. Australia. Port Lincoln, *R. Brown, Wilhelmi;* Murray river to St. Vincent's Gulf, *F. Mueller* and others; Kangaroo Island, *Waterhouse.*

W. Australia. Swan River, *Drummond, 1st Coll. also 4th Coll. n.* 216; Capel and Blackwood rivers, *Oldfield.*

Var. *squarrosa.* More hispid, with larger flower-heads and more numerous recurved outer involucral bracts.—*P. squarrosa*, Steetz in Pl. Preiss i. 488; Sond. in Linnæa, xxv. 529.—Port Phillip, *Gunn;* S. Australia, *F. Mueller* and others, and various parts of W. Australia, *Drummond, Preiss, n.* 108, *Oldfield, Maxwell.*

The species is very common in the northern hemisphere in the Old World, and has much spread with cultivation in other countries. It may therefore be an introduced plant in many of the Australian localities. In others however there is every probability of its being truly indigenous.

93. CREPIS, Linn.

(Youngia, *DC.*)

Involucre of a single row of nearly equal bracts, with a few small outer ones. Receptacle without scales. Florets all ligulate. Achenes oblong, cylindrical or scarcely flattened, striate, tapering at the top, but without a distinct beak. Pappus of numerous fine white soft simple bristles.—Annual or perennial herbs, usually branched. Leaves alternate or radical, mostly toothed or lobed. Flower-heads in loose irregular corymbs or panicles. Florets yellow.

A large genus, widely distributed over the temperate regions of the northern hemisphere, with a few subtropical species, the only Australian one extending into tropical and Eastern Asia.

1. **C. japonica,** *Benth. Fl. Hongk.* 194. An erect slender annual, 6 in. to near 1 ft. high, glabrous or slightly pubescent or hairy near the base. Leaves mostly radical, petiolate, varying from obovate, nearly entire and 1 to 2 in. long, to lyrate or pinnatifid, 2 to 4 in. long, with a large terminal toothed lobe. Stem-leaves few or in the Australian specimens usually none. Panicle slender, loosely corymbose. Flower-heads small, numerous. Involucres about 2½ lines long, containing 10 to 15 small yellow florets.—*Prenanthes japonica*, Linn.; *Youngia japonica*, DC. Prod. vii. 194, also *Y. Thunbergiana*, and some others of DC. See A. Gray in Mem. Amer. Acad. vi. 396.

Queensland. Shoalwater Bay, *R. Brown;* Dawson river, *F. Mueller;* Dunk Island, *M'Gillivray;* Rockhampton, *Thozet, Dallachy.*

N. S. Wales. Hunter's and Grose rivers, *R. Brown;* Macleay and Hastings rivers, *Beckler.*

The species is common in India, and extends on the one hand to Ceylon and the Mauritius, and on the other to China and Japan.

94. SONCHUS, Linn.

Involucre ovoid, with imbricate bracts, and usually becoming conical after flowering. Receptacle without scales. Florets all ligulate. Achenes flattened and striate, not beaked. Pappus of numerous fine bristles, usually soft and white.—Herbs either annual or in species not Australian perennial or shrubby. Leaves alternate, usually toothed or lobed. Flower-heads small or large in loose corymbs or panicles. Florets yellow or (in species sometimes separated from the genus) blue.

A considerable genus, ranging over the temperate regions of the northern hemisphere, the Australian species extending over the whole range of the genus, and introduced into almost every part of the world.

1. **S. oleraceus,** *Linn. Spec.* 1116. An erect annual, with a hollow stem, 1 to 3 or even 4 ft. high. Leaves thin, bordered by irregular acute or prickly teeth, otherwise either undivided or pinnatifid with a broad heart-shaped or triangular terminal lobe, the upper ones narrow and clasping the stem with short auricles. Flower-heads in a short corymbose terminal panicle, sometimes almost umbellate. Florets of a pale yellow.—Steetz in Pl. Preiss. i. 489; *S. asper*, Fuchs; Hook. f. Fl. Tasm. i. 227; Steetz in Pl. Preiss. i. 489; *S. ciliatus*, Lam., and *S. fallax*, Wallr., DC. Prod. vii. 185.

N. S. Wales. Port Jackson, *R. Brown, Woolls;* Hastings and Clarence rivers, *Beckler.*

Victoria. Near Skipton, *Whan.*

Tasmania. Port Dalrymple, *R. Brown;* common, but only near the sea and on the north shore of the island, *J. D. Hooker.*

S. Australia. Mount Gambier, Bugle Range, Torrens river, etc., *F. Mueller;* towards Spencer's Gulf, *Waterhouse.*

W. Australia. Swan River, *Drummond, n.* 75, 317, *Preiss, n.* 116, 117; Don river, *Maxwell;* Murchison river, *Oldfield.*

A weed of cultivation, probably indigenous to Europe and temperate Asia, but now distributed over the greater part of the globe and perhaps truly indigenous in Australia. There are two marked varieties; in the one (*S. asper*) the ribs of the achenes are perfectly smooth; in the other, for which the name of *S. oleraceus* is more specially retained, they are marked with transverse asperities. Both occur in Australia, as also a maritime variety with the flowers almost as large as in *S. arvensis,* but without the glandular hispid involucres of that species.

The following European *Cichoriaceæ* are amongst the introduced weeds in some of the Australian colonies.

Arnoseris pusilla, Gærtn.; DC. Prod. vii. 79 (*Lapsana pusilla,* Willd.). A small annual, glabrous or nearly so. Leaves all radical, obovate or oblong, toothed. Flower-stalks slightly branched, leafless, the erect branches or peduncles enlarged and hollow upwards, each bearing a small head of yellow florets, all ligulate. Involucre of few nearly equal bracts with some very small outer ones. Achenes crowned only by a minute raised border.—Tasmania, so much spread as to be apparently indigenous, *J. D. Hooker.*

Cichorium Intybus, Linn.; DC. Prod. vii. 84 (*Succory* or *Chicory*). A perennial, more or less hairy, erect with stiff spreading branches. Radical leaves spreading on the ground and lower stem-leaves pinnatifid with a large terminal lobe, upper leaves small and less cut. Flower-heads in closely sessile clusters of 2 or 3 along the branches, with large florets, all ligulate, of a bright blue. Involucre of few nearly equal bracts, with a few small outer ones. Achenes crowned by a ring of minute scales.—S. Australia, *F. Mueller.*

Leontodon hirtus, Linn. (*Thrincia hirta,* DC. Prod vii. 99). A small herb with a perennial stock. Leaves all radical, oblong or linear, toothed, sinuate or pinnatifid. Scapes seldom above 6 in. high, bearing a single head of rather small yellow florets, all ligulate. Involucre of about 10 to 12 nearly equal bracts with a very few or scarcely any small outer ones. Achenes slightly tapering at the top, those of the outer row with a very short scaly pappus, the others of brownish plumose bristles.—Near Paramatta, *Woolls.*

Tragopogon porrifolius, Linn.; DC. Prod. vii. 113 (*Salsify*). A glabrous biennial or perennial of 1 to 2 ft., with a taproot. Radical and lower leaves long and grass-like, entire, shortly dilated and sheathing at the base, the upper ones shorter and broader. Peduncles long, thickened at the summit, each with a single head of purple florets. Involucre of 8 to 12 nearly equal bracts longer than the florets. Achenes narrowed into a long beak, bearing a pappus of feathery bristles.—Now spreading in the cultivated fields near Camden, *Woolls.*

Lactuca saligna, Linn.; DC. Prod. vii. 136. An erect stiff glabrous annual or biennial of 2 or 3 ft. Leaves narrow, erect, clasping the stem with pointed auricles, entire or with few teeth or narrow lobes. Flower-heads rather small, clustered on the short branches of a simple panicle. Involucre narrow, 4 or 5 lines long, of a few imbricate bracts. Florets 6 to 10, of a pale yellow, all ligulate. Achenes much flattened, obovate-oblong, produced into a slender beak two or three times their own length and bearing a pappus of numerous white silky simple bristles.—Australia Felix, *F. Mueller.*

Taraxacum Dens-leonis, Desf.; DC. Prod. vii. 145 (including also perhaps the whole 25 supposed species of De Candolle's § 2) (*Dandelion*). A perennial with a thick bitter taproot. Leaves radical, varying from linear-lanceolate and almost entire to deeply pinnatifid, the lobes often curved downwards. Scapes leafless, rarely exceeding 6 in., bearing a single rather large flower-head. Involucre of several nearly equal bracts with some smaller outer imbricate ones often recurved. Florets yellow, all ligulate. Achenes scarcely compressed, striate, tapering into a slender beak two or three times their own length and bearing a pappus of numerous simple hairs.—Now common about Melbourne and in the plains of the Avon, *F. Mueller;* also in Tasmania, and in West Australia, *Drummond, n.* 367.

INDEX OF GENERA AND SPECIES.

The synonyms and species incidentally mentioned are printed in italics.

END OF VOL. III.

J. E. TAYLOR AND CO., PRINTERS,
LITTLE QUEEN STREET, LINCOLN'S INN FIELDS.

For EU product safety concerns, contact us at Calle de José Abascal, 56–1°,
28003 Madrid, Spain or eugpsr@cambridge.org.

www.ingramcontent.com/pod-product-compliance
Ingram Content Group UK Ltd.
Pitfield, Milton Keynes, MK11 3LW, UK
UKHW041844190726
13854UKWH00002B/704

* 9 7 8 1 1 0 8 0 3 7 4 0 2 *